AN INTRODUCTION TO CONTINUUM PHYSICS

Adopting a unified mathematical framework, this textbook gives a comprehensive derivation of the rules of continuum physics, describing how the macroscopic response of matter emerges from the underlying discrete molecular dynamics. Covered topics include elasticity and elastodynamics, electromagnetics, fluid dynamics, diffusive transport in fluids, capillary physics, and thermodynamics. By also presenting mathematical methods for solving boundary-value problems across this breadth of topics, readers develop understanding and intuition that can be applied to many important real-world problems within the physical sciences and engineering. A wide range of guided exercises are included, with accompanying answers, allowing readers to develop confidence in using the tools they have learned. This book requires an understanding of linear algebra and vector calculus and will be a valuable resource for undergraduate and graduate students in physics, chemistry, engineering, and geoscience.

STEVEN R. PRIDE is Adjunct Professor at the University of California, Berkeley, and Scientist at the Lawrence Berkeley National Laboratory. His primary research expertise is the physics of porous media. Over three decades, he has published extensively across the physical sciences from statistical physics to seismology.

AN INTRODUCTION TO CONTINUUM PHYSICS

STEVEN R. PRIDE
University of California, Berkeley

CAMBRIDGE
UNIVERSITY PRESS

CAMBRIDGE
UNIVERSITY PRESS

Shaftesbury Road, Cambridge CB2 8EA, United Kingdom

One Liberty Plaza, 20th Floor, New York, NY 10006, USA

477 Williamstown Road, Port Melbourne, VIC 3207, Australia

314–321, 3rd Floor, Plot 3, Splendor Forum, Jasola District Centre,
New Delhi – 110025, India

103 Penang Road, #05–06/07, Visioncrest Commercial, Singapore 238467

Cambridge University Press is part of Cambridge University Press & Assessment,
a department of the University of Cambridge.

We share the University's mission to contribute to society through the pursuit of
education, learning and research at the highest international levels of excellence.

www.cambridge.org
Information on this title: www.cambridge.org/9781108844611

DOI: 10.1017/9781108951982

When citing this work, please include a reference to the DOI 10.1017/9781108951982

First published 2025

Cover Image: Ali Kahfi / Digital Vision Vectors / Getty Images

A catalogue record for this publication is available from the British Library

A Cataloging-in-Publication data record for this book is available from the Library of Congress

ISBN 978-1-108-84461-1 Hardback

Contents

Preface

Various forms of the contents of this textbook have been taught by me to upper-division undergraduate students at the University of California (UC) at Berkeley every other year since 2005. These undergraduates have come from a range of physical-science departments including Earth sciences, most fields of engineering, chemistry, and physics. What is unusual about this text compared to other books on continuum mechanics is that the continuum rules describing macroscopic physics are derived by starting at the discrete scale of atomic trajectories and using an averaging procedure to arrive at the handful of macroscopic conservation laws that describe the dynamics of large numbers of atoms. An identical approach provides the rules of continuum electromagnetics and gravity. The various constitutive laws needed to complement the conservation and electromagnetic laws are shown to derive from the thermodynamics of either equilibrium or nonequilibrium systems. One advantage of this approach is that the student acquires from the outset a clear physical understanding of what each of the macroscopic response fields in continuum physics is representing. It also makes explicit how the mechanics of discrete particle systems, which is widely taught to undergraduates, transitions to continuum mechanics for the averaged response of the multitudes of molecules that surround each point in space. In this approach, we do not replace the actual system of discrete molecules with a space-filling nineteenth-century "continuous medium" that is supporting fields whose nature must be guessed at. Instead, we sum (average) over the actual molecules surrounding each point to define both our continuously distributed macroscopic fields and the differential equations that control them. The "theory of continuum physics" so obtained can be considered an exact theory, which is a novel perspective.

The book is meant to tie together the various loose threads of physics and mathematics that a typical United States upper-division undergraduate student in the physical sciences has been taught, while adding to this knowledge and packaging it in the form that physical scientists working at nonatomic scales actually use in practice. Although Earth-science applications, which is my field of research, at times motivate some of the examples and exercises, overall the book applies to any macroscopic body and no inherent geoscience knowledge is assumed of the reader. Although I have been teaching this material to undergraduates, this book would more normally be considered a graduate-level text in the United States. It is my hope that even practicing scientists further along in their career will find value in the presentations and proofs, many of which can only be found in this book. In revisiting the origins of continuum physics from a first-principles molecular perspective, it has been a pleasure to obtain fresh insights about the set of rules describing macroscopic response despite the topics having been taught in many cases for nearly two centuries.

Let me begin by explaining the genesis for this book, which traces through parts of my scientific biography in a somewhat self-indulgent manner. Perhaps these stories can benefit a student or young professor who, like me when I was starting out, is curious about the underpinnings of continuum mechanics and not always satisfied with the status quo of understanding or presentation as given in the classic texts.

When I was an undergraduate student in geophysics at UC Berkeley in the 1980s, we were required to take an upper-division class in the physics department called "classical mechanics." That class taught us how to calculate the movement of discrete masses connected to each other via springs, presumably as an analog for the dynamics of a crystal lattice. The dynamical equation controlling the movement $\boldsymbol{x}(t)$ of each discrete mass was the familiar Newton's law $m\ddot{\boldsymbol{x}} = \boldsymbol{F}$ along with Hooke's spring force $\boldsymbol{F} = -k\boldsymbol{x}$. The same semester, we also had to take an upper-division class in seismology that considered a different dynamics associated with the vibrations inside of the Earth excited by an earthquake or explosion. The dynamical equation controlling seismic waves, that was just written down by the professor teaching the class, looked vaguely like Newton's law in that there was a mass density times an acceleration, but the forcing causing the acceleration appeared to be quite distinct from the Hookean spring force causing the discrete masses to move even if it derived from something also called "Hooke's law." I went off to graduate school knowing there must be an explicit link between these two quite different perspectives of treating "mechanics."

In 1992, as a young professor (maître de conférences) at the University of Paris, specifically at the Institut de Physique du Globe de Paris (IPGP), I needed to prepare my first graduate-level class in continuum physics. For that purpose, I worked out for myself how to start at the discrete scale of molecular dynamics and obtain the macroscopic rules controlling the conservation of mass and momentum and electromagnetic response of vast numbers of molecules. I was also not satisfied at the time with how the classic texts on continuum mechanics first defined strain as an isolated kinematic topic on its own and then arbitrarily used this definition in Hooke's law for the stresses. I had a desire to see all constitutive laws as deriving from thermodynamics and so I worked out, using a simple averaging procedure, the definition of strain that naturally arises from the rate that work is performed in deforming a sample. The result for the strain tensor, though not new, was pleasing in its clear physical interpretation. Not long after, a famous seismology professor at IPGP called me into his office and asked me how, if I was given a rock sample with stress applied to it, the strain tensor for the sample should be measured. I showed him the simple answer I was presenting to the students, which he was pleased with. It was motivating to my pedagogical endeavors that a brilliant professor was not certain how something fundamental to his field, the strain tensor, is measured in practice.

I then decided to move with my young family to the University of Rennes, where I was required to teach the normal French course load of six classes per year that I had been insulated from while in Paris. A distinction about the French higher-education system as compared to the American system is that after the first two years when the students are required to select a major, the rest of their classes over their third and fourth years at university typically come only from the department associated with their major. The "upper-division" French students in geophysics still need a range of additional physics classes in order to complete their four-year university education so the four physics classes that I taught to our third- and fourth-year students were: (1) continuum mechanics, (2) fluid mechanics, (3) electromagnetics, and (4) a combined statistical mechanics and thermodynamics. Because this range of physics was taught each year to the same group of students by a single professor, me, from a perspective geared toward applications in a particular field, the students were able to see the mathematical similarities and connections across these various subject areas. Another great aspect of the French educational system is that it is frowned upon, or at least not common, to assign textbooks as is common in the United States. Instead, each professor in France is expected to write their own course content and distribute notes to the students. This exercise, done in my poor French and taking considerable time away from my research, forced me as a young professor to dig deep in developing my own understanding of each subject and not just lean on the crutch of an established text. That combined with a wholistic view of physics engendered by teaching a breadth of physics topics each year I believe has benefited my research over the many years and contributed to the nature of the present book.

In the early 2000s, I returned with my family to the United States to take over running a family business and obtained a scientist position at the Lawrence Berkeley National Laboratory and Adjunct Professor position at UC Berkeley in order to continue doing science at the same time. For some reason, continuum mechanics is not commonly taught in the United States to undergraduates in the physical sciences, which is strange because physical scientists not working at atomic or subatomic scales, which is most of us, use continuum mechanics to quantitatively model the macroscopic response of matter. I assume that undergraduates are not required to take continuum mechanics in the United States because tensors are needed to properly describe the macroscopic response of matter and tensors are not a normal part of the undergraduate curriculum. This seemed silly to me because tensor calculus follows smoothly, almost effortlessly, from the vector calculus that students are exposed to. So when I was asked to teach a class to our geophysics undergraduates at UC Berkeley, I decided to give them a full-blown class in continuum physics in my preferred way of starting at the discrete scale of atomic trajectories, which resulted in this textbook.

The assumed background of a student using this book is a course in first-year university physics, linear algebra, vector calculus, and ordinary-differential equations. The book is broken into two parts: Part I "Continuum Physics" and Part II "Mathematical Methods." I go through much, but certainly not all, of both parts of this text in a single-semester class. I have tried to only include topics in the book that I have used during my career, found useful and/or interesting and that should, in my opinion, be a part of the working-knowledge base of any physical scientist interested in modeling the macroscopic response of matter to various stimulations. It would be impossible for me to teach continuum physics to undergraduates without also teaching a considerable amount of mathematics at the same time with a particular focus on Fourier analysis, which is the reason for Part II. For example, in the thermodynamics of Part I, we derive a range of constitutive laws in which the response is instantaneous in the force. But to address delayed-linear response, you have to be versed in both Fourier analysis and contour-integration methods and so we cannot address delayed response until these mathematical topics have been covered. Although Part I and Part II can serve as their own stand-alone texts in continuum physics and the mathematical methods of physics, respectively, they are designed to work together. At the undergraduate level for physical-science students in the United States, there has been an increase in breadth requirements and a decrease in required major classes at most universities over the past two decades. My approach for dealing with this is to teach both continuum physics and associated topics in mathematical physics together in a single class.

Chapter 1 on tensor calculus is meant to be a student's first exposure to tensors, which are shown to be a direct extension from the vectors with which they are familiar. I am not aware of another text that shows students how to obtain the various types of important tensor-calculus product rules that we use repeatedly throughout all of continuum physics. This is an essential chapter that cannot be skipped over, containing material that cannot be found in other texts. I have never found that the introduction of tensor calculus as the first topic to address in continuum physics is either boring or difficult for the student, which is a traditional concern I have heard from many professors teaching continuum mechanics.

Continuum mechanics is developed in Chapter 2, and I always teach through most of this chapter, though skipping over the details of modeling discrete atomic trajectories. The basic averaging (adding) procedure for transitioning from atomic trajectories to the various continuum fields is laid out and the macroscopic statements of the conservation of mass and momentum are obtained from this perspective. In particular, this approach makes clear that the forces causing a collection of molecules to move are the forces that come from other molecules lying outside the collection, which leads to the definition of the stress tensor and body forces in terms of the molecular interactions. Chapter 3 derives the macroscopic Maxwell's equations in a way analogous to Jackson's (1975) treatment in his classic text. My treatment of the electromagnetic constitutive laws from the thermodynamic perspective of following the energetic changes within a mass element that is possibly in motion is unique, I believe, in the literature and is consistent with my handling of all the constitutive laws throughout the book. In my one-semester class, I usually have to skip Chapter 3 due to time constraints, which is regretful.

Chapter 4 is another chapter not to be skipped over that covers elasticity theory from a distinct, and I believe new-to-the-literature thermodynamics perspective. Strain as used in Hooke's law is not introduced, as it is in nearly all texts on continuum mechanics, as its own a priori stand-alone kinematic measure of deformation. Instead, we focus on the work of deforming a body and the measure of strain is forced upon us as the conjugate variable to stress in this work expression. Hooke's law follows directly and exactly from this thermodynamics perspective and in a form that has not hitherto been appreciated. The basics of linear elastodynamic response is covered including reflection, transmission, and evanescent waves at an interface. Several important theorems about the elastodynamic displacement field are also derived. Chapter 5 treats viscous fluid flow, giving some novel perspectives surrounding the condition for flow to be considered incompressible in addition to the usual introductory range of viscous-flow-related topics. I always cover viscous flow in my one-semester class. Other topics in continuum mechanics, such as plasticity and fracture mechanics, are not treated in this introductory text. The guided end-of-chapter exercises throughout Part I of the book treat problems that can, for the most part, be solved using simple integration.

The main purpose of Chapter 6 on continuum thermodynamics is to show where the constitutive laws of reversible processes come from. It is my experience that thermodynamics is not well understood by undergraduates in the physical sciences, and yet it provides core universally valid results in our modeling at macroscopic scales and is relatively easy to remember and use. Defining a fundamental function and extracting the information it contains by taking derivatives should be engrained in all undergraduates in the physical sciences, and I take the student through the process. An emphasis is placed on understanding the chemical potential, which is often given short shrift in other thermodynamics textbooks geared more to physics than chemistry. It is shown how the chemical potential controls first-order phase transitions, chemical reactions, solubility of minerals in liquids, and practical industrial processes such as reverse-osmosis filtering. A novel derivation is given for the functional form of the chemical potential in liquid solutions, which is widely used across disciplines. A proper formulation of the thermodynamics in solids is also given, which is unique in the literature. I have elected not to delve into ensemble-based statistical mechanics in this book.

Chapter 7 treats out-of-equilibrium fluid systems where transport processes are attempting to return the system to equilibrium. After a standard derivation of the energy conservation law in a fluid system, specific systems are then treated where the entropy production is identified and transport constitutive equations defined. We cover three types of systems in Chapter 7: (1) single-component (single molecular species) systems where only heat transport and fluid flow is occurring; (2) two-component "binary" systems involving solute and solvent molecules that are fully miscible in which both heat transport and solute diffusion are occurring along with fluid flow; and (3) two-component immiscible systems in which surface tension is important and the two fluids do not fully mix even in equilibrium. This last topic is covered both from the van der Waal's or Cahn-Hilliard perspective, which is a topic perhaps too advanced for presentation to undergraduates and thus skipped-over by me in my one-semester class, as well as from a more macroscopic perspective that is appropriate for undergraduates. In this chapter, I also show how to prove Onsager reciprocity (symmetry of the linear transport laws) both using a new-to-the-literature approach based on the self-adjoint nature of the macroscopic linear governing equations controlling transport processes and using the traditional statistical-mechanics viewpoint due to Onsager where time-reversal symmetry of the underlying discrete-particle mechanics is invoked. Typically in my one-semester class, the only part of Chapter 7 that I have time to teach is advective heat transport in single component systems.

For Part II of the book on mathematical methods, the content is fairly standard even if the presentation order and demonstrations are distinct from other books. The main theme is how to solve linear partial-differential equations analytically using a range of approaches. However, after introducing Fourier analysis and contour integration, we also use these tools to address a range of diverse topics such as linear-response constitutive equations when delay and loss processes are operating, distinctions between group velocity and phase velocity, the central-limit theorem and time-series analysis (data processing). We culminate in Chapter 12 with a treatment of the Green's functions for many of the macroscopic vectorial responses introduced in Part I of the book. For an undergraduate to understand Chapter 12, which is where the point source response of Part I continuum processes is obtained, one needs to cover most of the preceding chapters of Part II. I teach most, though not all, of Part II in my one-semester class. Generally speaking, the students find the first part of the class on deriving the rules of continuum physics to be more challenging and the second part on mathematical methods to be relatively easier.

Let me share a final math-related experience from my own undergraduate years at Berkeley, which had a big influence on my science career. As geophysics majors at UC Berkeley, we were required to take a mathematical methods of physics class (the usual topics such as contour integration, Fourier analysis, and partial-differential equations, similar to Part II of the present book) from a professor associated with the math department. Two weeks into the class, the professor had a heart attack and the class was canceled. Three of us geophysics majors went to the chair of undergraduate affairs in the math department at the time, Professor Rainer Sachs, and asked what we should do given that we needed to take the class in order to graduate the following semester. Ray asked to see the book we had been assigned, which was Eugene Butkov's (1968) *Mathematical Physics*, and he proposed the following: Each week, we would read on our own a chapter and work problems given at the end of the chapter that Ray would assign. We then would go to Ray's office once a week where he had each of us work the problems on his chalkboard in front of him. Only after we had done all that would Ray occasionally go to the chalkboard himself and share his own insights on the topics being covered. I loved the easy pedagogic style of Butkov's book and Ray's approach of having us self-teach ourselves knowing that each week we would be under his scrutiny solving problems at the chalkboard. Without this fabulous undergraduate experience, I am not sure I would have gone into theoretical research. If a student out there could have a similar experience of self-teaching themselves continuum physics and associated mathematical methods using the present book, I would be thrilled.

I would like to thank my wife Laurence and my now grown boys Thomas and Samuel who were patient over the many years, while I not only ran a business outside of academics but tried my best to continue doing innovative research with my colleagues and students and to work, at least occasionally, on the content of this book. I would also like to thank Kathy Hargreaves at Dangerous Curve LLC who took my copious hand-written class notes many years ago and converted them into Latex for me, which was a key step in this book coming into existence. I thank a long-time collaborator and friend Eirik Flekkøy for having read and commented on portions of Chapter 7 and Ben Gilbert for providing feedback on Chapter 6. Last, a special thank you to Raymond Jeanloz who provided greatly appreciated and insightful suggestions throughout the entire book.

Part I

Continuum Physics

1
An Introduction to Tensor Calculus

1.1 Overall Context

We will be developing the laws of continuum physics throughout the first part of this book. We will do so in an uncommon but pedagogic way by starting with the laws that describe the discrete movement of individual atoms and then summing over the molecular dynamics. The emergent continuum laws so obtained come in the form of partial-differential equations (PDEs) that determine how fields are changing in time at each point in space based on how the fields are varying in space in the immediate neighborhood of that point. The second part of the book will treat, for the most part, mathematical techniques for analytically solving the PDEs with a heavy dose of Fourier analysis and contour-integration methods. Students learning this material from me over the years have reported that the first part where the continuum rules are established is more difficult for them compared to the second part where math problems are solved. Perhaps this is analogous to how building a toy model is more challenging than playing with the toy once it is built.

In continuum mechanics in particular, the key field representing the underlying molecular-force interactions is a tensor (the "stress tensor") and across all of continuum physics, the material properties and constitutive laws are often only describable using tensors. In short, it is impossible to learn continuum physics properly without a solid foundation in tensors and tensor calculus, and this is why we begin the book with this foundational topic. What you learn in this first chapter, especially the tensor-calculus product-rule identities of Section 1.7, will be used at every step throughout our development of the rules of continuum physics. Fortunately, tensors and tensor calculus are a natural, even effortless, extension from the concepts of vectors and vector calculus that I assume you are familiar with. This chapter reviews the various types of spatial derivatives employed in continuum physics (the gradient, divergence, and curl), while allowing these spatial derivatives to act upon tensor fields, which is assumed to be new to the reader. It is my experience that even more senior research scientists can benefit from this chapter's survey of tensor calculus in preparation for the derivations in all the chapters that follow.

Throughout the book, our focus is on analytical understanding of the physics and mathematics and this involves pencil and paper work. You need to develop confidence in pushing the symbols around the page as you handle and ultimately solve the PDEs we will be deriving. The goal is to build intuition and hands-on familiarity with the physical processes being discussed. Simulating macroscopic experiments in the real world, often performed in complicated heterogeneous bodies of matter with irregular boundaries, is called the *forward problem* and usually needs to be performed numerically because analytical solutions of the governing equations are not possible. Recording the material response at places within a body during various types of experiments and minimizing the difference between the recorded data and simulations of the data with the goal of determining the physical properties throughout the body is called the *inverse problem* and is also a numerical exercise in nearly all cases. But we will not be addressing in this book numerical aspects of the forward and inverse problems posed in macroscopic bodies. Instead we content ourselves with first developing the PDEs that control basic processes of interest across many physical-science disciplines (Part I) and then solving simplified forms of the equations in simple geometries where analytical results are possible so that your physical intuition about the physics can be developed (Part II).

1.2 Some Actors

Any physical quantity continuously distributed over the space of some region is called a *field*. Continuum physics involves the study of fields. Fields can be *scalars, vectors,* or *tensors*.

Scalar Fields: A field quantity that has no intrinsic direction is called a scalar field. Examples include temperature, pressure, and various types of densities. In the nomenclature of tensors, a scalar can be called a zeroth-order tensor.

Vector Fields: A field quantity that has a direction associated with it is called a vector field. Examples include electric fields, fluid velocity, and gravitational acceleration. Vector fields are represented at each point in space by an arrow whose length denotes the amplitude of the vector field at that point. In the nomenclature of tensors, a vector can be called a first-order tensor. Vector fields as depicted in Fig. 1.1 can be written analytically in different ways:

$$\begin{aligned} \boldsymbol{r} &\mathrel{\widehat{=}} \text{position vector used to identify points in space} \\ &= x_1\hat{\boldsymbol{x}}_1 + x_2\hat{\boldsymbol{x}}_2 + x_3\hat{\boldsymbol{x}}_3 = x\hat{\boldsymbol{x}} + y\hat{\boldsymbol{y}} + z\hat{\boldsymbol{z}} \\ &= (x_1, x_2, x_3) = x_i\hat{\boldsymbol{x}}_i \quad \text{(summation over repeated indices)}, \end{aligned} \tag{1.1}$$

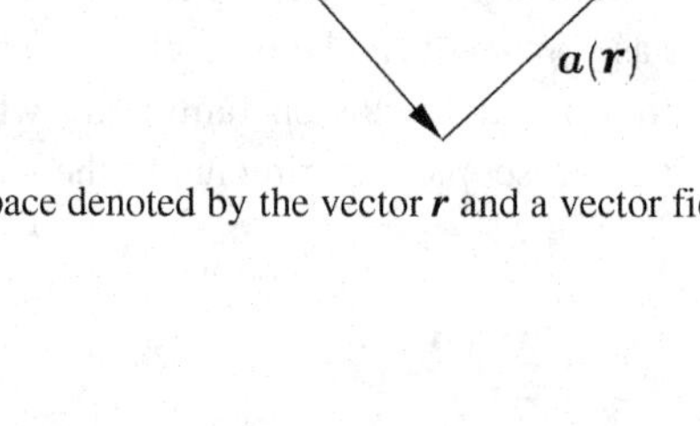

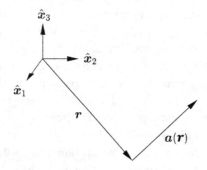

Figure 1.1 Points in space denoted by the vector $\boldsymbol{r}$ and a vector field $\boldsymbol{a}(\boldsymbol{r})$ at each point $\boldsymbol{r}$.

$$\begin{aligned}\boldsymbol{a}(\boldsymbol{r}) = \boldsymbol{a}(x_1, x_2, x_3) &\hat{=} \text{vector field defined at each point } \boldsymbol{r} \\ &= a_1\hat{\boldsymbol{x}}_1 + a_2\hat{\boldsymbol{x}}_2 + a_3\hat{\boldsymbol{x}}_3 = a_x\hat{\boldsymbol{x}} + a_y\hat{\boldsymbol{y}} + a_z\hat{\boldsymbol{z}} \\ &= (a_1, a_2, a_3) = a_i\hat{\boldsymbol{x}}_i.\end{aligned} \tag{1.2}$$

The caret symbol ^ placed above a vector means that vector is unitless and has an amplitude of 1, that is, $\hat{\boldsymbol{a}} \hat{=} \boldsymbol{a}/|\boldsymbol{a}|$, where $|\boldsymbol{a}| = \sqrt{(a_1^2 + a_2^2 + a_3^2)}$ denotes the amplitude of vector $\boldsymbol{a}$.

> IMPORTANT: Whenever an index appears twice in an expression, you always sum over that index. The index that is summed over is sometimes called a *dummy index* because the index does not survive the summation and could be given any name. For example, we have $a_ib_i = a_jb_j = a_nb_n = \sum_{n=1}^3 a_nb_n = \sum_{j=1}^3 a_jb_j = \sum_{i=1}^3 a_ib_i = a_1b_1 + a_2b_2 + a_3b_3$, where the i, j and n are examples of the dummy indices that we sum over. The summation over repeated indices in vectorial and tensorial expressions is called the *Einstein summation convention* and simply saves us from having to write the summation sign over and over.

Another type of vector is the vector operator that we call the *gradient operator* that is defined

$$\nabla = \hat{\boldsymbol{x}}\frac{\partial}{\partial x} + \hat{\boldsymbol{y}}\frac{\partial}{\partial y} + \hat{\boldsymbol{z}}\frac{\partial}{\partial z} \hat{=} \text{gradient operator (a vector operator).} \tag{1.3}$$

For example, if $\psi(\boldsymbol{r}) = \psi(x, y, z)$ is some scalar field, then the gradient of ψ is

$$\nabla\psi = \hat{\boldsymbol{x}}\frac{\partial\psi}{\partial x} + \hat{\boldsymbol{y}}\frac{\partial\psi}{\partial y} + \hat{\boldsymbol{z}}\frac{\partial\psi}{\partial z}. \tag{1.4}$$

and is a vector that we can also write $\nabla\psi = \hat{\boldsymbol{x}}_i\partial\psi/\partial x_i$ using the summation convention. The gradient vector $\nabla\psi$ is oriented in the direction that the scalar field ψ is increasing the most rapidly and the amplitude $|\nabla\psi|$ gives the rate of that maximum increase.

For a vector field $\boldsymbol{a} = a_i\hat{\boldsymbol{x}}_i$, we call the a_i the *scalar components of the vector* and call the unit vectors $\hat{\boldsymbol{x}}_i$ in each direction i the *base vectors*. Note that a vector at some point in space is an arrow with a length and is completely independent of the coordinate system we use to describe it. So $\boldsymbol{a}$ happily exists as the same arrow and does not change if we rotate the coordinate system. Note, however, that the scalar components of the vector a_i will change as we rotate our coordinate system (alter the orientations of the base vectors) or switch to another coordinate system such as cylindrical coordinates.

Some authors put an arrow above a symbol to denote that it is a vector field, i.e., $\vec{a}$. When working in typed text, we always use a bold-face symbol to denote a vector, i.e., $\boldsymbol{a}$. When writing by hand, we have elected not to use an arrow over a symbol but instead use a squiggly underscore, i.e., $\underset{\sim}{a}$.

You are free to develop your own vectorial and tensorial notation when writing by hand but using squiggly underscores for vectors and tensors has served me well over a long career. Note that if you do not use some type of notation to denote that a symbol is a vector or tensor, you will be in a constant state of confusion when manipulating the fields of continuum physics.

Second-Order Tensor Fields: A field quantity that acts as the proportionality between two vector fields that are related to each other at each point in space is called a *second-order tensor* field (can equivalently be called a "second-rank" tensor). Another word that is synonymous to second-order tensor is *dyad* or *dyadic*. We write a second-order tensor field as

$$\begin{aligned}
\boldsymbol{T}(\boldsymbol{r}) &\widehat{=} \text{a second-order tensor field defined at each point } \boldsymbol{r} \\
&= T_{xx}\hat{\boldsymbol{x}}\hat{\boldsymbol{x}} + T_{xy}\hat{\boldsymbol{x}}\hat{\boldsymbol{y}} + T_{xz}\hat{\boldsymbol{x}}\hat{\boldsymbol{z}} \\
&\quad + T_{yx}\hat{\boldsymbol{y}}\hat{\boldsymbol{x}} + T_{yy}\hat{\boldsymbol{y}}\hat{\boldsymbol{y}} + T_{yz}\hat{\boldsymbol{y}}\hat{\boldsymbol{z}} \qquad (1.5) \\
&\quad + T_{zx}\hat{\boldsymbol{z}}\hat{\boldsymbol{x}} + T_{zy}\hat{\boldsymbol{z}}\hat{\boldsymbol{y}} + T_{zz}\hat{\boldsymbol{z}}\hat{\boldsymbol{z}} \\
&= T_{ij}\,\hat{\boldsymbol{x}}_i\hat{\boldsymbol{x}}_j \quad \text{(summation over repeated indices assumed).} \qquad (1.6)
\end{aligned}$$

Just like the vector $\boldsymbol{a} = a_x\hat{\boldsymbol{x}} + a_y\hat{\boldsymbol{y}} + a_z\hat{\boldsymbol{z}}$ is the sum of three vectors in the three coordinate directions, so the second-order tensor $\boldsymbol{T}$ is the sum of nine second-order tensors as made explicit in Eq. (1.5). The T_{ij} are the scalar components of the second-order tensor and the various base vector pairs $\hat{\boldsymbol{x}}_i\hat{\boldsymbol{x}}_j$ for the various possible i and j are what we call second-order tensors. And just like we can write a vector in the array format $\boldsymbol{a} = (a_x, a_y, a_z)$, so can we write a second-order tensor as

$$\boldsymbol{T} = \begin{pmatrix} T_{xx} & T_{xy} & T_{xz} \\ T_{yx} & T_{yy} & T_{yz} \\ T_{zx} & T_{zy} & T_{zz} \end{pmatrix}. \qquad (1.7)$$

So a second-order tensor can be represented as a matrix. Much of what you learned about matrices in linear algebra applies to how we use second-order tensors. The main difference between a matrix and a second-order tensor is that although a matrix may have any dimension ($N \times M$) and corresponds to any proportionality between an M and N dimensioned vector (first-order) array, a second-order tensor is a field quantity distributed through three-dimensional space and is always a (3×3) matrix in three-dimensional space and is a physical field that is always the proportionality between two vector fields that each have clear physical meaning as will be demonstrated repeatedly throughout this book.

An example of a second-order tensor is two vector fields that sit side by side to each other in an expression without a scalar or vector product (that are defined in an upcoming section) between them:

$$\begin{aligned}
\boldsymbol{ab} &= \left(a_x\hat{\boldsymbol{x}} + a_y\hat{\boldsymbol{y}} + a_z\hat{\boldsymbol{z}}\right)\left(b_x\hat{\boldsymbol{x}} + b_y\hat{\boldsymbol{y}} + b_z\hat{\boldsymbol{z}}\right) \qquad (1.8) \\
&= a_xb_x\,\hat{\boldsymbol{x}}\hat{\boldsymbol{x}} + a_xb_y\,\hat{\boldsymbol{x}}\hat{\boldsymbol{y}} + a_xb_z\,\hat{\boldsymbol{x}}\hat{\boldsymbol{z}} \\
&\quad + a_yb_x\,\hat{\boldsymbol{y}}\hat{\boldsymbol{x}} + a_yb_y\,\hat{\boldsymbol{y}}\hat{\boldsymbol{y}} + a_yb_z\,\hat{\boldsymbol{y}}\hat{\boldsymbol{z}} \qquad (1.9) \\
&\quad + a_zb_x\,\hat{\boldsymbol{z}}\hat{\boldsymbol{x}} + a_zb_y\,\hat{\boldsymbol{z}}\hat{\boldsymbol{y}} + a_zb_z\,\hat{\boldsymbol{z}}\hat{\boldsymbol{z}} \\
&= a_ib_j\hat{\boldsymbol{x}}_i\hat{\boldsymbol{x}}_j \quad \text{(summation over repeated indices as always).} \qquad (1.10)
\end{aligned}$$

It is convenient to construct the 3×3 matrix representing $\boldsymbol{ab}$ as the matrix product between $\boldsymbol{a}$ written as a 3×1 array and $\boldsymbol{b}$ written as a 1×3 array, which corresponds to the multiplications of Eq. (1.8):

$$ab = \begin{pmatrix} a_x \\ a_y \\ a_z \end{pmatrix} \begin{pmatrix} b_x, b_y, b_z \end{pmatrix} = \begin{pmatrix} a_xb_x & a_xb_y & a_xb_z \\ a_yb_x & a_yb_y & a_yb_z \\ a_zb_x & a_zb_y & a_zb_z \end{pmatrix}. \tag{1.11}$$

When two vectors sit next to each other to form a second-order tensor, it is common to call that product the *tensor product* or *dyadic product*, even if we will not employ these words outside of this paragraph. Some authors in the engineering literature introduce a special symbol $\otimes$ to denote the tensor product, i.e., $\boldsymbol{a} \otimes \boldsymbol{b} \,\widehat{=}\, \boldsymbol{ab}$. So for the tensor product between the base vectors in any second-order or higher-order tensorial expression, these authors write, for example, $\hat{\boldsymbol{x}} \otimes \hat{\boldsymbol{y}}$ to represent what most authors write more simply as $\hat{\boldsymbol{x}}\hat{\boldsymbol{y}}$. The extra symbol $\otimes$ uses space on the page without providing any needed clarification, which is why we do not use it.

Another example of a second-order tensor is the gradient of a vector field. Working in Cartesian coordinates where derivatives of base vectors are zero, we have

$$\nabla \boldsymbol{a} = \left(\hat{\boldsymbol{x}}\frac{\partial}{\partial x} + \hat{\boldsymbol{y}}\frac{\partial}{\partial y} + \hat{\boldsymbol{z}}\frac{\partial}{\partial z}\right)(a_x\hat{\boldsymbol{x}} + a_y\hat{\boldsymbol{y}} + a_z\hat{\boldsymbol{z}}) \tag{1.12}$$

$$= \frac{\partial a_x}{\partial x}\hat{\boldsymbol{x}}\hat{\boldsymbol{x}} + \frac{\partial a_y}{\partial x}\hat{\boldsymbol{x}}\hat{\boldsymbol{y}} + \ldots = \frac{\partial a_j}{\partial x_i}\hat{\boldsymbol{x}}_i\hat{\boldsymbol{x}}_j, \tag{1.13}$$

which can again be written in array form as

$$\nabla \boldsymbol{a} = \begin{pmatrix} \partial_x \\ \partial_y \\ \partial_z \end{pmatrix} \begin{pmatrix} a_x, a_y, a_z \end{pmatrix} = \begin{pmatrix} \frac{\partial a_x}{\partial x} & \frac{\partial a_y}{\partial x} & \frac{\partial a_z}{\partial x} \\ \frac{\partial a_x}{\partial y} & \frac{\partial a_y}{\partial y} & \frac{\partial a_z}{\partial y} \\ \frac{\partial a_x}{\partial z} & \frac{\partial a_y}{\partial z} & \frac{\partial a_z}{\partial z} \end{pmatrix}. \tag{1.14}$$

We emphasize that we get these simple expressions for the components of $\nabla \boldsymbol{a}$ only in Cartesian coordinates where derivatives of the base vectors are zero because the base vectors in Cartesians are spatially uniform. When the components of $\nabla \boldsymbol{a}$ are written out in curvilinear coordinates (cylindrical, spherical, etc.) in which the base vectors themselves vary with position in space and thus have nonzero spatial derivatives, the result of performing $\nabla \boldsymbol{a}$ is more complicated. The expressions for $\nabla \boldsymbol{a}$ in arbitrary orthogonal curvilinear coordinates, cylindrical coordinates, and spherical coordinates are all given in Section 1.8.6.

Just like with a matrix, we can talk about the transpose of a second-order tensor $\boldsymbol{T} = T_{ij}\hat{\boldsymbol{x}}_i\hat{\boldsymbol{x}}_j$ and write

$$\begin{aligned} \boldsymbol{T}^T &\,\widehat{=}\, \text{the transpose of } \boldsymbol{T} \\ &= T_{ij}\hat{\boldsymbol{x}}_j\hat{\boldsymbol{x}}_i = T_{ji}\hat{\boldsymbol{x}}_i\hat{\boldsymbol{x}}_j. \end{aligned} \tag{1.15}$$

Thus to perform the transpose, we can either flip the indices on the scalar components $T_{ij} \rightarrow T_{ji}$ of the tensor or flip the position of the two base vectors as they sit side by side.

Note that like with a vector, a tensor $\boldsymbol{T}$ exists at a point and is independent of the coordinate system. If we rotate or change coordinate systems, $\boldsymbol{T}$ does not change. However, the scalar components of the tensor T_{ij} will change as we change the coordinates because the base vectors $\hat{\boldsymbol{x}}_i$ are changing. When working in typed text, we always denote a second-order tensor with bold type. When we write a second-order tensor by hand, we use two squiggly underscores $\boldsymbol{T} = \underset{\approx}{T}$.

Higher-Order Tensor Fields: The generalization to higher-order tensors is straightforward. A third-order tensor is written

$$ {}_3\boldsymbol{P} = P_{ijk}\hat{\boldsymbol{x}}_i\hat{\boldsymbol{x}}_j\hat{\boldsymbol{x}}_k \tag{1.16}$$

a fourth-order tensor as

$$ {}_4\boldsymbol{Q} = Q_{ijkl}\hat{\boldsymbol{x}}_i\hat{\boldsymbol{x}}_j\hat{\boldsymbol{x}}_k\hat{\boldsymbol{x}}_l \tag{1.17}$$

and so on for still higher-order tensors. Summation over each index is again assumed.

If, for example, a second-order tensor $\boldsymbol{A}$ happens to sit next to two vectors $\boldsymbol{a}$ and $\boldsymbol{b}$ we would have the fourth-order tensor

$$ \boldsymbol{Aab} = A_{ij}a_k b_l \hat{\boldsymbol{x}}_i\hat{\boldsymbol{x}}_j\hat{\boldsymbol{x}}_k\hat{\boldsymbol{x}}_l. \tag{1.18}$$

In general, we have $\boldsymbol{Aab} \neq \boldsymbol{Aba} \neq \boldsymbol{aAb} \neq \boldsymbol{bAa} \neq \boldsymbol{abA} \neq \boldsymbol{baA}$, so the order, from left to right, in which tensorial expressions sit next to each other to form higher-order tensors is very important.

The transpose of higher-order tensors must be specified by the way in which the base vectors are moved around relative to each other in the desired transpose operation. So, for example, for the fourth-order tensor ${}_4\boldsymbol{Q} = Q_{ijkl}\hat{\boldsymbol{x}}_i\hat{\boldsymbol{x}}_j\hat{\boldsymbol{x}}_k\hat{\boldsymbol{x}}_l$, we can define transpose operations such as

$$ {}_4\boldsymbol{Q}^{\overset{T}{2134}} = Q_{ijkl}\hat{\boldsymbol{x}}_j\hat{\boldsymbol{x}}_i\hat{\boldsymbol{x}}_k\hat{\boldsymbol{x}}_l = Q_{jikl}\hat{\boldsymbol{x}}_i\hat{\boldsymbol{x}}_j\hat{\boldsymbol{x}}_k\hat{\boldsymbol{x}}_l \tag{1.19}$$

$$ {}_4\boldsymbol{Q}^{\overset{T}{1243}} = Q_{ijkl}\hat{\boldsymbol{x}}_i\hat{\boldsymbol{x}}_j\hat{\boldsymbol{x}}_l\hat{\boldsymbol{x}}_k = Q_{ijlk}\hat{\boldsymbol{x}}_i\hat{\boldsymbol{x}}_j\hat{\boldsymbol{x}}_k\hat{\boldsymbol{x}}_l \tag{1.20}$$

$$ {}_4\boldsymbol{Q}^{\overset{T}{3412}} = Q_{ijkl}\hat{\boldsymbol{x}}_k\hat{\boldsymbol{x}}_l\hat{\boldsymbol{x}}_i\hat{\boldsymbol{x}}_j = Q_{klij}\hat{\boldsymbol{x}}_i\hat{\boldsymbol{x}}_j\hat{\boldsymbol{x}}_k\hat{\boldsymbol{x}}_l \tag{1.21}$$

and so on. There are $4! - 1 = 23$ such transposes for a fourth-order tensor, that is, there are $4! - 1$ different ways of placing the four base vectors next to each other that are different than in the nontransposed form. Similarly, an nth-order tensor would have $n! - 1$ different possible transpose operations; so a second-order tensor has only one way to write the transpose.

We write an nth-order tensor ${}_n\boldsymbol{Q}$ by hand as ${}_n\underset{\approx}{Q}$ for $n > 2$.

1.3 Some Acts

In tensor calculus, just like in vector calculus, we define two commonly employed types of products between vectors and tensors called the *scalar product* and the *vector product*.

Scalar Products: A scalar product between two vector fields $\boldsymbol{a}$ and $\boldsymbol{b}$ that have an angle θ between them at each point as depicted in Fig. 1.2 is the product of the amplitude of the two vectors after one of the two vectors is projected into the direction of the other vector. The scalar product $\boldsymbol{a}\cdot\boldsymbol{b}$ between two vectors is a scalar and is denoted with a dot sitting between the vectors and is defined by the following rule

$$\boldsymbol{a}\cdot\boldsymbol{b} = |\boldsymbol{a}||\boldsymbol{b}|\cos\theta, \quad \text{where } |\boldsymbol{a}| = \sqrt{a_x^2 + a_y^2 + a_z^2}. \tag{1.22}$$

So $\boldsymbol{a}\cdot\boldsymbol{b} = 0$ if $\boldsymbol{a}\perp\boldsymbol{b}$, which means that $\hat{\boldsymbol{x}}\cdot\hat{\boldsymbol{y}} = 0$ and $\hat{\boldsymbol{x}}\cdot\hat{\boldsymbol{z}} = 0$, but $\hat{\boldsymbol{x}}\cdot\hat{\boldsymbol{x}} = 1$, etc. Using these rules, we thus have

$$\begin{aligned}\boldsymbol{a}\cdot\boldsymbol{b} &= \left[a_x\hat{\boldsymbol{x}} + a_y\hat{\boldsymbol{y}} + a_z\hat{\boldsymbol{z}}\right]\cdot\left[b_x\hat{\boldsymbol{x}} + b_y\hat{\boldsymbol{y}} + b_z\hat{\boldsymbol{z}}\right]\\ &= a_xb_x + a_yb_y + a_zb_z \\ &= a_ib_i.\end{aligned} \tag{1.23}$$

The scalar product is also called the *dot product* or the *inner product*.

What if vector field $\boldsymbol{a}$ is related to vector field $\boldsymbol{b}$ at some point in space? How do you obtain $\boldsymbol{a}$ given $\boldsymbol{b}$? That is what a second-order tensor such as $\boldsymbol{T} = T_{ij}\hat{\boldsymbol{x}}_i\hat{\boldsymbol{x}}_j$ does for us once we introduce the scalar product:

$$\begin{aligned}\boldsymbol{a} &= \boldsymbol{T}\cdot\boldsymbol{b}\\ &= \begin{pmatrix} & T_{xx}\hat{\boldsymbol{x}}\hat{\boldsymbol{x}} & + & T_{xy}\hat{\boldsymbol{x}}\hat{\boldsymbol{y}} & + & T_{xz}\hat{\boldsymbol{x}}\hat{\boldsymbol{z}} \\ + & T_{yx}\hat{\boldsymbol{y}}\hat{\boldsymbol{x}} & + & T_{yy}\hat{\boldsymbol{y}}\hat{\boldsymbol{y}} & + & T_{yz}\hat{\boldsymbol{y}}\hat{\boldsymbol{z}} \\ + & T_{zx}\hat{\boldsymbol{z}}\hat{\boldsymbol{x}} & + & T_{zy}\hat{\boldsymbol{z}}\hat{\boldsymbol{y}} & + & T_{zz}\hat{\boldsymbol{z}}\hat{\boldsymbol{z}} \end{pmatrix}\cdot\left(b_x\hat{\boldsymbol{x}} + b_y\hat{\boldsymbol{y}} + b_z\hat{\boldsymbol{z}}\right) \qquad (1.24)\\ &= \left(T_{xx}\,b_x + T_{xy}\,b_y + T_{xz}\,b_z\right)\hat{\boldsymbol{x}} + \left(T_{yx}\,b_x + T_{yy}\,b_y + T_{yz}\,b_z\right)\hat{\boldsymbol{y}}\\ &\quad + \left(T_{zx}\,b_x + T_{zy}\,b_y + T_{zz}\,b_z\right)\hat{\boldsymbol{z}} \qquad (1.25)\\ &= \left(T_{ij}\hat{\boldsymbol{x}}_i\hat{\boldsymbol{x}}_j\right)\cdot\left(b_k\hat{\boldsymbol{x}}_k\right) = T_{ij}b_k\hat{\boldsymbol{x}}_i\left(\hat{\boldsymbol{x}}_j\cdot\hat{\boldsymbol{x}}_k\right) = T_{ij}b_j\hat{\boldsymbol{x}}_i, \qquad (1.26)\end{aligned}$$

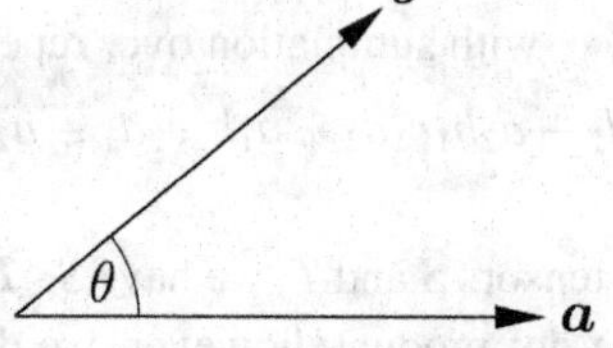

Figure 1.2 Two vectors $\boldsymbol{a}$ and $\boldsymbol{b}$ with an angle θ between them.

where in the last line we used that $\hat{\boldsymbol{x}}_j \cdot \hat{\boldsymbol{x}}_k$ requires $k=j$. Using the familiar matrix multiplication for the scalar product, this can be written

$$\begin{bmatrix} a_x \\ a_y \\ a_z \end{bmatrix} = \begin{bmatrix} T_{xx} & T_{xy} & T_{xz} \\ T_{yx} & T_{yy} & T_{yz} \\ T_{zx} & T_{zy} & T_{zz} \end{bmatrix} \begin{bmatrix} b_x \\ b_y \\ b_z \end{bmatrix}. \tag{1.27}$$

> IMPORTANT: Second-order tensor fields are always maps between two vectors that are physically related to each other at a point $\boldsymbol{r}$. You cannot visualize directly a second-order tensor (or higher-order tensors) using your 3D sense of perception. But you can picture in your mind's eye the two vectors (arrows) that are related to each other at a point and thus imagine there is a mapping (second-order tensor) that takes the one vector to the other.

Note that throughout this entire book, we work exclusively in orthogonal coordinates where dot products are zero between the different base vectors of a coordinate system. It is possible, for example, in crystallography, to want to work in *skew* coordinate systems where the base vectors are not orthogonal to each other. Complicating ideas such as covariant and contravariant base vectors arise and the reader interested in tensor calculus in skew coordinates is directed toward specialized texts (e.g., Lebedev et al., 2010).

We can also speak of the *double-dot product* : between tensors, that in this book is defined

$$\boldsymbol{ab} : \boldsymbol{cd} = (\boldsymbol{a} \cdot \boldsymbol{d})(\boldsymbol{b} \cdot \boldsymbol{c}) \tag{1.28}$$

$$= \left(a_i\hat{\boldsymbol{x}}_i \cdot d_j\hat{\boldsymbol{x}}_j\right)\left(b_k\hat{\boldsymbol{x}}_k \cdot c_l\hat{\boldsymbol{x}}_l\right) \tag{1.29}$$

$$= a_i d_j b_k c_l \left(\hat{\boldsymbol{x}}_i \cdot \hat{\boldsymbol{x}}_j\right)\left(\hat{\boldsymbol{x}}_k \cdot \hat{\boldsymbol{x}}_l\right). \tag{1.30}$$

Other authors define the double-dot product as $(\boldsymbol{a} \cdot \boldsymbol{c})(\boldsymbol{b} \cdot \boldsymbol{d})$. Either definition works if used consistently. We choose the convention of Eq. (1.28) so that when you see the : between vectors or base vectors, you perform the first dot product between the vectors that reside immediately on either side of the dot symbol and once that is done, perform the second dot product between the remaining vectors. This convention is the easiest to remember and is highly recommended. In writing out a tensorial expression such as given in Eq. (1.29), always use a different index for each base vector and associated coefficient. Because of the nature of the scalar product in orthogonal coordinate systems, we thus have $l=k$ and $j=i$ in Eq. (1.30) or

$$\boldsymbol{ab} : \boldsymbol{cd} = a_i b_k c_k d_i \quad \text{with summation over repeated indices} \tag{1.31}$$

$$= a_1 b_1 c_1 d_1 + a_2 b_1 c_1 d_2 + a_1 b_2 c_2 d_1 + a_2 b_2 c_2 d_2 \quad \text{in 2D.} \tag{1.32}$$

Note that for two second-order tensors $\boldsymbol{S}$ and $\boldsymbol{T}$, we have $\boldsymbol{S} \cdot \boldsymbol{T} = \left(\boldsymbol{T}^T \cdot \boldsymbol{S}^T\right)^T$ and that $\boldsymbol{S} \cdot \boldsymbol{T} \neq \boldsymbol{T} \cdot \boldsymbol{S}$ in general. For the double-dot product, however, we do have $\boldsymbol{S} : \boldsymbol{T} = \boldsymbol{T} : \boldsymbol{S}$ for any $\boldsymbol{S}$ and $\boldsymbol{T}$, where

$$\boldsymbol{S}:\boldsymbol{T} = S_{ij}\,\hat{\boldsymbol{x}}_i\hat{\boldsymbol{x}}_j : T_{kl}\,\hat{\boldsymbol{x}}_k\hat{\boldsymbol{x}}_l \tag{1.33}$$

$$= S_{ij}T_{kl}\left(\hat{\boldsymbol{x}}_j \cdot \hat{\boldsymbol{x}}_k\right)\left(\hat{\boldsymbol{x}}_i \cdot \hat{\boldsymbol{x}}_l\right) \quad \text{which requires } l = i \text{ and } k = j \tag{1.34}$$

$$= S_{ij}T_{ji} \quad \text{with summation over repeated indices.} \tag{1.35}$$

Renaming the dummy indices gives $\boldsymbol{S}:\boldsymbol{T} = S_{ij}T_{ji} = S_{ji}T_{ij} = T_{ij}S_{ji} = \boldsymbol{T}:\boldsymbol{S}$.

The second-order *identity tensor* $\boldsymbol{I}$ is defined $\boldsymbol{I} = \delta_{ij}\,\hat{\boldsymbol{x}}_i\hat{\boldsymbol{x}}_j$, where the δ_{ij} are called the *Kronecker coefficients* and are defined

$$\delta_{ij} = \begin{cases} 0 & \text{if } i \neq j \\ 1 & \text{if } i = j \end{cases} \quad \text{so that} \quad \boldsymbol{I} = \begin{pmatrix} 1 & 0 & 0 \\ 0 & 1 & 0 \\ 0 & 0 & 1 \end{pmatrix}. \tag{1.36}$$

Upon summing over the indices, we have $\boldsymbol{I} = \hat{\boldsymbol{x}}_1\hat{\boldsymbol{x}}_1 + \hat{\boldsymbol{x}}_2\hat{\boldsymbol{x}}_2 + \hat{\boldsymbol{x}}_3\hat{\boldsymbol{x}}_3 = \hat{\boldsymbol{x}}\hat{\boldsymbol{x}} + \hat{\boldsymbol{y}}\hat{\boldsymbol{y}} + \hat{\boldsymbol{z}}\hat{\boldsymbol{z}}$. The identity tensor works as follows: $\boldsymbol{A}\cdot\boldsymbol{I} = \boldsymbol{I}\cdot\boldsymbol{A} = \boldsymbol{A}$ for any second-order tensor $\boldsymbol{A}$. We further have that if the position vector is written $\boldsymbol{r} = x_j\hat{\boldsymbol{x}}_j$ in Cartesian coordinates, then $\boldsymbol{I} = \nabla\boldsymbol{r} = (\partial x_j/\partial x_i)\hat{\boldsymbol{x}}_i\hat{\boldsymbol{x}}_j = \delta_{ij}\hat{\boldsymbol{x}}_i\hat{\boldsymbol{x}}_j$.

A double-dot product with the identity tensor results in $\boldsymbol{A}:\boldsymbol{I} = A_{ij}\,\delta_{ji} = A_{ii} = \text{tr}\,\{\boldsymbol{A}\} = A_{11} + A_{22} + A_{33}$, which is called the *trace* of second-order tensor $\boldsymbol{A}$. The trace is the sum of the second-order tensor components along the diagonal, for example, $\boldsymbol{I}:\boldsymbol{I} = 3$ (in 3D). The double-dot product between two second-order tensors is the trace of the scalar (matrix) product of the two tensors, that is, $\boldsymbol{A}:\boldsymbol{B} = A_{ij}B_{ji} = \text{tr}\,\{\boldsymbol{A}\cdot\boldsymbol{B}\} = \text{tr}\,\{\boldsymbol{B}\cdot\boldsymbol{A}\}$.

We can extend the number of dot products we take between two higher-order tensors to as many as desired. So the *triple-dot product* $\overset{3}{:}$ between, say, two third-order tensors ${}_3\boldsymbol{S}$ and ${}_3\boldsymbol{T}$ can be defined

$$\begin{aligned} {}_3\boldsymbol{S} \overset{3}{:} {}_3\boldsymbol{T} &= S_{ijk}\hat{\boldsymbol{x}}_i\hat{\boldsymbol{x}}_j\hat{\boldsymbol{x}}_k \overset{3}{:} T_{lmn}\hat{\boldsymbol{x}}_l\hat{\boldsymbol{x}}_m\hat{\boldsymbol{x}}_n \\ &= S_{ijk}T_{lmn}(\hat{\boldsymbol{x}}_k\cdot\hat{\boldsymbol{x}}_l)(\hat{\boldsymbol{x}}_j\cdot\hat{\boldsymbol{x}}_m)(\hat{\boldsymbol{x}}_i\cdot\hat{\boldsymbol{x}}_n), \end{aligned}$$

which tells us that $n = i$, $m = j$ and $l = k$ so that the triple-dot product between two third-order tensors comes out to be

$${}_3\boldsymbol{S} \overset{3}{:} {}_3\boldsymbol{T} = S_{ijk}T_{kji} = \text{tr}\,\{{}_3\boldsymbol{S} : {}_3\boldsymbol{T}\} \tag{1.37}$$

with summation over repeated indices. We can extend such notation and definition to still higher-order dot products between still higher-order tensors.

Note that each dot product removes two base vectors from a tensorial expression. So without writing anything out, we know that a tensorial expression like ${}_8\boldsymbol{A} \overset{5}{:} {}_6\boldsymbol{B}$ is a fourth-order tensor, that is, the eighth-order tensor ${}_8\boldsymbol{A}$ contributes 8 base vectors to this expression and the sixth-order tensor ${}_6\boldsymbol{B}$ contributes 6 more base vectors but the 5 dot products remove 10 of those base vectors so that the result is a fourth-order tensor. As practice, we can write this lengthy example out to give

$$\begin{aligned} &{}_8\boldsymbol{A} \overset{5}{:} {}_6\boldsymbol{B} = \\ &A_{ijklmnop}B_{qrstuv}\hat{\boldsymbol{x}}_i\hat{\boldsymbol{x}}_j\hat{\boldsymbol{x}}_k(\hat{\boldsymbol{x}}_p\cdot\hat{\boldsymbol{x}}_q)(\hat{\boldsymbol{x}}_o\cdot\hat{\boldsymbol{x}}_r)(\hat{\boldsymbol{x}}_n\cdot\hat{\boldsymbol{x}}_s)(\hat{\boldsymbol{x}}_m\cdot\hat{\boldsymbol{x}}_t)(\hat{\boldsymbol{x}}_l\cdot\hat{\boldsymbol{x}}_u)\hat{\boldsymbol{x}}_v, \end{aligned} \tag{1.38}$$

which tells us that $q = p$, $r = o$, $s = n$, $t = m$, and $u = l$ so that we obtain the fourth-order tensor

$$ {}_8\boldsymbol{A} \overset{5}{:} {}_6\boldsymbol{B} = A_{ijklmnop} B_{ponmlv} \hat{\boldsymbol{x}}_i \hat{\boldsymbol{x}}_j \hat{\boldsymbol{x}}_k \hat{\boldsymbol{x}}_v \tag{1.39} $$

with summation implied over all the dummy (i.e., repeated) indices. In our development of continuum physics, we will not need to work with tensors higher than the sixth order or with more than three dot products between two tensors.

Vector Products: A vector product between two vectors $\boldsymbol{a}$ and $\boldsymbol{b}$ is a vector that is perpendicular to the two vectors as depicted in Fig. 1.3 and that has an amplitude equal to the area of the parallelogram formed with the two vectors as sides. We have

$$ \boldsymbol{c} = \boldsymbol{a} \times \boldsymbol{b}; \quad \boldsymbol{c} \perp \text{ to both } \boldsymbol{a} \text{ and } \boldsymbol{b} \tag{1.40} $$

$$ |\boldsymbol{c}| = |\boldsymbol{a}||\boldsymbol{b}| \sin\theta \tag{1.41} $$

$$ \boldsymbol{a} \times \boldsymbol{b} = 0 \quad \text{if } \boldsymbol{a} \,||\, \boldsymbol{b}. \tag{1.42} $$

Use the right-hand rule to determine the sense of $\boldsymbol{a} \times \boldsymbol{b}$. Note that $\boldsymbol{a} \times \boldsymbol{b} = -\boldsymbol{b} \times \boldsymbol{a}$. We can thus obtain

$$ \hat{\boldsymbol{x}} \times \hat{\boldsymbol{y}} = \hat{\boldsymbol{z}}; \quad \hat{\boldsymbol{x}} \times \hat{\boldsymbol{z}} = -\hat{\boldsymbol{y}}; \quad \hat{\boldsymbol{x}} \times \hat{\boldsymbol{x}} = 0 \tag{1.43} $$

and so forth for all the vector products between all base vectors. Using these rules, we can write

$$ \begin{aligned} \boldsymbol{a} \times \boldsymbol{b} &= \left(a_x\hat{\boldsymbol{x}} + a_y\hat{\boldsymbol{y}} + a_z\hat{\boldsymbol{z}}\right) \times \left(b_x\hat{\boldsymbol{x}} + b_y\hat{\boldsymbol{y}} + b_z\hat{\boldsymbol{z}}\right) \\ &= a_x b_y\, \hat{\boldsymbol{z}} - a_x b_z\, \hat{\boldsymbol{y}} - a_y b_x\, \hat{\boldsymbol{z}} + a_y b_z\, \hat{\boldsymbol{x}} + a_z b_x\, \hat{\boldsymbol{y}} - a_z b_y\, \hat{\boldsymbol{x}} \\ &= \left(a_y b_z - a_z b_y\right) \hat{\boldsymbol{x}} - \left(a_x b_z - a_z b_x\right) \hat{\boldsymbol{y}} + \left(a_x b_y - a_y b_x\right) \hat{\boldsymbol{z}}. \end{aligned} \tag{1.44} $$

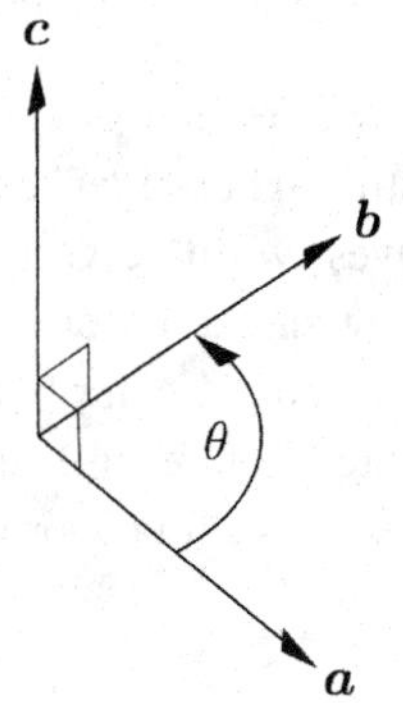

Figure 1.3 The vector product $\boldsymbol{c} = \boldsymbol{a} \times \boldsymbol{b}$ is a vector perpendicular to $\boldsymbol{a}$ and $\boldsymbol{b}$ as determined by the right-hand rule.

We can thus write the vector product using the matrix determinant in the following way:

$$\boldsymbol{a} \times \boldsymbol{b} = \begin{vmatrix} \hat{\boldsymbol{x}} & \hat{\boldsymbol{y}} & \hat{\boldsymbol{z}} \\ a_x & a_y & a_z \\ b_x & b_y & b_z \end{vmatrix}, \text{ where } |\cdots| \text{ denotes } \textit{taking the determinant}. \tag{1.45}$$

Note that the vector product is also called the *cross product*.

Doing Vector (Cross) Products with Scalar (Dot) Products: It is convenient for proving identities involving the cross product to write a cross product in a way that only involves dot products. To do so, we introduce the *alternating* or *permutation* or *antisymmetric* or *Levi–Civita* (these are all synonyms) third-order tensor ${}_3\boldsymbol{\epsilon}$ that is defined

$${}_3\boldsymbol{\epsilon} = \epsilon_{ijk}\,\hat{\boldsymbol{x}}_i\hat{\boldsymbol{x}}_j\hat{\boldsymbol{x}}_k \tag{1.46}$$

with scalar components called the *Levi–Civita coefficients* given by

$$\epsilon_{ijk} = \begin{cases} +1 & \text{for counterclockwise index positions: 123, 231, 312} \\ -1 & \text{for clockwise positions: 132, 321, 213} \\ 0 & \text{for every other index combination,} \end{cases} \tag{1.47}$$

where this cyclic ordering of the indices can be remembered using the mnemonic device of Fig. 1.4. To perform the cross product with the Levi–Civita tensor, we do the following

$$\begin{aligned} \boldsymbol{a} \times \boldsymbol{b} &= {}_3\boldsymbol{\epsilon} : \boldsymbol{ba} = -{}_3\boldsymbol{\epsilon} : \boldsymbol{ab} = -\epsilon_{ijk}\,\hat{\boldsymbol{x}}_i\,\hat{\boldsymbol{x}}_j\,\hat{\boldsymbol{x}}_k : a_l\,\hat{\boldsymbol{x}}_l\,b_m\,\hat{\boldsymbol{x}}_m \\ &= -\epsilon_{ijk}\,a_l b_m\,\hat{\boldsymbol{x}}_i\left(\hat{\boldsymbol{x}}_k \cdot \hat{\boldsymbol{x}}_l\right)\left(\hat{\boldsymbol{x}}_j \cdot \hat{\boldsymbol{x}}_m\right) \\ &= -\epsilon_{ijk}\,a_k b_j\,\hat{\boldsymbol{x}}_i \qquad (1.48) \\ &= \hat{\boldsymbol{x}}_1\left[a_2b_3 - a_3b_2\right] + \hat{\boldsymbol{x}}_2\left[a_3b_1 - a_1b_3\right] + \hat{\boldsymbol{x}}_3\left[a_1b_2 - a_2b_1\right] \qquad (1.49) \end{aligned}$$

which is identical to Eq. (1.44). Note that the double-dot product with the Levi–Civita tensor uses our double-dot convention of Eq. (1.28), which explains the minus sign in Eq. (1.48) in comparison to other authors who use the alternative but less-intuitive definition of the double-dot product.

If the Levi–Civita tensor is double dotted into the second-order tensor $\boldsymbol{A}$, we obtain the vector

$${}_3\boldsymbol{\epsilon} : \boldsymbol{A} = \hat{\boldsymbol{x}}_1\left(A_{32} - A_{23}\right) + \hat{\boldsymbol{x}}_2\left(A_{13} - A_{31}\right) + \hat{\boldsymbol{x}}_3\left(A_{21} - A_{12}\right). \tag{1.50}$$

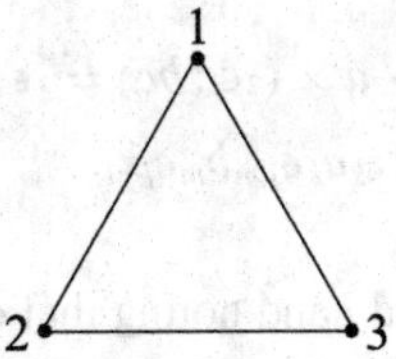

Figure 1.4 Mnemonic triangle showing the counterclockwise ordering of the three Levi–Civita indices

So if the tensor $\boldsymbol{A}$ is symmetric, then ${}_3\boldsymbol{\epsilon} : \boldsymbol{A} = 0$. Any second-order tensor can be separated into symmetric and antisymmetric portions $\boldsymbol{A} = \boldsymbol{A}^{(s)} + \boldsymbol{A}^{(a)}$, where

$$\boldsymbol{A}^{(s)} = \frac{1}{2}\left(\boldsymbol{A} + \boldsymbol{A}^T\right), \tag{1.51}$$

$$\boldsymbol{A}^{(a)} = \frac{1}{2}\left(\boldsymbol{A} - \boldsymbol{A}^T\right). \tag{1.52}$$

So the antisymmetric portion of any second-order tensor has zeroes along the diagonal and off-diagonal components that are "antisymmetric" $A_{ij}^{(a)} = -A_{ji}^{(a)}$. So the operation ${}_3\boldsymbol{\epsilon} : \boldsymbol{A} = {}_3\boldsymbol{\epsilon} : \boldsymbol{A}^{(a)}$ given by Eq. (1.50) involves only the antisymmetric portion of $\boldsymbol{A}$. This is because the Levi–Civita coefficients are anti-symmetric, that is, $\epsilon_{ijk} = -\epsilon_{ikj} = -\epsilon_{jik}$. You can prove as an end-of-chapter exercise that the double-dot product of any tensor that is antisymmetric in the last two base vectors with any tensor that is symmetric in the first two base vectors gives zero.

The antisymmetric nature of ${}_3\boldsymbol{\epsilon}$ can further be seen by dotting it into any vector $\boldsymbol{a}$ to obtain the second-order tensor

$${}_3\boldsymbol{\epsilon} \cdot \boldsymbol{a} = \epsilon_{ijk} a_k \hat{\boldsymbol{x}}_i \hat{\boldsymbol{x}}_j = \begin{pmatrix} 0 & a_3 & -a_2 \\ -a_3 & 0 & a_1 \\ a_2 & -a_1 & 0 \end{pmatrix}, \tag{1.53}$$

which is the definition of an antisymmetric second-order tensor. So the dot product of the third-order Levi–Civita tensor with any vector always produces an antisymmetric second-order tensor.

There is a useful identity involving the Levi–Civita coefficients

$$\epsilon_{ijk}\epsilon_{ilm} = \delta_{jl}\delta_{km} - \delta_{jm}\delta_{kl}, \tag{1.54}$$

where the δ_{ij} are the Kronecker coefficients. For the left-hand side to be nonzero, we need $j, k, l, m \neq i$ as well as both $j \neq k$ and $l \neq m$. If we take $j = l$ and $k = m$, then $\epsilon_{ijk}\epsilon_{ilm} = \epsilon_{ijk}\epsilon_{ijk} = 1$. If we take $j = m$ and $k = l$, then $\epsilon_{ijk}\epsilon_{ilm} = \epsilon_{ijk}\epsilon_{jik} = -1$. This suite of conditions is exactly satisfied by the right-hand side of Eq. (1.54). The identity of Eq. (1.54) allows us to prove relations that involve two cross products.

As an example, express the double-cross product between three vectors as

$$\boldsymbol{a} \times \boldsymbol{b} \times \boldsymbol{c} = -\boldsymbol{a} \times ({}_3\boldsymbol{\epsilon} : \boldsymbol{bc}) = {}_3\boldsymbol{\epsilon} : \boldsymbol{a}\,({}_3\boldsymbol{\epsilon} : \boldsymbol{bc}), \tag{1.55}$$

$$= \epsilon_{ijk} a_k \epsilon_{jlm} b_m c_l \hat{\boldsymbol{x}}_i. \tag{1.56}$$

Exchanging the dummy indices i and j and noting that $\epsilon_{jik} = -\epsilon_{ijk}$ gives

$$\boldsymbol{a} \times \boldsymbol{b} \times \boldsymbol{c} = -\epsilon_{ijk}\epsilon_{ilm} a_k b_m c_l \hat{\boldsymbol{x}}_j. \tag{1.57}$$

The identity of Eq. (1.54) then results in

$$\boldsymbol{a} \times \boldsymbol{b} \times \boldsymbol{c} = \left(\delta_{jm}\delta_{kl} - \delta_{jl}\delta_{km}\right) a_k b_m c_l \hat{\boldsymbol{x}}_j, \tag{1.58}$$

$$= \left(a_k b_j c_k - a_k b_k c_j\right) \hat{\boldsymbol{x}}_j, \tag{1.59}$$

$$= \boxed{(\boldsymbol{a} \cdot \boldsymbol{c})\, \boldsymbol{b} - (\boldsymbol{a} \cdot \boldsymbol{b})\, \boldsymbol{c}.} \tag{1.60}$$

The Levi–Civita third-order tensor allows us to rewrite cross products in terms of dot products, which simplifies obtaining vectorial and tensorial identities that involve the cross product.

1.4 The Integral Theorems

There are a variety of extremely useful theorems that involve the combined operations of differentiation and integration (the derivative and antiderivative). We will use these theorems over and over again in our development and manipulation of the rules of continuum physics.

Fundamental Theorem of 3D Calculus: For some volumetric region Ω bounded by the closed surface $\partial\Omega$ that has an outward normal $\boldsymbol{n}$ at each point of $\partial\Omega$ as depicted in Fig. 1.5, we have

$$\boxed{\int_{\Omega} \nabla\psi(\boldsymbol{r})\, \mathrm{d}^3\boldsymbol{r} = \int_{\partial\Omega} \boldsymbol{n}\psi(\boldsymbol{r})\, \mathrm{d}^2\boldsymbol{r},} \qquad \textit{fundamental theorem of 3D calculus} \tag{1.61}$$

where ψ is a scalar, vector, or tensor field of any order. If ψ is either a vector or tensor, it is necessary that it appears in the position given within the integrand on the right-hand side of Eq. (1.61). Although this theorem is used repeatedly in physics, for some peculiar reason it is rarely presented in vector calculus texts intended for the undergraduate level. As a guided exercise at the end of the chapter, you can prove Eq. (1.61) rather easily. Equation (1.61) is the 3D generalization of the *fundamental theorem of 1D calculus*

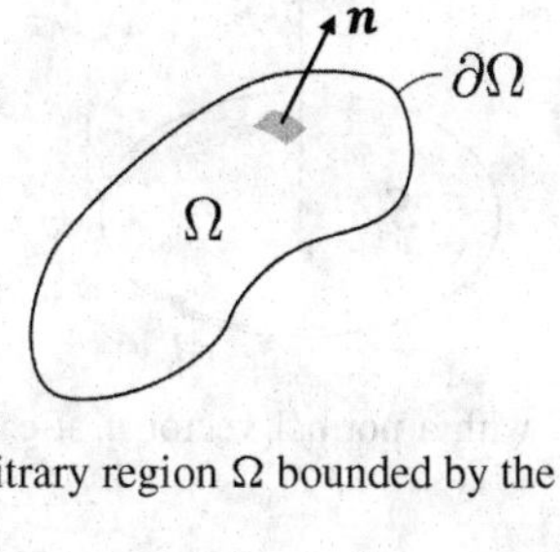

Figure 1.5 An arbitrary region Ω bounded by the closed surface $\partial\Omega$.

$$\int_a^b \frac{\mathrm{d}\psi(x)}{\mathrm{d}x}\,\mathrm{d}x = \psi(b) - \psi(a). \qquad \textit{fundamental theorem of 1D calculus} \tag{1.62}$$

In the above volume and surface integrals, we have employed the notation that $\mathrm{d}^3\boldsymbol{r} = \mathrm{d}x\,\mathrm{d}y\,\mathrm{d}z = \mathrm{d}V$ denotes a volume element and $\mathrm{d}^2\boldsymbol{r} = \mathrm{d}S$ denotes a surface element. With $\psi \to \boldsymbol{a}$ in the fundamental theorem of 3D calculus, where $\boldsymbol{a}$ is a vector or tensor field of any order, taking the trace over the two base vectors gives

$$\int_\Omega \nabla \cdot \boldsymbol{a}\,\mathrm{d}^3\boldsymbol{r} = \int_{\partial\Omega} \boldsymbol{n} \cdot \boldsymbol{a}\,\mathrm{d}^2\boldsymbol{r} \quad \textit{divergence theorem} \tag{1.63}$$

that is also called *Gauss' theorem.*

One may similarly obtain

$$\int_\Omega \nabla \times \boldsymbol{a}\,\mathrm{d}^3\boldsymbol{r} = \int_{\partial\Omega} \boldsymbol{n} \times \boldsymbol{a}\,\mathrm{d}^2\boldsymbol{r} \quad \textit{curl theorem} \tag{1.64}$$

which is not Stokes' theorem even if it involves the *curl* operation $\nabla \times \boldsymbol{a}$ defined in Section 1.6.

Stokes' Theorem: For some finite open (possibly curved) surface S having normal $\boldsymbol{n}$ at each point on S and bounded by a closed contour Γ as depicted in Fig. 1.6, we have

$$\int_S \boldsymbol{n} \cdot (\nabla \times \boldsymbol{a})\,\mathrm{d}S = \oint_\Gamma \boldsymbol{a} \cdot \mathrm{d}\boldsymbol{l} \qquad \textit{Stokes' theorem} \tag{1.65}$$

where $\mathrm{d}\boldsymbol{l}$ is the infinitesimal length vector tangent to points on Γ and $\oint_\Gamma$ means that we start the integral on the closed contour Γ at one point, go around the contour in the counterclockwise direction and finish the integral at that same point.

When applied to a plane, Stokes' theorem is equivalent to a theorem that is usually called *Green's theorem.* So taking the open surface S to reside in the x, y plane and bounded by the closed curve Γ and considering two functions of (x, y) that we call $P(x, y)$ and $Q(x, y)$, Green's theorem states

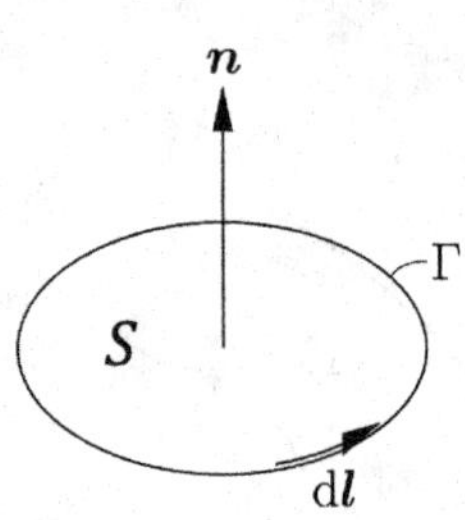

Figure 1.6 An open surface S with a normal vector $\boldsymbol{n}$ at each point and bounded by the closed contour Γ.

$$\int_S \left(\frac{\partial Q(x,y)}{\partial x} - \frac{\partial P(x,y)}{\partial y} \right) \mathrm{d}x\mathrm{d}y = \oint_\Gamma \left[P(x,y)\,\mathrm{d}x + Q(x,y)\,\mathrm{d}y \right]. \tag{1.66}$$

By substituting $P(x, y) = a_x(x, y)$ and $Q(x, y) = a_y(x, y)$, Green's theorem becomes Stokes' theorem. You can prove Green's theorem as an easy guided, end-of-chapter exercise. Green's theorem will also be used in Chapter 11 to prove Cauchy's theorem, which is the foundation for all contour-integration methods.

Differentiating under the Integral Sign: Another class of integral theorems involves time differentiation of spatial integrals when the limits of the integral domain are themselves variable in time.

In 3D, let's imagine a spatial integral domain $\Omega(t)$ whose enclosing boundary $\partial\Omega(t)$ is changing through time t because each point of the boundary is moving with a velocity $\boldsymbol{v}(\boldsymbol{r}, t)$. In this scenario, the time derivative of a volume integral over the time-variable domain $\Omega(t)$ is given by

$$\boxed{\frac{\mathrm{d}}{\mathrm{d}t} \int_{\Omega(t)} \psi(\boldsymbol{r}, t)\,\mathrm{d}^3\boldsymbol{r} = \int_{\Omega(t)} \frac{\partial \psi(\boldsymbol{r}, t)}{\partial t}\,\mathrm{d}^3\boldsymbol{r} + \int_{\partial\Omega(t)} \boldsymbol{n} \cdot \boldsymbol{v}(\boldsymbol{r}, t)\psi(\boldsymbol{r}, t)\,\mathrm{d}^2\boldsymbol{r},} \tag{1.67}$$

which is called the *Reynolds transport theorem*. Here, ψ can be a scalar, vector, or tensor field of any order. Applying the divergence theorem to the surface integral gives the alternative expression

$$\frac{\mathrm{d}}{\mathrm{d}t} \int_{\Omega(t)} \psi(\boldsymbol{r}, t)\,\mathrm{d}^3\boldsymbol{r} = \int_{\Omega(t)} \left\{ \frac{\partial \psi(\boldsymbol{r}, t)}{\partial t} + \nabla \cdot [\boldsymbol{v}(\boldsymbol{r}, t)\psi(\boldsymbol{r}, t)] \right\} \mathrm{d}^3\boldsymbol{r}. \tag{1.68}$$

If the field ψ is a vector or tensor, it is very important that it is placed after the velocity field in the tensorial expression that the divergence operates on. If the boundary $\partial\Omega$ is not moving with a velocity $\boldsymbol{v}$, then the time derivative passes through the volume integral and acts directly on the integrand ψ and there is no surface integral term or divergence term. If $\psi = 1$, then $V(t) = \int_{\Omega(t)} \mathrm{d}^3\boldsymbol{r}$ is the evolving volume of the domain $\Omega(t)$ and the Reynolds transport theorem gives $\mathrm{d}V(t)/\mathrm{d}t = \int_{\partial\Omega(t)} \boldsymbol{n} \cdot \boldsymbol{v}\,\mathrm{d}^2\boldsymbol{r}$, which is a self-evident fact and a result we will use in Chapter 4.

The Reynolds transport theorem of Eq. (1.67) is the 3D generalization of the rule for time differentiating 1D spatial integrals over a time-variable domain

$$\boxed{\frac{\mathrm{d}}{\mathrm{d}t} \int_{a(t)}^{b(t)} \psi(x, t)\,\mathrm{d}x = \int_{a(t)}^{b(t)} \frac{\partial \psi(x, t)}{\partial t}\,\mathrm{d}x + \frac{\mathrm{d}b(t)}{\mathrm{d}t}\psi(b(t), t) - \frac{\mathrm{d}a(t)}{\mathrm{d}t}\psi(a(t), t),} \tag{1.69}$$

which is called the *Leibniz rule*. Note that $\mathrm{d}b(t)/\mathrm{d}t$ is the velocity at which the domain limit $x = b$ is moving in the $+x$ direction and similarly for $\mathrm{d}a(t)/\mathrm{d}t$.

1.5 Divergence of Vector (and Tensor) Fields

The divergence operation is the dot product between the gradient operator and either a vector or tensor

$$\nabla \cdot \boldsymbol{a} = \left[\hat{x}\frac{\partial}{\partial x} + \hat{y}\frac{\partial}{\partial y} + \hat{z}\frac{\partial}{\partial z}\right] \cdot \left[a_x\hat{x} + a_y\hat{y} + a_z\hat{z}\right]. \tag{1.70}$$

In carrying out the products and derivatives, note that in Cartesian coordinates the base vectors are uniform constants and have the property that $\partial\hat{x}_j/\partial x_i = 0$. But this is not true for other base vectors in curvilinear coordinates (e.g., cylindrical and spherical). Since the derivatives of the base vectors are zero in Cartesian coordinates, we have

$$\nabla \cdot \boldsymbol{a} = \frac{\partial a_x}{\partial x} + \frac{\partial a_y}{\partial y} + \frac{\partial a_z}{\partial z} = \frac{\partial a_i}{\partial x_i} \tag{1.71}$$

with summation over the index i as always. In Section 1.8.6, we will investigate how this and other tensor-calculus operations involving the ∇ operator are different in orthogonal curvilinear coordinates. Our goal in the present section is specifically to provide physical understanding about what the divergence of vector (and tensor) fields is telling us about the field.

To do so, construct a region Ω around the point $\boldsymbol{r}$ that has a tiny volume $\delta V = \int_\Omega \mathrm{d}V$, where the δ means "tiny." The region is surrounded by the closed surface $\partial\Omega$. The divergence theorem says that:

$$\frac{1}{\delta V}\int_\Omega \nabla \cdot \boldsymbol{a}\,\mathrm{d}V = \frac{1}{\delta V}\int_{\partial\Omega} \boldsymbol{n} \cdot \boldsymbol{a}\,\mathrm{d}S. \tag{1.72}$$

In the limit as $\delta V \to 0$, $\nabla \cdot \boldsymbol{a} \approx$ constant in Ω so

$$\begin{aligned}\nabla \cdot \boldsymbol{a} &= \lim_{\delta V \to 0} \frac{1}{\delta V}\int_\Omega \nabla \cdot \boldsymbol{a}\,\mathrm{d}V \\ &= \lim_{\delta V \to 0} \frac{1}{\delta V}\int_{\partial\Omega} \boldsymbol{n} \cdot \boldsymbol{a}\,\mathrm{d}S,\end{aligned} \tag{1.73}$$

which represents the accumulation of the physical quantity carried by $\boldsymbol{a}$. Imagine the vector $\boldsymbol{a}$ as a flux of some type carrying a physical quantity with it. If the flux into a small volume element is different than the flux out of the element, which is what the surface integral of Eq. (1.73) quantifies, then the divergence of this flux is nonzero. The divergence of a vector field is thus associated with the idea of accumulation (if the divergence is negative) or depletion (if the divergence is positive) in a tiny region surrounding the point in question.

To conclude: *If $|\boldsymbol{n} \cdot \boldsymbol{a}|$ is larger on one side of the small element Ω compared to the other, the divergence is nonzero. If the divergence is positive, the physical quantity carried by* $\boldsymbol{a}$ *is depleting in Ω, while if the divergence is negative, the physical quantity is accumulating.* This is shown in Fig. 1.7.

There are two ways we commonly use to visualize a vector field as depicted in Fig. 1.8. In the first, we place a vector at each point in space. In the second, called *current lines* (or flow lines), the direction of the field at a point is tangent to the current line at that point

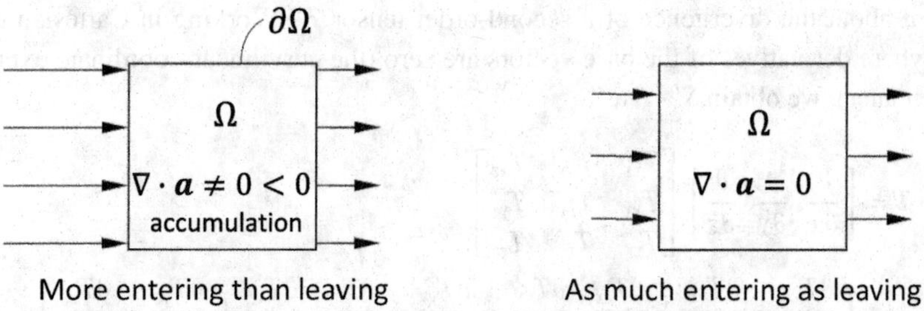

Figure 1.7 Figure showing that $\nabla \cdot \boldsymbol{a} < 0$ quantifies the accumulation of the physical quantity carried by the "flux" $\boldsymbol{a}$ due to more of the quantity fluxing into a volume than fluxing out. If the flux arrows leaving the volume are greater than those entering, the quantity is depleting and $\nabla \cdot \boldsymbol{a} > 0$. If the flux in equals the flux out, then there is no accumulation or depletion and $\nabla \cdot \boldsymbol{a} = 0$.

and the amplitude is given by the density of current lines in the neighborhood surrounding the point.

In the current line approach, the places where $\nabla \cdot \boldsymbol{a} \neq 0$ are always associated with the start of a new line as shown in Fig. 1.9. In particular, if we imagine the electric field $\boldsymbol{E}(\boldsymbol{r})$ around a point charge q, all field lines start at the charge location where $\nabla \cdot \boldsymbol{E} \neq 0$ but at all other points where there is no point charge, field lines are not being created and $\nabla \cdot \boldsymbol{E} = 0$.

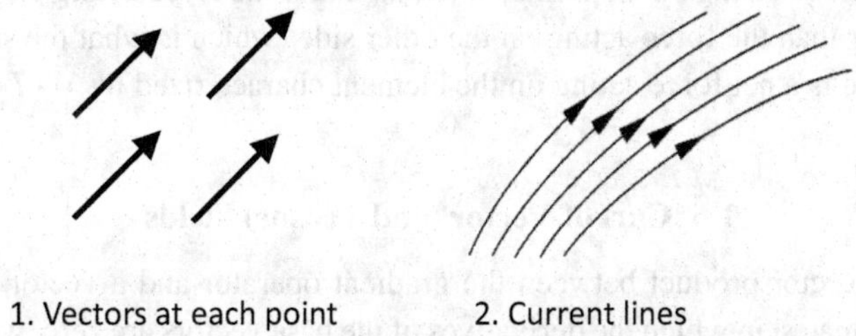

Figure 1.8 Two approaches that are commonly employed for picturing a vector field.

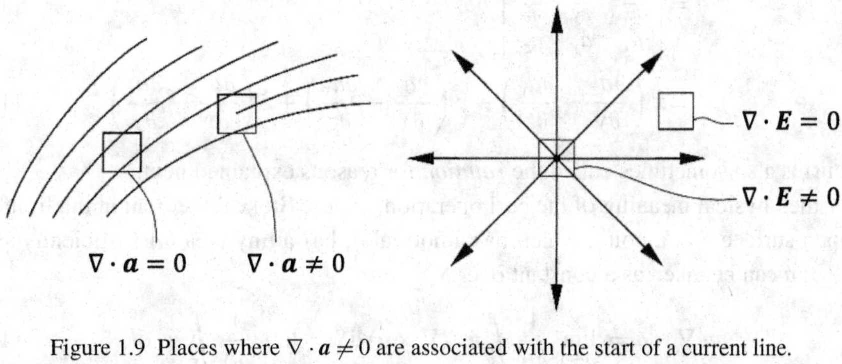

Figure 1.9 Places where $\nabla \cdot \boldsymbol{a} \neq 0$ are associated with the start of a current line.

What about the divergence of a second-order tensor $\boldsymbol{T}$? Working in Cartesian coordinates where derivatives of the base vectors are zero (the curvilinear-coordinate expression is given later), we obtain $\nabla \cdot \boldsymbol{T}$ as

$$\nabla \cdot \boldsymbol{T} = \left[\frac{\partial}{\partial x}, \frac{\partial}{\partial y}, \frac{\partial}{\partial z}\right] \begin{bmatrix} T_{xx} & T_{xy} & T_{xz} \\ T_{yx} & T_{yy} & T_{yz} \\ T_{zx} & T_{zy} & T_{zz} \end{bmatrix}$$

$$= \Bigg(\underbrace{\frac{\partial T_{xx}}{\partial x} + \frac{\partial T_{yx}}{\partial y} + \frac{\partial T_{zx}}{\partial z}}_{x \text{ component}}, \underbrace{\frac{\partial T_{xy}}{\partial x} + \frac{\partial T_{yy}}{\partial y} + \frac{\partial T_{zy}}{\partial z}}_{y \text{ component}}, \underbrace{\frac{\partial T_{xz}}{\partial x} + \frac{\partial T_{yz}}{\partial y} + \frac{\partial T_{zz}}{\partial z}}_{z \text{ component}} \Bigg) \tag{1.74}$$

$$= \left(\hat{\boldsymbol{x}}_i \frac{\partial}{\partial x_i}\right) \cdot \left(T_{jk}\, \hat{\boldsymbol{x}}_j \hat{\boldsymbol{x}}_k\right) = \frac{\partial T_{jk}}{\partial x_i} \left(\hat{\boldsymbol{x}}_i \cdot \hat{\boldsymbol{x}}_j\right) \hat{\boldsymbol{x}}_k = \frac{\partial T_{ik}}{\partial x_i} \hat{\boldsymbol{x}}_k \widehat{=} \nabla \cdot \boldsymbol{T}. \tag{1.75}$$

To visualize or intuit the meaning of the divergence of a second-order tensor, we can use our same device of considering a tiny element Ω of volume δV surrounding the point where the vector $\nabla \cdot \boldsymbol{T}$ is being evaluated and obtain (because δV is so small that $\nabla \cdot \boldsymbol{T}$ is uniform inside of Ω)

$$\nabla \cdot \boldsymbol{T} = \lim_{\delta V \to 0} \frac{1}{\delta V} \int_{\Omega} \nabla \cdot \boldsymbol{T}\, \mathrm{d}V = \lim_{\delta V \to 0} \frac{1}{\delta V} \int_{\partial\Omega} \boldsymbol{n} \cdot \boldsymbol{T}\, \mathrm{d}S. \tag{1.76}$$

Regardless of what the second-order tensor $\boldsymbol{T}$ actually represents, think of the vector $\boldsymbol{n} \cdot \boldsymbol{T}\, \mathrm{d}S$ as being a force acting on all points on $\partial\Omega$ so that if the force acting on one side of the element is larger than the force acting on the other side (which is what the surface integral allows for), there is a net force acting on the element characterized by $\nabla \cdot \boldsymbol{T} \neq 0$.

1.6 Curl of Vector (and Tensor) Fields

The curl is the vector product between the gradient operator and a vector (or tensor). In Cartesian coordinates, in which the derivatives of the base vectors are zero, we can calculate the curl of a vector field using the determinant rule

$$\nabla \times \boldsymbol{a} = \begin{vmatrix} \hat{\boldsymbol{x}} & \hat{\boldsymbol{y}} & \hat{\boldsymbol{z}} \\ \frac{\partial}{\partial x} & \frac{\partial}{\partial y} & \frac{\partial}{\partial z} \\ a_x & a_y & a_z \end{vmatrix}$$

$$= \hat{\boldsymbol{x}} \left(\frac{\partial a_z}{\partial y} - \frac{\partial a_y}{\partial z}\right) - \hat{\boldsymbol{y}} \left(\frac{\partial a_z}{\partial x} - \frac{\partial a_x}{\partial z}\right) + \hat{\boldsymbol{z}} \left(\frac{\partial a_y}{\partial x} - \frac{\partial a_x}{\partial y}\right). \tag{1.77}$$

The curl is also sometimes called the *rotation* for reasons explained next.

For the physical meaning of the curl operation, we use Stokes' theorem in the limit that the open surface S at a point, which has a normal $\boldsymbol{n}$, has a tiny area δA sufficiently small that $\nabla \times \boldsymbol{a}$ can be taken as a constant over S

$$\boldsymbol{n} \cdot \nabla \times \boldsymbol{a} = \lim_{\delta A \to 0} \frac{1}{\delta A} \int_S \boldsymbol{n} \cdot (\nabla \times \boldsymbol{a})\, \mathrm{d}S = \lim_{\delta A \to 0} \frac{1}{\delta A} \oint_{\Gamma} \boldsymbol{a} \cdot \mathrm{d}\boldsymbol{l}. \tag{1.78}$$

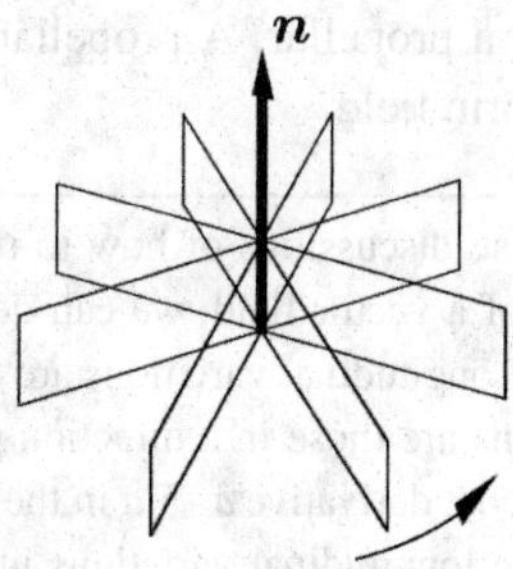

Figure 1.10 A waterwheel with flat blades and an axle in the $\boldsymbol{n}$ direction.

Thus, if the vector field $\boldsymbol{a}$ is tangent in places to the curve Γ and if this tangential component is larger on one side of S compared to the other so that integral over the closed line Γ is nonzero, then we will have $\nabla \times \boldsymbol{a} \neq 0$. We can associate this with the idea that the field has a rotation on S.

To visualize the curl, we use a *water wheel* consisting of an axle oriented in the direction $\boldsymbol{n}$ and having flat blades coming off the axle perpendicularly as depicted in Fig. 1.10. We immerse this wheel in our vector field $\boldsymbol{a}$ that we imagine to be the flow of water regardless of what $\boldsymbol{a}$ really corresponds to. We change the orientation of the axle of the wheel and observe how fast the wheel is moving if at all. The direction $\boldsymbol{n}$ of the axle at which the wheel turns the fastest in the counterclockwise direction gives the direction of $\nabla \times \boldsymbol{a}$, while the rate of rotation gives $|\nabla \times \boldsymbol{a}|$. The right-hand rule corresponds to the wheel rotating in the counterclockwise direction.

So to mentally investigate the curl (or rotation) associated with some field, we imagine the vector field to be a flow field and probe the field with our water wheel to see in what orientation the wheel turns the fastest if, indeed, it can turn at all, cf., Fig. 1.11. Note that

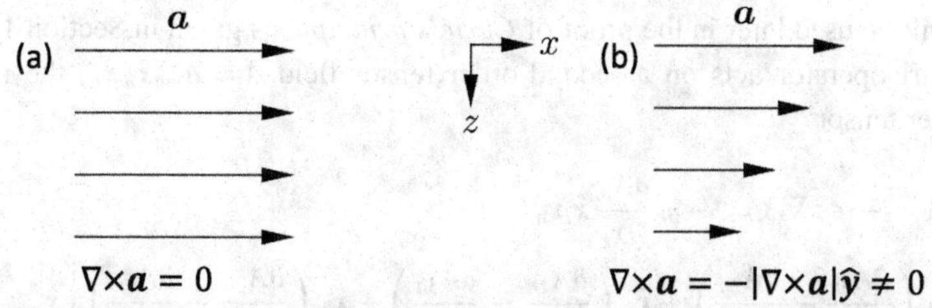

Figure 1.11 Use a water wheel to probe a field and see whether the wheel can turn: (a) a uniform vector field or a field with only longitudinal variation in the direction of the field cannot make a water wheel turn so that $\nabla \times \boldsymbol{a} = 0$; (b) a vector field that has spatial variation in a direction transverse to the field direction will always make a water wheel turn resulting in nonzero $\nabla \times \boldsymbol{a}$.

a water wheel is not the same as a propellar. A propellar would turn even in case (a) of Fig. 1.11 corresponding to a uniform field.

CONCLUSIONS: To conclude these discussions of how to understand and visualize nonzero values of the divergence and curl of a vector field, we can define *longitudinal* and *transverse* spatial variations of a vector field. Longitudinal variations are those in the same direction of the vector field and transverse variations are those in a direction perpendicular to the vector field. Longitudinal variations (the directional derivative $\hat{\boldsymbol{a}} \cdot \nabla \boldsymbol{a}$ in the direction of the vector field) lead to a nonzero divergence (unless the longitudinal variations in each direction sum to zero) but do not contribute to the curl. Similarly, transverse variations always lead to a nonzero curl but do not contribute to the divergence. The decomposition of a vector field into longitudinal and transverse variations will correspond to P-waves and S-waves, respectively, when we discuss elastic-wave propagation.

We can also perform the *curl* differential operation using the Levi–Civita tensor in the following way:

$$\nabla \times \boldsymbol{a} = -{}_3\boldsymbol{\epsilon} : \nabla \boldsymbol{a} = -\epsilon_{ijk}\, \hat{\boldsymbol{x}}_i \hat{\boldsymbol{x}}_j \hat{\boldsymbol{x}}_k : \hat{\boldsymbol{x}}_l \frac{\partial}{\partial x_l} a_m \hat{\boldsymbol{x}}_m = \boxed{-\epsilon_{ijk} \frac{\partial a_j}{\partial x_k} \hat{\boldsymbol{x}}_i} \tag{1.79}$$

$$= \hat{\boldsymbol{x}}_1 \left(\frac{\partial a_3}{\partial x_2} - \frac{\partial a_2}{\partial x_3} \right) + \hat{\boldsymbol{x}}_2 \left(\frac{\partial a_1}{\partial x_3} - \frac{\partial a_3}{\partial x_1} \right) + \hat{\boldsymbol{x}}_3 \left(\frac{\partial a_2}{\partial x_1} - \frac{\partial a_1}{\partial x_2} \right).$$

Note that if we had let the Levi–Civita tensor act on the transpose tensor $(\nabla \boldsymbol{a})^T$ we obtain

$${}_3\boldsymbol{\epsilon} : (\nabla \boldsymbol{a})^T = \nabla \times \boldsymbol{a}, \tag{1.80}$$

which shows that

$${}_3\boldsymbol{\epsilon} : \left[\nabla \boldsymbol{a} + (\nabla \boldsymbol{a})^T \right] = \nabla \times \boldsymbol{a} - \nabla \times \boldsymbol{a} = 0. \tag{1.81}$$

This fact will be used later in the proof of *Curie's principle* as given in Section 1.8.5.

If the curl operator acts on a second-order tensor field $\boldsymbol{A} = A_{mn} \hat{\boldsymbol{x}}_m \hat{\boldsymbol{x}}_n$, the result is a second-order tensor

$$\nabla \times \boldsymbol{A} = -{}_3\boldsymbol{\epsilon} : \nabla \boldsymbol{A} = -\epsilon_{ijk} \frac{\partial A_{jn}}{\partial x_k} \hat{\boldsymbol{x}}_i \hat{\boldsymbol{x}}_n \tag{1.82}$$

$$= \left[\hat{\boldsymbol{x}}_1 \left(\frac{\partial A_{3n}}{\partial x_2} - \frac{\partial A_{2n}}{\partial x_3} \right) + \hat{\boldsymbol{x}}_2 \left(\frac{\partial A_{1n}}{\partial x_3} - \frac{\partial A_{3n}}{\partial x_1} \right) + \hat{\boldsymbol{x}}_3 \left(\frac{\partial A_{2n}}{\partial x_1} - \frac{\partial A_{1n}}{\partial x_2} \right) \right] \hat{\boldsymbol{x}}_n. \tag{1.83}$$

Being able to do curl operations using the dot product makes it possible to prove many useful things about curl operations.

For example, in both fluid mechanics and electromagnetism, frequent use is made of the identity

$$\nabla \times \nabla \times \boldsymbol{u} = \nabla(\nabla \cdot \boldsymbol{u}) - \nabla^2 \boldsymbol{u}. \tag{1.84}$$

We can prove this identity using the Levi–Civita tensor as follows:

$$\begin{aligned}\nabla \times \nabla \times \boldsymbol{u} &= \epsilon_{ijk}\, \hat{\boldsymbol{x}}_i \hat{\boldsymbol{x}}_j \hat{\boldsymbol{x}}_k : \frac{\partial}{\partial x_l} \hat{\boldsymbol{x}}_l \left[\epsilon_{mno}\, \hat{\boldsymbol{x}}_m \hat{\boldsymbol{x}}_n \hat{\boldsymbol{x}}_o : \hat{\boldsymbol{x}}_p \frac{\partial}{\partial x_p} u_q \hat{\boldsymbol{x}}_q \right] \qquad (1.85)\\ &= \epsilon_{ijk}\, \epsilon_{mno} \frac{\partial^2 u_q}{\partial x_l\, \partial x_p}\, \hat{\boldsymbol{x}}_i \left(\hat{\boldsymbol{x}}_k \cdot \hat{\boldsymbol{x}}_l\right) \left(\hat{\boldsymbol{x}}_j \cdot \hat{\boldsymbol{x}}_m\right) \left(\hat{\boldsymbol{x}}_o \cdot \hat{\boldsymbol{x}}_p\right) \left(\hat{\boldsymbol{x}}_n \cdot \hat{\boldsymbol{x}}_q\right).\end{aligned}$$

So $l = k$, $m = j$, $p = o$, and $q = n$, to give

$$\nabla \times \nabla \times \boldsymbol{u} = \epsilon_{ijk}\, \epsilon_{jno} \frac{\partial^2 u_n}{\partial x_k\, \partial x_o} \hat{\boldsymbol{x}}_i. \tag{1.86}$$

As shown earlier, we have the identity that $\epsilon_{ijk}\epsilon_{ino} = \delta_{jn}\delta_{ko} - \delta_{jo}\delta_{kn}$. If we exchange the indices i and j and use that $\epsilon_{jik} = -\epsilon_{ijk}$ we then have the identity

$$\epsilon_{ijk}\, \epsilon_{jno} = \delta_{io}\delta_{kn} - \delta_{in}\delta_{ko}. \tag{1.87}$$

Using this in Eq. (1.86) gives the sought after result

$$\begin{aligned}\nabla \times \nabla \times \boldsymbol{u} &= \hat{\boldsymbol{x}}_i \frac{\partial}{\partial x_i} \left(\frac{\partial u_k}{\partial x_k} \right) - \frac{\partial^2 u_i}{\partial x_k^2} \hat{\boldsymbol{x}}_i \qquad (1.88)\\ &= \nabla(\nabla \cdot \boldsymbol{u}) - \nabla^2 \boldsymbol{u}. \qquad (1.89)\end{aligned}$$

Using ${}_3\boldsymbol{\epsilon}$, it is also straightforward to show

$$\nabla \times \nabla \alpha = 0 \tag{1.90}$$

$$\nabla \cdot (\nabla \times \boldsymbol{a}) = 0 \tag{1.91}$$

both of which are used throughout continuum physics.

1.7 Tensor-Calculus Product Rules

What if the gradient operator ∇ in an expression acts on several vectorial or tensorial terms with, possibly, various scalar products present between the vectors and tensors? What is the equivalent of the scalar "product rule" $\partial(\alpha\beta)/\partial x = \beta\partial\alpha/\partial x + \alpha\partial\beta/\partial x$ for various types of products involving vectors and tensors?

As an example, let's consider the specific expression

$$\nabla \cdot (\boldsymbol{ab} \cdot \boldsymbol{T}), \tag{1.92}$$

where $\boldsymbol{a}$ and $\boldsymbol{b}$ are both vectors and $\boldsymbol{T}$ is a second-order tensor, all of which vary in space. In this expression, $\boldsymbol{ab}$ is a second-order tensor and so is $\boldsymbol{ab} \cdot \boldsymbol{T}$ so that $\nabla \cdot (\boldsymbol{ab} \cdot \boldsymbol{T})$ is a vector. To distribute the derivative in this expression, we work in Cartesian coordinates in which the base vectors $\hat{\boldsymbol{x}}_i$ are uniform so that $\partial\hat{\boldsymbol{x}}_j/\partial x_i = 0$ for all base vectors and coordinate directions. But after we distribute the derivatives in Cartesians, we will write the resulting expression in the general bold-face notation that then applies to even curvilinear coordinates in which some of the base vectors can have nonzero derivatives. Any tensor identity expressed in bold-face notation is valid for any coordinate system and is the preferred way to express tensor-calculus product rules.

For each vectorial or tensorial term in the expression, we use a different set of Cartesian-coordinate indices to write out

$$\nabla\cdot(\boldsymbol{ab}\cdot\boldsymbol{T}) = \underbrace{\hat{\boldsymbol{x}}_i\frac{\partial}{\partial x_i}}_{\nabla}\cdot\left(\underbrace{a_j\hat{\boldsymbol{x}}_j\, b_k\hat{\boldsymbol{x}}_k}_{\boldsymbol{ab}}\cdot\underbrace{T_{lm}\hat{\boldsymbol{x}}_l\hat{\boldsymbol{x}}_m}_{\boldsymbol{T}}\right) \tag{1.93}$$

$$= \frac{\partial}{\partial x_i}\left(a_j b_k T_{lm}\right)\left(\hat{\boldsymbol{x}}_i\cdot\hat{\boldsymbol{x}}_j\right)\left(\hat{\boldsymbol{x}}_k\cdot\hat{\boldsymbol{x}}_l\right)\hat{\boldsymbol{x}}_m. \tag{1.94}$$

In passing from the first line to the second, we pulled out the scalar components and derivative while preserving the position of the base vectors and scalar products between the base vectors. Next, because of the nature of the scalar product in orthogonal coordinates, we have that $j = i$ and $l = k$. This allows us to write

$$\nabla\cdot(\boldsymbol{ab}\cdot\boldsymbol{T}) = \frac{\partial}{\partial x_i}\left(a_i b_k T_{km}\right)\hat{\boldsymbol{x}}_m \tag{1.95}$$

$$= \left[\left(\frac{\partial a_i}{\partial x_i}\right) b_k T_{km} + a_i\left(\frac{\partial b_k}{\partial x_i}\right) T_{km} + a_i b_k\frac{\partial T_{km}}{\partial x_i}\right]\hat{\boldsymbol{x}}_m. \tag{1.96}$$

In going from the first to second expression, we just employed the usual derivative product rule for scalar fields.

The final step is what students often find the most difficult. One must look at Eq. (1.96) and identify the equivalent expression in bold face. So you have to make identifications like $\partial a_i/\partial x_i = \nabla\cdot\boldsymbol{a}$ and $b_k T_{km}\hat{\boldsymbol{x}}_m = \boldsymbol{b}\cdot\boldsymbol{T}$. Carrying this out, we obtain at last

$$\nabla\cdot(\boldsymbol{ab}\cdot\boldsymbol{T}) = (\nabla\cdot\boldsymbol{a})\,\boldsymbol{b}\cdot\boldsymbol{T} + \boldsymbol{a}\cdot(\nabla\boldsymbol{b})\cdot\boldsymbol{T} + \boldsymbol{b}\cdot(\boldsymbol{a}\cdot\nabla\boldsymbol{T})\,. \tag{1.97}$$

Once we type the final expression in bold face (or write by hand the expression with squiggly underscores), it applies to any orthogonal curvilinear coordinates and is a generally valid identity not limited to Cartesian coordinates. Note that $\nabla\boldsymbol{T}$ is an example of a third-order tensor. We have thus determined how to distribute the ∇ operator onto the vectors and tensors in a multi-term expression involving scalar products. By using the earlier Levi–Civita alternating tensor, we can do the same for multi-term expressions involving vector products.

Using this ability, it is now straightforward to derive a long list of useful tensor-calculus product rules. In the following list, α is a scalar field, $\boldsymbol{a}$ and $\boldsymbol{b}$ are again vector fields, $\boldsymbol{A}$ and $\boldsymbol{B}$ are second-order tensor fields and ${}_4\boldsymbol{C}$ is a fourth-order tensor field:

$$\nabla \times (\alpha \boldsymbol{a}) = \alpha \nabla \times \boldsymbol{a} + (\nabla \alpha) \times \boldsymbol{a} \tag{1.98}$$

$$\nabla \times (\boldsymbol{a} \times \boldsymbol{b}) = \nabla \cdot (\boldsymbol{b}\boldsymbol{a} - \boldsymbol{a}\boldsymbol{b}) \tag{1.99}$$

$$\nabla \cdot (\boldsymbol{a}\boldsymbol{b}) = (\nabla \cdot \boldsymbol{a})\boldsymbol{b} + \boldsymbol{a} \cdot \nabla \boldsymbol{b} \tag{1.100}$$

$$\nabla \cdot (\boldsymbol{a} \times \boldsymbol{b}) = \boldsymbol{b} \cdot (\nabla \times \boldsymbol{a}) - \boldsymbol{a} \cdot (\nabla \times \boldsymbol{b}) \tag{1.101}$$

$$\nabla \cdot (\boldsymbol{A}\boldsymbol{B}) = (\nabla \cdot \boldsymbol{A})\boldsymbol{B} + \boldsymbol{A}^T \cdot \nabla \boldsymbol{B} \tag{1.102}$$

$$\nabla \cdot (\boldsymbol{A} \cdot \boldsymbol{B}) = (\nabla \cdot \boldsymbol{A}) \cdot \boldsymbol{B} + \boldsymbol{A}^T : \nabla \boldsymbol{B} \tag{1.103}$$

$$\nabla(\boldsymbol{a} \cdot \boldsymbol{b}) = (\nabla \boldsymbol{a}) \cdot \boldsymbol{b} + (\nabla \boldsymbol{b}) \cdot \boldsymbol{a} \tag{1.104}$$

$$= \boldsymbol{a} \cdot (\nabla \boldsymbol{b}) + \boldsymbol{b} \cdot (\nabla \boldsymbol{a}) + \boldsymbol{a} \times (\nabla \times \boldsymbol{b}) + \boldsymbol{b} \times (\nabla \times \boldsymbol{a}) \tag{1.105}$$

$$\nabla[(\nabla \alpha) \cdot \nabla \alpha] = 2(\nabla \alpha) \cdot \nabla\nabla \alpha = 2(\nabla\nabla \alpha) \cdot \nabla \alpha \tag{1.106}$$

$$\nabla (\boldsymbol{A} \cdot \boldsymbol{B}) = (\nabla \boldsymbol{A}) \cdot \boldsymbol{B} + [(\nabla \boldsymbol{B}) \cdot \boldsymbol{A}]^{\overset{T}{132}} \tag{1.107}$$

$$\nabla(\boldsymbol{A} \cdot \boldsymbol{a}) = (\nabla \boldsymbol{A}) \cdot \boldsymbol{a} + (\nabla \boldsymbol{a}) \cdot \boldsymbol{A}^T \tag{1.108}$$

$$\nabla(\boldsymbol{a} \cdot \boldsymbol{A}) = \left[\nabla \left(\boldsymbol{A}^T\right)\right] \cdot \boldsymbol{a} + (\nabla \boldsymbol{a}) \cdot \boldsymbol{A} \tag{1.109}$$

$$\nabla \cdot (\alpha \boldsymbol{A}) = \nabla \alpha \cdot \boldsymbol{A} + \alpha \nabla \cdot \boldsymbol{A} \tag{1.110}$$

$$\nabla \cdot (\boldsymbol{a} \cdot \boldsymbol{A}) = \nabla \boldsymbol{a} : \boldsymbol{A} + \left(\nabla \cdot \boldsymbol{A}^T\right) \cdot \boldsymbol{a} \tag{1.111}$$

$$\nabla \cdot (\boldsymbol{A} \cdot \boldsymbol{a}) = \boldsymbol{A} : (\nabla \boldsymbol{a})^T + (\nabla \cdot \boldsymbol{A}) \cdot \boldsymbol{a} \tag{1.112}$$

$$\nabla \cdot (\boldsymbol{a}\boldsymbol{A}) = (\nabla \cdot \boldsymbol{a})\boldsymbol{A} + \boldsymbol{a} \cdot \nabla \boldsymbol{A} \tag{1.113}$$

$$\nabla \cdot (\boldsymbol{A}\boldsymbol{a}) = (\nabla \cdot \boldsymbol{A})\, \boldsymbol{a} + \boldsymbol{A}^T \cdot \nabla \boldsymbol{a} \tag{1.114}$$

$$\nabla \cdot ({}_4\boldsymbol{C} : \boldsymbol{A}) = (\nabla \cdot {}_4\boldsymbol{C}) : \boldsymbol{A} + {}_4\boldsymbol{C}^{\overset{T}{2341}} \;\vdots\; \nabla \boldsymbol{A} \tag{1.115}$$

Note that in the last identity, the symbol $\vdots$ means "the triple dot product" as defined earlier. Having the above types of tensor-calculus product rules available to us for arbitrary curvilinear coordinates allows many results in continuum physics to be developed in the chapters that follow.

It may not seem obvious that tensor-calculus product rules proven in Cartesian coordinates, in which derivatives of the base vectors are zero, but written in bold-face notation after distributing the derivatives are in fact generally valid for all orthogonal curvilinear coordinate systems. To demonstrate this fact using an example, consider the identity $\nabla \cdot (\boldsymbol{A} \cdot \boldsymbol{a}) = \boldsymbol{A} : (\nabla \boldsymbol{a})^T + (\nabla \cdot \boldsymbol{A}) \cdot \boldsymbol{a}$ given in the above list as derived, for convenience, in Cartesian coordinates. To prove this identity is valid in say cylindrical coordinates, we simply carry out all the given operations using that $\partial \hat{\boldsymbol{r}}/\partial \phi = \hat{\boldsymbol{\phi}}$ and $\partial \hat{\boldsymbol{\phi}}/\partial \phi = -\hat{\boldsymbol{r}}$ with all other derivatives of the base vectors equal to zero. In cylindrical coordinates, the various terms are

$$\nabla \boldsymbol{a} = \begin{pmatrix} \partial a_r/\partial r & \partial a_\phi/\partial r & \partial a_z/\partial r \\ (\partial a_r/\partial \phi - a_\phi)/r & (\partial a_\phi/\partial \phi + a_r)/r & \partial a_z/\partial \phi \\ \partial a_r/\partial z & \partial a_\phi/\partial z & \partial a_z/\partial z \end{pmatrix}, \tag{1.116}$$

a result that will be obtained more formally in the upcoming Section 1.8.6 on orthogonal curvilinear coordinates. We then have that

$$
\begin{aligned}
\boldsymbol{A} : (\nabla \boldsymbol{a})^T &= \operatorname{tr}\left\{\boldsymbol{A} \cdot (\nabla \boldsymbol{a})^T\right\} \\
&= A_{rr}\frac{\partial a_r}{\partial r} + A_{r\phi}\frac{\partial a_\phi}{\partial r} + A_{rz}\frac{\partial a_z}{\partial r} \\
&\quad + \frac{A_{\phi r}}{r}\left(\frac{\partial a_r}{\partial \phi} - a_\phi\right) + \frac{A_{\phi\phi}}{r}\left(\frac{\partial a_\phi}{\partial \phi} + a_r\right) + A_{\phi z}\frac{\partial a_z}{\partial \phi} \\
&\quad + A_{zr}\frac{\partial a_r}{\partial z} + A_{z\phi}\frac{\partial a_\phi}{\partial z} + A_{zz}\frac{\partial a_z}{\partial z},
\end{aligned} \tag{1.117}
$$

$$
\begin{aligned}
\nabla \cdot \boldsymbol{A} &= \hat{\boldsymbol{r}}\left(\frac{1}{r}\frac{\partial (rA_{rr})}{\partial r} + \frac{1}{r}\frac{\partial A_{\phi r}}{\partial \phi} - \frac{A_{\phi\phi}}{r} + \frac{A_{zr}}{\partial z}\right) \\
&\quad + \hat{\boldsymbol{\phi}}\left(\frac{1}{r}\frac{\partial (rA_{r\phi})}{\partial r} + \frac{1}{r}\frac{\partial A_{\phi\phi}}{\partial \phi} + \frac{A_{\phi r}}{r} + \frac{A_{z\phi}}{\partial z}\right) \\
&\quad + \hat{\boldsymbol{z}}\left(\frac{1}{r}\frac{\partial (rA_{rz})}{\partial r} + \frac{1}{r}\frac{\partial A_{\phi z}}{\partial \phi} + \frac{A_{zz}}{\partial z}\right),
\end{aligned} \tag{1.118}
$$

and

$$
\begin{aligned}
\nabla \cdot (\boldsymbol{A} \cdot \boldsymbol{a}) &= \frac{1}{r}\frac{\partial}{\partial r}\left[r\left(A_{rr}a_r + A_{r\phi}a_\phi + A_{rz}a_z\right)\right] \\
&\quad + \frac{1}{r}\frac{\partial}{\partial \phi}\left(A_{\phi r}a_r + A_{\phi\phi}a_\phi + A_{\phi z}a_z\right) \\
&\quad + \frac{\partial}{\partial z}\left(A_{zr}a_r + A_{z\phi}a_\phi + A_{zz}a_z\right).
\end{aligned} \tag{1.119}
$$

Using these expressions, a final bit of algebra demonstrates that $\nabla \cdot (\boldsymbol{A} \cdot \boldsymbol{a}) = \boldsymbol{A} : (\nabla \boldsymbol{a})^T + (\nabla \cdot \boldsymbol{A}) \cdot \boldsymbol{a}$ is valid in cylindrical coordinates despite having been derived initially in Cartesian coordinates.

1.8 Additional Topics Involving Tensors

Each topic treated in this section is important because it will be used in our development of the rules of continuum physics. However, at this point, you have been exposed to enough tensor calculus that after first working through Section 1.9 on the Dirac delta function and working some end-of-chapter exercises to sharpen your skills, you can move ahead to Chapter 2 on continuum mechanics if you so choose. The topics treated in this section will be referred to each time they are needed in the chapters to follow. On the other hand, working through this section now will make you more proficient with tensors and tensor calculus, will show you some interesting uses and facts of tensors, and will better prepare you for the chapters that follow.

1.8.1 Taylor Series of Fields in Three-Dimensional Space

It is useful in various physics contexts to represent a scalar, vector, or tensor field near a particular point as a power series in the local coordinates adjacent to that point if the field and its spatial derivatives are known at that point. This is what the Taylor-series expansion does for us and, in three-dimensional space, requires the use of higher-order tensors and dot products, even for the representation of scalar fields.

But let's begin with the well-known example of a scalar function $\psi(x)$ in just one spatial dimension x that does not require the use of tensors. We expand $\psi(x)$ about a particular point x_0 as a power series that is called the Taylor series

$$\boxed{\psi(x) = \sum_{n=0}^{\infty} (x - x_0)^n a_n.} \tag{1.120}$$

Because $(x - x_0)^0 = 1$, the first coefficient a_0 is found by simply evaluating this series at $x = x_0$

$$\psi(x_0) = a_0. \tag{1.121}$$

Each successive coefficient is found by taking successive derivatives and evaluating at $x = x_0$

$$\left.\frac{\mathrm{d}\psi}{\mathrm{d}x}\right|_{x_0} = a_1, \quad \left.\frac{\mathrm{d}^2\psi}{\mathrm{d}x^2}\right|_{x_0} = (2)(1)a_2, \quad \text{and} \quad \left.\frac{\mathrm{d}^3\psi}{\mathrm{d}x^3}\right|_{x_0} = (3)(2)(1)a_3 \tag{1.122}$$

so that each coefficient in the Taylor series is given by the derivatives of the function at the point x_0 as

$$\boxed{a_n = \frac{1}{n!}\left.\frac{\mathrm{d}^n\psi}{\mathrm{d}x^n}\right|_{x_0} \quad \text{for} \quad n = 1, 2 \ldots \infty} \tag{1.123}$$

and where, again, $a_0 = \psi(x_0)$.

Next, for the Taylor series of any field (scalar, vector, or tensor) that is distributed in three-dimensional space, we again expand this field in a power series of the local coordinates about a particular point $\boldsymbol{r}_0$ but now each term n in the series involves coefficients that are tensors of (at least) order n. For a field written as ${}_t\boldsymbol{\psi}(\boldsymbol{r})$, where t denotes the tensorial order of the field (so $t = 0$ is a scalar field, $t = 1$ a vector field and so on), the Taylor series expansion for this field is written

$$
{}_t\psi(\boldsymbol{r}) = \sum_{n=0}^{\infty} (\boldsymbol{r}-\boldsymbol{r}_0)^n \overset{n}{\vdots}\, {}_{\{n+t\}}\boldsymbol{A}, \tag{1.124}
$$

where the notation is made clear by writing out the first few terms of the series

$$
\begin{aligned}
{}_t\psi(\boldsymbol{r}) = {}_t\boldsymbol{A} &+ (\boldsymbol{r}-\boldsymbol{r}_0)\cdot {}_{\{1+t\}}\boldsymbol{A} \\
&+ (\boldsymbol{r}-\boldsymbol{r}_0)\,(\boldsymbol{r}-\boldsymbol{r}_0) : {}_{\{2+t\}}\boldsymbol{A} \\
&+ (\boldsymbol{r}-\boldsymbol{r}_0)\,(\boldsymbol{r}-\boldsymbol{r}_0)\,(\boldsymbol{r}-\boldsymbol{r}_0) \overset{3}{\vdots}\, {}_{\{3+t\}}\boldsymbol{A} + \ldots .
\end{aligned} \tag{1.125}
$$

So for a field of tensorial order t in three-dimensional space, the Taylor-series coefficients at each n are tensors ${}_{\{n+t\}}\boldsymbol{A}$ of tensorial order $n+t$ (i.e., that have $n+t$ base vectors).

To find these tensorial coefficients, begin by evaluating the tensorial power series at $\boldsymbol{r}=\boldsymbol{r}_0$ to give the first tensorial coefficient as

$$
{}_t\psi(\boldsymbol{r}_0) = {}_t\boldsymbol{A}. \tag{1.126}
$$

The subsequent tensorial coefficients are obtained by taking successive gradients of the series and evaluating at $\boldsymbol{r}=\boldsymbol{r}_0$ beginning with

$$
\nabla\, ({}_t\psi)|_{\boldsymbol{r}_0} = \nabla\boldsymbol{r}\cdot {}_{\{t+1\}}\boldsymbol{A} = {}_{\{1+t\}}\boldsymbol{A}, \tag{1.127}
$$

$$
\nabla\nabla\, ({}_t\psi)|_{\boldsymbol{r}_0} = \nabla\nabla\, (\boldsymbol{r}\boldsymbol{r}) : {}_{\{2+t\}}\boldsymbol{A}, \tag{1.128}
$$

where the tensorial coefficients ${}_{\{n+t\}}\boldsymbol{A}$ are constants (the ∇ acting on them gives zero) and where in Eq. (1.127) we used the earlier result that $\nabla\boldsymbol{r}=\boldsymbol{I}$ is the second-order identity tensor. It is a straightforward exercise to show that $\nabla\nabla\,(\boldsymbol{r}\boldsymbol{r}) = \hat{\boldsymbol{x}}_i\hat{\boldsymbol{x}}_j\hat{\boldsymbol{x}}_j\hat{\boldsymbol{x}}_i + \hat{\boldsymbol{x}}_i\hat{\boldsymbol{x}}_j\hat{\boldsymbol{x}}_i\hat{\boldsymbol{x}}_j$ with summation over the repeated indices as always, which is a type of fourth-order identity-transpose tensor. To work with it, we double dot it into the tensorial coefficient ${}_{\{2+t\}}\boldsymbol{A} = A_{kl\ldots\alpha}\hat{\boldsymbol{x}}_k\hat{\boldsymbol{x}}_l\ldots\hat{\boldsymbol{x}}_\alpha$, where how many base vectors this tensorial coefficient has beyond the first two depends on the tensorial order t of the field being expanded. A scalar field ($t=0$) has no additional base vectors, a vector field ($t=1$) one additional base vector and so on. We have

$$
\nabla\nabla\,(\boldsymbol{r}\boldsymbol{r}) : {}_{\{2+t\}}\boldsymbol{A} = \left(\hat{\boldsymbol{x}}_i\hat{\boldsymbol{x}}_j\hat{\boldsymbol{x}}_j\hat{\boldsymbol{x}}_i + \hat{\boldsymbol{x}}_i\hat{\boldsymbol{x}}_j\hat{\boldsymbol{x}}_i\hat{\boldsymbol{x}}_j\right) : A_{kl\ldots\alpha}\hat{\boldsymbol{x}}_k\hat{\boldsymbol{x}}_l\ldots\hat{\boldsymbol{x}}_\alpha \tag{1.129}
$$

$$
= \left(A_{ij\ldots\alpha} + A_{ji\ldots\alpha}\right)\hat{\boldsymbol{x}}_i\hat{\boldsymbol{x}}_j\ldots\hat{\boldsymbol{x}}_\alpha. \tag{1.130}
$$

Because $\nabla\nabla(\boldsymbol{r}\boldsymbol{r})$ is symmetric in the last two base vectors, Eq. (1.128) requires the tensorial coefficient ${}_{\{2+t\}}\boldsymbol{A}$ to be symmetric in the first two base vectors, so that

$$
\nabla\nabla\, ({}_t\psi)|_{\boldsymbol{r}_0} = (2)\,{}_{\{2+t\}}\boldsymbol{A}. \tag{1.131}
$$

An identical analysis for the third-order coefficient that exploits the symmetry of $\nabla\nabla\nabla(\boldsymbol{r}\boldsymbol{r}\boldsymbol{r})$ yields

$$
\nabla\nabla\nabla\, ({}_t\psi)|_{\boldsymbol{r}_0} = (3)(2)\,{}_{\{3+t\}}\boldsymbol{A} \tag{1.132}
$$

with ${}_{\{3+t\}}\boldsymbol{A}$ having complete symmetry between the first three base vectors. Thus, the nth tensorial coefficient in the Taylor series of a tensorial field is given as

$$
{}_{\{n+t\}}\boldsymbol{A} = \frac{1}{n!}\,{}_{n}\nabla\,({}_{t}\boldsymbol{\psi})|_{\boldsymbol{r}_0} \quad \text{for} \quad n = 1, 2 \ldots \infty, \tag{1.133}
$$

where ${}_{n}\nabla = \nabla\nabla \ldots \nabla$ means n successive applications of the gradient operator acting on the tensor field ${}_{t}\boldsymbol{\psi}$ of order t before evaluating at the particular point in 3D space $\boldsymbol{r} = \boldsymbol{r}_0$. The leading $n = 0$ coefficient is again ${}_{t}\boldsymbol{A} = {}_{t}\boldsymbol{\psi}\,(\boldsymbol{r}_0)$.

So as an example from Chapter 2, we may wish to represent the electric field $\boldsymbol{E}(\boldsymbol{r})$ within a molecule whose center is located at $\boldsymbol{r}_0$ as an explicit function in the local coordinates $\boldsymbol{r} - \boldsymbol{r}_0$ within the molecule. We thus perform a Taylor-series expansion of the electric field about $\boldsymbol{r}_0$ to give

$$
\begin{aligned}
\boldsymbol{E}(\boldsymbol{r}) = {} & \boldsymbol{E}(\boldsymbol{r}_0) + (\boldsymbol{r} - \boldsymbol{r}_0) \cdot \nabla\boldsymbol{E}|_{\boldsymbol{r}_0} + \\
& + \frac{1}{2!}(\boldsymbol{r} - \boldsymbol{r}_0)\,(\boldsymbol{r} - \boldsymbol{r}_0) : \nabla\nabla\boldsymbol{E}|_{\boldsymbol{r}_0} + O\left(|\boldsymbol{r} - \boldsymbol{r}_0|^3\right).
\end{aligned} \tag{1.134}
$$

The notation $O\left(|\boldsymbol{r} - \boldsymbol{r}_0|^3\right)$ is called the "big-O" notation and is used to represent the part of the series that is being truncated in a certain limit such as $\boldsymbol{r} - \boldsymbol{r}_0 \to 0$. In this limit for this particular example, the amplitude $|\boldsymbol{r} - \boldsymbol{r}_0|^3$ is the largest part of what is being truncated in Eq. (1.134), which is what the notation $O\left(|\boldsymbol{r} - \boldsymbol{r}_0|^3\right)$ is saying. Let's write a power series in the parameter ϵ as $f(\epsilon) = a_0 + a_1\epsilon + a_2\epsilon^2 + a_3\epsilon^3 + \ldots$. If in the limit of $\epsilon \to 0$, we truncate the series after the first two terms, we use the big-O notation to write $f(\epsilon) = a_0 + a_1\epsilon + O(\epsilon^2)$, where the argument of $O(\epsilon^2)$ represents the size or "order" of what is being neglected in the limit, which in this case is $O(\epsilon^2) = a_2\epsilon^2[1 + (a_3/a_2)\epsilon + (a_4/a_2)\epsilon^2 + \ldots] \to a_2\epsilon^2$ as $\epsilon \to 0$.

1.8.2 Functions of Second-Order Tensors

It will arise that we want to consider a function whose argument is a second-order tensor, that is, $f(\boldsymbol{A})$ where both $\boldsymbol{A}$ and $f(\boldsymbol{A})$ are second-order tensors. What do we mean by this?

If $f(\alpha)$ is some function of a scalar α, we expand $f(\alpha)$ as a Taylor series about $\alpha = 0$ as

$$
\begin{aligned}
f(\alpha) = {} & f(0) + \frac{1}{1!}\left.\frac{\mathrm{d}f(\alpha)}{\mathrm{d}\alpha}\right|_{\alpha=0} \alpha \\
& + \frac{1}{2!}\left.\frac{\mathrm{d}^2 f(\alpha)}{\mathrm{d}\alpha^2}\right|_{\alpha=0} \alpha^2 + \frac{1}{3!}\left.\frac{\mathrm{d}^3 f(\alpha)}{\mathrm{d}\alpha^3}\right|_{\alpha=0} \alpha^3 + \ldots.
\end{aligned} \tag{1.135}
$$

Because α is a scalar, so is $f(\alpha)$. We now define the second-order tensor $f(\boldsymbol{A})$ as the operation or rule

$$
\boxed{
\begin{aligned}
f(\boldsymbol{A}) = & f(0)\boldsymbol{I} + \frac{1}{1!}\left.\frac{\mathrm{d}f(\alpha)}{\mathrm{d}\alpha}\right|_{\alpha=0}\boldsymbol{A} \\
& + \frac{1}{2!}\left.\frac{\mathrm{d}^2 f(\alpha)}{\mathrm{d}\alpha^2}\right|_{\alpha=0}\boldsymbol{A}\cdot\boldsymbol{A} + \frac{1}{3!}\left.\frac{\mathrm{d}^3 f(\alpha)}{\mathrm{d}\alpha^3}\right|_{\alpha=0}\boldsymbol{A}\cdot\boldsymbol{A}\cdot\boldsymbol{A} + \ldots,
\end{aligned}
} \tag{1.136}
$$

where $\boldsymbol{A}$ is a second-order tensor, $\boldsymbol{I}$ is the second-order identity tensor, and each term in the series is a second-order tensor.

Let's give some examples. Consider the specific scalar function $f_1(\alpha) = (1-\alpha)^{-1} = 1 + \alpha + \alpha^2 + \ldots$. We then can define the second-order tensor $f_1(\boldsymbol{A})$ operation as $f_1(\boldsymbol{A}) = (\boldsymbol{I} - \boldsymbol{A})^{-1}$, where $\boldsymbol{I}$ is again the second-order identity tensor. The operation $(\boldsymbol{I}-\boldsymbol{A})^{-1}$ is understood, through Eq. (1.136), to mean the expansion

$$
(\boldsymbol{I}-\boldsymbol{A})^{-1} = \boldsymbol{I} + \boldsymbol{A} + \boldsymbol{A}\cdot\boldsymbol{A} + \boldsymbol{A}\cdot\boldsymbol{A}\cdot\boldsymbol{A} + \ldots. \tag{1.137}
$$

We then expect that $(\boldsymbol{I}-\boldsymbol{A})\cdot(\boldsymbol{I}-\boldsymbol{A})^{-1} = \boldsymbol{I}$, which can be verified through explicit multiplication

$$
(\boldsymbol{I}-\boldsymbol{A})\cdot(\boldsymbol{I}-\boldsymbol{A})^{-1} = (\boldsymbol{I}-\boldsymbol{A})\cdot(\boldsymbol{I} + \boldsymbol{A} + \boldsymbol{A}\cdot\boldsymbol{A} + \boldsymbol{A}\cdot\boldsymbol{A}\cdot\boldsymbol{A} + \ldots) = \boldsymbol{I}. \tag{1.138}
$$

So a function of a second-order tensor, that is itself a second-order tensor, is coherently defined through the expansion of Eq. (1.136).

As another specific example, define a second-order tensor $\boldsymbol{B}$ as the function of another second-order tensor $\boldsymbol{A}$ through the operations

$$
\boldsymbol{B} = -\ln(\boldsymbol{I}-\boldsymbol{A}) = \boldsymbol{A} + \frac{1}{2}\boldsymbol{A}\cdot\boldsymbol{A} + \frac{1}{3}\boldsymbol{A}\cdot\boldsymbol{A}\cdot\boldsymbol{A} + \ldots. \tag{1.139}
$$

We can then take the exponential of $-\boldsymbol{B}$ by which we mean the expansion

$$
\begin{aligned}
\exp(-\boldsymbol{B}) &= \boldsymbol{I} - \frac{1}{1!}\boldsymbol{B} + \frac{1}{2!}\boldsymbol{B}\cdot\boldsymbol{B} - \frac{1}{3!}\boldsymbol{B}\cdot\boldsymbol{B}\cdot\boldsymbol{B} + \ldots \\
&= \exp(\ln(\boldsymbol{I}-\boldsymbol{A})) = \boldsymbol{I} - \boldsymbol{A}.
\end{aligned} \tag{1.140}
$$

This last relation then gives

$$
\boldsymbol{A} = \boldsymbol{I} - \exp(-\boldsymbol{B}) = \frac{1}{1!}\boldsymbol{B} - \frac{1}{2!}\boldsymbol{B}\cdot\boldsymbol{B} + \frac{1}{3!}\boldsymbol{B}\cdot\boldsymbol{B}\cdot\boldsymbol{B} + \ldots. \tag{1.141}
$$

If we then substitute the original definition of $\boldsymbol{B}$ from Eq. (1.139), we obtain

$$
\begin{aligned}
\boldsymbol{A} = & \frac{1}{1!}\left(\boldsymbol{A} + \frac{1}{2}\boldsymbol{A}\cdot\boldsymbol{A} + \frac{1}{3}\boldsymbol{A}\cdot\boldsymbol{A}\cdot\boldsymbol{A} + \ldots\right) \\
& - \frac{1}{2!}(\boldsymbol{A}\cdot\boldsymbol{A} + \boldsymbol{A}\cdot\boldsymbol{A}\cdot\boldsymbol{A} + \ldots) + \frac{1}{3!}(\boldsymbol{A}\cdot\boldsymbol{A}\cdot\boldsymbol{A} + \ldots) + \ldots
\end{aligned} \tag{1.142}
$$

$$
= \boldsymbol{A}. \tag{1.143}
$$

So these operations are internally consistent.

To conclude, the function of a second-order tensor is defined here to be another second-tensor tensor as calculated by the expansion of Eq. (1.136). We can also define the inverse function of a second-order tensor as we have shown in the above examples that we will see again later in the book.

1.8.3 Rotation of the Cartesian Coordinates

As we have emphasized, a tensor of any order, including a first-order tensor or vector, is a field that exists at each point in space independently of whatever coordinates we choose to work in. But it can arise that we want to work in a Cartesian coordinate system $\hat{\boldsymbol{x}}'_i$ that has been rotated from an initial system $\hat{\boldsymbol{x}}_i$ as shown in Fig. 1.12. We would like to know how the *scalar components* of the vectors and tensors change when we rotate the base vectors to have new orientations. It is sometimes stated that a tensor is defined by the rules derived below for how the Cartesian components of the tensor change with the changing orientation of the base vectors. However, we have already seen that tensors of any order are coherently defined without first having in place such "coordinate-rotation rules." The fundamental nature (and need) of tensors as used in continuum physics is again that they map, using dot products, one tensor (including vectors) into another tensor and such tensorial mappings exist independently of knowing how the Cartesian coefficients of a tensor change with the orientation of the coordinates.

A rotation of angle θ_1 about the $\hat{\boldsymbol{x}}_1$ axis is allowed for by the matrix operation (θ_1 positive is in a counterclockwise sense when $\hat{\boldsymbol{x}}_1$ is oriented toward the observer)

$$\begin{bmatrix} x'_1 \\ x'_2 \\ x'_3 \end{bmatrix} = \begin{bmatrix} 1 & 0 & 0 \\ 0 & \cos\theta_1 & \sin\theta_1 \\ 0 & -\sin\theta_1 & \cos\theta_1 \end{bmatrix} \begin{bmatrix} x_1 \\ x_2 \\ x_3 \end{bmatrix}, \tag{1.144}$$

which is easily confirmed by doing the trigonometry in Fig. 1.13.

We define the "rotation matrix" for rotations around the x_1 axis as

$$R^{(1)}_{ij}(\theta_1) = \begin{bmatrix} 1 & 0 & 0 \\ 0 & \cos\theta_1 & \sin\theta_1 \\ 0 & -\sin\theta_1 & \cos\theta_1 \end{bmatrix}. \tag{1.145}$$

Figure 1.12 Two Cartesian-coordinate systems that are rotated relative to each other.

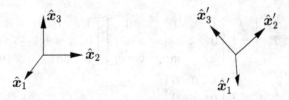

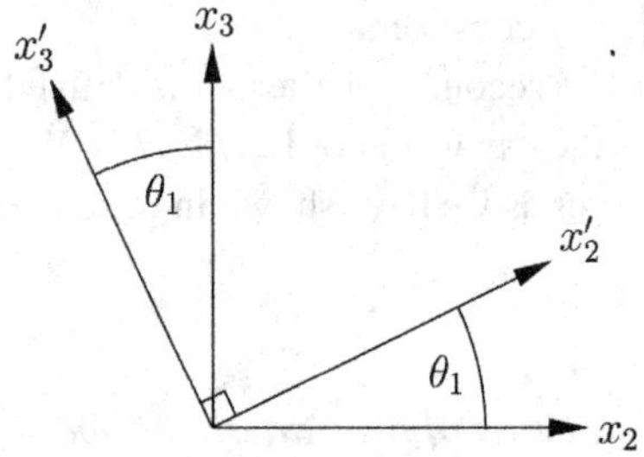

Figure 1.13 Rotating the coordinates counterclockwise by an angle θ_1 around the x_1 axis.

The components of a vector $\boldsymbol{a} = a_1\,\hat{\boldsymbol{x}}_1 + a_2\,\hat{\boldsymbol{x}}_2 + a_3\,\hat{\boldsymbol{x}}_3$ transform in the rotated coordinate system to $a_i' = R_{ij}^{(1)}(\theta_1)\,a_j$ with summation over the repeated index j being performed as the operation

$$\begin{bmatrix} a_1' \\ a_2' \\ a_3' \end{bmatrix} = \begin{bmatrix} 1 & 0 & 0 \\ 0 & \cos\theta_1 & \sin\theta_1 \\ 0 & -\sin\theta_1 & \cos\theta_1 \end{bmatrix} \begin{bmatrix} a_1 \\ a_2 \\ a_3 \end{bmatrix}. \tag{1.146}$$

We just have to carry out the matrix multiplication.

Similarly, a second-order tensor $\boldsymbol{T} = T_{ij}\,\hat{\boldsymbol{x}}_i\,\hat{\boldsymbol{x}}_j$ has Cartesian components T_{ij} that transform as

$$T_{ij}' = R_{ik}^{(1)}(\theta_1)\,R_{jl}^{(1)}(\theta_1)\,T_{kl} \quad \text{(sum over repeated indices)}, \tag{1.147}$$

where we have to apply the rotation matrix to each base vector. If we write the sums over repeated indices using matrix multiplication (the inner product), we must rearrange this expression as $T_{ij}' = R_{ik}^{(1)}\,T_{kl}\left(R_{jl}^{(1)}\right)^T$ so that the position of the indices correspond to the inner product and the matrix operation

$$\begin{bmatrix} T_{11}' & T_{12}' & T_{13}' \\ T_{21}' & T_{22}' & T_{23}' \\ T_{31}' & T_{32}' & T_{33}' \end{bmatrix} = \begin{bmatrix} 1 & 0 & 0 \\ 0 & \cos\theta_1 & \sin\theta_1 \\ 0 & -\sin\theta_1 & \cos\theta_1 \end{bmatrix} \begin{bmatrix} T_{11} & T_{12} & T_{13} \\ T_{21} & T_{22} & T_{23} \\ T_{31} & T_{32} & T_{33} \end{bmatrix} \begin{bmatrix} 1 & 0 & 0 \\ 0 & \cos\theta_1 & -\sin\theta_1 \\ 0 & \sin\theta_1 & \cos\theta_1 \end{bmatrix}. \tag{1.148}$$

Note that if $\boldsymbol{T}$ is proportional to the second-order identity tensor $\boldsymbol{T} = T\delta_{ij}\hat{\boldsymbol{x}}_i\hat{\boldsymbol{x}}_j$, we have

$$\begin{bmatrix} T_{11}' & T_{12}' & T_{13}' \\ T_{21}' & T_{22}' & T_{23}' \\ T_{31}' & T_{32}' & T_{33}' \end{bmatrix} = T \begin{bmatrix} 1 & 0 & 0 \\ 0 & \cos\theta_1 & \sin\theta_1 \\ 0 & -\sin\theta_1 & \cos\theta_1 \end{bmatrix} \begin{bmatrix} 1 & 0 & 0 \\ 0 & \cos\theta_1 & -\sin\theta_1 \\ 0 & \sin\theta_1 & \cos\theta_1 \end{bmatrix} \tag{1.149}$$

$$= T \begin{bmatrix} 1 & 0 & 0 \\ 0 & 1 & 0 \\ 0 & 0 & 1 \end{bmatrix}. \tag{1.150}$$

So the second-order identity tensor satisfies $T'_{ij} = T\delta'_{ij} = T_{ij} = T\delta_{ij}$ or $\delta'_{ij} = \delta_{ij}$ for any rotation of the coordinates about the x_1 axis.

Similarly, a fourth-order tensor has components that transform with coordinate rotations about the x_1 axis as

$$C'_{ijkl} = R^{(1)}_{im}(\theta_1)\, R^{(1)}_{jn}(\theta_1)\, R^{(1)}_{ko}(\theta_1)\, R^{(1)}_{lp}(\theta_1)\, C_{mnop} \tag{1.151}$$

with summation over the repeated indices *m*, *n*, *o*, and *p*. Higher-order tensors are handled in an analogous manner, using one rotation matrix for each base vector.

A rotation about the x_2 axis is accomplished using (θ_2 is again in the counterclockwise sense when $\hat{\boldsymbol{x}}_2$ is oriented toward the observer but note the sign change on the $\sin\theta_2$ relative to the other rotations)

$$R^{(2)}_{ij}(\theta_2) = \begin{bmatrix} \cos\theta_2 & 0 & -\sin\theta_2 \\ 0 & 1 & 0 \\ \sin\theta_2 & 0 & \cos\theta_2 \end{bmatrix} \tag{1.152}$$

and about the x_3 axis using

$$R^{(3)}_{ij}(\theta_3) = \begin{bmatrix} \cos\theta_3 & \sin\theta_3 & 0 \\ -\sin\theta_3 & \cos\theta_3 & 0 \\ 0 & 0 & 1 \end{bmatrix}. \tag{1.153}$$

Now, any conceivable rotation $(\theta_1, \theta_2, \theta_3)$ is accomplished using the rotation matrix

$$R_{ij}(\theta_1, \theta_2, \theta_3) = R^{(1)}_{ik}(\theta_1)\, R^{(2)}_{kl}(\theta_2)\, R^{(3)}_{lj}(\theta_3), \tag{1.154}$$

that is, just matrix multiply the three rotation matrices together, which you can do as an end-of-chapter exercise. Tensors of any order again have Cartesian components that transform according to the above rules using $R_{ij}(\theta_1, \theta_2, \theta_3)$ as the rotation matrix, using one rotation matrix for each base vector.

As an example, the second-order identity-tensor coefficients δ_{ij} transform as $R_{ik}(\theta_1, \theta_2, \theta_3) R_{jl}(\theta_1, \theta_2, \theta_3)\delta_{kl} = R_{ik}(\theta_1, \theta_2, \theta_3)\left[R_{jk}(\theta_1, \theta_2, \theta_3)\right]^T = \delta_{ij}$, which you can confirm through direct matrix multiplication as an end-of-chapter exercise. This means that the second-order identity tensor is *isotropic*, which means that $\delta'_{ij} = \delta_{ij}$ for arbitrary coordinate rotations.

It can be convenient for proving transformation identities involving higher-order tensors if we consider small rotations $\delta\theta_i$ around each axis so that $\cos\delta\theta_i = 1 + O(\delta\theta_i^2)$ and $\sin\delta\theta_i = \delta\theta_i\left[1 + O(\delta\theta_i^2)\right]$. Upon ignoring the $O(\delta\theta_i^2)$ terms in what follows, the rotation matrix becomes

$$R_{ij}(\delta\theta_1, \delta\theta_2, \delta\theta_3) = \begin{pmatrix} 1 & 0 & 0 \\ 0 & 1 & 0 \\ 0 & 0 & 1 \end{pmatrix} + \delta\theta_1 \begin{pmatrix} 0 & 0 & 0 \\ 0 & 0 & 1 \\ 0 & -1 & 0 \end{pmatrix} + \delta\theta_2 \begin{pmatrix} 0 & 0 & -1 \\ 0 & 0 & 0 \\ 1 & 0 & 0 \end{pmatrix} + \delta\theta_3 \begin{pmatrix} 0 & 1 & 0 \\ -1 & 0 & 0 \\ 0 & 0 & 0 \end{pmatrix}. \quad (1.155)$$

This can also be written as

$$R_{ij} = \delta_{ij} + \delta\theta_m \epsilon_{mij}, \quad (1.156)$$

where the ϵ_{mij} are the Levi–Civita coefficients. So, for example, if we want to find the isotropic second-order tensor that, by definition, has coefficients that satisfy $T'_{ij} = T_{ij}$, we write

$$\begin{aligned} T'_{ij} = T_{ij} &= R_{ik}R_{jl}T_{kl}, &(1.157)\\ &= (\delta_{ik} + \delta\theta_m\epsilon_{mik})\left(\delta_{jl} + \delta\theta_m\epsilon_{mjl}\right) T_{kl}, &(1.158)\\ &= \left[\delta_{ik}\delta_{jl} + \delta\theta_m\left(\epsilon_{mik}\delta_{jl} + \epsilon_{mjl}\delta_{ik}\right)\right] T_{kl}, &(1.159)\\ &= T_{ij} + \delta\theta_m\left(\epsilon_{mik}T_{kj} + \epsilon_{mjk}T_{ik}\right). &(1.160)\end{aligned}$$

This equation is satisfied if

$$\epsilon_{mik}T_{kj} = -\epsilon_{mjk}T_{ik}, \quad (1.161)$$

which has the solution $T_{ij} = T\delta_{ij}$, where T is any scalar, as can be shown through direct substitution. Thus we have that $\boldsymbol{T} = T\delta_{ij}\hat{\boldsymbol{x}}_i\hat{\boldsymbol{x}}_j = T\boldsymbol{I}$ is the form of the one and only isotropic second-order tensor.

If we want to find the isotropic third-order tensor that, by definition, has coefficients that satisfy $T'_{ijk} = T_{ijk}$, we write

$$\begin{aligned} T'_{ijk} = T_{ijk} &= R_{il}R_{jm}R_{kn}T_{lmn}, &(1.162)\\ &= \left(\delta_{il} + \delta\theta_p\epsilon_{pil}\right)\left(\delta_{jm} + \delta\theta_p\epsilon_{pjm}\right)\left(\delta_{kn} + \delta\theta_p\epsilon_{pkn}\right) T_{lmn}, &(1.163)\\ &= \left[\delta_{il}\delta_{jm}\delta_{kn} + \delta\theta_p\left(\epsilon_{pil}\delta_{jm}\delta_{kn} + \epsilon_{pjm}\delta_{il}\delta_{kn} + \epsilon_{pkn}\delta_{il}\delta_{jm}\right)\right] T_{lmn} &(1.164)\\ &= T_{ijk} + \delta\theta_p\left(\epsilon_{pil}T_{ljk} + \epsilon_{pjm}T_{imk} + \epsilon_{pkn}T_{ijn}\right), &(1.165)\end{aligned}$$

where terms of $O(\delta\theta_i^2)$ are again ignored. This equation is satisfied if

$$\epsilon_{pil}T_{ljk} + \epsilon_{pjm}T_{imk} + \epsilon_{pkn}T_{ijn} = 0. \quad (1.166)$$

The solution of this equation is $T_{ijk} = T\epsilon_{ijk}$, where again T is any scalar, as can be seen through substitution

$$\epsilon_{pil}\epsilon_{ljk} + \epsilon_{pjl}\epsilon_{ilk} + \epsilon_{pkl}\epsilon_{ijl} = 0. \quad (1.167)$$

Rewrite this as

$$\epsilon_{lpi}\epsilon_{ljk} - \epsilon_{lpj}\epsilon_{lik} + \epsilon_{lpk}\epsilon_{lij} = 0. \tag{1.168}$$

The identity of Eq. (1.54) can then be used to write each term on the left-hand side as

$$\epsilon_{lpi}\epsilon_{ljk} = \delta_{pj}\delta_{ik} - \delta_{pk}\delta_{ij}, \tag{1.169}$$

$$-\epsilon_{lpj}\epsilon_{lik} = -\delta_{pi}\delta_{jk} + \delta_{pk}\delta_{ij}, \tag{1.170}$$

$$\epsilon_{lpk}\epsilon_{lji} = -\delta_{pj}\delta_{ik} + \delta_{pi}\delta_{jk}, \tag{1.171}$$

which sum to zero when substituted into Eq. (1.168). Thus, we have shown that ${}_3\boldsymbol{T} = T\epsilon_{ijk}\hat{\boldsymbol{x}}_i\hat{\boldsymbol{x}}_j\hat{\boldsymbol{x}}_k = (T)\,({}_3\boldsymbol{\epsilon})$ is the form of the one and only third-order isotropic tensor with T some arbitrary scalar.

1.8.4 Isotropic Tensors of Any Order

As just seen, isotropic tensors are those tensors whose components do not change when we change the orientation of the Cartesian coordinates. Specifically, the second-order coefficients δ_{ij} and third-order coefficients ϵ_{ijk} do not change when changing the orientation of the coordinates. As such, even-ordered isotropic tensors have coefficients that are multiples of the Kronecker coefficients δ_{ij} and odd-ordered isotropic tensors have coefficients that involve the single presence of the Levi–Civita coefficients ϵ_{ijk} and additional multiples of the Kronecker coefficients that get to the desired (odd) tensorial order. So higher-order isotropic tensors involve multiples of Kronecker and Levi–Civita coefficients with numbers of indices that add up to the tensorial order (or rank) of interest.

We call a zeroth-order tensor a scalar and all scalars are, by definition, independent of the orientation of the axes. So all scalars are isotropic.

We call a first-order tensor a vector and all vectors of finite length have components that change when the axes are rotated. So an "isotropic vector" has zero length and does not exist.

As proven above, there is one fundamental *second-order isotropic tensor* ${}_2\boldsymbol{I}$, which is the second-order identity tensor $\boldsymbol{I}$,

$${}_2\boldsymbol{I} = \boldsymbol{I} = \delta_{ij}\hat{\boldsymbol{x}}_i\hat{\boldsymbol{x}}_j. \tag{1.172}$$

Similarly, we showed there is one fundamental *third-order isotropic tensor* ${}_3\boldsymbol{I}$, which is the third-order Levi–Civita alternating (or "antisymmetric" or "permutation") tensor,

$${}_3\boldsymbol{I} = {}_3\boldsymbol{\epsilon} = \epsilon_{ijk}\hat{\boldsymbol{x}}_i\hat{\boldsymbol{x}}_j\hat{\boldsymbol{x}}_k. \tag{1.173}$$

We can multiply these fundamental second-order and third-order isotropic tensors by scalars, and the result will also be an isotropic tensor.

Higher-order isotropic tensors have coefficients that involve additional multiples of the Kronecker coefficients. So, for example, there are multiple fundamental *fourth-order isotropic tensors* ${}_4\boldsymbol{I}^{(m)}$. If we define the first dummy index of these coefficients to always be i, there are three unique ways to place the remaining j, k, and l indices across the a, b, and c positions of $\delta_{ia}\delta_{bc}$, that is,

$$ {}_4\boldsymbol{I}^{(1)} = \delta_{il}\delta_{jk}\hat{\boldsymbol{x}}_i\hat{\boldsymbol{x}}_j\hat{\boldsymbol{x}}_k\hat{\boldsymbol{x}}_l, \tag{1.174} $$

$$ {}_4\boldsymbol{I}^{(2)} = \delta_{ik}\delta_{jl}\hat{\boldsymbol{x}}_i\hat{\boldsymbol{x}}_j\hat{\boldsymbol{x}}_k\hat{\boldsymbol{x}}_l, \tag{1.175} $$

$$ {}_4\boldsymbol{I}^{(3)} = \delta_{ij}\delta_{kl}\hat{\boldsymbol{x}}_i\hat{\boldsymbol{x}}_j\hat{\boldsymbol{x}}_k\hat{\boldsymbol{x}}_l = \boldsymbol{I}\boldsymbol{I}. \tag{1.176} $$

There are not more than these three fundamental fourth-order isotropic tensors due to the symmetry $\delta_{ij} = \delta_{ji}$. The rule for the number M of fundamental even-ordered $n = 4, 6, 8, \ldots$ isotropic tensors involving only the Kronecker coefficients is

$$ M = \prod_{i=1}^{n/2} (n - 2i + 1) \quad \text{for even } n. \tag{1.177} $$

So for $n = 4$, we have $M = (n-1)(n-3) = (3)(1) = 3$ as seen in Eq. (1.177).

The fundamental *fifth-order isotropic tensors* ${}_5\boldsymbol{I}^{(m)}$ have coefficients that involve a single multiplication between the Levi–Civita coefficients and the Kronecker coefficients. If we define the first dummy index of these fifth-order coefficients as i, there are six unique nonzero ways to place the remaining j, k, l, m indices across the a, b, c, d positions of $\epsilon_{iab}\delta_{cd}$ and four unique nonzero way to place the j, k, l, m across the a, b, c, d positions of $\delta_{ia}\epsilon_{bcd}$ to give

$$ {}_5\boldsymbol{I}^{(1)} = \epsilon_{ijk}\delta_{lm}\hat{\boldsymbol{x}}_i\hat{\boldsymbol{x}}_j\hat{\boldsymbol{x}}_k\hat{\boldsymbol{x}}_l\hat{\boldsymbol{x}}_m = ({}_3\boldsymbol{\epsilon})\,\boldsymbol{I}, \tag{1.178} $$

$$ {}_5\boldsymbol{I}^{(2)} = \epsilon_{ijl}\delta_{km}\hat{\boldsymbol{x}}_i\hat{\boldsymbol{x}}_j\hat{\boldsymbol{x}}_k\hat{\boldsymbol{x}}_l\hat{\boldsymbol{x}}_m, \tag{1.179} $$

$$ {}_5\boldsymbol{I}^{(3)} = \epsilon_{ijm}\delta_{kl}\hat{\boldsymbol{x}}_i\hat{\boldsymbol{x}}_j\hat{\boldsymbol{x}}_k\hat{\boldsymbol{x}}_l\hat{\boldsymbol{x}}_m, \tag{1.180} $$

$$ {}_5\boldsymbol{I}^{(4)} = \epsilon_{ikl}\delta_{jm}\hat{\boldsymbol{x}}_i\hat{\boldsymbol{x}}_j\hat{\boldsymbol{x}}_k\hat{\boldsymbol{x}}_l\hat{\boldsymbol{x}}_m, \tag{1.181} $$

$$ {}_5\boldsymbol{I}^{(5)} = \epsilon_{ikm}\delta_{jl}\hat{\boldsymbol{x}}_i\hat{\boldsymbol{x}}_j\hat{\boldsymbol{x}}_k\hat{\boldsymbol{x}}_l\hat{\boldsymbol{x}}_m, \tag{1.182} $$

$$ {}_5\boldsymbol{I}^{(6)} = \epsilon_{ilm}\delta_{jk}\hat{\boldsymbol{x}}_i\hat{\boldsymbol{x}}_j\hat{\boldsymbol{x}}_k\hat{\boldsymbol{x}}_l\hat{\boldsymbol{x}}_m, \tag{1.183} $$

$$ {}_5\boldsymbol{I}^{(7)} = \delta_{ij}\epsilon_{klm}\hat{\boldsymbol{x}}_i\hat{\boldsymbol{x}}_j\hat{\boldsymbol{x}}_k\hat{\boldsymbol{x}}_l\hat{\boldsymbol{x}}_m = \boldsymbol{I}\,({}_3\boldsymbol{\epsilon}), \tag{1.184} $$

$$ {}_5\boldsymbol{I}^{(8)} = \delta_{ik}\epsilon_{jlm}\hat{\boldsymbol{x}}_i\hat{\boldsymbol{x}}_j\hat{\boldsymbol{x}}_k\hat{\boldsymbol{x}}_l\hat{\boldsymbol{x}}_m, \tag{1.185} $$

$$ {}_5\boldsymbol{I}^{(9)} = \delta_{il}\epsilon_{jkm}\hat{\boldsymbol{x}}_i\hat{\boldsymbol{x}}_j\hat{\boldsymbol{x}}_k\hat{\boldsymbol{x}}_l\hat{\boldsymbol{x}}_m, \tag{1.186} $$

$$ {}_5\boldsymbol{I}^{(10)} = \delta_{im}\epsilon_{jkl}\hat{\boldsymbol{x}}_i\hat{\boldsymbol{x}}_j\hat{\boldsymbol{x}}_k\hat{\boldsymbol{x}}_l\hat{\boldsymbol{x}}_m, \tag{1.187} $$

There are not more than these 10 fundamental fifth-order isotropic tensors because $\delta_{ij} = \delta_{ji}$ and $\epsilon_{ijk} = -\epsilon_{ikj} = -\epsilon_{kij}$, that is, multiplying by -1 does not create a distinct fundamental isotropic tensor.

Proceeding like above gives the 15 *sixth-order isotropic tensors* ${}_6\boldsymbol{I}^{(m)}$ involving the Kronecker coefficients:

$$ {}_6\boldsymbol{I}^{(1)} = \delta_{in}\delta_{ml}\delta_{jk}\hat{\boldsymbol{x}}_i\hat{\boldsymbol{x}}_j\hat{\boldsymbol{x}}_k\hat{\boldsymbol{x}}_l\hat{\boldsymbol{x}}_m\hat{\boldsymbol{x}}_n, \tag{1.188} $$
$$ {}_6\boldsymbol{I}^{(2)} = \delta_{in}\delta_{mk}\delta_{jl}\hat{\boldsymbol{x}}_i\hat{\boldsymbol{x}}_j\hat{\boldsymbol{x}}_k\hat{\boldsymbol{x}}_l\hat{\boldsymbol{x}}_m\hat{\boldsymbol{x}}_n, \tag{1.189} $$
$$ {}_6\boldsymbol{I}^{(3)} = \delta_{in}\delta_{mj}\delta_{kl}\hat{\boldsymbol{x}}_i\hat{\boldsymbol{x}}_j\hat{\boldsymbol{x}}_k\hat{\boldsymbol{x}}_l\hat{\boldsymbol{x}}_m\hat{\boldsymbol{x}}_n, \tag{1.190} $$
$$ {}_6\boldsymbol{I}^{(4)} = \delta_{im}\delta_{nl}\delta_{jk}\hat{\boldsymbol{x}}_i\hat{\boldsymbol{x}}_j\hat{\boldsymbol{x}}_k\hat{\boldsymbol{x}}_l\hat{\boldsymbol{x}}_m\hat{\boldsymbol{x}}_n, \tag{1.191} $$
$$ {}_6\boldsymbol{I}^{(5)} = \delta_{im}\delta_{nk}\delta_{jl}\hat{\boldsymbol{x}}_i\hat{\boldsymbol{x}}_j\hat{\boldsymbol{x}}_k\hat{\boldsymbol{x}}_l\hat{\boldsymbol{x}}_m\hat{\boldsymbol{x}}_n, \tag{1.192} $$
$$ {}_6\boldsymbol{I}^{(6)} = \delta_{im}\delta_{nj}\delta_{kl}\hat{\boldsymbol{x}}_i\hat{\boldsymbol{x}}_j\hat{\boldsymbol{x}}_k\hat{\boldsymbol{x}}_l\hat{\boldsymbol{x}}_m\hat{\boldsymbol{x}}_n, \tag{1.193} $$
$$ {}_6\boldsymbol{I}^{(7)} = \delta_{il}\delta_{nm}\delta_{jk}\hat{\boldsymbol{x}}_i\hat{\boldsymbol{x}}_j\hat{\boldsymbol{x}}_k\hat{\boldsymbol{x}}_l\hat{\boldsymbol{x}}_m\hat{\boldsymbol{x}}_n = \left({}_4\boldsymbol{I}^{(1)}\right)\boldsymbol{I}, \tag{1.194} $$
$$ {}_6\boldsymbol{I}^{(8)} = \delta_{il}\delta_{nk}\delta_{jm}\hat{\boldsymbol{x}}_i\hat{\boldsymbol{x}}_j\hat{\boldsymbol{x}}_k\hat{\boldsymbol{x}}_l\hat{\boldsymbol{x}}_m\hat{\boldsymbol{x}}_n, \tag{1.195} $$
$$ {}_6\boldsymbol{I}^{(9)} = \delta_{il}\delta_{nj}\delta_{km}\hat{\boldsymbol{x}}_i\hat{\boldsymbol{x}}_j\hat{\boldsymbol{x}}_k\hat{\boldsymbol{x}}_l\hat{\boldsymbol{x}}_m\hat{\boldsymbol{x}}_n, \tag{1.196} $$
$$ {}_6\boldsymbol{I}^{(10)} = \delta_{ik}\delta_{nm}\delta_{jl}\hat{\boldsymbol{x}}_i\hat{\boldsymbol{x}}_j\hat{\boldsymbol{x}}_k\hat{\boldsymbol{x}}_l\hat{\boldsymbol{x}}_m\hat{\boldsymbol{x}}_n = \left({}_4\boldsymbol{I}^{(2)}\right)\boldsymbol{I} \tag{1.197} $$
$$ {}_6\boldsymbol{I}^{(11)} = \delta_{ik}\delta_{nl}\delta_{jm}\hat{\boldsymbol{x}}_i\hat{\boldsymbol{x}}_j\hat{\boldsymbol{x}}_k\hat{\boldsymbol{x}}_l\hat{\boldsymbol{x}}_m\hat{\boldsymbol{x}}_n, \tag{1.198} $$
$$ {}_6\boldsymbol{I}^{(12)} = \delta_{ik}\delta_{nj}\delta_{lm}\hat{\boldsymbol{x}}_i\hat{\boldsymbol{x}}_j\hat{\boldsymbol{x}}_k\hat{\boldsymbol{x}}_l\hat{\boldsymbol{x}}_m\hat{\boldsymbol{x}}_n, \tag{1.199} $$
$$ {}_6\boldsymbol{I}^{(13)} = \delta_{ij}\delta_{nm}\delta_{kl}\hat{\boldsymbol{x}}_i\hat{\boldsymbol{x}}_j\hat{\boldsymbol{x}}_k\hat{\boldsymbol{x}}_l\hat{\boldsymbol{x}}_m\hat{\boldsymbol{x}}_n = \left({}_4\boldsymbol{I}^{(3)}\right)\boldsymbol{I} = \boldsymbol{I}\left({}_4\boldsymbol{I}^{(3)}\right) = \boldsymbol{III}, \tag{1.200} $$
$$ {}_6\boldsymbol{I}^{(14)} = \delta_{ij}\delta_{nl}\delta_{km}\hat{\boldsymbol{x}}_i\hat{\boldsymbol{x}}_j\hat{\boldsymbol{x}}_k\hat{\boldsymbol{x}}_l\hat{\boldsymbol{x}}_m\hat{\boldsymbol{x}}_n = \boldsymbol{I}\left({}_4\boldsymbol{I}^{(2)}\right), \tag{1.201} $$
$$ {}_6\boldsymbol{I}^{(15)} = \delta_{ij}\delta_{nk}\delta_{lm}\hat{\boldsymbol{x}}_i\hat{\boldsymbol{x}}_j\hat{\boldsymbol{x}}_k\hat{\boldsymbol{x}}_l\hat{\boldsymbol{x}}_m\hat{\boldsymbol{x}}_n = \boldsymbol{I}\left({}_4\boldsymbol{I}^{(1)}\right). \tag{1.202} $$

Using the rule of Eq. (1.177) for this case of order $n = 6$, we have $M = (n-1)(n-3)(n-5) = (5)(3)(1) = 15$ as the number of fundamental sixth-order isotropic tensors involving only the Kronecker coefficients. To these can be added the sixteenth and final sixth-order isotropic tensor

$$ {}_6\boldsymbol{I}^{(16)} = \epsilon_{ijk}\epsilon_{lmn}\hat{\boldsymbol{x}}_i\hat{\boldsymbol{x}}_j\hat{\boldsymbol{x}}_k\hat{\boldsymbol{x}}_l\hat{\boldsymbol{x}}_m\hat{\boldsymbol{x}}_n = ({}_3\boldsymbol{\epsilon})\,({}_3\boldsymbol{\epsilon}) \tag{1.203} $$

with the ϵ_{ijk} the Levi–Civita coefficients.

If you want to determine the seventh-order isotropic tensors, arrange the j, k, l, m, n, o indices in the a, b, c, d, e, f positions of the following coefficients: $\epsilon_{iab}\delta_{cd}\delta_{ef}$, then $\delta_{ia}\epsilon_{bcd}\delta_{ef}$ and finally $\delta_{ia}\delta_{bc}\epsilon_{def}$. However, we will not carry out this exercise because the highest-order isotropic tensor we will encounter in our treatment of constitutive laws in this book is the sixth order as given in Section 4.1.5.

1.8.5 Curie's Principle for the Constitutive Laws of Isotropic Media

Curie's principle (Curie, 1894), as named after physicist Pierre Curie, has been a source of controversy over the years but will be taken here to be the noncontroversial statement (theorem, in fact, as will be demonstrated below) that says: "*in a constitutive law of an isotropic material, a generalized response has the same tensorial order as the generalized forces that create it.*" Curie's (1894) main point is actually the corollary statement that if a response and a force in a constitutive law are to have different tensorial orders, the material must possess *anisotropy*, that is, cannot be purely isotropic. Although we are getting ahead of ourselves in terms of the physics, we will clarify the meaning of these various words and prove the above italicized statement now, rather than in later applications, because the demonstration comes directly from the nature of the isotropic tensors that were just determined in Sections 1.8.3 and 1.8.4.

We focus here on constitutive laws associated with reversible processes, but Curie's principle also applies to irreversible processes for which the constitutive laws are called transport laws (Chapter 7). If a "force" (or "cause") is applied to an element of matter to create some "response" (or "effect"), the process is called *reversible* if when the force is returned to its initial value, the response returns to its initial value. Elastic deformation and electric and magnetic polarization are examples of reversible processes as will be developed from first principles in Chapters 3 and 4. For reversible processes, the constitutive laws are always temporal differential equations and, as developed in Chapter 6 on the thermodynamics of reversible processes, are derived by taking total time derivatives of a scalar "fundamental function" that we define here in generic form to be $u = u(\alpha, \boldsymbol{a}, \boldsymbol{A})$. This scalar function depends on time-variable forces that we represent here as a scalar $\alpha(t)$, a vector $\boldsymbol{a}(t)$, and a second-order tensor $\boldsymbol{A}(t)$ that is always symmetric. When we develop the physical nature of such a fundamental function in the chapters that follow, the function u will be seen to represent the internal energy of an element (defined later), while the scalar α can be representing entropy (defined later), the vector $\boldsymbol{a}$ can be representing the dielectric displacement or applied electric field (defined later) and the second-order tensor $\boldsymbol{A}$ is representing the elastic deformation or strain tensor (defined later) with $\boldsymbol{A}$ being symmetric. However, such physical interpretations are not required in our proof of Curie's principle that only requires a function $u = u(\alpha, \boldsymbol{a}, \boldsymbol{A})$ with $\boldsymbol{A}$ symmetric and knowledge about isotropic tensors.

Begin the proof by taking a total time derivative of the given fundamental function $u = u(\alpha, \boldsymbol{a}, \boldsymbol{A})$ to obtain

$$\frac{\mathrm{d}u}{\mathrm{d}t} = \left(\frac{\partial u}{\partial \alpha}\right)\frac{\mathrm{d}\alpha}{\mathrm{d}t} + \left(\frac{\partial u}{\partial \boldsymbol{a}}\right)\cdot\frac{\mathrm{d}\boldsymbol{a}}{\mathrm{d}t} + \left(\frac{\partial u}{\partial \boldsymbol{A}}\right):\frac{\mathrm{d}\boldsymbol{a}}{\mathrm{d}t}. \tag{1.204}$$

The partial derivatives in brackets are called the "responses" to which we give the symbolic names

$$\beta = \frac{\partial u}{\partial \alpha} \tag{1.205}$$

$$\boldsymbol{b} = \frac{\partial u}{\partial \boldsymbol{a}} \mathrel{\widehat{=}} \frac{\partial u}{\partial a_i}\hat{\boldsymbol{x}}_i, \tag{1.206}$$

$$\boldsymbol{B} = \frac{\partial u}{\partial \boldsymbol{A}} \mathrel{\widehat{=}} \frac{\partial u}{\partial A_{ij}}\hat{\boldsymbol{x}}_i\hat{\boldsymbol{x}}_j. \tag{1.207}$$

The second statements for both $\boldsymbol{b}$ and $\boldsymbol{B}$ define what it means to take a partial derivative when the independent variable is a vector or tensor. In later development, we will see that β is representing temperature if α is entropy, $\boldsymbol{b}$ is the total electric field that includes polarization if $\boldsymbol{a}$ is the applied electric field (dielectric displacement), $\boldsymbol{B}$ is the stress tensor if $\boldsymbol{A}$ is the strain tensor and Eq. (1.204) is the first law of thermodynamics. But again, such physical identifications are not required in this proof of Curie's principle, that simply posits the existence of a fundamental function $u = u(\alpha, \boldsymbol{a}, \boldsymbol{A})$ with $\boldsymbol{A}$ being symmetric.

The "constitutive laws" are the total time derivatives of the generalized responses $\beta, \boldsymbol{b}$, and $\boldsymbol{B}$:

$$\frac{d\beta}{dt} = \left(\frac{\partial^2 u}{\partial \alpha^2}\right)\frac{d\alpha}{dt} + \left(\frac{\partial^2 u}{\partial \boldsymbol{a}\partial \alpha}\right)\cdot\frac{d\boldsymbol{a}}{dt} + \left(\frac{\partial^2 u}{\partial \boldsymbol{A}\partial \alpha}\right):\frac{d\boldsymbol{A}}{dt}, \tag{1.208}$$

$$\frac{d\boldsymbol{b}}{dt} = \left(\frac{\partial^2 u}{\partial \alpha\partial \boldsymbol{a}}\right)\frac{d\alpha}{dt} + \left(\frac{\partial^2 u}{\partial \boldsymbol{a}\partial \boldsymbol{a}}\right)\cdot\frac{d\boldsymbol{a}}{dt} + \left(\frac{\partial^2 u}{\partial \boldsymbol{A}\partial \boldsymbol{a}}\right):\frac{d\boldsymbol{A}}{dt}, \tag{1.209}$$

$$\frac{d\boldsymbol{B}}{dt} = \left(\frac{\partial^2 u}{\partial \alpha\partial \boldsymbol{A}}\right)\frac{d\alpha}{dt} + \left(\frac{\partial^2 u}{\partial \boldsymbol{a}\partial \boldsymbol{A}}\right)\cdot\frac{d\boldsymbol{a}}{dt} + \left(\frac{\partial^2 u}{\partial \boldsymbol{A}\partial \boldsymbol{A}}\right):\frac{d\boldsymbol{A}}{dt}, \tag{1.210}$$

where the various double derivatives of the fundamental function having different tensorial orders are called "material properties" and can be given the symbolic names that possess the following symmetries:

$$\gamma = \frac{\partial\beta}{\partial\alpha} = \frac{\partial^2 u}{\partial\alpha^2}, \tag{1.211}$$

$$\boldsymbol{c} = \frac{\partial\beta}{\partial\boldsymbol{a}} = \frac{\partial^2 u}{\partial\boldsymbol{a}\partial\alpha} = \frac{\partial^2 u}{\partial\alpha\partial\boldsymbol{a}} = \frac{\partial\boldsymbol{b}}{\partial\alpha}, \tag{1.212}$$

$$\boldsymbol{D} = \frac{\partial\beta}{\partial\boldsymbol{A}} = \frac{\partial^2 u}{\partial\boldsymbol{A}\partial\alpha} = \frac{\partial^2 u}{\partial\alpha\partial\boldsymbol{A}} = \frac{\partial\boldsymbol{B}}{\partial\alpha} = \boldsymbol{D}^T, \tag{1.213}$$

$$\boldsymbol{E} = \frac{\partial\boldsymbol{b}}{\partial\boldsymbol{a}} = \frac{\partial^2 u}{\partial\boldsymbol{a}\partial\boldsymbol{a}} = \boldsymbol{E}^T, \tag{1.214}$$

$${}_3\boldsymbol{F} = \frac{\partial\boldsymbol{b}}{\partial\boldsymbol{A}} = \frac{\partial^2 u}{\partial\boldsymbol{A}\partial\boldsymbol{a}} = \left(\frac{\partial^2 u}{\partial\boldsymbol{a}\partial\boldsymbol{A}}\right)^{\overset{T}{231}} = {}_3\boldsymbol{F}^{\overset{T}{213}}, \tag{1.215}$$

$${}_3\boldsymbol{F}^{\overset{T}{312}} = \frac{\partial\boldsymbol{B}}{\partial\boldsymbol{a}} = \frac{\partial^2 u}{\partial\boldsymbol{a}\partial\boldsymbol{A}} = \left(\frac{\partial^2 u}{\partial\boldsymbol{A}\partial\boldsymbol{a}}\right)^{\overset{T}{312}} = {}_3\boldsymbol{F}^{\overset{T}{321}}, \tag{1.216}$$

$${}_4\boldsymbol{G} = \frac{\partial\boldsymbol{B}}{\partial\boldsymbol{A}} = \frac{\partial^2 u}{\partial\boldsymbol{A}\partial\boldsymbol{A}} = {}_4\boldsymbol{G}^{\overset{T}{3412}} = {}_4\boldsymbol{G}^{\overset{T}{2134}} = {}_4\boldsymbol{G}^{\overset{T}{1243}}. \tag{1.217}$$

So in terms of the material properties γ (a scalar), $\boldsymbol{c}$ (a vector), $\boldsymbol{D}$ (a second-order tensor), $\boldsymbol{E}$ (a second-order tensor), ${}_3\boldsymbol{F}$ (a third-order tensor), and ${}_4\boldsymbol{G}$ (a fourth-order tensor), the constitutive laws can be written

$$\frac{d\beta}{dt} = \gamma\frac{d\alpha}{dt} + \boldsymbol{c}\cdot\frac{d\boldsymbol{a}}{dt} + \boldsymbol{D}:\frac{d\boldsymbol{A}}{dt}, \tag{1.218}$$

$$\frac{d\boldsymbol{b}}{dt} = \boldsymbol{c}\frac{d\alpha}{dt} + \boldsymbol{E}\cdot\frac{d\boldsymbol{a}}{dt} + {}_3\boldsymbol{F}:\frac{d\boldsymbol{A}}{dt}, \tag{1.219}$$

$$\frac{d\boldsymbol{B}}{dt} = \boldsymbol{D}\frac{d\alpha}{dt} + {}_3\boldsymbol{F}^{\overset{T}{312}}\cdot\frac{d\boldsymbol{a}}{dt} + {}_4\boldsymbol{G}:\frac{d\boldsymbol{A}}{dt}. \tag{1.220}$$

Because such total differentials can be integrated reversibly, these laws correspond to "reversible processes." So a response of a given tensorial order can, in general, be generated by a force of different tensorial order.

For the material to be called "isotropic," the coefficients of each tensorial material property must be invariant to rotations of the coordinates, that is, each material property must involve a scalar times the fundamental isotropic tensor(s) having the same tensorial order as the material property. So in an isotropic material, we must have

$$\boldsymbol{c} = 0 \text{ (because there are no isotropic vectors)}, \tag{1.221}$$

$$\boldsymbol{D} = d\,\delta_{ij}\hat{\boldsymbol{x}}_i\hat{\boldsymbol{x}}_j = d\boldsymbol{I} \text{ (with } d \text{ a scalar)}, \tag{1.222}$$

$$\boldsymbol{E} = e\,\delta_{ij}\hat{\boldsymbol{x}}_i\hat{\boldsymbol{x}}_j = e\boldsymbol{I} \text{ (with } e \text{ a scalar)}, \tag{1.223}$$

$$_3\boldsymbol{F} = f\epsilon_{ijk}\hat{\boldsymbol{x}}_i\hat{\boldsymbol{x}}_j\hat{\boldsymbol{x}}_k = (f)\,({}_3\boldsymbol{\epsilon}) \text{ (with } f \text{ a scalar)}, \tag{1.224}$$

$$_3\boldsymbol{F}^{\overset{T}{312}} = {}_3\boldsymbol{F} \text{ (because } \epsilon_{kij} = \epsilon_{ijk}), \tag{1.225}$$

$$_4\boldsymbol{G} = g_1\left({}_4\boldsymbol{I}^{(1)} + {}_4\boldsymbol{I}^{(2)}\right) + (g_2)\,{}_4\boldsymbol{I}^{(3)} \text{ (with } g_1 \text{ and } g_2 \text{ scalars)}. \tag{1.226}$$

The three fundamental fourth-order isotropic tensors ${}_4\boldsymbol{I}^{(1)}$, ${}_4\boldsymbol{I}^{(2)}$, and ${}_4\boldsymbol{I}^{(3)}$ are given in Eqs (1.174)–(1.176). As shown in an end-of-chapter exercise, we have ${}_4\boldsymbol{I}^{(1)} : \mathrm{d}\boldsymbol{A}/\mathrm{d}t = \mathrm{d}\boldsymbol{A}/\mathrm{d}t$, ${}_4\boldsymbol{I}^{(2)} : \mathrm{d}\boldsymbol{A}/\mathrm{d}t = \mathrm{d}\boldsymbol{A}^T/\mathrm{d}t = \mathrm{d}\boldsymbol{A}/\mathrm{d}t$, and ${}_4\boldsymbol{I}^{(3)} : \mathrm{d}\boldsymbol{A}/\mathrm{d}t = \left[\mathrm{d}(\boldsymbol{I}:\boldsymbol{A})/\mathrm{d}t\right]\boldsymbol{I}$, where $\boldsymbol{I}$ is the second-order identity tensor.

Because ${}_3\boldsymbol{F}$ is proportional to the Levi–Civita tensor, when it is double dotted into the symmetric tensor $\mathrm{d}\boldsymbol{A}/\mathrm{d}t$ we get zero as proven earlier. When this same ${}_3\boldsymbol{F}$ is dotted into the vector $\mathrm{d}\boldsymbol{a}/\mathrm{d}t$, we obtain an antisymmetric tensor, also as proven earlier. However, $\mathrm{d}\boldsymbol{B}/\mathrm{d}t$ is a symmetric tensor because $\boldsymbol{A}$ is symmetric, which tells us that the scalar material property f must be zero. Last, we can decompose the second-order tensor $\boldsymbol{A}$ into so-called *isotropic* and *deviatoric* portions as

$$\boldsymbol{A} = \underbrace{\left(\frac{\boldsymbol{I}:\boldsymbol{A}}{\boldsymbol{I}:\boldsymbol{I}}\right)\boldsymbol{I}}_{\textit{Isotropic portion}} + \underbrace{\boldsymbol{A} - \left(\frac{\boldsymbol{I}:\boldsymbol{A}}{\boldsymbol{I}:\boldsymbol{I}}\right)\boldsymbol{I}}_{\textit{Deviatoric portion } \boldsymbol{A}^D}, \tag{1.227}$$

and similarly for $\boldsymbol{B} = (\boldsymbol{I}:\boldsymbol{B}/\boldsymbol{I}:\boldsymbol{I})\,\boldsymbol{I} + \boldsymbol{B}^D$. The deviatoric or "true-tensorial" portion of a second-order tensor has zero trace, that is, $\boldsymbol{I}:\boldsymbol{A}^D = 0 = \boldsymbol{I}:\boldsymbol{B}^D$ and continues to be symmetric if the second-order tensor being decomposed is symmetric. Note that if $\boldsymbol{B}$ is the stress tensor, as defined in Chapter 2, then $-(\boldsymbol{I}:\boldsymbol{B})/(\boldsymbol{I}:\boldsymbol{I})$ is the scalar pressure.

Thus, in an isotropic material, the constitutive laws contained within $u = u(\alpha, \boldsymbol{a}, \boldsymbol{A})$ are

$$\frac{\mathrm{d}\beta}{\mathrm{d}t} = \gamma\frac{\mathrm{d}\alpha}{\mathrm{d}t} + d\,\frac{\mathrm{d}(\boldsymbol{I}:\boldsymbol{A})}{\mathrm{d}t}, \tag{1.228}$$

$$\frac{\mathrm{d}}{\mathrm{d}t}\left(\frac{\boldsymbol{I}:\boldsymbol{B}}{\boldsymbol{I}:\boldsymbol{I}}\right) = d\,\frac{\mathrm{d}\alpha}{\mathrm{d}t} + \left(g_2 + \frac{2g_1}{\boldsymbol{I}:\boldsymbol{I}}\right)\frac{\mathrm{d}(\boldsymbol{I}:\boldsymbol{A})}{\mathrm{d}t}, \tag{1.229}$$

$$\frac{\mathrm{d}\boldsymbol{b}}{\mathrm{d}t} = e\frac{\mathrm{d}\boldsymbol{a}}{\mathrm{d}t}, \tag{1.230}$$

$$\frac{\mathrm{d}\boldsymbol{B}^D}{\mathrm{d}t} = g_1\frac{\mathrm{d}\boldsymbol{A}^D}{\mathrm{d}t}. \tag{1.231}$$

We see that the time rate of each "response" in these isotropic laws (the left-hand side) has the same tensorial order as the time rate of the conjugate "forces" that are creating it (the right-hand side) and that all the material properties are now simple scalars (γ, d, e, g_1, and g_2 with $f = 0$). This is the content of *Curie's principle of constitutive laws in isotropic media* that we have now demonstrated to be a theorem for the constitutive laws contained in the fundamental function $u = u(\alpha, \boldsymbol{a}, \boldsymbol{A})$ with $\boldsymbol{A}$ symmetric. So, for example, if a vectorial response $\boldsymbol{b}$ is created by a second-order tensor $\boldsymbol{A}$ as controlled by the third-order material property ${}_3\boldsymbol{F}$ (piezoelectricity is a classic example), the material cannot be isotropic, that is, it must possess *anisotropy* so that ${}_3\boldsymbol{F}$ is not isotropic, which is the main message Curie (1894) was conveying.

1.8.6 Tensor Calculus in Orthogonal Curvilinear Coordinates

We now present the detailed expressions for various common tensor-calculus operations involving the gradient operator in orthogonal curvilinear coordinates. The various tensor-calculus operations are also given explicitly in both cylindrical and spherical coordinates, which are the two most commonly employed curvilinear coordinates you will encounter and the only curvilinear coordinates used in this book. The treatment that follows is inspired by the fabulous treatment of orthogonal curvilinear coordinates in Appendix A of the fluid-mechanics text by Happel and Brenner (1983).

To begin, consider the differences between Cartesian coordinates and some arbitrary orthogonal curvilinear coordinate system as shown in Fig. 1.14. The distance vector in Cartesian coordinates is written in array format as $\boldsymbol{r} = (x_1, x_2, x_3)$. Because each coordinate in Cartesians measures linear distance along that coordinate, we have that the infinitesimal distance vector $\mathrm{d}\boldsymbol{r}$ between two positions in space $\mathrm{d}\boldsymbol{r} = \hat{\boldsymbol{x}}_1\, \mathrm{d}\ell_1 + \hat{\boldsymbol{x}}_2\, \mathrm{d}\ell_2 + \hat{\boldsymbol{x}}_3\, \mathrm{d}\ell_3$ is written

$$\mathrm{d}\boldsymbol{r} = \hat{\boldsymbol{x}}_1\, \mathrm{d}x_1 + \hat{\boldsymbol{x}}_2\, \mathrm{d}x_2 + \hat{\boldsymbol{x}}_3\, \mathrm{d}x_3 \tag{1.232}$$

because $\mathrm{d}\ell_i = \mathrm{d}x_i$ in each direction i. Thus, the gradient operator is simply

$$\nabla = \hat{\boldsymbol{x}}_1 \frac{\partial}{\partial x_1} + \hat{\boldsymbol{x}}_2 \frac{\partial}{\partial x_2} + \hat{\boldsymbol{x}}_3 \frac{\partial}{\partial x_3}, \tag{1.233}$$

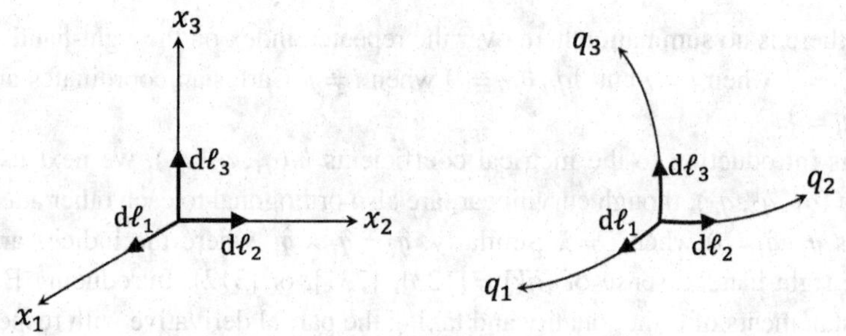

Figure 1.14 Cartesian coordinates on the left and some orthogonal curvilinear coordinate system on the right.

where we can employ the notation, if we so choose, that $\nabla = \partial/\partial \boldsymbol{r}$ with the Eq. (1.233) interpretation of what $\partial/\partial \boldsymbol{r}$ means to do.

The situation is different in orthogonal curvilinear coordinates because the coordinates in each orthogonal direction do not necessarily represent distance but instead can be represented by angles and because the base vectors vary their orientation, in general, as we move along a given coordinate. Although some point in space can again be represented in array format as $\boldsymbol{r} = (q_1, q_2, q_3)$, the infinitesimal distance between two points in space $\mathrm{d}\boldsymbol{r} = \hat{\boldsymbol{q}}_1\, \mathrm{d}\ell_1 + \hat{\boldsymbol{q}}_2\, \mathrm{d}\ell_2 + \hat{\boldsymbol{q}}_3\, \mathrm{d}\ell_3$ is now

$$\mathrm{d}\boldsymbol{r} = \hat{\boldsymbol{q}}_1 \frac{\mathrm{d}q_1}{h_1} + \hat{\boldsymbol{q}}_2 \frac{\mathrm{d}q_2}{h_2} + \hat{\boldsymbol{q}}_3 \frac{\mathrm{d}q_3}{h_3}, \tag{1.234}$$

that is, infinitesimal distance in each coordinate direction is given by

$$\mathrm{d}\ell_i = \frac{\mathrm{d}q_i}{h_i(q_1, q_2, q_3)}, \tag{1.235}$$

where the coefficients h_i are called the *metrical coefficients* for the particular orthogonal curvilinear coordinate system under consideration. These metrical coefficients convert change along a coordinate direction to change in distance along that coordinate and themselves will vary through space in general. *Specifying a curvilinear coordinate system amounts to specifying the functional dependence of the three metrical coefficients on the coordinates q_1, q_2, and q_3.*

The gradient operator in orthogonal curvilinear coordinates is then

$$\nabla = \hat{\boldsymbol{q}}_1\, h_1 \frac{\partial}{\partial q_1} + \hat{\boldsymbol{q}}_2\, h_2 \frac{\partial}{\partial q_2} + \hat{\boldsymbol{q}}_3\, h_3 \frac{\partial}{\partial q_3}. \tag{1.236}$$

A useful definition for the base vectors comes from combining $\mathrm{d}\boldsymbol{r} = \hat{\boldsymbol{q}}_1\, \mathrm{d}\ell_1 + \hat{\boldsymbol{q}}_2\, \mathrm{d}\ell_2 + \hat{\boldsymbol{q}}_3\, \mathrm{d}\ell_3$ with Eq. (1.235)

$$\hat{\boldsymbol{q}}_i = h_i \frac{\partial \boldsymbol{r}}{\partial q_i} \quad \text{where } i = 1, 2, \text{ or } 3 \tag{1.237}$$

and where there is no summation here over the repeated index on the right-hand side. Note that $\partial q_i/\partial q_j = 1$ when $i = j$ but $\partial q_i/\partial q_j = 0$ when $i \neq j$. Cartesian coordinates are defined by taking $h_i = 1$.

With this introduction to the metrical coefficients $h_i(q_1, q_2, q_3)$, we next use that the coordinates (q_1, q_2, q_3), though curvilinear, are also orthogonal to each other at each point. This means $\hat{\boldsymbol{q}}_i \cdot \hat{\boldsymbol{q}}_j = 0$ when $i \neq j$. Similarly, $\hat{\boldsymbol{q}}_i = \hat{\boldsymbol{q}}_j \times \hat{\boldsymbol{q}}_k$ where the indices are ordered here in the right-handed sense of $[ijk] = [123]$, $[231]$, or $[312]$. Introducing Eq. (1.237) into these statements of orthogonality and taking the partial derivative with respect to each coordinate q_i, one arrives, eventually, at the following results for the derivatives of the base vectors in an orthogonal curvilinear coordinate system

$$\frac{\partial \hat{\boldsymbol{q}}_j}{\partial q_i} = \hat{\boldsymbol{q}}_i h_j \frac{\partial}{\partial q_j}\left(\frac{1}{h_i}\right) \qquad \text{where } j \neq i \tag{1.238}$$

$$\frac{\partial \hat{\boldsymbol{q}}_i}{\partial q_i} = -\hat{\boldsymbol{q}}_j h_j \frac{\partial}{\partial q_j}\left(\frac{1}{h_i}\right) - \hat{\boldsymbol{q}}_k h_k \frac{\partial}{\partial q_k}\left(\frac{1}{h_i}\right) \qquad \text{where } j \neq k \neq i \tag{1.239}$$

and where there is no summation over repeated indices in these expressions. Again, in Cartesian coordinates, all of these derivatives are zero.

The second-order identity tensor $\boldsymbol{I}$ in orthogonal curvilinear coordinates is defined

$$\boldsymbol{I} = \nabla \boldsymbol{r} = \hat{\boldsymbol{q}}_1\, h_1 \frac{\partial \boldsymbol{r}}{\partial q_1} + \hat{\boldsymbol{q}}_2\, h_2 \frac{\partial \boldsymbol{r}}{\partial q_2} + \hat{\boldsymbol{q}}_3\, h_3 \frac{\partial \boldsymbol{r}}{\partial q_3}. \tag{1.240}$$

which from Eq. (1.237) is simply

$$\boldsymbol{I} = \hat{\boldsymbol{q}}_1\hat{\boldsymbol{q}}_1 + \hat{\boldsymbol{q}}_2\hat{\boldsymbol{q}}_2 + \hat{\boldsymbol{q}}_3\hat{\boldsymbol{q}}_3 = \delta_{ij}\hat{\boldsymbol{q}}_i\hat{\boldsymbol{q}}_j, \tag{1.241}$$

just like in Cartesian coordinates.

We also have that infinitesimal surface elements $\mathrm{d}S_i$ having a normal in the $\hat{\boldsymbol{q}}_i$ direction are given by

$$\mathrm{d}S_1 = \mathrm{d}\ell_2 \mathrm{d}\ell_3 = \frac{\mathrm{d}q_2 \mathrm{d}q_3}{h_2 h_3} \tag{1.242}$$

$$\mathrm{d}S_2 = \mathrm{d}\ell_1 \mathrm{d}\ell_3 = \frac{\mathrm{d}q_1 \mathrm{d}q_3}{h_1 h_3} \tag{1.243}$$

$$\mathrm{d}S_3 = \mathrm{d}\ell_1 \mathrm{d}\ell_2 = \frac{\mathrm{d}q_1 \mathrm{d}q_2}{h_1 h_2}. \tag{1.244}$$

Similarly, the infinitesimal volume element is

$$\mathrm{d}V = \mathrm{d}\ell_1 \mathrm{d}\ell_2 \mathrm{d}\ell_3 = \frac{\mathrm{d}q_1 \mathrm{d}q_2 \mathrm{d}q_3}{h_1 h_2 h_3}. \tag{1.245}$$

So given the metrical coefficients for an orthogonal curvilinear coordinate system, we can now calculate spatial derivatives of vectors and set up surface and volume integrals in those coordinates.

For example, to perform the divergence of a vector field $\nabla \cdot \boldsymbol{a}$, we write

$$\nabla \cdot \boldsymbol{a} = \hat{\boldsymbol{q}}_i\, h_i \frac{\partial}{\partial q_i} \cdot (\hat{\boldsymbol{q}}_j a_j) \tag{1.246}$$

$$= h_i \hat{\boldsymbol{q}}_i \cdot \left[\left(\frac{\partial \hat{\boldsymbol{q}}_j}{\partial q_i}\right) a_j + \hat{\boldsymbol{q}}_j \frac{\partial a_j}{\partial q_i}\right] \tag{1.247}$$

where now there is summation assumed over repeated indices and, as earlier, we use a different index for each vector in the tensor-calculus expression to be evaluated prior to performing any scalar products. We perform the explicit sum over repeated indices, insert Eqs (1.238) and (1.239) for the various derivatives of the base vectors and use the orthogonality condition that $\hat{\boldsymbol{q}}_i \cdot \hat{\boldsymbol{q}}_j = \delta_{ij}$, which is nonzero only if $j = i$, to obtain

$$\nabla \cdot \boldsymbol{a} = h_1 h_2 h_3 \left[\frac{\partial}{\partial q_1}\left(\frac{a_1}{h_2 h_3}\right) + \frac{\partial}{\partial q_2}\left(\frac{a_2}{h_1 h_3}\right) + \frac{\partial}{\partial q_3}\left(\frac{a_3}{h_1 h_2}\right) \right]. \tag{1.248}$$

This then yields the Laplacian $\nabla^2\psi \mathrel{\widehat{=}} \nabla \cdot \nabla\psi$ of any scalar field ψ to be

$$\nabla^2\psi = h_1 h_2 h_3 \times \left[\frac{\partial}{\partial q_1}\left(\frac{h_1 \partial\psi/\partial q_1}{h_2 h_3}\right) + \frac{\partial}{\partial q_2}\left(\frac{h_2 \partial\psi/\partial q_2}{h_1 h_3}\right) + \frac{\partial}{\partial q_3}\left(\frac{h_3 \partial\psi/\partial q_3}{h_1 h_2}\right) \right]. \tag{1.249}$$

Similar operations yield the curl in the form

$$\begin{aligned} \nabla \times \boldsymbol{a} = {} & \hat{\boldsymbol{q}}_1 h_2 h_3 \left[\frac{\partial}{\partial q_2}\left(\frac{a_3}{h_3}\right) - \frac{\partial}{\partial q_3}\left(\frac{a_2}{h_2}\right) \right] \\ & + \hat{\boldsymbol{q}}_2 h_1 h_3 \left[\frac{\partial}{\partial q_3}\left(\frac{a_1}{h_1}\right) - \frac{\partial}{\partial q_1}\left(\frac{a_3}{h_3}\right) \right] \\ & + \hat{\boldsymbol{q}}_3 h_1 h_2 \left[\frac{\partial}{\partial q_1}\left(\frac{a_2}{h_2}\right) - \frac{\partial}{\partial q_2}\left(\frac{a_1}{h_1}\right) \right]. \end{aligned} \tag{1.250}$$

The second-order tensor $\nabla\boldsymbol{a}$ in orthogonal curvilinear coordinates is defined (again with summation over repeated indices)

$$\nabla\boldsymbol{a} = \hat{\boldsymbol{q}}_i\, h_i \frac{\partial}{\partial q_i}(\hat{\boldsymbol{q}}_j a_j) = \hat{\boldsymbol{q}}_i\, h_i \left[\left(\frac{\partial \hat{\boldsymbol{q}}_j}{\partial q_i}\right) a_j + \hat{\boldsymbol{q}}_j \frac{\partial a_j}{\partial a_i} \right]. \tag{1.251}$$

So performing the explicit sum over repeated indices and employing Eqs (1.238) and (1.239) for the derivatives of the base vectors, we obtain the nine components of $\nabla\boldsymbol{a}$ as

$$\begin{aligned} \nabla\boldsymbol{a} = {} & \hat{\boldsymbol{q}}_1\hat{\boldsymbol{q}}_1 h_1 \left[\frac{\partial a_1}{\partial q_1} + h_2 a_2 \frac{\partial}{\partial q_2}\left(\frac{1}{h_1}\right) + h_3 a_3 \frac{\partial}{\partial q_3}\left(\frac{1}{h_1}\right) \right] \\ & + \hat{\boldsymbol{q}}_1\hat{\boldsymbol{q}}_2 h_1 \left[\frac{\partial a_2}{\partial q_1} - h_2 a_1 \frac{\partial}{\partial q_2}\left(\frac{1}{h_1}\right) \right] \\ & + \hat{\boldsymbol{q}}_1\hat{\boldsymbol{q}}_3 h_1 \left[\frac{\partial a_3}{\partial q_1} - h_3 a_1 \frac{\partial}{\partial q_3}\left(\frac{1}{h_1}\right) \right] \\ & + \hat{\boldsymbol{q}}_2\hat{\boldsymbol{q}}_1 h_2 \left[\frac{\partial a_1}{\partial q_2} - h_1 a_2 \frac{\partial}{\partial q_1}\left(\frac{1}{h_2}\right) \right] \\ & + \hat{\boldsymbol{q}}_2\hat{\boldsymbol{q}}_2 h_2 \left[\frac{\partial a_2}{\partial q_2} + h_3 a_3 \frac{\partial}{\partial q_3}\left(\frac{1}{h_2}\right) + h_1 a_1 \frac{\partial}{\partial q_1}\left(\frac{1}{h_2}\right) \right] \\ & + \hat{\boldsymbol{q}}_2\hat{\boldsymbol{q}}_3 h_2 \left[\frac{\partial a_3}{\partial q_2} - h_3 a_2 \frac{\partial}{\partial q_3}\left(\frac{1}{h_2}\right) \right] \\ & + \hat{\boldsymbol{q}}_3\hat{\boldsymbol{q}}_1 h_3 \left[\frac{\partial a_1}{\partial q_3} - h_1 a_3 \frac{\partial}{\partial q_1}\left(\frac{1}{h_3}\right) \right] \\ & + \hat{\boldsymbol{q}}_3\hat{\boldsymbol{q}}_2 h_3 \left[\frac{\partial a_2}{\partial q_3} - h_2 a_3 \frac{\partial}{\partial q_2}\left(\frac{1}{h_3}\right) \right] \\ & + \hat{\boldsymbol{q}}_3\hat{\boldsymbol{q}}_3 h_3 \left[\frac{\partial a_3}{\partial q_3} + h_1 a_1 \frac{\partial}{\partial q_1}\left(\frac{1}{h_3}\right) + h_2 a_2 \frac{\partial}{\partial q_2}\left(\frac{1}{h_3}\right) \right]. \end{aligned} \tag{1.252}$$

Last, the divergence of a second-order tensor $\boldsymbol{A}$ is then

$$\begin{aligned}
\nabla\cdot\boldsymbol{A} = \hat{\boldsymbol{q}}_1 \Bigg\{ & h_1h_2h_3\left[\frac{\partial}{\partial q_1}\left(\frac{A_{11}}{h_2h_3}\right)+\frac{\partial}{\partial q_2}\left(\frac{A_{21}}{h_1h_3}\right)+\frac{\partial}{\partial q_3}\left(\frac{A_{31}}{h_1h_2}\right)\right] \\
& + h_1h_1A_{11}\frac{\partial}{\partial q_1}\left(\frac{1}{h_1}\right) + h_1h_2A_{12}\frac{\partial}{\partial q_2}\left(\frac{1}{h_1}\right) + h_1h_3A_{13}\frac{\partial}{\partial q_3}\left(\frac{1}{h_1}\right) \\
& - h_1h_1A_{11}\frac{\partial}{\partial q_1}\left(\frac{1}{h_1}\right) - h_1h_2A_{22}\frac{\partial}{\partial q_1}\left(\frac{1}{h_2}\right) - h_1h_3A_{33}\frac{\partial}{\partial q_1}\left(\frac{1}{h_3}\right)\Bigg\} \\
+ \hat{\boldsymbol{q}}_2 \Bigg\{ & h_1h_2h_3\left[\frac{\partial}{\partial q_1}\left(\frac{A_{12}}{h_2h_3}\right)+\frac{\partial}{\partial q_2}\left(\frac{A_{22}}{h_1h_3}\right)+\frac{\partial}{\partial q_3}\left(\frac{A_{32}}{h_1h_2}\right)\right] \\
& + h_2h_1A_{21}\frac{\partial}{\partial q_1}\left(\frac{1}{h_2}\right) + h_2h_2A_{22}\frac{\partial}{\partial q_2}\left(\frac{1}{h_2}\right) + h_2h_3A_{23}\frac{\partial}{\partial q_3}\left(\frac{1}{h_2}\right) \\
& - h_2h_1A_{11}\frac{\partial}{\partial q_2}\left(\frac{1}{h_1}\right) - h_2h_2A_{22}\frac{\partial}{\partial q_2}\left(\frac{1}{h_2}\right) - h_2h_3A_{33}\frac{\partial}{\partial q_2}\left(\frac{1}{h_3}\right)\Bigg\} \\
+ \hat{\boldsymbol{q}}_3 \Bigg\{ & h_1h_2h_3\left[\frac{\partial}{\partial q_1}\left(\frac{A_{13}}{h_2h_3}\right)+\frac{\partial}{\partial q_2}\left(\frac{A_{23}}{h_1h_3}\right)+\frac{\partial}{\partial q_3}\left(\frac{A_{33}}{h_1h_2}\right)\right] \\
& + h_3h_1A_{31}\frac{\partial}{\partial q_1}\left(\frac{1}{h_3}\right) + h_3h_2A_{32}\frac{\partial}{\partial q_2}\left(\frac{1}{h_3}\right) + h_3h_3A_{33}\frac{\partial}{\partial q_3}\left(\frac{1}{h_3}\right) \\
& - h_3h_1A_{11}\frac{\partial}{\partial q_3}\left(\frac{1}{h_1}\right) - h_3h_2A_{22}\frac{\partial}{\partial q_3}\left(\frac{1}{h_2}\right) - h_3h_3A_{33}\frac{\partial}{\partial q_3}\left(\frac{1}{h_3}\right)\Bigg\}.
\end{aligned} \tag{1.253}$$

This can be compared to the same expression given in Cartesian coordinates

$$\begin{aligned}
\nabla\cdot\boldsymbol{A} = {} & \hat{\boldsymbol{x}}_1\left(\frac{\partial A_{11}}{\partial x_1}+\frac{\partial A_{21}}{\partial x_2}+\frac{\partial A_{31}}{\partial x_3}\right)+\hat{\boldsymbol{x}}_2\left(\frac{\partial A_{12}}{\partial x_1}+\frac{\partial A_{22}}{\partial x_2}+\frac{\partial A_{32}}{\partial x_3}\right) \\
& + \hat{\boldsymbol{x}}_3\left(\frac{\partial A_{13}}{\partial x_1}+\frac{\partial A_{23}}{\partial x_2}+\frac{\partial A_{33}}{\partial x_3}\right).
\end{aligned} \tag{1.254}$$

Using the above, the most pertinent expressions for the special cases of cylindrical and spherical coordinates follow.

Cylindrical Coordinates

In cylindrical coordinates $(q_1, q_2, q_3) \,\widehat{=}\, (r, \theta, z)$ with unit vectors $\hat{\boldsymbol{r}}$ (radial direction), $\hat{\boldsymbol{\theta}}$ (circumferential direction around the z axis), and $\hat{\boldsymbol{z}}$ (axial direction) that are orthogonal to each other at each point in space, the metrical coefficients are

$$\frac{1}{h_1}=1, \quad \frac{1}{h_2}=r \quad \text{and} \quad \frac{1}{h_3}=1. \tag{1.255}$$

This simply says that distance in the θ direction goes as θr. The mapping of the Cartesian base vectors into the cylindrical-coordinate base vectors is performed using the matrix multiplication

$$\begin{bmatrix}\hat{\boldsymbol{r}}\\ \hat{\boldsymbol{\theta}}\\ \hat{\boldsymbol{z}}\end{bmatrix} = \begin{bmatrix}\cos\theta & \sin\theta & 0\\ -\sin\theta & \cos\theta & 0\\ 0 & 0 & 1\end{bmatrix}\begin{bmatrix}\hat{\boldsymbol{x}}\\ \hat{\boldsymbol{y}}\\ \hat{\boldsymbol{z}}\end{bmatrix}. \tag{1.256}$$

Some further trigonometry gives $r = \sqrt{x^2 + y^2}$, $\theta = \tan^{-1}(y/x)$, and $z = z$ as well as $x = r\cos\theta$ and $y = r\sin\theta$.

The nonzero derivatives of the base vectors are given by Eqs (1.238) and (1.239) to be

$$\frac{\partial \hat{\boldsymbol{r}}}{\partial \theta} = \hat{\boldsymbol{\theta}} \quad \text{and} \quad \frac{\partial \hat{\boldsymbol{\theta}}}{\partial \theta} = -\hat{\boldsymbol{r}}. \tag{1.257}$$

Consider three fields: $\psi(\boldsymbol{r})$ a scalar, $\boldsymbol{a}(\boldsymbol{r}) = (a_r, a_\theta, a_z)$ a vector, and

$$\boldsymbol{A}(\boldsymbol{r}) = \begin{pmatrix} A_{rr} & A_{r\theta} & A_{rz} \\ A_{\theta r} & A_{\theta\theta} & A_{\theta z} \\ A_{zr} & A_{z\theta} & A_{zz} \end{pmatrix} \tag{1.258}$$

a second-order tensor. The various standard operations in cylindrical coordinates involving ∇ acting on the scalar and vector fields are

$$\nabla\psi = \hat{\boldsymbol{r}}\frac{\partial\psi}{\partial r} + \hat{\boldsymbol{\theta}}\frac{1}{r}\frac{\partial\psi}{\partial\theta} + \hat{\boldsymbol{z}}\frac{\partial\psi}{\partial z} \tag{1.259}$$

$$\nabla^2\psi = \frac{1}{r}\frac{\partial}{\partial r}\left(r\frac{\partial\psi}{\partial r}\right) + \frac{1}{r^2}\frac{\partial^2}{\partial\theta^2} + \frac{\partial^2\psi}{\partial z^2} \tag{1.260}$$

and

$$\nabla\cdot\boldsymbol{a} = \frac{1}{r}\frac{\partial}{\partial r}(ra_r) + \frac{1}{r}\frac{\partial a_\theta}{\partial\theta} + \frac{\partial a_z}{\partial z} \tag{1.261}$$

$$\nabla\times\boldsymbol{a} = \hat{\boldsymbol{r}}\left(\frac{1}{r}\frac{\partial a_z}{\partial\theta} - \frac{\partial a_\theta}{\partial z}\right) + \hat{\boldsymbol{\theta}}\left(\frac{\partial a_r}{\partial z} - \frac{\partial a_z}{\partial r}\right) + \hat{\boldsymbol{z}}\left(\frac{1}{r}\frac{\partial}{\partial r}(ra_\theta) - \frac{1}{r}\frac{\partial a_r}{\partial\theta}\right) \tag{1.262}$$

$$\nabla^2\boldsymbol{a} = \hat{\boldsymbol{r}}\left(\nabla^2 a_r - \frac{2}{r^2}\frac{\partial a_\theta}{\partial\theta} - \frac{a_r}{r^2}\right) + \hat{\boldsymbol{\theta}}\left(\nabla^2 a_\theta + \frac{2}{r^2}\frac{\partial a_r}{\partial\theta} - \frac{a_\theta}{r^2}\right) + \hat{\boldsymbol{z}}\nabla^2 u_z. \tag{1.263}$$

In this last expression, the Laplacian operator ∇^2 acting on the three scalar components of the vector $\boldsymbol{a}$ is given by Eq. (1.260). The two most common tensorial operations we will encounter are

$$\begin{aligned}\nabla\boldsymbol{a} = {} & \hat{\boldsymbol{r}}\hat{\boldsymbol{r}}\frac{\partial a_r}{\partial r} + \hat{\boldsymbol{r}}\hat{\boldsymbol{\theta}}\frac{\partial a_\theta}{\partial r} + \hat{\boldsymbol{r}}\hat{\boldsymbol{z}}\frac{\partial a_z}{\partial r} \\ & + \hat{\boldsymbol{\theta}}\hat{\boldsymbol{r}}\frac{1}{r}\left(\frac{\partial a_r}{\partial\theta} - a_\theta\right) + \hat{\boldsymbol{\theta}}\hat{\boldsymbol{\theta}}\frac{1}{r}\left(\frac{\partial a_\theta}{\partial\theta} + a_r\right) + \hat{\boldsymbol{\theta}}\hat{\boldsymbol{z}}\frac{1}{r}\frac{\partial a_z}{\partial\theta} \\ & + \hat{\boldsymbol{z}}\hat{\boldsymbol{r}}\frac{\partial a_r}{\partial z} + \hat{\boldsymbol{z}}\hat{\boldsymbol{\theta}}\frac{\partial a_\theta}{\partial z} + \hat{\boldsymbol{z}}\hat{\boldsymbol{z}}\frac{\partial a_z}{\partial z}\end{aligned} \tag{1.264}$$

and

$$\begin{aligned}\nabla\cdot\boldsymbol{A} = {} & \hat{\boldsymbol{r}}\left[\frac{1}{r}\frac{\partial}{\partial r}(rA_{rr}) + \frac{1}{r}\frac{\partial A_{\theta r}}{\partial\theta} + \frac{\partial A_{zr}}{\partial z} - \frac{A_{\theta\theta}}{r}\right] \\ & + \hat{\boldsymbol{\theta}}\left[\frac{1}{r}\frac{\partial}{\partial r}(rA_{r\theta}) + \frac{1}{r}\frac{\partial A_{\theta\theta}}{\partial\theta} + \frac{\partial A_{z\theta}}{\partial z} + \frac{A_{\theta r}}{r}\right] \\ & + \hat{\boldsymbol{z}}\left[\frac{1}{r}\frac{\partial}{\partial r}(rA_{rz}) + \frac{1}{r}\frac{\partial A_{\theta z}}{\partial\theta} + \frac{\partial A_{zz}}{\partial z}\right].\end{aligned} \tag{1.265}$$

Spherical Coordinates

In spherical coordinates (r, θ, ϕ), with θ now measuring latitude down from a "z axis" of revolution and ϕ measuring longitude around the z axis and with unit vectors $\hat{\boldsymbol{r}}$, $\hat{\boldsymbol{\theta}}$, and $\hat{\boldsymbol{\phi}}$ that are orthogonal to each other at each point in space, the metrical coefficients are

$$\frac{1}{h_1} = 1, \quad \frac{1}{h_2} = r, \quad \text{and} \quad \frac{1}{h_3} = r\sin\theta. \tag{1.266}$$

This says that at each latitude θ coming down from the z axis, distance in the longitudinal direction around the z axis goes as $\phi r \sin\theta$. The mapping of the Cartesian base vectors into the spherical-coordinate base vectors is performed using the matrix multiplication

$$\begin{bmatrix} \hat{\boldsymbol{r}} \\ \hat{\boldsymbol{\theta}} \\ \hat{\boldsymbol{\phi}} \end{bmatrix} = \begin{bmatrix} \sin\theta\cos\phi & \sin\theta\sin\phi & \cos\theta \\ \cos\theta\cos\phi & \cos\theta\sin\phi & -\sin\theta \\ -\sin\phi & \cos\phi & 0 \end{bmatrix} \begin{bmatrix} \hat{\boldsymbol{x}} \\ \hat{\boldsymbol{y}} \\ \hat{\boldsymbol{z}} \end{bmatrix}. \tag{1.267}$$

Some further trigonometry gives $r = \sqrt{x^2 + y^2 + z^2}$, $\theta = \cos^{-1}\left(z/\sqrt{x^2+y^2+z^2}\right)$, and $\phi = \tan^{-1}(y/x)$ as well as $x = r\sin\theta\cos\phi$, $y = r\sin\theta\sin\phi$, and $z = r\cos\theta$.

Of the nine possible derivatives of the base vectors, Eqs (1.238) and (1.239) give that five are nonzero

$$\frac{\partial\hat{\boldsymbol{r}}}{\partial\theta} = \hat{\boldsymbol{\theta}} \quad \text{and} \quad \frac{\partial\hat{\boldsymbol{\theta}}}{\partial\theta} = -\hat{\boldsymbol{r}} \tag{1.268}$$

as well as

$$\frac{\partial\hat{\boldsymbol{r}}}{\partial\phi} = \hat{\boldsymbol{\phi}}\sin\theta, \quad \frac{\partial\hat{\boldsymbol{\theta}}}{\partial\phi} = \hat{\boldsymbol{\phi}}\cos\theta, \quad \text{and} \quad \frac{\partial\hat{\boldsymbol{\phi}}}{\partial\phi} = -\hat{\boldsymbol{r}}\sin\theta - \hat{\boldsymbol{\theta}}\cos\theta. \tag{1.269}$$

Again consider three fields: $\psi(\boldsymbol{r})$ a scalar, $\boldsymbol{a}(\boldsymbol{r}) = (a_r, a_\theta, a_\phi)$ a vector, and

$$\boldsymbol{A}(\boldsymbol{r}) = \begin{pmatrix} A_{rr} & A_{r\theta} & A_{r\phi} \\ A_{\theta r} & A_{\theta\theta} & A_{\theta\phi} \\ A_{\phi r} & A_{\phi\theta} & A_{\phi\phi} \end{pmatrix} \tag{1.270}$$

a second-order tensor. The standard operations in spherical coordinates involving ∇ and the scalar and vector fields are

$$\nabla\psi = \hat{\boldsymbol{r}}\frac{\partial\psi}{\partial r} + \hat{\boldsymbol{\theta}}\frac{1}{r}\frac{\partial\psi}{\partial\theta} + \hat{\boldsymbol{\phi}}\frac{1}{r\sin\theta}\frac{\partial\psi}{\partial\phi} \tag{1.271}$$

$$\nabla^2\psi = \frac{1}{r^2}\left[\frac{\partial}{\partial r}\left(r^2\frac{\partial\psi}{\partial r}\right) + \frac{1}{\sin\theta}\frac{\partial}{\partial\theta}\left(\sin\theta\frac{\partial\psi}{\partial\theta}\right) + \frac{1}{\sin^2\theta}\frac{\partial^2\psi}{\partial\phi^2}\right] \tag{1.272}$$

and

$$\nabla\cdot\boldsymbol{a}=\frac{1}{r^2}\frac{\partial}{\partial r}\left(r^2a_r\right)+\frac{1}{r\sin\theta}\frac{\partial}{\partial\theta}\left(\sin\theta a_\theta\right)+\frac{1}{r\sin\theta}\frac{\partial a_\phi}{\partial\phi} \tag{1.273}$$

$$\begin{aligned}\nabla\times\boldsymbol{a}=&\hat{\boldsymbol{r}}\frac{1}{r\sin\theta}\left(\frac{\partial}{\partial\theta}\left(\sin\theta a_\phi\right)-\frac{\partial a_\theta}{\partial\phi}\right)+\hat{\boldsymbol{\theta}}\frac{1}{r}\left(\frac{1}{\sin\theta}\frac{\partial a_r}{\partial\phi}-\frac{\partial}{\partial r}\left(ra_\phi\right)\right)\\&+\hat{\boldsymbol{\phi}}\frac{1}{r}\left(\frac{\partial}{\partial r}\left(ra_\theta\right)-\frac{\partial a_r}{\partial\theta}\right)\end{aligned} \tag{1.274}$$

$$\begin{aligned}\nabla^2\boldsymbol{a}=&\hat{\boldsymbol{r}}\left[\nabla^2a_r-\frac{2}{r^2}\left(a_r+\frac{\partial a_\theta}{\partial\theta}+\frac{\cos\theta}{\sin\theta}a_\theta-\frac{1}{\sin\theta}\frac{\partial a_\phi}{\partial\phi}\right)\right]\\&+\hat{\boldsymbol{\theta}}\left[\nabla^2a_\theta+\frac{1}{r^2}\left(2\frac{\partial a_r}{\partial\theta}-\frac{a_\theta}{\sin^2\theta}-\frac{2\cos\theta}{\sin^2\theta}\frac{\partial a_\phi}{\partial\phi}\right)\right]\\&+\hat{\boldsymbol{\phi}}\left[\nabla^2a_\phi+\frac{1}{r^2}\left(\frac{2}{\sin\theta}\frac{\partial a_r}{\partial\phi}+\frac{2\cos\theta}{\sin^2\theta}\frac{\partial a_\theta}{\partial\phi}-\frac{a_\phi}{\sin^2\theta}\right)\right].\end{aligned} \tag{1.275}$$

In this last expression, the Laplacian operator ∇^2 acting on the three scalar components of the vector $\boldsymbol{a}$ is given by Eq. (1.272). The two most common tensorial operations involving ∇ are

$$\begin{aligned}\nabla\boldsymbol{a}=&\hat{\boldsymbol{r}}\hat{\boldsymbol{r}}\frac{\partial a_r}{\partial r}+\hat{\boldsymbol{r}}\hat{\boldsymbol{\theta}}\frac{\partial a_\theta}{\partial r}+\hat{\boldsymbol{r}}\hat{\boldsymbol{\phi}}\frac{\partial a_\phi}{\partial r}\\&+\hat{\boldsymbol{\theta}}\hat{\boldsymbol{r}}\frac{1}{r}\left(\frac{\partial a_r}{\partial\theta}-a_\theta\right)+\hat{\boldsymbol{\theta}}\hat{\boldsymbol{\theta}}\frac{1}{r}\left(\frac{\partial a_\theta}{\partial\theta}+a_r\right)+\hat{\boldsymbol{\theta}}\hat{\boldsymbol{\phi}}\frac{1}{r}\frac{\partial a_\phi}{\partial\theta}\\&+\hat{\boldsymbol{\phi}}\hat{\boldsymbol{r}}\frac{1}{r}\left(\frac{1}{\sin\theta}\frac{\partial a_r}{\partial\phi}-a_\phi\right)+\hat{\boldsymbol{\phi}}\hat{\boldsymbol{\theta}}\frac{1}{r}\left(\frac{1}{\sin\theta}\frac{\partial a_\theta}{\partial\phi}-\frac{\cos\theta}{\sin\theta}a_\phi\right)\\&+\hat{\boldsymbol{\phi}}\hat{\boldsymbol{\phi}}\frac{1}{r}\left(\frac{1}{\sin\theta}\frac{\partial a_\phi}{\partial\phi}+a_r+\frac{\cos\theta}{\sin\theta}a_\theta\right)\end{aligned} \tag{1.276}$$

and

$$\begin{aligned}\nabla\cdot\boldsymbol{A}=&\hat{\boldsymbol{r}}\left[\frac{1}{r^2}\frac{\partial}{\partial r}\left(r^2A_{rr}\right)+\frac{1}{r\sin\theta}\left(\frac{\partial}{\partial\theta}\left(\sin\theta A_{\theta r}\right)+\frac{\partial A_{\phi r}}{\partial\phi}\right)-\frac{\left(A_{\theta\theta}+A_{\phi\phi}\right)}{r}\right]\\&+\hat{\boldsymbol{\theta}}\left[\frac{1}{r^2}\frac{\partial}{\partial r}\left(r^2A_{r\theta}\right)+\frac{1}{r\sin\theta}\left(\frac{\partial}{\partial\theta}\left(\sin\theta A_{\theta\theta}\right)+\frac{\partial A_{\phi\theta}}{\partial\phi}\right)+\frac{A_{\theta r}}{r}-\frac{\cos\theta}{r\sin\theta}A_{\phi\phi}\right]\\&+\hat{\boldsymbol{\phi}}\left[\frac{1}{r^2}\frac{\partial}{\partial r}\left(r^2A_{r\phi}\right)+\frac{1}{r\sin\theta}\left(\frac{\partial}{\partial\theta}\left(\sin\theta A_{\theta\phi}\right)+\frac{\partial A_{\phi\phi}}{\partial\phi}\right)+\frac{A_{\phi r}}{r}-\frac{\cos\theta}{r\sin\theta}A_{\theta\phi}\right].\end{aligned} \tag{1.277}$$

1.9 The Dirac Delta Function

Throughout our development and implementation of continuum physics, we need to represent fields that are highly concentrated at a single point in space (or time, if considering time functions). The *Dirac delta* function $\delta(x-x_1)$ is used to represent a field highly concentrated at the point $x=x_1$ and is loosely, though insufficiently, defined as

$$\delta(x-x_1) = \begin{cases} \infty & \text{if } x = x_1 \\ 0 & \text{if } x \neq x_1 \end{cases} \tag{1.278}$$

and in such a way that if $a < x_1 < b$, then

$$\int_a^b \delta(x-x_1)\,\mathrm{d}x = 1 \quad \text{(no units).} \tag{1.279}$$

In some sense, we have that when $x = x_1$, then $\delta(0) = \mathrm{d}x^{-1}$ so that the *integral property of the Dirac delta function* of Eq. (1.279) is satisfied (indeed, when working numerically on discretized domains, this is one way to define the Dirac delta). Due to this integral property, we also have

$$\boxed{\int_a^b f(x)\,\delta(x-x_1)\,\mathrm{d}x = f(x_1),} \tag{1.280}$$

which is called the *sifting property*. In the integration process, the Dirac delta function samples f, where the argument of the Dirac goes to zero. Note as well that from Eq. (1.279) (or equivalently the sifting property), we necessarily have that $\delta(x)$ has physical of units of x^{-1} whatever the physical units of x are. This is important to remember in physics applications.

Because Eq. (1.278) is not a sufficient definition for a well-behaved differentiable function, we better define the Dirac delta function $\delta(x-x_1)$ as the limit of well-defined functions such as

$$\delta(x-x_1) = \lim_{\sigma\to 0} \frac{S(x-x_1+\sigma) - S(x-x_1-\sigma)}{2\sigma} \tag{1.281}$$

$$= \lim_{\sigma\to 0} \frac{\sin\left[(x-x_1)/\sigma\right]}{\pi(x-x_1)} \tag{1.282}$$

$$= \lim_{\sigma\to 0} \frac{1}{\sigma\pi\left[1+(x-x_1)^2/\sigma^2\right]} \tag{1.283}$$

$$= \lim_{\sigma\to 0} \frac{1}{\sigma\sqrt{2\pi}} \exp\left(\frac{-(x-x_1)^2}{2\sigma^2}\right), \tag{1.284}$$

where in the first example here, $S(x)$ is the unit step function defined to be 0 for $x < 0$ and 1 for $x > 0$. Each of these expressions for $\delta(x-x_1)$ satisfies the integral constraint of Eq. (1.279) for any value of σ. In the limit as the parameter σ (which has the same units as x and characterizes the width of the function) goes to zero, these also satisfy the sifting property of Eq. (1.280). The second example here is the scaled "sinc" function and will be shown to satisfy the required integral properties in Chapter 11 on contour integration. The last example is the familiar Gaussian function with standard deviation σ and will be shown to satisfy the required integral properties in Chapter 10 on Fourier analysis. We will use the Dirac to represent highly concentrated fields in nature; however, in each application

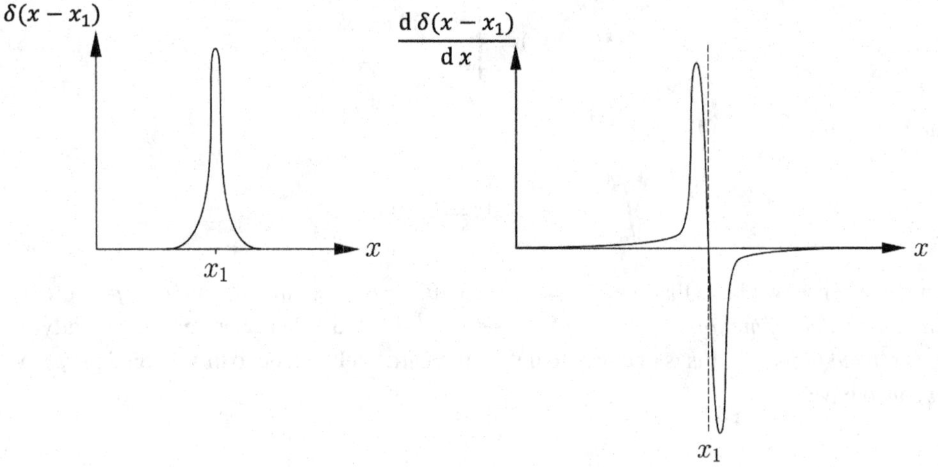

Figure 1.15 The Dirac delta and its derivative for small σ in Eqs (1.284) and (1.285).

to a physics problem, the Dirac delta function will be integrated over. Due to the defining integrals of Eqs (1.279) and (1.280), such integration leads to well-behaved finite results despite the fact that the Dirac function becomes very large when its argument is zero and as $\sigma \to 0$ (small but finite) in Eqs (1.281)–(1.284).

By representing the Dirac as a limit of a well-behaved function, one can take derivatives of $\delta(x - x_1)$. For example, for the Gaussian representation of $\delta(x - x_1)$, we have

$$\frac{\mathrm{d}\,\delta(x-x_1)}{\mathrm{d}x} = \lim_{\sigma\to 0} -\frac{(x-x_1)}{\sigma^3\sqrt{2\pi}} \exp\left(\frac{-(x-x_1)^2}{2\sigma^2}\right). \tag{1.285}$$

Visually, one can see the effect of taking the derivative of a Dirac delta by using Eqs (1.284) and (1.285) as shown in Fig. 1.15. One can also understand the derivative of the Dirac delta through the usual definition of the derivative, which is equivalent to taking the derivative of Eq. (1.281)

$$\frac{\mathrm{d}\delta(x-x_1)}{\mathrm{d}x} = \lim_{\sigma\to 0} \frac{\delta(x-x_1+\sigma) - \delta(x-x_1-\sigma)}{2\sigma}, \tag{1.286}$$

that is, as the sum of two Dirac functions of opposite sign that approach each other, also as depicted in Fig. 1.15.

To use the derivative of the Dirac delta function, note that if $a < x_1 < b$,

$$\int_a^b f(x)\,\frac{\mathrm{d}\,\delta(x-x_1)}{\mathrm{d}x}\,\mathrm{d}x = \int_a^b \left\{\frac{\mathrm{d}}{\mathrm{d}x}\Big[f(x)\,\delta(x-x_1)\Big] - \frac{\mathrm{d}f(x)}{\mathrm{d}x}\delta(x-x_1)\right\}\mathrm{d}x \tag{1.287}$$

$$= -\left.\frac{\mathrm{d}f(x)}{\mathrm{d}x}\right|_{x=x_1} \tag{1.288}$$

because $\delta(a-x_1)=0$ and $\delta(b-x_1)=0$. Equation (1.288) is called the *derivative-sifting property* and generalizes to an arbitrary number of derivatives n as

$$\int_a^b f(x)\,\frac{\mathrm{d}^n\delta(x-x_1)}{\mathrm{d}x^n}\,\mathrm{d}x=(-1)^n\left.\frac{\mathrm{d}^n f(x)}{\mathrm{d}x^n}\right|_{x=x_1}. \tag{1.289}$$

So taking the derivative of a Dirac delta function is legitimate at least when one then integrates with it, which you will see is always done in all applications. In Chapter 2, the Dirac delta is used to represent the position of single atoms. In this same context, the derivative of the Dirac is used to represent electric dipoles, in which a concentration of discrete positive charge is located a small distance σ away from a concentration of discrete negative charge.

We can define the *unit step function* $S(x-x_1)$ as the integral of the Dirac delta function

$$S(x-x_1)=\int_{-\infty}^{x}\delta(x_0-x_1)\,\mathrm{d}x_0\mathrel{\widehat{=}}\begin{cases}1 & x\geq x_1\\ 0 & x<x_1\end{cases}. \tag{1.290}$$

Note that the 1 here is unitless regardless of the physical units of x. The *ramp function* $R(x-x_1)$ is similarly defined as the integral of the step function

$$\begin{aligned}R(x-x_1)=(x-x_1)S(x-x_1)&=\int_{-\infty}^{x}\mathrm{d}x_0\int_{-\infty}^{x_0}\delta(x_2-x_1)\,\mathrm{d}x_2\\ &=\int_{-\infty}^{x}\mathrm{d}x_0\,S(x_0-x_1)\mathrel{\widehat{=}}\begin{cases}x-x_1 & x\geq x_1\\ 0 & x<x_1\end{cases}.\end{aligned} \tag{1.291}$$

Thus we also have that

$$\boxed{\delta(x-x_1)=\frac{\mathrm{d}}{\mathrm{d}x}S(x-x_1)=\frac{\mathrm{d}^2}{\mathrm{d}x^2}R(x-x_1).} \tag{1.292}$$

We will use these results for the step and ramp functions in the development of our elastodynamic Green's tensor in Chapter 12.

If we make the substitution of variables that $x\to ax'$ where a is some scalar, then $\mathrm{d}x=a\mathrm{d}x'$ and

$$\int_{-\infty}^{\infty}\delta(x)\,\mathrm{d}x=1=\int_{-\infty}^{\infty}\delta(ax')\,a\mathrm{d}x'. \tag{1.293}$$

If a is negative, the integral would be from $+\infty$ to $-\infty$ which, upon using that $\int_{+\infty}^{-\infty}\mathrm{d}x'=-\int_{-\infty}^{+\infty}\mathrm{d}x'$, yields the same result of 1. So we can conclude that

$$\delta(ax)=\frac{\delta(x)}{|a|}, \tag{1.294}$$

which is important to remember in some applications. For example, if a Dirac function is moving about as a wave response with wave speed c, this can be expressed $\delta(t-r/c)=\delta(r/c-t)=\delta((r-ct)/c)=c\delta(r-ct)$, which is good to know when comparing solutions of the wave equation obtained using different approaches.

One can further generalize to consider a Dirac delta function that is a function of another function, say $g(x)$, that has zeroes and ask about the nature of the integral

$$I = \int_{-\infty}^{\infty} \mathrm{d}x\, \delta[g(x)] f(x). \tag{1.295}$$

To treat this integral, we note that where $g(x) \neq 0$ the Dirac function is zero. So there is only contribution to the integral at the places where $x \to x_i$ where the x_i are the zeroes of $g(x)$ (i.e., the places where $g(x_i) = 0$). As $x \to x_i$ we can represent $g(x)$ as the lead term of the Taylor expansion of $g(x)$, which is $(x - x_i)g'(x_i)$ where $g'(x_i) \mathrel{\widehat{=}} \mathrm{d}g(x)/\mathrm{d}x|_{x=x_i}$. Thus, we can write

$$\delta[g(x)] = \sum_i \delta[(x - x_i)g'(x_i)] = \sum_i \frac{\delta(x - x_i)}{|g'(x_i)|} \tag{1.296}$$

so that we have

$$I = \int_{-\infty}^{\infty} \mathrm{d}x\, \delta[g(x)] f(x) = \sum_i \frac{f(x_i)}{|g'(x_i)|} \tag{1.297}$$

where the sum is over all the zeroes x_i of $g(x)$ found within the domain of integration.

To represent a field concentrated at a point $\boldsymbol{r}_1$ in 3D space using Cartesian coordinates, we use the notation

$$\delta(\boldsymbol{r} - \boldsymbol{r}_1) \mathrel{\widehat{=}} \delta(x - x_1)\, \delta(y - y_1)\, \delta(z - z_1). \tag{1.298}$$

Note that $\delta(\boldsymbol{r} - \boldsymbol{r}_1)$ has units of inverse length cubed, because $\delta(\psi)$ with ψ some scalar, has units of ψ^{-1} as Eq. (1.279) or Eq. (1.294) makes clear. In cylindrical coordinates (r, θ, z), we employ the notation

$$\delta(\boldsymbol{r} - \boldsymbol{r}_1) = \delta(r - r_1)\delta[r(\theta - \theta_1)]\delta(z - z_1) = \delta(r - r_1)\frac{\delta(\theta - \theta_1)}{r}\delta(z - z_1), \tag{1.299}$$

where we used the scaling property of Eq. (1.294). In this cylindrical-coordinate notation, the integrated result over the 3D whole space Ω_∞ is

$$\int_{\Omega_\infty} \delta(\boldsymbol{r} - \boldsymbol{r}_1)\, \mathrm{d}V = \int_0^{\infty} \mathrm{d}r \int_0^{2\pi} r\mathrm{d}\theta \int_{-\infty}^{\infty} \mathrm{d}z\, \delta(r - r_1)\frac{\delta(\theta - \theta_1)}{r}\delta(z - z_1) = 1, \tag{1.300}$$

where we used that $\mathrm{d}V = (\mathrm{d}r)(r\mathrm{d}\theta)(\mathrm{d}z)$ in cylindrical coordinates. Similarly, in spherical coordinates (r, θ, ϕ), we use

$$\delta(\boldsymbol{r} - \boldsymbol{r}_1) = \delta(r - r_1)\delta[r(\theta - \theta_1)]\delta[r\sin\theta(\phi - \phi_1)] = \frac{\delta(r - r_1)\delta(\theta - \theta_1)\delta(\phi - \phi_1)}{r^2 \sin\theta} \tag{1.301}$$

to represent a field concentrated at a point. Employing $dV = (dr)(rd\theta)(r\sin\theta d\phi)$ then gives again

$$\int_{\Omega_\infty} \delta(\boldsymbol{r}-\boldsymbol{r}_1)\,dV = \int_0^\infty dr \int_0^\pi r d\theta \int_0^{2\pi} r\sin\theta d\phi \, \frac{\delta(r-r_1)\delta(\theta-\theta_1)\delta(\phi-\phi_1)}{r^2\sin\theta} = 1 \tag{1.302}$$

as required.

In 3D space, we then have the sifting property for a point $\boldsymbol{r}_1$ lying within a volumetric region Ω

$$\int_\Omega \delta(\boldsymbol{r}-\boldsymbol{r}_1)\psi(\boldsymbol{r})\,dV = \psi(\boldsymbol{r}_1), \tag{1.303}$$

where the field ψ can be a scalar, vector, or tensor of any order. We also have the 3D version of the derivative sifting property

$$\int_\Omega [\nabla\delta(\boldsymbol{r}-\boldsymbol{r}_1)]\,\psi(\boldsymbol{r})\,dV = -\left.\nabla\psi\right|_{\boldsymbol{r}=\boldsymbol{r}_1} \tag{1.304}$$

and for multiple applications of the gradient operator

$$\int_\Omega [{}_n\nabla\delta(\boldsymbol{r}-\boldsymbol{r}_1)]\,\psi(\boldsymbol{r})\,dV = (-1)^n \, {}_n\nabla\psi|_{\boldsymbol{r}=\boldsymbol{r}_1}, \tag{1.305}$$

where ${}_n\nabla = \nabla\nabla\ldots\nabla$ represents n successive ∇ operations.

1.10 Exercises

1. Through direct calculation of the derivatives, demonstrate that

$$\nabla\cdot(\nabla\times\boldsymbol{a}) = 0 \tag{1.306}$$
$$\nabla\times(\nabla\psi) = 0. \tag{1.307}$$

Now, prove these same identities, again through direct calculation, using the Levi–Civita alternating third-order tensor.

2. Using the method demonstrated in Section 1.7, in which the tensorial expressions are first expressed in Cartesian coordinates, derivatives between the scalar components carried out using the usual product rule and dot products performed between base vectors prior to returning to the bold-face representation valid for all coordinate systems, prove all eleven of the identities in the list of Eqs (1.98)–(1.115) that do not involve a curl operation. For an even greater challenge, also prove the identities involving the curl by using the Levi–Civita tensor.

3. With $\boldsymbol{I} = \delta_{ij}\hat{\boldsymbol{x}}_i\hat{\boldsymbol{x}}_j$ being the identity tensor and $\boldsymbol{r} = x_i\hat{\boldsymbol{x}}_i$ the position vector in Cartesian coordinates, prove

$$\nabla\boldsymbol{r} = \boldsymbol{I} \tag{1.308}$$
$$\nabla\cdot\boldsymbol{r} = 3. \tag{1.309}$$

4. If $\boldsymbol{a}$ is some spatially variable vector field in a region Ω and $\boldsymbol{n}$ is the outward normal to the closed surface $\partial\Omega$ that surrounds Ω, show that

$$\int_{\partial\Omega} \boldsymbol{a}\boldsymbol{n}\, \mathrm{d}S = \int_{\Omega} (\nabla \boldsymbol{a})^T\, \mathrm{d}V. \tag{1.310}$$

5. *Prove the fundamental theorem of 3D calculus*: If $\boldsymbol{a}$ is any spatially variable and differentiable vector field $\boldsymbol{a}(x, y, z) = a_x(x, y, z)\hat{\boldsymbol{x}} + a_y(x, y, z)\hat{\boldsymbol{y}} + a_z(x, y, z)\hat{\boldsymbol{z}}$, use the fundamental theorem of 1D calculus to show that if region Ω is a cube with sides of length L then

$$\int_{\Omega} \nabla \boldsymbol{a}\, \mathrm{d}^3\boldsymbol{r} \,\hat{=} \int_0^L \mathrm{d}x \int_0^L \mathrm{d}y \int_0^L \mathrm{d}z\, \nabla \boldsymbol{a}(x, y, z) \tag{1.311}$$

$$\begin{aligned} &= \int_0^L \mathrm{d}y \int_0^L \mathrm{d}z\, \hat{\boldsymbol{x}}\boldsymbol{a}(L, y, z) - \int_0^L \mathrm{d}y \int_0^L \mathrm{d}z\, \hat{\boldsymbol{x}}\boldsymbol{a}(0, y, z) \\ &+ \int_0^L \mathrm{d}x \int_0^L \mathrm{d}z\, \hat{\boldsymbol{y}}\boldsymbol{a}(x, L, z) - \int_0^L \mathrm{d}x \int_0^L \mathrm{d}z\, \hat{\boldsymbol{y}}\boldsymbol{a}(x, 0, z) \\ &+ \int_0^L \mathrm{d}x \int_0^L \mathrm{d}y\, \hat{\boldsymbol{z}}\boldsymbol{a}(x, y, L) - \int_0^L \mathrm{d}x \int_0^L \mathrm{d}y\, \hat{\boldsymbol{z}}\boldsymbol{a}(x, y, 0). \end{aligned} \tag{1.312}$$

Then show that the six surface integrals on the right-hand side here are the contributions from each of the six cube faces coming from the surface integral

$$\int_{\partial\Omega} \boldsymbol{n}\boldsymbol{a}\, \mathrm{d}^2\boldsymbol{r}. \tag{1.313}$$

You thus obtain the fundamental theorem of 3D calculus $\int_{\Omega} \nabla \boldsymbol{a}\, \mathrm{d}^3\boldsymbol{r} = \int_{\partial\Omega} \boldsymbol{n}\boldsymbol{a}\, \mathrm{d}^2\boldsymbol{r}$ using the fundamental theorem of 1D calculus for the case of a cubic integration domain. For an arbitrarily shaped region Ω, you fill the region with tiny (approaching infinitesimal) cubes in a cubic packing. The sum of the volume integrals of $\nabla \boldsymbol{a}$ for each tiny cube adds up to the volume integral of $\nabla \boldsymbol{a}$ over all of Ω. For adjacent tiny cubes that share the same surface, the normal for each cube is oppositely directed and have surface integrals of $\boldsymbol{n}\boldsymbol{a}$ that cancel when summing over the cubes except for the surfaces that are coincident with $\partial\Omega$. The fundamental theorem of 3D calculus is thus proven for arbitrarily shaped regions.

6. Starting from each of the nine components represented by the second-order tensor integral identity

$$\int_{\Omega} \nabla \boldsymbol{a}\, \mathrm{d}^3\boldsymbol{r} = \int_{\partial\Omega} \boldsymbol{n}\boldsymbol{a}\, \mathrm{d}^2\boldsymbol{r}, \tag{1.314}$$

where $\boldsymbol{a}$ is some spatially variable vector field, work in Cartesian coordinates to demonstrate the divergence theorem

$$\int_{\Omega} \nabla \cdot \boldsymbol{a}\, \mathrm{d}^3\boldsymbol{r} = \int_{\partial\Omega} \boldsymbol{n} \cdot \boldsymbol{a}\, \mathrm{d}^2\boldsymbol{r}. \tag{1.315}$$

7. *Prove Green's theorem*: Green's theorem is the statement that on the (x, y) plane, two differentiable functions $P(x, y)$ and $Q(x, y)$ satisfy

$$\int_S \left(\frac{\partial Q(x, y)}{\partial x} - \frac{\partial P(x, y)}{\partial y} \right) \mathrm{d}x\, \mathrm{d}y = \oint_\Gamma \left[P(x, y)\, \mathrm{d}x + Q(x, y)\, \mathrm{d}y \right], \tag{1.316}$$

where S is any surface on the (x, y) plane that is bounded by the closed contour Γ and with the sense of the contour integral being counterclockwise. This can be proven rather trivially for the case of the rectangular surface shown in Fig. 1.16. To do so, simply integrate the left-hand side of Eq. (1.316) over the rectangle and use the 1D fundamental theorem of calculus to obtain

$$\int_{y_1}^{y_2} \mathrm{d}y \int_{x_1}^{x_2} \mathrm{d}x \frac{\partial Q(x, y)}{\partial x} = \int_{y_1}^{y_2} \mathrm{d}y \left[Q(x_2, y) - Q(x_1, y) \right] \tag{1.317}$$

$$-\int_{x_1}^{x_2} \mathrm{d}x \int_{y_1}^{y_2} \mathrm{d}y \frac{\partial P(x, y)}{\partial y} = \int_{x_1}^{x_2} \mathrm{d}x \left[P(x, y_1) - P(x, y_2) \right]. \tag{1.318}$$

Adding these together and identifying the right-hand side as the right-hand side of Eq. (1.316) proves Green's theorem for any rectangle.

To prove this for an arbitrary surface S bounded by the contour Γ such as depicted in Fig. 1.17, you fill the surface S with small squares as shown in Fig. 1.17 and apply Green's theorem as just proven to each such small square. The surface integral over

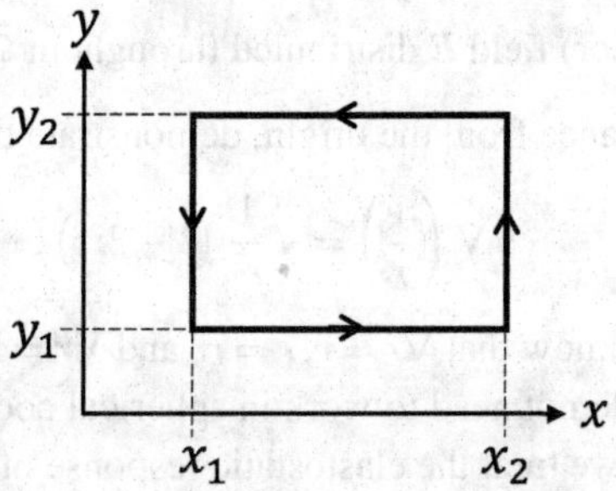

Figure 1.16 Simple rectangle used for proving Green's theorem.

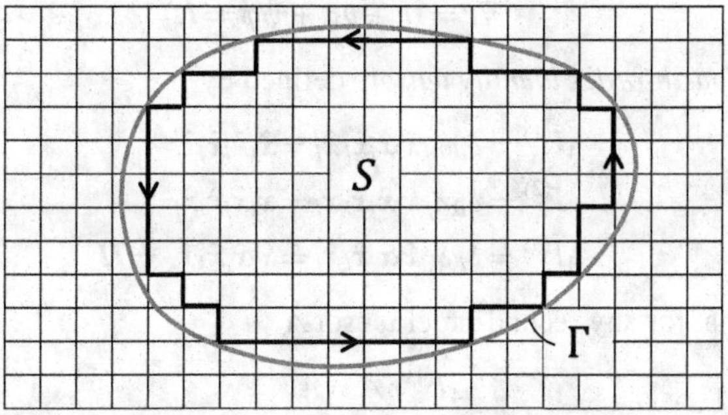

Figure 1.17 Surface S bounded by some contour Γ and filled with small squares made arbitrarily small.

any arbitrary S in Eq. (1.316) (the left-hand side) is obtained by adding the surface integrals over all the small squares together. After applying Green's theorem as proven above for each small square, adjacent squares have line integrals on their shared side that cancel such that the only contribution of the line integrals from the sum of small squares is the integral along the heavy jagged line shown in Fig. 1.17. In the limit as the small squares become quite small, the heavy jagged line becomes indistinguishable from the closed contour Γ and the theorem is proven for any S and not just rectangles.

8. Prove the following tensor identity involving the scalar field $\rho(\boldsymbol{r})$ (this identity is needed in Chapter 7 for the development of the differential rules that control how fluid density varies across a meniscus separating two distinct fluids)

$$\nabla \cdot \left[-\frac{1}{2} |\nabla \rho|^2 \boldsymbol{I} + (\nabla \rho)(\nabla \rho) \right] = (\nabla^2 \rho)(\nabla \rho), \tag{1.319}$$

where $\boldsymbol{I}$ is again the identity tensor and $|\nabla \rho|^2 = (\nabla \rho) \cdot (\nabla \rho)$. Do this using the method given in Section 1.7. With $\boldsymbol{E} = \nabla \rho$, this is also an identity developed in the proof of the Maxwell stress tensor of Chapter 3.

9. On a surface $\partial \Omega$ that surrounds some region of space Ω and has a normal vector $\boldsymbol{n}$, show that

$$\boldsymbol{n} \times \nabla \times \boldsymbol{E} = \boldsymbol{n} \cdot [\boldsymbol{I}(\nabla \cdot \boldsymbol{E}) - \nabla \boldsymbol{E}] \tag{1.320}$$

for some vector (or tensor) field $\boldsymbol{E}$ distributed throughout Ω and on $\partial \Omega$.

10. With r being radial distance from the origin, demonstrate that

$$\nabla \nabla \left(\frac{1}{r} \right) = -\frac{1}{r^3} \left(\boldsymbol{I} - 3\hat{\boldsymbol{r}}\hat{\boldsymbol{r}} \right), \tag{1.321}$$

where you will need to know that $\nabla r = \hat{\boldsymbol{r}}$, $\boldsymbol{r} = r\hat{\boldsymbol{r}}$ and $\nabla \boldsymbol{r} = \boldsymbol{I}$ (the second-order identity tensor). Note that you do not need to work in spherical coordinates to prove this. This is a needed result once we treat the elastostatic response of a solid.

11. In spherical coordinates, with the position vector defined as $\boldsymbol{r} = r\hat{\boldsymbol{r}}$, demonstrate that

$$\nabla \boldsymbol{r} = \hat{\boldsymbol{r}}\hat{\boldsymbol{r}} + \hat{\boldsymbol{\theta}}\hat{\boldsymbol{\theta}} + \hat{\boldsymbol{\phi}}\hat{\boldsymbol{\phi}} = \boldsymbol{I}. \tag{1.322}$$

12. For the three *fourth-order identity tensors* defined by

$$_4\boldsymbol{I}^{(1)} = \delta_{il}\delta_{jk}\hat{\boldsymbol{x}}_i\hat{\boldsymbol{x}}_j\hat{\boldsymbol{x}}_k\hat{\boldsymbol{x}}_l = \hat{\boldsymbol{x}}_i\hat{\boldsymbol{x}}_j\hat{\boldsymbol{x}}_j\hat{\boldsymbol{x}}_i \tag{1.323}$$

$$_4\boldsymbol{I}^{(2)} = \delta_{ik}\delta_{jl}\hat{\boldsymbol{x}}_i\hat{\boldsymbol{x}}_j\hat{\boldsymbol{x}}_k\hat{\boldsymbol{x}}_l = \hat{\boldsymbol{x}}_i\hat{\boldsymbol{x}}_j\hat{\boldsymbol{x}}_i\hat{\boldsymbol{x}}_j \tag{1.324}$$

$$_4\boldsymbol{I}^{(3)} = \delta_{ij}\delta_{kl}\hat{\boldsymbol{x}}_i\hat{\boldsymbol{x}}_j\hat{\boldsymbol{x}}_k\hat{\boldsymbol{x}}_l = \hat{\boldsymbol{x}}_i\hat{\boldsymbol{x}}_i\hat{\boldsymbol{x}}_k\hat{\boldsymbol{x}}_k = \boldsymbol{I}\boldsymbol{I} \tag{1.325}$$

demonstrate that for any second-order tensor $\boldsymbol{A}$

$$_4\boldsymbol{I}^{(1)} : \boldsymbol{A} = \boldsymbol{A} \tag{1.326}$$

$$_4\boldsymbol{I}^{(2)} : \boldsymbol{A} = \boldsymbol{A}^T \tag{1.327}$$

$$_4\boldsymbol{I}^{(3)} : \boldsymbol{A} = \text{tr}\,\{\boldsymbol{A}\}\,\boldsymbol{I}. \tag{1.328}$$

These three fourth-order identity tensors have components that are entirely independent of the coordinates being used and can also be called the fourth-order *isotropic* tensors. Of the 4! possible ways of distributing the indices i, j, k, and l over two Kronecker delta functions, there are only three unique ways as given in Eqs (1.323)–(1.325). So there are three and only three fourth-order isotropic tensors, and we will use them later when we derive the laws of elasticity in an isotropic solid.

13. Derive the fourth-order tensor identity

$$\nabla\nabla\,(\boldsymbol{rr}) = \hat{\boldsymbol{x}}_i\hat{\boldsymbol{x}}_j\hat{\boldsymbol{x}}_i\hat{\boldsymbol{x}}_j + \hat{\boldsymbol{x}}_i\hat{\boldsymbol{x}}_j\hat{\boldsymbol{x}}_j\hat{\boldsymbol{x}}_i = {}_4\boldsymbol{I}^{(2)} + {}_4\boldsymbol{I}^{(1)}. \tag{1.329}$$

To do so, write each $\nabla = \hat{\boldsymbol{x}}_i \partial/\partial x_i$ and $\boldsymbol{r} = x_i\hat{\boldsymbol{x}}_i$ in Cartesian coordinates with each vector having its own index and use the fact that $\partial x_i/\partial x_j = \delta_{ij}$. Similarly derive the sixth-order tensor identity

$$\begin{aligned}\nabla\nabla\nabla\,(\boldsymbol{rrr}) &= \hat{\boldsymbol{x}}_i\hat{\boldsymbol{x}}_j\hat{\boldsymbol{x}}_k\hat{\boldsymbol{x}}_i\hat{\boldsymbol{x}}_j\hat{\boldsymbol{x}}_k + \hat{\boldsymbol{x}}_i\hat{\boldsymbol{x}}_j\hat{\boldsymbol{x}}_k\hat{\boldsymbol{x}}_i\hat{\boldsymbol{x}}_k\hat{\boldsymbol{x}}_j \\ &+ \hat{\boldsymbol{x}}_i\hat{\boldsymbol{x}}_j\hat{\boldsymbol{x}}_k\hat{\boldsymbol{x}}_j\hat{\boldsymbol{x}}_i\hat{\boldsymbol{x}}_k + \hat{\boldsymbol{x}}_i\hat{\boldsymbol{x}}_j\hat{\boldsymbol{x}}_k\hat{\boldsymbol{x}}_j\hat{\boldsymbol{x}}_k\hat{\boldsymbol{x}}_i \\ &+ \hat{\boldsymbol{x}}_i\hat{\boldsymbol{x}}_j\hat{\boldsymbol{x}}_k\hat{\boldsymbol{x}}_k\hat{\boldsymbol{x}}_i\hat{\boldsymbol{x}}_j + \hat{\boldsymbol{x}}_i\hat{\boldsymbol{x}}_j\hat{\boldsymbol{x}}_k\hat{\boldsymbol{x}}_k\hat{\boldsymbol{x}}_j\hat{\boldsymbol{x}}_i\end{aligned} \tag{1.330}$$

by again writing each ∇ and $\boldsymbol{r}$ in Cartesian coordinates. Then show that for some third-order tensor ${}_3\boldsymbol{A} = A_{lmn}\hat{\boldsymbol{x}}_l\hat{\boldsymbol{x}}_m\hat{\boldsymbol{x}}_n$ that

$$[\nabla\nabla\nabla\,(\boldsymbol{rrr})] \stackrel{3}{\vdots}\, {}_3\boldsymbol{A} = \left(A_{ijk} + A_{ikj} + A_{jik} + A_{jki} + A_{kij} + A_{kji}\right)\hat{\boldsymbol{x}}_i\hat{\boldsymbol{x}}_j\hat{\boldsymbol{x}}_k \tag{1.331}$$

$$= \boldsymbol{A} + \boldsymbol{A}^{\overset{T}{132}} + \boldsymbol{A}^{\overset{T}{213}} + \boldsymbol{A}^{\overset{T}{231}} + \boldsymbol{A}^{\overset{T}{312}} + \boldsymbol{A}^{\overset{T}{321}}. \tag{1.332}$$

Such transpose identities are used in the derivation of the Taylor series coefficients for Taylor series of fields (scalars, vectors, or tensors) in three-dimensional space as described in Section 1.8.1.

14. Demonstrate through matrix multiplication that the Cartesian-coordinate rotation matrix for counterclockwise rotations θ_1, θ_2, and θ_3 about the Cartesian axes x_1, x_2, and x_3 is given by

$$R_{ij}(\theta_1,\theta_2,\theta_3) = R_{ik}(\theta_1)R_{kl}(\theta_2)R_{lj}(\theta_3)$$
$$= \begin{bmatrix} \cos\theta_2\cos\theta_3 & \cos\theta_2\sin\theta_3 & -\sin\theta_2 \\ \sin\theta_1\sin\theta_2\cos\theta_3 - \cos\theta_1\sin\theta_3 & \sin\theta_1\sin\theta_2\sin\theta_3 + \cos\theta_1\cos\theta_3 & \sin\theta_1\cos\theta_2 \\ \cos\theta_1\cos\theta_3\sin\theta_2 - \sin\theta_1\sin\theta_3 & \cos\theta_1\sin\theta_2\sin\theta_3 - \sin\theta_1\cos\theta_3 & \cos\theta_1\cos\theta_2 \end{bmatrix}$$

where $R_{ik}(\theta_1)$, $R_{kl}(\theta_2)$, and $R_{lj}(\theta_3)$ are given by Eqs (1.145), (1.152), and (1.153). Then show by direct matrix multiplication that for any rotation of the Cartesian coordinates, the identity tensor in the rotated coordinates is $R_{ik}(\theta_1,\theta_2,\theta_3)R_{jl}(\theta_1,\theta_2,\theta_3)\delta_{kl} = R_{ik}(\theta_1,\theta_2,\theta_3)\left[R_{jk}(\theta_1,\theta_2,\theta_3)\right]^T = \delta_{ij}$. So the second-order identity tensor is an isotropic second-order tensor, that is, a second-order tensor whose components do not change when we make arbitrary changes to the orientation of the axes.

15. For arbitrary orthogonal-curvilinear coordinates having metrical coefficients $h_1(q_1,q_2,q_3)$, $h_2(q_1,q_2,q_3)$, and $h_3(q_1,q_2,q_3)$ as well as unit base vectors $\hat{\boldsymbol{q}}_1$, $\hat{\boldsymbol{q}}_2$, and $\hat{\boldsymbol{q}}_3$, demonstrate that

$$\nabla\nabla\,(\boldsymbol{rr}) = \hat{\boldsymbol{q}}_i\hat{\boldsymbol{q}}_j\hat{\boldsymbol{q}}_i\hat{\boldsymbol{q}}_j + \hat{\boldsymbol{q}}_i\hat{\boldsymbol{q}}_j\hat{\boldsymbol{q}}_j\hat{\boldsymbol{q}}_i \tag{1.333}$$

using the ideas developed in Section 1.8.6.

16. Using the well-known definite integral $\int_{-\infty}^{\infty} e^{-u^2}\,du = \sqrt{\pi}$ that we will prove in Chapter 10, demonstrate that the representation of the Dirac delta function given by

$$\delta(x - x_1) = \lim_{\sigma\to 0} \frac{e^{-(x-x_1)^2/(2\sigma^2)}}{\sigma\sqrt{2\pi}} \tag{1.334}$$

indeed possesses the required property of the Dirac delta that

$$\int_{-\infty}^{\infty} \frac{e^{-(x-x_1)^2/(2\sigma^2)}}{\sigma\sqrt{2\pi}}\,\mathrm{d}x = 1 \tag{1.335}$$

for any σ. Taking $\sigma \to 0$ is what allows us to obtain the other key property of the Dirac delta that $\int_{-\infty}^{\infty} f(x)\delta(x-x_1)\,\mathrm{d}x = f(x_1)$, which is called the sifting property.

If we now define the Dirac delta to be some arbitrary power n of the above bell-shaped curve

$$\delta(x - x_1) = \lim_{\sigma\to 0} \frac{e^{-n(x-x_1)^2/(2\sigma^2)}}{c_\sigma}, \tag{1.336}$$

show that the normalization constant c_σ that allows this representation to possess the required property $\int_{-\infty}^{\infty} \delta(x-x_1)\,\mathrm{d}x = 1$ is

$$c_\sigma = \sigma\sqrt{\frac{2\pi}{n}}. \tag{1.337}$$

17. Given, say, a second-order tensor field $\boldsymbol{A}(\boldsymbol{r})$ (but this could also be a scalar or vector field) and a point $\boldsymbol{r}_s$ located somewhere within a volumetric region Ω and not on the boundary $\partial\Omega$, demonstrate the 3D gradient-sifting property of the 3D Dirac delta that states

$$\nabla \boldsymbol{A}(\boldsymbol{r})|_{\boldsymbol{r}=\boldsymbol{r}_s} = -\int_\Omega \mathrm{d}\boldsymbol{r}^3\ [\nabla\delta(\boldsymbol{r}-\boldsymbol{r}_s)]\ \boldsymbol{A}(\boldsymbol{r}). \tag{1.338}$$

To do so, you will need to use the fundamental theorem of 3D calculus and the sifting property of the 3D Dirac delta function.

18. Show that if you represent the 3D Dirac delta function using the Gaussian function in the limit as the standard of deviation becomes small, you can represent its gradient as

$$\nabla\delta(\boldsymbol{r}-\boldsymbol{r}_s) = \lim_{\sigma\to 0} -\frac{(\boldsymbol{r}-\boldsymbol{r}_s)}{\sigma^5(2\pi)^{3/2}} \exp\left(\frac{-|\boldsymbol{r}-\boldsymbol{r}_s|^2}{2\sigma^2}\right). \tag{1.339}$$

19. For the integral $I = \int_{-\infty}^{\infty} \mathrm{d}x\,\delta[g(x)]f(x)$, when $g(x) = \sin(\pi x/L)$ and $f(x) = e^{-x/L}$, show that

$$I = \frac{eL}{\pi(e-1)}. \tag{1.340}$$

HINT: it will prove useful to remember the binomial expansion $(1-u)^{-1} = \sum_{n=0}^{\infty} u^n$ for $|u| < 1$.

20. Demonstrate that for $a > 0$,

$$\delta(x^2 - a^2) = \frac{1}{2a}\left[\delta(x+a) + \delta(x-a)\right]. \tag{1.341}$$

21. For a symmetric second-order tensor given by $\boldsymbol{S} = \boldsymbol{ab} + \boldsymbol{ba}$ and a so called anti-symmetric second-order tensor given by $\boldsymbol{A} = \boldsymbol{cd} - \boldsymbol{dc}$, where $\boldsymbol{a}$, $\boldsymbol{b}$, $\boldsymbol{c}$, and $\boldsymbol{d}$ are vectors, show that $\boldsymbol{S} : \boldsymbol{A} = 0$.

2

Continuum Mechanics

Our goal throughout all the remaining chapters of Part I is to derive the macroscopic rules that control the collective movement of large numbers of atoms. We call this set of rules the laws of *continuum physics*. We can separate them into the rules that control the movement of matter called *continuum mechanics*, the rules of the macroscopic force fields called *continuum electromagnetism* and *continuum gravity* that are causing, at least in part, the movement of matter and the rules controlling how certain responses of matter are produced by certain forces that we call *constitutive laws*, which come from the ideas and postulates of equilibrium and nonequilibrium thermodynamics. In what follows, we take the nonconventional approach of establishing our macroscopic continuum laws on the underlying physics that controls how each individual atom is moving about.

2.1 System of Discrete Particles → Continuous Description

To model the mechanics of any macroscopic Earth system of interest (the crust, a sedimentary basin, the soils under a building, a rock in the laboratory, a grain in a rock, etc.) notice that:

- Just one gram of quartz or SiO_2, which fits within a cube about 7 mm on each side, contains:

$$\begin{aligned} N &= \text{number of molecules} \approx 1\text{ g} \left(\frac{1\text{ mole}}{60\text{ g}}\right)\left(\frac{6.0\times 10^{23}\text{ molecules}}{\text{mole}}\right) \\ &= 10^{22}\ \text{SiO}_2\text{ molecules.} \end{aligned}$$

  There are a total of 90 fundamental particles (electrons, neutrons, and protons) in each quartz molecule for a total of 9×10^{23} fundamental particles in each gram of quartz.
- It's neither feasible nor necessary to model the movement of each fundamental particle of each atom using the rules of quantum mechanics in order to describe the dynamics of macroscopic systems.
- At each point $\boldsymbol{r}$ within our macroscopic system, we need only model the collective averaged (or summed) movement of all the atoms in a volumetric neighborhood of size V_0 surrounding that point as depicted in Fig. 2.1. The average molecular behavior is what we measure at the macroscopic scale.

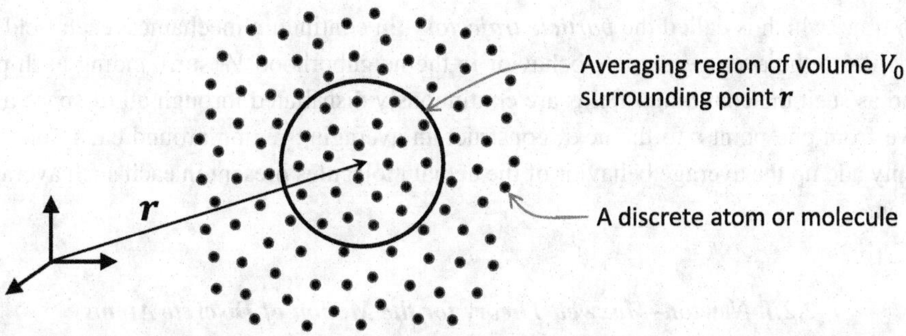

Figure 2.1 Around each point $\boldsymbol{r}$ within a system containing a vast number of interacting atoms or molecules, an averaging region of volume V_0 is constructed. The average atomic behavior in each such region surrounding each point $\boldsymbol{r}$ corresponds to the various fields of continuum mechanics.

In what follows, we may use the word *particles* to refer either to the *fundamental* electrons and the nuclei (the grouping of neutrons and protons at the center of each atom) or to the collective groupings of the electrons and nuclei into atoms or the grouping of atoms into molecules. But we should be aware, and it will be pointed out, that how the interaction forces between the particles are described will be different depending on whether we are describing electrons and nuclei or atoms or molecules.

Physicist Richard Feynman said in Volume 1 of his Feynman Lectures on Physics (Feynman, 1963) that if he had to distill all of science down to a single sentence, that sentence would be: "*all things are made of atoms – little particles that move around in perpetual motion, attracting each other when they are a little distance apart, but repelling upon being squeezed into one another.*" Despite not needing to specify the trajectory of each atom in order to make macroscopic predictions, all macroscale physical processes are controlled by the way the atoms in each V_0 are moving and how they interact with their surroundings.

We want our continuum laws and continuum fields to be founded on the reality of this underlying molecular dynamics. So in what follows, we derive our continuum laws by averaging over the molecular-scale physics. We do not replace the system of discrete particles by a space-filling, infinitely divisible "continuous mass" supporting macroscopic fields whose nature must be guessed at, which was the approach of Cauchy in the nineteenth century prior to the certainty of atoms. We obtain our macroscopic rules by simply averaging over the discrete molecules actually present, which can be considered an exact process. This approach provides unambiguous definition of the macroscopic fields that can also be used to average molecular simulations to obtain macroscopic fields. It provides knowledge not available when using the nineteenth-century Cauchy perspective.

2.2 Mechanics of Discrete Particle Systems

Discrete particle mechanics is distinct from the continuum mechanics we will derive because one models the location of each particle (or atom) and how that location changes with time, which is called the *particle trajectory*. In continuum mechanics, each field corresponds to the average particle behavior in the neighborhood V_0 surrounding each point $\boldsymbol{r}$ and as such the continuum fields are continuously distributed through all of space as we move from one point $\boldsymbol{r}$ to the next, construct an averaging region around each point, and simply add up the average behavior of the actual molecules present in each such averaging region.

2.2.1 Newton–Maxwell Theory for the Motion of Discrete Atoms

The Conservation of Linear Momentum

From the classical Newtonian perspective, each discrete atom p in the universe ($p = 1, 2, \ldots, N_\infty$) follows a trajectory $\boldsymbol{r}_p(t)$ controlled by Newton's second law (the conservation of linear momentum).

$$\frac{\mathrm{d}}{\mathrm{d}t}\left[m_p\,\dot{\boldsymbol{r}}_p(t)\right] = \boldsymbol{F}_p(t) \qquad \textit{conservation of linear momentum,} \tag{2.1}$$

where

$$\boldsymbol{r}_p(t) \mathrel{\widehat{=}} \text{position of } p \text{ at time } t \text{ that we call the } \textit{trajectory} \tag{2.2}$$

$$\dot{\boldsymbol{r}}_p(t) = \frac{\mathrm{d}\boldsymbol{r}_p(t)}{\mathrm{d}t} = \boldsymbol{v}_p(t) \mathrel{\widehat{=}} \text{velocity of } p \text{ at time } t \text{ (and position } \boldsymbol{r} = \boldsymbol{r}_p) \tag{2.3}$$

$$m_p \mathrel{\widehat{=}} \text{mass of } p \text{ (time independent)} \tag{2.4}$$

$$\boldsymbol{F}_p(t) \mathrel{\widehat{=}} \text{force acting on } p \text{ at time } t. \tag{2.5}$$

Given the initial conditions $\boldsymbol{r}_p(t=0)$ and $\boldsymbol{v}_p(t=0)$, the trajectory $\boldsymbol{r}_p(t)$ is the solution of a highly nonlinear multibody problem due to the fact that $\boldsymbol{F}_p$ depends on the positions of the other atoms.

If p represents a molecule (grouping of atoms) that has an orientation that influences the nature of the force fields emanating from (or acting upon) the molecule, we would also have to write down and solve the conservation of angular momentum to obtain not only the trajectory $\boldsymbol{r}_p(t)$ of the center of mass of the molecule but how its orientation changes through time. We will forego that complication in this pedagogic introduction to discrete particle mechanics and simply assume each particle is an isotropic atom that has a mass m_p, a net charge q_p (possibly zero), and can become both electrically and magnetically polarized.

The Force $\boldsymbol{F}_p$ Acting on Atom p

In this discrete treatment of the individual atomic trajectories, we assume that atom p may acquire an electric-dipole moment $\boldsymbol{p}_p$ corresponding to a charge separation $+\delta q_p$ and $-\delta q_p$ created across a distance $\boldsymbol{d}_p$ within the atom so that $\boldsymbol{p}_p(t) = |\delta q_p(t)|\boldsymbol{d}_p(t)$, where by convention $\boldsymbol{d}_p$ points from the negative charge toward the positive charge and is thus in the direction opposite to the associated electric field, which is defined by the direction a positive test charge moves. Such induced electric-dipole moments can occur either due to fluctuations of the electrons relative to the nucleus as described by quantum mechanics or due to an electric field acting on the atom that pushes more electrons to one side of the nucleus compared to the other.

Similarly, we assume that atom p may acquire a magnetic-dipole moment $\boldsymbol{\mu}_p$. The fundamental particles have no magnetic "charge" (no magnetic monopoles) but they do have magnetic-dipole moments called quantum *spin*. The magnetic-dipole moment of an entire atom $\boldsymbol{\mu}_p$ may be due to a coherent alignment of the quantum spins of the electrons and nucleus or, as will be assumed in the classic treatment given here, due to a circulation of the electrons around the nucleus. In the case of a circulation of electrons around the nucleus, the orientation of the dipole moment is determined using the right-hand rule by pointing your fingers in the direction that the electrons are circulating so that $\boldsymbol{\mu}_p$ is in the direction of your thumb. When a magnetic field from another atom acts on atom p, the electron circulation is altered and $\boldsymbol{\mu}_p$ is altered.

Classically, the forces $\boldsymbol{F}_p$ acting on p come from all the other atoms $n \neq p$ in the universe. But an atom does not exert a force on itself. These forces are given by

$$\boldsymbol{F}_p(t) = \int_{\Omega_\infty} \mathrm{d}^3\boldsymbol{r} \sum_{\substack{n=1 \\ n\neq p}}^{N_\infty} \Bigg\{ \underbrace{m_p\delta(\boldsymbol{r}-\boldsymbol{r}_p)\boldsymbol{g}_n(\boldsymbol{r},t)}_{\substack{\uparrow \\ \textit{gravity}}} + \underbrace{\left[q_p\delta(\boldsymbol{r}-\boldsymbol{r}_p) - \boldsymbol{p}_p\cdot\nabla\delta(\boldsymbol{r}-\boldsymbol{r}_p)\right]\boldsymbol{E}_n(\boldsymbol{r},t)}_{\substack{\uparrow \\ \textit{electrical}}}$$
$$+ \underbrace{\left[\left(q_p\dot{\boldsymbol{r}}_p + \dot{\boldsymbol{p}}_p\right)\delta(\boldsymbol{r}-\boldsymbol{r}_p) - \boldsymbol{\mu}_p \times \nabla\delta(\boldsymbol{r}-\boldsymbol{r}_p)\right] \times \boldsymbol{B}_n(\boldsymbol{r},t)}_{\substack{\uparrow \\ \textit{magnetic}}} \Bigg\} \tag{2.6}$$

with $\boldsymbol{g}_n$ the acceleration of gravity emanating from atom n and $\boldsymbol{E}_n$ and $\boldsymbol{B}_n$ the electric and magnetic fields due to n. The integral over all space and the insertion of the Dirac delta serves to evaluate the force fields $\boldsymbol{g}_n(\boldsymbol{r}, t)$, $\boldsymbol{E}_n(\boldsymbol{r}, t)$, and $\boldsymbol{B}_n(\boldsymbol{r}, t)$, that vary continuously through all of space, onto the atom p located at $\boldsymbol{r} = \boldsymbol{r}_p(t)$. Let's explain the mathematical form of the electric- and magnetic-dipole terms.

Note that from the derivative property of the Dirac delta function, we have that in the electric-dipole force term $\boldsymbol{F}_{pn}^{(ed)}$ between atoms p and n

$$\boldsymbol{F}_{pn}^{(ed)} = -\boldsymbol{p}_p(t)\cdot\int_{\Omega_\infty} \mathrm{d}^3\boldsymbol{r}\left[\nabla\delta(\boldsymbol{r}-\boldsymbol{r}_p(t))\right]\boldsymbol{E}_n(\boldsymbol{r},t) = \boldsymbol{p}_p(t)\cdot\nabla\boldsymbol{E}_n(\boldsymbol{r},t)\big|_{\boldsymbol{r}=\boldsymbol{r}_p(t)}, \tag{2.7}$$

where we are already putting to use our tensor and Dirac-delta calculus from Chapter 1 (i.e., $\nabla\boldsymbol{E}_n$ is a second-order tensor and we used the derivative sifting property of the Dirac delta as proven in Exercise 17 of Chapter 1). Let's say that the negatively charged tail of the dipole of atom p is located at $\boldsymbol{r}_p - \boldsymbol{d}_p/2$ and that the positively charged head is therefore at $\boldsymbol{r}_p + \boldsymbol{d}_p/2$. The electric-dipole force $\boldsymbol{F}_{pn}^{(ed)}$ can thus also be written

$$F_{pn}^{(ed)} = -\delta q_p E_n(r_p - d_p/2) + \delta q_p E_n(r_p + d_p/2) \tag{2.8}$$

$$= -\delta q_p E_n(r_p) - \frac{(-\delta q_p) d_p}{2} \cdot \nabla E_n(r)|_{r=r_p} + \delta q_p E_n(r_p) + \frac{\delta q_p d_p}{2} \cdot \nabla E_n(r)|_{r=r_p} + \dots \tag{2.9}$$

$$= p_p \cdot \nabla E_n(r)|_{r=r_p} + O\left(|d_p|^2/|r_{pn}|^2\right), \tag{2.10}$$

where we expanded each $E_n(r_p \pm d_p/2)$ in a Taylor series around the distance to the center of the atom r_p and assumed that $|d_p|$ is small relative to the distance $|r_{pn}| = |r_p - r_n|$ between atoms p and n over which ∇E_n is varying. We have thus shown why we allowed for this electrical force acting on a dipole in Eq. (2.6) by using the gradient of the Dirac delta function. This electric-dipole force $F_{pn}^{(ed)}$ between p and n operates even when atom p is uncharged $q_p = 0$. It is the source for the so-called *van der Waals* force interaction discussed below and is the dominant attractive force between two uncharged atoms that do not covalently share electrons.

Similarly, in the magnetic-dipole force term $F_{pn}^{(md)}$ between atom n and the magnetic-dipole moment of atom p we have

$$F_{pn}^{(md)} = -\mu_p(t) \times \int_{\Omega_\infty} d^3r \left[\nabla\delta(r - r_p(t))\right] \times B_n(r, t) = \mu_p(t) \times \nabla \times B_n(r, t)\Big|_{r=r_p(t)} \tag{2.11}$$

where we integrated the identity $\nabla \times (B_n\delta) = (\nabla \times B_n)\,\delta + (\nabla\delta) \times B_n$ over all space and used that $n \times B_n\delta(r - r_p)$ is zero on the surface $\partial\Omega_\infty$ at infinity.

If we carry out the integral over all of space given in Eq. (2.6) and use the properties of the Dirac delta function, we thus obtain

$$F_p(t) = \sum_{\substack{n=1 \\ n \neq p}}^{N_\infty} \Big[m_p g_n(r_p, t) + q_p E_n(r_p, t) + p_p \cdot \nabla E_n(r, t)|_{r=r_p} + \left(q_p \dot{r}_p + \dot{p}_p\right) \times B_n(r_p, t) + \mu_p \times \nabla \times B_n(r, t)|_{r=r_p} \Big]. \tag{2.12}$$

The force acting on atom p comes overwhelmingly from the electromagnetic (EM) force fields E_n and B_n emanating from each of the other atoms n, which makes the determination of the trajectory $r_p(t)$ a highly nonlinear multibody problem. To proceed, we need a theory for the determination of the gravitational acceleration and EM fields created by atom n.

Newtonian Gravity and Maxwellian Electromagnetism

We next give the rules for determining the three force fields $g_n(r, t)$, $E_n(r, t)$, and $B_n(r, t)$ emanating from each atom n.

The gravitational acceleration g_n created by atom n is given by the *Newtonian theory of gravity*:

$$\nabla \cdot \boldsymbol{g}_n(\boldsymbol{r}, t) = -4\pi G\, m_n\, \delta(\boldsymbol{r} - \boldsymbol{r}_n(t)) \tag{2.13}$$

$$\nabla \times \boldsymbol{g}_n(\boldsymbol{r}, t) = 0. \tag{2.14}$$

The fact that $\nabla \times \boldsymbol{g}_n = 0$ means that the vector $\boldsymbol{g}_n$ can be written as the gradient of a scalar potential $\boldsymbol{g}_n = -\nabla U_n$, where $U_n(\boldsymbol{r}, t)$ is called the *gravitational potential*. From the first law of gravity above, U_n must satisfy

$$\nabla^2 U_n = 4\pi G m_n \delta(\boldsymbol{r} - \boldsymbol{r}_n), \tag{2.15}$$

where $\delta(\boldsymbol{r} - \boldsymbol{r}_n) \mathrel{\widehat{=}}$ the 3D Dirac delta $= \delta(x - x_n)\, \delta(y - y_n)\, \delta(z - z_n)$ and where $G \mathrel{\widehat{=}}$ *universal gravitational constant* $= 6.67 \times 10^{-11}\,\mathrm{Nm^2/kg^2}$. In Part II of the book, you will learn how to solve the partial-differential equation (PDE) $\nabla^2\, U_n = 4\pi G\, m_n\, \delta(\boldsymbol{r} - \boldsymbol{r}_n)$ to obtain the 3D solution $U_n = -G\, m_n / |\boldsymbol{r} - \boldsymbol{r}_n|$. For the purpose of determining the trajectory $\boldsymbol{r}_p(t)$, the gravitational interaction $m_p\, \boldsymbol{g}_n\, (\boldsymbol{r} = \boldsymbol{r}_p, t)$ between particle p and n is entirely negligible compared to the EM forces as will be shown below.

The electric field $\boldsymbol{E}_n(\boldsymbol{r}, t)$ (units of V/m) and magnetic field $\boldsymbol{B}_n(\boldsymbol{r}, t)$ (units of T) produced by atom n are governed by *Maxwell's equations* in a vacuum with n moving about with trajectory $\boldsymbol{r}_n(t)$

$$\epsilon_0\, \nabla \cdot \boldsymbol{E}_n = q_n\, \delta(\boldsymbol{r} - \boldsymbol{r}_n) - \boldsymbol{p}_n \cdot \nabla \delta(\boldsymbol{r} - \boldsymbol{r}_n) \qquad \textit{Coulomb law} \tag{2.16}$$

$$\nabla \times \boldsymbol{E}_n = -\frac{\partial \boldsymbol{B}_n}{\partial t} \qquad \textit{Faraday's law} \tag{2.17}$$

$$\frac{1}{\mu_0} \nabla \times \boldsymbol{B}_n = \epsilon_0 \frac{\partial \boldsymbol{E}_n}{\partial t} + \left(q_n \dot{\boldsymbol{r}}_n + \dot{\boldsymbol{p}}_n\right)\, \delta(\boldsymbol{r} - \boldsymbol{r}_n) - \boldsymbol{\mu}_n \times \nabla \delta(\boldsymbol{r} - \boldsymbol{r}_n) \qquad \textit{Ampère's law} \tag{2.18}$$

$$\nabla \cdot \boldsymbol{B}_n = 0 \qquad \textit{Gauss' law} \tag{2.19}$$

where

$\epsilon_0 = 8.85 \times 10^{-12}$ F/m $\mathrel{\widehat{=}}$ electrical permittivity of vacuum

$\mu_0 = 4\pi \times 10^{-7}$ H/m $\mathrel{\widehat{=}}$ magnetic permeability of vacuum

$q_n \mathrel{\widehat{=}}$ net charge [units of C] on n (2.20)

$$\boldsymbol{p}_n(t) \mathrel{\widehat{=}} \text{electric-dipole moment of } n \text{ [units of Cm]} = \alpha_n \sum_{\substack{m=1\\ m \neq n}}^{N_\infty} \boldsymbol{E}_m\, (\boldsymbol{r}_n(t), t) \tag{2.21}$$

$$\boldsymbol{\mu}_n(t) \mathrel{\widehat{=}} \text{magnetic-dipole moment of } n \text{ [units of Cm}^2\text{/s]} = \beta_n \sum_{\substack{m=1\\ m \neq n}}^{N_\infty} \boldsymbol{B}_m\, (\boldsymbol{r}_n(t), t) \tag{2.22}$$

$\alpha_n \mathrel{\widehat{=}}$ electric polarizability of n [units of $\mathrm{C/(m^2V)}$] (2.23)

$\beta_n \mathrel{\widehat{=}}$ magnetic polarizability of n [units of $\mathrm{Cm^4/(s^2V)}$]. (2.24)

We assume the reader has been exposed to Maxwell's equations before. But in brief, a time-varying magnetic field will generate an electric field that circulates around the magnetic field lines, a process that is called *electromagnetic induction* and that is quantified by Faraday's law defined above. Similarly, a time-varying electric field will generate a magnetic field that circulates around the electric field lines as quantified by Ampère's law above. What we call Coulomb's law, which is the differential rule that results in the Coulomb force law for the electrical interaction between charged particles, is more commonly called "Gauss's law for the electric field" and what we call Gauss' law above is called "Gauss's law for the magnetic field." Note that if an atom n has a net charge q_n we call that atom an *ion*; a *cation* if there are more protons than electrons and an *anion* if there are more electrons than protons.

We sometimes introduce the *electromagnetic potentials* φ_n and $\boldsymbol{A}_n$ as

$$\boldsymbol{E}_n = -\nabla\varphi_n + \frac{\partial \boldsymbol{A}_n}{\partial t} \text{ and } \boldsymbol{B}_n = \nabla \times \boldsymbol{A}_n \text{ with } \nabla \cdot \boldsymbol{A}_n = 0. \tag{2.25}$$

Thus, for example, we have that for a nonpolarized atom bearing a net charge q_n (an ion), Coulomb's law can be rewritten

$$\nabla^2\varphi_n = -\frac{q_n}{\epsilon_0}\delta(\boldsymbol{r}-\boldsymbol{r}_n) \Rightarrow \varphi_n = \frac{q_n}{4\pi\,\epsilon_0\,|\boldsymbol{r}-\boldsymbol{r}_n|}. \tag{2.26}$$

Note the ratio of gravitational to electrical attraction energies between two oppositely charged ions is: $m_p U_{pn}/(q_p\varphi_{pn}) = 4\pi\epsilon_0 G m_p m_n/(q_p q_n) \approx 5 \times 10^{-34}$, where we put in numbers appropriate for a Na cation and Cl anion. So the gravitational interaction between two isolated ions is unimportant compared to the electrical interaction.

Even when an atom has no net charge, quantum-mechanical fluctuations of the electron positions create an instantaneous dipole moment that in turn induces a dipole moment in a neighboring atom n as quantified by the polarizability α_n and two aligned dipoles attract. This is called *van der Waals attraction* which has an interaction energy that falls off as r^{-6} with the distance of separation r (a factor of r^{-3} controls how the electric field from one dipole falls off with distance to create polarization on a neighboring atom located at distance r and another factor of r^{-3} is the energy of attraction between two aligned dipoles, which can be defined as the work required to separate to infinity two aligned dipoles initially a distance r apart). Van der Waals forces, which are always present, create attraction between two atoms and are also fantastically more important than gravitational forces between two atoms that are within a few atomic distances from each other. However, in the above classical formalism, no rules have yet been given for quantifying the fluctuating dipole moment of an atom. Also missing in the above description is a hard-sphere repulsive force that prevents two finite-width atoms from overlapping, details of which also have their origin in the quantum theory of electron interactions.

We can approximate the fluctuating atomic dipole moment by adding a randomly varying dipole-moment vector $\boldsymbol{\eta}_n(t)$ to Eq. (2.21) that acts as a source term for the van der Waals interaction. We can also create an approximate hard-sphere repulsion that does not require modeling the electron interactions between two atoms in proximity. But what changes in the above if we take the particles to be electrons and nuclei? For one thing, we would have $\alpha_n = 0$, $\boldsymbol{\eta}_n(t) = 0$ and thus $\boldsymbol{p}_n(t) = 0$ so that the particles do not polarize electrically. But more deeply, what prevents an electron from collapsing into and permanently residing within the positively charged nucleus? Is there some additional "quantum force" at work within and perhaps between atoms that allows the atoms to exist, not penetrate into each other and to possibly bond with each other and that is not yet present in the above classical Newtonian-Maxwell description?

2.2.2 Quantum Theory for the Motion of Discrete Fundamental Particles

Our purposes here are pedagogic. The goal is not to make actual quantum predictions of the discrete movement of the electrons within an atom but to give a rapid two-page presentation of quantum theory that elucidates ideas while neglecting the complicating realities of spin, the discretization of the EM fields and special relativity. Our primary goal is to compare such a bare-bones version of quantum mechanics to the classical Newtonian mechanics just presented and observe the distinctions.

The Schrödinger Equation

For a system of $p = 1, 2, \ldots, N_\infty$ fundamental particles (electrons and nuclei in what follows and not their subsequent grouping into atoms and molecules), we define the so-called *Schrödinger equation*

$$\boxed{i\hbar\frac{\partial\psi}{\partial t} = \sum_{p=1}^{N_\infty}\left\{-\frac{\hbar^2}{2m_p}\left|\nabla_p - \frac{i}{\hbar}\sum_{\substack{n=1\\n\neq p}}^{N_\infty} q_p\boldsymbol{A}_n\right|^2 + \sum_{\substack{n=1\\n\neq p}}^{N_\infty}\left(q_p\varphi_n + m_pU_n\right)\right\}\psi,} \tag{2.27}$$

where $\psi = \psi(\boldsymbol{r}_1, \boldsymbol{r}_2, \ldots \boldsymbol{r}_{N_\infty}, t)$ is called *the wave function* and $\hbar = 1.05 \times 10^{-34}$ J s $\widehat{=}$ *Planck's constant* and where we use the notation that $|\mathcal{L}|^2 = \mathcal{L} \cdot \mathcal{L}$ for a vector operator $\mathcal{L}$. The wave function is complex

$$\psi = Re^{iS/\hbar} = \psi_R + i\psi_I, \tag{2.28}$$

where R is the amplitude and $S/\hbar$ the phase. The wavefunction ψ depends on the coordinates $\boldsymbol{r}_1, \boldsymbol{r}_2, \ldots \boldsymbol{r}_{N_\infty}$ used to define each particle's possible position. So ψ is thus defined in a space of $3N_\infty$ dimensions plus time in which

$$\nabla_p = \frac{\partial}{\partial \boldsymbol{r}_p} \mathrel{\widehat{=}} \text{gradient with respect to the coordinates used to locate particle } p. \tag{2.29}$$

The EM potentials φ_n and $\boldsymbol{A}_n$ and gravitational potential U_n (which is negligible) are those defined earlier. The fundamental particles p are not entire atoms, and there are no EM polarization forces that need to be accounted for within the Schrödinger equation if we neglect quantum spin (magnetic dipole moments), which we will do. How electrons actually interact with other electrons and with the positively charged nuclei is more accurately described as a discrete exchange of photons in the discretized quantum version of electromagnetism called *quantum electrodynamics* that will not be considered here.

The basic postulate of quantum mechanics is that the *probability density* ρ for finding the N_∞ particles in a small neighborhood surrounding the possible positions $\boldsymbol{r}_1, \boldsymbol{r}_2, \ldots \boldsymbol{r}_{N_\infty}$ of the N_∞ particles is given by

$$\rho = R^2 = \psi\psi^*. \tag{2.30}$$

In other words, we have that

$$\rho(\boldsymbol{r}_1, \boldsymbol{r}_2, \ldots \boldsymbol{r}_{N_\infty})\, d^3\boldsymbol{r}_1\, d^3\boldsymbol{r}_2 \ldots d^3\boldsymbol{r}_{N_\infty} \;\widehat{=}\; \text{the probability that particle 1 is within a neighborhood } d^3\boldsymbol{r}_1 \text{ of } \boldsymbol{r}_1\text{, particle 2 is within a neighborhood } d^3\boldsymbol{r}_2 \text{ of } \boldsymbol{r}_2\text{, and so forth for all } N_\infty \text{ particles.} \tag{2.31}$$

The Quantum Trajectory

The basic "quantum recipe" for determining the trajectory of a particle p is to solve the Schrödinger equation to obtain the $3N_\infty + 1$ dimensioned complex function $\psi = \psi(\boldsymbol{r}_1, \boldsymbol{r}_2, \ldots \boldsymbol{r}_{N_\infty}, t)$, determine the probability density $\rho = \psi\psi^*$ and then make the best guess for the location of particle p by averaging the possible positions $\boldsymbol{r}_p$ as:

$$\boxed{\langle \boldsymbol{r}_p\rangle(t) = \int \rho(\boldsymbol{r}_1, \ldots, \boldsymbol{r}_{N_\infty}, t)\, \boldsymbol{r}_p\, \mathrm{d}^3\boldsymbol{r}_1 \mathrm{d}^3\boldsymbol{r}_2 \ldots \mathrm{d}^3\boldsymbol{r}_{N_\infty}.} \tag{2.32}$$

Quantum theory thus gives statistical best guesses for the trajectory $\langle \boldsymbol{r}_p\rangle(t)$ of particle p and appears quite distinct from the Newtonian description that produces deterministic trajectories $\boldsymbol{r}_p(t)$.

Distinctions between Newtonian and Quantum Theory

To better see the distinctions and similarities between the quantum and Newtonian descriptions, let's make an additional postulate due to physicist David Bohm (Bohm, 1952) that states that the momentum of particle p is deterministic and given by:

$$m_p \boldsymbol{v}_p = \nabla_p S - \sum_{\substack{n=1\\ n\neq p}}^{N_\infty} q_p \boldsymbol{A}_n \quad \text{with } p = 1, 2, \ldots, N_\infty, \tag{2.33}$$

where S is again the phase of the wave function and $\boldsymbol{v}_p = \dot{\boldsymbol{r}}_p$. In this formulation of quantum mechanics, the statistical uncertainty lies in the uncertainty of the initial conditions, the trajectories themselves being deterministic. There is ongoing debate and controversy surrounding the conceptual basis of quantum theory. But whether the Bohm momentum hypothesis is invoked or not, the statistical best guess of the trajectory is given by the $\langle \boldsymbol{r}_p\rangle(t)$ of Eq. (2.32) in quantum mechanics. The reason we are introducing Bohm's momentum postulate here is pedagogic; it allows us to see explicitly the distinctions and similarities between Newtonian and quantum predictions of the movement of discrete particles.

If we introduce $\psi = Re^{iS/\hbar}$ into the Schrödinger equation, separate into two equations corresponding to the real and imaginary parts, divide the real part by R, introduce the Bohm momentum postulate of Eq. (2.33) and then take the gradient ∇_p, we obtain exactly (as you can show in Exercise 1 at the end of the chapter):

$$\frac{\mathrm{d}}{\mathrm{d}t}(m_p\,\boldsymbol{v}_p) = -\nabla_p Q + \underbrace{\sum_{\substack{n=1\\ n\neq p}}^{N_\infty}\{m_p\,\boldsymbol{g}_n + q_p\boldsymbol{E}_n + q_p\boldsymbol{v}_p \times \boldsymbol{B}_n\}}_{\boldsymbol{F}_p\ \text{classical force}}, \tag{2.34}$$

where $\mathrm{d}\boldsymbol{v}_p/\mathrm{d}t \,\widehat{=}\, \partial \boldsymbol{v}_p/\partial t + \left(\sum_{n=1}^{N_\infty} \boldsymbol{v}_n \cdot \nabla_n\right)\boldsymbol{v}_p$ and where

$$Q = -\sum_{n=1}^{N_\infty} \frac{\hbar^2}{2m_n}\frac{\nabla_n^2 R}{R} \,\widehat{=}\, \textit{quantum potential}. \tag{2.35}$$

So the introduction of the Bohm momentum postulate implies that each particle trajectory is deterministic in nature as in the classical Newtonian description. However, there is a nonlocal force $-\nabla_p Q$ acting on particle p, the so-called *quantum force*, that is new in the quantum description and that has a strong influence on the particle's trajectory. The adjective *nonlocal* means that a particle can influence the other particles and the other particles can feed back on the first particle instantaneously, which makes quantum mechanics distinctly different from the Newton–Maxwell mechanics presented earlier. Due to the smoothing effects of the Laplacian operator in the definition of the quantum potential Q, with greater smoothing created by more intervening particles, the quantum force is most important in describing the interaction between small numbers of fundamental particles, such as those within a single atom or between electrons from two atoms that come into close proximity, and is less important in describing interactions between particles in separated atoms involving many intervening particles.

The statistical nature of the problem in this *Bohmian* perspective is controlled by the unknown initial conditions of the particle positions. The imaginary part of the Schrödinger equation produces

$$\frac{\partial \rho}{\partial t} = -\sum_{p=1}^{N_\infty} \nabla_p \cdot (\rho\,\boldsymbol{v}_p) \tag{2.36}$$

where again $\rho = R^2 = \psi\psi^*$ is the probability density for locating the particles in the system. This probability density ρ evolves according to Eq. (2.36) from some given initial distribution defined at $t = t_0$. The best guess of each deterministic particle trajectory given the uncertainty in the initial positions at $t = t_0$ is again given by Eq. (2.32). So invoking the Bohmian momentum postulate does not change the calculation of the most probable trajectory $\langle \boldsymbol{r}_p\rangle(t)$ for each particle p compared to the earlier quantum recipe. But it does make explicit that the distinction between a classical and quantum prediction of a particle trajectory is the presence of a new quantum force in the quantum description.

From the above perspective, it is the quantum force that keeps two particles from occupying the same position (the so-called exclusion principle) preventing, for example, electrons from permanently residing within the positive nucleus. It is the chaotic movement of the electrons as influenced by the quantum force that is responsible for each atom having a fluctuating electric-dipole moment that induces a fluctuating dipole moment in an adjacent atom resulting in an electrical attraction between the atoms (the van der Waals attraction force). The quantum force also contributes to repulsion that prevents electrons from one atom overlapping with electrons in an adjacent atom and to the various types of bonding that allow two atoms to be attached. In conclusion, and for the purpose of modeling the movement of large numbers of atoms, *quantum mechanics determines the attractive and repulsive forces between interacting atoms, while to an excellent approximation, Newton's law describes the center-of-mass trajectory of each individual atom or molecule under the influence of these forces.*

2.2.3 The Molecular Dynamics Model

The compromise model then combines Newtonian dynamics for the atomic trajectories with models of the atomic interaction forces that at least approximately allow for the quantum nature of the force interactions.

The force $\boldsymbol{f}_{pn}$ that an atom n exerts on another atom p depends on the distance r_{pn} between the atoms and can be formulated in terms of an interaction potential $U_{pn}(r_{pn})$ as

$$\boldsymbol{f}_{pn} = -\frac{\partial U_{pn}}{\partial r_{pn}} \hat{\boldsymbol{r}}_{pn}, \tag{2.37}$$

where in terms of the trajectories $r_{pn}(t) = |\boldsymbol{r}_p(t) - \boldsymbol{r}_n(t)|$ and $\hat{\boldsymbol{r}}_{pn}(t) = \left[\boldsymbol{r}_p(t) - \boldsymbol{r}_n(t)\right]/r_{pn}(t)$, which is directed from n toward p. We thus have $\boldsymbol{f}_{pn} = -\boldsymbol{f}_{np}$, which is commonly called *Newton's third law.*

Though limited in application, one well-known form for the interaction of uncharged isotropic atoms is the *Lennard-Jones* potential given by

$$U_{pn} = 4\varepsilon_{pn}\left[\left(\frac{\sigma_{pn}}{r_{pn}}\right)^{12} - \left(\frac{\sigma_{pn}}{r_{pn}}\right)^{6}\right], \tag{2.38}$$

where σ_{pn} is a characteristic atomic length, ε_{pn} has units of energy and both parameters depend on the specific atoms p and n involved in the interaction. The interaction-potential term proportional to r_{pn}^{-6} represents the dipolar electric attraction between p and n that we call *van der Waals attraction* as discussed earlier, while the term proportional to r_{pn}^{-12} approximates the *quantum repulsion force* that prevents overlap of the electron clouds from atoms p and n. A model for the interaction potential like Eq. (2.38) when combined with Eq. (2.37) approximately captures the interaction force between atoms, at least in this special case of electrically neutral, isotropic atoms (e.g., the Noble gas atoms). Such a model includes quantum effects, at least approximately.

So in what follows, we base our continuum physics on the "molecular dynamics" model:

$$\boxed{m_p \frac{\mathrm{d}^2 \boldsymbol{r}_p}{\mathrm{d}t^2} = \sum_{\substack{n=1 \\ n \neq p}}^{N_\infty} \boldsymbol{f}_{pn},} \tag{2.39}$$

where an appropriate modeling of $\boldsymbol{f}_{pn}$ is assumed to exist for the specific atoms and interactions (or bonding) that are present. If modeling interactions between anisotropic molecules instead of isotropic atoms, the conservation of angular momentum must additionally be allowed for to determine the orientation of each molecule in addition to its center-of-mass trajectory $\boldsymbol{r}_p(t)$ because the force potential now depends on the orientation of each molecule in addition to the separation distance r_{pn}. However, in our derivation that follows of the macroscopic stress tensor and macroscopic law of motion in continuum mechanics, we simply require that the interatomic forces $\boldsymbol{f}_{pn}$ exist and are short ranged. We will not need to quantify these forces with a model applicable to specific atoms or molecules.

That said, research on using quantum mechanics to determine the interatomic potentials for use in molecular-dynamics simulations is ongoing. In addition to the relatively weak van der Waal's interaction already discussed, other types of attractive atomic interactions include: *ionic bonds* in minerals (electrostatic interaction between charged atoms, where one atom in the mineral has donated one or more electrons to a neighbor), *covalent bonds* (neighboring atoms that share electrons), *hydrogen bonds* (dipolar electrostatic interaction between neighboring molecules in which each molecule has one or more hydrogen atoms covalently bonded to an anion), and *metallic bonds* (attraction between the nuclei of metal atoms and the free electrons in the conduction band of metals). First-principles quantum-mechanical modeling for such interaction scenarios and for specific minerals and fluids is both highly developed and rapidly evolving. The reader wishing to understand foundational ideas involved in such interatomic-force modeling, such as the *Born–Oppenheimer approximation* that allows the wave function of nuclei to be treated separately from the wave function of electrons due to the electrons being so much lighter than the nuclei or the *Hellmann–Feynman theorem* that says, in the present context, that interactions between nuclei and electrons or between one nucleus and another can be modeled as classical electrostatic–force interactions, might begin with textbooks such as Atkins and Friedman (2011), Weiner (2002), and Zhang (1999).

2.3 Averaging over the Molecular Dynamics

We now will show one way to average, or sum, over the discrete system of large numbers of atoms in order to arrive at the rules controlling the collective average movement of the atoms surrounding each point in space. To do so, we first need to introduce the notion of averaging a field. We then need to introduce a way to represent discrete atoms as continuously distributed fields. You will see that both tasks are quite simple.

2.3.1 The Volume Average

Around each point $\boldsymbol{r}$ of space, we now construct an averaging region $\Omega_0(\boldsymbol{r})$ that has a constant volume V_0. We can imagine $\boldsymbol{r}$ as being at the geometrical center of $\Omega_0(\boldsymbol{r})$ and that the shape of Ω_0 is the same for all points $\boldsymbol{r}$. We further imagine that within $\Omega_0(\boldsymbol{r})$ there is a scalar, vector, or tensor field $\psi(\boldsymbol{r}_0)$, where $\boldsymbol{r}_0$ are the local coordinates used to locate points within $\Omega_0(\boldsymbol{r})$. The average of $\psi(\boldsymbol{r}_0)$ is denoted with an overbar and defined as

$$\boxed{\overline{\psi}(\boldsymbol{r}) = \frac{1}{V_0}\int_{\Omega_0(\boldsymbol{r})} \psi(\boldsymbol{r}_0)\, \mathrm{d}^3\boldsymbol{r}_0.} \quad \textit{volume average} \tag{2.40}$$

Our continuum fields in what follows will be defined as volume averages.

2.3.2 *The Gradient of the Volume Average*

So what is $\nabla\overline{\psi}(\boldsymbol{r})$? The spatial derivative of any function in 3D including $\overline{\psi}$ is defined by $\mathrm{d}\boldsymbol{r}\cdot\nabla\overline{\psi} = \overline{\psi}(\boldsymbol{r}+\mathrm{d}\boldsymbol{r}) - \overline{\psi}(\boldsymbol{r})$, where $\mathrm{d}\boldsymbol{r}$ is some small change in position. In words, the spatial derivative in the direction $\mathrm{d}\boldsymbol{r}$ of a volume averaged function is the difference in the volume average when the averaging volume is located at $\boldsymbol{r}+\mathrm{d}\boldsymbol{r}$ and $\boldsymbol{r}$ or in symbols

$$\mathrm{d}\boldsymbol{r}\cdot\nabla\overline{\psi}(\boldsymbol{r}) = \frac{1}{V_0}\int_{\Omega_0(\boldsymbol{r}+\mathrm{d}\boldsymbol{r})} \psi(\boldsymbol{r}_0)\,\mathrm{d}V_0 - \frac{1}{V_0}\int_{\Omega_0(\boldsymbol{r})} \psi(\boldsymbol{r}_0)\,\mathrm{d}V_0. \tag{2.41}$$

After performing the difference, only the shaded regions in Fig. 2.2 contribute to $\mathrm{d}\boldsymbol{r}\cdot\nabla\overline{\psi}$. Note that volume elements in the two shaded regions can be written

$$\begin{aligned} \mathrm{d}V_+ &= \boldsymbol{n}\cdot\mathrm{d}\boldsymbol{r}\,\mathrm{d}^2\boldsymbol{r}_0 \\ \mathrm{d}V_- &= -\boldsymbol{n}\cdot\mathrm{d}\boldsymbol{r}\,\mathrm{d}^2\boldsymbol{r}_0, \end{aligned} \tag{2.42}$$

where $\mathrm{d}^2\boldsymbol{r}_0$ is a surface element on the boundary $\partial\Omega_0$ of each averaging region. So the difference in the volume integrals over the two shaded regions gives

$$\mathrm{d}\boldsymbol{r}\cdot\nabla\overline{\psi}(\boldsymbol{r}) = \mathrm{d}\boldsymbol{r}\cdot\left[\frac{1}{V_0}\int_{\partial\Omega_0(\boldsymbol{r})} \boldsymbol{n}\,\psi(\boldsymbol{r}_0)\,\mathrm{d}^2\boldsymbol{r}_0\right]. \tag{2.43}$$

Since this must hold true for any $\mathrm{d}\boldsymbol{r}$, we have our final result

$$\nabla\overline{\psi}(\boldsymbol{r}) = \frac{1}{V_0}\int_{\partial\Omega_0(\boldsymbol{r})} \boldsymbol{n}\,\psi(\boldsymbol{r}_0)\,\mathrm{d}^2\boldsymbol{r}_0, \tag{2.44}$$

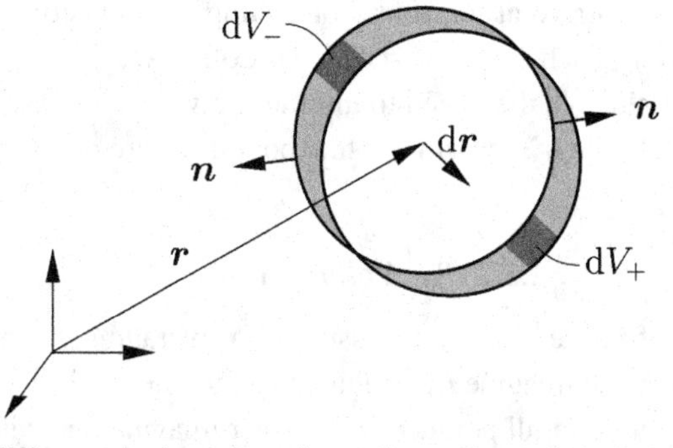

Figure 2.2 The difference in the volume integral over the regions $\Omega_0(\boldsymbol{r}+\mathrm{d}\boldsymbol{r})$ and $\Omega_0(\boldsymbol{r})$ as $\mathrm{d}\boldsymbol{r}\to 0$ defines the gradient of the average as a surface integral over $\partial\Omega_0(\boldsymbol{r})$.

where again $\overline{\psi}(\boldsymbol{r}) = V_0{}^{-1} \int_{\Omega_0(\boldsymbol{r})} \psi(\boldsymbol{r}_0)\, \mathrm{d}^3\boldsymbol{r}_0$. So $\nabla\overline{\psi}$ is nonzero when the local (nonaveraged) field ψ is larger on one side of the averaging volume compared to the other.

Note that from the *fundamental theorem of 3D calculus* given earlier as Eq. (1.61), we also have that $\nabla\overline{\psi} = \overline{\nabla_0\psi}$. If we consider the special case of $\psi(\boldsymbol{r}_0) = \psi_0$ (a spatial constant), because $\int_{\partial\Omega_0} \boldsymbol{n}\psi_0\, \mathrm{d}^2\boldsymbol{r}_0 = \int_{\Omega_0} \nabla_0\psi_0\, \mathrm{d}^3\boldsymbol{r}_0 = 0$, we can rewrite our gradient-of-the-volume-average result as

$$\boxed{\nabla\,\overline{\psi}(\boldsymbol{r}) = \frac{1}{V_0}\int_{\partial\Omega_0(\boldsymbol{r})} \boldsymbol{n}\,\left[\psi(\boldsymbol{r}_0) - \overline{\psi}(\boldsymbol{r})\right]\,\mathrm{d}^2\boldsymbol{r}_0.} \qquad \textit{gradient of a volume-averaged field} \tag{2.45}$$

This states that the gradient of an averaged field at some point $\boldsymbol{r}$ is independent of the average value of the field at that point. The gradient of the average is representing the average spatial variation of the local field from one side of the averaging region to the other, which is independent of the mean value in the region.

2.3.3 Discrete Molecules → Molecular Distribution Fields

In order to apply such averaging to our discrete molecularity, we assume that a molecule is so small that its spatial extent can be represented as a Dirac field $\delta\big(\boldsymbol{r} - \boldsymbol{r}_p(t)\big)$. This is the key idea for how we transform a discrete trajectory $\boldsymbol{r}_p(t)$ into a continuously distributed field $\delta\,\big(\boldsymbol{r} - \boldsymbol{r}_p(t)\big)$ that can be volume integrated or spatially differentiated.

Thus, we may define the *mass distribution field*

$$m(\boldsymbol{r}, t) = \sum_{p=1}^{N_\infty} m_p\, \delta\big(\boldsymbol{r} - \boldsymbol{r}_p(t)\big). \quad (\text{units of kg/m}^3) \tag{2.46}$$

The average value of m in Ω_0 gives the mass density $\rho(\boldsymbol{r}, t)$ (not to be confused with the quantum probability density defined earlier)

$$\boxed{\rho(\boldsymbol{r}, t) = \overline{m}(\boldsymbol{r}, t) = \frac{1}{V_0}\int_{\Omega_0(\boldsymbol{r})} \sum_{p=1}^{N_\infty} m_p\, \delta(\boldsymbol{r}_0 - \boldsymbol{r}_p(t))\, \mathrm{d}^3\boldsymbol{r}_0 = \sum_{p=1}^{N_0(\boldsymbol{r},t)} \frac{m_p}{V_0},} \tag{2.47}$$

where $N_0(\boldsymbol{r}, t)$ is the number of molecules in $\Omega_0(\boldsymbol{r})$ at a given instant in time. The *mass density* $\rho(\boldsymbol{r}, t)$ is our first example of a *continuum field* that represents the average (or summed) molecular behavior in a neighborhood $\Omega_0(\boldsymbol{r})$ surrounding each point $\boldsymbol{r}$.

We may similarly define a momentum distribution field

$$\boldsymbol{p}(\boldsymbol{r}, t) = \sum_{p=1}^{N_\infty} m_p\boldsymbol{v}_p(t)\,\delta\big(\boldsymbol{r} - \boldsymbol{r}_p(t)\big), \quad \text{where} \quad \boldsymbol{v}_p(t) \,\hat{=}\, \frac{\mathrm{d}\boldsymbol{r}_p(t)}{\mathrm{d}t} \tag{2.48}$$

that has an average value $\overline{\boldsymbol{p}}(\boldsymbol{r}, t) = V_0^{-1} \int_{\Omega_0(\boldsymbol{r})} \boldsymbol{p}(\boldsymbol{r}_0, t)\, d^3\boldsymbol{r}_0$ that defines the *momentum density*

$$\bar{\boldsymbol{p}}(\boldsymbol{r},t) = \frac{1}{V_0}\int_{\Omega_0(\boldsymbol{r})} \sum_{p=1}^{N_\infty} m_p\, \boldsymbol{v}_p(t)\, \delta\big(\boldsymbol{r}_0 - \boldsymbol{r}_p(t)\big)\, \mathrm{d}^3\boldsymbol{r}_0 = \frac{1}{V_0}\sum_{p=1}^{N_0(\boldsymbol{r},t)} m_p\, \boldsymbol{v}_p \mathrel{\widehat{=}} \rho\boldsymbol{v} \tag{2.49}$$

where we introduced the *center-of-mass velocity* as

$$\boxed{\boldsymbol{v}(\boldsymbol{r},t) = \frac{\bar{\boldsymbol{p}}}{\rho} = \frac{\sum_{p=1}^{N_0} m_p \boldsymbol{v}_p}{\sum_{p=1}^{N_0} m_p}.} \tag{2.50}$$

The field $\boldsymbol{v}(\boldsymbol{r},t)$ is another example of a *continuum field* that represents molecular behavior, that is, fields defined at each point $\boldsymbol{r}$ that represent the average discrete-particle behavior in a region Ω_0 surrounding that point. The molecular averaging being discussed here corresponds to simply adding up the contributions to a continuum field that comes from each discrete particle within an averaging region and makes no assumptions about the positions or interactions between the discrete particles in Ω_0.

2.4 The Continuum Concept

It is worth taking a pause to reflect on what we mean by the word *continuum*. It is common in nearly all texts of continuum mechanics to say that "the continuum" refers to a macroscopic body of matter free of voids that is being approximated as a continuous mass across all scales and not treated as a collection of discrete particles (electrons and nuclei or atoms or molecules). However, in our perspective, that is not the right way to understand the word continuum. The macroscopic body of matter is by definition a collection of discrete molecules. In formulating the rules of continuum physics, we are not replacing the actual material with an imagined continuous mass nor approximating the complicated molecular interactions. Although a modern understanding of solids and liquids could describe "the continuum" as referring to the continuously distributed electron density $\psi_e\psi_e^*$ with ψ_e the quantum wave function of the electrons in a condensed (nongaseous) medium, it is our perspective that the word continuum is referring to the continuity of the fields that are defined at each point $\boldsymbol{r}$ as the actual average molecular behavior in each volumetric region $\Omega_0(\boldsymbol{r})$ that surrounds each point $\boldsymbol{r}$. To repeat, *the continuously distributed macroscopic fields of continuum mechanics are defined by going to any point* **r** *and adding up (averaging) the molecular behavior in a volumetric region that surrounds that point.* A spatial derivative of such continuum fields is the difference in the average molecular behavior in two partially overlapping volumetric regions whose centers reside a small distance from each other in the direction of the spatial derivative.

In a formal sense, the word continuum should not be a noun referring to the body under consideration (i.e., "the continuum") but an adjective used to describe the way we are allowing for the average molecular behavior surrounding each point $\boldsymbol{r}$ (i.e., "continuum field" or "continuum description" or "continuum mechanics," etc.). Throughout the book, we use the adjectives "continuum" and "macroscopic" to mean the same thing.

In a heterogeneous material, the value of a continuum field at a point can depend on the size of the averaging volume as the averaging region becomes larger and additional molecules and molecular interactions are incorporated into the sum over the molecules. However, the mathematical form of the macroscopic governing equations derived in this and the following chapters does not depend on the size of the averaging volume. When we set up a macroscopic boundary-value problem (BVP) within a heterogeneous domain, we specify both the spatial distribution of the boundary values and initial conditions throughout the domain as well as the spatial distribution of the material properties. The resolution used in prescribing such spatial distributions implicitly corresponds to setting the size of the averaging volume, with the averaging volume needing to be somewhat smaller than the spatial resolution of the gradients of the fields. Additionally, if the macroscopic problem involves waves, the averaging volume must be smaller than the smallest wavelength being modeled.

Although the material properties specified throughout a heterogeneous material will change as we change the resolution of their given distribution, which implicitly corresponds to changing the size of the averaging volume, we do not need to set the size of the averaging volume when deriving the form of the macroscopic conservation and constitutive laws in what follows. A practical way to understand the implicit size of the averaging volume is to consider numerical simulations of a macroscopic BVP. The size of the grid cells used to discretize the domain corresponds to the size of the averaging volume that was used, again implicitly, in the derivation of the macroscopic governing equations.

Each point surrounded by each averaging region has two key aspects of the molecular response associated with it: (1) the average molecular behavior in the averaging volume (i.e., the value of the continuum field at that point) and (2) the spatial variation of the molecular behavior across the averaging volume (i.e., the gradient of the continuum field at that point). As we saw earlier in Eq. (2.45), these two aspects of the molecular behavior are distinct, for example, the average may be increasing through time at a point $\boldsymbol{r}$ while the gradient remains constant and vice versa.

We just saw two examples of continuum fields: (1) the mass density $\rho(\boldsymbol{r})$ is the total mass of the molecules in the volumetric region surrounding point $\boldsymbol{r}$ divided by the volume of the region and (2) the center-of-mass velocity $\boldsymbol{v}(\boldsymbol{r})$ is the total momentum of the molecules in the volumetric region surrounding $\boldsymbol{r}$ divided by the mass of all molecules in the volume. We will see many more examples of continuum fields as we develop the rules of continuum physics, all defined as averages of what is actually happening within the averaging volume surrounding a point in an actual collection of molecules. The word "continuum" means "averaged over the discrete molecularity."

2.5 Continuum Statement of Mass Conservation

We now obtain our first *continuum law*, or "rule," that relates changes through time of ρ to changes in space of $\boldsymbol{v}$ and that expresses the conservation of mass.

We focus on how the mass density ρ associated with a region Ω_0 changes in time as particles (atoms or molecules) enter and exit Ω_0 across $\partial\Omega_0$. Taking the time derivative of ρ gives:

$$\frac{\partial \rho(\boldsymbol{r},t)}{\partial t} = \frac{1}{V_0}\int_{\Omega_0(\boldsymbol{r})} \sum_{p=1}^{N_\infty} m_p \frac{\partial \delta\big(\boldsymbol{r}_0 - \boldsymbol{r}_p(t)\big)}{\partial t}\, \mathrm{d}^3\boldsymbol{r}_0. \tag{2.51}$$

The first question is how do you do the time derivative $\partial\delta\left(\boldsymbol{r}-\boldsymbol{r}_p(t)\right)/\partial t$?

To answer this, let's first consider any well-behaved function $f(x-x_p(t))$ of one spatial dimension x and determine how to do the time derivative

$$\frac{\partial f\left(x - x_p(t)\right)}{\partial t}. \tag{2.52}$$

Begin by letting $a(x,t) = x - x_p(t)$ represent the argument of $f(a)$ and use the chain rule in the two forms

$$\frac{\partial f}{\partial t} = \left(\frac{\mathrm{d}f}{\mathrm{d}a}\right)\frac{\partial a}{\partial t} \quad \text{and} \quad \frac{\partial f}{\partial x} = \left(\frac{\mathrm{d}f}{\mathrm{d}a}\right)\frac{\partial a}{\partial x}.$$

Now $\partial a/\partial x = 1$ so $\partial f/\partial x = \mathrm{d}f/\mathrm{d}a$ while $\partial a/\partial t = -\mathrm{d}x_p(t)/\mathrm{d}t$. Thus,

$$\frac{\partial f}{\partial t} = -\left(\frac{\partial f}{\partial x}\right)\frac{\mathrm{d}x_p(t)}{\mathrm{d}t} = -v_{px}(t)\frac{\partial f}{\partial x},$$

where $v_{px}(t) \widehat{=} \mathrm{d}x_p(t)/\mathrm{d}t$. We then can generalize to 3D by letting $f\left(\boldsymbol{r}-\boldsymbol{r}_p(t)\right) = f\left(x - x_p(t), y - y_p(t), z - z_p(t)\right)$, $\nabla = \hat{\boldsymbol{x}}\partial_x + \hat{\boldsymbol{y}}\partial_y + \hat{\boldsymbol{z}}\partial_z$ and $\boldsymbol{v}_p(t) = \hat{\boldsymbol{x}}v_{px}(t) + \hat{\boldsymbol{y}}v_{py}(t) + \hat{\boldsymbol{z}}v_{pz}(t)$ to obtain

$$\boxed{\frac{\partial}{\partial t} f\left(\boldsymbol{r}-\boldsymbol{r}_p(t)\right) = -\boldsymbol{v}_p(t)\cdot\nabla f\big(\boldsymbol{r}-\boldsymbol{r}_p(t)\big).} \tag{2.53}$$

Inserting the Dirac delta function then gives the needed result that $\partial\delta\left(\boldsymbol{r}-\boldsymbol{r}_p(t)\right)/\partial t = -\boldsymbol{v}_p(t)\cdot\nabla\delta\left(\boldsymbol{r}-\boldsymbol{r}_p(t)\right)$.

Given the above, we then have

$$\begin{aligned}
\frac{\partial \rho}{\partial t} &= -\frac{1}{V_0}\int_{\Omega_0}\sum_{p=1}^{N_\infty} m_p\, \boldsymbol{v}_p(t)\cdot\nabla_0\,\delta\big(\boldsymbol{r}_0 - \boldsymbol{r}_p(t)\big)\,\mathrm{d}^3\boldsymbol{r}_0 \\
&= -\frac{1}{V_0}\int_{\Omega_0}\sum_{p=1}^{N_\infty} \nabla_0\cdot\Big[m_p\, \boldsymbol{v}_p(t)\,\delta\big(\boldsymbol{r}_0 - \boldsymbol{r}_p(t)\big)\Big]\,\mathrm{d}^3\boldsymbol{r}_0 \\
&= -\frac{1}{V_0}\int_{\partial\Omega_0}\sum_{p=1}^{N_\infty} \boldsymbol{n}\cdot\Big[m_p\, \boldsymbol{v}_p(t)\,\delta\big(\boldsymbol{r}_0 - \boldsymbol{r}_p(t)\big)\Big]\,\mathrm{d}^2\boldsymbol{r}_0 \\
&= -\nabla\cdot\left[\frac{1}{V_0}\int_{\Omega_0}\sum_{p=1}^{N_\infty} m_p\, \boldsymbol{v}_p(t)\,\delta\big(\boldsymbol{r}_0 - \boldsymbol{r}_p(t)\big)\,\mathrm{d}^3\boldsymbol{r}_0\right] \\
&= -\nabla\cdot(\rho\boldsymbol{v}),
\end{aligned} \tag{2.54}$$

which says that density increases at a point if there is more mass flux $\rho \boldsymbol{v}$ into the volume surrounding the point than mass flux leaving the volume. To move from the first to the second line, we use the identity $\nabla \cdot [\boldsymbol{a}\beta(\boldsymbol{r})] = \boldsymbol{a} \cdot \nabla\beta(\boldsymbol{r})$ when vector $\boldsymbol{a}$ is independent of $\boldsymbol{r}$ and the only spatial dependence is in the scalar field $\beta(\boldsymbol{r})$. To go from the second to third line, we use the divergence theorem and to move from the third to fourth line we use our result above for the gradient of a volume average. Using the identity $\nabla \cdot (\rho \boldsymbol{v}) = \nabla\rho \cdot \boldsymbol{v} + \rho\nabla \cdot \boldsymbol{v}$, we then have

$$\boxed{\frac{\partial \rho}{\partial t} + \boldsymbol{v} \cdot \nabla\rho = -\rho\nabla \cdot \boldsymbol{v}.} \qquad \textit{continuum statement of the conservation of mass} \tag{2.55}$$

The operator $\partial/\partial t + \boldsymbol{v} \cdot \nabla$ appears often in continuum mechanics and can be given a physical interpretation.

2.6 The Eulerian and Lagrangian Perspectives

Imagine that the averaging region $\Omega_0(\boldsymbol{r})$ is not fixed in time with molecules entering and exiting across $\partial\Omega_0$. Instead imagine that $\boldsymbol{r}$ defines the center of mass of a fixed number of molecules N_0 and that $\Omega(\boldsymbol{r}, t)$ expands and contracts and translates through space and time to always contain this collection of N_0 molecules within it as the center of mass of the collection $\boldsymbol{r} = \boldsymbol{r}(t)$ trajects through space and time.

In this case, a continuum field like

$$\overline{\psi}(\boldsymbol{r}(t), t) = \frac{1}{V(\boldsymbol{r}(t))} \int_{\Omega(\boldsymbol{r}(t))} \psi(\boldsymbol{r}_0, t)\, \mathrm{d}^3\boldsymbol{r}_0 \tag{2.56}$$

has a *total time derivative* given by

$$\frac{\mathrm{d}}{\mathrm{d}t}\overline{\psi}(\boldsymbol{r}(t), t) = \frac{\partial\overline{\psi}}{\partial t} + \frac{\mathrm{d}\boldsymbol{r}}{\mathrm{d}t} \cdot \frac{\partial}{\partial \boldsymbol{r}}\overline{\psi} = \frac{\partial}{\partial t}\overline{\psi} + \boldsymbol{v} \cdot \nabla\overline{\psi}, \tag{2.57}$$

where $\boldsymbol{v} = \mathrm{d}\boldsymbol{r}(t)/\mathrm{d}t$ is the velocity of the center of mass of the collection of molecules.

Thus, the operator

$$\frac{\mathrm{d}}{\mathrm{d}t} = \frac{\partial}{\partial t} + \boldsymbol{v} \cdot \nabla \mathrel{\widehat{=}} \textit{total} \text{ or } \textit{material} \text{ time derivative} \tag{2.58}$$

denotes the changes in time of a property of an element that contains a fixed mass M of molecules as we follow the trajectory of that mass element through time and space. The time derivative $\mathrm{d}\overline{\psi}(\boldsymbol{r}(t), t)/\mathrm{d}t$ is taken from the perspective of an observer moving with the constant-mass element as the temporal changes are monitored of whatever $\overline{\psi}$ represents about the collection of molecules. The time derivative $\partial\overline{\psi}(\boldsymbol{r}, t)/\partial t + \boldsymbol{v}(\boldsymbol{r}, t) \cdot \nabla\overline{\psi}(\boldsymbol{r}, t)$ is taken from the perspective of a stationary observer watching the molecules move into and out of a stationary element whose center is located at $\boldsymbol{r}$ and that is used to define $\overline{\psi}(\boldsymbol{r}, t)$.

So from a moving frame in which we follow a constant mass M of molecules while the volume $V(t)$ containing it changes, we can rewrite our continuum statement of mass conservation as

$$\frac{1}{\rho(t)}\frac{\mathrm{d}\rho(t)}{\mathrm{d}t} = -\nabla\cdot\boldsymbol{v} = \frac{V(t)}{M}\frac{\mathrm{d}}{\mathrm{d}t}\left[\frac{M}{V(t)}\right] = -\frac{1}{V(t)}\frac{\mathrm{d}V(t)}{\mathrm{d}t}, \tag{2.59}$$

where $\rho(t) = M/V(t)$ is the mass density of the molecules. The identification $\nabla\cdot\boldsymbol{v} = V^{-1}\mathrm{d}V/\mathrm{d}t$ also provides the physical interpretation of $\nabla\cdot\boldsymbol{v}$ as the fractional volume change per unit time encompassing the constant mass M of N_0 particles.

When we look at the atomic trajectories from a stationary frame of reference and fix a constant volumetric region $\Omega_0(\boldsymbol{r})$ around each point $\boldsymbol{r}$ within a material and keep track of the particles and their properties as they enter and leave that volume through time, this is called an "Eulerian" description of continuum mechanics. This will be the perspective we take throughout the book and is always valid. When we imagine to follow each constant number N_0 of particles and imagine the region $\Omega(\boldsymbol{r}(t))$ that contains these N_0 particles to evolve in time and space to fit around these particles, we call this a "Lagrangian" description of continuum mechanics. Generally, a Lagrangian description is only possible for a solid because the atoms do not mix in this case as they do in a gas or liquid. There is much discussion in the literature surrounding how to implement Lagrangian dynamics in practice even if the concept is simple enough when applied to solid matter. Again, the operator $\partial/\partial t + \boldsymbol{v}\cdot\nabla$ that shows up in the stationary-frame Eulerian perspective just becomes a simple time derivative $\mathrm{d}/\mathrm{d}t$ in the moving-frame Lagrangian perspective.

When formulating the partial-differential conservation laws of continuum physics to be used for simulating various physical processes involving flux and accumulation of matter and energy, we will take the Eulerian perspective. However, as will be seen in later chapters, when formulating the accompanying constitutive laws of equilibrium thermodynamics in particular, it is natural to imagine samples (or elements) of constant mass as they traject through space and time and to keep track of the changes happening to such samples as we move along with the sample, which is the Lagrangian perspective. However, once we have established our constitutive laws of equilibrium thermodynamics using total time derivatives $\mathrm{d}\psi/\mathrm{d}t$ of thermodynamic continuum fields ψ in a frame moving with the sample, we simply substitute $\mathrm{d}\psi/\mathrm{d}t = \partial\psi/\partial t + \boldsymbol{v}\cdot\nabla\psi$ and work from the Eulerian, or stationary-frame, perspective that applies to all material types under all circumstances.

2.7 Continuum Statement of Linear-Momentum Conservation

We next obtain the macroscopic "law of motion" that represents the conservation of linear momentum and that controls the collective movement of the molecules in the neighborhood (averaging volume) surrounding each point in space. It is founded on the *molecular dynamics* model

$$m_p\frac{\mathrm{d}\boldsymbol{v}_p(t)}{\mathrm{d}t} = \sum_{\substack{n=1\\ n\neq p}}^{N_\infty}\boldsymbol{f}_{pn}, \quad \text{where} \quad \boldsymbol{v}_p(t) = \frac{\mathrm{d}\boldsymbol{r}_p(t)}{\mathrm{d}t}. \tag{2.60}$$

This is the law that controls the trajectory of each atom or molecule p under the influence of the forces $\boldsymbol{f}_{pn}$ emanating from the other $n \neq p$ atoms or molecules.

We focus on the time derivative of the *momentum density* $\rho\boldsymbol{v}$ defined earlier:

$$
\begin{aligned}
\frac{\partial}{\partial t}(\rho \boldsymbol{v}) &= \frac{1}{V_0}\int_{\Omega_0}\sum_{p=1}^{N_\infty} m_p \frac{\partial}{\partial t}\Big[\boldsymbol{v}_p(t)\,\delta\big(\boldsymbol{r}_0-\boldsymbol{r}_p(t)\big)\Big]\,\mathrm{d}^3\boldsymbol{r}_0 \\
&= \frac{1}{V_0}\int_{\Omega_0}\sum_{p=1}^{N_\infty} m_p \frac{\mathrm{d}\,\boldsymbol{v}_p(t)}{\mathrm{d}t}\,\delta\big(\boldsymbol{r}_0-\boldsymbol{r}_p(t)\big)\,\mathrm{d}^3\boldsymbol{r}_0 \\
&\quad -\frac{1}{V_0}\int_{\Omega_0}\sum_{p=1}^{N_\infty} m_p\,\boldsymbol{v}_p\boldsymbol{v}_p\cdot\nabla_0\,\delta\big(\boldsymbol{r}_0-\boldsymbol{r}_p(t)\big)\,\mathrm{d}^3\boldsymbol{r}_0
\end{aligned}
\tag{2.61}
$$

where we used the product rule of calculus to distribute the time derivative and then, in the second term so obtained, used Eq. (2.53) to perform the time derivative. Using our molecular dynamics model (Newton's second law), we can replace the acceleration of a particle $m_p\mathrm{d}\boldsymbol{v}_p/\mathrm{d}t$ with the force acting on it

$$
\frac{\partial}{\partial t}(\rho \boldsymbol{v}) = \frac{1}{V_0}\int_{\Omega_0}\sum_{p=1}^{N_\infty}\sum_{\substack{n=1\\ n\neq p}}^{N_\infty}\boldsymbol{f}_{pn}\,\delta\big(\boldsymbol{r}_0-\boldsymbol{r}_p(t)\big)\,\mathrm{d}^3\boldsymbol{r}_0 - \frac{1}{V_0}\int_{\Omega_0}\sum_{p=1}^{N_\infty}\nabla_0\cdot\Big[m_p\,\boldsymbol{v}_p\boldsymbol{v}_p\,\delta\big(\boldsymbol{r}_0-\boldsymbol{r}_p(t)\big)\Big]
\tag{2.62}
$$

and where again we used that $\boldsymbol{v}_p(t)$ has no spatial dependence (it is not a field) to move the gradient operator in the second term. Using the averaging theorem from earlier $\overline{\nabla_0\cdot\psi} = \nabla\cdot\overline{\psi}$, we then have

$$
\begin{aligned}
\frac{\partial}{\partial t}\left[\rho(\boldsymbol{r},t)\boldsymbol{v}(\boldsymbol{r},t)\right] &= \frac{1}{V_0}\sum_{p=1}^{N_0(\boldsymbol{r},t)}\sum_{n=N_0(\boldsymbol{r},t)+1}^{N_\infty}\boldsymbol{f}_{pn} \\
&\quad -\nabla\cdot\left[\frac{1}{V_0}\int_{\Omega_0(\boldsymbol{r})}\sum_{p=1}^{N_\infty} m_p\boldsymbol{v}_p(t)\boldsymbol{v}_p(t)\,\delta\big(\boldsymbol{r}_0-\boldsymbol{r}_p(t)\big)\,\mathrm{d}^3\boldsymbol{r}_0\right] && (2.63)\\
&= \frac{1}{V_0}\sum_{p=1}^{N_0(\boldsymbol{r},t)}\sum_{n=N_0(\boldsymbol{r},t)+1}^{N_\infty}\boldsymbol{f}_{pn} - \nabla\cdot\left[\frac{1}{V_0}\sum_{p=1}^{N_0(\boldsymbol{r},t)} m_p\boldsymbol{v}_p(t)\boldsymbol{v}_p(t)\right]. && (2.64)
\end{aligned}
$$

In the first line, we used $\boldsymbol{f}_{pn} = -\boldsymbol{f}_{np}$ (Newton's third law) to deduce that the force on molecules p within Ω_0 (i.e., molecules 1 to N_0) comes from the molecules n outside Ω_0 (i.e., molecules N_0+1 to N_∞). So Newton's third law requires that for p and n both within Ω_0, we have $\sum_{p=1}^{N_0}\sum_{n=1}^{N_0}\boldsymbol{f}_{pn} = 0$ with $\boldsymbol{f}_{pp} = 0$. We can define the macroscopic force density

$$
\boldsymbol{F} = \frac{1}{V_0}\sum_{p=1}^{N_0}\sum_{n=N_0+1}^{N_\infty}\boldsymbol{f}_{pn}
\tag{2.65}
$$

as the force that atoms or molecules outside of the averaging region exert on atoms inside the region and divided by the volume V_0 of the averaging region.

To analyze the term in brackets in Eq. (2.64), we decompose the molecular velocities $\boldsymbol{v}_p$ into a mean value $\boldsymbol{v}$ throughout Ω_0 (the center of mass velocity defined earlier) and random fluctuations $\delta\boldsymbol{v}_p$ about that mean, i.e.,

$$\boldsymbol{v}_p = \boldsymbol{v} + \delta\boldsymbol{v}_p, \quad \text{where} \sum_{p=1}^{N_0} m_p\,\delta\boldsymbol{v}_p = 0. \tag{2.66}$$

Using the fact that the fluctuations of the particle momenta sum to zero, we have

$$\frac{1}{V_0}\sum_{p=1}^{N_0} m_p\boldsymbol{v}_p\boldsymbol{v}_p = \rho\,\boldsymbol{v}\boldsymbol{v} + \frac{1}{V_0}\sum_{p=1}^{N_0} m_p\,\delta\boldsymbol{v}_p\delta\boldsymbol{v}_p. \tag{2.67}$$

Because the $\delta\boldsymbol{v}_p$ are random and uncorrelated in the various directions, we have

$$\sum_{p=1}^{N_0} \delta v_{px}\,\delta v_{py} = \sum_{p=1}^{N_0} \delta v_{px}\,\delta v_{pz} = \sum_{p=1}^{N_0} \delta v_{py}\,\delta v_{pz} = 0 \tag{2.68}$$

and

$$\sum_{p=1}^{N_0} m_p\,\delta v_{px}^2 = \sum_{p=1}^{N_0} m_p\,\delta v_{py}^2 = \sum_{p=1}^{N_0} m_p\,\delta v_{pz}^2 = \frac{1}{3}\sum_{p=1}^{N_0} m_p\,\delta\boldsymbol{v}_p\cdot\delta\boldsymbol{v}_p. \tag{2.69}$$

Thus, combining these results we have

$$\frac{1}{V_0}\sum_{p=1}^{N_0} m_p\,\delta\boldsymbol{v}_p\,\delta\boldsymbol{v}_p = \frac{1}{V_0}\sum_{p=1}^{N_0}\frac{m_p}{3}\left|\delta\boldsymbol{v}_p\right|^2\left(\hat{\boldsymbol{x}}\hat{\boldsymbol{x}} + \hat{\boldsymbol{y}}\hat{\boldsymbol{y}} + \hat{\boldsymbol{z}}\hat{\boldsymbol{z}}\right) = \frac{1}{V_0}\sum_{p=1}^{N_0}\frac{m_p}{3}\left|\delta\boldsymbol{v}_p\right|^2\boldsymbol{I}, \tag{2.70}$$

where $\boldsymbol{I} = \hat{\boldsymbol{x}}\hat{\boldsymbol{x}} + \hat{\boldsymbol{y}}\hat{\boldsymbol{y}} + \hat{\boldsymbol{z}}\hat{\boldsymbol{z}}$ is the identity tensor and $\delta\boldsymbol{v}_p\cdot\delta\boldsymbol{v}_p = |\delta\boldsymbol{v}_p|^2 = \delta v_{px}^2 + \delta v_{py}^2 + \delta v_{pz}^2$.

We now identify:

$$P_I(\boldsymbol{r}, t) = \frac{2}{3}\left[\frac{1}{V_0}\sum_{p=1}^{N_0}\frac{1}{2}m_p\left|\delta\boldsymbol{v}_p\right|^2\right] \;\widehat{=}\; \textit{internal (or “thermal”) pressure.} \tag{2.71}$$

This internal pressure corresponds to the kinetic energy of all the random thermal motion of the particles inside of Ω_0 and then divided by V_0 and multiplied by 2/3 so that the internal pressure has units of an energy density (J/m^3). As will be seen in Section 2.9, the total (or thermodynamic) pressure also has an external force contribution due to atoms just outside the averaging region exerting a force on atoms just inside the averaging region.

For an ideal gas (no force interactions between the particles), P_I is the total pressure. In this special case of noninteracting particles, a standard result of *statistical mechanics* is that temperature can be identified as

$$k_B T = \frac{2}{3}\left[\frac{1}{N_0}\sum_{p=1}^{N_0}\frac{1}{2}m_p\left|\delta\boldsymbol{v}_p\right|^2\right], \tag{2.72}$$

where $k_B \mathrel{\widehat{=}}$ Boltzmann's constant $= 1.38 \times 10^{-23}$ J/K. In words, $k_B T = 2/3$ "the average kinetic energy associated with the molecular fluctuations." Combining this expression for the thermal energy $k_B T$ with our expression Eq. (2.71) for the internal pressure yields the ideal gas law

$$P_I V_0 = N_0 k_B T, \tag{2.73}$$

where

$k_B N_A = R = 8.314$ J /(K mole) $\mathrel{\widehat{=}}$ the ideal gas constant
$N_A = 6.022 \times 10^{23} \mathrel{\widehat{=}}$ Avogadro's Number or the number of objects in one mole.

If we define $M_0 = N_0/N_A$ as the moles of particles inside of V_0, the ideal gas law is $P_I V_0 = M_0 R T$. But note that Eq. (2.71) for the internal pressure P_I is valid for any material including one with strong interactions between the atoms and molecules.

So P_I represents the random kinetic energy density of the molecules as they thermally fluctuate. Alternatively, we can also interpret P_I as a flux of momentum due to the random thermal motions across some surface. To do so, we use Eqs (2.70) and (2.71) to write

$$P_I \boldsymbol{I} = \frac{1}{V_0} \sum_{p=1}^{N_0} m_p \, \delta \boldsymbol{v}_p \, \delta \boldsymbol{v}_p \quad (\text{where } \boldsymbol{v}_p = \boldsymbol{v} + \delta \boldsymbol{v}_p) \tag{2.74}$$

so that

$$\boldsymbol{n}\, P_I = \frac{1}{V_0} \sum_{p=1}^{N_0} \left(\boldsymbol{n} \cdot \delta \boldsymbol{v}_p \right) \, m_p \delta \boldsymbol{v}_p \quad \text{and has units of } \frac{\text{kg m/s}}{\text{m}^2 \text{ s}} \tag{2.75}$$

where $\boldsymbol{n}$ is some direction that will be taken to be the unit vector perpendicular to a surface element having area ΔA in what follows. So P_I has the units of momentum per unit area per unit time.

We can define the length

$$d = \frac{1}{N_0} \sum_{p=1}^{N_0} \left| \boldsymbol{n} \cdot \delta \boldsymbol{v}_p \right| \, \Delta t \tag{2.76}$$

as the average distance that an atom moves in either the $+\boldsymbol{n}$ or $-\boldsymbol{n}$ direction in a time Δt. If we define the volume V_0 containing the N_0 atoms we are considering as $V_0 = d \Delta A$, Eq. (2.75) can be written

$$\boldsymbol{n}\, P_I = \frac{N_0 \langle m_p \delta \boldsymbol{v}_p \rangle_{\boldsymbol{n}}}{\Delta A \, \Delta t}, \tag{2.77}$$

where the average momentum of an atom in the $\boldsymbol{n}$ direction in angle brackets is defined

$$\langle m_p \delta \boldsymbol{v}_p \rangle_{\boldsymbol{n}} = \frac{\sum_{p=1}^{N_0} \left(\boldsymbol{n} \cdot \delta \boldsymbol{v}_p \right) m_p \left(\boldsymbol{n} \cdot \delta \boldsymbol{v}_p \right) \boldsymbol{n}}{\sum_{p=1}^{N_0} \left| \boldsymbol{n} \cdot \delta \boldsymbol{v}_p \right|}. \tag{2.78}$$

We wrote $\delta \boldsymbol{v}_p = (\boldsymbol{n} \cdot \delta \boldsymbol{v}_p)\boldsymbol{n} + (\hat{\boldsymbol{t}}_p \cdot \delta \boldsymbol{v}_p)\hat{\boldsymbol{t}}_p$ in Eq. (2.75) and used that the components of the thermal velocities that are transverse to the direction $\boldsymbol{n}$ sum to zero due to being random. This average atomic momentum as weighted over the N_0 random atomic velocities in the $\boldsymbol{n}$ direction is thus positive with atoms moving with components in either the $+\boldsymbol{n}$ or $-\boldsymbol{n}$ directions making positive contributions to the average.

A *flux* of some physical quantity is a vector representing the amount of said quantity that traverses a unit area element perpendicular to the flux direction in unit time. We thus can interpret Eqs (2.77) and (2.78) to mean

$$\boldsymbol{n}\, P_I \;\widehat{=}\; \textit{total momentum flux of the thermal fluctuations in the direction } \boldsymbol{n}, \tag{2.79}$$

which is always positive. So P_I can either be interpreted in terms of such a flux of the random thermal momentum across a surface or it can be interpreted as $2/3$ of the kinetic energy density of the random thermal motion. But it does not involve atom–atom interactions. If you imagine cupping air inside your two hands in an air-tight manner and somehow increasing the kinetic energy of the random thermal motions of the air molecules so that P_I increases, you will feel an increasing outward "pressure" as the random molecular trajectories of the air molecules crash into the molecules of your hands with increasing momentum transfer into your hands.

To sum up, we have $\nabla \cdot \left[V_0^{-1} \sum_{p=1}^{N_0} m_p \delta \boldsymbol{v}_p \delta \boldsymbol{v}_p\right] = \nabla \cdot (P_I \boldsymbol{I}) = \nabla P_I$ so that the continuum statement of the conservation of linear momentum as given in Eq. (2.64) can now be stated

$$\frac{\partial}{\partial t}(\rho \boldsymbol{v}) = -\nabla \cdot (\rho\, \boldsymbol{v}\boldsymbol{v}) - \nabla P_I + \boldsymbol{F}, \tag{2.80}$$

a first continuum statement of the conservation of linear momentum

where $\boldsymbol{F}$ is the force density corresponding to the force that molecules outside Ω_0 exert on the molecules inside Ω_0 as given by Eq. (2.65) and will be further analyzed in Section 2.8. The left side of Eq. (2.80) is quantifying the changes of momentum in time for the molecules contained in an averaging volume surrounding a given point in the material. This momentum may be changing either because there is a difference between molecular momentum fluxing into and out of an averaging volume (the divergence and gradient terms on the right-hand side corresponding, respectively, to the accumulation of the mean momentum and the random thermal momentum of the molecules) or because there is a net force $\boldsymbol{F}$ acting on the molecules in the averaging volume coming from the molecules outside the averaging volume. We now go on to analyze and model the nature of the external force $\boldsymbol{F}$.

2.8 The External Forces $\boldsymbol{F}$

The force created by the atoms outside our averaging volume on the atoms inside is again

$$\boldsymbol{F}(\boldsymbol{r},t) = \frac{1}{V_0}\int_{\Omega_0(\boldsymbol{r})} \sum_{p=1}^{N_\infty} \sum_{\substack{n=1 \\ n\neq p}}^{N_\infty} \boldsymbol{f}_{pn}\, \delta\big(\boldsymbol{r}_0 - \boldsymbol{r}_p(t)\big)\, \mathrm{d}^3\boldsymbol{r}_0 \tag{2.81}$$

$$= \frac{1}{V_0}\sum_{p=1}^{N_0} \sum_{n=N_0+1}^{N_\infty} \boldsymbol{f}_{pn} \quad (\text{because } \boldsymbol{f}_{pn} = -\boldsymbol{f}_{np} \text{ in } \Omega_0), \tag{2.82}$$

where at any instance t there are $N_0(\boldsymbol{r},t)$ atoms within our averaging region $\Omega_0(\boldsymbol{r})$ centered on the point $\boldsymbol{r}$ and $N_\infty - N_0$ atoms outside the averaging region. This force density $\boldsymbol{F}$ decomposes into two contributions

$$\boldsymbol{F} = \boldsymbol{F}_c + \boldsymbol{F}_b \tag{2.83}$$

called "contact" and "body" forces as laid out in this section. Due to the short-range nature of the molecular interaction forces $\boldsymbol{f}_{pn}$, the atoms immediately outside Ω_0 act on the atoms immediately inside to create a net macroscopic contact force $\boldsymbol{F}_c$ acting on Ω_0. The tensorial nature of stress as developed here is due entirely to such short-range atomic force interactions across the surface $\partial\Omega_0$. By contrast, the body forces $\boldsymbol{F}_b$ are due to accumulations or concentrations of moving atoms even at distances large relative to the extent of Ω_0. Such concentrations create macroscopic EM and gravitational forces that vary slowly across Ω_0 to produce the body force $\boldsymbol{F}_b$.

2.8.1 Contact Forces

The nature of $\boldsymbol{f}_{pn}$ is that it falls off rapidly with the distance r_{pn} separating any two p and n. For example, we have seen that the Lennard-Jones model for the interaction between uncharged isotropic atoms (e.g., the Noble gas atoms) is

$$\boldsymbol{f}_{pn} = -\frac{24\varepsilon}{\sigma}\Bigg[\underbrace{\left(\frac{\sigma}{r_{pn}}\right)^{7}}_{\uparrow\ \text{Attractive force}} - \underbrace{2\left(\frac{\sigma}{r_{pn}}\right)^{13}}_{\uparrow\ \text{Repulsive force}}\Bigg]\hat{\boldsymbol{r}}_{pn}, \tag{2.84}$$

where $\hat{\boldsymbol{r}}_{pn}$ is directed from atom n toward atom p, σ is a characteristic atomic size, and ε has units of energy and both are properties of the two atoms involved in the interaction. Compared to the van der Waals attractive force between neighboring uncharged atoms that are a distance $r_{pn} = \sigma$ apart, the attractive force between atoms twice removed ($r_{pn} \approx 2\sigma$) is 128 times smaller and the attractive force between atoms three times removed ($r_{pn} \approx 3\sigma$) is 2,187 times smaller. For the ionic interactions in a mineral, although an isolated cation p has a Coulombic force of attraction to an isolated anion n that varies slowly as r_{pn}^{-2}, because each cation in the mineral is surrounded by anions and each anion by cations, such screening causes the force of attraction between a cation and anion in the mineral to vary exponentially $\exp(-r_{pn}/\delta)$ with a decay length δ (skin depth) on the order of the interionic distances (cf., Israelachvili, 2011). So only *near-neighbor interactions* need be allowed for, which means only those atoms that are within a distance of a few σ of the surface $\partial\Omega_0$ surrounding our averaging volume will contribute to the contact force density $\boldsymbol{F}_c$.

We now define a *near-neighbor interaction tensor* $\boldsymbol{\sigma}$ (no relation to the above length σ in the Lennard-Jones model) that has units of force per unit area and that represents the near-neighbor force interactions across a surface element $dS_0 = (\Delta l)^2$ of the outer surface $\partial\Omega_0$ surrounding each averaging region Ω_0 that has the outward normal $\boldsymbol{n}$. This $\boldsymbol{\sigma}$ is required to satisfy the definition

$$dS_0\, \boldsymbol{n}\cdot\boldsymbol{\sigma} \mathrel{\widehat{=}} \text{net near-neighbor force that atoms on the positive } \boldsymbol{n} \text{ side of } dS_0 \text{ exert on atoms on the negative } \boldsymbol{n} \text{ side} \tag{2.85}$$

and we need a means to calculate this $\boldsymbol{\sigma}$ from knowledge of the forces $\boldsymbol{f}_{pn}$ between each pair of atoms.

Consider a small cubic volume element $\Delta V = (\Delta l)^3$ whose center lies at position $\boldsymbol{r}_0$ somewhere within an averaging element Ω_0 (i.e., $\Delta V \ll V_0$) as depicted in Fig. 2.3. The size Δl of this cube must be in the range $a \ll \Delta l \ll R$, where a is the size of the individual atoms ($\approx \sigma$ in the Lennard-Jones model) and R is the local radius of curvature of the surface $\partial\Omega_0$ surrounding the averaging volume. At a minimum, the cube used to define $\boldsymbol{\sigma}(\boldsymbol{r}_0)$ must contain a handful of atoms but would preferably contain thousands or more of atoms. With $(\Delta l)^2$ being the cross-sectional area of the cube, we identify the second-order tensor $\boldsymbol{\sigma}$ in terms of the atomic forces $\boldsymbol{f}_{pn}$ as

$$(\Delta l)^2\, \boldsymbol{\sigma}(\boldsymbol{r}_0) = \hat{\boldsymbol{x}} \sum_{\substack{p\\x<0}} \sum_{\substack{n\\x>0}} \langle \boldsymbol{f}_{pn} \rangle + \hat{\boldsymbol{y}} \sum_{\substack{p\\y<0}} \sum_{\substack{n\\y>0}} \langle \boldsymbol{f}_{pn} \rangle + \hat{\boldsymbol{z}} \sum_{\substack{p\\z<0}} \sum_{\substack{n\\z>0}} \langle \boldsymbol{f}_{pn} \rangle \tag{2.86}$$

where the angle brackets are denoting a temporal average over the fluctuating position of the jiggling atoms within $\Delta V = (\Delta l)^3$, details of which we will not elaborate upon.

So at every point $\boldsymbol{r}_0$ both within Ω_0 and on $\partial\Omega_0$, we allow three planes to intersect normally at that point, and perform the three sets of summations by adding up the net force that the atoms on one side of each plane of area $(\Delta l)^2$ exert on the atoms on the other side

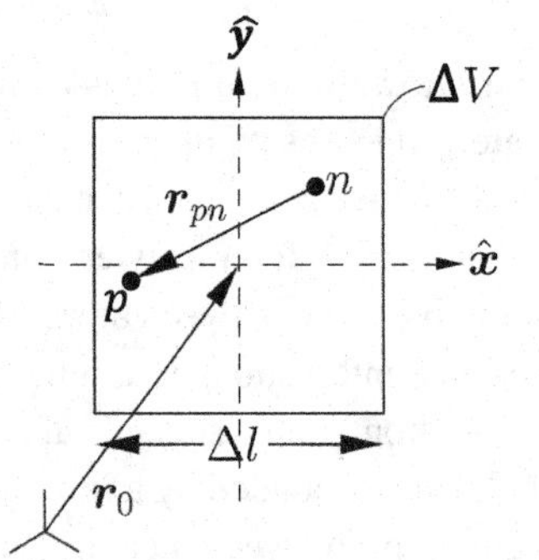

Figure 2.3 At each point $\boldsymbol{r}_0$ within $\Omega_0(\boldsymbol{r})$, construct three perpendicular planes and count the force that atoms n exert on atoms p across each such plane to determine $\boldsymbol{\sigma}$ of Eq. (2.86).

of that plane. In this way, the second-order near-neighbor-interaction tensor $\boldsymbol{\sigma}(\boldsymbol{r}_0)$ at each point $\boldsymbol{r}_0$ is allowing for all possible force interactions between pairs of atoms across all three normally intersecting planes.

For an averaging volume Ω_0 that is a rectangular prism, such as a cube, that possesses the six outward normals $\boldsymbol{n} = \pm\hat{\boldsymbol{x}}_i$ on $\partial\Omega_0$, Eq. (2.86) exactly satisfies the requirement of Eq. (2.85) for points $\boldsymbol{r}_0$ lying on $\partial\Omega_0$. However, if there are only a handful of atoms in each small element $(\Delta l)^3$, then for an $\boldsymbol{n}$ that is not in one of the three directions $\hat{\boldsymbol{x}}_i$, Eq. (2.86) will only approximately satisfy Eq. (2.85) due to the precise position of these handful of atoms even after temporal averaging. This can be handled in three ways: (1) we agree to always define our continuum fields using averaging elements Ω_0 that are rectangular prisms for which Eq. (2.86) satisfies exactly the requirement of Eq. (2.85), (2) if we insist on working with a macroscopic averaging surface $\partial\Omega_0$ that has curvature, we resolve the surface into tiny jagged incremental steps Δl in which $\boldsymbol{n}$ is always in one of the three directions $\hat{\boldsymbol{x}}_i$ used in defining $\boldsymbol{\sigma}$ (this was the approach you used for proving the "fundamental theorem of 3D calculus" as well as "Green's theorem" as exercises in Chapter 1) or (3) by increasing the size of Δl to contain a multitude of atoms while still remaining small relative to any possible curvature of $\partial\Omega_0$, Eq. (2.86) will become less sensitive to the position of any one interacting pair of atoms and will increasingly satisfy the requirement of Eq. (2.85) for arbitrarily oriented $\boldsymbol{n}$ on a curved $\partial\Omega_0$.

Summing up all these molecular-force interactions across all of $\partial\Omega_0$ gives the net *contact-force density* $\boldsymbol{F}_c$ acting on Ω_0 corresponding to the net force that atoms just outside the averaging volume Ω_0 exert on atoms just inside the volume

$$\boldsymbol{F}_c = \frac{1}{V_0}\int_{\partial\Omega_0} \boldsymbol{n}\cdot\boldsymbol{\sigma}(\boldsymbol{r}_0)\,\mathrm{d}S_0. \tag{2.87}$$

Upon using the definition of the gradient of a volume-averaged quantity as given earlier by Eq. (2.44), we end up with the result

$$\boxed{\boldsymbol{F}_c = \nabla\cdot\boldsymbol{S}(\boldsymbol{r}), \quad \text{where} \quad \boldsymbol{S}(\boldsymbol{r}) = \frac{1}{V_0}\int_{\Omega_0} \boldsymbol{\sigma}(\boldsymbol{r}_0)\,\mathrm{d}V_0 \mathrel{\widehat{=}} \textit{molecular-interaction tensor}.} \tag{2.88}$$

So $\nabla\cdot\boldsymbol{S}(\boldsymbol{r}) \neq 0$ if atoms on one side of Ω_0 are pushing in harder than atoms on the other side. The average near-neighbor forces accounted for in the second-order tensor $\boldsymbol{S}$ are a major contribution to the stress tensor $\boldsymbol{\tau}$ that we will soon define by adding to $\boldsymbol{S}$ the random kinetic energy density of the atoms that we identified as the isotropic tensor $-P_I\boldsymbol{I}$ in Section 2.7.

For some arbitrary macro surface element of area $\Delta\mathcal{A}$ located at the point $\boldsymbol{r}$ and having normal $\boldsymbol{n}$, we interpret

$$\Delta\mathcal{A}\,\boldsymbol{n}\cdot\boldsymbol{S}(\boldsymbol{r}) \mathrel{\widehat{=}} \begin{array}{l}\text{the averaged net force vector (in Newtons) that atoms on}\\ \text{the positive } \boldsymbol{n} \text{ side of } \Delta\mathcal{A} \text{ exert on atoms on the negative } \boldsymbol{n} \text{ side.}\end{array} \tag{2.89}$$

This force vector is what we feel when we try to push our hands into each other. If we change the orientation of $\Delta\mathcal{A}$ by changing the normal $\boldsymbol{n}$ while keeping the center of $\Delta\mathcal{A}$ at the same point $\boldsymbol{r}$ in the material, different near-neighbor atoms are acting across this different surface and we would end up, in general, with a different force vector. So to get a force vector, or *traction*, at a point $\boldsymbol{r}$, one must always also specify the orientation $\boldsymbol{n}$ of the surface $\Delta\mathcal{A}$ at that point. Note that from the way that $\boldsymbol{\sigma}$ is defined in Eq. (2.86), the outward component of traction $\boldsymbol{n}\cdot\boldsymbol{S}\cdot\boldsymbol{n}$ is positive when the atoms on the outside, or "positive $\boldsymbol{n}$" side, of the surface pull outward on the atoms just inside the surface and is negative when the outside atoms push inward on the atoms just inside the surface. So $\boldsymbol{n}\cdot\boldsymbol{S}\cdot\boldsymbol{n}$ is positive in tension (inside atoms attracted to the outside atoms) and negative in compression (inside atoms repulsed by the outside atoms).

In passing, we note that chemists often define the molecular-interaction tensor as $\boldsymbol{S} = \boldsymbol{W}/V_0$, where the second-order tensor $\boldsymbol{W}$ is called the *virial tensor* (units of energy) and is typically obtained from the atomic positions and forces using the prescription $\boldsymbol{W} = (1/2)\sum_p \sum_{n\neq p}\langle(\boldsymbol{r}_p - \boldsymbol{r}_n)\boldsymbol{f}_{pn}\rangle$ where the two summations are over all the atoms within the averaging region Ω_0. The factor of 1/2, is because $(\boldsymbol{r}_p - \boldsymbol{r}_n)\boldsymbol{f}_{pn} = (\boldsymbol{r}_n - \boldsymbol{r}_p)\boldsymbol{f}_{np}$. This may appear to be a distinct definition of $\boldsymbol{S}$ from our approach of deriving a local interaction tensor $\boldsymbol{\sigma}$ using Eq. (2.86) at each point of Ω_0 and then volume averaging $\boldsymbol{\sigma}$ over Ω_0. Our approach makes explicit that $\nabla\cdot\boldsymbol{S}$ represents exactly the net near-neighbor forces that atoms outside of Ω_0 exert on atoms inside of Ω_0, while it is less obvious that $\nabla\cdot(\boldsymbol{W}/V_0)$ satisfies this physical requirement. Further, if we pass a plane having a normal $\hat{\boldsymbol{x}}$ through the center of the collection of atoms in a cubic box Ω_0 of volume $V_0 = L_0^3$, it is not obvious that $\hat{\boldsymbol{x}}\cdot(\boldsymbol{W}/L_0^3) = L_0^{-2}\sum_p\sum_{n\neq p}\langle[(x_p - x_n)/L_0]\boldsymbol{f}_{pn}\rangle/2$ is the average force per unit area that atoms on the positive $\hat{\boldsymbol{x}}$ side of the plane exerts on atoms on the negative $\hat{\boldsymbol{x}}$ side, the dimensionless weight $(x_p - x_n)/L_0 < 1$ appears to throw off the sum. Nonetheless, it is widely assumed that the virial tensor does satisfy the interpretation of Eq. (2.89) with authors such as Weiner (2002) arguing for such an interpretation at least in the case of regular crystalline lattices. Again, our definition of $\boldsymbol{S}$ based on volume averaging the near-neighbor interaction tensor $\boldsymbol{\sigma}$ for the force interactions between atoms residing on either side of a plane to give the averaged interpretation of Eq. (2.89) guarantees that $\boldsymbol{F}_c = \nabla\cdot\boldsymbol{S}$ is exactly the net force that atoms just outside of Ω_0 exert on atoms just inside.

Last, the molecular-interaction tensor $\boldsymbol{S}$ is necessarily symmetric as will soon be demonstrated in Section 2.9.1.

2.8.2 Body Forces

In addition to the *contact-force density* $\boldsymbol{F}_c = \nabla\cdot\boldsymbol{S}$, the atoms outside an averaging volume can contribute an additional contribution to $\boldsymbol{F}$ called *body forces* $\boldsymbol{F}_b$ if there are large concentrations of mass, charge, or current in regions even far removed from Ω_0 as depicted in Fig. 2.4. Such concentrations create gravitational and/or EM force fields $\boldsymbol{g}$, $\boldsymbol{E}$, and $\boldsymbol{B}$ that fall off slowly with distance from the zone of concentration compared to the near-neighbor force interactions of the contact-force density. The body-force density is then defined using Eq. (2.6) as

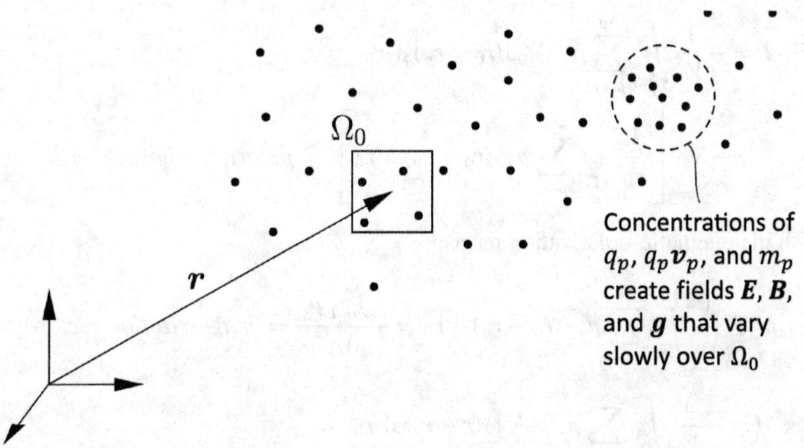

Figure 2.4 Slowly varying force fields $\boldsymbol{E}$, $\boldsymbol{B}$, and $\boldsymbol{g}$ created from particle concentrations outside Ω_0.

$$\boldsymbol{F}_b = \frac{1}{V_0}\int_{\Omega_0}\sum_{p=1}^{N_\infty}\{[q_p\boldsymbol{E} + (q_p\boldsymbol{v}_p + \dot{\boldsymbol{p}}_p) \times \boldsymbol{B} + m_p\,\boldsymbol{g}]\,\delta(\boldsymbol{r}_0 - \boldsymbol{r}_p)$$
$$- [\boldsymbol{p}_p \cdot \nabla_0\delta(\boldsymbol{r}_0 - \boldsymbol{r}_p)]\,\boldsymbol{E} - [\boldsymbol{\mu}_p \times \nabla_0\delta(\boldsymbol{r}_0 - \boldsymbol{r}_p)] \times \boldsymbol{B}\}\;\mathrm{d}^3\boldsymbol{r}_0, \tag{2.90}$$

where the $\boldsymbol{g}$, $\boldsymbol{E}$, and $\boldsymbol{B}$ vectors explicitly present here are taken as varying slowly across the integration domain Ω_0. Carrying out the integral in this case gives

$$\boldsymbol{F}_b = (\rho_e - \nabla \cdot \boldsymbol{P})\,\boldsymbol{E} + \left(\boldsymbol{J} + \frac{\partial \boldsymbol{P}}{\partial t} + \nabla \times \boldsymbol{M}\right) \times \boldsymbol{B} + \rho\boldsymbol{g} \mathrel{\widehat{=}} \textit{body-force density} \tag{2.91}$$

where the various densities conjugate to the force fields $\boldsymbol{E}$, $\boldsymbol{B}$, and $\boldsymbol{g}$ are

$$\rho_e = \frac{1}{V_0}\sum_{p=1}^{N_0} q_p \mathrel{\widehat{=}} \textit{free-charge density} \tag{2.92}$$

$$\boldsymbol{J} = \frac{1}{V_0}\sum_{p=1}^{N_0} q_p\;\boldsymbol{v}_p \mathrel{\widehat{=}} \textit{free-current density} \tag{2.93}$$

$$\rho = \frac{1}{V_0}\sum_{p=1}^{N_0} m_p \mathrel{\widehat{=}} \textit{mass density} \tag{2.94}$$

along with the electric-polarization terms

$$\boldsymbol{P} = \frac{1}{V_0}\int_{\Omega_0}\sum_{p=1}^{N_0}\boldsymbol{p}_p\delta(\boldsymbol{r}_0 - \boldsymbol{r}_p)\;\mathrm{d}^3\boldsymbol{r}_0 = \frac{\sum_{p=1}^{N_0}\boldsymbol{p}_p}{V_0} \mathrel{\widehat{=}} \textit{electrical-polarization vector} \tag{2.95}$$

$$-\nabla\cdot\boldsymbol{P} = -\frac{1}{V_0}\int_{\Omega_0}\sum_{p=1}^{N_0}\boldsymbol{p}_p\cdot\nabla_0\delta(\boldsymbol{r}_0-\boldsymbol{r}_p)\,\mathrm{d}^3\boldsymbol{r}_0$$

$$= -\nabla\cdot\left[\frac{1}{V_0}\int_{\Omega_0}\sum_{p=1}^{N_0}\boldsymbol{p}_p\delta(\boldsymbol{r}_0-\boldsymbol{r}_p)\,\mathrm{d}^3\boldsymbol{r}_0\right] \mathrel{\widehat{=}} \textit{polarized-charge density} \tag{2.96}$$

and with the magnetic-polarization terms

$$\boldsymbol{M} = \frac{1}{V_0}\int_{\Omega_0}\sum_{p=1}^{N_0}\boldsymbol{\mu}_p\delta(\boldsymbol{r}_0-\boldsymbol{r}_p)\,\mathrm{d}^3\boldsymbol{r}_0 = \frac{\sum_{p=1}^{N_0}\boldsymbol{\mu}_p}{V_0} \mathrel{\widehat{=}} \textit{magnetization vector} \tag{2.97}$$

$$\nabla\times\boldsymbol{M} = -\frac{1}{V_0}\int_{\Omega_0}\sum_{p=1}^{N_0}\boldsymbol{\mu}_p\times\nabla_0\delta(\boldsymbol{r}-\boldsymbol{r}_p)\,\mathrm{d}^3\boldsymbol{r}_0$$

$$= \nabla\times\left[\frac{1}{V_0}\int_{\Omega_0}\sum_{p=1}^{N_0}\boldsymbol{\mu}_p\delta(\boldsymbol{r}_0-\boldsymbol{r}_p)\,\mathrm{d}^3\boldsymbol{r}_0\right] \mathrel{\widehat{=}} \textit{bound-current density.} \tag{2.98}$$

In the final equation, we use the vector identity $\nabla\times(\boldsymbol{a}\beta) = -\boldsymbol{a}\times\nabla\beta$, where $\boldsymbol{a}$ is a vector that does not vary in space and β is a scalar that does vary in space.

The excess charge in Ω_0 associated with $-\nabla\cdot\boldsymbol{P}$ is created by the slow spatial longitudinal variation of the electric field $\boldsymbol{E}$ across the integration domain that results in more local dipole heads (positive charge) that stick into Ω_0 through $\partial\Omega_0$ on one side of the domain than there are dipole tails (negative charge) on the other side of the domain. So only if there is longitudinal spatial variation in $\boldsymbol{E}$ across an integration domain do we get a charge excess $-\nabla\cdot\boldsymbol{P}$. The body-force due to $\boldsymbol{E}$ then acting on the excess polarized-charge density $-\nabla\cdot\boldsymbol{P}$ in Ω_0 can be called the *electrostrictive force*. The effect of this electrostrictive body force generating stress and strain in a body as controlled by our macroscopic statement of conservation of momentum and Hooke's law of elasticity (as derived in Chapter 4) is called *electrostriction*. Similarly, the effective current density associated with $\nabla\times\boldsymbol{M}$ happens when the magnetic field $\boldsymbol{B}$ has a slow transverse spatial variation across the integration domain that creates more local magnetization on one side of the domain compared to the other. So only if there is a transverse spatial variation of $\boldsymbol{B}$ across the domain do we get the effective current density term $\nabla\times\boldsymbol{M}$ that the magnetic field $\boldsymbol{B}$ then acts upon to create a body force that can result in deformation that is called *magnetostriction*. We will show in Chapter 3, once we have derived the macroscopic or continuum form of the Maxwell equations, that the total effective EM body force can be equivalently and exactly expressed only in terms of the force fields $\boldsymbol{E}$ and $\boldsymbol{B}$ in a compact result known as the Maxwell stress tensor.

Again, an atom or molecule with a net charge q_p is called an ion. If the number of positive ions (cations) is the same as the number of negative ions (anions) in each element of a material, the free-charge density is $\rho_e = 0$, which is the most common situation in a material. But there are many important circumstances, for example, in a liquid electrolyte near the interface with a solid mineral, where $\rho_e \neq 0$ and has gradients even in equilibrium.

In a metal, we need to interpret the charged particles q_p as being the electrons and nuclei. The electrons in the conduction band of metallic atoms are free to move between the atoms of the metal and are associated with a relatively large $\boldsymbol{v}_p$, especially under the influence of an applied electric field, compared to the metallic nuclei which are heavy and locked in place and have, effectively, $\boldsymbol{v}_p \approx 0$. So current density in a metal is due to the free electrons in the conduction band moving under the influence of a macroscopic electric field.

Rules for determining the continuum force fields $\boldsymbol{E}$, $\boldsymbol{B}$, and $\boldsymbol{g}$ as well as the charge density ρ_e and the electric and magnetic polarization vectors $\boldsymbol{P}$ and $\boldsymbol{M}$ are derived in Chapter 3. To conclude here, the total force $\boldsymbol{F}$ that atoms outside of each averaging region Ω_0 exert on the atoms inside the region is

$$\boxed{\begin{aligned} \boldsymbol{F} &= \boldsymbol{F}_c + \boldsymbol{F}_b \\ &= \nabla \cdot \boldsymbol{S} + (\rho_e - \nabla \cdot \boldsymbol{P})\,\boldsymbol{E} + \left(\boldsymbol{J} + \frac{\partial \boldsymbol{P}}{\partial t} + \nabla \times \boldsymbol{M}\right) \times \boldsymbol{B} + \rho \boldsymbol{g}. \end{aligned}} \tag{2.99}$$

We have divided this force into a contact-force contribution $\nabla \cdot \boldsymbol{S}$ that corresponds to the near-neighbor force interaction that atoms just outside an averaging volume Ω_0 exert on atoms just inside Ω_0 and into a body-force contribution $(\rho_e - \nabla \cdot \boldsymbol{P})\,\boldsymbol{E} + (\boldsymbol{J} + \partial \boldsymbol{P}/\partial t + \nabla \times \boldsymbol{M}) \times \boldsymbol{B} + \rho \boldsymbol{g}$ that corresponds to the electrical, magnetic, and gravitational forces generated by concentrations of atoms at larger distance from the averaging volume resulting in force fields $\boldsymbol{E}$, $\boldsymbol{B}$, and $\boldsymbol{g}$ that can be taken as slowly varying across Ω_0.

2.9 The Stress Tensor

It is tradition to combine $-P_I \boldsymbol{I}$ and the molecular-interaction tensor $\boldsymbol{S}$ into a single second-order tensor $\boldsymbol{\tau}$ called the *stress tensor*

$$\boxed{\boldsymbol{\tau}(\boldsymbol{r}, t) = -P_I(\boldsymbol{r}, t)\,\boldsymbol{I} + \boldsymbol{S}(\boldsymbol{r}, t) \,\widehat{=}\, \textit{the stress tensor}.} \tag{2.100}$$

The divergence of the stress tensor is then the vector

$$\nabla \cdot \boldsymbol{\tau} = -\nabla P_I + \nabla \cdot \boldsymbol{S}, \tag{2.101}$$

which is a force density acting on the atoms in each averaging volume that we will call the *net traction-force density* and denote as $\boldsymbol{f} = \nabla \cdot \boldsymbol{\tau}$.

Any second-order tensor, including $\boldsymbol{\tau}$, can be separated into *isotropic* and *deviatoric* contributions as

$$\boldsymbol{\tau} = \underbrace{\left[\frac{\mathrm{tr}\{\boldsymbol{\tau}\}}{\mathrm{tr}\{\boldsymbol{I}\}}\boldsymbol{I}\right]}_{\text{Isotropic}} + \underbrace{\left[\boldsymbol{\tau} - \frac{\mathrm{tr}\{\boldsymbol{\tau}\}}{\mathrm{tr}\{\boldsymbol{I}\}}\boldsymbol{I}\right]}_{\text{Deviatoric}}, \tag{2.102}$$

where the trace of tensor $\boldsymbol{\tau}$ is defined $\mathrm{tr}\{\boldsymbol{\tau}\} = \boldsymbol{I} : \boldsymbol{\tau} = \tau_{ii} = \tau_{xx} + \tau_{yy} + \tau_{zz}$ (in 3D), which is the sum along the diagonal of the tensor. So the deviatoric portion of a second-order tensor has zero trace. The isotropic and deviatoric parts of the stress tensor each have their unique physical meaning and result in distinct independent deformations as will be seen when we consider the response of isotropic elastic materials in Chapter 4. In what follows, we further address the decomposition of the stress tensor into its isotropic and deviatoric contributions.

But we begin by determining the symmetry of the stress tensor, which because the identity tensor is symmetric, is equivalent to determining the symmetry of the molecular-interaction tensor $\boldsymbol{S}$.

2.9.1 Conservation of Angular Momentum and the Symmetry of $\boldsymbol{\tau}$

What we have shown thus far is that the macroscopic statement of the conservation of linear momentum is

$$\frac{\partial}{\partial t}(\boldsymbol{v}\rho) = -\nabla \cdot (\boldsymbol{v}\boldsymbol{v}\rho) + \nabla \cdot \boldsymbol{\tau} + \boldsymbol{F}_b, \tag{2.103}$$

where $\boldsymbol{v}\rho$ is the linear momentum density.

The angular momentum density is the vector $\boldsymbol{r} \times \boldsymbol{v}\rho$. For some arbitrary constant-in-time macroscopic region Ω that is surrounded by the surface $\partial\Omega$ and embedded within a still larger body, we can state the conservation of angular momentum for this subregion Ω as the requirement

$$\frac{\partial}{\partial t}\int_\Omega \boldsymbol{r} \times \boldsymbol{v}\rho \, \mathrm{d}V = -\int_{\partial\Omega} \boldsymbol{n} \cdot (\boldsymbol{v}\boldsymbol{r} \times \boldsymbol{v}\rho) \, \mathrm{d}S + \int_{\partial\Omega} \boldsymbol{r} \times (\boldsymbol{n} \cdot \boldsymbol{\tau}) \, \mathrm{d}S + \int_\Omega \boldsymbol{r} \times \boldsymbol{F}_b \, \mathrm{d}V. \tag{2.104}$$

The first surface integral on the right-hand side is the advective accumulation of angular momentum crossing into or out of Ω, while the second surface integral is the net force moment or *torque* created by the molecular-interaction forces $\boldsymbol{n} \cdot \boldsymbol{S}$ and the flux of random thermal momentum $\boldsymbol{n}P_I$ that act at the surface $\partial\Omega$. The final volume integral is the net force moment due to the body forces.

Use the divergence theorem to write the advective accumulation of angular momentum as

$$-\int_{\partial\Omega} \boldsymbol{n} \cdot (\boldsymbol{v}\boldsymbol{r} \times \boldsymbol{v}\rho) \, \mathrm{d}S = -\int_\Omega \nabla \cdot (\boldsymbol{v}\boldsymbol{r} \times \boldsymbol{v}\rho) \, \mathrm{d}V = +\int_\Omega \nabla \cdot [\boldsymbol{v}\,({}_3\boldsymbol{\epsilon} : \boldsymbol{r}\boldsymbol{v}\rho)] \, \mathrm{d}V, \tag{2.105}$$

where ${}_3\boldsymbol{\epsilon}$ is the third-order Levi–Civita tensor of Chapter 1, and we are using our double-dot product convention also as addressed after Eq. (1.48) in Chapter 1. Let's analyze the integrand by writing out its detailed components to give

$$\nabla \cdot [\boldsymbol{v}\,({}_3\boldsymbol{\epsilon} : \boldsymbol{r}\boldsymbol{v}\rho)] = \hat{\boldsymbol{x}}_i \frac{\partial}{\partial x_i} \cdot \left[v_j \hat{\boldsymbol{x}}_j \epsilon_{klm} \hat{\boldsymbol{x}}_k \hat{\boldsymbol{x}}_l \hat{\boldsymbol{x}}_m : x_n \hat{\boldsymbol{x}}_n v_o \hat{\boldsymbol{x}}_o \rho \right] \tag{2.106}$$

$$= \frac{\partial}{\partial x_i} \left[v_j \epsilon_{klm} x_n v_o \rho \right] (\hat{\boldsymbol{x}}_i \cdot \hat{\boldsymbol{x}}_j)(\hat{\boldsymbol{x}}_m \cdot \hat{\boldsymbol{x}}_n)(\hat{\boldsymbol{x}}_l \cdot \hat{\boldsymbol{x}}_o) \hat{\boldsymbol{x}}_k \tag{2.107}$$

$$= \epsilon_{klm} \left[v_m v_l \rho + x_m \frac{\partial}{\partial x_i} (v_i v_l \rho) \right] \hat{\boldsymbol{x}}_k \tag{2.108}$$

$$= -\boldsymbol{v} \times \boldsymbol{v}\rho - \boldsymbol{r} \times \nabla \cdot (\boldsymbol{v}\boldsymbol{v}\rho)\,. \tag{2.109}$$

Because $\boldsymbol{v} \times \boldsymbol{v} = 0$, we have so far shown that

$$\int_\Omega \boldsymbol{r} \times \left[\frac{\partial(\boldsymbol{v}\rho)}{\partial t} + \nabla \cdot (\boldsymbol{v}\boldsymbol{v}\rho) - \boldsymbol{F}_b \right] \mathrm{d}V = \int_{\partial\Omega} \boldsymbol{r} \times (\boldsymbol{n} \cdot \boldsymbol{\tau})\, \mathrm{d}S. \tag{2.110}$$

Using the Levi–Civita third-order tensor and the fact that $\boldsymbol{a} \cdot \boldsymbol{A} = \boldsymbol{A}^T \cdot \boldsymbol{a}$ for any vector $\boldsymbol{a}$ and second-order tensor $\boldsymbol{A}$, the surface integral of the torque acting at the surface may be written

$$\int_{\partial\Omega} \boldsymbol{r} \times (\boldsymbol{n} \cdot \boldsymbol{\tau})\, \mathrm{d}S = -\int_{\partial\Omega} {}_3\boldsymbol{\epsilon} : \boldsymbol{r}\,(\boldsymbol{n} \cdot \boldsymbol{\tau})\; \mathrm{d}S \tag{2.111}$$

$$= -\int_{\partial\Omega} \left[{}_3\boldsymbol{\epsilon} : \boldsymbol{r}\boldsymbol{\tau}^T \right] \cdot \boldsymbol{n}\, \mathrm{d}S \tag{2.112}$$

$$= -\int_{\partial\Omega} \boldsymbol{n} \cdot \left[{}_3\boldsymbol{\epsilon} : \boldsymbol{r}\boldsymbol{\tau}^T \right]^T \; \mathrm{d}S \tag{2.113}$$

$$= -\int_\Omega \nabla \cdot \left[{}_3\boldsymbol{\epsilon} : \boldsymbol{r}\boldsymbol{\tau}^T \right]^T \; \mathrm{d}V. \tag{2.114}$$

Writing out the integrand in its detailed components gives

$$-\nabla \cdot \left[{}_3\boldsymbol{\epsilon} : \boldsymbol{r}\boldsymbol{\tau}^T \right]^T = -\hat{\boldsymbol{x}}_i \frac{\partial}{\partial x_i} \cdot \left[\epsilon_{jkl} \hat{\boldsymbol{x}}_j \hat{\boldsymbol{x}}_k \hat{\boldsymbol{x}}_l : x_m \hat{\boldsymbol{x}}_m \tau_{on} \hat{\boldsymbol{x}}_n \hat{\boldsymbol{x}}_o \right]^T \tag{2.115}$$

$$= -\epsilon_{jkl} \frac{\partial}{\partial x_i} (x_l \tau_{ik})\, \hat{\boldsymbol{x}}_j \tag{2.116}$$

$$= {}_3\boldsymbol{\epsilon} : \boldsymbol{\tau} + \boldsymbol{r} \times \nabla \cdot \boldsymbol{\tau}. \tag{2.117}$$

With this, we now have the exact statement for the conservation of angular momentum in some region Ω

$$\int_\Omega \boldsymbol{r} \times \left[\frac{\partial(\boldsymbol{v}\rho)}{\partial t} + \nabla \cdot (\boldsymbol{v}\boldsymbol{v}\rho) - \nabla \cdot \boldsymbol{\tau} - \boldsymbol{F}_b \right] \mathrm{d}V = \int_\Omega {}_3\boldsymbol{\epsilon} : \boldsymbol{\tau}\, \mathrm{d}V. \tag{2.118}$$

The integrand on the left-hand side is exactly zero by the conservation of linear momentum. Because the subregion Ω is arbitrarily defined within the larger body, the integrand on the right-hand side must be zero:

$$ {}_3\boldsymbol{\epsilon} : \boldsymbol{\tau} = 0 \quad \text{or} \quad \boxed{\boldsymbol{\tau} = \boldsymbol{\tau}^T.} \tag{2.119}$$

We used the fact proven by Eq. (1.50) of Chapter 1 that the double dot product of the antisymmetric tensor ${}_3\boldsymbol{\epsilon}$ with a symmetric tensor is zero. Because $\boldsymbol{\tau} = -P_l\boldsymbol{I} + \boldsymbol{S}$, we have

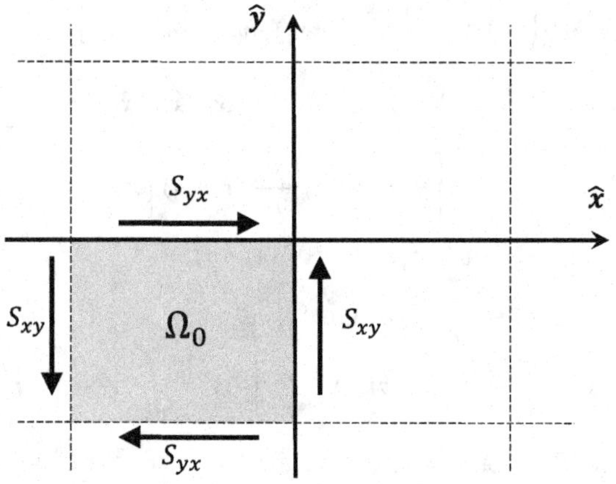

Figure 2.5 No net torque on each averaging element Ω_0 means $S_{xy} = S_{yx}$.

also proven that if $\boldsymbol{n} \cdot \boldsymbol{S}$ is the molecular interaction force across a surface as defined earlier by Eq. (2.89), and we assume our $\boldsymbol{S}$ satisfies this requirement, then $\boldsymbol{S} = \boldsymbol{S}^T$ is a requirement of angular-momentum conservation.

To visualize this, divide some large body under investigation into cubic voxels, where each voxel is an averaging element Ω_0. As seen in Fig. 2.5, the force condition $S_{ij} = S_{ji}$ results in no torque and no rotation of each cubic voxel about its center. The local near-neighbor-interaction tensor $\boldsymbol{\sigma}$ defined by Eq. (2.86) on each small element of volume $(\Delta l)^3$ contributing to an averaging volume Ω_0 (where $(\Delta l)^3$ may contain only a handful of molecules but would preferably contain thousands or more) may be locally nonsymmetric due to the specific location of the molecules within a particular $(\Delta l)^3$, even after temporal averaging over the thermal fluctuations. However, when $\boldsymbol{\sigma}$ is integrated over an entire averaging volume Ω_0 to give $\boldsymbol{S}$, such local asymmetry will average to produce a tensor $\boldsymbol{S}$ that is necessarily symmetric.

2.9.2 Energy Conservation and the Poynting Vector

Using the conservation of mass and the definition of the total derivative $\mathrm{d}/\mathrm{d}t = \partial/\partial t + \boldsymbol{v} \cdot \nabla$, the conservation of linear momentum can be written from a frame of reference that moves along with the velocity $\boldsymbol{v}$ of each mass element

$$\rho \frac{\mathrm{d}\boldsymbol{v}}{\mathrm{d}t} = \nabla \cdot \boldsymbol{\tau} + \boldsymbol{F}_b. \tag{2.120}$$

Dot multiplying this equation with $\boldsymbol{v}$ gives a statement of macroscopic energy conservation for mechanical response

$$\frac{\rho}{2} \frac{\mathrm{d}(\boldsymbol{v} \cdot \boldsymbol{v})}{\mathrm{d}t} + \boldsymbol{\tau} : \nabla \boldsymbol{v} = -\nabla \cdot (-\boldsymbol{\tau} \cdot \boldsymbol{v}) + \boldsymbol{F}_b \cdot \boldsymbol{v}, \tag{2.121}$$

where we used the identity $\nabla \cdot (\boldsymbol{\tau} \cdot \boldsymbol{v}) = \boldsymbol{\tau} : (\nabla \boldsymbol{v})^T + (\nabla \cdot \boldsymbol{\tau}) \cdot \boldsymbol{v}$, plus the fact that because $\boldsymbol{\tau} = \boldsymbol{\tau}^T$, we have $\boldsymbol{\tau} : \nabla \boldsymbol{v} = \boldsymbol{\tau} : (\nabla \boldsymbol{v})^T$. In later chapters, we will show that $\boldsymbol{\tau}$ is the sum of two terms, one representing elastic stress $\boldsymbol{\tau}^e$ and one representing viscous stress $\boldsymbol{\tau}^v$. It will be demonstrated that $\boldsymbol{\tau}^e : \nabla \boldsymbol{v}$ is the rate that energy per unit volume is being stored reversibly, while $\boldsymbol{\tau}^v : \nabla \boldsymbol{v}$ is the rate at which energy is being dissipated in the presence of viscosity. The term $\rho[\mathrm{d}(\boldsymbol{v} \cdot \boldsymbol{v})/\mathrm{d}t]/2$ is the rate at which the macroscopic kinetic energy density of the mass element is changing. The term $\boldsymbol{F}_b \cdot \boldsymbol{v}$ is that rate at which the body force is performing work on the mass element, which corresponds to energy imparted to the mass element. So the term $-\nabla \cdot (-\boldsymbol{\tau} \cdot \boldsymbol{v})$ represents the rate at which energy is flowing into or out of the mass element. The vector

$$\boxed{\boldsymbol{s} = -\boldsymbol{\tau} \cdot \boldsymbol{v} = -\boldsymbol{v} \cdot \boldsymbol{\tau}} \tag{2.122}$$

thus represents the flux of mechanical energy with $\boldsymbol{s}$ being called the *Poynting vector*. A Poynting vector will also be defined for electromagnetics in Chapter 3. We will calculate the Poynting vector in Chapter 4 for various types of elastodynamic waves.

Equation (2.121) says that if more mechanical energy is fluxing into an element than is fluxing out and/or if the body forces are doing work on the mass element, such energy change for the element goes into all of: (1) the macroscopic kinetic energy of the element, (2) energy being elastically stored in the element, and/or (3) energy being dissipated to heat in the element.

2.9.3 Isotropic Stress and the Total Pressure

The *total pressure* P (that we can also call the *thermodynamic pressure*) is defined as the negative of the isotropic portion of the stress tensor, which in 3D gives

$$\boxed{P(\boldsymbol{r}, t) \,\widehat{=}\, -\frac{\boldsymbol{I} : \boldsymbol{\tau}(\boldsymbol{r}, t)}{\boldsymbol{I} : \boldsymbol{I}} = -\frac{1}{3}\tau_{ii}(\boldsymbol{r}, t) = P_I(\boldsymbol{r}, t) - \frac{1}{3}S_{ii}(\boldsymbol{r}, t).} \tag{2.123}$$

Thus, the total pressure is due to both the random momentum flux of atoms (P_I) plus the atomic force interactions between atoms just outside of the averaging region and those just inside ($-S_{ii}/3$). In an ideal gas, there is only P_I because the molecular-interaction stress $\boldsymbol{S} = 0$. Note that from its Eq. (2.71) definition, P_I is always positive. But in liquids, solids, and real gases, both terms contribute to P. When the atoms outside an averaging element are pushing in with a net inward compressive force (corresponding to repulsion between the inside and outside atoms), the pressure is positive. But if there is enough outward attraction of the atoms inside the averaging element to the atoms outside, this isotropic outward tension $S_{ii}/3$ can become sufficiently large compared to P_I that the pressure P becomes negative. A negative total pressure will occur most easily in a solid but can occur in liquids as well. It will not occur in gases.

2.9.4 Deviatoric Stress Tensor

The *deviatoric stress tensor* is then defined

$$\boldsymbol{\tau}^D(\boldsymbol{r}, t) = \boldsymbol{\tau}(\boldsymbol{r}, t) + P(\boldsymbol{r}, t)\,\boldsymbol{I}, \tag{2.124}$$

which has zero trace and thus no isotropic or pressure component associated with it. More explicitly, if $\boldsymbol{\tau}$ is given in 3D Cartesian coordinates as

$$\boldsymbol{\tau} = \begin{pmatrix} \tau_{xx} & \tau_{xy} & \tau_{xz} \\ \tau_{xy} & \tau_{yy} & \tau_{yz} \\ \tau_{xz} & \tau_{yz} & \tau_{zz} \end{pmatrix}, \tag{2.125}$$

the deviatoric stress tensor is then

$$\boldsymbol{\tau}^D = \begin{pmatrix} (2\tau_{xx} - \tau_{yy} - \tau_{zz})/3 & \tau_{xy} & \tau_{xz} \\ \tau_{xy} & (2\tau_{yy} - \tau_{xx} - \tau_{zz})/3 & \tau_{yz} \\ \tau_{xz} & \tau_{yz} & (2\tau_{zz} - \tau_{xx} - \tau_{yy})/3 \end{pmatrix}. \tag{2.126}$$

So the deviatoric stress tensor is nonzero whenever any of the longitudinal-stress components τ_{xx}, τ_{yy}, and τ_{zz} along the main diagonal of $\boldsymbol{\tau}$ are different from each other or whenever any of the off-diagonal or shear-stress components τ_{xy}, τ_{xz}, and τ_{yz} are nonzero. Later in Chapter 4, we will see that when only deviatoric stress acts on a solid elastic sample, the sample's shape will change and, if the sample is elastically isotropic, its volume will not change.

2.9.5 Traction Vector

To visualize what the stress tensor $\boldsymbol{\tau}$ represents at some point in a material, one must always imagine a planar surface element that passes through that point, having a normal vector $\boldsymbol{n}$ and an area $\Delta\mathcal{A}$ as depicted in Fig. 2.6. The net force per unit area that atoms on the positive $\boldsymbol{n}$ side of the surface element exert on atoms on the negative $\boldsymbol{n}$ side when combined with the net flux of random atomic momentum across the surface element yields a vector $\boldsymbol{T}$ that is called the *traction vector*. It is the traction vector associated with each planar element at a point that can be visualized, not the stress tensor. The stress boundary conditions on a body are given by specifying the traction vector over the surface surrounding the body.

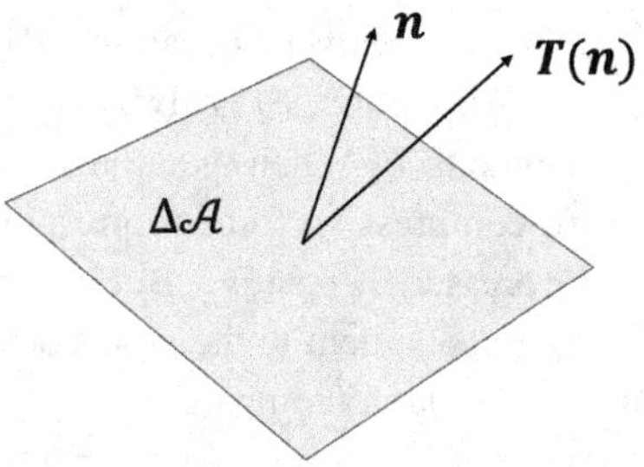

Figure 2.6 A surface element with unit normal $\boldsymbol{n}$ and area $\Delta\mathcal{A}$.

The traction vector is related to the unit normal $\boldsymbol{n}$ of the area element and the stress tensor $\boldsymbol{\tau}$ by the simple relation

$$\boldsymbol{T} = \boldsymbol{n} \cdot \boldsymbol{\tau} \mathrel{\widehat{=}} \textit{traction vector} \text{ associated with the surface element } \Delta\mathcal{A} \tag{2.127}$$

$$= -\boldsymbol{n}\, P_I + \boldsymbol{n} \cdot \boldsymbol{S} \quad \overset{\textit{UNITS}}{\left(\frac{\text{force}}{\text{area}} = \frac{\text{energy}}{\text{volume}} = \frac{\text{momentum}}{(\text{area})\,(\text{time})}\right)} \tag{2.128}$$

where

$$\boldsymbol{n}\, P_I \mathrel{\widehat{=}} \text{momentum flux vector associated with thermal fluctuations}$$

$$\boldsymbol{n} \cdot \boldsymbol{S} \mathrel{\widehat{=}} \text{force/area that atoms on the } +\boldsymbol{n} \text{ side of element exert on atoms on the } -\boldsymbol{n} \text{ side.}$$

So as promised in Chapter 1, the way a second-order stress tensor like $\boldsymbol{\tau}$ enters the continuum physics is as the proportionality between two vectors with clear physical interpretation like the traction vector $\boldsymbol{T}$ and the normal vector $\boldsymbol{n}$ to the surface element across which the traction is being determined. We have $\boldsymbol{T} = \boldsymbol{n} \cdot \boldsymbol{\tau} = \boldsymbol{\tau} \cdot \boldsymbol{n}$ because $\boldsymbol{\tau}$ is symmetric. If we remain at the same point in space and change the orientation of the surface element (i.e., change the direction of $\boldsymbol{n}$), there are different atoms involved in the nearest neighbor interaction across the differently oriented surface elements and the traction vector $\boldsymbol{T}$ will change most typically. However, the stress tensor $\boldsymbol{\tau}$ at a point does not change with the orientation of the surface element passing through that point and is a field that allows for all possible near-neighbor interactions and random momentum fluctuations present in the neighborhood surrounding each point. Note that the physical units of traction (or stress) can be written as (1) a force per unit area or (2) a density of energy or (3) a flux of momentum. Depending on the context, we will have (and have already had) occasion to take all three perspectives when interpreting stress.

The best way to visualize all six components of the stress tensor at once is to imagine a cubic element of solid being acted upon by the stress tensor. In your mind's eye, place the cubic sample in a reservoir of uniform stress tensor and visualize the traction vector acting on each of the six faces of the cube as shown in Fig. 2.7. Because the normal vector on opposing faces of the cube is oriented in opposite directions, the longitudinal stress components τ_{xx}, τ_{yy}, and τ_{zz} correspond to normal traction vectors acting on their corresponding cube face that are oriented in the opposite direction for any two opposing faces of the cube. So if a component τ_{yy} is pulling out on the face with a normal $\boldsymbol{n} = +\hat{\boldsymbol{y}}$, it is also pulling out on the opposing face on the other side of the cube with normal $\boldsymbol{n} = -\hat{\boldsymbol{y}}$. Similarly, the shear components of stress τ_{xy}, τ_{xz}, and τ_{yx} correspond to tangential traction vectors acting on their corresponding cube face and are also oriented in the opposite direction for any two opposing cube faces.

The stress tensor is defined in such a way that for a surface element with normal $\boldsymbol{n}$, the normal component of the traction vector $\boldsymbol{n} \cdot \boldsymbol{T} = \boldsymbol{n} \cdot \boldsymbol{\tau} \cdot \boldsymbol{n} = \boldsymbol{\tau} : \boldsymbol{nn}$ (that we may also call the *longitudinal* stress component for that surface element) is positive in tension. So when $\boldsymbol{n} \cdot \boldsymbol{T}$ is positive, there is a net force pulling outward on the surface in the direction of the normal $\boldsymbol{n}$. So a compressive traction, that corresponds to a force pushing in on the surface in the

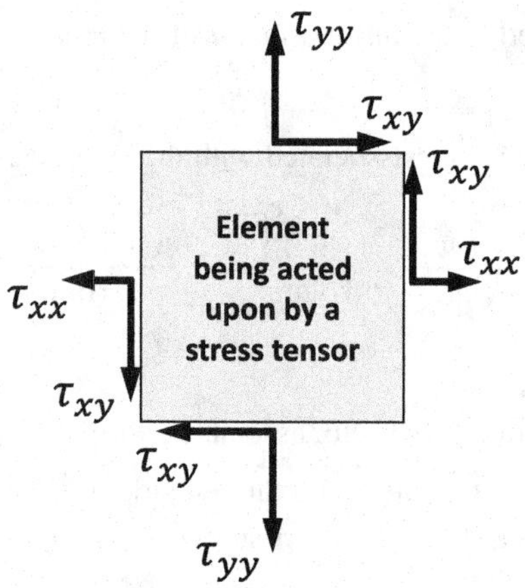

Figure 2.7 A cubic element immersed in a reservoir of uniform stress tensor $\boldsymbol{\tau}$ with the resulting traction vectors present on the six cube faces.

direction opposite to $\boldsymbol{n}$, will have $\boldsymbol{n} \cdot \boldsymbol{T} = \boldsymbol{n} \cdot \boldsymbol{\tau} \cdot \boldsymbol{n}$ negative. *Longitudinal stress is defined to be positive in tension and negative in compression.*

2.9.6 Continuity of the Traction Vector at an Interface

Across surfaces in a body, including the interface between two distinct materials as depicted in Fig. 2.8, the traction vector can be shown to be continuous across the interface.

To demonstrate this, we write the conservation of momentum as $\rho \, d\boldsymbol{v}/dt = \nabla \cdot \boldsymbol{\tau} + \boldsymbol{F}_b$ and integrate this statement over the shaded disc region Ω_ϵ shown in Fig. 2.8 in the limit as the half-width ϵ of the disc goes to zero. We have

$$\lim_{\epsilon \to 0} \left[\int_{\Omega_\epsilon} \rho \frac{d\boldsymbol{v}}{dt} \, dV = \int_{\partial\Omega_\epsilon} \boldsymbol{n} \cdot \boldsymbol{\tau} \, dS + \int_{\Omega_\epsilon} \boldsymbol{F}_b \, dV \right] \tag{2.129}$$

where the divergence theorem is used to rewrite the volume integral of $\nabla \cdot \boldsymbol{\tau}$ as a surface integral of the traction $\boldsymbol{n} \cdot \boldsymbol{\tau}$. In the limit as $\epsilon \to 0$, the two remaining volume integrals vanish and the surface integral over the circumferential face of the disc is also zero so that only the two surface integrals over the bounding faces $\Sigma_\pm$ survive

$$\int_{\Sigma_+} \boldsymbol{n} \cdot \boldsymbol{\tau}_1 \, dS - \int_{\Sigma_-} \boldsymbol{n} \cdot \boldsymbol{\tau}_2 \, dS = 0. \tag{2.130}$$

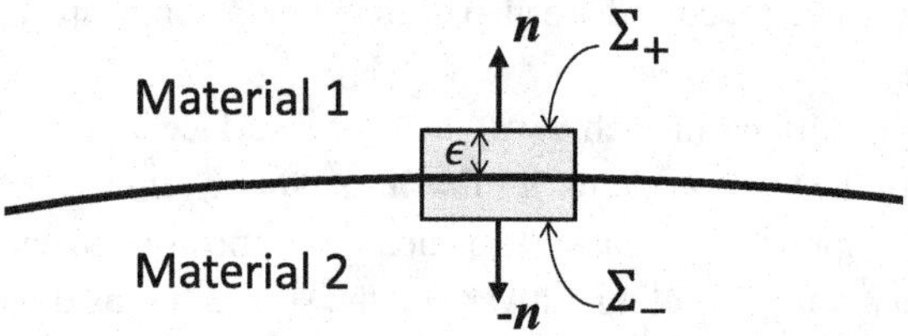

Figure 2.8 A disc Ω_ϵ straddling an interface.

Because this must hold for any definition of the bounding surfaces $\Sigma_\pm$, we must have that for any interface having normal $\boldsymbol{n}$

$$\boxed{\boldsymbol{n}\cdot\boldsymbol{\tau}_1 = \boldsymbol{n}\cdot\boldsymbol{\tau}_2;} \tag{2.131}$$

that is, the traction vector is continuous across any interface within a macroscopic body, including a material interface as shown in the figure. This is true both in the dynamic and static case. This can be thought of as a statement of momentum conservation; namely, all momentum fluxing into the interface fluxes out with no momentum accumulating on the interface.

An important exception to the above occurs when there is either a thin layer of excess free charge on the interface between materials or if there is an electrical-polarization vector $\boldsymbol{P}$ present having a normal component that is different on either side of the interface. As will be seen in Chapter 3, the electrical polarization will be different if material 1 and material 2 have different electrical permitivities and if there is, simultaneously an electric field present that has a normal component to the interface. If we allow for an electrical body force in the form $\boldsymbol{F}_b = (\rho_e - \nabla\cdot\boldsymbol{P})\boldsymbol{E}$ in Eq. (2.129) and divide that equation by the area S of both bounding surfaces $\Sigma_\pm$, the body-force term is

$$\frac{1}{S}\int_{\Omega_\epsilon}(\rho_e - \nabla\cdot\boldsymbol{P})\boldsymbol{E}\,\mathrm{d}S = \left[\lim_{\epsilon\to 0}\frac{1}{S}\int_{\Omega_\epsilon}\rho_e\,\mathrm{d}V - \boldsymbol{n}\cdot\boldsymbol{P}_1 + \boldsymbol{n}\cdot\boldsymbol{P}_2\right]\overline{\boldsymbol{E}}, \tag{2.132}$$

where we applied the divergence theorem to the polarization vector term and where if there is an excess of free charge on the interface giving rise to a surface-charge density q_f (Coulombs per meter squared), we have

$$q_f = \lim_{\epsilon\to 0}\frac{1}{S}\int_{\Omega_\epsilon}\rho_e\,\mathrm{d}V. \tag{2.133}$$

Further, as will be seen in Chapter 3, the horizontal component of the electric field is continuous across an interface while if there is a contrast in the dielectric constant or electrical conductivity across the interface, there will be a discontinuity in the normal component of the electric field across the interface. In this case, it is the average field across the interface that acts on the charge excess as given by

$$\overline{\boldsymbol{E}} = \frac{(\boldsymbol{n}\cdot\boldsymbol{E}_1 + \boldsymbol{n}\cdot\boldsymbol{E}_2)}{2}\boldsymbol{n} + \boldsymbol{n}\times\boldsymbol{E}\times\boldsymbol{n}, \tag{2.134}$$

where $\boldsymbol{n}\times\boldsymbol{E}\times\boldsymbol{n} = \boldsymbol{n}\cdot(\boldsymbol{nE} - \boldsymbol{En}) = \boldsymbol{E}\cdot(\boldsymbol{I} - \boldsymbol{nn})$ is the horizontal component of the electric field that is continuous across the interface (do the two cross products by applying the right-hand rule twice to quickly see this for yourself). Thus, when such electric fields and electrical permittivity and conductivity contrasts are present, there is a discontinuity in the traction vector given by

$$\boldsymbol{n}\cdot\boldsymbol{\tau}_1 - \boldsymbol{n}\cdot\boldsymbol{\tau}_2 = \left[q_f + \boldsymbol{n}\cdot\boldsymbol{P}_2 - \boldsymbol{n}\cdot\boldsymbol{P}_1\right]\left[\frac{(\boldsymbol{n}\cdot\boldsymbol{E}_1 + \boldsymbol{n}\cdot\boldsymbol{E}_2)}{2}\boldsymbol{n} + \boldsymbol{n}\times\boldsymbol{E}\times\boldsymbol{n}\right]. \tag{2.135}$$

Applying a time-varying electric field in this scenario will generate a time-varying stress discontinuity that will generate elastodynamic response. Because the excess charge on the interface is transiently created initially by more current entering the interface than leaving or due to a discontinuity in the electrical polarization (this will be addressed in Sections 3.9 and 3.10), the stress discontinuity is quadratic in the electric field present and is most often quite negligible.

An analogous argument holds for any of the other body forces present $(\boldsymbol{J} + \partial \boldsymbol{P}/\partial t + \nabla \times \boldsymbol{M}) \times \boldsymbol{B}$ or $\rho \boldsymbol{g}$. If there is an excess electric current in the plane of the interface in the presence of a magnetic field that has a component normal to the interface or if there is an excess mass on the interface in the presence of a gravitational acceleration, there also develops a discontinuity in the traction vector. Typically only the EM sources of traction-vector discontinuity are ever important.

2.9.7 Net Traction–Force Density

The net traction-force density $\boldsymbol{f} = \nabla \cdot \boldsymbol{\tau}$ represents the net force acting upon (and random thermal momentum accumulating within) the averaging region Ω_0 that surrounds the point where $\boldsymbol{f}$ is being defined. The development that follows serves as another more macroscopic way to help make this point explicit to you. Picture a rectangular averaging prism Ω_0 having volume $V_0 = \Delta x \, \Delta y \, \Delta z$ as depicted in Fig. 2.9. There are tractions $\boldsymbol{n} \cdot \boldsymbol{\tau}$ acting on all the faces of this prism including the hidden faces in the figure that have normals $\boldsymbol{n} = -\hat{\boldsymbol{x}}$ and $-\hat{\boldsymbol{y}}$ and $-\hat{\boldsymbol{z}}$. For each of the six faces of the rectangular prism, the traction vector $\boldsymbol{T} = \boldsymbol{n} \cdot \boldsymbol{\tau}$ can be written:

$$\text{On } \boldsymbol{n} = +\hat{\boldsymbol{x}}, \quad \boldsymbol{T}\left(\tfrac{\Delta x}{2}, 0, 0\right) = \hat{\boldsymbol{x}}\, \tau_{xx}\left(\tfrac{\Delta x}{2}, 0, 0\right) + \hat{\boldsymbol{y}}\, \tau_{xy}\left(\tfrac{\Delta x}{2}, 0, 0\right) + \hat{\boldsymbol{z}}\, \tau_{xz}\left(\tfrac{\Delta x}{2}, 0, 0\right), \tag{2.136}$$

$$\text{On } \boldsymbol{n} = -\hat{\boldsymbol{x}}, \quad \boldsymbol{T}\left(\tfrac{-\Delta x}{2}, 0, 0\right) = -\hat{\boldsymbol{x}}\, \tau_{xx}\left(\tfrac{-\Delta x}{2}, 0, 0\right) - \hat{\boldsymbol{y}}\, \tau_{xy}\left(\tfrac{-\Delta x}{2}, 0, 0\right) - \hat{\boldsymbol{z}}\, \tau_{xz}\left(\tfrac{-\Delta x}{2}, 0, 0\right), \tag{2.137}$$

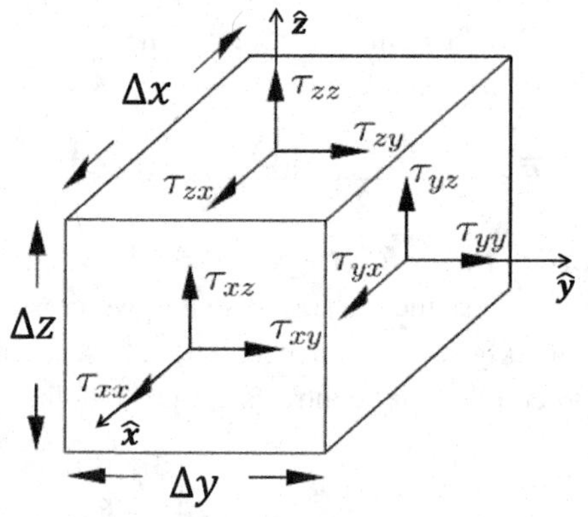

Figure 2.9 A prism Ω_0 subject to boundary tractions.

$$\text{On } \boldsymbol{n} = +\hat{\boldsymbol{y}}, \quad \boldsymbol{T}\left(0, \tfrac{\Delta y}{2}, 0\right) = \hat{\boldsymbol{x}}\, \tau_{yx}\left(0, \tfrac{\Delta y}{2}, 0\right) + \hat{\boldsymbol{y}}\, \tau_{yy}\left(0, \tfrac{\Delta y}{2}, 0\right) + \hat{\boldsymbol{z}}\, \tau_{yz}\left(0, \tfrac{\Delta y}{2}, 0\right), \tag{2.138}$$

$$\text{On } \boldsymbol{n} = -\hat{\boldsymbol{y}}, \quad \boldsymbol{T}\left(0, \tfrac{-\Delta y}{2}, 0\right) = -\hat{\boldsymbol{x}}\, \tau_{yx}\left(0, \tfrac{-\Delta y}{2}, 0\right) - \hat{\boldsymbol{y}}\, \tau_{yy}\left(0, \tfrac{-\Delta y}{2}, 0\right) - \hat{\boldsymbol{z}}\, \tau_{yz}\left(0, \tfrac{-\Delta y}{2}, 0\right), \tag{2.139}$$

$$\text{On } \boldsymbol{n} = +\hat{\boldsymbol{z}}, \quad \boldsymbol{T}\left(0, 0, \tfrac{\Delta z}{2}\right) = \hat{\boldsymbol{x}}\, \tau_{zx}\left(0, 0, \tfrac{\Delta z}{2}\right) + \hat{\boldsymbol{y}}\, \tau_{zy}\left(0, 0, \tfrac{\Delta z}{2}\right) + \hat{\boldsymbol{z}}\, \tau_{zz}\left(0, 0, \tfrac{\Delta z}{2}\right), \tag{2.140}$$

$$\text{On } \boldsymbol{n} = -\hat{\boldsymbol{z}}, \quad \boldsymbol{T}\left(0, 0, \tfrac{-\Delta z}{2}\right) = -\hat{\boldsymbol{x}}\, \tau_{zx}\left(0, 0, \tfrac{-\Delta z}{2}\right) - \hat{\boldsymbol{y}}\, \tau_{zy}\left(0, 0, \tfrac{-\Delta z}{2}\right) - \hat{\boldsymbol{z}}\, \tau_{zz}\left(0, 0, \tfrac{-\Delta z}{2}\right). \tag{2.141}$$

The total force (resultant or sum) of all these six tractions is

$$\begin{aligned} \Delta x\, \Delta y\, \Delta z \boldsymbol{f} = \Delta y\, \Delta z &\left[\boldsymbol{T}\left(\tfrac{\Delta x}{2}, 0, 0\right) + \boldsymbol{T}\left(\tfrac{-\Delta x}{2}, 0, 0\right)\right] \\ + \Delta x\, \Delta z &\left[\boldsymbol{T}\left(0, \tfrac{\Delta y}{2}, 0\right) + \boldsymbol{T}\left(0, \tfrac{-\Delta y}{2}, 0\right)\right] \\ + \Delta x\, \Delta y &\left[\boldsymbol{T}\left(0, 0, \tfrac{\Delta z}{2}\right) + \boldsymbol{T}\left(0, 0, \tfrac{-\Delta z}{2}\right)\right]. \end{aligned} \tag{2.142}$$

Using the definition of the derivative $\mathrm{d}\psi(x)/\mathrm{d}x = \left[\psi(x + \Delta x/2) - \psi(x - \Delta x/2)\right]/\Delta x$ along with the definition of all the traction vectors acting on the six cube faces given above, we then end up with

$$\begin{aligned} \boldsymbol{f} = &\hat{\boldsymbol{x}}\frac{\partial \tau_{xx}}{\partial x} + \hat{\boldsymbol{y}}\frac{\partial \tau_{xy}}{\partial x} + \hat{\boldsymbol{z}}\frac{\partial \tau_{xz}}{\partial x} \\ &+ \hat{\boldsymbol{x}}\frac{\partial \tau_{yx}}{\partial y} + \hat{\boldsymbol{y}}\frac{\partial \tau_{yy}}{\partial y} + \hat{\boldsymbol{z}}\frac{\partial \tau_{yz}}{\partial y} \\ &+ \hat{\boldsymbol{x}}\frac{\partial \tau_{zx}}{\partial z} + \hat{\boldsymbol{y}}\frac{\partial \tau_{zy}}{\partial z} + \hat{\boldsymbol{z}}\frac{\partial \tau_{zz}}{\partial z}. \end{aligned} \tag{2.143}$$

Using the definitions in Cartesian coordinates

$$\nabla = \hat{\boldsymbol{x}}\frac{\partial}{\partial x} + \hat{\boldsymbol{y}}\frac{\partial}{\partial y} + \hat{\boldsymbol{z}}\frac{\partial}{\partial z} = \hat{\boldsymbol{x}}_i\frac{\partial}{\partial x_i} \quad \text{and} \quad \boldsymbol{\tau} = \tau_{ij}\, \hat{\boldsymbol{x}}_i \hat{\boldsymbol{x}}_j$$

we have thus proven that

$$\boxed{\boldsymbol{f} = \nabla \cdot \boldsymbol{\tau}} = \frac{1}{V_0}\int_{\partial\Omega_0} \boldsymbol{T}\, \mathrm{d}S = -\nabla P_I + \boldsymbol{F}_c. \tag{2.144}$$

This exercise, that is redundant to what we have already shown, is to make explicit from a different perspective that if we add up the net force from all the tractions acting on the six surfaces of an averaging prism, we end up with a net force density $\boldsymbol{f} = \nabla \cdot \boldsymbol{\tau}$ acting on the prism. If $\boldsymbol{f}$ is nonzero and not balanced by body forces, the atoms in an averaging volume will experience an average acceleration in the direction of $\boldsymbol{f}$, which brings us back to our continuum statement of the conservation of momentum.

2.10 Summary of Continuum Mechanics (So Far)

Bringing together what we have derived so far, you can identify as a tensorial exercise at the end of the chapter that $\partial(\rho\boldsymbol{v})/\partial t + \nabla\cdot(\rho\boldsymbol{v}\boldsymbol{v}) = \rho(\partial\boldsymbol{v}/\partial t + \boldsymbol{v}\cdot\nabla\boldsymbol{v})$, which allows the continuum statement of Newton's second law (the conservation of linear momentum) to be stated for a stationary frame of reference as:

$$\rho\left(\frac{\partial\boldsymbol{v}}{\partial t} + \boldsymbol{v}\cdot\nabla\boldsymbol{v}\right) = \nabla\cdot\boldsymbol{\tau} + \boldsymbol{F}_b. \tag{2.145}$$

The conservation of angular momentum at the macroscopic level is the statement that

$$\boldsymbol{\tau} = \boldsymbol{\tau}^T. \tag{2.146}$$

Our derivation of Newton's law in a continuum description has been performed by averaging over the actual atomic behavior in an averaging volume surrounding each point and has defined each of the fields in terms of the average atomic behavior within the averaging volume. We have learned a lot about the nature of our continuum fields by taking this approach. For example, the velocity $\boldsymbol{v}$ represents the center-of-mass motion of the collection of atoms surrounding each point. The force density $\nabla\cdot\boldsymbol{\tau}$ represents both the net interatomic force that atoms outside the collection exerts on the atoms that are part of the collection as well as any accumulation of momentum due to the random thermal motions of the atoms being larger on one side of the collection compared to the other. The averaging over the molecules produces a stress tensor $\boldsymbol{\tau}$ consisting of one portion that allows for the random thermal energy of the molecules and one portion that allows for the near-neighbor molecular-force interactions. This decomposition tells us, for example, that the pressure can become negative at a point if the atoms outside the collection of molecules surrounding that point pull outward with an isotropic tension that is greater in magnitude than the random thermal-energy density of the collection. Our knowledge is increased when the rules of continuum physics are obtained by averaging over the actual molecules and interactions that are present as opposed to replacing the molecules with an infinitely divisible continuous mass that supports macroscopic fields whose nature must be guessed at.

When we use the fact that the stress tensor is symmetric, Newton's law in the continuum description provides three equations (each component of the equation) for nine unknowns (the three vector components v_i and the six tensor components τ_{ij}). If ρ is treated as an unknown field rather than a constant material property (though for liquids and solids we can usually approximate that ρ does not change as stress changes), we use the conservation of mass

$$\frac{\partial\rho}{\partial t} + \boldsymbol{v}\cdot\nabla\rho = -\rho\nabla\cdot\boldsymbol{v} \tag{2.147}$$

to determine ρ. We clearly need more equations, called constitutive equations, to begin solving problems in continuum mechanics.

Such constitutive equations derive from thermodynamic arguments as we demonstrate in the chapters to follow and also depend on the particular type of material to be treated (e.g., viscous fluid vs. elastic solid). For example, in an elastic solid, the thermodynamic energy analysis given in Chapter 4 yields relations between reversible changes in the stress tensor and reversible changes in a new quantity yet to be defined called the strain tensor. Until such constitutive laws are provided for a particular application in a particular material, the statements of the conservation of mass and momentum derived above cannot be solved to obtain the density (if treated as a variable), velocity and stress fields.

The body forces are given by

$$\boxed{\boldsymbol{F}_b = (\rho_e - \nabla \cdot \boldsymbol{P})\,\boldsymbol{E} + \left(\boldsymbol{J} + \frac{\partial \boldsymbol{P}}{\partial t} + \nabla \times \boldsymbol{M}\right) \times \boldsymbol{B} + \rho \boldsymbol{g}} \tag{2.148}$$

and need the additional field equations of macroscopic EM theory and Newtonian gravity to be derived in the next chapter. In Chapter 3, we will show that the EM body forces can be rewritten as the divergence of the so-called Maxwell stress tensor that only involves the total electric and magnetic fields $\boldsymbol{E}$ and $\boldsymbol{B}$ that contain within them any electric or magnetic polarization. In describing point sources for the generation of seismic waves, we often ascribe to the vector $\boldsymbol{F}_b$ the influence of an explosion or slip on a fault surface. We will show how to represent such seismic sources as an equivalent body-force density in Chapter 4.

Under static (or quasi-static) conditions where the inertial acceleration can be entirely neglected, we have the *equations of static equilibrium* that are given by

$$\nabla \cdot \boldsymbol{\tau} = -\boldsymbol{F}_b \tag{2.149}$$

or $\nabla \cdot \boldsymbol{\tau} = 0$ if body forces are absent. These equations are the force balance that allow us to solve problems under the influence of static (or quasi-static) loads in which the generation of seismic waves can be entirely neglected. *Quasistatic* means that if there is a time-varying applied force or boundary condition, the wavelengths of the seismic waves generated by such time-varying sources are much larger than the size of the system being studied so that such waves can be neglected, in which case Eq. (2.149) is the appropriate force balance. But again, before we can begin solving for the quasi-elastostatic response, we need to supplement Eq. (2.149) with a constitutive law between stress and strain as is described in Chapter 4.

Let us stand back and consider the averaging we have performed over the molecular dynamics to obtain the above macroscopic conservation laws and body-force identifications. This averaging does not involve a probability distribution and does not correspond to ensemble-based statistical mechanics; though that perspective can be taken when deriving the macroscopic fields of continuum mechanics as phase-space averages over the underlying molecularity (e.g., Weiner, 2002). Our averaging simply corresponds to adding up the molecular contributions to produce the macroscopic fields associated with each volume that surrounds each point. No assumptions of any kind are made as to how the atoms are arranged within the averaging region. There may be any kind of defects or dislocations or microcracks or other flaws in crystal lattices or other more amorphous solids. There can be any degree of heterogeneous mixtures or impurities in the solids, liquids, or gases being treated. For any atomic landscape found within an averaging region and for any assumed form of the atomic-force interactions, the above boxed equations of this summary are exact results. Although each continuum field in these equations is defined as the sum over molecular contributions, the form of the macroscopic laws in the boxes above are independent of the precise location or intermolecular-force details of the molecules in each finite averaging volume surrounding each point of a body. The form of the macroscopic laws is also independent of the size (volume) of the averaging region.

We will see throughout the upcoming chapters that the needed (but presently missing) "constitutive laws" for various reversible processes derive from the postulates of *equilibrium thermodynamics*, which provide macroscopic relations, usually in differential form, between the time-varying fields that do not appeal to the detailed atomic structure and can be considered exact if developed properly (which, historically, is not always the case as we will see). Approximations enter the theory of continuum physics only if we want to model the particular numerical value of a material property present in a constitutive law, perhaps along with its temperature, stress, and compositional dependence. Such modeling of the numerical values of material properties requires a detailed understanding of the structure and constituents and molecular interactions that are occurring within an averaging volume of given size and thus lies outside the macroscopic nature of thermodynamics, being the subject matter of statistical mechanics and not something to be addressed in this book. We will take the material properties in all that follows as determined from macroscopic experiments. For an irreversible (i.e., heat producing) process in which an equilibrating flux is linear to the force driving it, the postulates of *nonequilibrium thermodynamics* provide the form of the linear flux–force relations called *transport laws*, which are another type of macroscopic constitutive law involving material properties that are called *transport coefficients*. Finally, for certain irreversible processes in which the flux is nonlinear to the driving force, such as for the plastic response of solids or for other processes that may involve reaching a force threshold prior to occurring, the nature of the nonlinear constitutive laws is still the subject of fundamental research with mathematical forms that are often, though not always, more empirically motivated than theoretically derived. Welcome to the frontiers of materials-science research.

2.11 Exercises

As just stated in the summary, we cannot begin to solve problems in continuum mechanics until we have developed the constitutive laws that apply to a certain material and application. So other than for specialized stress problems, like problem 5 below, the exercises here cannot yet correspond to solutions of continuum-mechanics boundary-value problems (i.e., solving PDEs in a given domain subject to prescribed traction or velocity boundary conditions on the domain limit along with initial conditions throughout the domain for problems with temporal variations).

1. *From the Schrödinger equation to Newton's law*: Begin with the one particle Schrödinger equation for a particle of mass m moving about in a classical force potential field U:

$$i\hbar\frac{\partial\psi(\boldsymbol{r},t)}{\partial t}=-\frac{\hbar^2}{2m}\nabla^2\psi(\boldsymbol{r},t)+U(\boldsymbol{r},t)\psi(\boldsymbol{r},t). \tag{2.150}$$

The so-called wave function ψ is complex and can be written

$$\psi(\boldsymbol{r},t)=R(\boldsymbol{r},t)e^{iS(\boldsymbol{r},t)/\hbar}, \tag{2.151}$$

where both the amplitude R and phase function S are real functions. Introduce Eq. (2.151) into Eq. (2.150), carry out the time and space derivatives, separately group the real and imaginary terms in the resulting equation to obtain two equations, introduce the Bohm momentum postulate

$$m\boldsymbol{v}=\nabla S, \tag{2.152}$$

where $\boldsymbol{v}$ is the particle velocity, introduce the so-called *quantum potential* Q defined as

$$Q=-\frac{\hbar^2}{2m}\frac{\nabla^2 R}{R} \tag{2.153}$$

and finally use the definition of the probability density $\rho=R^2$ for the probability of finding the particle at position $\boldsymbol{r}$ at time t to obtain the following two equations (from the real and imaginary parts of the initial Schrödinger equation):

$$m\left(\frac{\partial\boldsymbol{v}}{\partial t}+\boldsymbol{v}\cdot\nabla\boldsymbol{v}\right)=-\nabla Q-\nabla U \tag{2.154}$$

$$\frac{1}{\rho}\left(\frac{\partial\rho}{\partial t}+\boldsymbol{v}\cdot\nabla\rho\right)=-\nabla\cdot\boldsymbol{v} \tag{2.155}$$

which contain the same information as the complex Schrödinger equation. The first of these is Newton's second law with the presence of the *quantum force* $-\nabla Q$, and the second is the rule for how the probability density varies through time and space given some initial probability distribution $\rho_0(\boldsymbol{r})=\rho(\boldsymbol{r},t=0)$ at $t=0$. HINTS: To demonstrate the above, you will need to know that $\mathrm{d}/\mathrm{d}x\left[f(x)e^{g(x)}\right]=(\mathrm{d}f/\mathrm{d}x+f\mathrm{d}g/\mathrm{d}x)\,e^{g}$ and that $\nabla(\boldsymbol{a}\cdot\boldsymbol{a})=2(\nabla\boldsymbol{a})\cdot\boldsymbol{a}$ which also equals $2\boldsymbol{a}\cdot\nabla\boldsymbol{a}$ for the special case where $\boldsymbol{a}=\nabla\alpha$ (so that $\nabla\boldsymbol{a}=\nabla\nabla\alpha$ is a symmetric tensor). Also, determine $\nabla\psi$ first and then from that result take the divergence to obtain $\nabla^2\psi=\nabla\cdot\nabla\psi$. You will also need to know an obvious thing like $\partial R^2/\partial t=2R\partial R/\partial t$.

2. Maxwell's equations in a vacuum (no charges or matter present) are as follows:

$$\nabla \times \boldsymbol{E} = -\frac{\partial \boldsymbol{B}}{\partial t}$$

$$\frac{1}{\mu_0}\nabla \times \boldsymbol{B} = \epsilon_0 \frac{\partial \boldsymbol{E}}{\partial t},$$

where we also have $\nabla \cdot \boldsymbol{E} = \nabla \cdot \boldsymbol{B} = 0$ and where ϵ_0 and μ_0 are universal constants given earlier. Combine these two first-order differential equations into one second-order differential equation involving only $\boldsymbol{E}$ and one second-order differential equation involving only $\boldsymbol{B}$. You will soon be learning that a PDE of the form $c^2\nabla^2\psi - \partial^2\psi/\partial t^2 = 0$ is called a *wave equation* for a wavefield ψ, where c is the wave speed. Given the above, as well as an identity demonstrated in Chapter 1, determine an expression for the speed of light in vacuum. HINT: Very easy problem meant to give you confidence in pushing symbols around the page and to prepare you for Chapter 3. Just start taking the curl of Maxwell's equations and see how it works out. The identity from Chapter 1 that you will need is $\nabla \times \nabla \times \boldsymbol{a} = \nabla(\nabla \cdot \boldsymbol{a}) - \nabla^2 \boldsymbol{a}$.

3. *The van der Waals attraction*: Random quantum-mechanical fluctuations of the electron positions in an atom create an electric dipole in that atom at any instant whose electric field induces an electric dipole in a neighboring atom and two aligned dipoles attract each other. This is called *van der Waals attraction* and is the reason for why two even uncharged atoms are attracted to each other. Earlier in the chapter, we stated without proof that the energy of attraction associated with this mechanism varies with the distance of separation r_{pn} between atom n and atom p as r_{pn}^{-6}. We stated that the electric field emanating from the dipole in atom n varies with r_{pn} as r_{pn}^{-3} and that the attraction energy between two aligned dipoles varies as r_{pn}^{-3}, which results in the r_{pn}^{-6} dependence of the van der Waals attraction energy between atoms n and p. In this exercise, you will prove these assertions.

   Start with the electrical attraction force between two aligned dipoles as depicted in Fig. 2.10 for which there are four charge centers that electrically interact. The Coulomb force interaction between any two charge centers, say 1 and 2, that are a distance r_{21} apart goes as $\boldsymbol{f}_{21} = q_1 q_2 \hat{\boldsymbol{r}}_{21}/(4\pi\epsilon_0 r_{21}^2)$. Because $\hat{\boldsymbol{r}}_{21} = -\hat{\boldsymbol{r}}_{12}$, we have that $\boldsymbol{f}_{21} = -\boldsymbol{f}_{12}$. For the charge-center interactions depicted in Fig. 2.10, if you use a truncated Taylor series expansion for a distance between charge centers such as

$$\frac{1}{r_{32}^2} = \frac{1}{[r_{pn} - (d_p + d_n)/2]^2} = \frac{1}{r_{pn}^2}\left[1 + \frac{(d_p + d_n)}{r_{pn}} + \frac{3}{4}\left(\frac{d_p + d_n}{r_{pn}}\right)^2 + \dots\right], \quad (2.156)$$

   show that the total force that atom n exerts on atom p through all four charge interactions associated with the two aligned dipoles is given by

$$\boldsymbol{f}_{pn} = -\frac{3 q_n q_p (d_p + d_n)^2}{8\pi\epsilon_0 r_{pn}^4}\hat{\boldsymbol{r}}_{pn}. \quad (2.157)$$

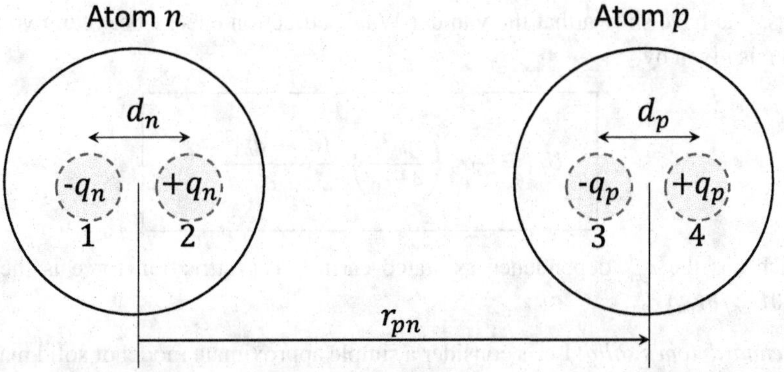

Figure 2.10 A spontaneously created dipole $q_n d_n$ in atom n induces a dipole $q_p d_p$ in atom p and two aligned dipoles attract. This is the mechanism of the van der Waals attraction between two atoms.

For the associated interaction energy U_{pn}, where $\boldsymbol{f}_{pn} = -\nabla U_{pn}$, show that

$$U_{pn} = -\frac{q_n q_p (d_p + d_n)^2}{8\pi \epsilon_0 r_{pn}^3}, \tag{2.158}$$

which does indeed have an r_{pn}^{-3} dependence as stated.

For the full spatial dependence of the attractive interaction, we next allow for how the dipole $q_n d_n$ of atom n polarizes atom p to create the dipole $q_p d_p$. The electric field $\boldsymbol{E}_n$ coming from the dipole of atom n is given as the sum of the two electric fields $\boldsymbol{E}_n = \boldsymbol{E}_{n+} + \boldsymbol{E}_{n-}$ coming from the positive and negative charge centers making up the dipole. The electric potential φ coming from a charge q located at $r = 0$ is the solution of the Poisson equation $\epsilon_0 \nabla^2 \varphi = -q\delta(r)$, which has the solution $\varphi(r) = q/(4\pi \epsilon_0 r)$ that then yields $\boldsymbol{E}(r) = -\nabla\varphi(r) = q/(4\pi \epsilon_0 r^2)\hat{\boldsymbol{r}}$. Use this result to show that along the line that goes through the two charge centers, the electric field coming from atom n is given by

$$\boldsymbol{E}_n = \frac{q_n \hat{\boldsymbol{r}}_{pn}}{4\pi \epsilon_0 (r - d_n/2)^2} - \frac{q_n \hat{\boldsymbol{r}}_{pn}}{4\pi \epsilon_0 (r + d_n/2)^2}. \tag{2.159}$$

The dipole induced in atom p is given by $q_p d_p \hat{\boldsymbol{r}}_{pn} = \alpha_p \, \boldsymbol{E}_n|_{r=r_{pn}}$, where α_p is the polarizability of atom p. If you perform another truncated Taylor expansion of the two terms in Eq. (2.159), show that the induced dipole moment of atom p is given by

$$q_p d_p = \frac{\alpha_p q_n d_n}{2\pi \epsilon_0 r_{pn}^3}, \tag{2.160}$$

which does indeed have an r_{pn}^{-3} dependence as stated.

Thus, you have shown that the van der Waals attraction energy between two atoms n and p is given by

$$U_{pn} = -\alpha_p \left(\frac{q_n d_n}{4\pi \epsilon_0}\right)^2 \frac{(d_p + d_n)^2}{d_p d_n} r_{pn}^{-6}, \tag{2.161}$$

which has the r_{pn}^{-6} dependence as stated earlier. The attraction force is then $\boldsymbol{f}_{pn} = -\left(\partial U_{pn}/\partial r_{pn}\right)\hat{\boldsymbol{r}}_{pn}$.

4. *A Lennard-Jones solid*: Let's consider a simple approximate model of solid matter that says we have a material composed of an isotropic atom arranged in a cubic packing. On average, there is one atom in each cube of size d^3. These atoms are assumed to interact with each other according to the Lennard-Jones force interaction of Eq. (2.84) as set by the parameters σ (atomic size) and ε (interaction energy). They also jiggle about their equilibrium position at the center of each cube with a random thermal velocity. These random thermal motions have an average kinetic energy as set by the temperature and an associated *internal or ideal* pressure P_I as given earlier by $P_I = k_B T n_0$, where $n_0 = N_0/V_0 = 1/d^3$ (the number density of atoms) in this simple cubic model of the atom positions.

   If we allow there to be eight such atoms in a local averaging cube of volume $(\Delta l)^3 = (2d)^3$ so that $\Delta l = 2d$, calculate the local near-neighbor interaction tensor $\boldsymbol{\sigma}$ for this cube of eight interacting atoms using Eqs (2.86) and (2.84) under the assumption that $d > \sigma$ so that only the attractive forces need to be allowed for. Show that $\boldsymbol{\sigma} = \sigma_{xx}\hat{\boldsymbol{x}}\hat{\boldsymbol{x}} + \sigma_{yy}\hat{\boldsymbol{y}}\hat{\boldsymbol{y}} + \sigma_{zz}\hat{\boldsymbol{z}}\hat{\boldsymbol{z}}$, where

$$\sigma_{xx} = \sigma_{yy} = \sigma_{zz} = \frac{24\varepsilon}{\sigma^3}\left(\frac{\sigma}{d}\right)^9 \left(1 + \frac{1}{2^3} + \frac{1}{3^4}\right) \tag{2.162}$$

$$= 24\varepsilon\sigma^6 \left(1 + \frac{1}{2^3} + \frac{1}{3^4}\right) n_0^3. \tag{2.163}$$

   Show that the equation of state for the total pressure P of this cubically packed Lennard-Jones solid is

$$P = k_B T n_0 - 24\varepsilon\sigma^6 \left(1 + \frac{1}{2^3} + \frac{1}{3^4}\right) n_0^3. \tag{2.164}$$

   HINT: Assign a number to each of the eight atoms in the local averaging cube of volume $(\Delta l)^3 = (2d)^3$, for example, as shown above. To use Eq. (2.86) to calculate the local near-neighbor interaction tensor $\boldsymbol{\sigma}$, note that for each of the four atoms n on the positive side of each of the intersecting planes of separation, there are four atoms p on the negative side of that plane. So there are a total of 16 force pairs between the atoms to calculate across each of the planes of separation. Note that the unit vector $\hat{\boldsymbol{r}}_{pn} = (\boldsymbol{r}_p - \boldsymbol{r}_n)/r_{pn}$ between each such pair points from atom n toward atom p. Begin with the separation plane $x=0$ that has $\hat{\boldsymbol{x}}$ as its unit normal. There are four pairs where the separation distance between the atomic centers is $r_{pn}=d$. Using the naming convention given in Fig. 2.11, these four pairs have force directions of $\hat{\boldsymbol{r}}_{15}=\hat{\boldsymbol{r}}_{26}=\hat{\boldsymbol{r}}_{37}=\hat{\boldsymbol{r}}_{48}=-\hat{\boldsymbol{x}}$. There are eight pairs where the atomic separation distance is $r_{pn}=\sqrt{2}d$ that have the force directions $\hat{\boldsymbol{r}}_{35}=\hat{\boldsymbol{r}}_{46}=(-\hat{\boldsymbol{x}}-\hat{\boldsymbol{z}})/\sqrt{2}$, $\hat{\boldsymbol{r}}_{17}=\hat{\boldsymbol{r}}_{28}=(-\hat{\boldsymbol{x}}+\hat{\boldsymbol{z}})/\sqrt{2}$, $\hat{\boldsymbol{r}}_{16}=\hat{\boldsymbol{r}}_{38}=(-\hat{\boldsymbol{x}}+\hat{\boldsymbol{y}})/\sqrt{2}$, and $\hat{\boldsymbol{r}}_{25}=\hat{\boldsymbol{r}}_{47}=(-\hat{\boldsymbol{x}}-\hat{\boldsymbol{y}})/\sqrt{2}$. Last, there are four pairs where the atomic separation distance is $r_{pn}=\sqrt{3}d$ that have the force directions $\hat{\boldsymbol{r}}_{27}=(-\hat{\boldsymbol{x}}-\hat{\boldsymbol{y}}+\hat{\boldsymbol{z}})/\sqrt{3}$, $\hat{\boldsymbol{r}}_{18}=(-\hat{\boldsymbol{x}}+\hat{\boldsymbol{y}}+\hat{\boldsymbol{z}})/\sqrt{3}$, $\hat{\boldsymbol{r}}_{36}=(-\hat{\boldsymbol{x}}+\hat{\boldsymbol{y}}-\hat{\boldsymbol{z}})/\sqrt{3}$, and $\hat{\boldsymbol{r}}_{45}=(-\hat{\boldsymbol{x}}-\hat{\boldsymbol{y}}-\hat{\boldsymbol{z}})/\sqrt{3}$. Adding up the 16 interaction pairs across the plane of separation having normal $\hat{\boldsymbol{x}}$ gives a contribution to $\boldsymbol{\sigma}$ of $\sigma_{xx}\hat{\boldsymbol{x}}\hat{\boldsymbol{x}}$ with σ_{xx} given by Eq. (2.162). Due to the perfect cubic symmetry, we will get the same result for the $\hat{\boldsymbol{y}}\hat{\boldsymbol{y}}$ and $\hat{\boldsymbol{z}}\hat{\boldsymbol{z}}$ components of $\boldsymbol{\sigma}$.

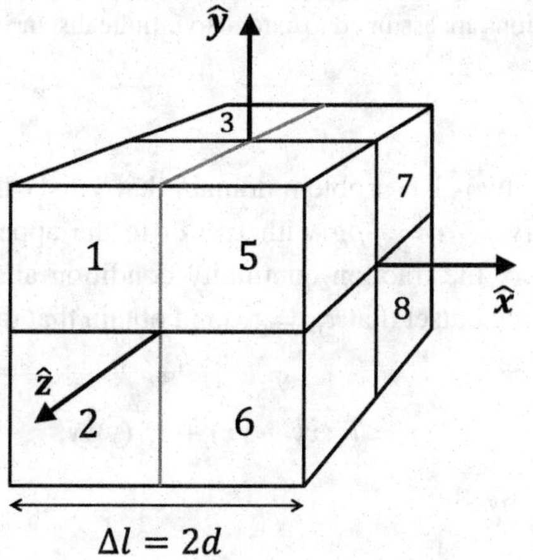

Figure 2.11 A cubic arrangement of eight atoms with an atom present at the center of each of the eight numbered cubes.

5. *Static stress distribution in a gravity field*: This is a very simple static-stress boundary-value problem where the traction vector is prescribed over the domain boundaries in such a special (and admittedly artificial) way as depicted in Fig. 2.12 that the static-force balance $\nabla\cdot\boldsymbol{\tau}=-\rho\boldsymbol{g}$ provides enough equations (with ρ and $\boldsymbol{g}$ given over the problem domain) to determine all the components of the stress tensor throughout the problem domain given $\boldsymbol{T}=\boldsymbol{n}\cdot\boldsymbol{\tau}$ on the domain boundaries that have outward normal $\boldsymbol{n}$.

The 2D problem domain is rectangular with two planar slabs of different solid material having densities ρ_1 and ρ_2 that sit one on top of the other with a welded interface at $y=0$. On the top and bottom boundaries, as designated by $y=+H$ and $y=-H$, the traction vector is uniform and given by $\boldsymbol{T}(x,y=+H)=T_x\hat{\boldsymbol{x}}$ and $\boldsymbol{T}(x,y=-H)=-T_x\hat{\boldsymbol{x}}+(\rho_1+\rho_2)gH\hat{\boldsymbol{y}}$ with T_x a given constant. The traction vector on the two lateral boundaries is also uniform and given by $\boldsymbol{T}(x=0,y)=-T_x\hat{\boldsymbol{y}}$ and $\boldsymbol{T}(x=L,y)=T_x\hat{\boldsymbol{y}}$ with T_x the same given constant. For gravity in the downward direction $\boldsymbol{g}=-g\hat{\boldsymbol{y}}$ with g given, find all the nonzero components of the stress tensor $\boldsymbol{\tau}$

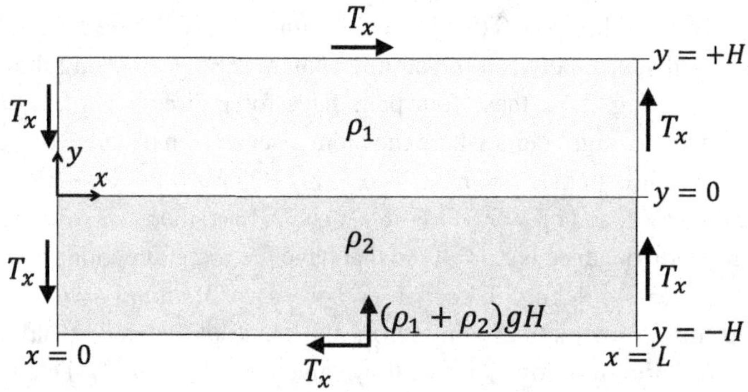

Figure 2.12 Two solid slabs of density ρ_1 and ρ_2 in a rectangular 2D domain with uniform traction boundary conditions prescribed over each of the four boundary faces. The slabs and uniform boundary tractions are assumed to extend to infinite distance in the third dimension coming out of the plane.

throughout the entire two-slab problem domain that satisfy the given traction boundary conditions, satisfy $\nabla \cdot \boldsymbol{\tau} = -\rho \boldsymbol{g}$ with $\rho = \rho_1$ in the upper slab and $\rho = \rho_2$ in the lower slab, and satisfy the traction continuity condition at the interface $y = 0$. You should assume from the outset that $\tau_{yx} = \tau_{xy}$ and obtain that the stress tensor satisfying all of these conditions is

$$\boldsymbol{\tau} = T_x(\hat{\boldsymbol{x}}\hat{\boldsymbol{y}} + \hat{\boldsymbol{y}}\hat{\boldsymbol{x}}) + \tau_{yy}(y)\hat{\boldsymbol{y}}\hat{\boldsymbol{y}}, \tag{2.165}$$

where $\tau_{yy}(y)$ is given by

$$\tau_{yy}(y) = \begin{cases} \rho_1 g y - \rho_1 g H, & 0 < y < H \\ \rho_2 g y - \rho_1 g H, & -H < y < 0. \end{cases} \tag{2.166}$$

What is "unrealistic" about this problem is that gravity alone would not produce the lower uniform traction in the $\hat{\boldsymbol{y}}$ direction when there is no normal component of traction $\pm \tau_{xx}\hat{\boldsymbol{x}}$ on the lateral boundaries. You would need to apply additional vertical pressure to the lower boundary as you get near the lateral boundaries at $x = 0, L$ in order to achieve a normal component of stress $(\rho_1 + \rho_2)gH\hat{\boldsymbol{y}}$ that is uniform over the entire $y = -H$ surface. But the simple solution you will find for the problem as stated with the given uniform traction boundary conditions is in fact unique and exact.

6. Using your tensorial expertise gained in Chapter 1, demonstrate the following two facts:

$$\frac{\partial}{\partial t}(\rho \boldsymbol{v}) + \nabla \cdot (\rho \boldsymbol{v}\boldsymbol{v}) = \rho \left(\frac{\partial \boldsymbol{v}}{\partial t} + \boldsymbol{v} \cdot \nabla \boldsymbol{v} \right)$$

$$\nabla \cdot \left[(\nabla \boldsymbol{v})^T \right] = \nabla (\nabla \cdot \boldsymbol{v}).$$

HINT: For the first of these, you will also need to use the continuum statement of the conservation of mass $\partial \rho / \partial t = -\nabla \cdot (\rho \boldsymbol{v})$ and any needed tensorial product-rule identities from the Table in Section 1.7 of Chapter 1. For the second, follow the approach given in Section 1.7 of Chapter 1.

7. A stress tensor $\boldsymbol{\tau}(x, y, z)$ in Cartesian coordinates has the explicit functional dependence

$$\boldsymbol{\tau} = \begin{pmatrix} \frac{a\ln(x/x_o)}{z/z_o} & b & \frac{c}{(x/x_o)(z/z_o)} \\ b & \frac{d}{(x/x_o)(z/z_o)} & e \\ \frac{c}{(x/x_o)(z/z_o)} & e & \frac{f\ln(z/z_o)}{x/x_o} \end{pmatrix},$$

where the constants a, b, c, d, e, and f all have units of stress and x_o and z_o are given lengths. Determine the analytical expression for the three components of the net traction density vector $\boldsymbol{f} = \nabla \cdot \boldsymbol{\tau}$. If there are no body forces present, determine the places in this material (points, lines, planes, etc.) where the acceleration ($\rho\, d\boldsymbol{v}/dt$) is exactly zero. As a check, this should occur on the line given in (x, y, z) coordinates as $(cz_o/f, y, cx_o/a)$.

8. In two dimensions, with y vertical on the page and x horizontal, show that the deviatoric stress tensor and pressure are defined

$$\boldsymbol{\tau}^D = \begin{pmatrix} (\tau_{xx} - \tau_{yy})/2 & \tau_{xy} \\ \tau_{xy} & (\tau_{yy} - \tau_{xx})/2 \end{pmatrix}$$

and

$$P = -\frac{(\tau_{xx} + \tau_{yy})}{2}.$$

For the special case of a square element subject to the stress tensor components τ_{xx} (a given positive value), $\tau_{yy} = 2\tau_{xx}$ and $\tau_{xy} = 0$, draw the traction vector on each of the four square faces that are created from the deviatoric-stress tensor (i.e., draw the vector $\boldsymbol{n} \cdot \boldsymbol{\tau}^D$ on each of the four faces) and then draw the traction vectors on the four faces that are created from the isotropic stress tensor (i.e., draw the vector $\boldsymbol{n} \cdot (-P\boldsymbol{I})$ for each of the four faces).

3

Continuum Theory of Electromagnetism and Gravity

In Chapter 2, we obtained the macroscopic continuum statement of the conservation of linear momentum, defining the nature of the contact forces and body forces in the process. The body forces are electromagnetic (EM) and gravitational in nature. In this chapter, we derive the continuum equations that control the macroscopic EM and gravity fields. We do so by averaging over the discrete molecularity in a manner analogous to what we did in deriving the continuum statements of the conservation of mass and momentum, that is, we use a Dirac delta to represent each discrete particle as a field and then volume average over the discrete particle behavior. This derivation of the continuum EM equations broadly follows the development given in the classic text by Jackson (1975), although the notation and details of the averaging procedure are somewhat different even if the results are the same. This rather involved tensorial exercise, perhaps the most difficult in the book, not only gives insight to the nature of the macroscopic electric and magnetic fields but is an excellent opportunity to test your tensor-calculus skills developed in Chapter 1.

Once we have the definition of the various EM fields and the associated field equations in a continuum description, we must derive the constitutive laws between the various fields, which is a thermodynamics exercise performed from a frame of reference that moves along with each possibly moving mass element under consideration. We will allow for how the EM constitutive laws depend explicitly on the material velocity $\boldsymbol{v}$ and will also develop conditions for the neglect of such material movement on the EM response.

Although in the overall organization of continuum mechanics, our motivation for developing the macroscopic EM and gravity equations is that they produce the body forces of continuum mechanics, we emphasize that the macroscopic EM and gravity fields are very important unto themselves. EM fields are used to transmit information across macroscopic distances, and both EM and gravity fields that probe a material can be used to garner information about that material (the inverse problem). However, this chapter does not develop such applications of EM and gravitational theory, focusing instead on the derivation of the pertinent macroscopic governing equations and boundary conditions and their solution for a limited number of pedagogic problems.

3.1 Derivation of the Continuum Maxwell's Equations

The fundamental subatomic particles whose motions influence our macroscopic EM fields are the electrons and protons within each atom. The protons are grouped with neutrons in the nucleus, which is a few 10^{-15} m in diameter. The electrons are in a cloud surrounding each nucleus to form an atom that is a couple 10^{-10} m in diameter. So compared to the dimensions of our averaging volume, which might be centimeters in diameter or larger depending on the desired resolution of a given macroscopic application, the positively charged nuclei can be modeled as point charges. The electrons are exactly point charges. To put these sizes in more human terms, if the nucleus is the size of a sand grain, the atom is the size of a large building and a 1-cm diameter averaging volume is the size of the sun. So the positively charged nucleus is indeed well approximated as a point charge when adding up the charge and current density inside of each averaging volume, which is the main task in what follows.

We will develop the governing equations of continuum EM using entirely classical ideas and will not attempt to describe the quantized nature of the magnetic spin of each subatomic particle. The molecular magnetization that will be allowed for is due to charge circulation about the center of each molecule but not that due to the spin of each subatomic particle. This neglect of quantum spin will not affect the form of the macroscopic equations that are derived. However, the contribution of spin to the net magnetization of each element is implicitly allowed for in the EM constitutive laws that are derived.

The local form of Maxwell's equations controlling the EM fields emanating from the multitude of discrete particles moving about in vacuum are

$$\nabla \cdot \boldsymbol{B} = 0 \quad \textit{Gauss' law} \tag{3.1}$$

$$\epsilon_0 \nabla \cdot \boldsymbol{E} = \alpha(\boldsymbol{r}, t) \quad \textit{Coulomb's law} \tag{3.2}$$

$$\nabla \times \boldsymbol{E} = -\frac{\partial \boldsymbol{B}}{\partial t} \quad \textit{Faraday's law} \tag{3.3}$$

$$\frac{1}{\mu_0}\nabla \times \boldsymbol{B} = \epsilon_0 \frac{\partial \boldsymbol{E}}{\partial t} + \boldsymbol{j}(\boldsymbol{r}, t) \quad \textit{Ampère's law} \tag{3.4}$$

where the local electric charge density α and electric current density $\boldsymbol{j}$ fields due to the nuclei and electrons are given by

$$\alpha(\boldsymbol{r}, t) = \sum_p q_p \delta(\boldsymbol{r} - \boldsymbol{r}_p(t)) \tag{3.5}$$

$$\boldsymbol{j}(\boldsymbol{r}, t) = \sum_p q_p \frac{\mathrm{d}\boldsymbol{r}_p(t)}{\mathrm{d}t} \delta(\boldsymbol{r} - \boldsymbol{r}_p(t)). \tag{3.6}$$

We emphasize that the index p here denotes a particular nucleus or electron and does not refer to their subsequent grouping into atoms and molecules, which will be performed systematically in what follows. The charge on each particle is q_p and each particle's trajectory is $\boldsymbol{r}_p(t)$ with a corresponding velocity given by $\boldsymbol{v}_p = \mathrm{d}\boldsymbol{r}_p/\mathrm{d}t$. The universal constants

$\epsilon_0 \approx 8.85 \times 10^{-12}$ F/m and $\mu_0 = 4\pi \times 10^{-7}$ H/m are the electrical permittivity and magnetic permeability of vacuum, where $c = 1/\sqrt{\epsilon_0 \mu_0} \approx 3.00 \times 10^8$ m/s is the speed of light in vacuum.

Using a fixed frame of reference with coordinates $\boldsymbol{r}$ that identify the center of each identical averaging volume $\Omega_0(\boldsymbol{r})$, we now volume average the local Maxwell's equations using the earlier averaging results from Chapter 2 for any field ψ of arbitrary tensorial order

$$\nabla \overline{\psi}(\boldsymbol{r}) = \frac{1}{V_0} \int_{\partial\Omega_0(\boldsymbol{r})} \boldsymbol{n}\psi(\boldsymbol{r}_0)\, \mathrm{d}^2\boldsymbol{r}_0 = \frac{1}{V_0} \int_{\Omega_0(\boldsymbol{r})} \nabla_0 \psi(\boldsymbol{r}_0)\, \mathrm{d}^3\boldsymbol{r}_0 = \overline{\nabla_0 \psi}, \tag{3.7}$$

where

$$\overline{\psi}(\boldsymbol{r}) = \frac{1}{V_0} \int_{\Omega_0(\boldsymbol{r})} \psi(\boldsymbol{r}_0)\, \mathrm{d}^3\boldsymbol{r}_0, \tag{3.8}$$

which results in the continuum form of Maxwell equations expressed in a fixed frame of reference

$$\nabla \cdot \overline{\boldsymbol{B}} = 0 \tag{3.9}$$

$$\epsilon_0 \nabla \cdot \overline{\boldsymbol{E}} = \overline{\alpha}(\boldsymbol{r}, t) \tag{3.10}$$

$$\nabla \times \overline{\boldsymbol{E}} = -\frac{\partial \overline{\boldsymbol{B}}}{\partial t} \tag{3.11}$$

$$\frac{1}{\mu_0} \nabla \times \overline{\boldsymbol{B}} = \epsilon_0 \frac{\partial \overline{\boldsymbol{E}}}{\partial t} + \overline{\boldsymbol{j}}(\boldsymbol{r}, t). \tag{3.12}$$

Clearly, anything new associated with the macroscopic or continuum form of Maxwell's equations resides within the average charge density $\overline{\alpha}$ and average current density $\overline{\boldsymbol{j}}$ and will be associated with polarization fields generated within the molecules.

Prior to carrying out the volume integration associated with $\overline{\alpha}$ and $\overline{\boldsymbol{j}}$, we first define molecular properties associated with each molecular grouping of nuclei and electrons. We also distinguish between those charges that are bound to particular molecules denoted by q_{mp} (the charge on particle p bound to molecule m) and those free electrons present in the conduction bands of metals that are free to move between the atoms of the metal under the influence of an electric field. To identify such free electrons, we use an index p_f and say there are a total N_f of such free electrons in the averaging region. In a metal, the nuclei and other electrons not in the conduction band are bound and have charges q_{mp}. It may turn out that a given molecule has a net total charge, in which case we call that molecule an ion. Such ions can move between the other molecules, with some resistive drag, under the influence of the forcing coming from the total electric and magnetic fields present. We will allow for that possibility as well, which is the only source of macroscopic current in nonmetals.

The position of a bound particle p within a molecule is written as $\boldsymbol{r}_p = \boldsymbol{r}_m + \boldsymbol{r}_{mp}$, where $\boldsymbol{r}_m$ defines the position of the center of mass of the molecule and $\boldsymbol{r}_{mp}$ is the vector from the molecule's center of mass to the position of particle p. The local charge density α associated with the M molecules in the averaging volume and the N_f free electrons not associated with any one molecule is given by

$$\alpha = \sum_{p_f=1}^{N_f} q_{p_f}\delta(\boldsymbol{r} - \boldsymbol{r}_{p_f}(t)) + \sum_{m=1}^{M}\sum_{p=1}^{N_m} q_{mp}\delta(\boldsymbol{r} - \boldsymbol{r}_m(t) - \boldsymbol{r}_{mp}(t)) \tag{3.13}$$

$$= \sum_{p_f=1}^{N_f} q_{p_f}\delta(\boldsymbol{r} - \boldsymbol{r}_{p_f}(t)) + \sum_{m=1}^{M}\sum_{p=1}^{N_m} q_{mp}\left[\delta(\boldsymbol{r} - \boldsymbol{r}_m) - \boldsymbol{r}_{mp}\cdot\nabla\delta(\boldsymbol{r} - \boldsymbol{r}_m)\right.$$
$$\left. + \frac{1}{2!}\boldsymbol{r}_{mp}\boldsymbol{r}_{mp} : \nabla\nabla\delta(\boldsymbol{r} - \boldsymbol{r}_m) + O\left(|\boldsymbol{r}_{mp}|^3\right)\right], \tag{3.14}$$

where N_m is the number of nuclei and nonfree electrons associated with each molecule m. In passing from the first to second lines, we have developed the 3D Dirac delta function in a Taylor series (cf., Section 1.8.1) about the center of mass of each molecule. As discussed in Chapter 1, the big-O notation $O\left(|\boldsymbol{r}_{mp}|^3\right)$ characterizes the amplitude of the part of the series that is being neglected in the limit as $|\boldsymbol{r}_{mp}|$ is tending to zero, that is, small relative to the size of the collection of M molecules.

The various coefficients in the Taylor expansion are now identified as the moments of the multipoles associated with each molecule m

$$q_m = \sum_{p=1}^{N_m} q_{mp} \,\widehat{=}\, \textit{molecular monopole moment} \text{ or } \textit{molecular charge} \tag{3.15}$$

$$\boldsymbol{p}_m = \sum_{p=1}^{N_m} q_{mp}\boldsymbol{r}_{mp} \,\widehat{=}\, \textit{molecular dipole-moment vector} \tag{3.16}$$

$$\boldsymbol{Q}_m = \frac{1}{2}\sum_{p=1}^{N_m} q_{mp}\boldsymbol{r}_{mp}\boldsymbol{r}_{mp} \,\widehat{=}\, \textit{molecular quadrupole-moment tensor} \tag{3.17}$$

and so on for the higher-order tensorial moments (octopoles, etc.) that are not shown. The quadrupole and higher-order moments are negligible in the macroscale response, but we retain the quadrupole moment in the development as representative of the higher-order moments. Again, if $q_m \neq 0$, molecule m is called an ion.

In terms of the molecular properties, the local charge density can thus be written

$$\alpha(\boldsymbol{r}) = \sum_{p_f}^{N_f} q_{p_f}\delta(\boldsymbol{r} - \boldsymbol{r}_{p_f}(t))$$
$$+ \sum_{m=1}^{M}\left[q_m\delta(\boldsymbol{r} - \boldsymbol{r}_m) - \boldsymbol{p}_m\cdot\nabla\delta(\boldsymbol{r} - \boldsymbol{r}_m) + \boldsymbol{Q}_m : \nabla\nabla\delta(\boldsymbol{r} - \boldsymbol{r}_m)\right]. \tag{3.18}$$

Because the dipole-moment vector $\boldsymbol{p}_m$ and symmetric quadrupole-moment second-order tensor $\boldsymbol{Q}_m$ are molecular properties that do not vary in space, we can write

$$\boldsymbol{p}_m \cdot \nabla \delta(\boldsymbol{r}-\boldsymbol{r}_m) = \nabla \cdot \left[\boldsymbol{p}_m \delta(\boldsymbol{r}-\boldsymbol{r}_m)\right] \tag{3.19}$$

$$\boldsymbol{Q}_m : \nabla\nabla \delta(\boldsymbol{r}-\boldsymbol{r}_m) = \nabla\nabla : \left[\boldsymbol{Q}_m^T \delta(\boldsymbol{r}-\boldsymbol{r}_m)\right] = \nabla \cdot \left\{\nabla \cdot \left[\boldsymbol{Q}_m \delta(\boldsymbol{r}-\boldsymbol{r}_m)\right]\right\}, \tag{3.20}$$

where we used that $\boldsymbol{Q}_m$ is symmetric.

With the above molecular charge moments defined, the local charge density is volume averaged and Eq. (3.7) used to give

$$\overline{\alpha}(\boldsymbol{r}) = \rho_e(\boldsymbol{r}) - \nabla \cdot \left[\boldsymbol{P}(\boldsymbol{r}) - \nabla \cdot \boldsymbol{Q}(\boldsymbol{r})\right]. \tag{3.21}$$

The excess free-charge density ρ_e is given by

$$\begin{aligned}\rho_e(\boldsymbol{r}) &= \frac{1}{V_0}\int_{\Omega_0(\boldsymbol{r})}\left[\sum_{p_f=1}^{N_f(\boldsymbol{r})} q_{p_f}\delta(\boldsymbol{r}-\boldsymbol{r}_{p_f}(t)) + \sum_{m=1}^{M(\boldsymbol{r})} q_m \delta(\boldsymbol{r}_0-\boldsymbol{r}_m)\right] \mathrm{d}^3\boldsymbol{r}_0 \\ &= \frac{\sum_{p_f=1}^{N_f(\boldsymbol{r})} q_{p_f} + \sum_{m=1}^{M(\boldsymbol{r})} q_m}{V_0} \mathrel{\widehat{=}} \textit{excess free electrical-charge density}.\end{aligned} \tag{3.22}$$

The polarized charge density $-\nabla \cdot \boldsymbol{P}$ is that due to dipole moments as discussed in Chapter 2, while the quadrupolar charge density $\nabla \cdot (\nabla \cdot \boldsymbol{Q})$ is that due to quadrupoles and is usually negative. We have

$$\boldsymbol{P}(\boldsymbol{r}) = \frac{1}{V_0}\int_{\Omega_0(\boldsymbol{r})}\sum_{m=1}^{M(\boldsymbol{r})} \boldsymbol{p}_m \delta(\boldsymbol{r}_0-\boldsymbol{r}_m)\,\mathrm{d}^3\boldsymbol{r}_0 = \frac{\sum_{m=1}^{M(\boldsymbol{r})}\boldsymbol{p}_m}{V_0} \mathrel{\widehat{=}} \textit{dipole-moment density} \tag{3.23}$$

$$\boldsymbol{Q}(\boldsymbol{r}) = \frac{1}{V_0}\int_{\Omega_0(\boldsymbol{r})}\sum_{m=1}^{M(\boldsymbol{r})} \boldsymbol{Q}_m \delta(\boldsymbol{r}_0-\boldsymbol{r}_m)\,\mathrm{d}^3\boldsymbol{r}_0 = \frac{\sum_{m=1}^{M(\boldsymbol{r})}\boldsymbol{Q}_m}{V_0} \mathrel{\widehat{=}} \textit{quadrupole-moment density} \tag{3.24}$$

and where

$$\nabla \cdot \boldsymbol{Q}(\boldsymbol{r}) = \frac{1}{V_0}\int_{\Omega_0(\boldsymbol{r})}\sum_{m=1}^{M(\boldsymbol{r})} \boldsymbol{Q}_m \cdot \nabla_0 \delta(\boldsymbol{r}_0-\boldsymbol{r}_m)\,\mathrm{d}^3\boldsymbol{r}_0. \tag{3.25}$$

The dipole-moment density vector $\boldsymbol{P}$ is often just called the *electrical polarization* field. The vectors $-\boldsymbol{P}$ and $\nabla \cdot \boldsymbol{Q}$ are the dipolar and quadrupolar contributions to the continuum-scale charge separation, or "polarization" or "displacement," within Ω_0 that is provoked when electric fields are applied to matter.

Given the above, the averaged form of Coulomb's law $\epsilon_0 \nabla \cdot \overline{\boldsymbol{E}} = \overline{\alpha}$ can be written

$$\nabla \cdot \left(\epsilon_o \overline{\boldsymbol{E}} + \boldsymbol{P} - \nabla \cdot \boldsymbol{Q}\right) = \rho_e. \tag{3.26}$$

The *dielectric displacement* vector $\boldsymbol{D}$ (units of C/m^2) is now introduced through the definition

$$\boldsymbol{D} = \epsilon_0 \overline{\boldsymbol{E}} + \boldsymbol{P} - \nabla \cdot \boldsymbol{Q} \widehat{=} \textit{dielectric displacement} \tag{3.27}$$

so that the macroscopic statement of Coulomb's law is compactly written

$$\nabla \cdot \boldsymbol{D} = \rho_e. \tag{3.28}$$

It is perhaps more informative to write the definition of Eq. (3.27) as a statement for the total averaged electric field within an averaging region

$$\overline{\boldsymbol{E}} = \frac{1}{\epsilon_0} (\boldsymbol{D} - \boldsymbol{P} + \nabla \cdot \boldsymbol{Q}), \tag{3.29}$$

which says that the total field is a combination of an applied field $\boldsymbol{D}$ due to any excess free charge or any induction processes present as well as the sum of the dipolar and quadrupolar polarization fields $-\boldsymbol{P} + \nabla \cdot \boldsymbol{Q}$ that are created due to the application of $\boldsymbol{D}$ to the material. We will develop in Section 3.2 the constitutive laws relating $\boldsymbol{D}$ to $\epsilon_0 \overline{\boldsymbol{E}}$. The name "dielectric displacement" is a remnant of the nineteenth-century origins of EM theory, and $\boldsymbol{D}/\epsilon_0$ would perhaps more pedagogically be named the "applied" electric field and $\overline{\boldsymbol{E}}$ the "total" electric field, the difference between the two being due to the polarization of the molecules as quantified by Eq. (3.29), so that total field = applied field + polarization fields.

We next turn to the average current density $\bar{\boldsymbol{j}} = V_0^{-1} \int_{\Omega_0} \sum_p q_p (\mathrm{d}\boldsymbol{r}_p/\mathrm{d}t) \delta(\boldsymbol{r}_0 - \boldsymbol{r}_p)\, \mathrm{d}^3\boldsymbol{r}_0$. Prior to performing this integral, we again distinguish between the free electrons in the conduction band if dealing with a metal and the other particles bound to molecules. We again write the position of each nucleus and electron as $\boldsymbol{r}_p = \boldsymbol{r}_m + \boldsymbol{r}_{mp}$ in order to develop molecular properties through the expansion of $\delta(\boldsymbol{r} - \boldsymbol{r}_m - \boldsymbol{r}_{mp})$ about the center-of-mass of each molecule

$$\boldsymbol{j} = \sum_{p_f=1}^{N_f} q_{p_f} \frac{\mathrm{d}\boldsymbol{r}_{p_f}}{\mathrm{d}t} \delta(\boldsymbol{r} - \boldsymbol{r}_{p_f}(t)) + \sum_{m=1}^{M} \sum_{p=1}^{N_m} q_{mp} \left(\frac{\mathrm{d}\boldsymbol{r}_m}{\mathrm{d}t} + \frac{\mathrm{d}\boldsymbol{r}_{mp}}{\mathrm{d}t} \right) \Big[\delta(\boldsymbol{r} - \boldsymbol{r}_m) - \boldsymbol{r}_{mp} \cdot \nabla \delta(\boldsymbol{r} - \boldsymbol{r}_m)$$
$$+ \frac{1}{2} \boldsymbol{r}_{mp} \boldsymbol{r}_{mp} : \nabla\nabla \delta(\boldsymbol{r} - \boldsymbol{r}_m) + O\left(|\boldsymbol{r}_{mp}|^3\right) \Big]. \tag{3.30}$$

Upon carrying out the term-by-term multiplications, using the earlier identifications for the molecular charge q_m and dipole moment $\boldsymbol{p}_m$, identifying both the molecular center-of-mass velocity as $\boldsymbol{v}_m = \mathrm{d}\boldsymbol{r}_m(t)/\mathrm{d}t$ and free electron velocity as $\boldsymbol{v}_{p_f} = \mathrm{d}\boldsymbol{r}_{p_f}(t)/\mathrm{d}t$, neglecting terms of $O\left(|\boldsymbol{r}_{mp}|^3\right)$ and averaging over Ω_0 we obtain

$$\bar{\boldsymbol{j}} = \frac{1}{V_0} \int_{\Omega_0} \Bigg\{ \sum_{p_f=1}^{N_f} q_{p_f} \boldsymbol{v}_{p_f} \delta(\boldsymbol{r}_0 - \boldsymbol{r}_{p_f}(t)) + \sum_{m=1}^{M} \Bigg[\left(q_m \boldsymbol{v}_m + \frac{\mathrm{d}\boldsymbol{p}_m}{\mathrm{d}t} \right) \delta(\boldsymbol{r}_0 - \boldsymbol{r}_m) - \boldsymbol{v}_m \boldsymbol{p}_m \cdot \nabla_0 \delta(\boldsymbol{r}_0 - \boldsymbol{r}_m)$$
$$+ \boldsymbol{v}_m \boldsymbol{Q}_m : \nabla_0 \nabla_0 \delta(\boldsymbol{r}_0 - \boldsymbol{r}_m) - \sum_{p=1}^{N_m} q_{mp} \frac{\mathrm{d}\boldsymbol{r}_{mp}}{\mathrm{d}t} \boldsymbol{r}_{mp} \cdot \nabla_0 \delta(\boldsymbol{r}_0 - \boldsymbol{r}_m) \Bigg] \Bigg\} \mathrm{d}^3\boldsymbol{r}_0. \tag{3.31}$$

We now proceed to identify these terms in terms of the time derivatives of the electric polarizations $\boldsymbol{P}$ and $\nabla \cdot \boldsymbol{Q}$ and in terms of two new macroscopic vectors $\boldsymbol{J}$ and $\boldsymbol{M}$ that we now define.

The macroscopic current density $\boldsymbol{J}$ (C/m^2/s) is defined

$$\boldsymbol{J} = \frac{1}{V_0} \int_{\Omega_0} \left[\sum_{p_f=1}^{N_f} q_{pf} \boldsymbol{v}_{p_f} \delta(\boldsymbol{r}_0 - \boldsymbol{r}_{p_f}) + \sum_{m=1}^{M} q_m \boldsymbol{v}_m \delta(\boldsymbol{r}_0 - \boldsymbol{r}_m) \right] \mathrm{d}^3 \boldsymbol{r}_0$$

$$= \frac{\sum_{p_f=1}^{N_f} q_{p_f} \boldsymbol{v}_{p_f} + \sum_{m=1}^{M} q_m \boldsymbol{v}_m}{V_0} \,\widehat{=}\, \textit{current density vector}. \tag{3.32}$$

We can say that the ionic velocities have three contributions $\boldsymbol{v}_m = \boldsymbol{v} + \boldsymbol{v}_{em} + \delta \boldsymbol{v}_m$ with $\delta \boldsymbol{v}_m$ being the random thermal motions that average to zero in their contribution to $\boldsymbol{J}$ and can be ignored. The overall average movement of all the molecules in the averaging volume is $\boldsymbol{v}$, which is the center-of-mass velocity we defined as one of our main continuum fields in Chapter 2. In addition to this, if an electric field acts on an ion, the ion will move with a speed that is different than the collective movement $\boldsymbol{v}$ of the surrounding molecules that are mostly not ions. So $\boldsymbol{v}_{em}$ is the velocity of the ions as measured in a frame of reference that is itself moving at velocity $\boldsymbol{v}$. Such electric-field-induced movement of the charge relative to the mean velocity of all the molecules is called *electromigration*. A similar decomposition of the free electron velocities can be made ($\boldsymbol{v}_{pf} = \boldsymbol{v} + \boldsymbol{v}_{epf} + \delta \boldsymbol{v}_{pf}$), which results in the definition of the total current density $\boldsymbol{J}$

$$\boldsymbol{J} = \rho_e \boldsymbol{v} + \boldsymbol{J}_{em}, \tag{3.33}$$

where $\rho_e \boldsymbol{v}$ is the charge flux if the material as a whole has a net free charge and net movement and $\boldsymbol{J}_{em}$ is the electromigration of the positively and negatively charged ions or the free electrons if dealing with a metal

$$\boldsymbol{J}_{em} = \frac{1}{V_0} \left(\sum_{p_f=1}^{N_f} q_{p_f} \boldsymbol{v}_{ep_f} + \sum_{m=1}^{M} q_m \boldsymbol{v}_{em} \right). \tag{3.34}$$

Later in Section 3.3.7, an appropriate macroscopic constitutive model for electromigration will be given in terms of the electric field that drives it.

In passing we note that the current density $\boldsymbol{J}_{em}$ can also be interpreted as a flux of electrical charge across a surface element that is moving with the mean velocity of all the molecules. Let's say that only ionic electromigration is occurring. With N_{ion} being the total number of ions in a volume V_0, we define a length d

$$d = \frac{1}{N_{\text{ion}}} \sum_{m=1}^{N_{\text{ion}}} |\boldsymbol{n} \cdot \boldsymbol{v}_{em}| \, \Delta t, \tag{3.35}$$

where the sum is over just the cations and anions in the volume V_0. So, d is the average distance traveled by the ions in the direction $\boldsymbol{n}$ and in a time Δt as observed from a frame of reference moving with the average velocity $\boldsymbol{v}$. The volume V_0 containing the N_{ion} ions is taken to be $V_0 = d\Delta A$, where ΔA is the area of a macroscopic surface element having a normal $\boldsymbol{n}$. In this case, our definition of Eq. (3.34) can be rewritten

$$\boldsymbol{n} \cdot \boldsymbol{J}_{em} = \frac{N_{\text{ion}} \langle q_m \rangle_{\boldsymbol{n}}}{\Delta A \Delta t}, \tag{3.36}$$

where $\langle q_m \rangle_{\boldsymbol{n}}$ is the average charge amplitude that each ion contributes to the charge flux and is given by

$$\langle q_m \rangle_{\boldsymbol{n}} = \frac{\sum_{m=1}^{N_{\text{ion}}} q_m \boldsymbol{n} \cdot \boldsymbol{v}_{em}}{\sum_{m=1}^{N_{\text{ion}}} |\boldsymbol{n} \cdot \boldsymbol{v}_{em}|}. \tag{3.37}$$

Because the $\boldsymbol{n} \cdot \boldsymbol{v}_{em}$ of anions has the opposite sign from that of cations, each ion makes a positive contribution to the velocity-weighted average charge traversing the surface $\langle q_m \rangle_{\boldsymbol{n}}$. Thus, $\boldsymbol{J}_{em}$ represents the flux of ionic charge of either sign (C/m^2/s) crossing a surface element having normal $\boldsymbol{n}$ that is moving at velocity $\boldsymbol{v}$.

We next introduce the molecular magnetic moment associated with molecule m as

$$\boldsymbol{\mu}_m = \frac{1}{2} \sum_{p=1}^{N_m} q_{mp} \boldsymbol{r}_{mp} \times \frac{\mathrm{d}\boldsymbol{r}_{mp}}{\mathrm{d}t} \tag{3.38}$$

which is a vector representing the magnetic field generated by angular charge circulations around the center of mass of molecule m. From this, we can define the continuum-scale magnetization vector

$$\boldsymbol{M} = \frac{1}{V_0} \int_{\Omega_0} \sum_{m=1}^{M} \boldsymbol{\mu}_m \delta(\boldsymbol{r}_0 - \boldsymbol{r}_m)\, \mathrm{d}^3\boldsymbol{r}_0 = \frac{\sum_{m=1}^{M} \boldsymbol{\mu}_m}{V_0} \,\hat{=}\, \textit{magnetization vector}. \tag{3.39}$$

We next take the curl of $\boldsymbol{M}$ and use Eq. (3.7) to obtain

$$\nabla \times \boldsymbol{M} = \frac{1}{V_0} \int_{\Omega_0} \sum_{m=1}^{M} \sum_{p=1}^{N_m} \frac{q_{mp}}{2} \nabla_0 \times \left[\boldsymbol{r}_{mp} \times \frac{\mathrm{d}\boldsymbol{r}_{mp}}{\mathrm{d}t} \delta(\boldsymbol{r}_0 - \boldsymbol{r}_m) \right] \mathrm{d}^3\boldsymbol{r}_0 \tag{3.40}$$

$$= \frac{1}{V_0} \int_{\Omega_0} \sum_{m=1}^{M} \sum_{p=1}^{N_m} \frac{q_{mp}}{2} [\nabla_0 \delta(\boldsymbol{r}_0 - \boldsymbol{r}_m)] \times \boldsymbol{r}_{mp} \times \frac{\boldsymbol{r}_{mp}}{\mathrm{d}t} \mathrm{d}^3\boldsymbol{r}_0 \tag{3.41}$$

$$= \frac{1}{V_0} \int_{\Omega_0} \sum_{m=1}^{M} \sum_{p=1}^{N_m} \frac{q_{mp}}{2} \left(\boldsymbol{r}_{mp} \frac{\mathrm{d}\boldsymbol{r}_{mp}}{\mathrm{d}t} - \frac{\mathrm{d}\boldsymbol{r}_{mp}}{\mathrm{d}t} \boldsymbol{r}_{mp} \right) \cdot \nabla_0 \delta(\boldsymbol{r}_0 - \boldsymbol{r}_m)\, \mathrm{d}^3\boldsymbol{r}_0 \tag{3.42}$$

where for a spatially constant vector $\boldsymbol{b}$ and spatially variable scalar field α, we used the identity $\nabla \times (\alpha \boldsymbol{b}) = \nabla \alpha \times \boldsymbol{b}$ in going from the first line to the second and used the identity valid for any three vectors that $\boldsymbol{a} \times (\boldsymbol{b} \times \boldsymbol{c}) = (\boldsymbol{bc} - \boldsymbol{cb}) \cdot \boldsymbol{a}$ in going from the second line to the third.

Next, the time derivatives of $\boldsymbol{P}$ and $\nabla \cdot \boldsymbol{Q}$ can be identified as

$$\frac{\partial \boldsymbol{P}(\boldsymbol{r}, t)}{\partial t} = \frac{1}{V_0} \int_{\Omega_0(\boldsymbol{r})} \sum_{m=1}^{M(\boldsymbol{r})} \frac{\partial}{\partial t} \left[\boldsymbol{p}_m(t)\, \delta(\boldsymbol{r}_0 - \boldsymbol{r}_m(t)) \right] \mathrm{d}^3\boldsymbol{r}_0 \tag{3.43}$$

$$= \frac{1}{V_0} \int_{\Omega_0(\boldsymbol{r})} \sum_{m=1}^{M(\boldsymbol{r})} \left[\frac{\mathrm{d}\boldsymbol{p}_m}{\mathrm{d}t} \delta(\boldsymbol{r}_0 - \boldsymbol{r}_m(t)) - \boldsymbol{p}_m(t) \boldsymbol{v}_m(t) \cdot \nabla_0 \delta(\boldsymbol{r}_0 - \boldsymbol{r}_m(t)) \right] \mathrm{d}^3\boldsymbol{r}_0 \tag{3.44}$$

and

$$-\frac{\partial}{\partial t} \nabla \cdot \boldsymbol{Q}(\boldsymbol{r}, t) = -\frac{1}{V_0} \int_{\Omega_0(\boldsymbol{r})} \sum_{m=1}^{M(\boldsymbol{r})} \sum_{p=1}^{N_m} \frac{\partial}{\partial t} \left[\frac{q_{mp}}{2} \boldsymbol{r}_{mp}(t) \boldsymbol{r}_{mp}(t) \cdot \nabla_0 \delta(\boldsymbol{r}_0 - \boldsymbol{r}_m(t)) \right] \mathrm{d}^3\boldsymbol{r}_0 \tag{3.45}$$

$$= \frac{1}{V_0} \int_{\Omega_0(\boldsymbol{r})} \sum_{m=1}^{M(\boldsymbol{r})} \sum_{p=1}^{N_m} \frac{q_{mp}}{2} \left[- \left(\frac{\mathrm{d}\boldsymbol{r}_{mp}(t)}{\mathrm{d}t} \boldsymbol{r}_{mp}(t) + \boldsymbol{r}_{mp}(t) \frac{\mathrm{d}\boldsymbol{r}_{mp}(t)}{\mathrm{d}t} \right) \cdot \nabla_0 \delta(\boldsymbol{r}_0 - \boldsymbol{r}_m(t)) \right.$$

$$\left. + \boldsymbol{r}_{mp}(t) \boldsymbol{r}_{mp}(t) \cdot \nabla_0 \nabla_0 \delta(\boldsymbol{r}_0 - \boldsymbol{r}_m(t)) \cdot \boldsymbol{v}_m \right] \mathrm{d}^3\boldsymbol{r}_0. \tag{3.46}$$

In each of these expressions, we distributed the time derivative in the first line and used the result that

$$\frac{\partial}{\partial t} \delta(\boldsymbol{r} - \boldsymbol{r}_m(t)) = -\nabla \delta(\boldsymbol{r} - \boldsymbol{r}_m(t)) \cdot \frac{\mathrm{d}\boldsymbol{r}_m(t)}{\mathrm{d}t} \tag{3.47}$$

to get to the second line.

Given the above, the various terms in Eq. (3.31) can be identified as terms within the expressions for $\boldsymbol{J}$, $\nabla \times \boldsymbol{M}$, $\partial \boldsymbol{P}/\partial t$, and $\partial \nabla \cdot \boldsymbol{Q}/\partial t$ to give the average of the local current density in the form

$$\bar{\boldsymbol{j}} = \boldsymbol{J} + \frac{\partial}{\partial t} (\boldsymbol{P} - \nabla \cdot \boldsymbol{Q}) + \nabla \times \boldsymbol{M}$$

$$+ \nabla \cdot \left[\frac{1}{V_0} \int_{\Omega_0} \sum_{m=1}^{M} (\boldsymbol{v}_m \boldsymbol{p}_m - \boldsymbol{p}_m \boldsymbol{v}_m)\, \delta(\boldsymbol{r}_0 - \boldsymbol{r}_m)\, \mathrm{d}^3\boldsymbol{r}_0 \right]$$

$$+ \frac{1}{V_0} \int_{\Omega_0} \sum_{m=1}^{M} \left[\boldsymbol{v}_m \boldsymbol{Q}_m : \nabla_0 \nabla_0 \delta(\boldsymbol{r}_0 - \boldsymbol{r}_m) - \boldsymbol{Q}_m \cdot \nabla_0 \nabla_0 \delta(\boldsymbol{r}_0 - \boldsymbol{r}_m) \cdot \boldsymbol{v}_m \right] \mathrm{d}^3\boldsymbol{r}_0. \tag{3.48}$$

The last two remaining integrals can be further manipulated.

To do so, we again write the molecular velocity as $\boldsymbol{v}_m = \boldsymbol{v} + \boldsymbol{v}_{em} + \delta \boldsymbol{v}_m$ and argue that only the average molecular velocity $\boldsymbol{v}$ will contribute significantly to the integrals of Eq. (3.48) because the $\boldsymbol{v}_{em}$ are either zero for molecules with no net charge or are both positive and negative depending on the ionic charge q_m, which cancel in their contribution to $\bar{\boldsymbol{j}}$. In this case, if we use the identity $\nabla \times (\boldsymbol{a} \times \boldsymbol{b}) = \nabla \cdot (\boldsymbol{ba} - \boldsymbol{ab})$, we can write

$$\nabla \times (\boldsymbol{P} \times \boldsymbol{v}) = \nabla \cdot (\boldsymbol{vP} - \boldsymbol{Pv}) = \nabla \cdot \left[\frac{1}{V_0} \int_{\Omega_0} \sum_{m=1}^{M} (\boldsymbol{v}\boldsymbol{p}_m - \boldsymbol{p}_m \boldsymbol{v})\, \delta(\boldsymbol{r}_0 - \boldsymbol{r}_m)\, \mathrm{d}^3\boldsymbol{r}_0 \right] \tag{3.49}$$

and

$$-\nabla\times\left[(\nabla\cdot\boldsymbol{Q})\times\boldsymbol{v})\right]=\nabla\cdot\left[(\nabla\cdot\boldsymbol{Q})\boldsymbol{v}-\boldsymbol{v}(\nabla\cdot\boldsymbol{Q})\right] \tag{3.50}$$

$$=\frac{1}{V_0}\int_{\Omega_0}\sum_{m=1}^{M}\left[\boldsymbol{v}\boldsymbol{Q}_m:\nabla_0\nabla_0\delta(\boldsymbol{r}_0-\boldsymbol{r}_m)-\boldsymbol{Q}_m\cdot\nabla_0\nabla_0\delta(\boldsymbol{r}_0-\boldsymbol{r}_m)\cdot\boldsymbol{v}\right]\,\mathrm{d}^3\boldsymbol{r}_0, \tag{3.51}$$

where we used that $\nabla\cdot\overline{\boldsymbol{a}}=\overline{\nabla_0\cdot\boldsymbol{a}}$, $\boldsymbol{Q}$ is symmetric and only the Dirac has spatial dependence.

Using $\boldsymbol{D}-\epsilon_0\overline{\boldsymbol{E}}=\boldsymbol{P}-\nabla\cdot\boldsymbol{Q}$, we thus obtain that

$$\bar{\boldsymbol{j}}=\boldsymbol{J}+\frac{\partial}{\partial t}\left(\boldsymbol{D}-\epsilon_0\overline{\boldsymbol{E}}\right)+\nabla\times\left[\boldsymbol{M}+\left(\boldsymbol{D}-\epsilon_0\overline{\boldsymbol{E}}\right)\times\boldsymbol{v}\right]. \tag{3.52}$$

So upon returning to the averaged form of Ampère's law $\mu_0^{-1}\nabla\times\overline{\boldsymbol{B}}=\epsilon_0\partial\overline{\boldsymbol{E}}/\partial t+\bar{\boldsymbol{j}}$, we obtain

$$\nabla\times\left[\frac{1}{\mu_0}\overline{\boldsymbol{B}}-\boldsymbol{M}-(\boldsymbol{D}-\epsilon_0\overline{\boldsymbol{E}})\times\boldsymbol{v}\right]=\frac{\partial\boldsymbol{D}}{\partial t}+\boldsymbol{J}. \tag{3.53}$$

The term in brackets on the left-hand side is now called $\boldsymbol{H}$ (units of C/s/m)

$$\boldsymbol{H}=\frac{1}{\mu_0}\overline{\boldsymbol{B}}-\boldsymbol{M}-(\boldsymbol{D}-\epsilon_0\overline{\boldsymbol{E}})\times\boldsymbol{v}\,\widehat{=}\,\textit{the magnetic field}. \tag{3.54}$$

As in the earlier relation between $\overline{\boldsymbol{E}}$ and $\boldsymbol{D}$, it is informative to rewrite this as an expression for $\overline{\boldsymbol{B}}$ (units of T = kg/C/s)

$$\overline{\boldsymbol{B}}=\mu_0\left[\boldsymbol{H}+\boldsymbol{M}+(\boldsymbol{P}-\nabla\cdot\boldsymbol{Q})\times\boldsymbol{v}\right]\,\widehat{=}\,\textit{the magnetic induction}. \tag{3.55}$$

Historically, and as indicated earlier, the name given to $\boldsymbol{H}$ is *the magnetic field* and that given to $\overline{\boldsymbol{B}}$ is *the magnetic induction* (units of T). If possible to rename these fields, in analogy to $\boldsymbol{D}$ and $\overline{\boldsymbol{E}}$, it would perhaps be better to call $\mu_0\boldsymbol{H}$ the applied magnetic field and call $\overline{\boldsymbol{B}}$ the total averaged magnetic field, which is the sum of the applied magnetic field and the various magnetic polarization fields.

So the final macroscopic form of Ampère's law is written compactly as

$$\nabla\times\boldsymbol{H}=\frac{\partial\boldsymbol{D}}{\partial t}+\boldsymbol{J} \tag{3.56}$$

and is seen to be a relation between the applied fields generated through induction by a current density $\boldsymbol{J}=\rho_e\boldsymbol{v}+\boldsymbol{J}_{em}$.

To conclude this section, we drop the overbars on $\overline{\boldsymbol{E}}$ and $\overline{\boldsymbol{B}}$ denoting averaging to give the continuum (or macroscopic) Maxwell's equations in a stationary frame of reference as

$$\nabla \cdot \boldsymbol{B} = 0 \tag{3.57}$$

$$\nabla \cdot \boldsymbol{D} = \rho_e \tag{3.58}$$

$$\nabla \times \boldsymbol{E} = -\frac{\partial \boldsymbol{B}}{\partial t} \tag{3.59}$$

$$\nabla \times \boldsymbol{H} = \frac{\partial \boldsymbol{D}}{\partial t} + \boldsymbol{J}_{em} + \rho_e \boldsymbol{v} \tag{3.60}$$

Equations (3.59) and (3.60) are two vector equations for the four unknown EM vectors $\boldsymbol{E}$, $\boldsymbol{D}$, $\boldsymbol{H}$, and $\boldsymbol{B}$ and the unknown electromigration current density $\boldsymbol{J}_{em}$. The unknown charge-excess density ρ_e is determined from Eq. (3.58) once $\boldsymbol{D}$ is known and finally Eq. (3.57) can be thought to be implicit within Eq. (3.59) when the fields are varying in time. To be able to solve the macroscopic Maxwell's equations we need three additional relations, called constitutive laws, between the vectors $\boldsymbol{E}, \boldsymbol{D}, \boldsymbol{H}, \boldsymbol{B}$, and $\boldsymbol{J}_{em}$

The difference between the electric fields $\boldsymbol{D}$ and $\boldsymbol{E}$ and the magnetic fields $\boldsymbol{H}$ and $\boldsymbol{B}$ are due to polarization of the material as the definitions

$$\epsilon_0 \boldsymbol{E} = \boldsymbol{D} - \boldsymbol{P} + \nabla \cdot \boldsymbol{Q} \tag{3.61}$$

$$\frac{1}{\mu_0} \boldsymbol{B} = \boldsymbol{H} + \boldsymbol{M} + (\boldsymbol{P} - \nabla \cdot \boldsymbol{Q}) \times \boldsymbol{v} \tag{3.62}$$

indicate. The induced polarization characterized by $-\boldsymbol{P}$ (dipolar electric polarization), $\nabla \cdot \boldsymbol{Q}$ (quadrupolar electric polarization), and $\boldsymbol{M}$ (dipolar magnetic polarization) are reversible processes that require modeling through equilibrium-thermodynamics constitutive laws. Further, the electromigration accounted for by $\boldsymbol{J}_{em}$ is an irreversible (heat generating) process that also needs its own distinct type of constitutive law as we consider in the sections that follow.

3.2 The Maxwell Stress Tensor

The macroscopic EM body force in the continuum theory of electromagnetism is given by

$$\boldsymbol{F}_b = \overline{\alpha} \boldsymbol{E} + \bar{\boldsymbol{j}} \times \boldsymbol{B}. \tag{3.63}$$

This is the force density (N m^{-3}) that the total electric and magnetic fields $\boldsymbol{E}$ and $\boldsymbol{B}$ exert on each element of a body. Using Eq. (3.21) for the macroscopic charge density $\overline{\alpha}$ and Eq. (3.52) for the macroscopic current density $\bar{\boldsymbol{j}}$ then gives

$$\begin{aligned} \boldsymbol{F}_b = & \left[\rho_e - \nabla \cdot (\boldsymbol{P} - \nabla \cdot \boldsymbol{Q})\right] \boldsymbol{E} \\ & + \left\{ \rho_e \boldsymbol{v} + \boldsymbol{J}_{em} + \frac{\partial (\boldsymbol{P} - \nabla \cdot \boldsymbol{Q})}{\partial t} + \nabla \times \left[\boldsymbol{M} + (\boldsymbol{P} - \nabla \cdot \boldsymbol{Q}) \times \boldsymbol{v}\right] \right\} \times \boldsymbol{B}. \end{aligned} \tag{3.64}$$

This expression generalizes the earlier development of Chapter 2 that did not include either the *convective bound-electric-current density* vector $\nabla \times \left[(\boldsymbol{P} - \nabla \cdot \boldsymbol{Q}) \times \boldsymbol{v}\right]$ or *quadrupolar electric-polarization* vector $\nabla \cdot \boldsymbol{Q}$. By combining Eqs (3.57)–(3.62), you can prove as an end-of-chapter exercise that these EM body forces can be written exactly using only the vectors $\boldsymbol{E}$ and $\boldsymbol{B}$ as

$$\boldsymbol{F}_b = \epsilon_0 (\nabla \cdot \boldsymbol{E})\boldsymbol{E} + \frac{1}{\mu_0}(\nabla \times \boldsymbol{B}) \times \boldsymbol{B} - \epsilon_0 \frac{\partial \boldsymbol{E}}{\partial t} \times \boldsymbol{B}. \tag{3.65}$$

If we further employ the tensor-product identities of Eqs (1.100) and (1.101) from Chapter 1, this expression can further be rewritten exactly as

$$\boxed{\boldsymbol{F}_b = \nabla \cdot \boldsymbol{\sigma}_M - \epsilon_0 \frac{\partial}{\partial t}(\boldsymbol{E} \times \boldsymbol{B}),} \tag{3.66}$$

where the symmetric second-order tensor $\boldsymbol{\sigma}_M$ is defined

$$\boxed{\boldsymbol{\sigma}_M = \epsilon_0 \left(\boldsymbol{E}\boldsymbol{E} - \frac{(\boldsymbol{E} \cdot \boldsymbol{E})\boldsymbol{I}}{2} \right) + \frac{1}{\mu_0} \left(\boldsymbol{B}\boldsymbol{B} - \frac{(\boldsymbol{B} \cdot \boldsymbol{B})\boldsymbol{I}}{2} \right)} \tag{3.67}$$

and is called the *Maxwell stress tensor*, where $\boldsymbol{I}$ is the second-order identity tensor. The divergence of the Maxwell stress tensor, along with the time derivative of $-\epsilon_0 \boldsymbol{E} \times \boldsymbol{B}$, is the most general compact expression for the EM body forces in our macroscopic continuum mechanics and allows for all conceivable electrical and magnetic polarization processes. That said, depending on the application, it may be more convenient to use the portions of Eq. (3.64) that apply to that application.

3.3 Electromagnetic Constitutive Laws

Constitutive laws always derive from the thermodynamics of either reversible processes (energy being stored or released reversibly through a sequence of equilibrium states) or irreversible processes (heat being generated irreversibly during out-of-equilibrium transport). Polarization of the molecules, either electric or magnetic, is well approximated as being a reversible process. If you turn off the applied fields, the molecules will return to their original unpolarized state with no change in the system's heat. As such, the constitutive laws that allow for the polarization of matter derive from equilibrium thermodynamics. Similarly, the electromigration of charge $\boldsymbol{J}_{em}$ relative to the collective movement $\boldsymbol{v}$ of the molecules derives from the postulates of nonequilibrium thermodynamics as will be shown below and generates heat irreversibly other than in some materials that exhibit electronic superconductivity at extremely low temperatures. If there is heterogeneity within an averaging element, a process like the electromigration of ions can result in ionic accumulation and depletion (i.e., polarization) at the scale of the heterogeneity so that both reversible and irreversible processes are occurring simultaneously. For now, we assume that polarization is purely reversible and that electromigration of charge is purely irreversible, in which case there is no time delay between a time-varying applied field and the time-varying response field. Later in Chapter 11, we will show how to allow for delays between a force and a response in the constitutive laws.

3.3.1 EM in Moving Frames of Reference

As a general rule, we formulate continuum thermodynamics in a frame of reference that moves along with each element of constant mass that is under consideration. The above continuum form of Maxwell's equations was derived in a frame of reference that is stationary. If instead we consider a frame of reference that moves with the center-of-mass velocity $\boldsymbol{v} = \mathrm{d}\boldsymbol{r}(t)/\mathrm{d}t$ of a mass element, it is assumed that the form of Maxwell's equations for the element in a uniformly moving frame is the same as that in the stationary frame, which is called *Galilean invariance*. However, the EM fields in the moving frame, $\boldsymbol{B}'(\boldsymbol{r}(t), t)$, $\boldsymbol{E}'(\boldsymbol{r}(t), t)$, $\boldsymbol{H}'(\boldsymbol{r}(t), t)$, and $\boldsymbol{D}'(\boldsymbol{r}(t), t)$, are possibly distinct from those in the stationary frame, $\boldsymbol{B}(\boldsymbol{r}, t)$, $\boldsymbol{E}(\boldsymbol{r}, t)$, $\boldsymbol{H}(\boldsymbol{r}, t)$, and $\boldsymbol{D}(\boldsymbol{r}, t)$, in a manner that we will derive below.

In the moving frame, in which spatial derivatives are taken with respect to $\boldsymbol{r}(t)$ at each instant, Galilean invariance says that Maxwell's equations have the form

$$\nabla \times \boldsymbol{E}'(\boldsymbol{r}(t), t) = -\frac{\mathrm{d}\boldsymbol{B}'(\boldsymbol{r}(t), t)}{\mathrm{d}t} \tag{3.68}$$

$$\nabla \times \boldsymbol{H}'(\boldsymbol{r}(t), t) = \frac{\mathrm{d}\boldsymbol{D}'(\boldsymbol{r}(t), t)}{\mathrm{d}t} + \boldsymbol{J}_{em}(\boldsymbol{r}(t), t). \tag{3.69}$$

As in Section 2.6, the time derivatives in this moving frame are given in terms of fields and derivatives in the stationary frame as

$$\frac{\mathrm{d}\boldsymbol{B}'(\boldsymbol{r}(t), t)}{\mathrm{d}t} = \frac{\partial \boldsymbol{B}(\boldsymbol{r}, t)}{\partial t} + \boldsymbol{v} \cdot \nabla \boldsymbol{B}(\boldsymbol{r}, t) \tag{3.70}$$

$$\frac{\mathrm{d}\boldsymbol{D}'(\boldsymbol{r}(t), t)}{\mathrm{d}t} = \frac{\partial \boldsymbol{D}(\boldsymbol{r}, t)}{\partial t} + \boldsymbol{v} \cdot \nabla \boldsymbol{D}(\boldsymbol{r}.t), \tag{3.71}$$

where $\boldsymbol{v}$ is the velocity of the moving frame as observed from the stationary frame. These identifications say that

$$\boxed{\boldsymbol{B}' = \boldsymbol{B} \quad \text{and} \quad \boldsymbol{D}' = \boldsymbol{D}.} \tag{3.72}$$

Also note that when moving with the mass element, one does not notice the advective current $\rho_e \boldsymbol{v}$ if the element has a net free charge and this is why it is not present in Eq. (3.69); one only notices the current $\boldsymbol{J}_{em}$ of the charge moving relative to $\boldsymbol{v}$, which is what we call electromigration. If the current density in the stationary frame is given by $\boldsymbol{J}(\boldsymbol{r}, t) = \boldsymbol{J}_{em}(\boldsymbol{r}(t), t) + \rho_e \boldsymbol{v}(\boldsymbol{r}, t)$, the current density in the moving frame is $\boldsymbol{J}'(\boldsymbol{r}(t), t) = \boldsymbol{J}(\boldsymbol{r}, t) - \rho_e \boldsymbol{v}(\boldsymbol{r}, t) = \boldsymbol{J}_{em}(\boldsymbol{r}(t), t)$.

In assuming Galilean invariance, we are assuming that the velocities of the other elements adjacent to the element being analyzed are also moving at $\boldsymbol{v}$, so that the velocity $\boldsymbol{v}$ can be taken as a uniform constant, at least in the neighborhood of each element. As such, when we write the vector identity from Chapter 1 that

$$\nabla \times (\boldsymbol{v} \times \boldsymbol{B}) = \nabla \cdot (\boldsymbol{B}\boldsymbol{v} - \boldsymbol{v}\boldsymbol{B}) \tag{3.73}$$

$$= (\nabla \cdot \boldsymbol{B})\boldsymbol{v} + \boldsymbol{B} \cdot \nabla\boldsymbol{v} - (\nabla \cdot \boldsymbol{v})\boldsymbol{B} - \boldsymbol{v} \cdot \nabla\boldsymbol{B}, \tag{3.74}$$

because $\boldsymbol{v}$ is locally uniform, we have $\nabla \cdot \boldsymbol{v} = 0$ and $\nabla\boldsymbol{v} = 0$. Further, Gauss' law requires that $\nabla \cdot \boldsymbol{B} = 0$ to give

$$\boldsymbol{v} \cdot \nabla\boldsymbol{B} = -\nabla \times (\boldsymbol{v} \times \boldsymbol{B}). \tag{3.75}$$

An identical manipulation on $\nabla \times (\boldsymbol{v} \times \boldsymbol{D})$ with the requirement from Coulomb's law that $\nabla \cdot \boldsymbol{D} = \rho_e$ results in

$$\boldsymbol{v} \cdot \nabla\boldsymbol{D} = -\nabla \times (\boldsymbol{v} \times \boldsymbol{D}) + \rho_e \boldsymbol{v}. \tag{3.76}$$

Using these expressions in Eqs (3.70) and (3.71) and inserting into Eqs (3.68) and (3.69) gives

$$\nabla \times \left[\boldsymbol{E}'(\boldsymbol{r}(t), t) - \boldsymbol{v} \times \boldsymbol{B}(\boldsymbol{r}, t)\right] = -\frac{\partial \boldsymbol{B}(\boldsymbol{r}, t)}{\partial t} \tag{3.77}$$

$$\nabla \times \left[\boldsymbol{H}'(\boldsymbol{r}(t), t) + \boldsymbol{v} \times \boldsymbol{D}(\boldsymbol{r}, t)\right] = \frac{\partial \boldsymbol{D}(\boldsymbol{r}, t)}{\partial t} + \rho_e \boldsymbol{v} + \boldsymbol{J}_{em}(\boldsymbol{r}(t), t). \tag{3.78}$$

If the electric and magnetic fields in the moving frame are identified as

$$\boldsymbol{E}'(\boldsymbol{r}(t), t) = \boldsymbol{E}(\boldsymbol{r}, t) + \boldsymbol{v} \times \boldsymbol{B}(\boldsymbol{r}, t) \tag{3.79}$$

$$\boldsymbol{H}'(\boldsymbol{r}(t), t) = \boldsymbol{H}(\boldsymbol{r}, t) - \boldsymbol{v} \times \boldsymbol{D}(\boldsymbol{r}, t), \tag{3.80}$$

we obtain Maxwell's equations in the stationary frame

$$\nabla \times \boldsymbol{E}(\boldsymbol{r}, t) = -\frac{\partial \boldsymbol{B}(\boldsymbol{r}, t)}{\partial t} \tag{3.81}$$

$$\nabla \times \boldsymbol{H}(\boldsymbol{r}, t) = \frac{\partial \boldsymbol{D}(\boldsymbol{r}, t)}{\partial t} + \rho_e \boldsymbol{v} + \boldsymbol{J}_{em}(\boldsymbol{r}(t), t), \tag{3.82}$$

which are identical to the derived laws of Eqs (3.59) and (3.60). In Eqs (3.79) and (3.80), we have identified how the electric and magnetic fields in the moving frame of reference (left-hand side) are related to the electric and magnetic fields in the stationary frame of reference (right-hand side). The above holds in the limit where the velocity $\boldsymbol{v}$ is much slower than the speed of light. When the frame of reference is moving at a speed approaching the speed of light, the Lorentz transformation and special relativity (e.g., Kong, 1986) must be invoked which results in still different field transformation rules that we will not consider.

If the local velocity field $\boldsymbol{v}$ with which the frame of reference is moving has local gradients that cannot be neglected, Galilean invariance is violated because the frame of reference has a spatially variable velocity. In this case, the macroscopic form of Maxwell's equations in the moving frame takes the different form

$$\nabla \times \boldsymbol{E}' = -\frac{\mathrm{d}\boldsymbol{B}'}{\mathrm{d}t} + \boldsymbol{B}' \cdot [\nabla \boldsymbol{v} - (\nabla \cdot \boldsymbol{v})\boldsymbol{I}] \tag{3.83}$$

$$\nabla \times \boldsymbol{H}' = \frac{\mathrm{d}\boldsymbol{D}'}{\mathrm{d}t} - \boldsymbol{D}' \cdot [\nabla \boldsymbol{v} - (\nabla \cdot \boldsymbol{v})\boldsymbol{I}] + \boldsymbol{J}_{em}. \tag{3.84}$$

If we again identify $\boldsymbol{D}' = \boldsymbol{D}$ and $\boldsymbol{B}' = \boldsymbol{B}$ and go through the same analysis as above but now with $\nabla \boldsymbol{v} - (\nabla \cdot \boldsymbol{v})\boldsymbol{I}$ not taken to be zero, we again arrive at $\boldsymbol{E}' = \boldsymbol{E} + \boldsymbol{v} \times \boldsymbol{B}$ and $\boldsymbol{H}' = \boldsymbol{H} - \boldsymbol{v} \times \boldsymbol{D}$ and the required form of the Maxwell equations in the stationary frame of reference.

3.3.2 The Thermodynamics of Electromagnetic Fields

Given the above formulation, we now address the EM energy balance of a mass element from a frame of reference that is moving with the mass element in a flow field $\boldsymbol{v}$ that is locally uniform. To do so, we form the following dot products:

$$\boldsymbol{H}' \cdot \left[\nabla \times \boldsymbol{E}' = -\frac{\mathrm{d}\boldsymbol{B}'}{\mathrm{d}t}\right] \tag{3.85}$$

$$\boldsymbol{E}' \cdot \left[\nabla \times \boldsymbol{H}' = \frac{\mathrm{d}\boldsymbol{D}'}{\mathrm{d}t} + \boldsymbol{J}_{em}\right]. \tag{3.86}$$

Upon subtracting the first line from the second and using the identity $\nabla \cdot (\boldsymbol{a} \times \boldsymbol{b}) = \boldsymbol{b} \cdot (\nabla \times \boldsymbol{a}) - \boldsymbol{a} \cdot (\nabla \times \boldsymbol{b})$ we obtain the EM energy balance in a frame that is moving with the mass element

$$\boxed{\frac{\mathrm{d}\tilde{u}_{em}}{\mathrm{d}t} + \boldsymbol{J}_{em} \cdot \boldsymbol{E}' = -\nabla \cdot (\boldsymbol{E}' \times \boldsymbol{H}'),} \tag{3.87}$$

where we identify

$$\frac{\mathrm{d}\tilde{u}_{em}}{\mathrm{d}t} \mathrel{\widehat{=}} \boldsymbol{E}' \cdot \frac{\mathrm{d}\boldsymbol{D}'}{\mathrm{d}t} + \boldsymbol{H}' \cdot \frac{\mathrm{d}\boldsymbol{B}'}{\mathrm{d}t} \tag{3.88}$$

as the rate at which energy per unit volume is being stored reversibly in the EM fields from a frame that is moving with the mass element. The energy density $\tilde{u}_{em}$ has units of energy per unit volume. In the present chapter, we will neglect mechanical deformation, even if we are allowing for the rigid-body translation $\boldsymbol{v}$ of each mass element, and will assume that the only source of heat change is the irreversible heating caused by the electromigration of charge as quantified by $\boldsymbol{J}_{em} \cdot \boldsymbol{E}'$, which is also called *Joule heating*.

In this context, the internal energy per unit reference volume $\tilde{u}$ of a mass element is changing as

$$\frac{\mathrm{d}\tilde{u}}{\mathrm{d}t} = T\frac{\mathrm{d}\tilde{s}}{\mathrm{d}t} + \frac{\mathrm{d}\tilde{u}_{em}}{\mathrm{d}t}, \tag{3.89}$$

which can be called the *first law of thermodynamics*, where T is temperature and $\tilde{s}$ is the entropy per unit reference volume as will be discussed at length in Chapter 6. Because mechanical deformation is not being allowed for here, the reference volume is simply the constant volume surrounding the mass element. But in Chapter 6, we will allow for both volumetric and shear deformation and show how such deformation couples to the EM fields. The Joule heating, or irreversible heat loss, is given by the statement

$$T\frac{\mathrm{d}\tilde{s}}{\mathrm{d}t} = \boldsymbol{J}_{em}\cdot\boldsymbol{E}' \geq 0, \tag{3.90}$$

which can be called the *second law of thermodynamics*. Here, the only source of irreversible heating given on the right-hand side of this law is the electromigration of charge; however, a host of other nonequilibrium transport processes can cause entropy to irreversibly increase as will be discussed in Chapter 7. Note that entropy can also be changing if more (or less) heat is flowing into an element than is flowing out but, in this chapter, we are neglecting that reason for entropy change as well, which can be either positive or negative depending on the spatial variation of the heat flow.

Because mechanical deformation is not being allowed for, the only way that internal energy is changing is if there is more EM energy fluxing into a mass element than fluxing out. We thus have that

$$\frac{\mathrm{d}\tilde{u}}{\mathrm{d}t} = -\nabla\cdot\left(\boldsymbol{E}'\times\boldsymbol{H}'\right), \tag{3.91}$$

which is identical to Eq. (3.87) after inserting Eqs (3.89) and (3.90). We call the vector $\boldsymbol{s} = \boldsymbol{E}'\times\boldsymbol{H}' = (\boldsymbol{E}+\boldsymbol{v}\times\boldsymbol{B})\times(\boldsymbol{H}-\boldsymbol{v}\times\boldsymbol{D})$ the *electromagnetic-energy flux vector* or *electromagnetic Poynting vector*. The Poynting vector represents the flux of energy across a surface perpendicular to its direction from a frame of reference that is moving with the surface. We saw the mechanical energy Poynting vector in Section 2.9.2 and will return to the Poynting vector again in Chapter 4.

3.3.3 The Reversible Processes of EM Polarization

Equation (3.88) for $\mathrm{d}\tilde{u}_{em}/\mathrm{d}t$ implies that $\tilde{u}_{em}$ has the functional dependence $\tilde{u}_{em} = \tilde{u}_{em}(\boldsymbol{D}', \boldsymbol{B}')$ because if we take the total time derivative, we obtain

$$\frac{\mathrm{d}\tilde{u}_{em}(\boldsymbol{D}',\boldsymbol{B}')}{\mathrm{d}t} = \frac{\partial\tilde{u}_{em}(\boldsymbol{D}',\boldsymbol{B}')}{\partial\boldsymbol{D}'}\cdot\frac{\mathrm{d}\boldsymbol{D}'}{\mathrm{d}t} + \frac{\partial\tilde{u}_{em}(\boldsymbol{D}',\boldsymbol{B}')}{\partial\boldsymbol{B}'}\cdot\frac{\mathrm{d}\boldsymbol{B}'}{\mathrm{d}t}, \tag{3.92}$$

which is the same as Eq. (3.88) so long as we identify $\boldsymbol{E}'$ and $\boldsymbol{H}'$ as

$$\boldsymbol{E}' = \frac{\partial\tilde{u}_{em}}{\partial\boldsymbol{D}'} \mathrel{\widehat{=}} \frac{\partial\tilde{u}_{em}}{\partial D_i'}\hat{\boldsymbol{x}}_i \quad\text{and}\quad \boldsymbol{H}' = \frac{\partial\tilde{u}_{em}}{\partial\boldsymbol{B}'} \mathrel{\widehat{=}} \frac{\partial\tilde{u}_{em}}{\partial B_i'}\hat{\boldsymbol{x}}_i \tag{3.93}$$

with summation over the index i. If we then take total time derivatives of $\boldsymbol{E}'$ and $\boldsymbol{H}'$, we end up with EM constitutive laws in the form

$$\frac{\mathrm{d}\boldsymbol{E}'}{\mathrm{d}t} = \boldsymbol{C}_{ED} \cdot \frac{\mathrm{d}\boldsymbol{D}'}{\mathrm{d}t} + \boldsymbol{C}_{EB} \cdot \frac{\mathrm{d}\boldsymbol{B}'}{\mathrm{d}t} \tag{3.94}$$

$$\frac{\mathrm{d}\boldsymbol{H}'}{\mathrm{d}t} = \boldsymbol{C}_{HD} \cdot \frac{\mathrm{d}\boldsymbol{D}'}{\mathrm{d}t} + \boldsymbol{C}_{HB} \cdot \frac{\mathrm{d}\boldsymbol{B}'}{\mathrm{d}t}, \tag{3.95}$$

where the various material properties, that implicitly allow for polarization, are symmetric second-order tensors defined as

$$\boldsymbol{C}_{ED} = \frac{\partial \boldsymbol{E}'}{\partial \boldsymbol{D}'} = \frac{\partial^2 \tilde{u}_{em}}{\partial \boldsymbol{D}' \partial \boldsymbol{D}'} = \boldsymbol{C}_{ED}^T \,\widehat{=}\, \text{electrical-impermittivity tensor} \tag{3.96}$$

$$\boldsymbol{C}_{HB} = \frac{\partial \boldsymbol{H}'}{\partial \boldsymbol{B}'} = \frac{\partial^2 \tilde{u}_{em}}{\partial \boldsymbol{B}' \partial \boldsymbol{B}'} = \boldsymbol{C}_{HB}^T \,\widehat{=}\, \text{magnetic-impermeability tensor} \tag{3.97}$$

$$\boldsymbol{C}_{HD} = \frac{\partial \boldsymbol{H}'}{\partial \boldsymbol{D}'} = \frac{\partial^2 \tilde{u}_{em}}{\partial \boldsymbol{B}' \partial \boldsymbol{D}'} = \left(\frac{\partial^2 \tilde{u}_{em}}{\partial \boldsymbol{D}' \partial \boldsymbol{B}'}\right)^T = \left(\frac{\partial \boldsymbol{E}'}{\partial \boldsymbol{B}'}\right)^T = \boldsymbol{C}_{EB}^T \,\widehat{=}\, \text{magnetoelectric tensor} \tag{3.98}$$

The electrical-impermittivity and magnetic-impermeability tensors being defined here are the inverse tensors of the electrical-permittivity and magnetic-permeability tensors.

If we now invoke the moving-frame field transformations of Section 3.3.1

$$\frac{\mathrm{d}\boldsymbol{E}'}{\mathrm{d}t} = \frac{\mathrm{d}\boldsymbol{E}}{\mathrm{d}t} + \boldsymbol{v} \times \frac{\mathrm{d}\boldsymbol{B}}{\mathrm{d}t} = \frac{\partial \boldsymbol{E}}{\partial t} + \boldsymbol{v} \cdot \nabla \boldsymbol{E} + \boldsymbol{v} \times \left(\frac{\partial \boldsymbol{B}}{\partial t} + \boldsymbol{v} \cdot \nabla \boldsymbol{B}\right), \tag{3.99}$$

$$\frac{\mathrm{d}\boldsymbol{H}'}{\mathrm{d}t} = \frac{\mathrm{d}\boldsymbol{H}}{\mathrm{d}t} - \boldsymbol{v} \times \frac{\mathrm{d}\boldsymbol{D}}{\mathrm{d}t} = \frac{\partial \boldsymbol{H}}{\partial t} + \boldsymbol{v} \cdot \nabla \boldsymbol{H} - \boldsymbol{v} \times \left(\frac{\partial \boldsymbol{D}}{\partial t} + \boldsymbol{v} \cdot \nabla \boldsymbol{D}\right), \tag{3.100}$$

$$\frac{\mathrm{d}\boldsymbol{D}'}{\mathrm{d}t} = \frac{\partial \boldsymbol{D}}{\partial t} + \boldsymbol{v} \cdot \nabla \boldsymbol{D}, \tag{3.101}$$

$$\frac{\mathrm{d}\boldsymbol{B}'}{\mathrm{d}t} = \frac{\partial \boldsymbol{B}}{\partial t} + \boldsymbol{v} \cdot \nabla \boldsymbol{B}, \tag{3.102}$$

the EM constitutive laws of Eqs (3.94) and (3.95) are given in a stationary frame of reference as

$$\frac{\partial \boldsymbol{E}}{\partial t} + \boldsymbol{v} \cdot \nabla \boldsymbol{E} + \boldsymbol{v} \times \left(\frac{\partial \boldsymbol{B}}{\partial t} + \boldsymbol{v} \cdot \nabla \boldsymbol{B}\right) = \boldsymbol{C}_{ED} \cdot \left(\frac{\partial \boldsymbol{D}}{\partial t} + \boldsymbol{v} \cdot \nabla \boldsymbol{D}\right) + \boldsymbol{C}_{EB} \cdot \left(\frac{\partial \boldsymbol{B}}{\partial t} + \boldsymbol{v} \cdot \nabla \boldsymbol{B}\right) \tag{3.103}$$

$$\frac{\partial \boldsymbol{H}}{\partial t} + \boldsymbol{v} \cdot \nabla \boldsymbol{H} - \boldsymbol{v} \times \left(\frac{\partial \boldsymbol{D}}{\partial t} + \boldsymbol{v} \cdot \nabla \boldsymbol{D}\right) = \boldsymbol{C}_{EB}^T \cdot \left(\frac{\partial \boldsymbol{D}}{\partial t} + \boldsymbol{v} \cdot \nabla \boldsymbol{D}\right) + \boldsymbol{C}_{HB} \cdot \left(\frac{\partial \boldsymbol{B}}{\partial t} + \boldsymbol{v} \cdot \nabla \boldsymbol{B}\right). \tag{3.104}$$

These constitutive laws for reversible polarization processes will be slightly different if mass density is changing through time and if coupling to elastic deformation and heat changes are allowed for as will be shown in Chapter 6. But when the only energy being accounted for is EM, these laws are exact statements, involving tensorial material properties $\boldsymbol{C}_{ED}$, $\boldsymbol{C}_{HB}$, and $\boldsymbol{C}_{EB}$ that must be measured experimentally. The inverse-permittivity or *impermittivity* tensor $\boldsymbol{C}_{ED}$, inverse-permeability or *impermeability* tensor $\boldsymbol{C}_{HB}$, and magnetoelectric polarization tensor $\boldsymbol{C}_{EB}$ are allowing for induced dipolar, quadrupolar, and higher-order electric polarization as well as induced magnetic polarization and circulation of polarized charge. In Section 3.7, we will show that the velocity terms in the constitutive equations are, most commonly, quite negligible.

Media that have nonzero magnetoelectric coupling as characterized by $\boldsymbol{C}_{EB}$ are called *bianisotropic*. Examples of bianisotropic media include chromium oxide and various iron oxides. However, for most media, such magnetoelectric coupling is entirely negligible and, as such, we will take $\boldsymbol{C}_{EB} = 0$ in what follows.

Further, if the remaining impermittivity and impermeability tensors are not changing significantly through time (i.e., are not varying with the time-varying EM fields), the constitutive laws of Eqs (3.94) and (3.95) can be time integrated to give $\boldsymbol{E}' = \boldsymbol{C}_{ED} \cdot \boldsymbol{D}'$ and $\boldsymbol{H}' = \boldsymbol{C}_{HB} \cdot \boldsymbol{B}'$, which are the EM constitutive laws written in so-called inverse form (for reasons discussed in Chapter 11 when delayed linear response is allowed for). Under these conditions, the constitutive laws can be written in their normal form and in the stationary frame of reference as

$$\boldsymbol{D} = \epsilon_0 \boldsymbol{K}_e \cdot (\boldsymbol{E} + \boldsymbol{v} \times \boldsymbol{B}) \tag{3.105}$$

$$\boldsymbol{B} = \mu_0 \boldsymbol{K}_m \cdot (\boldsymbol{H} - \boldsymbol{v} \times \boldsymbol{D}), \tag{3.106}$$

where the dimensionless second-order tensor $\boldsymbol{K}_e$ $(= \boldsymbol{C}_{ED}^{-1}/\epsilon_0)$ is called the *dielectric-constant* tensor and the dimensionless second-order tensor $\boldsymbol{K}_m$ $(= \boldsymbol{C}_{HB}^{-1}/\mu_0)$ is called the *relative magnetic-permeability* tensor. If the response is isotropic, meaning the vectors $\boldsymbol{D}$ and $\boldsymbol{E}$ (and $\boldsymbol{B}$ and $\boldsymbol{H}$) are always oriented in the same direction, the second-order tensors can be replaced by scalars $\boldsymbol{D} = \epsilon_0 \kappa_e (\boldsymbol{E} + \boldsymbol{v} \times \boldsymbol{B})$ and $\boldsymbol{B} = \mu_0 \kappa_m (\boldsymbol{H} - \boldsymbol{v} \times \boldsymbol{D})$, where κ_e is the scalar dielectric constant and κ_m the scalar relative magnetic permeability. We also commonly write $\varepsilon = \epsilon_o \kappa_e$ and $\mu = \mu_0 \kappa_m$ for the permittivity and permeability of an isotropic material.

The above formal manipulations, that remarkably are based only on the macroscopic form of Maxwell's equations as coupled to energy interpretations from thermodynamics, have yielded the form of the constitutive laws associated with the polarization processes in media moving with uniform velocity $\boldsymbol{v}$. But we have not gained much insight into the nature of polarization.

3.3.4 Qualitative Discussion of Electric Polarization

For our pedagogic aims here, we focus only on induced dipolar charge separation and ignore quadrupolar and higher-order electric polarization. We will also assume the velocity of the material is small enough that its influence on the polarization process can be ignored, conditions for which are derived in Section 3.7. To qualitatively understand dipolar electric polarization, we return to the relation between the total electric field $\boldsymbol{E}$, applied field $\boldsymbol{D}$, and the polarization field $\boldsymbol{P}$ due to charge becoming separated in each molecule

$$\underset{\substack{\uparrow \\ \text{total}}}{\epsilon_0 \boldsymbol{E}} = \underset{\substack{\uparrow \\ \text{applied}}}{\boldsymbol{D}} - \underset{\substack{\uparrow \\ \text{polarization}}}{\boldsymbol{P}}. \tag{3.107}$$

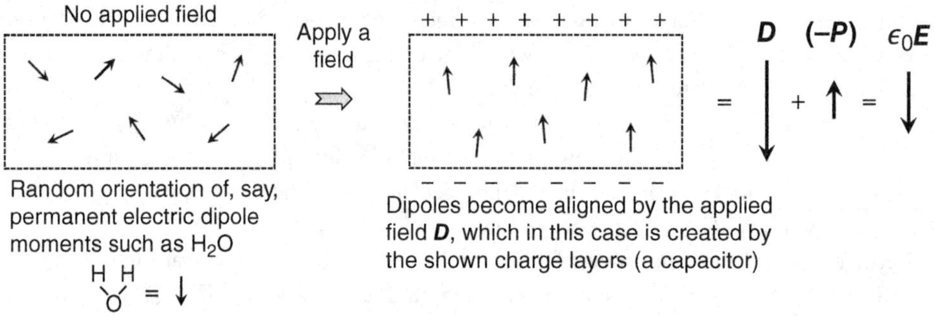

Figure 3.1 Schematic of the alignment of electric dipoles under an applied field $\boldsymbol{D}$ to create the polarization field $-\boldsymbol{P}$ that is in the direction opposite to $\boldsymbol{D}$. All arrows correspond to electric-field directions (which is the opposite of the dipole-moment direction in a polarized molecule).

When an electric field $\boldsymbol{D}$ is applied to a bunch of molecules, any permanent electric dipoles (like in a water molecule) become more or less aligned. Further, within each atom, electrons are pushed more to one side of the nucleus compared to the other. This charge separation in each atom or molecule coherently adds up under the influence of the applied field $\boldsymbol{D}$ to give an additional net electric field $-\boldsymbol{P}$ called the *polarization field* as depicted in Fig. 3.1.

Note that the direction of an electric field as denoted by the arrows in Fig. 3.1 is determined by the direction that a positive test charge would move. And as we already noted in Chapter 2, by convention the dipole moment created within each atom by the applied field (or due to permanent dipoles moments in molecules such as water) is in the direction opposite to the electric field between the separated charge centers. The polarization vector $\boldsymbol{P}$ is defined to be the sum of all the molecular dipole moments in an element (and then divided by the volume of the element) and is thus oriented in the direction opposite to the electric field associated with the net charge separation of the electrical polarization, which explains the negative sign on the polarization vector in $\epsilon_0 \boldsymbol{E} = \boldsymbol{D} - \boldsymbol{P}$.

The electric dipoles of the molecules (either permanent or induced by $\boldsymbol{D}$) become aligned by $\boldsymbol{D}$ and always act to reduce the size of $\boldsymbol{D}$ so that $|\boldsymbol{D}| > |\epsilon_0 \boldsymbol{E}|$, that is, the electric polarization field is always oriented in a direction opposite to the applied field. So the total electric field $\boldsymbol{E}$ is always smaller than the applied field $\boldsymbol{D}/\epsilon_0$. Applied fields are generated either from an overall charge separation across the volume under consideration (as depicted by the capacitor example in Fig. 3.1) or from a time-varying magnetic field.

Although individual molecules can have a permanent electric dipole moment (e.g., water), in the absence of an applied electric field to align them there is typically no permanent electric polarization for a large collection of permanent dipole moments, even in the frozen solid state at low temperature. Exceptions to this are when certain resins, waxes, and other polymer materials are heated to melting and then cooled to being solid in the presence of a large electric field, which allows a quasi-permanent polarization to remain even once the applied electric field is turned off. A material that exhibits this phenomenon is called an *electret*. Due to finite though quite small electrical conductivity (free charge mobility) within the electret, the polarization field eventually disappears entirely but this may take years. You might think that if you freeze liquid water into ice in the presence of a large applied electric field that aligns the dipolar molecules and then turn off the applied field, there would remain a permanent polarization. But the protons (H^+) residing on the hydrogen-bond bridges between any two oxygens in the structure of ice are quite mobile. Although a proton remains in the neighborhood of the oxygen to which it is initially attached, it can move across the hydrogen-bond bridge to a neighboring oxygen or move from one hydrogen-bond bridge to another on the same oxygen. Such local proton mobility quickly depolarizes the ice once the applied field is turned off.

Historically, for an isotropic material, the electric polarization field $\boldsymbol{P}$ is expressed in terms of the total electric field $\boldsymbol{E}$ as $\boldsymbol{P} = \epsilon_0 \chi_e \boldsymbol{E}$, where the material property characterizing the polarization process is called the *electrical susceptibility* χ_e, which is a positive dimensionless quantity. If the material were anisotropic and the polarization field could be in a direction different than $\boldsymbol{E}$, the electrical susceptibility would be a second-order tensor $\boldsymbol{\chi}_e$ and $\boldsymbol{P} = \epsilon_0 \boldsymbol{\chi}_e \cdot \boldsymbol{E}$. Again, in continuum physics, a second-order tensor is always the proportionality (mapping) between two vectors related to each other at a point.

The relation between the applied field $\boldsymbol{D}$ and the total field $\epsilon_0 \boldsymbol{E}$ is then obtained by combining $\boldsymbol{P} = \epsilon_0 \chi_e \boldsymbol{E}$ with $\epsilon_0 \boldsymbol{E} = \boldsymbol{D} - \boldsymbol{P}$ to give, as earlier, $\boldsymbol{D} = \epsilon_0\, \kappa_e \boldsymbol{E}$. The dielectric constant $\kappa_e = 1 + \chi_e$ thus obeys the constraint $\kappa_e > 1$ so that $\epsilon_0 |\boldsymbol{E}| < |\boldsymbol{D}|$ as required and with equality holding in a vacuum where polarization does not occur. For water molecules, that have permanent electric dipole moments that become aligned by the application of a static $\boldsymbol{D}$, we have $\kappa_e \approx 80$ which is a large value (lots of polarization) compared to most materials. For example, quartz has a static dielectric constant of $\kappa_e \approx 4$. For an anisotropic material, we have the earlier tensor relation $\boldsymbol{D} = \epsilon_0\, \boldsymbol{K}_e \cdot \boldsymbol{E}$, where $\boldsymbol{K}_e = \boldsymbol{I} + \boldsymbol{\chi}_e$ with $\boldsymbol{I}$ the identity tensor.

3.3.5 Instantaneous versus Delayed Electric Polarization

Throughout the above discussion, it is important to note that we have assumed the polarization takes place instantaneously and that past values of the applied field are not influencing the present value of the polarization field. If delayed polarization response is possible, we can write, for example, the electric-polarization constitutive law in the temporal convolution form

$$\boldsymbol{D}(t) = \epsilon_0 \int_{-\infty}^{t} \boldsymbol{K}_{te}(t-u) \cdot \boldsymbol{E}(u)\, \mathrm{d}u, \tag{3.108}$$

where the second-order tensor $\boldsymbol{K}_{te}(t)$ is now time dependent with units of inverse time and allows for delay between the applied and polarization fields. In Part II of the book, once we have learned about Fourier analysis and contour integration, we will show that such delayed polarization response can be allowed for using a complex frequency-dependent dielectric-constant tensor $\tilde{\boldsymbol{K}}_{te}(\omega)$, where ω is circular frequency in a frequency-dependent constitutive law $\tilde{\boldsymbol{D}}(\omega) = \epsilon_0 \tilde{\boldsymbol{K}}_{te}(\omega) \cdot \tilde{\boldsymbol{E}}(\omega)$. With increasing frequency of the applied field, polarization processes become weaker because they don't have enough time to occur in each temporal cycle, which results in $\tilde{\boldsymbol{K}}_{te}(\omega)$ being frequency dependent. In Chapter 11, the complex part of the susceptibility or dielectric constant will be shown to produce both a phase lag between the applied field and the polarization field and an irreversible loss of energy. In our introductory treatment given above, we have assumed implicitly that polarization occurs instantaneously with changes in the applied field, which corresponds to $\boldsymbol{K}_{te}(t-u) = \boldsymbol{K}_e\,\delta(t-u)$ and the neglect of how polarization decreases with increasing frequency. But if frequency-dependent polarization is measured and modeled, it can provide important diagnostic information about the material being polarized and the polarization process.

3.3.6 Qualitative Discussion of Magnetization

For our pedagogic aims here, we will focus only on the dipolar magnetization characterized by the magnetization vector $\boldsymbol{M}$ and ignore the circulation of polarized electric charge characterized by $\boldsymbol{v} \times (\boldsymbol{P} - \nabla \cdot \boldsymbol{Q})$. To qualitatively understand this *magnetic polarization* or "magnetization," we write

$$\underset{\substack{\uparrow \\ \text{total}}}{\frac{1}{\mu_0}\boldsymbol{B}} = \underset{\substack{\uparrow \\ \text{applied}}}{\boldsymbol{H}} + \underset{\substack{\uparrow \\ \text{induced} \\ \text{magnetization}}}{\boldsymbol{M}} + \underset{\substack{\uparrow \\ \text{permanent} \\ \text{magnetization}}}{\boldsymbol{M}_0} \quad . \tag{3.109}$$

Although there are no magnetic monopoles (no magnetic charges), there is a *quantum spin* (magnetic dipole) associated with each electron and nucleus. Additionally, as accounted for in our derivation of the continuum form of Maxwell's equations, current loops can be induced in molecules that also create effective magnetic dipoles. The applied field $\boldsymbol{H}$ can orient these magnetic dipoles to induce a net magnetization $\boldsymbol{M}$ that is usually in the same direction as $\boldsymbol{H}$, and we can write this relation for an isotropic material as $\boldsymbol{M} = \chi_m \boldsymbol{H}$, where χ_m is the dimensionless *magnetic susceptibility*. Somewhat rarely, magnetization can occur in the direction opposite to the applied field so that χ_m can be negative. However, in so-called *diamagnetic media* where this occurs (such as bismuth, ammonia, some living organisms, and superconductors), the effect is very weak and χ_m is only slightly smaller than 0, for example, ammonia has $\chi_m \approx -10^{-5}$. Most matter has $0 < \chi_m \ll 1$ and is called *paramagnetic*. Paramagnetic matter typically has χ_m on the order of 10^{-3} and $\boldsymbol{H}$-induced alignment of the magnetic dipoles and therefore χ_m decreases with increasing temperature. Media in which the magnetization is in a direction somewhat different than the applied field, which would necessitate the magnetic susceptibility to be a second-order tensor, are extremely rare so that magnetization can almost always be considered an isotropic response.

Some materials called *ferromagnetic* (e.g., solid iron) can have a permanent alignment of the electron spins frozen into them once the temperature drops below the so-called Curie temperature T_c, which for pure iron is $T_c = 1{,}043$ K. This results in a permanent magnetization $\boldsymbol{M}_0$ even when $\boldsymbol{H} = 0$. A material with $\boldsymbol{M}_0 \neq 0$ is called a *magnet*. The possible presence of a permanent magnetic polarization $\boldsymbol{M}_0 \neq 0$ even in the absence of an applied magnetic field is possible because there are no mobile magnetic charges (no magnetic monopoles) that can depolarize the magnet. Recall that in *electrets*, there are mobile electric charge carriers that eventually depolarize the material in the absence of an applied electric field.

So the relation between the total magnetic field, the applied magnetic field and the permanent magnetization can be written

$$\frac{1}{\mu_0}\boldsymbol{B} = \underbrace{\boldsymbol{M}_0 + \kappa_m \boldsymbol{H}}_{\text{ferromagnetic}} \quad \underbrace{(\text{or } = \kappa_m \boldsymbol{H})}_{\text{paramagnetic}}, \tag{3.110}$$

where $\kappa_m = 1 + \chi_m > 1$ is again the relative magnetic permeability (dimensionless) and $\mu = \mu_0 \kappa_m$ is the magnetic permeability. Only ferromagnetic media rich in iron, cobalt, or nickel have κ_m significantly greater than 1, approaching and exceeding 10^3 in some cases. So, very commonly, unless iron is present, we have $\boldsymbol{B} = \mu_0 \boldsymbol{H}$ in upper-crustal Earth materials (e.g., water and sand grains), which means magnetization (induced magnetic dipole alignment) can be totally ignored in such materials.

3.3.7 The Irreversible Process of Electromigration

The rate at which heat is irreversibly being generated by electromigration is $\boldsymbol{J}_{em} \cdot \boldsymbol{E}'$, which is the product of the flux of electrical charge $\boldsymbol{J}_{em}$ with the electrical force $\boldsymbol{E}'$ driving the flux, both being defined in the frame of reference that is moving with the mass element. The basic assumption of irreversible thermodynamics is that the flux trying to equilibrate a nonequilibrium situation is created by the force that appears in the expression for the irreversible heat (or entropy) generation. In our present case, this relation is $\boldsymbol{J}_{em} = \boldsymbol{J}_{em}(\boldsymbol{E}')$. For the electromigration of ions and free electrons relative to the mass-averaged velocity $\boldsymbol{v}$ of all the particles in a mass element, this relation is almost always linear, which allows us to state that

$$\boldsymbol{J}_{em} = \boldsymbol{\sigma} \cdot \boldsymbol{E}' = \boldsymbol{\sigma} \cdot (\boldsymbol{E} + \boldsymbol{v} \times \boldsymbol{B}), \tag{3.111}$$

where $\boldsymbol{\sigma}$ is the second-order electrical conductivity tensor (units of S/m), and in the second expression, we have expressed the driving force in terms of the fields in the stationary frame of reference [Eq. (3.79)]. This law of electromigration is called *Ohm's law*. In isotropic media, we have the simple expression $\boldsymbol{J}_{em} = \sigma(\boldsymbol{E} + \boldsymbol{v} \times \boldsymbol{B})$, where the electrical conductivity σ is now a scalar. We will not have anything more to say here about the conductivity σ for metals or semi-conductors where electrons are the charge carriers.

However, we will digress on an important source of electrical conductivity in the Earth, which is ionic electromigration in liquids. *Strong electrolytes* are defined as solutions in which salt crystals put into a solvent like liquid water completely dissolve into cations and anions. We will give here the expression for the electrical conductivity of a strong electrolyte. Define $\tilde{N}$ as the number of salt molecules per unit volume of solution. Each salt molecule dissolves into ν_+ cations and ν_- anions. Charge neutrality will be satisfied in the solution which requires $\nu_+ z_+ + \nu_- z_- = 0$ where the ionic valences z_+ and z_- are the number and sign of fundamental charges $e = 1.6 \times 10^{-19}$ C on each ion. One can write the dissolution of a salt molecule as the stoichiometric formula $C_{\nu_+}A_{\nu_-} = \nu_+ C^{z+} + \nu_- A^{z-}$. Some examples of strong electrolytes include sodium chloride $NaCl = Na^+ + Cl^-$

with $\nu_\pm = 1$ and $z_\pm = \pm 1$, magnesium sulfate $MgSO_4 = Mg^{+2} + SO_4^{-2}$ with $\nu_\pm = 1$ and $z_\pm = \pm 2$, and calcium chloride $CaCl_2 = Ca^{+2} + 2Cl^{-1}$ with $\nu_+ = 1$, $\nu_- = 2$, $z_+ = 2$, and $z_- = -1$.

We now appeal to Eq. (3.34) for the definition of the current density $\boldsymbol{J}_{em}$ of an ionic fluid. Treating a strong electrolyte in which $\tilde{N}$ salt molecules per unit volume have completely dissolved into cations and anions, the sum of Eq. (3.34) is performed and one obtains

$$\boldsymbol{J}_{em} = e(z_+ \nu_+ \boldsymbol{v}_+ + z_- \nu_- \boldsymbol{v}_-)\tilde{N}, \tag{3.112}$$

where the $\boldsymbol{v}_\pm$ are the average velocities of the cations and anions in excess of the collective velocity $\boldsymbol{v}$ and are due to the forcing provided by the field $\boldsymbol{E}'$. We define the mobility b of any object moving with resistance at a velocity $\boldsymbol{v}_r$ relative to the collective movement of the fluid $\boldsymbol{v}$ and under the influence of an applied force $\boldsymbol{F}$ from the relation $\boldsymbol{v}_r = b\boldsymbol{F}$. For our ions in solutions, the force on a cation is $ez_+(\boldsymbol{E} + \boldsymbol{v} \times \boldsymbol{B})$ and that on an anion is $ez_-(\boldsymbol{E} + \boldsymbol{v} \times \boldsymbol{B})$. So $\boldsymbol{v}_+$ carries positive charge in the direction of $\boldsymbol{E} + \boldsymbol{v} \times \boldsymbol{B}$ and $\boldsymbol{v}_-$ carries negative charge in the negative direction compared to $\boldsymbol{v}_+$ and both contributions to the electromigration are positive. We then obtain the expression for the electrolyte conductivity

$$\sigma = e^2(z_+^2 \nu_+ b_+ + z_-^2 \nu_- b_-)\tilde{N}. \tag{3.113}$$

If there are multiple types of strong (i.e., completely dissolving) molecules m present, that include any strong acids like HCl and strong bases like NaOH in addition to strong salts like NaCl, we have

$$\sigma = e^2 \sum_m \left(z_{m+}^2 \nu_{m+} b_{m+} + z_{m-}^2 \nu_{m-} b_{m-}\right) \tilde{N}_m, \tag{3.114}$$

where the $\tilde{N}_m$ are the number densities (number per unit volume) of the various salts, acids, and bases that have been added to the pure solvent to make the electrolyte. What to use for the ionic mobilities $b_{m\pm}$?

A simple model due to Einstein imagines the electromigrating ions as if they were spheres plowing through a sea of solvent molecules that can be treated with a continuum description using a shear viscosity η. When fluid mechanics is presented in Chapter 5, a reasonable approximation due to Stokes will be shown to give the so-called *Einstein–Stokes model* of ionic mobilities

$$b_\pm = \frac{1}{6\pi \eta R_\pm}, \tag{3.115}$$

where $R_\pm$ are the effective radii of the ions. Values to use for the radii of the ions for some typical ions in aqueous solution are given in Table 3.1. These listed effective radii are those that produce the mobilities in water and allow for the ions to be hydrated with water molecules. What makes the Einstein–Stokes model for the ionic mobilities useful is that it allows, at least approximately, for the temperature and nonlinear salt concentration dependence of the electrical conductivity of the solution through the way that viscosity η varies with temperature and ionic concentrations. Although such trends are only

Table 3.1 *The effective radii to use in the Einstein–Stokes ionic-mobility model of Eq. (3.115).*

Ion	Effective ionic radius $R_\pm$
H^+	0.23×10^{-10} m
Li^+	2.12×10^{-10} m
Na^+	1.63×10^{-10} m
K^+	1.12×10^{-10} m
Ag^+	1.33×10^{-10} m
Ca^{+2}	2.74×10^{-10} m
Mg^{+2}	3.08×10^{-10} m
OH^-	0.41×10^{-10} m
Cl^-	1.07×10^{-10} m
Br^-	1.05×10^{-10} m
NO_3^-	1.15×10^{-10} m
F^-	1.47×10^{-10} m
CO_3^{-2}	2.34×10^{-10} m
SO_4^{-2}	2.04×10^{-10} m

approximate and ultimately breakdown over very large ranges of temperature and ionic concentration, the Einstein–Stokes model with $\eta(T, \tilde{N}_m)$ dependence provides a reasonable first model of the ionic mobilities. The effective ionic radii in Table 3.1, although allowing for hydration effects, are usually approximated as being independent of temperature and salt concentration.

A final comment about the irreversible heat-producing process of electromigration is that we have been assuming the current to respond instantaneously to changes in the driving electric field. In analogy to the delayed polarization discussed in Section 3.3.5, if the electric field has a time-harmonic temporal variation, as the temporal frequency begins to increase, there can develop a phase lag between the current and the time-varying electric field. Such phase lags are associated with any reversible (non-heat-producing) polarization processes at work, resulting in the conductivity being representable as a complex and frequency-dependent material property $\tilde{\sigma}(\omega)$. We will ignore such frequency dependence in the present treatment of electromigration but will return to the general subject of frequency-dependent linear constitutive laws in Chapter 11 once the purely mathematical topics of Fourier analysis and contour integration have been presented.

3.4 Summary of Continuum EM in Moving Media

Let's bring together everything we have learned about the continuum form of Maxwell's equations when linear constitutive laws apply to a material with the following properties: (1) it does not possess a permanent magnetization ($\boldsymbol{M}_0 = 0$), (2) magnetoelectric polarization processes can be neglected ($\boldsymbol{C}_{EB} = 0$), and (3) it is moving with a velocity $\boldsymbol{v}$ without strong local gradients. The governing equations in this case are:

$$\nabla \cdot \boldsymbol{B} = 0 \quad \textit{Gauss' law} \tag{3.116a}$$

$$\nabla \cdot \boldsymbol{D} = \rho_e \quad \textit{Coulomb's law} \tag{3.116b}$$

$$\nabla \times \boldsymbol{E} = -\frac{\partial \boldsymbol{B}}{\partial t} \quad \textit{Faraday's law} \tag{3.116c}$$

$$\nabla \times \boldsymbol{H} = \frac{\partial \boldsymbol{D}}{\partial t} + \boldsymbol{J}_{em} + \rho_e \boldsymbol{v} + \boldsymbol{J}_a \quad \textit{Ampère's law} \tag{3.116d}$$

$$\boldsymbol{D} = \epsilon_0 \boldsymbol{K}_e \cdot (\boldsymbol{E} + \boldsymbol{v} \times \boldsymbol{B}) \quad \text{(units of C m}^{-2}\text{)} \tag{3.116e}$$

$$\boldsymbol{B} = \mu_0 \kappa_m (\boldsymbol{H} - \boldsymbol{v} \times \boldsymbol{D}) \quad \text{(units of Tesla = T = kg C}^{-1}\text{ s}^{-1}\text{)} \tag{3.116f}$$

$$\boldsymbol{J}_{em} = \boldsymbol{\sigma} \cdot (\boldsymbol{E} + \boldsymbol{v} \times \boldsymbol{B})\,. \quad \text{(units of C m}^{-2}\text{ s}^{-1}\text{).} \tag{3.116g}$$

The above laws are being expressed in a stationary frame of reference from which the material velocity $\boldsymbol{v}$ and EM fields are measured. The *applied* (or *antenna*) current density $\boldsymbol{J}_a$ in Ampère's law is a source term corresponding to injecting current into a problem domain through a metal electrode or by exciting induction fields using, say, an insulated current-loop antenna. Taking the divergence of Faraday's law, we have $\nabla \cdot \boldsymbol{B} =$ const (independent of time) at each point of space. Since at some time in the distant past we can assume $\boldsymbol{B} = 0$ in each region under consideration (could also be our initial condition), we can take $\nabla \cdot \boldsymbol{B} = 0$ as an implicit part of Faraday's law and do not need to include this equation (Gauss' law) when we solve Maxwell's equations for the time-varying fields; however, for static fields, we do need to write down $\nabla \cdot \boldsymbol{B} = 0$ explicitly. To determine the time-varying fields $\boldsymbol{E}$ and $\boldsymbol{B}$ in a heterogeneous but initially uncharged macroscopic body of matter, we use Eqs (3.116c) to (3.116g) alone and then use $\rho_e = \nabla \cdot \boldsymbol{D}$ after the fact to determine any accumulated free-charge density ρ_e once $\boldsymbol{D}$ is known. Locally, we can have $\nabla \cdot \boldsymbol{D} \neq 0$ (charge build up) when the fields are present in a heterogeneous material.

3.5 Conservation of Electric Charge

By taking the divergence of Ampère's law and introducing Coulomb's law, we obtain

$$\frac{\partial \rho_e}{\partial t} = -\nabla \cdot (\boldsymbol{v} \rho_e) - \nabla \cdot \boldsymbol{J}_{em}. \tag{3.117}$$

This is the statement of the conservation of electric charge from a stationary frame of reference that is built right into Maxwell's equations. Any excess charge density that is building up at a point is due to the advective and electromigration accumulations as accounted for by the two divergences on the right-hand side. From a frame of reference that is moving with the locally uniform velocity $\boldsymbol{v}$, Ampère's law is $\nabla \times \boldsymbol{H}' = \mathrm{d}\boldsymbol{D}'/\mathrm{d}t + \boldsymbol{J}_{em}$ and a total time derivative of Coulomb's law gives $\nabla \cdot \mathrm{d}\boldsymbol{D}'/\mathrm{d}t = \mathrm{d}\rho_e/\mathrm{d}t$. Taking the divergence of Ampère's law then gives the conservation of charge from the moving frame of reference $\nabla \cdot \boldsymbol{J}_{em} = -\mathrm{d}\rho_e/\mathrm{d}t$.

3.6 The Wave-Propagation, Diffusion, and Quasi-static Limits

The macroscopic form of Maxwell's equations contain three important limits that can be called the *wave-propagation*, *diffusion*, and *quasi-static* limits. In this section, we describe these three limits when the material velocity $\boldsymbol{v}$ can be neglected in the governing equations. The conditions for neglecting the material velocity in the constitutive laws will be given in Section 3.7 that follows. To make the analysis more simple and transparent without losing the pedagogic concept of defining these three limits, we further assume the material to be isotropic.

In all that follows, we assume the material has a finite electrical conductivity σ as it does throughout the Earth to which we imagine the macroelectromagnetic theory is being applied. Insulating materials for which the electrical conductivity is zero are called *dielectrics*. In a dielectric material, time-dependent response is in the wave-propagation limit and there is no diffusive response, which is caused by electrical conduction of free charge. The quasi-static response of a dielectric that holds when the wavelengths of a time-dependent response are much larger than the size of the domain being modeled is also distinct from the quasi-static response of a conductor that supports electromigration as will be pointed out below.

3.6.1 The Wave-Propagation Limit

If an EM disturbance is such that in Ampère's law,

$$\left|\frac{\partial \boldsymbol{D}}{\partial t}\right| \gg |\boldsymbol{J}_{em}|, \tag{3.118}$$

the EM disturbance will advance through space and time as a propagating wave. To quantify the condition for this limit to hold, the temporal variations are characterized by a period T or equivalently by a characteristic circular frequency $\omega = 2\pi/T$ so that the time derivative can be estimated as $|\partial/\partial t| \approx \omega$. For example, time-harmonic response that has the temporal dependence $e^{-i\omega t}$ has $\partial e^{-i\omega t}/\partial t = -i\omega e^{-i\omega t}$. With $\boldsymbol{J}_{em} = \sigma \boldsymbol{E}$ and $\boldsymbol{D} = \varepsilon \boldsymbol{E}$, where ε is the electrical permittivity of the material, the limit of Eq. (3.118) corresponds to the condition

$$\omega \gg \frac{\sigma}{\varepsilon}. \tag{3.119}$$

For a common near-surface Earth material having, say, $\sigma \approx 10^{-2}$ S/m and $\varepsilon \approx 10^{-10}$ F/m, wave propagation occurs when $f = \omega/(2\pi) \gg 10$ MHz. In the opposite limit of $|\partial \boldsymbol{D}/\partial t| \ll |\boldsymbol{J}_{em}|$, the EM disturbance will advance as a diffusion which will be treated after.

So in the wave-propagation limit for which $|\partial \boldsymbol{D}/\partial t| \gg |\boldsymbol{J}_{em}|$, Maxwell's equations can be written

$$\frac{1}{\mu}\nabla \times \boldsymbol{E} = -\frac{\partial \boldsymbol{H}}{\partial t} \tag{3.120}$$

$$\nabla \times \boldsymbol{H} = \varepsilon \frac{\partial \boldsymbol{E}}{\partial t} + \boldsymbol{J}_a, \tag{3.121}$$

where $\boldsymbol{J}_a$ represents say a current-loop antenna that acts as the source of the EM disturbance. Taking the curl of Faraday's law and then using Ampère's law gives a single wave equation

$$\nabla \times \left(\frac{1}{\mu}\nabla \times \boldsymbol{E}\right) + \varepsilon \frac{\partial^2 \boldsymbol{E}}{\partial t^2} = -\frac{\partial \boldsymbol{J}_a}{\partial t}. \tag{3.122}$$

In a nonferromagnetic material for which $\mu \approx \mu_0$ or in a uniform material of any type, $1/\mu$ can be taken outside of the curl operation on the left-hand side to obtain

$$c^2 \nabla \times \nabla \times \boldsymbol{E} + \frac{\partial^2 \boldsymbol{E}}{\partial t^2} = -\frac{1}{\varepsilon}\frac{\partial \boldsymbol{J}_a}{\partial t}, \tag{3.123}$$

where c is the wave speed

$$c = \frac{1}{\sqrt{\mu\varepsilon}}. \tag{3.124}$$

Further, in a homogeneous and uncharged material for which $\nabla \cdot \boldsymbol{E} = 0$, we have the identity $\nabla \times \nabla \times \boldsymbol{E} = \nabla(\nabla \cdot \boldsymbol{E}) - \nabla^2 \boldsymbol{E} = -\nabla^2 \boldsymbol{E}$, which allows Eq. (3.123) to be written as an explicit vector wave equation

$$\boxed{c^2 \nabla^2 \boldsymbol{E} - \frac{\partial^2 \boldsymbol{E}}{\partial t^2} = \frac{1}{\varepsilon}\frac{\partial \boldsymbol{J}_a}{\partial t}.} \tag{3.125}$$

For an EM wave of frequency f, the wavelength is $\lambda = c/f$.

In this wave-propagation limit, we can estimate the size of the fields relative to each other by estimating spatial derivatives as $\nabla \approx 2\pi/\lambda$ and temporal derivatives as $\partial/\partial t \approx 2\pi f$. From Faraday's law $|\nabla \times \boldsymbol{E}| = |\partial \boldsymbol{B}/\partial t|$ and Ampère's law $|\nabla \times \boldsymbol{H}| = |\partial \boldsymbol{D}/\partial t|$, we obtain the estimates

$$\frac{|\boldsymbol{B}|}{|\boldsymbol{E}|} \approx \frac{1}{c} \quad \text{and} \quad \frac{|\boldsymbol{D}|}{|\boldsymbol{H}|} \approx \frac{1}{c}. \tag{3.126}$$

An EM wave might be generated in the near-surface of the Earth by a loop antenna (as used in ground-penetrating radar equipment) creating an electric field on the order of say 10^{-2} V/m at a frequency of say 1 GHz. The EM velocity in the near surface is $c \approx 10^8$ m/s, so that the magnetic-field amplitude of this EM wave would be $|\boldsymbol{B}| \approx 10^{-6}$ T. For reference, the Earth's static field near the surface is on the order of 10^{-5} T and the natural electric fields, called telluric fields, coming from the atmosphere with a frequency of say $f = 1$ Hz might have amplitudes on the order of 10^{-6} V/m. The EM wavelength in this loop-antenna example would be $\lambda = 0.1$ m.

3.6.2 The Diffusion Limit

In the opposite limit in which $|\partial \boldsymbol{D}/\partial t| \ll |\boldsymbol{J}_{em}|$ or $\omega \ll \sigma/\varepsilon$, we have

$$\frac{1}{\mu}\nabla \times \boldsymbol{E} = -\frac{\partial \boldsymbol{H}}{\partial t} \tag{3.127}$$

$$\nabla \times \boldsymbol{H} = \sigma \boldsymbol{E} + \boldsymbol{J}_a, \tag{3.128}$$

which can be combined into a single diffusion equation as

$$\nabla \times \left(\frac{1}{\mu}\nabla \times \boldsymbol{E}\right) + \sigma \frac{\partial \boldsymbol{E}}{\partial t} = -\frac{\partial \boldsymbol{J}_a}{\partial t}. \tag{3.129}$$

As above, if μ is uniform in space or if the material is not ferromagnetic, we obtain

$$D_{em}\nabla \times \nabla \times \boldsymbol{E} + \frac{\partial \boldsymbol{E}}{\partial t} = -\frac{1}{\sigma}\frac{\partial \boldsymbol{J}_a}{\partial t}, \tag{3.130}$$

where the EM diffusivity D_{em} (units of m^2/s) is defined

$$D_{em} = \frac{1}{\sigma \mu}. \tag{3.131}$$

For a homogeneous and uncharged material in which $\nabla \cdot \boldsymbol{E} = 0$, we have $\nabla \times \nabla \times \boldsymbol{E} = -\nabla^2 \boldsymbol{E}$, which results in the explicit vector diffusion equation

$$\boxed{D_{em}\nabla^2 \boldsymbol{E} - \frac{\partial \boldsymbol{E}}{\partial t} = \frac{1}{\sigma}\frac{\partial \boldsymbol{J}_a}{\partial t}.} \tag{3.132}$$

For a disturbance having a circular frequency ω, the EM skin depth δ_{em} is defined from this diffusion equation using the conventions $\nabla \approx \sqrt{2}/\delta_{em}$ and $\partial/\partial t \approx \omega$ to give

$$\delta_{em} = \sqrt{\frac{2D_{em}}{\omega}} = \sqrt{\frac{2}{\sigma \mu \omega}} \;\widehat{=}\; \textit{EM skin depth} \tag{3.133}$$

and is the distance over which a diffusive response is significantly varying in space. In the first exercise at the end of the chapter, you will see quantitatively that for a planar EM diffusion occurring in the z direction, the envelope of the response decays spatially as $e^{-z/\delta_{em}}$ with the skin depth δ_{em} defined precisely as in Eq. (3.133) and explains the convention of using the approximation $\nabla \approx \sqrt{2}/\delta_{em}$ for making order-of-magnitude estimates in diffusion problems.

In the diffusive limit $\omega \ll \sigma/\varepsilon$, the amplitudes of the fields in Faraday's law $|\nabla \times \boldsymbol{E}| = |\partial \boldsymbol{B}/\partial t|$ and Ampère's law $|\nabla \times \boldsymbol{H}| = |\sigma \boldsymbol{E}|$ can be estimated by approximating derivatives as $\nabla \approx \sqrt{2}/\delta_{em}$ and $\partial/\partial t \approx \omega$ to obtain

$$\frac{|\boldsymbol{B}|}{|\boldsymbol{E}|} \approx \sqrt{\frac{\sigma\mu}{\omega}} \quad \text{and} \quad \frac{|\boldsymbol{D}|}{|\boldsymbol{H}|} \approx \varepsilon\sqrt{\frac{\mu\omega}{\sigma}}. \tag{3.134}$$

One may also use the constitutive laws to obtain $|\boldsymbol{D}|/|\boldsymbol{H}|$ from $|\boldsymbol{B}|/|\boldsymbol{H}|$. Thus, the amplitudes of the fields relative to each other have a completely different character in the diffusive limit compared to the wave-propagation limit treated above. If the near-surface electric field amplitude is again say $|\boldsymbol{E}| \approx 10^{-2}$ V/m but is now generated by a source $\boldsymbol{J}_a$ operated at $f = 10^4$ Hz typical of the diffusive regime used in exploration geophysics, the corresponding magnetic field would be $|\boldsymbol{B}| \approx 10^{-8}$ T. The skin depth in this example has the order of magnitude $\delta_{em} \approx 100$ m.

When both $|\partial \boldsymbol{D}/\partial t|$ and $|\sigma \boldsymbol{E}|$ are of comparable magnitude and one cannot neglect one in comparison to the other, then we have attenuated wave propagation with amplitude loss due to the Joule heating of electrical conduction. You can treat this case in the first exercise at the end of the chapter.

3.6.3 The Quasi-static Limit

The quasi-static limit occurs when the size of the domain of investigation L is much smaller than the diffusive skin depth δ_{em}

$$L \ll \delta_{em} = \sqrt{\frac{2}{\sigma\mu\omega}}. \tag{3.135}$$

When this occurs, the effects of induction are not important within the domain (induction corresponds to the time derivative in Faraday's) and the spatial distribution of the field lines are determined by neglecting all time derivatives in Maxwell's equations. So the governing equations of quasi-static response are

$$\nabla \cdot \boldsymbol{B} = 0, \tag{3.136}$$

$$\nabla \times \boldsymbol{E} = 0, \tag{3.137}$$

$$\nabla \times \boldsymbol{H} = \boldsymbol{J}_{em} + \boldsymbol{J}_a \tag{3.138}$$

$$\boldsymbol{J}_{em} = \sigma \boldsymbol{E} \tag{3.139}$$

$$\boldsymbol{B} = \mu \boldsymbol{H}. \tag{3.140}$$

In quasi-static problems, the applied current density $\boldsymbol{J}_a$ is often created by a conduction current I_a (C/s) injected into the conductive problem domain through an electrode having a nominal surface area S_e with the resulting magnitude $|\boldsymbol{J}_a| \approx I_a/S_e$.

Faraday's law $\nabla \times \boldsymbol{E} = 0$ is now satisfied by an electric field of the form $\boldsymbol{E} = -\nabla\varphi$, where φ is called the scalar electric potential. Taking the divergence of Ampère's law gives

$$\nabla \cdot (\sigma \nabla \varphi) = \nabla \cdot \boldsymbol{J}_a. \tag{3.141}$$

To solve for the quasi-static response, we take the known source term $(\nabla \cdot \boldsymbol{J}_a)_0$ to be constant in time though variable in space and solve the scalar Poisson problem Eq. (3.141) subject to the appropriate boundary conditions to obtain the truly static fields φ_0 and $\boldsymbol{E}_0 = -\nabla\varphi_0$. Now if the source term has a time dependence written as $\nabla \cdot \boldsymbol{J}_a = s(t)\,(\nabla \cdot \boldsymbol{J}_a)_0$, where $s(t)$ is a dimensionless time function, then so long as the skin depth δ_{em} is much larger than the problem domain, the solution for the time-varying or quasi-static electric field is simply $\boldsymbol{E} = s(t)\boldsymbol{E}_0$. This is what we mean by *quasi-static* response.

The quasi-static conduction currents flowing in the material will generate magnetic fields $\boldsymbol{H}$. Further, even if the source current is an insulated loop of current that is not generating any other conduction current throughout the domain so that $\boldsymbol{E} = 0$, the current loop $\boldsymbol{J}_a$ will generate a magnetic field. In order to satisfy $\nabla \cdot \boldsymbol{B} = 0$, we define a magnetostatic vector potential as $\boldsymbol{B} = \nabla \times \boldsymbol{A}$. In this case, the governing equation for $\boldsymbol{A}$ is

$$\nabla \times \left(\frac{1}{\mu}\nabla \times \boldsymbol{A}\right) = \sigma \boldsymbol{E} + \boldsymbol{J}_a, \tag{3.142}$$

where $\boldsymbol{E}$ is either assumed known from the electrostatic solution above or is zero so that the right-hand side is a known source term for the vector potential $\boldsymbol{A}$. By adding to $\boldsymbol{A}$ the gradient of some scalar potential, one may always define a vector potential that satisfies $\nabla \cdot \boldsymbol{A} = 0$. In this case, and when μ can be taken as spatially uniform as it is in nonferromagnetic media, the identity $\nabla \times \nabla \times \boldsymbol{A} = \nabla(\nabla \cdot \boldsymbol{A}) - \nabla^2\boldsymbol{A}$ allows the vector potential $\boldsymbol{A}$ to be the solution of the vector Poisson equation

$$\nabla^2\boldsymbol{A} = -\mu\,(\sigma \boldsymbol{E} + \boldsymbol{J}_a), \tag{3.143}$$

where the right-hand side consists entirely of known source terms that satisfy $\nabla \cdot (\sigma \boldsymbol{E} + \boldsymbol{J}_a) = 0$. We will not consider techniques for the solution of this problem at this time.

For those quasi-static problems where the source generates a conduction current that in turn generates a magnetic field, we have $\nabla \times \boldsymbol{B} = \mu(\sigma \boldsymbol{E} + \boldsymbol{J}_a)$. If we estimate the spatial derivative here as $\nabla \approx 1/L$, where L is some macroscopic length within our element related to the distance over which the magnetic field is varying with distance from places where conduction current is concentrated, we can obtain an estimate of the size of $|\boldsymbol{B}|$ relative to $|\boldsymbol{E}|$ as

$$\frac{|\boldsymbol{B}|}{|\boldsymbol{E}|} \approx \mu\sigma L. \tag{3.144}$$

An overestimate of $|\boldsymbol{B}|$ relative to $|\boldsymbol{E}|$ would take L to be the size of the system being investigated. If the source $\boldsymbol{J}_a$ has generated an electric field of amplitude of say $|\boldsymbol{E}| \approx 10^{-2}$ V/m at a distance L from the source electrode, we then have that for near-surface Earth properties, $|\boldsymbol{B}| \approx 10^{-7}$ T using $L = 10$ m.

If the material is a dielectric (an electrically insulating material that has effectively $\sigma \to 0$), the quasi-static limit is defined by the condition $L \ll \lambda$ (the wavelength of EM waves) and the quasi-static response is controlled by slightly different governing equations. Specifically, the electrostatic response is controlled by $\nabla \times \boldsymbol{E} = 0$ and $\nabla \cdot \boldsymbol{D} = \rho_e$ or with $\boldsymbol{E} = -\nabla\varphi$

$$\nabla \cdot (\varepsilon \nabla\varphi) = -\rho_e, \tag{3.145}$$

which can be compared to Eq. (3.141). In a dielectric, there are no electric currents in the steady state ($\boldsymbol{J}_{em} \to 0$) and the only source for steady-state magnetic fields is an applied magnetic field that generates magnetization throughout the body.

3.7 On Neglecting the Material-Velocity Contributions to the Constitutive Laws

In looking at the governing equations of Section 3.4, if the material-velocity $\boldsymbol{v}$ terms are important in the three constitutive laws $\boldsymbol{J}_{em} = \sigma(\boldsymbol{E} + \boldsymbol{v} \times \boldsymbol{B})$, $\boldsymbol{D} = \varepsilon(\boldsymbol{E} + \boldsymbol{v} \times \boldsymbol{B})$, and $\boldsymbol{B} = \mu(\boldsymbol{H} - \boldsymbol{v} \times \boldsymbol{D})$, the analysis of an EM problem becomes significantly more complicated.

Our goal in this section is to determine conditions for neglecting the material velocity in the EM problem description even though the material is moving during a certain application. The leading order contribution of $\boldsymbol{v}$ to an EM response is that in which the governing equations for $\boldsymbol{E}$ and $\boldsymbol{H}$ are linear in $\boldsymbol{v}$, which corresponds to

$$\nabla \times \boldsymbol{E} = -\mu \frac{\partial}{\partial t} (\boldsymbol{H} - \varepsilon \boldsymbol{v} \times \boldsymbol{E}) \tag{3.146}$$

$$\nabla \times \boldsymbol{H} = \sigma (\boldsymbol{E} + \mu \boldsymbol{v} \times \boldsymbol{H}) + \varepsilon \frac{\partial}{\partial t} (\boldsymbol{E} + \mu \boldsymbol{v} \times \boldsymbol{H}) + \boldsymbol{J}_a. \tag{3.147}$$

Here, $\boldsymbol{J}_a$ is again an applied current density for a given application of interest, though the source for the EM fields could also come from a boundary condition.

Let's assume there is a large static uniform magnetic field $\boldsymbol{B}_0$, such as the Earth's field, that is present prior to and during the EM application associated with the source term $\boldsymbol{J}_a$. There are also motions of the material $\boldsymbol{v}$ throughout the EM problem domain. In this context, we decompose the electric and magnetic fields throughout the domain as follows:

$$\boldsymbol{E} = \boldsymbol{E}_{v0} + \boldsymbol{E}_a + \boldsymbol{E}_{va} \tag{3.148}$$

$$\boldsymbol{H} = \frac{\boldsymbol{B}_0}{\mu} + \boldsymbol{H}_{v0} + \boldsymbol{H}_a + \boldsymbol{H}_{va}, \tag{3.149}$$

where $\boldsymbol{E}_{v0}$ and $\boldsymbol{H}_{v0}$ are the fields generated by (and linear in) the movement $\boldsymbol{v}$ in the large static background field $\boldsymbol{B}_0/\mu$, $\boldsymbol{E}_a$, and $\boldsymbol{H}_a$ are the fields generated by the applied source $\boldsymbol{J}_a$ in the absence of material movement and, finally, $\boldsymbol{E}_{va}$ and $\boldsymbol{H}_{va}$ are the fields generated by (and linear in) the movement $\boldsymbol{v}$ in the fields $\boldsymbol{E}_a$ and $\boldsymbol{H}_a$. From the perspective of understanding the recordings during an application associated with $\boldsymbol{J}_a$, the field $\boldsymbol{E}_{v0}$ represents a background noise and we want to determine here whether $|\boldsymbol{E}_{v0}|/|\boldsymbol{E}_a|$ is significant relative to one. The field $\boldsymbol{E}_{va}$ is the electric field due to material movement while $\boldsymbol{J}_a$ is being applied and we again want to determine whether $|\boldsymbol{E}_{va}|/|\boldsymbol{E}_a|$ is significant and therefore must be modeled as part of the response. Similar comments hold for the various $\boldsymbol{H}$ fields in the decomposition.

Assuming that the time dependence of $\boldsymbol{v}$ is such that the source term $\sigma \boldsymbol{v} \times \boldsymbol{B}_0$ generates quasi-static fields $\boldsymbol{E}_{v0}$ and $\boldsymbol{H}_{v0}$ in our problem domain, we have

$$\nabla \times \boldsymbol{E}_{v0} = 0 \tag{3.150}$$

$$\nabla \times \boldsymbol{H}_{v0} = \sigma \boldsymbol{E}_{v0} + \sigma \boldsymbol{v} \times \boldsymbol{B}_0. \tag{3.151}$$

The fields $\boldsymbol{E}_{v0}$ and $\boldsymbol{H}_{v0}$ can be solved for by introducing a scalar electric potential and a vector magnetic potential as in Section 6.3.3. The amplitude of $|\boldsymbol{E}_{v0}|$ is given by $|\boldsymbol{E}_{v0}| \approx |\boldsymbol{v}||\boldsymbol{B}_0|$. If $\boldsymbol{B}_0$ is identified as the Earth's field with an amplitude of $|\boldsymbol{B}_0| \approx 10^{-5}$ T, then $|\boldsymbol{E}_{v0}| \approx |\boldsymbol{v}|10^{-5}$ V/m. Whether the fields $\boldsymbol{E}_{v0}$ and $\boldsymbol{H}_{v0}$ in this case are negligible or important compared to the fields generated by $\boldsymbol{J}_a$ depends on the amplitudes of $|\boldsymbol{v}|$ and $|\boldsymbol{J}_a|$. Common mechanical noise sources such as motor vehicles on roads or trees blowing in wind can generate ground motion bounded as say $|\boldsymbol{v}| < 10^{-2}$ m/s and even much less, which generates an antenna response of $|\boldsymbol{E}_{v0}| < 10^{-7}$ V/m and likely considerably smaller. So scenarios where the $\boldsymbol{E}_{v0}$ and $\boldsymbol{H}_{v0}$ are comparable to or larger than the magnetotelluric fields coming from the Earth's atmosphere, that commonly have electric fields on the order 10^{-6} V/m, need material velocities significantly greater than the ambient background mechanical vibrations that are inevitably present in any EM application in the Earth.

The fields $\boldsymbol{E}_a$ and $\boldsymbol{H}_a$ generated by the source $\boldsymbol{J}_a$ in the absence of material movement satisfy

$$\nabla \times \boldsymbol{E}_a + \mu \frac{\partial \boldsymbol{H}_a}{\partial t} = 0 \tag{3.152}$$

$$\nabla \times \boldsymbol{H}_a - \varepsilon \frac{\partial \boldsymbol{E}_a}{\partial t} - \sigma \boldsymbol{E}_a = \boldsymbol{J}_a, \tag{3.153}$$

which means the fields $\boldsymbol{E}_{va}$ and $\boldsymbol{H}_{va}$ that are generated by the material movement and that are linear in $\boldsymbol{v}$ are governed by

$$\nabla \times \boldsymbol{E}_{va} + \mu \frac{\partial \boldsymbol{H}_{va}}{\partial t} = \varepsilon\mu \frac{\partial}{\partial t} (\boldsymbol{v} \times \boldsymbol{E}_a) \tag{3.154}$$

$$\nabla \times \boldsymbol{H}_{va} - \varepsilon \frac{\partial \boldsymbol{E}_{va}}{\partial t} - \sigma \boldsymbol{E}_{va} = \varepsilon\mu \frac{\partial}{\partial t} (\boldsymbol{v} \times \boldsymbol{H}_a) + \sigma\mu \boldsymbol{v} \times \boldsymbol{H}_a. \tag{3.155}$$

The right-hand side of these equations are the source terms generating the fields on the left-hand side. We will analyze the amplitudes of the various terms in these equations in the three limits of wave-propagation, diffusion, and quasi-static response as controlled by the frequency (temporal variations) of the source $\boldsymbol{J}_a$.

3.7.1 The Wave-Propagation Regime

In the higher frequency wave-propagation limit where $|\varepsilon \partial \boldsymbol{E}/\partial t| \gg |\sigma \boldsymbol{E}|$, we estimate derivatives in the retained terms of Eq. (3.155) by taking $\partial/\partial t \approx \omega = 2\pi f$ and $\nabla \approx 2\pi/\lambda$, where $\lambda = c/f$ and the wave speed c is again $c = 1/\sqrt{\varepsilon\mu}$. Because each retained term has roughly the same order-of-magnitude amplitude, we obtain

$$\frac{|\nabla \times \boldsymbol{H}_{va}|}{|\varepsilon\mu\partial\,(\boldsymbol{v} \times \boldsymbol{H}_a)\,/\partial t|} \approx 1 \tag{3.156}$$

or

$$\frac{|\boldsymbol{H}_{va}|}{|\boldsymbol{H}_a|} \approx \varepsilon\mu f\lambda|\boldsymbol{v}| = \frac{|\boldsymbol{v}|}{c}. \tag{3.157}$$

A similar handling of Eq. (3.154) gives equivalently

$$\frac{|\boldsymbol{E}_{va}|}{|\boldsymbol{E}_a|} \approx \frac{|\boldsymbol{v}|}{c}. \tag{3.158}$$

Thus, so long as the material velocity satisfies the condition

$$|\boldsymbol{v}| \ll c, \tag{3.159}$$

which it always does because $c \approx 10^8$ m/s in most Earth materials, we can use $\boldsymbol{B} = \mu\boldsymbol{H}$ and $\boldsymbol{D} = \varepsilon\boldsymbol{E}$ as our constitutive laws for the modeling of EM wave propagation.

3.7.2 The Diffusion Regime

In the lower-frequency diffusive limit where $|\varepsilon\partial\boldsymbol{E}/\partial t| \ll |\sigma\boldsymbol{E}|$, we estimate derivatives as $\partial/\partial t \approx \omega$ and $\nabla \approx \sqrt{2}/\delta_{em}$, where δ_{em} is the EM skin depth defined earlier as $\delta_{em} = \sqrt{2/(\sigma\mu\omega)}$. From Ampère's law in this diffusive limit, we have

$$\frac{|\nabla \times \boldsymbol{H}_{va}|}{|\sigma\mu\boldsymbol{v} \times \boldsymbol{H}_a|} \approx 1 \tag{3.160}$$

or

$$\frac{|\boldsymbol{H}_{va}|}{|\boldsymbol{H}_a|} \approx \sqrt{\frac{\sigma\mu}{\omega}}|\boldsymbol{v}|. \tag{3.161}$$

The limit $|\boldsymbol{H}_{va}|/|\boldsymbol{H}_a| \to 0$ means that we can take $\sigma(\boldsymbol{E} + \boldsymbol{v} \times \boldsymbol{B}) \to \sigma\boldsymbol{E}$ in Ampère's law. This limit corresponds to the material velocity satisfying the condition

$$|\boldsymbol{v}| \ll \sqrt{\frac{\omega}{\sigma\mu}} \tag{3.162}$$

For a near-surface Earth material with typical values of $\sigma \approx 10^{-2}$ S/m and $\mu \approx 10^{-6}$ H/m, we have the condition $|\boldsymbol{v}| \ll 10^4\sqrt{\omega}$ m/s. The lowest frequency we typically employ in magnetotelluric imaging of the Earth's interior is on the order 10^{-2} Hz. Thus, so long as $|\boldsymbol{v}| \ll 10^3$ m/s, as we can reasonably assume it will be in typical EM diffusion applications such as magnetotellurics, we can use $\boldsymbol{J}_{em} = \sigma\boldsymbol{E}$ as our constitutive law for modeling EM diffusion. However, in materials with sufficiently high electrical conductivity in applications with sufficiently small temporal frequencies, one may indeed need to retain the velocity term in $\boldsymbol{J}_{em} = \sigma(\boldsymbol{E} + \boldsymbol{v} \times \boldsymbol{B})$ as is the case in modeling problems of magnetohydrodynamics.

A similar handling of Faraday's law in the diffusive limit gives

$$\frac{|\boldsymbol{E}_{va}|}{|\boldsymbol{E}_a|} \approx \varepsilon \sqrt{\frac{\mu\omega}{\sigma}}\, |\boldsymbol{v}|. \tag{3.163}$$

The limit $|\boldsymbol{E}_{va}|/|\boldsymbol{E}_a| \to 0$ means that we can take $\mu(\boldsymbol{H} - \boldsymbol{v} \times \boldsymbol{D}) \to \mu \boldsymbol{H}$ in the time derivative of Faraday's law. This limit corresponds to the material velocity satisfying the condition

$$|\boldsymbol{v}| \ll \frac{1}{\varepsilon}\sqrt{\frac{\sigma}{\mu\omega}}. \tag{3.164}$$

For a near-surface Earth material with $\varepsilon = 10^{-10}$ F/m, $\sigma = 10^{-2}$ S/m, and $\mu = 10^{-6}$ H/m, the condition for the neglect of the velocity in Faraday's law becomes $|\boldsymbol{v}| \ll 10^{12}/\sqrt{\omega}$ m/s, which is always satisfied. So we may use $\boldsymbol{B} = \mu \boldsymbol{H}$ as the magnetic constitutive law in all EM wave and diffusion problems.

3.7.3 The Quasi-static Regime

Last, in the quasi-static limit, we have

$$\nabla \times \boldsymbol{E}_{va} = 0 \tag{3.165}$$

$$\nabla \times \boldsymbol{H}_{va} - \sigma \boldsymbol{E}_{va} = \sigma \mu \boldsymbol{v} \times \boldsymbol{H}_a \tag{3.166}$$

which with $\nabla \approx 1/L$ yields the estimate

$$\frac{|\boldsymbol{H}_{va}|}{|\boldsymbol{H}_a|} \approx \sigma \mu L\, |\boldsymbol{v}|. \tag{3.167}$$

Thus, so long as

$$v \ll \frac{1}{\sigma \mu L}, \tag{3.168}$$

which is normally the case, we can neglect the effect of material movement in a quasi-static EM response generated by the source $\boldsymbol{J}_a$. This said, there can be circumstances during fluid flow where even though the quasi-static magnetic field generated by the fluid movement in a large background field is small relative to the background field, the movement-induced field is influencing the fluid flow in a nonnegligible way. This is the subject of *magnetohydrodynamics* that we will return to in Chapter 5 on fluid mechanics.

3.8 Summary of Continuum EM in Practice

From the above analysis, if the material movement is to have an influence on the EM fields in any practical application of interest, it will only come through the constitutive law $\boldsymbol{J}_{em} = \sigma(\boldsymbol{E} + \boldsymbol{v} \times \boldsymbol{B})$. So in practice, for modeling the macroscopic EM fields in any

EM application, even when the material is moving, we only need consider the following governing equations:

$\nabla \cdot \boldsymbol{B} = 0$	*Gauss' law*	(3.169a)
$\nabla \cdot \boldsymbol{D} = \rho_e$	*Coulomb's law*	(3.169b)
$\nabla \times \boldsymbol{E} = -\dfrac{\partial \boldsymbol{B}}{\partial t}$	*Faraday's law*	(3.169c)
$\nabla \times \boldsymbol{H} = \dfrac{\partial \boldsymbol{D}}{\partial t} + \boldsymbol{J}_{em} + \rho_e \boldsymbol{v} + \boldsymbol{J}_a$	*Ampère's law*	(3.169d)
$\boldsymbol{D} = \epsilon_0 \boldsymbol{K}_e \cdot \boldsymbol{E}$	(units of C m^{-2})	(3.169e)
$\boldsymbol{B} = \mu_0 \kappa_m \boldsymbol{H}$	(units of Tesla = T = kg C^{-1} s^{-1})	(3.169f)
$\boldsymbol{J}_{em} = \boldsymbol{\sigma} \cdot (\boldsymbol{E} + \boldsymbol{v} \times \boldsymbol{B})$.	(units of C m^{-2} s^{-1}).	(3.169g)

Further, we only expect the velocity term in $\boldsymbol{J}_{em} = \sigma (\boldsymbol{E} + \boldsymbol{v} \times \boldsymbol{B})$ to be of importance when we are in the diffusive or low-frequency limit and when high-conductivity materials like liquid metals or plasmas are being treated, which can be called the *magnetohydrodynamics* regime.

3.9 Electromagnetic Continuity Conditions at an Interface

We now determine how the various EM fields transition across a material interface in a body that is possibly in motion. The interface may be arbitrarily defined as any imagined surface within an otherwise homogeneous or heterogenous body or, more interestingly, as an actual interface between two different materials and upon which an excess surface charge density typically accumulates when subjected to electric fields. There also sometimes resides a double layer of charge on the interface, corresponding to a layer of positive charge on one side of the interface that is balanced by a layer of negative charge on the immediate other side of the interface. This double layer is present prior to an EM application and although not contributing to the net charge on the interface can contribute to a distinct electrical current that is concentrated on the surface.

To derive such interface continuity (or discontinuity) conditions for the EM fields across an interface, we work from a frame of reference that is moving with the interface and integrate each of Maxwell's equations expressed in this frame of reference over the shaded disc-shaped region Ω_ϵ shown in Fig. 3.2 in the limit as the half-width ϵ of the disc is reduced to molecular distances. Beginning with Faraday's law, we therefore consider the following operation:

$$\lim_{\epsilon \to 0} \int_{\Omega_\epsilon} \left(\nabla \times \boldsymbol{E}' + \frac{\mathrm{d}\boldsymbol{B}'}{\mathrm{d}t} \right) \mathrm{d}V = 0. \tag{3.170}$$

This integral is evaluated by performing the following sequence of steps: (1) use the integral theorem of Chapter 1 that says $\int_\Omega \nabla \times \boldsymbol{a} \, \mathrm{d}V = \int_{\partial\Omega} \boldsymbol{n} \times \boldsymbol{a} \, \mathrm{d}S$, to rewrite the volume

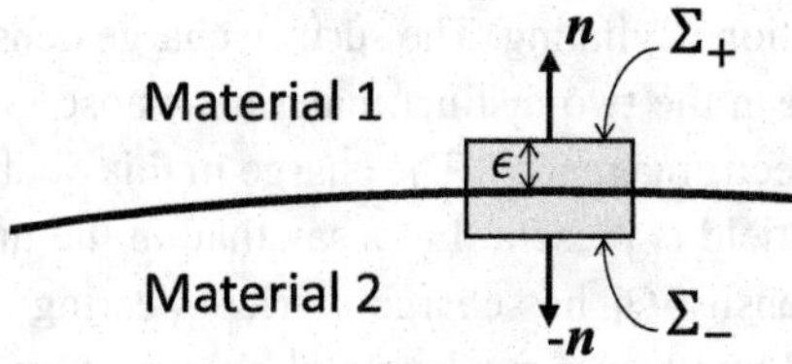

Figure 3.2 A disc Ω_ϵ straddling an interface in the limit as ϵ shrinks to molecular distances.

integral of the curl as a surface integral; (2) as $\epsilon \to 0$ (i.e., molecular distances), allow the volume integral of the total time derivative to vanish entirely; (3) also as $\epsilon \to 0$, ignore entirely the contribution in the surface integral from the circumferential or "ribbon" surface of the disc that resides between the two flat faces Σ^+ and Σ^- that are a distance ϵ apart; and last, (4) describe the fields in the moving frame of reference from the stationary frame $\boldsymbol{E}' = \boldsymbol{E} + \boldsymbol{v} \times \boldsymbol{B}$. This gives

$$\lim_{\epsilon \to 0} \int_{\partial\Omega_\epsilon} \boldsymbol{n} \times (\boldsymbol{E} + \boldsymbol{v} \times \boldsymbol{B})\, \mathrm{d}S = \int_{\Sigma+} \boldsymbol{n}_1 \times (\boldsymbol{E}_1 + \boldsymbol{v} \times \boldsymbol{B}_1)\, \mathrm{d}S + \int_{\Sigma-} \boldsymbol{n}_2\, (\boldsymbol{E}_2 + \boldsymbol{v} \times \boldsymbol{B}_2)\, \mathrm{d}S = 0. \tag{3.171}$$

If we take the two flat faces of the disc to be sufficiently small in lateral extent that the integrands of the surface integrals are constant over each flat surface of area $A = \int_{\Sigma^+} \mathrm{d}S = \int_{\Sigma^-} \mathrm{d}S$ and note that $\boldsymbol{n}_2 = -\boldsymbol{n}_1$ on these flat surfaces, we have the result that

$$\boldsymbol{n}_1 \times (\boldsymbol{E}_1 + \boldsymbol{v} \times \boldsymbol{B}_1 - \boldsymbol{E}_2 - \boldsymbol{v} \times \boldsymbol{B}_2) = 0. \tag{3.172}$$

As per Section 3.7, we only expect the velocity-dependent terms to be important for magnetohydrodynamics problems. In all other applications, we can neglect the velocity terms and arrive at the standard interface condition

$$\boxed{\boldsymbol{n} \times (\boldsymbol{E}_1 - \boldsymbol{E}_2) = 0,} \tag{3.173}$$

which says that the horizontal component of the electric field is continuous even across a moving interface.

Let's next integrate Ampère's law in the moving frame of reference over the same disc straddling the interface with the same assumptions as above to give

$$\lim_{\epsilon \to 0} \int_{\Omega_\epsilon} \left(\nabla \times \boldsymbol{H}' - \frac{\mathrm{d}\boldsymbol{D}'}{\mathrm{d}t} - \boldsymbol{J}_{em} \right) \mathrm{d}V = 0. \tag{3.174}$$

If the interface is between two contrasting media or phases, there sometimes develops a charge separation across the interface that can be characterized as a *double layer* of electrical charge. Charged sites on one side of the interface may react with ions on the other side resulting in a layer of excess charge Q_1 on one side (Coulombs per meter squared of a given sign) that has the opposite sign from a layer $Q_2 = -Q_1$ on the other side of the interface. These two charge layers reside within molecular distances ϵ of each other and are present prior to a particular EM application beginning. The surface charge densities are commonly set by chemical reactions between the two distinct materials whose modeling lies outside of our theory of macroscopic electromagnetics. The charge in this double layer can be free to move if a tangential electric field is present. Let's say that on the material 1 side of the interface, the surface-charge density Q_1 has charge carriers bearing a charge $z_1 e$, where the valence z_1 sets the sign and number of fundamental charges e on each charge carrier and has the same sign as Q_1. Further, we can assign a mobility b_1 to these charge carriers that has units of velocity divided by force. If there were multiple species of charge carriers contributing to the net charge density, each with a distinct mobility, we would need to sum over these various species. If say the charge layer Q_1 corresponds to the ionized mineral surface sites of a solid surface, then $b_1 = 0$ while if they correspond to free ions, we have $b_1 \neq 0$. Immediately across the interface is a layer of charge density $Q_2 = -Q_1$ that has charge carriers $z_2 = -z_1$ with their own distinct mobility b_2. When the tangential component of an electric field $\boldsymbol{E}_t$ is present at the interface

$$\boldsymbol{E}_t = \boldsymbol{E} - (\boldsymbol{n} \cdot \boldsymbol{E})\boldsymbol{n} = (\boldsymbol{I} - \boldsymbol{n}\boldsymbol{n}) \cdot \boldsymbol{E} = \boldsymbol{n} \times \boldsymbol{E} \times \boldsymbol{n} \tag{3.175}$$

it acts on each of these two charge layers at the surface and creates a surface current $\boldsymbol{j}_s$ (C/s/m) given by

$$\boldsymbol{j}_s = \frac{\lim_{\epsilon \to 0} \int_{\Omega_\epsilon} \boldsymbol{J}_{em} \, \mathrm{d}V}{A} = e z_1 Q_1 \, (b_1 + b_2) \, \boldsymbol{E}_t, \tag{3.176}$$

where $z_1 Q_1 = z_2 Q_2$ is always positive.

Taking the same steps for Eq. (3.174) as in our handling above of the integration over the shrinking disc of Faraday's law, including neglecting the velocity terms which is justified in this case for all EM regimes including the magnetohydrodynamics regime, gives

$$\boxed{\boldsymbol{n}_1 \times (\boldsymbol{H}_1 - \boldsymbol{H}_2) = \boldsymbol{j}_s.} \tag{3.177}$$

If in a given scenario either $Q_1 = 0$ or $b_1 = b_2 = 0$ so that $\boldsymbol{j}_s = 0$, there is no distinct surface current on the interface and the tangential component of the magnetic field is continuous

$$\boldsymbol{n}_1 \times (\boldsymbol{H}_1 - \boldsymbol{H}_2) = 0, \tag{3.178}$$

which is the most common scenario. In practice, if the surface current density $\boldsymbol{j}_s$ (C/s/m) corresponds to a current $\ell|\boldsymbol{j}_s|$ passing normally across a line of length ℓ on the surface and if the volume current density $\boldsymbol{J}_{em}$ (C/s/m^2) producing the magnetic field $\boldsymbol{H}$ corresponds to a current $\epsilon\ell|\boldsymbol{J}_{em}|$, where ϵ is the molecular-scale width of the zone that carries the surface current, then so long as $|\boldsymbol{j}_s| \ll \epsilon|\boldsymbol{J}_{em}|$, Eq. (3.178) can be used with accuracy.

Integrating the conservation of charge in a frame moving with the interface, $\nabla \cdot \boldsymbol{J}_{em} + \mathrm{d}\rho_e/\mathrm{d}t = 0$, and taking the same series of steps in treating the shrinking disc while allowing for a surface current $\boldsymbol{j}_s$ in the plane of the interface, one obtains

$$\boldsymbol{n}_1 \cdot (\boldsymbol{\sigma}_1 \cdot \boldsymbol{E}_1 - \boldsymbol{\sigma}_2 \cdot \boldsymbol{E}_2) + \nabla_t \cdot \boldsymbol{j}_s = -\frac{\partial q_f}{\partial t} \tag{3.179}$$

where

$$q_f = \frac{\lim_{\epsilon \to 0} \int_{\Omega_\epsilon} \rho_e \, dV}{A} \tag{3.180}$$

is the excess free-charge density q_f (C/m^2) on the surface that accumulates through time if there is more current coming into a patch of the interface than leaving. We see that there are two contributions to this accumulation of excess free charge: (1) if there is more current coming into the surface from a direction normal to the interface than is leaving normally from the other side, or (2) if the surface currents, which are confined to the interface, have a nonzero divergence on the surface $\nabla_t \cdot \boldsymbol{j}_s$, where ∇_t corresponds to tangential spatial derivatives in the plane of the interface. Formally, if there is a tangential electrical field $\boldsymbol{E}_t$ present at the surface, it acts on the accumulating free charge q_f so that the surface current should more completely be written using a rule such as

$$\boldsymbol{j}_s = e\left[z_1 Q_1 (b_1 + b_2) + z_f q_f b_f\right] \boldsymbol{E}_t, \tag{3.181}$$

where z_f is the valence (sign and number of fundamental charges) associated with the charge carriers represented by q_f and where b_f is the mobility of these charge carriers. In practice, because the free-charge density that builds on the surface q_f is proportional to the electric fields in a given application and because the electric fields then act back on this accumulated charge q_f to create a surface current, allowing for this source of surface current is a nonlinear effect and is often entirely negligible. So we can allow for q_f to be generated on a surface in a given application involving conductors (metal or ionic) and simultaneously neglect the associated surface current so that the transverse component of the magnetic fields are continuous across the interface as in Eq. (3.178), assuming that the double layer of charge present before the start of the EM application is also negligible. This approximation also corresponds to neglecting $\nabla_t \cdot \boldsymbol{j}_s$ in Eq. (3.179). We will put numbers to these notions in Section 3.10.2.

By integrating Coulomb's law $\nabla \cdot \boldsymbol{D}' = \rho_e$ over the same shrinking disc, we additionally obtain

$$\epsilon_0 \boldsymbol{n}_1 \cdot (\boldsymbol{K}_{e1} \cdot \boldsymbol{E}_1 - \boldsymbol{K}_{e2} \cdot \boldsymbol{E}_2) = q_f. \tag{3.182}$$

Equations (3.179) and (3.182) together form a time differential equation for $q_f(t)$ that can be solved under the initial condition that $q_f(0) = 0$ at the start of an EM application. In solving a particular problem to obtain solutions for the EM fields in a region having a material interface, one must use all of the above continuity conditions. As will be seen in Section 3.10.1, the time Δt it takes for the free charge to accumulate to a steady state on the interface goes as the ratio of permittivity to conductivity ($\Delta t \approx \varepsilon/\sigma$). If T is a characteristic time period of the source or boundary condition that is exciting the time-varying fields, so long as $T \gg \Delta t$ we can take $\partial q_f/\partial t = 0$ in Eq. (3.179) and solve for all the EM fields. Once the electric fields are known, we can determine q_f (the excess free charge per unit area on the surface) if so desired using Eq. (3.182).

For quasi-static problems characterized by an electric potential $\varphi(\boldsymbol{r}, t)$ and magnetic vector potential $\boldsymbol{A}(\boldsymbol{r}, t)$ that is solenoidal $\nabla \cdot \boldsymbol{A} = 0$, the continuity of the horizontal component of the electric field is equivalent to the condition that the electric potential is continuous across the interface, that is, $\varphi_1 = \varphi_2$ so that upon taking the tangential derivative of this continuity condition, we obtain that the tangential component of the electric field is continuous as required. In a magnetostatic problem, the continuity of the normal component of the magnetic induction is obtained by integrating Gauss' law over a disc straddling the interface to give $\boldsymbol{n} \cdot (\mu_1 \boldsymbol{H}_1 - \mu_2 \boldsymbol{H}_2) = 0$ or $\boldsymbol{n} \cdot (\nabla \times \boldsymbol{A}_1 - \nabla \times \boldsymbol{A}_2) = 0$, which is satisfied if $\boldsymbol{A}_1 = \boldsymbol{A}_2$ along the interface. The tangential component of the magnetostatic field satisfies Eq. (3.177) derived above $\boldsymbol{n} \times \left[(\nabla \times \boldsymbol{A}_1)/\mu_1 - (\nabla \times \boldsymbol{A}_2)/\mu_2\right] = \boldsymbol{j}_s$. We will not further consider magnetostatic problems in this book.

Earth materials always have a nonzero electrical conductivity. But we can imagine problems posed in perfect dielectrics where there are no free-charge carriers and the material can only become polarized. We might have some free charge that acts as the source for the electric fields and results in nonzero values of the free-charge density ρ_e where that charge resides, but there are no steady-state electric currents in a dielectric (i.e., no electromigration of free charge). For the interface between two dielectrics, the integration of $\nabla \cdot \boldsymbol{D} = \rho_e$ over a disc results in

$$\boldsymbol{n} \cdot (\boldsymbol{D}_1 - \boldsymbol{D}_2) = 0, \tag{3.183}$$

where the free-charge density on the interface between dielectrics is zero $q_f = 0$. For electrostatic problems, for which $\nabla \times \boldsymbol{E} = 0$ and $\boldsymbol{E} = -\nabla\phi$, you use the continuity condition of Eq. (3.183) stated as $\kappa_1 \boldsymbol{n} \cdot \nabla\varphi_1 = \kappa_2 \boldsymbol{n} \cdot \nabla\varphi_2$ for an isotropic dielectric where the κ_i are the dielectric constants of the two dielectrics. You use this condition for a dielectric in place of the condition $\sigma_1 \boldsymbol{n} \cdot \nabla\varphi_1 = \sigma_2 \boldsymbol{n} \cdot \nabla\varphi_2$ that holds for steady-state problems when there is finite isotropic conductivity present (and negligible surface current on the interface).

Finally, in problems where there is both electromigration and electric polarization taking place, as in Earth materials, the total excess surface charge q_s on the interface is the sum of the free-charge density q_f and a bound-charge density q_b on the surface. The free-charge surface density is defined from the interface condition

$$\boldsymbol{n} \cdot (\boldsymbol{D}_1 - \boldsymbol{D}_2) = \epsilon_0 \boldsymbol{n} \cdot (\kappa_1 \boldsymbol{E}_1 - \kappa_2 \boldsymbol{E}_2) = q_f, \tag{3.184}$$

while the bound-charge surface density q_b is created by polarization differences across the interface. You can think of the excess bound charge q_b as being due to more (or less) charge from material-2 dipole heads on the interface compared to the charge from material-1 dipole tails. The total surface charge density q_s is obtained by integrating $\epsilon_0 \nabla \cdot \boldsymbol{E} = \nabla \cdot \boldsymbol{D} - \nabla \cdot \boldsymbol{P}$ over the same shrinking disc that straddles the interface to obtain

$$\epsilon_0 \boldsymbol{n} \cdot (\boldsymbol{E}_1 - \boldsymbol{E}_2) = q_s = q_f + q_b. \tag{3.185}$$

If the free-charge excess q_f for isotropic materials is given by Eq. (3.184) and $q_s = q_f + q_b$ is given by Eq. (3.185), then the bound surface-charge density is given by

$$-\boldsymbol{n} \cdot (\boldsymbol{P}_1 - \boldsymbol{P}_2) = \epsilon_0 \boldsymbol{n} \cdot [(1 - \kappa_1)\boldsymbol{E}_1 - (1 - \kappa_2)\boldsymbol{E}_2] = q_b. \tag{3.186}$$

So the polarized excess surface charge q_b at an interface is controlled by the extent to which the dielectric constants are greater than one. If the interface is between two dielectrics, then we have $q_f = 0$ so that $q_s = q_b$ in Eq. (3.185). For the case of two dielectrics, q_s is calculated (if desired) using Eq. (3.185) once the electric fields are determined using the two continuity conditions $\boldsymbol{n} \cdot (\boldsymbol{D}_1 - \boldsymbol{D}_2) = 0$ and $\boldsymbol{n} \times (\boldsymbol{E}_1 - \boldsymbol{E}_2) = 0$, which for electrostatic problems in isotropic dielectrics are equivalent to $\kappa_1 \boldsymbol{n} \cdot \nabla \varphi_1 = \kappa_2 \boldsymbol{n} \cdot \nabla \varphi_2$ and $\varphi_1 = \varphi_2$.

3.10 Example Problems That Use the Continuity Conditions

3.10.1 The Time to Build Up Charge on an Interface

As an example that uses the above interface conditions to obtain both the accumulating free-charge density $q_f(t)$ on an interface and the fields themselves, consider two infinite planar slabs of different isotropic materials that interface along the surface $x = 0$ as shown in Fig. 3.3. The problem is modeled here as being purely quasi-electrostatic $\boldsymbol{E} = -\nabla \varphi$, that is, induction effects are ignored, which means that the EM skin depth δ_{em} of the transient EM response is assumed to be much larger than the distance between the two electrodes. The quasi-static electric potential $\varphi(x, t)$ satisfies the boundary conditions on the bounding electrodes that $\varphi(d, t) = +V$ at $x = d$ and $\varphi(-d, t) = -V$ at $x = -d$. For $t < 0$ prior to the voltages being applied, the interface at $x = 0$ has no surface-charge density. At $t = 0$, the voltages are applied and it takes a finite time for the electric fields to achieve a steady state as free surface-charge density $q_f(t)$ accumulates on the interface $x = 0$.

The electric potentials in each material $\varphi_1(x, t)$ and $\varphi_2(x, t)$ are solutions of the Laplace equation $\nabla^2 \varphi = 0$, which in this simple 1D problem have the solution

$$\varphi_i(x, t) = A_i(t) - E_i(t)x, \tag{3.187}$$

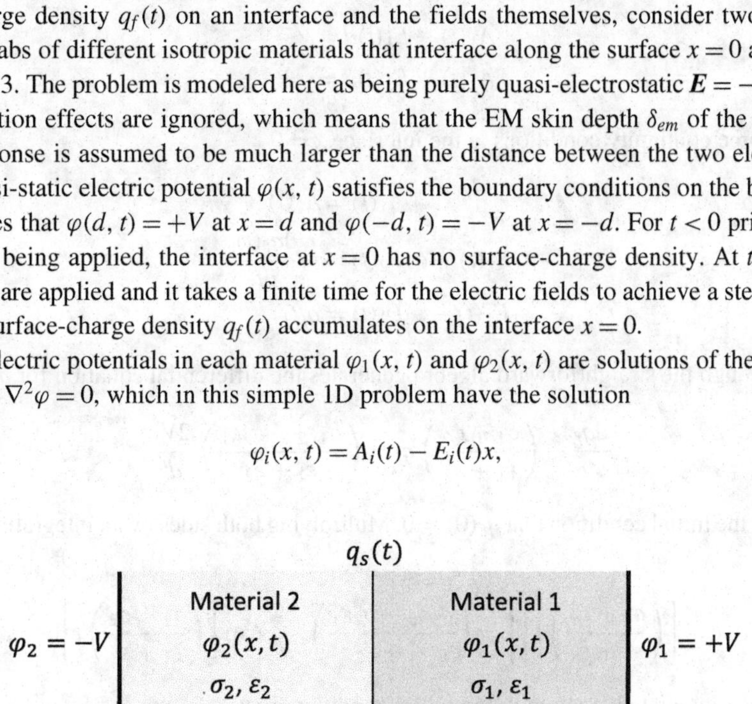

Figure 3.3 Potentials $\varphi = \pm V$ are suddenly applied at $t = 0$ to the electrodes at $x = \pm d$, where V is a constant in time once applied leading to an electric field $E_x(x, t)$ across the system and a building charge density $q_s(t) = q_f(t) + q_b(t)$ on $x = 0$ prior to steady state.

where $i = 1, 2$. We thus have five time functions to solve for: $A_1(t)$, $A_2(t)$, $E_1(t)$, $E_2(t)$, and $q_f(t)$. These five time functions are obtained from the five conditions on the various interfaces:

$$\varphi_1(+d, t) = +V \tag{3.188}$$

$$\varphi_2(-d, t) = -V \tag{3.189}$$

$$\varphi_1(0, t) = \varphi_2(0, t) \tag{3.190}$$

$$-\sigma_1 \left.\frac{\partial \varphi_1}{\partial x}\right|_{x=0} + \sigma_2 \left.\frac{\partial \varphi_2}{\partial x}\right|_{x=0} = -\frac{\mathrm{d}q_f(t)}{\mathrm{d}t} \tag{3.191}$$

$$-\varepsilon_1 \left.\frac{\partial \varphi_1}{\partial x}\right|_{x=0} + \varepsilon_2 \left.\frac{\partial \varphi_2}{\partial x}\right|_{x=0} = q_f(t), \tag{3.192}$$

where $\varepsilon_i = \epsilon_0 \kappa_i$ if you prefer to use the dielectric constant for each material. In Eq. (3.191), we can neglect the surface current term $\nabla_t \cdot \boldsymbol{j}_s$ exactly because all current in this problem is normal to the various interfaces.

Inserting the form of the solutions for the electric potentials then gives the two voltage boundary conditions on the bounding electrodes

$$A_1(t) - E_1(t)d = V \tag{3.193}$$

$$A_2(t) + E_2(t)d = -V \tag{3.194}$$

and the three continuity conditions at the interface $x = 0$

$$A_1(t) = A_2(t) \tag{3.195}$$

$$\sigma_1 E_1(t) - \sigma_2 E_2(t) = -\frac{\mathrm{d}q_f(t)}{\mathrm{d}t} \tag{3.196}$$

$$\varepsilon_1 E_1(t) - \varepsilon_2 E_2(t) = q_f(t). \tag{3.197}$$

Going through the straightforward algebra generates the differential equation for $q_f(t)$

$$\frac{\mathrm{d}q_f}{\mathrm{d}t} + \left(\frac{\sigma_1 + \sigma_2}{\varepsilon_1 + \varepsilon_2}\right) q_f = \left(\frac{\sigma_1 \varepsilon_2 - \sigma_2 \varepsilon_1}{\varepsilon_1 + \varepsilon_2}\right) \frac{2V}{d} \tag{3.198}$$

subject to the initial condition that $q_f(0) = 0$. Multiplying both sides by an integrating factor gives

$$\frac{\mathrm{d}}{\mathrm{d}t}\left\{q_f \exp\left[\left(\frac{\sigma_1 + \sigma_2}{\varepsilon_1 + \varepsilon_2}\right) t\right]\right\} = \left(\frac{\sigma_1 \varepsilon_2 - \sigma_2 \varepsilon_1}{\varepsilon_1 + \varepsilon_2}\right) \frac{2V}{d} \exp\left[\left(\frac{\sigma_1 + \sigma_2}{\varepsilon_1 + \varepsilon_2}\right) t\right], \tag{3.199}$$

which when integrated along with the initial condition gives

$$\boxed{q_f(t) = \left(\frac{\sigma_1 \varepsilon_2 - \sigma_2 \varepsilon_1}{\sigma_1 + \sigma_2}\right) \frac{2V}{d} \left\{1 - \exp\left[-\left(\frac{\sigma_1 + \sigma_2}{\varepsilon_1 + \varepsilon_2}\right) t\right]\right\}.} \tag{3.200}$$

We then obtain the final results for the electric potentials and electric fields as

$$
\begin{aligned}
-\frac{\partial \varphi_1(x,t)}{\partial x} = E_1(t) = &-\left(\frac{\sigma_2}{\sigma_1+\sigma_2}\right)\frac{2V}{d} \\
&-\left(\frac{\sigma_1\varepsilon_2-\sigma_2\varepsilon_1}{(\sigma_1+\sigma_2)(\varepsilon_1+\varepsilon_2)}\right)\frac{2V}{d}\exp\left[-\left(\frac{\sigma_1+\sigma_2}{\varepsilon_1+\varepsilon_2}\right)t\right] \qquad (3.201)\\
-\frac{\partial \varphi_2(x,t)}{\partial x} = E_2(t) = &-\left(\frac{\sigma_1}{\sigma_1+\sigma_2}\right)\frac{2V}{d} \\
&+\left(\frac{\sigma_1\varepsilon_2-\sigma_2\varepsilon_1}{(\sigma_1+\sigma_2)(\varepsilon_1+\varepsilon_2)}\right)\frac{2V}{d}\exp\left[-\left(\frac{\sigma_1+\sigma_2}{\varepsilon_1+\varepsilon_2}\right)t\right] \qquad (3.202)\\
A_1(t) = A_2(t) = &\left(\frac{\sigma_1-\sigma_2}{\sigma_1+\sigma_2}\right)V \\
&-2\left(\frac{\sigma_1\varepsilon_2-\sigma_2\varepsilon_1}{(\sigma_1+\sigma_2)(\varepsilon_1+\varepsilon_2)}\right)V\exp\left[-\left(\frac{\sigma_1+\sigma_2}{\varepsilon_1+\varepsilon_2}\right)t\right]. \qquad (3.203)
\end{aligned}
$$

The total excess charge density q_s on the interface $x = 0$ is the sum of the excess free-charge density q_f that builds through time and the excess bound polarized charge density q_b and is given by $q_s = \epsilon_o\,[E_1(t) - E_2(t)]$ or

$$
q_s(t) = \epsilon_o\left\{\left(\frac{\sigma_1-\sigma_2}{\sigma_1+\sigma_2}\right)\frac{2V}{d} - \left(\frac{\sigma_1\varepsilon_2-\sigma_2\varepsilon_1}{(\sigma_1+\sigma_2)(\varepsilon_1+\varepsilon_2)}\right)\frac{4V}{d}\exp\left[-\left(\frac{\sigma_1+\sigma_2}{\varepsilon_1+\varepsilon_2}\right)t\right]\right\}. \qquad (3.204)
$$

When the voltage drop is first turned on at $t = 0$, this expression becomes

$$
q_s(0) = \epsilon_0\left(\frac{\varepsilon_1-\varepsilon_2}{\varepsilon_1+\varepsilon_2}\right)\frac{2V}{d}. \qquad (3.205)
$$

So the charge on the interface at $t = 0$ is due only to the contrast in polarization across $x = 0$, which occurs "instantaneously" once the voltage drop is turned on in the quasi-electrostatic limit we are treating here.

After a transition time of a few $\Delta t = (\varepsilon_1 + \varepsilon_2)/(\sigma_1 + \sigma_2)$, all the terms in these solutions that contain the exponential disappear and all of the time functions $A_1 = A_2$, E_1, E_2, q_f, and q_s become steady state constants. We have the order-of-magnitude estimate for near-surface Earth properties that $\Delta t = (\varepsilon_1 + \varepsilon_2)/(\sigma_1 + \sigma_2) \approx 10^{-8}$ s, so that steady state is achieved very rapidly after applying the voltage drop across the two slabs. The quasi-static limit treated above requires that the skin depth $\delta_{em} = \sqrt{2/(\sigma\mu\omega)}$ with $\omega \approx 2\pi/\Delta t$ satisfies $\delta_{em} \gg 2d$.

Once in the steady state, the potentials depend only on the conductivities σ_1 and σ_2 and are independent of the permittivities. The steady-state electric-current density $J_{x\infty}$ in the x direction is

$$
J_{x\infty} = \sigma_1 E_1 = \sigma_2 E_2 = -\left(\frac{2}{1/\sigma_1 + 1/\sigma_2}\right)\frac{V}{d} \qquad (3.206)
$$

being a uniform constant throughout the two-slab system. So the effective electrical conductivity of this specific two slab system is $\sigma_e = 2/(1/\sigma_1 + 1/\sigma_2)$, which is called a harmonic average.

3.10.2 A Sphere Immersed in a Uniformly Applied Electric Field

Consider a sphere of conductivity σ_2 and permittivity ε_2 that is embedded in an infinite medium having conductivity σ_1 and permittivity ε_1. There is a uniform electric field $\boldsymbol{E} = E_x\hat{\boldsymbol{x}}$ applied throughout the infinite medium 1 in the x direction that has been present for a very long time so that the problem is defined in the steady state. All this is depicted in Fig. 3.4. One of the primary goals here is to solve this problem first by making the approximation that the surface currents $\boldsymbol{j}_s$ due to free charge q_f that accumulates on the surface are negligible and then to calculate this $\boldsymbol{j}_s$, which is quadratic in the applied electric field, to see whether the approximation is valid.

The fields in this problem are static and controlled by the static governing equations

$$\nabla \times \boldsymbol{E} = 0 \tag{3.207}$$

$$\nabla \times \boldsymbol{H} = \sigma \boldsymbol{E}. \tag{3.208}$$

The first of these (Faraday's law) is satisfied by an electric field $\boldsymbol{E} = -\nabla\varphi$ with φ the electric potential and the second of these (Ampère's law) provides the governing equation for the magnetic field as well as providing the equation for the electric potential after taking the divergence. Because the electrical conductivity is uniform inside of and outside of the sphere, if we take the divergence of Ampère's law, the electric potentials φ_1 (outside the sphere) and φ_2 (inside the sphere) are seen to be solutions of the Laplace equation

$$\nabla^2 \varphi_{1,2} = 0. \tag{3.209}$$

Being a second-order differential equation whose solution requires, at least implicitly, two spatial integrations that generate two unknown integration constants, there are two conditions on the electric potential φ_1 that will determine the two constants in the region $r < a$

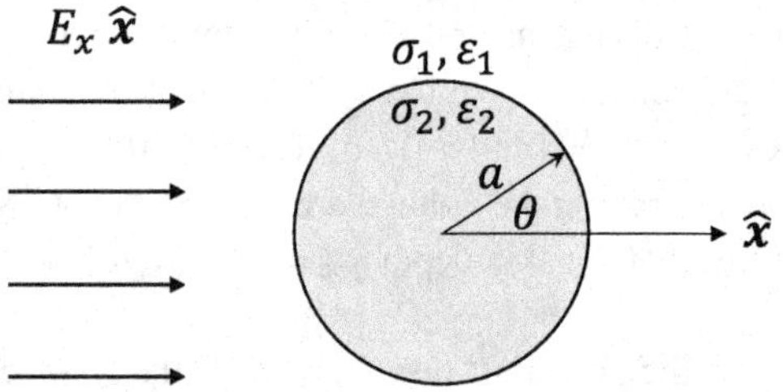

Figure 3.4 An applied electric field $\boldsymbol{E} = E_x\hat{\boldsymbol{x}}$ is uniform throughout an infinite medium 1 having electrical properties σ_1 and ε_1. This electric field is incident on a sphere having electrical properties σ_2 and ε_2, which creates excess surface charge on the sphere surface that is positive in some places and negative in others and makes the total electric fields both inside and outside the sphere be different than the uniform applied field.

and two conditions on φ_2 that determine the two constants in $r > a$. Two of these four needed conditions are the continuity conditions given on $r = a$, one is given at $r = 0$ and another is given at $r = \infty$.

One condition on the sphere surface is that the tangential component of the electric field is continuous across the surface, which is satisfied if the electric potential is continuous

$$\varphi_1 = \varphi_2 \quad \text{on } r = a. \tag{3.210}$$

The other condition is that as much electric current is flowing in normally to each patch on the sphere surface as is flowing out for the fields to be in a steady state and for charge not to be actively accumulating on $r = a$. In stating this condition, we make the approximation that the surface current density $\boldsymbol{j}_s$ is entirely negligible so that the condition of Eq. (3.179) becomes

$$\sigma_1 \frac{\partial \varphi_1}{\partial r}\bigg|_{r=a} \approx \sigma_2 \frac{\partial \varphi_2}{\partial r}\bigg|_{r=a} \tag{3.211}$$

in the steady state. Neglecting the divergence of the surface current in this condition makes the problem linear in the applied field and is the one approximation being made here. This approximation will be justified after the fact. The condition at $r = 0$ is that the electric fields must remain finite while the condition at $r = \infty$ is that the only electric field is the given uniform applied field $\boldsymbol{E} = E_x \hat{\boldsymbol{x}}$ in the x direction.

Given the above, the goal here is to find the functions $\varphi_1(r, \theta)$ in $r > a$ and $\varphi_2(r, \theta)$ in $r < a$ working in spherical coordinates with an origin at the sphere center. Once these are known, we can determine the electric fields $\boldsymbol{E}_{1,2}(r, \theta)$ as well as the amount of free charge (charge per unit area) q_f that the applied field creates on the sphere surface, which allows the actual electric fields to be different from the uniform applied field $\varphi_x = -E_x x = -E_x r\cos\theta$, where $x = r\cos\theta$ in spherical coordinates. The formula derived earlier is that $q_f = \boldsymbol{n} \cdot \boldsymbol{D}_1 - \boldsymbol{n} \cdot \boldsymbol{D}_2$ which for our spherical surface at $r = a$ is given by

$$q_f = -\,\varepsilon_1 \frac{\partial \varphi_1}{\partial r}\bigg|_{r=a} + \varepsilon_2 \frac{\partial \varphi_2}{\partial r}\bigg|_{r=a}, \tag{3.212}$$

where we used that in spherical coordinates $\boldsymbol{n} = \hat{\boldsymbol{r}}$, and $\boldsymbol{D} = -\varepsilon\left(\hat{\boldsymbol{r}}\partial\varphi/\partial r + \hat{\boldsymbol{\theta}} r^{-1}\partial\varphi/\partial\theta\right)$. Last, an integration of Ampère's law then gives the magnetic field driven by the current in the problem.

Prior to developing more formal techniques for such boundary-value problems (BVPs) in Part II, we begin with the Laplace equation expressed in spherical coordinates when the potential is independent of the longitudinal coordinate ϕ as in the present problem

$$\nabla^2\varphi(\boldsymbol{r}) = \frac{1}{r^2}\left[\frac{\partial}{\partial r}\left(r^2\frac{\partial\varphi(r,\theta)}{\partial r}\right) + \frac{1}{\sin\theta}\frac{\partial}{\partial\theta}\left(\sin\theta\frac{\partial\varphi(r,\theta)}{\partial\theta}\right)\right] = 0. \tag{3.213}$$

First, note by direct substitution that the applied field $\varphi_x(r, \theta) = -E_x r\cos\theta$ exactly satisfies this Laplace equation in spherical coordinates. We then suppose that an even more general solution capable of allowing for the induced charge distribution over the sphere surface is of the form

$$\varphi(r,\theta) = -cE_x r\cos\theta\left(\frac{a}{r}\right)^n, \tag{3.214}$$

where c is some constant and n must be found by substituting this expression into Eq. (3.213) and taking derivatives. You should do this and show that

$$-cE_x r\cos\theta\left(\frac{a}{r}\right)^n[(1-n)(-n)+2(1-n)-2]=0. \tag{3.215}$$

So the n that provides a solution to Laplace's equation in spherical coordinates is given by the equation $n(n-3)=0$. Thus, either $n=0$ or $n=3$ in Eq. (3.214) provide solutions to Laplace's equation in spherical coordinates when there is no longitudinal (or ϕ) dependence.

So outside the sphere, we propose a solution of the form

$$\varphi_1(r,\theta) = -c_1E_x r\cos\theta - c_2E_x r\cos\theta\left(\frac{a}{r}\right)^3, \tag{3.216}$$

where c_1 and c_2 are constants that are determined from the boundary conditions. Because as $r\to\infty$ we must have $\varphi_1\to\varphi_x=-E_x r\cos\theta$, we therefore must take $c_1=1$. Similarly, inside the sphere, we propose a solution of the same form

$$\varphi_2(r,\theta) = -c_3E_x r\cos\theta - c_4E_x r\cos\theta\left(\frac{a}{r}\right)^3 \tag{3.217}$$

where, because the fields must remain finite at $r=0$, requires that $c_4=0$. There are therefore two remaining constants to find, c_2 and c_3, which are found from the two continuity conditions on $r=a$.

From Eq. (3.210) on $r=a$, we have that

$$-E_x a\cos\theta - c_2E_x a\cos\theta = -c_3E_x a\cos\theta \tag{3.218}$$

or $c_3=1+c_2$. Using this in Eq. (3.211) then gives

$$\sigma_1\left(-E_x\cos\theta+2c_2E_x\cos\theta\right) = -\sigma_2(1+c_2)E_x\cos\theta \tag{3.219}$$

as the equation that determines c_2. Thus, the two electric potentials are given by

$$\varphi_1(r,\theta) = -E_x r\cos\theta\left[1+\left(\frac{\sigma_1-\sigma_2}{2\sigma_1+\sigma_2}\right)\left(\frac{a}{r}\right)^3\right], \tag{3.220}$$

$$\varphi_2(r,\theta) = -E_x r\cos\theta\,\frac{3\sigma_1}{2\sigma_1+\sigma_2}. \tag{3.221}$$

The associated electric fields $\boldsymbol{E} = -\hat{\boldsymbol{r}}\partial\varphi/\partial r - \hat{\boldsymbol{\theta}}r^{-1}\partial\varphi/\partial\theta$ are then

$$\boldsymbol{E}_1(r,\theta) = E_x\cos\theta\left[1 - 2\left(\frac{\sigma_1-\sigma_2}{2\sigma_1+\sigma_2}\right)\left(\frac{a}{r}\right)^3\right]\hat{\boldsymbol{r}}$$

$$- E_x\sin\theta\left[1 + \left(\frac{\sigma_1-\sigma_2}{2\sigma_1+\sigma_2}\right)\left(\frac{a}{r}\right)^3\right]\hat{\boldsymbol{\theta}}, \tag{3.222}$$

$$\boldsymbol{E}_2(r,\theta) = E_x\cos\theta\frac{3\sigma_1}{2\sigma_1+\sigma_2}\hat{\boldsymbol{r}} - E_x\sin\theta\frac{3\sigma_1}{2\sigma_1+\sigma_2}\hat{\boldsymbol{\theta}}. \tag{3.223}$$

Note that when $\theta = \pi/2$, we have $\hat{\boldsymbol{\theta}} = -\hat{\boldsymbol{x}}$. The free-charge density (charge per unit area) on the sphere's surface is then

$$q_f = 3E_x\cos\theta\left(\frac{\varepsilon_1\sigma_2 - \varepsilon_2\sigma_1}{2\sigma_1+\sigma_2}\right). \tag{3.224}$$

The total excess charge density on the sphere's surface is the sum of q_f and the bound charge q_b due to the contrast in polarization across $r = a$ and is given by $q_s = -\,\epsilon_0\,(\partial\varphi_1/\partial r - \partial\varphi_2/\partial r)|_{r=a}$ or

$$q_s = 3E_x\cos\theta\epsilon_0\left(\frac{\sigma_2-\sigma_1}{2\sigma_1+\sigma_2}\right). \tag{3.225}$$

Thus, as θ increases from 0 to π, the downfield side of the sphere has a q_s of one sign and the upfield side a q_s of the opposite sign. The total charge on the sphere surface is zero, but the two sides of opposite charge on the sphere surface make the sphere act like a dipole for observation points $r \gg a$.

To obtain the magnetic field $\boldsymbol{H}$ that is created by the current in this problem, note that the current is in the $\hat{\boldsymbol{r}}$ and $\hat{\boldsymbol{\theta}}$ directions so that the only nonzero component of the magnetic field is perpendicular to these current components and thus $\boldsymbol{H} = H_\phi(r,\theta)\hat{\boldsymbol{\phi}}$. To obtain this magnetic field, write Ampère's law in spherical coordinates using the curl expression from Chapter 1 along with $H_\theta = H_r = 0$ to obtain in $r > a$ that the $\hat{\boldsymbol{\theta}}$ component of Ampère's law is

$$-\frac{1}{r}\frac{\partial}{\partial r}(rH_{1\phi}) = -\sigma_1 E_x\sin\theta\left[1 + \left(\frac{\sigma_1-\sigma_2}{2\sigma_1+\sigma_2}\right)\left(\frac{a}{r}\right)^3\right]. \tag{3.226}$$

Integrating this over r after multiplying through by r gives the solution

$$H_{1\phi}(r,\theta) = \frac{1}{2}\sigma_1 E_x r\sin\theta\left[1 - 2\left(\frac{\sigma_1-\sigma_2}{2\sigma_1+\sigma_2}\right)\left(\frac{a}{r}\right)^3\right] + \frac{c_{1\theta}(\theta)}{r}, \tag{3.227}$$

where the integration constant $c_{1\theta}(\theta)$ is some function of θ to be determined. The $\hat{\boldsymbol{r}}$ component of Ampère's law is

$$\frac{1}{r\sin\theta}\left[\frac{\partial}{\partial\theta}\left(\sin\theta H_{1\phi}\right)\right]=\sigma_1 E_x\cos\theta\left[1-2\left(\frac{\sigma_1-\sigma_2}{2\sigma_1+\sigma_2}\right)\left(\frac{a}{r}\right)^3\right]. \tag{3.228}$$

Multiplying through by $r\sin\theta$ and integrating over θ then gives

$$H_{1\phi}(r,\theta)=\frac{1}{2}\sigma_1 E_x r\sin\theta\left[1-2\left(\frac{\sigma_1-\sigma_2}{2\sigma_1+\sigma_2}\right)\left(\frac{a}{r}\right)^3\right]+\frac{c_{1r}(r)}{\sin\theta}, \tag{3.229}$$

where the integration constant $c_{1r}(r)$ is some function of r. We must have $c_{1\theta}/r=c_{1r}/\sin\theta$ or $c_{1\theta}=c/\sin\theta$ and $c_{1r}=c/r$, where c is some constant independent of θ and r. Because $\sin\theta=0$ at $\theta=0$ (and π), which is in our domain of interest, we must take $c=0$ for the magnetic field to be finite at $\theta=0$. So the magnetic field in $r>a$ is

$$\boxed{\boldsymbol{H}_1(r,\theta)=H_{1\phi}(r,\theta)\hat{\boldsymbol{\phi}}=\frac{1}{2}\sigma_1 E_x r\sin\theta\left[1-2\left(\frac{\sigma_1-\sigma_2}{2\sigma_1+\sigma_2}\right)\left(\frac{a}{r}\right)^3\right]\hat{\boldsymbol{\phi}}.} \tag{3.230}$$

As $r\to\infty$ there is an infinite amount of current flowing in the x direction, so there is a linear in r divergence to the magnetic field that is inherent to the unusual nature of having a uniformly applied field (and electric current) extending out to infinity.

For the inside of the sphere $r<a$, the $\hat{\boldsymbol{\theta}}$ component of Ampère's law is

$$-\frac{1}{r}\frac{\partial}{\partial r}\left(rH_{2\phi}\right)=-\sigma_2 E_x\sin\theta\frac{3\sigma_1}{2\sigma_1+\sigma_2}, \tag{3.231}$$

while the $\hat{\boldsymbol{r}}$ component is

$$\frac{1}{r\sin\theta}\frac{\partial}{\partial\theta}\left(\sin\theta H_{2\phi}\right)=\sigma_2 E_x\cos\theta\frac{3\sigma_1}{2\sigma_1+\sigma_2}. \tag{3.232}$$

Integrating the θ-component equation after multiplying by r gives

$$H_{2\phi}(r,\theta)\hat{\boldsymbol{\phi}}=\frac{3}{2}E_x r\sin\theta\frac{\sigma_1\sigma_2}{2\sigma_1+\sigma_2}+\frac{c_{2\theta}(\theta)}{r}. \tag{3.233}$$

Because the magnetic field remains finite at $r=0$, the integration constant is zero $c_{2\theta}(\theta)=0$. Integrating the r-component equation after multiplying through be $r\sin\theta$ gives the same result. So we have

$$\boxed{\boldsymbol{H}_2=H_{2\phi}(r,\theta)\hat{\boldsymbol{\phi}}=\frac{3}{2}E_x r\sin\theta\frac{\sigma_1\sigma_2}{2\sigma_1+\sigma_2}\hat{\boldsymbol{\phi}}} \tag{3.234}$$

and we see that on $r=a$, the tangential component of the magnetic field is continuous across $r=a$: $H_{1\phi}(a,\theta)=H_{2\phi}(a,\theta)$ as it must be because we neglected the surface current in writing the condition of Eq. (3.217). In this example, the excess free-charge density q_f is present on the surface and is acted upon by a nonzero electric field that is tangential to the surface $\boldsymbol{E}_t = -E_x \sin\theta 3\sigma_1/(2\sigma_1+\sigma_2)\hat{\boldsymbol{\theta}}$, but we have assumed that the resulting surface current density, which is quadratic in E_x, is negligible despite being nonzero.

To justify this approximation, let's insert some numbers. Take material 1 to be a liquid electrolyte having a salt concentration of M moles per liter (i.e., ionic conduction in material 1) so that the charges that build on the sphere surface are ions. The condition for neglecting the surface current $\boldsymbol{j}_s$ (C/s/m) is that $|\boldsymbol{j}_s| \ll \epsilon|\boldsymbol{J}_{em}|$, where ϵ is the molecular-scale width of the interface where surface current is occurring. The surface current can be modeled as $\boldsymbol{j}_s = ezbq_f\boldsymbol{E}_t$, where b is the mobility of the ions that accumulate on the surface and z the valence. Using approximate values we have $q_f \approx \epsilon_0 E_x$, where $\epsilon_0 = 8.85\times10^{-12}$ F/m and $|\boldsymbol{E}_t| \approx E_x$ so that $|\boldsymbol{j}_s| \approx eb\epsilon_0 E_x^2$. Similarly, the ionic electromigration (C/s/m^2) in the electrolyte goes as $\boldsymbol{J}_{em} = \sigma_1\boldsymbol{E}$ or $|\boldsymbol{J}_{em}| \approx e^2 bN_A(10^3 \text{ liters/m}^3)ME_x$, where $N_A = 6\times10^{23}$ is Avogadro's number and the 10^3 is how many liters are in a cubic meter. Thus, the condition for the surface current to be negligible is that

$$E_x \ll \frac{\epsilon e N_A 10^3 M}{\epsilon_0} \approx \frac{(10^{-9})(10^{-19})(10^{24})(10^3)}{10^{-11}}M = 10^{10}M \tag{3.235}$$

in volts per meter. The smallest possible molarity of an aqueous electrolyte is 10^{-7} moles per liter (pure water) while $M = 10$ moles per liter is approaching salt saturation. Thus, unless the applied electric field is extraordinarily large, the nonlinear surface current is negligible in most applications as was assumed here from the beginning. If the sphere is metallic with $\sigma_1/\sigma_2 \to 0$, then $\boldsymbol{E}_t = -E_x\sin\theta 3\sigma_1/\sigma_2\hat{\boldsymbol{\theta}} \to 0$ and surface currents are zero. The charge induced on a metal surface is $q_f = 3E_x\varepsilon_1\cos\theta$, while the electric field inside and on a metallic sphere is zero per the above expressions. The steady-state (dc) real dielectric constant of a metal is not experimentally measured and is commonly taken to be 1, i.e., $\varepsilon_2 = \epsilon_0$ for a metal and plays no role in the above problem in the metallic-sphere limit where $\sigma_1/\sigma_2 \to 0$.

If the problem had been posed as a "metallic sphere embedded in insulating air," where, effectively, $\sigma_1 \to 0$, $\varepsilon_1 \to \epsilon_0$, and $\sigma_2 \to \infty$ (i.e., fantastically greater than σ_1), the above expressions reduce to

$$\varphi_1(r,\theta) = -E_x r\cos\theta\left[1-\left(\frac{a}{r}\right)^3\right], \tag{3.236}$$

$$\varphi_2(r,\theta) = 0, \tag{3.237}$$

$$\boldsymbol{E}_1(r,\theta) = E_x\cos\theta\left[1+2\left(\frac{a}{r}\right)^3\right]\hat{\boldsymbol{r}} - E_x\sin\theta\left[1-\left(\frac{a}{r}\right)^3\right]\hat{\boldsymbol{\theta}}, \tag{3.238}$$

$$\boldsymbol{E}_2(r,\theta) = 0, \tag{3.239}$$

$$q_f = 3E_x\epsilon_0\cos\theta, \tag{3.240}$$

$$\boldsymbol{H}_1 = \boldsymbol{H}_2 = 0. \tag{3.241}$$

So there is no electric field inside of the metal conductor. The surface charge excess q_f corresponds to mobile electrons in the conduction band of the metal but because the tangential electric field on the surface of the sphere is exactly zero, there truly is no surface current or indeed any current in this problem, which is why the magnetic fields are zero. As an end-of-chapter exercise, you can solve the similar problem when the sphere and surrounding material are perfect dielectrics.

3.11 Electromagnetic Boundary Conditions and Initial Conditions

We now derive the nature of the boundary conditions on the closed surface $\partial\Omega$ that surrounds a modeling domain Ω as well as the nature of the initial conditions throughout Ω. Such boundary conditions on the limits of our system domain are distinct from the interface continuity conditions on a material or other interface residing within the domain that was the focus of Sections 3.9 and 3.10. Given a linearized form of the EM governing equations, these initial and boundary conditions are those that permit a unique solution throughout the domain.

Let's imagine that in a heterogeneous and anisotropic domain Ω, there are two sets of EM fields, $\boldsymbol{E}_1$ and $\boldsymbol{H}_1$ and $\boldsymbol{E}_2$ and $\boldsymbol{H}_2$, where each set satisfies the same boundary conditions on $\partial\Omega$, the same initial conditions at $t=0$ throughout Ω and are generated by the same source term $\boldsymbol{J}_a$. If we can choose the boundary and initial conditions such that the difference fields $\delta\boldsymbol{E}=\boldsymbol{E}_1-\boldsymbol{E}_2$ and $\delta\boldsymbol{H}=\boldsymbol{H}_1-\boldsymbol{H}_2$ are everywhere zero, we say the solution is *unique*. A problem that gives unique solutions is said to be *well posed*. So our goal here is to find the nature of the boundary and initial conditions such that an EM BVP is well posed and the solution within the domain is guaranteed to be unique.

If we subtract the EM governing equations for $\boldsymbol{E}_1$ and $\boldsymbol{H}_1$ from those controlling $\boldsymbol{E}_2$ and $\boldsymbol{H}_2$, the difference fields therefore satisfy

$$\nabla\times\delta\boldsymbol{E}=-\boldsymbol{\mu}\cdot\frac{\partial\delta\boldsymbol{H}}{\partial t} \tag{3.242}$$

$$\nabla\times\delta\boldsymbol{H}=\boldsymbol{\varepsilon}\cdot\frac{\partial\delta\boldsymbol{E}}{\partial t}+\boldsymbol{\sigma}\cdot\delta\boldsymbol{E}. \tag{3.243}$$

The source term $\boldsymbol{J}_a$ is absent because it is common to both sets of possible EM fields. For generality, we assume that $\boldsymbol{\mu}$, $\boldsymbol{\varepsilon}$, and $\boldsymbol{\sigma}$ are all symmetric, positive-definite, real, second-order tensors and have arbitrary spatial variability throughout the domain Ω. A symmetric second-order tensor $\boldsymbol{A}$ is said to be *positive definite* if the scalar $\boldsymbol{a}\cdot\boldsymbol{A}\cdot\boldsymbol{a}\geq 0$ for all real vectors $\boldsymbol{a}$. That $\boldsymbol{\mu}$ and $\boldsymbol{\varepsilon}$ are symmetric and positive-definite tensors is a result of equilibrium thermodynamics and is due to these material properties being the second-derivatives of a fundamental energy function (which guarantees their symmetry) as well as to the concept of stability (which guarantees they are positive definite). These ideas will be developed in Chapter 6 that is devoted to equilibrium thermodynamics. That $\boldsymbol{\sigma}$ is a symmetric and positive-definite tensor is a result of nonequilibrium thermodynamics and is due to the time reversibility of the underlying molecular dynamics (which guarantees symmetry) and the second law of thermodynamics (which guarantees positive definiteness). These ideas will be developed and derived in Chapter 7 that is devoted to nonequilibrium thermodynamics. Again, if we can show that $\delta\boldsymbol{E}$ and $\delta\boldsymbol{H}$ are everywhere zero throughout the domain for all time, the solution to our EM problem is unique and well posed.

Dot multiply Eq. (3.243) with $\delta\boldsymbol{E}$ and Eq. (3.242) with $\delta\boldsymbol{H}$, subtract, and integrate over all of space and time to give

$$-\int_0^t \mathrm{d}t' \int_\Omega \nabla\cdot(\delta\boldsymbol{E}\times\delta\boldsymbol{H})\,\mathrm{d}V$$

$$= \int_0^t \mathrm{d}t' \int_\Omega \left[\frac{1}{2}\frac{\partial}{\partial t}(\delta\boldsymbol{E}\cdot\boldsymbol{\varepsilon}\cdot\delta\boldsymbol{E}+\delta\boldsymbol{H}\cdot\boldsymbol{\mu}\cdot\delta\boldsymbol{H})+\delta\boldsymbol{E}\cdot\boldsymbol{\sigma}\cdot\delta\boldsymbol{E}\right]\mathrm{d}V. \tag{3.244}$$

Using the divergence theorem on the left-hand side and carrying out the time integral of the time derivative term on the right-hand side then gives

$$-\int_0^t \mathrm{d}t' \int_{\partial\Omega} \boldsymbol{n}\cdot(\delta\boldsymbol{E}\times\delta\boldsymbol{H})\,\mathrm{d}S$$

$$= \frac{1}{2}\int_\Omega \Bigg[\delta\boldsymbol{E}(t)\cdot\boldsymbol{\varepsilon}\cdot\delta\boldsymbol{E}(t)+\delta\boldsymbol{H}(t)\cdot\boldsymbol{\mu}\cdot\delta\boldsymbol{H}(t)-\delta\boldsymbol{E}(0)\cdot\boldsymbol{\varepsilon}\cdot\delta\boldsymbol{E}(0)-\delta\boldsymbol{H}(0)\cdot\boldsymbol{\mu}\cdot\delta\boldsymbol{H}(0)$$

$$+\int_0^t \delta\boldsymbol{E}(t')\cdot\boldsymbol{\sigma}\cdot\boldsymbol{E}(t')\,\mathrm{d}t'\Bigg]\,\mathrm{d}V. \tag{3.245}$$

We now impose the following conditions on our BVP:

1. **Boundary Conditions**: Require that $\boldsymbol{n}\cdot(\delta\boldsymbol{E}\times\delta\boldsymbol{H}) = -\delta\boldsymbol{E}\cdot(\boldsymbol{n}\times\delta\boldsymbol{H}) = \delta\boldsymbol{H}\cdot(\boldsymbol{n}\times\delta\boldsymbol{E}) = 0$ so that the left-hand side of Eq. (3.245) is zero. This can occur either by specifying the tangential components of $\boldsymbol{E}$ over the boundary so that $\boldsymbol{n}\times\delta\boldsymbol{E}=0$ or by specifying the tangential components of $\boldsymbol{H}$ so that $\boldsymbol{n}\times\delta\boldsymbol{H}=0$. Note as well that if $\boldsymbol{n}\times\boldsymbol{E}$ is known for all time over the boundary, then after performing time differentiation in Ampère's law, $\boldsymbol{n}\times\nabla\times\boldsymbol{H}$ is also known on the boundary. Similarly, if $\boldsymbol{n}\times\boldsymbol{H}$ is known for all time over the boundary, then after performing time differentiation in Faraday's law, $\boldsymbol{n}\times\nabla\times\boldsymbol{E}$ is known on the boundary. Thus, we can say equivalently that when either $\boldsymbol{n}\times\boldsymbol{E}$ or $\boldsymbol{n}\times\nabla\times\boldsymbol{E}$ are given at each point of $\partial\Omega$, the left-hand side of Eq. (3.245) is zero.
2. **Initial Conditions**: Require that both $\boldsymbol{E}(\boldsymbol{r},0)$ and $\boldsymbol{H}(\boldsymbol{r},0)$ are specified at time zero throughout all of Ω. This results in $\delta\boldsymbol{E}(0)=0$ and $\delta\boldsymbol{H}(0)=0$ on the right-hand side of Eq. (3.245). Note that Ampère's law evaluated at $t=0$ gives $\nabla\times\boldsymbol{H}(\boldsymbol{r},0) = \varepsilon\,(\partial\boldsymbol{E}(\boldsymbol{r},t)/\partial t)|_{t=0}+\sigma\boldsymbol{E}(\boldsymbol{r},0)$. So requiring $\boldsymbol{E}(\boldsymbol{r},0)$ and $\boldsymbol{H}(\boldsymbol{r},0)$ to be specified at $t=0$ is equivalent to requiring that $\boldsymbol{E}(\boldsymbol{r},0)$ and $(\partial\boldsymbol{E}(\boldsymbol{r},t)/\partial t)|_{t=0}$ are both specified.

Using these boundary conditions and initial conditions allows Eq. (3.245) to be written

$$\int_\Omega \left[\delta\boldsymbol{E}(t)\cdot\boldsymbol{\varepsilon}\cdot\delta\boldsymbol{E}(t)+\delta\boldsymbol{H}(t)\cdot\boldsymbol{\mu}\cdot\delta\boldsymbol{H}(t)+2\int_0^t \delta\boldsymbol{E}(t')\cdot\boldsymbol{\sigma}\cdot\delta\boldsymbol{E}(t')\,\mathrm{d}t'\right]\mathrm{d}V=0. \tag{3.246}$$

Each term in these integrals is always positive definite which means the only way for this equation to be satisfied is if $\delta\boldsymbol{E}=0$ and $\delta\boldsymbol{H}=0$ throughout all of Ω and for all time. We have thus shown that if we select the boundary conditions and initial conditions as specified above, we have a well-posed BVP that results in unique solutions of the governing equations. Later in Chapter 12, we will show how given nonzero values of the above boundary and initial conditions act as source terms for generating EM response and how, if we know the so-called Green's tensor of EM response, we can obtain solutions for this response.

So the tangential components $\boldsymbol{n} \times \boldsymbol{E}$ or $\boldsymbol{n} \times \boldsymbol{H}$ (or $\boldsymbol{n} \times \nabla \times \boldsymbol{E}$) must be specified over the domain boundary $\partial\Omega$ and the initial fields $\boldsymbol{E}(\boldsymbol{r}, 0)$ and $\boldsymbol{H}(\boldsymbol{r}, 0)$ (or $(\partial \boldsymbol{E}(\boldsymbol{r}, t)/\partial t)|_{t=0}$) must be specified through Ω for an EM problem to be well posed and thus possess a unique solution. For an electrostatic problem involving the electric potential φ, knowing the tangential component of $\boldsymbol{E}$ is equivalent to prescribing φ on $\partial\Omega$. We are not constrained by what these boundary and initial values are, only that they must be given. If, for example, we instead specify only the normal component of the electric or magnetic fields on the system boundary, unique solutions for the electric and magnetic fields throughout the domain cannot be guaranteed and the problem is not well posed.

3.12 Conclusions about Macroscopic Electromagnetism

We have completed our treatment of how to set up well-posed macroscopic BVPs for continuum-EM response that include the processes of electrical and magnetic polarization and electromigration (electrical current). Starting from the underlying molecularity, we derived the form of the macroscopic Maxwell's equations. An energy analysis of these macroscopic governing equations coupled with statements from equilibrium and nonequilibrium thermodynamics allowed us to extract the EM constitutive laws that govern polarization and electromigration processes. To perform this constitutive law analysis exactly requires taking total time derivatives and allowing for the EM effect of the material to be moving if it is moving. But we went on to show that for all practical EM problems of interest, the effect of the moving material can be ignored except in regimes corresponding to low-frequency diffusive behavior in a material with a large electrical conductivity that we call the *magnetohydrodynamic* regime. We also showed how to employ the EM continuity equations in solving problems that have a material interface within the domain of inquiry. Finally, we demonstrated what aspects of the electric and magnetic fields must be specified on the boundary of a system and as initial conditions so that unique solutions of the continuum Maxwell's equations can be obtained. In Chapter 12, we derive the so-called *electromagnetic representation theorem* that shows how nonzero values of these boundary and initial conditions act as source terms for the creation of EM fields within a domain.

You will gain practice and intuition by working through the guided end-of-chapter exercises as you are encouraged to do. In Exercise 1, you can treat the important topic of EM plane waves. For the reflection and transmission of plane EM waves, you can treat that topic as an exercise at the end of Chapter 4 after we treat the analogous problem of the reflection and transmission of elastodynamic (seismic) waves in Chapter 4. There are a large number of classic texts that develop the multitude of applications of the macroscopic EM theory derived in this chapter. Older examples include Stratton (1941), Jackson (1975), Landau and Lifshitz (1984), and Kong (1986), which were the specific four texts I used in developing my own understanding of continuum electromagnetics and its applications.

3.13 Newtonian Gravity

3.13.1 Derivation of the Continuum Laws of Newtonian Gravity

The local laws of gravity for a multitude of discrete particles p having mass m_p take the form

$$\nabla \cdot \boldsymbol{g}(\boldsymbol{r}, t) = -4\pi G \sum_p m_p \delta(\boldsymbol{r} - \boldsymbol{r}_p(t)) \tag{3.247}$$

$$\nabla \times \boldsymbol{g}(\boldsymbol{r}, t) = 0, \tag{3.248}$$

where $\boldsymbol{g}$ is the acceleration of gravity and $G = 6.67 \times 10^{-11} \mathrm{m}^3\ \mathrm{kg}^{-1}\ \mathrm{s}^{-2}$ is the universal gravitational constant. By averaging these equations over an averaging volume, one obtains the continuum form of gravity as

$$\nabla \cdot \overline{\boldsymbol{g}}(\boldsymbol{r}, t) = -4\pi G \bar{\rho}(\boldsymbol{r}, t) \tag{3.249a}$$

$$\nabla \times \overline{\boldsymbol{g}}(\boldsymbol{r}, t) = 0, \tag{3.249b}$$

where

$$\overline{\boldsymbol{g}}(\boldsymbol{r}, t) = \frac{1}{V_0} \int_{\Omega_0} \boldsymbol{g}(\boldsymbol{r}', t)\, \mathrm{d}^3\boldsymbol{r}' \tag{3.250}$$

$$\bar{\rho}(\boldsymbol{r}, t) = \frac{1}{V_0} \int_{\Omega_0} \sum_p m_p \delta(\boldsymbol{r}' - \boldsymbol{r}_p(t))\, \mathrm{d}^3\boldsymbol{r}' = \frac{1}{V_0} \sum_{p=1}^{N_0(\boldsymbol{r},t)} m_p \tag{3.251}$$

are the average acceleration of gravity and mass density of each averaging region $\Omega_0(\boldsymbol{r})$ of volume V_0 that surrounds each point $\boldsymbol{r}$ in a body as particles enter and leave the averaging region over time with $N_0(\boldsymbol{r}, t)$ particles present in the averaging region at each instant t.

These laws for $\overline{\boldsymbol{g}}$ should be compared to Maxwell's equations for $\boldsymbol{E}$ to see the similarities and differences, which indicates that there is no gravitational induction or polarization phenomena. The laws of gravity are directly analogous to the laws of electrostatics in nonpolarizable dielectrics. In the theory of continuum gravity, there are no "gravitational constitutive laws."

As we learned in our vector-calculus exercises in Chapter 1, if $\nabla \times \overline{\boldsymbol{g}} = 0$, we can write $\overline{\boldsymbol{g}} = -\nabla U$, where U is called the gravitational potential. In this case, determining the gravitational fields comes down to solving the Poisson problem

$$\nabla^2 U = 4\pi G \bar{\rho} \tag{3.252}$$

for some spatial distribution of $\bar{\rho}$ and under the condition that U be finite throughout a body being analyzed. With U so determined, we can then determine the acceleration of gravity $\overline{\boldsymbol{g}} = -\nabla U$ and the gravitational body-force density $\boldsymbol{F}_g = \bar{\rho}\overline{\boldsymbol{g}}$.

3.13.2 The Gravitational Force between Discrete Point Masses

Consider a mass m_1 that is sufficiently concentrated in space relative to the radial distances r from the mass under consideration that the mass can be treated as a point source. So the acceleration of gravity due to this point source is controlled by

$$\nabla \cdot \boldsymbol{g}(\boldsymbol{r}) = -4\pi G m_1 \delta(\boldsymbol{r}). \tag{3.253}$$

By spherical symmetry, we have $\boldsymbol{g}(\boldsymbol{r}) = -g_1(r)\hat{\boldsymbol{r}}_1$, where r is radial distance from the mass and $\hat{\boldsymbol{r}}_1$ is the unit radial vector directed away from m_1. To find the scalar function $g_1(r)$, we integrate Eq. (3.253) over a spherical region Ω_r of radius r that surrounds m_1 and apply the divergence theorem to obtain

$$\int_{\partial\Omega_r} \hat{\boldsymbol{r}}_1 \cdot \boldsymbol{g}\, \mathrm{d}^2\boldsymbol{r}' = -4\pi r^2 g_1(r) = -4\pi G m_1. \tag{3.254}$$

Thus the acceleration of gravity from this point mass decreases with distance as

$$g_1(r) = \frac{G m_1}{r^2}. \tag{3.255}$$

If there is a second point mass m_2 a distance r_{12} from m_1, the force $\boldsymbol{f}_{21}$ (Newtons) that m_1 exerts on m_2 is

$$\boldsymbol{f}_{21} = -\frac{G m_1 m_2}{r_{12}^2}\hat{\boldsymbol{r}}_{12}, \tag{3.256}$$

where $\hat{\boldsymbol{r}}_{12}$ is directed from 1 toward 2. If we had taken the opposite perspective and considered the acceleration of gravity coming from m_2 that then acts on m_1, the force that m_2 exerts on m_1 is

$$\boldsymbol{f}_{12} = -\frac{G m_1 m_2}{r_{12}^2}\hat{\boldsymbol{r}}_{21}. \tag{3.257}$$

Because on the line between the two masses we have $\hat{\boldsymbol{r}}_{21} = -\hat{\boldsymbol{r}}_{12}$, we have hat $\boldsymbol{f}_{12} = -\boldsymbol{f}_{21}$ (Newton's third law).

3.13.3 The Gravitational Force within a Spherical Body

We now treat the problem of how the acceleration of gravity $\boldsymbol{g}(\boldsymbol{r})$ (with the overbars dropped indicating that the macroscopic acceleration of gravity and mass density are averaged quantities) varies throughout a spherical macroscopic body that has a uniform mass density ρ and a radius R. The gravitational acceleration satisfies $\nabla \cdot \boldsymbol{g} = -4\pi G\rho$. Due to the spatial symmetry of the uniform spherical body, we can write $\boldsymbol{g}(\boldsymbol{r}) = -g_r(r)\hat{\boldsymbol{r}}$, where $g_r(r)$ is the acceleration of gravity a distance r from the sphere center. Working in spherical coordinates, the acceleration of gravity satisfies the equation

$$\frac{1}{r^2}\frac{\mathrm{d}}{\mathrm{d}r}\left[r^2 g_r(r)\right] = 4\pi G\rho. \tag{3.258}$$

Multiplying by r^2, integrating, and then dividing by r^2 gives

$$g_r(r) = \frac{4\pi G\rho}{3} r + \frac{c}{r^2}, \tag{3.259}$$

where c is an integration constant. Because the acceleration of gravity remains finite or zero at $r = 0$, we must take $c = 0$. Alternatively, if we integrate $\nabla \cdot \mathbf{g} = -4\pi G\rho$ over the sphere of radius r and use the divergence theorem, we obtain $4\pi r^2 g_r(r) = 4\pi G\rho(4\pi r^3/3)$, which gives the same result of $g_r(r) = 4\pi G\rho r/3$.

This result for the acceleration of gravity within a uniform spherical body can be written in a few equivalent ways

$$\boxed{g_r(r) = \frac{4\pi G\rho}{3} r = g_R \frac{r}{R} = \frac{G m_r}{r^2},} \tag{3.260}$$

where g_R is the acceleration of gravity on the surface of the spherical body given by

$$g_R = \frac{4\pi G\rho}{3} R \tag{3.261}$$

and m_r is the mass contained within a sphere of radius r within the larger spherical body of radius R

$$m_r = \frac{4\pi r^3 \rho}{3}. \tag{3.262}$$

So the acceleration of gravity within a spherical body at a distance r from the sphere center is identical to if we take the mass m_r and place it as a point mass at the center of the sphere, cf., Eq. (3.255). So the mass residing in the spherical shell between r and R is not contributing to $g_r(r)$. To get a feel for the amplitude of the acceleration of gravity, take $\rho = 10^3$ kg/m^3 and $R = 10^6$ m (the radius of a moon) to obtain $g_R = 0.28$ m/s^2, which can be compared to the acceleration of gravity at the surface of the Earth $g_E = 9.81$ m/s^2 ($\overline{\rho}_E = 5.51 \times 10^3$ kg/m^3 and $R_E = 6.37 \times 10^6$ m) and at the surface of the sun $g_S = 274$ m/s^2 ($\overline{\rho}_S = 1.41 \times 10^3$ kg/m^3 and $R_S = 696 \times 10^6$ m).

The acceleration of gravity at the center of a spherically symmetric mass-density distribution is zero. If we consider a body of uniform density having infinite extent, the acceleration of gravity is everywhere zero, that is, at any one point, there are mass elements that uniformly surround that point resulting in zero net acceleration. Equivalently stated, each point in a uniform infinite body is surrounded by a spherically symmetric mass distribution and therefore has zero gravitational acceleration.

3.14 Coriolis Force and Centrifugal Force

An additional macroscopic force arises in continuum mechanics when a body under consideration is rotating around an axis with an angular velocity Ω (radians per second). This force is continuously distributed through the rotating body and is quite analogous to the gravitational force just treated in that it is defined as the local mass density times an acceleration $\mathrm{d}^2\boldsymbol{r}(t)/\mathrm{d}t^2$, where $\boldsymbol{r}(t)$ is the location of each mass element rotating about the axis as defined from a frame of reference that is rotating with the body. As derived below, this inertial force $\rho\, \mathrm{d}^2\boldsymbol{r}(t)/\mathrm{d}t^2$ attributable to the uniformly rotating frame separates into two contributions called the *Coriolis* and *centrifugal* forces. These inertial forces are sometimes called "fictitious forces" because they do not arise through electromagnetism or gravity or other force interactions between molecules (i.e., stress or "contact" forces), but for a mass element rotating about an axis, they are quite real forces that push on that element producing either deformation or flow depending on the state of matter. It is the force you feel when you are being pushed outward while riding a merry-go-round or when you are pushed into the car door (or, hopefully, seat belt) when the car you're riding in goes around a corner.

Define the axis of rotation to be the z axis. We can derive the expressions for the Coriolis and centrifugal forces in any coordinate system, but will do so first in Cartesian coordinates and then give the corresponding expressions in cylindrical and spherical coordinates. The coordinates of a mass element in a fixed frame (or, equivalently, in a frame moving with uniform linear velocity, which is called an *inertial* frame) are denoted by $\boldsymbol{r}' = (x', y', z')$ while those in the rotating frame are depicted in Fig. 3.5 and given by $\boldsymbol{r} = (x, y, z)$ where

$$x(t) = \cos(\Omega t)\, x'(t) + \sin(\Omega t)\, y'(t) \tag{3.263}$$

$$y(t) = -\sin(\Omega t)\, x'(t) + \cos(\Omega t)\, y'(t) \tag{3.264}$$

$$z(t) = z'(t). \tag{3.265}$$

Taking a time derivative and using $\boldsymbol{v} = \mathrm{d}\boldsymbol{r}/\mathrm{d}t$ for the mass-element's velocity described in the rotating frame and $\boldsymbol{v}' = \mathrm{d}\boldsymbol{r}'/\mathrm{d}t$ in the stationary frame gives

$$v_x(t) = -\Omega\left[\sin(\Omega t)\, x'(t) - \cos(\Omega t)\, y'(t)\right] + \cos(\Omega t)\, v'_x(t) + \sin(\Omega t)\, v'_y(t) \tag{3.266}$$

$$v_y(t) = -\Omega\left[\cos(\Omega t)\, x'(t) + \sin(\Omega t)\, y'(t)\right] - \sin(\Omega t)\, v'_x(t) + \cos(\Omega t)\, v'_y(t) \tag{3.267}$$

$$v_z(t) = v'_z(t) \tag{3.268}$$

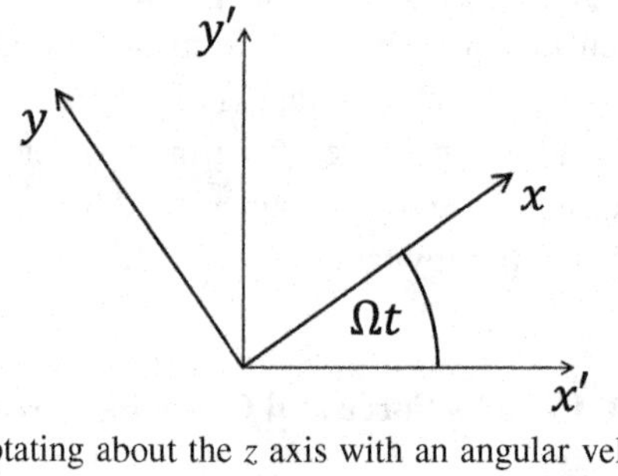

Figure 3.5 A body is rotating about the z axis with an angular velocity Ω. The Cartesian coordinates in a stationary frame are x', y', z' while the Cartesian coordinates that are rotating with the body are x, y, $z = z'$. We wish to express the dynamics of continuum mechanics in the frame that is rotating.

or equivalently after using Eqs (3.263) and (3.264),

$$v_x(t) = \Omega\, y(t) + \cos(\Omega t)\, v'_x(t) + \sin(\Omega t)\, v'_y(t) \tag{3.269}$$

$$v_y(t) = -\Omega\, x(t) - \sin(\Omega t)\, v'_x(t) + \cos(\Omega t)\, v'_y(t) \tag{3.270}$$

$$v_z(t) = v'_z(t). \tag{3.271}$$

Taking a time derivative of these expressions and using $\boldsymbol{a}(t) = \mathrm{d}^2\boldsymbol{r}/\mathrm{d}t^2$ for the mass element's acceleration in the rotating frame and $\boldsymbol{a}'(t) = \mathrm{d}^2\boldsymbol{r}'/\mathrm{d}t^2$ for the acceleration in the stationary frame and using Eqs (3.269) and (3.270) gives

$$a_x(t) = \cos(\Omega t)\, a'_x(t) + \sin(\Omega t)\, a'_y(t) + 2\Omega\, v_y(t) + \Omega^2 x \tag{3.272}$$

$$a_y(t) = -\sin(\Omega t)\, a'_x(t) + \cos(\Omega t)\, a'_y(t) - 2\Omega\, v_x(t) + \Omega^2 y \tag{3.273}$$

$$a_z(t) = a'_z(t). \tag{3.274}$$

These results can equivalently be written in array form as

$$\begin{pmatrix} a_x \\ a_y \\ a_z \end{pmatrix} = \begin{pmatrix} \cos\Omega t & \sin\Omega t & 0 \\ -\sin\Omega t & \cos\Omega t & 0 \\ 0 & 0 & 1 \end{pmatrix} \begin{pmatrix} a'_x \\ a'_y \\ a'_z \end{pmatrix} + \begin{pmatrix} 2\Omega v_y \\ -2\Omega v_x \\ 0 \end{pmatrix} + \begin{pmatrix} \Omega^2 x \\ \Omega^2 y \\ 0 \end{pmatrix}. \tag{3.275}$$

In Section 1.8.3, the 3×3 matrix here multiplying the acceleration components in the stationary frame is called the *rotation matrix*. But because the angle of rotation is time dependent and because the acceleration involves two time derivatives of the position vector, we end up with the two additional terms in Eq. (3.275).

We now focus on the dynamics of interest. If the force balance at each point in the body is written in the stationary frame as $\rho\boldsymbol{a}' = \boldsymbol{f}'$, where $\boldsymbol{f}'$ is some combination of EM, gravity, or traction forces as described in Chapter 2, the force components in the rotating frame are given in terms of the components in the stationary frame by multiplying with the rotation matrix

$$\begin{pmatrix} f_x \\ f_y \\ f_z \end{pmatrix} = \begin{pmatrix} \cos\Omega t & \sin\Omega t & 0 \\ -\sin\Omega t & \cos\Omega t & 0 \\ 0 & 0 & 1 \end{pmatrix} \begin{pmatrix} f'_x \\ f'_y \\ f'_z \end{pmatrix}. \tag{3.276}$$

If the body-force density in the stationary frame is not uniform or depends on the velocity of the body, then the x', y', z' dependence and v'_x, v'_y, v'_z dependence of f'_x, f'_y, f'_z should be expressed using Eqs (3.263), (3.264), (3.269), and (3.270) so that the f_x, f_y, f_z of Eq. (3.276) is given entirely in terms of the rotating coordinates.

If we multiply $\rho(a'_x, a'_y, a'_z)^T = (f'_x, f'_y, f'_z)^T$ by the rotation matrix and use Eq. (3.275), we obtain the vector force balance in the frame that is rotating with the body

$$\rho\boldsymbol{a} = \boldsymbol{f} + \underbrace{2\rho\Omega\left(v_y\hat{\boldsymbol{x}} - v_x\hat{\boldsymbol{y}}\right)}_{\substack{\uparrow \\ \text{Coriolis force density}}} + \underbrace{\rho\Omega^2\left(x\hat{\boldsymbol{x}} + y\hat{\boldsymbol{y}}\right)}_{\substack{\uparrow \\ \text{centrifugal force density}}}, \tag{3.277}$$

which identifies the *Coriolis force density* and *centrifugal force density* in the rotating Cartesian frame of reference. The centrifugal force is independent of whether each mass element is moving in the rotating frame and is directed radially away from the rotation axis. If we define the angular-velocity vector as $\boldsymbol{\Omega} = \Omega\hat{\boldsymbol{z}}$, the force balance in the rotating frame can be expressed in vector form as

$$\rho \boldsymbol{a} = \boldsymbol{f} - 2\rho\boldsymbol{\Omega} \times \boldsymbol{v} - \rho\boldsymbol{\Omega} \times (\boldsymbol{\Omega} \times \boldsymbol{r}). \tag{3.278}$$

In cylindrical coordinates that rotate with the body, these force densities become

$$\textit{Coriolis}: \quad -2\rho\boldsymbol{\Omega} \times \boldsymbol{v} = 2\rho\Omega\left(v_\theta\hat{\boldsymbol{r}} - v_r\hat{\boldsymbol{\theta}}\right) \tag{3.279}$$

$$\textit{centrifugal}: \quad -\rho\boldsymbol{\Omega} \times (\boldsymbol{\Omega} \times \boldsymbol{r}) = \rho\Omega^2 r\hat{\boldsymbol{r}}, \tag{3.280}$$

while in spherical coordinates that rotate with the body they are

$$\textit{Coriolis}: \quad -2\rho\boldsymbol{\Omega} \times \boldsymbol{v} = 2\rho\Omega\left[v_\phi \sin\theta\hat{\boldsymbol{r}} + v_\phi\cos\theta\hat{\boldsymbol{\theta}} + (v_r\sin\theta - v_\theta\cos\theta)\,\hat{\boldsymbol{\phi}}\right] \tag{3.281}$$

$$\textit{centrifugal}: \quad -\rho\boldsymbol{\Omega} \times (\boldsymbol{\Omega} \times \boldsymbol{r}) = \rho\Omega^2 r\sin\theta\left(\sin\theta\hat{\boldsymbol{r}} + \cos\theta\hat{\boldsymbol{\theta}}\right). \tag{3.282}$$

Again, the centrifugal force is always directed radially outward from the rotation axis and linearly increases with increasing distance from the rotation axis. The Coriolis force is only zero throughout the rotating body when the velocity of each mass element is parallel with the rotation axis.

To determine whether the Coriolis or centrifugal forces are important in some problem involving a rotation $\boldsymbol{\Omega}$ of a system (or body) about an axis, we can write the force balance of continuum mechanics when gravity is operational and our coordinates are rotating with the body as

$$\rho\left(\frac{\partial\boldsymbol{v}}{\partial t} + \boldsymbol{v}\cdot\nabla\boldsymbol{v}\right) = \nabla\cdot\boldsymbol{\tau} + \rho\boldsymbol{g} - 2\rho\boldsymbol{\Omega}\times\boldsymbol{v} - \rho\boldsymbol{\Omega}\times(\boldsymbol{\Omega}\times\boldsymbol{r}). \tag{3.283}$$

For problems near the surface of the Earth where gravity is important, the Earth's rotation gives $\Omega_E = 2\pi/[8.64 \times 10^4 \text{ s}] = 7.3 \times 10^{-5}$ rad/s, while $g_E = 9.81$ m/s^2 on the Earth's surface at $r = R_E = 6.37 \times 10^6$ m. The centrifugal force at the equator is directed radially outward from the center of the Earth in the direction opposite to gravity. We have $|\rho\boldsymbol{g}|/|\rho\boldsymbol{\Omega}\times(\boldsymbol{\Omega}\times\boldsymbol{r})| = g_E/(\Omega_E^2 R_E) \approx 290$. So the centrifugal force at the surface of the Earth due to the Earth's rotation is 290 times smaller than the force of gravity and can often be neglected in a first approximation.

This can be compared to a problem where somewhere near the surface of the Earth, a cylindrical body of radius $r = 1$ m is rotating about its axis at an angular velocity of $\Omega = 2\pi$ rad/s (one revolution per second). In this case, we have $g_E/(\Omega^2 r) \approx 0.25$ and the centrifugal force is four times larger than the force of gravity inside the cylinder and can no longer be neglected when modeling the deformation or flow inside the rotating cylinder.

Let's next consider the importance of the Coriolis force on the large-scale movement of the atmosphere. We define a dimensionless number Ro called the *Rossby number* as

$$\mathrm{Ro} = \frac{|\rho \boldsymbol{v} \cdot \nabla \boldsymbol{v}|}{|2\rho \boldsymbol{\Omega} \times \boldsymbol{v}|} = \frac{|\boldsymbol{v}|}{2\Omega \ell}, \tag{3.284}$$

where ℓ is the scale over which the fluid velocity $\boldsymbol{v}$ is significantly changing. For large-scale circulation of the atmosphere having $\ell = 10^5$ m at $|\boldsymbol{v}| = 10$ m/s, we have $\mathrm{Ro} = 0.7$ and the rotation of the Earth and the associated Coriolis force is what creates large-scale circulation cyclones. Indeed, as $\cos\theta$ changes sign from the Northern Hemisphere to the Southern Hemisphere in Eq. (3.281), the cyclones in the Southern Hemisphere are observed to rotate in the clockwise direction while those in Northern Hemisphere rotate in the counterclockwise direction. However, at the scale of a kitchen sink or bathtub where the Earth-rotation Coriolis force is sometimes imagined to set the direction of the water vortex at the drain, the Coriolis force is utterly negligible. The direction of the vortex at the drain in either hemisphere is instigated by some other larger-amplitude perturbation, such as somebody touching the water as it begins to drain.

3.15 Exercises

1. *An electromagnetic plane wave in an electrolyte*: Assume that a uniform electrolyte having permittivity ε, magnetic permeability μ and conductivity σ occupies the half-space $z > 0$ and is bounded by a plate at $z = 0$ that has no surface charge density on it and that has a given uniform time-harmonic electric field acting in it given by $\boldsymbol{E}_0 = E_0 \cos(\omega t)\hat{\boldsymbol{x}}$. This is depicted in Fig. 3.6. With the given electric field in the plate acting as a boundary condition at $z = 0$ that excites the EM fields throughout the electrolyte and with the requirement that $\boldsymbol{E} \to 0$ as $z \to \infty$ (the electrical conductivity attenuates the field from the plate as it propagates toward infinity), determine the electric field $\boldsymbol{E}(z, t)$ and magnetic field $\boldsymbol{H}(z, t)$ everywhere in space and time throughout the electrolyte. Assume that the time-harmonic electric field in the plate has been acting for a very long time so that at $t = 0$ the nonzero fields in the electrolyte extend out to infinity (there is no need to specify initial conditions).

   Further, show that if the half-space were occupied by a dielectric having permittivity ε, magnetic permeability μ and, by definition, $\sigma = 0$, your solution for $E_x(z, t)$ in the electrolyte as $\sigma \to 0$ satisfies a different boundary condition at $z = \infty$ given by

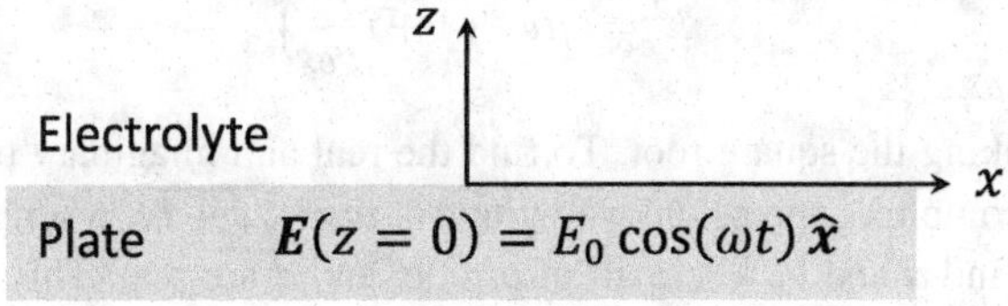

Figure 3.6 An infinite uniform electrolyte is bounded by a plate at $z = 0$ that has a uniform time-harmonic electric field present in it $\boldsymbol{E}(z = 0) = E_o \cos(\omega t)\hat{\boldsymbol{x}}$ that is the boundary condition for the fields throughout the electrolyte.

$$\left(\frac{\partial E_x}{\partial z}+\sqrt{\varepsilon\mu}\frac{\partial E_x}{\partial t}\right)\bigg|_{z=\infty}=0, \tag{3.285}$$

that is called an *outward-radiation condition*. In Chapter 8 of Part II of the book, radiation conditions will be shown to result in unique solutions to linear wave problems.

As this is your first exercise in macroscopic electromagnetics, some guidance will be provided that leads you to the solution. The starting place is the governing equations, which are Ampère's law and Faraday's law

$$\nabla\times\boldsymbol{H}=\varepsilon\frac{\partial\boldsymbol{E}}{\partial t}+\sigma\boldsymbol{E} \tag{3.286}$$

$$\nabla\times\boldsymbol{E}=-\mu\frac{\partial\boldsymbol{H}}{\partial t}. \tag{3.287}$$

Because the plate at $z=0$ is not charged, there is no free counter charge $\rho_e=0$ in the adjacent electrolyte so you have $\nabla\cdot\boldsymbol{E}=0$ throughout the electrolyte. Combine Ampère's and Faradays' laws using $\nabla\cdot\boldsymbol{E}=0$ and, by symmetry, the idea that only the x component of the electric field is excited by the boundary condition and that E_x can only vary in the z direction to obtain the electric-field governing equation

$$\frac{\partial^2 E_x(z,t)}{\partial z^2}=\varepsilon\mu\frac{\partial^2 E_x(z,t)}{\partial t^2}+\sigma\mu\frac{\partial E_x(z,t)}{\partial t}. \tag{3.288}$$

Begin by proposing a complex solution

$$E_x(z,t)=\tilde{E}_x(z,\omega)e^{-i\omega t}, \tag{3.289}$$

whose real part will be the actual solution of the problem that satisfies the boundary condition at $z=0$. Insert this complex time-harmonic form into the PDE for E_x and obtain

$$\frac{\mathrm{d}^2\tilde{E}_x(z,\omega)}{\mathrm{d}z^2}=-\left(\varepsilon\mu\omega^2 2+i\sigma\mu\omega\right)\tilde{E}_x(z,\omega). \tag{3.290}$$

This equation has a complex solution of the form $\tilde{E}_x(z,\omega)=A_k e^{ikz}$, where the complex parameter k, that we call the *wavenumber*, is obtained by inserting this form into the ODE for $\tilde{E}_x$ to obtain

$$k^2=\varepsilon\mu\omega^2\left(1+i\frac{\sigma}{\varepsilon\omega}\right). \tag{3.291}$$

Obtain k by taking the square root. To find the real and imaginary parts of the square root of some complex expression, we write $a+ib=\sqrt{c+id}$, where c and d are given and we must find a and b, where all of a, b, c, and d are real. This is done by taking the square of both sides to obtain $a^2-b^2+i2ab=c+id$ and then separately equating the real and imaginary parts. Perform this algebraic exercise and show that we have $k=\pm k(\omega)$ with

$$k(\omega) = \omega\sqrt{\varepsilon\mu}\left[\sqrt{\frac{1+\sqrt{1+\gamma^2}}{2}} + i\frac{\gamma}{\sqrt{2\left(1+\sqrt{1+\gamma^2}\right)}}\right], \tag{3.292}$$

and with the real, dimensionless, and frequency-dependent parameter γ defined as

$$\gamma = \frac{\sigma}{\varepsilon\omega}. \tag{3.293}$$

We recognize γ as the parameter that defines the high-frequency wave propagation regime when $\gamma \ll 1$ and the low-frequency diffusion regime when $\gamma \gg 1$. For the time being, do not make any assumptions about the magnitude of γ. Of our two solutions $\tilde{E}_x(z, \omega) = A_+ e^{+ik(\omega)z} + A_- e^{-ik(\omega)z}$, because the imaginary part of $k(\omega)$ is positive, the second solution diverges exponentially with increasing z rather than decreasing as it must, which requires $A_- = 0$. Thus, you have shown that

$$E_x(z, t) = A_+ \exp\left[i\omega\left(\sqrt{\varepsilon\mu}\sqrt{\frac{1+\sqrt{1+\gamma^2}}{2}}z - t\right)\right] \exp\left[\frac{-\omega\gamma\sqrt{\varepsilon\mu}\,z}{\sqrt{2\left(1+\sqrt{1+\gamma^2}\right)}}\right]. \tag{3.294}$$

Show that the real part of this expression, when evaluated at $z = 0$, requires that $A_+ = E_0$ from the given boundary condition on $z = 0$. Thus, the real solution that satisfies exactly all of the differential equations and boundary conditions and is, therefore, the one unique solution to this given problem is

$$\boxed{E_x(z, t) = E_0 \cos\left[\omega\left(\frac{z}{c(\omega)} - t\right)\right] \exp\left(-\frac{z}{\delta_{em}(\omega)}\right),} \tag{3.295}$$

where the frequency-dependent phase velocity is given by

$$c(\omega) = \sqrt{\frac{2}{\varepsilon\mu\left(1+\sqrt{1+\gamma^2}\right)}}, \tag{3.296}$$

while the frequency-dependent skin depth (or inverse attenuation coefficient) is

$$\delta_{em}(\omega) = \frac{\sqrt{2(1+\sqrt{1+\gamma^2})}}{\omega\gamma\sqrt{\varepsilon\mu}}. \tag{3.297}$$

So Eq. (3.295) is the exact solution of this EM problem valid for any value of $\gamma = \sigma/(\varepsilon\omega)$ and corresponds to an attenuated plane wave that loses amplitude exponentially with distance z propagated.

The magnetic field can either be obtained from Ampère's law or Faraday's law using the derived expression for $E_x(z, t)$. Using Faraday's law and the expression for the curl of a vector field, show that

$$\frac{\partial H_y(z,t)}{\partial t} = -\frac{1}{\mu}\frac{\partial E_x(z,t)}{\partial z}. \tag{3.298}$$

By first differentiating $E_x(z, t)$ with respect to z and then integrating over time, show that

$$\boxed{\begin{aligned} H_y(z,t) = \frac{E_0}{\mu}&\left\{\frac{1}{c(\omega)}\cos\left[\omega\left(\frac{z}{c(\omega)}-t\right)\right]\right. \\ &\left.-\frac{1}{\omega\delta_{em}(\omega)}\sin\left[\omega\left(\frac{z}{c(\omega)}-t\right)\right]\right\}\exp\left(-\frac{z}{\delta_{em}(\omega)}\right). \end{aligned}} \tag{3.299}$$

Show that you get this same result for $H_y(z, t)$ if you begin with Ampère's law, calculate $\varepsilon\partial E_x(z, t)/\partial t + \sigma E_x(z, t)$ and then integrate over z to get $H_y(z, t)$. But you will see that the Ampère's law approach is a lot more work even if the algebra eventually works out to give Eq. (3.299) as it must.

In the limit where $\gamma \to 0$ (pure wave propagation), show that

$$E_x(z,t) = E_0 \cos\left[\omega\left(\frac{z}{c}-t\right)\right], \tag{3.300}$$

where the frequency-independent phase velocity c is given by the well-known expression

$$c = \frac{1}{\sqrt{\varepsilon\mu}}, \tag{3.301}$$

and is the result you would have obtained for any ω if you had set $\sigma = 0$ from the beginning. Show that this solution satisfies the outward-radiation condition at $z = \infty$ (or, indeed, at any z).

In the opposite limit where $\gamma \to \infty$ (pure time-harmonic diffusion), show that

$$E_x(z,t) = E_0 \cos\left(\frac{z}{\delta_{em}} - \omega t\right)\exp\left(-\frac{z}{\delta_{em}}\right), \tag{3.302}$$

where the skin depth in this low-frequency diffusive limit is defined

$$\delta_{em} = \sqrt{\frac{2}{\sigma\mu\omega}}, \tag{3.303}$$

which is the expression given earlier in Section 3.6.2 when defining the diffusion limit of electromagnetism. The wavelength λ of pure time-harmonic diffusion is thus given by $\lambda = 2\pi\delta_{em}$. Make a sketch of $E_x(z, 0)/E_0 = \cos(z/\delta_{em})e^{-z/\delta_{em}}$ for this time-harmonic diffusional response at $t = 0$.

2. *The Maxwell stress tensor*: Fill in all the vectorial and tensorial steps in getting from Eq. (3.63) through to Eq. (3.67).

3. *An estimate for the solubility of salt in water*: Assume that we have a strong aqueous electrolyte of NaCl in which each Na^+ cation and each chlorine Cl^- anion are

separated from each other by an average distance r as the ions move about randomly with a certain amount of kinetic energy. A simple estimate for the number density of NaCl in water (number of NaCl per meter cubed of solution) at saturation, which is when additional NaCl molecules added to the solution no longer dissociate into ions, is obtained when the average distance between free ions $r = d$ is such that the electrostatic attraction energy $U_e(d)$ that is bringing the ions into association just equals the average kinetic energy $3kT/2$ (a result from statistical mechanics called the "equipartition of energy") of the random thermal motion of each molecule that is keeping the ions apart. To use this simple model for the saturation of a salt solution, we first need to determine the electrostatic interaction energy U_e between two ions separated by an average distance $r \geq d$.

The ions are solute particles within a sea of solvent water molecules characterized by an electrical permittivity $\varepsilon = \epsilon_0 \kappa_e$, where the dielectric constant of pure water is roughly $\kappa_e = 78$ at 25°C (and atmospheric pressure). If $E_r(r)$ is the radial electric field coming from the cation, this field acts on the anion of charge $-e$ to produce an attraction force between the ions of $F_r = -eE_r$ (we could also estimate this force as the electric field coming from the anion and acting on the cation). The attraction energy can be defined from $dU_e = F_r \, \mathrm{d}r = -eE_r(r) \, \mathrm{d}r$. To integrate this to obtain $U_e(r)$, we need to know $E_r(r)$. This is obtained from Coulomb's law $\nabla \cdot \boldsymbol{D} = \rho_e$, where the charge density associated with the cation can be written $\rho_e = +e\delta(\boldsymbol{r})$ and where the dielectric displacement is related to the electric field as $\boldsymbol{D} = \epsilon_0 \kappa_e \boldsymbol{E}$. Show that integrating both sides of Coulomb's law over a sphere $\Omega(r)$ of radius r gives

$$\epsilon_0 \kappa_e \int_{\Omega(r)} \nabla_o \cdot \boldsymbol{E}(\boldsymbol{r}_o) \, \mathrm{d}^3\boldsymbol{r}_o = \epsilon_0 \kappa_e \int_{\partial\Omega(r)} \hat{\boldsymbol{r}}_o \cdot \boldsymbol{E} \, \mathrm{d}^2\boldsymbol{r}_o = 4\pi \epsilon_0 \kappa_e r^2 E_r(r) = +e \quad (3.304)$$

or

$$E_r(r) = \frac{e}{4\pi \epsilon_0 \kappa_e r^2}. \quad (3.305)$$

Then by integrating $\mathrm{d}U_e = F_r \, \mathrm{d}r$ show that

$$U_e(r) = \frac{e^2}{4\pi \epsilon_0 \kappa_e r}. \quad (3.306)$$

At the distance $r = d$ at which the electrostatic attraction energy just equals the average molecular kinetic energy $U_e(d) \approx 3kT/2$, show that

$$d \approx \frac{e^2}{6\pi \epsilon_0 \kappa_e kT}, \quad (3.307)$$

which is sometimes called the Bjerrum length (formally, we have $d = 2\lambda_B/3$, where λ_B is the Bjerrum length). In other words, at the solubility limit there is one Na ion in each cube of size d^3 or one dissociated NaCl salt molecule in each $2d^3$. Define the molarity M_{sol} of the saline solution (moles of NaCl molecules per liter of solution) at the solubility limit as $M_{\text{sol}} = 10^{-3}/(N_A 2d^3)$, where $N_A = 6.02 \times 10^{23}$ is Avogadro's number and the factor 10^{-3} converts meters cubed into liters. For a solution at

$T = 298$ K (25°C), with the universal constants given by $\epsilon_0 = 8.85 \times 10^{-12}$ F/m, $k = 1.38 \times 10^{-23}$ J/K, and $e = 1.60 \times 10^{-19}$ C, show that we obtain $M_{\text{sol}} \approx$ 7.6 mol/liter for the room-temperature solubility of NaCl in water, that is, the saturation limit beyond which additional NaCl molecules will no longer dissociate into solute ions. The actual measured solubility of NaCl in an aqueous electrolyte is 6.2 mol/liter at 298 K. Given the simplicity of the model, our estimate of 7.6 mol/liter is surprisingly close.

Note that the larger the dielectric constant, the stronger will be the polarization of the intervening water, the smaller the electric field and the smaller the force of attraction between the cation and anion. The large number of $\kappa_e = 78$ used in the above estimate was for pure water at 298 K and atmospheric pressure. Because the Na^+ and Cl^- ions are hydrated (have water molecules attached to them that are already polarized), the dielectric constant of the water residing between the Na^+ and Cl^- is smaller than that for pure water resulting in a smaller solubility estimate. Further, at depth in the Earth's crust, at large enough temperature and pressure, the water molecules that are largely grouped tetrahedrally into groups of five molecules at lower temperature become disaggregated with larger numbers of single molecule water molecules (monomers) being present. This causes a large drop in the dielectric constant of water that allows the cations and anions in the saline pore waters to associate into neutral salt pairs that no longer contribute to the ionic electrical conductivity. So the electrical conductivity of saline-solution-filled rocks can actually decrease with increasing depth due to this effect of the decreasing dielectric constant of water causing an increase in electrostatic attraction between the cations and anions with an associated decrease in the solubility (salt saturation) of the saline solution.

4. For the same exercise treated in Section 3.10.1 but now with the voltage electrodes located at $x = +d_1$ and $x = -d_2$ as depicted in Fig. 3.7, find the electric potential throughout $-d_2 \leq x \leq d_1$, and show that the effective electrical conductivity of this two slab system is $\sigma_e = (d_1 + d_2)/(d_1/\sigma_1 + d_2/\sigma_2)$.

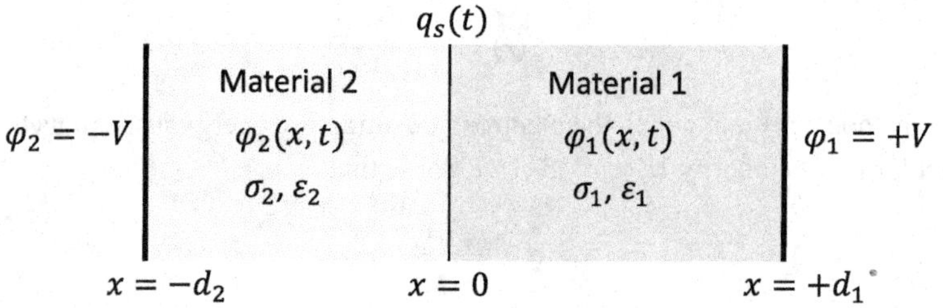

Figure 3.7 Same problem as treated in Section 3.10.1 but now with the voltages applied at $x = +d_1$ and $x = -d_2$.

5. If the sphere treated in Section 3.10.2 had been a perfect dielectric having permittivity ε_2 and if the infinite material into which the sphere is embedded had been a perfect dielectric having permittivity ε_1 with a uniform electric field applied to it, show that the resulting electrostatic fields in this case are given by

$$\varphi_1(r,\theta) = -E_x r\cos\theta\left[1+\left(\frac{\varepsilon_1-\varepsilon_2}{2\varepsilon_1+\varepsilon_2}\right)\left(\frac{a}{r}\right)^3\right], \tag{3.308}$$

$$\varphi_2(r,\theta) = -E_x r\cos\theta\frac{3\varepsilon_1}{2\varepsilon_1+\varepsilon_2}, \tag{3.309}$$

$$\begin{aligned}\boldsymbol{E}_1(r,\theta) &= E_x\cos\theta\left[1-2\left(\frac{\varepsilon_1-\varepsilon_2}{2\varepsilon_1+\varepsilon_2}\right)\left(\frac{a}{r}\right)^3\right]\hat{\boldsymbol{r}}\\ &\quad - E_x\sin\theta\left[1+\left(\frac{\varepsilon_1-\varepsilon_2}{2\varepsilon_1+\varepsilon_2}\right)\left(\frac{a}{r}\right)^3\right]\hat{\boldsymbol{\theta}},\end{aligned} \tag{3.310}$$

$$\boldsymbol{E}_2(r,\theta) = E_x\cos\theta\frac{3\varepsilon_1}{2\varepsilon_1+\varepsilon_2}\hat{\boldsymbol{r}} - E_x\sin\theta\frac{3\varepsilon_1}{2\varepsilon_1+\varepsilon_2}\hat{\boldsymbol{\theta}}, \tag{3.311}$$

$$q_s = 3\epsilon_0 E_x\cos\theta\left(\frac{\varepsilon_2-\varepsilon_1}{2\varepsilon_1+\varepsilon_2}\right), \tag{3.312}$$

where the surface-charge density $q_s = q_b$ in this case is due entirely to bound polarized charge excess at the interface. There are no magnetic fields in this problem involving dielectrics and an applied electric field.

6. Solve the same exercise as given in Section 3.10.2 and as depicted in Fig. 3.4 but now with the circular object with surface $r = a$ being a cylinder extending out of plane in the z direction and not a sphere as in Section 3.10.2. Follow the exact same steps as outlined in Exercise 4 except now working in cylindrical coordinates (i.e., find the n that solves the Laplace equation in cylindrical coordinates, etc.). Neglect the surface currents on the cylindrical surface and show that

$$\varphi_1(r,\theta) = -E_x r\cos\theta\left[1+\left(\frac{\sigma_1-\sigma_2}{\sigma_1+\sigma_2}\right)\left(\frac{a}{r}\right)^2\right], \quad r>a; \tag{3.313}$$

$$\varphi_2(r,\theta) = -E_x r\cos\theta\frac{2\sigma_1}{\sigma_1+\sigma_2}, \quad r<a; \tag{3.314}$$

$$\begin{aligned}\boldsymbol{E}_1(r,\theta) &= E_x\cos\theta\left[1-2\left(\frac{\sigma_1-\sigma_2}{\sigma_1+\sigma_2}\right)\left(\frac{a}{r}\right)^2\right]\hat{\boldsymbol{r}}\\ &\quad - E_x\sin\theta\left[1+\left(\frac{\sigma_1-\sigma_2}{\sigma_1+\sigma_2}\right)\left(\frac{a}{r}\right)^2\right]\hat{\boldsymbol{\theta}}, \quad r>a;\end{aligned} \tag{3.315}$$

$$\boldsymbol{E}_2(r,\theta) = E_x\cos\theta\frac{2\sigma_1}{\sigma_1+\sigma_2}\hat{\boldsymbol{r}} - E_x\sin\theta\frac{2\sigma_1}{\sigma_1+\sigma_2}\hat{\boldsymbol{\theta}}, \quad r<a; \tag{3.316}$$

$$q_s = q_f + q_b = 2E_x\cos\theta\epsilon_0\left(\frac{\sigma_2-\sigma_1}{\sigma_1+\sigma_2}\right), \quad \text{on } r=a; \tag{3.317}$$

$$\begin{aligned}\boldsymbol{H}_1(r,\theta) &= H_{1z}(r,\theta)\hat{z}\\ &= \sigma_1 E_x r\sin\theta\left[1-\left(\frac{\sigma_1-\sigma_2}{\sigma_1+\sigma_2}\right)\left(\frac{a}{r}\right)^2\right]\hat{z}, \quad r>a;\end{aligned} \tag{3.318}$$

$$\boldsymbol{H}_2(r,\theta) = H_{2z}(r,\theta)\hat{z} = \sigma_2 E_x r\sin\theta\frac{2\sigma_1}{\sigma_1+\sigma_2}\hat{z}, \quad r<a. \tag{3.319}$$

7. *The far-field electric field of an electric dipole*: An electric dipole corresponds to a charge $+q$ situated a small distance d from a charge $-q$. If the distance vector from the negative charge toward the positive charge is $\boldsymbol{d}$, the dipole-moment vector is defined $\boldsymbol{p} = q\boldsymbol{d}$. For a dielectric material extending out to infinity and having a uniform permittivity ε, find the static electric potential and electric field for a dipole $\boldsymbol{p}$ at distances large compared to the size d of the dipole.

   HINTS: One approach is to begin with a single point charge q at the origin for which the governing equations are $\nabla \cdot \boldsymbol{D}(\boldsymbol{r}) = q\delta(\boldsymbol{r})$ and $\nabla \times \boldsymbol{E} = 0$. With $\boldsymbol{E} = -\nabla\varphi$, the electric potential is therefore a solution of

$$\nabla^2\varphi(\boldsymbol{r}) = -\frac{q}{\varepsilon}\delta(\boldsymbol{r}). \tag{3.320}$$

   To solve this equation, work in spherical coordinates in which the angle θ measures distance from the axis defined by the direction $\boldsymbol{d}$ and note that by symmetry, there is no ϕ or θ dependence of this electric potential $\varphi(r)$. Using expressions for the Laplacian and 3D Dirac delta function in spherical coordinates given in Chapter 1, we have

$$\frac{1}{r^2}\frac{\partial}{\partial r}\left(r^2\frac{\partial\varphi(r)}{\partial r}\right) = -\frac{q}{\varepsilon}\frac{\delta(r)\delta(\theta)\delta(\phi)}{r^2\sin\theta}. \tag{3.321}$$

   Integrate this expression over a sphere of radius r (where in spherical coordinates, $\mathrm{d}V = r^2\sin\theta\,\mathrm{d}r\,\mathrm{d}\theta\,\mathrm{d}\phi$)

$$\int_0^r \mathrm{d}r_o\frac{\partial}{\partial r_o}\left(r_o^2\frac{\partial\varphi(r_o)}{\partial r_o}\right)\int_0^\pi \mathrm{d}\theta\,\sin\theta\int_0^{2\pi}\mathrm{d}\phi = -\frac{q}{\varepsilon}\int_0^r \mathrm{d}r_o\delta(r_o)\int_0^\pi \mathrm{d}\theta\,\delta(\theta)\int_0^{2\pi}\mathrm{d}\phi\,\delta(\phi) \tag{3.322}$$

   and show that

$$\frac{\partial\varphi(r)}{\partial r} = -\frac{q}{4\pi\varepsilon r^2} \tag{3.323}$$

   or, after integrating over r,

$$\varphi(r) = \frac{q}{4\pi\varepsilon r} = \frac{q}{4\pi\varepsilon|\boldsymbol{r}|}. \tag{3.324}$$

   With this as the solution for the potential from a single point charge, the potential from a dipole is exactly

$$\varphi(\boldsymbol{r}) = \frac{q}{4\pi\varepsilon}\left(\frac{1}{|\boldsymbol{r}-\boldsymbol{d}/2|} - \frac{1}{|\boldsymbol{r}+\boldsymbol{d}/2|}\right). \tag{3.325}$$

   In the limit as distances r become large compared to the size of the dipole $d \ll r$ where $d = \sqrt{\boldsymbol{d}\cdot\boldsymbol{d}}$, use the big-O notation to show that

$$\frac{1}{|\boldsymbol{r}+\boldsymbol{d}/2|} = \frac{1}{\sqrt{r^2[1+(d/(2r))^2]+\boldsymbol{r}\cdot\boldsymbol{d}}} = \frac{1}{r}\left[1-\frac{\boldsymbol{r}\cdot\boldsymbol{d}}{2r^2}+O\left(\left(\frac{d}{r}\right)^2\right)\right]. \tag{3.326}$$

Thus, show that when d/r is small, the far-field electric potential for the dipole $\boldsymbol{p} = q\boldsymbol{d}$ is

$$\boxed{\varphi(\boldsymbol{r}) = \frac{\boldsymbol{r}\cdot\boldsymbol{p}}{4\pi\varepsilon r^3} = \frac{p\cos\theta}{4\pi\varepsilon r^2}} \tag{3.327}$$

and that the associated far-field electric field is

$$\boxed{\boldsymbol{E} = \frac{p}{4\pi\varepsilon r^3}\left[2\cos\theta\hat{\boldsymbol{r}} + \sin\theta\hat{\boldsymbol{\theta}}\right]}. \tag{3.328}$$

If you are interested in the electric potential and electric field near to or between the two charges of the dipole, for which r can no longer be taken as large compared to d, Eq. (3.325) provides the exact solution at any distance.

8. *Point charge near a conducting metal plane*: Let's say that a point charge q is placed in a dielectric of permittivity ε at a distance $d/2$ from an infinitely extended metal plane. The metal conductor cannot support an electric field so it's surface is maintained at a constant electric potential. If we say that the sheet is grounded, that potential can be taken as zero. What is the solution for the electric potential throughout the dielectric? Take the normal to the conducting plane to be in the $\hat{z}$ direction. So you are looking for the electric potential $\varphi(x, y, z)$ created by the charge $+q$ in the half space $z \geq 0$ and satisfying the boundary conditions $\varphi(x, y, 0) = 0$ and $\varphi(x, y, \infty) = 0$.

   Using coordinates with an origin O as depicted in Fig. 3.8, the trick is to solve for the potential from a charge $+q$ located at $\boldsymbol{d} = (d/2)\hat{z}$ within a whole space of permittivity ε and to add to this the potential from a second charge $-q$, that is called an *image charge*, located at $\boldsymbol{d} = -(d/2)\hat{z}$ that also resides in the same whole space. Show that the sum

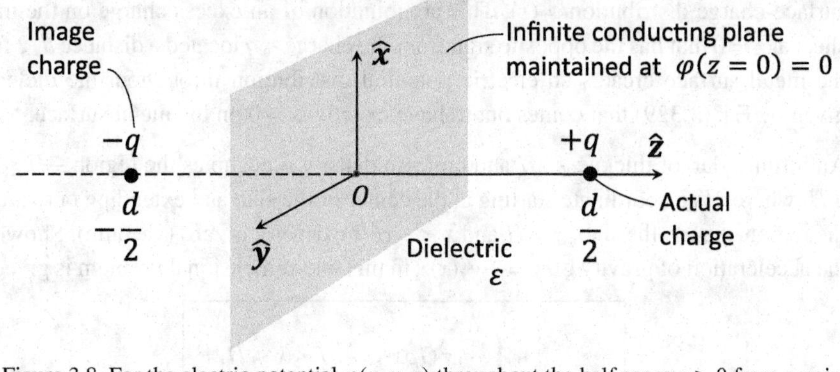

Figure 3.8 For the electric potential $\varphi(x, y, z)$ throughout the half space $z \geq 0$ from a point charge $+q$ located a distance $d/2$ from the plane of a conductor grounded to be at zero potential, the solution is the sum of the actual charge located at $z = d/2$ in a whole space and an image charge $-q$ located at $z = -d/2$ in the same whole space. This trick is called the *method of images*.

of these two potentials satisfies $\varphi = 0$ on the plane $z = 0$ and is given by Eq. (3.325) above, which can be expressed

$$\varphi(x, y, z) = \frac{q}{4\pi\varepsilon}\left(\frac{1}{\sqrt{x^2 + y^2 + z^2 - zd + d^2/4}} - \frac{1}{\sqrt{x^2 + y^2 + z^2 + zd + d^2/4}}\right) \tag{3.329}$$

and applies only to the region $z \geq 0$. One of the satisfying consequences of Section 3.11 on the uniqueness of the solution for linear EM problems is that if we find a solution that satisfies both the differential equations and the boundary conditions, by whatever guess or trick that is available to us as in the present circumstance, then we know this is the one and only solution of the problem.

Last, the charge $+q$ induces excess electrons on the metal surface $z = 0$ that have a surface-charge density q_f (measured in C/m^2). Show that this free charge induced on the metal surface is

$$q_f(r) = \frac{-qd}{4\pi\left(r^2 + d^2/4\right)^{3/2}}, \tag{3.330}$$

where $r^2 = x^2 + y^2$. Show that by integrating this surface-charge density over the entire metal surface at $z = 0$ that (working in polar coordinates)

$$\int_0^\infty \mathrm{d}r\, r q_f(r) \int_0^{2\pi} \mathrm{d}\theta = -q. \tag{3.331}$$

So the total charge induced on the metal surface is exactly the charge $-q$ of the fictitious image charge.

So in truth what is happening is that the electric field created by the actual point charge $+q$ creates a surface-charge density $q_f(r)$ on the metal surface of the opposite sign as given by Eq. (3.330). Once we place the charge $+q$ at a distance $d/2$ from the metal surface, electrons flow onto the surface (or off if $+q$ is negative) through the grounding wire that maintains the sheet at zero potential, thus creating the excess surface-charge distribution $q_f(r)$. This combination of an excess charge on the metal sheet at $z = 0$ that has the opposite sign from the charge $+q$ located a distance $d/2$ from the metal surface creates an electric potential distribution throughout the dielectric given by Eq. (3.329) that comes out to have exactly $\varphi = 0$ on the metal surface.

9. An infinite slab of thickness $2H$ and uniform density ρ occupies the region $-H \leq x \leq +H$, where x is a coordinate starting at the center of the slab and extending outward. In the region outside the slab, $x > H$ and $x < -H$ the density is zero (vacuum). Show that the acceleration of gravity $\boldsymbol{g}(\boldsymbol{r}) = -g_x(x)\hat{\boldsymbol{x}}$ in this one-dimensional problem is given by

$$\boxed{g_x(x) = \begin{cases} -4\pi G\rho x & -H \leq x \leq H, \\ -4\pi G\rho H & x > H, \\ +4\pi G\rho H & x < -H. \end{cases}} \tag{3.332}$$

4

Elasticity and Elastodynamics

Our next task is to understand how stress applied to a macroscopic sample of material causes the sample to deform. The idea of *elasticity* is that as stress is applied, the sample will deform to some different shape and/or volume, but when the stress is removed, the sample will return to its original shape and volume. If a sample does not return to its original shape and volume upon removing the applied stress, there is *plasticity* at work, which is an interesting process that has multiple causes, and is an especially complicated process in brittle materials like rocks that crack under applied stress, but that we will not treat in this book. The key to quantifying the relationship between applied stress and the elastic, or reversible, changes in shape and/or volume of a sample that is called *strain* is to focus on the work (energy) of deforming a sample. In other words, we take a thermodynamics perspective for deriving both our measure of strain and the associated stress–strain relation of elasticity called *Hooke's law*. We will return again to the theme of thermodynamics providing the constitutive laws of continuum physics in Chapter 6 that is exclusively dedicated to equilibrium thermodynamics and reversible processes such as elastic strain.

Our approach of defining strain through the energy being elastically stored in a sample is not common in texts on continuum mechanics. The usual approach is to imagine a kinematic definition of strain on its own terms based on the way that two points diverge to new points as the material deforms. Such a definition has no inherent connection to the energetics of deforming a sample. One then postulates that this kinematic definition of strain is proportional to stress through a linear phenomenological relation called Hooke's law. We prefer to avoid such speculation and let the work of deformation both define strain and produce Hooke's law, which for large strain produces different results compared to the standard phenomenological approach.

4.1 The Elastodynamic Governing Equations

In this section, we present the expression for the rate at which work is performed in deforming a sample and use this to derive the functional definitions of strain rate, strain, and the stress–strain elasticity relation called Hooke's law. In so doing, the entire suite of elastodynamic governing equations are obtained.

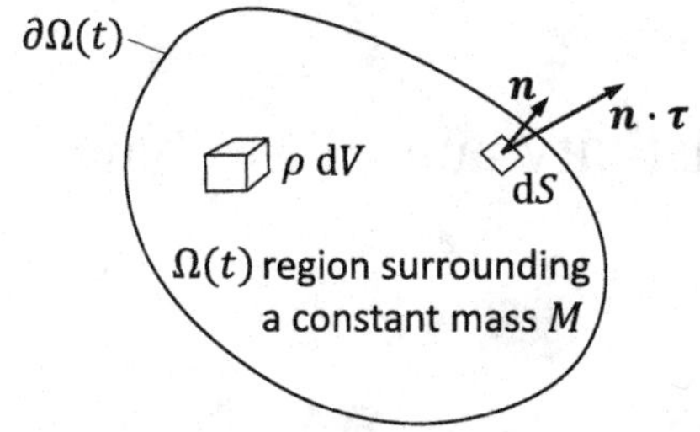

Figure 4.1 A solid sample $\Omega(t)$ that deforms and displaces to always enclose the same molecules.

4.1.1 The Internal Elastic Energy of a Sample

Consider a macroscopic chunk of matter, like a laboratory sample of rock. The region occupied by such a large macroscopic sample is denoted Ω, and each such sample is surrounded by a closed surface $\partial\Omega$ as depicted in Fig. 4.1. In deriving our measure of strain and the stress–strain relation for such a sample, we take the perspective that as the sample deforms, the region Ω evolves to always contain the same N molecules that have a constant total mass M. For the purposes of deriving the rules of elasticity, we are thus imagining this sample to be a solid. Once we establish the rules for the elastic response of a solid in this chapter, we will consider in Chapter 5 the simple changes to the rules if the sample is a fluid confined within Ω.

The sample is now immersed in a reservoir characterized by a uniform stress tensor. As the components of the stress tensor are varied in time to achieve a given level, the sample will deform through a sequence of equilibrium configurations. By focusing on the surface $\partial\Omega(t)$ surrounding the deforming sample, we can calculate how much work is performed on the sample by the stress acting on $\partial\Omega(t)$ and observe how the shape and/or volume of the sample changes, which results in our definition of strain rate and strain. We divide the evolving region $\Omega(t)$ into small elements dV that are filled with atoms and that possess a local value of the continuum fields $\boldsymbol{v}$ (center-of-mass velocity) and $\boldsymbol{\tau}$ (the stress tensor). There are also tractions $\boldsymbol{n} \cdot \boldsymbol{\tau}$ acting on each surface element dS of $\partial\Omega(t)$ that are causing movement $\boldsymbol{v}$ of those surface elements which corresponds to a rate of work $\boldsymbol{n} \cdot \boldsymbol{\tau} \cdot \boldsymbol{v}\, dS$ in moving each surface element on the evolving surface $\partial\Omega(t)$. Throughout what follows, we use the terms *mass element* and *sample* interchangeably to mean the region $\Omega(t)$ surrounding the constant mass M of N molecules that is deforming due to the applied stress.

In anticipation of Chapter 6 on thermodynamics, we define U as the internal energy of the mass element, which represents all accessible forms of energy within the element but excludes the net translational kinetic energy of the element as a whole and excludes the inaccessible nuclear energy stored in the nuclei of the atoms. As developed in Chapter 6, the internal energy of some element changes when either heat enters, leaves, or is created in the element or when work is performed on (or by) the element. In the present chapter, we assume that heat is neither entering or leaving nor being created within our elastically deforming mass element $\Omega(t)$, so that the internal energy of the mass element is only changing because of work being performed by the applied tractions, which is the statement

$$\frac{\mathrm{d}U}{\mathrm{d}t} = \int_{\partial\Omega(t)} \boldsymbol{n} \cdot \boldsymbol{\tau} \cdot \boldsymbol{v}\, \mathrm{d}S. \tag{4.1}$$

Dividing through by the constant mass M of the element defines the *specific internal energy* $\hat{u} = U/M$

$$\frac{\mathrm{d}\hat{u}}{\mathrm{d}t} = \frac{1}{\bar{\rho}} \left[\frac{1}{V(t)} \int_{\partial\Omega(t)} \boldsymbol{n} \cdot \boldsymbol{\tau} \cdot \boldsymbol{v}\, \mathrm{d}S \right], \tag{4.2}$$

where $\bar{\rho}$ is the total, or averaged, mass density of the element defined as

$$\bar{\rho} = \frac{M}{V(t)} = \frac{1}{V(t)} \int_{\Omega(t)} \rho\, \mathrm{d}V, \tag{4.3}$$

where a bar over a symbol denotes a volume average over $\Omega(t)$ at each instant and where $V(t) = \int_{\Omega(t)} \mathrm{d}V$ is the evolving volume of the sample. A process occurring with no net heat entering or leaving a mass element, as is assumed in the identification of Eq. (4.2), is called *adiabatic*. So in this chapter, we are modeling elastic deformation as being adiabatic. The specific energy $\hat{u}$ defined under these assumptions corresponds to the energy stored reversibly in the deforming atomic bonds within $\Omega(t)$, that we call *elastic-strain energy*. As work is performed by the stresses (tractions) applied to $\partial\Omega(t)$, all of that energy goes into strain energy that can be recovered once the stresses are returned to their initial values.

For the purpose of quantifying the change in internal energy, the sample is assumed to reside in a reservoir having a uniform stress tensor that can be identified as the average stress tensor $\bar{\boldsymbol{\tau}}$ of the sample $\Omega(t)$. To prove this common-sense assertion, we begin with the tensor identity $\nabla \cdot (\boldsymbol{\tau}\boldsymbol{r}) = (\nabla \cdot \boldsymbol{\tau})\,\boldsymbol{r} + \boldsymbol{\tau} \cdot \nabla\boldsymbol{r} = \boldsymbol{\tau}$, where we used that at each point within the possibly heterogeneous sample, the local stress field is in equilibrium $\nabla \cdot \boldsymbol{\tau} = 0$ for the specific purpose at hand of calculating the stored elastic energy and that $\nabla\boldsymbol{r} = \boldsymbol{I}$ is the identity tensor. Upon naming the uniform reservoir stress tensor as $\boldsymbol{\tau}_R$, averaging the statement $\boldsymbol{\tau} = \nabla \cdot (\boldsymbol{\tau}\boldsymbol{r})$ throughout Ω at each instance in time, using the divergence theorem and using the fact that $\boldsymbol{\tau}$ is symmetric, we obtain

$$\bar{\boldsymbol{\tau}} \mathrel{\widehat{=}} \frac{1}{V(t)} \int_{\Omega(t)} \boldsymbol{\tau}\, \mathrm{d}V = \frac{1}{V(t)} \int_{\Omega(t)} \nabla \cdot (\boldsymbol{\tau}\boldsymbol{r})\, \mathrm{d}V \tag{4.4}$$

$$= \frac{1}{V(t)} \int_{\partial\Omega(t)} \boldsymbol{n} \cdot \boldsymbol{\tau}_R \boldsymbol{r}\, \mathrm{d}S = \boldsymbol{\tau}_R \cdot \left[\frac{1}{V(t)} \int_{\partial\Omega(t)} \boldsymbol{n}\boldsymbol{r}\, \mathrm{d}S \right] = \boldsymbol{\tau}_R \cdot \left[\frac{1}{V(t)} \int_{\Omega(t)} \nabla\boldsymbol{r}\, \mathrm{d}V \right] = \boldsymbol{\tau}_R. \tag{4.5}$$

This proves the intuitive claim that the uniform stress tensor of the reservoir $\boldsymbol{\tau}_R$ is exactly the average stress tensor $\bar{\boldsymbol{\tau}}$ throughout the possibly heterogeneous sample in equilibrium. Stated differently, allowing for how the average stress tensor varies through time for a given mass element $\Omega(\boldsymbol{r}, t)$ that resides within a still larger system is equivalent to placing this mass element in a reservoir of uniform stress tensor and allowing this uniform reservoir stress to vary through time while the mass element passes through a sequence of evolving equilibrium configurations with elastic-strain energy being reversibly stored as quantified by $\hat{u}$.

So with the reservoir stress $\boldsymbol{\tau}_R = \overline{\boldsymbol{\tau}}$ being both uniform over $\partial\Omega(t)$ at each moment in time and symmetric, it also can be taken outside the surface integral of Eq. (4.2) to give the sought after expression for the rate at which elastic energy is being stored in the mass element

$$\frac{\mathrm{d}\hat{u}}{\mathrm{d}t} = \frac{\overline{\boldsymbol{\tau}}}{\overline{\rho}} : \left[\frac{1}{V(t)} \int_{\partial\Omega(t)} \boldsymbol{n}\boldsymbol{v}\, \mathrm{d}S \right]. \tag{4.6}$$

The tensor in square brackets quantifies the macroscopic rate of deformation of the sample.

For a mass element embedded in a larger body experiencing a heterogeneous stress and deformation distribution, in addition to temporal changes in the average stress tensor $\overline{\boldsymbol{\tau}}$ of the element, there may also be spatial variations of stress across the mass element resulting in $\nabla \cdot \overline{\boldsymbol{\tau}}$ being nonzero for that sample. When the sum of $\nabla \cdot \overline{\boldsymbol{\tau}}$ and any body forces present is nonzero, the mass element as a whole will accelerate and there will be a net (average) kinetic energy associated with that movement. Such macroscopic kinetic energy is accounted for separately in a macroscopic statement of mechanical energy conservation (e.g., as seen earlier in Section 2.9.2) and is not part of the adiabatic internal elastic-strain energy that $\hat{u}$ is accounting for. There is no inconsistency to modeling elastic deformation (strain) as a sequence of equilibrium configurations that is created only by the changing mean-stress tensor $\overline{\boldsymbol{\tau}}$ acting on a sample, while simultaneously allowing $\nabla \cdot \overline{\boldsymbol{\tau}}$ and any body forces present to create out-of-equilibrium conditions that result in acceleration of the sample as a whole.

4.1.2 Thermodynamic Strain Rate and Thermodynamic Strain

We now focus on the second-order tensor contained in the brackets on the right-hand side of Eq. (4.6). That tensor is, in general, not symmetric and is involved with a double-dot product with the symmetric tensor $\overline{\boldsymbol{\tau}}$. Let us briefly consider the general nature of the double-dot product involving symmetric tensors.

Any second-order tensor $\boldsymbol{S}$ can be written as the sum of symmetric and antisymmetric portions

$$\boldsymbol{S} = \underbrace{\frac{1}{2}\left[\boldsymbol{S} + \boldsymbol{S}^T\right]}_{symmetric} + \underbrace{\frac{1}{2}\left[\boldsymbol{S} - \boldsymbol{S}^T\right]}_{antisymmetric} \tag{4.7}$$

$$= \frac{1}{2}\begin{bmatrix} S_{11} + S_{11} & S_{12} + S_{21} \\ S_{21} + S_{12} & S_{22} + S_{22} \end{bmatrix} + \frac{1}{2}\begin{bmatrix} 0 & S_{12} - S_{21} \\ -(S_{12} - S_{21}) & 0 \end{bmatrix},$$

where we write the components in 2D to save time and space. An anti-symmetric tensor has zeroes along the diagonal and has off-diagonal components that satisfy $S_{ij} = -S_{ji}$.

Further, the double dot product of a symmetric tensor $\boldsymbol{A}$ and an antisymmetric tensor $\boldsymbol{B}$ is zero:

$$\boldsymbol{A}:\boldsymbol{B} = \mathrm{tr}\,\{\boldsymbol{A}\cdot\boldsymbol{B}\} = \mathrm{tr}\left\{\begin{pmatrix} A_{11} & A_{12} \\ A_{12} & A_{22}\end{pmatrix}\begin{pmatrix} 0 & B_{12} \\ -B_{12} & 0\end{pmatrix}\right\} \tag{4.8}$$

$$= \mathrm{tr}\left\{\begin{matrix} -A_{12}\,B_{12} & +A_{11}\,B_{12} \\ -A_{22}\,B_{12} & +A_{12}\,B_{12}\end{matrix}\right\} = 0.$$

So if we return to the second-order tensor defined by the bracketed term in Eq. (4.6), only the symmetric part of this tensor contributes to the work of deformation because the stress tensor is symmetric.

We therefore define the symmetric second-order *thermodynamic strain-rate tensor* as

$$\boxed{\frac{\mathrm{d}\boldsymbol{e}}{\mathrm{d}t} = \frac{1}{2V(t)}\int_{\partial\Omega(t)} [\boldsymbol{n}\boldsymbol{v} + \boldsymbol{v}\boldsymbol{n}]\;\mathrm{d}S.} \tag{4.9}$$

This rate can be experimentally measured by monitoring the velocity at which the sample's surface is moving. Using our Chapter 2 theorem for the gradient of a volume-averaged field as given in Eqs (2.40)–(2.44), we also have that

$$\frac{\mathrm{d}\boldsymbol{e}}{\mathrm{d}t} = \frac{1}{2}\left(\nabla\bar{\boldsymbol{v}} + (\nabla\bar{\boldsymbol{v}})^T\right), \tag{4.10}$$

where

$$\bar{\boldsymbol{v}} = \frac{1}{V(t)}\int_{\Omega(t)} \boldsymbol{v}\,\mathrm{d}V \tag{4.11}$$

is the volume-averaged velocity associated with $\Omega(t)$.

If we now return to Eq. (4.6) for how the specific internal elastic-strain energy of a sample is changing through time we have

$$\frac{\mathrm{d}\hat{u}}{\mathrm{d}t} = \frac{\bar{\boldsymbol{\tau}}}{\bar{\rho}} : \frac{\mathrm{d}\boldsymbol{e}}{\mathrm{d}t} \tag{4.12}$$

with the thermodynamic strain rate $\mathrm{d}\boldsymbol{e}/\mathrm{d}t$ given either by Eq. (4.9) or (4.10). This implies that $\hat{u} = \hat{u}(\boldsymbol{e})$ because upon taking a total time derivative

$$\frac{\mathrm{d}\hat{u}}{\mathrm{d}t} = \frac{\partial\hat{u}}{\partial\boldsymbol{e}} : \frac{\mathrm{d}\boldsymbol{e}}{\mathrm{d}t}, \tag{4.13}$$

we obtain Eq. (4.12) after identifying the mean stress tensor to be $\bar{\boldsymbol{\tau}} = \bar{\rho}\,\partial\hat{u}/\partial\boldsymbol{e}$ or equivalently

$$\boxed{\frac{\partial\hat{u}}{\partial\boldsymbol{e}} = \frac{\partial\hat{u}}{\partial e_{ij}}\hat{\boldsymbol{x}}_i\hat{\boldsymbol{x}}_j = \frac{\bar{\boldsymbol{\tau}}}{\bar{\rho}}.} \tag{4.14}$$

So the elastic-strain energy $\hat{u}$ depends on the *thermodynamic strain tensor* $\boldsymbol{e}$ that derives from gradients in the volume-averaged velocity $\bar{\boldsymbol{v}}$. Hooke's law will soon be obtained by taking a total derivative of Eq. (4.14) and thus also depends on gradients of the volume-averaged velocities. However, the inertial acceleration in our macroscopic theory of motion depends on the mass-averaged velocity field as we now quantify.

4.1.3 Volume-Averaged and Mass-Averaged Velocities in Elastodynamics

In Chapter 2, we averaged over the molecules to obtain the conservation of momentum in the form

$$\frac{\partial}{\partial t}(\boldsymbol{v}\rho) = -\nabla \cdot (\boldsymbol{v}\boldsymbol{v}\rho) + \nabla \cdot \boldsymbol{\tau} + \boldsymbol{F}_b, \tag{4.15}$$

where $\boldsymbol{v}$ is the center-of-mass velocity of the molecules surrounding each point and ρ is the mass of those molecules divided by the volume containing the molecules at each instant. This equation holds at every point $\boldsymbol{r}'$ throughout our sample $\Omega(t)$. Taking a volume average of this law of motion and noting that $\nabla' = \partial/\partial \boldsymbol{r}'$ gives

$$\frac{1}{V(t)} \int_{\Omega(t)} \left[\frac{\partial}{\partial t}(\boldsymbol{v}\rho) + \nabla' \cdot (\boldsymbol{v}\boldsymbol{v}\rho) \right] \mathrm{d}^3\boldsymbol{r}' = \frac{1}{V(t)} \int_{\Omega(t)} \left(\nabla' \cdot \boldsymbol{\tau} + \boldsymbol{F}_b \right) \mathrm{d}^3\boldsymbol{r}'. \tag{4.16}$$

From the Reynolds transport theorem of Eq. (1.68), we have

$$\frac{1}{V(t)} \int_{\Omega(t)} \left[\frac{\partial}{\partial t}(\boldsymbol{v}\rho) + \nabla' \cdot (\boldsymbol{v}\boldsymbol{v}\rho) \right] \mathrm{d}^3\boldsymbol{r}' = \frac{M}{V(t)M} \frac{\mathrm{d}}{\mathrm{d}t} \int_{\Omega(t)} \boldsymbol{v}\rho \, \mathrm{d}^3\boldsymbol{r}' = \bar{\rho} \frac{\mathrm{d}\langle \boldsymbol{v} \rangle}{\mathrm{d}t}, \tag{4.17}$$

where $\bar{\rho} = M/V(t)$ is again the average mass density of the collection of N molecules always contained within $\Omega(t)$ and $\langle \boldsymbol{v} \rangle$ is the mass-averaged velocity of these N molecules

$$\langle \boldsymbol{v} \rangle = \frac{1}{M} \int_{\Omega(t)} \boldsymbol{v}\rho \, \mathrm{d}^3\boldsymbol{r}' \tag{4.18}$$

$$= \frac{1}{M} \sum_i \left(\frac{1}{M_i} \sum_{p=N_{i-1}+1}^{N_i} m_p \frac{\mathrm{d}\boldsymbol{r}_p(t)}{\mathrm{d}t} \right) \frac{M_i}{\Delta V_i} \Delta V_i = \frac{1}{M} \sum_{p=1}^{N} m_p \frac{\mathrm{d}\boldsymbol{r}_p(t)}{\mathrm{d}t}. \tag{4.19}$$

In going from Eqs (4.18) to (4.19), the integral is performed by dividing $\Omega(t)$ into tiny voxels of volume ΔV_i where the index i denotes the name of each small voxel that contains $N_i - N_{i-1}$ molecules (where $N_0 = 0$) and that has a mass $M_i = \sum_{p=N_{i-1}+1}^{N_i} m_p$, a density $\rho_i = M_i/\Delta V_i$, and a center of mass velocity $\boldsymbol{v}_i = M_i^{-1} \sum_{p=N_{i-1}+1}^{N_i} m_p \mathrm{d}\boldsymbol{r}_p(t)/\mathrm{d}t$, where m_p is the mass of each molecule and $\mathrm{d}\boldsymbol{r}_p(t)/\mathrm{d}t$ the velocity. So $\langle \boldsymbol{v} \rangle$ is the center-of-mass velocity of the collection of N molecules. We further have that

$$\nabla \cdot \bar{\boldsymbol{\tau}} = \frac{1}{V(t)} \int_{\partial\Omega(t)} \boldsymbol{n} \cdot \boldsymbol{\tau} \, \mathrm{d}^2\boldsymbol{r}' = \frac{1}{V(t)} \int_{\Omega(t)} \nabla' \cdot \boldsymbol{\tau} \, \mathrm{d}^3\boldsymbol{r}', \tag{4.20}$$

$$\bar{\boldsymbol{F}}_b = \frac{1}{V(t)} \int_{\Omega(t)} \boldsymbol{F}_b \, \mathrm{d}^3\boldsymbol{r}', \tag{4.21}$$

where the theorem for the gradient of a volume-averaged field is used to obtain Eq. (4.20).

So from a stationary frame of reference for which $\mathrm{d}\langle \boldsymbol{v}\rangle/\mathrm{d}t = \partial\langle \boldsymbol{v}\rangle/\partial t + \langle \boldsymbol{v}\rangle \cdot \nabla\langle \boldsymbol{v}\rangle$, our macroscopic law of motion for each collection of N molecules whose center of mass is located at $\boldsymbol{r}(t) = M^{-1}\sum_{p=1}^{N} m_p \boldsymbol{r}_p(t)$ and moving with velocity $\langle \boldsymbol{v}\rangle = \mathrm{d}\boldsymbol{r}/\mathrm{d}t$ is

$$\bar{\rho}\left(\frac{\partial\langle \boldsymbol{v}\rangle}{\partial t} + \langle \boldsymbol{v}\rangle \cdot \nabla\langle \boldsymbol{v}\rangle\right) = \nabla \cdot \bar{\boldsymbol{\tau}} + \overline{\boldsymbol{F}}_b. \tag{4.22}$$

Using the conservation of mass $\partial\bar{\rho}/\partial t = -\nabla \cdot (\langle \boldsymbol{v}\rangle\bar{\rho})$ as first derived in Eq. (2.54), this can be rewritten in the exact same form as Eq. (4.15). What we've shown is that the macroscopic velocity involved in the inertial acceleration of our macroscopic law of motion is $\langle \boldsymbol{v}\rangle$, while the thermodynamic strain-rate tensor, and therefore $\bar{\boldsymbol{\tau}}/\bar{\rho}$, depends on $\nabla\bar{\boldsymbol{v}}$. Our next goal is to understand how $\nabla\langle \boldsymbol{v}\rangle$ is related to $\nabla\bar{\boldsymbol{v}}$.

To do so, begin with the alternative Lagrangian statement of mass conservation given by Eq. (2.59)

$$\frac{1}{\bar{\rho}}\frac{\mathrm{d}\bar{\rho}}{\mathrm{d}t} = -\nabla \cdot \langle \boldsymbol{v}\rangle = -\frac{1}{V(t)}\frac{\mathrm{d}V(t)}{\mathrm{d}t}, \tag{4.23}$$

where $\bar{\rho} = M/V(t)$. Independently from this statement, the rate at which the volume $V(t)$ surrounding the N molecules is changing is given by the purely kinematic statement

$$\frac{1}{V(t)}\frac{\mathrm{d}V(t)}{\mathrm{d}t} = \frac{1}{V(t)}\frac{\mathrm{d}}{\mathrm{d}t}\int_{\Omega(t)} \mathrm{d}^3\boldsymbol{r}' = \frac{1}{V(t)}\int_{\partial\Omega(t)} \boldsymbol{n}\cdot\boldsymbol{v}\,\mathrm{d}^2\boldsymbol{r}' = \nabla\cdot\bar{\boldsymbol{v}}, \tag{4.24}$$

where the second equality is again the Reynolds transport theorem. Equating Eqs (4.23) and (4.24) yields the requirement

$$\nabla\cdot\bar{\boldsymbol{v}} = \nabla\cdot\langle \boldsymbol{v}\rangle. \tag{4.25}$$

Although we have, in general, that $\langle \boldsymbol{v}\rangle \neq \bar{\boldsymbol{v}}$ for each collection of N molecules, unless for example each of the molecules in the collection has the same mass, we always have that $\nabla\cdot\bar{\boldsymbol{v}} = \nabla\cdot\langle \boldsymbol{v}\rangle$ exactly.

Let's now restate the theorem for the gradient of a volume-averaged field as a theorem for the gradient of a mass-averaged field. So for some local field $\psi(\boldsymbol{r}')$ that can be a tensor of any order (including the scalar "zeroth order" or vector "first order"), we define the mass average at each instance of time as

$$\langle\psi\rangle(\boldsymbol{r}) = \frac{1}{M}\int_{\Omega(\boldsymbol{r})} \psi(\boldsymbol{r}')\rho(\boldsymbol{r}')\,\mathrm{d}^3\boldsymbol{r}'. \tag{4.26}$$

The gradient of the mass average, at the same instant in time, is then obtained from the definition of the spatial derivative as

$$\mathrm{d}\boldsymbol{r}\cdot\nabla\langle\psi\rangle(\boldsymbol{r}) = \frac{1}{M}\left[\int_{\Omega(\boldsymbol{r}+\mathrm{d}\boldsymbol{r})} \psi(\boldsymbol{r}')\rho(\boldsymbol{r}')\,\mathrm{d}^3\boldsymbol{r}' - \int_{\Omega(\boldsymbol{r})} \psi(\boldsymbol{r}')\rho(\boldsymbol{r}')\,\mathrm{d}^3\boldsymbol{r}'\right] \tag{4.27}$$

in the limit as $|\mathbf{d}\boldsymbol{r}| \to 0$ and the difference in the two volume integrals becomes a surface integral. With $\mathrm{d}^3\boldsymbol{r}' = +(\mathbf{d}\boldsymbol{r}\cdot\boldsymbol{n})\,\mathrm{d}^2\boldsymbol{r}'$ on the one side of the element and $\mathrm{d}^3\boldsymbol{r}' = -(\mathbf{d}\boldsymbol{r}\cdot\boldsymbol{n})\,\mathrm{d}^2\boldsymbol{r}'$ on the other side and for arbitrary direction of $\mathbf{d}\boldsymbol{r}$, the gradient of the mass average at each instant of time is thus given by

$$\nabla\langle\psi\rangle(\boldsymbol{r},t) = \frac{1}{M}\int_{\partial\Omega(\boldsymbol{r}(t))} \boldsymbol{n}(\boldsymbol{r}',t)\psi(\boldsymbol{r}',t)\rho(\boldsymbol{r}',t)\,\mathrm{d}^2\boldsymbol{r}', \tag{4.28}$$

where $\partial\Omega\,(\boldsymbol{r}(t))$ is what we have been calling $\partial\Omega(t)$. Applying Eq. (4.28) to the velocity field gives

$$\nabla\langle\boldsymbol{v}\rangle = \frac{1}{M}\int_{\partial\Omega(t)} \boldsymbol{n}\boldsymbol{v}\rho\,\mathrm{d}^2\boldsymbol{r}'. \tag{4.29}$$

Writing the local density over the surface of $\partial\Omega(t)$ as $\rho(\boldsymbol{r}') = \bar{\rho} + \delta\rho(\boldsymbol{r}')$ then gives

$$\nabla\langle\boldsymbol{v}\rangle = \frac{1}{V(t)}\int_{\partial\Omega(t)} \boldsymbol{n}\boldsymbol{v}\left(1+\frac{\delta\rho}{\bar{\rho}}\right)\mathrm{d}^2\boldsymbol{r}' = \nabla\bar{\boldsymbol{v}} + \frac{1}{V(t)}\int_{\partial\Omega(t)} \boldsymbol{n}\boldsymbol{v}\frac{\delta\rho}{\bar{\rho}}\,\mathrm{d}^2\boldsymbol{r}'. \tag{4.30}$$

We proved above using the conservation of mass that $\nabla\cdot\langle\boldsymbol{v}\rangle = \nabla\cdot\bar{\boldsymbol{v}}$, which is the statement that

$$\frac{1}{V(t)}\int_{\partial\Omega(t)} \boldsymbol{n}\cdot\boldsymbol{v}\frac{\delta\rho}{\bar{\rho}}\,\mathrm{d}^2\boldsymbol{r}' = 0. \tag{4.31}$$

For this to be true for the normal component of $\boldsymbol{v}$ on $\partial\Omega(t)$, it will also be true for the tangential component because the velocities involved in reversibly storing elastic energy do not depend on the mass densities

$$\frac{1}{V(t)}\int_{\partial\Omega(t)} \boldsymbol{n}\boldsymbol{v}\frac{\delta\rho}{\bar{\rho}}\,\mathrm{d}^2\boldsymbol{r}' = 0. \tag{4.32}$$

We can then conclude that

$$\boxed{\nabla\bar{\boldsymbol{v}} = \nabla\langle\boldsymbol{v}\rangle.} \tag{4.33}$$

This identification allows us to use the center-of-mass velocity as our one and only velocity variable in the macroscopic theory of elastic response and says that inertia is not involved in the definition of elastic strain.

4.1.4 Hooke's Law of Nonlinear Elastic Deformation

To obtain Hooke's law, simply take a total derivative of Eq. (4.14) to obtain the relation between stress-change rate and strain-change rate

$$\frac{\mathrm{d}}{\mathrm{d}t}\left(\frac{\overline{\boldsymbol{\tau}}}{\bar{\rho}}\right) = \frac{\partial(\overline{\boldsymbol{\tau}}/\bar{\rho})}{\partial\boldsymbol{e}} : \frac{\mathrm{d}\boldsymbol{e}}{\mathrm{d}t}. \tag{4.34}$$

We now define the fourth-order *elastic stiffness tensor* ${}_4\boldsymbol{C}$ as

$$\frac{{}_4\boldsymbol{C}}{\bar{\rho}} = \frac{\partial(\overline{\boldsymbol{\tau}}/\bar{\rho})}{\partial \boldsymbol{e}} \mathrel{\hat{=}} \frac{\partial(\overline{\tau}_{kl}/\bar{\rho})}{\partial e_{ij}}\hat{\boldsymbol{x}}_i\hat{\boldsymbol{x}}_j\hat{\boldsymbol{x}}_k\hat{\boldsymbol{x}}_l = \frac{\partial^2 \hat{u}}{\partial e_{ij}\partial e_{kl}}\hat{\boldsymbol{x}}_i\hat{\boldsymbol{x}}_j\hat{\boldsymbol{x}}_k\hat{\boldsymbol{x}}_l \tag{4.35}$$

or in terms of the scalar components alone

$$C_{ijkl} = \bar{\rho}\frac{\partial^2 \hat{u}}{\partial e_{ij}\partial e_{kl}}. \tag{4.36}$$

The $3^4 = 81$ constants represented by C_{ijkl} are material properties called *elastic moduli* or *elastic stiffnesses* (SI units of Pa) that are defined for each macroscopic sample Ω of material surrounding each point $\boldsymbol{r}$. We will show in an upcoming section that at most 21 of these 81 elastic moduli are independent and that in an isotropic material there are only two independent stiffnesses.

Distributing the total derivative on the left-hand side of Eq. (4.34) gives Hooke's law in the form

$$\frac{\mathrm{d}\overline{\boldsymbol{\tau}}}{\mathrm{d}t} - \frac{\overline{\boldsymbol{\tau}}}{\bar{\rho}}\frac{\mathrm{d}\bar{\rho}}{\mathrm{d}t} = {}_4\boldsymbol{C} : \frac{\mathrm{d}\boldsymbol{e}}{\mathrm{d}t}. \tag{4.37}$$

The additional term on the left-hand side $\bar{\rho}^{-1}\overline{\boldsymbol{\tau}}\mathrm{d}\bar{\rho}/\mathrm{d}t$ is not standard in most texts that derive Hooke's law from a thermodynamics perspective (e.g., Landau and Lifshitz, 1986 or Nye, 1957). The reason is that in identifying the equivalent of our volume averages over each evolving region $\Omega(t)$, these authors incorrectly normalize the volume integrals by the initial volume $V(0)$ rather than the current volume $V(t)$ of $\Omega(t)$. Comparing the amplitude of the two terms on the left-hand side of Eq. (4.37)

$$\frac{|\overline{\boldsymbol{\tau}}\bar{\rho}^{-1}\mathrm{d}\bar{\rho}/\mathrm{d}t|}{|\mathrm{d}\overline{\boldsymbol{\tau}}/\mathrm{d}t|} \approx \frac{|\mathrm{d}\bar{\rho}|}{\bar{\rho}} = \frac{|\mathrm{d}V|}{V}, \tag{4.38}$$

we see that if the volumetric strain $|\mathrm{d}V|/V$ is small relative to one in a given application of Hooke's law, which is quite common in practice, the additional term can be neglected. This will be the perspective we take in analyzing linear elastodynamic response later in the chapter.

It should be emphasized that Hooke's law as given by Eq. (4.37) represents the complete nonlinear elastic strain response of a material being subjected to applied stress changes. One source of nonlinearity is the advective derivative $\bar{\boldsymbol{v}} \cdot \nabla\overline{\boldsymbol{\tau}}$ when we write out the total derivative $\mathrm{d}\overline{\boldsymbol{\tau}}/\mathrm{d}t = \partial\overline{\boldsymbol{\tau}}/\partial t + \bar{\boldsymbol{v}} \cdot \nabla\overline{\boldsymbol{\tau}}$ in the stationary-frame Eulerian perspective that we inevitably work from. Another source of nonlinearity is the additional nonstandard term $\bar{\rho}^{-1}\overline{\boldsymbol{\tau}}\,\mathrm{d}\bar{\rho}/\mathrm{d}t$ discussed above. Typically, the most important source of nonlinearity in real materials, especially if they contain flaws and cracks, is that the elastic stiffnesses ${}_4\boldsymbol{C}$ change as strain is changing. If this is occurring, additional rules are required for how the evolving stiffnesses depend on the evolving strain, as will be derived next in Section 4.1.5.

Using the tensorial identity that $\nabla \cdot [\langle \boldsymbol{v}\rangle\, \overline{\boldsymbol{\tau}}] = [\nabla \cdot \langle \boldsymbol{v}\rangle]\, \overline{\boldsymbol{\tau}} + \langle \boldsymbol{v}\rangle \cdot \nabla \overline{\boldsymbol{\tau}}$, along with the conservation of mass $\bar{\rho}^{-1}\mathrm{d}\bar{\rho}/\mathrm{d}t = -\nabla \cdot \langle \boldsymbol{v}\rangle$ and the definition of the strain-rate tensor, we can write Eq. (4.37) as

$$\frac{\partial \overline{\boldsymbol{\tau}}}{\partial t} + \nabla \cdot [\langle \boldsymbol{v}\rangle\, \overline{\boldsymbol{\tau}}] = \frac{{}_4\boldsymbol{C}}{2} : \left[\nabla \langle \boldsymbol{v}\rangle + (\nabla \langle \boldsymbol{v}\rangle)^T\right]. \tag{4.39}$$

This is Hooke's law in its most complete nonlinear form when expressed in a stationary reference frame.

4.1.5 The Macroscopic Rules of Nonlinear Elastodynamics

For a stationary frame of reference, we now gather together all we have learned about the macroscopic laws of conservation of mass and linear momentum along with the general form of Hooke's law as just derived:

$$\frac{\partial \bar{\rho}}{\partial t} = -\nabla \cdot [\langle \boldsymbol{v}\rangle \bar{\rho}] \qquad \textit{conservation of mass} \tag{4.40}$$

$$\frac{\partial}{\partial t}[\langle \boldsymbol{v}\rangle \bar{\rho}] = -\nabla \cdot [\langle \boldsymbol{v}\rangle\langle \boldsymbol{v}\rangle \bar{\rho}] + \nabla \cdot \overline{\boldsymbol{\tau}} + \overline{\boldsymbol{F}}_b \qquad \textit{conservation of linear momentum} \tag{4.41}$$

$$\frac{\partial \overline{\boldsymbol{\tau}}}{\partial t} = -\nabla \cdot [\langle \boldsymbol{v}\rangle \overline{\boldsymbol{\tau}}] + \frac{{}_4\boldsymbol{C}}{2} : \left[\nabla \langle \boldsymbol{v}\rangle + (\nabla \langle \boldsymbol{v}\rangle)^T\right]. \qquad \textit{Hooke's law} \tag{4.42}$$

There is a pleasing visual symmetry to these exact statements of nonlinear elastic response in which the temporal change of each varying field at a point (the left-hand side) is due to an advective accumulation of that particular field in the immediate neighborhood of the point (the first term on the right-hand side) and various source terms for each particular field (the additional terms on the right-hand side).

Given the fourth-order elastic stiffness tensor ${}_4\boldsymbol{C}$, these rules now provide enough equations to determine the fields $\overline{\rho}$, $\langle \boldsymbol{v}\rangle$, and $\overline{\boldsymbol{\tau}}$. For now, we will retain the cumbersome symbols that denote these fields as being averaged over each region $\Omega(t)$ surrounding each point of the material in a still larger macroscopic system being studied, but such averaging symbols will be dropped in later sections.

If the elastic stiffnesses ${}_4\boldsymbol{C}$ are themselves changing with time, taking the total derivative $\mathrm{d}\left({}_4\boldsymbol{C}/\bar{\rho}\right)/\mathrm{d}t$, then gives the differential constitutive law controlling this nonlinear effect

$$\bar{\rho}\frac{\mathrm{d}}{\mathrm{d}t}\left(\frac{{}_4\boldsymbol{C}}{\bar{\rho}}\right) = {}_6\boldsymbol{D} : \frac{\mathrm{d}\boldsymbol{e}}{\mathrm{d}t}, \tag{4.43}$$

where the sixth-order tensor ${}_6\boldsymbol{D}$ (same units as ${}_4\boldsymbol{C}$ which is Pa or force per unit area) is defined

$${}_6\boldsymbol{D} = \bar{\rho}\frac{\partial}{\partial \boldsymbol{e}}\left(\frac{{}_4\boldsymbol{C}}{\bar{\rho}}\right) = \bar{\rho}\frac{\partial^3 \hat{u}}{\partial e_{ij}\partial e_{kl}\partial e_{mn}}\hat{\boldsymbol{x}}_i\hat{\boldsymbol{x}}_j\hat{\boldsymbol{x}}_k\hat{\boldsymbol{x}}_l\hat{\boldsymbol{x}}_m\hat{\boldsymbol{x}}_n \tag{4.44}$$

and is the material property controlling how ${}_4\boldsymbol{C}$ is changing with strain. In a stationary frame of reference, this constitutive law can be written exactly as

$$\frac{\partial({}_4\boldsymbol{C})}{\partial t} = -\nabla \cdot [\langle \boldsymbol{v} \rangle \, ({}_4\boldsymbol{C})] + \frac{{}_6\boldsymbol{D}}{2} : \left[\nabla \langle \boldsymbol{v} \rangle + (\nabla \langle \boldsymbol{v} \rangle)^T\right], \tag{4.45}$$

which is the law for how the elastic stiffnesses change in proportion to the strain rate and has the same exactitude as Hooke's law of Eq. (4.42).

Equation (4.45) has a generalized "force" that is the second-order strain-rate tensor and a generalized "response" that is the time rate of change of the fourth-order stiffness tensor. You might be curious what happens to this equation in an isotropic material where both ${}_4\boldsymbol{C}$ and ${}_6\boldsymbol{D}$ must be isotropic tensors, that is, must have coefficients that do not change if the Cartesian coordinates are rotated. In an isotropic material, the "change moduli" ${}_6\boldsymbol{D}$ must be proportional to the 16 isotropic sixth-order tensors given by Eqs (1.188)–(1.203), while the stiffness tensor ${}_4\boldsymbol{C}$ must be proportional to the three fourth-order isotropic tensors, which can be written as

$$C_{ijkl} = G\left(\delta_{il}\delta_{jk} + \delta_{ik}\delta_{jl} - \frac{2}{3}\delta_{ij}\delta_{kl}\right) + K\delta_{ij}\delta_{kl}. \tag{4.46}$$

The scalar G is called the *shear modulus* and the scalar K the *bulk modulus*. Equation (4.46) will be obtained more deliberately in Section 4.4.3. You can show as a guided end-of-chapter exercise that for an isotropic material, Eq. (4.45) becomes the two scalar equations

$$\frac{\partial G}{\partial t} = -\nabla \cdot [\langle \boldsymbol{v} \rangle G] - D_G \nabla \cdot \langle \boldsymbol{v} \rangle, \tag{4.47}$$

$$\frac{\partial K}{\partial t} = -\nabla \cdot [\langle \boldsymbol{v} \rangle K] - D_K \nabla \cdot \langle \boldsymbol{v} \rangle. \tag{4.48}$$

So in an isotropic material, the shear modulus and bulk modulus change due to volumetric change alone. This is an example of Curie's principle for the constitutive laws of an isotropic material as introduced in Section 1.8.5. In writing these laws for how G and K change, we have assumed that G and K both increase with decreasing sample volume so that the isotropic change moduli D_G and D_K as given are both positive. If a process like cracks closing with decreasing sample volume is causing the elastic stiffnesses G and K to increase, which is the most common reason in rocks for elastic moduli to increase significantly with decreasing isotropic strain, the change moduli D_G and D_K are commonly measured (in Pa) to be significantly larger than either G or K. From a moving Lagrangian frame of reference, we can also write Eqs (4.47) and (4.48) using the total time derivative $\mathrm{d}/\mathrm{d}t = \partial/\partial t + \langle \boldsymbol{v} \rangle \cdot \nabla$ as

$$\frac{\mathrm{d}G}{\mathrm{d}t} = -(G + D_G)\nabla \cdot \langle \boldsymbol{v} \rangle, \tag{4.49}$$

$$\frac{\mathrm{d}K}{\mathrm{d}t} = -(K + D_K)\nabla \cdot \langle \boldsymbol{v} \rangle. \tag{4.50}$$

As another end-of-chapter exercise, you can integrate these exact differentials under the assumption of constant D_G and D_K to obtain how G and K vary with volumetric strain. We will not attempt to further quantify the nature of how the elastic stiffnesses change with changing strain.

The above rules allow for arbitrarily large levels of elastic strain to accumulate in each mass element of a large body as accounted for by the thermodynamic strain tensor $\boldsymbol{e}$, which can be updated through time if so desired in a stationary frame of reference as

$$\frac{\partial \boldsymbol{e}}{\partial t} = -\langle \boldsymbol{v} \rangle \cdot \nabla \boldsymbol{e} + \frac{1}{2}\left[\nabla \langle \boldsymbol{v} \rangle + (\nabla \langle \boldsymbol{v} \rangle)^T\right]. \tag{4.51}$$

The dependent variables that must be found (updated) at each instant of time in the theory of nonlinear elasticity and that are intrinsically coupled to each other are the mass density, stress tensor and the center-of-mass velocity associated with each element. No mention of the "displacement" field has been made in obtaining these nonlinear laws (displacement is defined in the next section). When strain is large, the displacement is calculated using one additional equation (given in the next section) that is solved simultaneously with the above Eqs (4.40)–(4.42). In Section 4.10.2, we show how to solve large-strain elastostatic problems using Eqs (4.40)–(4.42).

Although any level of strain can be modeled in this manner, it must be emphasized that the response will remain elastic (i.e., reversible) only up to some strain threshold where plasticity sets in. At higher temperatures, plasticity is typically due to movement of *dislocations* in the crystal lattice which are stress-induced lines of defects in the crystal. Movement of a dislocation line through the crystal is a thermally activated process that takes place at constant stress and is associated with lines of atoms that repeatedly debond, displace a unit lattice spacing, and rebond across a *glide plane*, which results in an irreversible or plastic strain of the crystal lattice. At lower temperature in materials having local heterogeneity (e.g., solids consisting of jumbles of different grains), plasticity is mainly due to brittle cracking. We have elected not to treat plasticity, either dislocation glide or brittle fracture, in this introductory textbook.

4.2 The Displacement $\boldsymbol{E}$ and Thermodynamic $\boldsymbol{e}$ Strain Tensors

In this section, we introduce the *displacement field*, which is the time integral of the center-of-mass velocity, and define another strain tensor that we call the displacement strain tensor $\boldsymbol{E}$, which is distinct from the thermodynamic strain tensor $\boldsymbol{e}$ that the stored elastic energy $\hat{u}$ and average stress $\bar{\boldsymbol{\tau}}$ depend on. We then find the exact analytical nonlinear relation between $\boldsymbol{e}$ and $\boldsymbol{E}$.

4.2.1 Displacement and the Displacement Strain Tensor

Define the macroscopic *displacement* $\langle \boldsymbol{u} \rangle$ of a mass element $\Omega(t)$ as the translocation of its center of mass. In terms of the molecular trajectories, the displacement is thus defined

$$\langle \boldsymbol{u} \rangle = \boldsymbol{r}(t) - \boldsymbol{r}(0) = \frac{1}{M} \sum_{p=1}^{N} m_p \left[\boldsymbol{r}_p(t) - \boldsymbol{r}_p(0) \right], \tag{4.52}$$

where $\boldsymbol{r}(t)$ is the trajectory of the center of mass of $\Omega(t)$ and $\boldsymbol{r}(0)$ is the location of the center of mass prior to the start of deformation. Because M and N are constants, we have exactly

$$\frac{\mathrm{d}\langle \boldsymbol{u} \rangle}{\mathrm{d}t} = \langle \boldsymbol{v} \rangle. \tag{4.53}$$

This result can also be derived from an alternative perspective using the mass-averaged definition

$$\langle \boldsymbol{u} \rangle = \frac{1}{M} \int_{\Omega(t)} \boldsymbol{u} \rho \, \mathrm{d}V, \tag{4.54}$$

where $\boldsymbol{u}$ is the local displacement of each small mass element $\rho \, \mathrm{d}V$ throughout the sample. Taking a total time derivative of this expression and using the Reynolds transport theorem of Eq. (1.68) gives

$$\frac{\mathrm{d}\langle \boldsymbol{u} \rangle}{\mathrm{d}t} = \frac{1}{M} \frac{\mathrm{d}}{\mathrm{d}t} \int_{\Omega(t)} \boldsymbol{u} \rho \, \mathrm{d}V = \frac{1}{M} \int_{\Omega(t)} \left[\frac{\partial (\boldsymbol{u} \rho)}{\partial t} + \nabla \cdot (\boldsymbol{v} \boldsymbol{u} \rho) \right] \mathrm{d}V \tag{4.55}$$

$$= \frac{1}{M} \int_{\Omega(t)} \left[\left(\frac{\partial \rho}{\partial t} \right) \boldsymbol{u} + \rho \left(\frac{\partial \boldsymbol{u}}{\partial t} \right) + [\nabla \cdot (\boldsymbol{v} \rho)] \boldsymbol{u} + \rho \boldsymbol{v} \cdot \nabla \boldsymbol{u} \right] \mathrm{d}V \tag{4.56}$$

$$= \frac{1}{M} \int_{\Omega(t)} \rho \left(\frac{\partial \boldsymbol{u}}{\partial t} + \boldsymbol{v} \cdot \nabla \boldsymbol{u} \right) \mathrm{d}V = \langle \boldsymbol{v} \rangle. \tag{4.57}$$

If you want to determine the displacement field $\langle \boldsymbol{u} \rangle (\boldsymbol{r}, t)$ from a stationary frame of reference and when strain is large, you must add to the governing Eqs (4.40)–(4.42) for $\bar{\rho}$, $\langle \boldsymbol{v} \rangle$, and $\overline{\boldsymbol{\tau}}$ the update equation for displacements

$$\frac{\partial \langle \boldsymbol{u} \rangle}{\partial t} = \langle \boldsymbol{v} \rangle \cdot (\boldsymbol{I} - \nabla \langle \boldsymbol{u} \rangle). \tag{4.58}$$

This is the perspective taken for solving problems in nonlinear elastostatics as described in Section 4.10.2.

In terms of the macroscopic gradients of the center-of-mass displacement field $\langle \boldsymbol{u} \rangle$, we define a different strain tensor $\boldsymbol{E}$ that we call the *displacement strain tensor* as

$$\boldsymbol{E} = \frac{1}{2} \left[\nabla \langle \boldsymbol{u} \rangle + (\nabla \langle \boldsymbol{u} \rangle)^T \right], \tag{4.59}$$

which is exactly equivalent to the statement

$$\boldsymbol{E} = \frac{1}{2M} \int_{\partial \Omega(t)} [\boldsymbol{n}(\boldsymbol{u} \rho) + (\boldsymbol{u} \rho) \boldsymbol{n}] \, \mathrm{d}S \tag{4.60}$$

given the definition of $\langle \boldsymbol{u} \rangle$ of Eq. (4.54). However, just as we argued that $\nabla \bar{\boldsymbol{v}} = \nabla \langle \boldsymbol{v} \rangle$, so will we have that $\nabla \bar{\boldsymbol{u}} = \nabla \langle \boldsymbol{u} \rangle$, where $\bar{\boldsymbol{u}} = V(t)^{-1} \int_{\Omega(t)} \boldsymbol{u}\, \mathrm{d}V$ is the volume-averaged displacement so that

$$\boldsymbol{E} = \frac{1}{2V(t)} \int_{\partial\Omega(t)} [\boldsymbol{n}\boldsymbol{u} + \boldsymbol{u}\boldsymbol{n}]\, \mathrm{d}S, \tag{4.61}$$

which can be used to experimentally measure the displacement strain tensor $\boldsymbol{E}$ by monitoring the local displacements over the sample surface and performing the integral.

However, we cannot identify the displacement strain tensor $\boldsymbol{E}$ to be the same as the thermodynamic strain tensor $\boldsymbol{e}$ except in the limit of small deformation. This can be seen by integrating the thermodynamic strain-rate tensor $\mathrm{d}\boldsymbol{e}/\mathrm{d}t$ to obtain $\boldsymbol{e}$ (and using the equivalence of mass-averaged and volume-averaged velocities for this strain-rate tensor)

$$\boldsymbol{e} = \int_0^{\boldsymbol{e}} \mathrm{d}\boldsymbol{e}' = \frac{1}{2M} \int_0^t \mathrm{d}t' \int_{\partial\Omega(t')} \left[\boldsymbol{n}\boldsymbol{v}(t')\rho(t') + \boldsymbol{v}(t')\rho(t')\boldsymbol{n}\right] \mathrm{d}S. \tag{4.62}$$

To address the displacement strain tensor $\boldsymbol{E}$, we note that if $\mathrm{d}\boldsymbol{u}(t')/\mathrm{d}t' = \boldsymbol{v}(t')$ is integrated from $t' = 0$ to $t' = t$, where $\boldsymbol{u}(0) = 0$, we have $\boldsymbol{u}(t) = \int_0^t \boldsymbol{v}(t')\, \mathrm{d}t'$ so that Eq. (4.60) can be written

$$\boldsymbol{E} = \frac{1}{2M} \int_{\partial\Omega(t)} \left[\boldsymbol{n} \left(\int_0^t \boldsymbol{v}(t')\mathrm{d}t'\right) \rho(t) + \left(\int_0^t \boldsymbol{v}(t')\mathrm{d}t'\right) \rho(t)\boldsymbol{n}\right] \mathrm{d}S. \tag{4.63}$$

Comparing Eqs (4.62) and (4.63), only when the deformation is sufficiently small that the integration domain limit $\partial\Omega(t)$ can be approximated as being constant in time $\partial\Omega(t) \to \partial\Omega_0$ when performing the integral do we have $\boldsymbol{e} = \boldsymbol{E}$. In fact, it is possible to find the exact nonlinear relation between $\boldsymbol{e}$ and $\boldsymbol{E}$ as we now show.

4.2.2 The Nonlinear Relation between e and E

In what follows, we first quantify the relation between the local displacements $\boldsymbol{u}$ on the surface $\partial\Omega(t)$ of a sample and the thermodynamic strain tensor $\boldsymbol{e}$. We then use the defining integral for $\boldsymbol{E}$ given by Eq. (4.61) both to determine the explicit functional relation between $\boldsymbol{E}$ and $\boldsymbol{e}$ and to show that $\boldsymbol{u} = \boldsymbol{r} \cdot \boldsymbol{E}$, where $\boldsymbol{r}$ is now the distance to points on the evolving surface $\partial\Omega(t)$, which we were calling $\boldsymbol{r}'$ earlier. To visualize what $\boldsymbol{E}$ or $\boldsymbol{e}$ is telling us about the deformation, we look at how points on a sample's surface displace due to $\boldsymbol{E}$.

We begin with the following relation between the local velocity $\boldsymbol{v}(\boldsymbol{r}, t)$ of points on the evolving surface $\partial\Omega(t)$ and the average thermodynamic strain-rate tensor $\mathrm{d}\boldsymbol{e}/\mathrm{d}t$ that is a uniform constant throughout the element

$$\boldsymbol{v}(\boldsymbol{r}, t) = \frac{\mathrm{d}\boldsymbol{u}(\boldsymbol{r}, t)}{\mathrm{d}t} = \boldsymbol{r} \cdot \frac{\mathrm{d}\boldsymbol{e}}{\mathrm{d}t} \quad \text{for points } \boldsymbol{r}(t) \text{ residing on } \partial\Omega(t). \tag{4.64}$$

The validity of this identification is proven by inserting it into the earlier definition of the thermodynamic strain-rate tensor Eq. (4.9) to give

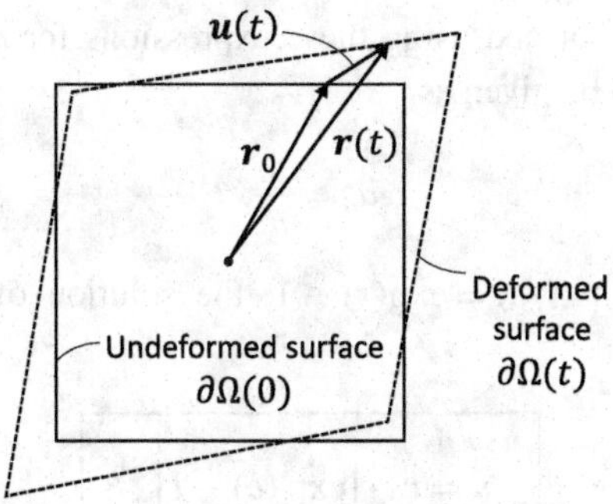

Figure 4.2 Displacement $\boldsymbol{u}(t)$ of a point $\boldsymbol{r}_0$ on the initial undeformed surface to the new position $\boldsymbol{r}(t)$ on the deformed surface.

$$\frac{\mathrm{d}\boldsymbol{e}}{\mathrm{d}t} = \frac{1}{2V(t)} \int_{\partial\Omega(t)} \left[\boldsymbol{n} \left(\boldsymbol{r} \cdot \frac{\mathrm{d}\boldsymbol{e}}{\mathrm{d}t} \right) + \left(\boldsymbol{r} \cdot \frac{\mathrm{d}\boldsymbol{e}}{\mathrm{d}t} \right) \boldsymbol{n} \right] \mathrm{d}S \tag{4.65}$$

$$= \frac{1}{2V(t)} \left[\left(\int_{\Omega(t)} \nabla \boldsymbol{r} \, \mathrm{d}V \right) \cdot \frac{\mathrm{d}\boldsymbol{e}}{\mathrm{d}t} + \frac{\mathrm{d}\boldsymbol{e}}{\mathrm{d}t} \cdot \left(\int_{\Omega(t)} (\nabla \boldsymbol{r})^T \, \mathrm{d}V \right) \right] = \frac{\mathrm{d}\boldsymbol{e}}{\mathrm{d}t}, \tag{4.66}$$

where we used that the strain-rate tensor is symmetric and that $\nabla \boldsymbol{r} = \boldsymbol{I}$ is the identity tensor.

As shown in Fig. 4.2, the distance to points on the evolving surface $\partial\Omega(t)$ of the mass element is given by $\boldsymbol{r}(t) = \boldsymbol{r}_0 + \boldsymbol{u}(t)$, where $\boldsymbol{r}_0$ is the time-independent distance to a surface element on the undeformed sample surface that later displaces by an amount $\boldsymbol{u}(t)$ to its new position $\boldsymbol{r}(t)$. Thus, from Eq. (4.64) we have that the local displacement $\boldsymbol{u}$ to the evolving sample surface is related to the evolving thermodynamic strain tensor $\boldsymbol{e}(t)$ of the entire mass element through the differential equation

$$\frac{\mathrm{d}\boldsymbol{u}}{\mathrm{d}t} = (\boldsymbol{r}_0 + \boldsymbol{u}) \cdot \frac{\mathrm{d}\boldsymbol{e}}{\mathrm{d}t}, \tag{4.67}$$

which we can write

$$\frac{\mathrm{d}\boldsymbol{u}}{\mathrm{d}t} - \boldsymbol{u} \cdot \frac{\mathrm{d}\boldsymbol{e}}{\mathrm{d}t} = \boldsymbol{r}_0 \cdot \frac{\mathrm{d}\boldsymbol{e}}{\mathrm{d}t}. \tag{4.68}$$

One way to solve this differential equation for $\boldsymbol{u}(\boldsymbol{e})$ is to note the initial condition that $\boldsymbol{u} = 0$ when $\boldsymbol{e} = 0$ at $t = 0$ and propose a series solution of the form

$$\boldsymbol{u} = \boldsymbol{r}_0 \cdot (a_1 \boldsymbol{e} + a_2 \boldsymbol{e} \cdot \boldsymbol{e} + a_3 \boldsymbol{e} \cdot \boldsymbol{e} \cdot \boldsymbol{e} + a_4 \boldsymbol{e} \cdot \boldsymbol{e} \cdot \boldsymbol{e} \cdot \boldsymbol{e} + \ldots), \tag{4.69}$$

where the constants a_n are found from the differential equation. Taking the time derivative gives

$$\frac{\mathrm{d}\boldsymbol{u}}{\mathrm{d}t} = \boldsymbol{r}_0 \cdot (a_1 \boldsymbol{I} + 2a_2 \boldsymbol{e} + 3a_3 \boldsymbol{e} \cdot \boldsymbol{e} + 4a_4 \boldsymbol{e} \cdot \boldsymbol{e} \cdot \boldsymbol{e} + \ldots) \cdot \frac{\mathrm{d}\boldsymbol{e}}{\mathrm{d}t}, \tag{4.70}$$

where $\boldsymbol{I}$ is the identity tensor and using these expressions for $\boldsymbol{u}$ and $d\boldsymbol{u}/dt$ in Eq. (4.68) requires the coefficients to be given as

$$a_n = \frac{1}{n!}. \tag{4.71}$$

From the expansion $\sum_{n=1}^{\infty} x^n/n! = \exp(x) - 1$, the solution of Eq. (4.68) can then be written

$$\boxed{\boldsymbol{u} = \boldsymbol{r}_0 \cdot \left[\exp(\boldsymbol{e}) - \boldsymbol{I}\right],} \tag{4.72}$$

where the second-order tensor $\exp(\boldsymbol{e})$ is shorthand for the multiplications (cf., Section 1.8.2)

$$\exp(\boldsymbol{e}) = \boldsymbol{I} + \frac{\boldsymbol{e}}{1!} + \frac{\boldsymbol{e}\cdot\boldsymbol{e}}{2!} + \frac{\boldsymbol{e}\cdot\boldsymbol{e}\cdot\boldsymbol{e}}{3!} + \frac{\boldsymbol{e}\cdot\boldsymbol{e}\cdot\boldsymbol{e}\cdot\boldsymbol{e}}{4!} + \dots. \tag{4.73}$$

Substituting the result of Eq. (4.72) into Eq. (4.68) shows the validity of this result for how the local displacement $\boldsymbol{u}(\boldsymbol{r}_0, t)$ of the external surface of a sample is related nonlinearly to the components of the thermodynamic strain tensor $\boldsymbol{e}(t)$ for that sample.

Using $\boldsymbol{r}_0 = \boldsymbol{r} - \boldsymbol{u}$ in Eq. (4.72), the local displacements over the evolving surface $\partial\Omega(t)$ have the alternate identification $\boldsymbol{u} = \boldsymbol{r} \cdot \left[\boldsymbol{I} - \exp(-\boldsymbol{e})\right]$ and if this expression is used in the integral definition of displacement strain $\boldsymbol{E}$ given by Eq. (4.61), we obtain the nonlinear relations between $\boldsymbol{e}$ and $\boldsymbol{E}$

$$\boldsymbol{E} = \boldsymbol{I} - \exp(-\boldsymbol{e}) \tag{4.74}$$

$$= \boldsymbol{e} - \frac{1}{2!}\boldsymbol{e}\cdot\boldsymbol{e} + \frac{1}{3!}\boldsymbol{e}\cdot\boldsymbol{e}\cdot\boldsymbol{e} - \dots \tag{4.75}$$

or equivalently

$$\boldsymbol{e} = -\ln(\boldsymbol{I} - \boldsymbol{E}) \tag{4.76}$$

$$= \boldsymbol{E} + \frac{1}{2}\boldsymbol{E}\cdot\boldsymbol{E} + \frac{1}{3}\boldsymbol{E}\cdot\boldsymbol{E}\cdot\boldsymbol{E} + \dots. \tag{4.77}$$

So if an experimentalist measures $\boldsymbol{E}$ using the operation of Eq. (4.61), the thermodynamic strain tensor $\boldsymbol{e}$ that the specific internal energy $\hat{u}(\boldsymbol{e})$ and stress depend on is obtained by performing the matrix multiplications (inner products) of Eq. (4.77). So long as the components of $\boldsymbol{E}$ are less than one, though maybe not greatly so for very large deformation, the expansion of Eq. (4.77) that yields $\boldsymbol{e}$ will converge. For components of $\boldsymbol{E}$ that are quite small relative to one, we have the small-strain approximation $\boldsymbol{e} \approx \boldsymbol{E}$ that is very often quite accurate in many applications.

The relation $\boldsymbol{u} = \boldsymbol{r} \cdot \boldsymbol{E}$ for points $\boldsymbol{r}(t)$ on $\partial\Omega(t)$ is actually a nonlinear relation between $\boldsymbol{u}$ and $\boldsymbol{E}$ because $\boldsymbol{r} = \boldsymbol{r}_0 + \boldsymbol{u}$. Indeed, using the nonlinear relations between $\boldsymbol{e}$ and $\boldsymbol{E}$, we can rewrite Eq. (4.72) as the relation

$$\boldsymbol{u} = \boldsymbol{r}_0 \cdot (\boldsymbol{I} - \boldsymbol{E})^{-1} \cdot \boldsymbol{E} = \boldsymbol{r}_0 \cdot (\boldsymbol{E} + \boldsymbol{E} \cdot \boldsymbol{E} + \boldsymbol{E} \cdot \boldsymbol{E} \cdot \boldsymbol{E} + \ldots), \tag{4.78}$$

which is the nonlinear relation between $\boldsymbol{u}(\boldsymbol{r}_0, t)$ and $\boldsymbol{E}(t)$. The nonlinear relations between $\boldsymbol{e}$ and $\boldsymbol{E}$ further give that

$$\frac{\mathrm{d}\boldsymbol{e}}{\mathrm{d}t} = (\boldsymbol{I} - \boldsymbol{E})^{-1} \cdot \frac{\mathrm{d}\boldsymbol{E}}{\mathrm{d}t} = (\boldsymbol{I} + \boldsymbol{E} + \boldsymbol{E} \cdot \boldsymbol{E} + \boldsymbol{E} \cdot \boldsymbol{E} \cdot \boldsymbol{E} + \ldots) \cdot \frac{\mathrm{d}\boldsymbol{E}}{\mathrm{d}t}. \tag{4.79}$$

Additionally, the local material velocity $\boldsymbol{v}$ over $\partial\Omega(t)$ is

$$\boldsymbol{v} = \frac{\mathrm{d}\boldsymbol{u}}{\mathrm{d}t} = \boldsymbol{r}_0 \cdot \exp(\boldsymbol{e}) \cdot \frac{\mathrm{d}\boldsymbol{e}}{\mathrm{d}t} = \boldsymbol{r} \cdot \frac{\mathrm{d}\boldsymbol{e}}{\mathrm{d}t}, \tag{4.80}$$

where we again used $\boldsymbol{r} = \boldsymbol{r}_0 + \boldsymbol{u} = \boldsymbol{r}_0 \cdot \exp(\boldsymbol{e})$. The relation $\boldsymbol{v} = \boldsymbol{r} \cdot \mathrm{d}\boldsymbol{e}/\mathrm{d}t$ on $\partial\Omega(t)$ is where we began this analysis into the nonlinear relations between $\boldsymbol{u}$ on $\partial\Omega(t)$ and either $\boldsymbol{e}$ or $\boldsymbol{E}$, so we have come full circle.

4.3 Visualizing Strain and Rotation

We can decompose the macroscopic displacement gradient $\nabla\bar{\boldsymbol{u}} = \nabla\langle\boldsymbol{u}\rangle$ as

$$\nabla\bar{\boldsymbol{u}} = \frac{1}{V(t)} \int_{\partial\Omega(t)} \boldsymbol{n}\boldsymbol{u}\, \mathrm{d}S = \frac{1}{2V(t)} \int_{\partial\Omega(t)} (\boldsymbol{n}\boldsymbol{u} + \boldsymbol{u}\boldsymbol{n})\, \mathrm{d}S + \frac{1}{2V(t)} \int_{\partial\Omega(t)} (\boldsymbol{n}\boldsymbol{u} - \boldsymbol{u}\boldsymbol{n})\, \mathrm{d}S, \tag{4.81}$$

where the evolving surface that contains the sample is again denoted $\partial\Omega(t)$. The symmetric displacement strain tensor $\boldsymbol{E}$ and the antisymmetric rotation tensor $\boldsymbol{\omega}$ are thus defined as $(\nabla\bar{\boldsymbol{u}} = \boldsymbol{E} + \boldsymbol{\omega})$

$$\boldsymbol{E} = \frac{1}{2V(t)} \int_{\partial\Omega(t)} (\boldsymbol{n}\boldsymbol{u} + \boldsymbol{u}\boldsymbol{n})\, \mathrm{d}S \tag{4.82}$$

$$\boldsymbol{\omega} = \frac{1}{2V(t)} \int_{\partial\Omega(t)} (\boldsymbol{n}\boldsymbol{u} - \boldsymbol{u}\boldsymbol{n})\, \mathrm{d}S. \tag{4.83}$$

Because $\overline{\boldsymbol{\tau}} = \overline{\boldsymbol{\tau}}^T$, we have $\overline{\boldsymbol{\tau}} : \boldsymbol{\omega} = 0$ and the rotation tensor $\boldsymbol{\omega}$ does not enter into our macroscopic thermodynamic rules that account for elastic deformation. Nonetheless, in this section, our goal is to visualize what both $\boldsymbol{E}$ and $\boldsymbol{\omega}$ are telling us about how an element is changing due to $\nabla\bar{\boldsymbol{u}}$, so we will consider $\boldsymbol{\omega}$ a bit further below. We note in passing that the second-order tensor $\boldsymbol{\omega}$ is denoting a distinct response from the curl or rotation vectorial operation $\nabla \times \bar{\boldsymbol{u}}$ as Fig. 4.3 demonstrates.

For our visualization purposes of understanding how the components of the strain tensor control the changing shape and volume of a sample, we assume the strain is small $|\boldsymbol{e}| \ll 1$ which gives

$$\boldsymbol{u} \approx \boldsymbol{r}_0 \cdot \boldsymbol{e} \quad \text{and} \quad \boldsymbol{e} \approx \boldsymbol{E}. \tag{4.84}$$

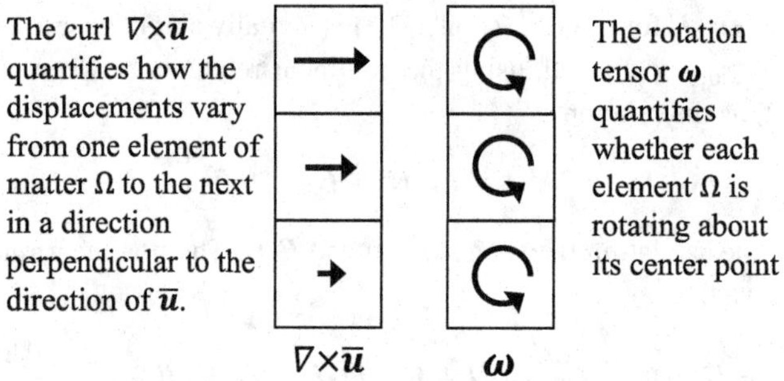
The curl $\nabla\times\bar{u}$ quantifies how the displacements vary from one element of matter Ω to the next in a direction perpendicular to the direction of $\bar{u}$.
The rotation tensor $\boldsymbol{\omega}$ quantifies whether each element Ω is rotating about its center point
$\nabla\times\bar{u}$
$\boldsymbol{\omega}$

Figure 4.3 The difference between the displacements associated with the curl and the rotation tensor.

So in what follows, to imagine the deformation associated with the second-order tensor $\boldsymbol{e}$ (or $\boldsymbol{E}$), we simply dot the distance vector to the undeformed surface $\boldsymbol{r}_0$ into the given $\boldsymbol{e}$ and observe the associated displacements $\boldsymbol{u}$ from the undeformed sample surface. As promised in Chapter 1, a second-order tensor that arrives in our description of continuum physics, like the strain tensor $\boldsymbol{e}$, is always the proportionality between two vectors that have clear physical definition, like $\boldsymbol{u}$ and $\boldsymbol{r}_0$.

4.3.1 Visualizing the Effect of the Rotation Tensor

To keep things simple and confined to the plane of the page, we work in 2D and write the rotation tensor as

$$\boldsymbol{\omega} = \begin{pmatrix} 0 & \omega \\ -\omega & 0 \end{pmatrix}. \tag{4.85}$$

The small-strain displacement on the surface of a 2D box $\boldsymbol{u} = \boldsymbol{r}_0 \cdot \boldsymbol{\omega}$ with $\boldsymbol{r}_0 = (x_1, x_2)$ in 2D has the two components

$$(u_1, u_2) = (x_1, x_2) \begin{pmatrix} 0 & \omega \\ -\omega & 0 \end{pmatrix} = (-x_2\omega, +x_1\omega). \tag{4.86}$$

If this displacement field is plotted around the outside surface of a square box, we obtain the rotated element as shown in Fig. 4.4. So indeed, the anti-symmetric rotation tensor creates a pure rigid-body rotation of the square.

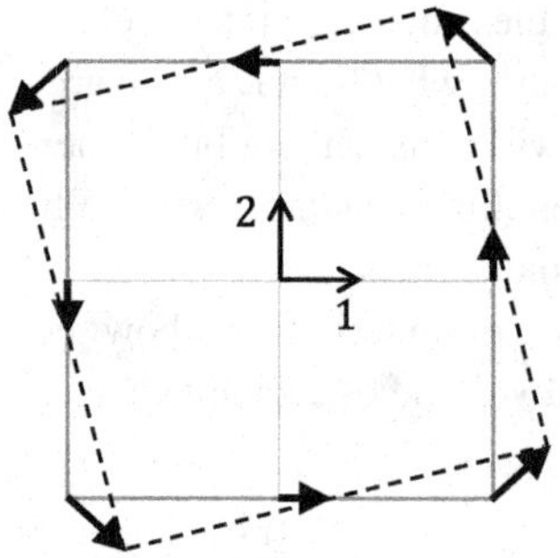
2
1

Figure 4.4 The rigid-body rotation produced by the rotation tensor $\boldsymbol{\omega}$.

4.3.2 Visualizing the Effect of the Strain Tensor

Any second-order tensor, including $\boldsymbol{e}$ and $\boldsymbol{\tau}$, can be decomposed into so-called *isotropic* and *deviatoric* portions

$$\boldsymbol{e} = \frac{\mathrm{tr}\{\boldsymbol{e}\}}{\mathrm{tr}\{\boldsymbol{I}\}}\boldsymbol{I} + \left[\boldsymbol{e} - \frac{\mathrm{tr}\{\boldsymbol{e}\}}{\mathrm{tr}\{\boldsymbol{I}\}}\boldsymbol{I}\right] \tag{4.87}$$

$$= \underbrace{\frac{\nabla\cdot\bar{\boldsymbol{u}}}{\mathrm{tr}\{\boldsymbol{I}\}}\boldsymbol{I}}_{\substack{\text{Isotropic strain}\\ \text{or pure dilation}}} + \underbrace{\frac{1}{2}\left[\nabla\bar{\boldsymbol{u}} + (\nabla\bar{\boldsymbol{u}})^T - 2\frac{\nabla\cdot\bar{\boldsymbol{u}}}{\mathrm{tr}\{\boldsymbol{I}\}}\boldsymbol{I}\right]}_{\substack{\text{Deviatoric strain}\\ \text{or pure shear}}}, \tag{4.88}$$

where $\mathrm{tr}\{\boldsymbol{e}\} = \nabla\cdot\bar{\boldsymbol{u}}$ and where $\mathrm{tr}\{\boldsymbol{I}\} = 3$ in 3D and $\mathrm{tr}\{\boldsymbol{I}\} = 2$ in 2D.

If we take the trace of the isotropic and deviatoric portions of $\boldsymbol{e}$ in either 2D or 3D, which means to sum along the diagonal of these two portions of the strain tensor, we obtain

$$\mathrm{tr}\left\{\frac{\nabla\cdot\bar{\boldsymbol{u}}}{\mathrm{tr}\{\boldsymbol{I}\}}\boldsymbol{I}\right\} = \nabla\cdot\bar{\boldsymbol{u}} \tag{4.89}$$

$$\mathrm{tr}\left\{\frac{1}{2}\left[\nabla\bar{\boldsymbol{u}} + (\nabla\bar{\boldsymbol{u}})^T - 2\frac{\nabla\cdot\bar{\boldsymbol{u}}}{\mathrm{tr}\{\boldsymbol{I}\}}\boldsymbol{I}\right]\right\} = \frac{1}{2}\left[\nabla\cdot\bar{\boldsymbol{u}} + \nabla\cdot\bar{\boldsymbol{u}} - 2\nabla\cdot\bar{\boldsymbol{u}}\right] = 0. \tag{4.90}$$

We will see that the deviatoric strain (or pure shear) produces no volume change and is only associated with shape change of the element. To describe deviatoric strain, a second-order tensor is required. By contrast, the isotropic strain (or pure dilation), is characterized by the scalar $\nabla\cdot\bar{\boldsymbol{u}}$ that is called the *dilatation*, which produces no shape change of an element, only a scalar volume change as we now go on to observe.

Isotropic Strain and Volume Change

To visualize the effect of the isotropic portion of the strain tensor (or pure dilation) $\boldsymbol{e} = (\nabla\cdot\bar{\boldsymbol{u}}/3)\boldsymbol{I}$, we consider the local small-strain displacement field in 3D given by

$$\boldsymbol{u} = \boldsymbol{r}_0\cdot\left[\frac{1}{3}\nabla\cdot\bar{\boldsymbol{u}}\,\boldsymbol{I}\right] = \left(\frac{\nabla\cdot\bar{\boldsymbol{u}}}{3}\right)\boldsymbol{r}_0. \tag{4.91}$$

For example, for a cube (3D) with $\boldsymbol{r}_0$ representing distance from the center of the cube to points on the surface or for a sphere of radius a, we use Eq. (4.91) to determine the displacements from the initial surface as depicted in Fig. 4.5. We see that a pure dilation increases (or decreases) the volume of a region while exactly preserving the shape. We could also describe this as a *self-similar* expansion (or contraction).

To understand how the volume of either the cube or the sphere changes with the dilatation $\nabla\cdot\bar{\boldsymbol{u}}$, we can take two approaches below that give identical exact results for arbitrary (not just small) isotropic-strain amplitude. In one approach, we integrate the conservation of mass $\mathrm{d}\bar{\rho}/\mathrm{d}t = -\bar{\rho}\nabla\cdot\bar{\boldsymbol{v}} = -\bar{\rho}\nabla\cdot\langle\bar{\boldsymbol{v}}\rangle$ written as

$$\frac{\mathrm{d}\bar{\rho}}{\bar{\rho}} = -\nabla\cdot\bar{\boldsymbol{v}}\,\mathrm{d}t \tag{4.92}$$

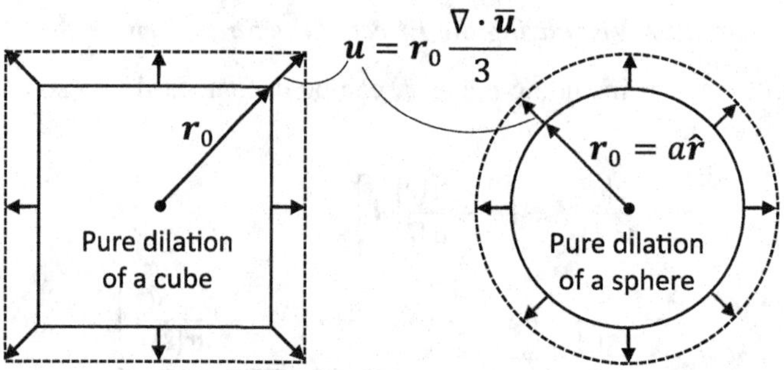

Figure 4.5 Pure volumetric expansion caused by a positive $\nabla \cdot \bar{u}$.

to obtain that

$$\ln\left(\frac{\bar{\rho}(t)}{\bar{\rho}(0)}\right) = -\nabla \cdot \bar{u}, \tag{4.93}$$

where $\bar{\rho}(0) = M/V(0)$ is the density in the undeformed state $\nabla \cdot \bar{u} = 0$ and $\bar{\rho}(t) = M/V(t)$ is the density in the current state of dilation. Expressing this in terms of the volume surrounding the mass M, we have

$$\boxed{V(t) = V(0)\exp\left(\nabla \cdot \bar{u}\right).} \tag{4.94}$$

This is the general exact nonlinear relation between the volume of a sample that initially has volume $V(0)$ and the dilatation $\nabla \cdot \bar{u}$. At the small strain of Eq. (4.91), this becomes

$$\nabla \cdot \bar{u} \approx \frac{[V(t) - V(0)]}{V(0)}, \tag{4.95}$$

which holds when $|\nabla \cdot \bar{u}| \ll 1$. The interpretation of $\nabla \cdot \bar{u}$ as being the fractional change in volume of a mass element is commonly found in continuum mechanics texts but is only strictly true in the limit of small dilatation, while Eq. (4.94) holds for any level of dilatation.

We can also obtain the same result of Eq. (4.94) from the alternative perspective provided by Eq. (4.72). In the case of a sphere with initial radius a subjected to isotropic strain for which $\boldsymbol{u} = u_r\hat{\boldsymbol{r}}$, $\boldsymbol{r}_0 = a\hat{\boldsymbol{r}}$, and $\boldsymbol{e} = (\nabla \cdot \bar{u}/3)\boldsymbol{I}$, Eq. (4.72) becomes

$$u_r = a\left[\exp\left(\frac{\nabla \cdot \bar{u}}{3}\right) - 1\right]. \tag{4.96}$$

So the radius of the dilated sphere is $a + u_r = a\exp(\nabla \cdot \bar{u}/3)$, and the volume is given by the cube of this radius as

$$V(t) = V(0)\exp\left(\nabla \cdot \bar{u}\right). \tag{4.97}$$

This is the same result as Eq. (4.94), which was obtained by integrating the conservation of mass. This is a demonstration that $\nabla \cdot \bar{u} = \nabla \cdot \langle \boldsymbol{u} \rangle$ because the right-hand side of Eq. (4.93) is formally $\nabla \cdot \langle \boldsymbol{u} \rangle$. To conclude, isotropic strain produces a volume change without changing the shape of the initial undeformed sample.

Deviatoric Strain and Shape Change

Let's explicitly denote the deviatoric strain tensor as

$$\boldsymbol{e}^D \mathrel{\widehat{=}} \text{deviatoric strain} = \frac{1}{2}\left[\nabla\bar{\boldsymbol{u}} + (\nabla\bar{\boldsymbol{u}})^T - \frac{2\nabla\cdot\bar{\boldsymbol{u}}}{\mathrm{tr}\{\boldsymbol{I}\}}\boldsymbol{I}\right]. \tag{4.98}$$

For visualization purposes, we again work in 2D so that

$$\boldsymbol{e}^D = \frac{1}{2}\left[\nabla\bar{\boldsymbol{u}} + (\nabla\bar{\boldsymbol{u}})^T - \nabla\cdot\bar{\boldsymbol{u}}\,\boldsymbol{I}\right] \text{ with the 2D identity tensor } \boldsymbol{I} = \begin{pmatrix}1 & 0\\ 0 & 1\end{pmatrix}. \tag{4.99}$$

Writing out the various components of $\boldsymbol{e}^D$ in matrix notation, we have (in 2D)

$$\boldsymbol{e}^D = \begin{bmatrix} \dfrac{\partial u_1}{\partial x_1} - \dfrac{1}{2}\left(\dfrac{\partial u_1}{\partial x_1} + \dfrac{\partial u_2}{\partial x_2}\right) & \dfrac{1}{2}\left(\dfrac{\partial u_1}{\partial x_2} + \dfrac{\partial u_2}{\partial x_1}\right) \\ \dfrac{1}{2}\left(\dfrac{\partial u_1}{\partial x_2} + \dfrac{\partial u_2}{\partial x_1}\right) & \dfrac{\partial u_2}{\partial x_2} - \dfrac{1}{2}\left(\dfrac{\partial u_1}{\partial x_1} + \dfrac{\partial u_2}{\partial x_2}\right) \end{bmatrix}, \tag{4.100}$$

which can be rewritten as

$$\boldsymbol{e}^D = \begin{bmatrix} e_d & e_{12} \\ e_{12} & -e_d \end{bmatrix}, \quad \text{where} \quad \begin{cases} e_d = \dfrac{1}{2}\left(\dfrac{\partial u_1}{\partial x_1} - \dfrac{\partial u_2}{\partial x_2}\right) \\ e_{12} = \dfrac{1}{2}\left(\dfrac{\partial u_1}{\partial x_2} + \dfrac{\partial u_2}{\partial x_1}\right) \end{cases} \tag{4.101}$$

$$= \begin{bmatrix} e_d & 0 \\ 0 & -e_d \end{bmatrix} + \begin{bmatrix} 0 & e_{12} \\ e_{12} & 0 \end{bmatrix}. \tag{4.102}$$

The two tensors in the decomposition of Eq. (4.102) represent the two ways that pure shear (or deviatoric) strain can be represented and visualized.

In the first way, we take $e_{12} = 0$ and $e_d \neq 0$ to obtain the small-strain displacement field given by

$$[u_1, u_2] = [x_1, x_2]\begin{bmatrix} e_d & 0 \\ 0 & -e_d \end{bmatrix} \tag{4.103}$$

or when written out

$$u_1 = x_1 e_d \quad \text{and} \quad u_2 = -x_2 e_d. \tag{4.104}$$

Drawing these displacements on the surface surrounding a square region gives a pictorial representation of pure shear as shown in Fig. 4.6. In this form of pure shear, the initial square is pushed in on two opposing sides and pulled out on the two other opposing sides without any change in volume.

In the second way to visualize pure shear, we take $e_d = 0$ and $e_{12} \neq 0$ to obtain the displacement field

$$[u_1, u_2] = [x_1, x_2]\begin{bmatrix} 0 & e_{12} \\ e_{12} & 0 \end{bmatrix} \tag{4.105}$$

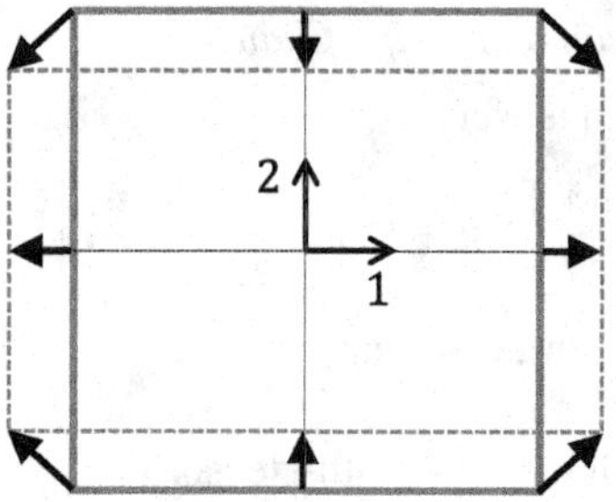

Figure 4.6 One mode of pure shear has two sides approach each other and two sides recede.

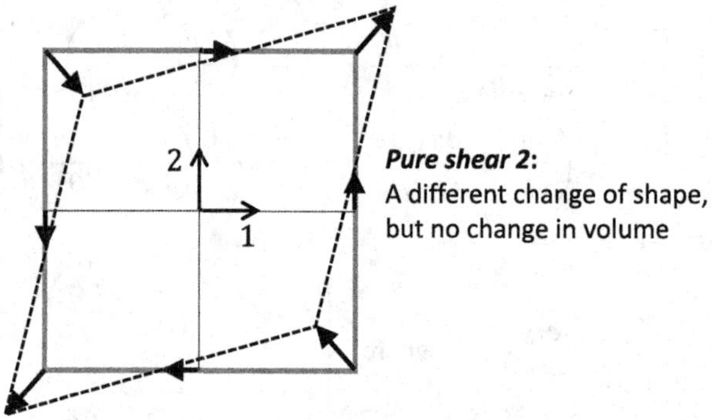

Figure 4.7 A second mode of pure shear has two corners approach each other and the other two corners recede, which is the same as the first mode after the coordinates are rotated to the left by 45°.

or when written out

$$u_1 = x_2\, e_{12} \quad \text{and} \quad u_2 = x_1\, e_{12}. \tag{4.106}$$

Drawing these displacements on the surface surrounding a square gives a different pictorial representation of pure shear as shown in Fig. 4.7. In general, an arbitrary deviatoric deformation can be a combination of both ways of representing pure shear. By rotating the coordinates by 45°, the "first way" of expressing pure shear becomes identical to the "second way."

To show this explicitly, let's begin with

$$e_{ij}^{D} = \begin{pmatrix} 0 & e_d \\ e_d & 0 \end{pmatrix} \tag{4.107}$$

and use the 2D rotation matrix from Chapter 1 that rotates a 2D vector by θ degrees

$$R_{ij} = \begin{pmatrix} \cos\theta & \sin\theta \\ -\sin\theta & \cos\theta \end{pmatrix} \tag{4.108}$$

to obtain the rotated components of the second-order tensor in 2D

$$\left(e_{ij}^{D}\right)' = R_{ik}\, R_{jl}\, e_{kl}^{D} = \begin{pmatrix} \cos\theta & \sin\theta \\ -\sin\theta & \cos\theta \end{pmatrix} \begin{pmatrix} 0 & e_d \\ e_d & 0 \end{pmatrix} \begin{pmatrix} \cos\theta & -\sin\theta \\ \sin\theta & \cos\theta \end{pmatrix}, \tag{4.109}$$

where, as in Chapter 1, we need to transpose the matrix acting from the right in order for the matrix multiplication to correspond to the expression written in terms of sums over indices. We obtain

$$\left(e_{ij}^{D}\right)' = \begin{pmatrix} e_d \sin\theta & e_d \cos\theta \\ e_d \cos\theta & -e_d \sin\theta \end{pmatrix} \begin{pmatrix} \cos\theta & -\sin\theta \\ \sin\theta & \cos\theta \end{pmatrix}$$
$$= \begin{bmatrix} 2e_d \cos\theta\, \sin\theta & e_d\left(\cos^2\theta - \sin^2\theta\right) \\ e_d\left(\cos^2\theta - \sin^2\theta\right) & -2e_d \cos\theta\, \sin\theta \end{bmatrix}. \tag{4.110}$$

For $\theta = \pi/4$ (a rotation by 45°), we then obtain

$$\left(e_{ij}^{D}\right)' = \begin{bmatrix} e_d & 0 \\ 0 & -e_d \end{bmatrix}, \quad \text{where} \quad \cos\frac{\pi}{4} = \sin\frac{\pi}{4} = \frac{1}{\sqrt{2}} \tag{4.111}$$

which is the "first way" of expressing a pure shear as given earlier. Again, if you look at Fig. 4.7 for the "second way" of representing pure shear and rotate your head counter clockwise by 45°, you can see the "first way" of representing pure shear.

To conclude, a deviatoric (or pure shear) strain causes: (1) a straight line (or plane in 3D) on the undeformed surface to remain a straight line on the deformed surface, just stretched, displaced, and rotated relative to its initial position; (2) parallel lines (or planes) on the undeformed surface to remain parallel on the deformed surface, just stretched, displaced, and rotated; (3) a circle (or sphere) to deform into an ellipse (or ellipsoid); and (4) no change in volume to occur.

4.4 Linear Elasticity

From here on, we drop the notation indicating that the fields $\bar{\boldsymbol{\tau}}$, $\langle \boldsymbol{v} \rangle$ (or $\bar{\boldsymbol{v}}$), $\langle \boldsymbol{u} \rangle$ (or $\bar{\boldsymbol{u}}$), and $\bar{\rho}$ are averaged, it now being implicitly understood that $\boldsymbol{\tau}$, $\boldsymbol{v}$, $\boldsymbol{u}$, and ρ represent the average over each macroscopic sample $\Omega(\boldsymbol{r})$ that surrounds each point $\boldsymbol{r}$ of a still larger region under investigation.

We earlier derived Hooke's law in the general nonlinear form

$$\frac{\mathrm{d}}{\mathrm{d}t}\left(\frac{\tau_{ij}}{\rho}\right) = \frac{C_{ijkl}}{\rho}\frac{\mathrm{d}e_{kl}}{\mathrm{d}t} \quad \text{or} \quad \frac{\mathrm{d}}{\mathrm{d}t}\left(\frac{\boldsymbol{\tau}}{\rho}\right) = \frac{{}_4\boldsymbol{C}}{\rho} : \frac{\mathrm{d}\boldsymbol{e}}{\mathrm{d}t}, \tag{4.112}$$

where the fourth-order stiffness tensor ${}_4\boldsymbol{C} = C_{ijkl}\,\hat{\boldsymbol{x}}_i\hat{\boldsymbol{x}}_j\hat{\boldsymbol{x}}_k\hat{\boldsymbol{x}}_l$ has a possibility of $3^4 = 81$ components but due to some symmetries that will be discussed below, there are a maximum of 21 independent components.

If the change with time of ${}_4\boldsymbol{C}/\rho$ is small and can be ignored, which corresponds to the strain components being small in some sense relative to one, Hooke's law can be time-integrated to give $\boldsymbol{\tau} - \boldsymbol{\tau}_I = {}_4\boldsymbol{C} : (\boldsymbol{e} - \boldsymbol{e}_I)$, where $\boldsymbol{\tau}_I$ and $\boldsymbol{e}_I$ are the initial stress and strain tensors related as $\boldsymbol{\tau}_I = {}_4\boldsymbol{C} : \boldsymbol{e}_I$. So the integrated Hooke's law can be written in component form as

$$\boxed{\tau_{ij} = C_{ijkl}\, e_{kl} \,\widehat{=}\, \text{stress–strain relation of linear elasticity.}} \tag{4.113}$$

We also have that $e_{ij} = E_{ij} = \left(\partial u_j/\partial x_i + \partial u_i/\partial x_j\right)/2$ in the small-strain limit where this linear form of Hooke's law applies.

But if the components C_{ijkl}/ρ are themselves varying with strain in a nonnegligible way, which corresponds to *nonlinear elasticity*, we must use the rate equation form of Eq. (4.112) with strain rate given in terms of gradients of the material velocity and work with the explicit rule of Eq. (4.45) for the way that C_{ijkl}/ρ changes. Throughout what follows, we restrict ourselves to the linear elasticity of Eq. (4.113).

4.4.1 Material-Independent Symmetries of the C_{ijkl} Stiffness Tensor

Note the following symmetries of C_{ijkl}

$$\begin{aligned} C_{ijkl} = C_{klij} &\quad \text{because } \rho\frac{\partial^2 \hat{u}}{\partial e_{ij}\,\partial e_{kl}} = \rho\frac{\partial^2 \hat{u}}{\partial e_{kl}\,\partial e_{ij}} \\ = C_{jikl} &\quad \text{because } \tau_{ij} = \tau_{ji} \\ = C_{ijlk} &\quad \text{because } e_{kl} = e_{lk}. \end{aligned} \tag{4.114}$$

These symmetries reduce the number of independent components from 81 to 21. The number of independent stiffnesses can be reduced further, depending on the geometrical symmetries of the material found within Ω. For a so-called isotropic material, which means that the components C_{ijkl} do not change as the coordinates are rotated into any orientation, there are only two independent stiffnesses as will be elaborated upon below.

4.4.2 The Stiffness Tensor in Matrix Notation

To see how the various symmetries $C_{ijkl} = C_{jikl}$, $C_{ijlk} = C_{ijkl}$, and $C_{ijkl} = C_{klij}$ reduce the number of independent constants from 81 to 21, as well as to obtain a convenient way to represent the unwieldy Hooke's law $\tau_{ij} = C_{ijkl}\, e_{lk}$, consider the following tensor-to-matrix index map

$$\begin{pmatrix} 11 & 12 & 13 \\ 21 & 22 & 23 \\ 31 & 32 & 33 \end{pmatrix} \Longrightarrow \begin{pmatrix} 1 & 6 & 5 \\ 6 & 2 & 4 \\ 5 & 4 & 3 \end{pmatrix}. \tag{4.115}$$

We can use this $ij \to k$ map to write the 3×3 symmetric tensors τ_{ij} and e_{kl} as six-component first-order arrays

$$\begin{pmatrix} \tau_{11} & \tau_{12} & \tau_{13} \\ & \tau_{22} & \tau_{23} \\ & & \tau_{33} \end{pmatrix} \Longrightarrow \begin{bmatrix} T_1 \\ T_2 \\ T_3 \\ T_4 \\ T_5 \\ T_6 \end{bmatrix} \quad \text{and} \quad \begin{pmatrix} e_{11} & e_{12} & e_{13} \\ & e_{22} & e_{23} \\ & & e_{33} \end{pmatrix} \Longrightarrow \begin{bmatrix} E_1 \\ E_2 \\ E_3 \\ E_4 \\ E_5 \\ E_6 \end{bmatrix}. \tag{4.116}$$

We can then write $\tau_{ij} = C_{ijkl}\, e_{lk}$ using a simple matrix representation $T_i = C_{ij}\, E_j$, where i and $j = 1, 2, 3, 4, 5, 6$, that is,

$$\begin{bmatrix} T_1 \\ T_2 \\ T_3 \\ T_4 \\ T_5 \\ T_6 \end{bmatrix} = \begin{bmatrix} C_{11} & C_{12} & C_{13} & C_{14} & C_{15} & C_{16} \\ & C_{22} & C_{23} & C_{24} & C_{25} & C_{26} \\ & & C_{33} & C_{34} & C_{35} & C_{36} \\ & & & C_{44} & C_{45} & C_{46} \\ & \text{symmetric} & & & C_{55} & C_{56} \\ & & & & & C_{66} \end{bmatrix} \begin{bmatrix} E_1 \\ E_2 \\ E_3 \\ 2E_4 \\ 2E_5 \\ 2E_6 \end{bmatrix} \tag{4.117}$$

The need for the 2 in front of E_4, E_5, and E_6 is due to the symmetry of the strain tensor as can be seen by writing out τ_{ij} explicitly

$$\begin{aligned} \tau_{ij} &= C_{ij11}\, e_{11} + C_{ij22}\, e_{22} + C_{ij33}\, e_{33} \\ &\quad + C_{ij12}\, e_{21} + C_{ij21}\, e_{12} + C_{ij23}\, e_{32} + C_{ij32}\, e_{23} + C_{ij13}\, e_{31} + C_{ij31}\, e_{13} \end{aligned} \tag{4.118}$$

or because of the symmetry $e_{ij} = e_{ji}$

$$T_k = C_{k1}\, E_1 + C_{k2}\, E_2 + C_{k3}\, E_3 + 2\, C_{k4}\, E_4 + 2\, C_{k5}\, E_5 + 2\, C_{k6}\, E_6. \tag{4.119}$$

Because the number of independent constants in C_{ij} is $\sum_{i=1}^{6} i = 21$, we see why there are at most 21 independent constants in C_{ijkl}. Representing Hooke's law, that is the proportionality between two symmetric second-order tensors, as a matrix multiplication between two first-order arrays of dimension 6 as given by Eq. (4.117) is sometimes called the *Voigt representation* of Hooke's law.

The above 6×6 matrix form of Hooke's law makes it particularly easy to see how to measure the various elastic constants C_{ij}. For example, to measure $C_{13} = C_{1133}$ you would apply a strain $E_3 = e_{33}$ to a sample keeping all other strain components at zero. You would then measure $T_1 = \tau_{11}$ under these strain conditions and obtain $C_{13} = C_{1133} = \tau_{11}/e_{33}$. In this same applied strain experiment, by measuring τ_{22} and τ_{33} you also obtain $C_{23} = C_{2233} = \tau_{22}/e_{33}$ and $C_{33} = C_{3333} = \tau_{33}/e_{33}$. Note that applying e_{33} to a cubic sample of size L^3 means to uniformly displace the two opposing faces with normals in the $\hat{\boldsymbol{x}}_3$ and $-\hat{\boldsymbol{x}}_3$ direction by an amount $\Delta u_3/2$ in the inward direction on both faces to obtain $e_{33} = -|\Delta u_3|/L$. As the two faces in the $\hat{\boldsymbol{x}}_3$ direction are pushed inward, in order to keep the two faces in the $\pm\hat{\boldsymbol{x}}_1$ direction from displacing outward and creating strain in the $\hat{\boldsymbol{x}}_1$ direction, we have to apply an inward force $-\tau_{11}L^2$ on the opposing faces $\pm\hat{\boldsymbol{x}}_1$. Similar comments hold for the 2 direction. By measuring the needed τ_{11} and τ_{22} for $e_{11} = e_{22} = 0$, we thus obtain the moduli $C_{1133} = \tau_{11}/e_{33}$ and $C_{2233} = \tau_{22}/e_{33}$. Recall that normal stress components, that we also call the longitudinal stress components, are defined to be positive in tension (when you pull out on a surface) and negative in compression (when you push in on the surface, which is the scenario being described here). Equivalent strain experiments, where you apply one component of strain while keeping all others at zero, allow all the C_{ijkl} to be measured in this manner.

4.4.3 Elastic Anisotropy and Isotropy

Any internal geometrical symmetries due to the material or crystal structure found inside a sample can reduce the number of independent constants C_{ijkl} for the sample to less than 21.

A material is elastically *anisotropic* if when we rotate the coordinates $(x_1, x_2, x_3) \to (x_1', x_2', x_3')$, some components of C_{ijkl} change to new values C_{ijkl}'. Rules for finding the new components in terms of the old components and by how much each coordinate direction is rotated were given in Section 1.8.3 of Chapter 1.

There can be confusion between the ideas of anisotropy and heterogeneity. In continuum physics, to define the material properties at a point, we consider the material in a finite region that surrounds the point and that contains a given amount of mass. We call this region a "mass element," "material element," or "sample." If all points in a large body are surrounded by mass elements that have the same material properties (elastic constants in the present context), we say the medium is *homogeneous*. If not, the medium is *heterogeneous*.

To describe anisotropy, consider first a generalized force vector acting on a mass element that produces a generalized response vector for that element. If the component of the response vector in the direction of the force vector has different amplitudes for different orientations of a constant-amplitude force vector applied to the element, or if the response vector ever develops a component in a direction different than the force vector, that element is *anisotropic*. In this case, we are required to use a second-order tensor, and not a simple scalar, as the material property that is the proportionality between the response and the force vectors for that mass element. This second-order tensor that accounts for the anisotropy will have scalar components that change when the base vectors are rotated to new orientations. For elasticity, the generalized force is the second-order stress tensor and the generalized response is the second-order strain tensor. If changing the orientation of the applied stress components results in the corresponding strain components having different amplitudes, the response is anisotropic. We derived Hooke's law of linear response in the "inverse form" (this language will be explained in Chapter 11) where the strain tensor $\boldsymbol{e}$ produces a stress tensor $\boldsymbol{\tau}$ after multiplication with the fourth-order stiffness tensor ${}_4\boldsymbol{C}$, that is, $\boldsymbol{\tau} = {}_4\boldsymbol{C} : \boldsymbol{e}$. If when we change the orientation of the coordinates, the components of stiffness change so that ${}_4\boldsymbol{C}' \neq {}_4\boldsymbol{C}$, that mass element is said to be elastically anisotropic.

Pure mineral single crystals always exhibit elastic anisotropy; for example, feldspar, quartz, calcite, and olivine are common anisotropic minerals found in Earth materials. However, if a mass element consists of a jumble of single-crystal mineral grains that are all oriented in different directions, although each grain when analyzed on its own is anisotropic, it is possible that the jumble of grains comprising the mass element is isotropic.

So the possible confusion between heterogeneity and anisotropy is that if we look at the material within a mass element, the reason for that element being anisotropic or isotropic is usually due to the local heterogeneity found within that particular element. So in terms of the local coordinates $\boldsymbol{r}_o$ locating points within the mass element, each $\mathrm{d}V_o$ associated with each $\boldsymbol{r}_o$ might have different local elastic constants or other material properties. Imagine that the material inside a mass element is layered. If each layer has different material properties, the layering, which represents *local* heterogeneity within the mass element, causes that particular element to be anisotropic. Next imagine the solid material inside a mass element is a jumble of differently oriented mineral grains that result in that solid being well-approximated as isotropic but that there is also a population of cracks distributed throughout this isotropic solid material. If the cracks have an average common orientation, the mass element will be anisotropic while if the cracks are randomly oriented the mass element will be isotropic. Whether each local dV_o within a mass element, that may contain many randomly oriented grains, has a crack or not changes the local elastic properties of that dV_o (this local dV_o will be anisotropic with a crack and isotropic without a crack) and both the local heterogeneity and orientation of the cracks throughout the sample determines whether the sample as a whole is isotropic or anisotropic.

As an example of the nomenclature, if in a large system composed of many elements, each mass element surrounding each point in the system has the exact same anisotropic material property, we would say that this anisotropic material is homogeneous throughout the large system even though the reason for each mass element being anisotropic may be local heterogeneity found *within* each mass element. Anisotropy is usually present to some degree for samples of most Earth materials, but we will largely ignore anisotropy in the introductory treatment that follows.

If $C_{ijkl} = C'_{ijkl}$ for any and all coordinate changes at a given point, the material is called *isotropic* at that point. Isotropic tensors are tensors whose components do not change when the orientation of the axes are changed. As we showed in Section 1.8.4, there are three and only three fourth-order isotropic tensors that when multiplied by scalars and added together produce the most general form of a fourth-order isotropic tensor. These three fourth-order isotropic tensors are again

$$ {}_4\boldsymbol{I}^{(1)} = \delta_{il}\delta_{jk}\hat{\boldsymbol{x}}_i\hat{\boldsymbol{x}}_j\hat{\boldsymbol{x}}_k\hat{\boldsymbol{x}}_l = \hat{\boldsymbol{x}}_i\hat{\boldsymbol{x}}_j\hat{\boldsymbol{x}}_j\hat{\boldsymbol{x}}_i \tag{4.120} $$

$$ {}_4\boldsymbol{I}^{(2)} = \delta_{ik}\delta_{jl}\hat{\boldsymbol{x}}_i\hat{\boldsymbol{x}}_j\hat{\boldsymbol{x}}_k\hat{\boldsymbol{x}}_l = \hat{\boldsymbol{x}}_i\hat{\boldsymbol{x}}_j\hat{\boldsymbol{x}}_i\hat{\boldsymbol{x}}_j \tag{4.121} $$

$$ {}_4\boldsymbol{I}^{(3)} = \delta_{ij}\delta_{kl}\hat{\boldsymbol{x}}_i\hat{\boldsymbol{x}}_j\hat{\boldsymbol{x}}_k\hat{\boldsymbol{x}}_l = \hat{\boldsymbol{x}}_i\hat{\boldsymbol{x}}_i\hat{\boldsymbol{x}}_j\hat{\boldsymbol{x}}_j = \boldsymbol{I}\boldsymbol{I}, \tag{4.122} $$

where, in the last expression, $\boldsymbol{I}$ is the second-order isotropic (or identity) tensor. The first isotropic form is an identity tensor in that if it is double-dot multiplied into a second-order tensor $\boldsymbol{A} = A_{ij}\hat{\boldsymbol{x}}_i\hat{\boldsymbol{x}}_j$ one obtains

$$ {}_4\boldsymbol{I}^{(1)} : \boldsymbol{A} = \delta_{il}\delta_{jk}\hat{\boldsymbol{x}}_i\hat{\boldsymbol{x}}_j\hat{\boldsymbol{x}}_k\hat{\boldsymbol{x}}_l : A_{mn}\hat{\boldsymbol{x}}_m\hat{\boldsymbol{x}}_n \tag{4.123} $$

$$ = \delta_{il}\delta_{jk}A_{mn}\hat{\boldsymbol{x}}_i\hat{\boldsymbol{x}}_j(\hat{\boldsymbol{x}}_l \cdot \hat{\boldsymbol{x}}_m)(\hat{\boldsymbol{x}}_k \cdot \hat{\boldsymbol{x}}_n) \tag{4.124} $$

$$ = \delta_{il}\delta_{jk}A_{lk}\hat{\boldsymbol{x}}_i\hat{\boldsymbol{x}}_j = A_{ij}\hat{\boldsymbol{x}}_i\hat{\boldsymbol{x}}_j = \boldsymbol{A}. \tag{4.125} $$

The second isotropic form is another type of identity tensor that produces the transpose

$$ {}_4\boldsymbol{I}^{(2)} : \boldsymbol{A} = \delta_{ik}\delta_{jl}A_{lk}\hat{\boldsymbol{x}}_i\hat{\boldsymbol{x}}_j = A_{ji}\hat{\boldsymbol{x}}_i\hat{\boldsymbol{x}}_j = \boldsymbol{A}^T \tag{4.126} $$

and the third isotropic form produces a purely isotropic second-order tensor ${}_4\boldsymbol{I}^{(3)} : \boldsymbol{A} = \boldsymbol{I}(\boldsymbol{I} : \boldsymbol{A}) = \mathrm{tr}\{\boldsymbol{A}\}\boldsymbol{I}$.

So the most general form for the fourth-order isotropic stiffness tensor is a linear combination of the above three forms or

$$ {}_4\boldsymbol{C} = \lambda^{(1)}\,{}_4\boldsymbol{I}^{(1)} + \lambda^{(2)}\,{}_4\boldsymbol{I}^{(2)} + \lambda^{(3)}\,{}_4\boldsymbol{I}^{(3)}. \tag{4.127} $$

The double-dot product of the first term and the second term onto the strain tensor $\boldsymbol{e}$ produces $\lambda^{(1)}\boldsymbol{e}$ and $\lambda^{(2)}\boldsymbol{e}^T = \lambda^{(2)}\boldsymbol{e}$ because the strain tensor is symmetric. We thus can take $\lambda^{(2)} = \lambda^{(1)}$ since these terms are contributing the same tensorial form to the stress tensor. The double-dot product of the third term onto the strain tensor produces $\lambda^{(3)}\mathrm{tr}\{\boldsymbol{e}\}\boldsymbol{I} = \lambda^{(3)}\nabla\cdot\boldsymbol{u}\,\boldsymbol{I}$ which is a purely isotropic stress. We thus can rewrite Eq. (4.127) as

$$ {}_4\boldsymbol{C} = \lambda^{(1)}\left({}_4\boldsymbol{I}^{(1)} + {}_4\boldsymbol{I}^{(2)} - \frac{2}{3}\,{}_4\boldsymbol{I}^{(3)}\right) + \left(\lambda^{(3)} + \frac{2\lambda^{(1)}}{3}\right){}_4\boldsymbol{I}^{(3)} \tag{4.128} $$

so that the first term, when double dotted into the strain tensor, leads to purely deviatoric stress (zero trace) and the second term to purely isotropic stress. We call $\lambda^{(1)} \mathrel{\widehat{=}} G$ the *shear modulus* of the isotropic material (units of stress or Pa) and the combination $\lambda^{(3)} + 2\lambda^{(1)}/3 \mathrel{\widehat{=}} K$ the *bulk modulus*. To conclude, the scalar components of the *fourth-order isotropic stiffness tensor* are most meaningfully written

$$ \boxed{C_{ijkl} = K\delta_{ij}\delta_{kl} + G\left(\delta_{ik}\delta_{jl} + \delta_{il}\delta_{jk} - \frac{2}{3}\delta_{ij}\delta_{kl}\right),} \tag{4.129} $$

which was more rapidly arrived at in the earlier Eq. (4.46). The first term involving K is responsible for the isotropic or compressional or volume-change response of the material and the second term involving G is responsible for the deviatoric or shear or shape-change response. But don't worry, in practice it is rarely necessary to remember this expression with all of its detailed indices.

This is because if you use the isotropic form of C_{ijkl} in Hooke's law $\tau_{ij} = C_{ijkl}e_{kl}$ you obtain (again, because $e_{ij} = e_{ji}$)

$$ \tau_{ij} = Ke_{kk}\delta_{ij} + 2G\left(e_{ij} - \frac{1}{3}e_{kk}\delta_{ij}\right), \tag{4.130} $$

which we can equivalently write in bold-face notation (again, $e_{kk} = \mathrm{tr}\{\boldsymbol{e}\} = \nabla\cdot\boldsymbol{u}$)

$$ \boldsymbol{\tau} = K\nabla\cdot\boldsymbol{u}\,\boldsymbol{I} + 2G\left(\boldsymbol{e} - \frac{\nabla\cdot\boldsymbol{u}}{3}\boldsymbol{I}\right). \tag{4.131} $$

If we introduce the definition of the strain tensor $\boldsymbol{e} = \left[\nabla\boldsymbol{u} + (\nabla\boldsymbol{u})^T\right]/2$, we then have the useful form of Hooke's law in a linear isotropic material that relates stress to gradients of displacement

$$ \boxed{\tau_{ij} = \underbrace{K\,\nabla\cdot\boldsymbol{u}\,\delta_{ij}}_{\text{Pure compression}} + \underbrace{G\left(\frac{\partial u_j}{\partial x_i} + \frac{\partial u_i}{\partial x_j} - \frac{2}{3}\nabla\cdot\boldsymbol{u}\,\delta_{ij}\right)}_{\text{Pure shear}}.} \tag{4.132} $$

This expression of the linear form of Hooke's law in an isotropic material is what you will more typically remember, not $\tau_{ij} = C_{ijkl}e_{kl}$ with C_{ijkl} given by Eq. (4.129).

If we work with the velocities $\boldsymbol{v}$ instead of the displacements $\boldsymbol{u}$, the exact nonlinear form of Hooke's law in an isotropic material is then

$$\frac{\partial \tau_{ij}}{\partial t} + \frac{\partial}{\partial x_k}\left(v_k \tau_{ij}\right) = K\nabla \cdot \boldsymbol{v}\, \delta_{ij} + G\left(\frac{\partial v_j}{\partial x_i} + \frac{\partial v_i}{\partial x_j} - \frac{2}{3}\nabla \cdot \boldsymbol{v}\, \delta_{ij}\right). \tag{4.133}$$

This form of Hooke's law is valid even if K and G are changing with strain as quantified through the earlier rules of Eqs (4.47) and (4.48). For linear (small strain) response, we ignore how K and G are changing and drop the nonlinear convective term on the left-hand side of Eq. (4.133) to produce the linear form of Hooke's law as expressed in terms of velocities. This is generally the best form to use when evaluating linear elastodynamic response numerically.

Recall our definition of the *total pressure*

$$-P = \frac{1}{3}\,\mathrm{tr}\{\boldsymbol{\tau}\} = \frac{\tau_{ii}}{3} = \frac{\tau_{xx} + \tau_{yy} + \tau_{zz}}{3}. \tag{4.134}$$

Taking the trace of the isotropic Hooke's law then gives

$$\boxed{-P = K\,\nabla \cdot \boldsymbol{u}.} \tag{4.135}$$

So the isotropic part of Hooke's law is a purely scalar relation that serves to define the bulk modulus K. This then leaves the deviatoric (or tensorial) part of Hooke's law as

$$\boxed{\tau_{ij}^D = G\left[\frac{\partial u_j}{\partial x_i} + \frac{\partial u_i}{\partial x_j} - \frac{2}{3}\nabla \cdot \boldsymbol{u}\, \delta_{ij}\right],} \tag{4.136}$$

which similarly serves to define the shear modulus G. In words, the second-order deviatoric (or shear) strain tensor is related to the second-order deviatoric (or shear) stress tensor by the scalar G in an isotropic material.

The above decomposition $\tau_{ij} = -P\delta_{ij} + \tau_{ij}^D$ shows that in an isotropic material, the generalized responses and forces have the same tensorial order. So the scalar P is produced only by the scalar $\nabla \cdot \boldsymbol{u}$ in an isotropic material. Similarly, the second-order deviatoric stress tensor $\boldsymbol{\tau}^D$ is produced only by the second-order deviatoric strain tensor $\boldsymbol{e}^D$ in an isotropic material. This is an example of the general result known as *Curie's principle* that we proved in Section 1.8.5 and that states: *in the constitutive laws of an isotropic material, "responses" (the left-hand side) have the same tensorial order as the "forces" (the right-hand side) that generate them and the material properties are simple scalars.*

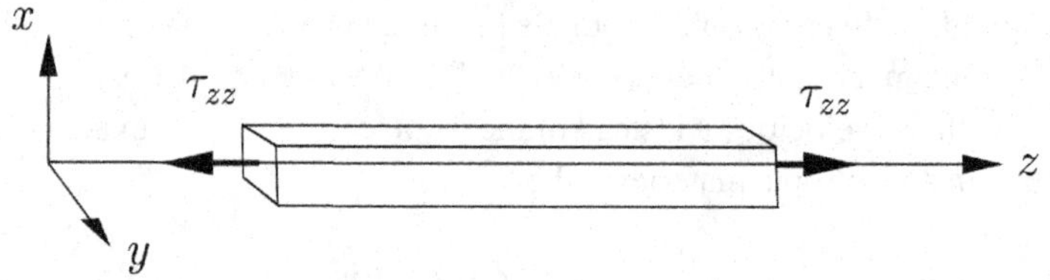

Figure 4.8 A long bar has a tensile stress τ_{zz} applied to its ends and no stress applied to its lateral boundaries. This experiment is used to define the elastic constants E and ν.

4.4.4 Different Forms of Hooke's Law in an Isotropic Medium

It takes two elastic constants to describe the elastic stress produced by an arbitrary deformation in an isotropic elastic material. Above, we used K and G but the two elastic constants of an isotropic material can be defined in other ways. For example, we can write the linear Hooke's law as

$$\tau_{ij} = \lambda \nabla \cdot \boldsymbol{u}\, \delta_{ij} + G \left[\frac{\partial u_j}{\partial x_i} + \frac{\partial u_i}{\partial x_j} \right], \tag{4.137}$$

where the elastic modulus λ is therefore defined in terms of K and G as

$$\lambda = K - 2G/3 \mathrel{\widehat{=}} \textit{the Lamé modulus}. \tag{4.138}$$

We also note parenthetically that many authors denote the shear modulus G with the symbol μ. We do not make this choice so that we may use μ to represent other material properties such as the magnetic permeability or coefficient of friction (i.e., in this book, G is only used to represent the elastic shear modulus).

We can also define the two elastic moduli of isotropic elasticity from an experiment performed on a long bar of material as depicted in Fig. 4.8. We pull on the ends of the rod (or bar) with a positive tensile stress $\tau_{zz} > 0$, while keeping $\tau_{xx} = \tau_{yy} = \tau_{xy} = \tau_{xz} = \tau_{yz} = 0$. We then measure $e_{xx} = \partial u_x/\partial x = \partial u_y/\partial y = e_{yy}$ and $e_{zz} = \partial u_z/\partial z$. The bar will elongate in the axial z direction so that e_{zz} is positive, and it will contract in the radial (or lateral) directions x and y so that $e_{xx} = e_{yy}$ is negative. From this "bar experiment," we define two constants:

$$E \mathrel{\widehat{=}} \text{Young's modulus} = \frac{\tau_{zz}}{e_{zz}} \quad \left(\text{when } \tau_{xx} = \tau_{yy} = \tau_{xy} = \tau_{xz} = \tau_{yz} = 0\right) \tag{4.139}$$

$$\nu \mathrel{\widehat{=}} \text{Poisson's ratio} = -\frac{e_{xx}}{e_{zz}} = -\frac{e_{yy}}{e_{zz}} \quad \left(\text{when } \tau_{xx} = \tau_{yy} = \tau_{xy} = \tau_{xz} = \tau_{yz} = 0\right). \tag{4.140}$$

A simple application of the isotropic Hooke's law in terms of K and G under the above stress conditions of the bar experiment allows you to prove as an exercise that:

$$\boxed{E = \frac{9GK}{3K + G} \quad \text{and} \quad \nu = \frac{1}{2}\left(\frac{3K - 2G}{3K + G}\right).} \tag{4.141}$$

Many authors in engineering only use E and ν as their two isotropic elastic moduli.

As implied earlier, it is often convenient to write out even the isotropic form of Hooke's law in the 6×6 matrix or Voigt form to give the relation between strain and stress as

$$\begin{bmatrix} \tau_{xx} \\ \tau_{yy} \\ \tau_{zz} \\ \tau_{yz} \\ \tau_{xz} \\ \tau_{xy} \end{bmatrix} = \begin{bmatrix} K+4G/3 & K-2G/3 & K-2G/3 & 0 & 0 & 0 \\ K-2G/3 & K+4G/3 & K-2G/3 & 0 & 0 & 0 \\ K-2G/3 & K-2G/3 & K+4G/3 & 0 & 0 & 0 \\ 0 & 0 & 0 & G & 0 & 0 \\ 0 & 0 & 0 & 0 & G & 0 \\ 0 & 0 & 0 & 0 & 0 & G \end{bmatrix} \begin{bmatrix} e_{xx} \\ e_{yy} \\ e_{zz} \\ 2e_{yz} \\ 2e_{xz} \\ 2e_{xy} \end{bmatrix}. \tag{4.142}$$

Using the above, it is then easy to obtain the relation between stress and strain

$$\begin{bmatrix} e_{xx} \\ e_{yy} \\ e_{zz} \\ 2e_{yz} \\ 2e_{xz} \\ 2e_{xy} \end{bmatrix} = \frac{1}{E} \begin{bmatrix} 1 & -\nu & -\nu & 0 & 0 & 0 \\ -\nu & 1 & -\nu & 0 & 0 & 0 \\ -\nu & -\nu & 1 & 0 & 0 & 0 \\ 0 & 0 & 0 & 2(1+\nu) & 0 & 0 \\ 0 & 0 & 0 & 0 & 2(1+\nu) & 0 \\ 0 & 0 & 0 & 0 & 0 & 2(1+\nu) \end{bmatrix} \begin{bmatrix} \tau_{xx} \\ \tau_{yy} \\ \tau_{zz} \\ \tau_{yz} \\ \tau_{xz} \\ \tau_{xy} \end{bmatrix}; \tag{4.143}$$

that is, $2(1+\nu)/E = 1/G$. You should verify that you can obtain both of these matrix forms starting from the initial form of the isotropic Hooke's law given by Eq. (4.130). The Hooke's-law form of Eq. (4.142) that determines the stress tensor given the strain tensor can be called the *stiffness form*. The form of Eq. (4.143) that determines the strain tensor given the stress tensor is called the *compliance form*. Later, in Chapter 11, for subtle reasons due to the possibility of a delay between the stress and strain response, we call the compliance form the "normal" form of Hooke's law and the stiffness form the "inverse" form.

4.5 Isotropic Elastodynamics and Linear Wave Propagation

For an isotropic elastic material, our laws of motion and deformation are given by:

$$\frac{\partial \rho}{\partial t} = -\nabla \cdot (\boldsymbol{v}\rho) \tag{4.144}$$

$$\frac{\partial (\boldsymbol{v}\rho)}{\partial t} = -\nabla \cdot (\boldsymbol{v}\boldsymbol{v}\rho) + \nabla \cdot \boldsymbol{\tau} + \boldsymbol{F}_b \tag{4.145}$$

$$\frac{\partial \boldsymbol{\tau}}{\partial t} = -\nabla \cdot (\boldsymbol{v}\boldsymbol{\tau}) + K\,\nabla \cdot \boldsymbol{v}\boldsymbol{I} + G\left[\nabla \boldsymbol{v} + \nabla \boldsymbol{v}^T - \frac{2}{3}\nabla \cdot \boldsymbol{v}\boldsymbol{I}\right]. \tag{4.146}$$

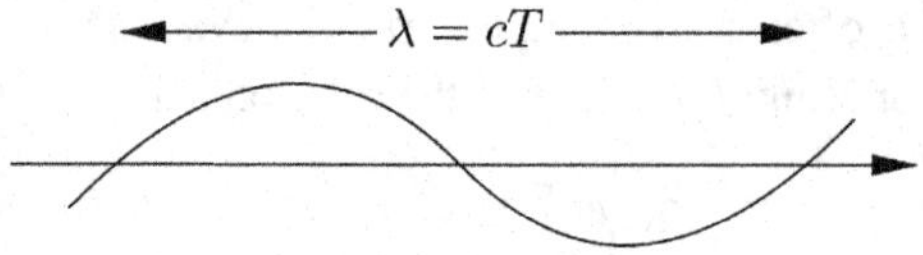

Figure 4.9 The relation between wavelength λ, wave period $T = 1/f$, and wavespeed c.

These exact laws for isotropic elastodynamic response are nonlinear due to the accumulation terms (first terms on the right-hand side) as well as due to any nonnegligible variations through time of K and/or G as modeled using Eqs (4.47) and (4.48). Again, only if the solid material contains a significant population of cracks will K or G change appreciably with changing strain.

To simplify these laws for application to wave propagation, let's estimate the amplitude of the accumulation terms and the density changes for a propagating wave. Consider a wave having wavelength λ and period T as depicted in Fig. 4.9. An order-of-magnitude estimate of the various terms in the elastodynamic governing equations is made using the identifications

$$\nabla \approx 1/\lambda, \qquad \lambda \mathrel{\widehat{=}} \text{wavelength}, \qquad f \mathrel{\widehat{=}} \text{wave frequency},$$
$$\partial/\partial t \approx 1/T = f, \qquad T \mathrel{\widehat{=}} \text{wave period}, \qquad c \mathrel{\widehat{=}} \text{wave speed} = \lambda/T = \lambda f.$$

Thus, for an elastic wave, an estimate of the magnitude of the accumulation terms relative to the partial time-derivative terms in each of Eqs (4.144)–(4.146) is

$$\frac{|\nabla \cdot (\boldsymbol{v}\rho)|}{|\partial\rho/\partial t|} \approx \frac{|\nabla \cdot (\boldsymbol{v}\boldsymbol{v}\rho)|}{|\partial(\boldsymbol{v}\rho)/\partial t|} \approx \frac{|\nabla \cdot (\boldsymbol{v}\boldsymbol{\tau})|}{|\partial\boldsymbol{\tau}/\partial t|} \approx \frac{|\boldsymbol{v}|}{c}, \tag{4.147}$$

where $|\boldsymbol{v}|$ is a characteristic amplitude of the particle velocities.

Now for a wave advancing in the x direction, we have $\boldsymbol{u} = \boldsymbol{u}(x - ct)$ which is perhaps the single most important characteristic of a wave response as will be elaborated upon below. Thus, in the usual manner, we write $\boldsymbol{u} = \boldsymbol{u}(a)$ where the argument is $a = x - ct$ so that $\partial\boldsymbol{u}/\partial x = (\mathrm{d}\boldsymbol{u}(a)/\mathrm{d}a)\partial a/\partial x = \mathrm{d}\boldsymbol{u}(a)/\mathrm{d}a$ and $\partial\boldsymbol{u}/\partial t = (\mathrm{d}\boldsymbol{u}(a)/\mathrm{d}a)\partial a/\partial t = -(\mathrm{d}\boldsymbol{u}(a)/\mathrm{d}a)c = -(\partial\boldsymbol{u}/\partial x)\, c$, which then gives

$$\frac{\partial \boldsymbol{u}}{\partial t} \approx \boldsymbol{v} = -\frac{\partial \boldsymbol{u}}{\partial x} c \quad \text{or} \quad \frac{|\boldsymbol{v}|}{c} \approx \left|\frac{\partial \boldsymbol{u}}{\partial x}\right| \approx \text{strain}. \tag{4.148}$$

So when strain levels satisfy $|\partial\boldsymbol{u}/\partial x| \ll 1$, the accumulation terms are negligible and the wave propagation can be considered linear. In lightly consolidated sedimentary rocks, for example, strain levels generally need to be less than 10^{-6} for rubbing between grain surfaces (a nonlinear friction effect we will not discuss) to be negligible and for wave propagation to be considered linear. So it does not always take a lot of strain to be present before nonlinear response becomes important. Assuming the small-strain condition to hold and the wave-propagation to be linear, the conservation of mass is approximated as $\rho(\boldsymbol{r}, t) \approx \rho_0(\boldsymbol{r})$, where ρ_0 is the mass density that holds prior to wave arrival.

We will also take K and G to be constant with time so that Hooke's law may be time integrated. Our governing equations for such linear wave propagation in an isotropic but possibly heterogeneous medium become

$$\rho \frac{\partial^2 \boldsymbol{u}}{\partial t^2} = \nabla \cdot \boldsymbol{\tau} + \boldsymbol{F}_b \tag{4.149}$$

$$\boldsymbol{\tau} = K \nabla \cdot \boldsymbol{u} \boldsymbol{I} + G\left[\nabla \boldsymbol{u} + \nabla \boldsymbol{u}^T - \frac{2}{3}\nabla \cdot \boldsymbol{u} \boldsymbol{I}\right], \tag{4.150}$$

where all of ρ, K, and G are now treated as time-independent material properties. If we further assume that ρ, K, and G are uniform through space, a simple tensor-calculus exercise (see the end of the chapter) gives

$$\nabla \cdot \boldsymbol{\tau} = \left(K + \frac{G}{3}\right) \nabla \nabla \cdot \boldsymbol{u} + G \nabla^2 \boldsymbol{u}. \tag{4.151}$$

Under these conditions, we obtain the *elastodynamic wave equation*

$$\rho \frac{\partial^2 \boldsymbol{u}}{\partial t^2} = \left(K + \frac{G}{3}\right) \nabla \nabla \cdot \boldsymbol{u} + G \nabla^2 \boldsymbol{u} + \boldsymbol{F}_b, \tag{4.152}$$

which holds for small-strain (linear) wave propagation in an isotropic and homogeneous body. We next examine the plane-wave response that is controlled by this equation.

4.6 Longitudinal and Transverse Plane Waves

A plane wave has the functional dependence

$$\boldsymbol{u}(\boldsymbol{r}, t) = \boldsymbol{u}\left(t - \frac{\hat{\boldsymbol{k}} \cdot \boldsymbol{r}}{c}\right), \tag{4.153}$$

where c is the wave speed and $\hat{\boldsymbol{k}}$ is a unit normal to the planar front of the wave as shown in Fig. 4.10. For any point $\boldsymbol{r}$ on the wave front, the distance $\hat{\boldsymbol{k}} \cdot \boldsymbol{r}$ represents distance in

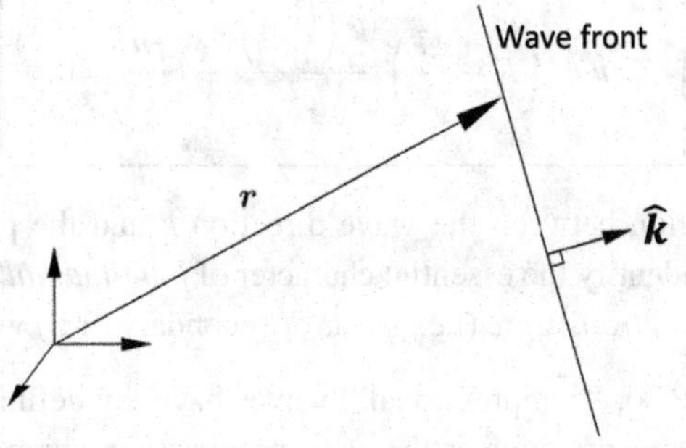

Figure 4.10 The distance propagated by a plane wave is $\hat{\boldsymbol{k}} \cdot \boldsymbol{r}$, where $\hat{\boldsymbol{k}}$ is the unit vector perpendicular to the front.

the direction of propagation so that $\hat{\boldsymbol{k}} \cdot \boldsymbol{r}/c$ is the time it takes the front to advance the distance $\hat{\boldsymbol{k}} \cdot \boldsymbol{r}$.

We further define the polarization of the wave by the direction $\hat{\boldsymbol{u}}$ (unit vector) of the displacements. We may write the general response of an elastodynamic plane wave as

$$\boldsymbol{u}(\boldsymbol{r}, t) = U\left(t - \frac{\hat{\boldsymbol{k}} \cdot \boldsymbol{r}}{c}\right) \hat{\boldsymbol{u}}. \tag{4.154}$$

The scalar function $U(t)$ is the displacement amplitude time function along the plane of points $\hat{\boldsymbol{k}} \cdot \boldsymbol{r} = 0$. That the functional dependence on t and $\boldsymbol{r}$ is as given in Eq. (4.154) for the displacements satisfying the elastodynamic wave equation will be proven below and is the single most important fact about seismic-wave propagation if forced to choose just one.

If we now insert this plane-wave response into the elastodynamic wave equation, we obtain (putting the body force to zero):

$$\rho \frac{\partial^2 U}{\partial t^2} \hat{\boldsymbol{u}} = \left(K + \frac{G}{3}\right) \frac{\hat{\boldsymbol{k}}\left(\hat{\boldsymbol{k}} \cdot \hat{\boldsymbol{u}}\right)}{c^2} \frac{\partial^2 U}{\partial t^2} + G \frac{\left(\hat{\boldsymbol{k}} \cdot \hat{\boldsymbol{k}}\right) \hat{\boldsymbol{u}}}{c^2} \frac{\partial^2 U}{\partial t^2}. \tag{4.155}$$

We have used the facts that

$$\nabla U\left(t - \frac{\hat{\boldsymbol{k}} \cdot \boldsymbol{r}}{c}\right) = -\frac{\partial U}{\partial t} \frac{\hat{\boldsymbol{k}}}{c} \tag{4.156}$$

$$\nabla \cdot \nabla U\left(t - \frac{\hat{\boldsymbol{k}} \cdot \boldsymbol{r}}{c}\right) = +\frac{\partial^2 U}{\partial t^2} \frac{\hat{\boldsymbol{k}} \cdot \hat{\boldsymbol{k}}}{c^2}, \tag{4.157}$$

where for plane-wave response, one has $\nabla = -\hat{\boldsymbol{k}} c^{-1} \partial/\partial t$. Note that these results are obtained as earlier by first writing the displacement amplitude as $U(a)$, where the argument is $a = t - \hat{\boldsymbol{k}} \cdot \boldsymbol{r}/c$ so that $\nabla U = (dU/da)\ \nabla a$, noting that $\mathrm{d}U/da = (\partial U/\partial t)/(\partial a/\partial t) = \partial U/\partial t$ and that $\nabla a = -\nabla\left(\hat{\boldsymbol{k}} \cdot \boldsymbol{r}/c\right) = -\hat{\boldsymbol{k}} \cdot \nabla \boldsymbol{r}/c$ and then recalling that $\nabla \boldsymbol{r} = \boldsymbol{I}$ (the identity tensor) so that for plane-wave response $\nabla U = -\hat{\boldsymbol{k}} c^{-1} \partial U/\partial t$. We next eliminate $\partial^2 U/\partial t^2$ and note that $\hat{\boldsymbol{k}} \cdot \hat{\boldsymbol{k}} = 1$ to obtain a vector identity for elastodynamic plane waves:

$$\boxed{\rho\, \hat{\boldsymbol{u}} - \left(K + \frac{G}{3}\right) \frac{\hat{\boldsymbol{k}}\left(\hat{\boldsymbol{k}} \cdot \hat{\boldsymbol{u}}\right)}{c^2} - G \frac{\hat{\boldsymbol{u}}}{c^2} = 0.} \tag{4.158}$$

By analyzing the relation between the wave direction $\hat{\boldsymbol{k}}$ and the polarization of the displacements $\hat{\boldsymbol{u}}$, we can identify the essential character of *longitudinal* (i.e., compressional or primary or P) waves and *transverse* (i.e., shear or secondary or S) waves.

- **Longitudinal** (or "P" or "compressional") waves have, by definition, the displacement polarization in the same direction as the wave propagation which means

$$\hat{\boldsymbol{u}} = \hat{\boldsymbol{k}} \quad \text{so that} \quad \hat{\boldsymbol{k}} \cdot \hat{\boldsymbol{u}} = 1.$$

Thus, from Eq. (4.158) we obtain

$$\left[\rho - \left(K + \frac{4G}{3}\right)/c_p^2\right]\hat{\boldsymbol{k}} = 0 \tag{4.159}$$

or

$$c_p = \sqrt{\frac{K + 4G/3}{\rho}} \quad \text{wave speed of longitudinal waves.} \tag{4.160}$$

The longitudinal (or P) plane-wave displacement and stress response for propagation in the $\hat{\boldsymbol{k}}$ direction is then

$$\boldsymbol{u}_p(\boldsymbol{r}, t) = \hat{\boldsymbol{k}}\, U_p\left(t - \frac{\hat{\boldsymbol{k}} \cdot (\boldsymbol{r} - \boldsymbol{r}_o)}{c_p}\right) \tag{4.161}$$

$$\boldsymbol{\tau}_p(\boldsymbol{r}, t) = -\frac{1}{c_p}\left[\left(K - \frac{2G}{3}\right)\boldsymbol{I} + 2G\,\hat{\boldsymbol{k}}\hat{\boldsymbol{k}}\right]\frac{\partial}{\partial t}U_p\left(t - \frac{\hat{\boldsymbol{k}} \cdot (\boldsymbol{r} - \boldsymbol{r}_o)}{c_p}\right), \tag{4.162}$$

where the scalar function $U_p(t)$ is the displacement in the $\hat{\boldsymbol{k}}$ direction recorded at an observation point $\boldsymbol{r}_o$. Note the presence of the shear modulus in the expression for the longitudinal wavespeed c_p. This is because a longitudinal wave is associated with a uniaxial deformation of the material, which is a deformation in which a rectangle with two of the edges parallel with the wave front will have those two edges approach and move away from each other while the other edges perpendicular to the wavefront will not move at all. Such uniaxial deformation corresponds to a shape change of the rectangle and is governed by the uniaxial elastic modulus H defined as

$$H = K + \frac{4G}{3}. \tag{4.163}$$

So compressional waves generate deviatoric shape change of elements in addition to volumetric changes which explains why both the shear modulus and bulk modulus are involved in the "compressional" (or P or longitudinal or uniaxial) wavespeed c_p.

The Poynting vector $\boldsymbol{s}_p = -\boldsymbol{v}_p \cdot \boldsymbol{\tau}_p$ represents the direction that energy is fluxing in the compressional wave. With $\boldsymbol{v}_p = \hat{\boldsymbol{k}}\partial U_p/\partial t$ and using the expression above for $\boldsymbol{\tau}_p$, we obtain

$$\boldsymbol{s}_p = -\boldsymbol{v}_p \cdot \boldsymbol{\tau}_p = \frac{H}{c_p}\left(\frac{\partial U_p}{\partial t}\right)^2 \hat{\boldsymbol{k}}, \tag{4.164}$$

which says that the energy flux for a plane compressional wave in an isotropic material is entirely in the direction $\hat{\boldsymbol{k}}$ of wave propagation.

- **Transverse** (or "S" or "shear") waves have, by definition, the displacement polarization in a direction perpendicular to the wave direction

$$\hat{\boldsymbol{u}} = \hat{\boldsymbol{t}}, \quad \text{where} \quad \hat{\boldsymbol{t}} \perp \hat{\boldsymbol{k}} \quad \text{so that} \quad \hat{\boldsymbol{t}} \cdot \hat{\boldsymbol{k}} = 0.$$

Thus, we obtain from Eq. (4.158) that

$$\left[\rho - \frac{G}{c_s^2}\right] \hat{\boldsymbol{t}} = 0 \tag{4.165}$$

or

$$c_s = \sqrt{\frac{G}{\rho}} \quad \text{wave speed of transverse waves.} \tag{4.166}$$

The transverse (or S) plane-wave displacement and stress response is then

$$\boldsymbol{u}_s(\boldsymbol{r}, t) = \hat{\boldsymbol{t}}\, U_s\left(t - \frac{\hat{\boldsymbol{k}} \cdot (\boldsymbol{r} - \boldsymbol{r}_o)}{c_s}\right) \tag{4.167}$$

$$\boldsymbol{\tau}_s(\boldsymbol{r}, t) = -\frac{G}{c_s}\left(\hat{\boldsymbol{k}}\hat{\boldsymbol{t}} + \hat{\boldsymbol{t}}\hat{\boldsymbol{k}}\right) \frac{\partial}{\partial t} U_s\left(t - \frac{\hat{\boldsymbol{k}} \cdot (\boldsymbol{r} - \boldsymbol{r}_o)}{c_s}\right), \tag{4.168}$$

where $U_s(t)$ is the shear-wave displacement in the $\hat{\boldsymbol{t}} \perp \hat{\boldsymbol{k}}$ direction recorded at $\boldsymbol{r}_o$. Note that for a shear wave

$$\nabla \boldsymbol{u}_s = -\frac{1}{c_s} \hat{\boldsymbol{k}}\hat{\boldsymbol{t}} \frac{\partial}{\partial t} U_s\left(t - \frac{\hat{\boldsymbol{k}} \cdot (\boldsymbol{r} - \boldsymbol{r}_o)}{c_s}\right) \tag{4.169}$$

$$\nabla \cdot \boldsymbol{u}_s = \operatorname{tr}\{\nabla \boldsymbol{u}_s\} = -\frac{1}{c_s}(\hat{\boldsymbol{k}} \cdot \hat{\boldsymbol{t}}) \frac{\partial}{\partial t} U_s\left(t - \frac{\hat{\boldsymbol{k}} \cdot (\boldsymbol{r} - \boldsymbol{r}_o)}{c_s}\right) = 0 \tag{4.170}$$

because $\hat{\boldsymbol{k}} \cdot \hat{\boldsymbol{t}} = 0$, which means the volume of the material does not change during shear-wave propagation in an isotropic material. Similarly, the pressure in a shear wave is given by $P_s = -\operatorname{tr}\{\boldsymbol{\tau}_s\}/3 = 0$. As such, the bulk modulus K is not involved in the expression for the shear (or S or transverse) wavespeed c_s.

With $\boldsymbol{v}_s = \hat{\boldsymbol{t}}\partial U_s/\partial t$, the Poynting vector for a plane shear wave in an isotropic material is then

$$\boldsymbol{s}_s = -\boldsymbol{v}_s \cdot \boldsymbol{\tau}_s = \frac{G}{c_s}\left(\frac{\partial U_s}{\partial t}\right)^2 \hat{\boldsymbol{k}}, \tag{4.171}$$

which again says that the energy flux for a plane shear wave in an isotropic material is entirely in the direction $\hat{\boldsymbol{k}}$ of wave propagation.

4.7 Reflection and Transmission of Plane Waves

We next consider what happens when a plane wave, either longitudinal (P) or transverse (S), is incident at a plane boundary that separates two materials that have distinct values of any of ρ, K, or G. For the transverse waves, we decompose the displacement polarization into one part whose component is normal to the interface and another part whose component is entirely parallel with the plane of the interface pointed in and out of the page. The transverse polarization with a normal component to the interface is called SV (shear, vertical) and the transverse polarization that is entirely parallel with the interface is called SH (shear, horizontal). Where a plane P wave is incident on the interface with an incident angle $\theta_i > 0$ relative to the normal from the interface, there are generated two reflected plane waves (P and SV) and two transmitted plane waves (P and SV). Similarly, where a plane SV wave is incident on the interface, there are generated two reflected waves (SV and P) and two transmitted waves (SV and P). However, where a plane SH wave is incident on the interface, there is only generated one reflected and one transmitted SH wave with no coupling to the P-SV polarization. The goal in what follows is to find both the direction $\hat{\boldsymbol{k}}$ of each reflected and transmitted plane wave and the amplitude of each wave relative to the amplitude of the incident wave. We treat one case of P-SV polarization and another case of SH polarization.

4.7.1 P-SV Polarization with an Incident P Wave

As shown in Fig. 4.11, we consider an incident P-wave with an angle θ_i from the vertical that generates a reflected and transmitted P wave and a reflected and transmitted SV wave. An analogous analysis to what follows holds when an SV-wave is incident on the interface. With a z coordinate pointing down and normal to the horizontal interface, the SV polarization is in the plane (x, z) of the page with no polarization in the y direction coming out of the page. The direction of the five waves are characterized in terms of their angles from the vertical direction by the relations

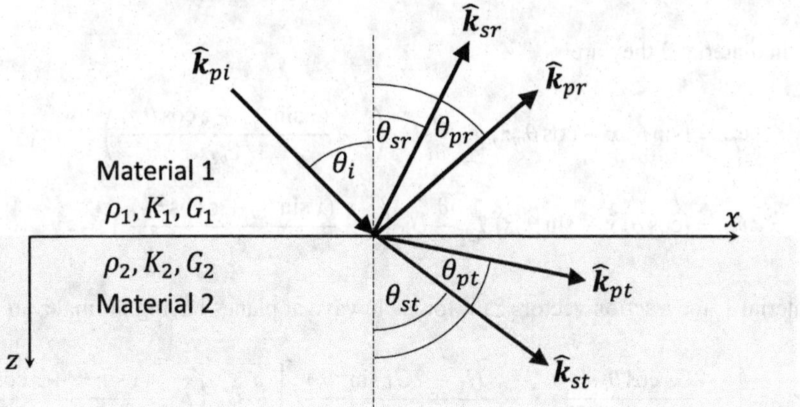

Figure 4.11 An incident P wave on $z = 0$ that generates, simultaneously and with the same time dependence, two reflected waves (P and SV) and two transmitted waves (P and SV).

$$\hat{\boldsymbol{k}}_{pi} = \sin\theta_i\hat{\boldsymbol{x}} + \cos\theta_i\hat{\boldsymbol{z}} \tag{4.172}$$

$$\hat{\boldsymbol{k}}_{pr} = \sin\theta_{pr}\hat{\boldsymbol{x}} - \cos\theta_{pr}\hat{\boldsymbol{z}} \tag{4.173}$$

$$\hat{\boldsymbol{k}}_{sr} = \sin\theta_{sr}\hat{\boldsymbol{x}} - \cos\theta_{sr}\hat{\boldsymbol{z}} \tag{4.174}$$

$$\hat{\boldsymbol{k}}_{pt} = \sin\theta_{pt}\hat{\boldsymbol{x}} + \cos\theta_{pt}\hat{\boldsymbol{z}} \tag{4.175}$$

$$\hat{\boldsymbol{k}}_{st} = \sin\theta_{st}\hat{\boldsymbol{x}} + \cos\theta_{st}\hat{\boldsymbol{z}} \tag{4.176}$$

and we must find the four angles θ_{pr}, θ_{ps}, θ_{pt}, and θ_{st} relative to the vertical in terms of the angle of incidence θ_i of the incident P wave. The two transverse waves created by reflection and transmission have polarizations $\hat{\boldsymbol{t}}$ perpendicular to the direction of propagation, that is, $\hat{\boldsymbol{t}} = \hat{\boldsymbol{y}} \times \hat{\boldsymbol{k}}$ with $\hat{\boldsymbol{y}}$ coming out of the page or

$$\hat{\boldsymbol{t}}_{sr} = -\cos\theta_{sr}\hat{\boldsymbol{x}} - \sin\theta_{sr}\hat{\boldsymbol{z}} \tag{4.177}$$

$$\hat{\boldsymbol{t}}_{st} = \cos\theta_{st}\hat{\boldsymbol{x}} - \sin\theta_{st}\hat{\boldsymbol{z}}. \tag{4.178}$$

By definition, the two longitudinal waves created by reflection and transmission have polarizations parallel with their direction of propagation.

We now use these facts in the plane-wave response for longitudinal and transverse waves as derived in the previous section. In material 1, the material velocity fields for each wave are

$$\boldsymbol{v}_{pi} = \left(\sin\theta_i\hat{\boldsymbol{x}} + \cos\theta_i\hat{\boldsymbol{z}}\right)\frac{\partial}{\partial t}U_p\left(t - \frac{(x\sin\theta_i + z\cos\theta_i)}{c_{p1}}\right) \tag{4.179}$$

$$\boldsymbol{v}_{pr} = \left(\sin\theta_{pr}\hat{\boldsymbol{x}} - \cos\theta_{pr}\hat{\boldsymbol{z}}\right)R_{pp}\frac{\partial}{\partial t}U_p\left(t - \frac{(x\sin\theta_{pr} - z\cos\theta_{pr})}{c_{p1}}\right) \tag{4.180}$$

$$\boldsymbol{v}_{sr} = -\left(\cos\theta_{sr}\hat{\boldsymbol{x}} + \sin\theta_{sr}\hat{\boldsymbol{z}}\right)R_{ps}\frac{\partial}{\partial t}U_p\left(t - \frac{(x\sin\theta_{sr} - z\cos\theta_{sr})}{c_{s1}}\right) \tag{4.181}$$

while in material 2 they are

$$\boldsymbol{v}_{pt} = \left(\sin\theta_{pt}\hat{\boldsymbol{x}} + \cos\theta_{pt}\hat{\boldsymbol{z}}\right)T_{pp}\frac{\partial}{\partial t}U_p\left(t - \frac{(x\sin\theta_{pt} + z\cos\theta_{pt})}{c_{p2}}\right) \tag{4.182}$$

$$\boldsymbol{v}_{st} = \left(\cos\theta_{st}\hat{\boldsymbol{x}} - \sin\theta_{st}\hat{\boldsymbol{z}}\right)T_{ps}\frac{\partial}{\partial t}U_p\left(t - \frac{(x\sin\theta_{st} + z\cos\theta_{st})}{c_{s2}}\right). \tag{4.183}$$

In material 1, the traction vectors $\hat{\boldsymbol{z}}\cdot\boldsymbol{\tau}$ for each wave at planes having normal $\hat{\boldsymbol{z}}$ are

$$\hat{\boldsymbol{z}}\cdot\boldsymbol{\tau}_{pi} = \left[-\frac{2G_1\cos\theta_i\sin\theta_i}{c_{p1}}\hat{\boldsymbol{x}} - \frac{(H_1 - 2G_1\sin^2\theta_i)}{c_{p1}}\hat{\boldsymbol{z}}\right]\frac{\partial}{\partial t}U_p\left(t - \frac{(x\sin\theta_i + z\cos\theta_i)}{c_{p1}}\right) \tag{4.184}$$

$$\hat{z}\cdot\boldsymbol{\tau}_{pr} = \left[\frac{2G_1\cos\theta_{pr}\sin\theta_{pr}}{c_{p1}}\hat{\boldsymbol{x}} - \frac{(H_1 - 2G_1\sin^2\theta_{pr})}{c_{p1}}\hat{z}\right] \times R_{pp}\frac{\partial}{\partial t}U_p\left(t - \frac{(x\sin\theta_{pr} - z\cos\theta_{pr})}{c_{p1}}\right) \quad (4.185)$$

$$\hat{z}\cdot\boldsymbol{\tau}_{sr} = -\frac{G_1}{c_{s1}}\left[(\cos^2\theta_{sr} - \sin^2\theta_{sr})\hat{\boldsymbol{x}} + 2\cos\theta_{sr}\sin\theta_{sr}\hat{z}\right] \times R_{ps}\frac{\partial}{\partial t}U_p\left(t - \frac{(x\sin\theta_{sr} - z\cos\theta_{sr})}{c_{s1}}\right), \quad (4.186)$$

where $H = K + 4G/3$ in both material 1 and 2 is the P-wave or "uniaxial" elastic modulus. The traction vectors in material 2 are

$$\hat{z}\cdot\boldsymbol{\tau}_{pt} = \left[-\frac{2G_2\cos\theta_{pt}\sin\theta_{pt}}{c_{p2}}\hat{\boldsymbol{x}} - \frac{(H_2 - 2G_2\sin^2\theta_{pt})}{c_{p2}}\hat{z}\right] \times T_{pp}\frac{\partial}{\partial t}U_p\left(t - \frac{(x\sin\theta_{pt} + z\cos\theta_{pt})}{c_{p2}}\right) \quad (4.187)$$

$$\hat{z}\cdot\boldsymbol{\tau}_{st} = -\frac{G_2}{c_{s2}}\left[(\cos^2\theta_{st} - \sin^2\theta_{st})\hat{\boldsymbol{x}} + 2\cos\theta_{st}\sin\theta_{st}\hat{z}\right] \times T_{ps}\frac{\partial}{\partial t}U_p\left(t - \frac{(x\sin\theta_{st} + z\cos\theta_{st})}{c_{s2}}\right). \quad (4.188)$$

The incident wave has a material-velocity time function measured at $x = 0$ and $z = 0$ to be $\partial U_p(t)/\partial t$, while the two reflected waves have amplitudes of R_{pp} and R_{ps} relative to the incident wave that are called *reflection coefficients*. Similarly, the two transmitted waves have amplitudes of T_{pp} and T_{ps} relative to the incident wave that are called *transmission coefficients*. We must find the reflection and transmission coefficients.

Where the incident wave is present at the interface $z = 0$, we require that all five waves possess the same time function $\partial U_p(t, x)/\partial t$ because each wave is generated without delay and with the same time dependence as the incident wave, which requires

$$p \mathrel{\widehat{=}} \frac{\sin\theta_i}{c_{p1}} = \frac{\sin\theta_{pr}}{c_{p1}} = \frac{\sin\theta_{sr}}{c_{s1}} = \frac{\sin\theta_{pt}}{c_{p2}} = \frac{\sin\theta_{st}}{c_{s2}}. \quad (4.189)$$

This result is called *Snell's law* and allows the angle that each wave makes relative to the vertical to be determined in terms of the incidence angle θ_i. The wave speeds in material $n = 1, 2$ are given by $c_{pn} = \sqrt{H_n/\rho_n}$ and $c_{sn} = \sqrt{G_n/\rho_n}$. The parameter p that is the same for all five waves is called the *horizontal slowness* (the inverse of a wavespeed is called a "slowness"). So Snell's law says that each of the five plane waves are coincident and moving horizontally along the interface $z = 0$ at the same speed $1/p$.

The four reflection and transmission coefficients are found from the four continuity conditions, which are (1) the welded-interface conditions that require material 1 and material 2 not to penetrate or separate from each other at the interface requiring that the two components of the material velocity be continuous across the interface and (2) the two continuity of stress conditions derived in Section 2.9.6. These four conditions at $z=0$ are stated

$$\hat{z}\cdot\left(\boldsymbol{v}_{pi}+\boldsymbol{v}_{pr}+\boldsymbol{v}_{sr}\right)=\hat{z}\cdot\left(\boldsymbol{v}_{pt}+\boldsymbol{v}_{st}\right) \tag{4.190}$$

$$\hat{x}\cdot\left(\boldsymbol{v}_{pi}+\boldsymbol{v}_{pr}+\boldsymbol{v}_{sr}\right)=\hat{x}\cdot\left(\boldsymbol{v}_{pt}+\boldsymbol{v}_{st}\right) \tag{4.191}$$

$$\hat{z}\cdot\left(\boldsymbol{\tau}_{pi}+\boldsymbol{\tau}_{pr}+\boldsymbol{\tau}_{sr}\right)\cdot\hat{z}=\hat{z}\cdot\left(\boldsymbol{\tau}_{pt}+\boldsymbol{\tau}_{st}\right)\cdot\hat{z} \tag{4.192}$$

$$\hat{z}\cdot\left(\boldsymbol{\tau}_{pi}+\boldsymbol{\tau}_{pr}+\boldsymbol{\tau}_{sr}\right)\cdot\hat{x}=\hat{z}\cdot\left(\boldsymbol{\tau}_{pt}+\boldsymbol{\tau}_{st}\right)\cdot\hat{x}. \tag{4.193}$$

Thus, upon substituting the expressions above for each of these plane-wave responses at the interface $z=0$ and using Snell's law gives four equations for the four unknowns R_{pp}, R_{ps}, T_{pp}, and T_{ps}. The two velocity-continuity conditions are written

$$\begin{bmatrix}\cos\theta_i & \sin\theta_{sr}\\ -\sin\theta_i & \cos\theta_{sr}\end{bmatrix}\begin{bmatrix}R_{pp}\\ R_{ps}\end{bmatrix}+\begin{bmatrix}\cos\theta_{pt} & -\sin\theta_{st}\\ \sin\theta_{pt} & \cos\theta_{st}\end{bmatrix}\begin{bmatrix}T_{pp}\\ T_{ps}\end{bmatrix}=\begin{bmatrix}\cos\theta_i\\ \sin\theta_i\end{bmatrix}, \tag{4.194}$$

while the two traction-continuity conditions are

$$\begin{bmatrix}-1 & -\gamma_1\cos\theta_{sr}\\ \cos\theta_i\sin\theta_{sr} & \sin^2\theta_{sr}-1/2\end{bmatrix}\begin{bmatrix}R_{pp}\\ R_{ps}\end{bmatrix}$$
$$+\begin{bmatrix}\eta_2 c_{p1}/c_{p2} & \gamma_2\sin\theta_i\cos\theta_{st}\\ \sin\theta_{sr}\cos\theta_{pt}G_2/G_1 & (1/2-\sin^2\theta_{st})G_2c_{s1}/(G_1c_{s2})\end{bmatrix}\begin{bmatrix}T_{pp}\\ T_{ps}\end{bmatrix}=\begin{bmatrix}1\\ \sin\theta_{sr}\cos\theta_i\end{bmatrix}, \tag{4.195}$$

where

$$\gamma_{1,2}=\frac{2G_{1,2}\sin\theta_i}{H_1-2G_1\sin^2\theta_i}\quad\text{and}\quad\eta_2=\frac{H_2-2G_2\sin^2\theta_{pt}}{H_1-2G_1\sin^2\theta_i}. \tag{4.196}$$

We will not write out the straightforward but complicated algebraic results for these reflection and transmission coefficients.

If the incident wave had an SV polarization, there would again be two reflected waves, SV and P, and two transmitted waves, SV and P, but the incident wave would have different material velocity and traction components at the interface compared to the above incident P-wave case, which would result in altered expressions in the above continuity conditions. But otherwise, the analysis would proceed directly as above.

4.7.2 SH Polarization

In the case of an incident SH plane wave with polarization in the $\hat{y}$ direction out of the plane of the page, there is only one reflected SH wave and one transmitted SH wave with no coupling to P waves. The analysis is therefore greatly simplified and we will give the relatively compact analytical expressions for R, the SH-to-SH reflection coefficient and T, the SH-to-SH transmission coefficient.

In this case, the incident, reflected, and transmitted SH waves make angles of θ_i, θ_r, and θ_t relative to the vertical and have material velocities given by

$$\boldsymbol{v}_i = \hat{\boldsymbol{y}} \frac{\partial U_s}{\partial t}\left(t - \frac{(x\sin\theta_i + z\cos\theta_i)}{c_{s1}}\right) \tag{4.197}$$

$$\boldsymbol{v}_r = \hat{\boldsymbol{y}} R \frac{\partial U_s}{\partial t}\left(t - \frac{(x\sin\theta_r - z\cos\theta_r)}{c_{s1}}\right) \tag{4.198}$$

$$\boldsymbol{v}_t = \hat{\boldsymbol{y}} T \frac{\partial U_s}{\partial t}\left(t - \frac{(x\sin\theta_t + z\cos\theta_t)}{c_{s2}}\right) \tag{4.199}$$

and traction vectors given by

$$\hat{\boldsymbol{z}} \cdot \boldsymbol{\tau}_i = -\hat{\boldsymbol{y}} \frac{G_1 \cos\theta_i}{c_{s1}} \frac{\partial U_s}{\partial t}\left(t - \frac{(x\sin\theta_i + z\cos\theta_i)}{c_{s1}}\right) \tag{4.200}$$

$$\hat{\boldsymbol{z}} \cdot \boldsymbol{\tau}_r = \hat{\boldsymbol{y}} \frac{G_1 \cos\theta_r}{c_{s1}} R \frac{\partial U_s}{\partial t}\left(t - \frac{(x\sin\theta_r - z\cos\theta_r)}{c_{s1}}\right) \tag{4.201}$$

$$\hat{\boldsymbol{z}} \cdot \boldsymbol{\tau}_t = -\hat{\boldsymbol{y}} \frac{G_2 \cos\theta_t}{c_{s2}} T \frac{\partial U_s}{\partial t}\left(t - \frac{(x\sin\theta_t + z\cos\theta_t)}{c_{s2}}\right). \tag{4.202}$$

So that at the $z = 0$ interface, all three waves have the same time dependence along x, we obtain Snell's law in the form

$$p = \frac{\sin\theta_i}{c_{s1}} = \frac{\sin\theta_r}{c_{s1}} = \frac{\sin\theta_t}{c_{s2}} \tag{4.203}$$

so that

$$\theta_r = \theta_i \quad \text{and} \quad \theta_t = \sin^{-1}\left(\frac{c_{s2}}{c_{s1}}\sin\theta_i\right). \tag{4.204}$$

The continuity of the material velocity and traction vectors across the interface $z = 0$ then gives the two equations

$$\begin{bmatrix} 1 & -1 \\ G_1\cos\theta_i/c_{s1} & G_2\cos\theta_t/c_{s2} \end{bmatrix} \begin{bmatrix} R \\ T \end{bmatrix} = \begin{bmatrix} -1 \\ G_1\cos\theta_i/c_{s1} \end{bmatrix}. \tag{4.205}$$

Using $c_s^2 = G/\rho$ then yields the solution for the reflection and transmission coefficients

$$R = \frac{\rho_1 c_{s1}\cos\theta_i - \rho_2 c_{s2}\cos\theta_t}{\rho_1 c_{s1}\cos\theta_i + \rho_2 c_{s2}\cos\theta_t} \tag{4.206}$$

$$T = \frac{2\rho_1 c_{s1}\cos\theta_i}{\rho_1 c_{s1}\cos\theta_i + \rho_2 c_{s2}\cos\theta_t}, \tag{4.207}$$

where

$$\cos\theta_t = \sqrt{1 - \left(\frac{c_{s2}}{c_{s1}}\right)^2 \sin^2\theta_i}. \tag{4.208}$$

The material property that influences reflection and transmission is thus the mass density times the wave speed, which is given the name *seismic impedance*. Note that by the way we have defined R and T, we always have $T = 1 + R$.

4.8 Evanescent Interface Waves

We now consider plane waves that propagate parallel with the interface but that have amplitudes that decrease exponentially with perpendicular distance from the interface along the plane of the wave. We demonstrate how such waves arise and how they are modeled. If the plane interface has a normal in the z direction, such interface waves are not associated with any time-averaged energy flux in the z direction, which along with their exponential fall off in amplitude with distance z from the interface, is why these waves are called *evanescent* (or disappearing). However, they are actual waves that have peak amplitudes at the interface and energy flux in the direction of the wave propagation, which is parallel with the interface.

In addition to the evanescent waves treated in this section, there can be guided waves that are trapped within low-velocity layers. For example, *Love waves* are a type of surface wave in which SH waves are trapped (guided) within a low-velocity layer that resides between a stress-free surface and an underlying material that has a faster wave speed. We have chosen not to address such guided waves.

4.8.1 Total Internal Reflection and Evanescent Transmitted Waves

One way that evanescent interface waves are created is by an incident wave arriving at the interface at a sufficiently large incident angle. At a certain critical incident angle $\theta_i = \theta_{cr}$, and under the condition that the transmitted wave velocity is larger than the incident wave velocity, the transmitted wave will have an angle of propagation relative to the vertical that is $\pi/2$ so that the transmitted wave is propagating in a direction parallel with the interface and is now called a *head wave*. We call this situation *total internal reflection* and will consider here the nature of what is happening with the reflected and transmitted plane waves when $\theta_i \geq \theta_{cr}$.

Let's begin with the case of transmission and reflection of SH waves just treated above. When the shear velocity in the transmitted domain 2 is larger than the shear velocity in the incident domain 1, we define the critical angle of incidence $\theta_i = \theta_{cr}$ from the condition $\theta_t = \pi/2$, which from Snell's law gives

$$\frac{\sin\theta_{cr}}{c_{s1}} = \frac{\sin(\pi/2)}{c_{s2}} = \frac{1}{c_{s2}}. \tag{4.209}$$

We thus have $\theta_{cr} = \sin^{-1}(c_{s1}/c_{s2})$. For $\theta_i \geq \theta_{cr}$, the propagation of the transmitted wave is in the x direction and controlled by the horizontal slowness $\sin\theta_t/c_{s2} = \sin\theta_i/c_{s1}$. The incident and reflected waves also have x components that propagate at this same horizontal slowness per Snell's law. However, when $\theta_i \geq \theta_{cr}$, there is no longer propagation of the transmitted wave in the vertical or z direction. From Eq. (4.208) we have

$$\cos\theta_t = \sqrt{1-\left(\frac{c_{s2}}{c_{s1}}\right)^2 \sin^2\theta_i} = i\sqrt{\left(\frac{c_{s2}}{c_{s1}}\right)^2 \sin^2\theta_i - 1} \,\hat{=}\, i\chi_s, \tag{4.210}$$

so that for angles of incidence greater than the critical angle we have that $\sin\theta_i > \sin\theta_{cr} = c_{s1}/c_{s2}$ so that the dimensionless parameter χ_s defined in Eq. (4.210) is real and positive, that is, $\cos\theta_t = i\chi_s$ becomes purely imaginary when $\theta_i > \theta_{cr}$.

To observe the implications of the transmitted wave's vertical slowness becoming purely imaginary, we appeal to the case that the incident wave's particle velocity is time harmonic with time dependence $\cos\omega t$, where ω is a given real circular frequency. To treat this, we take the time function $\partial U_s(s)/\partial t$ to have the complex form

$$\frac{\partial U_s(t)}{\partial t} = A_s e^{-i\omega t}, \tag{4.211}$$

where A_s is the real amplitude of the incident SH wave. Using this in the above expressions for the particle velocities and taking the real part to give the actual plane-wave response, we have

$$\boldsymbol{v}_i = \mathrm{Re}\left\{\hat{\boldsymbol{y}} \exp\left[-i\omega\left(t - \frac{(x\sin\theta_i + z\cos\theta_i)}{c_{s1}}\right)\right]\right\} = \hat{\boldsymbol{y}}\cos\left[\omega\left(t - \frac{x\sin\theta_i}{c_{s1}} - \frac{z\cos\theta_i}{c_{s1}}\right)\right] \tag{4.212}$$

$$\boldsymbol{v}_r = \mathrm{Re}\left\{\hat{\boldsymbol{y}} A_s R \exp\left[-i\omega\left(t - \frac{(x\sin\theta_i - z\cos\theta_i)}{c_{s1}}\right)\right]\right\} \tag{4.213}$$

$$\boldsymbol{v}_t = \mathrm{Re}\left\{\hat{\boldsymbol{y}} A_s T \exp\left[-i\omega\left(t - \frac{x\sin\theta_i}{c_{s1}}\right)\right] e^{-\omega\chi_s z/c_{s2}}\right\}. \tag{4.214}$$

Because $\cos\theta_t = i\chi_s$ is purely imaginary when $\theta_i > \theta_{cr}$, the earlier expressions for the reflection and transmission coefficients become complex at postcritical incidence. With $R = R_R + iR_I$ and $T = T_R + iT_I$, we have

$$R_R = \frac{(\rho_1 c_{s1}\cos\theta_i)^2 - (\rho_2 c_{s2}\chi_s)^2}{(\rho_1 c_{s1}\cos\theta_i)^2 + (\rho_2 c_{s2}\chi_s)^2} \tag{4.215}$$

$$R_I = T_I = -\frac{2\rho_1 c_{s1}\rho_2 c_{s2}\cos\theta_i \chi_s}{(\rho_1 c_{s1}\cos\theta_i)^2 + (\rho_2 c_{s2}\chi_s)^2} \tag{4.216}$$

$$T_R = \frac{2(\rho_1 c_{s1}\cos\theta_i)^2}{(\rho_1 c_{s1}\cos\theta_i)^2 + (\rho_2 c_{s2}\chi_s)^2}, \tag{4.217}$$

which still satisfies $T = 1 + R$. So the actual, real, time-harmonic particle velocities of the reflected and transmitted plane waves at postcritical incidence are

$$\boldsymbol{v}_r(t,x,z) = \hat{\boldsymbol{y}} A_s \left\{R_R \cos\left[\omega\left(t - \frac{x\sin\theta_i}{c_{s1}} + \frac{z\sin\theta_i}{c_{s1}}\right)\right] - R_I \sin\left[\omega\left(t - \frac{x\sin\theta_i}{c_{s1}} + \frac{z\sin\theta_i}{c_{s1}}\right)\right]\right\} \tag{4.218}$$

$$\boldsymbol{v}_t(t,x,z) = \hat{\boldsymbol{y}} A_s e^{-z/\delta_s} \left\{T_R \cos\left[\omega\left(t - \frac{x\sin\theta_i}{c_{s1}}\right)\right] - R_I \sin\left[\omega\left(t - \frac{x\sin\theta_i}{c_{s1}}\right)\right]\right\}, \tag{4.219}$$

where we define the postcritical-incidence transmitted-amplitude skin depth δ_s as

$$\delta_s = \frac{c_{s2}}{\omega \chi_s} = \frac{c_{s2}}{\omega\sqrt{c_{s2}^2 \sin^2\theta_i / c_{s1}^2 - 1}}. \tag{4.220}$$

For all $\theta_i > \theta_{cr}$, the transmitted plane wave propagates only in the x direction (horizontally) but as one proceeds down this vertically oriented plane into material 2 ($z > 0$), the amplitude of this plane wave decays exponentially. The larger that θ_i becomes and the larger the frequency, the more rapid is this exponential fall off in amplitude. When $\theta_i = \theta_{cr}$, there is no such fall off so that $\delta_s = \infty$, $\chi_s = 0$, $R_R = 1$ (total internal reflection), $R_I = 0$, and $T_R = 2$. Note that the transmitted wave is only defined in the region $z \geq 0$ just like the incident and reflected waves are only defined in the region $z \leq 0$. As with any angle of incidence, Snell's law requires that all three waves are coincident at the interface $z = 0$ and move horizontally along the interface at the same horizontal slowness $\sin\theta_i / c_{s1}$.

We now analyze the z-component of the Poynting vector associated with the transmitted wave at postcritical incidence and show that the time-averaged energy transmitting vertically through the interface into the underlying material 2 is exactly zero in this circumstance. To demonstrate this, let's begin by considering a stress tensor having the complex time-harmonic form $\boldsymbol{\tau}(\boldsymbol{r}, \omega)e^{-i\omega t}$, where $\boldsymbol{\tau}(\boldsymbol{r}, \omega) = \boldsymbol{\tau}_R(\boldsymbol{r}, \omega) + i\boldsymbol{\tau}_I(\boldsymbol{r}, \omega)$ is complex. The spatial dependence here need not be limited to the plane waves we are currently treating in this section. Similarly, we have a complex time-harmonic particle velocity given by $\boldsymbol{v}(\boldsymbol{r}, \omega)e^{-i\omega t}$, where $\boldsymbol{v}(\boldsymbol{r}, \omega) = \boldsymbol{v}_R(\boldsymbol{r}, \omega) + i\boldsymbol{v}_I(\boldsymbol{r}, \omega)$ is also complex. We then form a vector called the *complex Poynting vector* defined as

$$\boldsymbol{s}(\boldsymbol{r}, \omega) = -\boldsymbol{\tau}(\boldsymbol{r}, \omega) \cdot \boldsymbol{v}^*(\boldsymbol{r}, \omega) = -\left[\boldsymbol{\tau}_R \cdot \boldsymbol{v}_R + \boldsymbol{\tau}_I \cdot \boldsymbol{v}_I + i\left(\boldsymbol{\tau}_I \cdot \boldsymbol{v}_R - \boldsymbol{\tau}_R \cdot \boldsymbol{v}_I\right)\right], \tag{4.221}$$

where, per the usual definition, the complex conjugate is defined $\boldsymbol{v}^*(\boldsymbol{r}, \omega) = \boldsymbol{v}_R - i\boldsymbol{v}_I$.

The real time-harmonic stress tensor $\boldsymbol{\tau}(\boldsymbol{r}, t)$, particle velocity $\boldsymbol{v}(\boldsymbol{r}, t)$, and Poynting vector $\boldsymbol{s}(\boldsymbol{r}, t) = -\boldsymbol{\tau}(\boldsymbol{r}, t) \cdot \boldsymbol{v}(\boldsymbol{r}, t)$ are defined

$$\boldsymbol{\tau}(\boldsymbol{r}, t) = \mathrm{Re}\left\{\boldsymbol{\tau}(\boldsymbol{r}, \omega)e^{-i\omega t}\right\} = \boldsymbol{\tau}_R(\boldsymbol{r}, \omega)\cos\omega t + \boldsymbol{\tau}_I(\boldsymbol{r}, \omega)\sin\omega t, \tag{4.222}$$

$$\boldsymbol{v}(\boldsymbol{r}, t) = \mathrm{Re}\left\{\boldsymbol{v}(\boldsymbol{r}, \omega)e^{-i\omega t}\right\} = \boldsymbol{v}_R(\boldsymbol{r}, \omega)\cos\omega t + \boldsymbol{v}_I(\boldsymbol{r}, \omega)\sin\omega t, \tag{4.223}$$

$$\boldsymbol{s}(\boldsymbol{r}, t) = -\left[\boldsymbol{\tau}_R \cdot \boldsymbol{v}_R \cos^2\omega t + \boldsymbol{\tau}_I \cdot \boldsymbol{v}_I \sin^2\omega t + (\boldsymbol{\tau}_R \cdot \boldsymbol{v}_I + \boldsymbol{\tau}_I \cdot \boldsymbol{v}_R)\sin\omega t\cos\omega t\right]. \tag{4.224}$$

We now average the real Poynting vector $\boldsymbol{s}(\boldsymbol{r}, t)$ over a single time cycle and use the well-known integrals $(2\pi)^{-1}\int_0^{2\pi}\cos^2 u\, du = (2\pi)^{-1}\int_0^{2\pi}\sin^2 u\, du = 1/2$ and $(2\pi)^{-1}\int_0^{2\pi}\cos u \sin u\, du = 0$ to obtain

$$\langle \boldsymbol{s}(\boldsymbol{r}, t)\rangle = \frac{1}{2\pi}\int_0^{2\pi} \mathrm{d}(\omega t)\boldsymbol{s}(\boldsymbol{r}, t) = -\frac{1}{2}\left(\boldsymbol{\tau}_R \cdot \boldsymbol{v}_R + \boldsymbol{\tau}_I \cdot \boldsymbol{v}_I\right) \tag{4.225}$$

$$= \frac{1}{2}\mathrm{Re}\left\{\boldsymbol{s}(\boldsymbol{r}, \omega)\right\} = -\frac{1}{2}\mathrm{Re}\left\{\boldsymbol{\tau}(\boldsymbol{r}, \omega) \cdot \boldsymbol{v}^*(\boldsymbol{r}, \omega)\right\}. \tag{4.226}$$

So the time-averaged flux of any real time-harmonic elastodynamic energy is equivalently given by half the real part of the complex Poynting vector. This is a general result of time-harmonic response that is independent of specific applications.

Returning to our particular focus on the amount of time-harmonic plane-wave energy being transmitted normally at postcritical incidence, we have the complex forms obtained earlier (taking $\partial U_s(t)/\partial t = e^{-i\omega t}$)

$$\hat{z} \cdot \boldsymbol{\tau}_t(\boldsymbol{r}, \omega) = -\hat{y}\frac{iG_2\chi_s}{c_{s2}} T \exp\left(\frac{i\omega \sin\theta_i x}{c_{s1}}\right) e^{-z/\delta_s}, \tag{4.227}$$

$$\boldsymbol{v}_t(\boldsymbol{r}, \omega) = \hat{y} T \exp\left(\frac{i\omega \sin\theta_i x}{c_{s1}}\right) e^{-z/\delta_s}, \tag{4.228}$$

$$\boldsymbol{v}_t^*(\boldsymbol{r}, \omega) = \hat{y} T^* \exp\left(\frac{-i\omega \sin\theta_i x}{c_{s1}}\right) e^{-z/\delta_s}, \tag{4.229}$$

so that the time-averaged energy fluxing normally through the interface into material 2 at postcritical incidence is

$$\langle \hat{z} \cdot \boldsymbol{s}_t(\boldsymbol{r}, t)\rangle = -\frac{1}{2}\text{Re}\left\{\hat{z} \cdot \boldsymbol{\tau}_t(\boldsymbol{r}, \omega) \cdot \boldsymbol{v}_t^*(\boldsymbol{r}, \omega)\right\} = \frac{1}{2}\text{Re}\left\{TT^* e^{-2z/\delta_s}\frac{iG_2\chi_s}{c_{s2}}\right\} = 0, \tag{4.230}$$

because what is in brackets is purely imaginary. So we have proven that when $\theta_i > \theta_{cr}$, there is no energy fluxing normally through the interface into a transmitted wave. At all postcritical-incidence angles, the transmitted plane wave exists in the form of a vertical plane in $z \geq 0$ and propagates in the x (horizontal) direction with an amplitude on the plane that decreases exponentially with depth z. This postcritical transmitted plane wave is called an *evanescent* wave due to no energy feeding vertically into it even if it possesses energy that propagates horizontally that is a maximum at $z = 0_+$. Another term sometimes used for an evanescent wave is *inhomogeneous wave*. Such a postcritical transmitted wave is also called the *critically refracted wave* and in the case of elastodynamics the *head wave*.

We have quantified above the specific case of how an SH transmitted waves becomes an evanescent interface wave at postcritical incidence. Following an identical analysis, it is straightforward, though algebraically intensive due to the complicated nature of R_{pp}, R_{ps}, T_{ps}, and T_{pp}, to treat the case of total-internal reflection of plane P or SV incident waves. Any transmitted wave, P or SV or SH, that has a velocity greater than the velocity of the incident wave will, at large enough angle of incidence, go through critical refraction ($\theta_t = \pi/2$), resulting in evanescent waves at postcritical incidence that can be treated in the same way that we handled SH waves above.

In the case of SV waves incident at the interface, the reflected P wave has a larger velocity than the incident SV wave and, at large enough angle of incidence, has a vertical slowness that becomes purely imaginary. So for an incident SV wave, the reflected P wave can also become an evanescent interface wave at large enough angle of incidence. As an end-of-chapter exercise, you can treat the special case of SV plane waves incident at a stress-free surface like the surface of the Earth.

4.8.2 Rayleigh Waves and Stonely Waves

Another class of evanescent waves that propagate along the interface are generated by an excitation of the interface itself (an impact or explosion or fracture/slip event at the material interface) and are not created from either critical transmission or critical reflection of an incident wave as treated and discussed above. When an interface-generated wave propagates at the stress-free interface between a solid half-space and a gas or vacuum, we call it a *Rayleigh wave*. When the interface is between two solid media, or between a solid medium and a liquid medium, we call this interface-generated wave a *Stonely wave*. We will only model here the case of plane Rayleigh waves propagating along the surface of a solid elastic half space. Plane Stonely waves are handled in the same way but involve more algebra. A Rayleigh wave generated by a point source on the free surface will have an amplitude that falls off with distance propagated due to the circular spreading of the wavefront on the surface, but this surface wave will move at the same Rayleigh wave speed obtained below for plane waves. At great distance from the point source on the surface, plane-wave response becomes a reasonable approximation to spherical waves, at least in the vicinity of the surface.

The new thinking here is to pose the question: might there be a wave that satisfies the elastodynamic wave equation and the zero stress condition on $z=0$ and that propagates along the interface as a linear combination of P and SV response but that has its own distinct wave speed given by c_R that we must find? This c_R is required to satisfy $c_R < c_s < c_p$. The horizontal slowness for this surface wave is $p = 1/c_R$.

To answer this question, let's return to our expressions for plane P and plane SV waves, written for complex time-harmonic response as

$$\boldsymbol{v}_p = \left(\sin\theta_p \hat{\boldsymbol{x}} + \cos\theta_p \hat{\boldsymbol{z}}\right) A_p \exp\left[-i\omega\left(t - \frac{\sin\theta_p x}{c_p} - \frac{\cos\theta_p z}{c_p}\right)\right], \tag{4.231}$$

$$\boldsymbol{v}_s = \left(\cos\theta_s \hat{\boldsymbol{x}} - \sin\theta_s \hat{\boldsymbol{z}}\right) A_s \exp\left[-i\omega\left(t - \frac{\sin\theta_s x}{c_s} - \frac{\cos\theta_s z}{c_s}\right)\right], \tag{4.232}$$

where the A_p and A_s are the amplitudes of each response and we need to find A_s/A_p. From the requirement that the horizontal slowness p of both of these responses is the same, we have

$$p = \frac{1}{c_R} = \frac{\sin\theta_p}{c_p} = \frac{\sin\theta_s}{c_s}, \tag{4.233}$$

or, equivalently, $\sin\theta_p = pc_p$ and $\sin\theta_s = pc_s$. Because $1/c_R = p > 1/c_s > 1/c_p$ by supposition, the relation $\cos^2\theta = 1 - \sin^2\theta$ gives

$$\cos\theta_p = i\chi_p \mathrel{\widehat{=}} i\sqrt{(pc_p)^2 - 1} \quad \text{and} \quad \cos\theta_s = i\chi_s \mathrel{\widehat{=}} i\sqrt{(pc_s)^2 - 1}, \tag{4.234}$$

both of which are purely imaginary. There is a possibility of the $\chi_{p,s}$ being either positive or negative but we have chosen the positive sign so that the response has exponential decay, and not exponential growth, with distance z from the surface into the elastic half-space.

The complex particle velocity is then given as the evanescent response

$$\boldsymbol{v}_p = \left(pc_p\hat{\boldsymbol{x}} + i\chi_p\hat{\boldsymbol{z}}\right) A_p e^{-z/\delta_p} e^{-i\omega(t-px)}, \tag{4.235}$$

$$\boldsymbol{v}_s = \left(i\chi_s\hat{\boldsymbol{x}} - pc_s\hat{\boldsymbol{z}}\right) A_s e^{-z/\delta_s} e^{-i\omega(t-px)}, \tag{4.236}$$

and the total complex particle velocity of this proposed surface wave is $\boldsymbol{v}_R = \boldsymbol{v}_p + \boldsymbol{v}_s$ or

$$\boldsymbol{v}_R = A_p\left[\left(pc_p e^{-z/\delta_p} + i\chi_s e^{-z/\delta_s}(A_s/A_p)\right)\hat{\boldsymbol{x}} + \left(i\chi_p e^{-z/\delta_p} - pc_s e^{-z/\delta_s}(A_s/A_p)\right)\hat{\boldsymbol{z}}\right] e^{-i\omega(t-px)}. \tag{4.237}$$

The skin depths associated with the P and SV portions of the response are defined

$$\delta_p = \frac{c_p}{\omega\chi_p} \quad \text{and} \quad \delta_s = \frac{c_s}{\omega\chi_s}. \tag{4.238}$$

Once we find A_s/A_p, the actual real time-harmonic Rayleigh wave is given by $\text{Re}\{\boldsymbol{v}_R\}$.

The horizontal slowness $p = 1/c_R$ of this wave along with the amplitude ratio A_s/A_p are determined from the free surface condition that says that on $z = 0$, $\hat{\boldsymbol{z}}\cdot(\boldsymbol{\tau}_p + \boldsymbol{\tau}_s) = 0$. Using our earlier results, we find that on $z = 0$

$$\hat{\boldsymbol{z}}\cdot\boldsymbol{\tau}_p = -\frac{H}{c_p}\left[i2\chi_p p\frac{c_s^2}{c_p}\hat{\boldsymbol{x}} + (1 - 2p^2c_s^2)\hat{\boldsymbol{z}}\right] A_p e^{-i\omega(t-px)}, \tag{4.239}$$

$$\hat{\boldsymbol{z}}\cdot\boldsymbol{\tau}_s = -\frac{G}{c_s}\left[(1 - 2p^2c_s^2)\hat{\boldsymbol{x}} - i2\chi_s pc_s\hat{\boldsymbol{z}}\right] A_s e^{-i\omega(t-px)}, \tag{4.240}$$

where we used $H/G = c_p^2/c_s^2$. Adding these together and requiring the x and z components to both go to zero gives (after some algebraic rearrangement)

$$-\left(p^2 - \frac{1}{2c_s^2}\right) A_s + ip\frac{c_p}{c_s}\sqrt{p^2 - \frac{1}{c_p^2}} A_p = 0, \tag{4.241}$$

$$ip\frac{c_s}{c_p}\sqrt{p^2 - \frac{1}{c_s^2}} A_s + \left(p^2 - \frac{1}{2c_s^2}\right) A_p = 0. \tag{4.242}$$

Dividing through by A_p, we obtain the two equations

$$\frac{A_s}{A_p} = \frac{ip(c_p/c_s)\sqrt{p^2 - 1/c_p^2}}{p^2 - 1/(2c_s^2)} = -\frac{[p^2 - 1/(2c_s^2)]}{ip(c_s/c_p)\sqrt{p^2 - 1/c_s^2}}. \tag{4.243}$$

The first equality gives the relative amplitude A_s/A_p in the combined P-SV response of the Rayleigh wave. The second equality provides the equation that determines $p = 1/c_R$

$$\left(p^2 - \frac{1}{2c_s^2}\right)^2 = p^2\sqrt{p^2 - \frac{1}{c_p^2}}\sqrt{p^2 - \frac{1}{c_s^2}}, \tag{4.244}$$

which is called the Rayleigh equation. Making the substitution $r = (c_R/c_s)^2 = (pc_s)^{-2}$ and squaring both sides gives a cubic equation for r after some rearrangement

$$r^3 - 8r^2 + 16\left[\frac{3}{2} - \left(\frac{c_s}{c_p}\right)^2\right] r - 16\left[1 - \left(\frac{c_s}{c_p}\right)^2\right] = 0. \tag{4.245}$$

Thermodynamic stability requires that $K \geq 0$ and $G \geq 0$ so that $(c_s/c_p)^2 = G/(K + 4G/3)$ must lie within the range $0 \leq (c_s/c_p)^2 \leq 3/4$ over the wide range of positive elastic properties $0 \leq G/K \leq \infty$. In practice, nearly all materials lie in the narrower range of $0 \leq G/K \leq 3/2$, which results in the more realistic range

$$0 \leq \left(\frac{c_s}{c_p}\right)^2 \leq \frac{1}{2}. \tag{4.246}$$

The one real positive root of the cubic equation for r when $(c_s/c_p)^2$ lies in this range results in Rayleigh-wave speeds within the range

$$0.874c_s \leq c_R \leq 0.995c_s. \tag{4.247}$$

So over a wide range of material properties, the Rayleigh-wave speed is close to, but smaller than, the shear-wave velocity of the isotropic elastic half space.

The final real particle velocity associated with a time-harmonic plane Rayleigh wave traveling in the x direction is

$$\begin{aligned} \boldsymbol{v}_R(x, z, t) = \\ A_R \Bigg[\left(\frac{c_R}{c_s} e^{-z/\delta_p} - \frac{2\sqrt{1 - (c_R/c_s)^2}\sqrt{1 - (c_s/c_p)^2 (c_R/c_s)^2}\, e^{-z/\delta_s}}{2 - (c_R/c_s)^2} \right) \cos\left(\omega \left(t - \frac{x}{c_R}\right)\right) \hat{\boldsymbol{x}} \\ + \sqrt{1 - \left(\frac{c_s}{c_p}\right)^2 \left(\frac{c_R}{c_s}\right)^2} \left(e^{-z/\delta_p} - \frac{2e^{-z/\delta_s}}{2 - (c_R/c_s)^2} \right) \sin\left(\omega \left(t - \frac{x}{c_R}\right)\right) \hat{\boldsymbol{z}} \Bigg] \end{aligned} \tag{4.248}$$

where A_R is the real amplitude we assign to the Rayleigh wave. By taking the time integral of this expression, we obtain the particle displacements. On the surface $z = 0$ for a Rayleigh wave propagating in the $+x$ direction (i.e., to the right), the particle motions so obtained are counterclockwise elliptical (neither back-and-forth linear in the wave direction as for a P wave nor back-and-forth linear in the perpendicular direction as for an SV wave) and become exponentially smaller with depth. At sufficient depth, due to the different skin depths $\delta_p < \delta_s$, the particle motion becomes clockwise elliptical. It is straightforward to verify from the above complex expressions for the time-harmonic response that the time-averaged vertical energy flux associated with a Rayleigh wave is $\langle \hat{\boldsymbol{z}} \cdot \boldsymbol{s}_R \rangle = (1/2)\text{Re}\left\{\hat{\boldsymbol{z}} \cdot \boldsymbol{\tau}_R(\boldsymbol{r}, \omega) \cdot \boldsymbol{v}_R^*(\boldsymbol{r}, \omega)\right\} = 0$. Although the instantaneous Poynting vector quantifies a vertical energy flux that goes from positive to negative as we proceed through a harmonic cycle, the time-averaged vertical flux of energy is zero for a Rayleigh wave as it is for all evanescent waves on a horizontal interface.

A nearly identical analysis can be performed at an interface between two elastic media. One asks whether there is an interface response that is made up of four responses that are the P and SV plane waves in material 1 and material 2. All four plane waves are required to propagate horizontally with the same horizontal wave speed that must be smaller than the speed of all four P and SV plane waves. Using the particle velocity and stress continuity conditions at the interface, one finds both the wave speed of this surface wave as well as the three amplitude ratios A_{s1}/A_{p1}, A_{p2}/A_{p1}, and A_{s2}/A_{p1}. This surface wave is called a *Stonely wave* and is also evanescent. For the case of a liquid elastic half-space above and a solid elastic half-space below, the Stonely wave will be a linear sum of three horizontally propagating planes waves all advancing at the same speed along the interface: a P-wave response in the liquid and both a P-wave and SV-wave response in the solid. The continuity conditions again provide the speed of this Stonely wave and the two relative amplitudes A_{p2}/A_{p1} and A_{s2}/A_{p1}.

4.9 The Linear-Seismic Displacement Theorems

In this section, we treat several interrelated topics associated with seismic (i.e., elastodynamic) wave propagation due to a range of different forcing scenarios including the response from both distributed and point sources. The results to follow require the wave propagation to be linear, which requires the material properties ${}_4\boldsymbol{C}$ and ρ not to vary in time. But in all the theorems derived in this section, the stiffness tensor ${}_4\boldsymbol{C}$ can be arbitrarily anisotropic and both ${}_4\boldsymbol{C}$ and ρ can be arbitrarily heterogeneous throughout the body supporting the linear wave propagation. Further, even if the stress and strain response are delayed relative to each other, which as will be seen in Chapter 11 means there are dissipative or attenuative processes present causing the waves to lose energy with distance propagated, once we learn about the Fourier transform we can take Fourier transforms of the linear governing equations and work with a complex frequency-dependent stiffness tensor that allows for attenuation and all the theorems of this section remain.

In what follows, we write the elastodynamic wave equation in the linear form

$$\rho \frac{\partial^2 \boldsymbol{u}}{\partial t^2} = \nabla \cdot \boldsymbol{\tau} + \boldsymbol{F}(\boldsymbol{r}, t) \quad \text{with} \quad \boldsymbol{\tau} = {}_4\boldsymbol{C} : \nabla \boldsymbol{u}. \tag{4.249}$$

Note that because ${}_4\boldsymbol{C}$ is symmetric in the third and fourth base vectors, we have the identity ${}_4\boldsymbol{C} : \nabla \boldsymbol{u} = {}_4\boldsymbol{C} : [\nabla \boldsymbol{u} + (\nabla \boldsymbol{u})^T]/2 = {}_4\boldsymbol{C} : \boldsymbol{e} = \boldsymbol{\tau}$. The source for the seismic waves in what follows is assumed to be the effective body force $\boldsymbol{F}(\boldsymbol{r}, t)$ whose nature for many common types of seismic sources will be addressed as a theorem showing how localized displacement-discontinuity (or "slip") events across fault surfaces or localized stress-discontinuity events across impact surfaces translate exactly for linear elastodynamics into an effective body-force representation $\boldsymbol{F}$ of such sources.

Our goal for the rest of the chapter is not to solve a long list of specific linear-seismic boundary-value and initial-condition wave problems. Instead, the results that follow are general integral relations that apply to all linear seismic wave problems that show how sources for seismic waves produce a distant response in any domain. In Chapter 12, once we have learned the methods of Fourier analysis and contour integration, we will obtain the point-source response for seismic waves in a whole space that is also known as the elastodynamic Green's tensor. Such specific results from Chapter 12 complement the more general development that follows here.

4.9.1 Well-Posed Elastodynamic BVPs and Boundary and Initial Conditions

Given the above linear form of the elastodynamic governing equations, we seek to pose a BVP in a modeling domain Ω such that the solution for the displacement field $\boldsymbol{u}$ is unique. A linear BVP possessing unique solutions is called *well posed*. We show here that a well-posed problem requires boundary conditions to be specified over $\partial\Omega$ on the displacement and/or traction fields as well as certain initial conditions to be specified throughout Ω. This analysis is analogous to that for defining the conditions for a well-posed electromagnetics problem as treated in Section 3.11.

To obtain the conditions for uniqueness of solution, consider two fields $\boldsymbol{u}_1$ and $\boldsymbol{u}_2$ that both satisfy the same boundary and initial conditions throughout Ω and that both come from the same source $\boldsymbol{F}$. If we define the difference field $\delta\boldsymbol{u} = \boldsymbol{u}_1 - \boldsymbol{u}_2$ and if we can identify boundary and initial conditions such that $\delta\boldsymbol{u} = 0$ throughout Ω and for all time, then the solution is unique and the BVP is well posed.

The governing equation for $\delta\boldsymbol{u}$ is obtained by subtracting the equation for $\boldsymbol{u}_1$ from that for $\boldsymbol{u}_2$, which eliminates the common source term $\boldsymbol{F}$ to give

$$\nabla\cdot({}_4\boldsymbol{C}:\nabla\delta\boldsymbol{u}) = \rho\frac{\partial^2\delta\boldsymbol{u}}{\partial t^2}. \tag{4.250}$$

Dot multiply $\partial\delta\boldsymbol{u}/\partial t$ into this equation, use the identity $\nabla\cdot(\boldsymbol{a}\cdot\boldsymbol{A}) = \boldsymbol{a}\cdot\left(\nabla\cdot\boldsymbol{A}^T\right) + \nabla\boldsymbol{a}:\boldsymbol{A}$ and note that $\delta\boldsymbol{\tau} = {}_4\boldsymbol{C}:\nabla\delta\boldsymbol{u} = \delta\boldsymbol{\tau}^T$ to obtain

$$\nabla\cdot\left(\delta\boldsymbol{\tau}\cdot\frac{\partial\delta\boldsymbol{u}}{\partial t}\right) = \frac{\rho}{2}\frac{\partial}{\partial t}\left(\frac{\partial\delta\boldsymbol{u}}{\partial t}\cdot\frac{\partial\delta\boldsymbol{u}}{\partial t}\right) + \frac{1}{2}\frac{\partial}{\partial t}\left(\nabla\delta\boldsymbol{u}:{}_4\boldsymbol{C}:\nabla\delta\boldsymbol{u}\right). \tag{4.251}$$

Integrate over all time and space and use the divergence theorem to obtain

$$\int_0^t \mathrm{d}t'\int_{\partial\Omega}\boldsymbol{n}\cdot\delta\boldsymbol{\tau}\cdot\frac{\partial\delta\boldsymbol{u}}{\partial t'}\,\mathrm{d}S = \frac{1}{2}\int_\Omega\left(\rho\left.\frac{\partial\delta\boldsymbol{u}}{\partial t'}\cdot\frac{\partial\delta\boldsymbol{u}}{\partial t'}\right|_{t'=0}^{t'=t} + \left.\nabla\delta\boldsymbol{u}:{}_4\boldsymbol{C}:\nabla\delta\boldsymbol{u}\right|_{t'=0}^{t'=t}\right)\mathrm{d}V. \tag{4.252}$$

We now impose the following conditions on our BVP:

1 **Boundary conditions**: Require that either $\boldsymbol{n}\cdot\boldsymbol{\tau}$ or $\boldsymbol{u}$ are specified, patch by patch, over the system boundary $\partial\Omega$ for all time. In those places on $\partial\Omega$ where $\boldsymbol{u}$ is specified, $\partial\boldsymbol{u}/\partial t$ is also known so that $\partial\delta\boldsymbol{u}/\partial t = 0$ on those patches. On the other portions where $\boldsymbol{n}\cdot\boldsymbol{\tau}$ is given, we have $\boldsymbol{n}\cdot\delta\boldsymbol{\tau} = 0$, so that the surface integral on the left-hand side of Eq. (4.252) is zero for these boundary conditions.

2 **Initial conditions**: Require that both $\boldsymbol{u}(0)$ and $\partial\boldsymbol{u}/\partial t(0)$ are specified at time zero throughout all of Ω. If at $t = 0$, $\boldsymbol{u}$ is known throughout Ω then so is $\nabla\boldsymbol{u}$, which means that the terms on the right-hand side of Eq. (4.252) evaluated at $t = 0$ are both zero.

Using these boundary conditions and initial conditions in Eq. (4.252) then results in

$$\int_\Omega\left(\rho\frac{\partial\delta\boldsymbol{u}}{\partial t}\cdot\frac{\partial\delta\boldsymbol{u}}{\partial t} + \nabla\delta\boldsymbol{u}:{}_4\boldsymbol{C}:\nabla\delta\boldsymbol{u}\right)\mathrm{d}V = 0. \tag{4.253}$$

From ideas given in Chapter 6, stability of an equilibrium state requires that the stiffness tensor is positive definite meaning that $\nabla\delta\boldsymbol{u} : {}_4\boldsymbol{C} : \nabla\delta\boldsymbol{u} \geq 0$. Since each term in the integrand is positive definite, we must have that $\delta\boldsymbol{u} = 0$ for all space and time.

So if the boundary conditions and initial conditions are as specified above, we are guaranteed to have a unique solution and a well-posed BVP. The boundary conditions that result in a well-posed problem only require that either $\boldsymbol{n} \cdot \boldsymbol{\tau}$ or $\boldsymbol{u}$ is specified point by point over the surface but are independent of what these specific boundary values are. Similarly the initial conditions that result in a well-posed problem only require that both $\boldsymbol{u}$ and $\partial\boldsymbol{u}/\partial t$ are specified at all points throughout Ω at $t = 0$ but are independent of what these specific initial conditions are.

4.9.2 Seismic Reciprocity

As the next topic, we consider the way that the elastodynamic (or seismic) response at point a in space due to a source at point b is related to the seismic response at point b due to a source at point a. Such relations are known as *seismic reciprocity* and require nothing more than the elastodynamic laws to be linear.

First consider a more general problem in which we consider two distinct body forces $\boldsymbol{F}_a(\boldsymbol{r}, t)$ and $\boldsymbol{F}_b(\boldsymbol{r}, t)$ that have different distributions through space and time. These forces act throughout some possibly heterogeneous and anisotropic body Ω_∞. We will assume the spatial distribution of the forces is compact compared to the spatial extent of the body and that these forces act for a finite duration of time to generate seismic waves that propagate outward from the forced regions into the rest of the body. The body Ω_∞ can be taken to extend to such large distances compared to the extent of the forcing that the seismic wavefields are zero on the surface $\partial\Omega_\infty$ surrounding the body. We will consider such an infinity-body limit as a special case that gives interesting results. Prior to the sources turning on there are no seismic wave fields in the body.

Due to the assumed linearity, each source distribution generates its own set of seismic waves characterized by displacements $\boldsymbol{u}_a$ and $\boldsymbol{u}_b$ that each satisfy the linearized equations

$$\rho \frac{\partial^2 \boldsymbol{u}_a}{\partial t'^2} = \nabla \cdot \boldsymbol{\tau}_a + \boldsymbol{F}_a(\boldsymbol{r}, t') \quad \text{with} \quad \boldsymbol{\tau}_a = {}_4\boldsymbol{C} : \nabla \boldsymbol{u}_a \tag{4.254}$$

$$\rho \frac{\partial^2 \boldsymbol{u}_b}{\partial t'^2} = \nabla \cdot \boldsymbol{\tau}_b + \boldsymbol{F}_b(\boldsymbol{r}, t') \quad \text{with} \quad \boldsymbol{\tau}_b = {}_4\boldsymbol{C} : \nabla \boldsymbol{u}_b \tag{4.255}$$

with the only difference between the two partial-differential equations (PDEs) being the spatial and temporal nature of the force densities $\boldsymbol{F}_a$ and $\boldsymbol{F}_b$ that generate the two sets of wavefields $\boldsymbol{u}_a$ and $\boldsymbol{u}_b$. Again, the density ρ and stiffness tensor ${}_4\boldsymbol{C}$ can have arbitrary spatial distribution throughout the body Ω_∞ but are both taken as independent of time in the linearized theory.

The statement of reciprocity is obtained by dot multiplying the time-reversed and time-shifted wavefield $\boldsymbol{u}_b(t - t')$ into Eq. (4.254) controlling $\boldsymbol{u}_a(t')$ and similarly dot multiplying the time-reversed and time-shifted wavefield $\boldsymbol{u}_a(t - t')$ into Eq. (4.255) controlling $\boldsymbol{u}_b(t')$. Writing out the first equation so obtained, we have

$$\rho\left(\frac{\partial}{\partial t'}\left[\boldsymbol{u}_b(t-t')\cdot\frac{\partial \boldsymbol{u}_a(t')}{\partial t'}\right]-\frac{\partial \boldsymbol{u}_b(t-t')}{\partial t'}\cdot\frac{\partial \boldsymbol{u}_a(t')}{\partial t'}\right)$$
$$=\nabla\cdot\left[\boldsymbol{u}_b(t-t')\cdot\boldsymbol{\tau}_a(t')\right]-\nabla\boldsymbol{u}_b(t-t'):{}_4\boldsymbol{C}:\nabla\boldsymbol{u}_a(t')+\boldsymbol{u}_b(t-t')\cdot\boldsymbol{F}_a(t'), \quad (4.256)$$

where we used that the stress tensor is symmetric along with the tensor-product identity from Chapter 1 that $\left(\nabla\cdot\boldsymbol{\tau}^T\right)\cdot\boldsymbol{u}=\nabla\cdot(\boldsymbol{u}\cdot\boldsymbol{\tau})-\nabla\boldsymbol{u}:\boldsymbol{\tau}$. Identical manipulations on the second equation yields

$$\rho\left(\frac{\partial}{\partial t'}\left[\boldsymbol{u}_a(t-t')\cdot\frac{\partial \boldsymbol{u}_b(t')}{\partial t'}\right]-\frac{\partial \boldsymbol{u}_a(t-t')}{\partial t'}\cdot\frac{\partial \boldsymbol{u}_b(t')}{\partial t'}\right)$$
$$=\nabla\cdot\left[\boldsymbol{u}_a(t-t')\cdot\boldsymbol{\tau}_b(t')\right]-\nabla\boldsymbol{u}_a(t-t'):{}_4\boldsymbol{C}:\nabla\boldsymbol{u}_b(t')+\boldsymbol{u}_a(t-t')\cdot\boldsymbol{F}_b(t'). \quad (4.257)$$

We next subtract these two equations and integrate over all of Ω_∞ and over all of time $-\infty<t'<\infty$ to obtain

$$\rho\int_{\Omega_\infty}\left\{\left[\boldsymbol{u}_b(t-t')\cdot\frac{\partial \boldsymbol{u}_a(t')}{\partial t'}-\boldsymbol{u}_a(t-t')\cdot\frac{\partial \boldsymbol{u}_b(t')}{\partial t'}\right]_{-\infty}^{+\infty}\right.$$
$$\left.+\int_{-\infty}^{+\infty}\left(\frac{\partial \boldsymbol{u}_b(t')}{\partial t'}\cdot\frac{\partial \boldsymbol{u}_a(t-t')}{\partial t'}-\frac{\partial \boldsymbol{u}_b(t-t')}{\partial t'}\cdot\frac{\partial \boldsymbol{u}_a(t')}{\partial t'}\right)\mathrm{d}t'\right\}\mathrm{d}V$$
$$=\int_{-\infty}^{+\infty}\mathrm{d}t'\int_{\partial\Omega_\infty}\boldsymbol{n}\cdot\left[\boldsymbol{u}_b(t-t')\cdot\boldsymbol{\tau}_a(t')-\boldsymbol{u}_a(t-t')\cdot\boldsymbol{\tau}_b(t')\right]\mathrm{d}S$$
$$+\int_{-\infty}^{+\infty}\mathrm{d}t'\int_{\Omega_\infty}\left[\nabla\boldsymbol{u}_b(t-t'):{}_4\boldsymbol{C}:\nabla\boldsymbol{u}_a(t')-\nabla\boldsymbol{u}_b(t'):{}_4\boldsymbol{C}:\nabla\boldsymbol{u}_a(t-t')\right]\mathrm{d}V$$
$$+\int_{-\infty}^{+\infty}\mathrm{d}t'\int_{\Omega_\infty}\left[\boldsymbol{u}_b(t-t')\cdot\boldsymbol{F}_a(t')-\boldsymbol{u}_a(t-t')\cdot\boldsymbol{F}_b(t')\right]\mathrm{d}V, \quad (4.258)$$

where we used the divergence theorem to rewrite the first volume integral on the right-hand side as a surface integral over $\partial\Omega_\infty$ that has the outward normal $\boldsymbol{n}$. Because $C_{ijkl}=C_{klij}$, we used that $\nabla\boldsymbol{u}_a:{}_4\boldsymbol{C}:\nabla\boldsymbol{u}_b=\nabla\boldsymbol{u}_b:{}_4\boldsymbol{C}:\nabla\boldsymbol{u}_a$. Further, because all fields evaluated in the infinite past or infinite future are zero, the first term in this equation also vanishes.

As will be discussed in greater detail in Chapter 10 on Fourier analysis, the *convolution* operation $f*g$ of two functions $f(t)$ and $g(t)$ is the integral operation defined as

$$f*g\,\hat{=}\int_{-\infty}^{+\infty}f(t-t')g(t')\,\mathrm{d}t'. \quad (4.259)$$

By making the substitution $t''=t-t'$ with $\mathrm{d}t''=-\mathrm{d}t'$, the convolution is seen to have the commutative property that

$$f*g=\int_{-\infty}^{+\infty}f(t-t')g(t')\,\mathrm{d}t'=-\int_{+\infty}^{-\infty}f(t'')g(t-t'')\,\mathrm{d}t''=\int_{-\infty}^{+\infty}f(t')g(t-t')\,\mathrm{d}t'=g*f. \quad (4.260)$$

Using this fact in Eq. (4.258) allows a first statement of reciprocity to be written as

$$\begin{aligned}
&\int_{-\infty}^{\infty} \mathrm{d}t' \left\{ \int_{\partial\Omega_\infty} \boldsymbol{n} \cdot \left[\boldsymbol{u}_b(\boldsymbol{r}, t-t') \cdot \boldsymbol{\tau}_a(\boldsymbol{r}, t')\right] \mathrm{d}S \right. \\
&\quad \left. + \int_{\Omega_\infty} \boldsymbol{u}_b(\boldsymbol{r}, t-t') \cdot \boldsymbol{F}_a(\boldsymbol{r}, t')\, \mathrm{d}V \right\} \\
&= \int_{-\infty}^{\infty} \mathrm{d}t' \left\{ \int_{\partial\Omega_\infty} \boldsymbol{n} \cdot \left[\boldsymbol{u}_a(\boldsymbol{r}, t-t') \cdot \boldsymbol{\tau}_b(\boldsymbol{r}, t')\right] \mathrm{d}S \right. \\
&\quad \left. + \int_{\Omega_\infty} \boldsymbol{u}_a(\boldsymbol{r}, t-t') \cdot \boldsymbol{F}_b(\boldsymbol{r}, t')\, \mathrm{d}V \right\},
\end{aligned} \tag{4.261}$$

which makes no assumptions about the fields on the surface $\partial\Omega_\infty$ and is a generally valid result of linear elastodynamics. This is a first most general statement of seismic reciprocity that shows how the fields a and b emanating from force terms a and b are related to each other.

As a special case, assume the surface $\partial\Omega_\infty$ is sufficiently removed from the spatial extent of the forcing that the wavefields on $\partial\Omega_\infty$ can be taken to be zero in the limit. Alternatively, we may be dealing with a body such as the Earth that has a free-surface boundary condition $\boldsymbol{n} \cdot \boldsymbol{\tau}_{a,b} = 0$. In these cases, the surface integrals are zero in the reciprocity relation of Eq. (4.261). But there certainly exist other scenarios for less extended bodies with given inhomogeneous (nonzero) boundary conditions, in which the surface integrals of Eq. (4.261) need to be retained.

To more clearly see the content of this reciprocity statement, consider forces so concentrated in time and space that they can be represented using Dirac delta functions as

$$\boldsymbol{F}_a(\boldsymbol{r}, t) = \boldsymbol{P}_a\, \delta(\boldsymbol{r} - \boldsymbol{r}_a)\delta(t - t_a) \tag{4.262}$$

$$\boldsymbol{F}_b(\boldsymbol{r}, t) = \boldsymbol{P}_b\, \delta(\boldsymbol{r} - \boldsymbol{r}_b)\delta(t - t_b). \tag{4.263}$$

The impulse vectors $\boldsymbol{P}_a$ and $\boldsymbol{P}_b$ can be in different directions are independent of time and space, and have physical units of Ns. Using these Dirac body-force vectors in the time and space reciprocity integrals of Eq. (4.261) with the surface integrals set to zero (e.g., a body of infinite extent) then gives

$$\boldsymbol{u}_b(\boldsymbol{r}_a, t - t_a) \cdot \boldsymbol{P}_a = \boldsymbol{u}_a(\boldsymbol{r}_b, t - t_b) \cdot \boldsymbol{P}_b. \tag{4.264}$$

Consider the situation shown in Fig. 4.12. The vertical point force (or impulse) $\boldsymbol{P}_a$ located at the point $\boldsymbol{r} = \boldsymbol{r}_a$ generates a displacement field $\boldsymbol{u}_a$ throughout the body, while the horizontal point impulse $\boldsymbol{P}_b$ at the point $\boldsymbol{r} = \boldsymbol{r}_b$ generates a displacement field $\boldsymbol{u}_b$. For this scenario, the reciprocity relation of Eq. (4.264) says that the horizontal component of $\boldsymbol{u}_a$ evaluated at $\boldsymbol{r} = \boldsymbol{r}_b$ as multiplied by $|\boldsymbol{P}_b|$ is equal to the vertical component of $\boldsymbol{u}_b$ evaluated at $\boldsymbol{r} = \boldsymbol{r}_a$ as multiplied by $|\boldsymbol{P}_a|$. Again, this somewhat remarkable result is true regardless of how heterogeneous or anisotropic the body is. It only requires that the response be linear in

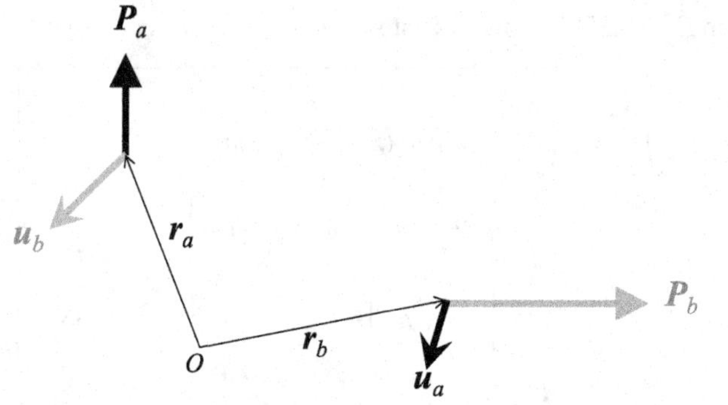

Figure 4.12 Displacement response $\boldsymbol{u}_a(\boldsymbol{r}_b)$ recorded at $\boldsymbol{r}_b$ generated by an impulse force $\boldsymbol{P}_a$ located at $\boldsymbol{r}_a$ is related to the displacement response $\boldsymbol{u}_b(\boldsymbol{r}_a)$ recorded at $\boldsymbol{r}_a$ generated by an impulse force $\boldsymbol{P}_b$ located at $\boldsymbol{r}_b$ by the reciprocity relation of Eq. (4.264).

the force. Although we did not formulate Hooke's law as a more general linear-response convolution in which past values of strain contribute to the present values of stress, which once we have covered contour integration in Chapter 11 will be shown to be associated with the irreversible generation of heat (attenuation), if we had done so, the same reciprocity relation of Eq. (4.264) would result.

It is possible to imagine scenarios, or "paradoxes," in which it seems questionable, at least qualitatively, that the reciprocity relation of Eq. (4.264) holds. For example, in Fig. 4.12, you can imagine placing a curved reflector of seismic waves around the point $\boldsymbol{r}_b$ that causes the seismic waves $\boldsymbol{u}_a$ propagating from the point $\boldsymbol{r}_a$ to converge and concentrate onto the point $\boldsymbol{r}_b$ to some degree. If there is no equivalent such curved reflector surrounding point $\boldsymbol{r}_a$, it may seem doubtful that the reciprocity relation holds. But keep in mind that the waves leaving the source at $\boldsymbol{r}_b$ would reflect off the reflector and converge to some degree at point $\boldsymbol{r}_a$. In every such "paradox" that has been numerically and experimentally analyzed in terms of linear wave propagation, reciprocity is observed to hold as it must.

As a final reciprocity topic, we introduce the second-order *Green's tensor* response as being that coming from point sources of the form of Eqs (4.262) and (4.263). The vectorial displacements associated with the Green's tensor response can be represented

$$\boldsymbol{u}_a(\boldsymbol{r}, t) = \boldsymbol{G}(\boldsymbol{r}, t \,|\, \boldsymbol{r}_a, t_a) \cdot \boldsymbol{P}_a \tag{4.265}$$

$$\boldsymbol{u}_b(\boldsymbol{r}, t) = \boldsymbol{G}(\boldsymbol{r}, t \,|\, \boldsymbol{r}_b, t_b) \cdot \boldsymbol{P}_b, \tag{4.266}$$

where the notation of the Green's tensor $\boldsymbol{G}(\boldsymbol{r}, t \,|\, \boldsymbol{r}', t')$ means there is a directed impulse source applied at point $\boldsymbol{r}'$ and time t' resulting in a response that is later observed at point $\boldsymbol{r}$ and time t. For $t < t'$, we define the Green's tensor to be zero. Furthermore, for reasons developed in Section 4.9.3, the Green's tensors are always defined so that either

$$\boldsymbol{G}(\boldsymbol{r}, t \,|\, \boldsymbol{r}', t') = 0 \quad \text{or} \quad \boldsymbol{n} \cdot {}_4\boldsymbol{C} : \nabla \boldsymbol{G}(\boldsymbol{r}, t \,|\, \boldsymbol{r}', t') = 0 \tag{4.267}$$

for point $\boldsymbol{r}$ residing on the boundaries $\partial\Omega$ of the domain Ω under consideration and as such the surface integrals in Eq. (4.261) are exactly zero for the Green's tensor response.

Inserting the Green's tensor representation of Eqs (4.265) and (4.266) into Eq. (4.264) means that for all imaginable directions for $\boldsymbol{P}_a$ and $\boldsymbol{P}_b$, the Green's tensor satisfies the symmetry relation

$$\boxed{\boldsymbol{G}(\boldsymbol{r}_a, t - t_a \,|\, \boldsymbol{r}_b, t_b) = \boldsymbol{G}^T(\boldsymbol{r}_b, t - t_b \,|\, \boldsymbol{r}_a, t_a),} \tag{4.268}$$

which is another way to express reciprocity. The Green's tensor $\boldsymbol{G}(\boldsymbol{r}, t \,|\, \boldsymbol{r}', t')$ is defined by the partial-differential equation

$$\rho \frac{\partial^2 \boldsymbol{G}}{\partial t^2} = \nabla \cdot ({}_4\boldsymbol{C} : \nabla \boldsymbol{G}) + \boldsymbol{I}\delta(\boldsymbol{r} - \boldsymbol{r}')\delta(t - t') \tag{4.269}$$

in which $\boldsymbol{I}$ is the identity tensor and $\boldsymbol{G}(\boldsymbol{r}, t \,|\, \boldsymbol{r}', t') = 0$ for $t - t' < 0$. The Green's tensor accounts for all possible components of displacement response at a receiver position $\boldsymbol{r}$ at time t due to an isotropic point source $\boldsymbol{I}$ located at $\boldsymbol{r}'$ that imparts a sharp impulse at time t'. For a homogeneous and isotropic wholespace, we will obtain the explicit expression for the elastodynamic Green's tensor in Chapter 12, which will indeed be seen to satisfy the symmetry property of Eq. (4.268).

Specific Green's tensor reciprocity relations correspond to taking particular cases for t, t_a, and t_b. For example, taking $t_a = t_b = 0$ (both impulse sources a and b are excited at the same time) results in

$$\boldsymbol{G}(\boldsymbol{r}_a, t \,|\, \boldsymbol{r}_b, 0) = \boldsymbol{G}^T(\boldsymbol{r}_b, t \,|\, \boldsymbol{r}_a, 0). \tag{4.270}$$

Taking $t_a = t'$, $t_b = -t''$, and $t = 0$ (along with $\boldsymbol{r}_a = \boldsymbol{r}'$ and $\boldsymbol{r}_b = \boldsymbol{r}''$) results in

$$\boldsymbol{G}(\boldsymbol{r}', -t' \,|\, \boldsymbol{r}'', -t'') = \boldsymbol{G}^T(\boldsymbol{r}'', t'' \,|\, \boldsymbol{r}', t'), \tag{4.271}$$

which is a symmetry, or reciprocity relation, that we will be using Section 4.9.3.

4.9.3 Representation of Displacement Using the Green's Tensor

We next determine how to use the Green's tensor to represent the linear displacements within some domain Ω that has sourcing coming from some combination of: (1) a given body-force distribution $\boldsymbol{F}$ throughout the domain; (2) given nonzero displacements or tractions over portions of the closed boundary $\partial\Omega$ around the domain; and (3) given nonzero initial conditions throughout the domain.

To do so, let's begin by writing the defining PDE for the Green's tensor in Cartesian coordinates and in scalar component form (where $\boldsymbol{G} = G_{jn}\hat{\boldsymbol{x}}_j\hat{\boldsymbol{x}}_n$ and $\boldsymbol{r} = x_i\hat{\boldsymbol{x}}_i$)

$$\rho \frac{\partial^2}{\partial t^2} G_{jn}(\boldsymbol{r}, t \,|\, \boldsymbol{r}', t') = \frac{\partial}{\partial x_i}\left[C_{ijkl}\frac{\partial}{\partial x_l} G_{kn}(\boldsymbol{r}, t \,|\, \boldsymbol{r}', t')\right] + \delta_{jn}\delta(\boldsymbol{r} - \boldsymbol{r}')\delta(t - t'). \tag{4.272}$$

Now, let's rewrite this equation by making the substitutions $\boldsymbol{r} \leftrightarrow \boldsymbol{r}'$ and $t \leftrightarrow -t'$ to obtain

$$\rho \frac{\partial^2}{\partial t'^2} G_{jn}(\boldsymbol{r}', -t' \,|\, \boldsymbol{r}, -t) = \frac{\partial}{\partial x_i'} \left[C_{ijkl} \frac{\partial}{\partial x_l'} G_{kn}(\boldsymbol{r}', -t' \,|\, \boldsymbol{r}, -t) \right] + \delta_{jn} \delta(\boldsymbol{r}' - \boldsymbol{r}) \delta(t - t'). \tag{4.273}$$

In Eq. (4.271) of the reciprocity section (Section 4.9.2), we proved that $G_{jn}(\boldsymbol{r}', -t' \,|\, \boldsymbol{r}, -t) = G_{nj}(\boldsymbol{r}, t \,|\, \boldsymbol{r}', t')$ so that we can write equivalently

$$\rho \frac{\partial^2}{\partial t'^2} G_{nj}(\boldsymbol{r}, t \,|\, \boldsymbol{r}', t') = \frac{\partial}{\partial x_i'} \left[C_{ijkl} \frac{\partial}{\partial x_l'} G_{nk}(\boldsymbol{r}, t \,|\, \boldsymbol{r}', t') \right] + \delta_{jn} \delta(\boldsymbol{r}' - \boldsymbol{r}) \delta(t - t'). \tag{4.274}$$

The differential equation for the displacements is written in component form as

$$\rho \frac{\partial^2}{\partial t'^2} u_j(\boldsymbol{r}', t') = \frac{\partial}{\partial x_i'} \left[C_{ijkl} \frac{\partial}{\partial x_l'} u_k(\boldsymbol{r}', t') \right] + F_j(\boldsymbol{r}', t'). \tag{4.275}$$

We now multiply Eq. (4.275) by $G_{nj}(\boldsymbol{r}, t \,|\, \boldsymbol{r}', t')$ and Eq. (4.274) by $u_j(\boldsymbol{r}', t')$, use the product rule $\partial(fg)/\partial x = f\partial g/\partial x + g\partial f/\partial x$ and subtract to obtain

$$\begin{aligned} \rho \frac{\partial}{\partial t'} \left(G_{nj} \frac{\partial u_j}{\partial t'} - u_j \frac{\partial G_{nj}}{\partial t'} \right) &= \frac{\partial}{\partial x_i'} \left(C_{ijkl} \frac{\partial u_k}{\partial x_l'} G_{nj} - C_{ijkl} u_j \frac{\partial G_{nk}}{\partial x_l'} \right) \\ &\quad - \frac{\partial G_{nj}}{\partial x_i'} C_{ijkl} \frac{u_k}{\partial x_l'} + \frac{\partial G_{nk}}{\partial x_l'} C_{ijkl} \frac{\partial u_j}{\partial x_i'} + G_{nj}(\boldsymbol{r}, t \,|\, \boldsymbol{r}', t') F_j(\boldsymbol{r}', t') \\ &\quad - u_n(\boldsymbol{r}', t') \delta(\boldsymbol{r}' - \boldsymbol{r}) \delta(t - t'). \end{aligned} \tag{4.276}$$

By renaming the dummy indices ($j \leftrightarrow k$ and $i \leftrightarrow l$), we have that

$$\frac{\partial G_{nk}}{\partial x_l'} C_{ijkl} \frac{\partial u_j}{\partial x_i'} = \frac{\partial G_{nj}}{\partial x_i'} C_{lkji} \frac{\partial u_k}{\partial x_l'}. \tag{4.277}$$

We also have the symmetries $C_{ijkl} = C_{klij} = C_{lkji}$ which means the first two terms of the second line in Eq. (4.276) vanish.

Integrating over t' from the initial condition at $t' = 0$ to $t' = \infty$ and over $\boldsymbol{r}'$ throughout the entire domain Ω, we obtain

$$\begin{aligned} &u_n(\boldsymbol{r}, t) \\ &= \int_0^\infty \mathrm{d}t' \int_\Omega \mathrm{d}^3\boldsymbol{r}' G_{nj}(\boldsymbol{r}, t \,|\boldsymbol{r}', t') F_j(\boldsymbol{r}', t') \\ &\quad + \int_0^\infty \mathrm{d}t' \int_{\partial\Omega} \mathrm{d}^2\boldsymbol{r}' n_i(\boldsymbol{r}') C_{ijkl}(\boldsymbol{r}') \left[\frac{\partial u_k(\boldsymbol{r}', t')}{\partial x_l'} G_{nj}(\boldsymbol{r}, t \,|\boldsymbol{r}', t') - \frac{\partial G_{nk}(\boldsymbol{r}, t \,|\boldsymbol{r}', t')}{\partial x_l'} u_j(\boldsymbol{r}', t') \right] \\ &\quad - \int_\Omega \mathrm{d}^3\boldsymbol{r}' \, \rho \left(G_{nj}(\boldsymbol{r}, t|\boldsymbol{r}', t') \frac{\partial u_j(\boldsymbol{r}', t')}{\partial t'} - u_j(\boldsymbol{r}', t') \frac{\partial G_{nj}(\boldsymbol{r}, t|\boldsymbol{r}', t')}{\partial t'} \right) \Bigg|_{t'=0}^{t'=\infty}. \end{aligned} \tag{4.278}$$

As usual, we used the divergence theorem to replace the volume integral with the surface integral. In the last term, we have that $G_{ij}(\boldsymbol{r}, t|\boldsymbol{r}', t') = 0$ for $t - t' < 0$, which means that when $t' = \infty$ both G_{ij} and $\partial G_{ij}/\partial t'$ are zero and that it is not necessary to integrate beyond $t' = t$. We thus have the final result that

$$
\begin{aligned}
u_n(\boldsymbol{r}, t) &\\
= \int_0^t \mathrm{d}t' &\int_\Omega \mathrm{d}^3\boldsymbol{r}' G_{nj}(\boldsymbol{r}, t\,|\boldsymbol{r}', t') F_j(\boldsymbol{r}', t') \\
+ \int_0^t \mathrm{d}t' &\int_{\partial\Omega} \mathrm{d}^2\boldsymbol{r}' n_i(\boldsymbol{r}') C_{ijkl}(\boldsymbol{r}') \left[\frac{\partial u_k(\boldsymbol{r}', t')}{\partial x'_l} G_{nj}(\boldsymbol{r}, t\,|\boldsymbol{r}', t') - \frac{\partial G_{nk}(\boldsymbol{r}, t\,|\boldsymbol{r}', t')}{\partial x'_l} u_j(\boldsymbol{r}', t') \right] \\
+ &\int_\Omega \mathrm{d}^3\boldsymbol{r}' \, \rho \left(G_{nj}(\boldsymbol{r}, t|\boldsymbol{r}', t') \frac{\partial u_j(\boldsymbol{r}', t')}{\partial t'} - u_j(\boldsymbol{r}', t') \frac{\partial G_{nj}(\boldsymbol{r}, t|\boldsymbol{r}', t')}{\partial t'} \right) \Bigg|_{t'=0}, \qquad (4.279)
\end{aligned}
$$

which is often called the *seismic-representation theorem.* This result shows that the source for elastodynamic displacements is any of: (1) a nonzero body force (the first line), (2) a nonzero boundary condition (the second line), or (3) a nonzero initial condition. Any combination of these three sources of seismic displacements can be occurring simultaneously.

To obtain unique solutions, we must have boundary conditions of either the displacements $\boldsymbol{u}$ being given on portions of $\partial\Omega$ or the traction $\boldsymbol{n} \cdot \boldsymbol{\tau}$ being given. On those portions of $\partial\Omega$, say $\partial\Omega_u$, where the displacements are given, we require the Green's tensor to vanish. Similarly, on those portions of $\partial\Omega$, say $\partial\Omega_t$, where the tractions are given, we require $\boldsymbol{n} \cdot {}_4\boldsymbol{C} : \nabla \boldsymbol{G}^T$ to vanish. As such, with $\partial\Omega = \partial\Omega_u + \partial\Omega_t$ we have the representation theorem given in perhaps its most useful bold-face form

$$
\begin{aligned}
\boldsymbol{u}(\boldsymbol{r}, t) &\\
= \int_0^t \mathrm{d}t' &\int_\Omega \mathrm{d}^3\boldsymbol{r}' \boldsymbol{F}(\boldsymbol{r}', t') \cdot \boldsymbol{G}^T(\boldsymbol{r}, t\,|\boldsymbol{r}', t') \\
+ \int_0^t \mathrm{d}t' &\int_{\partial\Omega_t} \mathrm{d}^2\boldsymbol{r}' \, \boldsymbol{n}(\boldsymbol{r}') \cdot \boldsymbol{\tau}(\boldsymbol{r}', t') \cdot \boldsymbol{G}^T(\boldsymbol{r}, t\,|\boldsymbol{r}', t') \\
- \int_0^t \mathrm{d}t' &\int_{\partial\Omega_u} \mathrm{d}^2\boldsymbol{r}' \, \boldsymbol{u}(\boldsymbol{r}', t') \boldsymbol{n}(\boldsymbol{r}') : {}_4\boldsymbol{C}(\boldsymbol{r}') : \nabla' \boldsymbol{G}^T(\boldsymbol{r}, t\,|\boldsymbol{r}', t') \\
+ &\int_\Omega \mathrm{d}^3\boldsymbol{r}' \rho \left(\frac{\partial \boldsymbol{u}(\boldsymbol{r}', t')}{\partial t'} \cdot \boldsymbol{G}^T(\boldsymbol{r}, t|\boldsymbol{r}', t') - \boldsymbol{u}(\boldsymbol{r}', t') \cdot \frac{\boldsymbol{G}^T(\boldsymbol{r}, t|\boldsymbol{r}', t')}{\partial t'} \right) \Bigg|_{t'=0}, \qquad (4.280)
\end{aligned}
$$

where we used that $\boldsymbol{A} \cdot \boldsymbol{a} = \boldsymbol{a} \cdot \boldsymbol{A}^T$ for any vector $\boldsymbol{a}$ and second-order tensor $\boldsymbol{A}$. So given a Green's tensor for a particular problem, that is required to satisfy homogeneous boundary conditions as just stated earlier, we can calculate the displacements through time and space by performing the above integrals over $\boldsymbol{r}'$ and t'. The Green's tensor is what propagates the three types of sourcing from a source point $\boldsymbol{r}'$ to an observation point $\boldsymbol{r}$. Finding the Green's tensor for complex boundaries and heterogeneous bodies is almost always a numerical exercise. However, for scenarios involving a homogeneous body with simple boundaries, analytical solutions can sometimes be found as will be seen in Chapter 12.

4.9.4 Representing Seismic Sources as Body-Force Equivalents

In the reciprocity section (Section 4.9.2), we considered the representation of a concentrated and directed seismic point source using an expression for the force density of the form $\boldsymbol{F}(\boldsymbol{r}, t) = \boldsymbol{N}(t)\delta(\boldsymbol{r})$, where $\boldsymbol{N}(t)$ is the directed force measured in N and $\delta(\boldsymbol{r}) = \delta(x)\delta(y)\delta(z)$ is the 3D Dirac delta having units of m^{-3}. As a special example, we considered $\boldsymbol{N}(t) = \boldsymbol{P}\delta(t)$ with $\boldsymbol{P}$ the impulse vector having units of N s. However, this type of point source is neither the only way nor the most common way that seismic waves are generated within a body.

In solids such as the Earth, a common source of seismic waves is a break that occurs across a plane that we call a fault or fracture plane. For example, when a shear-stress threshold is reached on a given fault plane, the material will suddenly slip by displacing parallel to the surface on one side of the plane and displacing in the opposite direction on the other side. Such a lateral displacement discontinuity across a fault generates seismic waves. A standard criterion for the critical shear traction at which slip is initiated in the direction $\hat{\boldsymbol{t}}$ parallel to a fault having normal $\boldsymbol{n}$ is called *Coulomb's friction law*, which can be stated

$$\boldsymbol{n} \cdot \boldsymbol{\tau} \cdot \hat{\boldsymbol{t}} \geq -\mu\, \boldsymbol{n} \cdot \boldsymbol{\tau} \cdot \boldsymbol{n}, \tag{4.281}$$

where the dimensionless coefficient μ is called the *friction coefficient* which is positive; although μ is most commonly less than 1, there are materials such as rubber on rubber that have $\mu > 1$. Recall that the traction component normal to the surface $\boldsymbol{n} \cdot \boldsymbol{\tau} \cdot \boldsymbol{n}$ is defined to be positive when in tension and negative when in compression, which explains the negative sign in Coulomb's law. To get a tangible feel for Coulomb's law, lightly place your hands together and then slide them relative to each other. Now push your hands together harder (i.e., increase the normal compressive traction $-\boldsymbol{n} \cdot \boldsymbol{\tau} \cdot \boldsymbol{n}$). To slide your hands in this case, you now have to use more muscular force to increase the shear traction $\boldsymbol{n} \cdot \boldsymbol{\tau} \cdot \hat{\boldsymbol{t}}$ on your hand surfaces (the fault) prior to getting slip. At the fine scale of the roughness of your skin, the enhanced normal traction locks the two rough surfaces together more tightly, requiring enhanced shear stress to initiate slip. An alternative type of seismic source is that at a critical tension stress given by $\boldsymbol{n} \cdot \boldsymbol{\tau} \cdot \boldsymbol{n} > \tau_c$, where τ_c is the positive *tensile strength* of the fracture interface, a fracture plane pulls apart in tension and in a direction normal to the fracture plane, which creates a normal component of displacement discontinuity across the plane and associated seismic waves. Still another type of source for seismic waves is an explosion or a vibrating piezoelectric body.

We can use the representation theorem of the previous section to show explicitly how to represent these various types of sources of elastodynamic waves as an effective or "equivalent" body force $\boldsymbol{F}_e$ to be used in the elastodynamic wave equation. We will show that a body-force equivalent of the form $\boldsymbol{F}_e = \boldsymbol{N}(t)\delta(\boldsymbol{r})$ corresponds to a source in which the stress is discontinuous and the displacement continuous across a surface whose size is small enough relative to the generated wavelengths that it is modeled as a point. A body-force equivalent of the form $\boldsymbol{F}_e = \boldsymbol{M}(t) \cdot \nabla\delta(\boldsymbol{r})$, where the second-order tensor $\boldsymbol{M}(t)$ is called the *moment tensor*, corresponds to a source in which the displacement is discontinuous and the stress is continuous across a surface or "fault" whose size is again small relative to wavelengths. Scenarios where the body-force equivalent is characterized by a directed impulse $\boldsymbol{N}(t)$ include impacts such as a hammer hitting a plate on the surface of the Earth. Scenarios where the body-force equivalent is characterized by a moment tensor $\boldsymbol{M}(t)$ include displacement discontinuities across small failure or slip or fault surfaces as well as explosions.

The Body-Force Equivalent for Slip Events and Opening Events

Consider a surface Σ_u across which slip (a displacement discontinuity) occurs with traction continuous across Σ_u. We enclose this slip surface within a small disc-shaped region $\Omega_{\epsilon u}$ of half-width ϵ in the limit as $\epsilon \to 0$ as shown in Fig. 4.13. Applying the displacement representation theorem of Eq. (4.280) to the source region $\Omega_{\epsilon u}$ and assuming that just prior to the slip event the displacements and particle velocities were zero throughout the medium, we have

$$\begin{aligned} \boldsymbol{u}(\boldsymbol{r}, t) = & \int_0^t \mathrm{d}t' \int_{\Omega_{\epsilon u}} \mathrm{d}^3\boldsymbol{r}' \, \boldsymbol{F}(\boldsymbol{r}', t') \cdot \boldsymbol{G}^T(\boldsymbol{r}, t \,|\boldsymbol{r}', t') \\ & + \int_0^t \mathrm{d}t' \int_{\partial\Omega_{\epsilon u}} \mathrm{d}^2\boldsymbol{r}' \, \boldsymbol{n}(\boldsymbol{r}') \cdot \boldsymbol{\tau}(\boldsymbol{r}', t') \cdot \boldsymbol{G}^T(\boldsymbol{r}, t \,|\boldsymbol{r}', t') \\ & - \int_0^t \mathrm{d}t' \int_{\partial\Omega_{\epsilon u}} \mathrm{d}^2\boldsymbol{r}' \, \boldsymbol{u}(\boldsymbol{r}', t') \, \boldsymbol{n}(\boldsymbol{r}') : {}_4\boldsymbol{C}(\boldsymbol{r}') : \nabla' \boldsymbol{G}^T(\boldsymbol{r}, t \,|\boldsymbol{r}', t'). \end{aligned} \tag{4.282}$$

In the limit as $\epsilon \to 0$, the surface integral around the circumferential face of $\partial\Omega_{\epsilon u}$ (the "ribbon" surface wrapping around the radial extent of the disc shown in Fig. 4.13) vanishes leaving only the contributions across the two flat faces $\partial\Omega_{u\pm}$ that bound Σ_u above and below. Let $[\boldsymbol{u}] = \boldsymbol{u}_+ - \boldsymbol{u}_-$ be the jump (slip) in the displacement vector across the surface Σ_u and let the jump in traction be zero $[\boldsymbol{n} \cdot \boldsymbol{\tau}] = 0$ across Σ_u. In the limit as $\epsilon \to 0$ so that $\partial\Omega_{u+} \to \Sigma_u$ from above with a normal $+\boldsymbol{n}$ and $\partial\Omega_{u-} \to \Sigma_u$ from below with a normal $-\boldsymbol{n}$, Eq. (4.282) becomes

$$\begin{aligned} \boldsymbol{u}(\boldsymbol{r}, t) = & \lim_{\epsilon \to 0} \int_0^t \mathrm{d}t' \int_{\Omega_{\epsilon u}} \mathrm{d}^3\boldsymbol{r}' \, \boldsymbol{F}(\boldsymbol{r}', t') \cdot \boldsymbol{G}^T(\boldsymbol{r}, t \,|\boldsymbol{r}', t') \\ & - \int_0^t \mathrm{d}t' \int_{\Sigma_u} \mathrm{d}^2\boldsymbol{r}_{su} \, [\boldsymbol{u}(\boldsymbol{r}_{su}, t')] \, \boldsymbol{n}(\boldsymbol{r}_{su}) : {}_4\boldsymbol{C}(\boldsymbol{r}_{su}) : \nabla' \boldsymbol{G}^T(\boldsymbol{r}, t \,|\boldsymbol{r}', t')\big|_{\boldsymbol{r}' = \boldsymbol{r}_{su}}, \end{aligned} \tag{4.283}$$

where $\boldsymbol{r}_{su}$ are points belonging to the slip surface Σ_u. We now use the 3D Dirac delta $\delta\left(\boldsymbol{r}' - \boldsymbol{r}_{su}\right)$ to write the gradient of the Green's tensor evaluated on the slip surface as a

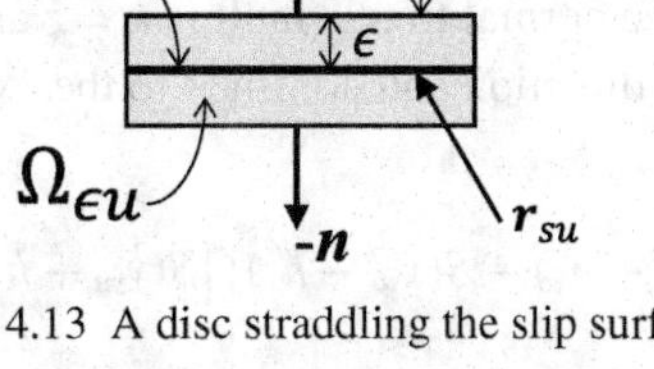

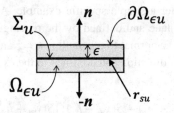

Figure 4.13 A disc straddling the slip surface Σ_u.

volume integral. So we integrate $\nabla'(\delta \boldsymbol{G}^T) = (\nabla'\delta)\,\boldsymbol{G}^T + \delta\,(\nabla'\boldsymbol{G}^T)$ over $\Omega_{\epsilon u}$, use the fundamental theorem of 3D calculus, use that $\delta\,(\boldsymbol{r}' - \boldsymbol{r}_{su}) = 0$ for $\boldsymbol{r}'$ on the bounding surface $\partial\Omega_{\epsilon u}$, and use the sifting property of the 3D Dirac to obtain

$$\nabla'\boldsymbol{G}^T(\boldsymbol{r}, t\,|\boldsymbol{r}', t')\big|_{\boldsymbol{r}'=\boldsymbol{r}_{su}} = -\int_{\Omega_{\epsilon u}} \mathrm{d}^3\boldsymbol{r}'\left[\nabla'\delta(\boldsymbol{r}' - \boldsymbol{r}_{su})\right]\boldsymbol{G}^T(\boldsymbol{r}, t\,|\boldsymbol{r}', t'). \tag{4.284}$$

Using this in the representation for $\boldsymbol{u}$ then gives

$$\begin{aligned}\boldsymbol{u}(\boldsymbol{r}, t) = \lim_{\epsilon\to 0}\int_0^t \mathrm{d}t'\int_{\Omega_{\epsilon u}} \mathrm{d}^3\boldsymbol{r}'\,\boldsymbol{F}(\boldsymbol{r}', t')\cdot\boldsymbol{G}^T(\boldsymbol{r}, t\,|\boldsymbol{r}', t')\\ + \int_0^t \mathrm{d}t'\int_{\Omega_{\epsilon u}} \mathrm{d}^3\boldsymbol{r}'\left(\int_{\Sigma_u} \mathrm{d}^2\boldsymbol{r}_{su}\,[\boldsymbol{u}(\boldsymbol{r}_{su}, t')]\,\boldsymbol{n}(\boldsymbol{r}_{su}) : {}_4\boldsymbol{C}(\boldsymbol{r}_{su})\cdot\nabla'\delta(\boldsymbol{r}' - \boldsymbol{r}_{su})\right)\cdot\boldsymbol{G}^T(\boldsymbol{r}, t\,|\boldsymbol{r}', t').\end{aligned} \tag{4.285}$$

We thus see that an equivalent way to model the effect of slip (or indeed any displacement discontinuity) on a surface is to add to any true body forces present $\boldsymbol{F}$ an effective body-force density $\boldsymbol{F}_e$ given by

$$\boxed{\boldsymbol{F}_e(\boldsymbol{r}, t) = \int_{\Sigma_u} \mathrm{d}^2\boldsymbol{r}_{su}\,[\boldsymbol{u}(\boldsymbol{r}_{su}, t)]\,\boldsymbol{n}(\boldsymbol{r}_{su}) : {}_4\boldsymbol{C}(\boldsymbol{r}_{su})\cdot\nabla\delta(\boldsymbol{r} - \boldsymbol{r}_{su}),} \tag{4.286}$$

which is also called a *body-force equivalent*. The term $[\boldsymbol{u}(\boldsymbol{r}_{su}, t)]\,\boldsymbol{n}(\boldsymbol{r}_{su}) : {}_4\boldsymbol{C}(\boldsymbol{r}_{su})$ that is distributed over the fault surface Σ_u is a second-order tensor that is sometimes called the *moment-density tensor*. It is symmetric because the stiffness tensor is symmetric in the third and fourth base vectors, i.e., $C_{ijkl} = C_{ijlk}$. Pulling out the ∇ operator that only acts on the $\boldsymbol{r}$ in the 3D Dirac delta, we can write

$$\boldsymbol{F}_e(\boldsymbol{r}, t) = \nabla\cdot\left[\int_{\Sigma_u} \mathrm{d}^2\boldsymbol{r}_{su}\,[\boldsymbol{u}(\boldsymbol{r}_{su}, t)]\,\boldsymbol{n}(\boldsymbol{r}_{su}) : {}_4\boldsymbol{C}(\boldsymbol{r}_{su})\delta(\boldsymbol{r} - \boldsymbol{r}_{su})\right], \tag{4.287}$$

where the second-order tensor inside of the square brackets has units of stress. Note that the operation of "pulling out the ∇" involved using the identity $\boldsymbol{A}^T\cdot\nabla\alpha = \nabla\cdot(\boldsymbol{A}\alpha) - \alpha\nabla\cdot\boldsymbol{A}$ for any second-order tensor field $\boldsymbol{A}(\boldsymbol{r})$ and scalar field $\alpha(\boldsymbol{r})$. The second-order tensor in Eq. (4.286) is the symmetric moment-density tensor that is independent of $\boldsymbol{r}$, which is how we get from Eqs (4.286) to (4.287).

To see how to use Eq. (4.287) to obtain an explicit expression for the body-force equivalent $\boldsymbol{F}_e(\boldsymbol{r}, t)$ of slip on a finite-sized fault (i.e., a fault that is not small relative to wavelengths), let's consider a simple specific example. Assume the fault surface Σ_u is confined to the (x_{su}, y_{su}) plane and defined by the rectangular domain $-h_x \le x_{su} \le h_x$, $-h_y \le y_{su} \le h_y$ and $z_{su} = 0$. The normal to this fault is $\boldsymbol{n} = \hat{\boldsymbol{z}}$ and the slip event is a uniform break with slip purely in the x direction and confined to the extent of this rectangular fault, that is,

$$\left[\boldsymbol{u}(x_{su}, y_{su}, t)\right] = s(t)\,[S(x_{su} + h_x) - S(x_{su} - h_x)]\left[S(y_{su} + h_y) - S(y_{su} - h_y)\right]\hat{\boldsymbol{x}}, \tag{4.288}$$

where $S(x)$ is the unit step function (defined to be 0 for $x<0$ and to be 1 for $x>0$) and $s(t)$ represents the magnitude and timing of the uniform slip over the rectangular fault. Assume the material to be isotropic so that ${}_4\boldsymbol{C}=\left[K\delta_{ij}\delta_{kl}+G(\delta_{ik}\delta_{jl}+\delta_{il}\delta_{jk}-2\delta_{ij}\delta_{kl}/3)\right]\hat{\boldsymbol{x}}_i\hat{\boldsymbol{x}}_j\hat{\boldsymbol{x}}_k\hat{\boldsymbol{x}}_l$, which gives

$$\hat{\boldsymbol{x}}\hat{\boldsymbol{z}}:{}_4\boldsymbol{C}=G\left(\hat{\boldsymbol{x}}\hat{\boldsymbol{z}}+\hat{\boldsymbol{z}}\hat{\boldsymbol{x}}\right). \tag{4.289}$$

Carrying out the integral over the fault plane in Eq. (4.287) using the sifting property of the 3D Dirac delta $\delta(\boldsymbol{r}-\boldsymbol{r}_{su})=\delta(x-x_{su})\delta(y-y_{su})\delta(z)$ and assuming the shear modulus is uniform, we obtain

$$\boldsymbol{F}_e(\boldsymbol{r},t)=s(t)G\nabla\cdot\left\{\left(\hat{\boldsymbol{x}}\hat{\boldsymbol{z}}+\hat{\boldsymbol{z}}\hat{\boldsymbol{x}}\right)\left[S(x+h_x)-S(x-h_x)\right]\left[S(y+h_y)-S(y-h_y)\right]\delta(z)\right\}. \tag{4.290}$$

Distributing the spatial derivatives then gives the desired result

$$\begin{aligned}\boldsymbol{F}_e(\boldsymbol{r},t)=s(t)G\Bigg\{&\left[S(x+h_x)-S(x-h_x)\right]\left[S(y+h_y)-S(y-h_y)\right]\frac{\mathrm{d}\delta(z)}{\mathrm{d}z}\hat{\boldsymbol{x}}\\&+\left[\delta(x+h_x)-\delta(x-h_x)\right]\left[S(y+h_y)-S(y-h_y)\right]\delta(z)\hat{\boldsymbol{z}}\Bigg\}.\end{aligned} \tag{4.291}$$

The first term in the $\hat{\boldsymbol{x}}$ direction corresponds to a uniform distribution of force dipoles over the entire fault plane, with the z derivative of the Dirac producing a positive force in the $\hat{\boldsymbol{x}}$ direction immediately on one side of the fault surface and a negative force in the $-\hat{\boldsymbol{x}}$ direction immediately on the other side. The difference in the Dirac terms in the second term is due to the abruptness of how the slip event starts at $x=-h_x$ and ends at $x=h_x$. There is created a lobe of compression adjacent to a lobe of dilation where the fault starts and ends, which produces a line of force along the fault limits at $x=\pm h_x$ that is in the $\hat{\boldsymbol{z}}$ direction at $x=-h_x$ and in the $-\hat{\boldsymbol{z}}$ direction at $x=+h_x$. This uniform slip event is on a rectangular surface of finite area $A_s=4h_xh_y$ that need not be small relative to the wavelengths that are generated. Using the $\boldsymbol{F}_e$ of Eq. (4.291) for this finite-sized slip event in the elastodynamic wave equation $\rho\partial^2\boldsymbol{u}/\partial t^2=\nabla\cdot({}_4\boldsymbol{C}:\nabla\boldsymbol{u})+\boldsymbol{F}_e$ allows us to determine the displacement field $\boldsymbol{u}(\boldsymbol{r},t)$ created by the rupture.

If we now take the limit where the rectangular fault dimensions h_x and h_y are much smaller than the wavelengths, we can use $\mathrm{d}S(x)/\mathrm{d}x=\delta(x)$ to write

$$S(x+h_x)-S(x-h_x)=\left(S(x)+h_x\frac{\mathrm{d}S(x)}{\mathrm{d}x}+\ldots\right)-\left(S(x)-h_x\frac{\mathrm{d}S(x)}{\mathrm{d}x}+\ldots\right) \tag{4.292}$$

$$\rightarrow 2h_x\delta(x) \tag{4.293}$$

and similarly

$$S(y+h_y)-S(y-h_y)\rightarrow 2h_y\delta(y). \tag{4.294}$$

Further, the difference in the Dirac functions can be expressed as the derivative of the Dirac

$$\frac{\delta(x+h_x)-\delta(x-h_x)}{2h_x}\rightarrow\frac{\mathrm{d}\delta(x)}{\mathrm{d}x}. \tag{4.295}$$

Writing again $A_s = 4h_x h_y$, Eq. (4.291) becomes

$$\boldsymbol{F}_e(x, y, z, t) = A_s s(t) G \left[\delta(x)\delta(y)\frac{\mathrm{d}\delta(z)}{\mathrm{d}z}\hat{\boldsymbol{x}} + \frac{\delta(x)}{\mathrm{d}x}\delta(y)\delta(z)\hat{\boldsymbol{z}} \right]. \tag{4.296}$$

Each of the two vector dipoles inside the brackets is called a *couple* and this body-force equivalent for slip on a small surface is therefore called a *double couple*. Intuitively, you may think that only the force dipole in the $\hat{\boldsymbol{x}}$ direction is needed for simulating slip in the $\hat{\boldsymbol{x}}$ direction, but the other force dipole in the $\hat{\boldsymbol{z}}$ direction is equally important and required for the simulation of slip on a fault that is small relative to wavelengths.

We can rewrite Eq. (4.296) in the form

$$\boldsymbol{F}_e(\boldsymbol{r}, t) = \boldsymbol{M}(t) \cdot \nabla\delta(\boldsymbol{r}), \tag{4.297}$$

where the second-order tensor $\boldsymbol{M}(t)$ is called the *moment tensor* and has units of energy (Joules). For our specific example here of slip $s(t)$ in the $\hat{\boldsymbol{x}}$ direction on a small fault having area A_s and normal $\hat{\boldsymbol{z}}$ that is embedded in a uniform isotropic material having a shear modulus G, the moment tensor is

$$\boldsymbol{M}(t) = A_s s(t) G \left(\hat{\boldsymbol{x}}\hat{\boldsymbol{z}} + \hat{\boldsymbol{z}}\hat{\boldsymbol{x}} \right). \tag{4.298}$$

Note that if slip is in the $\hat{\boldsymbol{z}}$ direction on a fault having normal $\hat{\boldsymbol{x}}$, repeating the above steps produces identically the same body-force equivalent. So the waves generated by either type of slip event are identically the same. Using recorded seismic waves to infer the orientation of a small slip surface always has such a 90° ambiguity with regard to the orientation of the slip surface that generated the waves.

So the moment tensor is only defined in the limit that the slip surface (or fault) Σ_u is so small relative to the wavelengths that it may be approximated as a point located either at the origin as we have been assuming above or at some other position $\boldsymbol{r}_s$. From either Eq. (4.286) or Eq. (4.287), the body-force equivalent for a fault so small that it can be approximated as a point is

$$\boxed{\boldsymbol{F}_e(\boldsymbol{r}, t) = \boldsymbol{M}(t) \cdot \nabla\delta(\boldsymbol{r} - \boldsymbol{r}_s) = \nabla \cdot [\boldsymbol{M}(t)\delta(\boldsymbol{r} - \boldsymbol{r}_s)],} \tag{4.299}$$

where the moment tensor corresponding to slip $[\boldsymbol{u}(t)]$ on the "point" fault of area A_s and normal $\boldsymbol{n}$ is

$$\boxed{\boldsymbol{M}(t) = A_s[\boldsymbol{u}(t)]\,\boldsymbol{n} : {}_4\boldsymbol{C}.} \tag{4.300}$$

This result for the moment tensor applies to an anisotropic material as well as for nonslip situations where $[\boldsymbol{u}(t)]$ is in the $\boldsymbol{n}$ direction, which corresponds to a tensile opening event or *tension crack* that again satisfies $[\boldsymbol{n} \cdot \boldsymbol{\tau}] = 0$ across the crack surface and will be considered next. Because the third and fourth base vectors of ${}_4\boldsymbol{C}$ are symmetric, so is the moment tensor $\boldsymbol{M}(t)$.

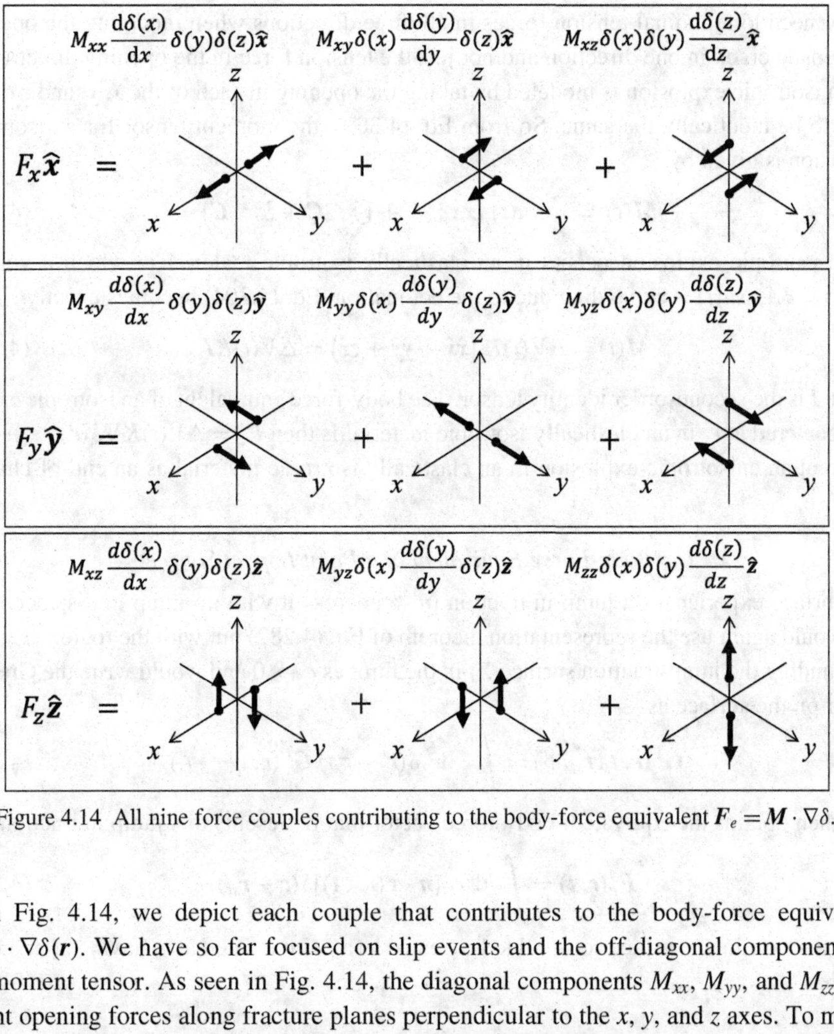

Figure 4.14 All nine force couples contributing to the body-force equivalent $\boldsymbol{F}_e = \boldsymbol{M} \cdot \nabla\delta$.

In Fig. 4.14, we depict each couple that contributes to the body-force equivalent $\boldsymbol{M}(t) \cdot \nabla\delta(\boldsymbol{r})$. We have so far focused on slip events and the off-diagonal components of the moment tensor. As seen in Fig. 4.14, the diagonal components M_{xx}, M_{yy}, and M_{zz} represent opening forces along fracture planes perpendicular to the x, y, and z axes. To model a small tension crack of area A_s that has a normal $\boldsymbol{n} = \hat{z}$ and is opening in the z direction as described by $[\boldsymbol{u}(t)] = \Delta u(t)\hat{z}$, we use Eq. (4.300) for the moment tensor to give

$$\boldsymbol{M}(t) = A_s \Delta u(t)\hat{z}\hat{z} : {}_4\boldsymbol{C}. \tag{4.301}$$

If the material is isotropic, we have

$$\hat{z}\hat{z} : {}_4\boldsymbol{C} = (K + 4G/3)\hat{z}\hat{z} + (K - 2G/3)\left(\hat{x}\hat{x} + \hat{y}\hat{y}\right), \tag{4.302}$$

which results in the body-force equivalent

$$\boldsymbol{F}_e = A_s \Delta u(t) \left[\left(K - \frac{2G}{3}\right)\left(\hat{x}\frac{d\delta(x)}{dx}\delta(y)\delta(z) + \hat{y}\delta(x)\frac{d\delta(y)}{dy}\delta(z)\right) + \left(K + \frac{4G}{3}\right)\hat{z}\delta(x)\delta(y)\frac{d\delta(z)}{dz}\right]. \tag{4.303}$$

So we need longitudinal tension forces in all three directions when modeling the opening of a tensile crack in one direction and not just the tension force in the opening direction.

An isotropic explosion is modeled by taking the opening in each of the x, y, and z directions to be identically the same. So from Eq. (4.300), the moment tensor for an isotropic explosion is given by

$$\boldsymbol{M}(t) = A_s \Delta u(t) \left(\hat{x}\hat{x} : {}_4\boldsymbol{C} + \hat{y}\hat{y} : {}_4\boldsymbol{C} + \hat{z}\hat{z} : {}_4\boldsymbol{C} \right). \tag{4.304}$$

If the isotropic explosion occurs in an elastically isotropic material in which a volume $\Delta V(t) = 3A_s\Delta u(t)$ is expanding due to the detonation, Eq. (4.304) becomes exactly

$$\boldsymbol{M}(t) = \Delta V(t) K \left(\hat{x}\hat{x} + \hat{y}\hat{y} + \hat{z}\hat{z} \right) = \Delta V(t) K \boldsymbol{I}, \tag{4.305}$$

where $\boldsymbol{I}$ is the second-order identity tensor. The body-force equivalent of an isotropic explosion centered at $\boldsymbol{r}_s$ in an elastically isotropic material is then $\boldsymbol{F}_e = \Delta V(t) K \nabla \delta(\boldsymbol{r} - \boldsymbol{r}_s)$. You can treat an anisotropic explosion in an elastically isotropic material as an end-of-chapter exercise.

The Body-Force Equivalent of a Point Impact Source

If a surface experiences a jump in traction $[\boldsymbol{n} \cdot \boldsymbol{\tau}]$ across it with no jump in displacement, one would again use the representation theorem of Eq. (4.282) but with the region $\Omega_{\epsilon t}$ now surrounding the jump-traction surface Σ_t in the limit as $\epsilon \to 0$ and would write the Green's tensor on the surface as

$$\boldsymbol{G}^T(\boldsymbol{r}, t \,|\, \boldsymbol{r}_{st}, t') = \int_{\Omega\epsilon t} \mathrm{d}\boldsymbol{r}' \, \delta(\boldsymbol{r}' - \boldsymbol{r}_{st}) \, \boldsymbol{G}^T(\boldsymbol{r}, t \,|\, \boldsymbol{r}', t'). \tag{4.306}$$

One then obtains the equivalent body-force vector that represents this jump in traction as

$$\boldsymbol{F}_e(\boldsymbol{r}, t) = \int_{\Sigma_t} \mathrm{d}^2\boldsymbol{r}_{st} [\boldsymbol{n} \cdot \boldsymbol{\tau}(\boldsymbol{r}_{st}, t)] \delta(\boldsymbol{r} - \boldsymbol{r}_{st}). \tag{4.307}$$

For a traction-jump surface so small to be treated as a point located at $\boldsymbol{r}_s$ having area A_s one then has

$$\boldsymbol{F}_e(\boldsymbol{r}, t) = A_s [\boldsymbol{n} \cdot \boldsymbol{\tau}(t)] \delta(\boldsymbol{r} - \boldsymbol{r}_s) = \boldsymbol{N}(t) \delta(\boldsymbol{r} - \boldsymbol{r}_s) \tag{4.308}$$

where we have identified the directed point force (measured in N)

$$\boxed{\boldsymbol{N}(t) = A_s [\boldsymbol{n} \cdot \boldsymbol{\tau}(t)].} \tag{4.309}$$

One fairly common seismic source where this description is appropriate is a hammer source that corresponds to a weight falling with a given acceleration and hitting a plate which then imparts a sharp momentum transfer into the body that the plate is attached to. If the plate of area A_s has a normal $\hat{z}$ with $\hat{z}$ in the direction of the weight drop and if the weight of mass m falls (accelerates) from a height h above the plate with an acceleration of ng, where n is the multiplier of the acceleration of gravity $g = 9.8$ m s^{-2} and impacts as a perfect impulse at a time $t = t_0$, a simple calculation gives $[\boldsymbol{n} \cdot \boldsymbol{\tau}(t)] = \hat{z} m \sqrt{ngh}\, \delta(t - t_0)/A_s$, where $\sqrt{ngh}$

is the velocity at impact and $[\boldsymbol{n} \cdot \boldsymbol{\tau}(t)]$ represents the flux of momentum imparted to the body due to the impact. The body-force equivalent to be used in the conservation of linear momentum is then $\boldsymbol{F}_e = m\sqrt{ngh}\,\delta(t - t_0)\delta(\boldsymbol{r} - \boldsymbol{r}_s)\,\hat{\boldsymbol{z}}$. Instead of a Dirac impulse in time, one could use another time function $s(t)$ that obeys $\int_{-\infty}^{\infty} s(t)\,\mathrm{d}t = 1$ but that is less sharp than the Dirac.

4.10 Elastostatics

4.10.1 Linear Elastostatics

Everything developed in the above section on the linear-seismic (i.e., linear-elastodynamic) displacement theorems applies to linear elastostatics, for which the forces are in balance and the material is not accelerating anywhere. Linear elastostatic problems involve finding solutions of

$$0 = \nabla \cdot \boldsymbol{\tau}(\boldsymbol{r}) + \boldsymbol{F}(\boldsymbol{r}) \quad \text{with} \quad \boldsymbol{\tau} = {}_4\boldsymbol{C} : \nabla \boldsymbol{u} \tag{4.310}$$

throughout a domain Ω that satisfy given values of either the traction $\boldsymbol{n} \cdot \boldsymbol{\tau}$ or displacement $\boldsymbol{u}$ on the boundary $\partial\Omega$. Numerically, these equations can be solved as two coupled first-order PDEs to give, simultaneously, the stress components and displacement components (nine unknowns in 3D). Alternatively, Hooke's law can be placed in the force balance to create a second-order PDE for the displacements. Once the displacements are determined, the stresses can be determined through Hooke's law. In Chapter 12, we will obtain the linear elastostatic point-source response, or Green's tensor, for a homogeneous isotropic wholespace.

4.10.2 Nonlinear Elastostatics

For elastostatic problems that are not linear in the applied displacements (or applied tractions or applied force densities) and that necessarily involve significant amounts of deformation, numerical solutions of the pertinent governing equations are required in general. Even posing the governing equations for the nonlinear elastostatic problem is not trivial. In the way we have obtained the rules of nonlinear elastic response in this chapter, the time-varying fields throughout some body of interest are the velocities $\boldsymbol{v}(\boldsymbol{r}, t)$, stress tensor $\boldsymbol{\tau}(\boldsymbol{r}, t)$, mass density $\rho(\boldsymbol{r}, t)$, and, possibly, elastic stiffnesses ${}_4\boldsymbol{C}(\boldsymbol{r}, t)$. If our problem is nonlinear elastodynamic wave propagation inside of a body with given boundary conditions and a given compact source for the seismic waves expressed as an effective body force, it is clear that our derived governing equations as given in Eqs (4.40)–(4.42) along with Eq. (4.45) for how the stiffnesses change, if they change, during wave propagation, are naturally suited to modeling this elastodynamic problem. To be sure, the numerical details of solving such nonlinear wave response are challenging, but it is clear what equations need to be solved for given boundary and initial conditions.

Let's now imagine a particular elastostatic problem in which a given (known) amount of displacement is applied to the exterior surface of some body so that the body is significantly, but elastically, deformed in the final static state of interest. Starting with the given undeformed body and given the displacement vector that takes each point of the undeformed external surface to the final deformed surface, we want to know how much traction is present at each point of the deformed surface and throughout the entire body to achieve the given surface displacement. How do we use Eqs (4.40)–(4.42) to solve this elastostatic problem?

There is no one single answer but one approach is to break the deformation into N increments. Given the total displacement $\boldsymbol{u}(\boldsymbol{r})$ from each point of the given initial surface $\partial\Omega_0$, the surface displacement in the nth increment is, say, $\delta\boldsymbol{u}_n(\boldsymbol{r}) = \boldsymbol{u}(\boldsymbol{r})/N$ as applied to the external surface $\partial\Omega_{n-1}$ of the body after $n-1$ increments. At increment n, one applies a velocity to the surface $\partial\Omega_{n-1}$ for a time T. This velocity on $\partial\Omega_{n-1}$ is, say, $\delta\boldsymbol{v}_n(\boldsymbol{r}, t) = (\pi/T)\delta\boldsymbol{u}_n(\boldsymbol{r})\sin^2(\pi t/T)$, where the time T is taken to be sufficiently long that the implicit elastodynamic wavelengths generated by $\delta\boldsymbol{v}_n(\boldsymbol{r}, t)$ on the surface $\partial\Omega_{n-1}$ are much larger than the body itself. This velocity applied to the surface smoothly increases from zero at $t = 0$, hits a maximum at $t = T/2$, and then smoothly returns to zero at $t = T$, resulting in a total displacement of each point of the surface given by $\int_0^T \delta\boldsymbol{v}_n(\boldsymbol{r}, t)\,\mathrm{d}t = \delta\boldsymbol{u}_n(\boldsymbol{r})$ as required. The density and stress-tensor fields in each deformation increment are written

$$\rho_n(\boldsymbol{r}, t) = \rho_n(\boldsymbol{r}, 0) + \delta\rho_n(\boldsymbol{r}, t) \tag{4.311}$$

$$\boldsymbol{\tau}_n(\boldsymbol{r}, t) = \boldsymbol{\tau}_n(\boldsymbol{r}, 0) + \delta\boldsymbol{\tau}_n(\boldsymbol{r}, t), \tag{4.312}$$

where $\rho_n(\boldsymbol{r}, 0) = \rho_{n-1}(\boldsymbol{r}, T)$ and $\boldsymbol{\tau}_n(\boldsymbol{r}, 0) = \boldsymbol{\tau}_{n-1}(\boldsymbol{r}, T)$, that is, the starting values of these fields at each deformation increment come from the final values of these fields at the previous increment after some rediscretization discussed below. The process is initialized with $\rho_0(\boldsymbol{r})$ and $\boldsymbol{\tau}_0(\boldsymbol{r})$ being known throughout the initial undeformed body Ω_0.

One then solves Eqs (4.40)–(4.42) in increment n throughout the body Ω_{n-1}. Such update equations are written in the form

$$\frac{\partial\delta\rho_n}{\partial t} = -\nabla\cdot(\delta\boldsymbol{v}_n\rho_n) \tag{4.313}$$

$$\frac{\partial\delta\boldsymbol{v}_n}{\partial t} = -\frac{1}{\rho_n}\nabla\cdot(\delta\boldsymbol{v}_n\delta\boldsymbol{v}_n\rho_n) + \frac{1}{\rho_n}\nabla\cdot\delta\boldsymbol{\tau}_n, \tag{4.314}$$

$$\frac{\partial\delta\boldsymbol{\tau}_n}{\partial t} = -\nabla\cdot(\delta\boldsymbol{v}_n\boldsymbol{\tau}_n) + {}_4\boldsymbol{C} : \nabla\delta\boldsymbol{v}_n \tag{4.315}$$

and possibly Eq. (4.45) or Eqs (4.47) and (4.48) if the elastic stiffnesses are allowed to change. The displacement field in each deformation increment comes from the definition $\delta\boldsymbol{v}_n \,\widehat{=}\, \mathrm{d}\delta\boldsymbol{u}_n/\mathrm{d}t = \partial\delta\boldsymbol{u}_n/\partial t + \delta\boldsymbol{v}_n\cdot\nabla\delta\boldsymbol{u}_n$ that yields the update equation

$$\frac{\partial\delta\boldsymbol{u}_n}{\partial t} = \delta\boldsymbol{v}_n\cdot(\boldsymbol{I} - \nabla\delta\boldsymbol{u}_n)\,. \tag{4.316}$$

In writing the above, we assume that at the start and end of each deformation increment, the fields are in static equilibrium so that $\nabla\cdot\boldsymbol{\tau}_n(\boldsymbol{r}, 0) + \boldsymbol{F}_n = 0$ and there is no acceleration occurring. One can solve these first-order PDEs efficiently using explicit finite-difference approximations in time and space, where the updates through time of each field at point $\boldsymbol{r}$ are given only in terms of the spatial derivatives at that point. One must be careful to use a sufficiently small time increment that the updates are stable.

At the end of each deformation increment n, one moves each grid point $\boldsymbol{r}_{n-1}$ throughout the domain Ω_{n-1} to a new position $\boldsymbol{r}_n = \boldsymbol{r}_{n-1} + \delta\boldsymbol{u}_n(\boldsymbol{r}_{n-1}, T)$ and reassigns the field values from the end of the nth increment to the start of the $(n+1)$th increment according to the mapping $\rho_{n+1}(\boldsymbol{r}_n, 0) = \rho_n(\boldsymbol{r}_{n-1}, T)$ and $\boldsymbol{\tau}_{n+1}(\boldsymbol{r}_n, 0) = \boldsymbol{\tau}_n(\boldsymbol{r}_{n-1}, T)$ and equivalently for the stiffness field if the stiffnesses are varying. The new collection of grid points $\boldsymbol{r}_n$ so created that make up the body Ω_n after the nth deformation increment is typically an irregular array of points. One likely should interpolate the fields given on this irregular grid onto a regular grid for solving the fields in the $(n+1)$th increment of deformation. By taking the number of deformation increments N large enough that the displacements $\delta\boldsymbol{u}_n$ at each increment are smaller than the grid spacing, tangling of the grid is avoided at each increment and the reinterpolation onto a regular grid is unambiguous.

The scheme sketched here is only meant to show how Eqs (4.40) and (4.42) can be used to solve a nonlinear elastostatics problem. Another problem that can be treated in this manner would be: given an elastically deformed body with given tractions $\boldsymbol{n} \cdot \boldsymbol{\tau}_0(\boldsymbol{r})$ on the surface, what is the size and shape of the body in the undeformed state where the applied tractions are zero? One would begin with a discretized grid in the deformed body and in each deformation increment allow the applied traction boundary condition to smoothly decrease by an amount $-\boldsymbol{n} \cdot \boldsymbol{\tau}_0(\boldsymbol{r})/N$ over a time T so that at the end of N increments the body surface is not subjected to any applied stress. The calculation of the displacement field and the rediscretization of the domain after each increment would proceed as described earlier. The final collection of grid points $\boldsymbol{r}_N = \boldsymbol{r}_{N-1} + \delta\boldsymbol{u}_N(\boldsymbol{r}_{N-1}, T)$ would give the size and shape of the undeformed body.

4.11 Overall Summary

We have developed the rules of elasticity using a novel approach that begins with a statement for the rate that work is performed on a mass element that is subjected to stress from a surrounding environment. From this one statement of the rate at which internal elastic energy is reversibly stored in a sample, we obtain the measures of strain rate and strain valid for any level of strain as well as the nonlinear form of Hooke's law that is also valid for any level of strain and a rule for how the elastic moduli change due to changing strain levels. We showed that the macroscopic rules of nonlinear elastic response as given in Section 4.1.5 involve a material velocity field that is the center-of-mass velocity of the molecules contained in each mass element. We also showed that the macroscopic gradient of the center-of-mass velocity field is the same as the macroscopic gradient of the volume-averaged velocity field. This thermodynamic approach for deriving the constitutive laws of elasticity theory is the same that we take for deriving any of the constitutive laws controlling reversible response across the breadth of physics and yields exact results without the use of postulates.

With this general foundation in nonlinear elastodynamics established, the chapter then focuses on linearized forms of the governing equations, establishing the basics of elastodynamic plane-wave response and general theorems about the creation of seismic wavefields, including how to represent various sources of seismic waves as effective body forces, called body-force equivalents. We will return to spherical elastodynamic response emanating from point sources in Chapter 12. The reader is encouraged to work through the exercises that follow to develop intuition and familiarity with the ideas and equations of elasticity theory.

4.12 Exercises

1. Using the conditions for defining the so-called bar moduli as depicted in Fig. 4.15 where $\tau_{zz} > 0$ and all other stress-tensor components are zero (i.e., $\tau_{xx} = \tau_{yy} = \tau_{xz} = \tau_{xy} = \tau_{yz} = 0$), imagine that we measure both $e_{xx} = \partial u_x/\partial x = \partial u_y/\partial y = e_{yy}$ and $e_{zz} = \partial u_z/\partial z$ and obtain the "bar moduli"

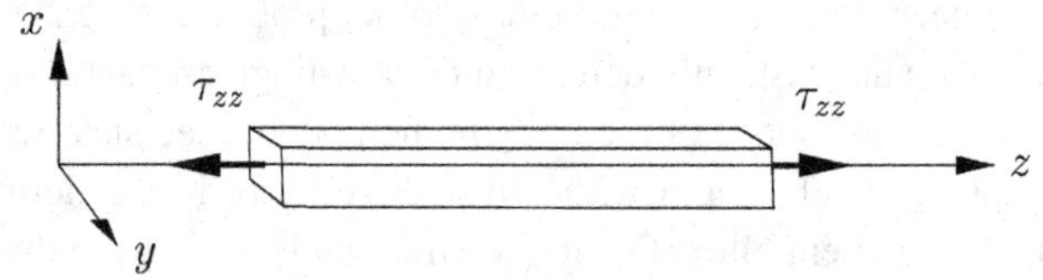

Figure 4.15 The bar moduli E and ν are defined by applying axial stress τ_{zz} to the ends of a long bar under the condition that no stress is applied to the lateral sides of the long bar.

$$E \mathrel{\widehat{=}} \text{Young's Modulus} = \frac{\tau_{zz}}{e_{zz}}$$

$$\nu \mathrel{\widehat{=}} \text{Poisson's Ratio} = -\frac{e_{xx}}{e_{zz}}.$$

Using any isotropic form of Hooke's law that you find convenient, prove that E and ν are given in terms of K and G as

$$\boxed{E = \frac{9GK}{3K+G} \quad \text{and} \quad \nu = \frac{1}{2}\left(\frac{3K-2G}{3K+G}\right).}$$

2. Insert the isotropic form for the stiffness components

$$C_{ijkl} = K\delta_{ij}\delta_{kl} + G\left(\delta_{ik}\delta_{jl} + \delta_{il}\delta_{jk} - \frac{2}{3}\delta_{ij}\delta_{kl}\right)$$

into Hooke's Law $\tau_{ij} = C_{ijkl}e_{kl}$ and obtain the relation between the stress components and the strain components

$$\boxed{\tau_{ij} = Ke_{kk}\delta_{ij} + 2G\left(e_{ij} - \frac{1}{3}e_{kk}\delta_{ij}\right).}$$

Recall that $\delta_{ij} = 0$ if $i \neq j$ and $\delta_{ij} = 1$ if $i = j$.

3. For the isotropic form of Hooke's Law

$$\boldsymbol{\tau} = K\nabla \cdot \boldsymbol{u}\boldsymbol{I} + G\left(\nabla\boldsymbol{u} + (\nabla\boldsymbol{u})^T - \frac{2}{3}\nabla \cdot \boldsymbol{u}\boldsymbol{I}\right)$$

show that in a medium where K and G are uniform spatial constants we have

$$\boxed{\nabla \cdot \boldsymbol{\tau} = \left(K + \frac{G}{3}\right)\nabla\nabla \cdot \boldsymbol{u} + G\nabla^2\boldsymbol{u}.}$$

4. For a uniaxial compression experiment in which, by definition, $e_{xx} = e_{yy} = 0$ (along with all off-diagonal components of $\boldsymbol{e}$) but $e_{zz} \neq 0$ (i.e., no movement at all in the x and y directions, only movement in the z direction), show that

$$\boxed{H \text{ (the uniaxial-strain modulus)} \;\hat{=}\; \frac{\tau_{zz}}{e_{zz}} = K + \frac{4G}{3}.}$$

For such uniaxial deformation, are τ_{xx} and τ_{yy} different from zero and, if so, what are τ_{xx}/τ_{zz} and τ_{yy}/τ_{zz}? Make a sketch of how a cube (square in 2D) changes shape during P-wave propagation in the z direction.

5. For the uniform and isotropic flat Earth model present in Fig. 4.16, assume that gravity $g_o\hat{z}$ alone is compressing the Earth normal to the plane surface $z = 0$ that is a free surface with zero traction (i.e., you can assume atmospheric pressure is zero).

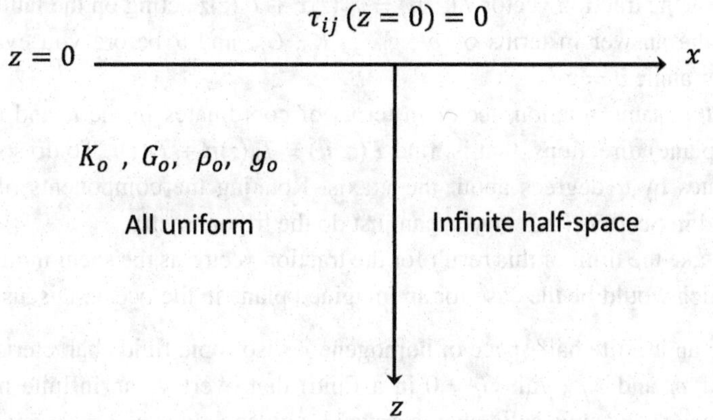

Figure 4.16 A flat Earth model with a surface $z = 0$ having zero traction and with all stress variation with depth z due to gravity alone.

Determine all six components of the stress tensor as a function of depth: $\tau_{xx}(z)$, $\tau_{yy}(z)$, $\tau_{zz}(z)$, $\tau_{xy}(z)$, $\tau_{xz}(z)$, and $\tau_{yz}(z)$. Similarly, determine all six components of the strain tensor as a function of depth $e_{ij}(z)$. HINT: Use the three components of the vector equation $\nabla \cdot \boldsymbol{\tau} = \rho_o g_o \hat{z}$ along with your ideas for whether there can be lateral variations (nonzero spatial derivatives with respect to x and y) of the stress or displacement/strain components in this homogeneous isotropic half space forced only in the z direction. You will also need to use an isotropic form of Hooke's Law. Your answer will be expressed only in terms of ρ_o, g_o, z, K_o, and G_o.

6. In the same homogeneous and isotropic Earth model of the previous exercise, determine the traction vector $\boldsymbol{T}(z, \boldsymbol{n}) = \boldsymbol{n} \cdot \boldsymbol{\tau}$ acting on a potential slip plane (fault) at depth z that is at an angle θ from the $\hat{z}$ axis and that has a unit normal $\boldsymbol{n}$ defined to be positive when its z-component is directed toward the surface. This is depicted in Fig. 4.17.

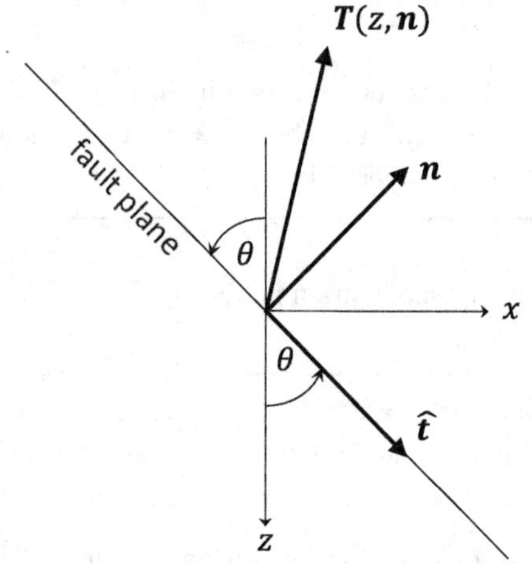

Figure 4.17 Details of a possible slip surface that has a traction vector acting on it within the flat Earth model of Exercise 6.

Express the traction vector $\boldsymbol{T}(z, \boldsymbol{n}) = T_x(z)\hat{\boldsymbol{x}} + T_z(z)\hat{\boldsymbol{z}}$ acting on the fault plane. First express the answer in terms of ρ_o, g_o, z, K_o, G_o, and θ before you evaluate at the particular angle $\theta = \pi/4$.

Find this same traction vector in terms of coordinates in the $\boldsymbol{n}$ and $\hat{\boldsymbol{t}}$ (tangent to the slip plane) directions, that is, find $\boldsymbol{T}(z, \boldsymbol{n}) = T_n(z)\boldsymbol{n} + T_t(z)\hat{\boldsymbol{t}}$. To do so, rotate your coordinates by θ degrees about the y axis. Rotating the components of a vector is discussed in Section 1.8.3 or you can just do the trigonometry.

Last, take the limit of this result for the traction vector as the shear modulus tends to zero, which would be the case for an imagined plane in the ocean. Discuss this result.

7. We have an infinite half-space of homogeneous isotropic fluid characterized by given values of ρ_f and K_f (with $G_f = 0$ in a fluid) that overlays an infinite half-space of homogeneous isotropic solid characterized by given values of ρ_s, K_s, and G_s. A plane compressional wave is incident on the plane interface located at $z = 0$ with a direction of propagation $\hat{\boldsymbol{k}}$ that is at an angle θ (as measured counterclockwise) from the vertical. Determine the reflection and transmission coefficients of all the waves created at the interface in a manner analogous to how we handled the interface between two solids.

8. *SV plane wave incident at a free surface*: Consider the situation depicted in Fig. 4.18 in which a plane SV wave is incident at a free surface, which generates a reflected

plane SV wave and reflected P wave as shown. The amplitude of the reflected SV wave relative to the amplitude of the SV incident wave is R_s and the amplitude of the reflected P wave relative to the incident wave is R_p. Find R_s and R_p.

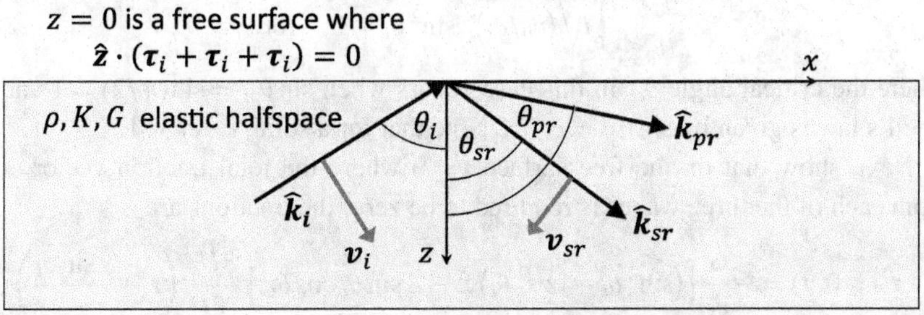

Figure 4.18 An SV wave is incident at the stress-free surface which generates a reflected SV wave and a reflected P wave. Because $c_p > c_s$, at large enough angle of incidence, the reflected P wave becomes critically reflected and evanescent. The arrows in gray give the direction of the SV particle motions.

HINTS: The incident SV plane wave has particle displacements given by $\boldsymbol{u}_i(\boldsymbol{r}, t) = \hat{\boldsymbol{u}}_i U_s(t - \hat{\boldsymbol{k}}_i \cdot \boldsymbol{r}/c_s)$, where $U_s(t)$ is some given time function. Show that the direction of the particle displacements (and therefore particle velocities) for each wave are

$$\hat{\boldsymbol{u}}_i = \cos\theta_i \hat{\boldsymbol{x}}_i + \sin\theta_i \hat{\boldsymbol{z}} \tag{4.317}$$

$$\hat{\boldsymbol{u}}_{sr} = -\cos\theta_{sr} \hat{\boldsymbol{x}}_i + \sin\theta_{sr} \hat{\boldsymbol{z}} \tag{4.318}$$

$$\hat{\boldsymbol{u}}_{pr} = \cos\theta_{pr} \hat{\boldsymbol{x}}_i + \sin\theta_{pr} \hat{\boldsymbol{z}} \tag{4.319}$$

and that the direction of each wave is

$$\hat{\boldsymbol{k}}_i = \sin\theta_i \hat{\boldsymbol{x}}_i - \cos\theta_i \hat{\boldsymbol{z}} \tag{4.320}$$

$$\hat{\boldsymbol{k}}_{sr} = \sin\theta_{sr} \hat{\boldsymbol{x}}_i + \cos\theta_{sr} \hat{\boldsymbol{z}} \tag{4.321}$$

$$\hat{\boldsymbol{k}}_{pr} = \hat{\boldsymbol{u}}_{pr}. \tag{4.322}$$

Taking a time derivative of these plane-wave displacements and requiring that on $z = 0$, the waves are in phase with each other so that Snell's law is

$$p = \frac{\sin\theta_i}{c_s} = \frac{\sin\theta_{sr}}{c_s} = \frac{\sin\theta_{pr}}{c_p}, \tag{4.323}$$

show that

$$\boldsymbol{v}_i(x, z, t) = (\cos\theta_i \hat{\boldsymbol{x}} + \sin\theta_i \hat{\boldsymbol{z}}) \frac{\partial U_s}{\partial t}\left(t - \frac{x\sin\theta_i}{c_s} + \frac{z\cos\theta_i}{c_s}\right), \tag{4.324}$$

$$\boldsymbol{v}_{sr}(x, z, t) = (-\cos\theta_i \hat{\boldsymbol{x}} + \sin\theta_i \hat{\boldsymbol{z}}) R_s \frac{\partial U_s}{\partial t}\left(t - \frac{x\sin\theta_i}{c_s} - \frac{z\cos\theta_i}{c_s}\right), \tag{4.325}$$

$$\boldsymbol{v}_{pr}(x, z, t) = \left(\frac{c_p}{c_s}\sin\theta_i \hat{\boldsymbol{x}} + \cos\theta_{pr} \hat{\boldsymbol{z}}\right) R_p \frac{\partial U_s}{\partial t}\left(t - \frac{x\sin\theta_i}{c_s} - \frac{z\cos\theta_{pr}}{c_p}\right). \tag{4.326}$$

It is convenient to represent all angular dependence in terms of $\sin\theta_i = pc_s$ so that, $\cos\theta_i = \sqrt{1-\sin^2\theta_i}$ and

$$\cos\theta_{pr} = \begin{cases} \sqrt{1-\left(c_p/c_s\right)^2\sin^2\theta_i} & \text{for } \theta_i < \theta_{cr} \\ i\sqrt{\left(c_p/c_s\right)^2\sin^2\theta_i - 1} & \text{for } \theta_i > \theta_{cr} \end{cases}, \tag{4.327}$$

where the critical angle θ_{cr} in this case occurs when $\sin\theta_{pr} = \sin(\pi/2) = 1$ and from Snell's law is given by $\sin\theta_{cr} = c_s/c_p$. Note that for a solid, $c_s/c_p < 1$.

Next, show that on the free surface $z=0$ where the total traction vector coming from each of the three waves is required to be zero, the tractions are

$$\hat{z}\cdot\boldsymbol{\tau}_i(x,0,t) = -\frac{G}{c_s}\left[\left(\sin^2\theta_i - \cos^2\theta_i\right)\hat{\boldsymbol{x}} - 2\sin\theta_i\cos\theta_i\hat{z}\right]\frac{\partial U_s}{\partial t}\left(t - \frac{x\sin\theta_i}{c_s}\right) \tag{4.328}$$

$$\hat{z}\cdot\boldsymbol{\tau}_{sr}(x,0,t) = -\frac{G}{c_s}\left[\left(\sin^2\theta_i - \cos^2\theta_i\right)\hat{\boldsymbol{x}} + 2\sin\theta_i\cos\theta_i\hat{z}\right]R_s\frac{\partial U_s}{\partial t}\left(t - \frac{x\sin\theta_i}{c_s}\right) \tag{4.329}$$

$$\hat{z}\cdot\boldsymbol{\tau}_{pr}(x,0,t) = -\frac{G}{c_s}\left[2\cos\theta_{pr}\sin\theta_i\hat{\boldsymbol{x}} - \frac{c_p}{c_s}\left(\sin^2\theta_i - \cos^2\theta_i\right)\hat{z}\right]R_p\frac{\partial U_s}{\partial t}\left(t - \frac{x\sin\theta_i}{c_s}\right). \tag{4.330}$$

Next, show that by equating to zero the x and z components of the free-surface condition $\hat{z}\cdot\left(\boldsymbol{\tau}_i + \boldsymbol{\tau}_{sr} + \boldsymbol{\tau}_{pr}\right) = 0$, you obtain two equations for R_s and R_p that are solved to give

$$R_s = \frac{\left(c_p/c_s\right)\left(\sin^2\theta_i - \cos^2\theta_i\right) - 4\cos\theta_{pr}\sin^2\theta_i\cos\theta_i}{\left(c_p/c_s\right)\left(\sin^2\theta_i - \cos^2\theta_i\right) + 4\cos\theta_{pr}\sin^2\theta_i\cos\theta_i} \tag{4.331}$$

$$R_p = \frac{4\sin\theta_i\cos\theta_i\left(\sin^2\theta_i - \cos^2\theta_i\right)}{\left(c_p/c_s\right)\left(\sin^2\theta_i - \cos^2\theta_i\right) + 4\cos\theta_{pr}\sin^2\theta_i\cos\theta_i}. \tag{4.332}$$

These expressions are real until $\theta_i > \theta_R$ at which time $\cos\theta_{pr}$ becomes purely imaginary. You will treat the case of postcritical incidence of an SV wave at a free surface in the next problem, where we specialize to the time-harmonic case.

9. *Postcritical incidence of an SV wave at a free surface*: We now consider that the SV incident plane wave of the previous problem is time harmonic so that

$$\frac{\partial U_s}{\partial t} = A_s e^{-i\omega t}. \tag{4.333}$$

For large enough incident angle that $\sin\theta_i > \sin\theta_{cr} = c_s/c_p$, we have $\cos\theta_{pr} = i\chi_p$ where

$$\chi_p = \sqrt{(c_p/c_s)^2\sin^2\theta_i - 1}. \tag{4.334}$$

In this case, show that the real particle velocities associated with the critically reflected P wave is given by

$$\boxed{\begin{aligned} \boldsymbol{v}_{pr}(x,z,t) = A_s \Bigg\{ \frac{c_p}{c_s} \sin\theta_i \Bigg[& R_{pR} \cos\left(\omega\left(t - \frac{x\sin\theta_i}{c_s}\right)\right) \\ & + R_{pI} \sin\left(\omega\left(t - \frac{x\sin\theta_i}{c_s}\right)\right)\Bigg] \hat{\boldsymbol{x}} \\ & - \chi_p \Bigg[R_{pI} \cos\left(\omega\left(t - \frac{x\sin\theta_i}{c_s}\right)\right) \\ & + R_{pR} \sin\left(\omega\left(t - \frac{x\sin\theta_i}{c_s}\right)\right)\Bigg] \hat{\boldsymbol{z}} \Bigg\} e^{-\omega\chi_p z/c_p}. \end{aligned}} \quad (4.335)$$

Because $\cos\theta_{pr} = i\chi_p$ is purely imaginary at postcritical incidence, the reflection coefficients R_p and R_s both become complex. So for the critically reflected evanescent P wave, we have $R_p = R_{pR} + iR_{pI}$, where

$$R_{pR} = \frac{4(c_p/c_s)\sin\theta_i\cos\theta_i(\sin^2\theta_i - \cos^2\theta_i)^2}{(c_p/c_s)^2(\sin^2\theta_i - \cos^2\theta_i)^2 + 16\chi_p^2\sin^4\theta_i\cos^2\theta_i} \quad (4.336)$$

$$R_{pI} = \frac{-16\chi_p\sin^3\theta_i\cos^2\theta_i\left(\sin^2\theta_i - \cos^2\theta_i\right)}{(c_p/c_s)^2(\sin^2\theta_i - \cos^2\theta_i)^2 + 16\chi_p^2\sin^4\theta_i\cos^2\theta_i}. \quad (4.337)$$

For such a critically reflected horizontally propagating P wave, the particle motion is clockwise elliptical.

Electromagnetic Plane Waves

In our treatment of electromagnetics in Chapter 3, we did not address the topic of the reflection and transmission of plane EM waves. This is because the reflection and transmission of EM waves is handled identically to how we treat the reflection and transmission of elastodynamic waves and we chose to present the elastodynamic case first. In the exercises that follow, you can now treat EM waves in the same way.

Return to the Maxwell equations of Chapter 3 that apply to an isotropic dielectric characterized by a uniform electrical permittivity ε and magnetic permeability μ

$$\nabla \times \boldsymbol{E} = -\mu\frac{\partial \boldsymbol{H}}{\partial t} \quad (4.338)$$

$$\nabla \times \boldsymbol{H} = \varepsilon\frac{\partial \boldsymbol{E}}{\partial t}. \quad (4.339)$$

To simplify, we will assume the material has zero conductivity. In Exercise 1 from Chapter 3, you determined the expression for a time-harmonic plane wave that applies when both electromigration (that attenuates wave energy) and electric and magnetic polarization are occurring simultaneously. You now will treat EM plane waves without electromigration and determine how they reflect and transmit at an interface.

10. *Electromagnetic plane-wave response in a uniform isotropic dielectric*: For a plane electromagnetic wave as controlled by Eqs (4.338) and (4.339), the electric and magnetic fields have the form

$$\boldsymbol{E}(\boldsymbol{r},t)=E\left(t-\frac{\hat{\boldsymbol{k}}\cdot\boldsymbol{r}}{c}\right)\hat{\boldsymbol{e}}, \tag{4.340}$$

$$\boldsymbol{H}(\boldsymbol{r},t)=H\left(t-\frac{\hat{\boldsymbol{k}}\cdot\boldsymbol{r}}{c}\right)\hat{\boldsymbol{h}}. \tag{4.341}$$

Using Eqs (4.338) and (4.339), you need to find the relation between the scalar functions $E(t)$ and $H(t)$, the relation between the plane-wave direction of propagation $\hat{\boldsymbol{k}}$ and the direction (polarization) of the electric-field vector $\hat{\boldsymbol{e}}$ and magnetic-field vector $\hat{\boldsymbol{h}}$ and finally the wave speed c. The unit vectors characterizing the wave's polarization $\hat{\boldsymbol{e}}$ and $\hat{\boldsymbol{h}}$ are uniform constants in the homogeneous medium under consideration.

To obtain these results, insert the plane-wave responses directly into Eqs (4.338) and (4.339) and use the vector identity $\nabla\times(\alpha\boldsymbol{a})=(\nabla\alpha)\times\boldsymbol{a}+\alpha\nabla\times\boldsymbol{a}$ along with the plane-wave identities $\nabla E=-(\partial E/\partial t)\hat{\boldsymbol{k}}/c$ and $\nabla H=-(\partial H/\partial t)\hat{\boldsymbol{k}}/c$ as proven earlier in Eq. (4.156). Show that the two Maxwell equations become

$$-\frac{1}{c}\frac{\partial E}{\partial t}\hat{\boldsymbol{k}}\times\hat{\boldsymbol{e}}=\mu\frac{\partial H}{\partial t}\hat{\boldsymbol{h}}, \tag{4.342}$$

$$-\frac{1}{c}\frac{\partial H}{\partial t}\hat{\boldsymbol{k}}\times\hat{\boldsymbol{h}}=\varepsilon\frac{\partial E}{\partial t}\hat{\boldsymbol{e}}. \tag{4.343}$$

First, consider the possibility that the electromagnetic response is longitudinal with $\hat{\boldsymbol{k}}\cdot\hat{\boldsymbol{e}}=1$ and, therefore, $\hat{\boldsymbol{k}}\times\hat{\boldsymbol{e}}=0$. Equation (4.342) can then be satisfied only if $\partial H/\partial t=0$, which results in $\partial E/\partial t=0$ from Eq. (4.343), and so there is no electromagnetic plane wave with longitudinal response of the electric field. The same conclusion is reached if the magnetic field is assumed to be longitudinal. You thus have shown that the electromagnetic wave under consideration is transverse with both the electric-field direction $\hat{\boldsymbol{e}}$ and magnetic-field direction $\hat{\boldsymbol{h}}$ being perpendicular to the wave direction $\hat{\boldsymbol{k}}$.

Equations (4.342) and (4.343) are satisfied by a transverse wave response that satisfies simultaneously the polarization conditions

$$\hat{\boldsymbol{h}}=\hat{\boldsymbol{k}}\times\hat{\boldsymbol{e}}, \tag{4.344}$$

$$\hat{\boldsymbol{e}}=-\hat{\boldsymbol{k}}\times\hat{\boldsymbol{h}}, \tag{4.345}$$

and the electric and magnetic scalar time-function relations

$$\frac{\partial E}{\partial t}=c\mu\frac{\partial H}{\partial t}, \tag{4.346}$$

$$\frac{\partial H}{\partial t}=c\varepsilon\frac{\partial E}{\partial t}. \tag{4.347}$$

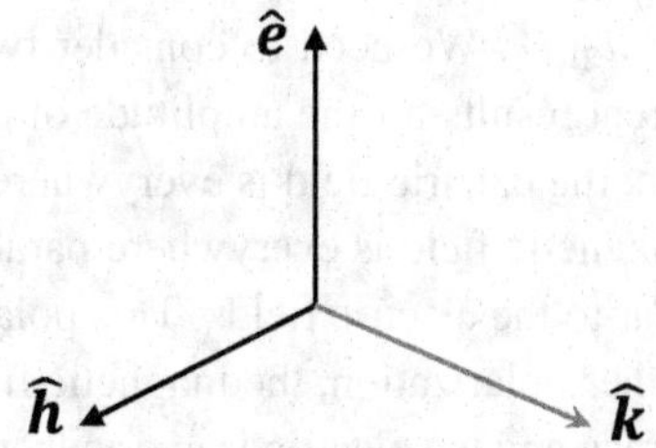

Figure 4.19 The unit vectors of electromagnetic waves $\hat{e}$, $\hat{h}$, $\hat{k}$ form a right-handed orthogonal set as shown.

Show that the two polarization conditions here require $\hat{e} \times \hat{h} = \hat{k}$ as depicted in Fig. 4.19. By combining the two time-function relations, show that $\partial H/\partial t = c\varepsilon \partial E/\partial t = c^2 \varepsilon\mu \partial H/\partial t$ or

$$\boxed{c = \frac{1}{\sqrt{\varepsilon\mu}}.} \tag{4.348}$$

Upon time-integrating Eqs (4.346) and (4.347) and requiring that if the magnetic field is zero, so is the electric field (which puts the integration constant to zero), you have shown $H(t) = \sqrt{\varepsilon/\mu}E(t)$. Thus for a given transverse polarization of the electric field in some electromagnetic plane-wave response

$$\boldsymbol{E}(\boldsymbol{r}, t) = E\left(t - \frac{\hat{\boldsymbol{r}} \cdot \hat{\boldsymbol{k}}}{c}\right)\hat{\boldsymbol{e}}, \tag{4.349}$$

the associated magnetic field is

$$\boxed{\boldsymbol{H}(\boldsymbol{r}, t) = \sqrt{\frac{\varepsilon}{\mu}}E\left(t - \frac{\hat{\boldsymbol{r}} \cdot \hat{\boldsymbol{k}}}{c}\right)\hat{\boldsymbol{k}} \times \hat{\boldsymbol{e}}.} \tag{4.350}$$

Show that the Poynting vector, which is a vector representing the flux of electromagnetic energy, is given by

$$\boxed{\boldsymbol{s} = \boldsymbol{E} \times \boldsymbol{H} = \sqrt{\frac{\varepsilon}{\mu}}\left[E\left(t - \frac{\hat{\boldsymbol{r}} \cdot \hat{\boldsymbol{k}}}{c}\right)\right]^2 \hat{\boldsymbol{k}}} \tag{4.351}$$

for an electromagnetic plane wave propagating in the $\hat{\boldsymbol{k}}$ direction at wave speed $c = 1/\sqrt{\varepsilon\mu}$.

11. *Electromagnetic reflection and transmission*: We now allow a plane electromagnetic wave to be incident at a planar interface separating one material having μ_1, ε_1 from

a second material having μ_2, ε_2. We need to consider two polarizations of the incident wave that give different results for the amplitude of the reflected and transmitted waves. In one polarization, the electric field is everywhere in a plane perpendicular to the interface, while the magnetic field is everywhere parallel to the plane of the interface and also perpendicular to the electric fields. This polarization is called *transverse magnetic* or TM. In the other polarization, the magnetic field is everywhere in a plane perpendicular to the interface and the electric is everywhere parallel to the plane of the interface. This polarization is called *transverse electric* or TE. The two situations are depicted in Fig. 4.20.

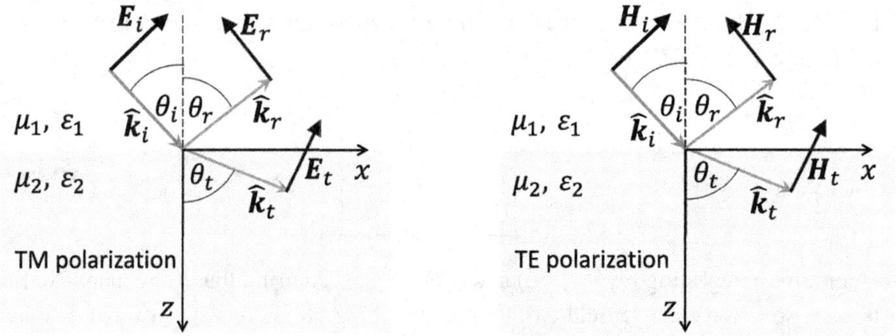

Figure 4.20 In TM polarization on the left, the magnetic field is everywhere pointing out of the page in the $+\hat{y}$ direction. In TE polarization on the right, the electric field is everywhere pointing into the page in the $-\hat{y}$ direction.

TM polarization: In this case, the various unit vectors associated with the incident, reflected, and transmitted waves are expressed in terms of the angle each wave direction makes relative to the vertical as

$$\hat{\boldsymbol{k}}_i = \sin\theta_i\hat{\boldsymbol{x}} + \cos\theta_i\hat{\boldsymbol{z}}, \tag{4.352}$$

$$\hat{\boldsymbol{k}}_r = \sin\theta_r\hat{\boldsymbol{x}} - \cos\theta_r\hat{\boldsymbol{z}}, \tag{4.353}$$

$$\hat{\boldsymbol{k}}_i = \sin\theta_t\hat{\boldsymbol{x}} + \cos\theta_t\hat{\boldsymbol{z}}, \tag{4.354}$$

$$\hat{\boldsymbol{e}}_i = \cos\theta_i\hat{\boldsymbol{x}} - \sin\theta_i\hat{\boldsymbol{z}}, \tag{4.355}$$

$$\hat{\boldsymbol{e}}_r = -\cos\theta_r\hat{\boldsymbol{x}} - \sin\theta_r\hat{\boldsymbol{z}}, \tag{4.356}$$

$$\hat{\boldsymbol{e}}_t = \cos\theta_t\hat{\boldsymbol{x}} - \sin\theta_t\hat{\boldsymbol{z}}, \tag{4.357}$$

$$\hat{\boldsymbol{h}}_i = \hat{\boldsymbol{h}}_r = \hat{\boldsymbol{h}}_t = +\hat{\boldsymbol{y}}. \tag{4.358}$$

The sum of the incident and reflected wave is the response in material 1, while the transmitted wave is the response in material 2. Across the interface $z = 0$, we proved earlier that the horizontal components of the electric and magnetic fields are continuous, that is, $\hat{\boldsymbol{z}} \times (\boldsymbol{E}_i + \boldsymbol{E}_r) = \hat{\boldsymbol{z}} \times \boldsymbol{E}_t$ and $\hat{\boldsymbol{z}} \times (\boldsymbol{H}_i + \boldsymbol{H}_r) = \hat{\boldsymbol{z}} \times \boldsymbol{H}_t$. Show that these continuity conditions on $z = 0$ are thus

$$\hat{z} \times \left[\hat{e}_i E\left(t - \frac{x \sin\theta_i}{c_1}\right) + \hat{e}_r R_{tm} E\left(t - \frac{x \sin\theta_r}{c_1}\right) \right] = \hat{z} \times \hat{e}_t T_{tm} E\left(t - \frac{x \sin\theta_t}{c_2}\right), \tag{4.359}$$

$$\sqrt{\frac{\varepsilon_1}{\mu_1}} \hat{z} \times \left[\hat{y} E\left(t - \frac{x \sin\theta_i}{c_1}\right) + \hat{y} R_{tm} E\left(t - \frac{x \sin\theta_r}{c_1}\right) \right] = \sqrt{\frac{\varepsilon_2}{\mu_2}} \hat{z} \times \hat{y} T_{tm} E\left(t - \frac{x \sin\theta_t}{c_2}\right), \tag{4.360}$$

which provide two equations for the two unknowns R_{tm} and T_{tm}.

As with the elastodynamic response, the incident, reflected, and transmitted waves are coincident at any x along the interface and thus each has identically the same time dependence on $z = 0$. For this to be true, we arrive at Snell's law that says each wave has the same horizontal slowness p

$$p \mathrel{\widehat{=}} \frac{\sin\theta_i}{c_1} = \frac{\sin\theta_r}{c_1} = \frac{\sin\theta_t}{c_2}, \tag{4.361}$$

so that $\theta_r = \theta_i$ and $\theta_t = \sin^{-1}(pc_2) = \sin^{-1}((c_2/c_1)\sin\theta_i) > \theta_i$ if $c_2 > c_1$.

Eliminating the time dependence that is the same in each term of Eqs (4.359) and (4.360) by Snell's law and performing the cross products, show that these continuity equations can be written

$$\begin{bmatrix} \cos\theta_i & \cos\theta_t \\ -\sqrt{\varepsilon_1/\mu_1} & \sqrt{\varepsilon_2/\mu_2} \end{bmatrix} \begin{bmatrix} R_{tm} \\ T_{tm} \end{bmatrix} = \begin{bmatrix} \cos\theta_i \\ \sqrt{\varepsilon_1/\mu_1} \end{bmatrix}, \tag{4.362}$$

which have the solution

$$R_{tm} = \frac{\sqrt{\varepsilon_2/\mu_2}\cos\theta_i - \sqrt{\varepsilon_1/\mu_1}\cos\theta_t}{\sqrt{\varepsilon_2/\mu_2}\cos\theta_i + \sqrt{\varepsilon_1/\mu_1}\cos\theta_t} \tag{4.363}$$

$$T_{tm} = \frac{2\sqrt{\varepsilon_1/\mu_1}\cos\theta_i}{\sqrt{\varepsilon_2/\mu_2}\cos\theta_i + \sqrt{\varepsilon_1/\mu_1}\cos\theta_t}. \tag{4.364}$$

TE polarization: In this polarization, the roles of the magnetic and electric field are reversed and the incident, reflected and transmitted plane waves have the electric-field and magnetic-field polarizations given by

$$\hat{\boldsymbol{h}}_i = \cos\theta_i \hat{\boldsymbol{x}} - \sin\theta_i \hat{\boldsymbol{z}}, \tag{4.365}$$

$$\hat{\boldsymbol{h}}_r = -\cos\theta_r \hat{\boldsymbol{x}} - \sin\theta_r \hat{\boldsymbol{z}}, \tag{4.366}$$

$$\hat{\boldsymbol{h}}_t = \cos\theta_t \hat{\boldsymbol{x}} - \sin\theta_t \hat{\boldsymbol{z}}, \tag{4.367}$$

$$\hat{\boldsymbol{e}}_i = \hat{\boldsymbol{e}}_r = \hat{\boldsymbol{e}}_t = -\hat{\boldsymbol{y}}. \tag{4.368}$$

In this case, go through the exact same procedure as used above for the case of TM polarization and show that for TE polarization

$$R_{te} = \frac{\sqrt{\varepsilon_1/\mu_1}\cos\theta_i - \sqrt{\varepsilon_2/\mu_2}\cos\theta_t}{\sqrt{\varepsilon_1/\mu_1}\cos\theta_i + \sqrt{\varepsilon_2/\mu_2}\cos\theta_t} \tag{4.369}$$

$$T_{te} = \frac{2\sqrt{\varepsilon_1/\mu_1}\cos\theta_i}{\sqrt{\varepsilon_1/\mu_1}\cos\theta_i + \sqrt{\varepsilon_2/\mu_2}\cos\theta_t}. \tag{4.370}$$

Uniaxial Problems in Linear Elastostatics

The equations of linear elastostatic response of an isotropic solid are $\nabla\cdot\boldsymbol{\tau} + \rho\boldsymbol{g} = 0$ with $\boldsymbol{\tau} = K\nabla\cdot\boldsymbol{u}\boldsymbol{I} + G\left(\nabla\boldsymbol{u} + (\nabla\boldsymbol{u})^T - 2\nabla\cdot\boldsymbol{u}\boldsymbol{I}/3\right)$. The body force here $\rho\boldsymbol{g}$ could be the force of gravity but can also be a placeholder for centrifugal or other forces.

12. *Best form for solving uniaxial problems in homogeneous isotropic media*: Show that these equations can be written for any isotropic homogeneous body as

$$H\nabla\nabla\cdot\boldsymbol{u} - G\nabla\times\nabla\times\boldsymbol{u} = -\rho\boldsymbol{g}, \tag{4.371}$$

where $H = K + 4G/3$ is the uniaxial or longitudinal or P-wave modulus. This form is optimal for solving uniaxial or, synonymously, longitudinal deformation problems in which the displacement varies spatially only in the direction that the displacement is directed. For such problems, $\nabla\times\boldsymbol{u} = 0$ and Eq. (4.371) is in a form that is immediately integrable with the only elastic constant being the uniaxial modulus H.

13. *Radial deformation of spheres*: If both the displacements and body forces in a spherical geometry are purely radial $\boldsymbol{u}(\boldsymbol{r}) = u_r(r)\hat{\boldsymbol{r}}$ and $\boldsymbol{g}(\boldsymbol{r}) = -g_r(r)\hat{\boldsymbol{r}}$, show that $\nabla\times\boldsymbol{u} = 0$ and that in spherical coordinates the uniaxial displacements are governed by

$$\frac{\mathrm{d}}{\mathrm{d}r}\left[\frac{1}{r^2}\frac{\mathrm{d}}{\mathrm{d}r}\left(r^2 u_r(r)\right)\right] = \frac{\rho}{H}g_r(r). \tag{4.372}$$

Further, show that the only nonzero stress components in this case are given by

$$\tau_{rr} = (H-2G)\left(\frac{du_r}{dr} + \frac{2u_r}{r}\right) + 2G\frac{du_r}{dr}, \tag{4.373}$$

$$\tau_{\theta\theta} = \tau_{\phi\phi} = (H-2G)\left(\frac{du_r}{dr} + \frac{2u_r}{r}\right) + 2G\frac{u_r}{r}. \tag{4.374}$$

Note that in spherical coordinates $\nabla\cdot\boldsymbol{u} = r^{-2}\mathrm{d}(r^2\mathrm{d}u_r/dr) = \mathrm{d}u_r/\mathrm{d}r + 2u_r/r$.

14. *Deformation of a self-gravitating sphere of approximately uniform density*: Obtain the radial displacement field due to gravity in a uniform self-gravitating sphere of radius R subject to the boundary condition that $\tau_{rr} = 0$ on the surface of the sphere. Make the

approximation that the response is linear in the force of gravity, which means taking $\rho(r) \approx \rho_0$ (a uniform constant) in the gravitational force density. As shown at the end of Chapter 3, the acceleration of gravity in a uniform sphere is given by $g_r(r) = g_R r/R$, where g_R is the acceleration of gravity at the surface of the sphere $g_R = 4\pi G\rho_0 R/3$, where G is the universal gravitational constant. Integrate Eq. (4.372) to obtain

$$\frac{1}{r^2}\frac{\mathrm{d}}{\mathrm{d}r}\left(r^2 u_r\right) = \frac{\rho_0 g_R}{2HR} r^2 + c_1, \tag{4.375}$$

where c_1 is an integration constant to be found from the free surface condition. Multiplying by r^2 and integrating again gives

$$u_r = \frac{\rho_0 g_R}{2HR}\frac{r^3}{5} + \frac{c_1 r}{3} + \frac{c_2}{r^2}. \tag{4.376}$$

Find the constants c_1 and c_2 from the conditions that the displacements are finite (indeed zero) at the sphere center $r = 0$, while the radial component of stress satisfies $\tau_{rr} = 0$ on $r = R$. Show that

$$\boxed{u_r(r) = -\frac{\rho_0 g_R R^2}{10H}\left[\left(\frac{5H-4G}{3H-4G}\right)\frac{r}{R} - \left(\frac{r}{R}\right)^3\right].} \tag{4.377}$$

Using this result, we can also obtain how the pressure varies with depth $P = -K\nabla\cdot\boldsymbol{u}$ and, from our earlier result, how the density varies $\rho(r) = \rho_0 \exp(-\nabla\cdot\boldsymbol{u})$ due to the displacement field. Show that in spherical coordinates with our purely radial displacement field $\boldsymbol{u} = u_r(r)\hat{\boldsymbol{r}}$ that $\nabla\cdot\boldsymbol{u} = \mathrm{d}u_r(r)/\mathrm{d}r + 2u_r/r$ so that

$$P(r) = \frac{\rho_0 g_R RK}{10H}\left[3\left(\frac{5H-4G}{3H-4G}\right) - 5\left(\frac{r}{R}\right)^2\right] \tag{4.378}$$

$$\rho(r) = \rho_0 \exp\left\{\frac{\rho_0 g_R R}{10H}\left[3\left(\frac{5H-4G}{3H-4G}\right) - 5\left(\frac{r}{R}\right)^2\right]\right\}. \tag{4.379}$$

For the case of an isolated sphere having $R = 10^6$ m, $\rho_0 = 2\times 10^3$ kg/m^3, and $K = 4G/3 = 10^{10}$ Pa, show that the pressure at the center of the sphere is $P(0) = 3.9 \times 10^8$ Pa, while the density at the center and surface of the sphere are $\rho(0)/\rho_0 \approx 1.04$ and $\rho(R)/\rho_0 \approx 1.01$. For this particular spherical body, the approximation of neglecting the variation of density in the force of gravity has resulted in an error of a couple percent.

15. *Deformation of a spherical shell*: A spherical shell of inner radius R_1 and outer radius R_2 is made of a uniform isotropic elastic solid having given K and G. The cavity within the spherical shell $r < R_1$ is a gas maintained at constant pressure P_1. The atmosphere surrounding the shell $r > R_2$ is also a gas that has an initial pressure of P_1 when the shell is in the undeformed state but is then increased to P_2. What is the resulting deformation and stress throughout the shell $R_1 \le r \le R_2$ once $P_2 > P_1$ under the assumption of linear elasticity?

Again, the displacement field in this case is purely radial $\boldsymbol{u}(\boldsymbol{r}) = u_r(r)\hat{\boldsymbol{r}}$ so that $\nabla \times \boldsymbol{u} = 0$. Because gravity is being ignored (the isolated sphere is too small to have a significant acceleration of gravity), Eq. (4.372) informs us that

$$\frac{1}{r^2}\frac{\mathrm{d}}{\mathrm{d}r}\left[r^2 u_r(r)\right] = c_1, \tag{4.380}$$

where c_1 is a constant found from the boundary conditions. Show that multiplying by r^2 and integrating gives

$$u_r(r) = \frac{c_1 r}{3} + \frac{c_2}{r^2}, \tag{4.381}$$

where c_2 is an integration constant that also must be found from the boundary conditions. The two boundary conditions for c_1 and c_2 are $\tau_{rr}(R_2) = -P_2$ and $\tau_{rr}(R_1) = -P_1$, which from Eq. (4.373) allows you to obtain the final results for all nonzero components of displacement and stress

$$u_r(r) = -\frac{(P_2R_2^3 - P_1R_1^3)}{3K(R_2^3 - R_1^3)}r - \frac{(P_2 - P_1)R_1^3R_2^3}{4G(R_2^3 - R_1^3)}\frac{1}{r^2}, \tag{4.382}$$

$$\tau_{rr}(r) = \frac{-P_2R_2^3 + P_1R_1^3 + (P_2 - P_1)R_1^3R_2^3/r^3}{R_2^3 - R_1^3}, \tag{4.383}$$

$$\tau_{\theta\theta}(r) = \tau_{\phi\phi}(r) = -\frac{(P_2R_2^3 - P_1R_1^3)}{R_2^3 - R_1^3} - \frac{(P_2 - P_1)R_1^3R_2^3}{2(R_2^3 - R_1^3)r^3}. \tag{4.384}$$

16. *Deformation of a rotating cylinder*: An isolated isotropic solid cylinder of radius R is rotating about its axis with a steady angular velocity of Ω rad/s, which creates an outward (radial) centrifugal force throughout the cylinder that induces deformation of the cylinder compared to when the cylinder is not rotating. Find the displacement field and all nonzero stress components under the condition that the normal stress on the cylinder surface is zero.

As shown at the end of Chapter 3, if we work in cylindrical coordinates that rotate with the cylinder, there is generated a radial centrifugal-force density given by $\rho\Omega^2 r\hat{\boldsymbol{r}}$. By symmetry, the only component of displacement induced by this centrifugal force is itself radial, that is, $\boldsymbol{u}(\boldsymbol{r}) = u_r(r)\hat{\boldsymbol{r}}$. In cylindrical coordinates, show that Eq. (4.371) becomes

$$\frac{\mathrm{d}}{\mathrm{d}r}\left[\frac{1}{r}\frac{\mathrm{d}}{\mathrm{d}r}(ru_r(r))\right] = -\frac{\rho\Omega^2}{H}r. \tag{4.385}$$

Show that the nonzero stress components for purely radial displacements in cylindrical coordinates are

$$\tau_{rr} = (H - 2G)\left(\frac{du_r}{dr} + \frac{u_r}{r}\right) + 2G\frac{du_r}{dr}, \tag{4.386}$$

$$\tau_{\theta\theta} = (H - 2G)\left(\frac{\mathrm{d}u_r}{\mathrm{d}r} + \frac{u_r}{r}\right) + 2G\frac{u_r}{r} \tag{4.387}$$

$$\tau_{zz} = (H - 2G)\left(\frac{\mathrm{d}u_r}{\mathrm{d}r} + \frac{u_r}{r}\right). \tag{4.388}$$

By integrating the equation for u_r and using the boundary conditions that $\tau_{rr}(R) = 0$ and that displacements remain finite at $r = 0$, show that

$$u_r(r) = -\frac{\rho\Omega^2 r}{8H}\left[r^2 - \left(\frac{2H - G}{H - G}\right)R^2\right], \tag{4.389}$$

which has its maximum value and is positive (i.e., outward) at $r = R$, that is, $u_r(R) = \rho\Omega^2 R^3/[8(H - G)]$. Further show that the stress components are

$$\tau_{rr}(r) = -\frac{\rho\Omega^2(2H - G)}{4H}\left[r^2 - R^2\right], \tag{4.390}$$

$$\tau_{\theta\theta}(r) = -\frac{\rho\Omega^2}{4H}\left[(2H - 3G)r^2 - (2H - G)\frac{R^2}{2}\right], \tag{4.391}$$

$$\tau_{zz}(r) - \frac{\rho\Omega^2(H - 2G)}{4H}\left[2r^2 - \left(\frac{2H - G}{H - G}\right)\frac{R^2}{2}\right]. \tag{4.392}$$

So at the center of the cylinder $r = 0$, the traction is outward $\tau_{rr} > 0$, corresponding to a state of outward tension. For a cylinder of radius $R = 1$ m rotating at one revolution per second $\Omega = 2\pi$ rad/s and made of a material having $K = 4/3G = 10^{10}$ Pa with density $\rho = 2 \times 10^3$ kg/m^3, the displacement at $r = R$ is $u_r(R)/R = 1.01 \times 10^{-5}$.

17. *Deformation of a cylindrical shell*: A cylindrical shell of inner radius R_1 and outer radius R_2 is made of a uniform isotropic elastic solid having given K and G. The cavity within the cylindrical shell $r < R_1$ is a gas with a pressure set at P_1. The atmosphere surrounding the shell $r > R_2$ is also a gas with a pressure set at P_2. What is the deformation and stress throughout the shell $R_1 \le r \le R_2$ due to P_1 and P_2 under the assumption of linear elasticity?

    In direct analogy to the treatment above of the spherical shell, work in cylindrical coordinates and show

$$u_r(r) = -\frac{(P_2R_2^2 - P_1R_1^2)r}{2(H - G)(R_2^2 - R_1^2)} - \frac{(P_2 - P_1)R_1^2R_2^2}{2G(R_2^2 - R_1^2)r}, \tag{4.393}$$

$$\tau_{rr}(r) = \frac{-(P_2R_2^2 - P_1R_1^2) + (P_2 - P_1)R_1^2R_2^2/r^2}{R_2^2 - R_1^2}, \tag{4.394}$$

$$\tau_{\theta\theta}(r) = \frac{-(P_2R_2^2 - P_1R_1^2) - (P_2 - P_1)R_1^2R_2^2/r^2}{R_2^2 - R_1^2}, \tag{4.395}$$

$$\tau_{zz}(r) = -\left(\frac{H - 2G}{H - G}\right)\left(\frac{P_2R_2^2 - P_1R_1^2}{R_2^2 - R_1^2}\right). \tag{4.396}$$

18. *Double-couple slip event*: A moment tensor source that represents slip in say the x direction on a small fault that has a normal $\hat{\boldsymbol{y}}$ has the form

$$\boldsymbol{M} = \begin{pmatrix} 0 & M_{xy} & 0 \\ M_{xy} & 0 & 0 \\ 0 & 0 & 0 \end{pmatrix}. \tag{4.397}$$

The effective body force is $\boldsymbol{F} = \boldsymbol{M}(t) \cdot \nabla\delta(\boldsymbol{r})$ for a slip surface centered at the origin, which results in the double-couple forcing $\boldsymbol{F} = \hat{\boldsymbol{x}} M_{xy}\delta(x)\delta(z)\, \mathrm{d}\delta(y)/\mathrm{d}y + \hat{\boldsymbol{y}} M_{xy}\delta(y)\delta(z)\, \mathrm{d}\delta(x)/\mathrm{d}x$. If we had assumed that slip was in the y direction on a fault having normal $\hat{\boldsymbol{x}}$, we obtain identically the same body-force equivalent. Show that if you rotate the coordinates by 45° around the z axis to give new coordinates $(x, y, z) \rightarrow (x', y', z)$, this same double-couple slip event has moment-tensor components in the rotated "primed" coordinates given by

$$\boldsymbol{M} = \begin{pmatrix} M_{xy} & 0 & 0 \\ 0 & -M_{xy} & 0 \\ 0 & 0 & 0 \end{pmatrix}. \tag{4.398}$$

The forcing in these rotated coordinates is $\boldsymbol{F} = \hat{\boldsymbol{x}}' M_{xy}\delta(y')\delta(z)\, \mathrm{d}\delta(x')/\mathrm{d}x' - \hat{\boldsymbol{y}}' M_{xy}\delta(x')\,\delta(z)\, \mathrm{d}\delta(y')/\mathrm{d}y'$, which now corresponds to tensile forcing on one plane ($x' = 0$) and compressive forcing on the perpendicular plane ($y' = 0$) and with the slip surface being at a 45° angle relative to these planes. Make a sketch of the forcing vectors that correspond to this slip event both in the unrotated and rotated coordinates. Can you visualize why both sets of vectors correspond to the same slip event?

19. *An anisotropic explosion*: If a "point" explosion takes place in an elastically isotropic material but the opening in each direction is distinct

$$[\boldsymbol{u}(t)] = \Delta u_x(t)\hat{\boldsymbol{x}} + \Delta u_y(t)\hat{\boldsymbol{y}} + \Delta u_z(t)\hat{\boldsymbol{z}},$$

the explosion will generate a spherically propagating compressional wave that has variable amplitude over its surface. Show that the moment tensor corresponding to this explosion is

$$\begin{aligned} \boldsymbol{M}(t) = \Delta V(t) K \Bigg[&\left(1 + \frac{2G}{3K}\left[\frac{2\Delta u_x - \Delta u_y - \Delta u_z}{\Delta u_x + \Delta u_y + \Delta u_z}\right]\right)\hat{\boldsymbol{x}}\hat{\boldsymbol{x}} \\ &+ \left(1 + \frac{2G}{3K}\left[\frac{2\Delta u_y - \Delta u_x - \Delta u_z}{\Delta u_x + \Delta u_y + \Delta u_z}\right]\right)\hat{\boldsymbol{y}}\hat{\boldsymbol{y}} \\ &+ \left(1 + \frac{2G}{3K}\left[\frac{2\Delta u_z - \Delta u_x - \Delta u_y}{\Delta u_x + \Delta u_y + \Delta u_z}\right]\right)\hat{\boldsymbol{z}}\hat{\boldsymbol{z}}\Bigg], \end{aligned} \tag{4.399}$$

where $\Delta V = A_s(\Delta u_x + \Delta u_y + \Delta u_z)$ is the opening volume. So, for example, if we have a cylindrically shaped explosive device that is positioned vertically in the elastically isotropic material such that the horizontal opening $\Delta u_x = \Delta u_y = \Delta u_H$ is greater than the vertical (z direction) opening $\Delta u_z = \Delta u_V$, show that the moment tensor becomes

$$M(t) = \Delta V(t) K \left[\left(1 + \frac{2G}{3K} \left[\frac{\Delta u_H - \Delta u_V}{2\Delta u_H + \Delta u_V} \right] \right) (\hat{x}\hat{x} + \hat{y}\hat{y}) + \left(1 - \frac{4G}{3K} \left[\frac{\Delta u_H - \Delta u_V}{2\Delta u_H + \Delta u_V} \right] \right) \hat{z}\hat{z} \right], \tag{4.400}$$

where $\Delta V(t) = A_s \left[2\Delta u_H(t) + \Delta u_V(t) \right]$.

20. *Isotropic moduli changes*: If a material is elastically isotropic, show that Eq. (4.45) for how the elastic moduli change with changing strain can be written in Cartesian coordinates as

$$\left\{ \left[\frac{\partial G}{\partial t} + \nabla \cdot (\boldsymbol{v} G) \right] \left(\delta_{il}\delta_{jk} + \delta_{ik}\delta_{jl} - \frac{2}{3}\delta_{ij}\delta_{kl} \right) + \left[\frac{\partial K}{\partial t} + \nabla \cdot (\boldsymbol{v} K) \right] \delta_{ij}\delta_{kl} \right\} \hat{\boldsymbol{x}}_i \hat{\boldsymbol{x}}_j \hat{\boldsymbol{x}}_k \hat{\boldsymbol{x}}_l = \sum_{m=1}^{16} \lambda_m \, {}_6\boldsymbol{I}^{(m)} : \frac{\mathrm{d}\boldsymbol{e}}{\mathrm{d}t}, \tag{4.401}$$

where $\mathrm{d}\boldsymbol{e}/\mathrm{d}t = \left[\nabla \boldsymbol{v} + (\nabla \boldsymbol{v})^T \right]/2$ is the strain-rate tensor, the ${}_6\boldsymbol{I}^{(m)}$ are the 16 sixth-order isotropic tensors derived in Chapter 1 as given by Eqs (1.188)–(1.203) and the λ_m are proportionality constants having units of Pa. Show that this equation can be written in terms of the components alone as

$$\begin{aligned} &\left[\frac{\partial G}{\partial t} + \nabla \cdot (\boldsymbol{v} G) \right] \left(\delta_{il}\delta_{jk} + \delta_{ik}\delta_{jl} - \frac{2}{3}\delta_{ij}\delta_{kl} \right) + \left[\frac{\partial K}{\partial t} + \nabla \cdot (\boldsymbol{v} K) \right] \delta_{ij}\delta_{kl} \\ &= \lambda_7 \delta_{il}\delta_{jk} \frac{\mathrm{d}e_{mm}}{\mathrm{d}t} + \lambda_{10} \delta_{ik}\delta_{jl} \frac{\mathrm{d}e_{mm}}{\mathrm{d}t} + \lambda_{13} \delta_{ij}\delta_{kl} \frac{\mathrm{d}e_{mm}}{\mathrm{d}t} \\ &+ (\lambda_1 + \lambda_4)\delta_{jk} \frac{\mathrm{d}e_{il}}{\mathrm{d}t} + (\lambda_2 + \lambda_5)\delta_{jl} \frac{\mathrm{d}e_{ik}}{\mathrm{d}t} + (\lambda_{14} + \lambda_{15})\delta_{ij} \frac{\mathrm{d}e_{kl}}{\mathrm{d}t} \\ &+ (\lambda_8 + \lambda_9)\delta_{il} \frac{\mathrm{d}e_{jk}}{\mathrm{d}t} + (\lambda_{11} + \lambda_{12})\delta_{ik} \frac{\mathrm{d}e_{jl}}{\mathrm{d}t} + (\lambda_3 + \lambda_6)\delta_{kl} \frac{\mathrm{d}e_{ij}}{\mathrm{d}t}. \end{aligned} \tag{4.402}$$

Note that the proportionality constant λ_{16} is not present because the double-dot product of an antisymmetric tensor with a symmetric tensor is zero. Show that this equation can only be satisfied by the isotropic part of e_{ij} (the left hand side cannot be proportional to the deviatoric part of the strain-rate tensor) and that this is equivalent to taking $\lambda_7 = \lambda_{10} = -D_G$, $\lambda_{13} = -D_K + 2D_G/3$ and all other $\lambda_i = 0$ (or, equivalently, taking $\lambda_7 + \lambda_1 + \lambda_4 + \lambda_8 + \lambda_9 = -D_G$ and $\lambda_{10} + \lambda_2 + \lambda_5 + \lambda_{11} + \lambda_{12} = -D_G$ and $\lambda_{13} + \lambda_{14} + \lambda_{15} + \lambda_3 + \lambda_6 = -D_K + 2D_G/3$) to give the two scalar equations

$$\frac{\partial G}{\partial t} = -\nabla \cdot [\boldsymbol{v} G] - D_G \nabla \cdot \boldsymbol{v}, \tag{4.403}$$

$$\frac{\partial K}{\partial t} = -\nabla \cdot [\boldsymbol{v} K] - D_K \nabla \cdot \boldsymbol{v}, \tag{4.404}$$

where $\nabla \cdot \boldsymbol{v} = \mathrm{d}e_{mm}/\mathrm{d}t$. These are the results quoted earlier as Eqs (4.47) and (4.48).

21. Show that by using the total differential $\mathrm{d}/\mathrm{d}t = \partial/\partial t + \boldsymbol{v} \cdot \nabla$, Eqs (4.403) and (4.404) can be integrated under the assumption of constant D_G and D_K to give the relation

between the shear and bulk moduli and the dilatation $\nabla \cdot \boldsymbol{u}$ (where $\boldsymbol{u}$ is the displacement field)

$$G = (G_0 + D_G)\, e^{-\nabla\cdot \boldsymbol{u}} - D_G \tag{4.405}$$

$$K = (K_0 + D_K)\, e^{-\nabla\cdot \boldsymbol{u}} - D_K. \tag{4.406}$$

Here, G_0 and K_0 are the shear and bulk moduli at zero dilatation (zero volumetric strain). These relations show that when the dilatation is negative (sample volume getting smaller due to isotropic strain), the isotropic elastic moduli increase exponentially with $|\nabla \cdot \boldsymbol{u}|$. If for the isotropic material we use $P = -K\nabla \cdot \boldsymbol{u}$ to replace the dilatation with applied pressure P, we have

$$G = (G_0 + D_G)\, e^{P/K} - D_G \tag{4.407}$$

$$K = (K_0 + D_K)\, e^{P/K} - D_K, \tag{4.408}$$

which are nonlinear equations that can be solved to give how $G(P)$ and $K(P)$ increase with increasing applied pressure P, which is what is more commonly measured in a laboratory. For solids having G_0 and K_0 on the order of 10^{10} Pa, the D_G and D_K are typically about a factor of 10 to 100 larger than K_0 and G_0 in cracked solids, with greater values corresponding to greater crack densities. The slope of such $G(P)$ and $K(P)$ curves are largest at small P and gradually decrease with increasing P. Feel free to solve for $G(P)$ and $K(P)$ numerically and make a plot. So long as $P/K < 1$, you can also approximate the exponential, solve for $K(P)$ and show that

$$\frac{\mathrm{d}K}{\mathrm{d}P} = \frac{D_K/K_0}{\sqrt{1 + 4D_K P/K_0^2}}, \tag{4.409}$$

where the slope has its maximum value of D_K/K_0 at $P = 0$ and then decreases with increasing P. Observing this slope experimentally allows the change modulus D_K to be measured.

5

Fluid Dynamics

Fluid dynamics (or synonymously *fluid mechanics*) is a broad subject that incorporates a wide range of subdisciplines. In this chapter, we focus mainly on the slow flow of an incompressible viscous fluid, such as liquid water, through various types of conduits or around solid objects. An Earth-science application of this aspect of fluid mechanics is understanding the flow of fluids through the pores of rocks. When coupled with the laws for how temperature is distributed, this chapter also applies to the convection of Earth's highly viscous mantle. Later, in Chapter 7, after we learn some thermodynamics in Chapter 6, we will cover three additional topics of fluid mechanics: (1) capillary physics, which deals with the simultaneous flow of two distinct partially immiscible fluids that do not perfectly mix together even in equilibrium (e.g., air and water); (2) solute diffusion, which deals with how one molecular component (the solute) within a liquid diffuses and convects and changes its concentration during the flow of that fluid; and (3) heat flow, which deals with how temperature diffuses and convects during fluid flow. Other topics of fluid mechanics, such as ideal fluid flow and turbulence (flows at large velocity where viscous stress becomes negligible), shock waves, and plasma physics, will not be treated at all in the introductory treatment that follows but are covered in the classic text of Landau and Lifshitz (1987).

5.1 The Laws of Viscous Fluid Movement

In fluid mechanics, our conservation of mass and momentum laws continue to control the dynamics

$$\frac{1}{\rho}\left(\frac{\partial \rho}{\partial t} + \boldsymbol{v} \cdot \nabla \rho\right) = -\nabla \cdot \boldsymbol{v} \tag{5.1}$$

$$\rho\left(\frac{\partial \boldsymbol{v}}{\partial t} + \boldsymbol{v} \cdot \nabla \boldsymbol{v}\right) = \nabla \cdot \boldsymbol{\tau} + \boldsymbol{F}_b, \tag{5.2}$$

which can also be written in forms that emphasize even more explicitly the accumulation of mass and linear momentum, that is, $\partial\rho/\partial t = -\nabla \cdot (\boldsymbol{v}\rho)$ and $\partial(\boldsymbol{v}\rho)/\partial t = -\nabla \cdot (\boldsymbol{v}\boldsymbol{v}\rho - \boldsymbol{\tau}) + \boldsymbol{F}_b$. However, in a fluid, the constitutive laws for how $\boldsymbol{\tau}$ is related to $\nabla \boldsymbol{v}$ are different from an elastic solid.

The defining characteristic of a fluid is that it cannot sustain a permanent static shear stress when it is not flowing which means that the shear modulus $G = 0$ in a fluid. Further, fluids are generally isotropic (liquid crystals are a well-known counter example not treated here), so Hooke's law of elasticity becomes the simple scalar relation

$$-\frac{\partial P}{\partial t} - \nabla \cdot (\boldsymbol{v}P) = K \nabla \cdot \boldsymbol{v}, \tag{5.3}$$

where P is the elastic fluid pressure (minus one third the trace of the elastic stress tensor) and K is the elastic bulk modulus of the fluid. The stress tensor in a fluid may then be written:

$$\boldsymbol{\tau} = -P\boldsymbol{I} + \boldsymbol{\tau}^{v}, \tag{5.4}$$

where $-P\boldsymbol{I}$ is the elastic stress tensor associated with reversible (i.e., elastic) compressions, and $\boldsymbol{\tau}^{v}$ are "viscous" stresses associated with momentum transfer through molecular collisions.

To understand the viscous stresses $\boldsymbol{\tau}^{v}$, we interpret the stress tensor as representing the flux of linear momentum. From the conservation of linear momentum $\partial(\rho\boldsymbol{v})/\partial t = -\nabla \cdot (\rho\boldsymbol{v}\boldsymbol{v} - \boldsymbol{\tau}) + \boldsymbol{F}_b$, it is clear that $-\boldsymbol{\tau}$, and not $+\boldsymbol{\tau}$, represents positive momentum flux. So for a fixed surface element having a normal $\boldsymbol{n}$, the positive flux of momentum (in the direction of $\boldsymbol{n}$) across the element is $-\boldsymbol{n} \cdot \boldsymbol{\tau}$ if the components of $\boldsymbol{\tau}$ are positive.

With this interpretation of the sign of momentum flux established, to understand the viscous stresses $\boldsymbol{\tau}^{v}$, consider the flow in the z direction depicted in Fig. 5.1. Random thermal fluctuations result in the faster molecules in one "layer" colliding with slower molecules in the neighboring layer. The slower molecules speed up and the faster molecules slow

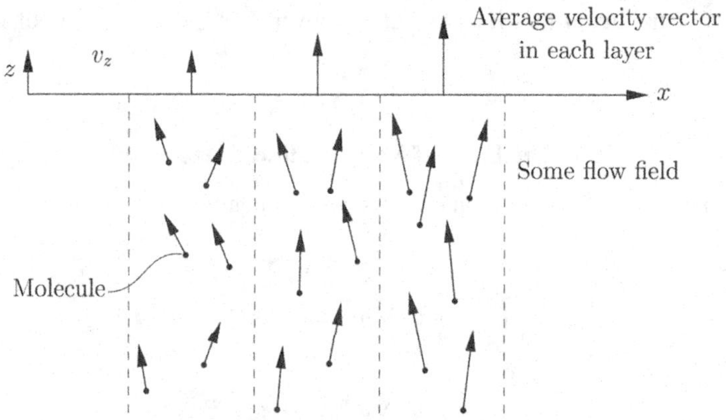

Figure 5.1 Flow in the z direction has a gradient in the positive x direction resulting in a momentum flux in the $-x$ direction created by molecular collisions between the faster and slower moving molecules.

down. In this example, there is thus a flux in the $-\hat{\boldsymbol{x}}$ direction of vertical ($\hat{z}$ direction) momentum or

$$\hat{\boldsymbol{x}} \cdot (-\boldsymbol{\tau}^v) \propto -\frac{\partial v_z}{\partial x}\,\hat{\boldsymbol{z}}, \tag{5.5}$$

where $\partial v_z/\partial x$ is positive, which corresponds to a flux of vertical momentum in the $-\hat{\boldsymbol{x}}$ direction. We can also think of $\hat{\boldsymbol{x}} \cdot \boldsymbol{\tau}^v$ as a shear force (per unit area of surface that is perpendicular to the x direction) that the molecules in the faster-moving layers exert on the molecules in the slower-moving layers due to collisions.

In general, and for either interpretation of $\boldsymbol{n} \cdot \boldsymbol{\tau}^v$ as representing either a negative momentum flux or a force per unit area on a surface element having a normal unit vector $\boldsymbol{n}$, we can write

$$\boldsymbol{n} \cdot \boldsymbol{\tau}^v \propto \boldsymbol{n} \cdot \nabla \boldsymbol{v} \tag{5.6}$$

or since $\boldsymbol{n}$ is arbitrary

$$\boldsymbol{\tau}^v \propto \nabla \boldsymbol{v}. \tag{5.7}$$

So the viscous stress is proportional to strain rate. This is distinct from the elastic stress in which the time derivative of elastic stress is proportional to strain rate.

As usual, we write $\nabla \boldsymbol{v}$ in symmetric and antisymmetric portions

$$\nabla \boldsymbol{v} = \tfrac{1}{2}\left[\nabla \boldsymbol{v} + (\nabla \boldsymbol{v})^T\right] + \tfrac{1}{2}\underbrace{\left[\nabla \boldsymbol{v} - (\nabla \boldsymbol{v})^T\right]}_{\text{Rotations}} \tag{5.8}$$

and because viscous force moments within each fluid element can be neglected so that rigid rotations of the fluid about the center of each averaging element are not present (and do not involve work even if present), we then arrive at the isotropic viscous stress law for a flowing fluid that is known as *Newtonian viscosity*:

$$\boxed{\boldsymbol{\tau}^v = \eta\underbrace{\left[\nabla \boldsymbol{v} + (\nabla \boldsymbol{v})^T - \frac{2}{3}\nabla \cdot \boldsymbol{v}\,\boldsymbol{I}\right]}_{\text{Purely deviatoric}} + \underbrace{\kappa\,\nabla \cdot \boldsymbol{v}\,\boldsymbol{I}}_{\text{Purely compressional}},} \tag{5.9}$$

where the material property η is called the *shear viscosity* and κ the *bulk viscosity* and both have units of Pa s. Typically, both η and κ have comparable orders of magnitude for most fluids but because most fluid-flow problems involve incompressible flow (i.e., $\nabla \cdot \boldsymbol{v} = 0$ as will be discussed later), we rarely see the bulk viscosity used in practice.

So our complete laws of fluid mechanics are

$$
\boxed{
\begin{aligned}
\frac{1}{\rho}\left(\frac{\partial \rho}{\partial t} + \boldsymbol{v}\cdot\nabla\rho\right) &= -\nabla\cdot\boldsymbol{v} && (5.10)\\
\rho\left(\frac{\partial \boldsymbol{v}}{\partial t} + \boldsymbol{v}\cdot\nabla\boldsymbol{v}\right) &= -\nabla P + \nabla\cdot\boldsymbol{\tau}^{v} + \boldsymbol{F}_b && (5.11)\\
\frac{\partial P}{\partial t} + \nabla\cdot(\boldsymbol{v}P) &= -K\,\nabla\cdot\boldsymbol{v} && (5.12)\\
\boldsymbol{\tau}^{v} = \eta\Big(\nabla\boldsymbol{v} + (\nabla\boldsymbol{v})^{T} &- \tfrac{2}{3}\nabla\cdot\boldsymbol{v}\,\boldsymbol{I}\Big) + \kappa\,\nabla\cdot\boldsymbol{v}\,\boldsymbol{I}. && (5.13)
\end{aligned}
}
$$

Fortunately, we rarely need to model all of these coupled nonlinear first-order equations simultaneously. Throughout each section of this chapter, we will make a range of standard approximations.

5.2 Linear Acoustics

The propagation of sound in fluids is the subject known as acoustics. In *linear acoustics*, the wave amplitudes are sufficiently small that we can neglect the accumulation terms in the above laws, which for the conservation of mass and momentum is equivalent to neglecting the advective derivatives $\boldsymbol{v}\cdot\nabla$. For a wavefield ψ, it was shown in Chapter 4 that the accumulation term $\nabla\cdot(\boldsymbol{v}\psi)$ has an amplitude relative to the time derivative $\partial\psi/\partial t$ given by $|\nabla\cdot(\boldsymbol{v}\psi)|/|\partial\psi/\partial t| = |\boldsymbol{v}|/c_p =$ wave strain and the wave strain is assumed to be quite small relative to one for a linear wave. In this expression, c_p is the acoustic wave speed.

Further, if we decompose the fluid density into a constant and fluctuating portion

$$
\rho(\boldsymbol{r}, t) = \rho_0 + \delta\rho(\boldsymbol{r}, t), \tag{5.14}
$$

the conservation of mass gives

$$
\frac{\partial\,\delta\rho}{\partial t} + \boldsymbol{v}\cdot\overset{\text{Small}}{\cancel{\nabla}}\,\delta\rho = -\rho_0\left(1 + \frac{\delta\rho}{\rho_0}\right)\nabla\cdot\boldsymbol{v}. \tag{5.15}
$$

To make order-of-magnitude estimates of the time and space derivatives in Eq. (5.15), we consider a wave having speed c_p, period T, and wavelength $\lambda = c_p T$ and identify $\partial/\partial t \approx 1/T$ and $|\nabla| \approx 1/\lambda$ so that Eq. (5.15) can be rewritten in the order-of-magnitude form

$$
\frac{|\delta\rho|/\rho_0}{(1 + |\delta\rho|/\rho_0)} \approx \frac{T}{\lambda}|\boldsymbol{v}| = \frac{|\boldsymbol{v}|}{c_p} \approx \text{wave strain} \ll 1. \tag{5.16}
$$

So density is well approximated as being constant during wave propagation. In passing, although for an acoustic wave, we can interpret $|\boldsymbol{v}|/c_p$ as the wave strain, for any problem in fluid mechanics, the ratio of fluid velocity to sound speed is given the special name

$$
M = \frac{|\boldsymbol{v}|}{c_p}, \tag{5.17}
$$

where the dimensionless number M is called the *Mach number*. For all flow problems we will consider in this chapter, and not just acoustics, the Mach number M can be considered to be small; however, in shock waves, which represents nonlinear response, the Mach number is no longer negligible.

With advective derivatives neglected and density taken to be time invariant due to the small wave-strain assumption, the laws of linear acoustics for a homogeneous fluid are

$$\rho_0 \frac{\partial \boldsymbol{v}}{\partial t} = -\nabla P + \left(\kappa + \frac{\eta}{3}\right) \nabla \nabla \cdot \boldsymbol{v} + \eta \nabla^2 \boldsymbol{v} + \boldsymbol{F}_b \tag{5.18}$$

$$\frac{\partial P}{\partial t} = -K \nabla \cdot \boldsymbol{v}. \tag{5.19}$$

For a homogeneous fluid, take the divergence of Eq. (5.18) and substitute in Eq. (5.19) to give

$$-\frac{\rho_0}{K} \frac{\partial^2 P}{\partial t^2} = -\nabla^2 P - \frac{(\kappa + 4\eta/3)}{K} \nabla^2 \frac{\partial P}{\partial t} + \nabla \cdot \boldsymbol{F}_b. \tag{5.20}$$

Identifying $c_p = \sqrt{K/\rho_0}$ then yields

$$\boxed{\nabla^2 \Bigg[P + \underbrace{\frac{(\kappa + 4\eta/3)}{K} \frac{\partial P}{\partial t}}_{\text{Viscous loss term}} \Bigg] - \frac{1}{c_p^2} \frac{\partial^2 P}{\partial t^2} = \nabla \cdot \boldsymbol{F}_b.} \tag{5.21}$$

When both the Hooke's law of Eq. (5.19) and the law of Newtonian viscosity are being allowed for, as they are in the acoustics of a viscous fluid, we say the wave response is *viscoelastic*.

To address the importance of the viscous "loss" contribution to the propagation of acoustic waves, consider the dimensionless ratio

$$\left| \frac{(\kappa + 4\eta/3)\, \partial P/\partial t}{KP} \right| \approx f \frac{(\kappa + 4\eta/3)}{K}, \tag{5.22}$$

where $f = 1/T \mathrel{\widehat{=}}$ wave frequency. Typical values in liquids are

$$\kappa + 4\eta/3 \approx 10^{-3} \text{ Pa s} \tag{5.23}$$

$$K \approx 10^9 \text{ Pa}. \tag{5.24}$$

Thus, so long as $f \ll 10^{12}$ Hz which is always the case (meaning our acoustic wavelengths $\lambda = c_p/f$ are much larger than 10^{-9} m which is a requirement for a continuum description to hold), we can neglect the viscosity term in the above wave equation that produces wave attenuation, and the law for acoustics is

$$\boxed{\nabla^2 P - \frac{1}{c_p^2} \frac{\partial^2 P}{\partial t^2} = \nabla \cdot \boldsymbol{F}_b,} \quad \text{where} \quad c_p = \sqrt{\frac{K}{\rho_0}}. \tag{5.25}$$

This is the equation that controls sound moving through air or any fluid and is a classic scalar wave equation as will be studied at length in Part II of this book. Note that in an ideal gas, a standard exercise in statistical mechanics shows that the adiabatic bulk modulus of the gas is $K = \gamma P$, where P is the fluid pressure and the dimensionless parameter γ is $\gamma = 1.4$ for air (this γ will be addressed in Chapter 6 on thermodynamics). However, we can again go through the exercise of writing the pressure field as $P(\boldsymbol{r}, t) = P_0 + \delta P(\boldsymbol{r}, t)$ and obtain that the wave-induced pressure perturbations satisfy $|\delta P|/P_0 \approx$ strain $\ll 1$ so that to leading order we have $c_p = \sqrt{\gamma P_0/\rho_0}$ in an ideal gas (which is also a nice approximation for most gases).

In a liquid like water, that has $\rho_0 = 10^3$ kg/m^3 and $K = 2.2 \times 10^9$ Pa, we have $c_p = 1483$ m/s. In a gas like air, that has $\rho_0 = 1.23$ kg/m^3 and $K \approx \gamma P_{\text{atm}} \approx 1.4 \times 10^5$ Pa, we have $c_p \approx 337$ m/s.

5.3 Incompressible Viscous Flow

We now consider situations where a viscous fluid is moving under the influence of a pressure gradient or gravity or the movement of a solid boundary. The most important simplification for solving this class of problems is that we can almost always assume the flow to be incompressible.

5.3.1 Conditions for Flow to Be Incompressible

Incompressible flow means we can neglect density changes in moving flow elements, that is,

$$-\frac{1}{\rho}\frac{\mathrm{d}\rho}{\mathrm{d}t} = \boxed{\nabla \cdot \boldsymbol{v} = 0.} \tag{5.26}$$

Despite pressure varying in space and time, $\nabla \cdot \boldsymbol{v} = 0$ means that the density changes that are also occurring can be neglected. The assumption of incompressible flow is always an approximation. Flows for which $\nabla \cdot \boldsymbol{v} \approx 0$ are called *incompressible* or *solenoidal*. Although $\nabla \cdot \boldsymbol{v} = 0$ is often called the *continuity condition*, we will call it the *incompressible-flow condition*.

To obtain conditions for this approximation to be valid, we start with the equation of motion

$$\rho\frac{\mathrm{d}\boldsymbol{v}}{\mathrm{d}t} = -\nabla P + \nabla \cdot \boldsymbol{\tau}^v \tag{5.27}$$

and take a total time derivative ($\mathrm{d}/\mathrm{d}t = \partial/\partial t + \boldsymbol{v} \cdot \nabla$) to give

$$\frac{\mathrm{d}\rho}{\mathrm{d}t}\frac{\mathrm{d}\boldsymbol{v}}{\mathrm{d}t} + \rho\frac{\mathrm{d}^2\boldsymbol{v}}{\mathrm{d}t^2} = -\nabla\frac{\mathrm{d}P}{\mathrm{d}t} + \nabla \cdot \frac{\mathrm{d}\boldsymbol{\tau}^v}{\mathrm{d}t}. \tag{5.28}$$

If we can show that the dimensionless number S defined as

$$S \mathrel{\widehat{=}} \left|\frac{\mathrm{d}\rho}{\mathrm{d}t}\frac{\mathrm{d}\boldsymbol{v}}{\mathrm{d}t}\right| \Big/ \left|\nabla\frac{\mathrm{d}P}{\mathrm{d}t}\right| \ll 1, \tag{5.29}$$

then changes in the density of a flow element may be neglected in the force balance despite changes in pressure being nonnegligible so that $\nabla \cdot \boldsymbol{v} \approx 0$ (incompressible flow).

Hooke's law states that

$$\nabla \cdot \boldsymbol{v} = -\frac{1}{\rho}\frac{\mathrm{d}\rho}{\mathrm{d}t} = -\frac{1}{K}\frac{\mathrm{d}P}{\mathrm{d}t}, \tag{5.30}$$

which expresses the dilemma surrounding the assumption of incompressible flow: How can we ignore the density variations in time and/or space while at the same time allow for the pressure variations? The answer is that we are ignoring the density variations relative to the pressure variations *within the force balance equation* of Eq. (5.27), which is what the dimensionless parameter S addresses specifically.

From Hooke's law we can write

$$\nabla \frac{\mathrm{d}P}{\mathrm{d}t} = \nabla\left[\frac{K}{\rho}\frac{\mathrm{d}\rho}{\mathrm{d}t}\right] = \nabla\left[c_p^2 \frac{\mathrm{d}\rho}{\mathrm{d}t}\right], \tag{5.31}$$

where $c_p = \sqrt{K/\rho}$ is again the speed of sound. Thus, we can write the dimensionless ratio S as

$$S \mathrel{\widehat{=}} \frac{\left|\mathrm{d}\rho/\mathrm{d}t\right| \left|\mathrm{d}\boldsymbol{v}/\mathrm{d}t\right|}{\left|\nabla(\mathrm{d}P/\mathrm{d}t)\right|} = \frac{\left|\mathrm{d}\rho/\mathrm{d}t\right| \left|\partial \boldsymbol{v}/\partial t + \boldsymbol{v} \cdot \nabla \boldsymbol{v}\right|}{\left|\nabla(c_p^2 \mathrm{d}\rho/\mathrm{d}t)\right|}. \tag{5.32}$$

Now if fields are varying over a distance ℓ and/or fields are changing with a frequency f (where $f = 1/T$ with $T \mathrel{\widehat{=}}$ the time period of change), we can approximate the differential operators as $\nabla \approx 1/\ell$ and $\partial/\partial t \approx f$ to obtain the order-of-magnitude estimate

$$S \approx \frac{f\ell|\boldsymbol{v}|}{c_p^2} + \frac{|\boldsymbol{v}|^2}{c_p^2} = M\left(\frac{\ell}{\lambda} + M\right), \tag{5.33}$$

where $\lambda = c_p/f$ is the wavelength of sound induced by any temporal variations occurring at a frequency f. The second condition that $|\boldsymbol{v}|^2/c_p^2 = M^2 \ll 1$ (where M is again the Mach number) is always satisfied for all subsonic flow scenarios. The first condition that $M\ell/\lambda \ll 1$ is always satisfied for all steady-flow scenarios ($f = 0$) and usually satisfied even when $f \neq 0$ because $\lambda > \ell$ and $M \ll 1$. So long as the wavelength of any induced sound is larger than the characteristic length ℓ over which fields are varying, which, for example, may be the pore size in a porous material, the flow can be taken as incompressible even when $f \neq 0$. We can conclude that in nearly all practical flow conditions where $|\boldsymbol{v}| \ll c_p$, we have

$$S \ll 1 \quad \text{and, therefore,} \quad \boxed{\nabla \cdot \boldsymbol{v} = 0} \tag{5.34}$$

and so flow can indeed be approximated as being incompressible in the solution of subsonic flow problems.

5.3.2 Incompressible Flow and Pressure Variations in Space and Time

Under steady-state conditions, we can understand that when $\nabla \cdot \boldsymbol{v} = 0$, as much fluid is entering each volume element as is leaving so that the density of the element is not changing. Imagine steady flow taking place in a tube under the influence of a pressure drop between the two ends of the tube. If there is a constriction in the tube, the condition $\nabla \cdot \boldsymbol{v} = 0$ means that the flow speed must increase as we enter the constricted zone and decrease as we leave the constricted zone so that the same volume of fluid is passing across each tube cross section per unit time. For the flow speed to be greater in the constricted zone, the local pressure gradient that is driving flow must be higher across the constricted zone compared to the average pressure gradient down the entire length of the tube. The condition $\nabla \cdot \boldsymbol{v} = 0$ is what allows us to calculate the locally increasing and decreasing pressure gradient on the upstream and downstream sides of the constriction.

Now, let's begin to vary in time the pressure in the upstream reservoir that feeds the inlet into the tube. The time-variable pressure in the upstream reservoir will work its way down the tube of length L with $\nabla \cdot \boldsymbol{v} = -\rho^{-1} \mathrm{d}\rho/\mathrm{d}t = -K^{-1} \mathrm{d}P/\mathrm{d}t \neq 0$, that is, as a *compressible flow*. As an exercise at the end of the chapter, you can show that the average flow across a cross section of a cylindrical pipe of radius a goes as

$$\overline{v}_z = -\frac{a^2}{8\eta} \frac{\partial P}{\partial z}, \tag{5.35}$$

where z is axial distance down the pipe and where the overbar denotes an average across each cross section of the tube. Using Hooke's law $-\partial P/\partial t = K \partial \overline{v}_z/\partial z$ for the compressions, the pressure satisfies a diffusion equation

$$D_p \frac{\partial^2 P}{\partial z^2} - \frac{\partial P}{\partial t} = 0, \quad \text{where} \quad D_p \mathrel{\widehat{=}} \text{pressure diffusivity} = \frac{K a^2}{8\eta}. \tag{5.36}$$

If the temporal variations in the upstream reservoir are time harmonic (say $\cos \omega t$) with a circular frequency $\omega = 2\pi f$, we have already seen in Chapter 3 that the distance diffused by a time-harmonic diffusion is characterized by the pressure-diffusion skin depth δ_p, which is defined, as for all time-harmonic diffusions, as

$$\delta_p = \sqrt{\frac{2 D_p}{\omega}}. \tag{5.37}$$

For sufficiently low frequencies that $\delta_p \gg L$, where L is the length of the tube, the pressure gradient down the length of the tube is in a quasi-static steady state characterized by $\partial^2 P/\partial z^2 = 0$, which from Eq. (5.35) says that $\partial \overline{v}_z/\partial z = 0$ (incompressible flow) and $\partial P/\partial z = (\Delta P/L) \cos \omega t$ (quasi-static forcing).

A bit less precisely, if we estimate derivatives in the diffusion equation using $\partial/\partial z \approx 1/L$ and $\partial/\partial t \approx f$, where f is the frequency in Hertz, the diffusion equation gives the incompressible-flow condition of $\partial^2 P/\partial z^2 = 0$ to be

$$f \ll \frac{D_p}{L^2}, \tag{5.38}$$

which is equivalent to $\delta_p \gg L$. If our pipe has a length of $L = 1$ m and a radius $a = 10^{-3}$ m, then because water has $K \approx 10^9$ Pa and $\eta = 10^{-3}$ Pa s, we find that $D_p/L^2 \approx 10^5$ Hz. So long as the temporal pressure variations in the reservoir feeding the tube are much smaller than 10^5 Hz in this example, the flow can be considered incompressible along the length of the tube and the evolving pressure gradient can be assumed to be established instantaneously ($\nabla \cdot \boldsymbol{v} = 0$).

5.3.3 Navier–Stokes Equations

Given that $\nabla \cdot \boldsymbol{v} = 0$ for low Mach-number (subsonic) flows or for time-harmonic diffusive flow scenarios at sufficiently low frequencies, we obtain the equations of incompressible viscous flow called the *Navier–Stokes equations*

$$\rho\left(\frac{\partial \boldsymbol{v}}{\partial t} + \boldsymbol{v}\cdot\nabla\boldsymbol{v}\right) = -\nabla P + \eta\nabla^2\boldsymbol{v} + \boldsymbol{F}_b \tag{5.39}$$

$$\nabla \cdot \boldsymbol{v} = 0. \tag{5.40}$$

In these equations, ρ is taken as a constant-in-time material property and the unknowns are the fluid pressure P and the flow velocity $\boldsymbol{v}$. You can think of the force balance (conservation of momentum) as providing three equations for the three components of $\boldsymbol{v}$ and the incompressible-flow condition $\nabla \cdot \boldsymbol{v} = 0$ as providing the additional equation for P.

The body forces, as always, are still given by

$$\boldsymbol{F}_b = (\rho_e - \nabla\cdot\boldsymbol{P})\boldsymbol{E} + \left(\boldsymbol{J} + \frac{\partial \boldsymbol{P}}{\partial t} + \nabla\times\boldsymbol{M}\right)\times\boldsymbol{B} + \rho\boldsymbol{g} \tag{5.41}$$

and are either electromagnetic or gravitational in nature. In the later section on magnetohydrodynamics, we will solve a flow problem that includes the effect of a magnetic force $\boldsymbol{J}\times\boldsymbol{B}$ and as an exercise at the end of the chapter, you can solve an electro-osmotic problem that includes the effect of an electric force $\rho_e\boldsymbol{E}$.

To include the effects of a gravitational acceleration $\boldsymbol{g} = g\hat{\boldsymbol{z}}$, the Navier–Stokes equations are written

$$\rho\left(\frac{\partial \boldsymbol{v}}{\partial t} + \boldsymbol{v}\cdot\nabla\boldsymbol{v}\right) = \eta\nabla^2\boldsymbol{v} - \nabla P + \rho\boldsymbol{g} \quad \text{and} \quad \nabla\cdot\boldsymbol{v} = 0. \tag{5.42}$$

It is common that a flow domain is sufficiently small relative to the size of the Earth that the acceleration of gravity $\boldsymbol{g}$ can be taken as a uniform constant throughout the domain. If we further assume the fluid is homogeneous so that ρ is also a uniform constant, the pressure field can be written in a way that includes the gravitational effects

$$P(\boldsymbol{r}, t) = P_0 + \rho\boldsymbol{g}\cdot\boldsymbol{r} + P_v(\boldsymbol{r}, t) \tag{5.43}$$

where the *viscous pressure* field $P_v(\boldsymbol{r}, t)$ is that required to overcome the resistance provided by viscosity in the flowing fluid. The Navier–Stokes equation then becomes

$$\rho\left(\frac{\partial \boldsymbol{v}}{\partial t}+\boldsymbol{v}\cdot\nabla\boldsymbol{v}\right)=\eta\nabla^2\boldsymbol{v}-\nabla P_v, \tag{5.44}$$

where ∇P_v is called the *viscous pressure gradient*. The viscous pressure gradient can be applied through boundary conditions on say a flow conduit to create flow down the conduit and/or created by the viscous flow around curved walls in order that $\nabla\cdot\boldsymbol{v}=0$.

For static (stationary) fluids where $P_v=0$ or for flows in the x, y directions that are perpendicular to the direction of gravity $\boldsymbol{g}=g\hat{z}$ for which $v_z=0$ and $\partial P_v/\partial z=0$ as required by the z-component of Navier Stokes, the pressure distribution in the domain is determined entirely by

$$P(x, y, z)=P_0+P_v(x, y)+\rho g z, \tag{5.45}$$

where $P_v(x, y)$ is independent of z. So pressure varies with depth z in a fluid layer as $\rho g z$ even when there is horizontal flow perpendicular to the z direction. At the bottom of a horizontally flowing ocean with, say, $z=1{,}000$ m, we have $\rho\approx 10^3$ kg/m^3 and $g\approx 10$ m/s^2. Atmospheric pressure is $P_0=1$ atm ≈ 1 bar $=10^5$ Pa. So we find, $P(z=1{,}000\text{ m})=10^5\text{ Pa}+10^7\text{ Pa}\approx 10$ MPa, which is roughly 100 times more than atmospheric pressure.

5.3.4 Vorticity and Pressure

When the nonlinear advective acceleration term $\rho\boldsymbol{v}\cdot\nabla\boldsymbol{v}$ is negligible, conditions for which are given in the next section, the Navier–Stokes equations are

$$\rho\frac{\partial \boldsymbol{v}}{\partial t}=-\nabla P+\eta\nabla^2\boldsymbol{v} \tag{5.46}$$

along with $\nabla\cdot\boldsymbol{v}=0$. For convenience in this discussion, we have dropped the body force term in the original Navier–Stokes equation.

Taking the divergence of this linear Navier–Stokes equation and using the incompressibility condition gives

$$\nabla^2 P=0. \tag{5.47}$$

So the pressure satisfies a Laplace equation for linear viscous-incompressible-flow problems.

If we take the curl of Eq. (5.46) and use that $\nabla\times\nabla P=0$ and take η to be a spatial constant, we obtain

$$D_v\nabla^2\boldsymbol{\omega}-\frac{\partial\boldsymbol{\omega}}{\partial t}=0 \tag{5.48}$$

where the *vorticity* vector field is defined

$$\boldsymbol{\omega}=\nabla\times\boldsymbol{v} \tag{5.49}$$

and the *viscous diffusivity* D_v, also called the kinematic viscosity and usually denoted as ν, is defined

$$D_v = \frac{\eta}{\rho} \mathrel{\widehat{=}} \nu. \tag{5.50}$$

So the vorticity of a viscous-linear-incompressible flow that has explicit time dependence satisfies a diffusion equation. At steady state, the vorticity, like the pressure field, satisfies a Laplace equation.

It may appear at first sight that we have decoupled the solution for the pressure P from that for the vorticity $\boldsymbol{\omega}$. We will consider boundary conditions in Section 5.5 and will see that the flow velocity must go to zero at solid walls. So the vorticity is always largest right at solid walls and falls off with distance away from the wall. Because the flow velocity and therefore any flow acceleration is zero right at the wall, the Navier–Stokes equation at the wall is $\eta\nabla^2\boldsymbol{v} = \nabla P$. Using the Chapter 1 identity $\nabla^2\boldsymbol{v} = \nabla(\nabla\cdot\boldsymbol{v}) - \nabla\times\nabla\times\boldsymbol{v}$ along with $\nabla\cdot\boldsymbol{v} = 0$ and $\boldsymbol{\omega} = \nabla\times\boldsymbol{v}$, the Navier–Stokes equation at the wall is $\eta\nabla\times\boldsymbol{\omega} = -\nabla P$. We can write this as the boundary condition

$$\eta\boldsymbol{n}\times\nabla\times\boldsymbol{\omega} = -\boldsymbol{n}\times\nabla P, \tag{5.51}$$

where $\boldsymbol{n}$ is the unit normal vector of the fluid–solid interface. So the pressure and vorticity fields are coupled to each other at the solid wall and the diffusion problem for vorticity is coupled to the Laplace problem for the pressure.

Last, if we integrate the Navier–Stokes equation at the solid wall $0 = -\eta\nabla\times\boldsymbol{\omega} - \nabla P$ over a thin disc that straddles the fluid–solid interface, use the fundamental theorem of 3D calculus from Chapter 1 and take the limit as the disc thickness shrinks to 0, we have that the viscous fluid exerts a force per unit area $\boldsymbol{n}\cdot\boldsymbol{\tau}_s$ on the solid surface given by

$$\boldsymbol{n}\cdot\boldsymbol{\tau}_s = -\eta\boldsymbol{n}\times\boldsymbol{\omega} - \boldsymbol{n}P, \tag{5.52}$$

where $\boldsymbol{n}$ is directed from solid to fluid here. The stress $-\eta\boldsymbol{n}\times\boldsymbol{\omega}$ is the tangential viscous-shear stress acting on the wall while $-\boldsymbol{n}P$ is the normal stress acting on the wall.

5.4 Dimensional Analysis

When a flow problem is posed, one should begin by determining the importance of the various terms in the equations and specifically any nonlinear terms such as $\boldsymbol{v}\cdot\nabla\boldsymbol{v}$ in Navier–Stokes. If a nonlinear term is determined to be negligible in comparison to other terms, the solution, either analytical or numerical, is greatly simplified. If a nonlinear term is small but nonnegligible, we can iteratively include its effect through what is called a *perturbation analysis*. As a general matter for solving any set of complicated nonlinear partial-differential equations (PDEs), and in particular when our solution is based on a perturbation analysis, it is good to take the physical dimensions out of the problem. When we do so, we can also establish a rule about flows in self-similar geometries called *similarity*. These interrelated topics are all treated in this section.

5.4.1 Direct Term-to-Term Comparison

In the Navier–Stokes equations

$$\rho\left(\frac{\partial \boldsymbol{v}}{\partial t}+\boldsymbol{v}\cdot\nabla\boldsymbol{v}\right)=\eta\,\nabla^2\boldsymbol{v}-\nabla P \quad \text{and} \quad \nabla\cdot\boldsymbol{v}=0, \tag{5.53}$$

if we compare the amplitude of the nonlinear convective (also called advective) acceleration $\rho\boldsymbol{v}\cdot\nabla\boldsymbol{v}$ to the amplitude of the viscous shearing term $\eta\nabla^2\boldsymbol{v}$ we obtain a dimensionless number Re, called the *Reynolds number*, that is defined physically as the ratio

$$\text{Re} \mathrel{\widehat{=}} \frac{|\rho\,\boldsymbol{v}\cdot\nabla\boldsymbol{v}|}{|\eta\,\nabla^2\boldsymbol{v}|}=\frac{\text{inertial convection}}{\text{viscous shearing}}. \tag{5.54}$$

To estimate Re, if $v_c=|\boldsymbol{v}|$ is a characteristic flow velocity and if ℓ is a characteristic distance over which $\boldsymbol{v}$ varies, we can estimate derivatives using $\nabla\sim 1/\ell$ to give

$$\text{Re}=\frac{\rho\,v_c^2/\ell}{\eta\,v_c/\ell^2}=\frac{\rho v_c\ell}{\eta}. \tag{5.55}$$

When Re $\ll 1$, the nonlinear term may be neglected. In the steady-state limit of a flow problem, when the forcing and flow fields are not varying in time, the low-Reynolds number form of Navier–Stokes is

$$\boxed{0=-\nabla P+\eta\nabla^2\boldsymbol{v} \quad \text{and} \quad \nabla\cdot\boldsymbol{v}=0.} \quad \textit{Stokes flow} \tag{5.56}$$

Flow fields satisfying these equations are called *Stokes flow*. We can assume Stokes flow even for flow past curved solid surfaces, for which the nonlinear $\boldsymbol{v}\cdot\nabla\boldsymbol{v}$ term is formally not exactly zero, so long as Re $\ll 1$. In such cases, the flow near a solid surface is parallel with the possibly curved surface and regularly increases with increasing distance from the surface without developing eddies or ripples. We refer to such regularly varying flow as being *laminar*. So Stokes flow is always laminar.

When Re > 1, the nonlinear term dominates, eddies near curved solid surfaces develop and turbulent flow eventually emerges at large enough Re. When Re $\gg 1$, the viscous stressing term can be dropped altogether in the Navier–Stokes equation and we say that the fluid flow is *ideal*. That said, near a stationary solid surface where the flow velocity goes to zero, viscosity will dominate in a thin *viscous boundary layer* whose thickness decreases with increasing flow velocity. High Reynolds number flow dominates our everyday human experience. For example, as we are walking down the street with our arms swinging at roughly 1 m/s, the air flow created by our swinging arms has a Reynolds number on the order of 10^4 so viscous stressing is completely negligible in the turbulent air flow created as we walk.

In a problem with temporal variations taking place with a characteristic time t_c so that the time derivative can be estimated as $\partial/\partial t\sim 1/t_c$, if we compare the amplitude of the temporal acceleration term to the viscous shearing term, we obtain another nondimensional number Rk called the *Roshko number*

$$\mathrm{Rk} \,\widehat{=}\, \frac{\rho|\partial \boldsymbol{v}/\partial t|}{\eta|\nabla^2 \boldsymbol{v}|} = \frac{\rho \ell^2}{\eta t_c}. \tag{5.57}$$

Note that we don't use Ro for the Roshko number because we have already identified Ro as the *Rossby number*, which is the dimensionless ratio of the inertial-advection force to the Coriolis force in a rotating body. If you have been exposed to fluid mechanics prior to this reading, you likely have heard of the Reynolds number but will never have heard of the Roshko number. A first reason is that many flow problems are posed in the steady state where $t_c \to \infty$ and $\mathrm{Rk} = 0$. Further, if the time variations in the problem are due to viscous diffusion as a result of an initial condition that creates transitory flow over a characteristic time scale t_c prior to achieving a steady state with respect to the imposed boundary conditions or applied forcing, the characteristic time can be identified as $t_c = \ell^2/D_v$, where the viscous diffusivity is again defined $D_v = \eta/\rho$, which results in a Roshko number of $\mathrm{Rk} = 1$ (i.e., if you want to treat this transitory diffusive flow in the passage toward a steady state, you have to include the time-derivative term in Navier Stokes). Only in scenarios where the temporal variations are due to applied time-varying forces or oscillating boundary walls having a characteristic frequency f_c, in which we identify $t_c = 1/f_c$, do we get a Roshko number different than 0 or 1 as given by $\mathrm{Rk} = \rho\ell^2 f_c/\eta$. If Rk is very small compared to 1, we can drop the temporal acceleration term altogether. Problems in which there is time-varying forcing or time-varying boundary conditions with a negligible Roshko number (and therefore negligible time derivative in the Navier–Stokes equation) are called *quasistatic*. As a limit in the time-varying problem treated in Section 5.6.5, we will give an example of quasistatic response. If $\mathrm{Re} \ll \mathrm{Rk}$, we might keep the temporal acceleration but should drop the nonlinear advective acceleration.

5.4.2 Nondimensional Governing Equations and Similarity

Another way to accomplish the above, that has additional advantages, is to restate the governing equations in entirely nondimensional form. We denote dimensionless fields, operators, coordinates, and time using a subscript n (for nondimensional) and use the characteristic values v_c, P_c, ℓ, and t_c to write

$$\boldsymbol{v} = v_c \boldsymbol{v}_n, \quad P = P_c P_n, \quad \nabla = \frac{1}{\ell}\nabla_n, \quad \frac{\partial}{\partial t} = \frac{1}{t_c}\frac{\partial}{\partial t_n}, \quad \boldsymbol{r} = \ell \boldsymbol{r}_n, \quad \text{and} \quad t = t_c t_n. \tag{5.58}$$

The characteristic pressure P_c will be defined momentarily. Introducing this change of variables into the Navier–Stokes equations, we get

$$\rho\left(\frac{v_c}{t_c}\frac{\partial \boldsymbol{v}_n}{\partial t_n} + \frac{v_c^2}{\ell}\boldsymbol{v}_n \cdot \nabla_n \boldsymbol{v}_n\right) = \frac{\eta v_c}{\ell^2}\nabla_n^2 \boldsymbol{v}_n - \frac{P_c}{\ell}\nabla_n P_n. \tag{5.59}$$

If we next identify the characteristic pressure as

$$P_c = \frac{\eta v_c}{\ell} \tag{5.60}$$

and divide through by $\eta v_c/\ell^2$ we obtain

$$\left(\frac{\rho \ell^2}{\eta t_c}\right)\frac{\partial \boldsymbol{v}_n}{\partial t_n} + \left(\frac{\rho v_c \ell}{\eta}\right)\boldsymbol{v}_n \cdot \nabla_n \boldsymbol{v}_n = \nabla_n^2 \boldsymbol{v}_n - \nabla_n P_n. \tag{5.61}$$

Introducing the Roshko number $\text{Rk} = \rho\ell^2/(\eta t_c)$ and Reynolds number $\text{Re} = \rho v_c \ell/\eta$ yields the nondimensional form of the Navier–Stokes equations

$$\boxed{\text{Rk}\,\frac{\partial \boldsymbol{v}_n}{\partial t_n} + \text{Re}\,\boldsymbol{v}_n \cdot \nabla \boldsymbol{v}_n = \nabla_n^2 \boldsymbol{v}_n - \nabla_n P_n \quad \text{with} \quad \nabla_n \cdot \boldsymbol{v}_n = 0.} \tag{5.62}$$

All symbols in the problem, including distance and time, are without physical dimension and other than for Rk and Re are of $O(1)$ in amplitude. This is an excellent form from which to begin numerical solution of the Navier–Stokes equations.

This exercise shows that a solution of these nondimensional equations are

$$\boldsymbol{v}_n = \boldsymbol{v}_n(\text{Re}, \text{Rk}, \boldsymbol{r}_n, t_n) \quad \text{and} \quad P_n = P_n(\text{Re}, \text{Rk}, \boldsymbol{r}_n, t_n). \tag{5.63}$$

This functional dependence is called *similarity* for reasons described next.

Consider two different flow domains that are simply scaled-up or scaled-down versions of each other. Geometries that look the same other than for a uniform inflation (or deflation) of the domain are called *self-similar*. The solution for the flow fields in each of these differently sized but self-similar domains will be identical to each other after a trivial scaling of the spatial and temporal units so long as each flow problem has the same dimensionless numbers Re and Rk. For the experiment at scale 1, let's say we have obtained the flow solutions with all symbols having their usual physical dimensions (meters, seconds, Pascals, etc.)

$$\boldsymbol{v}_1 = \boldsymbol{v}_1(\text{Re}_1, \text{Rk}_1, \boldsymbol{r}, t) \quad \text{and} \quad P_1 = P_1(\text{Re}_1, \text{Rk}_1, \boldsymbol{r}, t). \tag{5.64}$$

The solution for a self-similar version of the same problem at scale 2, again with all symbols having their usual physical dimensions, is given by the simple *similarity* scaling of the experimental fields obtained at scale 1

$$\boldsymbol{v}_2(\text{Re}_1, \text{Rk}_1, \boldsymbol{r}, t) = \frac{v_{c2}}{v_{c1}}\boldsymbol{v}_1\left(\text{Re}_1, \text{Rk}_1, \frac{\ell_2}{\ell_1}\boldsymbol{r}, \frac{t_{c2}}{t_{c1}}t\right) \tag{5.65}$$

$$P_2(\text{Re}_1, \text{Rk}_1, \boldsymbol{r}, t) = \frac{\eta_2 v_{c2}\ell_1}{\eta_1 v_{c1}\ell_2}P_1\left(\text{Re}_1, \text{Rk}_1, \frac{\ell_2}{\ell_1}\boldsymbol{r}, \frac{t_{c2}}{t_{c1}}t\right). \tag{5.66}$$

The requirement for this similarity scaling to hold is that the Reynolds and Roshko numbers at scale 2 are identical to their value at scale 1, that is, we must use a v_{c2}, ρ_2, η_2, and t_{c2} at scale 2 such that

$$\frac{\rho_2 \ell_2^2}{\eta_2 t_{c2}} = \frac{\rho_1 \ell_1^2}{\eta_1 t_{c1}} \quad \text{and} \quad \frac{\rho_2 v_{c2}\ell_2}{\eta_2} = \frac{\rho_1 v_{c1}\ell_1}{\eta_1}. \tag{5.67}$$

Laboratory experimentalists, for example, can exploit this by modeling a given flow on their benchtop that is supposed to represent a larger (or smaller) flow domain somewhere outside their lab. To get that Re and Rk are the same in the outside and benchtop flow problems, the experimentalists have to choose their flow speeds, applied temporal variations and fluid properties carefully for their analog laboratory experiments. Taking dimensions out of physics problems to obtain similarity scalings can be used for any type of problem.

5.4.3 Perturbation Analysis

Consider a steady-flow problem for which $\text{Rk} = 0$. We want to solve for the flow in a regime where Re is small but not entirely negligible. This typically involves flow near curved solid surfaces. How can we solve the nonlinear Navier–Stokes equation in this case?

One approach is to first identify $\epsilon = \text{Re}$ to be a small parameter, called a *perturbation parameter*. Using the dimensionless form of the Navier–Stokes equation given above but with the cumbersome subscript n indicating "nondimensional" dropped, one introduces into it the following *perturbation series* for the dimensionless flow fields

$$\boldsymbol{v}(\boldsymbol{r}) = \boldsymbol{v}_0(\boldsymbol{r}) + \epsilon\, \boldsymbol{v}_1(\boldsymbol{r}) + \epsilon^2\, \boldsymbol{v}_2(\boldsymbol{r}) + \ldots = \sum_{m=0}^{\infty} \epsilon^m\, \boldsymbol{v}_m(\boldsymbol{r}) \tag{5.68}$$

$$P(\boldsymbol{r}) = P_0(\boldsymbol{r}) + \epsilon\, P_1(\boldsymbol{r}) + \epsilon^2\, P_2(\boldsymbol{r}) + \ldots = \sum_{m=0}^{\infty} \epsilon^m\, P_m(\boldsymbol{r}). \tag{5.69}$$

If we then group the various terms by their common power of ϵ, the dimensionless Navier–Stokes equation becomes

$$\begin{aligned} 0 = {} & \epsilon^0 \left(-\nabla P_0 + \nabla^2 \boldsymbol{v}_0\right) \\ & + \epsilon^1 \left(-\nabla P_1 + \nabla^2 \boldsymbol{v}_1 - \boldsymbol{v}_0 \cdot \nabla \boldsymbol{v}_0\right) \\ & + \epsilon^2 \left(-\nabla P_2 + \nabla^2 \boldsymbol{v}_2 - \boldsymbol{v}_0 \cdot \nabla \boldsymbol{v}_1 - \boldsymbol{v}_1 \cdot \nabla \boldsymbol{v}_0\right) + O\left(\epsilon^3\right). \end{aligned} \tag{5.70}$$

This equation is satisfied if each grouping of terms is zero, which gives the sequential hierarchy of problems

$$\underline{\epsilon^0}: \quad -\nabla p_0 + \nabla^2 \boldsymbol{v}_0 = 0 \quad \text{with} \quad \nabla \cdot \boldsymbol{v}_0 = 0, \tag{5.71}$$

$$\underline{\epsilon^1}: \quad -\nabla p_1 + \nabla^2 \boldsymbol{v}_1 = \boldsymbol{v}_0 \cdot \nabla \boldsymbol{v}_0 \quad \text{with} \quad \nabla \cdot \boldsymbol{v}_1 = 0, \tag{5.72}$$

$$\underline{\epsilon^2}: \quad -\nabla p_2 + \nabla^2 \boldsymbol{v}_2 = \boldsymbol{v}_0 \cdot \nabla \boldsymbol{v}_1 + \boldsymbol{v}_1 \cdot \nabla \boldsymbol{v}_0 \quad \text{with} \quad \nabla \cdot \boldsymbol{v}_2 = 0, \tag{5.73}$$

$$\underline{\epsilon^3}: \quad \text{and so on.} \tag{5.74}$$

The zeroth-order problem is Stokes flow which always is laminar. The higher-order fields are solutions of the Stokes-flow equation but now with source terms on the right-hand side that come from the lower-order flow fields. These source terms can cause the higher-order fields to develop nonlaminar features such as eddies. Solving a few iterations of this hierarchy and combining them according to the ansatz of Eqs (5.68) and (5.69) and then reinstating dimensions gives the desired solution for flow past a particular curved object. This flow will no longer be perfectly laminar.

This approach, which is called a *perturbation method*, can be applied to any nonlinear PDE in nondimensional form when the nonlinearity as characterized by a perturbation parameter ϵ is small. There will always be a nondimensional number in front of the nonlinear term that is to be used as the perturbation (or expansion) parameter. If there are multiple dimensionless numbers in the partial-differential equations, one usually identifies one of them as the perturbation parameter and then, appropriate to the problem being solved, determines the value of the other dimensionless numbers in terms of powers of the perturbation parameter. So for a problem of time-varying viscous flow past a curved solid surface, one would estimate how large the Roshko number Rk is in terms of powers of Re, that is, we would write $\text{Rk} = \alpha\epsilon^n$. So if, for example, we put in numbers and determined that $\epsilon = \text{Re} = 0.2$ and $\text{Rk} = 0.005$, then in our asymptotic perturbation approach, we would take $\text{Rk} = 5\epsilon^3/8$ (i.e., $\alpha = 5/8$ and $n = 3$) and use this in the perturbation expansion. If $\text{Rk} = 1$, we would use this value for Rk (i.e., $\alpha = 1$ and $n = 0$).

In some problems where the flow domain extends to large distances compared to the size of the objects around which the flow occurs, if we compare the size of terms in the Navier–Stokes equation, it may be that Reynolds number as defined by Eq. (5.54) is small when derivatives of the flow fields are estimated close to a solid surface but becomes large when derivatives are evaluated at distances large compared to the curvature of the surface. In this case, the *method of matched asymptotic expansions*, which is another type of perturbation method, is important. In this approach, one perturbation expansion is performed for the fields in the inner region where fields are varying rapidly in space, which is expressed by making the change of spatial coordinates $\boldsymbol{r} = \epsilon\boldsymbol{r}'$ or $\nabla = \nabla'/\epsilon$ and solving for the fields in the $\boldsymbol{r}'$ coordinates. The inner solution is required to satisfy the boundary conditions on the solid wall but will not satisfy the boundary condition at large distance from the solid wall. You then perform a separate expansion for the fields in the outer region where fields are varying more slowly so that the spatial coordinates are not scaled by ϵ. The outer solution will satisfy the boundary condition at large distance from the solid walls. You then link the two solutions together using a matching criterion, which is the requirement that an inner expansion (i.e., with the scaling $\boldsymbol{r} = \epsilon\boldsymbol{r}'$) of the outer solution equals an outer expansion (i.e., with no ϵ scaling of $\boldsymbol{r}$) of the inner solution. You then add the inner and outer solutions together and subtract either the outer expansion of the inner solution or, what is identical, the inner expansion of the outer solution. Van Dyke (1964) and Nayfeh (1973) describe using such perturbation methods in fluid-mechanics applications.

For flow past solid bodies, when the Reynolds number is significantly greater than 1 so that the appropriate small dimensionless number is $\epsilon = 1/\text{Re}$, a *boundary layer* will always develop near the solid body where the velocity must decrease rapidly to zero, while outside the boundary layer the fluid can be modeled as being inviscid (i.e., ideal). In this case, the dimensionless Navier–Stokes equation with the characteristic pressure $P_c = \rho v_c^2$ becomes $\boldsymbol{v} \cdot \nabla \boldsymbol{v} = \epsilon \nabla^2 \boldsymbol{v} - \nabla P$, and the method of matched asymptotic expansion is applied to this form when treating viscous boundary layers at high Reynolds number.

5.5 Boundary Conditions

An essential part of any flow problem is specifying the boundary conditions that the flow fields must satisfy on the limits of the flow domain. Below, we begin by considering what boundary and initial conditions must be satisfied in a linear flow problem (i.e., one with negligible Reynolds number) in order for the flow field to be unique. We then consider the specific boundary-conditions that typically apply in most flow problems.

5.5.1 Well-Posed Linear Flow Problems

As we did with the linear forms of the macroscopic governing equations of electromagnetism and elastodynamics, we imagine that there are two sets of flow fields, say $\boldsymbol{v}_1, P_1$ and $\boldsymbol{v}_2, P_2$ that each satisfy the same linear governing equations of fluid flow with the same given body forces and within the same possibly heterogeneous flow domain (i.e., a domain in which ρ and η have arbitrary spatial variations). What is established below are the nature of the boundary and initial conditions that must be specified for the flow domain in order for the difference fields $\delta\boldsymbol{v} = \boldsymbol{v}_1 - \boldsymbol{v}_2$ and $\delta P = P_1 - P_2$ to be everywhere zero so that the solution of the linear flow problem is unique. When we have posed proper boundary and initial conditions that lead to unique solutions of the linear flow problem, we say that the flow problem is *well posed*.

After subtracting the linear governing equations (low Reynolds number) for each flow field in the same domain, because the given body forces and fluid heterogeneity are common to each flow problem we obtain the governing equations for the difference fields to be

$$\rho\frac{\partial \delta\boldsymbol{v}}{\partial t} = -\nabla\delta P + \nabla\cdot\delta\boldsymbol{\tau}^v \tag{5.75}$$

$$\delta\boldsymbol{\tau}^v = \eta\left[\nabla\delta\boldsymbol{v} + (\nabla\delta\boldsymbol{v})^T\right] \tag{5.76}$$

$$\nabla\cdot\delta\boldsymbol{v} = 0. \tag{5.77}$$

Although we are focusing on incompressible flow problems here, the same results hold for compressible fluid problems (visco-acoustics) that are linear in which case $\nabla\cdot\boldsymbol{v} = -K^{-1}\partial P/\partial t$ with both K and ρ independent of time. Dot multiplying Eq. (5.75) with $\delta\boldsymbol{v}$ and using the two identities that (1) $\delta\boldsymbol{v}\cdot(\nabla\cdot\delta\boldsymbol{\tau}^v) = \nabla\cdot(\delta\boldsymbol{v}\cdot\delta\boldsymbol{\tau}^v) - \nabla\delta\boldsymbol{v} : \delta\boldsymbol{\tau}^v$ where we used that $\delta\boldsymbol{\tau}^v$ is symmetric and (2) $\delta\boldsymbol{v}\cdot\nabla\delta P = \nabla\cdot(\delta\boldsymbol{v}\delta P) - \delta P\nabla\cdot\delta\boldsymbol{v} = \nabla\cdot(\delta\boldsymbol{v}\delta P)$, we have

$$\nabla\cdot\left[\delta\boldsymbol{v}\cdot(-P\boldsymbol{I} + \delta\boldsymbol{\tau}^v)\right] = \frac{\rho}{2}\frac{\partial}{\partial t}(\delta\boldsymbol{v}\cdot\delta\boldsymbol{v}) + \nabla\delta\boldsymbol{v} : \delta\boldsymbol{\tau}^v. \tag{5.78}$$

Because $\delta\boldsymbol{\tau}^v$ is symmetric, only the symmetric part of $\nabla\delta\boldsymbol{v}$ that double dots $\boldsymbol{\tau}^v$ makes a nonzero contribution in the last term. As such we have

$$\nabla\cdot\left[\delta\boldsymbol{v}\cdot(-\delta P\boldsymbol{I} + \delta\boldsymbol{\tau}^v)\right] = \frac{\rho}{2}\frac{\partial}{\partial t}(\delta\boldsymbol{v}\cdot\delta\boldsymbol{v}) + \frac{\eta}{2}\left[\nabla\delta\boldsymbol{v} + (\nabla\delta\boldsymbol{v})^T\right] : \left[\nabla\delta\boldsymbol{v} + (\nabla\delta\boldsymbol{v})^T\right]. \tag{5.79}$$

Integrating this equation over time and the entire fluid domain Ω that is bounded by the closed surface $\partial\Omega$ and using the divergence theorem gives

$$\int_0^t \mathrm{d}t' \int_{\partial\Omega} \boldsymbol{n}\cdot(-\delta P\boldsymbol{I}+\delta\boldsymbol{\tau}^v)\cdot\delta\boldsymbol{v}\,\mathrm{d}S = \int_\Omega \frac{\rho}{2}\left[\delta\boldsymbol{v}(t)\cdot\delta\boldsymbol{v}(t)\right]\mathrm{d}V - \int_\Omega \frac{\rho}{2}\left[\delta\boldsymbol{v}(0)\cdot\delta\boldsymbol{v}(0)\right]\mathrm{d}V$$
$$+\int_0^t \mathrm{d}t' \int_\Omega \frac{\eta}{2}\left[\nabla\delta\boldsymbol{v}+(\nabla\delta\boldsymbol{v})^T\right]:\left[\nabla\delta\boldsymbol{v}+(\nabla\delta\boldsymbol{v})^T\right]\mathrm{d}V. \tag{5.80}$$

We now impose the following conditions on our flow problem:

(i) **Boundary conditions**: Require that either $\boldsymbol{v}$ or $\boldsymbol{n}\cdot(-P\boldsymbol{I}+\boldsymbol{\tau}^v)$ are specified over each patch of the entire surface $\partial\Omega$ that encloses the flow domain. This results in the surface integral on the left-hand side of Eq. (5.80) being zero.
(ii) **Initial conditions**: Require that $\boldsymbol{v}(0)$ is specified at time zero throughout all of Ω. This results in $\delta\boldsymbol{v}(0)=0$ on the right-hand side of Eq. (5.80).

For these boundary and initial conditions, Eq. (5.80) becomes

$$0=\int_\Omega \frac{\rho}{2}\left[\delta\boldsymbol{v}(t)\cdot\delta\boldsymbol{v}(t)\right]\mathrm{d}V+\int_0^t \mathrm{d}t'\int_\Omega \frac{\eta}{2}\left[\nabla\delta\boldsymbol{v}+(\nabla\delta\boldsymbol{v})^T\right]:\left[\nabla\delta\boldsymbol{v}+(\nabla\delta\boldsymbol{v})^T\right]\mathrm{d}V. \tag{5.81}$$

Both of these volume integrals have integrands that are everywhere positive throughout all of Ω so that satisfaction of this equation is only possible if $\delta\boldsymbol{v}=0$ throughout all of Ω, which means that the flow that satisfies these boundary and initial conditions is unique.

The boundary values for either $\boldsymbol{v}$ or $\boldsymbol{n}\cdot(-P\boldsymbol{I}+\boldsymbol{\tau}^v)$ on $\partial\Omega$ are not determined from this analysis. All that this analysis says is that if boundary values of either of these vectors are specified on each patch of $\partial\Omega$, we are guaranteed that there is a unique flow field in the flow domain. So by whatever means, if we stumble across a flow field that both satisfies the linear governing equations as well as these boundary (and initial conditions in the case of time-dependent problems), we are guaranteed that this flow field is the one and only solution to the flow problem.

5.5.2 Specific Boundary Conditions

Knowing that we need to specify either $\boldsymbol{v}$ or $\boldsymbol{n}\cdot(-P\boldsymbol{I}+\boldsymbol{\tau}^v)$ on $\partial\Omega$, we next turn to what these boundary values must be for two commonly encountered flow-domain boundaries.

Where the flow domain is bounded by a solid wall, the fluid molecules are always adsorbed to the solid surface through either van der Waals forces or other chemical binding forces. So there is no slip or separation that develops between the fluid and the solid at a solid wall. If the solid wall is moving with a given velocity $\boldsymbol{v}_s$, the boundary condition for the flow velocity $\boldsymbol{v}$ is

$$\boxed{\boldsymbol{v}=\boldsymbol{v}_s.} \quad \textit{The no-slip boundary condition at a solid wall} \tag{5.82}$$

In the exercises at the end of the chapter, we will consider scenarios where $\boldsymbol{v}_s=0$, in which case the fluid movement in the flow domain is created by some type of applied force, and scenarios where $\boldsymbol{v}_s\neq 0$, which on its own can result in fluid movement throughout the flow domain.

In places where the limit $\partial\Omega$ of the flow domain is a free surface in which a liquid flow domain makes planar contact with a stationary gas having a given pressure P_{gas}, we have that

$$\boldsymbol{n}\cdot(-P\boldsymbol{I}+\boldsymbol{\tau}^{v}) = -\boldsymbol{n}P_{gas}, \qquad \textit{The free-planar-surface condition} \tag{5.83}$$

which for flow parallel with the planar free surface at the free surface becomes the two distinct conditions that $P = P_{gas}$ and $\boldsymbol{n}\cdot\boldsymbol{\tau}^{v} = 0$. If the free surface has some curvature to it, there is an additional term, called the capillary pressure, that is present on the right-hand side that depends on the amount of curvature and the surface tension as will be treated later in Chapter 7.

5.6 Analytical Solutions Involving Incompressible Viscous Flow

As you will see below, and as you can treat in the exercises, there are a handful of flow problems in various types of pipes or next to simple planar solid walls that yield to explicit exact solution. These solutions have, for the most part, been available since the nineteenth century but remain instructive for building intuition about incompressible viscous flow profiles next to solid walls where vorticity and viscosity are important. But for treating the local flow through, say, the pores of a rock if you could know the detailed pore geometry from, for example, X-ray imaging of a rock sample, one must rely on numerical solution of the Navier–Stokes equation, which is a topic we will not discuss here.

5.6.1 Steady Flow in a Planar Duct

We first solve for how a viscous fluid like water flows in an infinite planar duct of width $w = 2h$ as depicted in Fig. 5.2 under the influence of a uniform applied pressure gradient in the $\hat{z}$ direction that we denote as $\nabla P = (\Delta P/L)\,\hat{z}$, where $\Delta P/L$ is a given constant. In truth, the duct, or channel, must be connected to fluid reservoirs and if we assume that the terminus reservoirs differ in pressure by ΔP and are a distance L apart, there will be a uniform pressure gradient $\Delta P/L$ down the length of the duct other than for a small region

Figure 5.2 An infinite planar flow channel of width $w = 2h$ with a uniform pressure gradient driving flow.

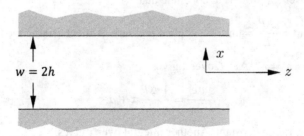

near the entrance and exit that can be neglected for sufficiently long ducts. Because the applied pressure gradient is not driving flow into the solid walls, we have by symmetry that $v_x = v_y = 0$. Using this one piece of intuition in the incompressible flow condition gives

$$\nabla \cdot \boldsymbol{v} = \frac{\partial v_z}{\partial z} = 0 \Rightarrow \boxed{v_z = v_z(x).} \tag{5.84}$$

To evaluate $\boldsymbol{v} \cdot \nabla \boldsymbol{v}$, we have:

$$\boldsymbol{v} = v_z(x)\,\hat{\boldsymbol{z}} \tag{5.85}$$

$$\nabla \boldsymbol{v} = \frac{\partial v_z(x)}{\partial x}\,\hat{\boldsymbol{x}}\hat{\boldsymbol{z}} \tag{5.86}$$

and so

$$\boldsymbol{v} \cdot \nabla \boldsymbol{v} = v_z(x)\frac{\partial v_z(x)}{\partial x}(\hat{\boldsymbol{z}} \cdot \hat{\boldsymbol{x}})\hat{\boldsymbol{z}} = 0 \tag{5.87}$$

because $\hat{\boldsymbol{z}} \cdot \hat{\boldsymbol{x}} = 0$. Thus, the two components of the Navier–Stokes equations become

$$\hat{\boldsymbol{x}}\!: 0 = -\frac{\partial P}{\partial x} + \eta\left(\frac{\partial^2}{\partial x^2} + \frac{\partial^2}{\partial z^2}\right)\overset{0}{\cancel{v_x}} = -\frac{\partial P}{\partial x} = 0 \Rightarrow \boxed{P = P(z)} \tag{5.88}$$

$$\hat{\boldsymbol{z}}\!: 0 = -\frac{\partial P}{\partial z} + \eta\frac{\partial^2 v_z}{\partial x^2} \Rightarrow \boxed{\frac{\partial^2 v_z}{\partial x^2} = \frac{1}{\eta}\frac{\partial P}{\partial z}.} \tag{5.89}$$

Taking the derivative of the z-component equation with respect to z and noting that $v_z(x)$ has no z dependence gives

$$\frac{\partial^2}{\partial x^2}\overset{0}{\cancel{\frac{\partial v_z}{\partial z}}} = 0 = \frac{1}{\eta}\frac{\partial^2 P}{\partial z^2}. \tag{5.90}$$

Integrating this over z allows us to conclude that

$$\frac{\partial P}{\partial z} = \text{constant} = \frac{\Delta P}{L} \mathrel{\widehat{=}} \text{uniform applied pressure gradient driving the flow.} \tag{5.91}$$

Integrating $\partial^2 v_z/\partial x^2 = \eta^{-1}\Delta P/L$ over x gives

$$\frac{\partial v_z}{\partial x} = \frac{1}{\eta}\frac{\Delta P}{L}x + a, \tag{5.92}$$

where a is an integration constant. Another integral over x gives

$$v_z(x) = \frac{1}{\eta}\frac{\Delta P}{L}\frac{x^2}{2} + ax + b, \tag{5.93}$$

where b is yet another integration constant. To find the constants a and b, we use the two no-slip boundary conditions on the two solid walls that

$$v_z(x=+h)=0=\frac{1}{\eta}\frac{\Delta P}{L}\frac{h^2}{2}+ah+b \tag{5.94}$$

$$v_z(x=-h)=0=\frac{1}{\eta}\frac{\Delta P}{L}\frac{h^2}{2}-ah+b. \tag{5.95}$$

Thus, we find that $a=0$ and

$$b=-\frac{1}{\eta}\frac{\Delta P}{L}\frac{h^2}{2} \tag{5.96}$$

so that we may write our solution of this flow problem in the final form

$$\boxed{v_z(x)=\frac{1}{2\eta}\frac{\Delta P}{L}\left(x^2-h^2\right).} \tag{5.97}$$

The parabolic flow profile as shown in Fig. 5.3 is a common characteristic of slow, steady, laminar flow through channels, pipes, and ducts.

The average fluid velocity across the duct is

$$\bar{v}_z=\frac{1}{2h}\int_{-h}^{+h}v_z(x)\,\mathrm{d}x=-\frac{h^2}{3\eta}\frac{\Delta P}{L} \tag{5.98}$$

This is a special case of *Darcy's law* that relates the average flow in a pore space to the macroscopic fluid-pressure gradient across the pore space. The local parabolic flow can then be expressed in terms of the average flow in the duct as $v_z(x)=(3\bar{v}_z/2)\left[1-(x/h)^2\right]$.

The frictional shear force acting on the walls of the flow channel is

$$\boldsymbol{n}\cdot\boldsymbol{\tau}^v=\hat{\boldsymbol{x}}\cdot\hat{\boldsymbol{\tau}}^v=\hat{\boldsymbol{x}}\cdot\left[\eta\frac{\partial v_z}{\partial x}\hat{\boldsymbol{x}}\hat{\boldsymbol{z}}\right]_{x=h}=\frac{1}{\eta}\frac{\Delta P}{L}h\hat{\boldsymbol{z}}. \tag{5.99}$$

You can treat problems similar to the above as exercises at the end of the chapter.

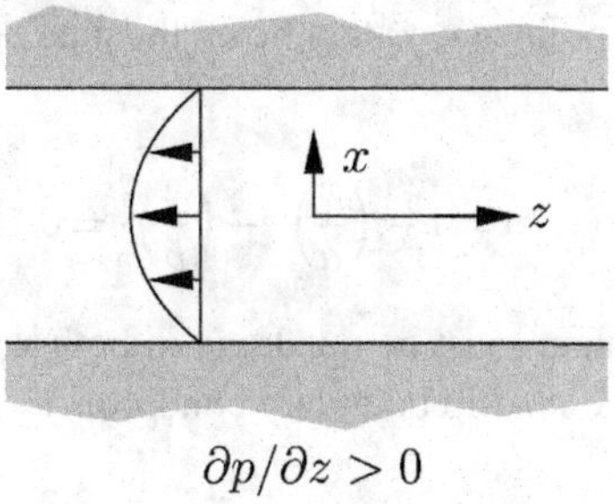

Figure 5.3 Parabolic flow profile in a planar duct.

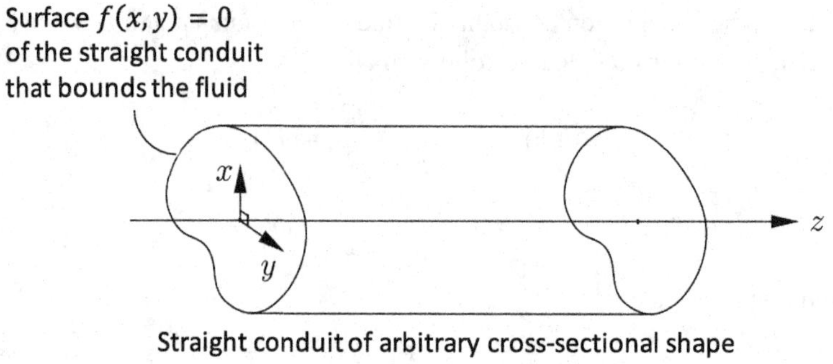

Figure 5.4 A straight conduit of arbitrary cross section.

5.6.2 Steady Flow in Straight Conduits of Arbitrary Cross Section

Assume that steady flow is driven by a uniform pressure gradient $(\Delta P/L)\,\hat{z}$ acting parallel with the axis of the conduit. The straight conduit has an arbitrary cross-sectional shape, which is the same along the entire length of the conduit as depicted in Fig. 5.4. Points (x, y) that reside at the interface between the fluid and the solid conduit are defined by the equation $f(x, y) = 0$ as shown in the figure. From symmetry, we have $v_x = v_y = 0$, and from $\nabla \cdot \boldsymbol{v} = \partial v_z/\partial z = 0,\ v_z = v_z(x, y)$. The advective acceleration in this case is

$$\boldsymbol{v} \cdot \nabla \boldsymbol{v} = \left(v_z \hat{z}\right) \cdot \nabla \left(v_z \hat{z}\right) = v_z \frac{\partial v_z}{\partial z} \hat{z} = 0. \tag{5.100}$$

Further, from the x and y components of Navier–Stokes, we have $\partial P/\partial x = 0$ and $\partial P/\partial y = 0$ so that $P = P(z)$ (only a function of length along the conduit). Further, taking the z derivative of the z component of Navier–Stokes gives $\partial^2 P/\partial z^2 = 0$ or upon integrating over z, $\partial P/\partial z =$ constant $= \Delta P/L$. Thus, the velocity is controlled by

$$\left(\frac{\partial^2}{\partial x^2} + \frac{\partial^2}{\partial y^2}\right) v_z(x, y) = \frac{1}{\eta}\frac{\Delta P}{L}, \quad \text{where } v_z = 0 \text{ on } f(x, y) = 0 \tag{5.101}$$

which is the boundary-value problem (BVP) for flow in straight pipes of arbitrary cross section.

We now consider the special case of a pipe having elliptical cross section as depicted in Fig. 5.5 where the position of points (x, y) that lie on the pipe boundary is determined from the equation for an ellipse

$$f(x, y) = 1 - \left(\frac{x}{a}\right)^2 - \left(\frac{y}{b}\right)^2 = 0. \tag{5.102}$$

The solution of the flow problem must be quadratic in both x and y. A quadratic function of x and y that also satisfies the no-slip boundary condition $v_z|_{f(x,y)=0} = 0$ is

$$v_z(x, y) = c\left[1 - \left(\frac{x}{a}\right)^2 - \left(\frac{y}{b}\right)^2\right]. \tag{5.103}$$

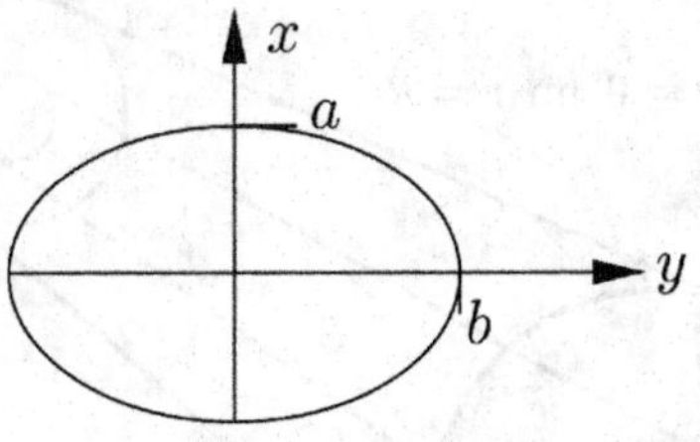

Figure 5.5 A pipe having elliptical cross section.

To find c, we note that $\partial^2 v_z(x, y)/\partial x^2 = -2c/a^2$ and $\partial^2 v_z(x, y)/\partial y^2 = -2c/b^2$ so that the Navier–Stokes equation of Eq. (5.101) gives

$$-2c\left(\frac{1}{a^2} + \frac{1}{b^2}\right) = \frac{\Delta P}{\eta L} \tag{5.104}$$

or

$$\boxed{v_z(x, y) = -\frac{1}{2\eta}\frac{\Delta P}{L}\frac{a^2 b^2}{a^2 + b^2}\left[1 - \left(\frac{x}{a}\right)^2 - \left(\frac{y}{b}\right)^2\right].} \tag{5.105}$$

Unfortunately, $v_z = cf(x, y)$ is only a solution of Navier–Stokes for special pipe boundaries $f(x, y) = 0$ that are quadratic in x and y.

For this case of a straight pipe with elliptical cross section, the average flow in the pipe can be obtained by averaging over a single quadrant of cross-sectional area $\pi ab/4$ as follows:

$$\overline{v}_z = \frac{4}{\pi ab}\int_0^a \mathrm{d}x \int_0^{b\sqrt{1-(x/a)^2}} \mathrm{d}y \left[-\frac{1}{2\eta}\frac{\Delta P}{L}\frac{a^2 b^2}{a^2 + b^2}\left(1 - \left(\frac{x}{a}\right)^2 - \left(\frac{y}{b}\right)^2\right)\right] \tag{5.106}$$

$$= -\frac{1}{4\eta}\left(\frac{a^2 b^2}{a^2 + b^2}\right)\frac{\Delta P}{L} \tag{5.107}$$

which is another special case of Darcy's law.

5.6.3 Steady Flow within Coaxial Pipes

We now consider flow within a concentric cylinder geometry as depicted in Fig. 5.6. This flow is not too important from an Earth-sciences perspective but is of interest mainly because it is exactly soluble. The inner core $0 < r < R_1$ is solid as is the region $r > R_2$ so that fluid resides and moves in the annular region $R_1 < r < R_2$. By symmetry, we can assume that $v_r = 0$.

When $v_r = 0$, the three components of the *Navier–Stokes equations* in cylindrical coordinates are

$$\hat{\boldsymbol{r}}: \quad -\rho\frac{v_\phi^2}{r} = -\frac{\partial P}{\partial r} - \eta\frac{2}{r^2}\frac{\partial v_\phi}{\partial \phi} \tag{5.108}$$

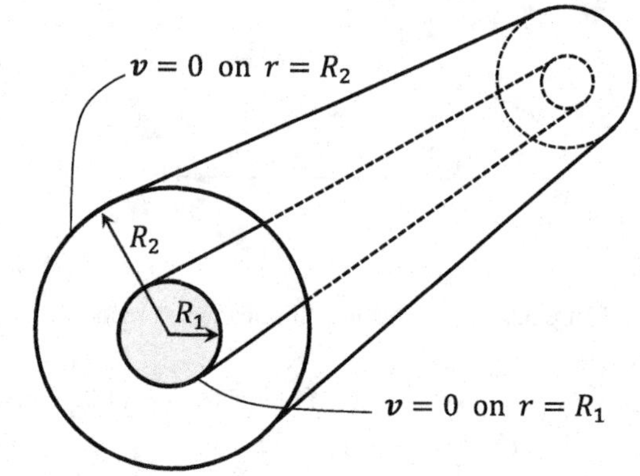

Figure 5.6 Coaxial cylinder geometry with viscous fluid residing in the annulus between the solid surface at $r = R_1$ and the other solid surface at $r = R_2$.

$$\hat{\boldsymbol{\phi}}: \quad \rho\left[\frac{v_\phi}{r}\frac{\partial v_\phi}{\partial \phi} + v_z\frac{\partial v_\phi}{\partial z}\right] = -\frac{1}{r}\frac{\partial P}{\partial \phi} + \eta\left[\nabla^2 v_\phi - \frac{v_\phi}{r^2}\right] \tag{5.109}$$

$$\hat{\boldsymbol{z}}: \quad \rho\left[\frac{v_\phi}{r}\frac{\partial v_z}{\partial \phi} + v_z\frac{\partial v_z}{\partial z}\right] = -\frac{\partial P}{\partial z} + \eta\nabla^2 v_z \tag{5.110}$$

and we have as well that

$$\nabla\cdot\boldsymbol{v} = 0 = \frac{1}{r}\frac{\partial v_\phi}{\partial \phi} + \frac{\partial v_z}{\partial z} \tag{5.111}$$

$$\nabla^2\psi = \frac{1}{r}\frac{\partial}{\partial r}\left(r\frac{\partial \psi}{\partial r}\right) + \frac{1}{r^2}\frac{\partial^2 \psi}{\partial \phi^2} + \frac{\partial^2 \psi}{\partial z^2}, \tag{5.112}$$

where ψ is any scalar field. So if flow in a coaxial pipe is driven in the z direction by a steady uniform pressure gradient $\nabla P = (\Delta P/L)\hat{z}$, we have that $v_\phi = 0$ and there is no variation of any field in the ϕ direction. Thus, the governing equations are

$$\nabla\cdot\boldsymbol{v} = \frac{\partial v_z}{\partial z} = 0 \quad \text{which says that} \quad v_z = v_z(r), \tag{5.113}$$

$$-\frac{\partial P}{\partial r} = 0 \quad \text{which says that} \quad P = P(z), \tag{5.114}$$

$$\frac{1}{r}\frac{\mathrm{d}}{\mathrm{d}r}\left(r\frac{\mathrm{d}v_z}{\mathrm{d}r}\right) = \frac{1}{\eta}\frac{\mathrm{d}P(z)}{\mathrm{d}z} = \frac{1}{\eta}\frac{\Delta P}{L}. \tag{5.115}$$

Integrating this differential equation once after multiplying through by r gives

$$\frac{\mathrm{d}v_z(r)}{\mathrm{d}r} = \frac{1}{\eta}\frac{\Delta P}{L}\frac{r}{2} + \frac{c_1}{r}. \tag{5.116}$$

Integrating again gives

$$v_z(r) = \frac{1}{\eta}\frac{\Delta P}{L}\frac{r^2}{4} + c_1\ln r + c_2, \tag{5.117}$$

where the integration constants c_1 and c_2 are found by applying the two boundary conditions that $v_z(R_1) = 0$ and $v_z(R_2) = 0$. The final result is

$$\boxed{v_z(r) = \frac{1}{4\eta}\frac{\Delta P}{L}\left[R_2^2 - r^2 + \frac{(R_2^2 - R_1^2)}{\ln(R_2/R_1)}\ln\left(\frac{r}{R_2}\right)\right].} \quad (5.118)$$

Additional problems in coaxial geometries are given in the exercises at the end of the chapter.

5.6.4 Steady Creeping Flow around a Solid Sphere

We now consider a problem where, *a priori*, due to the geometry of the flow domain alone, one cannot set to zero the nonlinear advection term $\boldsymbol{v} \cdot \nabla \boldsymbol{v}$ in the Navier–Stokes equation. The problem involves viscous flow around an isolated single solid sphere of radius a. The problem can be stated as one where the sphere is falling under the influence of a gravitational force in the z direction when the fluid at infinity is at rest. In this case, due to the balance between the gravitational force pushing the sphere in the z direction and the fluid pressure and viscous-shearing stress providing a resistive force to the motion, a steady state emerges in which the sphere is moving at a constant velocity, say U_z, with the fluid flowing around the sphere as it falls with no slip at the fluid–sphere interface. This problem is equivalent (other than for an additive constant and sign change) to a stationary sphere and a uniform fluid flow at infinity in the z direction given by $\boldsymbol{v} = U_z\hat{\boldsymbol{z}}$ with zero fluid velocity at the sphere surface. We will start with the latter perspective of flow around a stationary sphere.

Due to the curved surface of the sphere, $\boldsymbol{v} \cdot \nabla \boldsymbol{v}$ is not intrinsically zero and its amplitude relative to the viscous term in Navier–Stokes is given by Reynolds number $\text{Re} = \rho a U_z/\eta$. If we have a scenario in which $\text{Re} < 1$, we can use a perturbation method to solve this problem iteratively as described in Section 5.4.3. The leading-order Stokes-flow problem corresponding to $\text{Re} \ll 1$ is the starting point and will be our only focus here.

Stokes Flow Around a Sphere

The standard method for solving Stokes (or *creeping*) flow past an axisymmetric obstacle like a solid sphere is to introduce a so-called *stream function* ψ as Stokes (1851) originally did in his treatment of this problem that our own presentation here closely follows. In spherical coordinates (r, θ, ϕ) with latitudinal angle θ measured from the z direction and in which the flow fields are all independent of ϕ so that $\psi = \psi(r, \theta)$, the stream function is related to the flow velocity as

$$\boldsymbol{v} = -\nabla \times \left(\frac{\psi}{r \sin\theta}\hat{\boldsymbol{\phi}}\right), \quad (5.119)$$

which automatically satisfies $\nabla \cdot \boldsymbol{v} = 0$. Carrying out the curl in spherical coordinates gives the two components of flow as

$$v_r = -\frac{1}{r^2 \sin\theta}\frac{\partial \psi}{\partial \theta} \quad \text{and} \quad v_\theta = \frac{1}{r\sin\theta}\frac{\partial \psi}{\partial r}. \tag{5.120}$$

We are trying to solve the Stokes flow equation $-\nabla P + \eta \nabla^2 \boldsymbol{v} = 0$. By using $\nabla^2 \boldsymbol{v} = \nabla\nabla \cdot \boldsymbol{v} - \nabla \times \nabla \times \boldsymbol{v} = -\nabla \times \boldsymbol{\omega}$, where $\boldsymbol{\omega} = \nabla \times \boldsymbol{v}$ is the vorticity, the Stokes equation becomes $\nabla \times \boldsymbol{\omega} + \nabla P/\eta = 0$. Taking the curl under the assumption that the viscosity is spatially uniform then shows the vorticity to satisfy $\nabla \times \nabla \times \boldsymbol{\omega} = 0$.

For the boundary conditions on the surface of the sphere, we must have $v_r = 0$, which means that $\partial\psi/\partial\theta = 0$ all along the sphere surface or $\psi(a,\theta) = \text{constant}$ and since the constant does not enter the formalism we can require $\psi(a,\theta) = 0$. Similarly, $v_\theta = 0$ on the sphere surface requires that $\partial\psi/\partial r|_{r=a} = 0$. As $r \to \infty$, we have that $\boldsymbol{v} \to U_z \hat{\boldsymbol{z}} = U_z \cos\theta\, \hat{\boldsymbol{r}} - U_z \sin\theta\, \hat{\boldsymbol{\theta}}$. This requires that

$$\lim_{r\to\infty} \psi \to -\frac{1}{2} U_z r^2 \sin^2\theta \tag{5.121}$$

because one then has from Eq. (5.120) that $v_r \to U_z \cos\theta$ and $v_\theta \to -U_z \sin\theta$ as required.

Taking the curl of Eq. (5.119), we obtain the vorticity in terms of the stream function as

$$\boldsymbol{\omega} = \frac{\hat{\boldsymbol{\phi}}}{r\sin\theta} E^2 \psi \tag{5.122}$$

where in spherical coordinates the operator E^2 is defined

$$E^2 = \frac{\partial^2}{\partial r^2} + \frac{\sin\theta}{r^2}\frac{\partial}{\partial\theta}\left(\frac{1}{\sin\theta}\frac{\partial}{\partial\theta}\right). \tag{5.123}$$

Taking a curl of the vorticity gives

$$\nabla \times \boldsymbol{\omega} = \frac{1}{r^2 \sin\theta}\frac{\partial}{\partial\theta}(E^2\psi)\hat{\boldsymbol{r}} - \frac{1}{r\sin\theta}\frac{\partial}{\partial r}(E^2\psi)\hat{\boldsymbol{\theta}} = -\frac{\nabla P}{\eta} = -\frac{1}{\eta}\left(\frac{\partial P}{\partial r}\hat{\boldsymbol{r}} + \frac{1}{r}\frac{\partial P}{\partial\theta}\hat{\boldsymbol{\theta}}\right) \tag{5.124}$$

so that

$$\frac{\partial P}{\partial r} = -\frac{\eta}{r^2\sin\theta}\frac{\partial}{\partial\theta}\left(E^2\psi\right) \quad \text{and} \quad \frac{\partial P}{\partial\theta} = \frac{\eta}{\sin\theta}\frac{\partial}{\partial r}\left(E^2\psi\right). \tag{5.125}$$

Taking the curl of $\nabla \times \boldsymbol{\omega}$ while assuming η is a uniform constant then gives

$$\nabla \times \nabla \times \boldsymbol{\omega} = -\frac{\hat{\boldsymbol{\phi}}}{r\sin\theta} E^4 \psi = 0. \tag{5.126}$$

So the stream function is a solution of

$$E^4 \psi = 0 \tag{5.127}$$

that satisfies the earlier boundary conditions on $r = a$ and as $r \to \infty$.

To solve this equation, assume that ψ separates as

$$\psi(r,\theta) = \sin^2\theta\, f(r). \tag{5.128}$$

We apply the operator E^2 to obtain

$$E^2\psi = \sin^2\theta\left(\frac{d^2 f(r)}{dr^2} - \frac{2}{r^2}f(r)\right) \,\hat{=}\, \sin^2\theta\, g(r) \tag{5.129}$$

and then after an additional operation with E^2

$$E^4\psi = \sin^2\theta\left(\frac{d^2 g(r)}{dr^2} - \frac{2}{r^2}g(r)\right) = 0. \tag{5.130}$$

This is solved by $g(r) = c_1 r^2 + c_2/r$ so that $d^2 f/dr^2 - (2/r^2)f = c_1 r^2 + c_2/r$, which then gives

$$\psi = \sin^2\theta\left(\frac{c_1}{10}r^4 - \frac{c_2}{2}r + c_3 r^2 + \frac{c_4}{r}\right). \tag{5.131}$$

The constants are determined from the above boundary conditions. In the limit as $r \to \infty$, we find $c_1 = 0$ and $c_3 = -U_z/2$, while from the no-slip condition on $r = a$ that $\partial\psi/\partial r = \partial\psi/\partial\theta = 0$, we find $c_2 = -3aU_z/2$ and $c_4 = -a^3 U_z/4$, which gives

$$\psi = -\frac{U_z a^2}{4}\sin^2\theta\left(\frac{a}{r} - 3\frac{r}{a} + 2\frac{r^2}{a^2}\right). \tag{5.132}$$

By taking the derivatives of Eq. (5.120), we at last have the flow fields determined as

$$v_r = U_z\cos\theta\left(1 - \frac{3a}{2r} + \frac{a^3}{2r^3}\right) \quad \text{and} \quad v_\theta = -U_z\sin\theta\left(1 - \frac{3a}{4r} - \frac{a^3}{4r^3}\right). \tag{5.133}$$

This laminar flow field smoothly connects the boundary conditions on $r = a$ to the uniform flow field in the far field.

To obtain the associated fluid pressure, we first note that from Eq. (5.129)

$$E^2\psi = -\frac{3}{2}aU_z\frac{\sin^2\theta}{r}. \tag{5.134}$$

The total derivative of the fluid pressure is

$$dP = \frac{\partial P}{\partial r}dr + \frac{\partial P}{\partial\theta}d\theta = \eta\frac{3}{2}aU_z\left(\frac{2\cos\theta}{r^3}dr + \frac{\sin\theta}{r^2}d\theta\right) = -\eta\frac{3}{2}aU_z d\left(\frac{\cos\theta}{r^2}\right), \tag{5.135}$$

where we combined Eqs (5.134) and (5.125) to perform the partial derivatives with respect to r and θ. A final integration then yields

$$P = P_\infty - \frac{3}{2}\eta aU_z\frac{\cos\theta}{r^2}. \tag{5.136}$$

So $P - P_\infty$ is positive on the upstream polar cap of the sphere and negative on the downstream polar cap as it needs to be to create the viscous pressure gradients and associated flow around the sphere.

Net Force Balance on a Migrating Sphere: Stokes Mobility Law

We now imagine that the sphere is moving under the influence of an applied force F_z acting in the z direction. This force could be gravitational or electrical if the sphere has a net charge on it. In steady state, if the sphere is migrating with speed U_z in the direction of the applied F_z and the fluid is stationary at infinity, this corresponds to subtracting the uniform flow at infinity from the flow fields of the previous (stationary sphere) section and taking the previous velocity to have the opposite sign $-U_z$.

So the flow field around a migrating sphere from a frame of reference at rest with the fluid at infinity is

$$v_r = U_z \cos\theta \left(\frac{3a}{2r} - \frac{a^3}{2r^3}\right) \quad \text{and} \quad v_\theta = -U_z \sin\theta \left(\frac{3a}{4r} + \frac{a^3}{4r^3}\right) \tag{5.137}$$

with the pressure now reading

$$P = P_\infty + \frac{3}{2}\eta a U_z \frac{\cos\theta}{r^2}. \tag{5.138}$$

The force balance between F_z and the viscous forces resisting the sphere migration are

$$F_z = -\int_{\partial\Omega} \boldsymbol{n} \cdot (-P\boldsymbol{I} + \boldsymbol{\tau}^v) \cdot \hat{z}\, \mathrm{d}S, \tag{5.139}$$

where $\partial\Omega$ is the sphere surface. The nonzero components of viscous shear stress in the flow field are

$$\tau^v_{rr} = 2\eta \frac{\partial v_r}{\partial r} \quad \text{and} \quad \tau^v_{r\theta} = \eta \left(\frac{1}{r}\frac{\partial v_r}{\partial \theta} + \frac{\partial v_\theta}{\partial r} - \frac{v_\theta}{r}\right), \tag{5.140}$$

which when evaluated on the sphere surface using the derived flow field are

$$\tau^v_{rr} = 0 \quad \text{and} \quad \tau^v_{r\theta} = \frac{3}{2a}\eta U_z \sin\theta. \tag{5.141}$$

With $\boldsymbol{n} = \hat{\boldsymbol{r}}$, $\hat{\boldsymbol{r}} \cdot \hat{z} = \cos\theta$, $\hat{\boldsymbol{\theta}} \cdot \hat{z} = -\sin\theta$, and $\mathrm{d}S = a^2 \sin\theta\, \mathrm{d}\theta \mathrm{d}\phi$, we perform the integral over $r = a$ and obtain

$$F_z = -2\pi \int_0^\pi \mathrm{d}\theta\, \sin\theta\, a^2 \left(P_\infty \cos\theta + \frac{3}{2a}\eta U_z \cos^2\theta - \frac{3}{2a}\eta U_z \sin^2\theta\right) = 6\pi\eta a U_z. \tag{5.142}$$

We can rewrite this result as the often-used *Stokes mobility law*

$$\boxed{U_z = bF_z, \quad \text{where} \quad b = \frac{1}{6\pi\eta a} \mathrel{\widehat{=}} \textit{the mobility} \text{ (units of m/s/N).}} \tag{5.143}$$

As we saw earlier in our discussion of electromigration in Chapter 3, Einstein assumed that ions of effective radius a migrating under the influence of an electric field in an electrolyte of viscosity η have mobilities b that can be approximated by the Stokes mobility formula that we have now derived.

Using a perturbation expansion, refinements can be made for the mobility of a sphere migrating through a fluid that account for the nonlinear advective term $\boldsymbol{v} \cdot \nabla \boldsymbol{v}$ in Navier–Stokes (that was completely neglected in the Stokes flow treatment given above). Such refinement will render the relation between sphere velocity U_z and the applied force F_z nonlinear. We direct the interested reader to texts such as Van Dyke (1964) for such development.

5.6.5 Time-Varying Flow

The previous problems have involved steady forcing so that $\partial \boldsymbol{v}/\partial t = 0$. Let's now apply, for example, a uniform pressure gradient to the planar duct of width $w = 2h$ considered earlier in Section 5.6.1 but assume the pressure gradient driving the flow to vary sinusoidally in time as

$$\nabla P = \frac{\Delta P}{L} \cos \omega t \, \hat{z}. \tag{5.144}$$

The Roshko number for this problem is $\mathrm{Rk} = \rho h^2 \omega/(2\pi\eta)$, which we assume to be non-negligible so that the term $\partial \boldsymbol{v}/\partial t$ must be allowed for in the Navier–Stokes equation. It is always best when dealing with such time-harmonic forcing (i.e., sinusoidal time dependence) in problems involving linear physics to first work with the complex exponential time dependence

$$e^{-i\omega t} = \cos \omega t - i \sin \omega t \tag{5.145}$$

and then take the real part of the solution as the final answer if the forcing has $\cos \omega t$ time dependence or minus the imaginary part if the forcing has $\sin \omega t$ dependence.

So, for the planar duct problem, with the coordinate z representing distance down the duct and x distance across the duct, we have again that $v_z(x, t)$ is the only nonzero velocity component, a form that also satisfies the incompressible flow condition $\nabla \cdot \boldsymbol{v} = 0$. In this case, the Navier–Stokes equation and boundary conditions combine to form the time-dependent BVP

$$\rho \frac{\partial v_z(x, t)}{\partial t} = \eta \frac{\partial^2 v_z(x, t)}{\partial x^2} - \frac{\Delta P}{L} e^{-i\omega t}$$

with

$$v_z(x, t) = 0 \quad \text{on } x = \pm h.$$

This is a diffusion equation for v_z with a diffusivity given by $D_v = \eta/\rho$. After we solve this BVP for the complex function $v_z(x, t)$, the final step is to take the real part of v_z to obtain the solution of the original problem with $\cos \omega t$ time dependence.

We thus propose a solution of the form

$$v_z(x, t) = \tilde{v}_z(x, \omega)\, e^{-i\omega t} \tag{5.146}$$

and substitute into the Navier–Stokes equation to obtain

$$\eta \frac{\partial^2 \tilde{v}_z(x,\omega)}{\partial x^2} + i\,\omega\rho\,\tilde{v}_z(x,\omega) = \frac{\Delta P}{L}, \tag{5.147}$$

where we have canceled the common factor $e^{-i\omega t}$ on all terms.

A particular solution of this equation (i.e., one that satisfies the differential equation but not the boundary conditions) is

$$\tilde{v}_z^P(x,\omega) = \frac{1}{i\,\omega\rho}\frac{\Delta P}{L}. \tag{5.148}$$

To this we can add the general solution of the homogeneous equation

$$\frac{\partial^2 \tilde{v}_z^H(x,\omega)}{\partial x^2} + \frac{i\,\omega\rho}{\eta}\tilde{v}_z^H(x,\omega) = 0. \tag{5.149}$$

An exponential function is the natural choice for the solution here

$$\tilde{v}_z^H(x,\omega) = A_k e^{ikx}. \tag{5.150}$$

Substituting into the differential equation, we determine the k that satisfies the differential equation

$$\left[(ik)^2 + \frac{i\,\omega\rho}{\eta}\right] A_k\, e^{ikx} = 0 \tag{5.151}$$

or

$$k^2 = \frac{i\,\omega\rho}{\eta}. \tag{5.152}$$

Thus, there are two possible k values that allow the differential equation to be solved

$$k = +\sqrt{\frac{i\,\omega\rho}{\eta}} \quad \text{and} \quad k = -\sqrt{\frac{i\,\omega\rho}{\eta}}. \tag{5.153}$$

As an aside, what is $\sqrt{i}$? This is answered by considering that $\sqrt{i} = a + ib$ is a complex number so that by taking the square we have $i = a^2 - b^2 + i2ab$ which requires that $a = b$ and $a = \pm 1/\sqrt{2}$ or

$$\boxed{\sqrt{i} = \pm\frac{(1+i)}{\sqrt{2}}.} \tag{5.154}$$

The square root of a complex number is always multivalued, that is, can be either plus or minus. Another way of addressing $\sqrt{i}$ is to consider a unit circle on the complex plane and write $i = e^{i\pi/2}$ so that $\sqrt{i} = i^{1/2} = e^{i\pi/4} = (1+i)/\sqrt{2}$. But we also have that $i = e^{i(\pi/2+2\pi)}$ so that $\sqrt{i} = e^{i\pi/4}e^{i\pi} = -e^{i\pi/4} = -(1+i)/\sqrt{2}$. This ambiguity as to the sign of $\sqrt{i}$ does not bother us in the present problem because we get the same two roots for k regardless of which sign convention is used. We will use $\sqrt{i} = (1+i)/\sqrt{2}$.

The homogeneous solution of the differential equation is then

$$\tilde{v}_z^H(x,\omega) = A_+\, e^{i\sqrt{\frac{i\omega\rho}{\eta}}x} + A_-\, e^{-i\sqrt{\frac{i\omega\rho}{\eta}}x} \tag{5.155}$$

We find A_+ and A_- from the boundary conditions that

$$\tilde{v}_z = \tilde{v}_z^H + \tilde{v}_z^P = 0 \quad \text{at } x = \pm h \tag{5.156}$$

on the two solid walls. In practice, the algebra for finding A_+ and A_- is a bit tedious, though quite straightforward.

We can save time by noting that the final answer for $\tilde{v}_z(x,\omega)$ must be symmetric in x because of the symmetric boundary conditions on $x = \pm h$. Thus we can use the symmetric form of the solution

$$\tilde{v}_z^H(x,\omega) = A\cos kx, \qquad \text{where} \quad k = \sqrt{\frac{i\omega\rho}{\eta}} \tag{5.157}$$

since $\tilde{v}_z^P$ is already symmetric (it is a constant independent of spatial position). We find A from the boundary condition

$$\tilde{v}_z\,(x = \pm h) = \frac{1}{i\,\omega\rho}\frac{\Delta P}{L} + A\cos kh = 0, \tag{5.158}$$

which gives

$$A = -\frac{1}{i\,\omega\rho}\frac{\Delta P/L}{\cos kh} \tag{5.159}$$

and

$$\boxed{\tilde{v}_z(x,\omega) = +\frac{1}{i\,\omega\rho}\frac{\Delta P}{L}\left[1 - \frac{\cos kx}{\cos kh}\right].} \tag{5.160}$$

The final answer is given by

$$v_z(x,t) = \mathrm{Re}\left\{\tilde{v}_z(x,\omega)\, e^{-i\omega t}\right\}. \tag{5.161}$$

Due to the fact that $k = (1+i)\sqrt{\omega\rho/(2\eta)}$ is complex, determining the real part is a rather tedious, though straightforward, algebraic exercise. Going through this exercise yields

$$v_z(x,t) = -\frac{\delta_v^2}{2\eta}\frac{\Delta P}{L}\{[1 - f(x,\omega)]\sin\omega t + g(x,\omega)\cos\omega t\} \tag{5.162}$$

where

$$f(x, \omega) = \frac{\cosh(x/\delta_v)\cos(x/\delta_v)\cosh(h/\delta_v)\cos(h/\delta_v) + \sinh(x/\delta_v)\sin(x/\delta_v)\sinh(h/\delta_v)\sin(h/\delta_v)}{\cosh^2(h/\delta_v)\cos^2(h/\delta_v) + \sinh^2(h/\delta_v)\sin^2(h/\delta_v)}, \tag{5.163}$$

$$g(x, \omega) = \frac{\cosh(x/\delta_v)\cos(x/\delta_v)\sinh(h/\delta_v)\sin(h/\delta_v) - \sinh(x/\delta_v)\sin(x/\delta_v)\cosh(h/\delta_v)\cos(h/\delta_v)}{\cosh^2(h/\delta_v)\cos^2(h/\delta_v) + \sinh^2(h/\delta_v)\sin^2(h/\delta_v)}, \tag{5.164}$$

and

$$\delta_v = \sqrt{\frac{2\eta}{\omega\rho}} \mathrel{\widehat{=}} \textit{viscous skin depth} \text{ (units of length).} \tag{5.165}$$

This is an example where obtaining the exact analytical answer does not immediately allow you to gain intuition into the physics. The skin depth δ_v, that gets increasingly smaller with increasing frequency, characterizes the distance next to the walls where the velocity field is varying spatially (has vorticity).

To obtain an intuitive feel for what is happening, we consider the two limits of high and low frequencies. At high-frequencies where $\delta_v/h \ll 1$, the leading-order behavior in the limit as $\delta_v/h \to 0$ (so that $\cosh z/\delta_v \approx \sinh z/\delta_v \approx e^{z/\delta_v}/2$) is

$$\begin{aligned}\lim_{\delta_v \to 0} v_z(x, t) = -\frac{1}{\omega\rho}\frac{\Delta P}{L}\Bigg\{&\left[1 - \exp\left(\frac{x-h}{\delta_v}\right)\cos\left(\frac{x-h}{\delta_v}\right)\right]\sin\omega t \\ &+ \exp\left(\frac{x-h}{\delta_v}\right)\sin\left(\frac{x-h}{\delta_v}\right)\cos\omega t\Bigg\} \quad \text{for} \quad 0 \le x \le h.\end{aligned} \tag{5.166}$$

This expression, which applies to the upper part of the duct where $x > 0$, does indeed show that because of the rapid fall off of the exponential when we are not right at $x = h$, that all spatial variation is within a distance of a couple δ_v next to the walls and the solution is spatially uniform, with a $\sin\omega t$ time dependence, when we are more than a couple δ_v distance away from the walls. For the high-frequency solution valid across the entire duct, simply replace x with $|x|$ in Eq. (5.166).

In the opposite limit as $\delta_v/h \to \infty$ (low frequencies), the various trigonometric and hyperbolic functions are expanded using $\sinh u = u + u^3/6 + \ldots$, $\sin u = u - u^3/6 + \ldots$, $\cosh u = 1 + u^2/2 + \ldots$ and $\cos u = 1 - u^2/2 + \ldots$ for $u \ll 1$. A quick inspection shows the leading order behavior is given by

$$\lim_{\delta_v \to \infty} v_z(x, t) = \frac{1}{2\eta}\frac{\Delta P}{L}\left(x^2 - h^2\right)\cos\omega t, \tag{5.167}$$

which is the same result we obtained earlier for this problem of flow in a straight duct when the pressure gradient was constant in time. So when $\delta_v/h \gg 1$ (viscous skin depth much larger than the duct half-width), the flow profile is the same as in the static forcing case except that there is a cos ωt time dependence. We call this type of low-frequency response the *quasi-static* response, which corresponds to a finite Roshko number Rk that is nonetheless much smaller than 1 so that the time derivative in the Navier–Stokes equation is negligible. If we further let $\omega \to 0$, which corresponds to Rk = 0, the time dependence disappears altogether and we have the static-forcing flow case, called Poiseuille flow, as obtained earlier.

5.7 Magnetohydrodynamics

For any of the viscous flow scenarios modeled so far in this chapter, if there is a magnetic field present in the flow domain that has a component normal to the flow direction and if the fluid is an electrical conductor, an electrical current will be created (unless impeded in the steady state by an electrically insulating boundary) that is perpendicular to both the magnetic field and the flow direction and this electrical current will, in turn, create an additional magnetic field through induction that contributes to the forcing on the fluid. This coupled interaction between electromagnetics and fluid flow in a background magnetic field is called *magnetohydrodynamics*.

The Earth's magnetic field, that helps protect life on our planet by deflecting the charged particles in the solar wind, is created and maintained through a complicated nonlinear interaction of the viscous flow in the rotating outer core with the magnetic field that is present there. How the Earth acquired a magnetic field in the first place and how the magnetic field continues to evolve through time is a challenging problem of magnetohydrodynamics amenable only to numerical analysis. Our goal here is quite modest in comparison, aiming only to understand how magnetic fields and a flow field are coupled to each other in the simplest conceivable geometry of steady viscous flow in a planar channel driven by a pressure gradient in the presence of a magnetic field applied perpendicular to the flow channel.

5.7.1 The Governing Equations

The pertinent Maxwell's equations are those in which the only influence of the fluid velocity is through the current density $\boldsymbol{J}_{em} = \sigma\,(\boldsymbol{E} + \boldsymbol{v} \times \boldsymbol{B})$ and are thus given as in Section 3.8

$$\nabla \cdot \boldsymbol{B} = 0, \tag{5.168}$$

$$\nabla \cdot \boldsymbol{D} = 0, \tag{5.169}$$

$$\boldsymbol{B} = \mu \boldsymbol{H}, \tag{5.170}$$

$$\boldsymbol{D} = \varepsilon \boldsymbol{E} \tag{5.171}$$

$$\nabla \times \boldsymbol{E} = -\frac{\partial \boldsymbol{B}}{\partial t}, \tag{5.172}$$

$$\nabla \times \boldsymbol{H} = \frac{\partial \boldsymbol{D}}{\partial t} + \sigma\,(\boldsymbol{E} + \boldsymbol{v} \times \boldsymbol{B})\,. \tag{5.173}$$

We are assuming here that the fluid is both isotropic and uncharged ($\rho_e = 0$). For the limited scope of the following treatment, we will only focus on the quasi-static or emergent steady state that follows the start of a flow problem and will thus drop the time derivatives in Faraday's and Ampère's laws.

The coupling of the fluid flow to the electromagnetics is confined to Ampère's law. Assuming that the fluid has uniform μ and σ, we take the curl of Ampére's law and use Faraday's law $\nabla \times \boldsymbol{E} = 0$ to obtain

$$\nabla \times \nabla \times \boldsymbol{H} = \sigma \mu \nabla \times (\boldsymbol{v} \times \boldsymbol{H}) . \tag{5.174}$$

This equation determines the magnetic field $\boldsymbol{H}$ that is created by and coupled to the fluid flow $\boldsymbol{v}$.

The steady-state Navier–Stokes equations for incompressible viscous flow in the presence of a magnetic body-force density given by $\boldsymbol{F}_b = \boldsymbol{J}_{em} \times \boldsymbol{B} = \sigma \mu (\boldsymbol{E} + \mu \boldsymbol{v} \times \boldsymbol{H}) \times \boldsymbol{H} = \mu (\nabla \times \boldsymbol{H}) \times \boldsymbol{H} = -\mu \boldsymbol{H} \times (\nabla \times \boldsymbol{H})$ is then

$$\rho \boldsymbol{v} \cdot \nabla \boldsymbol{v} = -\nabla P + \eta \nabla^2 \boldsymbol{v} - \mu \boldsymbol{H} \times (\nabla \times \boldsymbol{H}) \tag{5.175}$$

$$\nabla \cdot \boldsymbol{v} = 0. \tag{5.176}$$

Confining ourselves to low-Reynolds number so that the inertial acceleration is negligible on the left-hand side and using the vector calculus identity from Chapter 1 that $2\boldsymbol{H} \times (\nabla \times \boldsymbol{H}) = \nabla (\boldsymbol{H} \cdot \boldsymbol{H}) - 2\boldsymbol{H} \cdot \nabla \boldsymbol{H}$ gives the governing equations for determining P and $\boldsymbol{v}$ as

$$\eta \nabla^2 \boldsymbol{v} = \nabla \left(P + \frac{\mu}{2} \boldsymbol{H} \cdot \boldsymbol{H} \right) - \mu \boldsymbol{H} \cdot \nabla \boldsymbol{H} \tag{5.177}$$

$$\nabla \cdot \boldsymbol{v} = 0. \tag{5.178}$$

Equation (5.174) for $\boldsymbol{H}$ can be rewritten using the vector calculus identities from Chapter 1 that

$$\nabla \times (\boldsymbol{v} \times \boldsymbol{H}) = \nabla \cdot (\boldsymbol{H}\boldsymbol{v} - \boldsymbol{v}\boldsymbol{H}) \tag{5.179}$$

$$= (\nabla \cdot \boldsymbol{H})\boldsymbol{v} + \boldsymbol{H} \cdot \nabla \boldsymbol{v} - (\nabla \cdot \boldsymbol{v})\boldsymbol{H} - \boldsymbol{v} \cdot \nabla \boldsymbol{H} \tag{5.180}$$

$$= \boldsymbol{H} \cdot \nabla \boldsymbol{v} - \boldsymbol{v} \cdot \nabla \boldsymbol{H} \tag{5.181}$$

and $\nabla \times \nabla \times \boldsymbol{H} = \nabla (\nabla \cdot \boldsymbol{H}) - \nabla^2 \boldsymbol{H} = -\nabla^2 \boldsymbol{H}$ to give

$$\nabla^2 \boldsymbol{H} - \mu \sigma \boldsymbol{v} \cdot \nabla \boldsymbol{H} + \mu \sigma \boldsymbol{H} \cdot \nabla \boldsymbol{v} = 0. \tag{5.182}$$

The boxed equations above are the coupled suite of three governing equations for steady-state magnetohydrodynamics in the low-Reynolds number and incompressible viscous flow regime. They permit the solution of $\boldsymbol{v}$, P, and $\boldsymbol{H}$ given appropriate boundary conditions. Given the $\boldsymbol{H}$ and $\boldsymbol{v}$ determined from these boxed equations, the electric field is calculated from Ampére's law as $\boldsymbol{E} = (\nabla \times \boldsymbol{H}) / \sigma - \mu \boldsymbol{v} \times \boldsymbol{H}$.

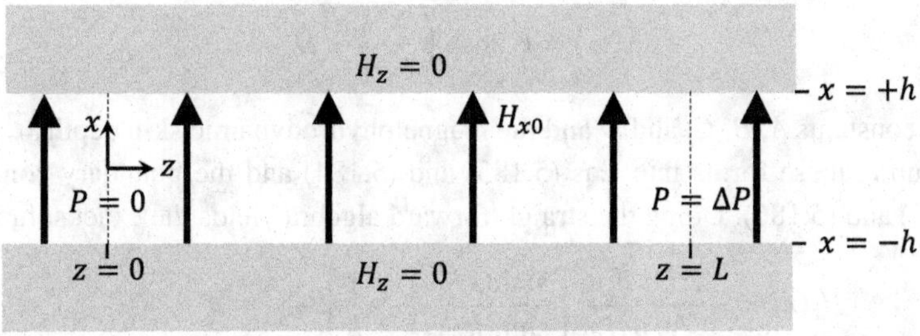

Figure 5.7 A planar channel of width $w = 2h$ with viscous flow driven by a uniform pressure gradient $\Delta P/L\hat{z}$ in the presence of a uniform vertical (x direction) applied magnetic field $H_{x0}\hat{x}$.

5.7.2 Steady Magnetohydrodynamic Flow in a Planar Channel

As our one and only example of solving a magnetohydrodynamics problem, we focus on the problem treated earlier of viscous flow in a planar channel of width $w = 2h$ as driven by the uniform applied pressure gradient $\partial P/\partial z = \Delta P/L$ but now in the presence of a steady magnetic field $H_{x0}\hat{x}$ applied through the solid walls across the channel. In the solid bounding the viscous fluid, we will assume there is no magnetic field parallel with the channel but it is easy to allow for that possibility. The problem is depicted in Fig. 5.7.

The viscous flow driven by the pressure gradient is in the z direction $\boldsymbol{v} = v_z(x)\hat{z}$, a form that satisfies the incompressible flow condition $\nabla \cdot \boldsymbol{v} = 0$. Due to the presence of the uniform vertical magnetic field across the channel $H_{x0}\hat{x}$, there is a uniform electrical current in the y direction that generates a secondary magnetic field $H_z(x)\hat{z}$. So we have $\boldsymbol{H} = H_{x0}\hat{x} + H_z(x)\hat{z}$, and we must find $H_z(x)$ subject to the boundary conditions that $H_z(\pm h) = 0$. Applying these constraints to the z-components of Eqs (5.182) and (5.177) gives the coupled equations for $H_z(x)$ and $v_z(x)$

$$\frac{\mathrm{d}^2 H_z(x)}{\mathrm{d}x^2} = -\mu\sigma H_{x0}\frac{\mathrm{d}v_z(x)}{\mathrm{d}x} \tag{5.183}$$

$$\eta\frac{\mathrm{d}^2 v_z(x)}{\mathrm{d}x^2} = \frac{\mathrm{d}P(z)}{\mathrm{d}z} - \mu H_{x0}\frac{\mathrm{d}H_z(x)}{\mathrm{d}x} \tag{5.184}$$

that are subject to the boundary conditions

$$H_z(x) = 0 \quad \text{on} \quad x = \pm h \tag{5.185}$$

$$v_z(x) = 0 \quad \text{on} \quad x = \pm h \tag{5.186}$$

as well as $P(0) = 0$ and $P(L) = \Delta P$. If the background magnetic field applied through the bounding walls has a nonzero component H_{z0} tangential to the walls, we would simply use $H_z(\pm h) = H_{z0}$ as the boundary condition for $H_z(x)$. Taking an additional z derivative of Eq. (5.184) gives $\mathrm{d}^2P(z)/\mathrm{d}z^2 = 0$ or $P(z) = \Delta P z/L$ and $\mathrm{d}P(z)/\mathrm{d}z = \Delta P/L$. Solutions for $H_z(x)$ and $v_z(x)$ are of the form

$$H_z(x) = A\sinh\left(\frac{x}{\delta_m}\right) + Bx \tag{5.187}$$

$$v_z(x) = C \cosh\left(\frac{x}{\delta_m}\right) + D, \tag{5.188}$$

where the constants A, B, C, and D and the magnetohydrodynamic skin depth δ_m are found by substituting these forms into Eqs (5.183) and (5.184) and the boundary conditions of Eqs (5.185) and (5.186). Doing the straightforward algebra yields the exact solutions

$$H_z(x) = -\frac{\Delta P}{L}\frac{h}{\mu H_{x0}}\left[\frac{\sinh(x/\delta_m)}{\sinh(h/\delta_m)} - \frac{x}{h}\right] \tag{5.189}$$

$$v_z(x) = -\frac{\Delta P}{L}\frac{h\delta_m}{\eta\sinh(h/\delta_m)}\left[\cosh\left(\frac{h}{\delta_m}\right) - \cosh\left(\frac{x}{\delta_m}\right)\right], \tag{5.190}$$

where the skin depth is found to be

$$\delta_m = \frac{1}{\mu\,|H_{x0}|}\sqrt{\frac{\eta}{\sigma}} \mathrel{\widehat{=}} \textit{magnetohydrodynamic skin depth.} \tag{5.191}$$

Note that $\mu H_{x0} = B_{x0}$ is the total magnetic field in the x direction as measured in Tesla. Although $v_z(x)$ is influenced only by the amplitude of the background field $|H_{x0}|$, the induced magnetic field $H_z(x)$ changes sign when H_{x0} changes sign.

The electric field is then obtained as

$$\boldsymbol{E} = \frac{\nabla\times\boldsymbol{H}}{\sigma} - \mu\boldsymbol{v}\times\boldsymbol{H} = -\left[\frac{1}{\sigma}\frac{\mathrm{d}H_z(x)}{\mathrm{d}x} + \mu H_{x0}v_z(x)\right]\hat{\boldsymbol{y}} \tag{5.192}$$

$$= \frac{\Delta P}{L}\frac{1}{\sigma\mu H_{x0}}\left[\frac{h}{\delta_m}\frac{\cosh(h/\delta_m)}{\sinh(h/\delta_m)} - 1\right]\hat{\boldsymbol{y}}, \tag{5.193}$$

which is a uniform field pointing out of the page when both $H_{x0} > 0$ and $\Delta P > 0$ (i.e., flow in the negative z direction) because $x\coth x > 1$ for all $x \geq 0$. So E_y also changes sign as H_{x0} changes sign. All three of the nonzero response fields $H_z(x)$, $v_z(x)$, and E_y (a spatial constant satisfying $\nabla\times\boldsymbol{E} = 0$) are linear in the driving pressure gradient $\Delta P/L$ that is creating them. To better understand the nature of this response, it is useful to explore the two limits in which $h/\delta_m \to 0$ (a vanishingly small applied field amplitude $|H_{x0}|$) and $h/\delta_m \to \infty$ (a large applied field amplitude $|H_{x0}|$).

In the limit as the applied magnetic field gets small so that $h/\delta_m \to 0$, we expand the hyperbolic sinusoids with small arguments to obtain to leading order in w/δ_m (and therefore x/δ_m)

$$H_z(x) = \frac{\Delta P}{L}\frac{x}{6\mu H_{x0}}\left[\left(\frac{h}{\delta_m}\right)^2 - \left(\frac{x}{\delta_m}\right)^2\right] \tag{5.194}$$

so that $H_z(x) \to 0$ as $h/\delta_m \to 0$ and $x/\delta_m \to 0$, that is, as H_{x0} vanishes. To leading order in the same limit, the flow field becomes

$$v_z(x) = -\frac{\Delta P}{L}\frac{h^2}{2\eta}\left[\left(1 - \frac{x^2}{h^2}\right)\left(1 - \frac{(h/\delta_m)^2}{6}\right) + \frac{(h/\delta_m)^2 - (x/\delta_m)^2}{12}\right]. \tag{5.195}$$

When $h/\delta_m = 0$ (no applied magnetic field), this becomes $v_z(x) = -(\Delta P/L)(h^2 - x^2)/(2\eta)$, which is the Poiseuille flow found earlier. Last, the induced electric field in the limit of a small applied magnetic field is

$$E_y = \frac{\Delta P}{L}\frac{2(h/\delta_m)^2}{3\sigma\mu H_{x0}}, \tag{5.196}$$

which vanishes when there is no applied magnetic field.

In the opposite limit of very large applied magnetic fields so that $h/\delta_m \to \infty$, we have

$$H_z(x) = -\frac{\Delta P}{L}\frac{h^2}{\mathrm{sgn}(H_{x0})}\sqrt{\frac{\sigma}{\eta}}\frac{1}{(h/\delta_m)}\left[\exp\left(-\frac{(h-|x|)}{\delta_m}\right) - \frac{x}{h}\right], \tag{5.197}$$

$$v_z(x) = -\frac{\Delta P}{L}\frac{h^2}{\eta}\frac{1}{(h/\delta_m)}\left[1 - \exp\left(-\frac{(h-|x|)}{\delta_m}\right)\right], \tag{5.198}$$

$$E_y = -\frac{\Delta P}{L}\frac{h}{\mathrm{sgn}(H_{x0})\sqrt{\sigma\eta}}. \tag{5.199}$$

The notation $\mathrm{sgn}(a)$ is called the "signum" function and simply extracts the sign of the argument a. So right at the walls in this limit of small δ_m (large applied magnetic field), the flow field increases from zero much more rapidly compared to when there is no magnetic field present. But as we head to the center of the channel and $(h - |x|)/\delta_m \gg 1$, the flow profile quickly flattens out across the channel width with an amplitude that decreases with increasing $|H_{x0}|$. Analogous comments hold for $H_z(x)$. The electric field in this limit of large $|H_{x0}|$ becomes a constant independent of $|H_{x0}|$.

5.8 Exercises

1. *Plane-wave acoustics*: Show that in a homogeneous, infinite fluid of negligible viscosity, if the recorded pressure of a plane sound wave propagating in the x direction at the point $x = 0$ is $P(x = 0, t) = f(t)$, then the pressure of the plane wave everywhere in the fluid is (with $\nabla \cdot \boldsymbol{F}_b = 0$)

$$\boxed{P(x, t) = f\left(t - \frac{x}{c_p}\right), \quad \text{where } c_p = \text{sound speed.}} \tag{5.200}$$

2. *Spherical-wave acoustics*: For spherical sound waves emanating from a point $r = 0$ in the same homogenous fluid, show that if the pressure recorded at some point $r = r_0$ is

$$P(r = r_0, t) = f(t) \tag{5.201}$$

that the pressure field in all of space and time (with $\nabla \cdot \boldsymbol{F}_b = 0$) is

$$P(r,t) = \frac{f\left(t - (r - r_0)/c_p\right)}{r/r_0}. \tag{5.202}$$

Note that in spherical coordinates and with the notation $\partial_r = \partial/\partial r$, the Laplacian operator is $\nabla^2 \psi(r) = r^{-2}\partial_r \left[r^2 \partial_r \psi(r)\right]$.

The most important fact about solutions of the wave equation (when c_p is uniform in space) is that the argument of the wave function has the common-sense time and space dependence $t - x/c_p$ (which we can equivalently write $x - c_p t$).

Viscous-Flow Problems with Exact Solutions

The number of analytically tractable viscous-flow problems in simple geometries that yield exact solutions is rather limited. Below are a handful of problems that will sharpen your intuition and that are not very difficult. You should try to solve all of them.

3. *Flow along an inclined slope*: As depicted in Fig. 5.8, liquid resides between the planes $x = 0$ and $x = H$. The region $x < 0$ is solid and $x > H$ is the atmosphere. You will assume that $H =$ constant along the flowing layer of liquid, which is an approximation built into the problem statement even if the solution to the problem as stated is exact. In the final exercise below, you can treat a problem where the layer of liquid has an evolving profile. By symmetry in this problem, $v_z = v_z(x)$ and $v_x = v_y = 0$. Also, $\boldsymbol{F}_b = \rho \boldsymbol{g}$ where $\boldsymbol{g}$ is directed down. Given the boundary conditions $P = P_{\text{atm}}$ on $x = H$, $\eta \partial v_z/\partial x = 0$ on $x = H$ and $v_z = 0$ on $x = 0$, show that

$$P = P(x) = P_{\text{atm}} + \rho g \cos\theta \,(H - x) \tag{5.203}$$

$$v_z(x) = \frac{\rho g}{2\eta} \sin\theta \,(2Hx - x^2). \tag{5.204}$$

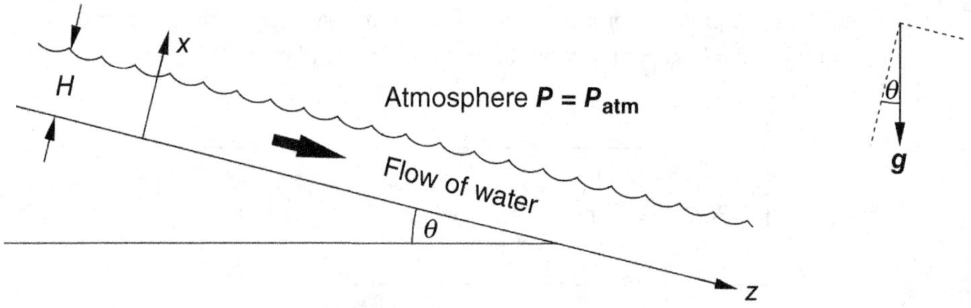

Figure 5.8 Viscous-fluid layer of constant thickness H flowing down a slope under the influence of gravity.

Sketch the flow field so that you can visualize this result. *Intuition in physics comes from treating simple systems where the laws of physics can be solved analytically and then obtaining a clear mental image of the mathematical solution so obtained.* So drawing a sketch in order to obtain a mental image of a mathematical result is as important for building intuition as is obtaining the mathematical solution itself.

4. *Steady flow in a cylindrical pipe*: Assume that steady flow is driven by a uniform pressure gradient of amplitude $\Delta P/L$ that acts throughout the cylindrical pipe in the z direction. Working in cylindrical coordinates [e.g., Eq. (5.110)], obtain the function $v_z(r)$ for a pipe of radius a and draw a sketch of this flow field across the diameter of the pipe. HINT: the solution must be everywhere finite which requires one of the integration constants to be zero.

   Go on to show that the average flow across the cross section of the cylindrical pipe is

$$\overline{v}_z = -\frac{a^2}{8\eta}\frac{\Delta P}{L}, \quad \text{where} \quad a \,\widehat{=}\, \text{pipe radius.} \tag{5.205}$$

   Again, this result for the average flow is another statement of *Darcy's law*. Be careful in how you perform the area integral over the circular cross section of the pipe (you should work in polar coordinates to perform the integral). Finally, verify that your result for the flow throughout the cylindrical pipe can be written $v_z(r) = 2\overline{v}_z\left[1-(r/a)^2\right]$.

Flows in a Coaxial Cylindrical Pipe

In Section 5.6.3, we gave the governing equations and solved an example problem for flow in a coaxial pipe. The next two problems are also posed in a coaxial pipe.

5. *Flow between rotating cylinders or "Couette flow"*: A viscous fluid resides in the annulus $R_1 \le r \le R_2$ between two solid rotating cylinders as depicted in Fig. 5.9, where $\Omega_2 =$ angular velocity (rad/s) of the outer cylinder and $\Omega_1 =$ angular velocity (rad/s) of the inner cylinder. The boundary conditions are that:

$$\boldsymbol{v}(r=R_1) = R_1\,\Omega_1\,\hat{\boldsymbol{\phi}} \tag{5.206}$$

$$\boldsymbol{v}(r=R_2) = R_2\,\Omega_2\,\hat{\boldsymbol{\phi}}. \tag{5.207}$$

   No pressure gradients are applied along the axis of the cylinders. In this case, $v_z = 0$ and $v_r = 0$. Under these conditions, show that

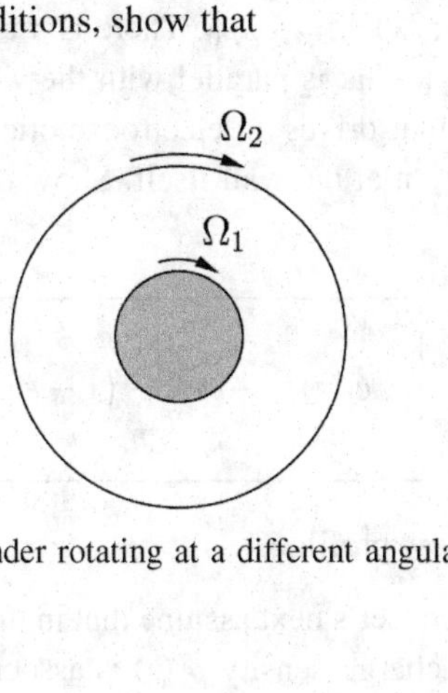

Figure 5.9 An inner solid cylinder rotating at a different angular speed Ω_1 than an outer solid cylinder Ω_2.

$$v_\phi(r) = \frac{\left(\Omega_2 R_2^2 - \Omega_1 R_1^2\right)}{R_2^2 - R_1^2} r + \frac{(\Omega_1 - \Omega_2) R_1^2 R_2^2}{R_2^2 - R_1^2} \frac{1}{r}. \tag{5.208}$$

HINT: Look for solutions of the form $v_\phi = A_n r^n$ and determine which n works.

Also, determine $P(r)$, which is easy, but nontrivial. To do so, look at the $\hat{\boldsymbol{r}}$ component of the Navier–Stokes equation and assume you know the pressure at $P(r = R_1)$. Eventually, at large enough angular velocity and, therefore, large enough Reynolds number, Couette flow becomes turbulent due to the nonlinear v_ϕ^2 term in the radial component of Navier Stokes.

6. *Flow created by axial movement of the inner cylinder*: In this case, the boundary conditions are that:

$$\boldsymbol{v}(r = R_1) = v_0 \hat{\boldsymbol{z}} \tag{5.209}$$

$$\boldsymbol{v}(r = R_2) = 0. \tag{5.210}$$

No pressure gradients are applied to the fluid in this problem. Show that

$$v_z(r) = v_0 \frac{\ln(r/R_2)}{\ln(R_1/R_2)}. \tag{5.211}$$

Electroosmotic Flow

Flows driven by an electric field acting on excess charge in the fluid are called "electroosmotic." The next two problems involve electroosmotic flow.

7. *An exponential charge distribution near a solid wall*: A planar solid wall with a unit normal in the $\hat{z}$ direction and located at $z = 0$ is bounded by a half-space of viscous electrolyte (a liquid having uniform viscosity η). The solid wall bears a surface charge per unit area σ_e (Coulombs per meter squared) that is balanced by an excess electric charge density in the viscous electrolyte that varies with distance as $\rho_e(z) = \rho_0 \exp(-z/\lambda)$, where λ is a given length and ρ_0 is related to σ_e as $\sigma_e = -\int_0^\infty \rho_0 \exp(-z/\lambda)\, \mathrm{d}z = -\lambda\rho_0$. There is a uniform applied electric field $E_x \hat{\boldsymbol{x}}$ throughout the electrolyte that is parallel with the wall and that acts on $\rho_e(z)$ to give a body force $\rho_e(z) E_x \hat{\boldsymbol{x}}$ that drives an electroosmotic flow in the x direction that satisfies the no slip condition at the wall itself. Show that the flow field throughout the electrolyte is

$$v_x(z) = -\frac{\sigma_e \lambda E_x}{\eta}\left(1 - e^{-z/\lambda}\right). \tag{5.212}$$

Make a sketch of this flow profile.

8. *The electroviscous effect*: Let's next assume that in the same problem geometry as just treated in 7, the excess charge density $\rho_e(z)$ is associated with an enhanced viscosity

compared to neutral electrolyte, in which $\eta(z) = \eta_\infty + \Delta\eta \exp(-z/\lambda)$, where η_∞ is the viscosity at great distance from the wall and $\eta_\infty + \Delta\eta$ is the enhanced viscosity right at $z = 0$. This enhanced viscosity where there is an excess charge density is called the *electroviscous effect*. First show that when $\eta = \eta(z)$ is spatially variable in z and for a flow field of the form $\boldsymbol{v} = v_x(z)\hat{\boldsymbol{x}}$, then

$$\nabla \cdot \boldsymbol{\tau}^v = \frac{\mathrm{d}}{\mathrm{d}z}\left(\eta(z)\frac{\mathrm{d}v_x(z)}{\mathrm{d}z}\right)\hat{\boldsymbol{x}}. \tag{5.213}$$

Then, using this result, show that the flow field induced by the same uniform electric field E_x acting on the excess charge density as in Problem 7 is now

$$\boxed{v_x(z) = \frac{\sigma_e \lambda E_x}{\Delta\eta} \ln\left[\frac{1 + (\Delta\eta/\eta_\infty)\exp(-z/\lambda)}{1 + \Delta\eta/\eta_\infty}\right].} \tag{5.214}$$

Show that in the limit that $\Delta\eta/\eta_\infty \to 0$, this solution is identical to that in Eq. (5.212).

9. *Nonsteady flow near a planar wall*: Solve for the fluid motion next to a solid plane that oscillates in time with the velocity $u_0 \cos\omega t\hat{\boldsymbol{x}}$ as depicted in Fig. 5.10. The fluid resides in the region $z > 0$, that is, the solid surface $z = 0$ that bounds the fluid is oscillating in time. The boundary conditions are that

$$\boldsymbol{v}(z = 0) = u_0 \cos\omega t\,\hat{\boldsymbol{x}} \tag{5.215}$$
$$\boldsymbol{v}(z = \infty) = 0. \tag{5.216}$$

The appropriate differential equation is the 1D diffusion equation

$$\frac{\eta}{\rho}\frac{\partial^2 v_x}{\partial z^2} - \frac{\partial v_x}{\partial t} = 0. \tag{5.217}$$

Show that

$$\boxed{v_x(z, t) = u_0\, e^{-z/\delta} \cos\left(\frac{z}{\delta} - \omega t\right), \quad \text{where} \quad \delta = \sqrt{\frac{2\eta}{\rho\,\omega}}} \tag{5.218}$$

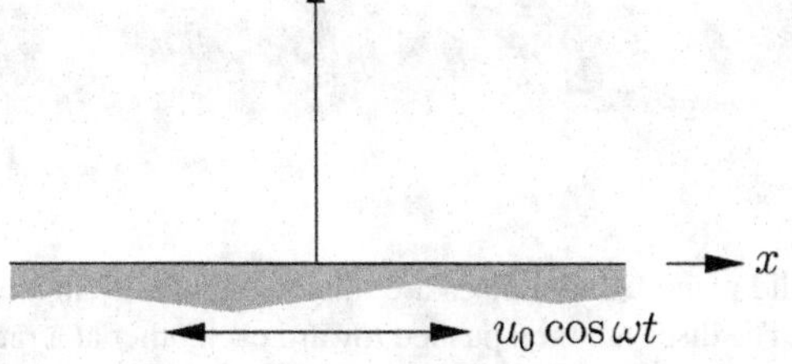

Figure 5.10 A solid plane bounded by a viscous fluid in $z > 0$ is oscillating in its plane.

and sketch a flow profile. Note that the length $\delta \mathrel{\widehat{=}}$ *skin depth* is common to all problems involving time-harmonic diffusion and always comes out as $\delta = \sqrt{2D/\omega}$, where D is the diffusivity. In the present problem where viscous shear is diffusing away from the moving wall, we have the viscous diffusivity $D = \eta/\rho$ (physical units of m^2/s). Later in the course, you will treat the problem of how the diurnal temperature fluctuations of the atmosphere penetrate into the Earth's subsurface which is a mathematically identical problem.

If you approximate your hand as a solid plane and place it on the surface of a liquid body and move it back and forth parallel with the surface, you will feel increasing resistance with increasing velocity that is quantified through the viscous shear stress $\hat{z} \cdot \boldsymbol{\tau}^v$. Please calculate this viscous shear stress acting on the solid plane. In truth, due to the 3D shape of your hand partially submerged in the water and your impatience with moving your hand slowly enough, your hand's movement will create a high Reynolds number flow and the resistance you feel will be more inertial than purely viscous.

The Lubrication Approximation

Sometimes an exact solution of a given flow through a duct or conduit that is driven by a fluid-pressure gradient or gravity is not possible to find analytically due to geometrical complexities but an excellent approximation can nonetheless be found in certain practical limits. The following are four such examples. They invoke the so-called *lubrication approximation*, which states that when the length of the conduit is large compared to its width, the force locally driving the flow can be approximated as varying only in the direction of the average fluid flux through the conduit and does not have a component in the direction perpendicular to that average flux. For flows involving a pressure gradient, this means the pressure field is approximated as only varying in the direction of the average fluid flux.

10. *Viscous flow between two converging plane-parallel solid discs*: Consider two planar circular discs of radius R that are separated by a gap of time-variable width $w(t)$ as

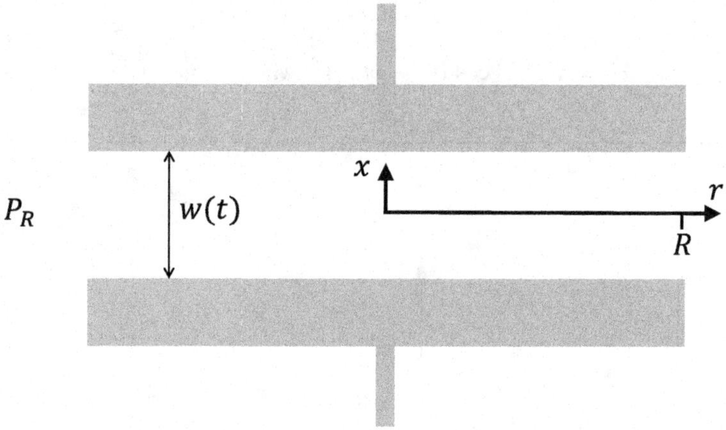

Figure 5.11 Two solid plane-parallel discs are immersed initially in a reservoir of viscous fluid at pressure P_R. The discs are then pushed toward each other at a rate $dw(t)/dt$ causing a pressure change and radial viscous flow out of the closing gap.

shown in Fig. 5.11. These plane-parallel discs are immersed in a liquid reservoir of viscosity η held at a constant fluid pressure P_R, such that the gap is filled with this same liquid. The discs are then pushed toward each other so that the gap closes at a given rate $\mathrm{d}w(t)/\mathrm{d}t < 0$. Determine the flow velocity and fluid pressure in the gap as $w(t)$ varies through time when $w(t) \ll R$. Assume that the rate of closure $\mathrm{d}w(t)/\mathrm{d}t$ is sufficiently small that the radial pressure gradient created between the center of the gap and the bounding fluid reservoir is fully established at each moment in time, which means that (cf., Section 5.3.2) the rate of change satisfies $w^{-1}\,|\mathrm{d}w/\mathrm{d}t| \ll D_p/R^2$, where D_p is the fluid-pressure diffusivity in the gap that goes roughly as $D_p = w^2K/\eta$, where $K \approx 10^9$ Pa is the bulk modulus of the liquid.

The flow problem is expressed in cylindrical coordinates and, by symmetry, has no angular dependence

$$\frac{\partial^2 v_r}{\partial r^2} + \frac{1}{r}\frac{\partial v_r}{\partial r} - \frac{v_r}{r^2} + \frac{\partial^2 v_r}{\partial x^2} = \frac{1}{\eta}\frac{\partial P}{\partial r} \quad r\text{-component of Navier Stokes} \tag{5.219}$$

$$\frac{\partial^2 v_x}{\partial r^2} + \frac{1}{r}\frac{\partial v_x}{\partial r} + \frac{\partial^2 v_x}{\partial x^2} = \frac{1}{\eta}\frac{\partial P}{\partial x} \quad x\text{-component of Navier Stokes} \tag{5.220}$$

$$\frac{\partial v_r}{\partial r} + \frac{v_r}{r} + \frac{\partial v_x}{\partial x} = 0 \quad \text{incompressible-flow condition} \tag{5.221}$$

and is subject to the boundary conditions that

$$v_r(r, x = \pm w/2) = 0 \tag{5.222}$$

$$v_x(r, x = \pm w/2) = \pm\frac{1}{2}\frac{\mathrm{d}w(t)}{\mathrm{d}t} \tag{5.223}$$

$$P(r = R, x) = P_R. \tag{5.224}$$

Formally, all of $P(r, x)$, $v_r(r, x)$, and $v_x(r, x)$ are functions of both r and x (and time) and are difficult to find in an exact way. However, in the limit as $w \ll R$, we can obtain a reasonable estimate throughout most of the gap by ignoring the diverging flow lines as the flow leaves the gap into the reservoir.

Thus, as an approximation, we ignore the x dependence in the pressure field $P(r, x) \approx P(r)$ so that $\partial P(r)/\partial x \approx 0$. This is called the *lubrication approximation*. The x-component of the Navier–Stokes equation is then solved by

$$\boxed{v_x(r, x) = \frac{\mathrm{d}w}{\mathrm{d}t}\frac{x}{w},} \tag{5.225}$$

which exactly satisfies the boundary conditions of Eq. (5.223).

With this leading-order approximation for $v_x(x)$, a radial derivative of the incompressible-flow condition yields

$$\frac{\partial^2 v_r}{\partial r^2} + \frac{1}{r}\frac{\partial v_r}{\partial r} - \frac{v_r}{r^2} = -\frac{\partial^2 v_x}{\partial r \partial x} = 0. \tag{5.226}$$

Using this in the radial component of the Navier–Stokes equation then yields

$$\frac{\partial^2 v_r(r,x)}{\partial x^2} = \frac{1}{\eta}\frac{dP(r)}{dr}. \tag{5.227}$$

Two integrations over x and use of the boundary conditions then gives

$$v_r(r,x) = -\frac{1}{2\eta}\frac{dP(r)}{dr}\left[\left(\frac{w}{2}\right)^2 - x^2\right]. \tag{5.228}$$

To find the radial pressure gradient, we can use the conservation of mass for incompressible flow in the form

$$-\pi r^2\frac{dw}{dt} = 2\pi r w \bar{v}_r(r), \tag{5.229}$$

where $\bar{v}_r(r)$ is the average radial velocity defined at each radial distance r as

$$\bar{v}_r(r) = \frac{2}{w}\int_0^{w/2} v_r(r,x)\,dx. \tag{5.230}$$

Equation (5.229) says that for incompressible flow, as the solid discs approach each other with $dw/dt < 0$ being given, the rate of change in gap volume within a radius r as given on the left-hand side goes entirely into radial volumetric flow across the radius r as given by the right-hand side. Upon carrying out the averaging integral, we find that

$$\frac{dP(r)}{dr} = \frac{6\eta}{w^3}\frac{dw}{dt}r, \tag{5.231}$$

so that

$$\boxed{v_r(r,x) = -\frac{3r}{w^3}\frac{dw}{dt}\left[\left(\frac{w}{2}\right)^2 - x^2\right].} \tag{5.232}$$

Multiply both sides of the pressure gradient by dr, integrate and use the pressure boundary condition, to give the pressure field as

$$\boxed{P(r) = P_R - \frac{3\eta}{w^3}\frac{dw}{dt}\left(R^2 - r^2\right).} \tag{5.233}$$

So as the discs are brought toward each other with dw/dt negative, the pressure is positive and has its largest value at the center $r=0$ and then parabolically decreases to the value P_R once $r=R$.

The above boxed solutions to this problem are an approximation based on the simple assumption $P(r,x) \approx P(r)$ that generates purely radial flow for each value of $w(t)$. Fill in all the missing steps to obtain each of the boxed equations. Show that this solution does not satisfy the local incompressible-flow condition of $\partial v_r/\partial r + v_r/r + \partial v_x/\partial x = 0$ but that it does satisfy the averaged incompressible-flow condition

$$\frac{\partial \bar{v}_r}{\partial r} + \frac{\bar{v}_r}{r} = -\frac{1}{w}\frac{dw}{dt}, \tag{5.234}$$

where the right-hand side is $-\partial v_x/\partial x$. Show that you could equivalently have found the same pressure gradient as Eq. (5.231) using this averaged form of the incompressible-flow condition in place of our conservation of incompressible-fluid mass argument of Eq. (5.229).

11. *Steady viscous flow in a variable-width duct*: A similar type of approximation can be made for pressure-driven flow between walls that have a variable in space, but constant in time, width $w(z) = h_+(z) + h_-(z)$ as shown in Fig. 5.12. The profiles of the upper and lower solid walls as characterized by the half-width functions $h_+(z)$ and $h_-(z)$ are taken to be known and possibly distinct. The flow is driven by a pressure drop down the length of the variable-width duct in the z direction and is assumed to be low-Reynolds number in nature. The 2D flow problem (no flow into or out of the page in the y direction) is expressed in cartesian coordinates as

$$\frac{\partial^2 v_z}{\partial z^2} + \frac{\partial^2 v_z}{\partial x^2} = \frac{1}{\eta}\frac{\partial P}{\partial z} \tag{5.235}$$

$$\frac{\partial^2 v_x}{\partial z^2} + \frac{\partial^2 v_x}{\partial x^2} = \frac{1}{\eta}\frac{\partial P}{\partial x} \tag{5.236}$$

$$\frac{\partial v_x}{\partial x} + \frac{\partial v_z}{\partial z} = 0, \tag{5.237}$$

subject to the boundary conditions that

$$v_x(x = h_+, z) = 0 \quad \text{and} \quad v_x(x = -h_-, z) = 0 \tag{5.238}$$

$$v_z(x = h_+, z) = 0 \quad \text{and} \quad v_z(x = -h_-, z) = 0 \tag{5.239}$$

$$P(x, z = 0) = P_R + \Delta P \tag{5.240}$$

$$P(x, z = L) = P_R, \tag{5.241}$$

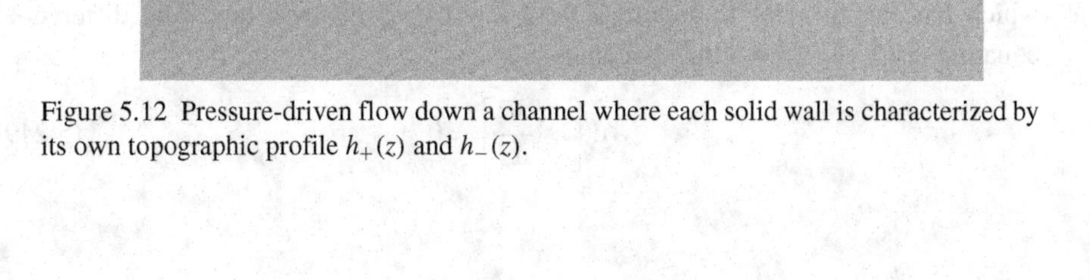

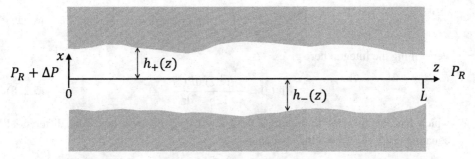

Figure 5.12 Pressure-driven flow down a channel where each solid wall is characterized by its own topographic profile $h_+(z)$ and $h_-(z)$.

where P_R is the pressure at the point $z = L$ of outflow. Formally, for arbitrary variations of the width $w(z) = h_+(z) + h_-(z)$, the flow fields and pressure are all functions of both x and z, that is, $v_x(x, z)$, $v_z(x, z)$ and $P(x, z)$, which for arbitrary variations in duct width $w(z)$, cannot be found analytically in an exact way even at low Reynolds number.

However, when the width only varies mildly with distance z and is small relative to the length L of the duct, we can attempt the *lubrication approximation* as made in the previous problem; namely, $P(x, z) \approx P(z)$. In this case, the x component of the Navier–Stokes equation and the boundary condition on the two walls are exactly satisfied by the leading-order approximation

$$v_x(x, z) = 0. \tag{5.242}$$

Using this in the incompressible-flow condition $\partial v_z/\partial z = 0$ so that $\partial^2 v_z/\partial z^2 = 0$, the z component of Navier Stokes becomes

$$\frac{\partial^2 v_z(x, z)}{\partial x^2} = \frac{1}{\eta}\frac{\mathrm{d}P(z)}{\mathrm{d}z}. \tag{5.243}$$

The quadratic function in x that satisfies this equation and also satisfies the no-slip boundary conditions on each wall $x = \pm h_\pm(z)$ is

$$v_z(x, z) = -\frac{1}{2\eta}\frac{\mathrm{d}P(z)}{\mathrm{d}z}\left\{\left(\frac{h_+(z) + h_-(z)}{2}\right)^2 - \left[x - \left(\frac{h_+(z) - h_-(z)}{2}\right)\right]^2\right\}. \tag{5.244}$$

To obtain the pressure gradient and pressure as a function of z, we then use the incompressible flow condition in the specific form that says the volumetric fluid flux across each width

$$w(z) = h_+(z) + h_-(z) \tag{5.245}$$

of the duct is the same, which is the statement

$$\frac{\mathrm{d}}{\mathrm{d}z}\left[w(z)\bar{v}_z(z)\right] = 0, \tag{5.246}$$

where $\bar{v}_z(z)$ is the average fluid flux across each width defined by

$$\bar{v}_z(z) = \frac{1}{w(z)}\int_{-h_-(z)}^{h_+(z)} v_z(x, z)\,\mathrm{d}x. \tag{5.247}$$

Performing the integral here gives

$$\bar{v}_z(z) = -\frac{w^2(z)}{12\eta}\frac{\mathrm{d}P(z)}{\mathrm{d}z} \tag{5.248}$$

which has the form of the average-flow law called Darcy's law. The differential equation $\mathrm{d}[w(z)\bar{v}_z(z)] = 0$ then becomes

$$\mathrm{d}\left[w^3\frac{\mathrm{d}P}{\mathrm{d}z}\right] = 0 \tag{5.249}$$

which integrates to give

$$\frac{dP(z)}{dz} = cw^{-3}(z). \tag{5.250}$$

The integration constant c is obtained by integrating from 0 to L to give the pressure gradient in final form as

$$\boxed{\frac{dP(z)}{dz} = -\frac{w^{-3}(z)\Delta P}{\int_0^L w^{-3}(z')\,dz'}.} \tag{5.251}$$

So where the width $w(z)$ is small, the pressure gradient is large in order to keep the volumetric flux along the length of the duct constant. The pressure field is then obtained after a final integration

$$\boxed{P(z) = P_R + \Delta P\left(1 - \frac{\int_0^z w^{-3}(z')\,dz'}{\int_0^L w^{-3}(z')\,dz'}\right).} \tag{5.252}$$

The flow velocity is written finally as

$$\boxed{v_z(x, z) = \frac{1}{2\eta}\frac{w^{-3}(z)\Delta P}{\int_0^L w^{-3}(z')\,dz'}\left\{\left(\frac{w(z)}{2}\right)^2 - \left[x - \left(\frac{h_+(z) - h_-(z)}{2}\right)\right]^2\right\},} \tag{5.253}$$

where again $w(z) = h_+(z) + h_-(z)$. Although this $v_z(x, z)$ and $v_x(x, z) = 0$ do not satisfy the local incompressible-flow condition $\partial v_x/\partial x + \partial v_z/\partial z = 0$, the total volumetric flux across each cross section of the duct is a constant along the duct which is an averaged statement of incompressible flow. In the special case of a planar duct where $w(z) = 2h$ is a constant, the above boxed expressions reduce to those given earlier in Section 5.6.1.

Fill in all the steps between the governing equations and the above boxed equations.

12. *Steady viscous flow between two plates sliding past each other*: Problems in which solid plates that are separated by a viscous fluid slip past each other is why these types of approximate solutions for flow in variable-width ducts or conduits bear the name *lubrication approximations*. The problem addressed here again involves two solid surfaces with a variable-width $w(z)$ fluid-filled gap between them. In this problem, we assume the lower surface is both infinite and planar and moving at a constant slipping velocity V_z relative to the upper surface that is curved and held stationary as depicted in Fig. 5.13. This is a slightly more general version of the problem treated by Batchelor (1967) in his classic text, in which a planar upper plate is held fixed at a small angle θ

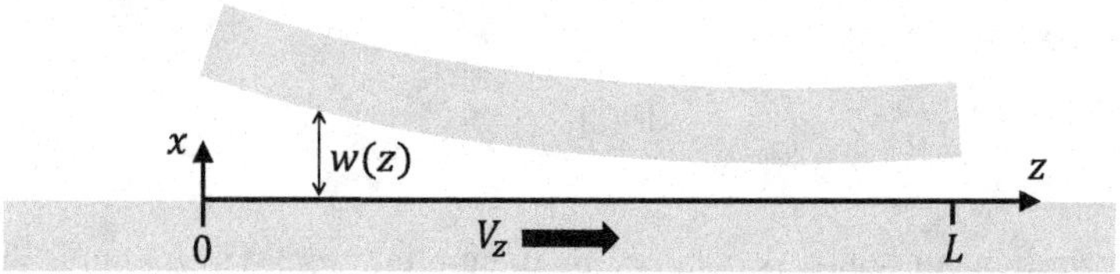

Figure 5.13 A planar solid lower plate is moving with a velocity V_z and a curved solid upper plate is held stationary.

above another infinitely long planar plate that is slipping past the upper plate in the z direction at the constant velocity V_z.

In the problem formulation, the (x, z) coordinates will be held stationary with the upper plate. We want to determine the flow velocity and fluid pressure throughout the gap created by this steady slipping motion between the plates when the spatial variation of the width $w(z)$ is small. This arrangement of having a viscous fluid lubricate the sliding between two plates when there is a variable-width gap that separates the plates arises in industrial and engineering practice. The governing equations are again

$$\frac{\partial^2 v_z}{\partial z^2} + \frac{\partial^2 v_z}{\partial x^2} = \frac{1}{\eta}\frac{\partial P}{\partial z} \tag{5.254}$$

$$\frac{\partial^2 v_x}{\partial z^2} + \frac{\partial^2 v_x}{\partial x^2} = \frac{1}{\eta}\frac{\partial P}{\partial x} \tag{5.255}$$

$$\frac{\partial v_x}{\partial x} + \frac{\partial v_z}{\partial z} = 0, \tag{5.256}$$

subject to the boundary conditions that

$$v_x(x = w(z), z) = 0 \tag{5.257}$$

$$v_x(x = 0, z) = 0 \tag{5.258}$$

$$v_z(x = w(z), z) = 0 \tag{5.259}$$

$$v_z(x = 0, z) = V_z \tag{5.260}$$

$$P(x, z = 0) = P_R \tag{5.261}$$

$$P(x, z = L) = P_R \tag{5.262}$$

where P_R is the constant pressure of the bounding fluid reservoir into which this slide-plate assembly is inserted. Formally, the flow and pressure fields are all functions of (x, z). But when $w(z)/L \ll 1$ and if we ignore the diverging flow lines at the entrance and exit of the gap, it is reasonable to invoke the *lubrication approximation* that $P(x, z) \approx P(z)$ so that $\partial P(x, z)/\partial x = 0$.

In this case, the x component of the Navier–Stokes equation is satisfied, along with the boundary conditions, by

$$v_x(x, z) = 0. \tag{5.263}$$

From the incompressible-flow condition, we then have $\partial^2 v_z/\partial z^2 \approx 0$ so that the z-component of the Navier–Stokes equation and the boundary conditions are satisfied by

$$v_z(x, z) = -\frac{1}{2\eta}\frac{\mathrm{d}P(z)}{\mathrm{d}z} x\,[w(z) - x] + V_z\left(1 - \frac{x}{w(z)}\right). \tag{5.264}$$

The second term is what drives the flow as the moving lower plate drags the viscous fluid along with it. The first term has a variable in z pressure gradient that increases or decreases the viscous flow rate at each z so that the volumetric flow rate through each cross section of the gap is the same.

As in all of our other examples of the lubrication approximation, these fields do not satisfy the local condition for incompressible flow but they do satisfy the averaged statement of incompressible flow that

$$\frac{\mathrm{d}}{\mathrm{d}z}\left[w(z)\bar{v}_z(z)\right] = 0, \tag{5.265}$$

where

$$w(z)\bar{v}_z(z) = \int_0^{w(z)} v_z(x, z)\,\mathrm{d}x = -\frac{1}{12\eta}\frac{\mathrm{d}P(z)}{\mathrm{d}z} w^3(z) + \frac{V_z}{2} w(z). \tag{5.266}$$

So taking the derivative and setting to zero gives

$$\mathrm{d}\left(w^3 \frac{\mathrm{d}P}{\mathrm{d}z}\right) = 6\eta V_z \mathrm{d}w, \tag{5.267}$$

which integrates to give

$$\frac{\mathrm{d}P}{\mathrm{d}z} = \frac{c}{w^3} + \frac{6\eta V_z}{w^2}. \tag{5.268}$$

The integration constant c is found by integrating from 0 to L and using that $P(0) = P(L)$ to give

$$c = -6\eta V_z \frac{\int_0^L w^{-2}(z')\,\mathrm{d}z'}{\int_0^L w^{-3}(z')\,\mathrm{d}z'}. \tag{5.269}$$

The pressure gradient is then given by

$$\frac{\mathrm{d}P(z)}{\mathrm{d}z} = \frac{6\eta V_z}{w^3(z)}\left[w(z) - \frac{\int_0^L w^{-2}(z')\,\mathrm{d}z'}{\int_0^L w^{-3}(z')\,\mathrm{d}z'}\right]. \tag{5.270}$$

Integrating this from $z = 0$ where $P(0) = P_R$ then gives

$$\boxed{P(z) = P_R + 6\eta V_z\left[\int_0^z w^{-2}(z')\,\mathrm{d}z' - \frac{\int_0^L w^{-2}(z')\,\mathrm{d}z'}{\int_0^L w^{-3}(z')\,\mathrm{d}z'}\int_0^z w^{-3}(z')\,\mathrm{d}z'\right].} \tag{5.271}$$

Last, the flow velocity may be written

$$v_z(x, z) = V_z\left(1 - \frac{x}{w(z)}\right)\left[1 - \frac{3x}{w(z)}\left(1 - \frac{1}{w(z)}\frac{\int_0^L w^{-2}(z')\,dz'}{\int_0^L w^{-3}(z')\,dz'}\right)\right]. \quad (5.272)$$

So we have solved this lubrication-layer problem for any given width function $w(z)$ between a sliding planar lower plate and a possibly slowly undulating in space but stationary upper plate.

For the specific lubrication problem addressed by Batchelor (1967) for which both plates are perfect planes and the upper plate is oriented at an angle θ relative to the lower plate, the width function is

$$w(z) = w_0 - z\tan\theta \quad \text{where} \quad w_0 = w_L + L\tan\theta, \quad (5.273)$$

with w_L the smallest gap width at the exit point $z = L$. In this case, we find

$$\frac{dP(z)}{dz} = \frac{6\eta V_z}{w^3(z)}\left[w(z) - \frac{2w_0 w_L}{w_0 + w_L}\right], \quad (5.274)$$

$$P(z) = P_R + \frac{6\eta V_z}{\tan\theta}\frac{[w_0 - w(z)][w(z) - w_L]}{w^2(z)(w_0 + w_L)} \quad (5.275)$$

and

$$v_z(x, z) = V_z\left(1 - \frac{x}{w(z)}\right)\left[1 - \frac{3x}{w(z)}\left(1 - \frac{2w_0 w_L}{w(z)(w_0 + w_L)}\right)\right]. \quad (5.276)$$

So in this problem, the pressure is P_R at $z = 0$, then increases with increasing z until it attains a smooth maximum at the position z_m where $w(z_m) = 2w_0 w_L/(w_0 + w_L)$. It then smoothly descends until it is again P_R at $z = L$. So the viscous pressure gradient retards the flow velocity through the entry portion of the gap and then after the smooth pressure maximum increases the velocity through the exit portion in order for the volumetric fluid flux to be constant all along the gap.

Fill in all the missing detailed steps between the governing equations and all of the boxed expressions above.

13. *Thinning of a vertical liquid film on a plate:* As a final problem invoking a lubrication approximation, consider that there is a thin film of viscous liquid having thickness H_0 that is, initially, residing horizontally on a solid plate and that is overlain by air. We then rotate the plate to be vertical so that gravity then drives a flow of the film downward in the z direction. This causes the film to have an evolving thickness $H(z, t)$

that is thinner near the top as the liquid drains out compared to the initial thickness H_0. We assume that the contact line between solid, liquid, and air is pinned and resides at $z = 0$ as depicted in Fig. 5.14. Justification for the assumption of a pinned contact line will be given later in Section 7.6.3 when we address capillary physics.

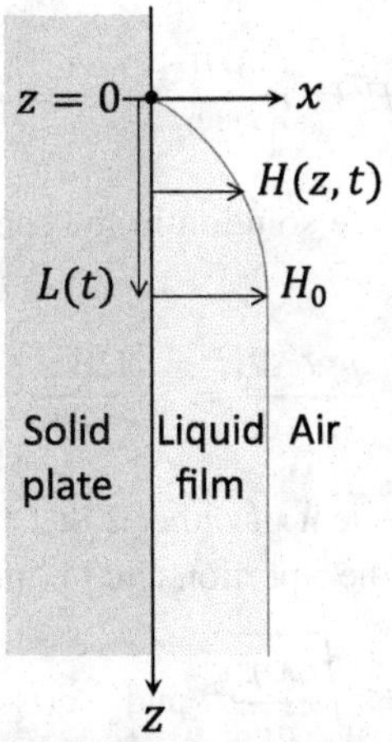

Figure 5.14 A vertically oriented thin film flows downward by gravity so that it has a thinner thickness $H(z, t)$ near the top compared to its initial thickness H_0. The contact line between the solid, liquid film, and air is assumed to be pinned at $z = 0$ (the black dot).

To find the evolving thickness $H(z, t)$ of this thin film, which is the goal of the exercise, we make another type of lubrication approximation that says that the flow even in the thinning part of the film is purely in the vertical z direction. The governing equation for the flow of the viscous liquid in this approximation is then

$$\eta \frac{\partial^2 v_z(x, z, t)}{\partial x^2} + \rho g = 0, \tag{5.277}$$

subject to the condition that there is no viscous shear stress at the liquid-air interface $x = H(z, t)$ and no slip on the solid surface $x = 0$, that is,

$$\left. \frac{\partial v_z}{\partial x} \right|_{x=H} = 0 \quad \text{and} \quad v_z|_{x=0} = 0. \tag{5.278}$$

Show that by integrating this equation twice and using the two boundary conditions, the flow field is given by

$$v_z(x, z, t) = \frac{\rho g}{\eta} \left(H(z, t)x - \frac{x^2}{2} \right) \tag{5.279}$$

and that film flow averaged across each width of the film is

$$\boxed{\bar{v}_z(z, t) = \frac{1}{H(z, t)} \int_0^{H(z,t)} v_z(x, z, t)\, \mathrm{d}x = \frac{\rho g}{3\eta} H^2(z, t).} \tag{5.280}$$

To find $H(z, t)$ we then express the conservation of mass of the incompressible flow where the film is thinning by the volume balance of the incompressible fluid

$$\frac{\partial}{\partial z}\left[H(z,t)\bar{v}_z(z,t)\right] = -\frac{\partial H(z,t)}{\partial t} \tag{5.281}$$

or

$$\frac{\rho g}{\eta} H^2(z,t)\frac{\partial H(z,t)}{\partial z} = -\frac{\partial H(z,t)}{\partial t}. \tag{5.282}$$

To solve this equation, write the solution in the separated form $H(z, t) = \zeta(z)\tau(t)$ to obtain

$$\frac{\rho g \zeta}{\eta}\frac{d\zeta}{dz} = -\frac{1}{\tau^3}\frac{d\tau}{dt} = \lambda, \tag{5.283}$$

where λ is a constant independent of either z or t that you will see drops out of the analysis. Integrate separately the z portion and t portion of these equations to obtain

$$\zeta(t) = \sqrt{\frac{2\lambda\eta z}{\rho g}} \quad \text{and} \quad \tau(t) = \sqrt{\frac{1}{2\lambda t}}. \tag{5.284}$$

Thus, show that you can express the thinning of the liquid film as it flows vertically under gravity as

$$H(z,t) = \begin{cases} \sqrt{\eta z/(\rho g t)} & \text{when } z < L(t) \\ H_0 & \text{when } z > L(t), \end{cases} \tag{5.285}$$

where $L(t)$ is the length of the tapered portion of the film defined by the condition $H(L, t) = H_0$ or

$$L(t) = \frac{H_0^2 \rho g t}{\eta}, \tag{5.286}$$

which increases linearly with time as the film thins.

6
Equilibrium Thermodynamics

Each mass element of material that contains a multitude of jiggling molecules has an average energy and an average volume associated with the given number of molecules as well as various thermodynamic state functions such as temperature, pressure, and chemical potential. Thermodynamics is the subject that describes the relations between these macroscopic variables once the concept of entropy is introduced. We can generally separate thermodynamics into two subject areas:

1. *Equilibrium thermodynamics* focuses on the relations and properties of systems in *equilibrium*. Equilibrium is defined when the average state of a system, as defined by its energy, volume, and number of molecules, is not changing over time. If we perturb a system by allowing energy, volume, or molecules to be added or subtracted, equilibrium thermodynamics addresses the new equilibrium state that emerges eventually following the perturbation. An important part of equilibrium thermodynamics, and perhaps our primary focus toward our goal of establishing the rules of macroscopic response, is defining the constitutive laws that control how changes in the various thermodynamic variables influence each other. The changes are between one equilibrium state to another. Entropy is the key new macroscopic variable that must be introduced that allows us to establish these general relations between different equilibrium states.
2. *Nonequilibrium thermodynamics* focuses on what is happening at the macroscopic level (i.e., not molecularly resolved) while a system is passing from one equilibrium state to another. The fluxes (movement of energy and matter) that allows a perturbed system to reequilibrate into a new equilibrium state will irreversibly create entropy until the entropy is a maximum, subject to constraints, in the new equilibrium state that emerges. The relations between the fluxes and the gradients in the various thermodynamic state functions that hold while a system is out of equilibrium are called *transport laws*, which are another class of constitutive laws that are distinct in character from the constitutive laws of equilibrium thermodynamics addressed in this chapter.

Hooke's law of elasticity is an example we have already seen of a constitutive law of equilibrium thermodynamics, while the law of Newtonian viscosity is an example we have already seen of a transport law of nonequilibrium thermodynamics. This chapter focuses exclusively on equilibrium thermodynamics that describe reversible processes from a continuum (averaged) perspective. Nonequilibrium thermodynamics and the approach for obtaining transport laws is treated in Chapter 7.

6.1 Survey of the Fundamentals of Equilibrium Thermodynamics

6.1.1 The Thermodynamic Independent Variables in a Fluid

The equilibrium state of a macroscopic element of a fluid containing a single molecular species is characterized by the variables:

$$U = \textit{internal energy} \text{ (total accessible energy) of the element,} \tag{6.1}$$

$$V = \textit{volume} \text{ of the element,} \tag{6.2}$$

$$N = \textit{number} \text{ of molecules in the element.} \tag{6.3}$$

We are envisioning each element of fluid under discussion to be surrounded by other elements as part of a larger macroscopic body being studied.

The internal energy of a fluid element represents all accessible forms of energy within the element. These include the random-motion thermal (kinetic) energy of the center-of-mass motion of the molecules and the internal modes of vibration of the molecules that together are known as *heat* energy, as well as the energy stored in the interaction fields between the molecules within an element and the energy stored in the interaction fields between the molecules inside the element and those outside the element, which are associated with the *work* required to add or extract a molecule from the element or to change the volume and/or shape of an element. The internal energy does not include the inaccessible nuclear energy stored in the nucleus of atoms nor the kinetic energy $\mathcal{M}\boldsymbol{v} \cdot \boldsymbol{v}/2$ associated with the macroscopic movement of the entire element of mass $\mathcal{M}$ that is moving with a center-of-mass velocity $\boldsymbol{v}$.

The thermodynamic independent variables U, V, and N are called *extensive* because if the size of an initial element and its associated mass $\mathcal{M}$ is increased by a factor β from $\mathcal{M}$ to $\beta\mathcal{M}$ by connecting together β identical versions of the initial element, all of U, V, and N for the composite will increase by a factor β.

In a continuum description of a single-component fluid that is, possibly, flowing, we imagine following a mass element about as work is either done on it or done by it or as the random thermal energy (i.e., heat) of the N molecules either accumulates or decreases in the element depending on whether more or less heat is flowing into than out of the element or whether there are fluxes across the element that are generating heat within the element. The entire macroscopic body consisting of many such elements may be out of equilibrium but a fundamental assumption of our continuum thermodynamics is that each element immediately adapts to external perturbations and can always be treated as being in equilibrium with itself unless chemical reactions or phase changes are occurring within the element. Equilibrium thermodynamics quantifies the changes to the equilibrium state of each element as the element is subjected to externally applied perturbations.

In continuum thermodynamics, it is common to normalize the extensive variables U, V, and N of an element by the mass of the element to create so-called *specific* variables. It is common practice to place a caret symbol over a quantity to designate that it has been normalized by mass but we elect in this chapter to simply use a lower case letter: $u = U/\mathcal{M}$, $v = V/\mathcal{M} = 1/\rho$, and $n = N/\mathcal{M}$. So in a single-component system, $n = 1/m$ is a constant where m is the mass of a single molecule, which is independent of the thermodynamic state of the mass element. One can also define continuum thermodynamic extensive variables as *densities*, which means the extensive thermodynamic variables are normalized by the volume of the mass element surrounding each element. However, because this normalizing volume typically varies during the course of thermodynamic changes, it is a less convenient normalization to consider. In Sections 6.6 and 6.7, where we generalize our treatment to include deviatoric stress and strain in a solid as well as electromagnetic polarization processes, we will normalize the thermodynamic extensive variables by a constant reference volume defined as the volume when the element has zero strain. This is not essential but is made to provide consistency with historic treatment of solids. To transition from a specific variable that is normalized by a constant mass to a density that is normalized by the constant reference volume, one simply multiplies by the constant mass density in the reference state.

Situations where each element contains a range of different molecular species are also very important, especially in chemistry, and will be addressed in some detail in Section 6.5. But for pedagogic simplicity in laying out the formalism of thermodynamics, we proceed for now under the assumption that each fluid element is comprised of N identical molecules.

6.1.2 Entropy

The quintessential concept of thermodynamics is *entropy*, which is the variable that allows changes in equilibrium due to external perturbations to be quantified. It is also the thermodynamic variable that is the most challenging to define. We will follow the postulational approach for defining entropy as originally devised by Herbert Callen in his classic and still highly recommended text on thermodynamics (Callen, 1985).

There exists a function of the thermodynamic independent variables called the entropy S

$$S = S(U, V, N) \mathrel{\widehat{=}} \textit{entropy} \tag{6.4}$$

that obeys the following three postulates:

1. *Entropy is a maximum in the equilibrium state.* Consider two systems (U_1, V_1, N_1) and (U_2, V_2, N_2) that are brought together and allowed to touch along a wall of separation.

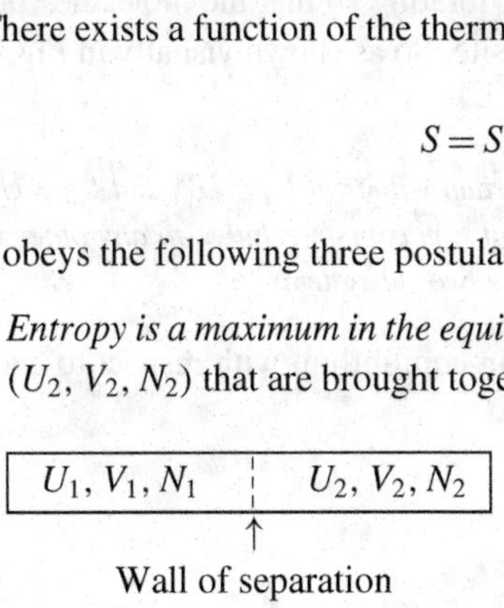

For the composite total system, we have $U = U_1 + U_2$, $V = V_1 + V_2$, and $N = N_1 + N_2$. While the totals for the composite U, V, and N are constants during equilibration across the wall, the entropy $S = S_1 + S_2$ is not a constant.

U_1	U_2	$S = S_1 + S_2$
U	0	$S_1(U) + S_2(0)$
$U - \Delta U$	$+\Delta U$	$S_1(U - \Delta U) + S_2(\Delta U)$
$\vdots$	$\vdots$	$\vdots$
U_1^{eq}	U_2^{eq}	$S_{\max} = S_1\left(U_1^{eq}\right) + S_2\left(U_2^{eq}\right)$
$\vdots$	$\vdots$	$\vdots$
ΔU	$U - \Delta U$	$S_1(\Delta U) + S_2(U - \Delta U)$
0	U	$S_1(0) + S_2(U)$

Ways to partition the energy between the two systems — Value of $S = S_1 + S_2$ for the various ways to partition energy

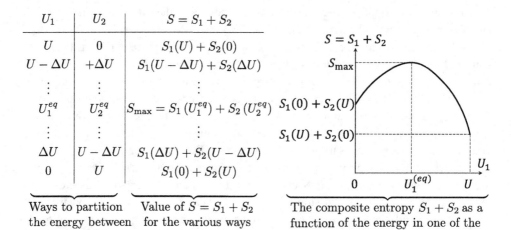

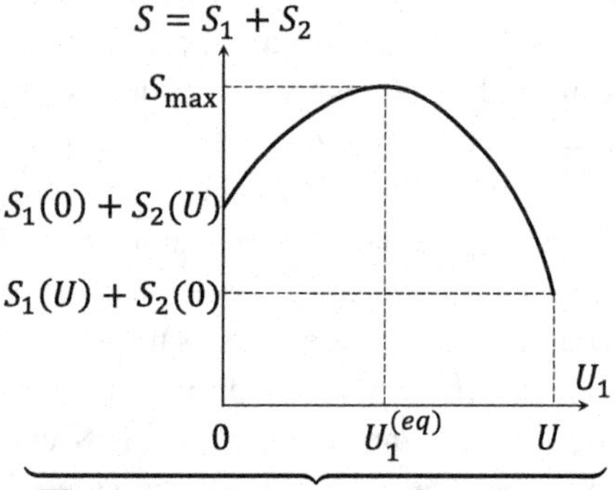

The composite entropy $S_1 + S_2$ as a function of the energy in one of the subsystems U_1 where $0 \le U_1 \le U$.

Figure 6.1 The heat energy in two subsystems separated by a diathermal wall flows until the entropy of the composite is at a maximum.

Let the wall of separation allow any of (i) energy to flow across the wall, (ii) the wall to displace, or (iii) molecules to pass through the wall. Such equilibration processes are what allow the composite at some time after the initial subsystems are brought into contact to achieve a new equilibrium state.

The first postulate concerning entropy is that the value of the composite entropy $S = S_1 + S_2$ in the new equilibrium state will be a maximum relative to all possible nonequilibrium ways of distributing the U_i, V_i, and N_i between the two subsystems $i = 1, 2$. To describe what this means let's consider a specific example. If the wall of separation allows the molecular kinetic and vibrational energy known as heat to flow across it (a so-called diathermal wall) but is immobile and impermeable to the molecules, heat will flow between the two subsystems until equilibrium is obtained as depicted in Fig. 6.1. Having zero energy in a subsystem in this example means that the molecules in that subsystem are not moving or vibrating so that the molecules in the other subsystem has all the energy in the composite. So as shown visually in Fig. 6.1, the entropy-maximum postulate states that:

When the energy in each subsystem takes the equilibrium values of $U_1 = U_1^{eq}$ and $U_2 = U_2^{eq} = U - U_1^{eq}$, the composite entropy $S = S_1 + S_2$ will be at a maximum relative to any other non-equilibrium way of partitioning the energy between the two subsystems.

Mathematically, when the composite system is in equilibrium with respect to energy transfers between the subsystems, we have

$$\left(\frac{\partial S}{\partial U_1}\right)_U = 0 \tag{6.5}$$

$$\left(\frac{\partial^2 S}{\partial U_1^2}\right)_U < 0. \tag{6.6}$$

Equation (6.5) says that entropy is an extremum at equilibrium and Eq. (6.6) says that this extremum is a maximum.

2. *Entropy is a monotonically increasing function of* U, V, and N:

$$\frac{\partial S}{\partial U} > 0, \quad \frac{\partial S}{\partial V} > 0, \quad \text{and} \quad \frac{\partial S}{\partial N} > 0. \tag{6.7}$$

This postulate allows the entropy function to be inverted. Specifically, we can always obtain

$$U = U(S, V, N) \,\hat{=}\, \textit{internal energy} \text{ function.} \tag{6.8}$$

The information contained in $U = U(S, V, N)$ is equivalent to that contained in $S = S(U, V, N)$.

3. *The physical meaning of entropy* is provided by *statistical mechanics* and is related to the concept of "order" by the postulate

$$S = -k_B \sum_{j=1}^{\Omega} p_j \ln p_j \quad \textit{statistical definition of entropy}, \tag{6.9}$$

where $k_B \approx 1.381 \times 10^{-23}$ J/K is Boltzmann's constant and

p_j = probability that the atoms of the element are in a state "j" given any constraints on the system (such as there being a given U, V, and N).

From a classical molecular-dynamics perspective, each "state" j corresponds to a list of $6N$ numbers when there are N atoms in the element that correspond to the position and momentum of each atom at some instant in time. As time proceeds, the N atoms pass from one state j to the next. From a quantum-mechanical perspective, the state is given by the list of discrete quantum numbers for each atom that set, among other things, the discrete energy level of each atom. In this case, there are a total of Ω different states available to a system corresponding to the different ways of distributing quantum numbers between the atoms so that the atoms and interaction energies present in each state add up to a total energy U. This total state count Ω is fantastically large but is an actual number and not infinity. Note that if the atoms always stay in one particular state $j = J$ so that $p_J = 1$ and $p_j = 0$ for $j \neq J$, which is a good definition of total order, then Eq. (6.9) says that $S = 0$ in a system that is perfectly ordered. If, however, for an element with given U, V, and N, each state is equally likely so that $p_j = 1/\Omega =$ constant, which is a good definition of total disorder as the system jumps from one state to another with equal probability under the constraint that U, V, and N are fixed, then Eq. (6.9) says that

$$S(U, V, N) = k_B \ln \Omega(U, V, N), \tag{6.10}$$

which also happens to be the maximum possible value for S given by Eq. (6.9). So entropy as defined by Eq. (6.9) quantifies the qualitative concept of disorder, as was independently demonstrated by Shannon (1948). As the number of possible states $\Omega = \Omega(U, V, N)$ increases with increasing U, V, or N, it is assumed that each state is equally probable so that the entropy also increases as given by Eq. (6.10). Ludwig Boltzmann made the postulate of Eq. (6.10) in 1877, that is called *Boltzmann's entropy formula*, thus giving entropy a clear statistical definition that can be used to theoretically calculate the function $S = S(U, V, N)$ for simple systems using molecularly resolved counting arguments. Statistical mechanics is the subject that allows macroscopic thermodynamic "fundamental functions," such as $S(U, V, N)$, to be determined based on specific models of the underlying molecularity, while thermodynamics is the subject that provides general relations between the macroscopic variables without appealing to any particular molecular landscape.

An important consequence of Eq. (6.10) is that entropy S, like the other thermodynamics variables U, V, and N, is extensive. If we add two identical equilibrium systems together that each have Ω states available to them, the number of states available to the new larger composite system is Ω^2, that is, when subsystem 1 is in state 1, subsystem 2 can be in Ω states and so on which results in Ω^2 states available to the larger composite system. Upon taking the logarithm of Ω^2, the entropy of the composite is seen to be double that of each subsystem. So the entropy function satisfies the extensive scaling rule $\beta S(U, V, N) = S(\beta U, \beta V, \beta N)$.

Like with the internal energy, volume, and particle numbers, we can normalize entropy by the mass of the fluid element it pertains to and define the specific entropy $s = S/\mathcal{M}$ that is a function of the specific energy and specific volume $s = s(u, v)$ for single-molecular-species elements.

6.1.3 Four Facts about Partial Derivatives

Before getting on with the main business of determining how the thermodynamic variables change from one equilibrium state to the next when a change to U, V, or N takes place, we need to establish four useful facts concerning partial derivatives that are used repeatedly in thermodynamics.

Consider first some thermodynamic function of two independent variables $f = f(x, y)$. Our thermodynamic functions are monotonic and invertible so that we may obtain both $y = y(x, f)$ and $x = x(y, f)$. Take a total derivative of f to obtain

$$\mathrm{d}f = \left(\frac{\partial f}{\partial x}\right)_y \mathrm{d}x + \left(\frac{\partial f}{\partial y}\right)_x \mathrm{d}y. \tag{6.11}$$

Although it is redundant notation to put the large parentheses around the partial derivatives and indicate with a subscript which variable is being held constant, it is clarifying to do so in thermodynamics given the myriad of variables and the practice is recommended. Next take the total derivative of say $y = y(x, f)$ to obtain

$$\mathrm{d}y = \left(\frac{\partial y}{\partial x}\right)_f \mathrm{d}x + \left(\frac{\partial y}{\partial f}\right)_x \mathrm{d}f. \tag{6.12}$$

Substituting in the expression for df then gives

$$\mathrm{d}y = \left(\frac{\partial y}{\partial x}\right)_f \mathrm{d}x + \left(\frac{\partial y}{\partial f}\right)_x \left[\left(\frac{\partial f}{\partial x}\right)_y \mathrm{d}x + \left(\frac{\partial f}{\partial y}\right)_x \mathrm{d}y\right], \tag{6.13}$$

which upon rearranging gives

$$0 = \left[\left(\frac{\partial y}{\partial f}\right)_x \left(\frac{\partial f}{\partial x}\right)_x + \left(\frac{\partial y}{\partial x}\right)_f\right] \mathrm{d}x + \left[\left(\frac{\partial y}{\partial f}\right)_x \left(\frac{\partial f}{\partial y}\right)_x - 1\right] \mathrm{d}y. \tag{6.14}$$

Since dx and dy are arbitrary, each coefficient in brackets must be zero. Setting the dy coefficient to zero gives the first useful fact about partial derivatives

$$\left(\frac{\partial f}{\partial y}\right)_x = \frac{1}{(\partial y/\partial f)_x}. \quad \textbf{Fact 1} \tag{6.15}$$

Setting the dx coefficient to zero then yields the second useful fact that

$$\left(\frac{\partial y}{\partial x}\right)_f = -\left(\frac{\partial y}{\partial f}\right)_x \left(\frac{\partial f}{\partial x}\right)_y, \tag{6.16}$$

which upon using the first fact can be rewritten as

$$\left(\frac{\partial y}{\partial x}\right)_f = -\frac{(\partial f/\partial x)_y}{(\partial f/\partial y)_x}. \quad \textbf{Fact 2} \tag{6.17}$$

This second fact in particular will be used over and over again in thermodynamics.

Two more facts can be established if we now consider two functions of two independent variables $f(x, y)$ and $g(x, y)$ both of which are invertible. So we may obtain, for example, the function $y = y(x, g)$ which when substituted into $f(x, y)$ means that f can also be expressed as a function of x and g, that is, $f = f(x, g)$. To see this more explicitly, again take a total derivative of $f = f(x, y)$

$$\mathrm{d}f = \left(\frac{\partial f}{\partial x}\right)_y \mathrm{d}x + \left(\frac{\partial f}{\partial y}\right)_x \mathrm{d}y \tag{6.18}$$

and then take a total derivative of $y = y(x, g)$

$$\mathrm{d}y = \left(\frac{\partial y}{\partial x}\right)_g \mathrm{d}x + \left(\frac{\partial y}{\partial g}\right)_x \mathrm{d}g. \tag{6.19}$$

Inserting the expression for dy into that for df then gives

$$\mathrm{d}f = \left(\frac{\partial f}{\partial x}\right)_y \mathrm{d}x + \left(\frac{\partial f}{\partial y}\right)_x \left[\left(\frac{\partial y}{\partial x}\right)_g \mathrm{d}x + \left(\frac{\partial y}{\partial g}\right)_x \mathrm{d}g\right] \tag{6.20}$$

$$= \left[\left(\frac{\partial f}{\partial x}\right)_y + \left(\frac{\partial f}{\partial y}\right)_x \left(\frac{\partial y}{\partial x}\right)_g\right] \mathrm{d}x + \left(\frac{\partial f}{\partial y}\right)_x \left(\frac{\partial y}{\partial g}\right)_x \mathrm{d}g. \tag{6.21}$$

This last expression tells us what we said earlier that $f = f(x, g)$ because upon taking the total derivative we have

$$\mathrm{d}f = \left(\frac{\partial f}{\partial x}\right)_g \mathrm{d}x + \left(\frac{\partial f}{\partial g}\right)_x \mathrm{d}g, \tag{6.22}$$

which is of identical form to the previous equation. So upon equating the coefficients of dx and dg in these last two expressions, we obtain the next two facts

$$\left(\frac{\partial f}{\partial g}\right)_x = \left(\frac{\partial f}{\partial y}\right)_x \left(\frac{\partial y}{\partial g}\right)_x \quad \textbf{Fact 3} \text{ (also called the } \textit{chain rule}\text{)} \tag{6.23}$$

and

$$\left(\frac{\partial f}{\partial x}\right)_g = \left(\frac{\partial f}{\partial x}\right)_y + \left(\frac{\partial f}{\partial y}\right)_x \left(\frac{\partial y}{\partial x}\right)_g . \quad \textbf{Fact 4} \tag{6.24}$$

Throughout the rest of this chapter and the next, as well as in the exercises at the end of this chapter, you will employ these facts.

6.1.4 Fundamental Functions and the Hierarchy of Information They Contain

Either $S = S(U, V, N)$ or its inverse relation $U = U(S, V, N)$ are called *fundamental functions* because they contain all conceivable thermodynamic information about the element of fluid (or system) they pertain to. We will focus in this section on $U = U(S, V, N)$ with $S = S(U, V, N)$ treated as an end-of-chapter exercise.

The thermodynamic information contained in $U(S, V, N)$ is extracted by taking derivatives. So let's take a total derivative:

$$\mathrm{d}U = \left(\frac{\partial U}{\partial S}\right)_{V,N} \mathrm{d}S + \left(\frac{\partial U}{\partial V}\right)_{S,N} \mathrm{d}V + \left(\frac{\partial U}{\partial N}\right)_{S,V} \mathrm{d}N. \tag{6.25}$$

Define the so-called *state functions* as the first derivatives of the fundamental function U:

$$T(S, V, N) = \frac{\partial U(S, V, N)}{\partial S} \mathrel{\widehat{=}} \text{temperature} \tag{6.26}$$

$$P(S,V,N) = -\frac{\partial U(S,V,N)}{\partial V} \widehat{=} \text{pressure} \tag{6.27}$$

$$\mu(S,V,N) = \frac{\partial U(S,V,N)}{\partial N} \widehat{=} \text{chemical potential.} \tag{6.28}$$

We thus obtain the important statement called the *first law of thermodynamics* (or conservation of energy)

$$\mathrm{d}U = \underset{\substack{\uparrow \\ \text{Heat} \\ \text{energy} \\ \text{changes}}}{T\,\mathrm{d}S} - \underset{\substack{\uparrow \\ \text{Work} \\ \text{energy} \\ \text{changes}}}{P\,\mathrm{d}V} + \underset{\substack{\uparrow \\ \text{Chemical} \\ \text{energy} \\ \text{changes}}}{\mu\,\mathrm{d}N}\,. \tag{6.29}$$

Note that in the partial derivatives that define T, P, and μ, because U, S, V, and N are all extensive (their value is doubled if two identical elements are combined into a new larger element that has twice the mass), the values of T, P, and μ are independent of the size of the system. A thermodynamic variable whose value is independent of the size of the element is called *intensive*. So T, P, and μ are the thermodynamic *intensive* variables. Let's discuss them.

We have an intuitive feel based on our human sensory experience what the temperature and pressure represent. Place your hand into a higher-temperature liquid than your body temperature and then into a lower-temperature liquid. For the higher-temperature bath, heat energy fluxes into your hand causing the molecules in your hand to acquire kinetic energy, which your senses notice as the bath being warm or hot. For the lower-temperature bath, heat energy fluxes out of your hand causing the molecules in your hand to lose kinetic energy, which your senses notice as the bath being cool or cold. *Temperature differences between a system (or element) and a surrounding reservoir is what controls whether heat energy flows into or out of the system, which is the essential character of temperature.* Heat will flow from regions of higher temperature toward regions of lower temperature.

Similarly, go to the bottom of a swimming pool to feel a higher-pressure liquid compared to the lower-pressure environment when you are near the surface and you feel the difference due to the pool-water atoms pushing in on your body with greater force at the bottom of the pool causing the volume containing your body mass to become reduced. *Pressure differences between a system and a surrounding reservoir is what controls whether the system volume increases or decreases, which is the essential character of pressure.* An element of fluid will have its volume reduced if the outside pressure of the reservoir is larger than the inside pressure of the element and will have its volume increased if the outside pressure is less than the inside pressure.

However, the chemical potential μ is less intuitive, with the *first law* given by Eq. (6.29) informing us that it quantifies by how much the energy of an element is increased at constant volume and entropy when one additional molecule is inserted into the element ($\mathrm{d}N = 1$). We will return to the chemical potential in later sections where we discuss all of (1) the state of equilibrium in single component fluids, (2) the chemical potential in multi-component (multiple molecular species) fluids, and (3) the chemical potential in solids. It will be demonstrated that entropy being a maximum in equilibrium requires that an element with a larger μ compared to its surroundings will want to give up molecules to the surroundings while if it has a smaller μ than the surroundings it will want to acquire molecules. *Chemical potential differences between a system and a surrounding reservoir is what controls whether molecules flow into or out of the system, which is the essential character of chemical potential.* Matter will flow from regions where the chemical potential is higher toward regions where the chemical potential is lower.

Next, we obtain the so-called *constitutive laws* by taking derivatives of the state functions, which involves second derivatives of the fundamental function. When presenting the constitutive laws in the sections to follow, we continue to make the simplifying assumption that our constant-mass element contains a single molecular species so that the number of such molecules is not changing in the element. If we normalize all the extensive variables by the mass of the element to obtain u, s, v, and n where again $n = N/\mathcal{M} = 1/m$, where m is the constant mass of a single molecule, we have $u = u(s, v)$ and $\mathrm{d}u = T\mathrm{d}s - P\mathrm{d}v$ with $\mathrm{d}n = 0$. In Section 6.5, we relax this assumption of a single molecular species and allow for a range of molecular species to be present and evolving in number within each element.

With the fundamental function $u = u(s, v)$ and state functions $T(s, v) = (\partial u/\partial s)_v$ and $P(s, v) = -(\partial u/\partial v)_s$, we now take a total derivative of both T and P

$$\mathrm{d}T = \frac{\partial T(s, v)}{\partial s}\,\mathrm{d}s + \frac{\partial T(s, v)}{\partial v}\,\mathrm{d}v \tag{6.30}$$

$$\mathrm{d}P = \frac{\partial P(s, v)}{\partial s}\,\mathrm{d}s + \frac{\partial P(s, v)}{\partial v}\,\mathrm{d}v. \tag{6.31}$$

Note that we have the so-called *Maxwell relation*

$$\frac{-\partial P(s, v)}{\partial s} = \frac{\partial^2 u(s, v)}{\partial s\,\partial v} = \frac{\partial^2 u(s, v)}{\partial v\,\partial s} = \frac{\partial T(s, v)}{\partial v} \tag{6.32}$$

because the order that we perform the two partial derivatives of $u(s, v)$ is interchangeable. We next define the so-called *material properties:*

$$T\left(\frac{\partial s}{\partial T}\right)_v = c_v \mathrel{\widehat{=}} \text{isochoric specific heat} \tag{6.33}$$

$$\frac{1}{v}\left(\frac{\partial v}{\partial T}\right)_s = \alpha_s \mathrel{\widehat{=}} \text{adiabatic thermal expansion coefficient} \tag{6.34}$$

$$-v\left(\frac{\partial P}{\partial v}\right)_s = K_s \mathrel{\widehat{=}} \text{adiabatic bulk modulus.} \tag{6.35}$$

The adjective *isochoric* means a process occurring at constant volume while *adiabatic* means that the reversible process is occurring without heat entering or leaving the mass element (system). The formally precise term to use in the context of Eqs (6.34) and (6.35) is *isentropic*, meaning "constant entropy," but adiabatic is more commonly employed for describing the material properties of reversible processes when measured at constant entropy. One can imagine an adiabatic process (one occurring with no heat flow into or out of an element) that generates entropy. For example, a rapid application of strain to a sample surrounded by an adiabatic (heat insulating) wall can set off viscous mechanisms that irreversibly generate entropy and therefore heat. Such a process is not isentropic but because the generated heat cannot escape, we would say that the process is adiabatic. For reversible processes that take place sufficiently slowly that entropy is not irreversibly generated, adiabatic and isentropic are synonymous. The isochoric specific heat c_v (units of J kg^{-1} K^{-1}) is telling us how much the heat in a kilogram of fluid $T\,\delta s$ increases when temperature is raised by δT at constant volume. The adiabatic thermal expansion coefficient α_s tells us how much volumetric strain $\delta v/v$ increases when δT is increased at constant entropy. Last, the adiabatic bulk modulus K_s tells us how much pressure increases δP when we decrease volume by an amount $\delta v/v$ at constant entropy.

Using Eq. (6.15) (or Fact 1) about partial derivatives as derived earlier, we can write $\partial T/\partial s = 1/(\partial s/\partial T) = T/c_v$ and $\partial T/\partial v = 1/(\partial v/\partial T) = 1/(\alpha_s v)$ to obtain the *constitutive laws* when (s, v) are the independent variables:

$$\mathrm{d}T = \frac{T}{c_v}\,\mathrm{d}s + \frac{1}{\alpha_s}\frac{\mathrm{d}v}{v} \tag{6.36}$$

$$-\mathrm{d}P = \frac{1}{\alpha_s v}\,\mathrm{d}s + K_s\frac{\mathrm{d}v}{v}. \tag{6.37}$$

This is a key result for modeling the response in a continuum. Note that $v = V/\mathcal{M} = 1/\rho$ where ρ is the mass density of the element and so v may be replaced by $1/\rho$ in these constitutive laws if desired.

So for each element surrounding each point $\boldsymbol{r}$ of a large system, the temporal changes of the state functions from one equilibrium state to the next are controlled by Eqs (6.36) and (6.37). We have obtained these constitutive laws by taking two derivatives of our fundamental function $u = u(s, v)$. We will later treat the additions that we need to make to these laws when dealing with an anisotropic solid for which permanent elastic shape change of each mass element is possible or when dealing with a multicomponent liquid. Note that Eq. (6.37) is the generalized form of Hooke's law for a fluid when (s, v) are taken as the independent thermodynamic variables. Within the present thermodynamic framework for describing changes within an element of fluid, we see that pressure can also vary due to heat exchanges $T\mathrm{d}s$ between the element and its surroundings in addition to the volumetric dilation $\mathrm{d}v/v$ that was allowed for in Chapter 4 on elasticity (with the shear modulus $G = 0$ for the fluid under consideration).

Note that we can divide the constitutive laws by a time increment $\mathrm{d}t$ and thus replace the differential operator d in the above with $\mathrm{d}/\mathrm{d}t$. If we are moving along with a mass element whose center of mass is moving through space with a velocity $\boldsymbol{v}(\boldsymbol{r}, t)$, the temporal changes of the thermodynamic variables that we observe in the element from this moving frame of reference are quantified by the time derivative $\mathrm{d}/\mathrm{d}t$. However, if our frame of reference is stationary as the element is moving about, we showed in Chapter 2 that the $\mathrm{d}/\mathrm{d}t$ in the above constitutive laws is replaced with

$$\frac{\mathrm{d}}{\mathrm{d}t} = \frac{\partial}{\partial t} + \boldsymbol{v}\cdot\nabla, \tag{6.38}$$

which is what we called the Eulerian, or stationary frame, perspective in Chapter 2. So when we develop the constitutive laws of equilibrium thermodynamics, the total derivative $\mathrm{d}/\mathrm{d}t$

quantifies the temporal changes of a thermodynamic variable as we move along with the mass element as it moves with a velocity $\boldsymbol{v}$. However, when we set up boundary-value and initial-value problems for large systems consisting of many mass elements, we typically work in a fixed frame of reference and use Eq. (6.38) when expressing the constitutive laws at each point in space and time.

6.1.5 The Euler Equation

There is an additional relation between the thermodynamic variables that is important to establish. It is based on the extensive character of an element's total energy $U = u\mathcal{M}$, total entropy $S = s\mathcal{M}$, total volume $V = v\mathcal{M}$, and total molecular number $N = n\mathcal{M}$. If we enhance the size of the element by adding together (bringing into contact) β identical elements, extensivity means that $\beta U(S, V, N) = U(\beta S, \beta V, \beta N)$. Taking a derivative of this expression with respect to β gives

$$\begin{aligned} U(S, V, N) &= \frac{\partial U(\beta S, \beta V, \beta N)}{\partial(\beta S)}\frac{\partial(\beta S)}{\partial \beta} + \frac{\partial U(\beta S, \beta V, \beta N)}{\partial(\beta V)}\frac{\partial(\beta V)}{\partial \beta} \\ &\quad + \frac{\partial U(\beta S, \beta V, \beta N)}{\partial(\beta N)}\frac{\partial(\beta N)}{\partial \beta} \\ &= \frac{\partial U(\beta S, \beta V, \beta N)}{\partial(\beta S)}S + \frac{\partial U(\beta S, \beta V, \beta N)}{\partial(\beta V)}V + \frac{\partial U(\beta S, \beta V, \beta N)}{\partial(\beta N)}N. \end{aligned} \quad (6.39)$$

We can now let $\beta = 1$ and use the definitions $T = \partial U/\partial S$, $-P = \partial U/\partial V$, and $\mu = \partial U/\partial N$ to obtain the so-called *Euler equation*

$$\boxed{U = TS - PV + \mu N.} \quad (6.40)$$

Dividing through by the mass $\mathcal{M}$ of the system then gives the equivalent statement in terms of the specific (or continuum) variables

$$u = Ts - Pv + \mu n. \quad (6.41)$$

Extending to multimolecular systems composed of M different molecular species simply requires replacing μN with $\sum_{m=1}^{M} \mu_m N_m$. One thing that the Euler equation tells us is that if we know the three state functions $T(S, V, N)$, $P(S, V, N)$, and $\mu(S, V, N)$, we can then obtain the fundamental function $U(S, V, N)$ from Eq. (6.40). So knowing all three state functions provides all possible thermodynamic information associated with an element and is equivalent to knowing the fundamental function.

6.1.6 The Gibbs–Duhem Equation

If we take a total derivative of the Euler equation

$$dU = (dT)S + TdS - (dP)V - PdV + (d\mu)N + \mu dN \quad (6.42)$$

and combine with the first law of thermodynamics $dU = TdS - PdV + \mu dN$, we end up with

$$\boxed{d\mu = \frac{V}{N}dP - \frac{S}{N}dT,} \tag{6.43}$$

which is an equilibrium relation known as the *Gibbs–Duhem equation*. This says that a change in the chemical potential is not independent from a change in temperature or pressure. The coefficient V/N is the average volume associated with a single fluid molecule in the element, while S/N is the average entropy that each molecule contributes to the element.

For a single-component fluid system (say pure water) that has uniform temperature $\nabla T = 0$, a gradient in fluid pressure that drives a fluid flow when the system is out of equilibrium as treated in Chapter 5 can be identified through the Gibbs–Duhem relation as a gradient in chemical potential that drives fluid molecules from regions with large chemical potential to regions with smaller chemical potential, that is, $\nabla\mu = V_1 \nabla P$ where V_1 is the volume of a single fluid molecule in the system.

6.1.7 Example of Moving from One Equilibrium State to Another

With the above thermodynamic formalism in place, let's play with it a bit and determine the mechanical-work and heat-energy changes as an element of constant mass moves between equilibrium states. We assume the material consists of a single molecular species so that n is a constant during all thermodynamic change, where $1/n$ is the mass of a single molecule of the material.

Consider the example shown in Fig. 6.2 as we pass from an initial equilibrium state a to a final state c by taking various paths on the $P - v$ plane that correspond to mechanical

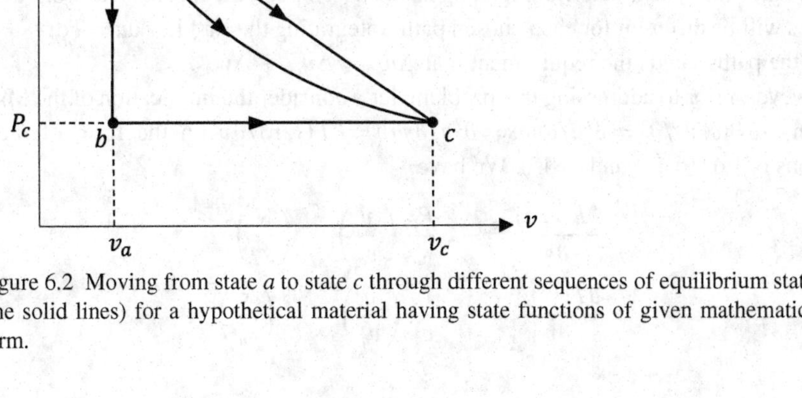

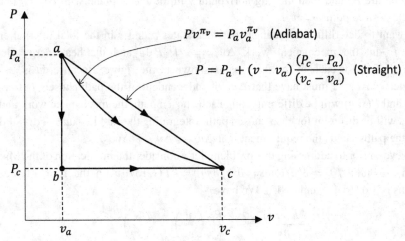

Figure 6.2 Moving from state a to state c through different sequences of equilibrium states (the solid lines) for a hypothetical material having state functions of given mathematical form.

work being performed on (or by) the system while heat is simultaneously being let into (or out of) the system in various ways. We could have chosen other thermodynamic planes such as $T - v$, $T - s$ or $P - s$ for this exercise but choose $P - v$ here due to its intuitive mechanical nature. For this particular material, let's say that the two state functions have been experimentally measured to have the forms

$$P(s, v) = P_a \left(\frac{v}{v_a}\right)^{-\pi_v} \left(\frac{s}{s_a}\right)^{\pi_s} \tag{6.44}$$

$$T(s, v) = T_a \left(\frac{v}{v_a}\right)^{-\tau_v} \left(\frac{s}{s_a}\right)^{\tau_s}, \tag{6.45}$$

where the various dimensionless exponents π_v, π_s, τ_v, and τ_s are all positive and have been experimentally determined. We will see below that Maxwell's relation provides relations between these exponents. The initial state characterized by v_a, P_a, and T_a is assumed known and, as will be seen, we can determine the corresponding entropy s_a as part of the analysis that follows.

Beginning in state a, we can take various paths to arrive at state c. For the particular example given in Fig. 6.2, we assume that state c is connected to state a by an adiabat $s = s_a$ of constant entropy as shown in the figure and for which the pressure and temperature along this adiabatic path have the form

$$P_{adi}(v) = P_a \left(\frac{v}{v_a}\right)^{-\pi_v} \tag{6.46}$$

$$T_{adi}(v) = T_a \left(\frac{v}{v_a}\right)^{-\tau_v}. \tag{6.47}$$

If our terminal state was at a point d not on the adiabat $s = s_a$ and given by $P = P_d$ and $v = v_d$, one could get to that point by first following the $s = s_a$ adiabat until we get to the given pressure P_d and then moving horizontally in the $P - v$ plane with $P = P_d$ until we get to the desired point $v = v_d$.

We want to determine the changes between states a and c in the total internal energy $\Delta u_{ac} = \int_{ac} du$, the mechanical work $\Delta w_{ac} = -\int_{ac} P\, dv$ and the heat-energy changes $\Delta q_{ac} = \int_{ac} T\, ds$ for each of the three paths shown in the figure. Because du is an exact (total) derivative, we must have that Δu_{ac} is independent of the path chosen, but because $T(s, v)$ and $P(s, v)$ will be different along each chosen path, the mechanical work and heat changes will be different for each chosen path. Integrating the first law $du = T ds - P dv$ for any of the paths yields the requirement that $\Delta u_{ac} = \Delta q_{ac} + \Delta w_{ac}$.

However, prior to addressing this problem, let's consider the implication of the Maxwell relation $-\partial P(s, v)/\partial s = \partial^2 u/\partial v \partial s = \partial^2 u/\partial s \partial v = \partial T(s, v)/\partial v$ on the form of the state functions of Eqs (6.44) and (6.45). We have

$$-\frac{\partial P(s, v)}{\partial s} = -\pi_s \frac{P_a}{s_a} \left(\frac{v}{v_a}\right)^{-\pi_v} \left(\frac{s}{s_a}\right)^{\pi_s - 1} \tag{6.48}$$

$$\frac{\partial T(s, v)}{\partial v} = -\tau_v \frac{T_a}{v_a} \left(\frac{v}{v_a}\right)^{-\tau_v - 1} \left(\frac{s}{s_a}\right)^{\tau_s}, \tag{6.49}$$

which upon equating these two expressions by Maxwell's relation results in

$$\frac{\pi_s}{\tau_v} \frac{P_a v_a}{T_a s_a} = \left(\frac{v}{v_a}\right)^{\pi_v - \tau_v - 1} \left(\frac{s}{s_a}\right)^{\tau_s - \pi_s + 1}. \tag{6.50}$$

The left-hand side is a constant independent of v and s and the right-hand side must hold true for all v and s. The only way for this to be possible is if the exponents are given by

$$\tau_v = \pi_v - 1 \quad \text{and} \quad \tau_s = \pi_s - 1, \tag{6.51}$$

which then results in

$$\frac{T_a s_a}{\pi_s} = \frac{P_a v_a}{\pi_v - 1}. \tag{6.52}$$

This expression gives the specific entropy in state a as $s_a = \pi_s P_a v_a / [T_a(\pi_v - 1)]$, which can be considered known if all of π_s, π_v, P_a, v_a, and T_a are known as is postulated. We emphasize to the student that this particular relation between s_a, P_a, v_a, and T_a only holds for the particular imagined material that has state functions of the form Eqs (6.44) and (6.45).

Next, let's consider the constitutive laws for this particular material type. We first determine the material properties c_v (isochoric specific heat), α_s (adiabatic thermal expansion coefficient), and K_s (adiabatic bulk modulus) using their earlier definitions and the above form of the state functions as

$$\frac{1}{c_v} \mathrel{\hat{=}} \frac{1}{T}\left(\frac{\partial T}{\partial s}\right)_{v,n} = \frac{\pi_s - 1}{s} \tag{6.53}$$

$$\frac{1}{\alpha_s} \mathrel{\hat{=}} v\left(\frac{\partial T}{\partial v}\right)_{s,n} = -(\pi_v - 1)T \tag{6.54}$$

$$K_s \mathrel{\hat{=}} -v\left(\frac{\partial P}{\partial v}\right)_{s,n} = \pi_v P. \tag{6.55}$$

Using these expressions for the material properties in the constitutive laws results in

$$\frac{\mathrm{d}T}{T} = (\pi_s - 1)\frac{\mathrm{d}s}{s} - (\pi_v - 1)\frac{\mathrm{d}v}{v} \tag{6.56}$$

$$\frac{\mathrm{d}P}{P} = (\pi_v - 1)\frac{T_a s_a}{P_a v_a}\frac{\mathrm{d}s}{s} - \pi_v \frac{\mathrm{d}v}{v}. \tag{6.57}$$

We just showed in Eq. (6.52) that Maxwell's relation requires $\pi_s = (\pi_v - 1)T_a s_a/(P_a v_a)$. Using this, the constitutive laws corresponding to our particular material can be integrated from the initial state s_a, v_a to a present state s, v to give the state functions

$$\frac{T}{T_a} = \left(\frac{s}{s_a}\right)^{\pi_s - 1} \left(\frac{v}{v_a}\right)^{1 - \pi_v} \tag{6.58}$$

$$\frac{P}{P_a} = \left(\frac{s}{s_a}\right)^{\pi_s} \left(\frac{v}{v_a}\right)^{-\pi_v} \tag{6.59}$$

where again T_a and P_a are assumed to be known. These are the state functions with the implications of Maxwell's relation built into them. What we have shown here is that working with the constitutive laws with the known state dependence of the material properties is equivalent to knowing the state functions themselves given knowledge of T_a and P_a in some initial state a. We often have constitutive equations available to us for some modeling application in a particular material but don't always begin with known state functions.

With the above established, we return to the specific question of how to determine Δu_{ac} between states a and c in the above figure. First, choose the adiabatic path $s = s_a$ for which $\Delta q_{ac} = 0$ and for which the pressure varies with specific volume as $P_{adi}(v)$ as given by Eq. (6.46), to give $\Delta w_{ac} = -\int_{v_a}^{v_c} P_{adi}(v)\, dv$. From the integrated form of the first law that says $\Delta u_{ac} = \Delta q_{ac} + \Delta w_{ac}$ we then have that for the adiabatic path

$$\Delta u_{ac} = \Delta w_{ac} = -P_a \int_{v_a}^{v_c} \left(\frac{v}{v_a}\right)^{-\pi_v} dv = -\frac{P_a v_a}{\pi_v - 1}\left[1 - \left(\frac{v_a}{v_c}\right)^{\pi_v - 1}\right]. \tag{6.60}$$

The result is negative because work is done by the mass element on the surroundings (stored energy is lost from the mass element) as it expands from a to c. Again, this result for Δu_{ac} must hold true for any path between points a and c due to du being an exact differential in the first law $du = T(s, v)ds - P(s, v)dv$.

Next, let's consider the linear path on the $P - v$ plane in which the pressure varies with volume as

$$P_{lin}(v) = P_a + (v - v_a)\left(\frac{P_c - P_a}{v_c - v_a}\right). \tag{6.61}$$

Integrating this gives that the mechanical work when passing from a to c along the linear pressure path is given by

$$\Delta w_{lin} = -\int_{v_a}^{v_c} P_{lin}(v)\, dv = -\left(\frac{P_a + P_c}{2}\right)(v_c - v_a), \tag{6.62}$$

which can also be observed graphically on the figure as the area under the linear pressure path between v_a and v_c. From $\Delta u_{ac} = \Delta q_{lin} + \Delta w_{lin}$, the heat added to the mass element along the linear path between a and c on the $P - v$ plane is

$$\Delta q_{lin} = \frac{(P_a + P_c)}{2} v_c \left(1 - \frac{v_a}{v_c}\right) - \frac{P_a v_a}{\pi_v - 1}\left[1 - \left(\frac{v_a}{v_c}\right)^{\pi_v - 1}\right], \tag{6.63}$$

which is necessarily positive, that is, the area under the linear path between a and c is greater than the area under the adiabatic (no heat change) path as seen in the figure above. So as anticipated, we see that the heat change and mechanical work involved in going from one equilibrium state to another depends on the path taken between the two states, while the change in the internal energy is path independent.

As the first exercise at the end of the chapter, you can show that Eq. (6.63) is also obtained by integrating Tds from state a to state c along the linear path. To do so, first equate Eqs (6.61) and (6.59) to obtain $s_{lin}(v)$ along the linear path, which when used in

Eq. (6.58) gives $T_{lin}(v)$ along the linear path. You then obtain Δq_{lin} for the linear path as the integral $\Delta q_{lin} = \int_{v_a}^{v_c} T_{lin}(v)\,(\mathrm{d}s_{lin}(v)/\mathrm{d}v)\,\mathrm{d}v$, which gives Eq. (6.63).

We can also go from state a to c by decreasing pressure at constant volume to arrive at state b as shown on the figure and then by increasing volume at constant pressure to arrive at state c. The first leg from a to b on this path is achieved by letting heat escape the mass element at constant volume which decreases the pressure. All mechanical work on this path is associated with the leg from b to c. So we have that

$$\Delta w_{abc} = \Delta w_{bc} = -P_c \int_{v_a}^{v_c} \mathrm{d}v = -P_c v_c \left(1 - \frac{v_a}{v_c}\right) = -P_a v_a \left(1 - \frac{v_a}{v_c}\right)\left(\frac{v_a}{v_c}\right)^{\pi_v - 1}. \tag{6.64}$$

The heat lost from the mass element is then given as $\Delta q_{abc} = \Delta u_{ac} - \Delta w_{abc}$ or

$$\Delta q_{abc} = -\frac{P_a v_a}{(\pi_v - 1)}\left[1 + (\pi_v - 1)\left(\frac{v_a}{v_c}\right)^{\pi_v} - \pi_v \left(\frac{v_a}{v_c}\right)^{\pi_v - 1}\right]. \tag{6.65}$$

For this path from a to b to c, we can also calculate directly this change in heat as $\Delta q_{abc} = \int_{ab} T\,\mathrm{d}s + \int_{bc} T\,\mathrm{d}s$. To do so for the leg from a to b, we use $v = v_a$ in Eq. (6.58) for $T(s, v)$ to obtain

$$\Delta q_{ab} = \int_{s_a}^{s_b} T\,\mathrm{d}s = T_a \int_{s_a}^{s_b} \left(\frac{s}{s_a}\right)^{\pi_s - 1} \mathrm{d}s = \frac{T_a s_a}{\pi_s}\left[\left(\frac{s_b}{s_a}\right)^{\pi_s} - 1\right]. \tag{6.66}$$

We have from Eq. (6.59) for $P(s, v)$ that

$$\frac{P_c}{P_a} = \left(\frac{s_b}{s_a}\right)^{\pi_s} = \left(\frac{v_a}{v_c}\right)^{\pi_v}. \tag{6.67}$$

Further, we have the earlier Maxwell-relation result that $T_a s_a/\pi_s = P_a v_a/(\pi_v - 1)$ so that

$$\Delta q_{ab} = -\frac{P_a v_a}{\pi_v - 1}\left[1 - \left(\frac{v_a}{v_c}\right)^{\pi_v}\right], \tag{6.68}$$

which is seen to be negative because we had to let heat out of the element to decrease the pressure at constant volume.

For the second leg from b to c, we have $P = P_c$, which from Eq. (6.59) for $P(s, v)$, allows us to determine that

$$\left(\frac{v}{v_a}\right)^{1-\pi_v} = \left(\frac{v_a}{v_c}\right)^{1-\pi_v}\left(\frac{s}{s_a}\right)^{(1-\pi_v)\pi_s/\pi_v}. \tag{6.69}$$

As such, along the path from b to c, the temperature varies as

$$T(s) = T_a \left(\frac{v_a}{v_c}\right)^{\pi_v - 1}\left(\frac{s}{s_a}\right)^{\pi_s/\pi_v - 1}. \tag{6.70}$$

Thus the heat gained by the element along this path is (note that the specific entropy at point c, which is on the adiabat, is $s_c = s_a$)

$$\Delta q_{bc} = \int_{s_b}^{s_c} T\,\mathrm{d}s = T_a \left(\frac{v_a}{v_c}\right)^{\pi_v - 1} \int_{s_b}^{s_a} \left(\frac{s}{s_a}\right)^{\pi_s/\pi_v - 1} \mathrm{d}s = \frac{P_a v_a}{(\pi_v - 1)}\pi_v \left(\frac{v_a}{v_c}\right)^{\pi_v - 1}\left(1 - \frac{v_a}{v_c}\right), \tag{6.71}$$

which is seen to be positive. Putting the results from the two legs together then give

$$\Delta q_{abc} = \Delta q_{ab} + \Delta q_{bc} = -\frac{P_a v_a}{\pi_v - 1}\left[1 + (\pi_v - 1)\left(\frac{v_a}{v_c}\right)^{\pi_v} - \pi_v\left(\frac{v_a}{v_c}\right)^{\pi_v - 1}\right], \quad (6.72)$$

which is identical to Eq. (6.65) as it must be.

The above pedagogical example focused, somewhat arbitrarily, on paths on the $P - v$ plane and other thermodynamic planes could have been considered. If instead we were given only constitutive laws, one must integrate the constitutive laws to find the relations between the thermodynamic variables as we change from one equilibrium state to another. If given the state functions, the relation between the changing thermodynamic variables are given by the state functions themselves. In the above exercise, we integrated these relations to obtain how the various energy changes (total internal energy, mechanical work, or heat) depend on the path (sequence of equilibrium states) taken to get from one equilibrium state to another.

6.2 Changing Variables in Thermodynamics

So the basic fundamental function containing all thermodynamic information of a fluid mass element containing a single chemical species is:

$$u = u(s, v) \mathrel{\widehat{=}} \text{internal energy/unit mass.} \quad (6.73)$$

This is the starting point for problems in which entropy and volume are convenient independent variables. However, if in an application we are controlling or keeping constant other variables such as temperature T or pressure P, can we redefine our fundamental function so that it is not a function of (s, v) but of other variables such as (T, v) or (s, P) or (T, P)? This is what the Legendre transform does as will be discussed below. But first we consider an example of when it might be convenient to change variables.

6.2.1 Example of Adiabatic or Isothermal Compressional Waves

If we are modeling a so-called *adiabatic process* in which s does not significantly change, then it is convenient to work with the independent variables (s, v) or possibly (s, P) so that there is only one independent variable (either v or P) changing during the process. For the same reason, if we are modeling a so-called *isothermal process* in which T does not significantly change, then it is convenient to work with the independent variables (T, v) or (T, P).

Let's consider the specific case of acoustic or longitudinal seismic-wave propagation depicted in Fig. 6.3 and ask, "Is acoustic wave propagation an isothermal or adiabatic process?" This question is answered by posing another question: "In each wave period t_p, how much heat is transferred from a compression to a dilation by thermal conduction?" A compression is nominally associated with a temperature increase and a dilation with a temperature decrease resulting in a flow of heat (thermal conduction) from the compression to the dilation. If a negligible amount of heat is conducted from the compression to the

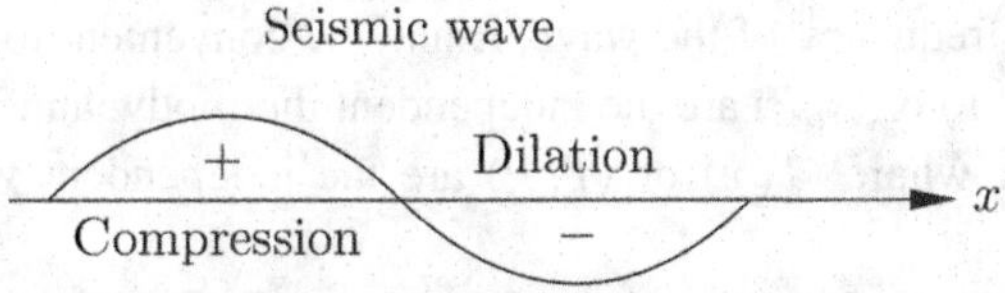

Figure 6.3 Compressions (increased temperature) and dilations (decreased temperature) of an acoustic wave that generate heat flow from the compressions toward the dilations.

dilation in a wave period t_p, acoustic waves can be modeled as being *adiabatic* (negligible amounts of heat transfer). If instead there is significant heat transfer from a compression to a dilation such that the temperature differences are equilibrated rapidly within each wave period, acoustic waves can be modeled as being *isothermal*.

From the diffusion equation of thermal conduction that will be derived in Chapter 7, we have

$$D_t \frac{\partial^2 T}{\partial x^2} - \frac{\partial T}{\partial t} = 0, \tag{6.74}$$

where D_t is the thermal diffusivity that will be shown to have the form $D_t = \lambda_t/(\rho c_v)$ where λ_t is the thermal conductivity, ρ the mass density, and c_v the isochoric specific heat. The time τ required to just equilibrate the temperature differences over an acoustic wavelength λ can be estimated from the thermal diffusion equation if we approximate derivatives as $\partial/\partial x \approx 1/\lambda$ and $\partial/\partial t \approx 1/\tau$ to give

$$\frac{D_t}{\lambda^2} \approx \frac{1}{\tau} \quad \text{or} \quad \tau = \frac{\lambda^2}{D_t}. \tag{6.75}$$

If $t_p \ll \tau$, then there is not enough time in a wave period to have much heat flow from a compression to a dilation and the wave propagation can be considered *adiabatic*. If $t_p \gg \tau$, then total heat exchange has time to occur so the temperature in compressions and dilations is the same and the wave propagation can be considered *isothermal*.

Now, for a wave of frequency f, we have

$$f = \frac{1}{t_p} \quad \text{and} \quad \lambda = \frac{c}{f} \quad \text{where } c \mathrel{\widehat{=}} \text{wave speed}. \tag{6.76}$$

The condition for adiabatic wave propagation is that

$$f \gg \frac{D_t f^2}{c^2} \quad \text{or} \quad \boxed{f \ll \frac{c^2}{D_t}.} \tag{6.77}$$

Similarly, the condition for isothermal wave propagation is that

$$f \ll \frac{D_t f^2}{c^2} \quad \text{or} \quad \boxed{f \gg \frac{c^2}{D_t}.} \tag{6.78}$$

So depending on the frequency of the wave, it may be convenient to work in a representation either where (s, v) or (s, P) are the independent thermodynamic variables (adiabatic wave propagation) or where (T, v) or (T, P) are the independent variables (isothermal wave propagation).

It turns out that for most liquids and solids (and therefore rocks) we have the order of magnitude values

$$D_t \mathrel{\widehat{=}} \frac{\lambda_t}{\rho\, c_v} \approx \frac{1(\mathrm{W/m/^\circ C})}{\left(10^3\,\mathrm{kg/m^3}\right)\left(10^3\,\mathrm{J/kg/^\circ C}\right)} \approx 10^{-6}\,\frac{\mathrm{m^2}}{\mathrm{s}} \tag{6.79}$$

$$c^2 = \left(10^3\,\frac{\mathrm{m}}{\mathrm{s}}\right)^2 \approx 10^6\,\frac{\mathrm{m^2}}{\mathrm{s^2}} \tag{6.80}$$

so

$$\frac{c^2}{D_t} \approx 10^{12}\,\mathrm{Hz} \gg \text{frequency of waves in the Earth.} \tag{6.81}$$

Note that the wavelength of a 10^{12} Hz compressional wave is of the order $\lambda = 10^{-9}$ m and since all wavelengths of interest are much larger than this in a continuum description of matter, compressional wave propagation really is *adiabatic* in practice and either (s, v) or (s, P) are the convenient thermodynamic variables to use for describing wave propagation.

6.2.2 The Legendre Transform

So we can imagine processes or experiments where other pairs of independent variables such as (s, P) or (T, v) or (T, P) are more convenient for expressing the first law of thermodynamics and the constitutive laws. Defining a new fundamental function that allows this to occur is what the Legendre transform treats.

Say we have a monotonically increasing function $y(x)$. It may arise that we would like to represent the information in $y(x)$ not as a function of x, but as a function of the slope $p = \mathrm{d}y/\mathrm{d}x$. We ask ourselves: "What function $\psi(p)$ can be used to uniquely reconstruct $y(x)$ and is, therefore, equivalent to $y(x)$?" Graphically, the answer comes from representing $y(x)$ as the envelope bounding all the tangent lines that touch the surface as depicted in Fig. 6.4. The slope p at each point x along $y(x)$ is different and related to the function $y(x)$ and the *tangent line intercept* $\psi(p)$ as

$$p = \frac{y(x) - \psi(p)}{x} \tag{6.82}$$

or

$$\boxed{\psi(p) = y(x) - xp, \quad \text{where} \quad p = \frac{\mathrm{d}y}{\mathrm{d}x}.} \tag{6.83}$$

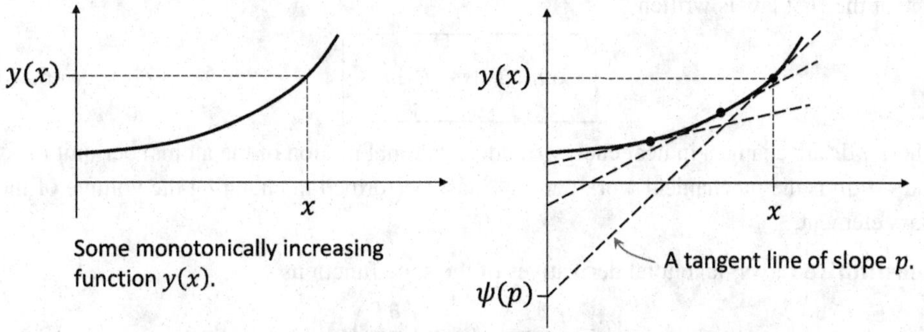

Figure 6.4 On the left, a monotonically increasing function $y(x)$. On the right, $y(x)$ represented by the envelope of tangent lines. Given the y-axis intercepts $\psi(p)$ for each tangent line of slope p, where $\psi(p)$ is called the *Legendre transform* of $y(x)$, the monotonic function $y(x)$ can be reconstructed exactly.

The function $\psi(p)$ is called the *Legendre transform* of $y(x)$. From each given $\psi(p)$ value along the y axis, we draw the corresponding straight line having slope p and the envelope of all these tangent lines is exactly the original function $y(x)$. So we see that $\psi(p)$ contains all the information contained in $y(x)$. The Legendre transform is what allows us to replace the fundamental function $u(s, v)$ by other fundamental functions that have other independent variables defined as the slope, or first derivatives, of u with respect to either s or v.

6.3 The Various Representations of Thermodynamics

For mass elements in which the number of molecules is not changing, we now use the Legendre transform of $u(s, v)$ to work out the thermodynamic relations in the various possible representations that have independent variables (s, p) or (T, v) or (T, p). To see the pattern at work, we begin by repeating the standard "internal energy" development where (s, v) are the independent variables.

6.3.1 Standard Internal-Energy Representation (s, v)

Fundamental Function: $u = u(s, v) \widehat{=}$ *internal energy / unit mass.*

Euler Equation: $u = Ts - Pv + \mu n$.

First Law: take a total derivative of the fundamental function to give

$$du = \left(\frac{\partial u}{\partial s}\right)_v ds + \left(\frac{\partial u}{\partial v}\right)_s dv. \tag{6.84}$$

State Functions: define the first derivatives to be the state functions in this representation

$$T(s, v) = \left(\frac{\partial u}{\partial s}\right)_v \quad \text{and} \quad P(s, v) = -\left(\frac{\partial u}{\partial v}\right)_s \tag{6.85}$$

so that the first law is written

$$\boxed{\mathrm{d}u = T\,\mathrm{d}s - P\,\mathrm{d}v,} \tag{6.86}$$

where $T\mathrm{d}s$ are changes in heat energy (random thermal motion of the atoms) per unit mass and $-P\mathrm{d}v$ is the mechanical work per unit mass performed in changing the volume of the mass element.

Constitutive Laws: take total derivatives of the state functions

$$\mathrm{d}T = \left(\frac{\partial T}{\partial s}\right)_v \mathrm{d}s + \left(\frac{\partial T}{\partial v}\right)_s \mathrm{d}v \tag{6.87}$$

$$\mathrm{d}P = \left(\frac{\partial P}{\partial s}\right)_v \mathrm{d}s + \left(\frac{\partial P}{\partial v}\right)_s \mathrm{d}v. \tag{6.88}$$

We have the so-called *Maxwell relation*

$$-\left(\frac{\partial P}{\partial s}\right)_v = \frac{\partial^2 u}{\partial s\,\partial v} = \frac{\partial^2 u}{\partial v\,\partial s} = \left(\frac{\partial T}{\partial v}\right)_s. \tag{6.89}$$

The first derivatives of the state functions (second derivatives of the fundamental function) are identifiable in terms of the following *material properties*

$$T\left(\frac{\partial s}{\partial T}\right)_v = c_v \mathrel{\widehat{=}} \textit{isochoric}\text{ specific heat}(\mathrm{J\,K^{-1}\,kg^{-1}}) \tag{6.90}$$

$$\frac{1}{v}\left(\frac{\partial v}{\partial T}\right)_s = \alpha_s \mathrel{\widehat{=}} \textit{adiabatic}\text{ thermal expansion coefficient}(\mathrm{K^{-1}}) \tag{6.91}$$

$$-v\left(\frac{\partial P}{\partial v}\right)_s = K_s \mathrel{\widehat{=}} \textit{adiabatic}\text{ bulk modulus}(\mathrm{Pa}). \tag{6.92}$$

So the constitutive laws in the internal-energy representation $(s,\, v)$ can be written

$$\mathrm{d}T = \frac{T}{c_v}\,\mathrm{d}s + \frac{1}{\alpha_s}\frac{\mathrm{d}v}{v} \tag{6.93}$$

$$-\mathrm{d}P = \frac{1}{\alpha_s\, v}\,\mathrm{d}s + K_s\frac{\mathrm{d}v}{v}. \tag{6.94}$$

Note how we can manipulate these constitutive laws. For example, if we consider a volume change at constant entropy, divide through by $\mathrm{d}v/v$ and take $\mathrm{d}s/\mathrm{d}v = 0$, we obtain

$$v\left.\frac{\mathrm{d}T}{\mathrm{d}v}\right|_{s=\mathrm{const}} \mathrel{\widehat{=}} v\left(\frac{\partial T}{\partial v}\right)_s = \frac{1}{\alpha_s} \tag{6.95}$$

$$-v\left.\frac{\mathrm{d}P}{\mathrm{d}v}\right|_{s=\mathrm{const}} \mathrel{\widehat{=}} -v\left(\frac{\partial P}{\partial v}\right)_s = K_s, \tag{6.96}$$

which are the same as Eqs (6.91) and (6.92).

6.3.2 Helmholtz Free-Energy Representation (*T*, *v*)

We now put the Legendre transform to work in defining a new fundamental function that has (T, v) as the independent variables.

Fundamental Function: $f = f(T, v) = u - Ts \,\widehat{=}\,$ *Helmholtz free energy / unit mass*, which is the Legendre transform of $u(s, v)$ that replaces s with T, that is, $f(T, v) = u(s, v) - (\partial u/\partial s)_v\, s = u - Ts$.

Euler Equation (combine the Euler equation for u with the Legendre transform): $f = -Pv + \mu n$.

First Law: take a total derivative of the fundamental function (Legendre transform)

$$\begin{aligned} \mathrm{d}f &= \mathrm{d}u - s\,\mathrm{d}T - T\,\mathrm{d}s \\ &= \cancel{T\mathrm{d}s} - P\,\mathrm{d}v - s\,\mathrm{d}T - \cancel{T\mathrm{d}s} \end{aligned} \tag{6.97}$$

or

$$\boxed{\mathrm{d}f = -s\,\mathrm{d}T - P\,\mathrm{d}v,} \tag{6.98}$$

which makes explicit that $f = f(T, v)$.

State Functions: define the first derivatives of the fundamental function to be the state functions in this representation

$$s(T, v) = -\left(\frac{\partial f}{\partial T}\right)_v \quad \text{and} \quad P(T, v) = -\left(\frac{\partial f}{\partial v}\right)_T. \tag{6.99}$$

Constitutive Laws: take total derivatives of the state functions

$$\mathrm{d}s = \left(\frac{\partial s}{\partial T}\right)_v \mathrm{d}T + \left(\frac{\partial s}{\partial v}\right)_T \mathrm{d}v \tag{6.100}$$

$$\mathrm{d}P = \left(\frac{\partial P}{\partial T}\right)_v \mathrm{d}T + \left(\frac{\partial P}{\partial v}\right)_T \mathrm{d}v. \tag{6.101}$$

The Maxwell relation for this representation is

$$-\left(\frac{\partial P}{\partial T}\right)_v = \frac{\partial^2 f}{\partial T\,\partial v} = \frac{\partial^2 f}{\partial v\,\partial T} = -\left(\frac{\partial s}{\partial v}\right)_T. \tag{6.102}$$

Define the three material properties in this representation as:

$$T\left(\frac{\partial s}{\partial T}\right)_v = c_v \,\widehat{=}\, \text{isochoric specific heat} \tag{6.103}$$

$$\frac{1}{v}\left(\frac{\partial v}{\partial T}\right)_P = \alpha_p \,\widehat{=}\, \text{isobaric thermal expansion} \tag{6.104}$$

$$-v\left(\frac{\partial P}{\partial v}\right)_T = K_T \,\widehat{=}\, \text{isothermal bulk modulus.} \tag{6.105}$$

To identify either $(\partial P/\partial T)_v$ or $(\partial s/\partial v)_T$, we must use the facts about partial derivatives derived earlier in Section 6.1.3.

Using Fact 2, we have

$$\left(\frac{\partial P}{\partial T}\right)_v = -\frac{(\partial v/\partial T)_p}{(\partial v/\partial P)_T} = K_T\,\alpha_p = \left(\frac{\partial s}{\partial v}\right)_T. \tag{6.106}$$

So the constitutive laws in the Helmholtz-energy representation (T, v) can be written:

$$T\mathrm{d}s = c_v\,\mathrm{d}T + T\,\alpha_p\,K_T\,\mathrm{d}v \tag{6.107}$$

$$-\mathrm{d}P = -\alpha_p\,K_T\,\mathrm{d}T + K_T\,\frac{\mathrm{d}v}{v}. \tag{6.108}$$

Note that we also have from the earlier Facts 1 and 2 about partial derivatives

$$\left(\frac{\partial s}{\partial v}\right)_T = -\frac{(\partial T/\partial v)_s}{(\partial T/\partial s)_v} = -\frac{vT\,(\partial s/\partial T)_v}{vT\,(\partial v/\partial T)_s} = -\frac{c_v}{vT\,\alpha_s} \tag{6.109}$$

so that we have proven the nontrivial relation between material properties that

$$K_T = -\frac{c_v}{vT\,\alpha_p\alpha_s}. \tag{6.110}$$

That we can obtain such generally valid relations between the material properties is one of the powerful strengths of the thermodynamic formalism. We will show in Section 6.4 on thermodynamic equilibrium and stability that $\alpha_p\alpha_s < 0$ so that $K_T > 0$. Having a fundamental function and extracting the information it contains by taking total derivatives and identifying the meaning of the partial derivatives that arrive is what allows a relation like Eq. (6.110) to be derived.

As another example of such general relations between the material properties, use the relation of Eq. (6.110) to rewrite the constitutive laws above as

$$T\mathrm{d}s = c_v\mathrm{d}T - \frac{c_v}{\alpha_s}\frac{\mathrm{d}v}{v} \tag{6.111}$$

$$-\mathrm{d}P = \frac{c_v}{v\alpha_s}\frac{\mathrm{d}T}{T} + K_T\frac{\mathrm{d}v}{v}. \tag{6.112}$$

The earlier form of the constitutive laws in the (s, v) representation are

$$\mathrm{d}T = \frac{T}{c_v}\mathrm{d}s + \frac{1}{\alpha_s}\frac{\mathrm{d}v}{v} \tag{6.113}$$

$$-\mathrm{d}P = \frac{1}{\alpha_s v}\mathrm{d}s + K_s\frac{\mathrm{d}v}{v}. \tag{6.114}$$

Equations (6.111) and (6.113) are identical. Equating Eqs (6.112) and (6.114) gives

$$\frac{c_v}{v\alpha_s}\frac{\mathrm{d}T}{T} + K_T\frac{\mathrm{d}v}{v} = \frac{1}{\alpha_s v}\mathrm{d}s + K_s\frac{\mathrm{d}v}{v}. \tag{6.115}$$

Substituting into this Eq. (6.113) for $\mathrm{d}T/T$ gives

$$\frac{c_v}{v\alpha_s}\left(\frac{\mathrm{d}s}{c_v}+\frac{1}{T\alpha_s}\frac{\mathrm{d}v}{v}\right)+K_T\frac{\mathrm{d}v}{v}=\frac{1}{\alpha_s v}\mathrm{d}s+K_s\frac{\mathrm{d}v}{v}, \tag{6.116}$$

which after rearrangement gives

$$\left(K_T+\frac{c_v}{vT\alpha_s^2}\right)\frac{\mathrm{d}v}{v}=K_s\frac{\mathrm{d}v}{v}. \tag{6.117}$$

We thus find that

$$\boxed{K_s=K_T+\frac{c_v}{vT\alpha_s^2},} \tag{6.118}$$

which is a general relation showing how the adiabatic bulk modulus is different from the isothermal bulk modulus.

6.3.3 Enthalpy Representation (s, P)

Fundamental Function: $h=h(s,P)=u-(-Pv)\,\widehat{=}\,$ *enthalpy / unit mass*, which is the Legendre transform of $u(s,v)$ that replaces v with $-P$, i.e., $h(s,P)=u(s,v)-(\partial u/\partial v)_s\, v=u+Pv$.

Euler Equation: $h=Ts+\mu n$.

First Law: take a total derivative of the fundamental function (Legendre transform)

$$\begin{aligned}\mathrm{d}h&=\mathrm{d}u+v\,\mathrm{d}P+P\,\mathrm{d}v\\&=T\,\mathrm{d}s-P\cancel{\mathrm{d}v}+v\,\mathrm{d}P+P\cancel{\mathrm{d}v}\end{aligned}$$

$$\boxed{\mathrm{d}h=T\,\mathrm{d}s+v\,\mathrm{d}P,} \tag{6.119}$$

which makes explicit that $h=h(s,P)$.

State Functions: define the first derivatives of the fundamental function to be the state functions in this representation

$$T(s,P)=\left(\frac{\partial h}{\partial s}\right)_P \quad \text{and} \quad v(s,P)=\left(\frac{\partial h}{\partial P}\right)_s. \tag{6.120}$$

Constitutive Laws: take total derivatives of the state functions

$$\mathrm{d}T=\left(\frac{\partial T}{\partial s}\right)_P\mathrm{d}s+\left(\frac{\partial T}{\partial P}\right)_s\mathrm{d}P \tag{6.121}$$

$$\mathrm{d}v=\left(\frac{\partial v}{\partial s}\right)_P\mathrm{d}s+\left(\frac{\partial v}{\partial P}\right)_s\mathrm{d}P. \tag{6.122}$$

The Maxwell relation in this representation is

$$\left(\frac{\partial T}{\partial P}\right)_s = \left(\frac{\partial v}{\partial s}\right)_p = \left(\frac{\partial v}{\partial T}\right)_p \left(\frac{\partial T}{\partial s}\right)_p = \frac{vT\,\alpha_p}{c_p}, \tag{6.123}$$

where the two new material properties are defined

$$T\left(\frac{\partial s}{\partial T}\right)_p = c_p \mathrel{\widehat{=}} \text{isobaric specific heat} \tag{6.124}$$

$$-v\left(\frac{\partial P}{\partial v}\right)_s = K_s \mathrel{\widehat{=}} \text{adiabatic bulk modulus.} \tag{6.125}$$

So the constitutive laws in the enthalpy-representation (s, P) can be written:

$$\frac{\mathrm{d}T}{T} = \frac{1}{c_p}\,\mathrm{d}s + \frac{v\,\alpha_p}{c_p}\,\mathrm{d}P \tag{6.126}$$

$$\frac{\mathrm{d}v}{v} = \frac{\alpha_p T}{c_p}\,\mathrm{d}s - \frac{1}{K_s}\,\mathrm{d}P. \tag{6.127}$$

We can now prove another relation between the material properties. Let's analyze the derivative $-(\partial P/\partial s)_v$ first by using the chain rule (Fact 3) to give

$$-\left(\frac{\partial P}{\partial s}\right)_v = -\left(\frac{\partial P}{\partial T}\right)_v \left(\frac{\partial T}{\partial s}\right)_v = -\frac{T}{c_v}\left(\frac{\partial P}{\partial T}\right)_v. \tag{6.128}$$

We can then use Fact 2 to obtain

$$-\left(\frac{\partial P}{\partial T}\right)_v = \frac{(\partial v/\partial T)_p}{(\partial v/\partial P)_T} = \alpha_p K_T \tag{6.129}$$

so that

$$-\left(\frac{\partial P}{\partial s}\right)_v = \frac{\alpha_p K_T T}{c_v}. \tag{6.130}$$

Using Fact 2 and Fact 3, we also have

$$-\left(\frac{\partial P}{\partial s}\right)_v = \frac{(\partial v/\partial s)_p}{(\partial v/\partial P)_s} = \frac{K_s}{v}\left(\frac{\partial v}{\partial T}\right)_p \left(\frac{\partial T}{\partial s}\right)_p = \frac{\alpha_p K_s T}{c_p}. \tag{6.131}$$

By equating Eqs (6.130) and (6.131), we thus obtain the generally valid relation between material properties

$$\frac{K_s}{K_T} = \frac{c_p}{c_v}. \tag{6.132}$$

The ratio of specific heats is given the symbolic name $\gamma = c_p/c_v$ and $\gamma > 1$ for all materials (as proven in Section 6.4) so that $K_s > K_T$. For example, dry air at $T = 293$ K has $\gamma = 1.4$, while liquid water at $T = 293$ K has $\gamma \approx 1.007$. The difference between K_s and K_T is much more important in gases compared to liquids and solids.

6.3.4 Gibbs Free-Energy Representation (T, P)

For this free energy we have to apply two Legendre transforms.

Fundamental Function: $g = g(T, P) = u - Ts - (-Pv) \;\widehat{=}\;$ *Gibbs free energy / unit mass*, which is the Legendre transform of $u(s, v)$ that replaces s with T and v with $-P$, that is, $g(T, P) = u(s, v) - (\partial u/\partial s)_v\, s - (\partial u/\partial v)_s\, v = u - TS + Pv$.

Euler Equation: $g = \mu n$. Because $n = 1/m$ is constant with m the mass of a single molecule, the Gibbs free energy g plays the same role as the chemical potential of the element due to the relation $\mu = mg$. So in what follows, you can everywhere replace g with μn (or μ/m) with $n = 1/m$ being a constant.

First Law: take a total derivative of the fundamental function (Legendre transform)

$$\begin{aligned} \mathrm{d}g &= \mathrm{d}u - s\,\mathrm{d}T - T\,\mathrm{d}s + v\,\mathrm{d}P + P\,\mathrm{d}v \\ &= T\,\mathrm{d}s - P\mathrm{d}v - s\,\mathrm{d}T - T\,\mathrm{d}s + v\,\mathrm{d}P + P\,\mathrm{d}v \end{aligned}$$

or

$$\boxed{\mathrm{d}g = -s\,\mathrm{d}T + v\,\mathrm{d}P.} \tag{6.133}$$

Note that with the substitution $\mathrm{d}g = n\mathrm{d}\mu$, the first law in the Gibbs representation is identically the Gibbs–Duhem equation.

State Functions: define the first derivatives of the fundamental function to be the state functions

$$s(T, P) = -\left(\frac{\partial g}{\partial T}\right)_p = -n\left(\frac{\partial \mu}{\partial T}\right)_p \quad \text{and} \quad v(T, P) = \left(\frac{\partial g}{\partial P}\right)_T = n\left(\frac{\partial \mu}{\partial P}\right)_T. \tag{6.134}$$

Constitutive Laws: take total derivatives of the state functions

$$\mathrm{d}s = \left(\frac{\partial s}{\partial T}\right)_p \mathrm{d}T + \left(\frac{\partial s}{\partial P}\right)_T \mathrm{d}P \tag{6.135}$$

$$\mathrm{d}v = \left(\frac{\partial v}{\partial T}\right)_p \mathrm{d}T + \left(\frac{\partial v}{\partial P}\right)_T \mathrm{d}P. \tag{6.136}$$

The Maxwell relation in this representation is

$$-\left(\frac{\partial s}{\partial P}\right)_T = \frac{\partial^2 g}{\partial T \partial P} = \frac{\partial^2 g}{\partial P \partial T} = \left(\frac{\partial v}{\partial T}\right)_p. \tag{6.137}$$

Using the three previously defined material properties c_p, α_p, and K_T, the constitutive laws in the Gibbs-energy representation (T, P) are written

$$\mathrm{d}s = \frac{c_p}{T}\,\mathrm{d}T - v\,\alpha_p\,\mathrm{d}P \tag{6.138}$$

$$\mathrm{d}v = v\,\alpha_p\,\mathrm{d}T - \frac{v}{K_T}\,\mathrm{d}P. \tag{6.139}$$

Because T and P are the most convenient variables to experimentally control in laboratory experiments, the Gibbs representation is one of the most frequently utilized.

6.4 About Thermodynamic Equilibrium

As postulated, when a system with given energy U, volume V, and number of molecules N is in thermodynamic equilibrium, the entropy is at a maximum relative to all nonequilibrium ways of internally segregating the energy, volume and molecules between subregions within the system. We will show below that entropy being a maximum at fixed energy is equivalent to energy being a minimum at fixed entropy. We will also quantify the condition placed on the intensive thermodynamic variables for two systems in contact to be in equilibrium with each other; namely, that temperature, pressure, and chemical potential must be uniform through any compound system that is in thermodynamic equilibrium. We then use the fact that entropy is a maximum (or energy a minimum) to place constraints on the sign or amplitude of the various material properties. We conclude this section with the example of first-order phase transitions in which one phase of matter transitions to another. Equilibrium or "coexistence" between two phases of matter is quantified by the equality of the chemical potentials of each phase.

6.4.1 Energy Is a Minimum in the Equilibrium State

As considered earlier in Section 6.1.2, let's imagine again that two subsystems are brought into contact and that a certain extensive quantity (internal energy, volume, or number of molecules) can vary in the two subsystems until the composite system is in equilibrium. In Section 6.1.2, we took internal energy to be the locally varying extensive variable within the composite system and said that a fundamental characteristic of equilibrium is that the entropy S of the closed composite will increase during the equilibration process, achieving a maximum once the two subsystems have equilibrated with each other. In addition to the internal energy of the two-subsystem composite $U = U_1 + U_2$, let's consider a second extensive variable (either volume or number of molecules in our formulation of thermodynamics so far) that we represent generically as $X = X_1 + X_2$ for the composite. Again, although both the composite totals X and U are constant during equilibration, the entropy S of the composite increases during equilibration until it is a maximum when X_1 (and therefore $X_2 = X - X_1$) and U_1 (and therefore $U_2 = U - U_1$) are at their equilibrium values. This entropy maximum principle can be stated in terms of the variations of X_1 alone as

$$\left(\frac{\partial S}{\partial X_1}\right)_U = 0 \quad \text{entropy is an extrema in equilibrium} \tag{6.140}$$

$$\left(\frac{\partial^2 S}{\partial X_1^2}\right)_U < 0 \quad \text{entropy is a maximum in equilibrium.} \tag{6.141}$$

Using the first and second facts about partial derivatives given in Section 6.1.3, we can write the derivative of the total energy of the composite system occurring at constant entropy as

$$\left(\frac{\partial U}{\partial X_1}\right)_S = -\frac{(\partial S/\partial X_1)_U}{(\partial S/\partial U)_{X_1}} = -T\left(\frac{\partial S}{\partial X_1}\right)_U = 0 \tag{6.142}$$

where $T = (\partial U/\partial S)_{X_1}$ is the temperature of the composite at equilibrium. So the energy U also has an extrema at equilibrium so long as we have let entropy escape from the composite during the equilibration of the two subsystems so that the total entropy remains constant. To determine whether the energy extrema is a maximum or minimum, we can take another derivative to obtain

$$\left(\frac{\partial^2 U}{\partial X_1^2}\right)_S = -T\left(\frac{\partial^2 S}{\partial X_1^2}\right)_U + \left(\frac{\partial S}{\partial X_1}\right)_U \frac{\partial T}{\partial X_1} \tag{6.143}$$

$$= -T\left(\frac{\partial^2 S}{\partial X_1^2}\right)_U \tag{6.144}$$

$$> 0, \tag{6.145}$$

where we used the entropy maximization results that $(\partial S/\partial X_1)_U = 0$ and $\left(\partial^2 S/\partial X_1^2\right)_U < 0$ in equilibrium. *So the internal energy is a minimum for a system that has relaxed to equilibrium as an extensive internal variable, either local volume or local numbers of molecules, changes with entropy remaining constant.*

Similar arguments show that the free energies (Helmholtz and Gibbs) and enthalpy discussed earlier are also minima in the equilibrated composite system. The Helmholtz energy is a minimum in equilibrium for processes where an extensive internal parameter, again either volume or number of molecules, adjusts itself between the subsystems with temperature of the composite remaining constant. The Gibbs energy is a minimum in equilibrium for a process where the number of molecules adjusts itself between the subsystems with both temperature and pressure remaining uniform throughout the composite. Last, the enthalpy is a minimum in equilibrium for a process where the number of molecules adjusts itself between the subsystems with entropy and pressure remaining constant.

6.4.2 Achieving the State of Equilibrium

These extremum principles (entropy maximum or equivalently energy minimum) can be put to use to characterize the essential nature of equilibrium. We consider again two subsystems brought into contact and allowed to exchange heat energy while keeping their volume and molecule numbers constant. We again have that $U = U_1 + U_2$ is constant while

$$S = S_1(U_1, V_1, N_1) + S_2(U_2, V_2, N_2) \tag{6.146}$$

increases until it is at an extremum $\mathrm{d}S = 0$ at equilibrium. If we take a derivative we have

$$\mathrm{d}S = \left(\frac{\partial S_1}{\partial U_1}\right)_{V_1,N_1} \mathrm{d}U_1 + \left(\frac{\partial S_2}{\partial U_2}\right)_{V_2,N_2} \mathrm{d}U_2 \tag{6.147}$$

$$= \frac{1}{T_1}\, dU_1 + \frac{1}{T_2}\, \mathrm{d}U_2 \tag{6.148}$$

$$= \left(\frac{1}{T_1} - \frac{1}{T_2}\right) \mathrm{d}U_1, \tag{6.149}$$

where we used that $\mathrm{d}U_1 + \mathrm{d}U_2 = 0$ during the equilibration since U is constant. Now in equilibrium we have $\mathrm{d}S = 0$ so that we have deduced that

$$\boxed{T_1 = T_2} \tag{6.150}$$

as the terminal state of the thermal equilibration process.

To conclude that T is a uniform spatial constant throughout a system in equilibrium, we only have used that entropy is an extremum at equilibrium. The consequence of entropy being a maximum in equilibrium is that while the system is out of equilibrium, the entropy is increasing so that $\mathrm{d}S > 0$. In this case, Eq. (6.149) becomes the condition $(1/T_1 - 1/T_2)\mathrm{d}U_1 > 0$. If heat energy is leaving region 1 and going into region 2, we have $\mathrm{d}U_1 < 0$ which thus requires that $T_1 > T_2$. We thus deduce as a consequence of entropy maximization that *heat will flow from regions of high temperature to regions of low temperature until equilibrium is established.* So temperature controls the direction of heat flow, which provides a physical interpretation of temperature. When we touch an object that is at a colder temperature than our finger, we feel the heat lost from our finger into the object that results in the molecules in our finger moving with less random kinetic energy, which provides the physical sensation of cold.

Carrying through the same argument while allowing volume to also be exchanged between the subsystems while preventing molecular flow results in

$$\mathrm{d}S = \left(\frac{\partial S_1}{\partial U_1}\right)_{V_1,N_1} \mathrm{d}U_1 + \left(\frac{\partial S_1}{\partial V_1}\right)_{U_1,N_1} \mathrm{d}V_1 + \left(\frac{\partial S_2}{\partial U_2}\right)_{V_2,N_2} \mathrm{d}U_2 + \left(\frac{\partial S_2}{\partial V_2}\right)_{U_2,N_2} \mathrm{d}V_2 \tag{6.151}$$

$$= \left(\frac{1}{T_1} - \frac{1}{T_2}\right) \mathrm{d}U_1 + \left(\frac{P_1}{T_1} - \frac{P_2}{T_2}\right) \mathrm{d}V_1, \tag{6.152}$$

where from the facts of partial derivatives $(\partial S/\partial V)_U = -(\partial U/\partial V)_S(\partial S/\partial U)_V \widehat{=} P/T$ in both subregions. Because we have already shown that $T_1 = T_2$ in equilibrium $\mathrm{d}S = 0$, we must have that

$$\boxed{P_1 = P_2} \tag{6.153}$$

as the additional terminal condition to $T_1 = T_2$ once both mechanical and thermal equilibrium is established.

During the equilibration process, we again must have $dS > 0$ due to entropy being a maximum at equilibrium, so if $dU_1 < 0$ (energy leaving subregion 1 and going into subregion 2), we again must have that $T_1 > T_2$ as deduced above. If $dV_1 < 0$ (volume of subregion 1 getting smaller and subregion 2 getting larger during the equilibration process), Eq. (6.155) also requires $P_1/T_1 < P_2/T_2$ or $P_1 < (T_1/T_2)P_2$. For a subregion that is at smaller pressure than its surroundings when both regions are at the same temperature, the volume change of that subregion will be negative, until the equilibrium state $P_1 = P_2$ is attained. Thus, *pressure controls the direction of volume change, with regions having smaller pressures than the surroundings experiencing volume reductions.* This provides a physical interpretation of pressure.

If we finally allow molecules to be exchanged along with heat while keeping the volume constant in the two subsystems, we have that

$$dS = \left(\frac{\partial S_1}{\partial U_1}\right)_{V_1,N_1} dU_1 + \left(\frac{\partial S_1}{\partial N_1}\right)_{U_1,V_1} dN_1 + \left(\frac{\partial S_2}{\partial U_2}\right)_{V_2,N_2} dU_2 + \left(\frac{\partial S_2}{\partial N_2}\right)_{U_2,V_2} dN_2 \tag{6.154}$$

$$= \left(\frac{1}{T_1} - \frac{1}{T_2}\right) dU_1 - \left(\frac{\mu_1}{T_1} - \frac{\mu_2}{T_2}\right) dN_1, \tag{6.155}$$

where we again used the first and second facts about partial derivatives along with the definition of the chemical potential $\mu = (\partial U/\partial N)_{S,V}$ in each subregion to deduce that $(\partial S/\partial N)_U = -(\partial U/\partial N)_S(\partial S/\partial U)_N = -\mu/T$. Thus, because $T_1 = T_2$ in equilibrium we have that

$$\boxed{\mu_1 = \mu_2} \tag{6.156}$$

as the additional condition for chemical equilibrium or equilibrium with respect to the transport of matter.

During the equilibration process we again must have $dS > 0$ due to entropy being a maximum at equilibrium, so if $dU_1 < 0$ (energy leaving subregion 1 and going into subregion 2), we again must have that $T_1 > T_2$ as already established. This time, if $dN_1 < 0$ (molecules leaving subregion 1 and going into subregion 2), we must also have that $\mu_1/T_1 > \mu_2/T_2$ or $\mu_1 > (T_1/T_2)\mu_2$. For a subregion that is at larger chemical potential than its surroundings when both regions are at the same temperature, that subregion will lose molecules to the surroundings, until the equilibrium state $\mu_1 = \mu_2$ is attained. Thus, *chemical potential controls the direction of the transport of matter, with matter flowing from regions of larger chemical potential to regions of lower chemical potential.* This provides a physical interpretation of chemical potential.

So the terminal state of full equilibration within a single-component fluid system is that the intensive variables T, P, and μ are uniform throughout the system.

6.4.3 Stability and Associated Constraints on Material Properties

The consequence of the entropy being a maximum (or energy being a minimum) is related to the notion of thermodynamic stability, which places constraints on the material properties. Because we will be connecting to the constitutive laws given earlier, we revert back to the specific thermodynamic variables rendered per unit mass.

To understand how the maximum or minimum nature of the extremum is related to the notion of stability, we imagine a system that is in perfect equilibrium, that is, T, P, and μ uniform throughout any subsystem we define within the total system. Draw an imaginary partition down the middle of the system creating two identical subsystems each at equilibrium. If the composite in equilibrium has a specific entropy s_e (entropy per unit mass) and a specific energy u_e (energy per unit mass), each of the two identical subsystems also have s_e and u_e. Let's now transfer a bit of energy from subsystem 1 to subsystem 2 while keeping their volume and molecule numbers the same. This will cause the entropy in subsystem 1 to change from $s_1 = s(u_e) = s_e$ to $s_1 = s(u_e - \Delta u) \neq s_e$ and the entropy in subsystem 2 to change from $s_2 = s(u_e) = s_e$ to $s_2 = s(u_e + \Delta u) \neq s_e$. After the energy transfer, the composite system is no longer in equilibrium and processes will be set up that drive the system back toward equilibrium.

Let's now consider the consequences of the curvature of $s(u)$ on the equilibration process as depicted in Fig. 6.5. Consider first the solid black curve $s(u)$ in the graph called *stable* in which $\partial^2 s/\partial u^2 < 0$ (negative curvature). When we calculate the new entropy of the out-of-equilibrium composite after the energy transfer, we obtain

$$\frac{s(u_e + \Delta u) + s(u_e - \Delta u)}{2} < s(u_e); \tag{6.157}$$

that is, the new entropy for the out-of-equilibrium composite is less than the equilibrium entropy $s_e = s(u_e)$ due to the negative curvature of the curve $s(u)$ which is consistent

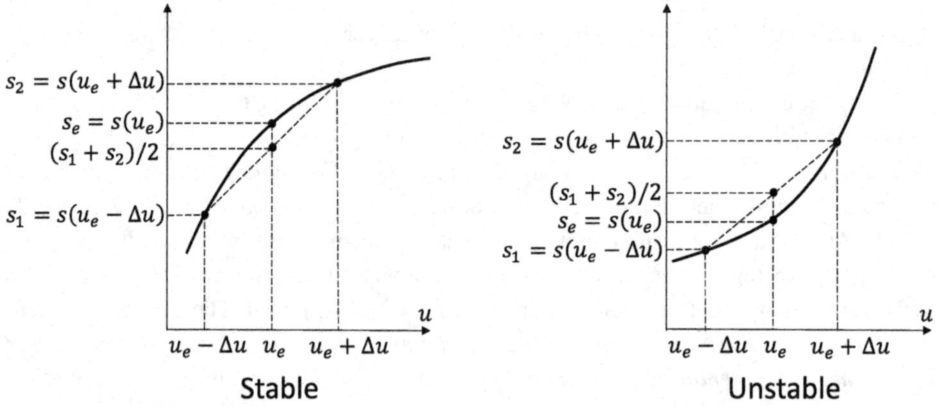

Figure 6.5 Possible monotonic dependence of entropy on energy $s(u)$. The system with negative curvature is stable, while positive curvature results in the system being unstable and therefore not realizable.

with our requirement that entropy is a maximum in equilibrium. When we calculate the slope of the curve $s(u)$ at the points $u_e + \Delta u$ and $u_e - \Delta u$, we obtain the temperatures $1/T_2 = \partial s(u)/\partial u|_{u_e+\Delta u} < \partial s(u)/\partial u|_{u_e-\Delta u} = 1/T_1$. So we have that $T_2 > T_1$. Heat energy will spontaneously flow from the higher-temperature and higher-energy subsystem 2 into the lower-temperature and lower-energy subsystem 1 until both subsystems are at the same temperature and both are at $s_e = s(u_e)$ which is the equilibrium state for the composite.

Thus, we see that when $s(u)$ has negative curvature, a transfer of energy from one subsystem to another, that has the effect to kick the composite out of equilibrium, will set up processes that drive the system back toward the equilibrium state. This is the definition of thermodynamic stability and we have learned that it requires that $\partial^2 s(u, v)/\partial^2 u < 0$. Similar conclusions hold for transfers of the specific volume between the two subsystem that also have the effect to kick the composite out of equilibrium, that is, we must also have that $\partial^2 s(u, v)/\partial v^2 < 0$ so that perturbations in mass density set up an equilibration process that drives the system back to equilibrium. This notion of thermodynamic stability, in which perturbations to equilibrium create internal "fluxes" that return a system to equilibrium, is called *Le Chatelier's principle*. While the internal fluxes, that we have also called equilibration processes, are occurring, entropy is increasing such that it is at a maximum in the final state of equilibrium.

If we had simultaneously transferred both energy and volume between the two subsystems so that the out-of-equilibrium entropy is characterized by $[s(u_e + \Delta u, v_e + \Delta v) + s(u_e - \Delta u, v_e - \Delta v, n_e)]/2 - s(u_e, v_e) \leq 0$, upon expanding to leading order in Δu and Δv, this inequality becomes

$$\frac{\partial^2 s}{\partial u^2}\Delta u^2 + 2\frac{\partial^2 s}{\partial u \partial v}\Delta u \Delta v + \frac{\partial^2 s}{\partial v^2}\Delta v^2 = \begin{bmatrix} \Delta u, & \Delta v \end{bmatrix} \begin{bmatrix} \partial^2 s/\partial u^2 & \partial^2 s/(\partial u\, \partial v) \\ \partial^2 s/(\partial u\, \partial v) & \partial^2 s/\partial v^2 \end{bmatrix} \begin{bmatrix} \Delta u \\ \Delta v \end{bmatrix} \leq 0. \tag{6.158}$$

This is called a *quadratic form*. A sufficient condition for a quadratic form involving a symmetric 2×2 matrix to be negative is that each diagonal element of the matrix must be negative and that the determinant of the matrix be positive, that is, the above form requires that

$$\frac{\partial^2 s}{\partial u^2} \leq 0, \quad \frac{\partial^2 s}{\partial v^2} \leq 0 \quad \text{and} \quad \left(\frac{\partial^2 s}{\partial u^2}\right)\left(\frac{\partial s^2}{\partial v^2}\right) - \left(\frac{\partial^2 s}{\partial u \partial v}\right)^2 > 0. \tag{6.159}$$

This has direct consequences for the sign of the material properties and their combinations as we will see.

Now if we imagine the curve $s(u)$ in the graph above called *unstable* in which the curvature is positive, we have that the perturbed composite entropy of the system out of equilibrium has an entropy $(s_1 + s_2)/2$ that is greater than the equilibrium entropy. Similarly, the higher energy subsystem 2 will have a temperature T_2 that is smaller than the temperature T_1 in the lower energy subsystem 1. This will cause heat energy to flow from the higher-temperature and lower-energy subsystem 1 to the lower-temperature and higher-energy subsystem 2 which drives the composite still further from equilibrium. This is the very notion of an unstable situation. We conclude that for thermodynamic equilibrium to be stable, the second derivatives of $s = s(u, v, n)$ with respect to the independent variables u, v, and n must all be negative while, simultaneously, the determinant of the matrix of second derivatives must be positive.

If we write the fundamental function as $u(s, v)$, we would conclude that the second derivatives of $u(s, v)$ with respect to the independent variables s and v must both be positive by virtue of the energy being a minimum when the system is in equilibrium. We also have that for perturbations simultaneously in both s and v that

$$\frac{\partial^2 u}{\partial s^2}\Delta s^2 + 2\frac{\partial^2 u}{\partial s \partial v}\Delta s \Delta v + \frac{\partial^2 u}{\partial v^2}\Delta v^2$$

$$= \begin{bmatrix} \Delta s, & \Delta v \end{bmatrix} \begin{bmatrix} \partial^2 u/\partial s^2 & \partial^2 u/(\partial s\,\partial v) \\ \partial^2 u/(\partial s\,\partial v) & \partial^2 u/\partial v^2 \end{bmatrix} \begin{bmatrix} \Delta s \\ \Delta v \end{bmatrix} \tag{6.160}$$

$$= \begin{bmatrix} \Delta s, & \Delta v \end{bmatrix} \begin{bmatrix} T/c_v & 1/(v\,\alpha_s) \\ 1/(v\,\alpha_s) & K_s/v \end{bmatrix} \begin{bmatrix} \Delta s \\ \Delta v \end{bmatrix} \tag{6.161}$$

$$= \begin{bmatrix} \Delta s, & \Delta v \end{bmatrix} \begin{bmatrix} \Delta T \\ -\Delta P \end{bmatrix} = \Delta s\,\Delta T - \Delta v\,\Delta P \geq 0. \tag{6.162}$$

In the above, we used our earlier definition of the constitutive laws for the *internal energy* representation $u(s, v)$. Because temperature is always positive, thermodynamic stability places the following constraints on the material properties

$$c_v > 0, \quad K_s > 0 \quad \text{and} \quad \alpha_s^2 - \frac{c_v}{v\,T\,K_s} > 0, \tag{6.163}$$

which allows for the possibility that the thermal expansion coefficient α_s could be negative.

Similar inequalities can be developed for the various Legendre transforms of $u(s, v)$ considered earlier. We now create quadratic forms for each form of the constitutive laws created in each representation.

In the *Helmholtz free energy* $f(T, v)$ representation of the constitutive laws we obtain the quadratic form

$$\begin{bmatrix} \Delta T, & \Delta v \end{bmatrix} \begin{bmatrix} c_v/T & \alpha_p\,K_T \\ -\alpha_p\,K_T & K_T/v \end{bmatrix} \begin{bmatrix} \Delta T \\ \Delta v \end{bmatrix} = \begin{bmatrix} \Delta T, & \Delta v \end{bmatrix} \begin{bmatrix} \Delta s \\ -\Delta P \end{bmatrix}$$

$$= \Delta s\,\Delta T - \Delta v\,\Delta P \geq 0. \tag{6.164}$$

For this quadratic form to be positive definite, we have the constraints that

$$c_v > 0, \quad K_T > 0, \quad \text{and} \quad \alpha_p^2 + \frac{c_v}{v\,T\,K_T} > 0. \tag{6.165}$$

The third inequality is a tautology because both terms are positive and thus provides no constraint on the material properties.

In the *enthalpy* $h(s,p)$ representation, we obtain the quadratic form

$$\begin{bmatrix} \Delta s, & -\Delta P \end{bmatrix} \begin{bmatrix} T/c_p & -v\,\alpha_p\,T/c_p \\ v\,\alpha_p\,T/c_p & v/K_s \end{bmatrix} \begin{bmatrix} \Delta s \\ -\Delta P \end{bmatrix} = \begin{bmatrix} \Delta s, & -\Delta P \end{bmatrix} \begin{bmatrix} \Delta T \\ \Delta v \end{bmatrix}$$
$$= \Delta s\,\Delta T - \Delta v\,\Delta P \geq 0. \tag{6.166}$$

For this quadratic form to be positive definite, we have the constraints that

$$c_p > 0, \quad K_s > 0, \quad \text{and} \quad \alpha_p^2 + \frac{c_p}{v\,T\,K_s} > 0. \tag{6.167}$$

Again, the third inequality provides no constraint on the material properties because both terms are positive.

In the *Gibbs free energy* $g(T,P)$ representation of the constitutive laws, we obtain the quadratic form

$$\begin{bmatrix} \Delta T, & -\Delta P \end{bmatrix} \begin{bmatrix} c_p/T & v\,\alpha_p \\ v\,\alpha_p & v/K_T \end{bmatrix} \begin{bmatrix} \Delta T \\ -\Delta P \end{bmatrix} = \begin{bmatrix} \Delta T, & -\Delta P \end{bmatrix} \begin{bmatrix} \Delta s \\ \Delta v \end{bmatrix}$$
$$= \Delta s\,\Delta T - \Delta v\,\Delta P \geq 0. \tag{6.168}$$

For this quadratic form to be positive definite, we have the constraints that

$$c_p > 0, \quad K_T > 0, \quad \text{and} \quad \frac{c_p}{v\,T\,K_T} - \alpha_p^2 > 0, \tag{6.169}$$

which provides a nontrivial constraint on α_p^2.

By using the relations between the partial derivatives, we have the following equalities derived either earlier or in the exercises:

$$\frac{1}{K_T} = \frac{1}{K_s} + \frac{v\,T\,\alpha_p^2}{c_p} \tag{6.170}$$

$$c_p = c_v + v\,T\,\alpha_p^2 K_T \tag{6.171}$$

$$\alpha_p\,\alpha_s = -\frac{c_v}{v\,T\,K_T}. \tag{6.172}$$

The last relation shows that either α_p or α_s must be negative in a particular material. Typically, it is α_s. A bit of algebra using these relations also shows that

$$\frac{K_s}{K_T} = \frac{c_p}{c_v} = 1 - \frac{\alpha_p}{\alpha_s} > 1. \tag{6.173}$$

As a special case, note that for an ideal gas, we give without proof the result from statistical mechanics that $K_T = P$ (the fluid pressure). As we saw earlier, an adiabatic bulk modulus is usually more appropriate for acoustics, which gives that $K_s = (c_p/c_v)P$ for an ideal gas.

The above represents all the constraints and relations we have between the various material properties c_p, c_v, K_T, K_s, α_p, and α_s of a single-component fluid. In particular, given either c_p or c_v (both of which are positive), either K_T or K_s (both of which are positive) and either α_p or α_s (one of which is positive and the other negative), we can calculate the other three properties using the above relations. In addition, we have the constraints coming from the internal-energy and Gibbs representation that $\alpha_s^2 > c_v/(v\,T\,K_s)$ and $\alpha_p^2 < c_p/(v\,T\,K_T)$.

6.4.4 First-Order Phase Transitions

Because we have been considering constant mass elements having a single molecular species, the chemical potential has not played much of a role in the thermodynamics so far because exchanges of matter between two elements in contact were not taking place. One topic where the chemical potential of a single component system plays a central role is at a first-order phase transition where the system transitions from one phase of matter to another, that is, from liquid to vapor (*vaporization*) or from solid to vapor (*sublimation*) or from solid to liquid (*fusion*, which is a word synonymous in this context to "melting").

As described below, when a single component system changes its phase due to a changing temperature and/or pressure, the chemical potential (or, equivalently, Gibbs free energy) is continuous across the transition while the derivatives of the chemical potential (or Gibbs free energy) with respect to both temperature and pressure have discontinuities corresponding to a release (or absorption) of heat that is called a *latent heat* and to a change in the mass density between the two phases. Because these first derivatives have a discontinuity across the phase transition, such phase changes are called either *first-order* phase transitions or, synonymously, *discontinuous* phase transitions. We will only treat here first-order phase transitions.

Another class of phase transitions have no discontinuity in the first derivatives of the Gibbs free energy but do have either discontinuities or singularities in the second derivatives (or material properties) and higher-order derivatives. These types of phase transitions are called either *second-order* phase transitions or, synonymously, either *continuous* phase transitions or *critical-point* phase transitions. The most well-known example is a ferromagnetic. As a ferromagnet cools from a sufficiently high starting temperature, it transitions from having no net magnetization to acquiring a net magnetization through the interaction and coherent alignment of the electron spins at a critical temperature ($T_{cr} = 1{,}044$ K for pure iron, which is called the *Curie temperature*, and can be compared to the melting or fusion temperature of pure iron that is 1,811 K). Other examples of second-order phase transitions include transitions to superconductivity or superfluidity at very low temperatures or the transition of a pure fluid to a supercritical fluid at sufficiently high temperature and pressure. The pressure and temperature point where a second-order transition occurs is called a *critical point*. The spatial correlation of an *order parameter*, such as the local magnetization in a ferromagnetic, diverges at the critical point, which is the inherent manifestation of a second-order transition. For a fluid such as pure water that approaches the critical point at high enough temperature and pressure ($T_{cr} = 647$ K and $P_{cr} = 21.8$ MPa for water), the order parameter is the local mass density and patches of larger and smaller density grow in spatial extent as the critical point is approached, causing light to scatter and the fluid to appear milky at the critical point, a phenomenon called *critical opalescence*.

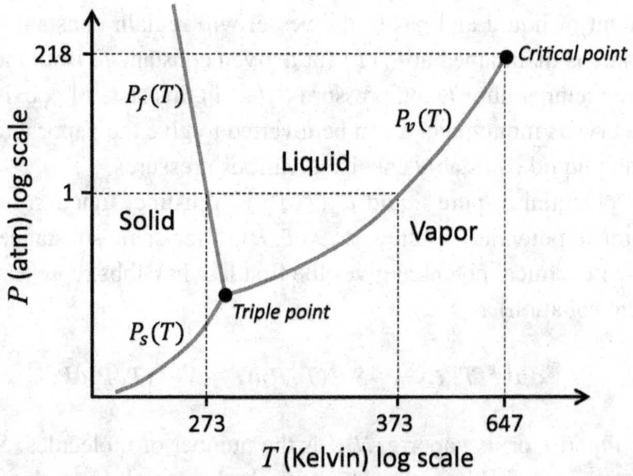

Figure 6.6 The phase diagram of equilibrium water showing at what P, T water stably exists as pure solid (ice), pure liquid, and pure vapor (gas). The lines separating the various pure phases are the curves of coexistence where the two phases on either side of the curves can stably coexist.

Critical opalescence also happens in binary liquids that are immiscible at low temperature but that become miscible above a critical-point temperature. We will not further consider second-order transitions.

Let's begin by presenting the *phase diagram* of pure water as given in Fig. 6.6. This diagram designates on the pressure–temperature plane those places where either pure solid, pure liquid, or pure vapor is the stable preferred phase for water. The curves separating the various pure phases are the *coexistence curves* that define the special P, T values where the phases on either side of the curve can stably coexist with each other without one phase growing at the expense of the other. The first-order phase transition separating each of the three phases of water is represented by these coexistence curves. One goal of what follows is to determine the $P(T)$ relation along the three coexistence curves. When one crosses a curve from one phase of matter to another, there is a jump in the state functions in the Gibbs representation, that is, the specific entropy $s(T, P) = -(\partial g(T, P)/\partial T)_P$ and specific volume (inverse mass density) $v(T, P) = (\partial g(T, P)/\partial P)_T$ both have discontinuities across the coexistence curves. However the Gibbs energy itself $g(T, P) = n\mu(T, P)$, where n^{-1} is the mass of a single molecule, is continuous across a coexistence curve.

At the triple point, all three phases (solid, liquid, and gas) of pure matter can stably coexist. At temperatures and pressures greater than the critical point, there ceases to be a distinction between liquid and vapor and the fluid is said to be in the *supercritical* state.

To proceed, let's focus on the coexistence curve that separates a liquid from its vapor (gas). The function $P_v(T)$ that defines this curve is called the *vapor pressure*. If there is a layer of liquid overlain by its gas in a closed vessel having a pressure $P_v(T)$ for some given T, the gas molecules will strike and stick to the liquid surface as often as the liquid molecules escape the liquid surface and become gas molecules. In this state of coexistence, the average amount of liquid and gas in the vessel will remain constant through time so long as the pressure is maintained at $P_v(T)$ for a given constant T. Both the liquid and gas will be at the same temperature T and pressure $P_v(T)$ in this state of coexistence. Because the coexistence curve is monotonic, it can be inverted to give the vapor temperature $T_v(P)$ at which vapor and liquid can stably coexist at various pressures.

The chemical potential in pure liquid $\mu^\ell(T, P)$ is a distinct function of T and P compared to the chemical potential of pure gas $\mu^g(T, P)$ of the same substance. Taking a total derivative of either chemical potential gives the first law in Gibbs representation (which is the Gibbs–Duhem equation) as

$$N^{\ell,g}\mathrm{d}\mu^{\ell,g}(T, P) = -S^{\ell,g}(T, P)\mathrm{d}T + V^{\ell,g}(T, P)\mathrm{d}P, \tag{6.174}$$

where for either liquid ℓ or its vapor g, $N^{\ell,g}$ is the number of molecules, $S^{\ell,g}$ the entropy, and $V^{\ell,g}$ the volume occupied by either phase inside the vessel. If we divide both sides of this equation by the mass of water in each phase, we get the statement for how chemical potential varies in specific form

$$n\mathrm{d}\mu^{\ell,g} = -s^{\ell,g}\mathrm{d}T + v^{\ell,g}\mathrm{d}P, \tag{6.175}$$

where the state functions of the liquid and its vapor are

$$s^{\ell,g}(T, P) = -n\left(\frac{\partial \mu^{\ell,g}(T, P)}{\partial T}\right)_P \quad \text{and} \quad v^{\ell,g}(T, P) = n\left(\frac{\partial \mu^{\ell,g}(T, P)}{\partial P}\right)_T \tag{6.176}$$

with $n = 1/m$ and m is the mass of a single molecule of the substance. The coexistence condition is that

$$\mu^\ell(T, P) = \mu^g(T, P), \tag{6.177}$$

which is what allows us to determine the coexistence curve $P_v(T)$ or, equivalently, $T_v(P)$.

The schematic trends with T and P of these various functions are depicted in Fig. 6.7. At a first-order phase transition like the liquid–gas transition, the chemical potential of each phase is continuous across the transition while the specific entropy and specific volume go through jumps. The qualitative trends depicted in Fig. 6.7 are consistent with those of an ideal gas $\mu(T, P) = \mu_0 T/T_0 - (1 + c)k_B T \ln(T/T_0) + k_B T \ln(P/P_0)$ for $T > T_0$ and $P > P_0$, where T_0, P_0 is some low temperature reference state (cf., Exercise 7 at the end of the chapter). However our gas, liquid, and solid need not be ideal for either these general trends or the present analysis to apply.

To find the coexistence curve $P_v(T)$ for vapor and liquid of the same pure substance contained in some vessel, we assume we know a particular point T_c, P_c on the coexistence curve. For water, this could be the well-known value of $P_c = 10^5$ Pa (atmospheric pressure) and $T_c = 373$ K. At this point we have $\mu^g(T_c, P_c) = \mu^\ell(T_c, P_c)$. As we move from this point in the T, P plane, the chemical potential of each phase will change according to Eq. (6.175) and so long as $\mathrm{d}\mu^g(T, P) = \mathrm{d}\mu^\ell(T, P)$ we will still be on the

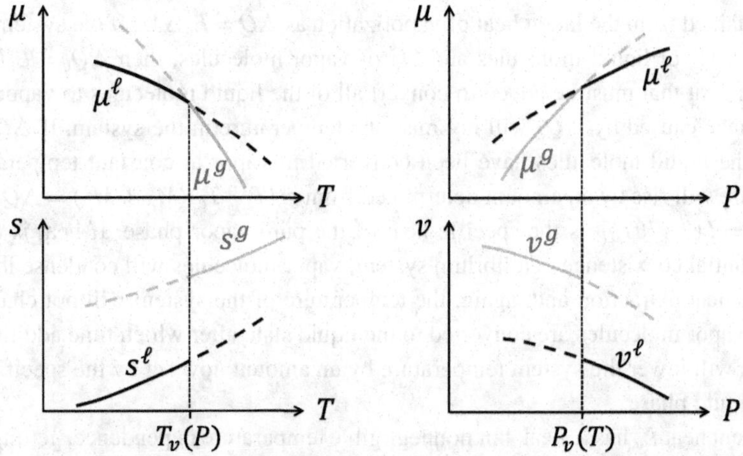

Figure 6.7 Schematic showing how the chemical potential as a function of P and T is continuous across the gas–liquid phase transition (the vertical dotted line) while both the specific entropy and specific volume, which are the first derivatives of the chemical potential $s = -n\,(\partial\mu/\partial T)_P$ and $v = n\,(\partial\mu/\partial P)_T$, go through jumps at the transition. The black curves are the liquid phase and the gray curves are the gas or vapor.

coexistence curve at this new T, P. We thus have that pressure variations along the coexistence curve are controlled by

$$(-s^g + s^\ell)\mathrm{d}T = (v^\ell - v_g)\mathrm{d}P_v \tag{6.178}$$

or upon rearranging

$$\boxed{\frac{\mathrm{d}P_v(T)}{\mathrm{d}T} = \frac{L_v}{T(v^g - v^\ell)},} \tag{6.179}$$

where L_v is called the *latent-heat of vaporization* and is given by

$$L_v = T(s^g - s^\ell). \tag{6.180}$$

It has units of Joules per kilogram and is always positive because the gas phase always has more disorder per unit mass (larger entropy) than does the liquid phase. The latent heat associated with vaporization or condensation works as follows.

If a system has liquid and vapor in stable coexistence at a particular T_c, P_c on the coexistence curve and if an amount of heat $\Delta Q > 0$ is injected into this equilibrium coexistent state, the added heat will cause liquid molecules to transition into the vapor state without raising the temperature in the system. The added heat gives the liquid molecules the additional energy they need to overcome the attractive forces holding them in the lower-entropy liquid state so they are liberated at constant temperature into the higher-entropy gaseous state. The mass ΔM of liquid molecules converted to vapor due to the injected heat is obtained from the latent heat of vaporization as $\Delta Q = L_v \Delta M$. If the system initially has a mass M_c^ℓ of liquid molecules and M_c^g of vapor molecules, then $\Delta Q_c = L_v M_c^\ell$ is the amount of heat that must be added to convert all of the liquid molecules to vapor and this phase change caused by ΔQ_c will not raise the temperature in the system. If $\Delta Q > \Delta Q_c$, once all the liquid molecules have been converted to vapor at constant temperature, the temperature will rise by an amount determined from $c_p^g(T - T_c)(M_c^\ell + M_c^g) = \Delta Q - L_v M_c^\ell$, where $c_p^g = T(\partial s^g/\partial T)_P$ is the specific heat of the pure vapor phase. If heat is extracted from the initial coexistence equilibrium system, vapor molecules will condense into liquid due to the heat extraction and, again, the temperature of the system will not change until all of the vapor molecules are converted to the liquid state after which time additional heat extraction will lower the system temperature by an amount now set by the specific heat of the pure liquid phase.

The latent heat L_v has a weak but nonnegligible temperature dependence, getting smaller with increasing temperature. Equation (6.179) is called the *Clapeyron* equation and once the temperature dependence of the latent heat and specific volumes are inserted, can be integrated to give the vapor pressure $P_v(T)$ of stable coexistence.

This same Clapeyron equation also controls the sublimation coexistence curve where solid transitions to gas $\mathrm{d}P_s(T)/\mathrm{d}T = L_s/[T(v^g - v^s)]$ with L_s the *latent heat of sublimation* and $v_s = 1/\rho_s$ the specific volume of the solid. Similarly, along the melting or fusion coexistence curve where solid transitions to liquid, we have $\mathrm{d}P_f(T)/\mathrm{d}T = L_f/[T(v^\ell - v^s)]$ with L_f called the *latent heat of fusion*. Water is one of the rare substances where the mass density in the liquid state is greater than in the solid state so that for water $v^\ell - v^s < 0$ (i.e., water expands upon freezing) and the slope of the melting coexistence curve $\mathrm{d}P_f(T)/\mathrm{d}T$ is negative as seen in Fig. 6.6.

If we continue to focus on the vaporization transition from liquid to gas, we have that $v^g \gg v^\ell$ so that v_ℓ can be ignored. If we further represent the gas as being ideal for which $1/v^g = \rho^g = mP/(k_B T)$ where, again, m is the mass of one molecule, we have

$$\frac{\mathrm{d}P_v}{P_v} = \frac{mL_v}{k_B}\frac{\mathrm{d}T}{T^2}, \tag{6.181}$$

which is called the *Clausius–Clapeyron* equation. If we say that we know a point on the coexistence curve T_c, P_c and know the latent heat at that point L_c, the decrease in latent-heat over some limited range of temperature that stays well below the critical point (where the temperature dependence becomes much stronger) can be approximated as being linear

$$L_v(T) = L_c + L^*\left(1 - \frac{T}{T_c}\right). \tag{6.182}$$

For water, for example, if we take $T_c = 373$ K when $P_c = 10^5$ Pa, we have $L_c = 2.256 \times 10^6$ J/kg and $L^* = 1.065 \times 10^6$ J/kg, which will make a reasonable prediction of the decrease in the latent heat for water vaporization from the triple point out to say no more than $T = 450$ K, always predicting the latent heat to be slightly greater than the actual latent heat except at the point $T = T_c = 373$ K. Beyond say 450 K, the latent heat of water begins to decrease much more rapidly than linearly.

Integrating the Clausius–Clapeyron equation with this linearly decreasing latent heat approximation gives

$$P_v(T) = P_c \left(\frac{T}{T_c}\right)^{-mL^*/(k_B T_c)} \exp\left[-\frac{m(L_c + L^*)}{k_B T}\left(1 - \frac{T}{T_c}\right)\right]. \quad (6.183)$$

This expression represents reasonably well the equilibrium pressure $P_v(T)$ of a gas in stable equilibrium with its liquid over a range of T that does not deviate too far from T_c and staying, say, 200 K less than the critical-point temperature in the case of pure water. We will not pause to further consider the coexistence curves for liquid freezing to solid $P_f(T)$ or for solid sublimating to gas $P_s(T)$.

With the above established, we can further understand perturbations to the state of equilibrium coexistence. Consider a closed rigid vessel with insulating walls that has a certain number of water molecules in the liquid state and a certain number in the gaseous state and that resides at a particular state of equilibrium T_c, P_c on the coexistence curve, that is, $P_c = P_v(T_c)$. There will be a layer of liquid water at the bottom of the vessel overlain by a layer of water vapor and the number of liquid and vapor molecules will not vary, on average, due to being in stable coexistence. Let's now perturb the system by inserting some additional water molecules having the vessel's starting temperature T_c, which will cause the pressure inside the vessel to increase. The vessel is now out of equilibrium and on the liquid side of the coexistence curve, with the gas phase now having a greater chemical potential than the liquid phase, cf., Fig. 6.7 for a pressure $P > P_v(T)$. Water-vapor molecules will therefore condense into the liquid state because matter always moves from the high-chemical-potential phase to the low-chemical-potential phase as the system attempts to return to equilibrium. As vapor condenses into liquid, latent heat will be released into the system as set by the latent heat of vaporization L_v. Because the walls of the system are insulated to heat flow, this latent-heat release will cause the temperature in the vessel to rise, which will drive the system back toward equilibrium and closer to the coexistence curve. Vapor molecules will continue to condense into liquid and the temperature will continue to rise due to the release of latent heat until the system is returned to an equilibrium state on the coexistence curve that is further up the $P_v(T)$ curve (i.e., closer to the critical point) and with now proportionally more of the water molecules in the liquid state than in the vapor state compared to the starting point at T_c, P_c prior to inserting water molecules. You can treat this scenario as an end-of-chapter exercise. Analogous comments hold for a perturbation caused by extraction of water molecules and a lowering of the pressure in the system which will cause liquid molecules to transition to vapor molecules. Heat will be absorbed by this vaporization transition as set by L_v and will cause the temperature in the insulated system to be lowered until the system resides at a new equilibrium coexistence state that is lower down on the coexistence curve $P_v(T)$. Such return to a stable state of equilibrium following a perturbation to the initial equilibrium state is the epitome of Le Chatelier's principle mentioned earlier. In this case, the process that returns our two-phase system to equilibrium following a pressure perturbation is a first-order phase transition and the associated release or absorption of latent heat.

Next, consider a cave having wet walls that are in equilibrium with humid air. Imagine warmer air coming into the cave through the cave entrance, which corresponds to heat being injected into the equilibrium cave. This added heat will cause the wet walls to release liquid water into the vapor state with no change in the cave temperature. Similarly, if cooler air comes into the cave, which corresponds to heat being extracted from the equilibrium cave, water vapor will condense onto the cave walls with no change in the cave temperature. The stable temperature of caves in the presence of externally applied heat perturbations (i.e., warmer or cooler air coming into the cave through the cave entrance) is due to first-order phase transitions induced by the heat perturbations. So the next time you are enjoying one of the world's fine wines that was aged in oak barrels in a humid cave of constant temperature, you can appreciate that it is the need of the chemical potential of the liquid and vapor states of water to be the same that is, at least in part, what is providing you with pleasure.

6.5 Multicomponent Fluid Solutions

To lay out the basic concepts of thermodynamics in as clear and unencumbered manner as possible, this chapter has so far focused on single-molecular-component fluids that cannot sustain static shear (shape change) and that are not subjected to macroscopic electric and magnetic fields. For the rest of the chapter, we show how to treat several generalizations to the continuum description of a simple isotropic single-component fluid.

We begin by considering a fluid composed of $M+1$ different molecular species labeled $m=0, 1, 2, \ldots M$. If one of the component species has much larger numbers than the other components, we call that species the *solvent* and label it $m=0$, while the other species are called the *solutes* and the collection of all the species inside of the element is called a *solution.* In what follows, we first determine the general functional form for each chemical potential in the multicomponent element and then consider the thermodynamic formalism in the Gibbs energy representation which is the most common and useful representation for describing multicomponent thermodynamics and chemical reactions. We conclude with determining the state of chemical equilibrium in a system where chemical reactions can occur between the constituents.

6.5.1 The Chemical Potentials of Fluid Solutions

When there are multiple molecular species in the fluid, the first law in the internal-energy representation has the form

$$dU = TdS - PdV + \sum_{j=0}^{M} \mu_j dN_j \tag{6.184}$$

and the state functions are defined

$$T = \left(\frac{\partial U}{\partial S}\right)_{V,N_j}, \quad -P = \left(\frac{\partial U}{\partial V}\right)_{S,N_j}, \quad \text{and} \quad \boxed{\mu_i = \left(\frac{\partial U}{\partial N_i}\right)_{S,V,N_j \neq N_i}}. \tag{6.185}$$

This definition of the chemical potential of species i in the internal-energy representation $\mu_i = (\partial U/\partial N_i)_{S,V,N_j}$ for $j \neq i$ is useful for obtaining an explicit functional form for the chemical potential that is widely used across the physical sciences.

In words, the chemical potential μ_i is the energy change when one molecule of species i is added with all other species held at their original numbers and under the condition that the volume and entropy of the element does not change. Now in adding a single species i molecule, the volume inherently increases since a single molecule has a finite volume v_i. Similarly, in adding a single molecule, the element now has more possible states available to it and the entropy increases. If we determine how the energy U changes when the volume and entropy increase due to the addition of species i molecules, we must subtract that energy from the total energy change in order that we consider only the energy change required to insert molecules at constant S and V. This is written mathematically as

$$\mu_i = \frac{\mathrm{d}U}{\mathrm{d}N_i} - \left(\frac{\partial U}{\partial V}\right)\frac{\partial V}{\partial N_i} - \left(\frac{\partial U}{\partial S}\right)\frac{\partial S}{\partial N_i} \tag{6.186}$$

$$= \frac{\mathrm{d}U}{\mathrm{d}N_i} + P\frac{\partial V}{\partial N_i} - T\frac{\partial S}{\partial N_i}. \tag{6.187}$$

Equation (6.187) is equivalent to Eq. (6.184) when $\mathrm{d}N_i \neq 0$ but $\mathrm{d}N_j = 0$ for $j \neq i$. The total energy change $\mathrm{d}U/\mathrm{d}N_i$ due to the addition of a single species-i molecule has two contributions: (1) an added molecule in equilibrium with the other molecules increases the system energy due to the kinetic and vibrational energy of the new molecule and (2) an inserted molecule requires work to be performed against the forces coming from the other molecules, which is a negative energy contribution when the intermolecular forces are attractive (e.g., van der Waals attractions between any atoms or electrostatic attractions between cations and anions).

The volume V of the entire element is $V = \sum_{j=0}^{M} v_j N_j$, where v_j is the volume of a single species j molecule so that

$$\frac{\partial V}{\partial N_i} = \frac{\partial}{\partial N_i}\sum_{j=0}^{M} v_j N_j = v_i \tag{6.188}$$

in Eq. (6.187). The volume v_i of a single molecule may depend on the pressure via the relation derived in Chapter 4 that says $v_i = v_i^{(0)} \exp\left[-(P-P_0)/K_i\right]$, where $v_i^{(0)}$ is the volume in a reference state having pressure P_0 and where K_i is the bulk modulus of the species i molecules defined $K_i = -v_i \partial P/\partial v_i$. We can take K_i to be the bulk modulus of an element composed only of species i molecules. Because K_i for liquids is typically measured in GPa and because the pressure range in many applications is far less than GPa, we can usually take $v_i \approx v_i^{(0)}$.

To address $\partial S/\partial N_i$, we need to know how the entropy of the element depends on the N_i. In the present special case of the internal-energy representation (which in statistical mechanics is given the intimidating name of the *microcanonical ensemble*), each conceivable configuration of the molecules in the element is equally probable per the Boltzmann postulate that leads to Eq. (6.10). If a system of volume V_i is occupied only by N_i molecules of species i, then the corresponding entropy is denoted $S_i(N_i)$ and will increase with increasing N_i. A collection of $M+1$ species randomly mixed together in equilibrium will have an entropy that is not only associated with the sum of the individual entropies $S_i(N_i)$ but also due to the mixing of the various species that by itself increases the entropy of the solution by an amount $S_{\text{mix}} = k_B \ln \Omega_{\text{mix}}$, which is called the *entropy of mixing*. We have that

$$S = \sum_{j=0}^{M} S_j(N_j) + k_B \ln \Omega_{\text{mix}}. \tag{6.189}$$

The state count Ω_{mix} is the number of distinct ways to place the molecules into the element, which because each such state is equally probable in the internal-energy representation, is a combinatorial calculation that does not depend on the energetic interactions between the molecules in each state. If the system were occupied only by species i molecules, then $S = S_i(N_i)$ and $\Omega_{\text{mix}} = 1$. We thus identify the entropy increase due to an added molecule as $\partial S/\partial N_i = \partial S_i/\partial N_i + \partial S_{\text{mix}}/\partial N_i$. In the internal-energy representation, the molecular interactions are accounted for in the term $\mathrm{d}U/\mathrm{d}N_i$ of Eq. (6.187) and, possibly, in the volume occupied by each molecule v_i.

To obtain Ω_{mix}, we begin by asking how many ways can N distinguishable objects be placed in a system? If we divide the system into N occupation bins, there are N possible objects that can be placed in the first bin, $N-1$ in the second bin and so on such that there are $N!$ possible ways to place the distinguishable objects in the system. Next, if $N = N_1 + N_2$ where one set of N_1 objects are all identical to each other and a different set of N_2 objects are all identical to each other, then not all of the $(N_1+N_2)!$ possible ways of inserting these objects are distinguishable; there will be a smaller number $\Omega(N_1, N_2)$ of distinguishable placements. To determine this Ω, we note that the total number of ways of placing the objects $(N_1+N_2)!$ is the product of the number of unique (distinguishable) ways $\Omega(N_1, N_2)$ multiplied by the number of different ways $N_1!$ of placing the N_1 identical objects for each of the Ω configurations as well as multiplied by the number of different ways $N_2!$ of placing the other set of identical objects in each unique configuration, such that $(N_1+N_2)! = \Omega(N_1, N_2)N_1!N_2!$ or

$$\Omega(N_1, N_2) = \frac{(N_1+N_2)!}{N_1!N_2!}. \tag{6.190}$$

We thus find that

$$\Omega_{\text{mix}} = \frac{\left(\sum_{j=0}^{M} N_j\right)!}{N_0!N_1!\ldots N_M!}. \tag{6.191}$$

by extending the argument to our element containing $M+1$ molecular species.

To evaluate $\ln \Omega_{\text{mix}}$, we first note that $\ln N! = \sum_{j=1}^{N} \ln j$ exactly. If the sum is multiplied by $\Delta j = 1$, it can be written as an integral, which is an approximation that becomes increasingly accurate for increasing N

$$\ln N! = \sum_{j=1}^{N} \ln j \, \Delta j \approx \int_1^N \ln x \, \mathrm{d}x = N \ln N - N + 1 \approx N \ln N - N. \tag{6.192}$$

This approximation $\ln N! \approx N \ln N - N$ is known as the *Sterling approximation*. For example, if we have $N = 10^3$ molecules of a particular species in our fluid element, the exact value is $\ln(10^3!) = 5{,}913$ (to the nearest integer) while the approximation gives $10^3 \ln 10^3 - 10^3 = 5{,}908$ (to the nearest integer), which is an error of only 0.08% that will continue to diminish with increasing N.

Using the Sterling approximation, we first write $N_t = \sum_{j=0}^{M} N_j$ for the total number of molecules in the element and obtain

$$S_{\text{mix}} = k_B \ln \Omega_{\text{mix}} = k_B \left(N_t \ln N_t - N_t - N_0 \ln N_0 + N_0 - N_1 \ln N_1 + N_1 - \ldots \right) \quad (6.193)$$

$$= k_B \left(N_t \ln N_t - \sum_{j=0}^{M} N_j \ln N_j \right) = -k_B \sum_{j=0}^{M} N_j \ln \left(\frac{N_j}{N_t} \right), \quad (6.194)$$

which is the well-known expression for the entropy of mixing in a multicomponent system. Taking the partial derivative of Eq. (6.194) with respect to N_i and using $\partial N_t / \partial N_i = 1$ then gives

$$\frac{\partial S_{\text{mix}}}{\partial N_i} = k_B \left(\ln N_t + \frac{N_t}{N_t} - \ln N_i - \frac{N_i}{N_i} \right) \quad (6.195)$$

$$= -k_B \ln \left(\frac{N_i}{N_t} \right). \quad (6.196)$$

The ratio $x_i = N_i / N_t$ is called the *number fraction* of species i in the multicomponent solution or, equivalently, the *mole fraction* if counting in terms of moles. Because $0 < x_i \leq 1$, we have $\partial S_{\text{mix}} / \partial N_i \geq 0$ as expected. This entropy change will be zero when species i is the only component in the element (there is only one distinguishable mixing state) and will increase with decreasing number fraction due to the increasing number of possible states compared to the pure state.

We have so far shown that

$$\mu_i = \frac{dU}{dN_i} - T \frac{\partial S_i}{\partial N_i} + v_i P + k_B T \ln \left(\frac{N_i}{N_t} \right). \quad (6.197)$$

It is convention to write the sum of the first two contributions as

$$\tilde{\mu}_i + k_B T \ln \gamma_i = \frac{dU}{dN_i} - T \frac{\partial S_i}{\partial N_i}, \quad (6.198)$$

where both $\tilde{\mu}_i$ and γ_i require physical interpretation. Although $\tilde{\mu}_i$ contributes to dU/dN_i and entirely accounts for $-T \partial S_i / \partial N_i$, the term $k_B T \ln \gamma_i$ only contributes to dU/dN_i. So the chemical potential of species-i molecules in a solution containing $N_t = \sum_{j=0}^{M} N_j$ total molecules is written

$$\boxed{\mu_i = \tilde{\mu}_i + v_i P + k_B T \ln \left(\gamma_i \frac{N_i}{N_t} \right),} \quad (6.199)$$

which is widely used across many different applications. The physical definitions given to $\tilde{\mu}_i$, that can be called the *reference-configuration chemical potential*, and to γ_i, that is called the *activity coefficient*, are coupled to each other and depend on whether the molecules in the system are electrically neutral or are charged ions.

If the molecular species are not charged and represent, for example, distinct liquids that fully mix together (e.g., water and ethanol), we interpret $\tilde{\mu}_i$ to correspond to an element containing only species i molecules $N_i/N_t = 1$. It thus represents the sum of three contributions: (1) the kinetic and vibrational energy of a single species i molecule, (2) the work performed against the intermolecular forces during the insertion of a molecule into the pure species-i system, which is negative when the interaction forces are attractive, and (3) the heat energy subtraction term $-T\partial S_i/\partial N_i$ that allows the insertion of a single molecule into the pure species i system to take place at constant entropy. So $\tilde{\mu}_i$ depends on temperature and pressure but is independent of the number fraction of species i in the solution. The term $k_BT \ln \gamma_i$ then represents the correction to the work performed against the intermolecular forces during insertion into the actual system containing all $M+1$ species. The activity coefficient γ_i depends on the average distance between molecules and thus has a strong number fraction dependence as well as a weaker pressure dependence that is often ignored and has the property that $\gamma_i \to 1$ as $N_i/N_t \to 1$. It gives the appearance in Eq. (6.199) to alter the effective number fraction of each species but is formally not a part of the entropy change $\partial S/\partial N_i$ and is instead a part of the work of insertion $\mathrm{d}U/\mathrm{d}N_i$. Note that in the v_iP contribution to Eq. (6.199), depending on how closely the other molecules can pack in against a single species i molecule, v_i can change, at least weakly, with varying number fractions (composition) of the solution and with pressure.

Alternatively, if the solution is an electrolyte consisting of a solvent contributing N_0 uncharged molecules to the element (e.g., water) and M species of charged ions, each species contributing N_i ions ($i = 1, M$) to the element, the above interpretations of $\tilde{\mu}_i$ and γ_i become problematic for the ions because there is no such thing as a pure state having only a single ion species present.

The convention for an electrolyte is to define the ionic activity coefficients γ_i so that they are one in the infinitely dilute state containing only solvent. So the reference configuration is pure solvent, and the $\tilde{\mu}_i$ corresponds to adding either a single solvent molecule or a single species-i ion into pure solvent. For the solvent $i = 0$, we have $\tilde{\mu}_0 + k_BT \ln \gamma_0 = \mathrm{d}U/\mathrm{d}N_0 - T\partial S_0/\partial N_0$ while for the species-i ions we have $\tilde{\mu}_i + k_BT \ln \gamma_i = \mathrm{d}U/\mathrm{d}N_i$. So $\tilde{\mu}_i$ allows for the work of inserting an ion into pure solvent molecules while $k_BT \ln \gamma_i$ is the work required to insert a species-i ion against the interactions with the other ions, which is dominated by the work of inserting the ion into the attractive-force countercharge cloud of counter ions that surround it in equilibrium.

A strong salt S added to solvent dissociates completely into ionic solute as $S = C_{\nu_+}A_{\nu_-} \to \nu_+C^{z+} + \nu_-A^{z-}$ where the $\nu_\pm$ are the numbers of cations C and anions A in each neutral salt molecule $S = C_{\nu_+}A_{\nu_-}$ and the valences $z_\pm$ are the number and sign of fundamental charges $e = 1.6 \times 10^{-19}$ C on each ion. Charge neutrality of the electrolyte requires that $z_+\nu_+ + z_-\nu_- = 0$ for each type of salt (or neutral acid or base) that dissolves into the electrolyte. We assume there can be many such strong salts, acids, or bases present that upon dissolution generate a total of M different species of cations and anions, each having an ion density (numbers per meter cubed) given by $\bar{N}_i$ in the neutral (or bulk) electrolyte. Such number densities can be expressed as molarities M_i (moles per liter) for each species using $M_i = \bar{N}_i/(10^3 N_A)$ with $N_A = 6.022 \times 10^{23}$ being Avogadro's number and the units on the 10^3 being liters per meter cubed.

Each cation in the electrolyte is surrounded by a cloud of anions whose densities are largest right near the cation at the distance of closest approach (the radius of the cation plus the radius of the anion) but diminish to their bulk densities $\bar{N}_i$ with distance from the cation. Identical comments hold for an anion surrounded by a cloud of cations. In 1923, Debye and Hückel treated this problem approximately (Lyklema, 1991) and determined the enhanced electrostatic energy $k_B T \ln \gamma_i$ stored in the interaction between an ion i of effective radius a_i and the surrounding cloud of counter charge. Their model gives the activity coefficient of ion i to be

$$\gamma_i = \exp\left(\frac{-z_i^2 \lambda_B}{2(\lambda_D + a_i)}\right), \tag{6.200}$$

where λ_B is called the *Bjerrum length* and is given by

$$\lambda_B = \frac{e^2}{4\pi \epsilon_0 \kappa k_B T} \tag{6.201}$$

with ϵ_0 the permittivity of vacuum and κ the dielectric constant of the solvent in the vicinity of each central ion. As seen earlier in Exercise 3 of Chapter 3, the Bjerrum length is the distance between a cation and anion when their electrostatic interaction energy just equals the thermal energy $k_B T$ in the system. Further, λ_D is called the *Debye length* and is a measure of the distance that the cloud of enhanced counterion density extends away from the central ion as given by

$$\lambda_D = \sqrt{\frac{\epsilon_0 \kappa k_B T}{e^2 \sum_{j=1}^{M} z_j^2 \bar{N}_j}}. \tag{6.202}$$

Again, the $\bar{N}_j$ are the number densities of each ionic species in neutral electrolyte (numbers per meter cubed).

We will not derive Eq. (6.200) but it is based on continuum modeling of the counter-charge cloud in which diffusion of counter ions away from the central ion is balanced exactly by electromigration (electrostatic attraction) of the counter ions toward the central ion. We have presented this result not only because it can be useful but because it shows explicitly some characteristics of the ionic activity coefficients of an electrolyte. For example, as the electrolyte becomes dilute and $\bar{N}_j \to 0$ in Eqs (6.200) and (6.202), each $\gamma_i \to 1$. Also, because Eq. (6.200) predicts that $\ln \gamma_i$ is negative in an electrolyte (it is the work performed against an attractive force), the ionic activity coefficients are always in the range $0 < \gamma_i \leq 1$, which is distinct from when we mix together fully miscible liquids of uncharged molecules that can result in activity coefficients, which are corrections to the work of inserting a species-i molecule into a pure species-i element, that can be either larger or smaller than 1 depending on the strength of the van der Waals interactions between the various molecules.

Because only neutral salt crystals (or acids or bases) from which the ions dissociate can be added to an electrolyte experimentally, the chemical potentials of each ion species cannot be individually measured. What can be measured is the chemical potential μ_S of each added salt (S), acid, or base which is related to the theoretical chemical potentials of the dissociated cations and ions as $\mu_S = \nu_+ \mu_C + \nu_- \mu_A$, which results in the measured activity coefficient of an added salt γ_S given as the average $(\nu_+ + \nu_-) \ln \gamma_S = \nu_+ \ln \gamma_C + \nu_- \ln \gamma_A$ or

$$\gamma_S = \left(\gamma_C^{\nu_+} \gamma_A^{\nu_-}\right)^{1/(\nu_+ + \nu_-)}. \tag{6.203}$$

When the theoretical estimates of γ_C and γ_A using the Debye and Hückel theory of Eq. (6.200) are used in this expression for γ_S and compared to experimental measurements of γ_S, an excellent fit is obtained except at the very highest levels of salt concentration, corresponding to where the approximation made by Debye and Hückel in their continuum modeling of the countercharge cloud that surrounds each ion breaks down.

For an electrolyte, the terms $\tilde{\mu}_i$ are interpreted to be the energy change when one molecule of species i (either a solvent molecule or an ion) is added to pure solvent at constant entropy. When adding ions ($i = 1, M$), the $\tilde{\mu}_i$ implicitly allow for ion–solvent interactions but neglect the energy stored in the ion–ion electrostatic interactions $k_B T \ln \gamma_i$ that the Debye–Hückel theory of Eq. (6.200) has explicitly modeled. Further, the effective volume v_i of a single ion in solution is different (smaller) than when the ion is in the salt crystal because of the way solvent molecules and counter ions pack in next to it in the liquid phase. We will not address the challenging topic of how the v_i depend on what is called the *ionic strength* $I = \left(\sum_{j=1}^{M} z_j^2 \bar{N}_j\right)/2$ of the electrolyte but there is such a dependence as has been measured experimentally.

To conclude, we have derived the form $\mu_i = \tilde{\mu}_i + k_B T \ln \gamma_i + v_i P + k_B T \ln \left(\tilde{N}_i / \sum_{j=0}^{M} \tilde{N}_j\right)$ for the chemical potential of molecular species i when there are $M+1$ species in an element. This expression applies to all fluids including liquid solutions for which the energetics of molecular interactions are embedded within the three terms $\tilde{\mu}_i$, γ_i, and v_i. The molecular interactions contained within $\tilde{\mu}_i$ do not depend on the number fractions in the element because these interactions are with the molecules in one of the two reference configurations, either (1) pure species-i molecules in the case where the molecules are not charged or (2) neutral solvent molecules in the case where the solutes are ions (an electrolyte). However, γ_i does depend on the number fractions as does the volume v_i of a single species-i molecule due to how strongly the other molecules interact with and pack in against an added species-i molecule. It is common in the literature either to say that the derived form for μ_i is phenomenological when applied to liquids (i.e., a nonderived form with coefficients adjusted to fit data) or to derive the form only in the special case of an ideal gas. By deriving this expression for μ_i in the internal-energy representation where each molecular state is equally probable with the combinatorial calculation of how the entropy changes due to the addition of a molecule being independent of the energy of each molecular state, we have shown that Eq. (6.199) has a broader foundation appropriate to liquid mixtures.

6.5.2 Multicomponent Fluids in the Gibbs Free-Energy Representation

Although the internal-energy representation is convenient for deriving the general functional nature of the chemical potential as given by Eq. (6.199), in experimental practice, it is difficult to control the entropy and volume of an element and more straightforward to control the temperature and pressure. As such, the Gibbs free-energy representation is generally the most useful representation for applications involving the chemical potential.

Taking a total derivative of the double Legendre transform definition $G = U - TS + PV$ of the Gibbs free energy and using the first law in the internal-energy representation gives

$$dG = -SdT + VdP + \sum_{i=0}^{M} \mu_i dN_i, \tag{6.204}$$

which shows that $G = G(T, P, N_i)$. The state functions in the multicomponent Gibbs representation are defined

$$-S(T, P, N_i) = \frac{\partial G}{\partial T}, \quad V(T, P, N_i) = \frac{\partial G}{\partial P} \quad \text{and} \quad \mu_i(T, P, N_i) = \frac{\partial G}{\partial N_i}, \tag{6.205}$$

while the various cross-derivative Maxwell relations are

$$-\frac{\partial S}{\partial P} = \frac{\partial V}{\partial T}, \quad \frac{\partial \mu_i}{\partial T} = -\frac{\partial S}{\partial N_i}, \quad \frac{\partial \mu_i}{\partial P} = \frac{\partial V}{\partial N_i}, \quad \text{and} \quad \frac{\partial \mu_i}{\partial N_j} = \frac{\partial \mu_j}{\partial N_i}. \tag{6.206}$$

Using the Euler relation for the internal energy $U = TS - PV + \sum_{j=0}^{M} \mu_j N_j$ in the Legendre transform definition of $G = U - TS + PV$, we obtain the Euler relation for the Gibbs free energy

$$G = \sum_{j=0}^{M} \mu_j N_j. \tag{6.207}$$

Taking a total derivative

$$dG = \sum_{j=0}^{M} \left(\mu_j dN_j + N_j d\mu_j \right) \tag{6.208}$$

and using Eq. (6.204) gives

$$0 = SdT - VdP + \sum_{j=0}^{M} N_j d\mu_j \tag{6.209}$$

which is the Gibbs–Duhem equilibrium relation for multicomponent elements that is independent of the particular energy representation we work in. In the present Gibbs-energy representation, the entropy is a function of T and P and each possible state does not have the same probability as it did in the internal-energy representation, so that $(\partial S/\partial N_i)_{T,P,N_j} \neq (\partial S/\partial N_i)_{V,N_j} = \partial S_i/\partial N_i + \partial S_{\text{mix}}/\partial N_i$ and we cannot use the result of Eq. (6.196) involving the entropy of mixing in the Maxwell relation of Eq. (6.206). However, we do have that $\partial V/\partial N_i = v_i(T, P, N_j)$ where v_i is again the average volume occupied by a single species-i molecule within the multicomponent fluid in the Gibbs representation.

The constitutive laws are the total derivatives of the state functions

$$\mathrm{d}S = \left(\frac{\partial S}{\partial T}\right)\mathrm{d}T + \left(\frac{\partial S}{\partial P}\right)\mathrm{d}P + \sum_{j=0}^{M}\left(\frac{\partial S}{\partial N_j}\right)\mathrm{d}N_j \tag{6.210}$$

$$\mathrm{d}V = \left(\frac{\partial V}{\partial T}\right)\mathrm{d}T + \left(\frac{\partial V}{\partial P}\right)\mathrm{d}P + \sum_{j=0}^{M}\left(\frac{\partial V}{\partial N_j}\right)\mathrm{d}N_j \tag{6.211}$$

$$\mathrm{d}\mu_i = \left(\frac{\partial \mu_i}{\partial T}\right)\mathrm{d}T + \left(\frac{\partial \mu_i}{\partial P}\right)\mathrm{d}P + \sum_{j=0}^{M}\left(\frac{\partial \mu_i}{\partial N_j}\right)\mathrm{d}N_j, \tag{6.212}$$

which after introducing the earlier definitions of the isobaric specific heat c_p, isobaric thermal expansion coefficient α_p, isothermal bulk modulus K_t, and molecular volume $v_j = \partial V/\partial N_j$ along with the Maxwell relations gives

$$\mathrm{d}S = \frac{c_p}{T}\mathrm{d}T - V\alpha_p\mathrm{d}P - \sum_{j=0}^{M}\left(\frac{\mu_j}{\partial T}\right)\mathrm{d}N_j \tag{6.213}$$

$$\mathrm{d}V = \alpha_p V\mathrm{d}T - \frac{V\mathrm{d}P}{K_t} + \sum_{j=0}^{M} v_j\mathrm{d}N_j \tag{6.214}$$

$$\mathrm{d}\mu_i = \left(\frac{\partial \mu_i}{\partial T}\right)\mathrm{d}T + v_i\mathrm{d}P + \sum_{j=0}^{M}\left(\frac{\partial \mu_i}{\partial N_j}\right)\mathrm{d}N_j. \tag{6.215}$$

Using the form $\mu_i = \tilde{\mu}_i + v_i P + k_B T \ln\gamma_i + k_B \ln(N_i/N_t)$ along with $\partial N_i/\partial N_j = \delta_{ij}$ (Kronecker delta), $\partial N_t/\partial N_j = 1$ and the approximation that the v_i are independent of N_i but the γ_i are not, we have that

$$\sum_{j=0}^{M}\left(\frac{\partial \mu_i}{\partial N_j}\right)\mathrm{d}N_j = k_B T\frac{\mathrm{d}N_i}{N_i} + k_B T\sum_{j=0}^{M}\left(\frac{1}{\gamma_i}\frac{\partial \gamma_i}{\partial N_j} - \frac{1}{N_t}\right)\mathrm{d}N_j. \tag{6.216}$$

In many solute transport scenarios involving dilute solutions, we have $\sum_{j=0}^{M}\mathrm{d}N_j = \mathrm{d}N_t \approx 0$ and $\gamma_i \approx 1$ so that

$$\sum_{j=0}^{N}\left(\frac{\partial \mu_i}{\partial N_j}\right)\mathrm{d}N_j \approx k_B T\frac{\mathrm{d}N_i}{N_i} \tag{6.217}$$

is often a valid approximation. We will return to the constitutive laws for a two-component solution of miscible liquids when we model solute diffusion in Chapter 7.

Last, the chemical potential in the Gibbs representation is often written in the phenomenological form

$$\mu_i(T, P, N_i) = \tilde{\mu}_i(T, P) + k_B T \ln\left(\frac{\gamma_i N_i}{N_t}\right). \tag{6.218}$$

Commonly, γ_i is taken to be a function of the N_i but not a function of pressure. In this case, one can then expand the reference configuration chemical potential $\tilde{\mu}_i(T, P)$ about a reference state pressure P_0 (say atmospheric pressure) to give

$$\tilde{\mu}_i(T, P) = \tilde{\mu}_i(T, P_0) + \frac{\partial \tilde{\mu}_i}{\partial P}(P - P_0) + \frac{1}{2!}\frac{\partial^2 \tilde{\mu}_i}{\partial P^2}(P - P_0)^2 + \ldots \tag{6.219}$$

where the partial derivatives with respect to pressure are evaluated at the reference pressure P_0. From the Maxwell relation of Eq. (6.206), we have

$$\frac{\partial \tilde{\mu}_i}{\partial P} = \frac{\partial V}{\partial N_i} = v_i^{(0)}, \tag{6.220}$$

where $v_i^{(0)}$ is the volume of a single species-i molecule at the reference pressure P_0. We also have

$$\frac{\partial^2 \tilde{\mu}_i}{\partial P^2} = \frac{\partial v_i^{(0)}}{\partial P} = -\frac{v_i^{(0)}}{K_i}, \tag{6.221}$$

where K_i is either the bulk modulus of pure phase i if the molecules are not charged or the bulk modulus of pure solvent in the case of an electrolyte.

So the chemical potential in the Gibbs representation is often taken to have the form

$$\mu_i(T, P, N_i) = \tilde{\mu}_i(T, P_0) + v_i^{(0)}(P - P_0)\left[1 - \frac{(P - P_0)}{2K_i} + \ldots\right] + k_B T \ln\left(\frac{\gamma_i N_i}{N_t}\right). \tag{6.222}$$

In liquids, the bulk modulus is measured in GPa and it is very common to work at pressures much smaller than GPa so that $P - P_0 \ll K_i$ giving the form

$$\boxed{\mu_i(T, P, N_i) = \tilde{\mu}_i(T, P_0) + v_i^{(0)}(P - P_0) + k_B T \ln\left(\frac{\gamma_i N_i}{N_t}\right).} \tag{6.223}$$

This form can be considered the same as Eq. (6.199) derived earlier from the internal-energy perspective, but this form requires us to assume that γ_i, if different from one, and $\tilde{\mu}_i$ are both defined at the reference-state pressure P_0. So the reference state in this form is defined not only based on the reference configuration that depends on whether the molecules are uncharged or charged but also at a given reference-state pressure.

In one of the many applications of this form, let's imagine two elements coming into contact. Each element has the same temperature and pressure but they have different number fractions N_i/N_t of solute i. The element that has the larger N_i/N_t will have an entropy-of-mixing contribution to its μ_i that is larger (less negative) than the element that has the smaller N_i/N_t. So from our earlier results on thermodynamic equilibrium, a net transport of species i molecule will occur from the element with the larger μ_i (and larger N_i/N_t) to the element with the smaller μ_i (and smaller N_i/N_t) until both elements have the same μ_i and are in equilibrium with each other.

This equilibration process can be understood from a less formal and more descriptive "diffusion" perspective. Focus on the interface separating the two elements and imagine counting the number of times per second that species-i solute molecules are crossing the interface from either side. Let's define element 1 to be the element with the larger N_i and therefore larger μ_i and element 2 to have less species-i molecules and a smaller μ_i. The random thermal molecular motions will result in more species-i molecules crossing from element 1 to element 2 compared to species-i molecules passing from element 2 into element 1. This net transport flux across the interface separating the elements will continue until the number of species-i molecules passing from element 1 to 2 is the same as those passing from element 2 to 1 at which time the two elements will be in equilibrium with each other with both the N_i/N_t and μ_i of each element the same. Solute diffusion always occurs in the direction from regions of larger chemical potential to regions of lower chemical potential.

6.5.3 Chemical Reactions

As another important application of using the chemical potentials of a multicomponent fluid in the Gibbs representation, we address the state of chemical equilibrium when species can react with each other within an element. Specifically, what is the equilibrium relationship between the number fractions of the various chemical species within a reactive multicomponent fluid element?

As an example, let's say that two molecular species represented by the symbols A and B and called the *reactants*, can react to form two other chemicals species C and D that are called the *products*. This reaction can be written as the balanced stoichiometric equation

$$aA + bB \leftrightarrows c\,C + dD, \tag{6.224}$$

where the numbers a, b, c, and d are the numbers of each chemical species involved in the reaction. The double arrows indicate that in equilibrium there are as many reactants going into products as there are products going back into reactants. For convenience, we move the reactants to the right-hand side and write the stoichiometric equation as

$$0 \leftrightarrows \sum_i \nu_i A_i, \tag{6.225}$$

where the dimensionless numbers ν_i are called the *stochiometric coefficients* in which the number of reactant molecules involved in a reaction is always negative and the number of product molecules formed in a reaction is always positive. If the reactants or products are ions having valences z_i, charge balance during the reaction requires that $\sum_i \nu_i z_i = 0$. In connecting to the example of Eq. (6.224), we thus have $\nu_1 = c$, $\nu_2 = d$, $\nu_3 = -a$, and $\nu_4 = -b$ as well as the molecular symbols $A_1 = C$, $A_2 = D$, $A_3 = A$, and $A_4 = B$.

Working in the Gibbs-energy representation

$$dG = -SdT + VdP + \sum_i \mu_i dN_i, \tag{6.226}$$

we can imagine to count the number of times N_R (a dimensionless number) that reactants associate to form products, where $N_R(t)$ is called the *extent of reaction*. If there are initially N_{i0} species-i molecules present in a reacting fluid element, then after N_R reaction occurrences, the number of molecules present are $N_i(t) = N_{i0} + \nu_i N_R(t)$ or $dN_i = \nu_i dN_R$. We thus have that for a reaction occurring at constant temperature and constant pressure

$$\frac{dG}{dN_R} = \sum_i \nu_i \mu_i, \tag{6.227}$$

where $\sum_i \nu_i \mu_i$ is called either the *chemical affinity* $\mathcal{A}$ or the *Gibbs free energy of reaction* ΔG_R, which are synonyms, that is, $\Delta G_R \mathrel{\widehat{=}} \mathcal{A} \mathrel{\widehat{=}} dG/dN_R = \sum_i \nu_i \mu_i$.

While the reaction is occurring and the system is attempting to achieve the equilibrium state of $dG/dN_R = 0$, we have that $dG < 0$ if $dN_R > 0$ because equilibrium is a free-energy minimum so that the free energy is going down while the reaction takes place. This means that $dG/dN_R = \sum_i \nu_i \mu_i < 0$ during equilibration, which says that reactants have a higher chemical potential on average compared to the products. So just like with the physical flux of matter, a reaction proceeds in the direction from higher chemical potential to lower chemical potential. This is a consequence of energy minimization in the state of equilibrium. The more negative that $\sum_i \nu_i \mu_i$ is, the larger the change in dG for each increment dN_R and the more rapidly the reaction is driven toward equilibrium. However, just as with the rate of first-order phase transitions, we will not attempt in this book to delve into the topic of the rate of reactions (so-called chemical kinetics) other than to say that the amplitude of the negative quantity $\sum_i \nu_i \mu_i$ is an important part of what sets the reaction rate.

When the reaction is in equilibrium and the balance between reactants associating into products equals the number of products disassociating into reactants, we have that $dG/dN_R = 0$ (free energy is a minimum in chemical equilibrium), which results in the chemical equilibrium condition

$$\boxed{\sum_{i=1}^{R} \nu_i \mu_i = 0,} \tag{6.228}$$

where we are assuming there are a total of R species of reactants and products that are in chemical equilibrium with each other. Inserting the earlier form $\mu_i = \tilde{\mu}_i + v_i P + k_B T \ln(\gamma_i N_i / N_t)$ into this condition for equilibrium gives

$$-\sum_{i=1}^{R} \nu_i \tilde{\mu}_i - \left(\sum_{i=1}^{R} \nu_i v_i \right) P = k_B T \sum_{i=1}^{R} \nu_i \ln\left(\frac{\gamma_i N_i}{N_t} \right). \tag{6.229}$$

We define the volume change of reaction Δv as

$$\Delta v = \sum_{i=1}^{R} \nu_i v_i \tag{6.230}$$

and the energy change of reaction in the reference configuration $\Delta\tilde{\mu}$ as

$$\Delta\tilde{\mu} = \sum_{i=1}^{R} \nu_i \tilde{\mu}_i. \tag{6.231}$$

Dividing both sides of Eq. (6.229) by $k_B T$ and taking the exponential gives the so-called *mass-balance* law for the reaction

$$\boxed{\prod_{i=1}^{R} \left(\frac{\gamma_i N_i}{N_t}\right)^{\nu_i} = K_{\text{eq}}(T, P),} \tag{6.232}$$

where the dimensionless *equilibrium constant* K_{eq} of the reaction is defined

$$\boxed{K_{\text{eq}}(T, P) = \exp\left[-\frac{(\Delta\tilde{\mu} + \Delta v P)}{k_B T}\right].} \tag{6.233}$$

The equilibrium constant is independent of the numbers of the various species in the element. The mass-balance law of Eq. (6.232) along with a known value for the equilibrium constant is the condition for chemical equilibrium.

So for the example reaction of Eq. (6.224), we have the mass-balance condition for equilibrium

$$\frac{(\gamma_C N_C/N_t)^c\,(\gamma_D N_D/N_t)^d}{(\gamma_A N_A/N_t)^a\,(\gamma_B N_B/N_t)^b} = K_{\text{eq}}. \tag{6.234}$$

Given nonequilibrium starting numbers N_{A0}, N_{B0}, N_{C0} and N_{D0}, the relation $N_i = N_{i0} + \nu_i N_R$ provides three more equations

$$\frac{N_A - N_{A0}}{\nu_A} = \frac{N_B - N_{B0}}{\nu_B} = \frac{N_C - N_{C0}}{\nu_C} = \frac{N_D - N_{D0}}{\nu_D}, \tag{6.235}$$

which when combined with the mass-balance law and $N_t = N_{nr} + N_A + N_B + N_C + N_D$, where N_{nr} are the number of nonreactive molecules in the batch reactor, allows the final equilibrium numbers of all species to be determined.

As a specific example, consider pure liquid water where water molecules can weakly dissociate into hydronium ion H_3O^+ and hydroxide ion OH^- as

$$2H_2O \leftrightarrows H_3O^+ + OH^-, \tag{6.236}$$

which can be written in the equivalently balanced form

$$H_2O \leftrightarrows H^+ + OH^-, \tag{6.237}$$

where H^+ represents a single proton. The dimensionless equilibrium constant $K_{eq} = K_w$ for the ionization of pure water at $T = 298$ K and atmospheric pressure $P = 10^5$ Pa is measured to be $K_w = 3.28 \times 10^{-18}$. As will be shown, this corresponds to a very dilute solution of protons and hydroxide ions in water so that the activity coefficient of each species, that represents interactions with the solute species (the H^+ and OH^{-1} ions), can be taken to be one; $\gamma_{H+} \to 1$, $\gamma_{OH-} \to 1$, and $\gamma_{H2O} \to 1$. The mass-balance law is then

$$\frac{(N_{H+}/N_t)(N_{OH-}/N_t)}{N_{H2O}/N_t} = K_w = 3.28 \times 10^{-18} \tag{6.238}$$

at ambient temperature and pressure. To an accuracy of nine decimal places, we can take $N_{H2O}/N_t = 1$. Further, we must also have that $N_{H+}/N_t = N_{H+}/N_{H2O} = N_{OH-}/N_{H2O}$ to give $(N_{H+}/N_{H2O})^2 = 3.28 \times 10^{-18}$ or

$$\frac{N_{H+}}{N_{H2O}} = \sqrt{K_w} = 1.81 \times 10^{-9}. \tag{6.239}$$

So the numbers of free protons in pure liquid water is measured in parts per billion, which is indeed dilute. Chemists often express concentrations of liquid solutions in "moles per liter" or *molarity M*. In a liter of water we have that $N_{H+}/N_{H2O} = M_{H+}/M_{H2O}$. The molarity of the water molecules in pure water at ambient conditions is determined using a measured mass density of 997 grams per liter for pure water and a molar mass (grams per mole) for water of 18.014 grams per mole as read from the periodic table, to give

$$M_{H2O} = \frac{997 \text{ grams / liter}}{18.104 \text{ grams / mole}} = 55.3\ \frac{\text{moles}}{\text{liter}}. \tag{6.240}$$

We then have $M_{H+} = \sqrt{K_w} M_{H2O} = 1.81 \times 10^{-9} M_{H2O}$ at $T = 298$ K and $P = 10^5$ Pa or

$$\boxed{M_{H+} = 1.00 \times 10^{-7}\ \frac{\text{moles}}{\text{liter}}.} \tag{6.241}$$

The pH scale is defined as $\text{pH} = -\log_{10} M_{H+}$ so that for the free protons in pure water at ambient conditions we have the well-known value pH = 7. As temperature goes up from 298 K at atmospheric pressure, the pH of pure water goes down (more protons are released) as controlled by the increasing K_w of Eq. (6.233) under the caveat that $\Delta\tilde{\mu}$ increases mildly with temperature and that the water density gets smaller in its contribution to M_{H2O}. At atmospheric pressure, the pressure term in Eq. (6.233) for the ionization of water is negligible. At $T = 323K$, pure water is measured to have pH = 6.63. On the logarithmic scale of pH, each decrease of $\log_{10} 2 = 0.30$ in pH corresponds to a factor of 2 increase in the free-proton concentration.

6.5.4 Osmosis, Reverse Osmosis, and Osmotic Pressure

For another example of using the chemical potential in multicomponent systems to address an important industrial (and biological) problem, we consider two systems that are separated by a membrane (cell membranes in biology) that is permeable to solvent molecules like water but that is not permeable to solute molecules like hydrated ions or other solutes that are larger in diameter compared to the solvent molecules.

Let each system consist of the same two molecular components (the solvent and solute) but with one system having a larger concentration of solute than the other. For the systems to be in equilibrium when they are separated by a membrane that is permeable only to solvent molecules, a larger pressure must be applied to the larger-concentration system compared to the lower-concentration system. The pressure difference between the systems at equilibrium is called the *osmotic pressure*. If each system has the same pressure, the composite system is out of equilibrium and solvent will spontaneously flow across the membrane from the more dilute system to the more concentrated system in an attempt to reduce the difference in the solvent's chemical potential in the two systems. This spontaneous flow of solvent from the less concentrated system (larger solvent chemical potential) to the more concentrated system (smaller solvent chemical potential) is called *osmosis*.

If a pressure is applied to the more-concentrated system that results in a pressure difference between the systems that is larger than the equilibrium osmotic pressure, solvent molecules will flow in the opposite direction, from the more-concentrated system to the more-dilute system, which further concentrates the solute in the concentrated system and further dilutes the solute in the dilute system. This is called *reverse osmosis* and is one of the most widely used industrial practices for separating solvent from solute, which can be used, for example, to purify ocean water into drinking water. In reverse-osmosis scenarios where the goal is to separate solvent from solute, the more dilute system is often taken to be pure solvent.

To find the equilibrium osmotic pressure, we equate the solvent's chemical potential in system 1 and system 2 that each have different concentrations of solute. Because solute cannot flow across the membrane to achieve equilibrium, we do not consider the chemical potential of the solute, which in general will be different for each system in equilibrium when the systems are maintained at different pressures. Equating the two solvent chemical potentials gives

$$\tilde{\mu}_0 + v_0 P_1 + k_B T \ln\left(\gamma^{(1)}(1-x^{(1)})\right) = \tilde{\mu}_0 + v_0 P_2 + k_B T \ln\left(\gamma^{(2)}(1-x^{(2)})\right), \qquad (6.242)$$

where v_0 is the volume of a single solvent molecule and the number fractions of solute are defined $x^{(i)} = N_1^{(i)}/(N_0^{(i)} + N_1^{(i)})$ with $i = 1, 2$ designating each system, $N_0^{(i)}$ the number of solvent molecules in system i and $N_1^{(i)}$ the number of solute molecules. We thus obtain the expression for the osmotic pressure to be

$$\boxed{P_1 - P_2 = \frac{k_B T}{v_0} \ln\left(\frac{\gamma^{(2)}(1-x^{(2)})}{\gamma^{(1)}(1-x^{(1)})}\right).} \qquad (6.243)$$

This is the equation that must be satisfied for the two systems separated by the solvent-permeable membrane to be in thermodynamic equilibrium with each other. If system 1 has the larger solute concentration, we have that $\gamma^{(2)}(1-x^{(2)}) > \gamma^{(1)}(1-x^{(1)})$ so that the osmotic pressure $P_1 - P_2$ is positive, that is, you must increase the pressure on the system that has the greater solute concentration in order for two systems separated by a solvent-permeable membrane to be in equilibrium when they have different solute concentrations $x^{(1)} \neq x^{(2)}$.

In the limit that system 2 is pure solvent, we have $\ln\left[\gamma^{(2)}(1-x^{(2)})\right]=0$ and

$$P_1 - P_2 = -\frac{k_B T}{v_0} \ln\left(\gamma^{(1)}(1-x^{(1)})\right). \tag{6.244}$$

Because the argument of the logarithm is less than 1, the osmotic pressure is again seen to be positive. If the solute concentration in system 1 is sufficiently dilute that $\ln\left[\gamma^{(1)}(1-x^{(1)})\right] \approx -x^{(1)}$, we obtain the dilute-concentration limit

$$P_1 - P_2 = \frac{k_B T x^{(1)}}{v_0}, \tag{6.245}$$

which is a result known as the *van 't Hoff* equation. So if $x^{(1)} = 10^{-3}$, $v_0 = 3 \times 10^{-29}$ m^3 (a liquid water molecule) and $k_B T = 4 \times 10^{-21}$ J (room temperature), we obtain $P_1 - P_2 = 1.33 \times 10^5$ Pa, which is 33% greater than atmospheric pressure.

Finally, if both system 1 (larger initial solute concentration) and system 2 (smaller initial solute concentration) are contained in rigid vessels of constant volume and if the two systems begin at the same pressure $P_1 = P_2$, which corresponds to a nonequilibrium initial condition, system 2 would spontaneously supply solvent molecules to system 1 until P_2 is dropped sufficiently and P_1 increased sufficiently that the equilibrium condition of Eq. (6.243) is satisfied.

6.6 Equilibrium Thermodynamics of Elastic Solids

6.6.1 General Considerations

We begin by reviewing some key results from Chapter 4 associated with elastic deformation in solids. From the perspective of following changes to an element of elastic solid that has constant mass $\mathcal{M}$ but variable volume $V(t)$, we showed earlier in Section 4.1 that the rate that strain energy is changing in the element is given by

$$V(t)\, \boldsymbol{\tau} : \frac{\mathrm{d}\boldsymbol{e}}{\mathrm{d}t} \mathrel{\widehat{=}} \text{the rate of total mechanical work performed in elastically deforming the mass element of volume } V(t), \tag{6.246}$$

where $\boldsymbol{\tau}$ is the stress tensor initially defined in Chapter 2 and $\boldsymbol{e}$ the strain tensor defined in Chapter 4.

As shown in Section 4.1, the strain tensor $\boldsymbol{e}$, that we called the "thermodynamic strain tensor," has an exact nonlinear relation to a distinctly different strain tensor $\boldsymbol{E} = \left[\nabla\boldsymbol{u} + (\nabla\boldsymbol{u})^T\right]/2$ where $\boldsymbol{u}$ is the center-of-mass displacement that is defined in terms of molecular trajectories as $\boldsymbol{u} = \mathcal{M}^{-1}\sum_{p=1}^{N} m_p \left[\boldsymbol{r}_p(t) - \boldsymbol{r}_p(0)\right]$ with N being the constant number of molecules in the element, m_p the mass of molecule p, and $\boldsymbol{r}_p(t)$ the trajectory of molecule p. The relationship between $\boldsymbol{e}$ and $\boldsymbol{E}$ is $\boldsymbol{e} = -\ln(\boldsymbol{I} - \boldsymbol{E})$ as understood through expansion to mean $\boldsymbol{e} = \boldsymbol{E} + \boldsymbol{E}\cdot\boldsymbol{E}/2 + \boldsymbol{E}\cdot\boldsymbol{E}\cdot\boldsymbol{E}/3 + \ldots$. Alternatively, in terms of the center-of-mass velocity of the element $\boldsymbol{v} = \mathrm{d}\boldsymbol{u}/\mathrm{d}t$, we also have the exact relation

$\mathrm{d}\boldsymbol{e}/\mathrm{d}t = \left[\nabla \boldsymbol{v} + (\nabla \boldsymbol{v})^T\right]/2$. Total time derivatives, as always, are taken from the perspective of an observer moving with the mass element. We further have $\mathrm{d}\boldsymbol{e}/\mathrm{d}t = (\boldsymbol{I} - \boldsymbol{E})^{-1} \cdot \mathrm{d}\boldsymbol{E}/\mathrm{d}t$. Only when the components of $\boldsymbol{E}$ are small so that $|\boldsymbol{E}| \ll 1$ does $\boldsymbol{e} = \boldsymbol{E}$.

If we now render the rate of changing strain energy to be per unit volume in the undeformed reference configuration at $t = 0$ in which the mass element has volume V_0, we must divide Eq. (6.246) by the constant V_0 to give

$$\frac{V}{V_0}\boldsymbol{\tau} : \frac{\mathrm{d}\boldsymbol{e}}{\mathrm{d}t} \mathrel{\widehat{=}} \text{the rate of total mechanical work performed in deforming elastically the mass element as normalized by } V_0. \tag{6.247}$$

In Section 4.1, we arrived at the above interpretations of $\boldsymbol{\tau} : \mathrm{d}\boldsymbol{e}/\mathrm{d}t$ as being the rate of strain energy change per unit evolving volume by considering the actual work performed by the traction vector in deforming a mass element's surface. We can consider the interpretations of Eqs (6.246) and (6.247) to be exact statements. This point is being emphasized again here because it is common in the standard reference texts that treat the thermodynamics of solids (e.g., Landau and Lifshitz, 1986 or Nye, 1957) to interpret without proof that $\boldsymbol{\tau} : \mathrm{d}\boldsymbol{e}/\mathrm{d}t$ is the total work rate divided by the constant V_0. This is an approximation that is only valid to the extent that the prefactor in Eq. (6.247) can be taken to be 1, that is, when $V/V_0 \approx 1$. We will return to this point after we have derived the thermodynamic constitutive laws using the exact interpretation of $\boldsymbol{\tau} : \mathrm{d}\boldsymbol{e}/\mathrm{d}t$ given above.

Given this preamble about the rate of mechanical work in a sample of elastic solid having constant mass, we now start over and consider how to allow for shape changes of solid mass elements in addition to the volumetric changes we have allowed for up to now for fluid elements that cannot sustain shape changes. The total energy U of the mass element is taken to be a function of the extensive entropy S, the extensive strain $\boldsymbol{\mathcal{E}}$, and the extensive number of molecules or atoms in the element N, that is, $U = U(S, \boldsymbol{\mathcal{E}}, N)$. We define the second-order extensive-strain tensor as

$$\boldsymbol{\mathcal{E}} = V_0 \boldsymbol{e}, \tag{6.248}$$

which has the property that if the initial size of the mass element is doubled, so is $\boldsymbol{\mathcal{E}}$. Taking a total time derivative of the fundamental function $U = U(S, \boldsymbol{\mathcal{E}}, N)$ gives

$$\frac{\mathrm{d}U}{\mathrm{d}t} = \left(\frac{\partial U}{\partial S}\right)_{\boldsymbol{\mathcal{E}},N} \frac{\mathrm{d}S}{\mathrm{d}t} + \left(\frac{\partial U}{\partial \boldsymbol{\mathcal{E}}}\right)_{S,N} : \frac{\mathrm{d}\boldsymbol{\mathcal{E}}}{\mathrm{d}t} + \left(\frac{\partial U}{\partial N}\right)_{S,\boldsymbol{\mathcal{E}}} \frac{\mathrm{d}N}{\mathrm{d}t}. \tag{6.249}$$

We identify the intensive partial derivatives as

$$T = \left(\frac{\partial U}{\partial S}\right)_{\boldsymbol{\mathcal{E}},N}, \quad \tilde{v}\boldsymbol{\tau} = \left(\frac{\partial U}{\partial \boldsymbol{\mathcal{E}}}\right)_{S,N} \quad \text{and} \quad \mu = \left(\frac{\partial U}{\partial N}\right)_{S,\boldsymbol{\mathcal{E}}}, \tag{6.250}$$

where as usual T is temperature and μ is chemical potential and where

$$\tilde{v} = \frac{V}{V_0} = \frac{\rho_0}{\rho}, \tag{6.251}$$

where $\rho = \mathcal{M}/V$ and $\rho_0 = \mathcal{M}/V_0$. The extensive form of the first law of thermodynamics for a solid is then

$$\frac{\mathrm{d}U}{\mathrm{d}t} = T\frac{\mathrm{d}S}{\mathrm{d}t} + V\boldsymbol{\tau} : \frac{\mathrm{d}\boldsymbol{e}}{\mathrm{d}t} + \mu\frac{\mathrm{d}N}{\mathrm{d}t}. \tag{6.252}$$

If there are multiple atomic or molecular species present in the solid, this generalizes to

$$\frac{\mathrm{d}U}{\mathrm{d}t} = T\frac{\mathrm{d}S}{\mathrm{d}t} + V\boldsymbol{\tau} : \frac{\mathrm{d}\boldsymbol{e}}{\mathrm{d}t} + \sum_i \mu_i\frac{\mathrm{d}N_i}{\mathrm{d}t}. \tag{6.253}$$

The first term on the right-hand side accounts for heat energy changes in an element of constant shape, constant volume and constant mass, the second term is the mechanical work performed in deforming a solid element of constant mass and when heat cannot enter or leave the element and the last term is the chemical work required to insert or remove molecules as performed against intermolecular forces for an element of constant shape and volume and when heat cannot enter or leave the element.

Next, we define densities by normalizing the extensive variables by the constant reference volume V_0 of the underformed state

$$\tilde{u} = \frac{U}{V_0}, \quad \tilde{s} = \frac{S}{V_0}, \quad \boldsymbol{e} = \frac{\boldsymbol{\mathcal{E}}}{V_0}, \quad \text{and} \quad \tilde{n} = \frac{N}{V_0}. \tag{6.254}$$

Dividing Eq. (6.252) by V_0 gives

$$\frac{\mathrm{d}\tilde{u}}{\mathrm{d}t} = T\frac{\mathrm{d}\tilde{s}}{\mathrm{d}t} + \tilde{v}\boldsymbol{\tau} : \frac{\mathrm{d}\boldsymbol{e}}{\mathrm{d}t} + \mu\frac{\mathrm{d}\tilde{n}}{\mathrm{d}t}. \tag{6.255}$$

If we now write the stress tensor in terms of isotropic and deviatoric portions $\boldsymbol{\tau} = -P\boldsymbol{I} + \boldsymbol{\tau}^D$ and similarly for the strain-rate tensor $\mathrm{d}\boldsymbol{e}/\mathrm{d}t = (\nabla \cdot \boldsymbol{v}/3)\boldsymbol{I} + \mathrm{d}\boldsymbol{e}^D/\mathrm{d}t$, Eq. (6.255) can be rewritten as

$$\frac{\mathrm{d}\tilde{u}}{\mathrm{d}t} = T\frac{\mathrm{d}\tilde{s}}{\mathrm{d}t} - \tilde{v}P\nabla \cdot \boldsymbol{v} + \tilde{v}\boldsymbol{\tau}^D : \frac{\mathrm{d}\boldsymbol{e}^D}{\mathrm{d}t} + \mu\frac{\mathrm{d}\tilde{n}}{\mathrm{d}t}. \tag{6.256}$$

Now, from the conservation of mass applied to an element of constant mass, we have $\nabla \cdot \boldsymbol{v} = -(\mathrm{d}\rho/\mathrm{d}t)/\rho = (\mathrm{d}V/\mathrm{d}t)/V = (\mathrm{d}\tilde{v}/\mathrm{d}t)/\tilde{v}$ so that the first law for an elastically deforming solid becomes

$$\boxed{\frac{\mathrm{d}\tilde{u}}{\mathrm{d}t} = T\frac{\mathrm{d}\tilde{s}}{\mathrm{d}t} - P\frac{\mathrm{d}\tilde{v}}{\mathrm{d}t} + \tilde{v}\boldsymbol{\tau}^D : \frac{\mathrm{d}\boldsymbol{e}^D}{\mathrm{d}t} + \mu\frac{\mathrm{d}\tilde{n}}{\mathrm{d}t}.} \tag{6.257}$$

This says that the internal energy as normalized by the constant V_0 has the functional dependence

$$\tilde{u} = \tilde{u}(\tilde{s}, \tilde{v}, \boldsymbol{e}^D, \tilde{n}). \tag{6.258}$$

We see that in a deforming solid, there are now two dimensionless strain variables $\tilde{v} = V/V_0$, which accounts for volumetric changes of the mass element, and $\boldsymbol{e}^D$, which accounts for shape changes with both such changes occurring for an element of constant mass. In a fluid, the deviatoric strain and deviatoric stress are zero. Last, the state functions in this representation are

$$T(\tilde{s}, \tilde{v}, \boldsymbol{e}^D, \tilde{n}) = \frac{\partial \tilde{u}}{\partial \tilde{s}} = \frac{\partial U}{\partial S} \tag{6.259}$$

$$-P(\tilde{s}, \tilde{v}, \boldsymbol{e}^D, \tilde{n}) == \frac{\partial \tilde{u}}{\partial \tilde{v}} = \frac{\partial U}{\partial V} \tag{6.260}$$

$$\tilde{v}\boldsymbol{\tau}^D(\tilde{s}, \tilde{v}, \boldsymbol{e}^D, \tilde{n}) = \frac{\partial \tilde{u}}{\partial \boldsymbol{e}^D} = \frac{\partial \tilde{u}}{\partial e^D_{ij}}\hat{\boldsymbol{x}}_i\hat{\boldsymbol{x}}_j = \frac{\partial U}{\partial (V_0\boldsymbol{e}^D)} \tag{6.261}$$

$$\mu(\tilde{s}, \tilde{v}, \boldsymbol{e}^D, \tilde{n}) = \frac{\partial \tilde{u}}{\partial \tilde{n}} = \frac{\partial U}{\partial N}. \tag{6.262}$$

We will return to these expressions momentarily when deriving the constitutive laws.

If we return to the extensive form of the first law when there are multiple constituents as given by Eq. (6.253) and insert the isotropic and deviatoric decompositions of the stress and strain tensors, we have

$$\frac{dU}{dt} = T\frac{dS}{dt} - P\frac{dV}{dt} + \tilde{v}\boldsymbol{\tau}^D : \frac{d(V_0\boldsymbol{e}^D)}{dt} + \sum_i \mu_i \frac{dN_i}{dt}. \tag{6.263}$$

This says that the total internal energy of our mass element has the functional dependence in extensive form of

$$U = U(S, V, V_0\boldsymbol{e}^D, N_1, N_2, \ldots). \tag{6.264}$$

We can now establish the Euler relation for a deforming solid by noting that the extensive internal energy satisfies the scaling $\beta U(S, V, V_0\boldsymbol{e}^D, N_1, N_2, \ldots) = U(\beta S, \beta V, \beta V_0\boldsymbol{e}^D, \beta N_1, \beta N_2, \ldots)$ for some arbitrary scale factor β of the initial size of the element. By taking a derivative with respect to β, using the above state-function identifications of the partial derivatives of U and then setting $\beta = 1$, Euler's equation for a solid is obtained as

$$\boxed{U = TS - PV + V\boldsymbol{\tau}^D : \boldsymbol{e}^D + \sum_i \mu_i N_i.} \tag{6.265}$$

In the way we are formulating the continuum rules in this section on the thermodynamics of solids, we normalize the Euler equation by the constant V_0 to obtain equivalently

$$\tilde{u} = T\tilde{s} - P\tilde{v} + \tilde{v}\boldsymbol{\tau}^D : \boldsymbol{e}^D + \sum_i \mu_i \tilde{n}_i. \tag{6.266}$$

Although a generally valid and important result for a deforming solid, we will not need this Euler relation in order to develop the constitutive laws, which is our primary goal here.

6.6.2 Constitutive Laws for an Elastically Deforming Solid

We now obtain the constitutive laws for the case when the solid element is composed of a single molecular constituent and when the mass of each solid element is constant through time. In this case, our fundamental function is $\tilde{u}(\tilde{s}, \tilde{v}, \boldsymbol{e}^D)$ and we need not consider derivatives with respect to $\tilde{n}$. To obtain the constitutive laws we simply take total time derivatives of each of the state functions given by Eqs (6.259)–(6.262). Let's perform the total derivative of the state function $\tilde{v}\boldsymbol{\tau}^D$ carefully. We have

$$\frac{d}{dt}\left(\tilde{v}\boldsymbol{\tau}^D\right) = \left(\frac{\partial(\tilde{v}\boldsymbol{\tau}^D)}{\partial\tilde{s}}\right)_{\tilde{v},\boldsymbol{e}^D}\frac{d\tilde{s}}{dt} + \left(\frac{\partial(\tilde{v}\boldsymbol{\tau}^D)}{\partial\tilde{v}}\right)_{\tilde{s},\boldsymbol{e}^D}\frac{d\tilde{v}}{dt} + \left(\frac{\partial(\tilde{v}\boldsymbol{\tau}^D)}{\partial\boldsymbol{e}^D}\right)_{\tilde{s},\tilde{v}} : \frac{d\boldsymbol{e}^D}{dt}. \tag{6.267}$$

Distribute the total derivative d/dt onto the two terms on the left-hand side. For those partial derivatives performed with $\tilde{v}$ held constant, the $\tilde{v}$ can be taken outside the partial derivative. For the partial derivative with respect to $\tilde{v}$, we have

$$\left(\frac{\partial(\tilde{v}\boldsymbol{\tau}^D)}{\partial\tilde{v}}\right)_{\tilde{s},\boldsymbol{e}^D} = \boldsymbol{\tau}^D + \tilde{v}\left(\frac{\partial\boldsymbol{\tau}^D}{\partial\tilde{v}}\right)_{\tilde{s},\boldsymbol{e}^D}. \tag{6.268}$$

We thus obtain that

$$\frac{d\boldsymbol{\tau}^D}{dt} = \left(\frac{\partial\boldsymbol{\tau}^D}{\partial\tilde{s}}\right)_{\tilde{v},\boldsymbol{e}^D}\frac{d\tilde{s}}{dt} + \left(\frac{\partial\boldsymbol{\tau}^D}{\partial\tilde{v}}\right)_{\tilde{s},\boldsymbol{e}^D}\frac{d\tilde{v}}{dt} + \left(\frac{\partial\boldsymbol{\tau}^D}{\partial\boldsymbol{e}^D}\right)_{\tilde{s},\tilde{v}} : \frac{d\boldsymbol{e}^D}{dt} \tag{6.269}$$

is the constitutive law controlling changes of the deviatoric-stress state function $\boldsymbol{\tau}^D = \boldsymbol{\tau}^D(\tilde{s}, \tilde{v}, \boldsymbol{e}^D)$.

The entire coupled suite of constitutive laws is obtained by taking total time derivatives of all the state functions to give

$$\frac{dT}{dt} = \left(\frac{\partial T}{\partial\tilde{s}}\right)_{\tilde{v},\boldsymbol{e}^D}\frac{d\tilde{s}}{dt} + \left(\frac{\partial T}{\partial\tilde{v}}\right)_{\tilde{s},\boldsymbol{e}^D}\frac{d\tilde{v}}{dt} + \left(\frac{\partial T}{\partial\boldsymbol{e}^D}\right)_{\tilde{s},\tilde{v}} : \frac{d\boldsymbol{e}^D}{dt} \tag{6.270}$$

$$\frac{dP}{dt} = \left(\frac{\partial P}{\partial\tilde{s}}\right)_{\tilde{v},\boldsymbol{e}^D}\frac{d\tilde{s}}{dt} + \left(\frac{\partial P}{\partial\tilde{v}}\right)_{\tilde{s},\boldsymbol{e}^D}\frac{d\tilde{v}}{dt} + \left(\frac{\partial P}{\partial\boldsymbol{e}^D}\right)_{\tilde{s},\tilde{v}} : \frac{d\boldsymbol{e}^D}{dt} \tag{6.271}$$

$$\frac{d\boldsymbol{\tau}^D}{dt} = \left(\frac{\partial\boldsymbol{\tau}^D}{\partial\tilde{s}}\right)_{\tilde{v},\boldsymbol{e}^D}\frac{d\tilde{s}}{dt} + \left(\frac{\partial\boldsymbol{\tau}^D}{\partial\tilde{v}}\right)_{\tilde{s},\boldsymbol{e}^D}\frac{d\tilde{v}}{dt} + \left(\frac{\partial\boldsymbol{\tau}^D}{\partial\boldsymbol{e}^D}\right)_{\tilde{s},\tilde{v}} : \frac{d\boldsymbol{e}^D}{dt}. \tag{6.272}$$

We have the following three Maxwell relations

$$-\frac{\partial P}{\partial\tilde{s}} = \frac{\partial^2\tilde{u}}{\partial\tilde{s}\partial\tilde{v}} = \frac{\partial^2\tilde{u}}{\partial\tilde{v}\partial\tilde{s}} = \frac{\partial T}{\partial\tilde{v}} \tag{6.273}$$

$$\tilde{v}\frac{\partial\boldsymbol{\tau}^D}{\partial\tilde{s}} = \frac{\partial^2\tilde{u}}{\partial\tilde{s}\partial\boldsymbol{e}^D} = \frac{\partial^2\tilde{u}}{\partial\boldsymbol{e}^D\partial\tilde{s}} = \frac{\partial T}{\partial\boldsymbol{e}^D} \tag{6.274}$$

$$\boldsymbol{\tau}^D + \tilde{v}\frac{\partial\boldsymbol{\tau}^D}{\partial\tilde{v}} = \frac{\partial^2\tilde{u}}{\partial\tilde{v}\partial\boldsymbol{e}^D} = \frac{\partial^2\tilde{u}}{\partial\boldsymbol{e}^D\partial\tilde{v}} = -\frac{\partial P}{\partial\boldsymbol{e}^D}. \tag{6.275}$$

We next define the scalar material properties as

$$T\left(\frac{\partial \tilde{s}}{\partial T}\right)_{\tilde{v},\boldsymbol{e}^D} = \tilde{c}_v \mathrel{\widehat{=}} \text{isochoric heat capacity (J m}^{-3}\,\text{K}^{-1}) \tag{6.276}$$

$$\frac{1}{\tilde{v}}\left(\frac{\partial \tilde{v}}{\partial T}\right)_{\tilde{s},\boldsymbol{e}^D} = \alpha_s \mathrel{\widehat{=}} \text{adiabatic thermal expansion (K}^{-1}) \tag{6.277}$$

$$-\tilde{v}\left(\frac{\partial P}{\partial \tilde{v}}\right)_{\tilde{s},\boldsymbol{e}^D} = K_s \mathrel{\widehat{=}} \text{adiabatic bulk modulus (Pa)}, \tag{6.278}$$

the second-order-tensor material properties as

$$\left(\frac{\partial T}{\partial \boldsymbol{e}^D}\right)_{\tilde{s},\tilde{v}} = \boldsymbol{A}_s = \boldsymbol{A}_s^T \mathrel{\widehat{=}} \text{adiabatic thermal-deviation tensor (K)} \tag{6.279}$$

$$-\left(\frac{\partial P}{\partial \boldsymbol{e}^D}\right)_{\tilde{s},\tilde{v}} = \boldsymbol{F}_s = \boldsymbol{F}_s^T \mathrel{\widehat{=}} \text{adiabatic pressure-deviation tensor (Pa)}, \tag{6.280}$$

and the fourth-order-tensor material property as

$$\left(\frac{\partial \boldsymbol{\tau}^D}{\partial \boldsymbol{e}^D}\right)_{\tilde{s},\tilde{v}} = \frac{1}{\tilde{v}}\frac{\partial^2 \tilde{u}}{\partial \boldsymbol{e}^D \partial \boldsymbol{e}^D} = {}_4\boldsymbol{G}_s = {}_4\boldsymbol{G}_s^{\overset{T}{3412}} = {}_4\boldsymbol{G}_s^{\overset{T}{2134}} = {}_4\boldsymbol{G}_s^{\overset{T}{1243}} \mathrel{\widehat{=}} \text{adiabatic shear moduli (Pa)}. \tag{6.281}$$

The various transpose symmetries in these material-property tensors are due to the strain tensor being symmetric. Using these material property definitions along with the Maxwell relations, the constitutive laws for an arbitrarily anisotropic solid can then be expressed

$$\frac{dT}{dt} = \frac{T}{\tilde{c}_v}\frac{d\tilde{s}}{dt} + \frac{1}{\alpha_s \tilde{v}}\frac{d\tilde{v}}{dt} + \boldsymbol{A}_s : \frac{d\boldsymbol{e}^D}{dt} \tag{6.282}$$

$$-\frac{dP}{dt} = \frac{1}{\alpha_s \tilde{v}}\frac{d\tilde{s}}{dt} + \frac{K_s}{\tilde{v}}\frac{d\tilde{v}}{dt} + \boldsymbol{F}_s : \frac{d\boldsymbol{e}^D}{dt} \tag{6.283}$$

$$\frac{d\boldsymbol{\tau}^D}{dt} + \frac{\boldsymbol{\tau}^D}{\tilde{v}}\frac{d\tilde{v}}{dt} = \frac{\boldsymbol{A}_s}{\tilde{v}}\frac{ds}{dt} + \frac{\boldsymbol{F}_s}{\tilde{v}}\frac{d\tilde{v}}{dt} + {}_4\boldsymbol{G}_s : \frac{d\boldsymbol{e}^D}{dt}. \tag{6.284}$$

If the solid is approximated as being isotropic, then Curie's principle as proven in Section 1.8.5 requires that $\boldsymbol{A}_s = 0$, $\boldsymbol{F}_s = 0$, and ${}_4\boldsymbol{G}_s = G_s(\delta_{ik}\delta_{jl} + \delta_{il}\delta_{jk} - 2\delta_{ij}\delta_{kl}/3)\hat{\boldsymbol{x}}_i\hat{\boldsymbol{x}}_j\hat{\boldsymbol{x}}_k\hat{\boldsymbol{x}}_l$, which results in

$$\frac{dT}{dt} = \frac{T}{\tilde{c}_v}\frac{d\tilde{s}}{dt} + \frac{1}{\alpha_s \tilde{v}}\frac{d\tilde{v}}{dt} \tag{6.285}$$

$$-\frac{dP}{dt} = \frac{1}{\alpha_s \tilde{v}}\frac{d\tilde{s}}{dt} + \frac{K_s}{\tilde{v}}\frac{d\tilde{v}}{dt} \tag{6.286}$$

$$\frac{d\boldsymbol{\tau}^D}{dt} + \frac{\boldsymbol{\tau}^D}{\tilde{v}}\frac{d\tilde{v}}{dt} = 2G_s\frac{d\boldsymbol{e}^D}{dt}, \tag{6.287}$$

where G_s is the adiabatic shear modulus (a scalar) of the material. For a fluid, in which the deviatoric stress and strain is zero, Eqs (6.285) and (6.286) are exactly the constitutive laws as given earlier in the chapter.

The term $\tilde{v}^{-1}\boldsymbol{\tau}^D \mathrm{d}\tilde{v}/\mathrm{d}t$ on the left-hand side of the differential equation for the deviatoric stress variations is nonstandard in most treatments of the thermodynamics of deforming solids but is a requirement that derives from our distinguishing between the evolving volume $V(t)$ of an element of constant mass and the constant reference volume V_0 of the mass element that held prior to the start of deformation. The amplitude of the term $\tilde{v}^{-1}\boldsymbol{\tau}^D \mathrm{d}\tilde{v}/\mathrm{d}t$ is given by

$$\frac{|\boldsymbol{\tau}^D \, \tilde{v}^{-1}\mathrm{d}\tilde{v}/\mathrm{d}t|}{|\mathrm{d}\boldsymbol{\tau}^D/\mathrm{d}t|} \approx \frac{|\mathrm{d}V|}{V}. \tag{6.288}$$

In applications where the volumetric strain $|\mathrm{d}V|/V \approx |\nabla \cdot \boldsymbol{u}|$ is small, which are quite common, this additional term is negligible and there is no distinction in the results obtained using the exact interpretation of $\boldsymbol{\tau} : \mathrm{d}\boldsymbol{e}/\mathrm{d}t$ versus the more common but approximate interpretation.

Note that in the above boxed constitutive laws, we are following a mass element and monitoring the changes through time that happen to this element from a frame of reference that moves along with the element. If we want to express the constitutive laws in the stationary frame of reference from which we normally work, we must make the substitution

$$\frac{\mathrm{d}}{\mathrm{d}t} = \frac{\partial}{\partial t} + \boldsymbol{v} \cdot \nabla \tag{6.289}$$

throughout the constitutive laws. From the fact that $\tilde{v}^{-1}\mathrm{d}\tilde{v}/\mathrm{d}t = \nabla \cdot \boldsymbol{v}$ and the identity that $\nabla \cdot (\boldsymbol{v}\boldsymbol{\tau}^D) = (\nabla \cdot \boldsymbol{v})\boldsymbol{\tau}^D + \boldsymbol{v} \cdot \nabla \boldsymbol{\tau}^D$ along with the definition of the total derivative $\mathrm{d}\boldsymbol{\tau}^D/\mathrm{d}t = \partial \boldsymbol{\tau}^D/\partial t + \boldsymbol{v} \cdot \nabla \boldsymbol{\tau}^D$, we can write the left-hand sides of Eqs (6.284) or (6.287) in the stationary frame as

$$\frac{\mathrm{d}\boldsymbol{\tau}^D}{\mathrm{d}t} + \frac{\boldsymbol{\tau}^D}{\tilde{v}}\frac{\mathrm{d}\tilde{v}}{\mathrm{d}t} = \frac{\partial \boldsymbol{\tau}^D}{\partial t} + \nabla \cdot \left(\boldsymbol{v}\boldsymbol{\tau}^D\right). \tag{6.290}$$

So for a stationary frame of reference, the constitutive laws for an isotropic solid (for example) are written

$$\frac{\partial T}{\partial t} + \boldsymbol{v} \cdot \nabla T = \frac{T}{\tilde{c}_v}\left(\frac{\partial \tilde{s}}{\partial t} + \boldsymbol{v} \cdot \nabla \tilde{s}\right) + \frac{1}{\alpha_s}\nabla \cdot \boldsymbol{v} \tag{6.291}$$

$$-\left(\frac{\partial P}{\partial t} + \boldsymbol{v} \cdot \nabla P\right) = \frac{1}{\alpha_s \tilde{v}}\left(\frac{\partial \tilde{s}}{\partial t} + \boldsymbol{v} \cdot \nabla \tilde{s}\right) + K_s \nabla \cdot \boldsymbol{v} \tag{6.292}$$

$$\frac{\partial \boldsymbol{\tau}^D}{\partial t} + \nabla \cdot \left(\boldsymbol{v}\boldsymbol{\tau}^D\right) = G_s\left(\nabla \boldsymbol{v} + (\nabla \boldsymbol{v})^T - \frac{2}{3}\nabla \cdot \boldsymbol{v}\boldsymbol{I}\right). \tag{6.293}$$

For small enough $\boldsymbol{v}$, the nonlinear advective terms can be dropped here as well.

If we want to change independent variables and work in other thermodynamic representations we again use the Legendre transform. For example, to change from using volume and deviatoric strain to using pressure and deviatoric stress as the independent variable, we would use a Legendre transform and define a new fundamental function, called the enthalpy, as $\tilde{h} \mathrel{\hat{=}} \tilde{u} + P\tilde{v} - \tilde{v}\boldsymbol{\tau}^D : \boldsymbol{e}^D$. We will not further discuss the various Legendre transforms that are possible for a solid that is deforming.

6.6.3 The Chemical Potential and Solubility of a Solid Mineral

A solid mineral (e.g., calcite, quartz, thousands of others) is built from cations and anions. When the solid mineral is in contact with a liquid solvent like water, the mineral's ions can partially dissolve into the solvent according to the weak dissolution reaction $C_{\nu_+}A_{\nu_-}(s) \leftrightarrows \nu_+ C^{z+}(l) + \nu_- A^{z-}(l)$ where the mineral molecule in solid form is $C_{\nu_+}A_{\nu_-}(s)$ and its ions as solute in liquid solution are $C^{z+}(l)$ and $A^{z-}(l)$. Charge neutrality always requires $\nu_+ z_+ + \nu_- z_- = 0$ for the dissolving mineral molecule. In order to determine how much mineral solute the liquid solution holds in equilibrium, we need to know both the chemical potential of the mineral molecule in the solid μ_s and the chemical potential of the mineral molecule in liquid solution $\mu_l = \nu_+ \mu_C(l) + \nu_- \mu_A(l)$. Once $\mu_s = \mu_l$, the solid–liquid system is in equilibrium and the liquid solution is saturated in the mineral solute. The concentration of mineral solute at saturation is called the *solubility* and is determined by specifying an equilibrium constant for the dissolution reaction that is called, in this context, the *solubility product* as determined below.

To obtain an expression for the chemical potential of a single-molecule solid (so not a solid solution), we work in the internal-energy representation and use the definition

$$\mu_s \mathrel{\hat{=}} \left(\frac{\partial U}{\partial N}\right)_{S,V,V_0\boldsymbol{e}^D} \tag{6.294}$$

$$= \frac{dU}{dN} - \left(\frac{\partial U}{\partial S}\right)\frac{\partial S}{\partial N} - \left(\frac{\partial U}{\partial V}\right)\frac{\partial V}{\partial N} - \left(\frac{\partial U}{\partial (V_0\boldsymbol{e}^D)}\right) : \frac{\partial (V_0\boldsymbol{e}^D)}{\partial N} \tag{6.295}$$

$$= \frac{dU}{dN} - T\frac{\partial S}{\partial N} + P\frac{\partial V}{\partial N} - \frac{V\boldsymbol{\tau}^D}{V_0} : \frac{\partial (V_0\boldsymbol{e}^D)}{\partial N}. \tag{6.296}$$

There is no entropy of mixing in this single component system, and it is convention to write

$$\tilde{\mu}_s = \frac{dU}{dN} - T\frac{\partial S}{\partial N}, \tag{6.297}$$

where dU/dN represents the random kinetic energy and vibrational energy of a single molecule as well as the work performed against intermolecular forces in inserting the molecule and $-T\partial S/\partial N$ is the heat that must be subtracted so that the insertion takes place at constant entropy. The remaining two mechanical work terms of Eq. (6.296) correspond to the amount of energy that must be subtracted from $\tilde{\mu}_s$ in order for the insertion to take place at constant volume and deviatoric strain.

The volume of the solid element under consideration goes as $V = v_s N$ where v_s is the volume of a single mineral molecule in the solid so that $\partial V/\partial N = v_s$. But the question of whether the deviatoric strain changes when a molecule is added is more subtle. Note that

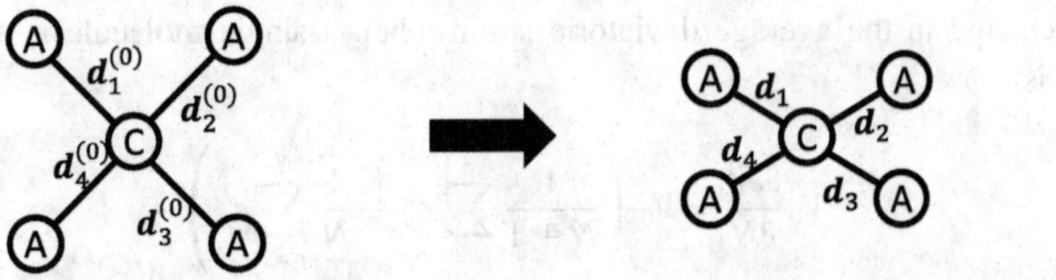

Figure 6.8 Schematic showing how molecular strain is defined in terms of changing bond lengths.

in the treatment that follows, no point defects in the mineral are being allowed for. Point defects (both additional atoms in the lattice called *interstitials* and missing atoms called *vacancies*) would be treated as additional species in a more complete analysis. Here we assume the entire mineral volume is filled uniformly with mineral molecules.

Consider first the strain $\boldsymbol{e}_p$ of a single mineral molecule p that is defined by the changes in the bond lengths as shown in Fig. 6.8. The bond-length vectors for each bond i of molecule p prior to the start of deformation are denoted $\boldsymbol{d}_{p,i}^{(0)}$. In Chapter 4, we show that any directed line segment, such as a bond vector within the mineral, changes with the changing strain tensor as

$$\boldsymbol{d}_{p,i} = \left[\exp\left(\boldsymbol{e}_p\right)\right] \cdot \boldsymbol{d}_{p,i}^{(0)} \tag{6.298}$$

$$= \left[\exp\left(\frac{\nabla \cdot \boldsymbol{u}_p}{3}\right) \exp\left(\boldsymbol{e}_p^D\right)\right] \cdot \boldsymbol{d}_{p,i}^{(0)}, \tag{6.299}$$

where the molecular strain tensor has been decomposed into isotropic (volume change) and deviatoric (shape change) portions

$$\boldsymbol{e}_p = (\nabla \cdot \boldsymbol{u}_p)\boldsymbol{I}/3 + \boldsymbol{e}_p^D \tag{6.300}$$

and where the notation exp $(\boldsymbol{e})$ involving a second-order tensor $\boldsymbol{e}$ means the operation

$$\exp\left(\boldsymbol{e}\right) = \boldsymbol{I} + \frac{\boldsymbol{e}}{1!} + \frac{\boldsymbol{e} \cdot \boldsymbol{e}}{2!} + \frac{\boldsymbol{e} \cdot \boldsymbol{e} \cdot \boldsymbol{e}}{3!} + \dots \tag{6.301}$$

So at vanishingly small strain $|\boldsymbol{e}_p| \to 0$ we have $\exp\left(\boldsymbol{e}_p\right) \to \boldsymbol{I}$ and there is no change to the bond vectors. It was also shown in Chapter 4 that the exact finite-strain relation between the dilatation and volume of a molecule (or any volumetric region) is given by $v_s = v_s^{(0)} \exp\left(\nabla \cdot \boldsymbol{u}_p\right)$ where $v_s^{(0)}$ is the molecular volume prior to the start of deformation.

We can define the average deviatoric strain tensor in the mineral element as the average over the molecular strain

$$\boldsymbol{e}^D = \frac{1}{N} \sum_{p=1}^{N} \boldsymbol{e}_p^D. \tag{6.302}$$

So the change in the average deviatoric strain when a single molecule is added to the mineral is

$$V_0 \frac{\partial \boldsymbol{e}^D}{\partial N} = V_0 \left(\frac{1}{N+1} \sum_{p=1}^{N+1} \boldsymbol{e}_p^D - \frac{1}{N} \sum_{p=1}^{N} \boldsymbol{e}_p^D \right). \tag{6.303}$$

Using the identity $1/(N+1) = 1/N - 1/[N(N+1)]$ we can then write

$$V_0 \frac{\partial \boldsymbol{e}^D}{\partial N} = \frac{V_0}{N} \left(\boldsymbol{e}_{N+1}^D - \frac{1}{N+1} \sum_{p=1}^{N+1} \boldsymbol{e}_p^D \right) \tag{6.304}$$

$$= v_s^{(0)} \left(\boldsymbol{e}_{N+1}^D - \boldsymbol{e}^D \right). \tag{6.305}$$

So only to the extent that a newly added molecule has a different molecular deviatoric strain $\boldsymbol{e}_{N+1}^D$ than the average molecular deviatoric strain throughout an element will deviatoric stress and strain make a nonzero contribution to the mineral's chemical potential.

We can equivalently formulate this derivative $V_0 \partial \boldsymbol{e}^D / \partial N$ as the average deviatoric strain change when a molecule is taken away. We will take the perspective that any one molecule within an element will have a deviatoric strain (or shape) that is the same as the average deviatoric strain (average shape) of the others, so that $\boldsymbol{e}_p^D = \boldsymbol{e}^D$ and $V_0 \partial \boldsymbol{e}^D / \partial N = 0$. Because the molecules being added or subtracted at the mineral surface might be in a different stress state due to the mineral-liquid boundary condition (see below) compared to molecules deeper inside the mineral, you could also argue that molecules at the surface will have an at least slightly different shape than molecules deeper inside. To model this, you need to know the stress and strain distribution inside the mineral. Again, we assume here that this effect is small and take $V_0 \partial \boldsymbol{e}^D / \partial N = 0$.

The chemical potential of a single-species solid mineral then takes the simple form

$$\boxed{\mu_s = \tilde{\mu}_s + v_s P} \tag{6.306}$$

regardless of how much deviatoric stress and strain are present in the element.

To determine the equilibrium constant for the dissolution reaction once the solid mineral is placed in liquid solvent, we also need the chemical potential of the mineral ions μ_C (the mineral cation) and μ_A (the mineral anion) in the liquid solution, which are given by Eq. (6.199) or Eq. (6.223) in the form

$$\mu_C = \tilde{\mu}_C + v_C P_l + k_B T \ln \left(\frac{\gamma_C N_C}{N_t} \right) \tag{6.307}$$

$$\mu_A = \tilde{\mu}_A + v_A P_l + k_B T \ln \left(\frac{\gamma_A N_A}{N_t} \right), \tag{6.308}$$

where P_l is the pressure in the liquid solution, the $\tilde{\mu}_{C,A}$ are the chemical potentials of the cations and ions in an infinitely dilute solution, and the $\gamma_{C,A}$ are the activity coefficients of the cations and anions in solution.

The stress state in the mineral is given by a stress tensor $\boldsymbol{\tau}$, whose components give the pressure in the solid as

$$P = -\frac{(\tau_{xx} + \tau_{yy} + \tau_{zz})}{3}. \tag{6.309}$$

If the solid and liquid are in contact at a planar interface having a normal in the x direction or $\boldsymbol{n} = \hat{\boldsymbol{x}}$, the stress boundary condition at this interface is that

$$\boldsymbol{n} \cdot \boldsymbol{\tau} = -P_l \boldsymbol{n} \tag{6.310}$$

or, equivalently, with $\boldsymbol{n} = \hat{\boldsymbol{x}}$

$$\tau_{xx} = -P_l \quad \text{and} \quad \tau_{xy} = \tau_{xz} = 0 \tag{6.311}$$

with all the other three stress components, τ_{yy}, τ_{zz}, and τ_{yz}, free to have whatever value they need to have in the solid element to be in mechanical equilibrium with the rest of the larger solid body. For example, there may be a distant boundary condition where stress is applied to the solid body that results in a nontrivial equilibrium stress distribution throughout the solid elements of this larger solid body. But for those elements right at the interface between solid and fluid having normal $\hat{\boldsymbol{x}}$, the solid stress components are given by Eq. (6.311) with τ_{yy}, τ_{zz}, and τ_{yz} not set by this boundary condition.

Chemical equilibrium between the solid and liquid occurs when $\mu_s = \nu_+ \mu_C + \nu_- \mu_A$ or

$$\tilde{\mu}_s - v_s \frac{(-P_l + \tau_{yy} + \tau_{zz})}{3} = \tilde{\mu}_l + v_l P_l + k_B T \ln\left[\left(\frac{\gamma_C N_C}{N_t}\right)^{\nu_+}\left(\frac{\gamma_A N_A}{N_t}\right)^{\nu_-}\right], \tag{6.312}$$

where $\tilde{\mu}_l = \nu_+ \tilde{\mu}_C + \nu_- \tilde{\mu}_A$ and $v_l = \nu_+ v_C + \nu_- v_A$. We thus define a dimensionless equilibrium constant, called the *solubility product* K_{sp}, that is independent of the mineral solute concentration, as

$$\boxed{K_{\text{sp}}(T, P_l, \tau_{yy}, \tau_{zz}) = \exp\left(\frac{\tilde{\mu}_s - \tilde{\mu}_l + (v_s/3 - v_l)P_l - v_s(\tau_{yy} + \tau_{zz})/3}{k_B T}\right),} \tag{6.313}$$

where, once again, we have defined the normal to the interface between the solid and liquid to be $\boldsymbol{n} = \hat{\boldsymbol{x}}$ so that τ_{yy} and τ_{zz} are the longitudinal stress components that are parallel with the interface and that are defined to be positive in tension and negative in compression. The condition $\mu_s = \mu_l$ of Eq. (6.312) then gives the mass-balance rule for the dissolution reaction at equilibrium

$$\boxed{\left(\frac{\gamma_C N_C}{N_t}\right)^{\nu_+}\left(\frac{\gamma_A N_A}{N_t}\right)^{\nu_-} = K_{\text{sp}}(T, P_l, \tau_{yy}, \tau_{zz}).} \tag{6.314}$$

To calculate the solubility as a number fraction, like N_C/N_t, we note that for each neutral mineral molecule that goes into solution there are ν_+ cations created and ν_- anions, which means that

$$\frac{N_C}{\nu_+} = \frac{N_A}{\nu_-}. \tag{6.315}$$

Thus, we define the solubility S for the dissolution as a number fraction given in terms of the solubility product K_{sp} and the activity coefficients $\gamma_{C,A}$ of the solute cations and anions as

$$S = \frac{N_C}{\nu_+ N_t} = \frac{N_A}{\nu_- N_t} = \left(\frac{K_{sp}}{(\nu_+ \gamma_C)^{\nu_+} (\nu_- \gamma_A)^{\nu_-}} \right)^{1/(\nu_+ + \nu_-)}. \tag{6.316}$$

In experimental practice, the calculation of a mineral-ion's concentration in an electrolyte solution in equilibrium with a solid mineral can become complicated due to electrolyte ions reacting with the dissolved mineral ions in solution. We will not delve into giving detailed examples of such solution chemistry.

Although we have defined the solubility as the dimensionless number fraction of the mineral ions in solution, most chemists define solubility as a molarity, that is, as the moles of mineral ions in a liter of solution. Converting from number fraction N_i/N_t for a species i to molarity M_i (moles per liter) is straightforward

$$M_i = \frac{\rho_t N_i/N_t}{\sum_i m_i (N_i/N_t)} \tag{6.317}$$

if given ρ_t the solution's mass density expressed as mass per liter and m_i the molar mass (mass per mole) of each constituent as reported in the periodic table. The mass-balance equation when expressed in terms of the molarity M_C of the mineral cation and the molarity M_A of the mineral anion is $(\gamma_C M_C)^{\nu_+} (\gamma_A M_A)^{\nu_-} = K^*_{sp}$, where the solubility product K^*_{sp} is what is most-commonly presented in tables for hundreds of minerals and has units of molarity raised to the $\nu_+ + \nu_-$ power, which will be different for molecules having different $\nu_+ + \nu_-$. The conversion from our dimensionless solubility product K_{sp} to the K^*_{sp} as reported in tables of experimental measurements is

$$K^*_{sp} = \left(\frac{\rho_t}{\sum_i m_i (N_i/N_t)} \right)^{(\nu_+ + \nu_-)} K_{sp}. \tag{6.318}$$

Our K_{sp} is inherently dimensionless because the entropy of mixing term in the solute-i chemical potential involves the logarithm of the dimensionless number fraction N_i/N_t and not the solute molarity.

We next consider a few special cases for determining the solubility product of Eq. (6.313).

1. *Negligible stress dependence*: If there is a sufficiently small liquid pressure and mineral stress such that $\left| v_s(-P_l + \tau_{yy} + \tau_{zz})/3 \right| \ll k_B T$ and $|v_l P_l| \ll k_B T$, the solubility product is approximated as $K_{sp}(T) = \exp\left[(\tilde{\mu}_s - \tilde{\mu}_l)/(k_B T) \right]$. For stresses on the order of atmospheric pressure 10^5 Pa and for an ionic molecular volume on the order 3×10^{-29} m^3, we have $v_s(-P_l + \tau_{yy} + \tau_{zz})/3 = 10^{-24}$ J which can be considered negligible relative to the thermal energy at room temperature $k_B T = 4 \times 10^{-21}$ J. Similar comments hold for $v_l P_l$ at atmospheric pressure. So for some mineral in contact with water at $T = 293$ K, if the solubility product is measured to be say $K_{sp}(293) = 10^{-10}$

(dimensionless), we would have $\tilde{\mu}_s - \tilde{\mu}_l = -9.3 \times 10^{-20}$ J. This being negative means that the attractive interaction energy between cations and anions in the solid $\tilde{\mu}_s$ is more negative (i.e., the work of insertion against an attractive force is negative) than the attractive interaction energy between a mineral ion and the pure liquid solvent molecules $\tilde{\mu}_l$ defined in the electrolyte's pure-solvent reference configuration.

2. *A suspension of isotropic solid grains*: If the solid phase is approximated as being isotropic and comes in the form of a suspension of grains in the liquid phase, such as sediments on the ocean floor, the stress tensor in the solid is then $\boldsymbol{\tau} = -P_l \boldsymbol{I}$ (i.e., $\tau_{xx} = \tau_{yy} = \tau_{zz} = -P_l$ everywhere in the solid grains) and the solubility product is

$$K_{\text{sp}}(T, P_l) = \exp\left(\frac{\tilde{\mu}_s - \tilde{\mu}_l + (v_s - v_l)P_l}{k_B T}\right). \tag{6.319}$$

Because the liquid solvent tends to pack in more tightly around mineral ions in the liquid than do the ions in solid, we have that $\Delta v = v_s - v_l > 0$ so that solubility is enhanced with increasing fluid pressure in the suspension of solid grains at constant temperature. If a suspension of solid grains has a solubility product of $K_{\text{sp}} = 10^{-10}$ at very low pressure and if, say, $v_s - v_l = 2 \times 10^{-29}$ m^3, then when the liquid pressure is increased to $P_l = 100$ MPa, the solubility product will be $K_{\text{sp}} = 1.7 \times 10^{-10}$ and there will be 170% more mineral solute in solution at equilibrium.

3. *Pressure solution*: Finally, if there is a significant component of compressive longitudinal stress parallel with the solid–liquid interface in the sense that $|v_s(\tau_{yy} + \tau_{zz})/3|$ is not negligible relative to $k_B T$, this compressive stress will always increase the solubility. In the Earth, this can lead to plastic deformation of rocks as mineral molecules in the rock grains dissolve from surfaces that have a large component of compressive stress parallel with the surface, diffuse away from that surface through the liquid-filled pore space and then precipitate on other rock grain surfaces that do not have such a large component of stress parallel with the surface. Such stress-induced mass redistribution inside of rocks is given the name *pressure solution* by Earth scientists. In practice, if two solid mineral surfaces having rough topography are pressed into each other, where two asperities are in contact, dissolution etch pits and/or microcracks will tend to form at the edge of the asperities in contact which grow into the minerals in an average direction that is perpendicular to the nominal plane of contact. Along such extending etch pit or microcrack surfaces, the large compressive stress parallel with those growing surfaces will create a stress-enhanced dissolution that eventually removes the two stress-bearing asperities in contact allowing other asperities to come into contact and begin to dissolve through a similar mechanism. This is a complicated process that has not yet been thoroughly modeled and understood.

6.7 Thermodynamics Allowing for Electromagnetic Polarization

We now augment the thermodynamics of solids just derived to include the energy changes associated with both electrical and magnetic polarization processes in the presence of macroscopic electric and magnetic fields. Our primary purpose here is to obtain the thermodynamic constitutive laws when electric and magnetic polarization is allowed for in addition to elastic deformation and heat changes.

Similar to the mechanical rate of work $\boldsymbol{\tau} : \mathrm{d}\boldsymbol{e}/\mathrm{d}t$ being a work rate normalized by the evolving volume V of a mass element, so the electromagnetic-polarization work rate $\boldsymbol{E}' \cdot \mathrm{d}\boldsymbol{D}'/\mathrm{d}t + \boldsymbol{H}' \cdot \mathrm{d}\boldsymbol{B}'/\mathrm{d}t$ obtained in Section 3.3.2 is also a density, in which the total electromagnetic work rate involved in polarizing a mass element is normalized by the evolving volume of the element. Note that in Section 3.3, we were assuming that volume would not change during the electromagnetic polarization. However, in this chapter, we are allowing for the more general case in which volume can change during polarization. Note that primes on the various electric and magnetic fields mean they are being defined in a frame of reference that is moving with the center-of-mass velocity $\boldsymbol{v}$ of the mass element. In Section 3.3.1, we identified these primed EM fields in the moving frame of reference in terms of the EM fields defined from a stationary frame of reference in which we are watching the mass element move with velocity $\boldsymbol{v}$.

Employing the formulation of Section 6.6, in which total energies and entropies of a single molecular component system are normalized by the constant reference volume V_0 of the mass element that holds prior to the start of deformation, the first law as augmented by the electromagnetic work of polarization is given by

$$\frac{\mathrm{d}\tilde{u}}{\mathrm{d}t} = T\frac{\mathrm{d}\tilde{s}}{\mathrm{d}t} + \tilde{v}\boldsymbol{\tau} : \frac{\mathrm{d}\boldsymbol{e}}{\mathrm{d}t} + \tilde{v}\left(\boldsymbol{E}' \cdot \frac{\mathrm{d}\boldsymbol{D}'}{\mathrm{d}t} + \boldsymbol{H}' \cdot \frac{\mathrm{d}\boldsymbol{B}'}{\mathrm{d}t}\right). \tag{6.320}$$

Inserting the isotropic and deviatoric decompositions of $d\boldsymbol{e}/\mathrm{d}t = (\nabla \cdot \boldsymbol{v}/3)\boldsymbol{I} + \boldsymbol{e}^D = \left[\nabla\boldsymbol{v} + (\nabla\boldsymbol{v})^T\right]/2$ and $\boldsymbol{\tau} = -P\boldsymbol{I} + \boldsymbol{\tau}^D$, the internal energy density has the functional dependence

$$\tilde{u} = \tilde{u}(\tilde{s}, \tilde{v}, \boldsymbol{e}^D, \boldsymbol{D}', \boldsymbol{B}') \tag{6.321}$$

and the five state functions in this representation are given by the partial derivatives

$$T = \frac{\partial \tilde{u}}{\partial \tilde{s}}, \quad -P = \frac{\partial \tilde{u}}{\partial \tilde{v}}, \quad \tilde{v}\boldsymbol{\tau}^D = \frac{\partial \tilde{u}}{\partial \boldsymbol{e}^D}, \quad \tilde{v}\boldsymbol{E}' = \frac{\partial \tilde{u}}{\partial \boldsymbol{D}'} \quad \text{and} \quad \tilde{v}\boldsymbol{H}' = \frac{\partial \tilde{u}}{\partial \boldsymbol{B}'}. \tag{6.322}$$

The constitutive laws are obtained as usual by taking total time derivatives of each of these state functions that we present, for convenience, in array format

$$\begin{bmatrix} \frac{\mathrm{d}T}{\mathrm{d}t} \\ \frac{\mathrm{d}P}{\mathrm{d}t} \\ \frac{\mathrm{d}\boldsymbol{\tau}^D}{\mathrm{d}t} \\ \frac{\mathrm{d}\boldsymbol{E}'}{\mathrm{d}t} \\ \frac{\mathrm{d}\boldsymbol{H}'}{\mathrm{d}t} \end{bmatrix} = \begin{bmatrix} \frac{\partial T}{\partial \tilde{s}} & \frac{\partial T}{\partial \tilde{v}} & \frac{\partial T}{\partial \boldsymbol{e}^D} : & \frac{\partial T}{\partial \boldsymbol{D}'} \cdot & \frac{\partial T}{\partial \boldsymbol{B}'} \cdot \\ \frac{\partial P}{\partial \tilde{s}} & \frac{\partial P}{\partial \tilde{v}} & \frac{\partial P}{\partial \boldsymbol{e}^D} : & \frac{\partial P}{\partial \boldsymbol{D}'} \cdot & \frac{\partial P}{\partial \boldsymbol{B}'} \cdot \\ \frac{\partial \boldsymbol{\tau}^D}{\partial \tilde{s}} & \frac{\partial \boldsymbol{\tau}^D}{\partial \tilde{v}} & \frac{\partial \boldsymbol{\tau}^D}{\partial \boldsymbol{e}^D} : & \frac{\partial \boldsymbol{\tau}^D}{\partial \boldsymbol{D}'} \cdot & \frac{\partial \boldsymbol{\tau}^D}{\partial \boldsymbol{B}'} \cdot \\ \frac{\partial \boldsymbol{E}'}{\partial \tilde{s}} & \frac{\partial \boldsymbol{E}'}{\partial \tilde{v}} & \frac{\partial \boldsymbol{E}'}{\partial \boldsymbol{e}^D} : & \frac{\partial \boldsymbol{E}'}{\partial \boldsymbol{D}'} \cdot & \frac{\partial \boldsymbol{E}'}{\partial \boldsymbol{B}'} \cdot \\ \frac{\partial \boldsymbol{H}'}{\partial \tilde{s}} & \frac{\partial \boldsymbol{H}'}{\partial \tilde{v}} & \frac{\partial \boldsymbol{H}'}{\partial \boldsymbol{e}^D} : & \frac{\partial \boldsymbol{H}'}{\partial \boldsymbol{D}'} \cdot & \frac{\partial \boldsymbol{H}'}{\partial \boldsymbol{B}'} \cdot \end{bmatrix} \begin{bmatrix} \frac{\mathrm{d}\tilde{s}}{\mathrm{d}t} \\ \frac{\mathrm{d}\tilde{v}}{\mathrm{d}t} \\ \frac{\mathrm{d}\boldsymbol{e}^D}{\mathrm{d}t} \\ \frac{\mathrm{d}\boldsymbol{D}'}{\mathrm{d}t} \\ \frac{\mathrm{d}\boldsymbol{B}'}{\mathrm{d}t} \end{bmatrix}. \tag{6.323}$$

These constitutive laws are supplemented by the following 10 Maxwell relations and associated material property names. For those material properties involving derivatives of the temperature, we have

$$-\frac{\partial P}{\partial \tilde{s}} = \frac{\partial^2 \tilde{u}}{\partial \tilde{s}\partial \tilde{v}} = \frac{\partial^2 \tilde{u}}{\partial \tilde{v}\partial \tilde{s}} = \frac{\partial T}{\partial \tilde{v}} \widehat{=} \frac{1}{\tilde{v}\alpha} \quad \text{inverse thermal-expansion coefficient} \tag{6.324}$$

$$\tilde{v}\frac{\partial \boldsymbol{\tau}^D}{\partial \tilde{s}} = \frac{\partial^2 \tilde{u}}{\partial \tilde{s}\partial \boldsymbol{e}^D} = \frac{\partial^2 \tilde{u}}{\partial \boldsymbol{e}^D\partial \tilde{s}} = \frac{\partial T}{\partial \boldsymbol{e}^D} \widehat{=} \boldsymbol{A} = \boldsymbol{A}^T \quad \text{thermal-deviation tensor} \tag{6.325}$$

$$\tilde{v}\frac{\partial \boldsymbol{E}'}{\partial \tilde{s}} = \frac{\partial^2 \tilde{u}}{\partial \tilde{s}\partial \boldsymbol{D}'} = \frac{\partial^2 \tilde{u}}{\partial \boldsymbol{D}'\partial \tilde{s}} = \frac{\partial T}{\partial \boldsymbol{D}'} \widehat{=} \boldsymbol{p} \quad \text{thermal-electric vector} \tag{6.326}$$

$$\tilde{v}\frac{\partial \boldsymbol{H}'}{\partial \tilde{s}} = \frac{\partial^2 \tilde{u}}{\partial \tilde{s}\partial \boldsymbol{B}'} = \frac{\partial^2 \tilde{u}}{\partial \boldsymbol{B}'\partial \tilde{s}} = \frac{\partial T}{\partial \boldsymbol{B}'} \widehat{=} \boldsymbol{q} \quad \text{thermal-magnetic vector,} \tag{6.327}$$

while for those involving derivatives of the pressure

$$\boldsymbol{\tau}^D + \tilde{v}\frac{\partial \boldsymbol{\tau}^D}{\partial \tilde{v}} = \frac{\partial^2 \tilde{u}}{\partial \tilde{v}\partial \boldsymbol{e}^D} = \frac{\partial^2 \tilde{u}}{\partial \boldsymbol{e}^D\partial \tilde{v}} = -\frac{\partial P}{\partial \boldsymbol{e}^D} \widehat{=} \boldsymbol{F} = \boldsymbol{F}^T \quad \text{pressure-deviation tensor} \tag{6.328}$$

$$\boldsymbol{E}' + \tilde{v}\frac{\partial \boldsymbol{E}'}{\partial \tilde{v}} = \frac{\partial^2 \tilde{u}}{\partial \tilde{v}\partial \boldsymbol{D}'} = \frac{\partial^2 \tilde{u}}{\partial \boldsymbol{D}'\partial \tilde{v}} = -\frac{\partial P}{\partial \boldsymbol{D}'} \widehat{=} \boldsymbol{s} \quad \text{pressure-electric vector} \tag{6.329}$$

$$\boldsymbol{H}' + \tilde{v}\frac{\partial \boldsymbol{H}'}{\partial \tilde{v}} = \frac{\partial^2 \tilde{u}}{\partial \tilde{v}\partial \boldsymbol{B}'} = \frac{\partial^2 \tilde{u}}{\partial \boldsymbol{B}'\partial \tilde{v}} = -\frac{\partial P}{\partial \boldsymbol{B}'} \widehat{=} \boldsymbol{t} \quad \text{pressure-magnetic vector,} \tag{6.330}$$

while for those involving derivatives of the deviatoric-stress tensor

$$\frac{\partial \boldsymbol{E}'}{\partial \boldsymbol{e}^D} = \frac{1}{\tilde{v}}\frac{\partial^2 \tilde{u}}{\partial \boldsymbol{e}^D\partial \boldsymbol{D}'} = \frac{1}{\tilde{v}}\left(\frac{\partial^2 \tilde{u}}{\partial \boldsymbol{D}'\partial \boldsymbol{e}^D}\right)^{\overset{T}{231}}$$

$$= \left(\frac{\partial \boldsymbol{\tau}^D}{\partial \boldsymbol{D}'}\right)^{\overset{T}{231}} \widehat{=} {}_3\boldsymbol{U}^{\overset{T}{231}} \quad \text{piezo-electric third-order tensor} \tag{6.331}$$

$$\frac{\partial \boldsymbol{H}'}{\partial \boldsymbol{e}^D} = \frac{1}{\tilde{v}}\frac{\partial^2 \tilde{u}}{\partial \boldsymbol{e}^D\partial \boldsymbol{B}'} = \frac{1}{\tilde{v}}\left(\frac{\partial^2 \tilde{u}}{\partial \boldsymbol{B}'\partial \boldsymbol{e}^D}\right)^{\overset{T}{231}}$$

$$= \left(\frac{\partial \boldsymbol{\tau}^D}{\partial \boldsymbol{B}'}\right)^{\overset{T}{231}} \widehat{=} {}_3\boldsymbol{W}^{\overset{T}{231}} \quad \text{piezo-magnetic third-order tensor} \tag{6.332}$$

and finally for those involving magnetoelectric coupling

$$\frac{\partial \boldsymbol{H}'}{\partial \boldsymbol{D}'} = \frac{1}{\tilde{v}}\frac{\partial^2 \tilde{u}}{\partial \boldsymbol{D}'\partial \boldsymbol{B}'} = \frac{1}{\tilde{v}}\left(\frac{\partial^2 \tilde{u}}{\partial \boldsymbol{B}'\partial \boldsymbol{D}'}\right)^T = \left(\frac{\partial \boldsymbol{E}'}{\partial \boldsymbol{B}'}\right)^T \widehat{=} \boldsymbol{C}_{EB}^T \quad \text{magnetoelectric tensor.} \tag{6.333}$$

We see that piezoelectric response, which is the generation of an electric-field vector from elastic deformation, is controlled by a material property that is a vector when generated by a scalar volume change and a third-order tensor when generated by deviatoric strain. Similar comments hold for piezomagnetic response, which is the generation of a magnetic-field vector from elastic deformation. Curie's principle as proven in Section 1.8.5 thus requires that both piezoelectric and piezomagnetic response can only occur in an anisotropic material. Thermal-electric and thermal-magnetic coupling will similarly only occur in an anisotropic material.

In addition to the above 10 off-diagonal material properties, the following on-diagonal properties are defined

$$\frac{1}{T}\frac{\partial T}{\partial \tilde{s}} \mathrel{\widehat{=}} \frac{1}{\tilde{c}} \quad \text{inverse volumetric heat capacity} \tag{6.334}$$

$$-\tilde{v}\frac{\partial P}{\partial \tilde{v}} \mathrel{\widehat{=}} K \quad \text{bulk modulus} \tag{6.335}$$

$$\frac{\partial \boldsymbol{\tau}^D}{\partial \boldsymbol{e}^D} \mathrel{\widehat{=}} {}_4\boldsymbol{G} = {}_4\overset{T}{\boldsymbol{G}}{}^{3412} = {}_4\overset{T}{\boldsymbol{G}}{}^{2134} = {}_4\overset{T}{\boldsymbol{G}}{}^{1243} \quad \text{shear stiffness fourth-order tensor} \tag{6.336}$$

$$\frac{\partial \boldsymbol{E}'}{\partial \boldsymbol{D}'} \mathrel{\widehat{=}} \boldsymbol{C}_{ED} = \boldsymbol{C}_{ED}^T \quad \text{inverse electrical-permittivity tensor} \tag{6.337}$$

$$\frac{\partial \boldsymbol{H}'}{\partial \boldsymbol{B}'} \mathrel{\widehat{=}} \boldsymbol{C}_{HB} = \boldsymbol{C}_{HB}^T \quad \text{inverse magnetic-permeability tensor.} \tag{6.338}$$

Many of the material properties in this particular internal-energy representation have no standard agreed-upon name. The inverse electrical permittivity can be called the *impermittivity* and the inverse magnetic permeability can be called the *impermeability*.

We can now write our constitutive laws in their most general form

$$\frac{\mathrm{d}T}{\mathrm{d}t} = \frac{T}{\tilde{c}}\frac{\mathrm{d}\tilde{s}}{\mathrm{d}t} + \frac{1}{\alpha\tilde{v}}\frac{\mathrm{d}\tilde{v}}{\mathrm{d}t} + \boldsymbol{A} : \frac{\mathrm{d}\boldsymbol{e}^D}{\mathrm{d}t} + \boldsymbol{p}\cdot\frac{\mathrm{d}\boldsymbol{D}'}{\mathrm{d}t} + \boldsymbol{q}\cdot\frac{\mathrm{d}\boldsymbol{B}'}{\mathrm{d}t} \tag{6.339}$$

$$-\frac{\mathrm{d}P}{\mathrm{d}t} = \frac{1}{\alpha\tilde{v}}\frac{\mathrm{d}\tilde{s}}{\mathrm{d}t} + \frac{K}{\tilde{v}}\frac{\mathrm{d}\tilde{v}}{\mathrm{d}t} + \boldsymbol{F} : \frac{\mathrm{d}\boldsymbol{e}^D}{\mathrm{d}t} + \boldsymbol{s}\cdot\frac{\mathrm{d}\boldsymbol{D}'}{\mathrm{d}t} + \boldsymbol{t}\cdot\frac{\mathrm{d}\boldsymbol{B}'}{\mathrm{d}t} \tag{6.340}$$

$$\frac{\mathrm{d}\boldsymbol{\tau}^D}{\mathrm{d}t} + \frac{\boldsymbol{\tau}^D}{\tilde{v}}\frac{\mathrm{d}\tilde{v}}{\mathrm{d}t} = \frac{\boldsymbol{A}}{\tilde{v}}\frac{\mathrm{d}\tilde{s}}{\mathrm{d}t} + \frac{\boldsymbol{F}}{\tilde{v}}\frac{\mathrm{d}\tilde{v}}{\mathrm{d}t} + {}_4\boldsymbol{G} : \frac{\mathrm{d}\boldsymbol{e}^D}{\mathrm{d}t} + {}_3\boldsymbol{U}\cdot\frac{\mathrm{d}\boldsymbol{D}'}{\mathrm{d}t} + {}_3\boldsymbol{W}\cdot\frac{\mathrm{d}\boldsymbol{B}'}{\mathrm{d}t} \tag{6.341}$$

$$\frac{\mathrm{d}\boldsymbol{E}'}{\mathrm{d}t} + \frac{\boldsymbol{E}'}{\tilde{v}}\frac{\mathrm{d}\tilde{v}}{\mathrm{d}t} = \frac{\boldsymbol{p}}{\tilde{v}}\frac{\mathrm{d}\tilde{s}}{\mathrm{d}t} + \frac{\boldsymbol{s}}{\tilde{v}}\frac{\mathrm{d}\tilde{v}}{\mathrm{d}t} + {}_3\overset{T}{\boldsymbol{U}}{}^{231} : \frac{\mathrm{d}\boldsymbol{e}^D}{\mathrm{d}t} + \boldsymbol{C}_{ED}\cdot\frac{\mathrm{d}\boldsymbol{D}'}{\mathrm{d}t} + \boldsymbol{C}_{EB}\cdot\frac{\mathrm{d}\boldsymbol{B}'}{\mathrm{d}t} \tag{6.342}$$

$$\frac{\mathrm{d}\boldsymbol{H}'}{\mathrm{d}t} + \frac{\boldsymbol{H}'}{\tilde{v}}\frac{\mathrm{d}\tilde{v}}{\mathrm{d}t} = \frac{\boldsymbol{q}}{\tilde{v}}\frac{\mathrm{d}\tilde{s}}{\mathrm{d}t} + \frac{\boldsymbol{t}}{\tilde{v}}\frac{\mathrm{d}\tilde{v}}{\mathrm{d}t} + {}_3\overset{T}{\boldsymbol{W}}{}^{231} : \frac{\mathrm{d}\boldsymbol{e}^D}{\mathrm{d}t} + \boldsymbol{C}_{EB}^T\cdot\frac{\mathrm{d}\boldsymbol{D}'}{\mathrm{d}t} + \boldsymbol{C}_{HB}\cdot\frac{\mathrm{d}\boldsymbol{B}'}{\mathrm{d}t}. \tag{6.343}$$

For the primed electromagnetic fields in the frame of reference moving with the center-of-mass velocity $\boldsymbol{v}$ of the mass element, we established in Section 3.3.1 that they are related to the unprimed electromagnetic fields in a stationary reference frame as

$$\frac{\mathrm{d}\boldsymbol{D}'}{\mathrm{d}t} = \frac{\partial \boldsymbol{D}}{\partial t} + \boldsymbol{v}\cdot\nabla\boldsymbol{D} \quad \text{and} \quad \frac{\mathrm{d}\boldsymbol{B}'}{\mathrm{d}t} = \frac{\partial \boldsymbol{B}}{\partial t} + \boldsymbol{v}\cdot\nabla\boldsymbol{B} \tag{6.344}$$

with similar definitions for $\mathrm{d}\boldsymbol{E}'/\mathrm{d}t$ and $\mathrm{d}\boldsymbol{H}'/\mathrm{d}t$. We further showed in Section 3.3.2 that

$$\boldsymbol{E}' = \boldsymbol{E} + \boldsymbol{v}\times\boldsymbol{B} \quad \text{and} \quad \boldsymbol{H}' = \boldsymbol{H} - \boldsymbol{v}\times\boldsymbol{D}. \tag{6.345}$$

But in Chapter 3, we also showed that in electromagnetic applications using the electromagnetic constitutive laws, the difference in the electromagnetic fields in a stationary and moving frame of reference can be entirely neglected. So on the left-hand side of the last three constitutive laws above, we can write

$$\frac{\mathrm{d}\boldsymbol{\tau}^D}{\mathrm{d}t} + \frac{\boldsymbol{\tau}^D}{\tilde{v}}\frac{\mathrm{d}\tilde{v}}{\mathrm{d}t} = \frac{\partial \boldsymbol{\tau}^D}{\partial t} + \nabla\cdot\left(\boldsymbol{v}\boldsymbol{\tau}^D\right), \tag{6.346}$$

$$\frac{\mathrm{d}\boldsymbol{E}'}{\mathrm{d}t} + \frac{\boldsymbol{E}'}{\tilde{v}}\frac{\mathrm{d}\tilde{v}}{\mathrm{d}t} = \frac{\partial \boldsymbol{E}}{\partial t} + \nabla\cdot(\boldsymbol{v}\boldsymbol{E}) \tag{6.347}$$

$$\frac{\mathrm{d}\boldsymbol{H}'}{\mathrm{d}t} + \frac{\boldsymbol{H}'}{\tilde{v}}\frac{\mathrm{d}\tilde{v}}{\mathrm{d}t} = \frac{\partial \boldsymbol{H}}{\partial t} + \nabla\cdot(\boldsymbol{v}\boldsymbol{H}). \tag{6.348}$$

Once allowance is made for all the material properties to depend on the independent thermodynamic variables $\tilde{s}$, $\tilde{v}$, $\boldsymbol{e}^D$, $\boldsymbol{D}$ and $\boldsymbol{B}$, the above boxed set of constitutive laws, along with the identifications of Eqs (6.344), (6.346)–(6.348), are the most general nonlinear constitutive laws that derive from the internal-energy representation of thermodynamics.

Through the Legendre transform, one can transform to other independent thermodynamic variables that may be preferable for certain applications. We will leave such exercises to the reader.

Last, in an isotropic material, all the various material properties must be proportional to fundamental isotropic tensors (see Section 1.8.4) which requires

$$\boldsymbol{p} = \boldsymbol{q} = \boldsymbol{s} = \boldsymbol{t} = 0, \quad \boldsymbol{A} = \boldsymbol{F} = 0, \quad {}_3\boldsymbol{U} = {}_3\boldsymbol{W} = 0, \tag{6.349}$$

$$\boldsymbol{C}_{ED} = C_{ED}\boldsymbol{I}, \quad \boldsymbol{C}_{EB} = C_{EB}\boldsymbol{I}, \quad \boldsymbol{C}_{HB} = C_{HB}\boldsymbol{I}, \tag{6.350}$$

$${}_4\boldsymbol{G} = G\left(\delta_{ik}\delta_{jl} + \delta_{il}\delta_{jk} - \frac{2}{3}\delta_{ij}\delta_{kl}\right)\hat{\boldsymbol{x}}_i\hat{\boldsymbol{x}}_j\hat{\boldsymbol{x}}_k\hat{\boldsymbol{x}}_l \tag{6.351}$$

and we can write the isotropic constitutive laws in a stationary frame of reference as

$$\frac{\partial T}{\partial t} + \boldsymbol{v}\cdot\nabla T = \frac{T}{\tilde{c}}\left(\frac{\partial \tilde{s}}{\partial t} + \boldsymbol{v}\cdot\nabla\tilde{s}\right) + \frac{1}{\alpha}\nabla\cdot\boldsymbol{v} \tag{6.352}$$

$$-\left(\frac{\partial P}{\partial t} + \boldsymbol{v}\cdot\nabla P\right) = \frac{1}{\alpha\tilde{v}}\left(\frac{\partial \tilde{s}}{\partial t} + \boldsymbol{v}\cdot\nabla\tilde{s}\right) + K\nabla\cdot\boldsymbol{v} \tag{6.353}$$

$$\frac{\partial \boldsymbol{\tau}^D}{\partial t} + \nabla\cdot\left(\boldsymbol{v}\boldsymbol{\tau}^D\right) = G\left(\nabla\boldsymbol{v} + (\nabla\boldsymbol{v})^T - \frac{2}{3}\nabla\cdot\boldsymbol{v}\boldsymbol{I}\right) \tag{6.354}$$

$$\frac{\partial \boldsymbol{E}}{\partial t} + \nabla\cdot(\boldsymbol{v}\boldsymbol{E}) = C_{ED}\left(\frac{\partial \boldsymbol{D}}{\partial t} + \boldsymbol{v}\cdot\nabla\boldsymbol{D}\right) + C_{EB}\left(\frac{\partial \boldsymbol{B}}{\partial t} + \boldsymbol{v}\cdot\nabla\boldsymbol{B}\right) \tag{6.355}$$

$$\frac{\partial \boldsymbol{H}}{\partial t} + \nabla\cdot(\boldsymbol{v}\boldsymbol{H}) = C_{EB}\left(\frac{\partial \boldsymbol{D}}{\partial t} + \boldsymbol{v}\cdot\nabla\boldsymbol{D}\right) + C_{HB}\left(\frac{\partial \boldsymbol{B}}{\partial t} + \boldsymbol{v}\cdot\nabla\boldsymbol{B}\right). \tag{6.356}$$

The various advective terms can be dropped for small enough $\boldsymbol{v}$. As discussed in Chapter 3, the cross-coupling coefficient C_{EB} between magnetic and electric field components is almost always negligible as well. We see that the piezoelectric and piezomagnetic effects disappear in an isotropic material.

6.8 Exercises

1. In the example material considered in Section 6.1.7, for the straight line path on the $P-v$ plane from state a to c, determine the change of heat Δq_{lin} not as $\Delta q_{lin} = \Delta u_{ac} - \Delta w_{lin}$ as we did in that section, but by integrating Tds along the linear path. Hint: Note that $s = s_a$ both in state a and state c. Begin with the relation $P_{lin}(v) = P(s, v)$ where P_{lin} is given by Eq. (6.61) and $P(s, v)$ for this material is given by Eq. (6.59), which then gives how the specific entropy $s_{lin}(v)$ varies with specific volume along the linear path. Use this in Eq. (6.58) for $T(s, v)$ to determine how temperature $T_{lin}(v)$ varies along the same linear path. The change in heat is then the integral

$$\Delta q_{lin} = \int_{v_a}^{v_c} T_{lin}(v) \frac{ds_{lin}(v)}{dv}\, dv.$$

Show that you get the same answer for Δq_{lin} using this approach as that given in Section 6.1.7.

2. For the constitutive laws in the s, v (or internal energy) representation

$$dT = \frac{T}{c_v}\, ds + \frac{1}{\alpha_s} \frac{dv}{v} \tag{6.357}$$

$$-dP = \frac{1}{\alpha_s v}\, ds + K_s \frac{dv}{v}, \tag{6.358}$$

it is experimentally determined that the material properties satisfy the following dependencies

$$\frac{1}{\alpha_s} = \frac{v}{v_o} T(s, v); \quad c_v = c_o \quad \text{and} \quad K_s = \frac{c_o v}{v_o^2} T(s, v), \tag{6.359}$$

where c_o is a constant. If the specific entropy and volume in a reference state are s_o and v_o and if the temperature and pressure in this reference state are given T_o and P_o, show that the state functions for this particular material are

$$T(s, v) = T_o e^{(s-s_o)/c_o} e^{(v/v_o - 1)} \tag{6.360}$$

$$P(s, v) = P_o + \frac{c_o}{v_o} \left[T_o - T(s, v)\right]. \tag{6.361}$$

3. *True density representation*: In this chapter, we formulated the thermodynamics of single-component fluid mass elements starting from the fundamental function in extensive form $U = U(S, V, N)$ from which we can derive Euler's relation $U = TS - PV + \mu N$ and for which we have the first law in the form $dU = TdS - PdV + \mu dN$, though if in a single component mass element the mass does not vary as it moves

around, we have $\mathrm{d}N = 0$. We then normalized the extensive variables by the constant mass $\mathcal{M}$ of the element and defined the *specific* variables $u = U/\mathcal{M}$, $s = S/\mathcal{M}$, $v = V/\mathcal{M} = 1/\rho$ and $n = N/\mathcal{M}$ which, trivially, leads to $u = u(s, v, n)$, the Euler relation $u = Ts - Pv + \mu n$ and the first law $\mathrm{d}u = T\mathrm{d}s - P\mathrm{d}v$ (since n is constant for our single component mass element).

Reformulate the description in terms of densities, that will be denoted in this problem with a bar over the symbol

$$\bar{u} = u\rho, \quad \bar{s} = s\rho, \quad \bar{n} = n\rho, \tag{6.362}$$

where $\rho = \mathcal{M}/V$ and the volume of the element V varies as the element moves around. Show that in terms of these variables, and due to n being a constant, we end up with a first law in the form

$$\mathrm{d}\bar{u} = T\mathrm{d}\bar{s} + \frac{\mu\bar{n}}{\rho}\mathrm{d}\rho. \tag{6.363}$$

Generally, it is much less convenient to formulate the relations of thermodynamics in terms of true densities (the extensive fields normalized by the volume V of the mass element that itself can vary) though it certainly can be done as shown here. An important example of a constitutive law that we have already seen that was derived starting with a true-density formulation (energy normalized by volume) was Hooke's law of Chapter 4.

In order to obtain a first law in the more recognizable form of a change in internal-energy density being equal to a change in heat-energy density plus a change in mechanical-energy density, replace the total derivative symbol d in the Eq. (6.363) by the total time derivative $\mathrm{d}/\mathrm{d}t = \partial/\partial t + \boldsymbol{v}\cdot\nabla$, use the appropriate Euler relation $\mu\bar{n} = \bar{u} - T\bar{s} + P$ and note that $\rho^{-1}\mathrm{d}\rho/\mathrm{d}t = -V^{-1}\mathrm{d}V/\mathrm{d}t = -\nabla\cdot\boldsymbol{v}$ to obtain

$$\frac{\mathrm{D}\bar{u}}{\mathrm{D}t} = T\frac{\mathrm{D}\bar{s}}{\mathrm{D}t} - P\nabla\cdot\boldsymbol{v}, \tag{6.364}$$

where we define a new type of material time derivative $\mathrm{D}/\mathrm{D}t$ as

$$\frac{\mathrm{D}\psi}{\mathrm{D}t} = \frac{1}{V}\frac{\mathrm{d}}{\mathrm{d}t}(V\psi) = \frac{\mathrm{d}\psi}{\mathrm{d}t} + \frac{\psi}{V}\frac{\mathrm{d}V}{\mathrm{d}t} \tag{6.365}$$

$$= \frac{\partial\psi}{\partial t} + \nabla\cdot(\boldsymbol{v}\psi), \tag{6.366}$$

where $\boldsymbol{v}$ is the center-of-mass velocity of the mass element under consideration. This derivative has the property that if $\psi = \psi_0$ is a constant in both space and time, then $\mathrm{D}\psi_0/\mathrm{D}t = \psi_0\nabla\cdot\boldsymbol{v}$ due to the changing volume surrounding a given mass. In general, when formulating thermodynamics in terms of densities, as we did in Section 6.6, it is standard to normalize extensive quantities by the constant volume of the mass element in the undeformed reference state V_0, which is analogous to normalizing the extensive quantities by the constant mass $\mathcal{M}$ of the mass element to form specific continuum variables.

Relations between Partial Derivatives

For the next five problems, you will use the relations between partial derivatives given in Section 6.1.3 as well as the form of the constitutive laws in the various thermodynamic representations as given throughout Section 6.3, to prove many nontrivial relations between the material properties or between various thermodynamic partial derivatives. These relations are general and apply to all materials and not to any one specific material. Worked examples of a similar nature within the chapter are given, for example, by Eqs (6.109) through (6.118) and by Eqs (6.128) through (6.132). The five exercises that follow build upon such worked examples.

4. Using the constitutive laws in both the (T, v) and (T, P) representations, prove that

$$c_v = c_p - vT\,\alpha_p^2\,K_T. \tag{6.367}$$

This is a generally valid relation between the specific heat at constant volume c_v and the specific heat at constant pressure c_p, and a great example of one of the quintessential thermodynamic facts one can demonstrate using the thermodynamic formalism.

5. Using Fact 2 about partial derivatives, prove that

$$\left(\frac{\partial P}{\partial T}\right)_s = \frac{c_p}{Tv\,\alpha_p}. \tag{6.368}$$

Then using the chain rule (Fact 3 about partial derivatives) for the same derivative, show that $K_s = -c_p/(Tv\alpha_p\alpha_s)$. We showed that as a consequence of a stable thermodynamics equilibrium, $\alpha_p\alpha_s < 0$ (usually due to $\alpha_s < 0$) so that $K_s > 0$, which is another requirement of stable thermodynamic equilibrium.

6. Using Fact 2 about partial derivatives, prove that

$$\left(\frac{\partial s}{\partial v}\right)_P = -K_s\,\alpha_s. \tag{6.369}$$

Then using the chain rule for the same derivative, again show that $K_s = -c_p/(Tv\alpha_p\alpha_s)$.

7. Using the constitutive laws in both (s, P) and (T, P) representations, prove that

$$\frac{1}{K_s} = \frac{1}{K_T} - \frac{vT\,\alpha_p^2}{c_p}. \tag{6.370}$$

This is a generally valid relation between the adiabatic bulk modulus K_s and the isothermal bulk modulus K_T and is another of the strikingly general results that can be obtained using thermodynamics.

8. In the definition of the isothermal bulk modulus

$$K_T = -v\left(\frac{\partial P}{\partial v}\right)_T,$$

use Fact 4 [Eq. (6.24)] as well as any other relations you can develop, to prove that

$$K_T = K_s - \frac{c_v}{vT\alpha_s^2}.$$

9. *Thermodynamics in the entropic representation*: Although rarely utilized, we can start with the fundamental function in the form $s = s(u, v)$ and take derivatives to obtain the first law, the definition of the state functions and the constitutive laws using the entropic representation when (u, v) are the independent variables. By taking derivatives, show that the first law in this representation (when particle numbers in the system of constant mass are not changing) is

$$\mathrm{d}s = \frac{1}{T}\mathrm{d}u + \frac{P}{T}\mathrm{d}v,$$

where T is temperature and P is pressure. By taking derivatives of the two state functions $1/T$ and P/T in this entropic representation, show that the constitutive laws when $(u.v)$ are the independent variables are

$$\mathrm{d}\left(\frac{1}{T}\right) = -\frac{1}{T^2 c_v}\,du - \frac{1}{T^2}\left(\frac{1}{v\alpha_s} + \frac{P}{c_v}\right)\,dv \tag{6.371}$$

$$\mathrm{d}\left(\frac{P}{T}\right) = -\frac{1}{T^2}\left(\frac{1}{v\alpha_s} + \frac{P}{c_v}\right)\,du - \frac{1}{T}\left(\frac{K_s}{v} + \frac{2P}{v\alpha_s T} + \frac{P^2}{c_v T}\right)\,\mathrm{d}v. \tag{6.372}$$

You should obtain this result by taking derivatives in the entropic representation, using the Maxwell relation $\partial^2 s/(\partial u \partial v) = \partial^2 s/(\partial v \partial u)$, the earlier Maxwell relation $(\partial P/\partial s)_v = -(\partial T/\partial v)_s$, and the four facts of partial derivatives including Fact 4 [Eq. (6.24)]. Once you have obtained the above form of the entropic constitutive laws, you should verify that they are identical to the earlier form of Eqs (6.36) and (6.37) given in the internal-energy representation once you make the substitution for ds given by $\mathrm{d}u = T\mathrm{d}s - P\mathrm{d}v$.

10. Using the Euler equation $ST = U + PV - \mu N$, show that the Gibbs–Duhem equation $\mathrm{d}\mu = (V/N)\mathrm{d}P - (S/N)\mathrm{d}T$ can be rewritten as

$$\mathrm{d}\left(\frac{\mu}{T}\right) = \frac{V}{N}\mathrm{d}\left(\frac{P}{T}\right) + \frac{U}{N}\mathrm{d}\left(\frac{1}{T}\right). \tag{6.373}$$

11. *Ideal gas*: Two equations of state for an ideal gas are usually written as

$$\frac{P}{T} = k_B\frac{N}{V} \quad \text{and} \quad \frac{1}{T} = ck_B\frac{N}{U},$$

where the dimensionless constant c takes the value $c = 3/2$ for monatomic ideal gases (gases consisting of single atoms moving about without interaction) and where $k_B = 1.381 \times 10^{-23}$ J/K is Boltzmann's constant. If these are the equations of state, then $S = S(U, V, N)$ must be the fundamental function.

- Find the third equation of state $\mu/T(U, V, N)$ by integrating the above Eq. (6.373) from some reference state in which μ_0, T_0, P_0, U_0, V_0, and N_0 are all given constants to obtain

$$\frac{\mu}{T}(U, V, N) = \frac{\mu_0}{T_0} - ck_B \ln\left(\frac{U}{U_0}\frac{N_0}{N}\right) - k_B \ln\left(\frac{V}{V_0}\frac{N_0}{N}\right).$$

  As an aside, if you insert the first two equations of state into this expression, you obtain $\mu(T, P)$ for a single-component ideal gas as

$$\mu(T,P) = \mu_0 \frac{T}{T_0} - (1+c)k_B T \ln\left(\frac{T}{T_0}\right) + k_B T \ln\left(\frac{P}{P_0}\right). \tag{6.374}$$

The qualitative trends of this function with T and P from some low temperature initial state T_0 and P_0 were depicted earlier in Fig. 6.7. A result from statistical mechanics is that for an ideal monatomic gas consisting of atoms of mass m, the chemical potential in the initial reference state is

$$\mu_0 = -k_B T_0 \ln\left[\left(\frac{2\pi m k_B T_0}{h^2}\right)^{3/2} \frac{V_0}{N_0},\right]$$

where $h = 6.626 \times 10^{-34}$ m^{2}kg/s is Planck's constant. Thus, show that

$$\boxed{\mu(T,P) = -k_B T \ln\left[\frac{V_0}{N_0}\left(\frac{2\pi m k_B T_0}{h^2}\right)^{3/2}\left(\frac{T}{T_0}\right)^{c+1}\frac{P_0}{P}\right]} \tag{6.375}$$

is the chemical potential of a single-species ideal gas as a function of T and P.

- Now use the Euler equation to obtain the fundamental function for an ideal gas $S = S(U, V, N)$ in the form

$$S(U,V,N) = N\sigma_0 + Nk_B \ln\left[\frac{V}{V_0}\left(\frac{U}{U_0}\right)^c \left(\frac{N}{N_0}\right)^{-(c+1)}\right], \tag{6.376}$$

where $\sigma_0 = k_B(c+1) - \mu_0/T_0$.

- Using the ideal gas equations of state, obtain the material properties as $K_T = P$, $\alpha_p = 1/T$, $c_v = ck_B/m$, $c_p = (c+1)k_B/m$, and $\alpha_s = -c/T$.
- Invert the above fundamental function for an ideal gas in the entropic representation $S = S(U, V, N)$ to obtain the fundamental function in the internal-energy representation $U = U(S, V, N)$ as

$$U(S,V,N) = U_0\left(\frac{V}{V_0}\right)^{-1/c}\left(\frac{N}{N_0}\right)^{(c+1)/c} \exp\left(\frac{S - \sigma_0 N}{ck_B N}\right).$$

- Next, obtain the fundamental function in the Helmholtz-free-energy representation $F = F(T, V, N) = U - TS$ in the form

$$F(T,V,N) = Nk_B T\left\{\frac{F_0}{N_0 k_B T_0} - \ln\left[\left(\frac{T}{T_0}\right)^c \frac{V}{V_0}\left(\frac{N}{N_0}\right)^{-1}\right]\right\},$$

where $T_0 = U_0/(ck_B N_0)$ and $F_0 = U_0 - \sigma_0 N_0 T_0$.

- An ideal gas is present in a cylinder that has closed end caps and that has an internal adiabatic and impermeable piston (wall) separating the cylinder into two chambers a and b. There are N_a atoms initially in chamber a and N_b atoms initially in chamber b. The initial volumes of chambers a and b are V_{a0} and V_{b0}. The pressure and chemical potential in the two chambers are different initially. The cylinder with its two chambers separated by the piston wall are in equilibrium with a surrounding reservoir that is maintained at constant temperature T.

We now release the piston wall and let it freely displace until the two chambers come to thermodynamic equilibrium (i.e., so that $P_a = P_b$ and $\mu_a = \mu_b$) all while the temperature T and particle numbers N_a and N_b are maintained as constants within each chamber. Show that equilibrium is attained when the volumes have changed to $V_a = V_{a0} + \Delta V_{eq}$ and $V_b = V_{b0} - \Delta V_{eq}$ where

$$\Delta V_{eq} = \frac{N_a V_{b0} - N_b V_{a0}}{N_a + N_b}.$$

Calculate the total mechanical work done in both chambers ΔW during the equilibration process

$$\Delta W = -\int_{V_{a0}}^{V_{a0}+\Delta V_{eq}} P_a(V_a)\,\mathrm{d}V_a - \int_{V_{b0}}^{V_{b0}-\Delta V_{eq}} P_b(V_b)\,\mathrm{d}V_b$$
$$= -k_B T\left[N_a \ln\left(\frac{1+V_{b0}/V_{a0}}{1+N_b/N_a}\right) + N_b \ln\left(\frac{1+V_{a0}/V_{b0}}{1+N_a/N_b}\right)\right].$$

Further argue that for this particular system, the entropies in each chamber are given by

$$S_a = N_a\sigma_0 + N_a k_B \ln\left(1+\frac{\Delta V}{V_{a0}}\right)$$
$$S_b = N_b\sigma_0 + N_b k_B \ln\left(1-\frac{\Delta V}{V_{b0}}\right),$$

where the ΔV is the volume change (either positive over negative) in each chamber as the piston wall is displaced from its initial position. Find the ΔV at which the composite cylinder entropy $S = S_a + S_b$ is a maximum (i.e., when $\partial S/\partial \Delta V = 0$) and compare to the ΔV_{eq} determined above. Show that the change in heat for the composite cylinder from the initial to the equilibrium state is given by $T\Delta S = T(S_{aeq} + S_{beq} - S_{a0} - S_{b0}) = -\Delta W$. So the work done in displacing the piston until equilibrium is reached comes entirely from heat fluxing into the cylinder from the surrounding heat reservoir, which is maintained at constant T. The total internal energy does not change during this particular equilibration process $\Delta U = U_{aeq} + U_{beq} - U_{a0} - U_{b0} = T\Delta S + \Delta W = 0$. But show that the total change in the Helmholtz free energy for the composite cylinder during the equilibration is $\Delta F = \Delta W$.

12. *Van der Waals gas*: The two equations of state for a van der Waals gas, in which the finite volume of each atom is accounted for through a parameter b and the finite force that the atoms inside a sample exert on an atom near the surface of a sample is accounted for through a parameter a, are

$$\frac{P}{T} = k_B\frac{n}{v}\left[\frac{1}{1-bn/v} - \frac{acn/v}{uv/n^2+a^2}\right] \quad \text{and} \quad \frac{1}{T} = \frac{ck_B v/n}{uv/n^2+a^2}.$$

Again $c = 3/2$ for a monatomic gas. For the van der Waals gas, obtain the equation of state for $\mu/T(u, v, n)$, the fundamental function $s = s(u, v, n)$, and the three material properties K_T, α_p, c_v and α_s.

13. *Pressure-induced first-order phase transition*: Consider a rigid chamber of volume 2×10^{-4} m^3 that is surrounded by insulating walls. Inside the chamber there is a single-molecular-species liquid layer at the bottom occupying half the chamber volume that is overlain by a layer of its vapor. The liquid and vapor are in equilibrium (stable coexistence) at the temperature $T_1 = 375$ K and vapor pressure $P_v(T_1) = P_1 = 10^5$ Pa. A small mass $\Delta M_l = 10^{-4}$ kg of liquid molecules at temperature T_1 is then injected through a tube traversing the chamber wall into the liquid layer, which increases the pressure in the chamber (the pressure adjustment can be assumed to occur instantaneously) and induces a first-order phase transition where vapor molecules condense into liquid molecules. This phase transition inside of the chamber that is surrounded by heat-insulating walls releases latent heat into the chamber as controlled by a latent heat of vaporization given by $L_v = 2.5 \times 10^6$ J/kg. The vapor pressure varies linearly with temperature in a small neighborhood surrounding the initial equilibrium point T_1, P_1 as given by $P_v(T) = P_1 + a(T - T_1)$ where $a = 4 \times 10^5$ Pa/K. The latent heat L_v can be modeled as a constant in this neighborhood of T_1, P_1.

Determine the mass of vapor (gas) molecules ΔM_{gl} that condenses into liquid so that the system resides at a new state of equilibrium coexistence T_2, P_2. Determine this T_2 using a specific heat given by $c_p = fc_p^g + (1-f)c_p^l$ where f is the volume fraction of the chamber occupied by vapor (gas) and $1-f$ the fraction occupied by liquid with $c_p^g = 10^3$ J/kg/K and $c_p^l = 4 \times 10^3$ J/kg/K which can be taken as constants in the neighborhood between T_1, P_1 and T_2, P_2. The liquid density in the initial state is $\rho_1^l = 10^3$ kg/m^3 so that the initial mass of liquid in the chamber is $M_1^l = 10^{-1}$ kg. In solving this problem, you can assume that the injected water molecules is a small fraction of the initial liquid molecules present, that is, $\Delta M^l/M_1^l = 10^{-3}$ and the induced pressure change can be approximated as being linear in $\Delta M^l/M_1^l$.

HINT: To solve this problem, you first need to estimate the pressure change for the isothermal injection of water molecules. This pressure change is controlled by the constitutive law $\mathrm{d}P = -K_T \mathrm{d}v/v$ when temperature is constant where K_T is the isothermal bulk modulus that is different in the liquid and gas; $K_T^l = 2 \times 10^9$ Pa and $K_T^g = 10^5$ Pa. The change in the liquid's specific volume is $\mathrm{d}v^l/v^l = (M^l/V^l)\mathrm{d}(V^l/M^l) = \mathrm{d}V^l/V^l - \mathrm{d}M^l/M^l$ where V^l is the volume of liquid in the chamber and M^l the mass of liquid. For the gas phase, prior to the phase transition but following the injection of liquid molecules, the change in the specific volume is due to a volume change alone $\mathrm{d}v^g = \mathrm{d}V^g/V^g$ because the mass of gas is not changing prior to the phase transition. You then can show that because $V_1^l = V_1^g$ prior to the injection of the water molecules into the rigid chamber of constant volume, the change in the liquid and gas pressures, which are the same, is

$$\Delta P = \frac{K_T^g}{1 + K_T^g/K_T^l} \frac{\Delta M_l}{M_1^l}. \tag{6.377}$$

Due to this perturbation in pressure that pushes the system off of the coexistence curve, vapor molecules will condense into liquid.

7

Nonequilibrium Diffusive Transport

We now put the equilibrium thermodynamics developed in Chapter 6 to use in the formulation of the continuum governing equations for problems involving the nonequilibrium processes of thermal diffusion and solute diffusion and for problems in which two fluid species flow in largely separate fluid domains while diffusing at least partially into each other across a transition layer whose position evolves in space and time. These problems involve fluid systems being out of equilibrium and will require us to develop the various pertinent transport laws appropriate to each situation. Diffusion and transport in solid-state systems, that is controlled by the point-defect structure in the solid, will not be treated. Borg and Dienes (1988) as well as Balluffi et al. (2005) provide introductions to the subject of solid-state diffusion for the reader interested in this topic.

As throughout the book, modeling the transport and equilibration at the continuum scale means that each point of a large out-of-equilibrium fluid body is surrounded by a macroscopic averaging element containing a multitude of atoms whose averaged, or summed, behavior defines the macroscopic response variables associated with that averaging element. Among these macroscopic response variables are the thermodynamic intensive parameters of temperature T, pressure P, and chemical potential μ. A key idea is that even though the system as a whole is out of equilibrium, the temporal changes of an averaging element's T, P, and μ and other thermodynamic variables are controlled by the rules and postulates of equilibrium thermodynamics as treated in Chapter 6. So an element is always assumed to be in thermodynamic equilibrium with itself as the intensive variables are changing through time with the only exceptions being when chemical reactions are taking place between the distinct molecular species within an element or when first-order phase transitions are taking place between distinct phases of matter within an element. We will not address the kinetics (i.e., rate laws) of either chemical reactions or first-order phase transitions, which involve their own unique considerations.

Nonequilibrium diffusive transport in fluid systems that is the focus of this chapter arises when neighboring elements have different values of T, P, and μ that produce equilibrating macroscopic flux that attempts to eliminate such macroscopic gradients in T, P, and μ. If there is greater flux of molecules, momentum, and/or energy into an element than leaving the element, the average values of T, P, and μ will change through time. The temporal change of an element's T, P, and μ is related to the temporal change of other thermodynamic fields at that element according to the constitutive laws of equilibrium thermodynamics. The transport or flux that is produced by the macroscopic gradients in T, P, and μ have their own distinct constitutive laws as addressed in this chapter that are called *transport laws.*

The continuum modeling of nonequilibrium fluid systems is based on the conservation of energy that we will derive in Section 7.1 using a conventional macroscopic (i.e., not molecularly resolved) perspective. To then use the energy-conservation law to address various diffusion problems, we must convert it to a statement of entropy conservation that is tailored to the specific type of system under consideration. Deriving an appropriate statement of entropy conservation allows us to develop the statement of irreversible entropy production for whatever nonequilibrium transport processes are being treated. From the statement of irreversible entropy production, we obtain the associated transport laws that are the flux-force constitutive laws in nonequilibrium diffusive situations. We then combine the conservation of entropy with appropriate equilibrium constitutive laws, as well as with the statements of conservation of molecular mass and conservation of momentum, to obtain the entire suite of governing equations for problems involving the spatial and temporal variations of temperature, concentration of chemical species, and viscous flow.

The specific systems to be treated represent the sectioning within this chapter. A first system is again a fluid made up of a single molecular species in which both the advection and diffusion of heat is occurring. In a second system, we consider a *binary fluid* made up of two molecular species, the *solute* and *solvent.* In this scenario, we want to solve simultaneously for the advection and diffusion of the solute concentration and temperature under the condition that in a final equilibrium static state, the solute concentration will be a uniform constant throughout the system. We describe this type of two-component system as being fully *miscible.* Finally, the third system to be treated is again a binary fluid but this time it is assumed that even in a final equilibrium state, there can be transition layers between uniform fluid-phase domains that are either mainly fluid 1 or mainly fluid 2. We describe this type of system as being either fully or partially *immiscible* and the overall name for the physics describing this third scenario is *capillary physics.*

We approach capillary physics first from the classical nineteenth-century perspective of van der Waals, as updated in the 1950s and 1960s by Cahn and Hilliard (e.g., Cahn and Hilliard, 1958), with the goal of generating the governing partial-differential equations that describe how two fluids with only partial miscibility flow together and at least partially diffuse into each other with the molecular interactions creating a capillary-stress tensor that acts throughout the transition layers. This "zoomed-in" Cahn–Hilliard approach resolves the transition layers between fluid domains explicitly but does not make clear some of the key macroscopic concepts of capillary physics. As such, we finish the chapter with a more macroscopic "zoomed-out" description of the interface between two immiscible fluids when the interface is modeled as a surface with a surface tension and not as a finite-width transition layer. We also treat contact lines where the two fluids and a solid boundary intersect, defining the wetting and nonwetting nature of the two fluids on the solid surface and the imbibition force that acts on the contact line to spontaneously pull the wetting fluid toward the nonwetting fluid when both fluids are in contact with a solid.

7.1 Energy Conservation in a Viscous Fluid

Consider a little volume element δV of viscous fluid that surrounds a point $\boldsymbol{r}$ and is fixed in time (the δ represents that this volume is small in size). We are taking a fixed-frame Eulerian perspective here where we watch fluid coming into and out of the element. There is an average velocity $\boldsymbol{v}$, a mass density ρ, and a specific internal energy (internal energy per unit mass) u associated with δV at any instant. In terms of these quantities, we can define the total mass of the element as $\delta M = \rho\, \delta V$ and the total internal energy of the element as $\delta U = \rho\, u\, \delta V$.

The total energy associated with this volume element is the sum of the total internal energy and the macroscopic kinetic energy

$$\delta E_{\text{total}} = \underbrace{\frac{1}{2}\, \delta M\, \boldsymbol{v} \cdot \boldsymbol{v}}_{\text{Kinetic energy}} + \underbrace{\delta U}_{\text{Internal energy}}. \tag{7.1}$$

The energy density associated with the element is then

$$\frac{\delta E_{\text{total}}}{\delta V} = \frac{1}{2}\, \rho\, \boldsymbol{v} \cdot \boldsymbol{v} + \rho u. \tag{7.2}$$

To establish the continuum statement of the conservation of energy, we imagine an arbitrary region Ω within the material that does not change with time and that is made up of all the little volume elements in the limit that $\delta V \to \mathrm{d}V$ as depicted in Fig. 7.1. The conservation of energy for this region requires that

$$\int_{\Omega} \frac{\partial}{\partial t}\left(\frac{1}{2}\, \rho\, \boldsymbol{v} \cdot \boldsymbol{v} + \rho u\right) \mathrm{d}V = -\int_{\partial\Omega} \boldsymbol{n} \cdot \boldsymbol{J}_E\, \mathrm{d}S, \tag{7.3}$$

where $\partial\Omega$ is the closed surface surrounding Ω and $\boldsymbol{J}_E$ is the *energy flux vector* (energy/time/area)

$$\boldsymbol{J}_E = \underbrace{\boldsymbol{v}\left(\frac{1}{2}\, \rho\, \boldsymbol{v} \cdot \boldsymbol{v} + \rho u\right)}_{\text{Energy advection}} + \underbrace{\boldsymbol{q}}_{\text{Heat flux}} - \underbrace{\boldsymbol{v} \cdot \boldsymbol{\tau}}_{\text{Mechanical work}}. \tag{7.4}$$

If we introduce $\boldsymbol{J}_E$ into the energy balance, apply the divergence theorem and use the fact that Ω is arbitrary, we obtain a first statement of energy conservation:

$$\frac{\partial}{\partial t}\left[\frac{1}{2}\, \rho\, \boldsymbol{v} \cdot \boldsymbol{v} + \rho u\right] = -\nabla \cdot \left[\boldsymbol{v}\left(\frac{1}{2}\, \rho\, \boldsymbol{v} \cdot \boldsymbol{v} + \rho u\right) + \boldsymbol{q} - \boldsymbol{v} \cdot \boldsymbol{\tau}\right]. \tag{7.5}$$

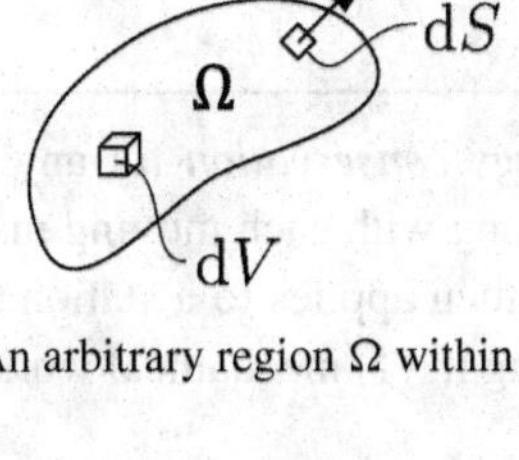

Figure 7.1 An arbitrary region Ω within a larger body.

The *total derivative* as first described in Sec. 2.6 is now introduced

$$\frac{\mathrm{d}\psi}{\mathrm{d}t} = \frac{\partial \psi}{\partial t} + \boldsymbol{v}\cdot\nabla\psi \mathrel{\widehat{=}} \text{the change in } \psi \text{ when we follow an element of fixed mass as it moves with velocity } \boldsymbol{v} \tag{7.6}$$

and the stress tensor in the fluid is decomposed into elastic and viscous contributions

$$\boldsymbol{\tau} = -P\boldsymbol{I} + \boldsymbol{\tau}^v, \tag{7.7}$$

where P is the fluid pressure that entirely accounts for elastic stress in the fluid and $\boldsymbol{\tau}^v$ is the viscous stress tensor as treated in Chapter 5 on fluid mechanics. If we were treating a solid here, instead of a fluid, we would need to use the full elastic stress tensor $\boldsymbol{\tau}^E$ in place of the purely isotropic elastic stress $-P\boldsymbol{I}$ appropriate to a fluid. Further, using the identity $\nabla\cdot(\boldsymbol{v}\,\psi) = \psi\,\nabla\cdot\boldsymbol{v} + \boldsymbol{v}\cdot\nabla\psi$, as well as $\nabla\cdot(\boldsymbol{v}\cdot\boldsymbol{\tau}) = \boldsymbol{v}\cdot(\nabla\cdot\boldsymbol{\tau}) + \boldsymbol{\tau}^T:\nabla\boldsymbol{v}$, we can rewrite Eq. (7.5) as

$$\begin{aligned}\frac{\mathrm{d}}{\mathrm{d}t}\left[\frac{1}{2}\rho\,\boldsymbol{v}\cdot\boldsymbol{v} + \rho u\right] = &-\left(\frac{1}{2}\rho\,\boldsymbol{v}\cdot\boldsymbol{v} + \rho u + P\right)\nabla\cdot\boldsymbol{v} \\ &- \boldsymbol{v}\cdot\nabla P + \boldsymbol{v}\cdot(\nabla\cdot\boldsymbol{\tau}^v) + \boldsymbol{\tau}^v:\nabla\boldsymbol{v} - \nabla\cdot\boldsymbol{q}.\end{aligned} \tag{7.8}$$

From the product rule of calculus, we further have

$$\frac{\mathrm{d}}{\mathrm{d}t}\left[\frac{1}{2}\rho\,\boldsymbol{v}\cdot\boldsymbol{v} + \rho u\right] = \left(\frac{1}{2}\boldsymbol{v}\cdot\boldsymbol{v} + u\right)\frac{\mathrm{d}\rho}{\mathrm{d}t} + \rho\,\boldsymbol{v}\cdot\frac{\mathrm{d}\boldsymbol{v}}{\mathrm{d}t} + \rho\frac{\mathrm{d}u}{\mathrm{d}t}. \tag{7.9}$$

Mass conservation states that

$$\frac{\mathrm{d}\rho}{\mathrm{d}t} = -\rho\,\nabla\cdot\boldsymbol{v} \tag{7.10}$$

while momentum conservation gives that

$$\boldsymbol{v}\cdot\left[\rho\frac{\mathrm{d}\boldsymbol{v}}{\mathrm{d}t} = -\nabla P + \nabla\cdot\boldsymbol{\tau}^v\right] \tag{7.11}$$

or

$$\rho\,\boldsymbol{v}\cdot\frac{\mathrm{d}\boldsymbol{v}}{\mathrm{d}t} = -\boldsymbol{v}\cdot\nabla P + \boldsymbol{v}\cdot(\nabla\cdot\boldsymbol{\tau}^v). \tag{7.12}$$

Thus, we obtain exactly

$$\boxed{\rho\frac{\mathrm{d}u}{\mathrm{d}t} = \frac{P}{\rho}\frac{\mathrm{d}\rho}{\mathrm{d}t} + \boldsymbol{\tau}^v:\nabla\boldsymbol{v} - \nabla\cdot\boldsymbol{q}} \tag{7.13}$$

as our general statement of *energy conservation* for any viscous fluid. It is written for a frame of reference that moves along with each moving mass element; however, upon substituting $\mathrm{d}/\mathrm{d}t \to \partial/\partial t + \boldsymbol{v}\cdot\nabla$, it then applies to a stationary frame. Equation (7.13) states that the internal energy is changing if: (1) mechanical work $(P/\rho)\mathrm{d}\rho/\mathrm{d}t$ is being performed that reversibly stores volumetric strain energy, (2) heat is irreversibly being generated due to viscous effects $\boldsymbol{\tau}^v : \nabla \boldsymbol{v}$, and (3) heat is accumulating due to more heat fluxing into the fluid element than fluxing out $-\nabla \cdot \boldsymbol{q}$.

7.2 Temperature Diffusion in a One-Component Fluid

To use the above statement of energy conservation in deriving the governing equations for various advection and diffusion scenarios, we must first transform it into a statement of entropy conservation which will require appealing to the specific type of system we want to treat. In this section, we will consider a viscous isotropic fluid made up of a single molecular species.

7.2.1 Entropy Conservation in a One-Component Fluid

We begin with the form of the first law of thermodynamics appropriate for a single-component fluid in equilibrium as discussed in Chapter 6 on thermodynamics

$$\frac{\mathrm{d}u}{\mathrm{d}t} = T\frac{\mathrm{d}s}{\mathrm{d}t} - P\frac{\mathrm{d}v}{\mathrm{d}t}, \tag{7.14}$$

where again

$$s \mathrel{\widehat{=}} \text{specific entropy} = \frac{\text{entropy}}{\text{unit mass}} \left(\frac{\mathrm{J}}{\mathrm{K\,kg}}\right)$$

and

$$v = \frac{1}{\rho} \mathrel{\widehat{=}} \text{specific volume} = \frac{\text{volume}}{\text{unit mass}}.$$

Thus, the first law $\rho\, \mathrm{d}u/\mathrm{d}t = \rho T\, \mathrm{d}s/\mathrm{d}t + (P/\rho)\mathrm{d}\rho/\mathrm{d}t$ when combined with Eq. (7.13) gives the heat-balance statement

$$\rho T\frac{\mathrm{d}s}{\mathrm{d}t} = -\nabla \cdot \boldsymbol{q} + \boldsymbol{\tau}^v : \nabla \boldsymbol{v}, \tag{7.15}$$

where each term has the following interpretation

$$\rho T\frac{\mathrm{d}s}{\mathrm{d}t} = \frac{\text{change in heat}}{\text{(unit volume) (unit time)}}$$

$$-\nabla \cdot \boldsymbol{q} = \frac{\text{accumulation of heat transported by conduction}}{\text{(unit volume) (unit time)}}$$

$$\boldsymbol{\tau}^v : \nabla \boldsymbol{v} = \frac{\text{irreversible production of heat by viscous flow}}{\text{(unit volume) (unit time)}}.$$

It is worth commenting again that this balance for how heat is changing applies to a fluid element that is out of equilibrium with its surroundings. Nonetheless, we have used the first law of thermodynamics to get to this statement and the first law is describing changes between equilibrium states. The presence of flux across a fluid element means there is transport trying to equilibrate differences in T and P between adjacent elements. So the gradients in T and P across an element are responsible for the nonequilibrium flux across an element. But the mean values of T and P in an element combine with the changing mean values of u, s, and v in the element as described by the rules of equilibrium thermodynamics. As stated in multiple places now, this is a fundamental concept in the modeling of nonequilibrium systems.

We next rewrite the heat balance explicitly as an *entropy balance* by dividing through by T

$$\rho \frac{\mathrm{d}s}{\mathrm{d}t} = -\frac{1}{T} \nabla \cdot \boldsymbol{q} + \frac{1}{T} \boldsymbol{\tau}^v : \nabla \boldsymbol{v} \tag{7.16}$$

$$= -\nabla \cdot \left(\frac{\boldsymbol{q}}{T} \right) + \frac{1}{T} \left[-\frac{\boldsymbol{q} \cdot \nabla T}{T} + \boldsymbol{\tau}^v : \nabla \boldsymbol{v} \right]. \tag{7.17}$$

So in terms of entropy accumulation, if there is more entropy flux $\boldsymbol{q}/T$ entering a volume element than leaving it, there will be an increase in entropy as quantified by $-\nabla \cdot (\boldsymbol{q}/T)$. Further, as heat flux is taking place in the material, there is an irreversible increase in entropy associated with this thermal-energy transport process as quantified by $-(\boldsymbol{q} \cdot \nabla T)/T^2$.

So we can identify the rate of irreversible entropy production σ for a one-component viscous fluid

$$\sigma = -\boldsymbol{q} \cdot \frac{\nabla T}{T^2} + \boldsymbol{\tau}^v : \frac{\nabla \boldsymbol{v}}{T} \mathrel{\widehat{=}} \frac{\text{the irreversible entropy production}}{\text{(unit volume)(unit time)}}. \tag{7.18}$$

The *second law of thermodynamics* is the statement that during the transport processes of equilibration, the entropy production is always positive, that is.,

$$\sigma \geq 0 \tag{7.19}$$

with equality holding only when no transport is occurring in an element that is in thermodynamic equilibrium with its surroundings. Again the transport processes that cause σ to be nonzero are also called *irreversible processes* due to the irreversible generation of heat, and we now go on to obtain the transport laws that are associated with them.

7.2.2 Transport Laws for a One-Component Fluid

In looking at the statement for irreversible entropy production σ, the various terms come in the form of a generalized flux dotted into the generalized force that is driving that flux. For the σ of Eq. (7.18), the first contribution identifies one flux as the heat flux $\boldsymbol{q}$ and this is conjugate to the driving force $-(\nabla T)/T^2$. In the second term, there is a momentum flux $\boldsymbol{\tau}^v$ that is conjugate to the driving force $(\nabla \boldsymbol{v})/T$.

The most general linear relation between these fluxes and forces are the rules

$$\begin{bmatrix} \boldsymbol{q} \\ \\ \boldsymbol{\tau}^{v} \end{bmatrix} = \begin{bmatrix} \boldsymbol{L}_q\cdot & {}_3\boldsymbol{L}_c : \\ \\ {}_3\boldsymbol{L}_c\cdot & {}_4\boldsymbol{L}_v : \end{bmatrix} \begin{bmatrix} -(\nabla T)/T^2 \\ \\ (\nabla \boldsymbol{v})/T \end{bmatrix}, \tag{7.20}$$

where the various transport material-property tensors $\boldsymbol{L}_q$, ${}_3\boldsymbol{L}_c$, and ${}_4\boldsymbol{L}_v$ are called *phenomenological constants.* That ${}_3\boldsymbol{L}_c$ is the same in both the heat-flow law and viscous-stress law and is symmetric across the first two indices ${}_3\boldsymbol{L}_c = {}_3\overset{T}{\boldsymbol{L}}{}_c^{213}$ is a fact known as *Onsager reciprocity* that we will demonstrate in Section 7.4. Onsager reciprocity requires the matrix of phenomenological constants in Eq. (7.20) to be symmetric, which also means that the second-order tensor $\boldsymbol{L}_q$ is symmetric as is the fourth-order viscosity tensor which satisfies the various symmetries ${}_4\boldsymbol{L}_v = {}_4\overset{T}{\boldsymbol{L}}{}_v^{3412} = {}_4\overset{T}{\boldsymbol{L}}{}_v^{2134} = {}_4\overset{T}{\boldsymbol{L}}{}_v^{1243}$ so that only the symmetric part of $\nabla \boldsymbol{v}$ need be considered.

However, the viscous fluid being treated is taken to be isotropic. From Curie's principle of Chapter 1, only fluxes and forces of the same tensorial order can be related to each other in an isotropic material. So the assumption of isotropy requires that ${}_e\boldsymbol{L}_c = \boldsymbol{L}_c\, {}_3\boldsymbol{\epsilon} = 0$ where ${}_3\boldsymbol{\epsilon}$ is the third-order antisymmetric Levi-Civita tensor. Further, for an isotropic fluid, we have the isotropic forms written in component form (e.g., $\boldsymbol{L}_q = L_{ij}^{q}\hat{\boldsymbol{x}}_i\hat{\boldsymbol{x}}_j$)

$$\frac{L_{ij}^{q}}{T^2} = \lambda_t \delta_{ij} \tag{7.21}$$

and

$$\frac{L_{ijkl}^{v}}{T} = \kappa\delta_{ij}\delta_{kl} + \eta\left(\delta_{ik}\delta_{jl} + \delta_{il}\delta_{jk} - \frac{2}{3}\delta_{ij}\delta_{kl}\right). \tag{7.22}$$

So the final transport laws for a one-component isotropic viscous fluid are the uncoupled equations

$$\boldsymbol{q} = -\lambda_t\, \nabla T; \quad \textit{Fourier's law of heat conduction} \tag{7.23}$$

$$\boldsymbol{\tau}^{v} = \kappa\ \nabla\cdot\boldsymbol{v}\boldsymbol{I} + \eta\left[\nabla\boldsymbol{v} + (\nabla\boldsymbol{v})^{T} - \frac{2}{3}\nabla\cdot\boldsymbol{v}\boldsymbol{I}\right]; \quad \textit{Newtonian viscosity.} \tag{7.24}$$

The material property λ_t is called the *thermal conductivity* (units of W m^{-1} K^{-1}) while the two viscosities are the bulk viscosity κ and the shear viscosity η as seen earlier in Chapter 5.

7.2.3 The Temperature-Diffusion Governing Equations

With the appropriate form for the transport laws established for an isotropic single-component fluid, we return to the statement of the conservation of entropy and replace $\mathrm{d}s/\mathrm{d}t$ with the proper statement for how temperature is changing. To do so, we need to employ our equilibrium-thermodynamics constitutive relations. Let's use the constitutive laws of a fluid in the Gibb's free-energy (T, P) representation which, for a single component fluid, we showed in Chapter 6 to have the form

$$T\frac{\mathrm{d}s}{\mathrm{d}t} = c_p\frac{\mathrm{d}T}{\mathrm{d}t} - v\,\alpha_p\,T\frac{\mathrm{d}P}{\mathrm{d}t} \tag{7.25}$$

$$\frac{1}{v}\frac{\mathrm{d}v}{\mathrm{d}t} = \alpha_p\frac{\mathrm{d}T}{\mathrm{d}t} - \frac{1}{K_T}\frac{\mathrm{d}P}{\mathrm{d}t}. \tag{7.26}$$

If the flow under consideration is being treated as incompressible, this means that pressure variations are not causing the density to vary appreciably. Under this approximation, we place the pressure term to zero in Eq. (7.26) to give

$$\frac{1}{v}\frac{\mathrm{d}v}{\mathrm{d}t} = \nabla\cdot\boldsymbol{v} = \alpha_P\,\frac{\mathrm{d}T}{\mathrm{d}t}. \tag{7.27}$$

This says that due to thermal expansion as the temperature is varying, the flow will not be solenoidal (i.e., divergence free). We will return to this in a moment.

To similarly neglect the pressure variations in the contribution to the heat of Eq. (7.25), an order-of-magnitude analysis requires that $v\alpha_p T|\Delta P| \ll c_p|\Delta T|$ or

$$|\Delta P| \ll \frac{c_p}{v\alpha_p}\frac{|\Delta T|}{T} \approx 10^{10}\,\mathrm{Pa}\,\frac{|\Delta T|}{T}, \tag{7.28}$$

where we inserted characteristic material properties for a liquid like water. This condition is assumed to be satisfied for the flow and temperature variations scenarios being modeled so that Eq. (7.25) gives

$$T\frac{\mathrm{d}s}{\mathrm{d}t} = c_p\frac{\mathrm{d}T}{\mathrm{d}t} \tag{7.29}$$

and the entropy (or heat) balance then becomes the equation controlling the temperature variations

$$\rho\,c_p\frac{\mathrm{d}T}{\mathrm{d}t} = -\nabla\cdot\boldsymbol{q} + \boldsymbol{\tau}^v : \nabla\boldsymbol{v}, \tag{7.30}$$

where $\boldsymbol{q} = -\lambda_t\nabla T$ from the Fourier heat law and $\boldsymbol{\tau}^v$ is given by the law of Newtonian viscosity.

For flow scenarios in the Earth (ground-water flow, atmospheric flow, mantle convection) the viscous heat production $\boldsymbol{\tau}^v : \nabla\boldsymbol{v}$ can be neglected. To see this, approximate $|\boldsymbol{\tau}^v : \nabla\boldsymbol{v}| \approx \eta|\boldsymbol{v}|^2/\ell^2$ and $|\nabla\cdot(\lambda_t\nabla T)| \approx \lambda_t T/\ell^2$ where ℓ is a characteristic length in the problem. Thus, when the dimensionless ratio defined by

$$\frac{|\boldsymbol{\tau}^v : \nabla\boldsymbol{v}|}{|\nabla\cdot(\lambda_t\nabla T)|} \approx \frac{\eta|\boldsymbol{v}|^2}{\lambda_t T} \tag{7.31}$$

is negligible, the viscous-shearing heat source may be neglected. This may be equivalently stated that when the flow velocity satisfies $|\boldsymbol{v}| \ll \sqrt{\lambda_t T/\eta}$, the heat of viscous shearing is negligible. For water at $T = 300$ K, this condition becomes $|\boldsymbol{v}| \ll 550$ m/s which is always the case for ground-water flow. If you are interested in temperature evolution during mantle convection, we have that $\lambda_t \approx 3\,\text{W m}^{-1}\,\text{K}^{-1}$ and $\eta \approx 10^{20}$ Pa s for mantle rock while $T \approx 3{,}000$ K so that the neglect of the heat of viscous shearing during mantle convection requires $|\boldsymbol{v}| \ll 10^{-8}$ m/s. It takes roughly 100 Myr for a complete convection cycle in the mantle, which corresponds to velocities on the order of $|\boldsymbol{v}| \approx 10^{-10}$ m/s. Thus, for mantle convection, even with the large viscosity of mantle rock, one can safely neglect the viscous-shearing heat-source term in the above equation for the temperature evolution.

So assuming the heat of viscous shearing is negligible, we then have

$$\rho\, c_p \left(\frac{\partial T}{\partial t} + \boldsymbol{v} \cdot \nabla T\right) = \nabla \cdot (\lambda_t \nabla T) \tag{7.32}$$

as the equation controlling temperature in a flowing fluid. As will be seen later, appropriate boundary conditions are that either T or $\boldsymbol{n} \cdot \boldsymbol{q}$ must be specified on the boundaries surrounding the region under study. Equation (7.32) is called an *advection–diffusion* equation and the advection term is often not negligible.

For the corresponding flow equations while temperature is evolving, we have formally that the compressibility (thermal expansion) constraint on the components of the flow velocity is that

$$\nabla \cdot \boldsymbol{v} = \alpha_P \left(\frac{\partial T}{\partial t} + \boldsymbol{v} \cdot \nabla T\right) \tag{7.33}$$

instead of the usual statement of $\nabla \cdot \boldsymbol{v} = 0$. So for such nonsolenoidal flow, the force balance controlling viscous flow becomes

$$\rho \left(\frac{\partial \boldsymbol{v}}{\partial t} + \boldsymbol{v} \cdot \nabla \boldsymbol{v}\right) = -\nabla P + \eta \nabla^2 \boldsymbol{v} + \left(\kappa + \frac{\eta}{3}\right) \nabla\nabla \cdot \boldsymbol{v} + \rho \boldsymbol{g}. \tag{7.34}$$

This can be complemented by an equation of state for $v(T) = 1/\rho(T)$ if density is changing appreciably with temperature during the advection and diffusion scenario.

However, for most advective heat transport problems, we can assume, as usual, that $\nabla \cdot \boldsymbol{v} \approx 0$, which in the context of fluid flow, is the statement that density changes in the force balance (Navier–Stokes) can be neglected. If we write the density and temperature as $\rho = \rho_0 + \Delta\rho$ and $T = T_0 + \Delta T$, where ρ_0 and T_0 are the initial density and temperature present prior to the start of advective heat transport, the density change due to the temperature change goes as $\Delta\rho \approx -\rho_0 \alpha_p \Delta T$ so that $\rho \approx \rho_0(1 - \alpha_p \Delta T)$. Because $\alpha_p \approx 10^{-4}\,\text{K}^{-1}$ for most liquids, so long as the temperature changes are less than a few hundred degrees Kelvin (or Celsius), the usual incompressibility condition of $\nabla \cdot \boldsymbol{v} = 0$ is sufficient for advective heat transport of a liquid. For a gas, we have $c_p \approx 1/T$ (measured in Kelvin) so that $\rho = \rho_0(1 - \Delta T/T)$, which for sufficiently large ΔT may require the use of Eqs (7.33) and (7.34).

Between Eqs (7.32)–(7.34), we have five equations for the five unknowns of T, P, and the three components of $\boldsymbol{v}$. We can then simultaneously solve these equations for these five unknowns subject to initial and boundary conditions. To make explicit that Eq. (7.33) provides the additional equation for determining P, we take the divergence of the force balance and insert the expression for $\nabla \cdot \boldsymbol{v}$ to get

$$\nabla^2 P = \left[\left(\kappa + \frac{4\eta}{3}\right)\nabla^2 - \rho\left(\frac{\partial}{\partial t} + \boldsymbol{v}\cdot\nabla\right)\right]\left[\alpha_p\left(\frac{\partial T}{\partial t} + \boldsymbol{v}\cdot\nabla T\right)\right] + \left(\frac{\partial \rho}{\partial T}\right)\nabla T \cdot \boldsymbol{g}. \tag{7.35}$$

So the pressure may be understood to satisfy this Poisson equation where the source term on the right-hand side varies with time as the temperature varies with time. For sufficiently small temperature changes such that $\nabla \cdot \boldsymbol{v} \approx 0$, we would have $\nabla^2 P = (\partial\rho/\partial T)\,\nabla T \cdot \boldsymbol{g}$.

Under scenarios where the thermal conductivity λ_t does not vary in space, we can rewrite the advection–diffusion equation of Eq. (7.32) as

$$\boxed{D_t\,\nabla^2 T - \frac{\partial T}{\partial t} - \boldsymbol{v}\cdot\nabla T = 0,} \tag{7.36}$$

where

$$D_t \,\hat{=}\, \textit{thermal diffusivity} = \frac{\lambda_t}{\rho c_p}. \tag{7.37}$$

To quantify whether the advection term is important relative to the diffusion term, we define the so-called Péclet number Pe as

$$\text{Pe} = \frac{|\boldsymbol{v}\cdot\nabla T|}{|D_t \nabla^2 T|} \approx \frac{|\boldsymbol{v}|\ell}{D_t}, \tag{7.38}$$

where ℓ is a characteristic length in the problem over which temperature is varying. When Pe $\ll 1$, advection can be neglected relative to diffusion and one ends up with

$$D_t\,\nabla^2 T - \frac{\partial T}{\partial t} = 0, \tag{7.39}$$

which is a scalar diffusion equation. In Part II of the book, we will be studying the scalar diffusion equation in detail. Equations (7.36) and (7.39) are the two equations most commonly used for the modeling of temperature distribution both when advection is important (Pe not negligible relative to one) and when advection is not important (Pe $\ll 1$).

7.3 Solute Diffusion in a Two-Component Miscible Fluid

We now will generalize the above to allow for a fluid made of two molecular species, one of which we call the solute species and the other the solvent species. Such two-component fluids are also called *binary fluids*. The goal of this section is to derive rules for how T and the concentration of solute are varying in space and time for a system out of equilibrium.

It is the assumption here that when the system finally attains full equilibrium, the concentration of solute will be a spatially uniform constant throughout the system; in this case, we say that the solute is *miscible* within the solvent.

We begin by deriving the statement of conservation of solute. We then obtain the law for the conservation of entropy for a binary fluid. Focus will be placed again on the irreversible entropy production, the expression of which allows the transport laws for coupled solute diffusion and heat diffusion to be written down. The entire suite of governing equations can then be presented in Section 7.3.4.

7.3.1 Conservation of Solute Mass

Let's begin by defining the concentration of solute as a mass ratio. If in each little volume element of the fluid ($\delta V \to dV$) there are N_1 solute molecules and N_2 solvent molecules and if each solute molecule has a mass m_1 and each solvent molecule has a mass m_2, then the total mass of the element is $M = m_1 N_1 + m_2 N_2$ and we define the mass concentration of solute c as the dimensionless mass ratio

$$c = \frac{m_1 N_1}{m_1 N_1 + m_2 N_2} = m_1 n_1. \tag{7.40}$$

Here, $n_i = N_i/M$ are the specific particle numbers (particles per unit total mass) of each species i.

If we now consider an arbitrary region Ω that is constant in time, that is surrounded by the closed surface $\partial\Omega$ and that is composed of a multitude of little volume elements dV that each contain a mass $\rho c dV$ of solute molecules, the law for the conservation of solute mass is the statement

$$\int_\Omega \frac{\partial(\rho c)}{\partial t}\, dV = -\int_{\partial\Omega} \boldsymbol{n} \cdot (\rho c \boldsymbol{v} + \boldsymbol{j})\, dS, \tag{7.41}$$

where the solute mass flux $\boldsymbol{j}$ is allowing for how solute molecules diffuse from regions of large concentration to regions of low concentration and $\rho c \boldsymbol{v}$ allows for how solute molecules convect due to fluid flow $\boldsymbol{v}$, where $\boldsymbol{v}$ is the center-of-mass velocity defined as always as the total momentum of all molecules (both species 1 and 2) within each element divided by the total mass of the element. The diffusive mass flux $\boldsymbol{j}$ is the process where through random thermal motions of the solute molecules in the presence of a concentration gradient ∇c, there will be more random jumps of solute molecules across a plane perpendicular to the gradient from the side of the plane that has the larger concentration of solute molecules compared to the number of jumps across the plane from the side that has the smaller concentration.

Applying the divergence theorem to Eq. (7.41) and arguing that the domain Ω is completely arbitrary then leads to the continuum statement of the conservation of solute molecules

$$\frac{\partial}{\partial t}(\rho c) + \nabla \cdot (\rho c \boldsymbol{v}) = -\nabla \cdot \boldsymbol{j}. \tag{7.42}$$

Distributing the derivatives on the left-hand side and using our earlier law for the conservation of all mass $\partial\rho/\partial t + \nabla\cdot(\rho\boldsymbol{v}) = 0$ then gives

$$\rho\left(\frac{\partial c}{\partial t} + \boldsymbol{v}\cdot\nabla c\right) = -\nabla\cdot\boldsymbol{j} \tag{7.43}$$

as the desired statement of solute mass conservation. Note that by introducing the usual definition of a total derivative as $\mathrm{d}c/\mathrm{d}t = \partial c/\partial t + \boldsymbol{v}\cdot\nabla c$, this law can also be written as $\rho\mathrm{d}c/\mathrm{d}t = -\nabla\cdot\boldsymbol{j}$. To proceed, we will need a constitutive law for the diffusive mass flux $\boldsymbol{j}$.

7.3.2 Entropy Conservation in a Two-Component Fluid

To obtain the statement of entropy conservation for a two-component fluid, we first write down a form of the first law appropriate to a two-component fluid and combine that with our earlier statement of energy conservation.

The fundamental function for a two-component fluid is the specific internal energy $u = u(s, v, n_1, n_2)$, where s is specific entropy, $v = 1/\rho$ is specific volume, and n_i is the specific number of particles of each species i. The first law is then of the form

$$\mathrm{d}u = T\mathrm{d}s - P\mathrm{d}v + \mu_1\mathrm{d}n_1 + \mu_2\mathrm{d}n_2. \tag{7.44}$$

To rewrite this in terms of changes in the solute mass ratio $c = m_1 n_1$, we note that the solvent mass ratio is $1 - c = m_2 n_2$ so that $\mathrm{d}n_1 = \mathrm{d}c/m_1$ and $\mathrm{d}n_2 = -\mathrm{d}c/m_2$. Thus, by identifying the chemical potential associated with solute concentration changes as

$$\mu = \frac{\mu_1}{m_1} - \frac{\mu_2}{m_2} \tag{7.45}$$

the first law can be rewritten in the time-differential form

$$\frac{\mathrm{d}u}{\mathrm{d}t} = T\frac{\mathrm{d}s}{\mathrm{d}t} + \frac{P}{\rho^2}\frac{\mathrm{d}\rho}{\mathrm{d}t} + \mu\frac{\mathrm{d}c}{\mathrm{d}t}, \tag{7.46}$$

that is, the internal-energy fundamental function is $u = u(s, \rho, c)$. Note that μ here has units of energy per unit mass and is physically representing the energy change per unit mass of a mass element of a binary fluid when the solute concentration changes at constant s and ρ, that is, $\mu = (\partial u/\partial c)_{s,\rho}$.

Let's multiply both sides of our law of conversation of solute mass by μ to get $\mu\rho\,\mathrm{d}c/\mathrm{d}t = -\mu\nabla\cdot\boldsymbol{j}$ which can be rewritten as

$$\rho\mu\frac{\mathrm{d}c}{\mathrm{d}t} + \nabla\cdot(\mu\boldsymbol{j}) - \boldsymbol{j}\cdot\nabla\mu = 0. \tag{7.47}$$

Our earlier statement of internal energy conservation is generally valid and given by $\rho\mathrm{d}u/\mathrm{d}t = (P/\rho)\mathrm{d}\rho/\mathrm{d}t - \nabla\cdot\boldsymbol{q} + \boldsymbol{\tau}^v : \nabla\boldsymbol{v}$ where $\boldsymbol{q}$ is the heat flux and $\boldsymbol{\tau}^v$ the viscous stress tensor. If we add to this the statement of Eq. (7.47) we obtain the energy conservation law applicable to a binary fluid

$$\rho \frac{\mathrm{d}u}{\mathrm{d}t} = \frac{P}{\rho}\frac{\mathrm{d}\rho}{\mathrm{d}t} + \rho\mu\frac{\mathrm{d}c}{\mathrm{d}t} - \nabla\cdot(\boldsymbol{q} - \mu\boldsymbol{j}) + \boldsymbol{\tau}^v : \nabla\boldsymbol{v} - \boldsymbol{j}\cdot\nabla\mu. \tag{7.48}$$

If we then introduce the first law of thermodynamics as given by Eq. (7.46) above and divide by T, we get the conservation of entropy in the form

$$\rho\frac{\mathrm{d}s}{\mathrm{d}t} = -\nabla\cdot\left(\frac{\boldsymbol{q} - \mu\boldsymbol{j}}{T}\right) + \frac{1}{T}\left[-\frac{(\boldsymbol{q} - \mu\boldsymbol{j})\cdot\nabla T}{T} - \boldsymbol{j}\cdot\nabla\mu + \boldsymbol{\tau}^v : \nabla\boldsymbol{v}\right]. \tag{7.49}$$

So $(\boldsymbol{q} - \mu\boldsymbol{j})/T$ corresponds to the entropy flux in a material when both heat flux and solute diffusion are occurring simultaneously and the term in square brackets corresponds to the irreversible entropy generation σ

$$\sigma = -(\boldsymbol{q} - \mu\boldsymbol{j})\cdot\frac{\nabla T}{T^2} - \boldsymbol{j}\cdot\frac{\nabla\mu}{T} + \boldsymbol{\tau}^v : \frac{\nabla\boldsymbol{v}}{T}. \tag{7.50}$$

Note that from the identity $(\nabla\mu)/T = \nabla(\mu/T) - \mu\nabla(1/T)$ we can rewrite this as

$$\boxed{\sigma = -\boldsymbol{q}\cdot\frac{\nabla T}{T^2} - \boldsymbol{j}\cdot\nabla\left(\frac{\mu}{T}\right) + \boldsymbol{\tau}^v : \frac{\nabla\boldsymbol{v}}{T}.} \tag{7.51}$$

The chemical potential in the Gibb's representation has an equation of state given by $\mu(T, P, c) = \mu_o(T, P) + k_B T \ln\left[c\,\gamma(c)\right]$ where the first term represents the energetic interaction between a solute molecule and the solvent molecules and the second term allows for both the entropy of mixing when an additional solute molecule is added to the solution and the interaction energy between solute molecules at sufficiently high concentration through a dimensionless *activity coefficient* $\gamma(c)$ that needs to be specified. In the limit where solute-solute interaction can be neglected at sufficiently low dilution, we have $\gamma(c) \to 1$. However, in obtaining the transport equations in what follows, we will not appeal to a specific equation of state for the chemical potential.

Although the second form of Eq. (7.51) may seem preferable for obtaining coupled transport laws for $\boldsymbol{q}$ and $\boldsymbol{j}$, we will work with the first form for σ given by Eq. (7.50) in what follows. However, either expression can be used and both lead to identical statements of the transport laws.

7.3.3 The Coupled Transport Laws of a Two-Component Fluid

We now use Eq. (7.50) for σ to identify how the fluxes $\boldsymbol{q} - \mu\boldsymbol{j}$, $\boldsymbol{j}$ and $\boldsymbol{\tau}$ are related to the generalized forces $(\nabla T)/T^2$, $(\nabla\mu)/T$ and $(\nabla\boldsymbol{v})/T$ that drive them. Assuming that our binary fluid is isotropic so that only fluxes and forces of the same tensorial order are coupled together and so that the transport coefficients are scalars and not tensors, the viscous stress tensor decouples from the gradients in T and μ and is related only to velocity gradients through the law of Newtonian viscosity as discussed earlier. A first form for the vectorial fluxes is then given by

$$q - \mu j = -L_t \left(\frac{\nabla T}{T^2}\right) - L_{t\mu}\left(\frac{\nabla \mu}{T}\right) \tag{7.52}$$

$$j = -L_{\mu t}\left(\frac{\nabla T}{T^2}\right) - L_{\mu}\left(\frac{\nabla \mu}{T}\right) \tag{7.53}$$

where from Onsager reciprocity we have $L_{t\mu} = L_{\mu t}$. From the transport law for j we can identify the gradient in the chemical potential as $\nabla\mu = -(T/L_\mu)j - (L_{t\mu}/L_\mu)\nabla T/T$ which allows the heat flux to be written

$$q = -\left(\frac{L_t}{T^2} - \frac{L_{t\mu}^2}{T^2 L_\mu}\right)\nabla T + \left(\mu + \frac{L_{t\mu}}{L_\mu}\right) j. \tag{7.54}$$

To address the solute mass flux law, we first express the chemical potential in the Gibb's representation $\mu = \mu(T, P, c)$ to obtain

$$\nabla\mu = \left(\frac{\partial\mu}{\partial T}\right)_{Pc}\nabla T + \left(\frac{\partial\mu}{\partial P}\right)_{Tc}\nabla P + \left(\frac{\partial\mu}{\partial c}\right)_{PT}\nabla c. \tag{7.55}$$

This then gives the solute-mass flux law as

$$\begin{aligned} j = &-\left[\frac{L_\mu}{T}\left(\frac{\partial\mu}{\partial T}\right)_{Pc} + \frac{L_{t\mu}}{T^2}\right]\nabla T - \frac{L_\mu}{T}\left(\frac{\partial\mu}{\partial P}\right)_{Tc}\nabla P \\ &- \frac{L_\mu}{T}\left(\frac{\partial\mu}{\partial c}\right)_{PT}\nabla c. \end{aligned} \tag{7.56}$$

We now give names to the various coefficients in these laws. When $j = 0$, we have that the heat flux must be given by the Fourier heat law $q = -\lambda_t \nabla T$ where λ_t is the thermal conductivity so that we must identify

$$\lambda_t = \frac{L_t}{T^2} - \frac{L_{t\mu}^2}{T^2 L_\mu}. \tag{7.57}$$

Further, when ∇T and ∇P are both zero, the solute-mass flux must be given by Fick's Law $j = -\rho D_c \nabla c$ where D_c is the diffusivity of solute molecules moving through the solution (units of $m^2\ s^{-1}$ and also called the diffusion coefficient or *solute diffusivity*). We thus must identify

$$\rho D_c = \frac{L_\mu}{T}\left(\frac{\partial\mu}{\partial c}\right)_{PT}. \tag{7.58}$$

For the other coefficients, we further identify

$$\begin{aligned} \frac{k_T}{T} &= \frac{(\partial\mu/\partial T)_{Pc}}{(\partial\mu/\partial c)_{PT}} + \frac{L_{t\mu}/L_\mu}{(\partial\mu/\partial c)_{PT}} = \frac{(\partial\mu/\partial T)_{Pc} + (\beta - \mu)/T}{(\partial\mu/\partial c)_{PT}} \\ &= \frac{(\beta-\mu)}{T}\left(\frac{\partial c}{\partial\mu}\right)_{PT} - \left(\frac{\partial c}{\partial T}\right)_{P\mu}, \end{aligned} \tag{7.59}$$

$$\xi_P = \frac{(\partial\mu/\partial P)_{Tc}}{(\partial\mu/\partial c)_{PT}} = \frac{(\partial v/\partial c)_{PT}}{(\partial\mu/\partial c)_{PT}}$$

$$= \left(\frac{\partial v}{\partial \mu}\right)_{PT} = -\left(\frac{\partial c}{\partial P}\right)_{T\mu} = -\left(\frac{\partial c}{\partial v}\right)_{T\mu}\left(\frac{\partial v}{\partial P}\right)_{T\mu} = \frac{v}{K_t}\left(\frac{\partial c}{\partial v}\right)_{T\mu}, \tag{7.60}$$

$$\beta = \mu + \frac{L_{t\mu}}{L_\mu} = \mu + k_T\left(\frac{\partial \mu}{\partial c}\right)_{PT} - T\left(\frac{\partial \mu}{\partial T}\right)_{Pc}. \tag{7.61}$$

In going between the various expressions given for these material parameters, we have used the four facts about partial derivatives from Chapter 6 as well as the Maxwell relation in the Gibb's representation $(\partial\mu/\partial P)_{Tc} = \partial^2 g/(\partial P\partial c) = \partial^2 g/(\partial c\partial P) = (\partial v/\partial c)_{TP}$. In terms of these parameters, the coupled transport laws can be written in final form as

$$\boldsymbol{q} = -\lambda_t \nabla T + \beta \boldsymbol{j} \tag{7.62}$$

$$\boldsymbol{j} = -\rho D_c\left(\frac{k_T}{T}\nabla T + \xi_P \nabla P + \nabla c\right). \tag{7.63}$$

One can of course substitute the law for $\boldsymbol{j}$ into that for $\boldsymbol{q}$. All of β (units of J kg^{-1}), ξ_P (units of Pa^{-1}) and k_T (dimensionless) can be experimentally measured as suggested by these laws. The coefficient k_T is called the *thermal-diffusion ratio*. Note that k_T and β are related through Eqs (7.59), (7.61) and an equation of state for the chemical potential. If an equation of state for μ is known, or if the specific volume can be assumed to be independent of pressure, which it can [see Eq. (7.79) below], then ξ_P can be determined from Eq. (7.60). However because $\xi_p \sim 1/K_t$ and because $P/K_t \to 0$ for nearly all flow scenarios, the force term $\xi_p \nabla P$ can safely be neglected in most flow and transport scenarios. If β is measured experimentally and an equation of state for μ is known, k_T can be determined from Eq. (7.59).

You can prove with guidance in the final end-of-chapter exercise a result due to Einstein for the solute diffusivity that is the relation $D_c = k_B T b$, where k_B is Boltzmann's constant and $b = v/F$ is the mobility of a solute molecule (the ratio of the terminal velocity v of a solute molecule in solution to the force F that is driving the velocity). As shown in Sec. 5.6.4, when a possibly hydrated solute molecule has an effective radius R that is larger than the size of the solvent molecules, the Stoke's approximation yields $b = 1/(6\pi\eta R)$, where η is the viscosity of the solution.

7.3.4 The Solute-Diffusion Governing Equations

We now bring together everything we have learned to give the governing partial-differential equations when all of solute diffusion, thermal diffusion and fluid flow are occurring.

In the Gibb's representation $g = g(T, P, c)$, the constitutive laws for reversible processes are

$$\frac{\mathrm{d}s}{\mathrm{d}t} = \frac{c_p}{T}\frac{\mathrm{d}T}{\mathrm{d}t} - v\alpha_p\frac{\mathrm{d}P}{\mathrm{d}t} - \left(\frac{\partial\mu}{\partial T}\right)\frac{\mathrm{d}c}{\mathrm{d}t}, \tag{7.64}$$

$$\frac{1}{v}\frac{\mathrm{d}v}{\mathrm{d}t} = \alpha_p\frac{\mathrm{d}T}{\mathrm{d}t} - \frac{1}{K_t}\frac{\mathrm{d}P}{\mathrm{d}t} + \frac{1}{v}\left(\frac{\partial v}{\partial c}\right)\frac{\mathrm{d}c}{\mathrm{d}t}, \tag{7.65}$$

$$\frac{\mathrm{d}\mu}{\mathrm{d}t} = \left(\frac{\partial\mu}{\partial T}\right)\frac{\mathrm{d}T}{\mathrm{d}t} + \left(\frac{\partial v}{\partial c}\right)\frac{\mathrm{d}P}{\mathrm{d}t} + \left(\frac{\partial\mu}{\partial c}\right)\frac{\mathrm{d}c}{\mathrm{d}t}. \tag{7.66}$$

The assumption we make here is that the flow is incompressible which means that pressure changes are not causing density to change. In the expression for the density (or specific volume) change $v^{-1}\mathrm{d}v/\mathrm{d}t = \nabla \cdot \boldsymbol{v}$ of Eq. (7.65), this means we can neglect the pressure changes so that density is only changing due to thermal expansion or compositional changes so that the incompressibility condition for the flow is that

$$\nabla \cdot \boldsymbol{v} = \alpha_p \frac{\mathrm{d}T}{\mathrm{d}t} + \frac{1}{v}\left(\frac{\partial v}{\partial c}\right)\frac{\mathrm{d}c}{\mathrm{d}t}. \tag{7.67}$$

Now it may turn out that the temperature and concentration changes are sufficiently small that we can indeed take $\nabla \cdot \boldsymbol{v} \approx 0$ to be the incompressibility condition but that requires additional justification as was given earlier for thermal advection in a one-component system.

Per the earlier discussion, if we again neglect the small pressure change contribution to the heat balance of Eq. (7.64) we have

$$T\frac{\mathrm{d}s}{\mathrm{d}t} = c_p \frac{\mathrm{d}T}{\mathrm{d}t} - T\left(\frac{\partial \mu}{\partial T}\right)\frac{\mathrm{d}c}{\mathrm{d}t}. \tag{7.68}$$

Similarly, in the transport laws, per the earlier discussion that $\xi_P \nabla P$ will force negligible solute flux compared to ∇c, which is equivalent to the statement that $|\Delta P| \ll K_t|\Delta c| \approx 10^{10}$ Pa $|\Delta c|$ which we assume to be satisfied, the transport laws can be rewritten so that

$$\boldsymbol{q} = -\tilde{\lambda}_t \nabla T - \rho D_c \beta \nabla c, \tag{7.69}$$

$$\boldsymbol{j} = -\rho D_c \frac{k_T}{T}\nabla T - \rho D_c \nabla c \tag{7.70}$$

where

$$\tilde{\lambda}_t = \lambda_t + \beta \rho D_c \frac{k_T}{T} \tag{7.71}$$

is the effective thermal conductivity when concentration gradients are absent. Equations (7.67) – (7.70) are the approximations related to the flow being incompressible, that is, the approximation that pressure changes are not causing the thermodynamic variables to change appreciably.

Returning to both the initial statement of the conservation of entropy $\rho T\mathrm{d}S/\mathrm{d}t = -\nabla \cdot \boldsymbol{q} + \mu \nabla \cdot \boldsymbol{j}$ and the statement of conservation of solute mass $\rho \mathrm{d}c/\mathrm{d}t = -\nabla \cdot \boldsymbol{j}$ we can finally write the suite of governing equations for the coupled advection and diffusion of temperature and solute concentration as

$$\rho c_p \left(\frac{\partial T}{\partial t} + \boldsymbol{v} \cdot \nabla T\right) = -\nabla \cdot \boldsymbol{q} + \tilde{\mu} \nabla \cdot \boldsymbol{j}, \tag{7.72}$$

$$\rho \left(\frac{\partial c}{\partial t} + \boldsymbol{v} \cdot \nabla c\right) = -\nabla \cdot \boldsymbol{j}, \tag{7.73}$$

$$\boldsymbol{q} = -\tilde{\lambda}_t \nabla T - \rho D_c \beta \nabla c, \tag{7.74}$$

$$\boldsymbol{j} = -\rho D_c \frac{k_T}{T}\nabla T - \rho D_c \nabla c \tag{7.75}$$

where

$$\tilde{\mu} = \mu - T\frac{\partial \mu}{\partial T}. \tag{7.76}$$

In addition to these four coupled first-order differential equations for T and c, we also must consider the flow equations. The compressibility condition due to thermal expansions and compositional changes is again

$$\nabla \cdot \boldsymbol{v} = \alpha_p \left(\frac{\partial T}{\partial t} + \boldsymbol{v} \cdot \nabla T\right) + \frac{1}{v}\left(\frac{\partial v}{\partial c}\right)\left(\frac{\partial c}{\partial t} + \boldsymbol{v} \cdot \nabla c\right) \tag{7.77}$$

and because $\nabla \cdot \boldsymbol{v} \neq 0$, the flow equations are again

$$\rho\left(\frac{\partial \boldsymbol{v}}{\partial t} + \boldsymbol{v} \cdot \nabla \boldsymbol{v}\right) = -\nabla P + \eta \nabla^2 \boldsymbol{v} + \left(\kappa + \frac{\eta}{3}\right) \nabla \nabla \cdot \boldsymbol{v} + \rho \boldsymbol{g}. \tag{7.78}$$

The density has a compositional dependence given by

$$\rho = \rho_1 c + \rho_2 (1 - c) = \rho_2 - (\rho_2 - \rho_1) c \tag{7.79}$$

where ρ_2 is the density of the pure solvent when $c = 0$ and ρ_1 is the density that the solution asymptotes toward as $c \to 1$. These pure solvent and solute densities are functions of temperature due to thermal expansion but not of pressure under the incompressibility assumption so that $\rho = \rho(T, c)$.

Again, it is Eq. (7.77) for $\nabla \cdot \boldsymbol{v}$ that provides the additional independent relation that allows the pressure to be determined. Just as for solenoidal flow and the earlier single-component thermal advection–diffusion scenario, this can be seen by taking the divergence of Eq. (7.78) and using Eq. (7.77) for $\nabla \cdot \boldsymbol{v}$ to obtain an equation for pressure

$$\begin{aligned} \nabla^2 P = &\left[\left(\kappa + \frac{4\eta}{3}\right)\nabla^2 - \rho\left(\frac{\partial}{\partial t} + \boldsymbol{v} \cdot \nabla\right)\right](\nabla \cdot \boldsymbol{v}) \\ &+ \left[\left(\frac{\partial \rho}{\partial T}\right)\nabla T + \left(\frac{\partial \rho}{\partial c}\right)\nabla c\right] \cdot \boldsymbol{g}. \end{aligned} \tag{7.80}$$

So the pressure satisfies a Poisson equation when both thermal expansion and compositional change cause $\nabla \cdot \boldsymbol{v} \neq 0$. The entire suite of Eqs (7.72) through (7.79) are to be solved simultaneously subject to appropriate initial and boundary conditions.

If all spatial and temporal variations of ρ are negligible so that $\nabla \cdot \boldsymbol{v} = -\rho^{-1} \mathrm{d}\rho/\mathrm{d}t = 0$, fluid flow is governed by this and the usual Navier–Stokes equation $\rho(\partial \boldsymbol{v}/\partial t + \boldsymbol{v} \cdot \nabla \boldsymbol{v}) = -\nabla P + \eta \nabla^2 \boldsymbol{v} + \rho \boldsymbol{g}$. If we further assume that all spatial variations of the transport coefficients are negligible so that they can be taken outside of $\nabla \cdot \boldsymbol{q}$ and $\nabla \cdot \boldsymbol{j}$, then the two governing equations for T and c are

$$\frac{\partial T}{\partial t} + \boldsymbol{v} \cdot \nabla T = \tilde{D}_t \nabla^2 T + \frac{D_c(\beta - \tilde{\mu})}{c_p} \nabla^2 c, \tag{7.81}$$

$$\frac{\partial c}{\partial t} + \boldsymbol{v} \cdot \nabla c = D_c \frac{k_T}{T} \nabla^2 T + D_c \nabla^2 c \tag{7.82}$$

where

$$\tilde{D}_t = \frac{\tilde{\lambda}_t}{\rho c_p} - \frac{\tilde{\mu} D_c}{c_p} \frac{k_T}{T} \tag{7.83}$$

is the effective thermal diffusivity of the fluid.

Last, if temperature variations are absent, the solute concentration obeys an advection–diffusion equation

$$\boxed{\frac{\partial c}{\partial t} + \boldsymbol{v} \cdot \nabla c - D_c \nabla^2 c = 0.} \tag{7.84}$$

As in the modeling of temperature, we can define the dimensionless Péclet number that quantifies whether advection is important relative to solute diffusion

$$\text{Pe} = \frac{|\boldsymbol{v}|\ell}{D_c}, \tag{7.85}$$

where ℓ is again a characteristic distance over which solute concentration is varying. If $\text{Pe} \ll 1$, the advection–diffusion equation for the solute concentration becomes the simple diffusion equation that will be studied extensively in Part II of the book. When both advection and solute diffusion are simultaneously occurring in some application at nonnegligible Péclet number, such transport is often given the name *dispersion*.

7.3.5 Taylor Dispersion in a Cylindrical Flow Tube

Solving advection–diffusion problems analytically is, most typically, a mathematical challenge. You can solve a simple example, though rather contrived, as an end-of-chapter exercise. Here, we address an example of advection–diffusion corresponding to how solute diffuses in a cylindrical tube in which solvent is flowing. The initial distribution of the solute is not essential. You can think of pure solvent flowing in the tube with its usual parabolic flow profile and then at a certain moment in time, solute is added at one end of the tube at a given concentration c_a. The solute concentration at the center of the tube is pulled forward more rapidly by advection in comparison to the advection near the tube walls. There will then ensue radial diffusion of the solute in addition to the dominant diffusion in the axial flow direction. As initially suggested by Taylor (1953), the net effect, at least in narrow tubes with moderate nonturbulent flow speeds where radial diffusion equilibrates more rapidly than axial diffusion, is that the average solute concentration across each cross section of the tube will obey a purely axial advection–diffusion equation in the tube but with an effective diffusivity that is now no longer just D_c but is influenced by the Péclet number of the flow as $D_{\text{eff}} = D_c(1 + \text{Pe}^2/48)$.

To demonstrate this result, the analysis that follows employs a perturbation expansion as described in Chapter 5 on viscous fluid flow. This requires taking the dimensions out of the problem before carrying out the perturbation analysis. Let's work in a frame of reference that moves along with the average speed of the fluid flow $\bar{v}_z$ in the cylindrical tube of radius a so that the fluid velocity in the cylindrical tube from this moving frame of reference is $\boldsymbol{v}'(r', t') = \left\{2\bar{v}_z[1 - (r'/a)^2] - \bar{v}_z\right\}\hat{z}$, a result you proved in Exercise 4 of Chapter 5. Note that we are putting primes on variables that have their usual mks (SI) dimensions so that when we take out dimensions, the variables are unprimed and thus easier to work with. Note that $\bar{v}_z$ (m/s) is not a variable but a set constant for the advection and diffusion. The advection–diffusion equation for the solute concentration $c'(r', z', t')$ in the cylindrical tube from a frame of reference that moves with the average flow speed is then

$$\frac{\partial c'}{\partial t'} + \bar{v}_z\left[1 - 2\left(\frac{r'}{a}\right)^2\right]\frac{\partial c'}{\partial z'} - D_t\left[\frac{\partial^2 c'}{\partial z'^2} + \frac{1}{r'}\frac{\partial}{\partial r'}\left(r'\frac{\partial c'}{\partial r'}\right)\right] = 0. \tag{7.86}$$

This is an exact statement to the extent that the gradients of solute concentration do not influence the fluid flow. When capillary physics is treated in a later section, we show how gradients of solute concentration can influence the flow field.

We now take dimensions out of the problem through the substitutions (unprimed variables are now dimensionless)

$$t' = \frac{a^2}{D_t}t, \quad z' = az, \quad r' = ar, \quad c' = c_a c, \tag{7.87}$$

so that the advection–diffusion equation in a moving frame of reference takes the dimensionless form

$$\frac{\partial c}{\partial t} + \text{Pe}(1 - 2r^2)\frac{\partial c}{\partial z} - \frac{\partial^2 c}{\partial z^2} - \frac{1}{r}\left[\frac{\partial}{\partial r}\left(r\frac{\partial c}{\partial r}\right)\right] = 0, \tag{7.88}$$

where $\text{Pe} = a\bar{v}_z/D_c$ is the Péclet number, which is assumed to be nonnegligible. Note that c' was already dimensionless (it is defined as a mass ratio) but we normalized this variable so that the concentration at peak value prior to diffusion and advection is 1. This is not required to do however.

Let's now introduce the one approximation into the analysis that results in the desired analytical result. Assume that the flow is happening slowly enough and that the tube is narrow enough that the radial diffusion happens rapidly compared to the axial diffusion. We can impose this condition mathematically by saying that if we stretch the time axis by the substitution $t \to t/\epsilon$ where ϵ is a small dimensionless "perturbation parameter," then the axial coordinate will stretch due to diffusion as $z \to z/\sqrt{\epsilon}$ because $z \propto \sqrt{t}$ for diffusion. However, because the radial diffusion occurs rapidly, we do not stretch the r coordinate as time is being slowed by ϵ. The advection–diffusion equation then becomes

$$\epsilon\frac{\partial c}{\partial t}+\sqrt{\epsilon}\mathrm{Pe}(1-2r^2)\frac{\partial c}{\partial z}-\epsilon\frac{\partial^2 c}{\partial z^2}-\frac{1}{r}\left[\frac{\partial}{\partial r}\left(r\frac{\partial c}{\partial r}\right)\right]=0, \tag{7.89}$$

when radial-diffusion equilibration is more rapidly achieved than axial diffusion.

Into this we insert the *ansatz* (or "asymptotic expansion" or "perturbation expansion")

$$c=c_0+\sqrt{\epsilon}c_1+\epsilon c_2+\ldots \tag{7.90}$$

and group the terms in the advection–diffusion equation by common powers of $\sqrt{\epsilon}$

$$\begin{aligned}0=\epsilon^0&\left[\frac{1}{r}\frac{\partial}{\partial r}\left(r\frac{\partial c_0}{\partial r}\right)\right]\\ &+\sqrt{\epsilon}\left[\mathrm{Pe}(1-2r^2)\frac{\partial c_0}{\partial z}-\frac{1}{r}\frac{\partial}{\partial r}\left(r\frac{\partial c_1}{\partial r}\right)\right]\\ &+\epsilon\left[\frac{\partial c_0}{\partial t}+\mathrm{Pe}(1-2r^2)\frac{\partial c_1}{\partial z}-\frac{\partial^2 c_0}{\partial z^2}-\frac{1}{r}\frac{\partial}{\partial r}\left(r\frac{\partial c_2}{\partial r}\right)\right]+\ldots.\end{aligned} \tag{7.91}$$

For this to be valid for any $\epsilon \le 1$, each term in square brackets must equal zero. The radial boundary conditions for each of these solute concentration fields c_i is that there is no radial diffusion at the tube wall $\partial c_i/\partial r|_{r=1}=0$ (where the dimensionless tube radius is 1) and that the concentration remains finite at $r=0$. The overall logic of the above perturbation scheme is that by stretching time as t/ϵ and axial distance as $z/\sqrt{\epsilon}$ without stretching r at all, the time and space dependence of the solution can be separated in a way that allows radial diffusion to occur rapidly and axial diffusion more slowly due to the small radius of the tube compared to its length. Once the functions $c_0(r,z,t)$, $c_1(r,z,t)$, and $c_2(r,z,t)$ are determined from the above differential equations and boundary conditions, these expressions can be inserted into the ansatz of Eq. (7.90), which is the solution of the advection–diffusion equation, Eq. (7.89). We can then let $\epsilon\to 1$ (remove the coordinate stretching) so that the final solution is $c(r,z,t)=c_0(r,z,t)+c_1(r,z,t)+c_2(r,z,t)+\ldots$, which satisfies the advection–diffusion equation while having a rapid radial diffusion built into it.

Thus, the leading order problem for $c_0(r,z,t)$ is a solution of

$$\frac{1}{r}\frac{\partial}{\partial r}\left(r\frac{\partial c_0}{\partial r}\right)=0. \tag{7.92}$$

Multiplying by r, integrating, and dividing by r gives

$$\frac{\partial c_0}{\partial r}=\frac{C(z,t)}{r}, \tag{7.93}$$

where $C(z,t)$ is an integration constant that must be zero either by the no-flux radial boundary condition at $r=1$ or by the finiteness of this gradient at $r=0$. Another integration then shows that $c_0=c_0(z,t)$ is a function only of z and t and is entirely independent of r. As such, $c_0(z,t)=\bar{c}(z,t)$, where $\bar{c}$ is the average across each cross section of the tube. The function $c_0(z,t)$ will be found from the diffusion equation that eventually emerges below.

The next problem for $c_1(r, z, t)$ is

$$\frac{1}{r}\frac{\partial}{\partial r}\left(r\frac{\partial c_1}{\partial r}\right) = \mathrm{Pe}\left(1 - 2r^2\right)\frac{\partial c_0}{\partial z}. \tag{7.94}$$

Multiplying by r, integrating, and then dividing by r gives

$$\frac{\partial c_1}{\partial r} = \frac{\mathrm{Pe}}{2}\left(r - r^3\right)\frac{\partial c_0}{\partial z} + \frac{C(z,t)}{r}, \tag{7.95}$$

where the integration constant $C(z, t)$ must again be zero either by the no-flux boundary condition on $r = 1$ or finiteness at $r = 0$. A final integration then gives

$$c_1(r, z, t) = \frac{\mathrm{Pe}}{4}\left(r^2 - \frac{r^4}{2}\right)\frac{\partial c_0}{\partial z} + D(z, t). \tag{7.96}$$

Because we identified $c_0(z, t) = \bar{c}(z, t)$ as the cross-sectional average concentration, we must have $\bar{c}_1 = 0$. To perform the average, note that because the cylinder has a radius of 1 in dimensionless variables and a cross-sectional area of π, the average of a radial function $f(r)$ over a cross section is $\bar{f} = \pi^{-1}\int_0^{2\pi}\int_0^1 \mathrm{d}r\, rf(r) = 2\int_0^1 \mathrm{d}r\, rf(r)$, that is, the average is performed by multiplying $f(r)$ by $2r\,\mathrm{d}r$ and integrating from 0 to 1. Thus, multiplying $c_1(r, z, t)$ by $2r\,\mathrm{d}r$ and integrating from $r = 0$ to $r = 1$ gives

$$D(z, t) = -\frac{\mathrm{Pe}}{12}\frac{\partial c_0}{\partial z} \tag{7.97}$$

and

$$c_1(r, z, t) = \frac{\mathrm{Pe}}{4}\left(r^2 - \frac{r^4}{2} - \frac{1}{3}\right)\frac{\partial c_0}{\partial z}. \tag{7.98}$$

Using this result for $c_1(r, z, t)$ in the next problem for $c_2(r, z, t)$ gives

$$\frac{1}{r}\frac{\partial}{\partial r}\left(r\frac{\partial c_2}{\partial r}\right) = \frac{\partial c_0}{\partial t} - \frac{\partial^2 c_0}{\partial z^2} + \frac{\mathrm{Pe}^2}{4}\left(r^2 - \frac{r^4}{2} - \frac{1}{3}\right)(1 - 2r^2)\frac{\partial^2 c_0}{\partial z^2}. \tag{7.99}$$

We now stop these expansions and average this last expression over the cross section of the tube.

So multiplying through by $2r\,\mathrm{d}r$ and integrating from 0 to 1 gives

$$2r\frac{\partial c_2}{\partial r}\bigg|_0^1 = 0 = \frac{\partial c_0}{\partial t} - \frac{\partial^2 c_0}{\partial z^2} + \frac{\mathrm{Pe}^2}{2}\int_0^1\left(-\frac{r}{3} + \frac{5}{3}r^3 - \frac{5}{2}r^5 + r^7\right)\mathrm{d}r\,\frac{\partial^2 c_0}{\partial z^2}, \tag{7.100}$$

where the left-hand side is again zero due to the no flux boundary condition at the tube wall $r = 1$. Performing these last integrals then gives that $c_0(z, t) = \bar{c}(z, t)$ satisfies the diffusion equation

$$\frac{\partial c_0(z,t)}{\partial t} = \left(1 + \frac{\mathrm{Pe}^2}{48}\right)\frac{\partial c_0(z,t)}{\partial z^2}. \tag{7.101}$$

If we now restore dimensions (put the primes back on the variables) and move back to a stationary frame of reference, we have that the average concentration across each cross section of the tube satisfies the advection–diffusion equation

$$\frac{\partial \bar{c}'}{\partial t'} + \bar{v}_z \frac{\partial \bar{c}'}{\partial z'} = D_c \left(1 + \frac{\mathrm{Pe}^2}{48}\right) \frac{\partial^2 \bar{c}'}{\partial z'^2}. \tag{7.102}$$

So the conclusion is that the cross-sectional average of the solute concentration in the cylindrical tube satisfies the advection–diffusion equation with an effective solute diffusivity given by

$$\boxed{D_{\mathrm{eff}} = D_c \left(1 + \frac{\mathrm{Pe}^2}{48}\right).} \tag{7.103}$$

We will solve Eq. (7.102) in Part II of the book, specifically in Section 10.6.4. Because Eq. (7.102) involves the average concentration across the parabolic flow profile in the cylindrical pore, it is often assumed to apply to the average concentration in a porous-continuum description of advection and solute diffusion in a porous material (e.g., a packing of sand grains). However, we should anticipate that in the expression for the effective diffusivity, the geometric factor of 1/48 will be different in the sand pack compared to the cylinder.

Note that temperature follows identically the same advection–diffusion equation as does the solute concentration. If we consider the particular case of a cylindrical flow tube with walls that are insulated to heat flux, the temperature and concentration problems are identical. So for the special case of an insulating flow tube, if a fluid having one temperature is introduced at one end of a tube that is initially flowing with the same fluid at a different temperature, the average temperature across the tube cross section will again follow an axial advection–diffusion equation with an effective diffusivity given by $D_t(1 + \mathrm{Pe}^2/48)$.

For water flowing in a glass flow tube, however, the thermal conductivity of the glass and water are of comparable order of magnitude with the glass actually more conductive than the water. In this case, one needs to model the complete problem of advection and diffusion for the water within the tube as well as the temperature diffusion across the glass of the tube under the two boundary conditions that the outside of the glass tube is in contact with a gas of given constant temperature and that the radial heat flux from the water into the glass is continuous across the water–glass interface. We will not attempt to solve this more involved advection–diffusion problem.

7.4 Transport Coefficient Reciprocity

We would like to demonstrate the reciprocity in the linear transport laws by two distinct arguments: (1) a nonstandard continuum argument based on a straightforward manipulation of the conservation and transport laws in systems where advection is occurring and

(2) the standard statistical-mechanics argument called *Onsager reciprocity* in which the thermal fluctuations through time in the reservoirs bounding a sample being studied are time reversible due to the underlying time reversibility of the molecular dynamics. In the Onsager treatment, advection is not taking place and so is a bit less general than the continuum argument where advection can occur. We will begin with the continuum argument that I don't think exists elsewhere.

7.4.1 Continuum Demonstration of Reciprocity with Flow Reversal

To be general, we formulate coupled transport equations for any response fields g and h that satisfy standard conservation equations. These fields could be temperature or concentrations as in the previous sections or any physical quantity that is diffusing and advecting according to the conservation laws

$$\frac{\partial g}{\partial t} + \boldsymbol{v} \cdot \nabla g = -\nabla \cdot \boldsymbol{J}_g \tag{7.104}$$

$$\frac{\partial h}{\partial t} + \boldsymbol{v} \cdot \nabla h = -\nabla \cdot \boldsymbol{J}_h, \tag{7.105}$$

with coupling occurring in the diffusive-flux transport equations

$$\boldsymbol{J}_g = -L_{gg} \nabla g - L_{gh} \nabla h \tag{7.106}$$

$$\boldsymbol{J}_h = -L_{hg} \nabla g - L_{hh} \nabla h. \tag{7.107}$$

Our purpose in this section is to investigate the relation between L_{gh} and L_{hg}. We are assuming the transport is isotropic. For anisotropic systems in which the transport coefficients would be second-order tensors, a similar but somewhat more involved exercise shows that the matrix of transport coefficients satisfies the same type of symmetry as that obtained below. Although the proposed transport of g and h being described here has some restrictions, it nonetheless covers many linear transport scenarios of interest when advection is occurring.

We now consider some material, possibly heterogeneous, occupying a region Ω in which the above transport is occurring. For simplicity, we will take Ω to be a cube with sides of length ℓ and consider two distinct boundary-value problems (BVPs) on this cube. In Problem 1, there is a drop in the field g across the cube and no drop in h while in Problem 2, there is a drop in h across the cube but no drop in g. Let's say the drop takes place in the z direction (could be any direction due to the assumed isotropy). Assuming that the fields g and h represent the nonequilibrium deviations from the associated equilibrium (constant) values in the cube and bounding reservoirs, the boundary conditions for these two problems are

$$g_1 = \pm \Delta g/2 \quad \text{on} \quad z = \pm \ell/2 \tag{7.108}$$

$$h_1 = 0 \quad \text{on} \quad z = \pm \ell/2 \tag{7.109}$$

$$g_2 = 0 \quad \text{on} \quad z = \pm \ell/2 \tag{7.110}$$

$$h_2 = \pm \Delta h/2 \quad \text{on} \quad z = \pm \ell/2 \tag{7.111}$$

with homogeneous (i.e., equal to zero) boundary conditions for the fields on the other four faces of the cube in the x and y directions.

Simultaneously, we also consider the average transport throughout the entire possibly heterogeneous cube that satisfies transport equations of the form

$$\overline{J}_g = -L^e_{gg} \nabla \overline{g} - L^e_{gh} \nabla \overline{h} \tag{7.112}$$

$$\overline{J}_h = -L^e_{hg} \nabla \overline{g} - L^e_{hh} \nabla \overline{h}, \tag{7.113}$$

where the overbars denote volume averaging and the superscript e on the coefficients means they are the effective coefficients for the entire possibly heterogeneous cube. Because for our particular BVP the average gradients are given by $\nabla \overline{g} = (\Delta g/\ell)\hat{z}$ and $\nabla \overline{h} = (\Delta h/\ell)\hat{z}$, we can define the effective coefficients by averaging the local flux to give

$$L^e_{gg} \frac{\Delta g}{\ell} \widehat{=} \frac{1}{\ell^3} \int_\Omega \hat{z} \cdot \left(L_{gg} \nabla g_1 + L_{gh} \nabla h_1 \right) \, dV \tag{7.114}$$

$$L^e_{gh} \frac{\Delta h}{\ell} \widehat{=} \frac{1}{\ell^3} \int_\Omega \hat{z} \cdot \left(L_{gg} \nabla g_2 + L_{gh} \nabla h_2 \right) \, dV \tag{7.115}$$

$$L^e_{hg} \frac{\Delta g}{\ell} \widehat{=} \frac{1}{\ell^3} \int_\Omega \hat{z} \cdot \left(L_{hg} \nabla g_1 + L_{hh} \nabla h_1 \right) \, dV \tag{7.116}$$

$$L^e_{hh} \frac{\Delta h}{\ell} \widehat{=} \frac{1}{\ell^3} \int_\Omega \hat{z} \cdot \left(L_{hg} \nabla g_2 + L_{hh} \nabla h_2 \right) \, dV. \tag{7.117}$$

The volume integral over the region Ω can be written as an integral from $z = -\ell/2$ to $z = +\ell/2$ over slices perpendicular to the z direction. The diffusive flux of g or h across each such slice is identical in the steady state. We elect to write this fact, for purposes in the demonstration that follows, as the average over the two terminal faces $z = \pm \ell/2$

$$L^e_{gg} \widehat{=} \frac{1}{2\Delta g \ell} \left[\int_{z=+\ell/2} \hat{z} \cdot \left(L_{gg} \nabla g_1 + L_{gh} \nabla h_1 \right) \, dS + \int_{z=-\ell/2} \hat{z} \cdot \left(L_{gg} \nabla g_1 + L_{gh} \nabla h_1 \right) \, dS \right] \tag{7.118}$$

$$L^e_{gh} \widehat{=} \frac{1}{2\Delta h \ell} \left[\int_{z=+\ell/2} \hat{z} \cdot \left(L_{gg} \nabla g_2 + L_{gh} \nabla h_2 \right) \, dS + \int_{z=-\ell/2} \hat{z} \cdot \left(L_{gg} \nabla g_2 + L_{gh} \nabla h_2 \right) \, dS \right] \tag{7.119}$$

$$L^e_{hg} \widehat{=} \frac{1}{2\Delta g \ell} \left[\int_{z=+\ell/2} \hat{z} \cdot \left(L_{hg} \nabla g_1 + L_{hh} \nabla h_1 \right) \, dS + \int_{z=-\ell/2} \hat{z} \cdot \left(L_{hg} \nabla g_1 + L_{hh} \nabla h_1 \right) \, dS \right] \tag{7.120}$$

$$L^e_{hh} \widehat{=} \frac{1}{2\Delta h \ell} \left[\int_{z=+\ell/2} \hat{z} \cdot \left(L_{hg} \nabla g_2 + L_{hh} \nabla h_2 \right) \, dS + \int_{z=-\ell/2} \hat{z} \cdot \left(L_{hg} \nabla g_2 + L_{hh} \nabla h_2 \right) \, dS \right]. \tag{7.121}$$

Note that we have not specified the size of Ω. If Ω is large enough to include heterogeneity of both the material and flow fields, then we expect the effective coefficients of the sample to be distinct from a simple arithmetic average of the local coefficients. But if Ω is so small that $\boldsymbol{v}$ and the material is uniform within Ω, then we expect that $L^e_{ij} = L_{ij}$ which means that over these small scales, the gradients of g and h are uniform within Ω and given by $\nabla g_1 = (\Delta g/\ell)\hat{\boldsymbol{z}}$, $\nabla h_1 = 0$, $\nabla g_2 = 0$ and $\nabla h_2 = (\Delta h/\ell)\hat{\boldsymbol{z}}$.

With all the above preliminaries out of the way, the proof is straightforward. Seeking the symmetry in the steady state where the time derivatives are explicitly zero and $\nabla \cdot \boldsymbol{v} = 0$, we form the following products between local fields from the two steady-state BVPs posed in Ω

$$g_2 \left[\nabla \cdot \left(g_1 \boldsymbol{v} - L_{gg} \nabla g_1 - L_{gh}(+\boldsymbol{v}) \nabla h_1\right) = 0\right] \tag{7.122}$$

$$g_1 \left[\nabla \cdot \left(-g_2 \boldsymbol{v} - L_{gg} \nabla g_2 - L_{gh}(-\boldsymbol{v}) \nabla h_2\right) = 0\right]. \tag{7.123}$$

Note that we wrote $\boldsymbol{v} \cdot \nabla g = \nabla \cdot (g\boldsymbol{v})$ because the steady flow is incompressible $\nabla \cdot \boldsymbol{v} = 0$. Further, for the Problem 1 statement we used a flow field $+\boldsymbol{v}$ while in the Problem 2 statement we used $-\boldsymbol{v}$. This is key to obtaining the symmetry we will derive. We are distinguishing that the coefficient $L_{gh}(\boldsymbol{v})$ may be different than the similar coefficient when the flow field is reversed $L_{gh}(-\boldsymbol{v})$ in Problem 2.

Upon using the identity $\nabla \cdot (\alpha \boldsymbol{a}) = \alpha \nabla \cdot \boldsymbol{a} + \boldsymbol{a} \cdot \nabla \alpha$ these products become

$$\begin{aligned} 0 =& \nabla \cdot \left[g_2 \left(g_1 \boldsymbol{v} - L_{gg} \nabla g_1 - L_{gh}(+\boldsymbol{v}) \nabla h_1\right)\right] \\ &+ \nabla g_2 \cdot \left[g_1 \boldsymbol{v} - L_{gg} \nabla g_1 - L_{gh}(+\boldsymbol{v}) \nabla h_1\right] \end{aligned} \tag{7.124}$$

$$\begin{aligned} 0 =& \nabla \cdot \left[g_1 \left(-g_2 \boldsymbol{v} - L_{gg} \nabla g_2 - L_{gh}(-\boldsymbol{v}) \nabla h_2\right)\right] \\ &+ \nabla g_1 \cdot \left[-g_2 \boldsymbol{v} - L_{gg} \nabla g_2 - L_{gh}(-\boldsymbol{v}) \nabla h_2\right]. \end{aligned} \tag{7.125}$$

Next we subtract these two statements, integrate over Ω, use the divergence theorem and apply the boundary-conditions both on $z = \pm \ell/2$ and on the lateral faces

$$\begin{aligned} 0 =& \frac{\Delta g}{2} \left\{ \int_{z=\ell/2} \hat{\boldsymbol{z}} \cdot \left(L_{gg} \nabla g_2 + L_{gh}(-\boldsymbol{v}) \nabla h_2\right) \mathrm{d}S \right. \\ &+ \left. \int_{z=-\ell/2} \hat{\boldsymbol{z}} \cdot \left(L_{gg} \nabla g_2 + L_{gh}(-\boldsymbol{v}) \nabla h_2\right) \mathrm{d}S \right\} \\ &+ \int_\Omega \left[-L_{gh}(+\boldsymbol{v}) \nabla g_2 \cdot \nabla h_1 + L_{gh}(-\boldsymbol{v}) \nabla g_1 \cdot \nabla h_2 + \boldsymbol{v} \cdot (g_1 \nabla g_2 + g_2 \nabla g_1)\right] \mathrm{d}V. \end{aligned} \tag{7.126}$$

Using the definition of the effective coefficient L^e_{gh} given by Eq. (7.119) then gives

$$\begin{aligned} \ell \Delta g \Delta h L^e_{gh}(-\boldsymbol{v}) = \int_\Omega \Big[& L_{gh}(+\boldsymbol{v}) \nabla g_2 \cdot \nabla h_1 - L_{gh}(-\boldsymbol{v}) \nabla g_1 \cdot \nabla h_2 \\ & - \boldsymbol{v} \cdot \nabla (g_1 g_2) \Big] \mathrm{d}V. \end{aligned} \tag{7.127}$$

An identical manipulation starting from the products

$$h_2\left[\nabla\cdot\left(h_1\boldsymbol{v}-L_{hg}(+\boldsymbol{v})\nabla g_1-L_{hh}\nabla h_1\right)=0\right] \tag{7.128}$$
$$h_1\left[\nabla\cdot\left(-h_2\boldsymbol{v}-L_{hg}(-\boldsymbol{v})\nabla g_2-L_{hh}\nabla h_2\right)=0\right] \tag{7.129}$$

then yields

$$\ell\Delta g\Delta h L^e_{hg}(+\boldsymbol{v})=\int_\Omega\Big[L_{hg}(-\boldsymbol{v})\nabla g_2\cdot\nabla h_1-L_{hg}(+\boldsymbol{v})\nabla g_1\cdot\nabla h_2 \tag{7.130}$$
$$+\boldsymbol{v}\cdot\nabla(h_1h_2)\Big]\,\mathrm{d}V. \tag{7.131}$$

Note that $\nabla\cdot(\boldsymbol{v}g_1g_2)=\boldsymbol{v}\cdot\nabla(g_1g_2)$ due to $\nabla\cdot\boldsymbol{v}=0$. Thus in Eq. (7.127) we have

$$\int_\Omega\boldsymbol{v}\cdot\nabla(g_1g_2)\,\mathrm{d}V=\int_\Omega\nabla\cdot(\boldsymbol{v}g_1g_2)\,\mathrm{d}V=\int_{\partial\Omega}\boldsymbol{n}\cdot\boldsymbol{v}g_1g_2\,\mathrm{d}S=0, \tag{7.132}$$

where the boundary conditions were applied to get zero for this term. For the same reason, the equivalent term in Eq. (7.131) also integrates to zero.

Thus, upon subtracting Eq. (7.131) from Eq. (7.127) we obtain

$$\ell\Delta g\Delta h\left[L^e_{gh}(-\boldsymbol{v})-L^e_{hg}(+\boldsymbol{v})\right]+\int_\Omega\left[L_{gh}(-\boldsymbol{v})-L_{hg}(+\boldsymbol{v})\right]\nabla g_1\cdot\nabla h_2\,\mathrm{d}V$$
$$=\int_\Omega\left[L_{gh}(+\boldsymbol{v})-L_{hg}(-\boldsymbol{v})\right]\nabla g_2\cdot\nabla h_1\,\mathrm{d}V. \tag{7.133}$$

We first consider the limit that Ω is so small that the flow field and transport coefficients are uniform inside Ω. As already stated, in this limit we have $\nabla g_2\to 0$, $\nabla h_1\to 0$, $\nabla g_1\to(\Delta g/\ell)\hat{\boldsymbol{z}}$, $\nabla h_2\to(\Delta h/\ell)\hat{\boldsymbol{z}}$, and $L^e_{gh}\to L_{gh}$, that is, the effective coefficients become the same as the local coefficients. In this limit, Eq. (7.133) can be written $2\ell\Delta g\Delta h\left[L_{gh}(-\boldsymbol{v})-L_{hg}(+\boldsymbol{v})\right]=0$ or

$$\boxed{L_{gh}(-\boldsymbol{v})=L_{hg}(+\boldsymbol{v}).} \tag{7.134}$$

So the local transport coefficients are symmetric in this flow-reversal sense.

We now consider a macroscopic Ω within which the $\boldsymbol{v}$ and local L_{ij} are spatially variable. The local symmetry $L_{gh}(-\boldsymbol{v})=L_{hg}(+\boldsymbol{v})$ is satisfied for each $\mathrm{d}V$ within Ω, which by just changing the sign of $\boldsymbol{v}$ is the same as $L_{gh}(+\boldsymbol{v})=L_{hg}(-\boldsymbol{v})$. We then appeal to Eq. (7.133) to obtain

$$\boxed{L^e_{gh}(-\boldsymbol{v})=L^e_{hg}(+\boldsymbol{v})} \tag{7.135}$$

for the overall transport associated with a macroscopic region Ω. We call this symmetry, or reciprocity, in these transport coefficients a *flow-reversal* symmetry. Now it is possible that the L^e_{gh} and L^e_{hg} may have a negligible dependence on the flow field in which case we just have the usual symmetry. But if these coefficients have a nonnegligible dependence on the flow field, we have $L^e_{gh}(\boldsymbol{v}) \neq L^e_{hg}(\boldsymbol{v})$; equality only emerges by changing the direction of flow in either Problem 1 or Problem 2 as defined above.

Last, although we derived the flow-reversal symmetry in the steady state, it is reasonable to assume that during the transient phase of a transport process, the symmetry continues to hold.

7.4.2 Onsager Demonstration of Reciprocity

This subsection is a topic in statistical mechanics and is formally outside the scope of the book. But given the frequent reference to "Onsager reciprocity" and the relative ease of the demonstration that does not require special training in statistical mechanics, we have elected to include it. You can find the proof of Onsager reciprocity in many places, including in Onsager's (1931) original paper as well as in the straightforward demonstration of de Groot (1951) that is the basis for what follows.

Let's say you have done a continuum calculation of the rate of entropy production. In the continuum analysis just given in Section 7.4.1, this would be $\sigma = -\boldsymbol{J}_g \cdot \nabla g - \boldsymbol{J}_h \cdot \nabla h \geq 0$. There could be many such flux-force products depending on the complexity of the transport scenario. To keep things simple like in Section 7.3, we will treat a small, uniform, and isotropic system and only consider the fluxes and forces in one particular direction. For the scenario involving g and h fields, we could identify the fluxes and forces in direction z as the scalar components J_g, $F_g = -\Delta g/\ell$, J_h, and $F_h = -\Delta h/\ell$. If we generalize, as is customary, to n different types of flux-force pairs, the rate of entropy production becomes $\sigma = J_i F_i$ where we sum over i from 1 to n and the associated linear phenomenological transport equations can be written

$$J_i = L_{ik} F_k, \tag{7.136}$$

where we sum over k. Again, the goal is to analyze the symmetry between L_{ik} and L_{ki}. We will limit ourselves to the case where there is no background flow involved in the transport. Allowing for the nonequilibrium process of flow complicates the analysis that follows. The only treatment of advective transport using the Onsager argument that I am aware of is given by Flekkøy et al. (2017) to which the reader with a background in statistical mechanics is referred.

We imagine a small central cell where the steady-state transport occurs. This cell is bounded by closed reservoirs (other than for their contact with the cell), which are assumed to be large and in a state of internal equilibrium. Whatever is fluxing across the cell comes from (or goes into) the bounding reservoirs. For each type of flux J_i, we can identify a state variable a_i in the reservoir such that its time derivative $\dot{a}_i$ is the flux, that is,

$$J_i = \dot{a}_i. \tag{7.137}$$

For example, if J is a heat flux, we would take a to be proportional to temperature because as heat fluxes from the reservoir into the transport cell, the temperature of the reservoir would linearly decrease at a rate proportional to the heat flux. If J was a flux of solute particles, a would be proportional to the number of solute particles in the reservoir so that $\dot{a}$ is the rate J at which the solute is fluxing out of the reservoir. We define a_i to be the deviation (or fluctuation) of the state variable from its equilibrium value so that $a_i = 0$ in each reservoir defines equilibrium and no flux of type i occurring across the transport cell.

The total entropy of the cell and reservoirs together will be a maximum in equilibrium and will be less than this maximum when transport is occurring. This fact can be expressed by an entropy S having the quadratic form

$$S = -\frac{1}{2} g_{ik} a_i a_k \tag{7.138}$$

with summation over the indices from 1 to n. So the entropy is zero at equilibrium (alternatively, we could add a positive constant S_0 to this quadratic form and nothing in what follows changes) and is less than this maximum when transport is occurring. The real second-order matrix g_{ik} is constrained to be symmetric so that it has real eigenvalues that must all be positive so that S is negative definite $S \leq 0$. If we take a time derivative to give the rate of entropy production, we have

$$\sigma \,\widehat{=}\, \dot{S} = -\dot{a}_i g_{ik} a_k. \tag{7.139}$$

Because $J_i = \dot{a}_i$ and $\dot{S} = J_i F_i$, we can identify the force F_i driving flux i as

$$F_i = \partial S / \partial a_i = -g_{ik} a_k. \tag{7.140}$$

So in the case of a heat flux J, the driving force would be proportional to the deviation a of the temperature in the reservoir from equilibrium, that is, a temperature drop across the central cell is what drives heat transport across the cell. In the case of a solute flux J, the driving force would be proportional to the deviation a of the solute-molecule numbers from equilibrium. The entropy production can thus be stated in the earlier form $\dot{S} = J_i F_i = -\dot{a}_i g_{ik} a_k$.

We now come to the bit of statistical mechanics that must be inserted. Due to random thermal motion of the molecules, the state-variable deviations a_i in the reservoirs fluctuate around their equilibrium value of zero. For closed systems in equilibrium and experiencing such fluctuations, we are interested in the probability $P\, \mathrm{d}a_1 \ldots \mathrm{d}a_n$ of finding the state parameters between a_1 and $a_1 + \mathrm{d}a_1$, a_2 and $a_2 + \mathrm{d}a_2$, and so forth. According to Boltzmann, the probability density of these fluctuations is $P(a_1, a_2, \ldots a_n) = c e^{S(a_1, a_2, \ldots a_n)/k_B}$ where the constant c is determined from the requirement that $\int \ldots \int P\, \mathrm{d}a_1 \ldots \mathrm{d}a_n = 1$ so that

$$P\, \mathrm{d}a_1 \ldots \mathrm{d}a_n = \frac{\exp(S/k_B)\, \mathrm{d}a_1 \ldots \mathrm{d}a_n}{\int \ldots \int \exp(S/k_B)\, \mathrm{d}a_1 \ldots \mathrm{d}a_n} \tag{7.141}$$

where k_B is Boltzmann's constant. Equation (7.141) is the bit of statistical mechanics being dropped into the proof. If desired, the quadratic form of Eq. (7.138) can be inserted to obtain an explicit expression for this probability. The range to integrate each state variable over is their fluctuation range which may be taken as $-\infty$ to ∞. The probability for fluctuations close to equilibrium ($a_i = 0$) is much higher than for larger fluctuations as the quadratic form of Eq. (7.138) guarantees.

We will soon want to know the average value of the product $a_i F_j$ for some given i and j. This is calculated as

$$\langle a_i F_j \rangle \hat{=} \int \ldots \int a_i F_j P \, da_1 \ldots da_n. \tag{7.142}$$

To perform the integrals, let's take a partial derivative of Eq. (7.141) to give

$$\begin{aligned} \frac{\partial P}{\partial a_j} da_1 \ldots da_n &= \frac{1}{k_B} \frac{\partial S}{\partial a_j} \frac{\exp(S/k_B) da_1 \ldots da_n}{\int \ldots \int \exp(S/k_B)\, da_1 \ldots da_n} \\ &= \frac{1}{k_B} F_j P da_1 \ldots da_n \end{aligned} \tag{7.143}$$

where we used that $F_j = \partial S / \partial a_j$. Thus, our average can be written

$$\langle a_i F_j \rangle = k_B \int \ldots \int da_1 \ldots da_n \left[\int a_i \frac{\partial P}{\partial a_j} \, da_j \right] \tag{7.144}$$

where we pulled out and isolated the integral over a_j. Using the fact that $\partial(a_i P)/\partial a_j = a_i \partial P/\partial a_j + \delta_{ij} P$ and noting that $\int_{-\infty}^{\infty} \partial(a_i P)/\partial a_j \, da_j = a_i P|_{-\infty}^{\infty} = 0$ (the probability is zero to find a_j at the extremes of its range), we then obtain that

$$\langle a_i F_j \rangle = -k_B \int \ldots \int da_1 \ldots da_n \left[\int P \delta_{ij} \, da_j \right] = -k_B \delta_{ij}. \tag{7.145}$$

This is the extent of our use of statistical mechanics in the proof.

We next consider the time reversibility of the fluctuations in an equilibrium system which can be expressed by the following correlation statement

$$\langle a_i(t) a_j(t+\tau) \rangle = \langle a_i(t) a_j(t-\tau) \rangle \tag{7.146}$$

where τ is a temporal displacement. If we now jump the movie of $a(t)$ forward by an amount τ and begin watching the fluctuations $a(t+\tau)$, we cannot tell the difference from the earlier viewing $a(t)$. This means we can replace $t \to t+\tau$ and there is no change in the temporal correlations, or

$$\langle a_i(t) a_j(t+\tau) \rangle = \langle a_i(t+\tau) a_j(t) \rangle. \tag{7.147}$$

If we subtract $\langle a_i(t) a_j(t) \rangle$ from both sides, we end up with

$$\langle a_i(t) \left[a_j(t+\tau) - a_j(t) \right] \rangle = \langle a_j(t) \left[a_i(t+\tau) - a_i(t) \right] \rangle. \tag{7.148}$$

By considering a τ that is larger than the most rapid fluctuation but small compared to the time it takes for a larger fluctuation to decay back toward zero, we can identify the difference term here as the flux J_i according to

$$J_i = \dot{a}_i = \frac{a_i(t+\tau) - a_i(t)}{\tau} = L_{ik} F_k. \tag{7.149}$$

where we introduced again our phenomenological transport equations. So using this in Eq. (7.148) gives

$$\langle a_i L_{jk} F_k \rangle = \langle a_j L_{ik} F_k \rangle. \tag{7.150}$$

Unlike a_i and F_k, the transport coefficients L_{jk} are taken to be nonfluctuating constants and thus moved outside the averaging brackets. Upon using our earlier result that $\langle a_i F_k \rangle = -k_B \delta_{ik}$, we have

$$L_{jk}\delta_{ik} = L_{ik}\delta_{jk}, \tag{7.151}$$

which then gives at last

$$\boxed{L_{ji} = L_{ij}.} \tag{7.152}$$

The above is the standard proof for the symmetry of the transport coefficients as originally obtained by Onsager (1931) and has been demonstrated when there is no fluid flow occurring $\boldsymbol{v} = 0$ in the element being analyzed. As we showed in the continuum demonstration of reciprocity, if there is a flow field $\boldsymbol{v} \neq 0$ influencing the transport through advection, then $L_{ij}(\boldsymbol{v}) = L_{ji}(-\boldsymbol{v})$.

7.5 Capillary Physics from the Cahn–Hilliard Perspective

Capillary physics is the topic of what happens at the interface where two fluids comprised of different molecules meet and cannot completely mix together even in full static equilibrium; we say that the two fluids are either fully or partially *immiscible*. This occurs at the transition from say air to water or from oil to water. Even in equilibrium, there is a gradient in the concentration of each fluid type across a thin transition layer. So capillary physics is distinct from the earlier treatment of a miscible solute and solvent where, in static equilibrium, the concentration of the solute ultimately becomes a uniform constant. That earlier treatment can be considered a special case of what will be developed below.

In what follows, we treat the advection and diffusion of a partially immiscible binary (two molecular component) liquid using the approach developed by van der Waals first in his 1873 doctoral thesis and then later in an 1893 journal article that has been translated from the Dutch by Rowlinson (1979). Van der Waals treated a liquid in equilibrium with its vapor. His approach was dusted off, partially reformulated and applied to a wider range of problems by Cahn and Hilliard (e.g., Cahn and Hilliard, 1958, and other works). The derivation of the governing equations for immiscible binary-liquid flow places an emphasis on the free-energy density of the binary liquid in the presence of equilibrium concentration and density gradients. As such, in this section alone, we will distinguish "specific" thermodynamic continuum variables (i.e., those normalized by mass) by using a caret over the symbol, while thermodynamic densities (i.e., normalized by volume) will be written as lower-case roman letters. So for example, f will mean the Helmholtz free energy per unit volume of the binary liquid and $\hat{f}$ will mean the Helmholtz free energy per unit mass.

The overall dynamics of a partially immiscible binary fluid are controlled by the conservation laws

$$\frac{1}{\rho}\left(\frac{\partial \rho}{\partial t} + \boldsymbol{v}\cdot\nabla\rho\right) = -\nabla\cdot\boldsymbol{v} \tag{7.153}$$

$$\rho\left(\frac{\partial c}{\partial t} + \boldsymbol{v}\cdot\nabla c\right) = -\nabla\cdot\boldsymbol{j} \tag{7.154}$$

$$\rho\left(\frac{\partial \boldsymbol{v}}{\partial t} + \boldsymbol{v}\cdot\nabla\boldsymbol{v}\right) = -\nabla P + \nabla\cdot\boldsymbol{\tau}_\gamma + \nabla\cdot\boldsymbol{\tau}_v - \rho\nabla U. \tag{7.155}$$

The total density ρ, concentration of type-1 molecules c (defined as a mass ratio), flow velocity $\boldsymbol{v}$, and fluid pressure P are the unknowns to be determined at each point within the fluid. The force potential U is that associated with some externally applied force such as gravity. Our main task in what follows is to obtain the constitutive laws for the momentum flux density $\boldsymbol{\tau}_\gamma$ due to surface tension effects, the momentum flux density $\boldsymbol{\tau}_v$ due to viscous effects, and the mass flux of type-1 molecules $\boldsymbol{j}$.

One of the main ideas of a partially (or totally) immiscible fluid is that even in a state of complete equilibrium in which the various nonequilibrium fluxes are zero, there will be equilibrium structure in the $\rho(\boldsymbol{r})$ and $c(\boldsymbol{r})$ profiles that separate fluid-phase domains that are mainly type-1 molecules in one domain and mainly type-2 molecules in another domain.

We will make the assumption that each fluid type is incompressible which means that in the expression for the total fluid density

$$\rho = c\rho_1 + (1-c)\rho_2, \tag{7.156}$$

the pure fluid-1 density ρ_1 and the pure fluid-2 density ρ_2 are both constants. This means that $d\rho/dt = (\rho_1 - \rho_2)dc/dt$ or

$$\nabla\cdot\boldsymbol{v} = \frac{(\rho_1 - \rho_2)}{\rho^2}\nabla\cdot\boldsymbol{j} \tag{7.157}$$

is the incompressible flow condition when compositional changes can occur due to c variations.

7.5.1 Constitutive Laws

To obtain the constitutive laws for a binary liquid being acted upon by externally applied forces and that can have internal structure (transition layers) in equilibrium, we focus on the isothermal Helmholtz free energy F in a finite region Ω of the binary liquid. This free energy should depend on how both the density ρ and the type-1 molecule concentration c are distributed in space in equilibrium as well as on the potential U associated with the externally applied body force

$$F = \int_\Omega f(T, \rho, \nabla\rho, c, \nabla c, U)\, dV. \tag{7.158}$$

To keep the pedagogic focus on how ρ and c vary from one fluid-phase domain to another while the binary liquid is flowing and diffusing, we will treat the isothermal case in what follows to avoid the coupling to temperature gradients and thermal fluxes. Since everything is formulated in terms of a free-energy that depends on temperature, it is straight forward to repeat the following analysis with the full thermal coupling.

The simplest form for the isothermal free-energy density f that allows $\rho(\boldsymbol{r})$ and $c(\boldsymbol{r})$ to be nonuniform in space in equilibrium is

$$f = \rho U + f_o(\rho, c) + \frac{\kappa_\rho}{2}|\nabla\rho|^2 + \frac{\kappa_c}{2}|\nabla c|^2, \tag{7.159}$$

where κ_ρ and κ_c are constant parameters that characterize the molecular interactions in the transition zone that are entirely responsible for surface-tension effects (excess energy in the transition zones). The potential field U is uniformly varying across the system and is associated with an applied body force $\boldsymbol{F}_b = -\rho\nabla U$ acting throughout the domain. So in the case of gravity, the acceleration of gravity is $-\nabla U$. The free-energy density $f_o(\rho, c)$ is that within uniform regions of the binary liquid. Assuming that there is partial miscibility between the phases in equilibrium, there are two possible fluid phases that stably emerge; one that is mainly type-1 molecules with some type-2 molecules as characterized by the equilibrium type-1 saturation $c^{(1)}$ and another that is mainly type-2 molecules with some type-1 molecules as characterized by the equilibrium type-1 saturation $c^{(2)}$. We will give possible forms for $f_o(\rho, c)$ later.

Equilibrium and the Reversible-Process Constitutive Laws

The state of equilibrium and associated internal fluid structure is obtained by minimizing F subject to the constraints that

$$M = \int_\Omega \rho\, dV \tag{7.160}$$

$$M_1 = \int_\Omega c\rho\, dV, \tag{7.161}$$

where M is the total mass and M_1 the mass of type-1 molecules in Ω both of which are constants as we consider different profiles of $\rho(\boldsymbol{r})$ and $c(\boldsymbol{r})$.

To obtain differential equations that control the profiles $\rho(\boldsymbol{r})$ and $c(\boldsymbol{r})$ in equilibrium, the fact that F will be a minimum with respect to variations (or perturbations) $\delta\rho(\boldsymbol{r})$ and $\delta c(\boldsymbol{r})$ in the equilibrium profiles $\rho(\boldsymbol{r})$ and $c(\boldsymbol{r})$ means that the variational derivative of F vanishes in equilibrium, that is, $\delta F = 0$. A variational derivative of a functional $F(\rho, c)$ of the spatial functions $\rho(\boldsymbol{r})$ and $c(\boldsymbol{r})$ means the operation $\delta F = (\partial F/\partial\rho)\delta\rho + (\partial F/\partial c)\delta c$ where the partial derivatives are evaluated in the equilibrium state corresponding to $\delta F = 0$. Further, since M and M_1 are invariant constants, we also have $\delta M = \delta M_1 = 0$ as the equilibrium profiles are perturbed. Overall, we have that in equilibrium, $\delta F + \lambda_\rho \delta M + \lambda_{c\rho}\delta M_1 = 0$ as perturbations $\delta\rho(\boldsymbol{r})$ and $\delta c(\boldsymbol{r})$ are considered, where the parameters λ_ρ and $\lambda_{c\rho}$ are called Lagrangian multipliers and must be found, or defined, as part of the problem. This overall variational approach of obtaining differential equations for a system in equilibrium is called the *method of Lagrangian multipliers*.

By forming δF, δM, and δM_1 using Eqs (7.158), (7.160), and (7.161), multiplying by the Lagrangian multipliers and adding together to form $\delta F + \lambda_\rho \delta M + \lambda_{c\rho}\delta M_1 = 0$ gives the equilibrium statement

$$0 = \int_\Omega dV \left[\left(\frac{\partial f_o}{\partial \rho} - \kappa_\rho \nabla^2 \rho + \lambda_\rho + c\lambda_{c\rho} + U \right) \delta\rho \right.$$
$$\left. + \left(\frac{\partial f_o}{\partial c} - \kappa_c \nabla^2 c + \rho \lambda_{c\rho} \right) \delta c \right]. \tag{7.162}$$

We have used that $\delta\left(|\nabla\rho|^2/2\right) = (\nabla\delta\rho \cdot \nabla\rho + \nabla\rho \cdot \nabla\delta\rho)/2 = \nabla\rho \cdot \nabla\delta\rho = \nabla \cdot (\delta\rho\nabla\rho) - \delta\rho\nabla^2\rho$, integrated over Ω, applied the divergence theorem and assumed that $\delta\rho$ and δc are either zero or periodic on the boundary $\partial\Omega$ to eliminate the surface integral. Because the integral in Eq. (7.162) must be zero for any and all perturbations $\delta\rho$ and δc, we obtain two differential equations that define equilibrium

$$\frac{\partial f_o}{\partial \rho} - \kappa_\rho \nabla^2 \rho + \lambda_\rho + c\lambda_{c\rho} + U = 0 \tag{7.163}$$

$$\frac{\partial f_o}{\partial c} - \kappa_c \nabla^2 c + \rho \lambda_{c\rho} = 0. \tag{7.164}$$

To proceed, we need to identify the physical meaning of the Lagrange multipliers λ_ρ and $\lambda_{c\rho}$. To do so, note that Eqs (7.163) and (7.164) define equilibrium both in uniform fluid regions where there are no gradients in $\rho(\boldsymbol{r})$ and $c(\boldsymbol{r})$ so that $\nabla^2\rho = 0$ and $\nabla^2 c = 0$ as well as in the transition regions where $\nabla^2\rho \neq 0$ and $\nabla^2 c \neq 0$. In uniform and nonuniform regions the derivatives $\partial f_o/\partial\rho$ and $\partial f_o/\partial c$ will be different but the constants λ_ρ and $\lambda_{c\rho}$ will be the same. We thus can identify the physical meaning of the Lagrangian multipliers by considering the thermodynamics of a uniform system for which $\nabla^2\rho = 0$, $\nabla^2 c = 0$, and $U = 0$ in Eqs (7.163) and (7.164).

Under such uniform field conditions, $f_o(\rho, c)$ alone is the free energy density which is a nonstandard combination of dependent and independent thermodynamic variables. More familiar for a binary fluid is the specific free energy $\hat{f}(T, \hat{v}, \hat{n}_1, \hat{n}_2)$ where $\hat{f}$ is the Helmholtz free energy per unit mass, $\hat{v} = 1/\rho$ is the volume per unit mass, and the $\hat{n}_i$ are the number of the type-1 and type-2 molecules per unit mass of fluid. We have

$$d\hat{f} = -\hat{s}dT - Pd\hat{v} + \mu_1 d\hat{n}_1 + \mu_2 d\hat{n}_2, \tag{7.165}$$

where $dT = 0$ in the isothermal system of interest here and the chemical potentials μ_i represent the incremental change in the free energy when a type-i molecular mass is added at constant entropy and volume. Using $\hat{v} = 1/\rho$, introducing the concentration parameters $c_1 = m_1\hat{n}_1$ and $c_2 = m_2\hat{n}_2$ where m_i is the mass of a type-i molecule, noting that $c_1 + c_2 = 1$ and choosing to work with $c = c_1$ as earlier, we obtain

$$d\hat{f}(\rho, c) = \frac{P}{\rho^2} d\rho + \hat{\mu} dc, \tag{7.166}$$

where $\hat{\mu}_1 = \mu_1/m_1$, $\hat{\mu}_2 = \mu_2/m_2$ and $\hat{\mu} = \hat{\mu}_1 - \hat{\mu}_2$ as in the earlier section. This allows us to identify the pressure P and chemical potential $\hat{\mu}$ through the partial derivatives $P/\rho^2 =$

$\partial \hat{f}/\partial \rho$ and $\hat{\mu} = \partial \hat{f}/\partial c$. We further have the Euler equation for this system in the form $\hat{f} = -P/\rho + \mu_1 \hat{n}_1 + \mu_2 \hat{n}_2$ which can be expressed $\hat{f} = -P/\rho + c\hat{\mu} + \hat{\mu}_2$.

We now switch to the nonstandard density form of the free energy $f_o(\rho, c) = \rho \hat{f}(\rho, c)$ that is the focus in the Cahn–Hilliard approach. Taking a derivative gives (the pressure terms cancel each other) $df_o = \hat{f} d\rho + \rho d\hat{f} = (c\hat{\mu} + \hat{\mu}_2) d\rho + \rho \hat{\mu} dc$ which allows us to identify

$$c\hat{\mu} + \hat{\mu}_2 = \frac{\partial f_o(\rho, c)}{\partial \rho} \tag{7.167}$$

$$\rho \hat{\mu} = \frac{\partial f_o(\rho, c)}{\partial c}. \tag{7.168}$$

Last, the Euler relation is restated in the form

$$-P(\rho, c) = f_o(\rho, c) - \rho \left[c\hat{\mu}(\rho, c) + \hat{\mu}_2(\rho, c) \right] \tag{7.169}$$

which gives

$$-P = f_o - \rho \frac{\partial f_o}{\partial \rho} \tag{7.170}$$

or

$$\frac{\partial f_o}{\partial \rho} = \frac{P + f_o}{\rho}. \tag{7.171}$$

It is through Eq. (7.171) that pressure P will be introduced below.

With the above established, we can identify the Lagrange multipliers in the uniform version of Eqs (7.163) and (7.164) as

$$\lambda_\rho + c\lambda_{c\rho} = -\frac{\partial f_o}{\partial \rho} = -c\hat{\mu} - \hat{\mu}_2 \tag{7.172}$$

$$\rho \lambda_{c\rho} = -\frac{\partial f_o}{\partial c} = -\rho \hat{\mu}. \tag{7.173}$$

or

$$\lambda_\rho = -\hat{\mu}_2 \quad \text{and} \quad \lambda_{c\rho} = -\hat{\mu}. \tag{7.174}$$

We thus interpret $-\lambda_{c\rho} = \hat{\mu}$ to be the chemical potential of primary interest even in the nonuniform binary fluid and $-\lambda_\rho = \hat{\mu}_2$ to be the chemical potential of type-2 molecules even in the nonuniform system.

The equilibrium conditions in the nonuniform system can now be stated as the differential equations

$$\hat{\mu}_2 = \frac{\partial f_o}{\partial \rho} - \kappa_\rho \nabla^2 \rho - c\hat{\mu} + U \tag{7.175}$$

$$\rho \hat{\mu} = \frac{\partial f_o}{\partial c} - \kappa_c \nabla^2 c. \tag{7.176}$$

When the system is in equilibrium, we have that $\nabla\hat{\mu}=0$ and $\nabla\hat{\mu}_2=0$. By taking $\hat{\mu}$ and $\hat{\mu}_2$ as uniform constants, Eqs (7.175) and (7.176) allow the internal structure $\rho(\boldsymbol{r})$ and $c(\boldsymbol{r})$ to be determined for the nonuniform system in equilibrium. This equilibrium state allows for phase separations and diffuse transition layers.

To obtain the statement of mechanical equilibrium in the binary liquid, we multiply Eq. (7.175) for $\hat{\mu}_2$ by $\nabla\rho$. We then use the tensor identity

$$(\nabla^2\rho)(\nabla\rho)=\nabla\cdot\left[-\frac{1}{2}|\nabla\rho|^2\boldsymbol{I}+(\nabla\rho)(\nabla\rho)\right] \tag{7.177}$$

where $\boldsymbol{I}$ is the identity tensor, write $\hat{\mu}_2\nabla\rho=\nabla(\rho\hat{\mu}_2)-\rho\nabla\hat{\mu}_2$, use equilibrium $\nabla\hat{\mu}_2=0$, write $U\nabla\rho=\nabla(\rho U)-\rho\nabla U$ and assume that κ_ρ is a uniform constant, to give

$$\begin{aligned} c\hat{\mu}\nabla\rho = &-\rho\nabla U+\left(\frac{\partial f_o}{\partial\rho}\right)\nabla\rho \\ &-\nabla\cdot\left[\left(\rho\hat{\mu}_2-\rho U-\frac{\kappa_\rho}{2}|\nabla\rho|^2\right)\boldsymbol{I}+\kappa_\rho(\nabla\rho)(\nabla\rho)\right]. \end{aligned} \tag{7.178}$$

Similarly, we multiply Eq. (7.176) for $\rho\hat{\mu}$ by ∇c and use $\rho\hat{\mu}\nabla c=\nabla(\rho c\hat{\mu})-c\hat{\mu}\nabla\rho$ due to equilibrium $\nabla\hat{\mu}=0$ to obtain

$$-c\hat{\mu}\nabla\rho=\left(\frac{\partial f_o}{\partial c}\right)\nabla c-\nabla\cdot\left[\left(\rho c\hat{\mu}-\frac{\kappa_c}{2}|\nabla c|^2\right)\boldsymbol{I}+\kappa_c(\nabla c)(\nabla c)\right]. \tag{7.179}$$

Adding Eqs (7.178) and (7.179) and using $\nabla f_o=(\partial f_o/\partial\rho)\nabla\rho+(\partial f_o/\partial c)\nabla c$ gives

$$\begin{aligned} 0=\nabla\cdot&\left[\left(f_o-\rho(c\hat{\mu}+\hat{\mu}_2-U)+\frac{\kappa_\rho}{2}|\nabla\rho|^2+\frac{\kappa_c}{2}|\nabla c|^2\right)\boldsymbol{I}\right. \\ &\left.-\kappa_\rho(\nabla\rho)(\nabla\rho)-\kappa_c(\nabla c)(\nabla c)\right]-\rho\nabla U. \end{aligned} \tag{7.180}$$

To introduce the pressure P into this statement of mechanical equilibrium, we first combine Eqs (7.171) and (7.175) to give

$$c\hat{\mu}+\hat{\mu}_2-U=\frac{(f_o+P)}{\rho}-\kappa_\rho\nabla^2\rho. \tag{7.181}$$

This allows Eq. (7.180) to be written as the standard statement of mechanical equilibrium

$$-\nabla P+\nabla\cdot\boldsymbol{\tau}_\gamma-\rho\nabla U=0, \tag{7.182}$$

where the stress tensor $\boldsymbol{\tau}_\gamma$ is defined as

$$\boxed{\boldsymbol{\tau}_\gamma=\left(\kappa_\rho\rho\nabla^2\rho+\frac{\kappa_\rho}{2}|\nabla\rho|^2+\frac{\kappa_c}{2}|\nabla c|^2\right)\boldsymbol{I}-\kappa_\rho(\nabla\rho)(\nabla\rho)-\kappa_c(\nabla c)(\nabla c).} \tag{7.183}$$

It is entirely through this constitutive law for the tensor $\boldsymbol{\tau}_\gamma$, that is sometimes called the *Korteweg stress tensor* (or *capillary stress tensor*), that surface-tension effects on the fluid flow and phase separation are accounted for. If we consider a planar interface with $\rho(x)$ decreasing across a transition layer from say a pure liquid to its vapor (so $c=0$ in this monomolecular fluid with only a density decrease across the interface), at a point in the transition layer where $d^2\rho/dx^2 = 0$ (the inflection point of the transition) or at least where $|\rho \, d^2\rho/dx^2| \ll (d\rho/dx)^2$, the Korteweg stress tensor predicts that $\boldsymbol{\tau}_\gamma \approx (\kappa_\rho/2)(d\rho/dx)^2 \left(-\hat{\boldsymbol{x}}\hat{\boldsymbol{x}} + \hat{\boldsymbol{y}}\hat{\boldsymbol{y}} + \hat{\boldsymbol{z}}\hat{\boldsymbol{z}}\right)$. Recall that a longitudinal stress component is positive in tension and negative in compression. So at the interface that has normal $\hat{\boldsymbol{x}}$, the liquid transition layer is in enhanced longitudinal compression (attraction toward the bulk of the liquid) in the direction perpendicular to the interface and is in longitudinal tension within the plane of the interface. As will be seen in the upcoming qualitative discussion of Section 7.6.1, this is consistent with a descriptive understanding of surface tension at the interface.

Nonequilibrium and the Irreversible-Process Transport Laws

In the absence of temperature gradients, our earlier statement for the entropy production and associated mass flux law of type-1 molecules continues to apply and gives

$$\boldsymbol{j} = -\frac{L_\mu}{T}\nabla\hat{\mu}, \tag{7.184}$$

where $\hat{\mu}$ is given by Eq. (7.176). We earlier defined

$$\rho D_c = \frac{L_\mu}{T}\frac{\partial\hat{\mu}}{\partial c}, \tag{7.185}$$

where D_c is the type-1 molecular diffusivity (units of $\mathrm{m}^2\,\mathrm{s}^{-1}$) in a uniform system where Eq. (7.168) holds for $\hat{\mu}$. We thus have

$$\frac{L_\mu}{T} = \frac{\rho^2 D_c}{\partial^2 f_o(\rho, c)/\partial c^2} \tag{7.186}$$

so that our transport law can be given as

$$\boldsymbol{j} = -\frac{\rho^2 D_c}{\partial^2 f_o/\partial c^2}\nabla\hat{\mu}, \tag{7.187}$$

where $\hat{\mu} = \hat{\mu}(T, \rho, \nabla\rho, c, \nabla c, U)$ is given by Eq. (7.168) as

$$\hat{\mu} = \frac{1}{\rho}\left(\frac{\partial f_o}{\partial c} - \kappa_c\nabla^2 c\right). \tag{7.188}$$

So in the presence of stable gradients in concentration in the fluid and associated surface tension, the nature of the type-1 diffusional flux $\boldsymbol{j}$ is considerably different than in purely miscible systems.

Last, the final constitutive law is that for the viscous stresses in this Newtonian fluid written as

$$\boldsymbol{\tau}_v = \eta\left[\nabla\boldsymbol{v} + (\nabla\boldsymbol{v})^T - \frac{2}{3}\nabla\cdot\boldsymbol{v}\right] + \kappa\nabla\cdot\boldsymbol{v}\boldsymbol{I}, \tag{7.189}$$

where η and κ are the shear and bulk viscosities that also vary with composition in general. Again, if we use η_i to denote the shear viscosity in each pure phase i and κ_i to denote the bulk viscosity in each pure phase, the simplest possible model is likely the geometric mean or Arrhenius form for the shear viscosity and a harmonic mean for the bulk viscosity, that is,

$$\eta = \eta_1^c \, \eta_2^{(1-c)} \tag{7.190}$$

$$\frac{1}{\kappa} = \frac{c}{\kappa_1} + \frac{(1-c)}{\kappa_2}. \tag{7.191}$$

We will not further speculate on the appropriateness of such mixing rules for the viscosities.

7.5.2 Forms for $f_o(\rho, c)$

We now discuss the functional form to the function $f_o(\rho, c)$ that allows for both phase separation and a reasonable expression for pressure to be obtained. We write

$$f_o(\rho, c) = f_c(c) + f_{\rho c}(\rho, c), \tag{7.192}$$

where $f_c(c)$ allows for the phases to separate and should therefore have the form of a double-well potential with wells centered on the two phases characterized by the type-1 molecule concentrations $c^{(1)}$ and $c^{(2)}$ in each phase. We thus have

$$f_c(c) = \nu (c - c^{(1)})^2 (c - c^{(2)})^2, \tag{7.193}$$

where ν is a constant parameter.

For the function $f_{\rho c}(\rho, c)$, we should choose a form that allows for pressure to be determined from Eq. (7.170) by differentiation with respect to ρ and that allows, at least approximately, for how the pressure depends on concentration. Pressure depends mainly on the total solution density. If we assume that $f_{\rho c}(\rho, c) \approx f_\rho(\rho)$, then as will be seen in the concluding section where we gather the entire suite of governing equations together, there is no dependence in the governing equations on the function $f_\rho(\rho)$ other than implicitly in the pressure and we can calculate the pressure explicitly from the governing equations without ever appealing to a specific form for $f_\rho(\rho)$. This is the approach we will take when treating a binary liquid.

When treating a pure liquid and its vapor, it is the mass density alone that defines the two phases with say ρ_1 the liquid and ρ_2 the vapor and $f_o = \nu_\rho (\rho - \rho_1)^2 (\rho - \rho_2)^2$.

7.5.3 Equilibrium Phase Partition

Consider a wholespace initially filled with all type-2 molecules in the halfspace $x < 0$ and with all type-1 molecules in $x > 0$. Diffusion ensues and the system then comes to complete equilibrium. We want to know the concentration profile $c(x)$ in equilibrium for the binary liquid. For this pedagogic purpose, we will make the just-stated approximation that $f_{\rho c}(\rho, c) \approx f_\rho(\rho)$ and take $\hat{\mu} = 0$ throughout the system in equilibrium. Under these simplifying assumptions, the equation that controls the profile $c(x)$ is obtained by inserting

Eq. (7.193) into Eq. (7.176) to obtain

$$\frac{d^2c(x)}{dx^2} = \frac{\nu}{\kappa_c}(c - c^{(1)})(c - c^{(2)})\left(c - \frac{(c^{(1)} + c^{(2)})}{2}\right). \tag{7.194}$$

The solution of this so-called *Cahn–Hilliard profile equation* is

$$c(x) = \frac{(c^{(1)} + c^{(2)})}{2} + \frac{(c^{(1)} - c^{(2)})}{2} \tanh\left(\frac{x}{d_c}\right), \tag{7.195}$$

where

$$d_c = \frac{\sqrt{2\kappa_c/\nu}}{c^{(1)} - c^{(2)}} \tag{7.196}$$

is the measure of interface thickness inferred from $c(x)$. A plot of this equilibrium structure is given in Fig. 7.2. An equivalent closed form solution for $\rho(x)$ is not available unless in $f_o(\rho, c) = f_c(c) + f_\rho(\rho)$ we give an equation for $f_\rho(\rho)$. We will not go through that exercise.

Last, we define a surface tension γ parameter as the excess energy in the variable-concentration and variable-density fields across the transition layer

$$\gamma = \int_{-\infty}^{\infty} \left[\kappa_c \left(\frac{\mathrm{d}c(x)}{\mathrm{d}x}\right)^2 + \kappa_\rho \left(\frac{\mathrm{d}\rho(x)}{\mathrm{d}x}\right)^2\right] \mathrm{d}x. \tag{7.197}$$

You can show in a guided end-of-chapter exercise that this definition of surface tension, when combined with the statement of mechanical equilibrium $-\nabla P + \nabla \cdot \boldsymbol{\tau}_\gamma = 0$ yields the famous Young–Laplace equation that will be independently derived from more macroscopic considerations in Section 7.6.

Since for the mixture of two binary liquids, the strong gradients will be in c and only weak gradients will be present in the solution density, we can approximate γ as

$$\gamma \approx \int_{-\infty}^{\infty} \kappa_c \left(\frac{\mathrm{d}c(x)}{\mathrm{d}x}\right)^2 \mathrm{d}x = \frac{d_c\kappa_c[c^{(1)} - c^{(2)}]^2}{3} = \frac{[c^{(1)} - c^{(2)}]}{3}\sqrt{\frac{2\kappa_c^3}{\nu}}. \tag{7.198}$$

So the parameter κ_c that characterizes the energy of interaction between species 1 and species 2 molecules controls the surface tension at the liquid–liquid interface.

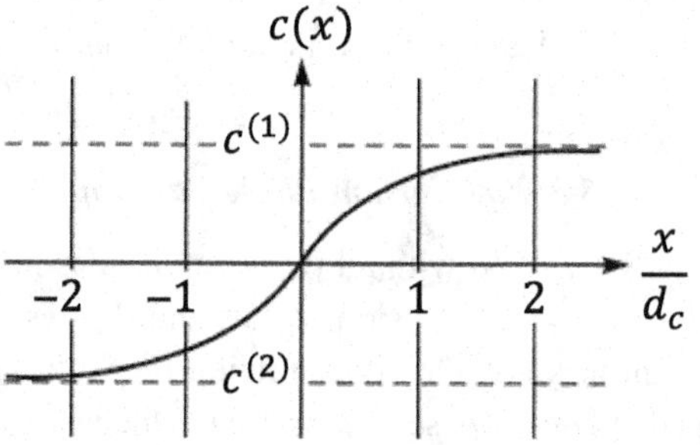

Figure 7.2 The equilibrium 1D structure of a diffuse interface as characterized by $c(x)$ of Eq. (7.195).

If we are dealing with a pure liquid 1 having a density ρ_1 in equilibrium with its vapor having a density ρ_2, the interface is a density transition

$$\rho(x) = \left(\frac{\rho_1 + \rho_2}{2}\right) + \left(\frac{\rho_1 - \rho_2}{2}\right) \tanh\left(\frac{x}{d_c}\right), \tag{7.199}$$

where d_c is now defined $d_c = \sqrt{2\kappa_\rho/\nu_\rho}/(\rho_1 - \rho_2)$ and the surface tension $\gamma = (\rho_1 - \rho_2)\sqrt{2\kappa_\rho^3/\nu_\rho}/3$.

7.5.4 Summary of Cahn–Hilliard Theory for a Binary Liquid

Finally, we summarize what was learned. In a binary liquid, the mass density goes as $\rho = \rho_1 c + \rho_2(1 - c)$, where ρ_1 and ρ_2 are the densities of pure fluid 1 and fluid 2 that are taken as constants in the isothermal and incompressible system being treated. The suite of governing equations are the rules of conservation of mass

$$\rho = \rho_1 c + \rho_2(1 - c) \tag{7.200}$$

$$\nabla \cdot \boldsymbol{v} = \frac{(\rho_1 - \rho_2)}{\rho^2} \nabla \cdot \boldsymbol{j} \tag{7.201}$$

$$\frac{\partial c}{\partial t} + \boldsymbol{v} \cdot \nabla c = -\frac{1}{\rho} \nabla \cdot \boldsymbol{j} \tag{7.202}$$

the rules of diffusive mass flux

$$\boldsymbol{j} = -\frac{\rho^2 D_c}{\partial^2 f_c/\partial c^2} \nabla \hat{\mu} \tag{7.203}$$

$$\hat{\mu} = \frac{1}{\rho}\left(\frac{\partial f_c(c)}{\partial c} - \kappa_c \nabla^2 c\right) \tag{7.204}$$

and the rules of conservation of momentum

$$\rho\left(\frac{\partial \boldsymbol{v}}{\partial t} + \boldsymbol{v} \cdot \nabla \boldsymbol{v}\right) = -\nabla P + \nabla \cdot \boldsymbol{\tau}_\gamma + \nabla \cdot \boldsymbol{\tau}_v - \rho \nabla U \tag{7.205}$$

$$\boldsymbol{\tau}_\gamma = \left(\kappa_\rho \rho \nabla^2 \rho + \frac{\kappa_\rho}{2}|\nabla \rho|^2 + \frac{\kappa_c}{2}|\nabla c|^2\right)\boldsymbol{I} - \kappa_\rho (\nabla \rho)(\nabla \rho) - \kappa_c (\nabla c)(\nabla c) \tag{7.206}$$

$$\boldsymbol{\tau}_v = \eta(\nabla \boldsymbol{v} + \nabla \boldsymbol{v}^T - \frac{2}{3}\nabla \cdot \boldsymbol{v}) + \kappa \nabla \cdot \boldsymbol{v} \boldsymbol{I} \tag{7.207}$$

along with the double-well expression for the function $f_c(c)$

$$f_c(c) = \nu[c - c^{(1)}]^2[c - c^{(2)}]^2 \tag{7.208}$$

and the rules given earlier for how η (shear viscosity) and κ (bulk viscosity) depend on c. In the force balance of Eq. (7.205), the surface tension effects come entirely from the Korteweg stress tensor $\boldsymbol{\tau}_\gamma$. As always, it is Eq. (7.201) for $\nabla \cdot \boldsymbol{v}$ that when combined with the force balance of Eq. (7.205) allows the pressure P to be determined. As such, and as promised, under the approximation that $f_o(\rho, c) \approx f_c(c) + f_\rho(\rho)$, the function f_ρ is not required in the above modeling. The above rules apply to a binary liquid flowing with velocity $\boldsymbol{v}$ and having a temporally and spatially varying concentration of liquid 1 molecules given by $c_1 = c$ (defined as a mass ratio) and a concentration of liquid 2 molecules given by $c_2 = 1 - c$. For a binary, incompressible liquid in which ρ_1 and ρ_2 are constants, the mass density is determined from the concentration field as $\rho = \rho_1 c + \rho_2(1-c)$ so that the gradient of the density is given by $\nabla\rho = (\rho_1 - \rho_2)\nabla c$ and the Laplacian as $\nabla^2\rho = (\rho_1 - \rho_2)\nabla^2 c$.

We have not yet said anything about the boundary conditions between this binary liquid and a bounding solid wall $\partial\Omega_s$ that has a normal vector $\boldsymbol{n}_s$. As always, the flow velocity must go to zero on solid walls, that is, $\boldsymbol{v} = 0$ on $\partial\Omega_s$. In Section 7.6, we introduce the macroscopic ideas of contact lines and contact angles where the two liquid phases and the solid come into contact with each other in the limit that $d_c \to 0$. Once those ideas are established, we leave it to you as a guided end-of-chapter exercise to show that the following boundary condition for the concentration field

$$\boldsymbol{n}_s \cdot \nabla c = -\frac{2(\gamma_{s2} - \gamma_{s1})}{d_c \gamma} \frac{(c - c^{(1)})(c - c^{(2)})}{(c^{(1)} - c^{(2)})} \quad \text{on} \quad \partial\Omega_s. \tag{7.209}$$

is how Cahn–Hilliard theory accounts for the coarser-resolution ideas of contact lines and contact angles, both moving and static. Here, γ is again the surface tension between the two liquids having fluid-1 molecule concentrations of $c^{(1)}$ and $c^{(2)}$, while γ_{s2} is the surface tension between the solid and the binary liquid having a fluid-1 molecule concentration of $c = c^{(2)}$ and γ_{s1} is the surface tension between the solid and the binary liquid having a concentration of $c = c^{(1)}$.

The above suite of nonlinear equations for the binary liquid allows for any of total miscibility, partial miscibility or total immiscibility of the two fluids. This is done by setting the equilibrium concentrations $c^{(1)}$ and $c^{(2)}$ of type-1 molecules in the two liquid phases present within the double-well potential $f_c(c)$ of Eq. (7.208). Setting $c^{(1)} = c^{(2)}$ corresponds to total miscibility, setting $c^{(1)} = 1$ and $c^{(2)} = 0$ corresponds to total immiscibility and taking $c^{(1)} \neq c^{(2)} \neq 0$ corresponds to partial miscibility. Note that even for totally miscible fluids, surface tension effects will be present as both diffusion and flow attempt to drive the system toward equilibrium. Further, even for totally immiscible fluids, there will be a finite-width transition zone of characteristic length $d_c = \sqrt{2\kappa_c/\nu}$ between the phases in equilibrium.

If we imagine a fluid channel between two plane-parallel solid walls (called a "Hele-Shaw cell") with flow in the z direction and initially filled with pure fluid 1 in $z > 0$ and pure fluid 2 in $z < 0$, as flow and diffusion ensues, we cannot assume that the flow field is independent of the concentration field due to the important role of the surface tension terms in Eq. (7.205). This makes the problem of the dynamics of a partially miscible binary liquid much more complicated than the Taylor-dispersion problem treated earlier in Section 7.3.5 where such surface-tension influences on the flow field due to ∇c and $\nabla\rho$ are neglected.

Last, it is straightforward to recast the above equations for the case of a pure liquid having density ρ_1 interacting with its vapor having density ρ_2 in which $\kappa_c = 0$ from the outset and $c = 0$ throughout the above with $f_o(\rho)$ having a double-well form. The conservation of mass laws are the statements of incompressible flow $\nabla \cdot \boldsymbol{v} = 0$ plus the accumulation of molecular mass due to advection and diffusion $\partial \rho / \partial t + \boldsymbol{v} \cdot \nabla \rho = -\nabla \cdot \boldsymbol{j}$ where the diffusive mass flux $\boldsymbol{j}$ in the variable density fluid is governed by

$$\boldsymbol{j} = -\frac{D_\rho}{\partial^2 f_0 / \partial \rho^2} \nabla \hat{\mu}, \tag{7.210}$$

$$\hat{\mu} = \frac{\partial f_0}{\partial \rho} - \kappa_\rho \nabla^2 \rho + U \tag{7.211}$$

with D_ρ the self-diffusivity of the fluid molecules in this single species system. Capillary effects are allowed for both through the Korteweg stress tensor

$$\boldsymbol{\tau}_\gamma = \left(\kappa_\rho \rho \nabla^2 \rho + \frac{\kappa_\rho}{2} |\nabla \rho|^2 \right) \boldsymbol{I} - \kappa_\rho (\nabla \rho)(\nabla \rho) \tag{7.212}$$

and the boundary condition on the fluid–solid surfaces

$$\boldsymbol{n}_s \cdot \nabla \rho = -\frac{2(\gamma_{s2} - \gamma_{s1})}{d_c \gamma} \frac{(\rho - \rho_1)(\rho - \rho_2)}{(\rho_1 - \rho_2)} \quad \text{on} \quad \partial \Omega_s, \tag{7.213}$$

which allows for the contact-line effects discussed in the following section.

7.6 Capillary Physics from a More Macroscopic Perspective

The above treatment for how a binary fluid flows and partially diffuses into each fluid domain up to a given saturation level is quite satisfying from a computational perspective if you are willing to discretize the fields to at least the resolution of the transition length d_c between the two phases. But it obfuscates some of the essential macroscopic physics at work where two largely immiscible fluids come into contact and are modeled from the lower-resolution perspective in which the interface is treated as a sharp step and not a finite-width transition layer. This section will derive many of the fundamental lower-resolution macroscopic laws of immiscible fluids that the zoomed-in Cahn–Hilliard partial-differential equations are allowing for.

7.6.1 A Qualitative Description of Surface Tension

Here we will describe what creates γ when, for example, a high-density liquid (or solid) is overlain by its low-density vapor (or a gas). Similar arguments describe the surface tension between two liquids.

As depicted in Fig. 7.3, we imagine drawing a plane at $z = 0$ through a liquid at some given pressure and somehow removing all the molecules that are above the plane and replacing them with a low-density vapor or gas at the same pressure. Prior to the $z > 0$ removal, as the molecules in the dense-liquid packing randomly jostle about due to their thermal energy, they are both attracted to one neighbor as they move away from that neighbor and are repulsed by another neighbor as they approach it. A molecule inside the liquid fluctuates about an equilibrium state where they are feeling more repulsion than attraction from a neighbor as set by the compressive liquid pressure. Once the dense packing

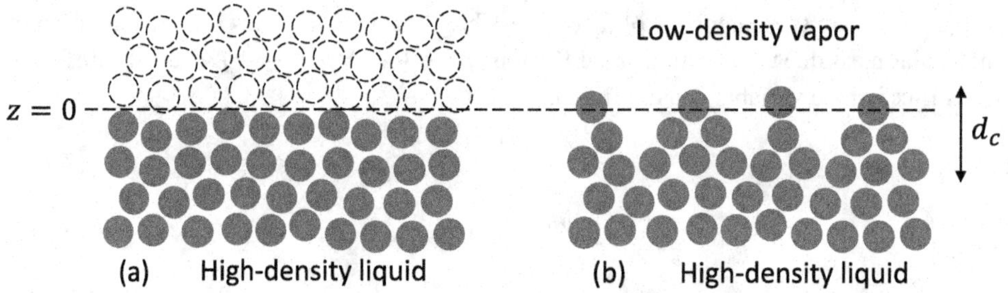

Figure 7.3 (a) We have a liquid in which we imagine removing instantaneously all the liquid molecules above a plane $z = 0$ as shown. (b) Once the dense liquid above $z = 0$ is gone and replaced by a low-density vapor at the same pressure, the molecules along the plane $z = 0$ have a much reduced upward pull from the missing dense packing of molecules. As such, these surface molecules are both pulled down into a tighter packing with the underlying liquid molecules, while also having some room above to wander more freely under their thermal motion without repulsive push back from the removed dense packing of molecules. The net effect of this thermal dance of molecules near the liquid surface is a gradient in molecular density from the vapor toward the dense liquid. The larger horizontal spacing between the molecules at the surface of the liquid compared to deeper within the liquid means these molecules are not subject to horizontal repulsion forces and are therefore in a state of heightened horizontal tension due to the purely attractive forces acting from both the left and right of each surface molecule. This is how surface tension is created. In the schematic snapshot of this thermal dance given here, the number of gray liquid molecules depicted in (a) and (b) are the same.

of molecules above $z = 0$ has been removed, the molecules along the plane have a much reduced force of attraction from above and are therefore pulled down into a more intimate coordinated packing with the underlying liquid molecules corresponding to a state of enhanced compression normal to the interface. Simultaneously, as these surface molecules move about with thermal energy, a random movement upward is no longer repulsed by molecules from above because the dense packing of molecules is no longer there. The net effect of such thermal movement near $z = 0$ is that although the average vertical distance between neighboring molecules is only slightly reduced due to ongoing repulsive push back from the underlying dense packing of molecules, the average horizontal distance between neighbors is increased more considerably with the liquid molecules the furthest from the underlying liquid packing having a larger average horizontal distance of separation compared to molecules further inside the liquid. This means that liquid molecules furthest from the underlying packing are in a state of horizontal tension due to attractive van der Waals forces pulling on the them both from the left and right with diminished horizontal repulsion force because the repulsion force falls with distance of separation much more rapidly than the attractive force. This is the inherent cause of surface tension and is consistent with the Korteweg stress tensor modeling of Cahn–Hilliard theory. Due to the vertical gradient in density over some distance d_c perpendicular to the surface of the liquid, with greater horizontal molecular distances for molecules at the surface compared to molecules deeper inside the liquid, there is an inherent nonzero horizontal tension always present on the first few layers of surface molecules that resists horizontal stretching of the surface layers. Qualitatively it is as if there is the skin of a balloon bounding the liquid at the surface $z = 0$.

It takes an amount of work $\gamma\delta A$ to change the gas–liquid surface area by an amount δA, which is elevated compared to changing the area of some arbitrary surface of molecules deep within the liquid by the same amount (e.g., by volumetric expansion). This distinct surface energy $\gamma\delta A$ must be added to $-P\delta V$ where P is the pressure if the system also experiences a change in volume δV. If one draws a line in the surface of the interface $z = 0$ using a curvilinear coordinate l_1 that is perpendicular to z, one can also identify γ as the tension force per unit line length acting normal to that line (say in a direction l_2) that resists strain of each surface element.

If the interface is between two immiscible liquids, one of the liquids will have a lower density and a larger average intermolecular distance compared to the other and will have distinct molecular polarizability properties that are involved in the van der Waals attraction. There again develops a gradient in density across a transition layer of thickness d_c and this surface region will again be in a state of enhanced tension due to horizontal attraction forces and diminished horizontal repulsion forces compared to deep within the denser and more polarizable liquid. The surface tension at a liquid–liquid interface will be measurably smaller than in the liquid-gas scenario described above. For example, at room temperature the liquid–liquid water–benzene interface has $\gamma = 35 \times 10^{-3}$ N/m while the water–air interface has $\gamma = 73 \times 10^{-3}$ N/m.

7.6.2 Capillary Pressure and the Young–Laplace Equation

If the surface separating two fluids is curved, there is a pressure jump across the interface due to the surface tension of the interface layer. This jump in pressure is called the *capillary pressure*. As an end-of-chapter exercise, you can obtain the nature of this pressure jump using Cahn–Hilliard theory but the result is more directly obtained as follows.

Consider a curved interface described by orthogonal curvilinear coordinates (l_1, l_2) lying parallel with the interface. Each point of the interface has a radius of curvature R_1 in the direction l_1 and R_2 in the perpendicular direction l_2. The amount of work performed in displacing the interface by a small amount δz perpendicular to each point of the interface is given by

$$\delta E = \int_S dl_1 dl_2 \, (-P_{in} + P_{ex})\delta z + \gamma \delta A. \tag{7.214}$$

Here, $-P_{in} dl_1 dl_2$ is the force pushing on an element $dl_1 dl_2$ of the interface in the direction of δz from the interior of the curved interface, while $P_{ex} dl_1 dl_2$ is the force pushing in the opposite direction on the element from the exterior side of the interface. The total change in surface area δA can be determined by how much each surface element changes area. As shown in Fig. 7.4, $dl_1 = R_1 d\theta$ and $dl_1' = (R_1 + \delta z)d\theta = dl_1(1 + \delta z/R_1)$ so that the stretched surface element $dl_1' dl_2'$ is given by

$$dl_1' dl_2' = dl_1 \left(1 + \frac{\delta z}{R_1}\right) dl_2 \left(1 + \frac{\delta z}{R_2}\right) \tag{7.215}$$

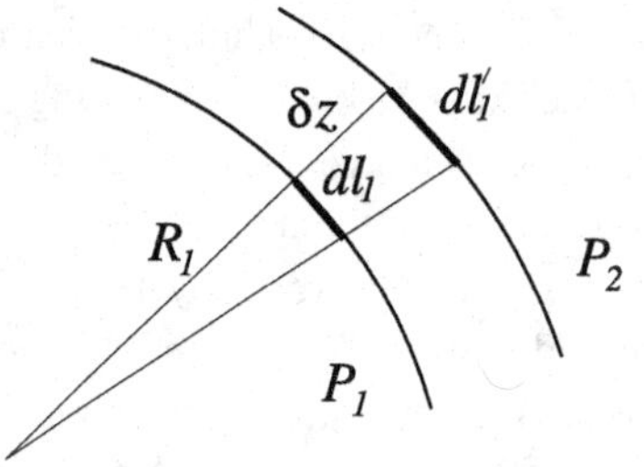

Figure 7.4 A sketch showing a displacement δz of a curved interface that is everywhere normal to the curved surface.

$$\approx dl_1 dl_2 \left[1 + \delta z \left(\frac{1}{R_1} + \frac{1}{R_2}\right)\right], \tag{7.216}$$

where the second line follows from the assumption that δz is small compared to the radii of curvature. The change in surface area is the difference in the surface integral $\delta A = \int_{S'} dl'_1 dl'_2 - \int_S dl_1 dl_2$ so that

$$\delta E = \int_S dl_1 dl_2 \, \delta z \left[P_{ex} - P_{in} + \gamma \left(\frac{1}{R_1} + \frac{1}{R_2}\right)\right]. \tag{7.217}$$

If we consider small but otherwise arbitrary displacements of the interface δz near equilibrium, we have that $\delta E = 0$ since the energy is at a minimum with respect to such perturbations at equilibrium. We are thus required to have

$$\boxed{\Delta P \hat{=} P_{in} - P_{ex} = \gamma \left(\frac{1}{R_1} + \frac{1}{R_2}\right)} \tag{7.218}$$

at equilibrium, which is the well-known *Young–Laplace equation* as originally described by Thomas Young and quantified by Pierre-Simon Laplace in separate work from the early 1800s. So the pressure jump across a curved interface separating two fluids $\Delta P = P_{in} - P_{ex}$, that is called the *capillary pressure*, is related to interface curvature through the surface-tension parameter γ. The pressure on the inside of any curved interface with an associated surface tension will always be larger than the pressure on the exterior side of the curved surface as quantified by Eq. (7.218). For a flat interface with no curvature ($R_1 = R_2 \to \infty$), $P_{in} = P_{ex}$ and there is no capillary-pressure drop.

7.6.3 Contact Lines

Now let two immiscible fluids be in simultaneous contact with an underlying solid as depicted in Fig. 7.5. The points of intersection between the two fluids and the solid form a line along the solid surface called the *contact line*. The two fluids have an interface called the *meniscus* that can be deformed more easily than the two fluid–solid interfaces.

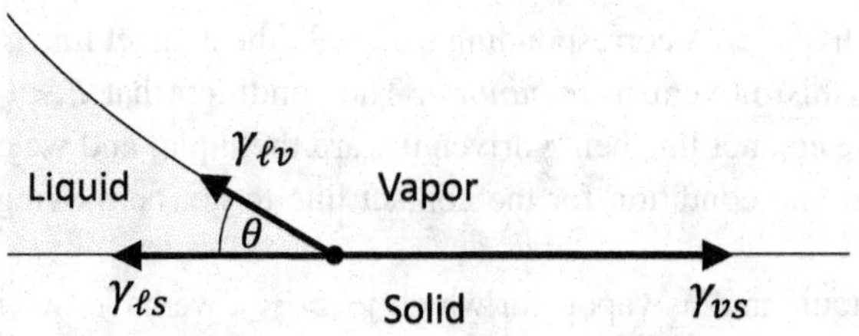

Figure 7.5 A contact line (the point of intersection between the three phases denoted by the black dot) is extending out of page. It is acted upon by the three interfacial tension forces. The force density acting on the contact line (per unit length extending out of page) in the plane of the solid is called the *imbibition-force density* and is given by $F = \gamma_{vs} - \gamma_{\ell s} - \gamma_{\ell v} \cos\theta$. When positive, the imbibition force is directed from the liquid toward the vapor. The contact angle θ is a macroscopic concept defined in the figure.

Contact Lines in Equilibrium

Each surface in Fig. 7.5 has a surface tension for the reasons described earlier. The vapor–solid interface has the tension-force density (N/m) γ_{vs}, the liquid–solid interface has the tension $\gamma_{\ell s}$ and the liquid–vapor interface the tension $\gamma_{\ell v}$. The contact line is thus acted upon by the three tensions shown in Fig. 7.5. When resolved onto the plane of the solid, the net tension-force density F (N/m) acting on the contact line is

$$F = \gamma_{vs} - \gamma_{\ell s} - \gamma_{\ell v} \cos\theta, \tag{7.219}$$

which is a force density parallel with the solid surface directed from the liquid toward the vapor phase when positive. This force acting normal to the contact line is called the *imbibition-force density*. If it is positive, the contact line will move toward the vapor while when it is negative it will move toward the liquid.

When $F = 0$, there is no net force acting on the contact line, which is therefore stationary. The condition $F = 0$ thus defines the equilibrium contact angle θ_c through the relation

$$\boxed{\gamma_{\ell v} \cos\theta_c = \gamma_{vs} - \gamma_{\ell s},} \tag{7.220}$$

which is another relation attributable to Thomas Young (1805). So the condition for a contact line not to be moving (i.e., to be in static equilibrium) is that $\theta = \theta_c$ with θ_c given by Eq. (7.220).

Contact Lines in Movement: Imbibition and Drainage

We can rewrite the imbibition force F that is acting normal to the contact line and in the plane of the solid surface as

$$F = \gamma_{\ell v} (\cos\theta_c - \cos\theta). \tag{7.221}$$

The angle θ that holds when $F \neq 0$ and the contact lines are in motion is called the *dynamic contact angle* and is usually time dependent in particular applications, as will be seen for the case of spontaneous imbibition in the example that follows. If the net force on the contact line is positive $F > 0$, corresponding to $\theta > \theta_c$, the contact line is driven toward the vapor and we call this movement *imbibition*. The condition that $F < 0$ corresponding to $\theta < \theta_c$ results in the contact line being driven toward the liquid and we call this movement *drainage*. So again, the condition for the contact line to not be moving is that $\theta = \theta_c$ (or $F = 0$).

If instead of a liquid and its vapor, for which $\gamma_{vs} > \gamma_{\ell s}$, we have two immiscible liquids and a solid meet at a contact line, the above argument repeats and we define the liquid that has the larger tension with the solid to be the *nonwetting* phase n and the liquid that has the smaller tension with the solid to be the *wetting* phase w. So the definition of wetting is that $\gamma_{ns} > \gamma_{ws}$. Because γ_{ns} is larger than γ_{ws}, the wetting phase has a greater affinity for the solid surface and so the solid surface attracts the contact line in the direction of the nonwetting phase (imbibition) unless there is a sufficiently large pressure difference applied across the meniscus to prevent such movement in the direction of the nonwetting phase as will be discussed below. To summarize, movement of the contact line in the direction of the wetting phase is called drainage and movement in the direction of the nonwetting phase is called imbibition. A nonzero imbibition-force density F (force per unit length of contact line acting normal to the contact line) is driving movement of the contact line in the plane of the solid surface and in a direction perpendicular to the contact line with $\theta \neq \theta_c$.

The simplest possible rule, or boundary condition or constitutive law, for the speed of imbibition v_c, which is the speed that the contact line is moving in the direction of imbibition under the influence of a positive F, is the linear relation $v_c = \alpha F/\eta_w$ where α is a dimensionless $O(1)$ "phenomenological coefficient" and η_w is the viscosity of the wetting fluid, i.e.,

$$\boxed{v_c = \alpha \frac{\gamma_{wn}}{\eta_w} \left(\cos\theta_c - \cos\theta\right).} \tag{7.222}$$

There is a growing literature surrounding the rules for the speed at which contact lines are moving and a review of the many proposed models and various experiments will not be attempted here. A range of experiments on contact lines advancing in imbibition suggest a relation of the form

$$v_c = \alpha(\theta_c) \frac{\gamma_{wn}}{\eta_w} \left(\cos\theta_c - \cos\theta\right)^\beta , \tag{7.223}$$

where, depending on the nature of the solid surface, the exponent β has been measured to lie in the range $1 \leq \beta \leq 2$. Molecularly smooth and clean solid surfaces tend to have an exponent $\beta \to 1$, which is the value that we will use in the two examples of imbibition dynamics considered in this chapter. To be consistent with our earlier handling of irreversible heat-generating processes, the relation between v_c and $F = \gamma_{wn}(\cos\theta_c - \cos\theta)$ would ideally be found through a careful analysis of the irreversible generation of heat per unit length of contact line $T\dot{S}$ (J/m/s) in the wedge-shaped "inner region" where the three phases come into contact and where wetting fluid is being pulled forward and bound to the solid surface while simultaneously experiencing a complicated viscous rotation that generates significant energy dissipation. If we obtain an expression for $T\dot{S}$ and make the identification $T\dot{S} = v_c F$ with F the positive imbibition force identified above, the phenomenological transport law for the speed of imbibition v_c can be identified. Simple analyses not entirely representing the complicated processes in the inner region of the three-phase contact predict that $T\dot{S} \propto v_c^2$, which is consistent with the linear relation of Eq. (7.222).

For spontaneous imbibition in conduits or droplet spreading on solid surfaces, the dynamic contact angle is observed to be time dependent $\theta(t)$ and we must find this evolving dynamic contact angle by combining the boundary condition of Eq. (7.222) with the modeling of the viscous flow as shown in the example that follows. Equation (7.222) is also assumed to hold in the case of slow drainage with both F and v_c now negative and with a distinct phenomenological coefficient α for drainage. However, as invasion speeds increase in drainage, the wetting fluid is not entirely replaced by nonwetting fluid and a thin film of wetting fluid remains on the solid walls. So "fast drainage" is often better described in terms of viscous flow alone without the use of Eq. (7.222) and evolving contact angles.

An Example of Spontaneous Imbibition

To illustrate how these ideas work in practice, consider a thin horizontal air-filled slit of width w (an air-filled rectangular duct) that resides between two solid walls of infinite extent. One side of the slit is brought into contact with a water reservoir maintained at atmospheric pressure P_a, while the other side of the slit is open to the atmosphere and therefore is also maintained at constant P_a. Because the air has such a small viscosity compared to the liquid water, we can ignore the viscous pressure drops in the air and take P_a to be a uniform constant in the slit between the meniscus and exit. The situation is depicted in Fig. 7.6.

Initially at $t = 0$, the meniscus separating the air from the water at the entrance to the water reservoir will be flat (no curvature) so that $P_w(z = 0, t = 0) = P_a$ and $\cos\theta(0) = 0$.

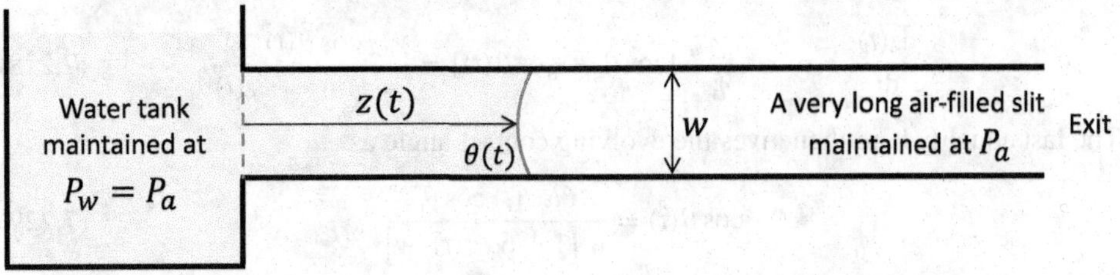

Figure 7.6 A long narrow slit is initially filled with air and brought into contact with a water reservoir on the left. Initially at $t = 0$, the meniscus is at $z = 0$ and has a contact angle of $\theta(0) = \pi/2$ (no curvature as depicted by the dashed line). Spontaneous imbibition then occurs in which the meniscus and viscous water is drawn into the slit causing the dynamic contact angle to monotonically decrease with increasing $z(t)$ within the range $\theta_c \leq \theta(t) \leq \pi/2$.

Due to the nonzero imbibition forces acting at the contact lines, the contact lines will be pulled into the slit bringing the viscous water with it. The invading water develops a parabolic flow profile behind the advancing meniscus that creates viscous resistance to the advancement. Resolving the details of the flow field near the contact line is complicated, especially in the *inner region* where the water and air and solid come into contact. We therefore consider only the average flow across the slit, which as derived in Chapter 5 on viscous flow is given by

$$\bar{v}_z = -\frac{w^2}{12\eta_w}\frac{\partial P_w}{\partial z}. \tag{7.224}$$

As the meniscus advances, it will develop an evolving curvature given by $1/R(t) = 2\cos\theta(t)/w$. To see this relation, draw a circle of radius $R(t) > w/2$ and use simple trigonometry to give $\cos\theta(t) = (w/2)/R(t)$. We next use the physics presented above to find this evolving contact angle $\theta(t)$.

Begin with the Young–Laplace result of Eq. (7.218) with $1/R_1 = 2\cos\theta(t)/w$ and $1/R_2 = 0$ for the rectangular duct to obtain the pressure jump across the advancing meniscus

$$P_w(z(t)) = P_a - \frac{2\gamma_{wa}\cos\theta(t)}{w} \tag{7.225}$$

so that the pressure gradient in the water is

$$\frac{\partial P_w}{\partial z} = \frac{P_w(z(t)) - P_a}{z(t)} = -\frac{2\gamma_{wa}\cos\theta(t)}{wz(t)}. \tag{7.226}$$

The average speed at which the water is advancing must be the same as the speed that the meniscus is advancing so that

$$\frac{\mathrm{d}z(t)}{\mathrm{d}t} = \bar{v}_z = \frac{\gamma_{wa}\cos\theta(t)}{6\eta}\frac{w}{z(t)}. \tag{7.227}$$

We also have the boundary condition of Eq. (7.222) for the speed v_c at which the contact lines are advancing

$$\frac{\mathrm{d}z(t)}{\mathrm{d}t} = v_c = \frac{\alpha\gamma_{wa}}{\eta_w}\left[\cos\theta_c - \cos\theta(t)\right] = \frac{\gamma_{wa}\cos\theta(t)}{6\eta}\frac{w}{z(t)}. \tag{7.228}$$

The last equality here then gives the evolving contact angle as

$$\cos\theta(t) = \frac{6\alpha\, z(t)\cos\theta_c}{w\left[1 + 6\alpha\, z(t)/w\right]}. \tag{7.229}$$

At $z(0) = 0$, this predicts that $\cos\theta(0) = 0$ or $\theta(0) = \pi/2$ as required. In the other limit as $z(t) \to \infty$, we then have the long-time equilibrium limit $\cos\theta(t) \to \cos\theta_c$.

Using this result for $\cos\theta(t)$ in Eq. (7.227) gives

$$\left(1 + 6\alpha\frac{z}{w}\right)\mathrm{d}z = \frac{\alpha\gamma_{wa}\cos\theta_c}{\eta_w}\mathrm{d}t, \tag{7.230}$$

which upon integration gives

$$3\alpha\left(\frac{z}{w}\right)^2+\frac{z}{w}-\frac{\alpha\gamma_{wa}\cos\theta_c}{w\eta_w}t=0. \tag{7.231}$$

This shows that once $z(t)/w \gg 1$, then $z^2 \propto t$, which is the experimentally observed result for spontaneous imbibition into either a single capillary or the air-filled pores of a porous material (cf., Feder et al., 2023). Solving for $z(t)/w$ then gives the final result for the advancement of the meniscus into the capillary slit

$$\boxed{\frac{z(t)}{w}=\frac{\sqrt{1+12\alpha^2\gamma_{wa}\cos\theta_c\, t/(w\eta_w)}-1}{6\alpha}.} \tag{7.232}$$

Taking the time derivative gives the evolving speed at which the meniscus spontaneously imbibes

$$\bar{v}_z(t)=v_c(t)=\frac{\mathrm{d}z(t)}{\mathrm{d}t}=\frac{\alpha\gamma_{wa}\cos\theta_c}{\eta_w\sqrt{1+12\alpha^2\gamma_{wa}\cos\theta_c\, t/(w\eta_w)}}. \tag{7.233}$$

So when $t=0$, we have Eq. (7.222) with $\theta(0)=\pi/2$ as required and as $t\to\infty$, we have $\mathrm{d}z(t)/\mathrm{d}t\to 0$ (the meniscus is no longer advancing) and $\cos\theta(t)\to\cos\theta_c$, which is the required condition of static equilibrium.

Static Equilibrium in Practice

To understand how static equilibrium is attained, other than by letting spontaneous imbibition play itself out over infinite time, consider a cylindrical capillary tube of radius a. The tube is connected at each end to a reservoir whose pressure can be independently set. If initially the meniscus between the two fluids inside the tube is flat and perpendicular to the surface of the solid (no capillary-pressure drop across the interface and therefore each reservoir maintained at the same pressure, say atmospheric), there will be an imbibition force F that drives the contact line toward the nonwetting fluid (spontaneous imbibition as just treated). If we want the contact line to be stationary, we must increase the pressure difference across the meniscus by either leaving the wetting side at atmospheric pressure and increasing the pressure in the reservoir on the nonwetting side or by leaving the nonwetting side at atmospheric pressure and lowering the pressure on the wetting side. This pressure difference will push on the interface in a direction opposite to the contact-line imbibition force F and will stretch and give curvature to the meniscus until the contact angle equals the equilibrium value θ_c at which time the contact line and fluid movement will be stationary in the capillary tube and the pressure in the wetting and nonwetting fluids are uniform constants between the reservoirs and the meniscus with a capillary-pressure drop across the stationary meniscus. The curvature $2/R=1/R+1/R$ of the meniscus at which this occurs is related to the cylindrical tube radius and equilibrium contact angle by $2/R=2\cos\theta_c/a$.

Thus, the Young–Laplace equation for a stationary meniscus in a cylindrical capillary tube of radius a is

$$\boxed{\Delta P_c \mathrel{\widehat{=}} P_{nc} - P_{wc} = \frac{2\gamma_{wn}\cos\theta_c}{a}.} \tag{7.234}$$

This defines the equilibrium capillary-pressure drop ΔP_c as the pressure drop condition at which $\theta = \theta_c$ and the contact lines are stationary. If the capillary were a rectangular slit of width w as in the previous subsection, then $\Delta P_c = 2\gamma_{wn}\cos\theta_c/w$. The fluid–fluid surface tension is denoted here as γ_{wn}.

If $P_n - P_w > \Delta P_c$, the pressure forces win out over the contact-line (or imbibition) force and the meniscus will be driven toward the wetting liquid (drainage) with $\theta(t) < \theta_c$. If $P_n - P_w < \Delta P_c$, the contact-line force wins out over the pressure forces and the meniscus will be driven toward the nonwetting fluid (imbibition) with $\theta(t) > \theta_c$.

The Simple Example of Capillary Rise

We now consider one of the oldest observations of capillary physics as depicted in Fig. 7.7. If an air-filled cylindrical glass capillary tube is brought into vertical contact with a body of water, the water will spontaneously rise into the tube. When the meniscus is at a height h above the surface of the body of water and is no longer moving, the contact angle will be at its equilibrium value θ_c. Here, we treat the final static equilibrium state only, relating h to θ_c. As a guided exercise at the end of the chapter, you can treat the dynamics of how the meniscus rises through time. As the meniscus rises in the tube, the contact angle decreases as does the imbibition speed until the meniscus is stationary at height h with contact angle θ_c.

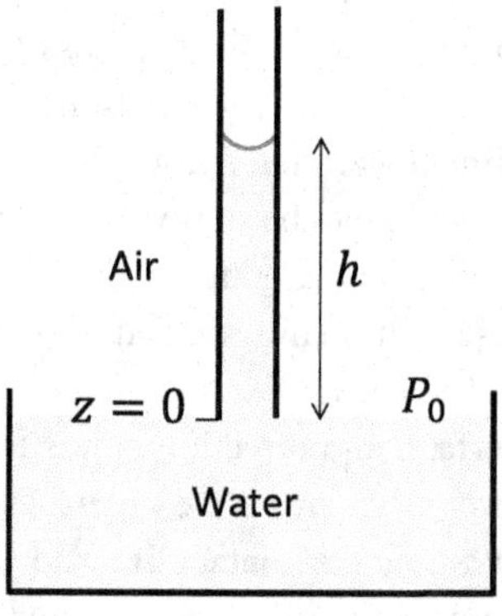

Figure 7.7 An air-filled capillary tube is brought into contact with a basin of water. Spontaneous imbibition draws the contact lines of the meniscus upward until the contact angle is at its equilibrium value θ_c and the meniscus is at a stationary height h above the surface of the water in the basin.

In the final static state after the rise has occurred, the pressure of the water and air on either side of the meniscus in the tube that is at a height h above the water surface is given by

$$P_w(h) = P_0 - \rho_w gh, \tag{7.235}$$

$$P_a(h) = P_0 - \rho_a gh, \tag{7.236}$$

where P_0 is the pressure at the surface of the water $z = 0$. The capillary-pressure drop across the stationary meniscus is then

$$P_w(h) - P_a(h) = -(\rho_w - \rho_a)\, gh \tag{7.237}$$

$$= -\frac{2\cos\theta_c \gamma_{wa}}{a}. \tag{7.238}$$

The capillary rise h is thus related to the equilibrium contact angle through the relation

$$\boxed{h = \frac{2\gamma_{wa}\cos\theta_c}{(\rho_w - \rho_a)ga}.} \tag{7.239}$$

This equation provides one way of measuring the surface tension of the air–water interface γ_{wa} by observing both the rise height h and the equilibrium contact angle θ_c of the stationary meniscus.

Static Equilibrium on Actual Rough and Dirty Surfaces

The preceding paragraphs are strictly valid only for molecularly smooth solid surfaces with no chemical variations over the surface. In practice, due to either rough solid surfaces or heterogeneity in the solid surface tensions over the solid surface, a contact line can become pinned and can tolerate a range of increasing or decreasing capillary-pressure drops before becoming unstuck. Note that for a perfectly clean surface where such pinning does not occur, the slightest deviation of the capillary-pressure drop from the equilibrium value corresponding to Eq. (7.234) will result in contact-line movement.

For example, consider the case shown in Fig. 7.8. A capillary tube of radius a has a small constriction (or bump) of height h_b in it on which the contact line has become pinned. In order for the meniscus to proceed to the right (drainage), one must increase the pressure drop so that the angle between the meniscus and the wall on the downslope of the bump is at least slightly greater than the equilibrium contact angle θ_c on that downslope as defined for these three phases. If the constrictive bump in the tube makes an angle θ_b relative to horizontal as shown in the figure, we have that the apparent equilibrium contact angle relative to the horizontal at which the meniscus can break free from the constriction in drainage is $\theta_a = \theta_c - \theta_b$. For imbibition, one would need to lower the pressure difference to an apparent contact angle of $\theta_a = \theta_c + \theta_b$ before the meniscus could move to the left. In this case, we have that over the range of pressure differences

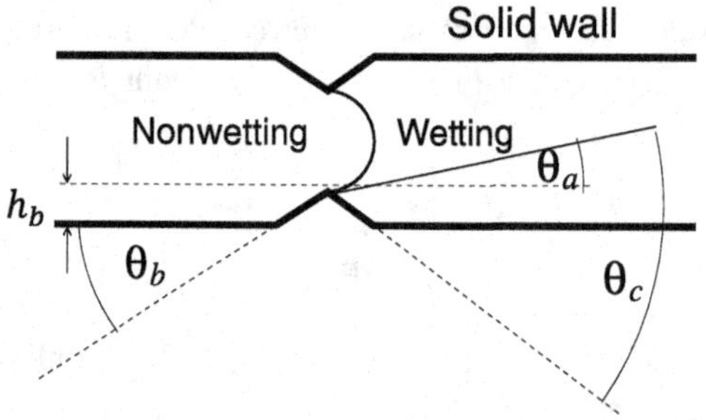

Figure 7.8 A capillary tube of radius a that has a constriction of height h_b making an angle θ_b relative to the horizontal. The equilibrium contact angle for these three phases is θ_c while θ_a is the apparent contact angle relative to the horizontal. As discussed in the text, such a constriction will allow the meniscus to remain pinned over a finite range of pressure differences $P_n - P_w$ as defined by Eq. (7.240).

$$\frac{2\gamma_{wn}\cos(\theta_c + \theta_b)}{(a - h_b)} < P_n - P_w < \frac{2\gamma_{wn}\cos(\theta_c - \theta_b)}{(a - h_b)} \tag{7.240}$$

the meniscus will remain pinned on the constriction.

Although the geometry in this illustrative example is somewhat contrived, the reality is that in the presence of roughness and/or chemical heterogeneity over the solid surface, menisci can remain pinned over a range of pressure differences and not just for a single value of pressure difference as implied by Eq. (7.234), which applies only to smooth and clean cylindrical surfaces. It is this effect that allows droplets of water to remain pinned to your windshield as you drive faster and faster in a car until they finally break free of the heterogeneity that was pinning their contact lines. Look for this the next time you are in a car when it is raining (while avoiding getting a traffic citation as you try to get the droplets to break free).

Visualizing Contact Line Movement

Given that both fluids must always satisfy the zero-flow boundary condition where they are in contact with the solid, how should we envision the contact line moving? This point of possible confusion is addressed in Fig. 7.9 where we present a schematic of what is going on in either imbibition or drainage.

It is perhaps useful to think of the interface between the two fluids as a stretched membrane of insect paper that has sticky glue on the side in contact with the nonwetting fluid. As the contact line is advancing to the right in imbibition with $\theta > \theta_c$, we can imagine laying down this insect paper onto the surface. The dots on the interface are meant to be markers that allow us to keep track of position along the fluid–fluid interface. Think of them as insects flying about in the nonwetting (e.g., vapor) phase that have become trapped in the glue and then get buried (or exposed) between the solid and the wetting fluid as the contact line advances in imbibition (or retreats in drainage). When the contact-line is moving to the left in drainage with $\theta < \theta_c$, it is like we are pulling up this insect paper. There have been experiments performed (Dussan and Davis, 1974) that actually put markers on the fluid–fluid interface and directly observe this type of motion.

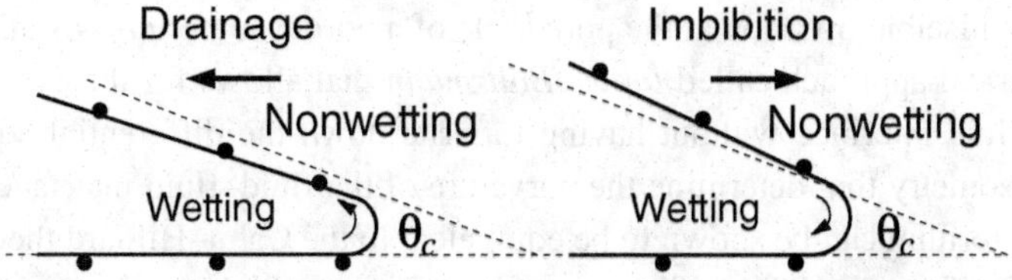

Figure 7.9 Schematic of how to understand the movement of a contact line and its relation to the no-slip boundary condition between the two fluids and the solid. Imagine that the fluid–fluid interface is a stretched piece of insect paper with glue on the side in contact with the nonwetting fluid. The dots are insects that have become trapped in the glue and that serve as markers. In drainage, we are pulling up the sticky paper and exposing previously buried insects. In imbibition, we are laying it down and burying insects.

7.6.4 Drainage and Imbibition in the Pores of a Porous Material

Imagine a drainage scenario with air entering one side of a packing of sand grains that is initially filled with water as the water is being extracted from the other side. We can imagine that the air moving into the sand pack is at atmospheric pressure at the entrance face and that the water is being extracted from the exit face by a pump working at a constant extraction rate, so that the smallest pressures in the system are at the exit face of the sand pack. As a meniscus bounding the invading air is advancing into the constriction between sand grains, the capillary-pressure drop across the meniscus must be increased by ongoing pumping that lowers the water pressure on the downstream side of the meniscus so that the radius of curvature of the meniscus becomes smaller. At a certain level of capillary-pressure drop, the meniscus has a small enough curvature that it can finally squeeze through the constriction and it will rapidly advance out the other side at a speed two or three orders of magnitude greater than the average rate of water withdrawal. While this rapid advancement is occurring, the single meniscus will separate into several menisci that enter into other constrictions between the sand grains until the capillary-pressure drops across these newly formed menisci have fallen back to sufficient levels that each meniscus is trapped again at the entrance of a constriction and must await the pressure drop to increase due to the steady withdrawal of water. This entire complicated, nonlinear drainage process of a meniscus going from being trapped in a constriction to advancing rapidly is called a *Haines jump* and the rules that govern the precise position and shape of the menisci and contact lines at all moments of a jump are not easily written down in a differential form that can then be solved numerically with enough resolution to observe the evolution of the menisci during the jumps. Imbibition in natural porous media is similarly complicated but now a meniscus is pulled rapidly through the constrictions by the imbibition force and becomes trapped at the entrance into the larger air-filled pores.

The rules of immiscible flow as specified earlier in this section require modeling the shape of the menisci that are no longer spherical in a natural sand pack as they are in a cylindrical capillary tube. The local curvature and local normal $\boldsymbol{n}$ of the fluid–fluid interface must be determined using differential geometry that accounts for the range of grain-surface shapes throughout the material. Due to the complexity of such formulations, the numerical solution of immiscible invasion at the pore scale of a porous material is usually performed using a numerical approach called *lattice Boltzmann* that allows for the evolving position of the fluid–fluid interface without having to write down the differential equations from differential geometry that determine the curvature of the fluid–fluid interface. The lattice-Boltzmann algorithm can be shown to be equivalent to the Cahn–Hilliard theory presented earlier (e.g., Swift et al., 1996) and is generally easier to implement in practice than finite-difference modeling of the Cahn–Hilliard governing equations. Working out general rules based on explicit modeling of the menisci curvature and contact line positions within sand packs or other realistic porous media is not practical.

It is also possible to derive differential rules for the average fluid speeds, fluid pressures, and saturation level in a porous material, where the average is over elements that are many grain diameters (or more) in linear dimension so that the differential rules are not resolving the details of the pores and the menisci shapes and positions. However, such macroscopic *porous-continuum* rules of immiscible flow are still evolving as the work of Breen et al. (2022) demonstrates.

7.7 Exercises

1. *Thermal diffusion into the Earth*: A simple practical problem is to determine how deep in the Earth should one place a root cellar or wine cave so that thermal diffusion from winter and summer does not penetrate into the underground storage area. The goal is for the storage area to remain at the year-round average temperature of the atmosphere above the cave or cellar and to not fluctuate with the seasons.

   The problem to be modeled is depicted in Fig. 7.10. The temperature in the Earth $T(z, t)$ as a function of depth z and time t is governed by the 1D diffusion equation

$$D_t \frac{\partial^2 T(z,t)}{\partial z^2} - \frac{\partial T(z,t)}{\partial t} = 0, \tag{7.241}$$

   where D_t is the thermal diffusivity of the Earth that is taken to be uniform. The boundary condition on the Earth's surface is

$$T(0,t) = T_0 + \Delta T \cos\left(\frac{2\pi t}{t_y}\right), \tag{7.242}$$

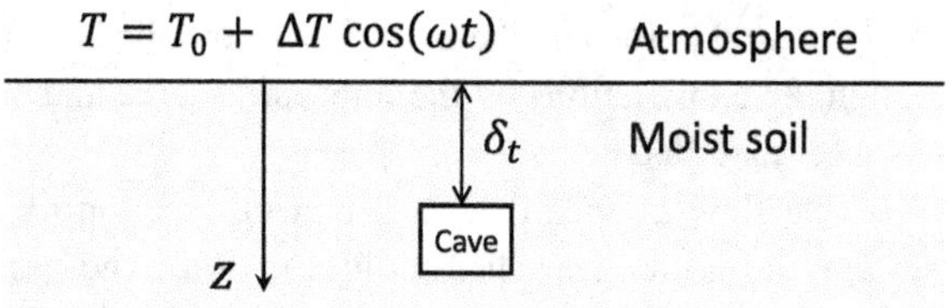

Figure 7.10 Sinusoidal temperature variations in the atmosphere penetrate a depth δ_t into the Earth.

where T_0 is the mean annual temperature of the atmosphere above this particular cave and ΔT is the temperature fluctuation between winter and summer. The temporal period is taken to be one year $t_y = 3.15 \times 10^7$ s. As $z \to \infty$, the fluctuations die out and $T \to T_0$.

To solve this BVP, assume a time-harmonic temporal response given in complex form as $T(z, t) = T_0 + \tilde{T}(z, \omega)e^{-i\omega t}$ and similarly for the spatial response $\tilde{T}(z, \omega) = A_k e^{ikz}$ where the temporal frequency is $\omega = 2\pi/t_y$ and where the "wave number" k must be found. Once all these terms are found, the actual temperature profile is taken as the real part of this form that will give the desired real boundary condition.

Show that inserting the complex time-harmonic form into the diffusion equation gives $k = \pm(1+i)\sqrt{\omega/(2D_t)}$ and that

$$T(z, t) = A_+ e^{i\left(\sqrt{\omega/(2D_t)}\, z - \omega t\right)} e^{-\sqrt{\omega/(2D_t)} z} + A_- e^{i\left(-\sqrt{\omega/(2D_t)}\, z - \omega t\right)} e^{+\sqrt{\omega/(2D_t)} z}, \tag{7.243}$$

which allows you to set $A_- = 0$ from the boundary condition at infinity. Show that the real part of this solution that satisfies the boundary condition at $z = 0$ is

$$\boxed{T(z, t) = T_0 + \Delta T \cos\left(\frac{z}{\delta_t} - \omega t\right) e^{-z/\delta_t},} \tag{7.244}$$

where the thermal skin depth is defined (like all skin depths associated with diffusion equations) as

$$\delta_t = \sqrt{\frac{2D_t}{\omega}}. \tag{7.245}$$

Temperature oscillations in the atmosphere produces a wave of temperature in the subsurface having a spatial wavelength $\lambda_t = 2\pi\delta_t$ and an amplitude that decays with depth as e^{-z/δ_t}.

So δ_t is a good measure of the depth of penetration of a sinusoidal temperature variation. Taking $\omega = 2\pi/t_y$ where $t_y = 3.15 \times 10^7$ s (one year) and a thermal diffusivity typical of moist soils $D_t = 6 \times 10^{-7}$ m^2/s, gives that $\delta_t = 2.5$ m. So a good depth to place your wine cave or root cellar is at δ_t or deeper so that the storage area remains at the average year-round temperature T_0. Note that if we take the time period to be one day $t_d = 8.64 \times 10^4$ s, the depth of penetration of the hot and cold of day and night is $\delta_t = 0.13$ m.

2. *A static temperature distribution*: When thermal diffusion has the opportunity to take place for a long time and the boundary conditions for some domain of interest do not change with time, the initial temperature distribution has time to equilibrate by diffusion and the temperature distribution in the terminal static state is governed by the Laplace equation $\nabla^2 T = 0$.

As an example of finding such a static temperature distribution, consider the case of a cylindrical glass tube of inner radius $r = a$ and outer radius $r = b$. The temperature of the gas inside the tube is maintained at a constant temperature T_a while the temperature in the gas outside the tube is maintained at a constant T_b. In the static state, find both the temperature distribution in the glass as well as the heat flux across the glass.

Working in cylindrical coordinates, the static temperature distribution in the glass is a solution of the differential equation

$$\frac{1}{r}\frac{\mathrm{d}}{\mathrm{d}r}\left(r\frac{\mathrm{d}T(r)}{\mathrm{d}r}\right) = 0, \tag{7.246}$$

subject to the BCs that $T(a) = T_a$ and $T(b) = T_b$. Integrate this equation and show that the static radial temperature distribution in the glass is

$$\boxed{T(r) = T_a + \frac{(T_b - T_a)}{\ln(b/a)} \ln\left(\frac{r}{a}\right),} \tag{7.247}$$

and that the radial heat flux $q_r(r)$ in the glass that has conductivity λ is

$$\boxed{q_r(r) = \frac{\lambda}{r}\frac{(T_a - T_b)}{\ln(b/a)}.} \tag{7.248}$$

3. *A moving boundary "Stefan problem"*: There are a range of diffusion problems in physics where, due to dissolution or a first-order phase transition (e.g., melting), one of the key boundary conditions in the domain of diffusion is specified on a surface that is changing position due to the dissolution or phase change taking place on that surface. These types of diffusion problems are often called *Stefan problems*.

   As an example, consider the interface between a solid and pure liquid when solid molecules can dissolve into the liquid at rates rapid relative to diffusion away from the solid–liquid interface and up to a solubility limit given by the concentration c_s. At $t = 0$, the liquid–solid interface is at $x = 0$ with solid occupying $x > 0$ and liquid $x < 0$. As time increases, solid molecules dissolve into the liquid and diffuse in the negative x direction to create an evolving concentration profile $c(x, t)$. The interface between the pure solid and liquid thus moves to the right and is located at the position $x = s(t)$. We'd like to find both $s(t)$ and $c(x, t)$ throughout the liquid $-\infty \le x \le s(t)$.

   We can formulate this Stefan problem as the BVP

$$\frac{\partial c}{\partial t} = D\frac{\partial^2 c}{\partial x^2}, \tag{7.249}$$

   subject to the boundary conditions that bound the liquid region $-\infty \le x \le s(t)$

$$c(-\infty, t) = 0 \quad \text{and} \quad c(s(t), t) = c_s. \tag{7.250}$$

   So the concentration begins at the saturation maximum of c_s at the evolving (increasing) interface position $x = s(t)$ and falls off to zero with distance to $x =$

$-\infty$ as controlled by the diffusion equation. The condition that allows $s(t)$ to be found is that as solid molecules enter the liquid, the interface position increases at the rate

$$c_s \frac{\mathrm{d}s(t)}{\mathrm{d}t} = D \frac{\partial c(x,t)}{\partial x}\bigg|_{x=s(t)}. \tag{7.251}$$

So all of Eqs (7.249) to (7.251) are coupled and together give $c(x, t)$ and $s(t)$.

A solution of the 1D diffusion equation takes the form

$$c(x,t) = A + B\int_0^{x/\sqrt{4Dt}} e^{-u^2}\,du, \tag{7.252}$$

where the integration constants A and B are found from the boundary conditions and must be independent of both x and t. To see that this is a solution of the 1D diffusion equation, just take derivatives using the Leibniz rule from Chapter 1 and show that

$$\frac{\partial c(x,t)}{\partial t} = -\frac{Bx}{4\sqrt{D}t^{3/2}} e^{-x^2/(4Dt)}, \tag{7.253}$$

$$D\frac{\partial c(x,t)}{\partial x} = \frac{B}{2}\sqrt{\frac{D}{t}} e^{-x^2/(4Dt)}, \tag{7.254}$$

$$D\frac{\partial^2 c(x,t)}{\partial x^2} = -\frac{Bx}{4\sqrt{D}t^{3/2}} e^{-x^2/(4Dt)}. \tag{7.255}$$

As an aside, the *error function* is defined $\mathrm{erf}(z) = (2/\sqrt{\pi})\int_0^z e^{-u^2}\,du$ and so we could have given the solution as $c(x,t) = A + B'\mathrm{erf}(x/\sqrt{4Dt})$. The normalization of the error function is because $\int_0^\infty e^{-u^2}\,du = \sqrt{\pi}/2$ so that $\mathrm{erf}(z) \to 1$ as $z \to \infty$.

To satisfy the boundary conditions, verify that $A = B' = B\sqrt{\pi}/2$ with A given by

$$A = \frac{c_s}{1 + \mathrm{erf}(s/\sqrt{4Dt})}. \tag{7.256}$$

But A must be independent of time which requires that

$$\boxed{s(t) = \beta\sqrt{4Dt},} \tag{7.257}$$

where β is a constant that is found from the condition of the moving interface

$$c_s\frac{\mathrm{d}s}{\mathrm{d}t} = D\,\frac{\partial c(x,t)}{\partial x}\bigg|_{x=s} = \frac{\sqrt{\pi}}{4}\sqrt{\frac{D}{t}}\frac{c_s e^{-\beta^2}}{1+\mathrm{erf}(\beta)}. \tag{7.258}$$

Inserting the required form of $s(t) = \beta\sqrt{4Dt}$ then yields the time-independent equation that determines β

$$\boxed{e^{-\beta^2} = \frac{4}{\sqrt{\pi}}\beta\,(1 + \mathrm{erf}(\beta)).} \tag{7.259}$$

The monotonic functions on the left-hand and right-hand side of this condition cross at a single positive real value of β that you can verify is $\beta \approx 0.30339$. With β so determined and the evolving position of the liquid–solid interface given by $s(t) = \beta\sqrt{4Dt}$, show that the concentration in the liquid domain $-\infty \leq x \leq s(t)$ evolves as

$$\boxed{c(x, t) = c_s \left[\frac{1 + \mathrm{erf}(x/\sqrt{4Dt})}{1 + \mathrm{erf}(\beta)} \right].} \tag{7.260}$$

4. *An exactly solvable advection–diffusion problem*: Earlier in the chapter, we used an approximate perturbation method to address the Taylor dispersion problem of the average advection and diffusion of solute in a cylindrical flow tube. Here, we consider an example of advection and diffusion of solute when an exact analytical solution is attainable due to the simple nature of the flow field.

At $t = 0$, we assume there is a concentration of solute in the binary solution that has a sinusoidal 1D variation $c(x, 0) = c_0 + \Delta c \cos kx$, where $k = 2\pi/\lambda$ and λ is the given wavelength of the spatial fluctuation. The advection is due to a simple uniform in space but sinusoidal in time flow given $v_x(t) = v_0 \cos \omega t$, where $\omega = 2\pi/T$ and T is the given period of the temporal oscillation. In this particular 1D spatial setting that is infinite in the x direction, the advection–diffusion equation for the solute concentration is

$$\frac{\partial c(x, t)}{\partial t} + v_0 \cos \omega t \frac{\partial c(x, t)}{\partial x} = D \frac{\partial^2 c(x, t)}{\partial x^2} \tag{7.261}$$

subject to the initial condition

$$c(x, 0) = c_0 + \Delta c \cos kx. \tag{7.262}$$

The problem is to find $c(x, t)$.

Given that the diffusion is created by the spatial variation of the initial condition, it is reasonable to propose a solution of the form

$$c(x, t) = c_0 + \mathrm{Re}\left\{\tau(t)e^{ikx}\right\} \tag{7.263}$$

for some complex time function $\tau(t)$ to be found. Putting this form into the advection–diffusion equation gives

$$\frac{\mathrm{d}\tau(t)}{\mathrm{d}t} = -\left(ikv_0 \cos \omega t + k^2 D\right) \tau(t), \tag{7.264}$$

which you can rewrite as

$$\frac{\mathrm{d}\tau}{\tau} = -ikv_0 \cos \omega t \, \mathrm{d}t - k^2 D \, \mathrm{d}t. \tag{7.265}$$

Integrate both sides and obtain

$$\tau(t) = \tau_0 e^{-k^2 Dt} e^{-i(kv_0/\omega)\sin\omega t}$$
$$= \tau_0 e^{-k^2 Dt}\left[\cos\left(\frac{kv_0}{\omega}\sin\omega t\right) - i\sin\left(\frac{kv_0}{\omega}\sin\omega t\right)\right]. \tag{7.266}$$

Show that the final solution for the real concentration field is then

$$c(x,t) = c_0 + \Delta c e^{-k^2 Dt}\left[\cos\left(\frac{kv_0}{\omega}\sin\omega t\right)\cos kx + \sin\left(\frac{kv_0}{\omega}\sin\omega t\right)\sin kx\right], \tag{7.267}$$

where you had to identify $\tau_0 = \Delta c$ so that the initial condition is exactly satisfied. So the initial concentration fluctuations in space decay to zero exponentially in time. Show that if the uniform flow field is steady $\omega = 0$, then $c(x,t) = c_0 + \Delta c e^{-k^2 t}\cos[k(x - v_0 t)]$ and corresponds to the steady translation of the frame of reference used to observe the otherwise pure diffusion.

5. *The Young–Laplace equation for capillary pressure from Cahn–Hilliard theory*: Let's assume we have a spherical bubble of one liquid that is surrounded by a second liquid and that the two fluids are partially miscible in each other. Using the Cahn–Hilliard formalism, obtain the Young–Laplace expression for the capillary pressure for this situation.

In the Cahn–Hilliard theory, there is assumed to be an equilibrium gradient of the concentration of liquid 1 molecules across a finite width layer of thickness d_c. For a large radius bubble R, in which we look at this transition layer from macroscopic scales, we have $d_c \ll R$ and the transition layer appears to be a sharp step. We assume for simplicity that the mass density variation across the transition layer between the two liquids is negligible. In this case, the surface tension parameter γ is defined from the Cahn–Hilliard molecular-interaction parameter κ_c as

$$\gamma = \int_{R-\Delta}^{R+\Delta} \kappa_c \left(\frac{\partial c}{\partial r}\right)^2 \mathrm{d}r, \tag{7.268}$$

where Δ is sufficiently large compared to d_c that $\partial c/\partial r|_{r=R\pm\Delta} = 0$. In this macroscopic limit, we thus can identify

$$1 = \int_{R-\Delta}^{R+\Delta} \frac{\kappa_c}{\gamma}\left(\frac{\partial c}{\partial r}\right)^2 \mathrm{d}r = \int_{R-\Delta}^{R+\Delta} \delta(r-R)\,\mathrm{d}r, \tag{7.269}$$

where $\delta(r-R) = (\kappa_c/\gamma)(\partial c/\partial r)^2$ emerges as the Dirac delta in the limit where $d_c/R \to 0$.

We next consider the condition of static equilibrium in Cahn–Hilliard theory $\nabla P = \nabla \cdot \boldsymbol{\tau}_\gamma$, where

$$\boldsymbol{\tau}_\gamma = \frac{\kappa_c}{2}(\nabla c \cdot \nabla c)\boldsymbol{I} - \kappa_c \nabla c \nabla c \tag{7.270}$$

is the Korteweg stress tensor when mass density variations are negligible. We now assume that $c(r)$ is purely radial as we transition from inside the bubble to outside across the spherical meniscus (transition layer). Working in spherical coordinates using the tensorial relations given in Chapter 1, and specifically Eq. (1.277), and noting that $\boldsymbol{I} = \hat{\boldsymbol{r}}\hat{\boldsymbol{r}} + \hat{\boldsymbol{\theta}}\hat{\boldsymbol{\theta}} + \hat{\boldsymbol{\phi}}\hat{\boldsymbol{\phi}}$, show that $\nabla P = \nabla \cdot \boldsymbol{\tau}_\gamma$ can be written

$$\frac{\mathrm{d}P(r)}{\mathrm{d}r} = -\frac{\kappa_c}{2}\frac{\mathrm{d}}{\mathrm{d}r}\left(\frac{\mathrm{d}c}{\mathrm{d}r}\right)^2 - \frac{2}{r}\kappa_c\left(\frac{\mathrm{d}c}{\mathrm{d}r}\right)^2. \tag{7.271}$$

Integrate both sides from $R - \Delta$ to $R + \Delta$ and use that $\mathrm{d}c(r)/\mathrm{d}r|_{r=R\pm\Delta} = 0$ to obtain

$$P_{ex} - P_{in} = -\int_{R-\Delta}^{R+\Delta} \frac{2}{r}\kappa_c\left(\frac{\mathrm{d}c}{\mathrm{d}r}\right)^2 \mathrm{d}r, \tag{7.272}$$

where P_{ex} is the pressure on the exterior of the liquid bubble and P_{in} the pressure on the inside. You can now use Eq. (7.269) to obtain

$$\boxed{P_{in} - P_{ex} = \frac{2\gamma}{R},} \tag{7.273}$$

which is precisely the Young–Laplace rule for the case of a spherical meniscus with $1/R_1 + 1/R_2 = 2/R$ separating two partially miscible liquids having a surface tension γ.

If we had considered a spherical droplet of liquid surrounded by its vapor, then it is the mass density variation across the transition layer that creates the macroscopic surface tension γ according to

$$\gamma = \int_{R-\Delta}^{R+\Delta} \kappa_\rho \left(\frac{\partial \rho}{\partial r}\right)^2 \mathrm{d}r, \tag{7.274}$$

where again Δ is sufficiently large compared to d_c that $\partial\rho/\partial r|_{r=R\pm\Delta} = 0$. In this macroscopic limit, we can again identify

$$1 = \int_{R-\Delta}^{R+\Delta} \frac{\kappa_\rho}{\gamma}\left(\frac{\partial \rho}{\partial r}\right)^2 \mathrm{d}r = \int_{R-\Delta}^{R+\Delta} \delta(r-R)\,\mathrm{d}r, \tag{7.275}$$

where $\delta(r-R)$ is again the Dirac delta. In this case of a spherical liquid droplet surrounded by its vapor in which the mass density varies radially across the transition layer $\rho(r)$, start with the definition

$$\boldsymbol{\tau}_\gamma = \kappa_\rho\left(\rho\nabla^2\rho + \frac{1}{2}\nabla\rho\cdot\nabla\rho\right)\boldsymbol{I} - \kappa_\rho\nabla\rho\nabla\rho \tag{7.276}$$

and go through the same steps as above in spherical coordinates to again obtain the macroscopic Young–Laplace rule for the capillary pressure $P_{in} - P_{ex}$.

6. *The static contact line condition* $\cos\theta_c = (\gamma_{s2} - \gamma_{s1})/\gamma_{12}$ *from Cahn–Hilliard theory*: We stated earlier that the macroscopic ideas of contact lines and contact angles are

accounted for in the Cahn–Hilliard theory of a binary liquid by using the boundary condition for the concentration c of fluid-1 molecules at a solid wall

$$\boldsymbol{n}_s \cdot \nabla c = -\frac{2(\gamma_{s2} - \gamma_{s1})}{d_c \gamma_{12}} \frac{(c - c^{(1)})(c - c^{(2)})}{(c^{(1)} - c^{(2)})}, \tag{7.277}$$

where $\boldsymbol{n}_s$ is the normal to the solid surface. Show that this boundary condition in static equilibrium produces Young's relation $\cos\theta_c = (\gamma_{s2} - \gamma_{s1})/\gamma_{12}$ for the equilibrium contact angle θ_c.

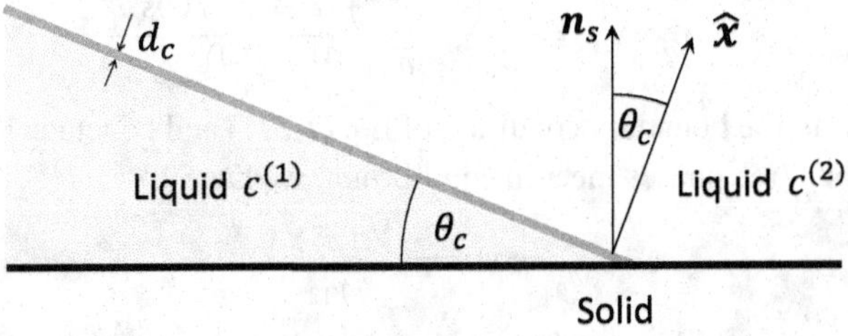

Figure 7.11 In the $d_c/R \ll 1$ limit where R is the radius of the meniscus and d_c the thickness of the transition region between the liquids (the gray layer in the figure that corresponds to the macroscopic concept of the meniscus), we can model the inner region as a wedge and the variations of concentration in the binary liquid as being perpendicular to the wedge in the linear direction x, where $\boldsymbol{n}_s \cdot \hat{\boldsymbol{x}} = \cos\theta_c$.

As shown in Fig. 7.11, assume that the thickness d_c of the transition layer separating the liquid with concentration $c^{(1)}$ from the liquid with concentration $c^{(2)}$ is much smaller than the radius of curvature of the meniscus. When we zoom in on the inner-region near the contact line in this case, we can assume that all concentration variations are in the linear x direction that is normal to the interface. In this case, when we are very close to the contact line, we have that

$$\boldsymbol{n}_s \cdot \nabla c(\boldsymbol{r}, t) = \boldsymbol{n}_s \cdot \hat{\boldsymbol{x}} \frac{\partial c(x, t)}{\partial x}. \tag{7.278}$$

This is true for both dynamic flowing conditions and static conditions. For static conditions, we have $\boldsymbol{n}_s \cdot \hat{\boldsymbol{x}} = \cos\theta_c$ where θ_c is the equilibrium contact angle and the equilibrium profile $c(x, t) = c(x)$ is a solution of the Cahn–Hilliard differential equation that was shown earlier to be

$$c(x) = \left(\frac{c^{(1)} + c^{(2)}}{2}\right) + \left(\frac{c^{(1)} - c^{(2)}}{2}\right) \tanh\left(\frac{x}{d_c}\right), \tag{7.279}$$

which can be inverted to give

$$\frac{x}{d_c} = \tanh^{-1}\left(\frac{c(x) - \left(c^{(1)} + c^{(2)}\right)/2}{\left(c^{(1)} - c^{(2)}\right)/2}\right). \tag{7.280}$$

Taking the derivative of the expression for the equilibrium profile $c(x)$ gives

$$\frac{\mathrm{d}c(x)}{\mathrm{d}x} = \left(\frac{c^{(1)} - c^{(2)}}{2d_c}\right) \operatorname{sech}^2\left(\frac{x}{d_c}\right). \tag{7.281}$$

The hyperbolic-function identity $\text{sech}^2 w = 1 - \tanh^2 w$ yields the identity $\text{sech}^2 \left(\tanh^{-1} u\right) = 1 - u^2$. Combine Eqs (7.281) and (7.280) and use this hyperbolic identity to obtain

$$\frac{dc(x)}{dx} = -\frac{2\left(c(x) - c^{(1)}\right)\left(c(x) - c^{(2)}\right)}{d_c\left(c^{(1)} - c^{(2)}\right)}. \tag{7.282}$$

So in static equilibrium and in the vicinity of the contact line when $d_c \ll R$ we have

$$\boldsymbol{n}_s \cdot \nabla c = -\frac{2\left(c - c^{(1)}\right)\left(c - c^{(2)}\right)\cos\theta_c}{d_c\left(c^{(1)} - c^{(2)}\right)}. \tag{7.283}$$

Compare this to the boundary condition of Eq. (7.277) and conclude that at the contact line where the three phases meet in equilibrium, we have

$$\cos\theta_c = \frac{\gamma_{s2} - \gamma_{s1}}{\gamma_{12}}, \tag{7.284}$$

which is Young's equilibrium force balance of the surface tensions acting normal to the contact line in the plane of the solid.

7. *The dynamics of capillary rise*: Consider again the case of a cylindrical capillary tube of radius a that is held vertically and allowed to make contact with a basin of water. In the chapter, we treated the terminal equilibrium state of the meniscus after it has risen into the capillary tube and is stationary at a height h above the surface of the basin. In this problem, you will fill in the details of the timing of how the meniscus rises until it reaches the static-equilibrium height h.

While the meniscus is rising and the viscous water is being pulled up into the capillary tube, parabolic flow in the water will occur with the creation of a viscous-pressure gradient $\partial P_w^v/\partial z$ in addition to the pressure variation due to gravity. We can continue to ignore the viscosity and viscous-pressure gradients in the low-viscosity air. So the pressure in the air and water on either side of the meniscus which is located at the height $z(t) \leq h$ is

$$P_a(z) = P_0 - \rho_a g z(t), \tag{7.285}$$

$$P_w(z) = P_0 - \rho_w g z(t) + \frac{\partial P_w^v}{\partial z} z(t) \tag{7.286}$$

so that the capillary pressure drop across the meniscus when it is at height $z(t)$ is

$$P_w(z) - P_a(z) = -(\rho_w - \rho_a) g z(t) + \frac{\partial P_w^v}{\partial z} z(t) = -\frac{2\gamma_{wa}\cos\theta(t)}{a}, \tag{7.287}$$

by the Young–Laplace relation for a cylindrical capillary tube. The second equality in Eq. (7.287) identifies the viscous pressure gradient in terms of the dynamic contact angle $\theta(t)$ as

$$\frac{\partial P_w^v}{\partial z} = -\frac{2\gamma_{wa}\cos\theta(t)}{a z(t)} + (\rho_w - \rho_a) g. \tag{7.288}$$

Further, the average value of the water velocity across the diameter of the tube $\bar{v}_z$ is, as was shown in Chapter 5 on viscous fluid flow, related to the viscous pressure gradient as

$$\frac{\mathrm{d}z(t)}{\mathrm{d}t} = \bar{v}_z = -\frac{a^2}{8\eta_w}\frac{\partial P_w^v}{\partial z}. \tag{7.289}$$

Additionally, the contact-line boundary condition of Eq. (7.222) sets the contact-line speed as

$$\frac{\mathrm{d}z(t)}{\mathrm{d}t} = v_c = \frac{\alpha\gamma_{wa}}{\eta_w}\left[\cos\theta_c - \cos\theta(t)\right]. \tag{7.290}$$

Combine Eqs (7.288)–(7.290) and show that the dynamic contact angle decreases with the increasing height of the meniscus as

$$\cos\theta(t) = \zeta(t)\left(\frac{\zeta(t)+\epsilon}{1+\epsilon}\right)\cos\theta_c, \tag{7.291}$$

where we define the scaled height of the meniscus as

$$\zeta(t) = \frac{z(t)}{h} \tag{7.292}$$

with

$$h = \frac{2\cos\theta_c\gamma_{wa}}{(\rho_w - \rho_a)ga} \tag{7.293}$$

being the final static equilibrium height of the meniscus (as derived earlier). We thus have $0 \leq \zeta(t) \leq 1$ during imbibition. The dimensionless parameter ϵ is defined

$$\epsilon = \frac{a}{4\alpha h} = \frac{(\rho_w - \rho_a)ga^2}{8\alpha\gamma_{wa}\cos\theta_c}. \tag{7.294}$$

To get a feel for the size of ϵ, consider the case of a glass capillary tube with a bore radius of $a = 10^{-3}$ m. Using $\alpha = 1$, $\gamma_{wa} = 7.3 \times 10^{-2}$ N/m (water-air) and $\cos\theta_c = 1$ (water-air-glass) we have $\epsilon = 1.7 \times 10^{-2}$. We will not set ϵ to zero here but could do so with essentially no error for glass capillary tubes having $a < 10^{-4}$ m. For this $a = 10^{-3}$ m glass-tube example, the static capillary rise is $h = 1.5 \times 10^{-2}$ m.

Using the above expression for $\cos\theta(t)$, show that $\zeta(t) = z(t)/h$ satisfies the differential equation

$$\frac{\mathrm{d}\zeta}{\mathrm{d}t} = -\frac{(\zeta^2 + \epsilon\zeta - 1 - \epsilon)}{(2+\epsilon)\tau_h} \tag{7.295}$$

subject to the initial condition $\zeta(0) = 0$ when $t = 0$ and where the characteristic rise time τ_h (measured in seconds) is given by

$$\tau_h = \left(\frac{1+\epsilon}{2+\epsilon}\right)\frac{\eta_w h}{\alpha\gamma_{wa}\cos\theta_c}. \tag{7.296}$$

For the above example with air-water-glass and a bore diameter of $a = 10^{-3}$ m, we have $\tau_h = 10^{-4}$ s. Rise times will become longer, however, for smaller radius capillary tubes (because $ha = \text{const} \approx 1.5 \times 10^{-5}$ m^2 for water-air-glass). This differential equation shows that $\mathrm{d}\zeta/\mathrm{d}t \to 0$ as $\zeta \to 1$ as it must.

Integrate the differential equation and obtain the result

$$\boxed{\zeta(t) = \frac{z(t)}{h} = (1+\epsilon)\left(\frac{1-e^{-t/\tau_h}}{1+\epsilon+e^{-t/\tau_h}}\right).} \tag{7.297}$$

At $t = 0$, we thus have $z(0) = 0$ as required and as $t \to \infty$, we have $z(t) \to h$ also as required. Although it formally takes an infinite amount of time for the meniscus to arrive precisely at h, the meniscus will effectively get to h after just a few τ_h, that is, after a few τ_h, the exponential functions here are effectively zero compared to one and $z(t) = h$.

Last, take a time derivative and obtain the imbibition speed

$$\frac{\mathrm{d}\zeta(t)}{\mathrm{d}t} = \frac{1}{h}\frac{\mathrm{d}z(t)}{\mathrm{d}t} = \frac{(1+\epsilon)(2+\epsilon)e^{-t/\tau_h}}{\tau_h\left(1+\epsilon+e^{-t/\tau_h}\right)^2}. \tag{7.298}$$

When $t = 0$, the speed is a maximum and given by $\mathrm{d}z(t)/\mathrm{d}t = h(1+\epsilon)/[(2+\epsilon)\tau_h]$, while as time advances there is an exponential slowing of the rise.

8. *The Einstein relation for the solute diffusivity D_c*: Imagine that a force $-\partial U/\partial x$ acts on the solute molecules of a liquid to drive a velocity $-b\partial U/\partial x$ and mass flux $-\rho c b\partial U/\partial x$ of solute molecules in the x direction, where b is the mobility of the solute molecules and ρc the mass of solute per unit volume. At a boundary, solute molecules accumulate and drive a counter diffusive mass flux of solute given by $-D_c\rho\partial c/\partial x$, such that in an equilibrium steady state the net flux is zero: $-\rho c b\partial U/\partial x - D_c\rho\partial c/\partial x = 0$. Such a steady state is also characterized by the force balance $-\partial\mu_1/\partial x - \partial U/\partial x = 0$ where μ_1 is the chemical potential of the solute that is approximated here in the dilute limit as $\mu_1 = \tilde{\mu}_1 + k_B T\ln(N_1/N_t)$. With N_t taken as constant in the dilute limit, integrate the force balance to obtain the much used "Boltzmann relation" $c = c_0\exp\left[-U/(k_B T)\right]$ that quantifies how concentrations or densities of solute molecules depend on a force potential U that is acting on them. Use the Boltzmann relation in the zero solute flux condition to obtain the desired result due to Einstein $D_c = k_B T b$.

Part II

Mathematical Methods

8
Partial-Differential Equations

We now change perspectives and develop mathematical techniques for solving a simplified subset of the various types of governing equations derived through Part I of the book. Having the appropriate partial-differential equations (PDEs) available to us for a given physics problem and understanding the restrictions on those equations due to having carefully derived them in Part I of the book is an important accomplishment. But to build intuition about the physics, one needs to solve the equations and visualize the nature of the fields so obtained. To do so analytically means we will be solving linearized forms of the equations in simple situations but that is precisely how you build intuition about the physics. The example problems and end-of-chapter exercises performed in Part I largely correspond to situations where an analytical solution can be obtained through integration alone. Our ultimate goal in this second part of the book is to get to the point that we can solve for the Green's function or Green's tensor of linearized versions of each physics problem we have encountered in Part I of the book. The Green's tensor is the response controlled by a vectorial set of PDEs due to an impulsive source that is applied at a point in space. The Green's tensor tells us a lot about the nature of the response controlled by a given set of PDEs and can be used to build the solution for nonimpulsive and spatially distributed sources of linear physics problems. To get there, however, we will develop a range of mathematical methods in the intervening chapters and treat a range of peripheral issues with these methods that the author has found useful over the years. The present Part II of the book serves as its own stand-alone introductory treatment of the subject generally known as *the mathematical methods of physics* but is also designed as the coherent continuation to Part I on the rules of continuum physics.

8.1 Overview of the Partial-Differential Equations to Be Treated

In the development of our laws of continuum physics, we saw on many occasions the following three PDEs:

$$\underbrace{D\,\nabla^2\psi - \frac{\partial\psi}{\partial t} = S_d}_{\substack{\text{Diffusion}\\ \text{equation}}}; \qquad \underbrace{c^2\,\nabla^2\,\psi - \frac{\partial^2\psi}{\partial t^2} = S_w}_{\substack{\text{Wave}\\ \text{equation}}}; \qquad \underbrace{\nabla^2\psi = S_p}_{\substack{\text{Potential}\\ \text{or “Poisson”}\\ \text{equation}}}. \tag{8.1}$$

To obtain these scalar equations in the above form, the material in the domain under consideration is assumed to be uniform and isotropic, that is, $D \widehat{=}$ diffusivity and $c \widehat{=}$ wave speed are given scalar constants and the response fields ψ are scalars.

For diffusion and potential problems in heterogeneous and anisotropic media, the physics can be represented

$$\nabla \cdot (\boldsymbol{\lambda} \cdot \nabla \psi) - \gamma \frac{\partial \psi}{\partial t} = S_d \quad \textit{diffusion equation} \tag{8.2}$$

$$\nabla \cdot (\boldsymbol{\lambda} \cdot \nabla \psi) = S_d, \quad \textit{potential equation} \tag{8.3}$$

where $\boldsymbol{\lambda}$ is a positive-definite second-order *conductivity* tensor (e.g., the heat-flux conductivity in a heat-flow problem) and γ is a positive scalar (e.g., the heat capacity in a heat-flow problem) whose precise definition depends on the type of diffusion occurring and that is sometimes 1. We see that potential problems can always be viewed as diffusion problems that have reached the steady state. By introducing a time-derivative term into the potential equation to create a diffusion equation, you obtain what is generally the fastest way to solve a potential problem numerically, that is, first solve the diffusion problem, where fields are being updated in time based only on the local values of the fields surrounding each point in space and then run that solution to steady state, where the time derivative term goes to zero, to obtain the solution of the potential problem.

For wave problems in heterogeneous media, we obtain different forms of wave equations depending on the physical context. For example, in an isotropic fluid we saw in Chapter 5 that the pressure response is governed by

$$K_f \nabla \cdot \left(\frac{1}{\rho} \nabla P \right) - \frac{\partial^2 P}{\partial t^2} = S_a, \quad \textit{acoustic wave equation} \tag{8.4}$$

where K_f (bulk modulus) and ρ (mass density) are both possibly variable in space (but not in time). In an elastic solid, we saw in Chapter 4 that the mechanical displacements $\boldsymbol{u}$ (a vector) are governed by

$$\nabla \cdot ({}_4\boldsymbol{C} : \nabla \boldsymbol{u}) - \rho \frac{\partial^2 \boldsymbol{u}}{\partial t^2} = -\boldsymbol{f}_a, \quad \textit{elastodynamic wave equation} \tag{8.5}$$

where ${}_4\boldsymbol{C}$ (the fourth-order stiffness tensor) and ρ (mass density) are possibly variable in space and $\boldsymbol{f}_a$ is an applied-force density creating the waves. For electromagnetic response in anisotropic and heterogeneous media we have

$$\nabla \times \boldsymbol{E} = -\boldsymbol{\mu} \cdot \frac{\partial \boldsymbol{H}}{\partial t} \tag{8.6}$$

$$\nabla \times \boldsymbol{H} = \boldsymbol{\epsilon} \cdot \frac{\partial \boldsymbol{E}}{\partial t} + \boldsymbol{\sigma} \cdot \boldsymbol{E} + \boldsymbol{J}_a, \tag{8.7}$$

where $\boldsymbol{\mu}$ (magnetic permeability), $\boldsymbol{\epsilon}$ (electrical permittivity), and $\boldsymbol{\sigma}$ (electrical conductivity) are all second-order tensors and $\boldsymbol{J}_a$ is the applied-current density creating the response.

When iron, nickel, and cobalt are not significantly present, we have $\boldsymbol{\mu} \approx \mu_0 \boldsymbol{I}$ where μ_0 is the permittivity of vacuum and

$$-\frac{1}{\mu_0}\nabla \times \nabla \times \boldsymbol{E} = \boldsymbol{\epsilon} \cdot \frac{\partial^2 \boldsymbol{E}}{\partial t^2} + \boldsymbol{\sigma} \cdot \frac{\partial \boldsymbol{E}}{\partial t} + \frac{\partial \boldsymbol{J}_a}{\partial t}. \quad \textit{Maxwell's wave equation} \tag{8.8}$$

Note that we have the vector differential identity proven in Chapter 1 that $\nabla \times \nabla \times \boldsymbol{E} = \nabla\nabla \cdot \boldsymbol{E} - \nabla^2 \boldsymbol{E}$ and if there is no net charge density throughout the material so that $\nabla \cdot \boldsymbol{E} = 0$, you have that $\nabla \times \nabla \times \boldsymbol{E} = -\nabla^2 \boldsymbol{E}$. For electromagnetic response, if $|\boldsymbol{\epsilon} \cdot \partial \boldsymbol{E}/\partial t| \gg |\boldsymbol{\sigma} \cdot \boldsymbol{E}|$ we have pure EM wave-propagation (a vector wave equation), while if $|\boldsymbol{\epsilon} \cdot \partial \boldsymbol{E}/\partial t| \ll |\boldsymbol{\sigma} \cdot \boldsymbol{E}|$ we have pure EM diffusion (a vector diffusion equation); for intermediate situations where both terms must be included, we have attenuated EM wave propagation. By assuming a homogeneous isotropic body and considering particular components of the vectors or by introducing potentials, these more complicated vector wave equations can be reduced in many cases to the form of the simple scalar wave equation.

If the source terms S_d, S_w, S_p, S_a, $\boldsymbol{f}_a$, and $\boldsymbol{J}_a$ above are zero, we say that the PDE itself is *homogeneous*, which is another distinct use of the word "homogeneous." If the source terms are nonzero, we say the PDE is *inhomogeneous*.

To define a well-posed *boundary-value problem* (BVP) within some domain Ω of interest, appropriate *boundary conditions* for the fields must be imposed on $\partial\Omega$ (the surface surrounding Ω) and appropriate *initial conditions* must be imposed at $t = 0$ (or at some other point in time) throughout all of Ω. A "well-posed problem" is one where the boundary and initial conditions permit *unique* solutions of the linearized physics as addressed in Section 8.2.

If the boundary conditions are constant in time and/or if the source terms in the PDEs are constant in time, solutions to the above PDEs can be obtained by the method of *separation of variables* as developed in this chapter.

Otherwise, the *Green's function method* works for the above linear BVPs, as does an approach based on orthogonal eigenfunctions such as the Fourier series (finite domains) and the Fourier transform (infinite or semi-infinite domains) as treated in Chapters 9–11.

8.2 Uniqueness of Solution for a Well-Posed BVP

We now address the type of boundary and initial conditions that must be imposed so that there is one and only one solution to the simple linear scalar forms of the PDEs given in Eq. (8.1) that will be the primary focus of the present and next few chapters. Generalization to problems involving heterogeneous and anisotropic systems as well as vectorial response is immediate and straightforward. For the case of electromagnetism, elastodynamics and viscous fluid dynamics, we have already given this "uniqueness-of-solution" analysis of the linearized physics in earlier chapters.

Imagine there are two fields, ψ_1 and ψ_2, that both satisfy the same PDE with the same coefficients and the same source terms and that both satisfy the same boundary and initial conditions. If we can show that the difference fields

$$\delta\psi = \psi_1 - \psi_2 \tag{8.9}$$

are everywhere zero in the domain Ω, we say that the solution is unique and that the BVP is well posed. The goal in the exercises that follow are to find the type of boundary and initial conditions that allow each type of BVP to have a unique solution, that is, to have $\delta\psi = 0$ throughout the domain and throughout time.

8.2.1 Conditions for Diffusion, Wave, and Potential BVPs to Be Well Posed

Each of these three basic responses involve subtle differences in the determination of what type of boundary and initial conditions are required to obtain unique solutions. Perhaps the main point of this analysis is to prove that if you stumble across a solution of a linear PDE that satisfies the type of boundary and initial conditions to be derived below, you can be certain that this solution is the one and only solution of the given BVP.

- **Diffusion Equation**: We write a diffusion equation for each possible field ψ_1 and ψ_2

$$D\,\nabla^2\psi_1 - \frac{\partial\psi_1}{\partial t} = S_d \tag{8.10}$$

$$D\,\nabla^2\psi_2 - \frac{\partial\psi_2}{\partial t} = S_d. \tag{8.11}$$

Subtracting the two equations gives

$$D\,\nabla^2\,\delta\psi - \frac{\partial\,\delta\psi}{\partial t} = 0, \tag{8.12}$$

where $\delta\psi = \psi_1 - \psi_2$. Next, multiply through by $\delta\psi$ to obtain

$$D\,\delta\psi\;\nabla^2\,\delta\psi - \delta\psi\,\frac{\partial\,\delta\psi}{\partial t} = 0. \tag{8.13}$$

Using $\nabla\cdot\left[(\nabla\alpha)\beta\right] = \beta\,\nabla^2\alpha + \nabla\alpha\cdot\nabla\beta$ and the fact that $\partial(\alpha\beta)/\partial t = \beta\partial\alpha/\partial t + \alpha\partial\beta/\partial t$, we obtain for the special case of $\alpha = \beta = \delta\psi$

$$D\,\nabla\cdot\left[(\nabla\,\delta\psi)\,\delta\psi\right] = \frac{1}{2}\frac{\partial}{\partial t}\,(\delta\psi)^2 + D\,\nabla\,\delta\psi\cdot\nabla\,\delta\psi. \tag{8.14}$$

Integrate both sides over Ω and over time:

$$D\int_0^t \mathrm{d}t' \int_\Omega \mathrm{d}V\;\nabla\cdot\left[(\nabla\,\delta\psi)\,\delta\psi\right] = \int_\Omega \mathrm{d}V \int_0^t \left[\frac{1}{2}\frac{\partial}{\partial t'}\,(\delta\psi)^2 + D\,\nabla\delta\psi\cdot\nabla\,\delta\psi\right] dt'. \tag{8.15}$$

Use the divergence theorem on the left-hand side to obtain

$$D\int_0^t \mathrm{d}t' \int_{\partial\Omega} \mathrm{d}S\,(\boldsymbol{n}\cdot\nabla\,\delta\psi)\;\delta\psi$$

$$= \int_\Omega \mathrm{d}V\left\{\frac{1}{2}\left[\delta\psi(t)\right]^2 - \frac{1}{2}\left[\delta\psi(t=0)\right]^2 + D\int_0^t \nabla\,\delta\psi\cdot\nabla\,\delta\psi\;\mathrm{d}t'\right\}. \tag{8.16}$$

We now impose the conditions that:

(i) **Boundary condition**: $\boldsymbol{n}\cdot\nabla\psi$ or ψ are specified over $\partial\Omega$ so that $(\boldsymbol{n}\cdot\nabla\,\delta\psi)\,\delta\psi=0$ on $\partial\Omega$, which results in the left-hand side of Eq. (8.16) being zero;
(ii) **Initial condition**: $\psi(\boldsymbol{r},t=0)$ is specified at $t=0$ so that $\delta\psi(\boldsymbol{r},t=0)=0$ throughout Ω.

Employing both of these conditions yields

$$\int_\Omega \mathrm{d}V\left\{\frac{1}{2}(\delta\psi)^2+D\int_0^t \mathrm{d}t'\,\nabla\,\delta\psi\cdot\nabla\,\delta\psi\right\}=0. \tag{8.17}$$

Since both terms are everywhere positive in both space and time, we must have $\delta\psi=0$ everywhere through space and time.

So, with the boundary and initial conditions imposed above, the solutions are unique and the BVP is well posed. This analysis does not say what the specific, possibly inhomogeneous boundary conditions for either ψ or $\boldsymbol{n}\cdot\nabla\psi$ must be, nor what the specific, possibly inhomogeneous initial condition for $\psi(t=0,\boldsymbol{r})$ must be. The analysis simply says that these particular boundary and initial conditions must be specified in order for a solution to be unique.

- **Wave equation**: The difference field in this case satisfies

$$c^2\,\nabla^2\,\delta\psi-\frac{\partial^2}{\partial t^2}\,\delta\psi=0 \tag{8.18}$$

Multiply through by $\partial\delta\psi/\partial t$ and manipulate to obtain

$$c^2\,\nabla\cdot\left[(\nabla\,\delta\psi)\frac{\partial\,\delta\psi}{\partial t}\right]-\frac{c^2}{2}\frac{\partial}{\partial t}\left[\nabla\,\delta\psi\cdot\nabla\,\delta\psi\right]=\frac{1}{2}\frac{\partial}{\partial t}\left(\frac{\partial\,\delta\psi}{\partial t}\right)^2. \tag{8.19}$$

Integrating over space and time yields

$$c^2\int_0^t \mathrm{d}t'\int_{\partial\Omega}(\boldsymbol{n}\cdot\nabla\,\delta\psi)\,\frac{\partial\,\delta\psi}{\partial t'}\,\mathrm{d}S=\int_\Omega \mathrm{d}V\left\{\frac{c^2}{2}\Big[\nabla\,\delta\psi(t)\cdot\nabla\,\delta\psi(t)\Big]_0^t+\frac{1}{2}\left(\frac{\partial\,\delta\psi}{\partial t}\right)^2\Bigg|_0^t\right\}. \tag{8.20}$$

We now impose the conditions:

(i) **Boundary condition**: $\boldsymbol{n}\cdot\nabla\psi$ or ψ are specified on $\partial\Omega$. Note that if ψ is given on $\partial\Omega$, then $\partial\psi/\partial t$ is also known on $\partial\Omega$. With these type of boundary conditions, we then have that $(\boldsymbol{n}\cdot\nabla\delta\psi)\partial\delta\psi/\partial t=0$ on the boundary $\partial\Omega$ and the left-hand side of Eq. (8.20) is zero.
(ii) **Initial conditions**: Both $\partial\psi/\partial t$ and ψ are specified at $t=0$. Note that if $\psi(\boldsymbol{r},t=0)$ is given throughout Ω at $t=0$, then $\nabla\psi|_{t=0}$ is also known throughout Ω at $t=0$.

Employing these conditions then yields

$$\int_\Omega \mathrm{d}V\left\{\frac{c^2}{2}\Big[\nabla\,\delta\psi\cdot\nabla\,\delta\psi\Big]+\frac{1}{2}\left(\frac{\partial\,\delta\psi}{\partial t}\right)^2\right\}=0, \tag{8.21}$$

which, due to both terms being everywhere positive, can only be satisfied by $\delta\psi$ being a uniform constant through all of space and time. But due to the initial condition on ψ, this uniform constant must be zero so that $\delta\psi = 0$ through space and time.

Thus, we learn that in addition to the same conditions required for a diffusion BVP to be well posed, we must additionally have the initial condition that $\partial\psi/\partial t$ be given throughout space at $t = 0$ in order for a wave BVP to be well posed. In passing, another common boundary condition for a wave problem that results in unique solutions is the mixed condition $\boldsymbol{n} \cdot \nabla\psi + c^{-1}\partial\psi/\partial t = 0$ that is called a *radiation condition*. This condition specifies that the waves present at the portion of the boundary where it applies are propagating out of Ω and are not incoming. To see this, consider an outward propagating wave at a boundary that has a normal $\hat{\boldsymbol{x}}$ which has the form $\psi(x - ct)$. Because $\partial\psi(x - ct)/\partial t = -c\partial\psi(x - ct)/\partial x$ for an outward radiating wave (x increasing with increasing t), we see that the radiation condition is satisfied by an outward propagating wave. This condition therefore adds a term to the left-hand side of Eq. (8.21) of the form $c^2 \int_0^t \mathrm{d}t' \int_{\partial\Omega} c^{-1}(\partial\delta\psi/\partial t')^2 \, \mathrm{d}S$ which is strictly positive and again results in $\delta\psi = 0$ and unique solutions.

- **Potential equation:** The difference field in this case satisfies

$$\nabla^2\delta\psi = 0. \tag{8.22}$$

If we multiply by $\delta\psi$ we have the identity that

$$\nabla \cdot [(\nabla\delta\psi)\,\delta\psi] = \nabla\delta\psi \cdot \nabla\delta\psi. \tag{8.23}$$

Integrate over Ω and use the divergence theorem to obtain

$$\int_{\partial\Omega} (\boldsymbol{n} \cdot \nabla\delta\psi)\,\delta\psi \, \mathrm{d}S = \int_{\Omega} \nabla\delta\psi \cdot \nabla\delta\psi \, \mathrm{d}V. \tag{8.24}$$

There are clearly no initial conditions to consider in this problem. Consider first the boundary condition that $\boldsymbol{n} \cdot \nabla\psi$ is a given function on $\partial\Omega$. This leads to $\boldsymbol{n} \cdot \nabla\delta\psi = 0$ and therefore

$$\int_{\Omega} \nabla\delta\psi \cdot \nabla\delta\psi \, \mathrm{d}V = 0, \tag{8.25}$$

which allows us to conclude that $\nabla\delta\psi = 0$ which means that $\delta\psi = \psi_1 - \psi_2 = \text{constant}$. So a boundary condition only involving $\boldsymbol{n} \cdot \nabla\psi$ in a potential problem only allows us to determine the solution uniquely up to an additive constant, that is, we can add a constant to whatever solution we find for a ψ satisfying this particular BC and we will still satisfy both the PDE and the BCs.

Now if at least at one point on $\partial\Omega$, the BC involves ψ itself, such as either ψ being given or $\boldsymbol{n} \cdot \nabla\psi = -\alpha\psi$ where $\alpha > 0$, then we conclude that $\psi_1 - \psi_2 = 0$ and we have an entirely unique solution.

8.2.2 More about Well-Posed BVPs

We obtained above the boundary and initial conditions that must be specified to obtain unique solutions of the scalar diffusion, wave and potential equations. As stated earlier, these particular equations that we will be studying over the present few chapters implicitly correspond to uniform and isotropic material properties throughout the domain Ω under study. However, the more general statement of say the diffusion equation is

$$-\nabla \cdot \boldsymbol{q} - \gamma \frac{\partial \psi}{\partial t} = S_d, \quad \text{where} \quad \boldsymbol{q} = -\boldsymbol{\lambda} \cdot \nabla \psi, \tag{8.26}$$

where $\boldsymbol{\lambda}$ is a second-rank tensor that is not purely isotropic (the material has anisotropy) and γ is a scalar material property (generalized susceptibility) that is positive due to thermodynamic stability as presented in Chapter 6 on equilibrium thermodynamics. It is straightforward to go through the same manipulations as just performed in the homogeneous and isotropic case (set up equations for the difference fields, multiply through by $\delta\psi$, integrate, etc.) and come to the exact same needed conditions for a well-posed diffusion problem. The only requirement in the anisotropic case is that $\nabla\psi \cdot \boldsymbol{\lambda} \cdot \nabla\psi > 0$, which corresponds to the irreversible heat production due to the conduction process and is always positive (the second law of thermodynamics as discussed in Chapter 7).

For all of diffusion, wave and potential equations and for either isotropic or anisotropic or homogeneous or heterogeneous material properties, the boundary conditions that must be specified on each patch of $\partial\Omega$ in order to obtain unique solutions must be of the following possible forms:

- **Dirichlet boundary conditions**: The field itself ψ is specified.
- **Neumann boundary conditions**: The normal derivative $\boldsymbol{n} \cdot \nabla\psi$ is specified. As just seen, a potential problem that only has Neumann conditions over all of $\partial\Omega$ is only known up to an additive uniform constant. But if at least one patch on $\partial\Omega$ has a Dirichlet or mixed condition (below) that is specified, this constant is zero.
- **Mixed boundary conditions**: Linear combinations of the field and its normal spatial derivative also result in unique solutions of the PDEs. Such mixed conditions come in the following forms for each of our PDEs under consideration:

$$\text{Diffusion equation}: \quad \boldsymbol{n} \cdot \nabla\psi + \alpha\psi = \text{given function on } \partial\Omega; \tag{8.27}$$

$$\text{Wave equation}: \quad \boldsymbol{n} \cdot \nabla\psi + \beta\frac{\partial \psi}{\partial t} = \text{given function on } \partial\Omega; \tag{8.28}$$

$$\text{Potential equation}: \quad \boldsymbol{n} \cdot \nabla\psi + \gamma\psi = \text{given function on } \partial\Omega. \tag{8.29}$$

In the mixed boundary condition for the wave equation, if we take $\beta = 1/c$ (the wave slowness) and set the right-hand side to zero, we have an outward radiation condition. Note that in using these mixed conditions, once we evaluate $\delta\psi = \psi_1 - \psi_2$, we obtain the difference fields on the boundary used on the left-hand side of Eqs (8.27)–(8.29).

$$\text{Diffusion equation}: \quad (\boldsymbol{n} \cdot \nabla\,\delta\psi)\,\delta\psi = -\alpha\,\delta\psi^2; \tag{8.30}$$

$$\text{Wave equation}: \quad (\boldsymbol{n}\cdot\nabla\,\delta\psi)\,\frac{\partial\delta\psi}{\partial t} = -\beta\left(\frac{\partial\delta\psi}{\partial t}\right)^2; \tag{8.31}$$

$$\text{Poisson equation}: \quad (\boldsymbol{n}\cdot\nabla\,\delta\psi)\,\delta\psi = -\gamma\,\delta\psi^2. \tag{8.32}$$

When these boundary values are inserted into the earlier development, we obtain $\delta\psi = 0$ throughout space and time (unique solutions) so long as $\alpha > 0$, $\beta > 0$, and $\gamma > 0$.

- **Periodic boundary conditions**: Imagine that the region Ω is a cube with the six faces located at $x_i = \pm L/2$ where $i = 1, 2, 3$. We can specify any of the above type of Dirichlet, Neumann, or mixed conditions on each face or at different parts of each face and obtain unique solutions. A final type of boundary condition is a so-called *periodic condition* in which for a given pair of opposing faces we have

$$\psi(x_i = +L/2) = \psi(x_i = -L/2). \tag{8.33}$$

  This means that if we consider two different possible solutions throughout the interior of Ω that are required to be equal to each other on the boundary, we have $\psi_1(x_i = +L/2) = \psi_2(x_i = +L/2) = \psi_1(x_i = -L/2) = \psi_2(x_i = L/2)$ or $\delta\psi = 0$ on each of the two opposing faces. With such periodic conditions we are then guaranteed that

$$\int_{\partial\Omega} (\boldsymbol{n}\cdot\nabla\,\delta\psi)\,\delta\psi\,\mathrm{d}S = 0, \tag{8.34}$$

  and we again have unique solutions. We would also obtain unique solutions if we had periodic conditions on just two opposing faces and some combination of Dirichlet, Neumann, and mixed conditions on the other four faces.

So any of Dirichlet, Neumann, mixed, or periodic conditions on the surface $\partial\Omega$ will result in unique solutions so long as the appropriate initial conditions are also specified in the case of diffusion and wave equations.

8.3 The Separation-of-Variables Method

We now lay out the general method of separation variables for diffusion and wave problems in 3D that have given inhomogeneous boundary and initial conditions. We do so in generality in this first section, without specifying the coordinate system or describing the specific boundary positions. Then, in Section 8.4, we carry out the method for specific problems with specified boundary positions and specified boundary and initial values.

The separation-of-variables approach has two significant constraints that limit its use: (1) the inhomogeneous boundary conditions cannot be a function of time and (2) if the PDE is inhomogeneous, which again means that there is a source term explicitly present on the right-hand side of the PDE, this source term must be independent of time. The eigenfunction approach of Chapter 9 does not have these limitations, nor does the Green's function method of Chapter 12.

8.3.1 Diffusion Equation BVPs

The problem is to find the $u(\boldsymbol{r}, t)$ that satisfies

$$D\,\nabla^2 u - \frac{\partial u}{\partial t} = 0 \quad \text{everywhere in } \Omega \tag{8.35}$$

subject to the initial condition

$$u(\boldsymbol{r}, t=0) = u_0(\boldsymbol{r}) \quad \text{throughout } \Omega$$

and either of the inhomogeneous boundary conditions

$$u(\boldsymbol{r}, t) = b_1(\boldsymbol{r}) \quad \text{or} \quad \boldsymbol{n}\cdot\nabla u(\boldsymbol{r}, t) = b_2(\boldsymbol{r}) \quad \text{for } \boldsymbol{r} \text{ on } \partial\Omega. \tag{8.36}$$

Note that over some parts of $\partial\Omega$, we could have $u(\boldsymbol{r}, t) = b_1(\boldsymbol{r})$ and on other parts $\boldsymbol{n}\cdot\nabla u(\boldsymbol{r}, t) = b_2(\boldsymbol{r})$. Note that these boundary values $b_1(\boldsymbol{r})$ and/or $b_2(\boldsymbol{r})$ are independent of time.

We begin by writing the solution in the separated form

$$u(\boldsymbol{r}, t) = \varphi(\boldsymbol{r}) + R(\boldsymbol{r})T(t), \tag{8.37}$$

where $\varphi(\boldsymbol{r})$ is the solution of the Laplace BVP

$$\nabla^2\,\varphi(\boldsymbol{r}) = 0 \quad \text{in } \Omega \tag{8.38}$$

that we introduce in order to satisfy the inhomogeneous and constant in time boundary conditions

$$\varphi(\boldsymbol{r}, t) = b_1(\boldsymbol{r}) \quad \text{or} \quad \boldsymbol{n}\cdot\nabla\varphi(\boldsymbol{r}, t) = b_2(\boldsymbol{r}) \quad \text{for } \boldsymbol{r} \text{ on } \partial\Omega. \tag{8.39}$$

We will return later to how to solve this Laplace BVP for specific BCs. This separation is made so that the spatial function $R(\boldsymbol{r})$ satisfies homogeneous boundary conditions on $\partial\Omega$.

Introducing $R(\boldsymbol{r})\,T(t)$ into the diffusion equation and dividing through by $DR(\boldsymbol{r})\,T(t)$ gives

$$\frac{1}{R}\,\nabla^2 R - \frac{1}{DT}\frac{\partial T}{\partial t} = 0. \tag{8.40}$$

For this to be true, we must have that

$$\frac{1}{R}\,\nabla^2 R = \lambda \quad \text{and} \quad \frac{1}{DT}\frac{\partial T}{\partial t} = \lambda \mathrel{\widehat{=}} \text{the separation constant} \tag{8.41}$$

where the separation constant λ is independent of space and time and will be found from the homogeneous boundary conditions.

The solution for $T(t)$ takes the form

$$T(t) = A\,e^{\lambda D t}, \quad \text{where} \quad A \mathrel{\widehat{=}} \text{some constant}. \tag{8.42}$$

For the time dependence not to blow up, we must have that λ is negative, which can be written $\lambda = -k^2$ so that

$$T(t) = A\, e^{-k^2 Dt}. \tag{8.43}$$

The space dependence $R(\boldsymbol{r})$ then satisfies the *Helmholtz equation*

$$\nabla^2 R(\boldsymbol{r}) = -k^2 R(\boldsymbol{r}) \quad \text{in } \Omega \tag{8.44}$$

subject to either of the homogeneous boundary conditions

$$R(\boldsymbol{r}) = 0 \quad \text{or} \quad \boldsymbol{n} \cdot \nabla R(\boldsymbol{r}) = 0 \quad \text{on } \partial\Omega. \tag{8.45}$$

For a finite domain Ω, this Helmholtz BVP only has solutions at discrete values k_n ($n = 0, 1, 2, \ldots$) of the separation constants (that are also called *eigenvalues* of the Laplacian operator) that will be determined from the homogeneous boundary conditions. For an infinite domain, there are a continuum of eigenvalues k. The consequence is that for a finite domain, the solution for $R(\boldsymbol{r})$ takes the form of a series (sum over n), while in an infinite domain, the solution has an integral representation (integral over all k). We focus for now on *finite domains*.

Corresponding to each discrete eigenvalue k_n, there is a corresponding eigenfunction ψ_n of the operator ∇^2 that satisfies

$$\nabla^2 \psi_n + k_n^2\, \psi_n = 0 \quad \text{in } \Omega \tag{8.46}$$

subject to the boundary conditions

$$\psi_n(\boldsymbol{r}) = 0 \quad \text{or} \quad \boldsymbol{n} \cdot \nabla \psi_n(\boldsymbol{r}) = 0 \quad \text{on } \partial\Omega. \tag{8.47}$$

The solution for $R(\boldsymbol{r})$ is the sum

$$R(\boldsymbol{r}) = \sum_n B_n\, \psi_n(\boldsymbol{r}), \tag{8.48}$$

where the constants B_n, when multiplied by the constant A above to produce other constants $A_n = AB_n$, will be found from the initial conditions. The eigenfunctions ψ_n possess the important *orthogonality* property that is what allows the constants $A_n (= AB_n)$ to be determined from the initial conditions and that can be stated

$$\int_\Omega \psi_n\, \psi_m\, \mathrm{d}V = \begin{cases} 0 & \text{if } n \neq m \\ \int_\Omega \psi_n^2\, dV \neq 0 & \text{if } n = m. \end{cases} \tag{8.49}$$

Proof. Construct the following products

$$\psi_m\, \nabla^2\, \psi_n + k_n^2\, \psi_m\, \psi_n = 0 \tag{8.50}$$

$$\psi_n\, \nabla^2\, \psi_m + k_m^2\, \psi_n\, \psi_m = 0. \tag{8.51}$$

Use the identity $\psi\,\nabla^2\psi = \nabla\cdot\left[(\nabla\psi)\,\psi\right] - \nabla\psi\cdot\nabla\psi$, subtract, integrate over Ω, use Green's theorem, and apply the homogeneous BCs to obtain

$$\left(k_n^2 - k_m^2\right)\int_\Omega \psi_m\,\psi_n\,\mathrm{d}V = 0, \tag{8.52}$$

which yields the orthogonality property of Eq. (8.49). □

Bringing it all together, the solution for $u(\boldsymbol{r}, t)$ in a *finite domain* takes the form

$$u(\boldsymbol{r}, t) = \sum_n A_n\, e^{-k_n^2 Dt}\,\psi_n(\boldsymbol{r}) + \varphi(\boldsymbol{r}), \tag{8.53}$$

where:

(i) $\varphi(\boldsymbol{r})$ is a solution of Laplace's equation that satisfies the inhomogeneous BCs (if there are such nonzero BCs) under the condition that the boundary values are independent of time and/or possibly a constant-in-time source term in the PDE itself.

(ii) k_n are the eigenvalues of the operator ∇^2, that are also called the separation constants, and are determined from the specific nature of the boundary $\partial\Omega$ as will be seen when we perform specific examples.

(iii) $\psi_n(\boldsymbol{r})$ are the eigenfunctions of the operator ∇^2 that satisfy homogeneous BCs and, as a result, can be shown to possess the all-important orthogonality property

$$\int_\Omega \psi_n(\boldsymbol{r})\,\psi_m(\boldsymbol{r})\,\mathrm{d}V = 0 \quad \text{if } m \neq n \tag{8.54}$$

$$\neq 0 \quad \text{if } m = n. \tag{8.55}$$

(iv) A_n are constants determined from the initial conditions and the orthogonality of the ψ_n as follows. Begin by writing down the initial condition remembering to include the time-independent $\varphi(\boldsymbol{r})$ if present

$$u(\boldsymbol{r}, 0) - \varphi(\boldsymbol{r}) = \sum_n A_n\,\psi_n(\boldsymbol{r}). \tag{8.56}$$

Multiply both sides by ψ_m, integrate over the problem domain and use orthogonality which states that out of the entire series in n, only the one particular term $n = m$ is non zero

$$\int_\Omega \left[u(\boldsymbol{r}, 0) - \varphi(\boldsymbol{r})\right]\psi_m\,\mathrm{d}V = A_m \int_\Omega \psi_m^2\,\mathrm{d}V, \tag{8.57}$$

which then gives (replace m with n)

$$A_n = \frac{\int_\Omega \left[u(\boldsymbol{r}, 0) - \varphi(\boldsymbol{r})\right]\psi_n\,\mathrm{d}V}{\int_\Omega \psi_n^2\,\mathrm{d}V}. \tag{8.58}$$

It may seem unsatisfactory, but in general, the analytical solution of a PDE in a finite domain will come in the form of an infinite series, each term corresponding to a particular discrete separation constant k_n. However, we will see that even just the first term of this series can provide physical insight into the problem.

8.3.2 Wave Equation BVPs

For wave problems in a finite domain, the BVP is stated

$$\nabla^2 u - \frac{1}{c^2}\frac{\partial^2 u}{\partial t^2} = 0 \qquad \text{throughout } \Omega \tag{8.59}$$

$$u(\boldsymbol{r}, 0) = u_0(\boldsymbol{r}) \qquad \text{throughout } \Omega \tag{8.60}$$

$$\left.\frac{\partial u(\boldsymbol{r}, t)}{\partial t}\right|_{t=0} = v_0(\boldsymbol{r}) \qquad \text{throughout } \Omega \tag{8.61}$$

along with either of the boundary conditions

$$u(\boldsymbol{r}, t) = b_1(\boldsymbol{r}) \quad \text{or} \quad \boldsymbol{n}\cdot\nabla u(\boldsymbol{r}, t) = b_2(\boldsymbol{r}) \quad \text{on } \partial\Omega. \tag{8.62}$$

We then separate variables $u(\boldsymbol{r}, t) = R(\boldsymbol{r})\,T(t) + \varphi(\boldsymbol{r})$

$$\frac{\nabla^2 R}{R} = \frac{1}{c^2 T}\frac{\partial^2 T}{\partial t^2} = -k^2 \tag{8.63}$$

so that

$$T(t) = A\,\sin kct + B\,\cos kct \tag{8.64}$$

and both $R(\boldsymbol{r})$ and $\varphi(\boldsymbol{r})$ are exactly as in the above diffusion problem.

Thus, in a finite domain (discrete k_n), we have:

$$\boxed{u(\boldsymbol{r}, t) = \sum_n \Big[A_n\,\sin k_n ct + B_n\,\cos k_n ct\Big]\,\psi_n(\boldsymbol{r}) + \varphi(\boldsymbol{r}),} \tag{8.65}$$

where the B_n are determined from the initial condition $u(\boldsymbol{r}, 0) = u_0(\boldsymbol{r})$ as

$$B_n = \frac{\int_\Omega \Big[u_0(\boldsymbol{r}) - \varphi(\boldsymbol{r})\Big]\psi_n\,\mathrm{d}V}{\int_\Omega \psi_n^2\,\mathrm{d}V}, \tag{8.66}$$

the A_n are determined from the other initial condition $\partial u/\partial t|_{t=0} = v_0(\boldsymbol{r})$ as

$$k_n\,c\,A_n = \frac{\int_\Omega v_0(\boldsymbol{r})\,\psi_n(\boldsymbol{r})\,\mathrm{d}V}{\int_\Omega \psi_n^2\,\mathrm{d}V} \tag{8.67}$$

and $\varphi(\boldsymbol{r})$ is again a solution of Laplace's equation that accounts for either a constant-in-time boundary condition or a constant-in-time source term in the PDE itself.

8.4 Examples of Using Separation of Variables

We now use the above formalism and treat a few specific problems to see how the method of separation of variables works in practice and to see how to find the eigenvalues (separation constants) k_n.

8.4.1 1D Diffusion

Consider an infinite slab of some material that has a field $P(x, t)$ that initially (at time $t=0$) has a uniform value P_0 throughout the slab and that equilibrates by diffusion with its surroundings that are maintained at $P=0$. The BVP is stated mathematically as

$$D\frac{\partial^2 P}{\partial x^2} - \frac{\partial P}{\partial t} = 0, \tag{8.68}$$

where D is the diffusivity and with the boundary conditions

$$P(x=a, t)=0 \quad \text{and} \quad P(x=0, t)=0 \tag{8.69}$$

and the initial condition

$$P(x, t=0) = P_0. \tag{8.70}$$

This is depicted in Fig. 8.1.

Separate variables in the form $P(x, t) = \psi(x)\,T(t)$ where there is no need for the solution of Laplace's equation φ because the BCs are homogeneous. Note that if the BCs above are not zero but are a given constant $P_b \neq P_0$ we could just take $\varphi(\boldsymbol{r}) = P_b$. But continuing with $P_b = 0$, we have that

$$\frac{1}{\psi(x)}\frac{\mathrm{d}^2\psi(x)}{\mathrm{d}x^2} = \frac{1}{D\,T(t)}\frac{\mathrm{d}T(t)}{\mathrm{d}t} = -k^2. \tag{8.71}$$

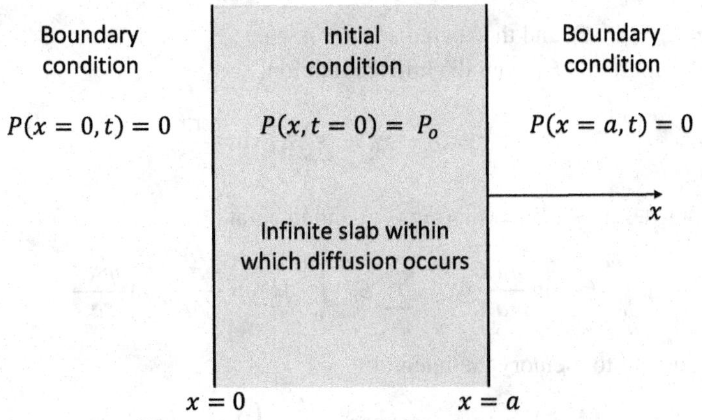

Figure 8.1 1D diffusion in an infinite slab with prescribed boundary conditions and a given initial condition.

Focusing on $\psi(x)$, we have

$$\frac{d^2\psi(x)}{dx^2} + k^2\,\psi(x) = 0 \quad \text{subject to } \psi(0) = \psi(a) = 0. \tag{8.72}$$

This DE, which is the 1D Helmholtz equation which is also the eigenvalue problem for the operator d^2/dx^2, has solutions

$$\psi(x) = A\,\cos kx + B\,\sin kx. \tag{8.73}$$

The BCs are satisfied if $A = 0$ and if the eigenvalues (separation constants) take the discrete form

$$k = k_n = \frac{n\pi}{a} \quad \text{for all } n = 1, 2, \ldots. \tag{8.74}$$

So we see how the BCs define the discrete nature of the eigenvalues in a finite domain. The corresponding eigenfunctions $\psi_n(x)$ are of the form

$$\psi_n(x) = B_n\,\sin\frac{n\pi x}{a}; \quad n = 1, 2, \ldots \tag{8.75}$$

For $T(t)$, we now have

$$\frac{dT_n}{dt} = -\left(\frac{n\pi}{a}\right)^2 D\,T_n \tag{8.76}$$

or

$$T_n(t) = C_n \exp\left[-\left(\frac{n\pi}{a}\right)^2 D\,t\right]; \quad n = 1, 2, \ldots \tag{8.77}$$

Putting all solutions together, we have

$$P(x,t) = \sum_{n=1}^{\infty} E_n\,e^{-(n\pi/a)^2 Dt}\,\sin\frac{n\pi x}{a}, \tag{8.78}$$

where we set $E_n = B_n\,C_n$ and then summed over n.

We find the constants E_n from the initial condition

$$P(x, t=0) = P_0 = \sum_{n=1}^{\infty} E_n\,\sin\frac{n\pi x}{a}. \tag{8.79}$$

Multiply both sides by $\psi_m(x) = \sin\,(m\pi x/a)$ and integrate

$$\int_0^a P_0\,\sin\frac{n\pi x}{a}\,dx = \sum_{n=1}^{\infty} E_n \int_0^a dx\,\sin\frac{m\pi x}{a}\,\sin\frac{m\pi x}{a}. \tag{8.80}$$

You should commit to memory the integral

$$\int_0^a \sin\left(\frac{n\pi x}{a}\right)\sin\left(\frac{m\pi x}{a}\right)\,dx = \begin{cases} 0 & \text{if } n \neq m \\ a/2 & \text{if } n = m. \end{cases} \tag{8.81}$$

This is easily proven from $\sin\theta = \left(e^{i\theta} - e^{-i\theta}\right)/2i$ which gives

$$\int_0^a \frac{\left(e^{in\pi x/a} - e^{-in\pi x/a}\right)}{2i} \frac{\left(e^{im\pi x/a} - e^{-im\pi x/a}\right)}{2i}\, \mathrm{d}x$$

$$= -\frac{1}{4}\int_0^a \left[e^{i(n+m)\pi x/a} + e^{-i(n+m)\pi x/a} - e^{i(n-m)\pi x/a} - e^{-i(n-m)\pi x/a}\right] \mathrm{d}x$$

$$= \frac{1}{2}\int_0^a \left[\cos\left(\frac{(n-m)\pi x}{a}\right) - \cos\left(\frac{(n+m)\pi x}{a}\right)\right] \mathrm{d}x$$

$$= \begin{cases} 0 & \text{if } n \neq m \\ a/2 & \text{if } n = m. \end{cases}$$

Though not needed here, a similar treatment yields both

$$\int_0^a \cos\left(\frac{n\pi x}{a}\right) \cos\left(\frac{m\pi x}{a}\right)) \,\mathrm{d}x = \begin{cases} 0 & \text{if } n \neq m \\ a/2 & \text{if } n = m. \end{cases} \tag{8.82}$$

and

$$\int_0^a \sin\left(\frac{n\pi x}{a}\right) \cos\left(\frac{m\pi x}{a}\right) \mathrm{d}x = 0. \tag{8.83}$$

We will also be using these facts in the next chapter when we utilize the Fourier series.

These integral results then give

$$\int_0^a P_0 \sin\frac{m\pi x}{a}\, \mathrm{d}x = \frac{a}{2} E_m \tag{8.84}$$

or, upon setting $m = n$,

$$E_n = \frac{2P_o}{a}\int_0^a \sin\frac{n\pi x}{a}\, \mathrm{d}x = -\frac{2P_0}{\pi n}\Big[\cos n\pi \; - 1\Big] = \begin{cases} 4P_0/(\pi n) & \text{for } n = \text{odd} \\ 0 & \text{for } n = \text{even}. \end{cases} \tag{8.85}$$

Thus, we have at last

$$\boxed{P(x,t) = \frac{4P_0}{\pi}\sum_{\substack{n \\ \text{odd}}} \frac{1}{n} \exp\left[-\left(\frac{n\pi}{a}\right)^2 Dt\right] \sin\frac{n\pi x}{a}.} \tag{8.86}$$

So writing out the first term gives

$$P(x,t) = \frac{4P_0}{\pi}\left[e^{-(\pi/a)^2 Dt} \sin\frac{\pi x}{a} + \cdots\right] \tag{8.87}$$

with all higher-order terms negligible when $t > a^2/(\pi^2 D) \mathrel{\widehat{=}} t_d$. This first term alone gives us insight into the nature of the solution after the diffusion in from the edges has had a chance to perturb the entire domain. Snapshots of the diffusion field in the slab at different t

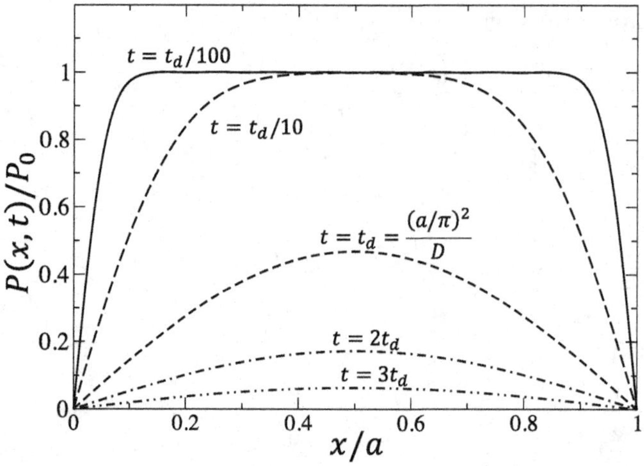

Figure 8.2 Plots of $P(x, t)$ for a few different t showing how diffusion in the presence of homogeneous boundary conditions is responsible for $P(x, t)$ evolving from $P(x, 0) = P_0$ at $t = 0$ to $P(x, \infty) = 0$ at infinite time.

(these were obtained by summing the first eight terms in the series) are plotted in Fig. 8.2. For $t > t_d$ as defined in the figure, just the single term given above involving $\sin(\pi x/a)$ is controlling the spatial distribution and the peak value of this half sinusoid is just decreasing in amplitude as t is increasing. Also, from the above results we can see that when $t \approx (a/2)^2/D$, the diffusion has penetrated all the way through to the midpoint of the slab $x = a/2$. *As seen earlier in the book, the relation between the time t to diffuse a distance x is roughly $t \approx x^2/D$, which is the most important single thing to remember about diffusion if forced to choose just one.*

8.4.2 1D Wave Problem

We next solve the exact same BVP as just treated but for the wave equation. The problem is described mathematically as

$$c^2 \frac{\partial^2 u}{\partial x^2} - \frac{\partial^2 u}{\partial t^2} = 0, \tag{8.88}$$

where c is the wave speed and with the boundary conditions

$$u(x = a, t) = 0 \quad \text{and} \quad u(x = 0, t) = 0 \tag{8.89}$$

and the initial conditions

$$u(x, t = 0) = u_0 \tag{8.90}$$

$$\left. \frac{\partial u(x, t)}{\partial t} \right|_{t=0} = 0. \tag{8.91}$$

Descriptively, this BVP corresponds to a string clamped on both ends that is pulled up initially at those two end points so that it resides at a uniform height u_0 above its equilibrium position $u(x, t) = 0$ prior to being released at $t = 0$ and executing vibrations $u(x, t)$ as controlled by the wave equation.

Separate variables as $u(x, t) = \psi(x)T(t)$ and put into the wave equation to obtain

$$\frac{c^2}{\psi(x)}\frac{\mathrm{d}^2\psi(x)}{\mathrm{d}x^2} = \frac{1}{T(t)}\frac{\mathrm{d}^2T(t)}{\mathrm{d}t^2} = -k^2. \tag{8.92}$$

Solve the equation for $\psi(x)$ to give

$$\psi(x) = A\sin\left(\frac{kx}{c}\right) + B\cos\left(\frac{kx}{c}\right) \tag{8.93}$$

subject to the BCs that $\psi(x = 0) = 0 = \psi(x = a)$, which require that $B = 0$ and the separation constant to be

$$k = k_n = \frac{n\pi c}{a} \quad \text{for } n = 1, 2, 3\ldots. \tag{8.94}$$

Similarly, the solution for $T(t)$ with these separation constants is

$$T(t) = C\sin\left(\frac{n\pi ct}{a}\right) + D\cos\left(\frac{n\pi ct}{a}\right). \tag{8.95}$$

The initial condition that $\partial u(x, t)/\partial t|_{t=0} = 0$ requires that $C = 0$. The solution is thus of the form

$$u(x, t) = \sum_{n=1}^{\infty} E_n \sin\left(\frac{n\pi x}{a}\right)\cos\left(\frac{n\pi ct}{a}\right). \tag{8.96}$$

The constants E_n are then found from the initial condition that $u(x, t = 0) = u_0$. Putting $t = 0$ into Eq. (8.96), using this initial condition, multiplying both sides by $\sin(m\pi x/a)$ and integrating both sides from $x = 0$ to $x = a$ yields

$$u_0\int_0^a \sin\left(\frac{m\pi x}{a}\right)\mathrm{d}x = \sum_{n=1}^{\infty} E_n \int_0^a \sin\left(\frac{n\pi x}{a}\right)\sin\left(\frac{m\pi x}{a}\right). \tag{8.97}$$

Performing the integrals on both sides identically as in the previous diffusion examples then yields

$$E_n = \frac{4u_0}{n\pi} \quad \text{for } n = 1, 3, 5\ldots \text{ (i.e., } n \text{ odd)} \tag{8.98}$$

and

$$u(x, t) = \frac{4u_0}{\pi}\sum_{\substack{n=1\\ \text{odd}}}^{\infty}\frac{1}{n}\sin\left(\frac{n\pi x}{a}\right)\cos\left(\frac{n\pi ct}{a}\right). \tag{8.99}$$

Using the identity that $\sin(x)\cos(y) = \left[\sin(x+y) + \sin(x-y)\right]/2$ then gives the solution in final form as

$$u(x,t) = \frac{2u_0}{\pi} \sum_{\substack{n=1 \\ \text{odd}}}^{\infty} \frac{1}{n} \left\{ \sin\left[\frac{n\pi}{a}(x+ct)\right] + \sin\left[\frac{n\pi}{a}(x-ct)\right] \right\}. \tag{8.100}$$

This corresponds to waves advancing through space and time according to the arguments $x+ct$ and $x-ct$ which for increasing t corresponds to waves moving both to the left on the string $(x+ct)$ and to the right $(x-ct)$.

8.4.3 2D Potential Problem

We now solve a potential problem in two spatial dimensions. The BVP is

$$\nabla^2 \varphi(x,y) = 0 \tag{8.101}$$

subject to the boundary conditions in a rectangular domain that

$$\varphi = 0 \quad \text{on } x = a,\ x = 0,\ y = b \tag{8.102}$$

$$\varphi = \Psi_0 \quad \text{on } y = 0 \tag{8.103}$$

where Ψ_0 is a given constant. The boundary values are perhaps more easily seen Fig. 8.3 To solve this BVP, we separate variables $\varphi(x,y) = X(x)\,Y(y)$ so that

$$\frac{1}{X(x)}\frac{d^2X(x)}{dx^2} + \frac{1}{Y(y)}\frac{d^2Y(y)}{dy^2} = 0. \tag{8.104}$$

This is satisfied if

$$\frac{1}{X(x)}\frac{d^2X(x)}{dx^2} = +\lambda \quad \text{and} \quad \frac{1}{Y(y)}\frac{d^2Y(y)}{dy^2} = -\lambda, \tag{8.105}$$

where the separation constant λ must be independent of x and y.

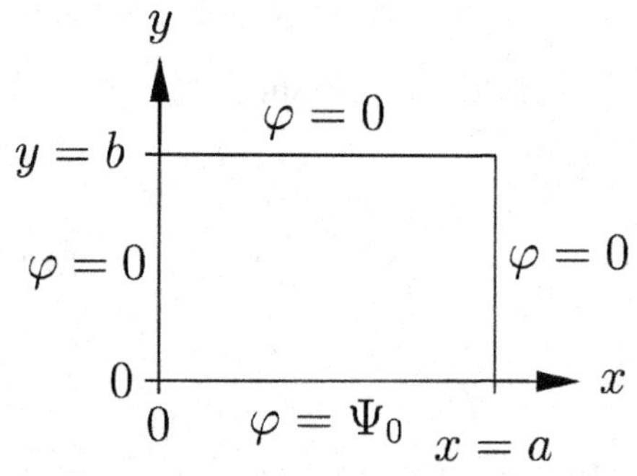

Figure 8.3 The potential boundary conditions in a simple rectangular domain.

So $d^2X(x)/dx^2 = \lambda X$ has solutions

$$\begin{aligned} X(x) &= A\ \cos\sqrt{-\lambda}\,x + B\ \sin\sqrt{-\lambda}\,x \quad \text{if } \lambda < 0 \\ &= Ax + B \quad \text{if } \lambda = 0 \\ &= A\ \cosh\sqrt{\lambda}\,x + B\ \sinh\sqrt{\lambda}\,x \quad \text{if } \lambda > 0. \end{aligned} \tag{8.106}$$

Depending on the nature of the boundary conditions, we have a choice to make for the sign of λ. In order to satisfy homogeneous BCs on $x = 0$, $x = a$, we want to have sines and cosines. Thus, we need $\lambda < 0$ or

$$\lambda = -k^2 \tag{8.107}$$

which then results in

$$X(x) = A\ \cos kx + B\ \sin kx \tag{8.108}$$

$$Y(y) = C\ \cosh ky + D\ \sinh ky. \tag{8.109}$$

To satisfy the BCs $X(0) = X(a) = 0$, we must have $A = 0$ as well as

$$k = k_n = \frac{n\pi}{a}, \quad \text{where } n = 1, 2, \ldots \tag{8.110}$$

so that

$$X_n(x) = B_n \sin\frac{n\pi x}{a}. \tag{8.111}$$

For $Y(0) \neq 0$ and $Y(b) = 0$ (the BCs in y), we take

$$Y_n(y) = D_n\ \sinh\frac{n\pi(b-y)}{a}. \tag{8.112}$$

Note that in writing $\sinh k_n(b-y)$ as the solution for Y_n, we are just taking a short cut to get a form of solution that satisfies the BCs. You can verify that it is just a linear combination of $\sinh k_n y$ and $\cosh k_n y$ and that it satisfies the ODE for $Y_n(y)$.

Putting it together, we have

$$\varphi(x, y) = \sum_{n=1}^{\infty} c_n\ \sin\frac{n\pi\, x}{a}\ \sinh\frac{n\pi(b-y)}{a}, \tag{8.113}$$

where we set $c_n = B_n D_n$. To find the c_n, we don't have an initial condition, but we do have the inhomogeneous BC on $y = 0$

$$\varphi(x, y=0) = \Psi_0 = \sum_n c_n\ \sin\frac{n\pi\, x}{a}\ \sinh\frac{n\pi\, b}{a}. \tag{8.114}$$

Multiply both sides by $\sin(m\pi\, x/a)$, integrate, and again use the orthogonality condition

$$\int_0^a dx\ \sin\frac{n\pi\, x}{a}\ \sin\frac{m\pi\, x}{a} = \begin{cases} 0 & \text{if } n \neq m \\ a/2 & \text{if } n = m \end{cases} \tag{8.115}$$

to find

$$
\begin{aligned}
c_m \sinh \frac{m\pi}{a}\frac{b}{2}\frac{a}{} &= \Psi_0 \int_0^a \sin \frac{m\pi\, x}{a}\, da \\
&= -\frac{a}{m\pi}\Psi_0 \cos \frac{m\pi\, x}{a}\Big|_0^a \\
&= \begin{cases} 2a\,\Psi_0/(m\pi) & \text{if } m = \text{odd} \\ 0 & \text{if } m = \text{even}. \end{cases}
\end{aligned}
\tag{8.116}
$$

Thus, we finally obtain the solution as

$$
\boxed{\varphi(x, y) = \sum_{n\ \text{odd}} \frac{4\Psi_0}{n\pi} \sin\left(\frac{n\pi\, x}{a}\right) \frac{\sinh n\pi\, (b-y)/a}{\sinh n\pi\, b/a}.}
\tag{8.117}
$$

So, in potential problems, it is either an inhomogeneous source term in the PDE or inhomogeneous boundary conditions that allow nontrivial (different from zero) solutions to emerge. In diffusion and wave problems with homogeneous boundary conditions and PDEs, it is the initial condition that results in nontrivial solutions.

8.4.4 More about Potential Problems $\nabla^2\varphi = 0$

Consider a simple finite-difference representation of $\nabla^2\,\varphi(x, y) = 0$ performed on a 2D square grid in Cartesian coordinates (to make the algebra simpler than in 3D) as shown in Fig. 8.4. The definition of the derivative as $\Delta \to 0$ gives

$$
\frac{\partial\varphi}{\partial x}\left(x + \frac{\Delta}{2}, y\right) = \frac{1}{\Delta}\Big[\varphi(x+\Delta, y) - \varphi(x, y)\Big]
\tag{8.118}
$$

$$
\frac{\partial\varphi}{\partial y}\left(x, y + \frac{\Delta}{2}\right) = \frac{1}{\Delta}\Big[\varphi(x, y+\Delta) - \varphi(x, y)\Big].
\tag{8.119}
$$

Similarly, taking another derivative gives

$$
\begin{aligned}
\frac{\partial^2\varphi}{\partial x^2}(x, y) &= \frac{1}{\Delta}\left[\frac{\partial\varphi}{\partial x}\left(x + \frac{\Delta}{2}, y\right) - \frac{\partial\varphi}{\partial x}\left(x - \frac{\Delta}{2}, y\right)\right] \\
&= \frac{1}{\Delta^2}\Big\{\varphi(x+\Delta, y) - \varphi(x, y) - \Big[\varphi(x, y) - \varphi(x-\Delta, y)\Big]\Big\} \\
&= \frac{1}{\Delta^2}\Big\{\varphi(x+\Delta, y) + \varphi(x-\Delta, y) - 2\,\varphi(x, y)\Big\}
\end{aligned}
\tag{8.120}
$$

$$
\frac{\partial^2\varphi}{\partial y^2}(x.y) = \frac{1}{\Delta^2}\big\{\varphi(x, y+\Delta) + \varphi(x, y-\Delta) - 2\varphi(x, y)\big\}.
\tag{8.121}
$$

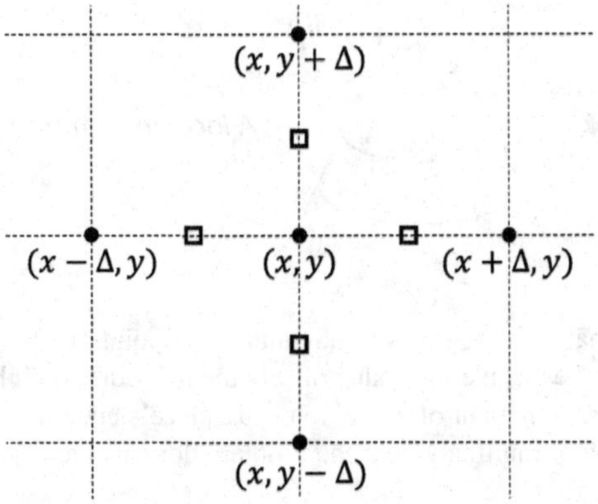

Figure 8.4 The grid points (black circles) surrounding a central point (x, y) in a 2D domain with the potential $\varphi(x, y)$ defined at each grid point. The squares indicate where the derivatives $\partial\phi/\partial x$ and $\partial\phi/\partial y$ are being calculated.

So in 2D, by adding together Eqs (8.120) and (8.121), we have shown that $\nabla^2\varphi = 0$ implies that

$$\varphi(x, y) = \frac{1}{4}\Big[\varphi(x + \Delta, y) + \varphi(x - \Delta, y) + \varphi(x, y + \Delta) + \varphi(x, y - \Delta)\Big]. \tag{8.122}$$

Thus we can conclude that the solution of Laplace's equation has the general and important interpretation

$$\boxed{\varphi(x, y) = \text{average value of } \varphi \text{ in the immediate. neighborhood surrounding } (x, y)} \tag{8.123}$$

This conclusion holds true in 3D and for any coordinate system.

Imagine that $\varphi(x, y)$ possesses a local maximum within some region Ω at a point (x_1, y_1) as shown in Fig. 8.5. Then the average value of $\varphi(x, y)$ in the neighborhood of (x_1, y_1) must be less than $\varphi(x_1, y_1)$. Thus, we must conclude that φ cannot possess a maximum anywhere within the problem domain. A similar visualization shows that φ cannot possess a minimum in the problem domain either.

So for $\nabla^2\varphi = 0$ in some region Ω, the potential $\varphi(\boldsymbol{r})$ has the following three key properties:

1. $\varphi(\boldsymbol{r})$ can possess no maxima or minima within the region Ω,
2. $\varphi(\boldsymbol{r})$ can only possess maxima and minima on the boundary $\partial\Omega$,
3. $\varphi(\boldsymbol{r})$ is the smoothest possible function in Ω that links the given boundary values of $\varphi(\boldsymbol{r})$ or $\boldsymbol{n} \cdot \nabla\varphi(\boldsymbol{r})$ or $\boldsymbol{n} \cdot \nabla\varphi(\boldsymbol{r}) + \alpha\,\varphi(\boldsymbol{r})$ with $\alpha > 0$ that are prescribed on each portion $\partial\Omega$ and that are the source (or data) producing nonzero values of $\varphi(\boldsymbol{r})$ throughout Ω.

These three facts are the most important general things to know about solutions of Laplace's equation.

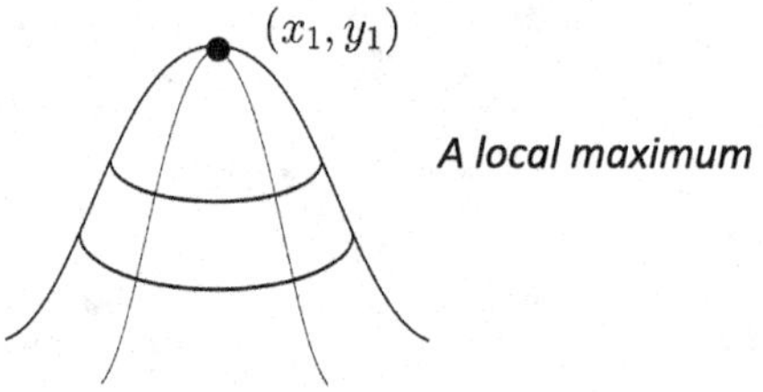

Figure 8.5 We see that if $\varphi(x_1, y_1)$ is a maximum at a point (x_1, y_1) within the problem domain, it cannot also be the average value of φ in the immediate neighborhood of (x_1, y_1), which is the derived requirement of a solution of Laplace's equation. Therefore, φ cannot possess either maxima or minima within the problem domain and also be a solution of the Laplace equation.

8.4.5 Essential Things to Remember about Waves, Diffusions, and Potentials

We briefly pause in our treatment of solving particular BVPs using the method of separation of variables to summarize the handful of most important things that we have already shown in our treatment of wave, diffusion and potential problems.

Waves: A solution of the wave equation in 1D has two possible forms $f(x - c\,t)$ and $f(x + c\,t)$ where c is the wavespeed. If we watch a wave move through space as t increases, we follow a particular point on the wavelet defined by amplitude $f(a) =$ constant where a is some constant. For $x - c\,t = a =$ constant, as t is increasing we must have that x is increasing so that $f(x - c\,t)$ corresponds to a wave propagating in the $+x$ direction at speed c. We call this an *outward-radiating* wave. Similarly, for $x + c\,t = a =$ constant, as t is increasing we must have x decreasing so that $f(x + c\,t)$ corresponds to a wave propagating in the $-x$ direction. We call this an *inward-radiating* wave. Again, the outward-radiating boundary condition at a boundary $x = L$ is written $(\partial f/\partial x + c^{-1}\partial f/\partial t)\big|_{x=L} = 0$. It is nearly trivial to say that waves have the time and space relation given by

$$t = x/c. \tag{8.124}$$

There is more to learn about how waves from line and point sources change amplitude with distance of propagation and how they scatter from objects and boundaries but the time and space relation is always simply $t = x/c$, which is the single most important thing to remember about a solution to a wave problem.

Diffusions: The time t it takes for diffusion controlled by a diffusivity D to advance a distance x can be estimated as

$$t \approx \frac{x^2}{D}. \tag{8.125}$$

We can use this order-of-magnitude estimate to get a quick feel for how much time it takes diffusion to advance a distance x. You can always remember this relation by simply writing down the diffusion equation $D\partial^2\psi(x, t)/\partial x^2 - \partial\psi(x, t)/\partial t = 0$ and estimating derivatives to give $(D/x^2 - 1/t)|\psi| \approx 0$, which is the same as Eq. (8.125). The relation $t \approx x^2/D$ for the estimate of the time t to diffuse a distance x is the single most important thing to remember about diffusion.

Potentials: As we just proved above, a solution of Laplace's equation within a given domain Ω cannot have a maximum or minimum within that domain and represents the smoothest possible way to connect the boundary values on the surrounding surface $\partial\Omega$, which are the single most important things to remember about potential problems. All maxima and minima in a potential problem must reside on the boundary $\partial\Omega$.

8.4.6 2D Diffusion

Getting back to solving specific BVPs using separation of variables, let's consider the diffusive equilibration of a rectangular domain with its surroundings. The BVP is defined

$$\left(\frac{\partial^2}{\partial x^2}+\frac{\partial^2}{\partial y^2}\right)P(x,y,t)-\frac{1}{D}\frac{\partial P(x,y,t)}{\partial t}=0, \tag{8.126}$$

subject to the BCs given in Fig. 8.6

$$P=0 \quad \text{on } x=0,\, x=a,\, y=0, \text{ and } y=b \tag{8.127}$$

as well as the initial condition that

$$P(x,y,t=0)=P_0. \tag{8.128}$$

Begin by separating the time dependence

$$P(x,y,t)=\psi(x,y)\,T(t) \tag{8.129}$$

to obtain

$$\frac{1}{\psi}\nabla^2\psi=\frac{1}{DT}\frac{dT}{dt}=-k^2, \tag{8.130}$$

where the separation constant was again chosen to be negative to obtain the proper time dependence given by $T(t)\propto e^{-Dk^2t}$. We then further separate the space dependence as

$$\psi(x,y)=X(x)\,Y(y) \tag{8.131}$$

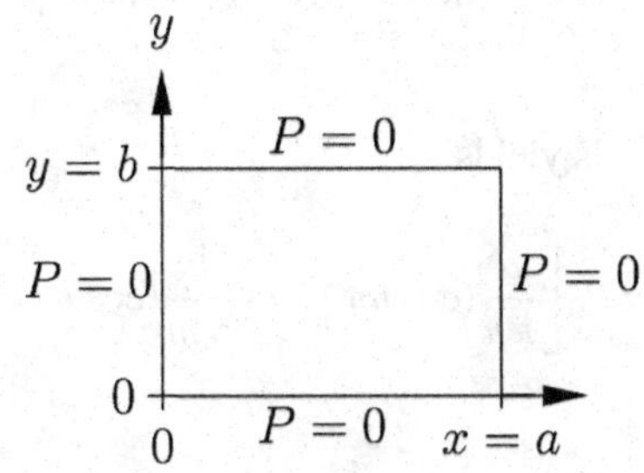

Figure 8.6 Boundary conditions for the given diffusion problem in a 2D rectangular domain.

to obtain

$$\frac{1}{Y}\frac{\mathrm{d}^2Y}{\mathrm{d}y^2} = -k_y^2 \quad \text{and} \quad \frac{1}{X}\frac{\mathrm{d}^2X}{\mathrm{d}x^2} = -k_x^2, \tag{8.132}$$

where we must have

$$k_x^2 + k_y^2 = k^2 \tag{8.133}$$

and where negative separation constants were chosen for both the x and y equations in order to get the sines and cosines that can satisfy homogeneous BCs. Specifically, to satisfy the BCs we take

$$X(x) = A_n \sin\frac{n\pi\, x}{a}; \qquad n = 1, 2, \ldots; \qquad k_{xn} = \frac{n\pi}{a} \tag{8.134}$$

$$Y(y) = B_m \sin\frac{m\pi\, y}{b}; \qquad m = 1, 2, \ldots; \qquad k_{ym} = \frac{m\pi}{b}. \tag{8.135}$$

Thus, we can identify

$$k^2 = k_{nm}^2 = k_{nx}^2 + k_{ym}^2 = \left(\frac{n\pi}{a}\right)^2 + \left(\frac{m\pi}{b}\right)^2 \tag{8.136}$$

and write

$$P(x, y, t) = \sum_{m=1}^{\infty}\sum_{n=1}^{\infty} A_{nm}\, e^{-\left[(n\pi/a)^2+(m\pi/b)^2\right]Dt} \sin\frac{n\pi\, x}{a} \sin\frac{m\pi\, y}{b}. \tag{8.137}$$

To obtain the constants A_{nm}, we evaluate this expression at time zero $P_0 = P(x, y, 0)$ (which is the initial condition), multiply both sides by $\sin(n'\pi\, x/a)\, \sin(m'\pi\, y/b)$, and integrate over the entire region

$$\begin{aligned} &P_0\left[\int_0^a \sin\frac{n'\pi\, x}{a}\,\mathrm{d}x\right]\left[\int_0^b \sin\frac{m'\pi\, y}{b}\,\mathrm{d}y\right] \\ &= \sum_{m,n} A_{nm}\left[\int_0^a \sin\frac{n'\pi\, x}{a}\sin\frac{n\pi\, x}{a}\,\mathrm{d}x\right]\left[\int_0^b \sin\frac{m'\pi\, y}{b}\sin\frac{m\pi\, y}{b}\,\mathrm{d}y\right]. \end{aligned} \tag{8.138}$$

Performing the various integrals yields

$$\frac{ab}{4}A_{nm} = P_0\left[\frac{a}{\pi n}(\cos n\pi - 1)\,\frac{b}{\pi m}(\cos m\pi - 1)\right] \tag{8.139}$$

or

$$A_{nm} = \frac{16\,P_0}{\pi^2 nm} \quad \text{with both } n \text{ and } m \text{ odd.} \tag{8.140}$$

Thus, we obtain the solution as

$$P(x, y, t) = \frac{16\,P_0}{\pi^2} \sum_{\substack{m,n \\ \text{odd}}} \frac{e^{-\left[(n\pi/a)^2+(m\pi/b)^2\right]Dt}}{nm} \sin\frac{n\pi x}{a} \sin\frac{m\pi y}{b}. \tag{8.141}$$

This is just the generalization to 2D of the 1D diffusion problem treated earlier and it is worth comparing the two solutions which have qualitatively similar behavior; namely, if a body (either an infinite slab in 1D or a rectangular element in the present 2D case) begins at a value $P(\boldsymbol{r}, 0) = P_0$ and then diffusively equilibrates with its surroundings, the boundary value $P = 0$ diffuses its way into the body until ultimately $P(\boldsymbol{r}, t) = 0$ throughout the body.

8.4.7 2D Vibration (Wave) Pattern

Consider the rectangular drumhead in Fig. 8.7. Displacements $u(x, y, t)$ normal to the stretched canvas satisfy the wave equation

$$\frac{\partial^2 u}{\partial x^2} + \frac{\partial^2 u}{\partial y^2} - \frac{1}{c^2}\frac{\partial^2 u}{\partial t^2} = 0 \tag{8.142}$$

subject to the BCs (clamped edges)

$$u = 0 \quad \text{on } x = 0,\, x = a,\, y = 0,\, y = b \tag{8.143}$$

and the initial conditions

$$u(x, y, t = 0) = u_0(x, y) \mathrel{\widehat{=}} \text{some given function} \tag{8.144}$$

$$\left.\frac{\partial u}{\partial t}(x, y, t)\right|_{t=0} = v_0(x, y) \mathrel{\widehat{=}} \text{another given function.} \tag{8.145}$$

Introduce the separated form

$$u(x, y, t) = X(x)\, Y(y)\, T(t) \tag{8.146}$$

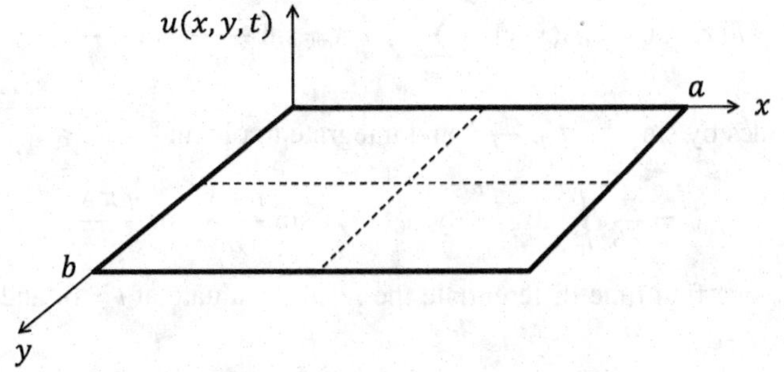

Figure 8.7 A vibrating rectangular drumhead (heavy black line) with two lower-order nodal lines $n = 2$, $m = 2$ shown as dashed lines.

and divide through by $X(x)Y(y)T(t)$ to get

$$\frac{1}{X}\frac{\mathrm{d}^2X}{\mathrm{d}x^2}+\frac{1}{Y}\frac{\mathrm{d}^2Y}{\mathrm{d}y^2}=\frac{1}{c^2T}\frac{\mathrm{d}^2T}{\mathrm{d}t^2}. \tag{8.147}$$

Separately require that

$$\frac{1}{X}\frac{\mathrm{d}^2X}{\mathrm{d}x^2}=-k_x^2 \tag{8.148}$$

$$\frac{1}{Y}\frac{\mathrm{d}^2Y}{\mathrm{d}y^2}=-k_y^2 \tag{8.149}$$

$$\frac{1}{c^2T}\frac{\mathrm{d}^2T}{\mathrm{d}t^2}=-k^2, \tag{8.150}$$

where $k^2=k_x^2+k_y^2$. We can introduce the "circular frequency" ω as $\omega^2=k^2c^2$ if we desire.

To satisfy the homogeneous BCs, we choose

$$X_n(x)=A_n\sin\frac{n\pi\, x}{a}\quad \text{or}\quad k_{xn}=\frac{n\pi}{a} \tag{8.151}$$

$$Y_m(y)=B_m\sin\frac{n\pi\, y}{b}\quad \text{or}\quad k_{ym}=\frac{m\pi}{b}. \tag{8.152}$$

Thus, the time solutions are of the form

$$T_{nm}(t)=A_{nm}\cos\omega_{nm}t+B_{nm}\sin\omega_{nm}t, \tag{8.153}$$

where

$$\omega_{nm}=\pi c\sqrt{\frac{n^2}{a^2}+\frac{m^2}{b^2}}. \tag{8.154}$$

Bringing it all together, we have

$$\boxed{u(x,y,t)=\sum_m\sum_n\left(A_{nm}\cos\omega_{nm}t+B_{nm}\sin\omega_{nm}t\right)\sin\frac{n\pi\, x}{a}\sin\frac{m\pi\, y}{a}.} \tag{8.155}$$

To find the A_{nm}, we use the initial condition at $t=0$

$$u(x,y,0)=u_0(x,y)=\sum_m\sum_n A_{nm}\sin\frac{n\pi\, x}{a}\sin\frac{m\pi\, y}{b}. \tag{8.156}$$

Multiply both sides by $\sin\frac{n'\pi\, x}{a}\sin\frac{m'\pi\, y}{b}$ and integrate to obtain

$$A_{nm}=\frac{4}{ab}\int_0^a \mathrm{d}x\int_0^b \mathrm{d}y\; u_0(x,y)\sin\frac{n\pi\, x}{a}\sin\frac{m\pi\, y}{b}. \tag{8.157}$$

To find the B_{nm}, we first time differentiate the result, evaluate at $t=0$, and use the other initial condition

$$\left.\frac{\partial u}{\partial t}(x,y,t)\right|_{t=0}=v_0(x,y)=\sum_m\sum_n \omega_{nm}B_{nm}\sin\frac{n\pi\, x}{a}\sin\frac{m\pi\, y}{b} \tag{8.158}$$

Similarly, this then allows the B_{nm} to be found as

$$B_{nm} = \frac{4}{ab\,\omega_{nm}} \int_0^a \mathrm{d}x \int_0^b \mathrm{d}y\ v_0(x, y)\ \sin\frac{n\pi\, x}{a}\ \sin\frac{m\pi\, y}{b}. \tag{8.159}$$

For specific given functions $u_0(x, y)$ and $v_0(x, y)$, the integrals are performed and the problem is solved. Each m, n pair is called a *"normal mode"* of vibration and has a specific frequency $\omega_{nm} = \pi c \sqrt{(n/a)^2 + (m/b)^2}$.

By using the suite of identities

$$\sin(\omega t + kr) = \sin(\omega t)\cos(kr) + \cos(\omega t)\sin(kr) \tag{8.160}$$

$$\sin(\omega t - kr) = \sin(\omega t)\cos(kr) - \cos(\omega t)\sin(kr) \tag{8.161}$$

$$\cos(\omega t + kr) = \cos(\omega t)\cos(kr) - \sin(\omega t)\sin(kr) \tag{8.162}$$

$$\cos(\omega t - kr) = \cos(\omega t)\cos(kr) + \sin(\omega t)\sin(kr), \tag{8.163}$$

we can express the above solution in the familiar form for the space and time dependence of a wave problem.

8.5 Conclusions

Generalization of the above problems to 3D (e.g., waves, diffusions, and potentials in rectangular prisms) is straightforward. In the exercises that follow, you can solve a range of problems that follow the general methodology laid out in the above examples of using the separation-of-variables methods. We have performed all of the above problems in Cartesian coordinates. We can perform separation of variables in cylindrical, spherical or any coordinate system that is convenient for the shape of the boundaries of a finite body under study. However, we elect to work only in Cartesian coordinates with our goal of getting to the Fourier transform as quickly as possible in this introductory treatment of linear PDEs.

8.6 Exercises

1. Show that

$$\sum_{n\ \mathrm{odd}} \frac{1}{n} \sin\left(\frac{n\pi x}{a}\right) = \frac{\pi}{2}. \tag{8.164}$$

HINT: one way to do this is to multiply both sides by $\sin(m\pi x/a)$ and integrate both sides from $x = 0$ to $x = a$.

2. *A 1D diffusion problem with an inhomogeneous boundary condition*: Find the diffusion field $P(x, t)$ in a slab that has width a in the x direction and that is infinite in the y and z directions and that is the solution of the well-posed problem:

$$\frac{\partial^2 P(x, t)}{\partial x^2} - \frac{1}{D}\frac{\partial P(x, t)}{\partial t} = 0 \tag{8.165}$$

subject to the boundary and initial conditions that

$$\left.\frac{\partial P}{\partial x}\right|_{x=a} = q \quad \text{given nonzero boundary value at } x = a \text{ that is independent of time} \tag{8.166}$$

$P(x=0,\, t) = 0$ boundary value at $x = 0$ (8.167)

$P(x,\, t=0) = P_0$ given initial value of field at $t = 0$ that is uniform throughout the slab. (8.168)

Note that this is a problem that has an inhomogeneous and constant in time BC. You allow for that using a time-independent field $\varphi(x)$ that is a solution of Laplace's equation and the above BCs. You should obtain

$$P(x,\, t) = qx + \sum_{\substack{n \\ \text{odd}}} \left(\frac{4P_0}{n\pi} - \frac{(-1)^{(n-1)/2} 8aq}{(n\pi)^2} \right) \exp\left[-\left(\frac{n\pi}{2a}\right)^2 Dt \right] \sin \frac{n\pi x}{2a}.$$

Using the first few terms of this series, make a sketch of this function for a few different t including $t = \infty$.

3. *The plucked string vibration problem*: Consider a string of length a pulled tight and held clamped at its two ends $x = 0$ and $x = a$ (think of a guitar string). We pull up on the string at a point $x = x_0$ and hold it there stationary. Then, at time $t = 0$, we let go of the string and the string begins to vibrate (it has waves move up and down its length). Find the displacement amplitude $w(x,\, t)$ that satisfies the scalar wave equation

$$\frac{\partial^2 w(x,\, t)}{\partial x^2} - \frac{1}{c^2}\frac{\partial^2 w(x,\, t)}{\partial t^2} = 0, \tag{8.169}$$

where c is the given wave speed of the string and that is subject to the BCs of being clamped at each end

$$w(x=0,\, t) = w(x=a,\, t) = 0 \tag{8.170}$$

and that is subject to the initial conditions that

$$w(x,\, t=0) = \begin{cases} b\,x/x_0; & x < x_0 \\ b(a-x)/(a-x_0); & x > x_0 \end{cases} \tag{8.171}$$

$$\left.\frac{\partial w}{\partial t}\right|_{t=0} = 0. \tag{8.172}$$

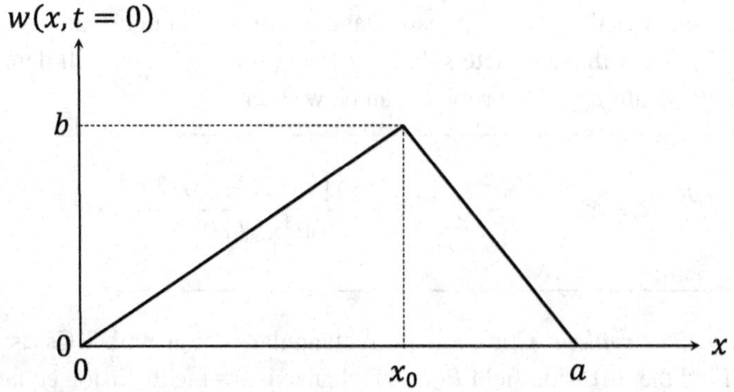

Figure 8.8 A string that has been pulled up to a given height b at the point $x = x_0$ just prior to its release at $t = 0$.

To visualize the initial condition just prior to letting the string go, see Fig. 8.8. The answer for the plucked-string problem that you will derive is

$$\boxed{w(x,t) = \frac{2ba^2}{\pi^2 x_0(a-x_0)} \sum_{n=1}^{\infty} \frac{1}{n^2} \sin\frac{n\pi x_0}{a} \sin\frac{n\pi x}{a} \cos\frac{n\pi ct}{a}.} \tag{8.173}$$

Use trigonometric identities to show that this has the "standard" form of waves moving up and down the string

$$w(x,t) = \frac{ba^2}{\pi^2 x_0(a-x_0)} \sum_{n=1}^{\infty} \frac{1}{n^2} \sin\frac{n\pi x_0}{a} \left[\sin\frac{n\pi(x-ct)}{a} + \sin\frac{n\pi(x+ct)}{a}\right]. \tag{8.174}$$

4. *A potential problem with a uniform source term*: Consider the potential problem that satisfies the Poisson equation (the inhomogeneous version of the Laplace equation)

$$\nabla^2\varphi(x,y) = -f_0, \tag{8.175}$$

where f_0 is a uniform constant throughout the rectangular domain of study that is bounded by the lines $x = 0$, $x = a$, $y = 0$, and $y = b$. The boundary conditions are homogeneous so it is the source term f_0 alone that is creating a nontrivial solution $\varphi(x,y)$. The BCs are specifically that

$$\varphi(x,y) = 0 \quad \text{on} \quad x = 0,\ x = a,\ y = 0,\ \text{and } y = b. \tag{8.176}$$

To solve this problem, you may note that a particular solution of the differential equation is

$$\varphi^p(x,y) = \frac{f_0}{2}x(a-x) \tag{8.177}$$

that indeed satisfies the Poisson equation and the BCs in x but not the BCs in y (so that φ^p is not, by itself, the complete solution of the problem). You will demonstrate that the complete solution to this problem can be written

$$\varphi(x, y) = \frac{f_0}{2}x(a-x) - \frac{4f_0a^2}{\pi^3}\sum_{n\ \mathrm{odd}}\frac{1}{n^3}\frac{\cosh\left[n\pi(2y-b)/(2a)\right]}{\cosh\left[n\pi b/(2a)\right]}\sin\frac{n\pi x}{a}. \tag{8.178}$$

5. *A 2D diffusion problem*: Consider the rectangular region with BCs as specified in Fig. 8.9. Find the diffusion field $P(x, y, t)$ that satisfies the diffusion equation

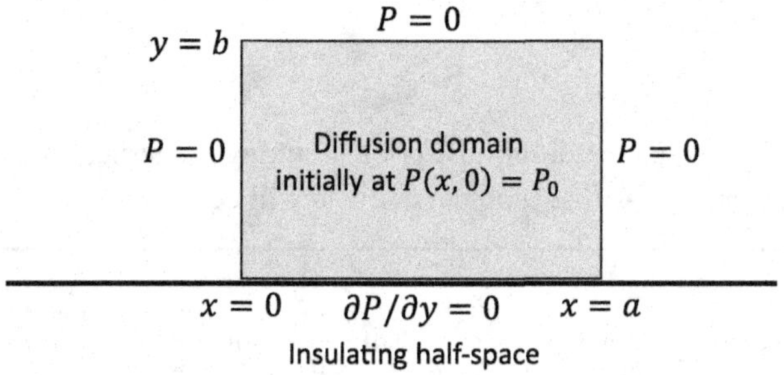

Figure 8.9 Simple rectangular domain in which diffusion modifies the initial condition in accord with the prescribed boundary conditions.

$$D\left(\frac{\partial^2 P(x, y, t)}{\partial x^2} + \frac{\partial^2 P(x, y, t)}{\partial y^2}\right) - \frac{\partial P(x, y, t)}{\partial t} = 0 \tag{8.179}$$

subject to the BCs that

$$\left.\frac{\partial P(x, y, t)}{\partial y}\right|_{y=0} = 0 \quad \text{and} \quad P(x=0, y, t) = P(x=a, y, t) = P(x, y=b, t) = 0 \tag{8.180}$$

and the initial condition that

$$P(x, y, t=0) = P_0. \tag{8.181}$$

You should obtain

$$P(x, y, t) = \frac{3P_0}{\pi^2}\sum_{n=1}^{\infty}\sum_{m=1}^{\infty}\frac{(-1)^{m+1}}{(2n-1)(2m-1)}\sin\left(\frac{\pi nx}{a}\right)\cos\left(\frac{\pi(2m-1)y}{2b}\right)$$
$$\times \exp\left\{-\left[\left(\frac{\pi n}{a}\right)^2 + \left(\frac{\pi(2m-1)}{2b}\right)^2\right]Dt\right\}.$$

9
The Fourier Series

The main purpose of this chapter is to derive and use the Fourier series for the purpose of solving boundary-value problems (BVPs) posed in Cartesian coordinates. We elect to derive the Fourier series as a special case of Sturm–Liouville theory that provides *eigenfunctions* that are a convenient basis for representing functions. We then show how to solve partial-differential equations (PDEs) posed in finite domains using the Fourier series. For context and contrast, we begin with the power series that you are familiar with from Chapter 1 as another way to represent functions distinct from the eigenfunction approach that is the focus of this chapter.

9.1 Power Series

As seen in Chapter 1, a power series representation of a differentiable function $f(x)$ of a single variable x is of the form

$$f(x) = \sum_{n=0}^{\infty} a_n\, x^n \tag{9.1}$$

$$= a_0 + a_1\, x + a_2\, x^2 + \ldots + a_{m-1}\, x^{m-1} + a_m\, x^m + a_{m+1}\, x^{m+1} + \ldots \tag{9.2}$$

To find the coefficients a_n, take m derivatives of $f(x)$. After taking m derivatives, all terms in the series with $n < m$ go to zero, the mth term in the series gives a constant and any remaining x dependence is from the terms $n \geq m+1$, that is,

$$\frac{\mathrm{d}^m f(x)}{\mathrm{d}x^m} = m!\, a_m + \frac{\mathrm{d}^m}{\mathrm{d}x^m} \sum_{n=m+1}^{\infty} a_n\, x^n \tag{9.3}$$

$$= m!\, a_m + \sum_{\ell=1}^{\infty} \frac{(m+\ell)!}{\ell!}\, a_{m+\ell}\, x^{\ell}. \tag{9.4}$$

Evaluate at $x = 0$ to remove the remaining x dependence and identify the a_n coefficient as

$$a_n = \frac{1}{n!} \left. \frac{\mathrm{d}^n f}{\mathrm{d}x^n} \right|_{x=0}. \tag{9.5}$$

A power series with these coefficients is called the *Taylor series* as treated in Section 1.8.1.

Such a representation is generally valid for $|x| < x_R$, where x_R is the *radius of convergence.* To find x_R, we can use the so-called *ratio test,* which says that for any series $S = \sum_n s_n$ to converge to a finite S, we must have

$$\frac{|s_{n+1}|}{|s_n|} < 1 \quad \text{as } n \to \infty. \tag{9.6}$$

Applying the ratio test to our Taylor series, we have

$$\frac{|a_{n+1}\, x^{n+1}|}{|a_n\, x^n|} < 1, \tag{9.7}$$

which identifies the radius of convergence for a Taylor-series expansion as

$$x_R = \frac{|a_n|}{|a_{n+1}|} \quad \text{as } n \to \infty. \tag{9.8}$$

We can apply this to a couple of familiar examples to calculate the radius of convergence. For example, consider the exponential function $f(x) = e^x$ that has Taylor-series coefficients given by

$$a_n = \frac{1}{n!}\frac{\mathrm{d}^n e^x}{\mathrm{d}x^n}\bigg|_{x=0} = \frac{1}{n!}. \tag{9.9}$$

The radius of convergence for the exponential function is then

$$x_R = \lim_{n\to\infty} \frac{(n+1)!}{n!} = \lim_{n\to\infty} (n+1) = \infty. \tag{9.10}$$

So the Taylor series of $f(x) = e^x$ converges for all $|x| < \infty$. Next consider, the function $f(x) = 1/(1-x)$ that has Taylor-series coefficients given by

$$a_n = \frac{1}{n!}\frac{\mathrm{d}^n (1-x)^{-1}}{\mathrm{d}x^n}\bigg|_{x=0} = \frac{(1)(2)...(n)}{n!} = 1. \tag{9.11}$$

We can use this to calculate the radius of convergence as

$$x_R = \lim_{n\to\infty} \frac{1}{1} = 1. \tag{9.12}$$

So the Taylor series of $f(x) = 1/(1-x)$ converges only for $|x| < 1$.

So given knowledge of a function and its derivatives at a point, the Taylor-series allows us to extrapolate that information into the neighborhood that surrounds the point out to a distance $|x| < x_R$, beyond which the power series will no longer converge.

For linear ordinary differential equations (ODEs) in the independent variable x in which the coefficients are themselves functions of x, one can represent both the dependent variable, say $y(x)$, as a power series $y(x) = \sum_{n=0}^{\infty} a_n x^n$ and the coefficients as Taylor series (unless the coefficients are already explicit polynomials in x), place such power series into the ODE and collect terms of similar powers of x. One then sets each such collection of terms associated with each power x^n to zero to obtain a recursion for the unknown coefficients a_n. We will not further elaborate on such a *power-series method* for solving ODEs.

9.2 Sturm–Liouville Theory

We now consider another way to represent some function $f(x)$ that provides the basis for Fourier analysis, which is a main focus throughout the rest of the book. Instead of representing the function $f(x)$ in terms of a sum of the powers x^n valid in a neighborhood $|x| < x_R$ surrounding the point where the derivatives are known, we instead use the representation

$$f(x) = \sum_{n=-\infty}^{\infty} a_n\, u_n(x), \tag{9.13}$$

where the so-called *eigenfunctions* $u_n(x)$ have to have certain properties that allow the coefficients a_n to be determined not by differentiation of $f(x)$ at a point but by integration of $f(x)$ over a given domain, say $a \le x \le b$. This series is convergent at all points x in the interval under the reasonable constraint that the function $f(x)$ be integrable over the domain $[a, b]$.

A broad class of eigenfunctions $u_n(x)$ that work well as the basis functions for the above representation of $f(x)$ are those defined as solutions of the so-called *Sturm–Liouville boundary-value problem*, which is

$$\mathcal{L}\, u_n(x) \mathrel{\widehat{=}} \frac{\mathrm{d}}{\mathrm{d}x}\left[p(x)\,\frac{\mathrm{d}\,u_n(x)}{\mathrm{d}x}\right] - q(x)\, u_n(x) = -\lambda_n\, r(x)\, u_n(x) \tag{9.14}$$

subject to the BCs on the interval $a \le x \le b$ that

$$u_n(a) + A\, u_n'(a) = 0 \tag{9.15}$$

$$u_n(b) + B\, u_n'(b) = 0, \tag{9.16}$$

where A and B are given constants and $u_n'(x) = \mathrm{d}u_n/\mathrm{d}x$. The coefficients $p(x)$ and $q(x)$ and the weighting-term $r(x)$ are all real, given functions of x, with the only constraint being that $r(x) > 0$. Another type of BC that works is *periodic* eigenfunctions

$$u_n(x) = u_n(x + b - a) \tag{9.17}$$

that produce the periodic boundary conditions $u_n(a) = u_n(b)$ and $u_n'(a) = u_n'(b)$.

The λ_n are called the *eigenvalues* of the Sturm–Liouville BVP. The eigenvalues allow solutions for the eigenfunctions $u_n(x)$ only at certain discrete values λ_n, with n corresponding to all positive and negative integers. We will see below, just as we did for the separation constants in the method of separation of variables, that it is the BCs that determine these discrete eigenvalues λ_n. If the interval over which the corresponding eigenfunctions $u_n(x)$ are defined is infinite, there is a continuum of eigenvalues λ that provide permissible solutions $u(\lambda, x)$ and the representation of $f(x)$ will be in the form of an integral $f(x) = \int_{-\infty}^{\infty} a(\lambda)\, u(\lambda, x)\, \mathrm{d}\lambda$ instead of as the discrete sum $\sum_{n=-\infty}^{\infty} a_n\, u_n(x)$.

The eigenfunctions possess the key property of *orthogonality* that is defined as

$$\boxed{\int_a^b \mathrm{d}x\, r(x)\, u_n(x)\, u_m^*(x) = 0 \quad \text{if } \lambda_n \neq \lambda_m.} \tag{9.18}$$

This property is what allows the coefficients a_n in the representation of $f(x)$ to be found.

To prove orthogonality of the eigenfunctions, consider for $m \neq n$

$$\mathcal{L}\, u_n(x) = -\lambda_n\, r(x)\, u_n(x) \tag{9.19}$$

$$\mathcal{L}\, u_m^*(x) = -\lambda_m^*\, r(x)\, u_m^*(x), \tag{9.20}$$

where the * denotes taking the complex conjugate if the $u_n(x)$ are complex, which they can be. The complex conjugate simply means replacing the imaginary unit i by $-i$ wherever it appears in a complex expression. We will prove below that the eigenvalues λ_n are real but do not assume so at this time. Note that per the terms of the Sturm–Liouville BVP, both the operator $\mathcal{L}$ and $r(x)$ are real. Multiply Eq. (9.19) by $u_m^*(x)$ and Eq. (9.20) by $u_n(x)$, subtract and integrate over (a, b) to obtain

$$\int_a^b \mathrm{d}x \left[u_m^*\, \mathcal{L}\, u_n - u_n\, \mathcal{L}\, u_m^* \right] = -\left(\lambda_n - \lambda_m^*\right) \int_a^b \mathrm{d}x\, r(x)\, u_n(x)\, u_m^*(x). \tag{9.21}$$

Using $\mathcal{L} \hat{=} \frac{\mathrm{d}}{\mathrm{d}x}\, p(x)\, \frac{\mathrm{d}}{\mathrm{d}x} - q(x)$, we find that

$$u_m^*\, \mathcal{L}\, u_n = \frac{\mathrm{d}}{\mathrm{d}x}\left[p\, u_m^* \frac{\mathrm{d}u_n}{\mathrm{d}x} \right] - p \frac{\mathrm{d}u_m^*}{\mathrm{d}x}\frac{\mathrm{d}u_n}{\mathrm{d}x} - q\, u_m^*\, u_n \tag{9.22}$$

$$u_n\, \mathcal{L}\, u_m^* = \frac{\mathrm{d}}{\mathrm{d}x}\left[p\, u_n \frac{\mathrm{d}u_m^*}{\mathrm{d}x} \right] - p \frac{\mathrm{d}u_m^*}{\mathrm{d}x}\frac{\mathrm{d}u_n}{\mathrm{d}x} - q\, u_m^*\, u_n \tag{9.23}$$

so that

$$\int_a^b \mathrm{d}x \left[u_m^*\, \mathcal{L}\, u_n - u_n\, \mathcal{L}\, u_m^* \right] = \left[p \left(u_m^* \frac{\mathrm{d}u_n}{\mathrm{d}x} - u_n \frac{\mathrm{d}u_m^*}{\mathrm{d}x} \right) \right]_a^b = 0 \tag{9.24}$$

after inserting the general BCs (insert them and see for yourself). We have thus demonstrated that

$$\left(\lambda_n - \lambda_m^*\right) \int_a^b \mathrm{d}x\, r(x) u_n(x)\, u_m^*(x) = 0, \tag{9.25}$$

which is the orthogonality statement when $m \neq n$. For $m = n$, the integral is strictly positive which thus requires that $\lambda_n = \lambda_n^*$ which demonstrates that the eigenvalues of the Sturm–Liouville BVP are purely real even if the associated eigenfunctions $u_n(x)$ can be complex. Again, when $n \neq m$ so that $\lambda_n - \lambda_m \neq 0$, the integral must be zero which is the desired statement of orthogonality of the eigenfunctions.

With the orthogonality of the eigenfunctions established, we can now represent any function $f(x)$ on the interval $a \leq x \leq b$ using

$$f(x) = \sum_n a_n\, u_n(x), \tag{9.26}$$

where the expansion coefficients a_n are determined using the orthogonality property by multiplying both sides by $r(x)\, u_m^*(x)$ and integrating

$$\int_0^a \mathrm{d}x \sum_n a_n\, u_n(x)\, u_m^*(x)\, r(x) = a_m \int_a^b r(x)\, u_m^*(x)\, u_m(x)\, \mathrm{d}x = \int_a^b f(x)\, r(x)\, u_m^*(x)\, \mathrm{d}x \tag{9.27}$$

or

$$a_n = \frac{\int_a^b \mathrm{d}x\, r(x) f(x)\, u_n^*(x)}{\int_a^b \mathrm{d}x\, r(x)\, u_n(x)\, u_n^*(x)}. \tag{9.28}$$

Note that the $u_n(x)$ are only known up to some arbitrary multiplicative constant. It is thus common to choose the constant so that

$$\int_a^b \mathrm{d}x\, r(x)\, u_n(x)\, u_n^*(x) = 1. \tag{9.29}$$

Finally, upon using such normalized eigenfunctions, we write:

$$f(x) = \sum_n a_n\, u_n(x) \tag{9.30}$$

$$= \sum_n \left[\int_a^b \mathrm{d}x'\, r(x')\, u_n^*(x') f(x') \right] u_n(x) \tag{9.31}$$

$$= \int_a^b \mathrm{d}x' f(x') \left[r(x') \sum_n u_n^*(x')\, u_n(x) \right], \tag{9.32}$$

which means that

$$r(x') \sum_n u_n^*(x')\, u_n(x) = \delta(x' - x) \mathrel{\widehat{=}} \text{the Dirac delta} \tag{9.33}$$

This result is often called the *completeness relation.*

A specific example of Sturm-Liouville theory that will be a focus throughout the remainder of the book occurs when $r(x) = p(x) = 1$ and $q(x) = 0$ which leads to the Fourier series when a and b are finite and the Fourier transform pair when $a = -\infty$ and $b = +\infty$. The continuum of frequencies ω (for time problems) and k (for spatial problems) of Fourier analysis in infinite domains correspond to the eigenvalues λ_n of Sturm–Liouville theory and are thus purely real as proven above. We will see that the Fourier eigenfunctions in infinite domains have the form e^{ikx} so that the completeness relation as just proven above becomes the result that $\int_{-\infty}^{\infty} e^{ik(x-x')}\, \mathrm{d}k/(2\pi) = \delta(x - x')$. Such Sturm–Liouville-theory results will be used in the chapters to follow.

When we use the method of separation of variables on the Helmholtz equation $\nabla^2 u(\boldsymbol{r}) + k^2 u(\boldsymbol{r}) = 0$ in various coordinate systems, we obtain a Sturm–Liouville BVP for each spatial coordinate with specific expressions emerging for the functions $p(x)$, $q(x)$, and $r(x)$. The corresponding eigenfunctions $u_n(x)$ defined on finite domains are called the *special functions* of mathematical physics, which include Bessel functions (that emerge for the radial dependence in cylindrical coordinates) as well as spherical harmonics and spherical Bessel functions (spherical coordinates). Other examples of orthogonal eigenfunctions that are solutions of specific Sturm–Liouville problems and that are used to represent functions over some interval include the various *orthogonal polynomials* such as Legendre polynomials, Hermite polynomials, Laguerre polynomials, Gegenbauer polynomials, and Tchebycheff polynomials. However, we will not consider finite-domain BVPs in curvilinear coordinates or use any of the many possible Sturm–Liouville eigenfunctions other than the sines and cosines of Fourier analysis for which $r(x) = p(x) = 1$ and $q(x) = 0$ in the Sturm–Liouville problem along with a periodic boundary condition.

To conclude, given functions for $r(x)$, $q(x)$, and $p(x)$ and boundary-condition constants A and B or assuming periodic boundary conditions in the Sturm–Liouville BVP, we obtain a complete set of orthogonal eigenfunctions $u_n(x)$ that can be used to represent any integrable function $f(x)$ on the interval $a \leq x \leq b$ as:

$$f(x) = \sum_n a_n\, u_n(x), \qquad \text{where} \quad a_n = \frac{\int_a^b \mathrm{d}x\, r(x) f(x)\, u_n^*(x)}{\int_a^b \mathrm{d}x\, r(x) \left[u_n(x)\right]^2}. \tag{9.34}$$

If you write down some functions for $r(x)$, $q(x)$, and $p(x)$ and define some boundary-condition constants A and B or assume periodic boundary conditions to create your own Sturm–Liouville BVP that has not previously been defined by somebody, once you find the associated orthogonal eigenfunctions $u_n(x)$ you can name them after yourself!

9.3 The Fourier Series

Perhaps the most important example of using Sturm–Liouville theory to represent functions is the Fourier series. The Fourier series is used in solving BVPs posed in finite domains and in Cartesian coordinates. As a matter of historical record, the Fourier series was first developed by Joseph Fourier as part of his 1822 book (Fourier, 1822) focused on heat-flow problems and this work then stimulated the more general theory of two other French mathematicians Charles-Francois Sturm and Joseph Liouville given above, which was first published in 1836 and 1837 (Sturm, 1836; Liouville and Sturm, 1837). That said, we prefer to think of the Fourier series as a special case of Sturm–Liouville theory.

To obtain the Fourier series, take $r(x) = 1$, $q(x) = 0$, $p(x) = 1$, and $\lambda_n = k_n^2$ to obtain the Sturm–Liouville problem defined by

$$\frac{\mathrm{d}^2 u_n}{\mathrm{d}x^2} = -k_n{}^2\, u_n \tag{9.35}$$

on an interval $a \leq x \leq a + L$ with the periodic condition

$$u_n(x) = u_n(x + L). \tag{9.36}$$

The solution is of the form $u_n(x) = e^{ik_n x}$ and $e^{-ik_n x}$. To satisfy the periodic condition

$$e^{ik_n x} = e^{ik_n (x+L)}, \tag{9.37}$$

we must choose the specific eigenvalues

$$k_n = \frac{2\pi n}{L} \quad \text{where} \quad n = 0, \pm 1, \pm 2, \dots \pm \infty \tag{9.38}$$

so that the eigenfunctions are

$$u_n(x) = e^{i2\pi nx/L} \quad \text{where} \quad n = 0, \pm 1, \pm 2, \dots \pm \infty, \tag{9.39}$$

which possess the orthogonality property

$$\int_a^{a+L} u_n^*(x)\, u_m(x)\, \mathrm{d}x = \int_a^{a+L} e^{i2\pi(m-n)x/L} \mathrm{d}x = \frac{L}{i2\pi(m-n)} e^{i2\pi(m-n)a/L}\left(e^{i2\pi(m-n)} - 1\right)$$
$$= L\,\delta_{nm} = \begin{cases} L & \text{if } n = m \\ 0 & \text{if } n \neq m. \end{cases} \tag{9.40}$$

Thus, we may represent $f(x)$ in the interval $a \le x \le a + L$ as

$$\boxed{f(x) = \sum_{n=-\infty}^{+\infty} c_n\, e^{i\,2\pi\, nx/L}} \tag{9.41}$$

or equivalently

$$\boxed{f(x) = \frac{a_0}{2} + \sum_{n=1}^{\infty}\left(a_n \cos\frac{2\pi n\,x}{L} + b_n \sin\frac{2\pi n\,x}{L}\right).} \tag{9.42}$$

We call the first boxed expression involving the c_n the *complex Fourier series* and the second boxed expression involving the a_n and b_n simply the *Fourier series*. Given the periodic nature of the trigonometric eigenfunctions, if we use the Fourier series to evaluate the function $f(x)$ outside the fundamental interval a to $a + L$, it too will be periodic with period L.

To get from the complex form to the usual form, use $e^{i\theta} = \cos\theta + i\sin\theta$ to write the complex form as

$$f(x) = c_0 + \sum_{n=1}^{\infty}\left[c_n e^{+i2\pi nx/L} + c_{-n} e^{-i2\pi nx/L}\right] \tag{9.43}$$

$$= c_0 + \sum_{n=1}^{\infty}\left[c_n\left(\cos\frac{2\pi nx}{L} + i\sin\frac{2\pi nx}{L}\right) + c_{-n}\left(\cos\frac{2\pi nx}{L} - i\sin\frac{2\pi nx}{L}\right)\right] \tag{9.44}$$

$$= c_0 + \sum_{n=1}^{\infty}\left[(c_n + c_{-n})\cos\frac{2\pi nx}{L} + i(c_n - c_{-n})\sin\frac{2\pi nx}{L}\right]. \tag{9.45}$$

The relation between the coefficients c_n and c_{-n} and the a_n and b_n can be read from the above as

$$a_n = c_n + c_{-n} \quad \text{and} \quad b_n = i(c_n - c_{-n}) \quad \text{for} \quad n = 1, 2, \ldots \infty. \tag{9.46}$$

We can also solve for the c_n as

$$c_n = \begin{cases} \frac{1}{2}\,(a_n - i\,b_n) & n > 0 \\ \frac{1}{2}\,(a_n + i\,b_n) & n < 0 \\ a_0/2 & n = 0. \end{cases} \tag{9.47}$$

For a real function $f(x)$, the coefficients a_n and b_n must be real and so the coefficients c_n will be complex.

The complex expansion coefficients c_n are obtained using the orthogonality property of the eigenfunctions. To do so, multiply both sides of the complex Fourier series by $e^{-i2\pi mx/L}$ (note that we are using the complex conjugate of the eigenfunction as was proven necessary in the Sturm–Liouville section) and then integrating over the interval a to $a+L$ to give

$$c_n = \frac{1}{L}\int_a^{a+L} f(x)\, e^{-i2\pi nx/L}\, \mathrm{d}x \quad \text{for} \quad -\infty < n < \infty. \tag{9.48}$$

The real coefficients a_n are obtained by multiplying both sides by $\cos 2\pi\, mx/L$ and integrating over the interval while the b_n are obtained by multiplying both sides by $\sin 2\pi\, mx/L$ and then integrating to give

$$a_n = \frac{2}{L}\int_a^{a+L} f(x)\, \cos\frac{2\pi n x}{L}\, \mathrm{d}x \quad \text{for} \quad n = 0, 1, 2, \ldots \infty \tag{9.49}$$

$$b_n = \frac{2}{L}\int_a^{a+L} f(x)\, \sin\frac{2\pi n x}{L}\, \mathrm{d}x \quad \text{for} \quad n = 1, 2, \ldots \infty. \tag{9.50}$$

Note that we used the orthogonality facts that you can demonstrate for yourself in the first exercise at the end of the chapter

$$\int_a^{a+L} \sin\left(\frac{2\pi nx}{L}\right) \sin\left(\frac{2\pi mx}{L}\right) dx = 0 \quad \text{when} \quad m \neq n \tag{9.51}$$

$$\int_a^{a+L} \cos\left(\frac{2\pi nx}{L}\right) \cos\left(\frac{2\pi mx}{L}\right) dx = 0 \quad \text{when} \quad m \neq n \tag{9.52}$$

$$\int_a^{a+L} \cos\left(\frac{2\pi nx}{L}\right) \sin\left(\frac{2\pi mx}{L}\right) dx = 0 \quad \text{for any } m \text{ and } n, \tag{9.53}$$

which, as usual, tells us that out of all the n only $n = m$ contributes. When $n = m$ and $n, m \geq 1$, we then have the integrals

$$\int_a^{a+L} \sin^2\left(\frac{2\pi n x}{L}\right) dx = \int_a^{a+L} \cos^2\left(\frac{2\pi n x}{L}\right) dx = \frac{L}{2} \tag{9.54}$$

$$\int_a^{a+L} \sin\left(\frac{2\pi n x}{L}\right) \cos\left(\frac{2\pi n x}{L}\right) dx = 0. \tag{9.55}$$

By using the above expressions to evaluate c_0 and a_0, we have

$$c_0 = \frac{1}{L}\int_a^{a+L} f(x)\, dx = \text{ average value of } f(x) \text{ over the interval } L \tag{9.56}$$

$$a_0 = \frac{2}{L}\int_a^{a+L} f(x)\, dx \tag{9.57}$$

so that we see $c_0 = a_0/2$ as given earlier.

9.4 Representing Functions Using the Fourier Series

Let's begin by evaluating the coefficients of a Fourier series representation for a few functions.

9.4.1 Example: A Quadratic Function

For example, to represent the function $g(x) = x^2$ as a Fourier series on the interval $-L/2 \leq x \leq L/2$, we write

$$x^2 = \frac{a_0}{2} + \sum_{n=1}^{\infty}\left[a_n \cos\left(\frac{2\pi nx}{L}\right) + b_n \sin\left(\frac{2\pi nx}{L}\right)\right]. \tag{9.58}$$

The goal is to find the constants a_0, a_n, and b_n for $n = 1, 2, \ldots \infty$. Because x^2 is even, we know that the $b_n = 0$ because $\sin(2\pi nx/L)$ is odd over the interval. You then use Eq. (9.49) to find the a_n.

This first time, however, let's begin from scratch and multiply both sides by $\cos(2\pi mx/L)$ and integrate over the interval

$$\begin{aligned}\int_{-L/2}^{L/2} x^2 \cos\left(\frac{2\pi mx}{L}\right) dx = {} & \frac{a_0}{2}\int_{-L/2}^{L/2} \cos\left(\frac{2\pi mx}{L}\right) dx \\ & + \sum_{n=1}^{\infty} a_n \int_{-L/2}^{L/2} \cos\left(\frac{2\pi mx}{L}\right) \cos\left(\frac{2\pi nx}{L}\right) dx \\ & + \sum_{n=1}^{\infty} b_n \int_{-L/2}^{L/2} \cos\left(\frac{2\pi mx}{L}\right) \sin\left(\frac{2\pi nx}{L}\right) dx. \end{aligned} \tag{9.59}$$

To pick up the coefficient a_0, we put $m = 0$, note that $\int_{-L/2}^{L/2} \cos(2\pi nx/L)\, dx = \int_{-L/2}^{L/2} \sin(2\pi nx/L)\, dx = 0$ and obtain

$$a_0 \frac{L}{2} = \int_{-L/2}^{L/2} x^2\, dx = \frac{1}{3}x^3\Big|_{-L/2}^{L/2} \tag{9.60}$$

which gives

$$a_0 = \frac{L^2}{6}. \tag{9.61}$$

For $m \geq 1$, the integral of $\cos(2\pi mx/L)\sin(2\pi nx/L)$ is zero for any m and n (including $m = n$) which, upon integrating the left-hand-side of Eq. (9.59) by parts twice, yields

$$a_m \frac{L}{2} = \int_{-L/2}^{L/2} x^2 \cos\left(\frac{2\pi mx}{L}\right) \mathrm{d}x \tag{9.62}$$

$$= \frac{x^2 L}{2\pi m} \sin\left(\frac{2\pi mx}{L}\right)\Bigg|_{-L/2}^{L/2} - \frac{L}{\pi m}\int_{-L/2}^{L/2} x \sin\left(\frac{2\pi mx}{L}\right) \mathrm{d}x \tag{9.63}$$

$$= -\frac{L}{\pi m}\left[-\frac{xL}{2\pi m}\cos\left(\frac{2\pi mx}{L}\right)\Bigg|_{-L/2}^{L/2} + \frac{L}{2\pi m}\int_{-L/2}^{L/2} \cos\left(\frac{2\pi mx}{L}\right) \mathrm{d}x\right] \tag{9.64}$$

$$= \frac{L^3}{2\pi^2 m^2}\cos(\pi m) = \frac{L^3}{2\pi^2 m^2}(-1)^m \tag{9.65}$$

so that

$$a_n = (-1)^n \left(\frac{L}{\pi n}\right)^2. \tag{9.66}$$

Again, $b_n = 0$, because x^2 is even and $\sin(2\pi mx/L)$ is odd over the given interval. You can also verify this by multiplying both sides of Eq. (9.58) by $\sin(2\pi mx/L)$, integrating over the interval and performing the integrals. Thus, the Fourier series representing $g(x) = x^2$ is

$$\boxed{g(x) = x^2 = \frac{L^2}{12} + \frac{L^2}{\pi^2}\sum_{n=1}^{\infty} \frac{(-1)^n}{n^2} \cos\left(\frac{2\pi nx}{L}\right)} \tag{9.67}$$

which is plotted in Fig. 9.1. As always with a function represented as a Fourier series over a given interval L, if we evaluate the Fourier series outside the interval it is periodic with period L.

9.4.2 Example: A Step Function

We next determine the Fourier series representation of the unit step function defined for the interval $0 \leq x \leq L$ as

$$f(x) = \begin{cases} 1 & 0 \leq x < h \\ 0 & h < x \leq L. \end{cases} \tag{9.68}$$

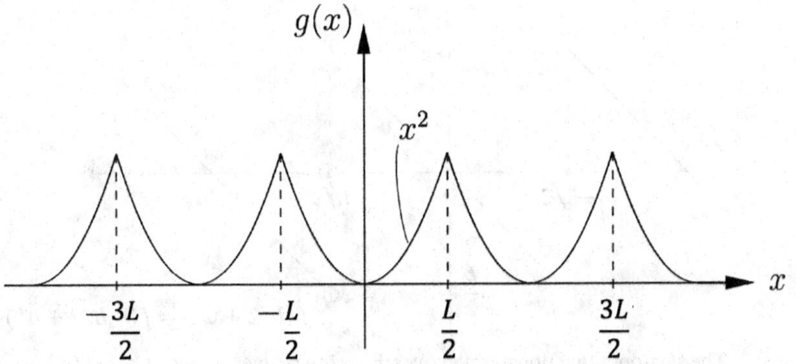

Figure 9.1 The function x^2 represented as a Fourier series over the fundamental interval $-L/2 \leq x \leq L/2$.

For practice and because the step function is neither odd nor even in the given interval which means that both the a_n and b_n are needed to represent the step function, we will do this example using the complex Fourier series

$$f(x) = \sum_{n=-\infty}^{+\infty} c_n \, e^{i2\pi nx/L}. \tag{9.69}$$

Thus, we multiply both sides of this by $e^{-i2\pi mx/L}$ and integrate over the interval $0 \leq x \leq L$ and determine that $n = m$ so that

$$c_n = \frac{1}{L}\int_0^L dx \, f(x)\, e^{-i2\pi nx/L} = \frac{1}{L}\int_0^h dx \, e^{-i2\pi nx/L} = \frac{\left(1 - e^{-i2\pi nh/L}\right)}{2i\pi n} \tag{9.70}$$

with $c_0 = h/L$.

Putting these into $\sum_{n=-\infty}^{\infty} c_n \, e^{i2\pi x/L}$ necessarily gives a real result. Writing the negative n part of the series separately from the positive n we have

$$f(x) = \frac{1}{\pi}\sum_{n=-\infty}^{+\infty} \frac{\left(e^{i2\pi nx/L} - e^{i2\pi n(x-h)/L}\right)}{2in} \tag{9.71}$$

$$= \frac{h}{L} + \frac{1}{\pi}\sum_{n=1}^{\infty}\frac{1}{n}\left[\frac{\left(e^{i2\pi nx/L} - e^{-i2\pi nx/L}\right)}{2i} - \frac{\left(e^{i2\pi n(x-h)/L} - e^{-i2\pi n(x-h)/L}\right)}{2i}\right]. \tag{9.72}$$

So the Fourier series of the step function on the interval $0 \leq x \leq L$ with the step at the point h within the interval is

$$f(x) = \text{step function} = \frac{h}{L} + \frac{1}{\pi}\sum_{n=1}^{\infty}\frac{1}{n}\left[\sin\left(\frac{2\pi nx}{L}\right) - \sin\left(\frac{2\pi n(x-h)}{L}\right)\right], \tag{9.73}$$

which periodically repeats outside the interval.

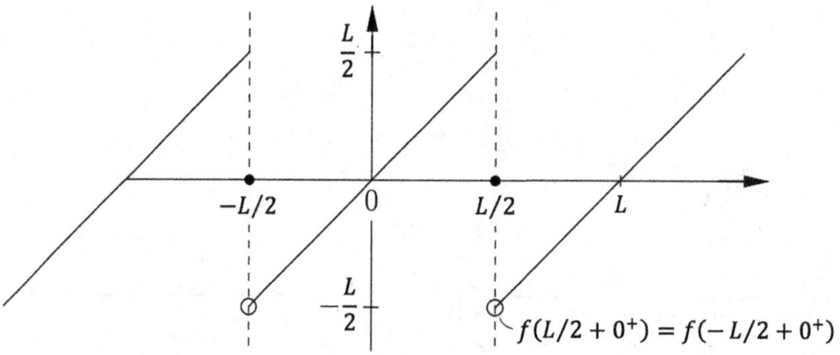

Figure 9.2 The sawtooth function defined over the primary interval $-L/2 \leq x \leq L/2$, where this function is zero precisely at the points $x = \pm nL/2$ for $n = 0, 1, 2, \ldots$.

9.4.3 Gibbs Phenomenon (Overshoot at Discontinuities)

Because of the periodic nature of functions being represented as a Fourier series, there are often discontinuities in the function being represented at the end of each interval. The step function just given is such an example.

Consider another function with discontinuities called the "sawtooth" function as shown in Fig. 9.2. The sawtooth function is defined

$$f(x) = \begin{cases} x & \text{on } -L/2 < x < L/2 \\ 0 & \text{at } x = L/2 \end{cases} \tag{9.74}$$

and is periodic outside the primary interval $[-L/2, L/2]$. In the figure, the notation $L/2 + 0^+$ is indicating the limit from the positive x side given by $L/2 + 0^+ = \lim_{\epsilon \to 0}(L/2 + \epsilon)$ for positive ϵ.

In general, the Fourier series can be shown to be "pointwise convergent" at all points x in the interval $a \leq x \leq a + L$. If $f_F(x)$ is the Fourier-series representation of the function $f(x)$, pointwise convergence means

$$f_F(x) = \lim_{\epsilon \to 0} \frac{1}{2}\Big[f(x + \epsilon) + f(x - \epsilon)\Big]. \tag{9.75}$$

In particular, at a discontinuity such as in the sawtooth function, this means

$$f_F(x = L/2) = \lim_{\epsilon \to 0} \frac{1}{2}\Big[f(L/2 + \epsilon) + f(L/2 - \epsilon)\Big] = 0 \tag{9.76}$$

and where $f(L/2 + \epsilon) = f(-L/2 + \epsilon)$ due to the periodicity. We will see below that although the Fourier series of the sawtooth function, and any other function with a discontinuity at the end of the primary interval, satisfies this condition right at the discontinuity $x = L/2$, it gets the function at $f(-L/2 + \epsilon)$ or $f(L/2 - \epsilon)$ wrong in the immediate neighborhood of the discontinuity even for very small ϵ.

The sawtooth function as defined here is an odd function, so the $a_n = 0$ and we have

$$f_F(x) = \sum_{n=1}^{\infty} b_n \sin\left(\frac{2\pi n x}{L}\right) \tag{9.77}$$

with the b_n given by

$$b_n = \frac{2}{L}\int_{-L/2}^{L/2} x \sin\left(\frac{2\pi n x}{L}\right) \mathrm{d}x = \frac{4}{L}\int_0^{L/2} x \sin\left(\frac{2\pi n x}{L}\right) \mathrm{d}x$$

$$= -\frac{L}{\pi n}\cos(\pi n) = \frac{L}{\pi n}(-1)^{n+1} \tag{9.78}$$

where the integral was performed by integrating by parts once. We thus have

$$\boxed{f_F(x) = \frac{L}{\pi}\sum_{n=1}^{\infty}\frac{(-1)^{n+1}}{n}\sin\left(\frac{2\pi n x}{L}\right).} \tag{9.79}$$

which is the Fourier-series representation of the sawtooth function $f(x)$.

To investigate the behavior of the Fourier series immediately adjacent to the discontinuity at $x = \pm L/2$, let's consider an approximation of the Fourier series $f_N(x) \approx f_F(x)$ obtained by including only a finite number of terms N in the series

$$f_N(x) = \frac{L}{\pi}\sum_{n=1}^{N}\frac{(-1)^{n+1}}{n}\sin\left(\frac{2\pi n x}{L}\right). \tag{9.80}$$

where $\lim_{N\to\infty} f_N(x) \to f_F(x)$. We can evaluate this function just short of the discontinuity at $x = L/2$ by taking N to be large in the expression

$$f_N\left(\frac{L}{2} - \frac{L}{2N}\right) = \frac{L}{\pi}\sum_{n=1}^{N}\frac{(-1)^{n+1}}{n}\sin\left(\pi n - \frac{\pi n}{N}\right) \tag{9.81}$$

$$= \frac{L}{\pi}\sum_{n=1}^{N}\frac{(-1)^{n+1}}{n}\left[\sin(\pi n)\cos\left(\frac{\pi n}{N}\right) - \cos(\pi n)\sin\left(\frac{\pi n}{N}\right)\right] \tag{9.82}$$

$$= \frac{L}{\pi}\sum_{n=1}^{N}\frac{1}{n}\sin\left(\frac{\pi n}{N}\right)\Delta n. \tag{9.83}$$

where we have written $\Delta n = 1$. In the limit as $N \to \infty$ which corresponds to the Fourier series, we can rewrite this sum as an integral using

$$x = \frac{\pi n}{N}; \quad \mathrm{d}x = \frac{\pi}{N}\Delta n; \quad x(n=1) = \frac{\pi}{N} \to 0; \quad x(n=N) = \pi \tag{9.84}$$

to give

$$\lim_{N\to\infty} f_N\left(\frac{L}{2} - \frac{L}{2N}\right) = \frac{L}{2}\left[\frac{1}{\pi}\int_0^{\pi}\mathrm{d}x\,\frac{\sin x}{x}\right] \tag{9.85}$$

$$= \frac{L}{2}\left[\frac{1}{\pi}\int_0^{\pi}\left(\frac{1}{1!} - \frac{x^2}{3!} + \frac{x^4}{5!} - \ldots\right)\mathrm{d}x\right] \tag{9.86}$$

$$= \frac{L}{2}\sum_{k=0}^{\infty}\frac{(-1)^k(\pi)^{2k}}{(2k+1)(2k+1)!} \approx 1.17\frac{L}{2}. \tag{9.87}$$

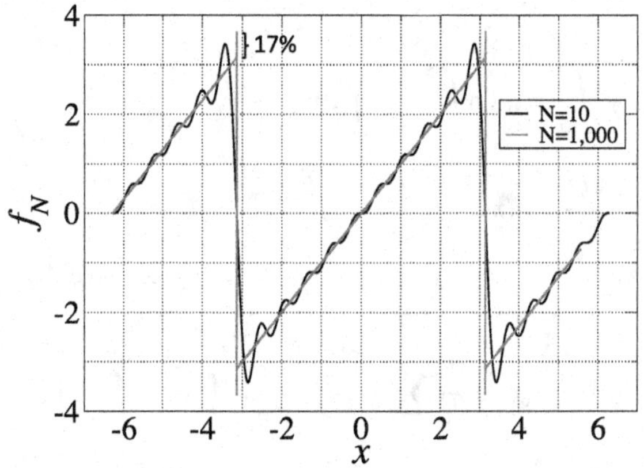

Figure 9.3 The 17% overshoot of the Fourier series $f_F(x) \approx f_N(x)$ at the discontinuity when $N = 10$ and $N = 1{,}000$. Making N larger removes the wiggles within the interval but does not remove the overshoot just shy of the end of the interval at $x = \pm L/2$. We took $L = 2\pi$ in this plot.

To perform the integral we expanded $\sin x$ in a Taylor series and integrated term by term. Summing over the first handful of terms gives $1.17L/2$ which can be confirmed with a hand calculator. So instead of going to $L/2$ in the limit as $N \to \infty$, the Fourier series gives a value for the sawtooth function at $x = L/2 - L/(2N)$ that is 17% too big even when $N \to \infty$.

Further, by an identical manipulation, we have

$$\lim_{N\to\infty} f_N\left(-\frac{L}{2} + \frac{L}{2N}\right) \approx -1.17\frac{L}{2}. \tag{9.88}$$

Thus, at each discontinuity, we have the correct behavior that $f_N(L/2(1 - 1/N)) + f_N(-L/2(1 + 1/N)) = 0$. However, right next to each discontinuity and just inside the primary interval, the Fourier series overshoots the exact value of the sawtooth function by 17%.

In Fig. 9.3, we show the approximation of the sawtooth function obtained using the first 10 terms and first 1,000 terms of the Fourier series where we can observe the overshoot in each approximation. This analysis shows that the 17% overshoot does not go away even as $N \to \infty$ (lots of terms in the series). Adding more terms to the series removes the little wiggles throughout the interval, but there is always a blip next to the discontinuity that is 17% too big. This is an intrinsic property of the Fourier series when applied to functions that have a discontinuity at the end of the fundamental period of the Fourier series. This overshoot at the discontinuity is called *Gibbs phenomenon*. It does not represent a serious problem though you should be aware that it exists if you need to evaluate a Fourier series right next to a discontinuity.

9.5 Using the Fourier Series to Solve BVPs

We now consider specific applications where we solve differential equations with associated boundary and initial conditions (i.e., boundary value problems) using the Fourier series.

9.5.1 Wave Equation with Initial Data

In the chapter on separation of variables, we have already solved the problem

$$c^2\frac{\partial^2\psi}{\partial x^2}-\frac{\partial^2\psi}{\partial t^2}=0 \text{ where } \psi(x=0)=\psi(x=L)=0 \tag{9.89}$$

subject to the initial data $\psi(x, 0)=u_0(x)$ and $\partial\psi/\partial t(x, 0)=v_0(x)$. This problem corresponds to a string of length L clamped at each end and given some initial conditions from which the vibrations of the string begin as controlled both by the wave equation and the boundary conditions.

We can also solve this BVP by using the Fourier series and assuming the solution has the form

$$\psi(x, t)=\sum_{n=1}^{\infty} b_n(t) \sin\left(\frac{n\pi x}{L}\right) \tag{9.90}$$

where we chose the sine basis due to the BC (the coefficients of the cosine basis are zero in order to satisfy the BCs). Note that we are taking the period for the Fourier series to be $2L$ even though the length of the vibrating string is L. This is so that the key lowest order $n=1$ mode starts at zero at $x=0$, monotonically increase to a maximum at $x=L/2$ and then decreases back to zero at $x=L$. If we had chosen the period to be L, then the lowest order mode would be zero at $x=L/2$ and we could not properly represent the physics.

Taking derivatives of this form of the solution gives

$$\frac{\partial^2\psi}{\partial x^2}=-\sum_{n=1}^{\infty}\left(\frac{n\pi}{L}\right)^2 b_n(t) \sin\left(\frac{n\pi x}{L}\right) \tag{9.91}$$

$$\frac{\partial^2\psi}{\partial t^2}=\sum_{n=1}^{\infty}\frac{\mathrm{d}^2 b_n(t)}{\mathrm{d}t^2} \sin\left(\frac{n\pi x}{L}\right) \tag{9.92}$$

Putting these second derivatives into the PDE then gives

$$\sum_{n=1}^{\infty}\left(\frac{\mathrm{d}^2 b_n}{\mathrm{d}t^2}+\left(\frac{n\pi c}{L}\right)^2 b_n\right) \sin\left(\frac{n\pi x}{L}\right)=0 \tag{9.93}$$

which, to be true for each x, requires

$$\frac{\mathrm{d}^2 b_n}{\mathrm{d}t^2}+\left(\frac{n\pi c}{L}\right)^2 b_n=0. \tag{9.94}$$

This ordinary differential equation has the solution

$$b_n(t)=A_n \cos\left(\frac{n\pi ct}{L}\right)+B_n \sin\left(\frac{n\pi ct}{L}\right) \tag{9.95}$$

or

$$\psi(x, t)=\sum_{n=1}^{\infty}\left[A_n \cos\left(\frac{n\pi ct}{L}\right)+B_n \sin\left(\frac{n\pi ct}{L}\right)\right] \sin\left(\frac{n\pi x}{L}\right) \tag{9.96}$$

Now, from the initial data,

$$u_0(x) = \sum_{n=1}^{\infty} A_n \sin\left(\frac{n\pi x}{L}\right) \tag{9.97}$$

$$v_0(x) = \sum_{n=1}^{\infty} B_n \frac{n\pi\, c}{L} \sin\left(\frac{n\pi\, x}{L}\right) \tag{9.98}$$

so that by the usual inversion method we obtain

$$A_n = \frac{2}{L}\int_0^L u_0(x) \sin\left(\frac{n\pi x}{L}\right) \mathrm{d}x \tag{9.99}$$

$$B_n = \frac{2}{n\pi\, c}\int_0^L v_0(x) \sin\left(\frac{n\pi x}{L}\right) \mathrm{d}x \tag{9.100}$$

where because we are integrating from 0 to L (see the first exercise at the end of the chapter) has no constraints placed the n.

So, to give a particular example, we take $u_0 = \text{constant} = C$ and $v_0 = 0$, which gives $B_n = 0$ and $A_n = -2C\left[\cos(n\pi) - 1\right]/(\pi n)$ so that $A_n = 4C/(\pi n)$ for $n = \text{odd}$ and $A_n = 0$ for n even. For this example, note as well that if we had integrated over a full period from 0 to $2L$ we would have obtained $A_n = 0$ for all n which is not a solution of the problem with the given initial conditions. We need to obtain the constants A_n by integrating over a half period. So for the special case of $u_0 = C$ and $v_0 = 0$ we have the solution of the BVP

$$\psi(x,t) = \frac{4C}{\pi}\sum_{\substack{n\\ \text{odd}}} \frac{1}{n} \cos\left(\frac{n\pi ct}{L}\right) \sin\left(\frac{n\pi x}{L}\right) \tag{9.101}$$

$$= \frac{2C}{\pi}\sum_{\substack{n\\ \text{odd}}} \frac{1}{n}\left[\sin\left(\frac{n\pi}{L}(x-ct)\right) + \sin\left(\frac{n\pi}{L}(x+ct)\right)\right].$$

which shows that the response is just a bunch of waves of different wavelengths (as set by n) moving to the right and left on the string clamped at each end (or corresponding to plane waves moving to the right and left within an infinite elastic slab of width L sandwiched between two rigid half-spaces).

9.5.2 Wave Equation with Time-Dependent BC

We know separation of variables does not work with time-dependent BCs. But a Fourier series approach has no problem. Consider

$$c^2\frac{\partial^2\psi(x,t)}{\partial x^2} - \frac{\partial^2\psi(x,t)}{\mathrm{d}t^2} = 0 \tag{9.102}$$

subject to $\psi(x=0) = 0$ and $\psi(x=L) = f(t)$ which corresponds to a string held fixed at $x = 0$ and shook at $x = L$. Assume $f(t)$ is periodic with period T and has been "turned on" for a very long time so that initial conditions are not pertinent.

To begin, represent $f(t)$ as a Fourier series

$$f(t) = \sum_{n=-\infty}^{+\infty} c_n\, e^{i2\pi nt/T} \tag{9.103}$$

$$c_n = \frac{1}{T}\int_{-T/2}^{+T/2} f(t)\, e^{-i2\pi nt/T}. \tag{9.104}$$

Given this time dependence, we represent the wave field with a similar time dependence

$$\psi(x, t) = \sum_{n=-\infty}^{+\infty} d_n(x)\, e^{i2\pi nt/T} \tag{9.105}$$

where we now must find the $d_n(x)$ that satisfy both the PDE and the BCs. Putting this representation of the wavefield into the wave equation, we have for each n

$$\frac{d^2 d_n(x)}{dx^2} + \frac{1}{c^2}\left(\frac{2\pi n}{T}\right)^2 d_n(x) = 0 \tag{9.106}$$

with $d_n(0) = 0$ and $d_n(L) = c_n$. Solutions are of the form (we can ignore the $\cos k_n x$ basis due to $d_n(0) = 0$)

$$d_n(x) = E_n \sin\left(\frac{2\pi n}{cT}x\right) \tag{9.107}$$

and we get the E_n from $d_n(L) = c_n$, or

$$E_n = \frac{c_n}{\sin(2\pi nL/(cT))} \tag{9.108}$$

Thus we find

$$\psi(x, t) = \sum_{n=-\infty}^{+\infty} \frac{c_n}{\sin(2\pi n L/(cT))} \sin\left(\frac{2\pi n x}{cT}\right) e^{\frac{i2\pi n}{T}t} \tag{9.109}$$

or

$$\psi(x, t) = \sum_{n=-\infty}^{+\infty} \frac{\sin(2\pi n x/(cT))}{\sin(2\pi n L/(cT))} \frac{1}{T}\int_{-T/2}^{+T/2} f(t')\, e^{-i\frac{2\pi n}{T}(t'-t)} \tag{9.110}$$

Note that if $f(t) = \cos 2\pi t/T = \cos\omega t$, then $c_{+1} = 1/2$, $c_{-1} = 1/2$ and all other $c_n = 0$. Thus upon using these two c_n in Eq. (9.109) we obtain

$$\boxed{\psi(x, t) = \frac{\sin(\omega x/c)}{\sin(\omega L/c)}\cos(\omega t) = \frac{(\sin[\omega(t - x/c)] + \sin[\omega(t + x/c)])}{2\sin(\omega L/c)}} \tag{9.111}$$

when $f(x) = \cos \omega t$ and $\omega \widehat{=} 2\pi/T$. Again, the method of separation of variables cannot work on this problem due to the time dependence of the boundary condition but the Fourier series approach used here has no problems.

9.5.3 Diffusion Equation with Time-Dependent BC

Consider the following diffusion boundary value problem

$$D\frac{\partial^2\psi(x,t)}{\partial x^2} - \frac{\partial\psi(x,t)}{\partial t} = 0 \tag{9.112}$$

subject to $\psi(x=0,t)=0$ and $\psi(x=L,t)=f(t)$. These are the same boundary conditions as in the previous wave problem involving a time-dependent inhomogeneous condition at $x=L$. But this time the diffusion equation is controlling the physics. We know that separation of variables will not work but a Fourier-series treatment as given in the previous problem will work just fine. Physically, this diffusion BVP with a time-dependent boundary condition could correspond to heat diffusing in a slab of width L with ψ corresponding to temperature and the temperature held fixed at a constant value at $x=0$ but with say a time-harmonic (e.g., the diurnal) temperature variation on the other side of the slab at $x=L$.

We again represent the periodic boundary condition $f(t)$ as a Fourier series

$$f(t) = \sum_{n=-\infty}^{+\infty} c_n\, e^{i2\pi nt/T} \quad \text{with} \quad c_n = \frac{1}{T}\int_{-T/2}^{+T/2} f(t)\, e^{-i2\pi nt/T}. \tag{9.113}$$

Given this time dependence, the temperature field has a similar time dependence

$$\psi(x,t) = \sum_{n=-\infty}^{+\infty} d_n(x)\, e^{i2\pi nt/T} \tag{9.114}$$

where we must find the $d_n(x)$ that satisfy both the PDE and the BCs. Putting this representation into the diffusion equation gives for each n

$$\frac{\mathrm{d}^2 d_n(x)}{\mathrm{d}x^2} + \frac{i2\pi n}{DT} d_n(x) = 0 \tag{9.115}$$

with $d_n(0)=0$ and $d_n(L)=c_n$. Solutions are of the form (we can ignore the $\cos k_n x$ basis due to $d_n(0)=0$)

$$d_n(x) = E_n \sin\left(\sqrt{\frac{i2\pi n}{DT}}x\right) \tag{9.116}$$

and we get the E_n from $d_n(L)=c_n$ or

$$E_n = \frac{c_n}{\sin\left(\sqrt{i2\pi n/(DT)}L\right)}. \tag{9.117}$$

Thus we find

$$\psi(x,t) = \sum_{n=-\infty}^{+\infty} c_n \frac{\sin\left(\sqrt{i2\pi n/(DT)}x\right)}{\sin\left(\sqrt{i2\pi n/(DT)}L\right)} \exp\left(\frac{i2\pi n}{T}\right). \tag{9.118}$$

The right-hand side works out to give a purely real result. This is particularly easy to see when the time function $f(t)$ in the boundary condition is even so that $c_n = c_{-n}$ for $n = 1, 2, 3 \ldots \infty$. Putting $n = -n$ in Eq. (9.118) is equivalent to taking the complex conjugate and a complex function (the sum $\sum_{n=1}^{\infty}$) when added to its complex conjugate (the sum $\sum_{n=-1}^{-\infty}$) is real.

Let's consider the simple time-harmonic boundary condition $\psi(x = L, t) = f(t) = a\cos\omega t$ so that we have $c_1 = a/2$, $c_{-1} = a/2$, $2\pi/T = \omega$ and all other $c_n = 0$. Recall that $\sqrt{i} = (1+i)/\sqrt{2}$. If we define the diffusive skin depth δ as $1/\delta = \sqrt{\omega/(2D)}$ and use the identity $\sin(a + ib) = \sin(a)\cosh(b) + i\cos(a)\sinh(b)$ when a and b are both real, going through the algebra yields the solution

$$\begin{aligned}\psi(x,t) &= a\cos\omega t\\ &\left(\frac{\sin(x/\delta)\sin(L/\delta)\cosh(x/\delta)\cosh(L/\delta) + \cos(x/\delta)\cos(L/\delta)\sinh(x/\delta)\sinh(L/\delta)}{\sin^2(L/\delta) + \sinh^2(L/\delta)}\right)\\ &- a\sin\omega t\\ &\left(\frac{\cos(x/\delta)\sin(L/\delta)\sinh(x/\delta)\cosh(L/\delta) - \sin(x/\delta)\cos(L/\delta)\cosh(x/\delta)\sinh(L/\delta)}{\sin^2(L/\delta) + \sinh^2(L/\delta)}\right),\end{aligned} \tag{9.119}$$

which is indeed purely real. This answer is seen to give $\psi(0, t) = 0$ and $\psi(L, t) = a\cos\omega t$ as required and is the exact solution of the given diffusion BVP with a time-harmonic boundary condition.

9.5.4 Uniform Load Applied to a Beam

Consider a uniform load q_o (force in the y direction per unit length in the x direction) applied to a beam of length L that is simply supported at its two end points as depicted in Fig. 9.4. The differential equation for small vertical deflections $y(x)$ is [we do not give the derivation of this equation, cf., Landau and Lifshitz, 1986, Eq. (20.4)]

$$\frac{d^4 y(x)}{dx^4} = \alpha\, q(x) \tag{9.120}$$

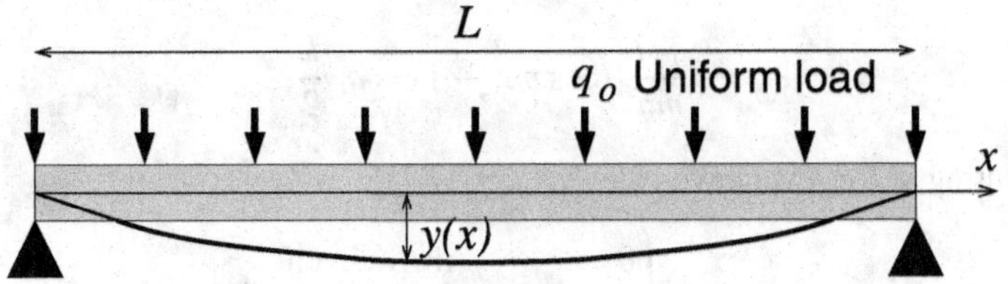

Figure 9.4 A beam is supported at its two ends and is uniformly loaded along its length.

where $q(x) = q_o$ is uniform along the beam length in this example and where α is the beam's elastic compliance (a given constant specific to the beam). The boundary conditions are that

$$y(0) = 0 \text{ and } y(L) = 0, \tag{9.121}$$

$$\left.\frac{d^2 y}{dx^2}\right|_{x=0} = 0 \text{ and } \left.\frac{d^2 y}{dx^2}\right|_{x=L} = 0. \tag{9.122}$$

We need four boundary conditions because the differential equation is fourth order and the second set of conditions is that the slope of the beam is uniform (the second derivatives are zero) as we approach the two support points at $x = 0$ and $x = L$ which is likely an approximation. The goal of this BVP is to find the deflection amplitude $y(x)$.

Due to the BCs, it is convenient to expand $y(x)$ in a Fourier sine series (the Fourier cosine series must have zero coefficients $a_n = 0$ because the cosine is nonzero at the two end points)

$$y(x) = \sum_{n=1}^{\infty} b_n \sin\frac{n\pi x}{L}. \tag{9.123}$$

This is a Fourier series representation of the response over a period $2L$. We again choose the fundamental interval to be $2L$ for the Fourier representation so that the lowest-order mode $n = 1$ starts at zero when $x = 0$, monotonically increases to a maximum at $x = L/2$ (the center of the beam) before returning to zero at $x = L$. If we had chosen an interval of L, this lowest order mode would be zero at the center of the beam and that is not physically correct. Upon taking four derivatives we obtain

$$\frac{d^4 y(x)}{dx^2} = \sum_{n=1}^{\infty} b_n \left(\frac{n\pi}{L}\right)^4 \sin\frac{n\pi x}{L}. \tag{9.124}$$

Also, expand $q(x) = q_0$ into a Fourier sine series (the coefficients of the cosine series are easily shown to be exactly zero)

$$q_0 = \sum_{n=1}^{\infty} q_n \sin\frac{n\pi x}{L}. \tag{9.125}$$

Find the q_n using orthogonality (sine inversion)

$$\int_0^L q_0 \sin\frac{m\pi x}{L}\,dx = q_m \int_0^L \sin^2\frac{m\pi x}{L}\,dx \tag{9.126}$$

or

$$-\frac{q_0 L}{m\pi}(\cos m\pi - 1) = q_m \frac{L}{2}. \tag{9.127}$$

Upon substituting m for n we have

$$q_n = \begin{cases} 0 & \text{if } n = \text{even} \\ 4q_0/(n\pi) & \text{if } n = \text{odd}. \end{cases} \tag{9.128}$$

The differential equation can thus be written

$$\sum_{n=1}^{\infty} b_n \left(\frac{n\pi}{L}\right)^4 \sin\frac{n\pi\, x}{L} = \alpha \sum_{\substack{n=1\\ \text{odd}}}^{\infty} \frac{4\, q_0}{n\pi} \sin\frac{n\pi\, x}{L}. \tag{9.129}$$

Equating coefficients for each n then gives

$$b_n = \begin{cases} 4\, L^4\, \alpha\, q_0/(n\pi)^5 & \text{if } n = \text{odd} \\ 0 & \text{if } n = \text{even} \end{cases} \tag{9.130}$$

which then yields the solution

$$\boxed{y(x) = \frac{4\,\alpha\, q_o\, L^4}{\pi^5} \sum_{\substack{n=1\\ \text{odd}}}^{\infty} \frac{1}{n^5} \sin\frac{n\pi\, x}{L}.} \tag{9.131}$$

Due to the n^5 in the denominator, the series rapidly converges and only the first few terms gives an excellent approximation to the solution.

As an aside, we can also solve this problem by assuming the solution is a polynomial of fourth order

$$y(x) = a_4 x^4 + a_3 x^3 + a_2 x^2 + a_1 x + a_0. \tag{9.132}$$

From the ODE, we have

$$a_4\, 4! = \alpha\, q_0 \quad \text{or} \quad a_4 = \frac{\alpha\, q_0}{24}. \tag{9.133}$$

From the BCs, we have

$$y(0) = 0 = a_0 \tag{9.134}$$

$$y(L) = 0 = \frac{\alpha\, q_0}{24} L^4 + a_3 L^3 + a_2 L^2 + a_1 L \tag{9.135}$$

$$\left.\frac{d^2y}{dx^2}\right|_{x=L} = 0 = \frac{\alpha\, a_0}{2} L^2 + 6a_3\, L + 2a_2 \tag{9.136}$$

$$\left.\frac{d^2y}{dx^2}\right|_{x=0} = 0 = 2a_2. \tag{9.137}$$

Thus, $a_0 = 0$, $a_2 = 0$, $a_3 = -\alpha\, q_0\, L/12$, $a_1 = \alpha\, q_0\, L^3/24$, and

$$\boxed{y(x) = \frac{\alpha\, q_0}{24}\left[x^4 - 2L\, x^3 + L^3\, x\right].} \tag{9.138}$$

To see whether this function is the same as that obtained using the Fourier series, let's represent this function as a Fourier series

$$y(x) = \sum_{n=1}^{\infty} b_n \sin \frac{n\pi x}{L}, \tag{9.139}$$

where the coefficients are given by

$$b_n = \frac{2}{L} \int_0^L \frac{\alpha q_0}{24} \left(x^4 - 2L x^3 + L^3 x\right) \sin \frac{n\pi x}{L} \, dx. \tag{9.140}$$

The various integrals are evaluated as follows:

$$\begin{aligned} \int_0^L x^4 \sin \frac{n\pi x}{L} &= -\frac{x^4}{(n\pi/L)} \cos \frac{n\pi x}{L}\bigg|_0^L + \frac{4}{(n\pi/L)} \left[\frac{\left(3(n\pi/L)^2 x^2 - 6\right)}{(n\pi/L)^4} \cos \frac{n\pi x}{L} \right]\Bigg|_0^L \\ &= -\frac{L^5}{n\pi} \cos n\pi + \frac{4 L^5}{(n\pi)^5} \left[\left(3(n\pi)^2 - 6\right) \cos n\pi + 6 \right] \\ &= \frac{24 L^5}{(n\pi)^5} - \frac{L^5 \cos n\pi}{n\pi} \left[1 - \frac{12}{(n\pi)^2} + \frac{24}{(n\pi)^4} \right]; \end{aligned} \tag{9.141}$$

$$\begin{aligned} \int_0^L x^3 \sin \frac{n\pi x}{L} \, dx &= -\frac{L^3}{(n\pi)^3} \left[\left(\frac{n\pi}{L}\right)^2 x^3 - 6x \right] \cos \frac{n\pi x}{L}\bigg|_0^L \\ &= -\frac{L^4}{(n\pi)^3} \left[(n\pi)^2 - 6 \right] \cos n\pi = -L^4 \left(\frac{1}{n\pi} - \frac{6}{(n\pi)^3} \right) \cos n\pi; \end{aligned} \tag{9.142}$$

and

$$\int_0^L x \sin \frac{n\pi x}{L} \, dx = -\frac{L^2}{n\pi} \cos n\pi. \tag{9.143}$$

Given these integrals, the coefficients b_n are then

$$\begin{aligned} b_n &= \frac{\alpha q_0}{12L} \left\{ \frac{24 L^5}{(n\pi)^5} - \frac{\cos n\pi \, L^5}{n\pi} \left[1 - \frac{12}{(n\pi)^2} + \frac{24}{(n\pi)^4} - 2 + \frac{12}{(n\pi)^2} + 1 \right] \right\} \\ &= \frac{\alpha q_0}{12L} \frac{24 L^5}{(n\pi)^5} (1 - \cos n\pi) \end{aligned} \tag{9.144}$$

or

$$b_n = \begin{cases} 4 L^4 \alpha q_0/(n\pi)^5 & \text{for } n = \text{odd} \\ 0 & \text{for } n = \text{even}. \end{cases} \tag{9.145}$$

Thus, we find that over $0 \leq x \leq L$

$$y(x) = \frac{\alpha\, q_0}{24}\left[x^4 - 2Lx^3 + L^3 x\right] = \frac{4\, q_0\, \alpha\, L^4}{\pi^5} \sum_{\substack{n=1 \\ \text{odd}}}^{\infty} \frac{1}{n^5} \sin \frac{n\pi\, x}{L}, \tag{9.146}$$

which is the result we found earlier.

In the case where $q(x) = q_0$, it was easier to have found the answer as a polynomial. When $q(x)$ is a more complicated function of x, the Fourier series approach can prove to be more effective, because the solution will no longer be a simple fourth order polynomial. In general, the solution for arbitrary $q(x)$ is of the form:

$$y(x) = \alpha \int_x \mathrm{d}x_3 \int_{x_3} \mathrm{d}x_2 \int_{x_2} \mathrm{d}x_1 \int_{x_1} \mathrm{d}x_0\, q(x_0) + a_3 x^3 + a_2 x^2 + a_1 x + a_0, \tag{9.147}$$

where the four constants are found from the four BCs. Whether it is faster to find the Fourier coefficients or to perform four embedded indefinite integrations of $q(x)$ will depend on the function $q(x)$.

9.5.5 2D Potential Problems

We can also use the Fourier-series "eigenfunction" approach to treat BVPs for potentials in 2D and 3D that are solutions of Laplace's equation. For a specific 2D potential BVP posed in a rectangular domain, consider

$$\frac{\partial^2 \varphi(x, y)}{\partial x^2} + \frac{\partial^2 \varphi(x, y)}{\partial y^2} = 0 \tag{9.148}$$

subject to the particular boundary conditions that

$$\left.\frac{\partial \varphi(x, y)}{\partial y}\right|_{y=0} = 0, \tag{9.149}$$

$$\varphi(x, b) = \varphi_0 \text{ (a given constant)}, \tag{9.150}$$

$$\varphi(0, y) = 0, \tag{9.151}$$

$$\varphi(a, y) = 0. \tag{9.152}$$

So the surface $y = 0$ of the rectangle is insulating.

To proceed with a Fourier-representation of the potential $\varphi(x, y)$, we use the same approach that we used for 1D wave and diffusion problems and write

$$\varphi(x, y) = \sum_{n=1}^{\infty} b_n(y) \sin\left(\frac{\pi n x}{a}\right) \tag{9.153}$$

that is a Fourier-series representation of the x dependence when the period of the x interval is $2a$ so that $n = 1$ is the proper lowest-order mode that starts at zero at $x = 0$, monotonically increases to a maximum at $x = a/2$ and then decrease down to zero at $x = a$. This representation satisfies the boundary conditions on $x = 0$ and $x = a$.

To determine the y dependence of the coefficients $b_n(y)$, we take two derivatives of this representation in both x and y and insert into the Laplace equation to obtain

$$\sum_{n=1}^{\infty}\left[\frac{\mathrm{d}^2 b_n(y)}{\mathrm{d}y^2}-\left(\frac{\pi n}{a}\right)^2 b_n(y)\right]\sin\left(\frac{\pi n x}{a}\right)=0. \tag{9.154}$$

The solutions for $b_n(y)$ that satisfy the Laplace equation for all x are then

$$b_n(y)=A_n\cosh\left(\frac{\pi n y}{a}\right)+B_n\sinh\left(\frac{\pi n y}{a}\right). \tag{9.155}$$

For our particular BVP with the insulating condition at $y=0$, we must set $B_n=0$ because after taking the y derivative, $\cosh(\pi n y/a)$ satisfies the insulating boundary condition at $y=0$.

We thus have that

$$\varphi(x, y)=\sum_{n=1}^{\infty}A_n\cosh\left(\frac{\pi n y}{a}\right)\sin\left(\frac{\pi n x}{a}\right) \tag{9.156}$$

and determine the A_n from the as yet unused condition on $y=b$ that $\varphi(x, b)=\varphi_0$. Multiplying both sides of this condition on $y=b$ by $\sin(\pi m x/a)$ and integrating from 0 to a then gives

$$\frac{a}{2}A_n\cosh\left(\frac{\pi n b}{a}\right)=\varphi_0\int_0^a\sin\left(\frac{\pi n x}{a}\right)\,\mathrm{d}x \tag{9.157}$$

$$=-\varphi_0\frac{a}{\pi n}\left[\cos(n\pi)-1\right] \tag{9.158}$$

or

$$A_n=\frac{4\varphi_0(-1)^n}{\pi n\cosh(\pi n b/a)}\quad\text{for}\quad n=1, 3, 5\ldots\infty\ (n\ \text{odd}). \tag{9.159}$$

We thus obtain the solution of the particular potential BVP as

$$\boxed{\varphi(x, y)=\frac{4\varphi_0}{\pi}\sum_{\substack{n=1\\ \text{odd}}}^{\infty}\frac{(-1)^n}{n}\frac{\cosh(\pi n y/a)}{\cosh(\pi n b/a)}\sin\left(\frac{\pi n x}{a}\right).} \tag{9.160}$$

As always, to express the fact that the sum is over odd n, we can also replace n by $2n-1$ throughout the right-hand side and sum from $n=1, 2, 3\ldots\infty$.

9.5.6 The Vibrating Rectangular Drumhead

We can also treat 2D and 3D wave and diffusion problems in finite-sized domains using the Fourier series.

As a specific example, we consider again the vibrating rectangular drumhead treated earlier using the method of separation of variables. The BVP for the displacements $u(x, y, t)$ of the stretched canvas of the drumhead is

$$\frac{\partial^2 u}{\partial x^2} + \frac{\partial^2 u}{\partial y^2} - \frac{1}{c^2}\frac{\partial^2 u}{\partial t^2} = 0 \tag{9.161}$$

along with the boundary condition that the edges of the drum are clamped

$$u = 0 \quad \text{on } x = 0, x = a, y = 0, y = b \tag{9.162}$$

and with the initial conditions

$$u(x, y, t = 0) = u_0(x, y) \;\widehat{=}\; \text{some given function} \tag{9.163}$$

$$\left.\frac{\partial u}{\partial t}(x, y, t)\right|_{t=0} = v_0(x, y) \;\widehat{=}\; \text{another given function.} \tag{9.164}$$

To solve this BVP using a Fourier-series "eigenfunction" approach, we represent the displacements using a double Fourier series for the spatial dependence

$$u(x, y, t) = \sum_{n=1}^{\infty}\sum_{m=1}^{\infty} b_{nm}(t) \sin\left(\frac{\pi n x}{a}\right) \sin\left(\frac{\pi m y}{b}\right), \tag{9.165}$$

which satisfies all the boundary conditions. Putting this representation into the wave equation gives

$$\sum_{n=1}^{\infty}\sum_{m=1}^{\infty} \left\{\left[\left(\frac{\pi n}{a}\right)^2 + \left(\frac{\pi m}{b}\right)^2\right] b_{nm}(t) + \frac{1}{c^2}\frac{\mathrm{d}^2 b_{nm}(t)}{\mathrm{d}t^2}\right\} \sin\left(\frac{\pi n x}{a}\right) \sin\left(\frac{\pi m y}{b}\right) = 0. \tag{9.166}$$

If we define the normal-mode circular frequencies

$$\omega_{nm} = \pi c \sqrt{\left(\frac{n}{a}\right)^2 + \left(\frac{m}{b}\right)^2}, \tag{9.167}$$

the $b_{nm}(t)$ that satisfy the wave equation for all x and y are

$$b_{nm}(t) = A_{nm}\cos(\omega_{nm} t) + B_{nm}\sin(\omega_{nm} t). \tag{9.168}$$

The constants A_{nm} and B_{nm} are then found from the initial conditions on the drumhead displacements $u_0(x, y)$ and velocities $v_0(x, y)$ as

$$A_{nm} = \frac{4}{ab}\int_0^a \mathrm{d}x \int_0^b \mathrm{d}y\; u_0(x, y) \sin\left(\frac{n\pi x}{a}\right) \sin\left(\frac{m\pi y}{b}\right). \tag{9.169}$$

$$B_{nm} = \frac{4}{ab\,\omega_{nm}}\int_0^a \mathrm{d}x \int_0^b \mathrm{d}y\; v_0(x, y) \sin\left(\frac{n\pi x}{a}\right) \sin\left(\frac{m\pi y}{b}\right). \tag{9.170}$$

So for example, if the initial condition for velocities is that $v_0 = 0$ so that the $B_{nm} = 0$, we obtain the solution as

$$u(x, y, t) = \sum_{n=1}^{\infty} \sum_{m=1}^{\infty} A_{nm} \cos(\omega_{nm} t) \sin\left(\frac{\pi n x}{a}\right) \sin\left(\frac{\pi m y}{b}\right), \tag{9.171}$$

where the A_{nm} are determined by the integral of Eq. (9.169) for some given initial stretching of the canvas drumhead $u_0(x, y)$.

9.6 Conclusions

These various examples show that any problem that we treated using the separation-of-variables method can also be handled with a Fourier-series or eigenfunction method. Further, the Fourier-series approach can handle time-dependent boundary conditions and time-dependent source terms in the PDE while the separation-of-variables method cannot. The way that we are implementing the Fourier-series method assumes a representation in separated form so it is not surprising that it is equivalent to, though more general than, the separation-of-variables method. If we had been working in cylindrical or spherical coordinates instead of Cartesian coordinates, our basis functions would not be just the sines and cosines of the Fourier-series approach. But such basis functions for use in cylindrical and spherical coordinates, such as Bessel functions, spherical Bessel functions, and spherical harmonics, are orthogonal eigenfunctions corresponding to a particular Sturm–Liouville BVP and the logic of how to use them to solve specific wave, diffusion, and potential BVPs posed in cylindrical and spherical coordinates is the same as our logic in using the Fourier-series eigenfunction approach as demonstrated through the above examples posed in Cartesian coordinates.

9.7 Exercises

1. Prove the following results that arise when working with the Fourier series:

$$\int_a^{a+L} \cos\left(\frac{2\pi n x}{L}\right) \cos\left(\frac{2\pi m x}{L}\right) \mathrm{d}x = \begin{cases} 0 & \text{if } n \neq m \\ L/2 & \text{if } n = m \text{ and } n, m \geq 1 \\ L & \text{if } n = m = 0 \end{cases} \tag{9.172}$$

$$\int_a^{a+L} \sin\left(\frac{2\pi n x}{L}\right) \sin\left(\frac{2\pi m x}{L}\right) \mathrm{d}x = \begin{cases} 0 & \text{if } n \neq m \text{ and } n, m \geq 1 \\ L/2 & \text{if } n = m \text{ and } n, m \geq 1 \\ 0 & \text{if } n = m = 0 \end{cases} \tag{9.173}$$

$$\int_a^{a+L} \sin\left(\frac{2\pi n x}{L}\right) \cos\left(\frac{2\pi m x}{L}\right) = 0 \quad \text{for any } n, m. \tag{9.174}$$

Show that the following are also true when the period is taken to be $2L$ instead of L as above

$$\int_a^{a+L} \cos\left(\frac{\pi n x}{L}\right)\cos\left(\frac{\pi m x}{L}\right)\,dx = \begin{cases} 0 & \text{if } a=0 \text{ and } n\neq m \\ 0 & \text{if } a\neq 0 \text{ and } n\neq m \text{ and } n,\, m \text{ even or } n,\, m \text{ odd} \\ L/2 & \text{if } n=m \text{ and } n,\, m\geq 1 \\ L & \text{if } n=m=0 \end{cases} \tag{9.175}$$

$$\int_a^{a+L} \sin\left(\frac{\pi n x}{L}\right)\sin\left(\frac{\pi m x}{L}\right)\,dx = \begin{cases} 0 & \text{if } a=0 \text{ and } n\neq m \\ 0 & \text{if } a\neq 0 \text{ and } n\neq m \text{ and } n,\, m \text{ even or } n,\, m \text{ odd} \\ L/2 & \text{if } n=m \text{ and } n,\, m\geq 1 \\ 0 & \text{if } n=m=0 \end{cases} \tag{9.176}$$

$$\int_a^{a+L} \sin\left(\frac{\pi n x}{L}\right)\cos\left(\frac{\pi m x}{L}\right) = \begin{cases} 0 & \text{if } a=-L/2 \text{ and } n\neq m \\ 0 & \text{all other } a,\ n\neq m \text{ and } n,\, m \text{ even or } n,\, m \text{ odd} \\ 0 & \text{if } n=m. \end{cases} \tag{9.177}$$

Be sure to verify Eqs (9.175) and (9.176) for the special case of $a=-L/2$.

2. *Diffusion equation with initial data*: Define a 1D diffusion boundary value problem in an infinite slab of width L

$$D\frac{\partial^2\psi(x,t)}{\partial x^2} - \frac{\partial\psi(x,t)}{\partial t} = 0 \quad \text{in the interval} \quad 0\leq x\leq L \tag{9.178}$$

with boundary conditions that

$$\psi(x=0,t)=0 \tag{9.179}$$

$$\left.\frac{\partial\psi(x,t)}{\partial x}\right|_{x=L} = 0, \tag{9.180}$$

and the initial condition that $\psi(x,t=0)=\psi_0$ (a spatial constant). These boundary conditions mean that our slab of width L is in contact at $x=0$ with a "well-stirred" reservoir maintained at $\psi=0$ and is insulated at $x=L$. Show that a Fourier-series approach analogous to that in Section 9.5.1 gives

$$\boxed{\psi(x,t) = \frac{4\psi_0}{\pi}\sum_{\substack{n=1 \\ \text{odd}}}^{\infty}\frac{1}{n}\exp\left[-\left(\frac{\pi n}{2L}\right)^2 Dt\right]\sin\left(\frac{\pi n x}{2L}\right)} \tag{9.181}$$

within the slab $0\leq x\leq L$. HINT: To account for the insulating boundary condition at $x=L$, you should work with a period of $4L$ in a Fourier-series representation

$\psi(x, t) = \sum_{n=1}^{\infty} b_n(t) \sin(\pi n x/(2L))$ and that when you use the orthogonality property for these basis functions $\sin(\pi n x/(2L))$ to account for the initial condition and find the $b_n(0)$, you should integrate from 0 to $2L$ and not 0 to L in order for orthogonality to emerge.

3. *The doubly plucked string vibration problem*: Consider a string of length a pulled tight and held clamped at its two end $x = 0$ and $x = a$ (think of a guitar string). We pull up on the string equally at two points $x = x_1$ and $x = x_2$ and hold the string level at a constant height b. Then, at time $t = 0$, we let go of the string and the string begins to vibrate (it has waves move up and down its length). Find the displacement amplitude $w(x, t)$ that satisfies the scalar wave equation

$$\frac{\partial^2 w(x, t)}{\partial x^2} - \frac{1}{c^2}\frac{\partial^2 w(x, t)}{\partial t^2} = 0, \tag{9.182}$$

where c is the given wave speed of the string and that is subject to the BCs of being clamped at each end

$$w(x = 0, t) = w(x = a, t) = 0 \tag{9.183}$$

and the initial conditions that

$$w(x, t = 0) = \begin{cases} b\,x/x_1 & \text{over } 0 \le x \le x_1 \\ b & \text{over } x_1 \le x \le x_2 \\ b(a - x)/(a - x_2) & \text{over } x_2 \le x \le a \end{cases} \tag{9.184}$$

and

$$\left.\frac{\partial w}{\partial t}\right|_{t=0} = 0. \tag{9.185}$$

To visualize the initial condition, see Fig. 9.5. So this is a wave BVP as presented in Section 9.5.1 excited by the initial condition. Using the appropriate Fourier-series representation for $w(x, t)$, show that the answer for the plucked-string problem is

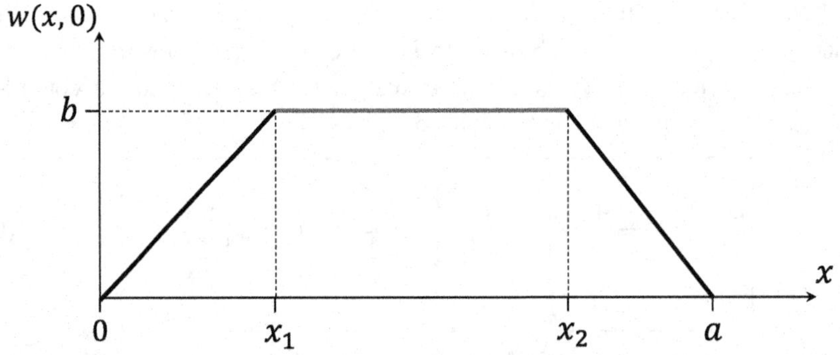

Figure 9.5 The initial steady position of the string held up at two points x_1 and x_2 prior to the release at $t = 0$.

$$w(x, t) = \frac{2ba}{\pi^2} \sum_{n=1}^{\infty} \left(\frac{\sin(\pi n x_1/a)}{x_1 n^2} + \frac{\sin(\pi n x_2/a)}{(a - x_2)n^2} \right) \sin\left(\frac{n\pi x}{a}\right) \cos\left(\frac{n\pi ct}{a}\right). \tag{9.186}$$

Then use trigonometric identities to show that this has the "standard" form of waves moving up and down the string

$$w(x, t) = \frac{ba}{\pi^2} \sum_{n=1}^{\infty} \left(\frac{\sin(\pi n x_1/a)}{x_1 n^2} + \frac{\sin(\pi n x_2/a)}{(a - x_2)n^2} \right) \times \left[\sin\left(\frac{n\pi(x - ct)}{a}\right) + \sin\left(\frac{n\pi(x + ct)}{a}\right) \right]. \tag{9.187}$$

4. *Forced string vibration problem*: Solve the 1D wave BVP with a spatially uniform forcing term $s(t)$ in the PDE that varies in time

$$c^2 \frac{\partial^2 u(x, t)}{\partial x^2} - \frac{\partial^2 u(x, t)}{\partial t^2} = s(t) \tag{9.188}$$

subject to boundary conditions that $u = 0$ at $x = 0, L$ (e.g., a clamped string of length L). The time function $s(t)$ is assumed to be an odd function that repeats with a period T (e.g., the sawtooth function treated in Section 9.4.3) that can be represented as the Fourier series

$$s(t) = \sum_{n=1}^{\infty} b_n \sin\left(\frac{2\pi nt}{T}\right). \tag{9.189}$$

You can assume you have calculated the b_n and they are known and a given. By representing the wavefield $u(x, t)$ in the form

$$u(x, t) = \sum_{n=1}^{\infty} f_n(x) \sin\left(\frac{2\pi nt}{T}\right) \tag{9.190}$$

find the $f_n(x)$ that not only satisfy the wave equation but also satisfy the two boundary conditions at $x = 0$ and $x = L$ and show that

$$u(x, t) = \left(\frac{T}{2\pi}\right)^2 \sum_{n=1}^{\infty} \frac{b_n}{n^2} \left\{ 1 - \cos\left(\frac{2\pi nx}{cT}\right) + \left[\frac{\cos[2\pi nL/(cT)] - 1}{\sin[2\pi nL/(cT)]} \right] \sin\left(\frac{2\pi nx}{cT}\right) \right\} \sin\left(\frac{2\pi nt}{T}\right). \tag{9.191}$$

HINT: you will find that $f_n(x)$ satisfies an inhomogeneous ordinary differential equation (ODE) in x. The solution of this can be written $f_n(x) = f_n^P(x) + f_n^H(x)$, where f_n^P is some particular solution of the inhomogeneous ODE that does not satisfy the boundary conditions while f_n^H is a solution of the homogeneous ODE that satisfies the boundary conditions that $f_n^H(0) = f_n(0) - f_n^P(0)$ and $f_n^H(L) = f_n(L) - f_n^P(L)$.

5. *Rectification:* When a flux is driven by a time-harmonic force, if the sign and amplitude of the flux does not have the sign of the force and is not simply proportional to the force throughout each temporal cycle, then we say *rectification* (a type of nonlinearity) is happening. So let's imagine that we have an alternating electric current $i(t) = A \sin \omega t$ that is passed through two types of rectifiers (there are many types of rectifiers).

   A half-wave rectifier only transmits current when it is flowing in the positive direction so that the output is

$$o(t) = A \sin \omega t \times \begin{cases} 1 & \text{when } 0 \leq t < \pi/\omega \\ 0 & \text{when } \pi/\omega \leq t \leq 2\pi/\omega. \end{cases} \tag{9.192}$$

Show that the Fourier series representation of this output over the interval $[0, 2\pi/\omega]$ is

$$o(t) = \frac{A}{\pi} + \frac{A}{2} \sin \omega t - \frac{2A}{\pi} \sum_{n=1 \text{ (odd)}}^{\infty} \frac{\cos(n+1)\omega t}{n(n+2)}. \tag{9.193}$$

   A full-wave rectifier transmits the instantaneous absolute value of the current which can be written

$$o(t) = A \sin \omega t \times \begin{cases} 1 & \text{when } 0 \leq t, \pi/\omega \\ -1 & \text{when } \pi/\omega \leq t \leq 2\pi/\omega. \end{cases} \tag{9.194}$$

Show that the Fourier series representation of this output over the interval $[0, 2\pi/\omega]$ is

$$o(t) = \frac{2A}{\pi} - \frac{4A}{\pi} \sum_{n=2 \text{ (even)}}^{\infty} \frac{\cos n\omega t}{n^2 - 1}. \tag{9.195}$$

6. Use the Fourier series to show that

$$\sinh\left(\frac{m\pi x}{b}\right) = \frac{2b^2}{\pi} \sum_{n=1}^{\infty} \left(\frac{(-1)^{n+1} n}{b^2 n^2 + a^2 m^2}\right) \sinh\left(\frac{m\pi a}{b}\right) \sin\left(\frac{n\pi x}{a}\right) \tag{9.196}$$

over the interval $-a \leq x \leq a$.

10

Fourier Analysis

Fourier analysis is one of the most important topics we address in Part II of the book. Understanding how spectra are calculated and what they tell us about a recorded response is used across many disciplines and applications. In this chapter, we derive the Fourier transform pair, discuss various conventions for defining the transform pairs, consider some examples of calculating Fourier transforms that do not require the ability to integrate on the complex plane (contour integration is treated in Chapter 11), consider some important general properties of Fourier transforms, use the Fourier transform to prove the most important theorems of probability theory, use the Fourier transform to solve both diffusion and wave initial-value problems and discuss some of the key aspects of time-series analysis that are strongly influenced by Fourier analysis. The presentation here is meant to be practical and direct. For the student seeking a more mathematical presentation filled with fascinating historical vignettes, the book by Körner (1988) is a recommended classic.

10.1 The Fourier Transform Pair

We begin by deriving the Fourier transform and discussing various conventions associated with its definition. The Fourier transform pair is simply the complex Fourier series in the limit where the period L tends to infinity so that the discrete frequencies $k_n = 2\pi n/L$ for integer n become a continuum of frequencies k and the summation over n becomes an integral over k.

10.1.1 Derivation from the Fourier Series

To derive the Fourier transform pair, we start with the complex form of the Fourier series and take the fundamental period to be from $a = -L/2$ to $a + L = L/2$ to obtain

$$f(x) = \sum_{n=-\infty}^{+\infty} c_n \, e^{i2\pi nx/L} \quad \text{where} \quad -\frac{L}{2} < x < \frac{L}{2} \tag{10.1}$$

and where the coefficients in the series are given by

$$c_n = \frac{1}{L} \int_{-L/2}^{+L/2} f(x) \, e^{-i2\pi nx/L} \, \mathrm{d}x. \tag{10.2}$$

We next consider what happens as $L \to \infty$. Let's identify the frequency k as

$$k = \frac{2\pi n}{L} \tag{10.3}$$

and the change in frequency Δk between n and $n + \Delta n$ (where $\Delta n = 1$) as

$$\Delta k = \frac{2\pi(n+1)}{L} - \frac{2\pi n}{L} = \frac{2\pi\,\Delta n}{L} \tag{10.4}$$

In this sense, we can think of the coefficients c_n as being the function $c(k)$. We elect to rewrite this function $c(k)$ as

$$\tilde{f}(k) = L\,c(k) \tag{10.5}$$

so the complex Fourier series may be rewritten as

$$f(x) = \frac{1}{2\pi} \sum_{kL/(2\pi)=-\infty}^{+\infty} \tilde{f}(k)\, e^{+ikx}\,\Delta k. \tag{10.6}$$

As $L \to \infty$, we take $2\pi\,\Delta n/L = \Delta k \to \mathrm{d}k$ to be an infinitesimal and replace the sum as an integral to give the pair of integrals

$$f(x) = \frac{1}{2\pi}\int_{-\infty}^{+\infty} \tilde{f}(k)\, e^{ikx}\,\mathrm{d}k \quad \textit{the inverse Fourier transform} \tag{10.7}$$

$$\tilde{f}(k) = \int_{-\infty}^{+\infty} f(x)\, e^{-ikx}\,\mathrm{d}x \quad \textit{the Fourier transform} \tag{10.8}$$

which is called the *Fourier-transform pair.* We call $\tilde{f}(k)$ the *Fourier transform* and call $f(x)$, that is obtained from the $\tilde{f}(k)$ by performing the integral above, the *inverse Fourier transform.* As with the Fourier series, we can still think of $f(x)$ as being the sum of a whole bunch of sines and cosines ($\sin kx$ and $\cos kx$) at different frequencies k, each weighted by the complex $\tilde{f}(k)$. As promised, to obtain the Fourier transform pair, we have just taken the limit of the Fourier series as the period becomes infinite, the frequencies become a continuum and the sum becomes an integral.

10.1.2 Description of the Forward and Inverse Fourier Transform

There are specific words and symbols used in discussing various aspects of the above transform pair. We begin with the following equivalent ways of writing the Fourier transform of a function $f(x)$

$$\tilde{f}(k) = \int_{-\infty}^{\infty} f(x)\, e^{-ikx}\,\mathrm{d}x = \mathrm{FT}\,\{f(x)\} = \tilde{f}_R(k) + i\tilde{f}_I(k) = A(k)e^{i\theta(k)} \mathrel{\widehat{=}} \textit{the spectrum}, \tag{10.9}$$

where $A(k) = \sqrt{\tilde{f}_R(k)^2 + \tilde{f}_I(k)^2}$ is the *amplitude spectrum* and $\theta(k) = \tan^{-1}\left(\tilde{f}_I(k)/\tilde{f}_R(k)\right)$ is the *phase spectrum* (both of which are real). In general, the Fourier transform or "spectrum" $\tilde{f}(k)$ is a complex function of k. So the amplitude spectrum $A(k)$ is telling us the amplitude to use for each sinusoid e^{ikx} at frequency k and the phase spectrum is telling us how each such sinusoid is shifted in x relative to $x = 0$, that is, $e^{i[kx+\theta(k)]} = e^{ik[x+\theta(k)/k]}$.

The inverse Fourier transform can also be called the *Fourier-integral representation* of the function $f(x)$. In the Fourier transform, the transform variable, or frequency, k is real as was proven in the section on Sturm-Liouville theory, cf., the discussion following Eq. (9.25). But in evaluating the inverse Fourier transform

$$f(x) = \frac{1}{2\pi}\int_{-\infty}^{\infty} \tilde{f}(k)\, e^{ikx}\, \mathrm{d}k = \mathrm{FT}^{-1}\left\{\tilde{f}(k)\right\} = \frac{1}{2\pi}\int_{-\infty}^{\infty} A(k)\, e^{ik[x+\theta(k)/k]}\, \mathrm{d}k, \tag{10.10}$$

we sometimes extend analysis onto the complex k plane in performing the above integral over real k as will be explained in Chapter 11 concerning contour integration. When we consider the Fourier transform, we say that we are analyzing a function "in the frequency domain" or "in the transform domain" or "in the spectral domain" (these are synonyms). When we come back from the transform domain to the "real domain" we say we are performing the "inverse transform."

If x represents distance in space, we often call the spatial circular frequency k the "wavenumber," where $k = 2\pi/\lambda$ with λ the wavelength associated with each wavenumber k (spatial circular frequency). If x were replaced by time t, we replace k by the temporal circular frequency ω where $\omega = 2\pi f$ with f frequency as measured in cycles per second and $f = 1/T$ where T is the time period associated with each ω (temporal circular frequency). So we can apply Fourier transforms to functions either in the space or time domains and we will return momentarily to some conventions associated with the space and time transforms. Note that both kx and ωt have units of radians just like the phase spectrum $\theta(k)$.

To conclude, the inverse Fourier transform says that we can represent some function $f(x)$ as a sum (integral in the limit) of sinusoids having different frequencies k with each sinusoid having an amplitude $A(k)$ and possibly shifted in space $e^{ik[x+\theta(k)/k]}$ as controlled by the phase spectrum $\theta(k)$. This is depicted in Fig. 10.1 for a handful of frequencies.

10.1.3 Condition for Validity

Whether the integral in the inverse Fourier transform converges to properly represent the function $f(x)$ depends on whether the Fourier transform $\tilde{f}(k)$ (the complex spectral coefficients associated with each sinusoid) is well behaved which in turn depends on the nature of the function $f(x)$. Because $|e^{ikx} f(x)| < |f(x)|$, a sufficient condition for the existence of the Fourier transform is that $\int_{-\infty}^{\infty} |f(x)|\, \mathrm{d}x$ is finite. Generally, a sufficient condition for this is that

$$\lim_{|x|\to\infty} f(x) \to 0, \tag{10.11}$$

that is, as we go to large positive x or to large negative x, if the function being transformed tends to zero as we head to infinity, the transform will exist and the Fourier representation of $f(x)$ is valid. We will assume, in general, that this is the case even if we have to imagine tapering a data set that we want to transform to zero as the argument gets large.

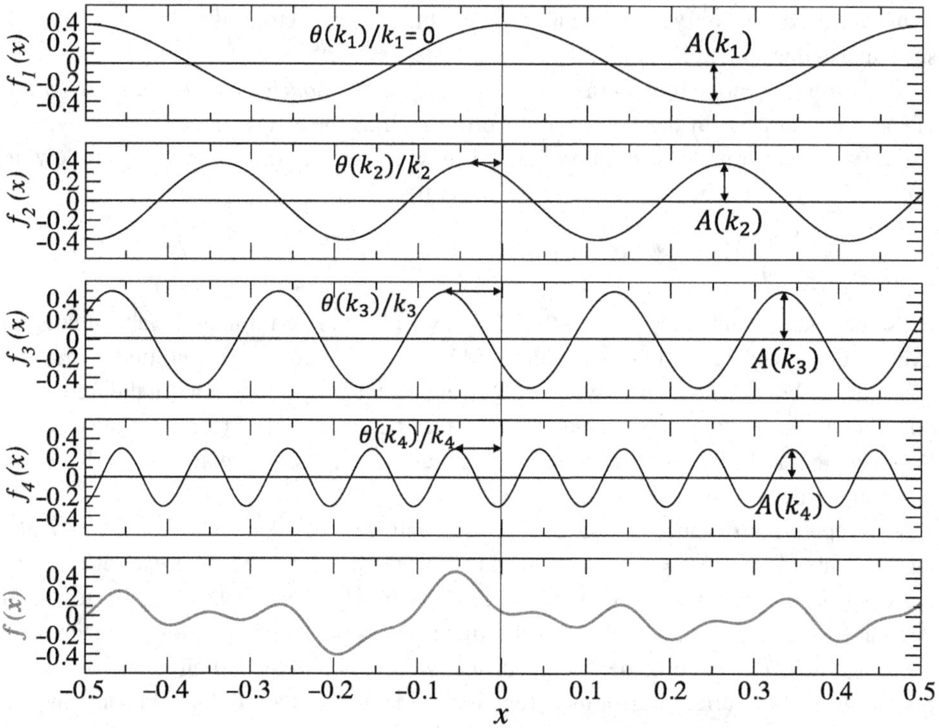

Figure 10.1 A real function $f(x)$ (shown in the lowest panel) is represented as four sinusoids (cosines in this example) having different frequencies k_n, different amplitudes $A(k_n)$ and shifted in space relative to $x = 0$ by an amount $\theta(k_n)/k_n$ (as shown by the horizontal arrows). These four cosines pictorially represent a spectrum given by $\tilde{f}(k) = \sum_{n=1}^{4} A(k_n)e^{i\theta(k_n)}\,[\delta(k-k_n) + \delta(k+k_n)]$ where $\theta(-k_n) = -\theta(k_n)$ (due to the fact that $f(x)$ is real as will be proven later). The inverse Fourier transform of this spectrum is $f(x) = \sum_{n=1}^{4} A(k_n)\left(e^{ik_n(x+\theta(k_n)/k_n)} + e^{-ik_n(x+\theta(k_n)/k_n)}\right)/(2\pi) = \sum_{n=1}^{4} A(k_n)$ $\cos\left[k_n\left(x+\theta(k_n)/k_n\right)\right]/\pi$. Thus, the function $f(x)$ in the lower panel is simply the sum of the four cosines depicted above $f(x) = \sum_{n=1}^{4} f_n(x)/\pi$. It is neither an odd or even function in x but it is purely real. Using enough different frequencies (formally a continuum of frequencies) with appropriate amplitudes and phases for the sinusoids at each frequency, any desired function can be represented using the inverse Fourier transform.

10.1.4 The Fourier-Integral Theorem

If we insert the expression for the Fourier transform $\tilde{f}(k)$ into the inverse Fourier transform, we obtain

$$f(x) = \frac{1}{2\pi}\int_{-\infty}^{\infty} \mathrm{d}k\, e^{ikx}\underbrace{\left[\int_{-\infty}^{\infty} f(u)\, e^{-iku}\, \mathrm{d}u\right]}_{\tilde{f}(k)} \tag{10.12}$$

$$= \frac{1}{2\pi}\int_{-\infty}^{\infty} \mathrm{d}u\, f(u) \int_{-\infty}^{\infty} \mathrm{d}k\, e^{ik(x-u)}. \tag{10.13}$$

This equation is called the *Fourier-integral theorem*. We first proved its validity on finite domains in Eq. (9.32) when we derived the Sturm-Liouville representation of functions. The *completeness relation*, also as first seen in the section on Sturm-Liouville theory, is the necessity that

$$\delta(x-u) = \frac{1}{2\pi}\int_{-\infty}^{\infty} e^{ik(x-u)}\mathrm{d}k. \tag{10.14}$$

We will independently prove the completeness relation below using the explicit case of the Fourier transform of the Gaussian function in the limit as the standard of deviation goes to zero and the Gaussian becomes the Dirac delta function. The completeness relation says that the inverse transform of a uniform spectrum that equals one for all k is the Dirac delta function. Using the completeness relation then gives

$$f(x) = \frac{1}{2\pi}\int_{-\infty}^{\infty} \mathrm{d}u\, f(u)\int_{-\infty}^{\infty} \mathrm{d}k\, e^{ik(x-u)} = \int_{-\infty}^{\infty} \mathrm{d}u\, f(u)\,\delta(x-u) = f(x). \tag{10.15}$$

So the transform pair is internally consistent which is the message of the Fourier-integral theorem.

10.1.5 Conventions on Signs, Symbols, and the 2π Factor

We could have adopted a different sign convention in the exponent of our transform pair; specifically,

$$f(x) = \frac{1}{2\pi}\int_{-\infty}^{\infty} \tilde{f}(k)\, e^{-ikx}\mathrm{d}k \tag{10.16}$$

$$\tilde{f}(k) = \int_{-\infty}^{\infty} f(x)\, e^{+ikx}\mathrm{d}x. \tag{10.17}$$

Running through the above proof of the Fourier-integral theorem with the change of variables $k \to -k$ and noting that $\delta(x-u) = \delta(u-x)$, we have

$$\int_{-\infty}^{\infty} e^{ik(x-u)}\mathrm{d}k = \int_{+\infty}^{-\infty} e^{i(-k)(x-u)}\mathrm{d}(-k) = \int_{-\infty}^{\infty} e^{-ik(x-u)}\mathrm{d}k = 2\pi\,\delta(x-u). \tag{10.18}$$

Thus we can adopt either sign convention in the exponentials of the transform pair and the Fourier-integral theorem is satisfied.

In physics, we use x to represent linear distance in space so that the transform variable, or wavenumber, k has units of "radians per meter." Similarly, we use t to represent time so that the transform variable, or circular frequency, ω has units of "radians per second." It is standard to adopt the following sign conventions for functions of space and time:

$$f(x) = \frac{1}{2\pi}\int_{-\infty}^{\infty} \tilde{f}(k)\, e^{+ikx}\,\mathrm{d}k \tag{10.19}$$

$$\tilde{f}(k) = \int_{-\infty}^{\infty} f(x)\, e^{-ikx}\,\mathrm{d}x \tag{10.20}$$

and

$$f(t) = \frac{1}{2\pi} \int_{-\infty}^{\infty} \tilde{f}(\omega)\, e^{-i\omega t}\, \mathrm{d}\omega \tag{10.21}$$

$$\tilde{f}(\omega) = \int_{-\infty}^{\infty} f(t)\, e^{+i\omega t}\, \mathrm{d}t. \tag{10.22}$$

If we have a function $\psi(x, t)$ of both space and time, we can perform a double transform over both space and time:

$$\psi(x, t) = \frac{1}{(2\pi)^2} \int_{-\infty}^{\infty} \mathrm{d}k \int_{-\infty}^{\infty} \mathrm{d}\omega\, \tilde{\tilde{\psi}}(k, \omega)\, e^{i(kx-\omega t)} \tag{10.23}$$

$$\tilde{\tilde{\psi}}(k, \omega) = \int_{-\infty}^{\infty} \mathrm{d}x \int_{-\infty}^{\infty} \mathrm{d}t\, \psi(x, t)\, e^{-i(kx-\omega t)}. \tag{10.24}$$

Later, we will perform quadruple transforms over each component of 3D space as well as time.

We also could have adopted any of the following different conventions on where to place the factor of 2π in the transform pair

$$f(x) = \int_{-\infty}^{\infty} \tilde{f}(k)\, e^{ikx}\, \mathrm{d}x \quad \text{and} \quad \tilde{f}(k) = \frac{1}{2\pi} \int_{-\infty}^{\infty} f(x)\, e^{-ikx}\, dx \tag{10.25}$$

or

$$f(x) = \frac{1}{\sqrt{2\pi}} \int_{-\infty}^{\infty} \tilde{f}(k)\, e^{ikx}\, \mathrm{d}x \quad \text{and} \quad \tilde{f}(k) = \frac{1}{\sqrt{2\pi}} \int_{-\infty}^{\infty} f(x)\, e^{-ikx}\, \mathrm{d}x \tag{10.26}$$

or

$$f(x) = \frac{1}{(2\pi)^{\alpha}} \int_{-\infty}^{\infty} \tilde{f}(k)\, e^{ikx}\, \mathrm{d}x \quad \text{and} \quad \tilde{f}(k) = \frac{1}{(2\pi)^{(1-\alpha)}} \int_{-\infty}^{\infty} f(x)\, e^{-ikx}\, \mathrm{d}x. \tag{10.27}$$

Nothing about our Fourier-integral theorem would change with any of these choices. Our choice in this book of putting the $(2\pi)^{-1}$ on the inverse transform is probably the most common across the physical sciences.

Some authors prefer to place the 2π in the exponentials and work with the transform pair

$$g(t) = \int_{-\infty}^{\infty} \tilde{g}(f)\, e^{-i2\pi ft}\, \mathrm{d}f \quad \text{and} \quad \tilde{g}(f) = \int_{-\infty}^{\infty} g(t)\, e^{+i2\pi ft}\, \mathrm{d}t, \tag{10.28}$$

where f is now "frequency" (and not "circular frequency" $\omega = 2\pi f$). The completeness relation is then (change the variables)

$$\int_{-\infty}^{\infty} e^{\pm i2\pi ft}\, \mathrm{d}f = \delta(t) \tag{10.29}$$

and the Fourier-integral theorem is again satisfied. Although our choice of putting the factor $(2\pi)^{-1}$ on the inverse transform is most common in the physical sciences, the choice of Eq. (10.28) is more symmetric and is arguably the "best" choice even though we choose not to adopt it.

So there are different conventions we can adopt for the sign to put in the exponential of the forward and inverse transform and for where to put the factors of 2π in the transform pair. All approaches give equivalent results in the sense that the Fourier-integral theorem is satisfied but it is important to choose a convention and keep with it through all of your work. The choices made here in Eqs (10.19) through (10.22) for spatial and temporal transforms are again the most common and will serve you well.

10.2 Some Examples of Calculating the Fourier Transform

10.2.1 Example 1: Gaussian Function

Consider the so-called Gaussian function

$$f(x) = \frac{1}{\sigma\sqrt{2\pi}} \exp\left(-\frac{x^2}{2\sigma^2}\right). \tag{10.30}$$

The normalizing factor of $\sigma\sqrt{2\pi}$ is there so that $\int_{-\infty}^{+\infty} f(x)\,\mathrm{d}x = 1$. Take the Fourier transform of the Gaussian to give

$$\tilde{f}(k) = \frac{1}{\sigma\sqrt{2\pi}} \int_{-\infty}^{\infty} e^{-x^2/(2\sigma^2) - ikx}\,\mathrm{d}x. \tag{10.31}$$

Complete the square in the exponent by writing

$$-\frac{x^2}{2\sigma^2} - ikx = -\left(\frac{x}{\sigma\sqrt{2}} + \frac{i\sigma\sqrt{2}k}{2}\right)^2 - \frac{\sigma^2 k^2}{2}. \tag{10.32}$$

Let $u = x/(\sigma\sqrt{2}) + i\sigma k/\sqrt{2}$ and $\mathrm{d}u = \mathrm{d}x/(\sigma\sqrt{2})$, so that

$$\tilde{f}(k) = \frac{1}{\sqrt{\pi}} e^{-\sigma^2 k^2/2} \int_{-\infty + ik\sigma/\sqrt{2}}^{\infty + ik\sigma/\sqrt{2}} e^{-u^2}\,\mathrm{d}u. \tag{10.33}$$

Now compared to ∞, we can neglect the relatively small imaginary part in the limits of the integral (this can also be justified using Cauchy's Theorem in Chapter 11) and obtain the integral as

$$\int_{-\infty}^{\infty} e^{-u^2}\,\mathrm{d}u = \sqrt{\pi}. \tag{10.34}$$

We thus have determined the Fourier transform of the Gaussian function to be

$$\boxed{\tilde{f}(k) = \exp\left(-\frac{\sigma^2 k^2}{2}\right).} \tag{10.35}$$

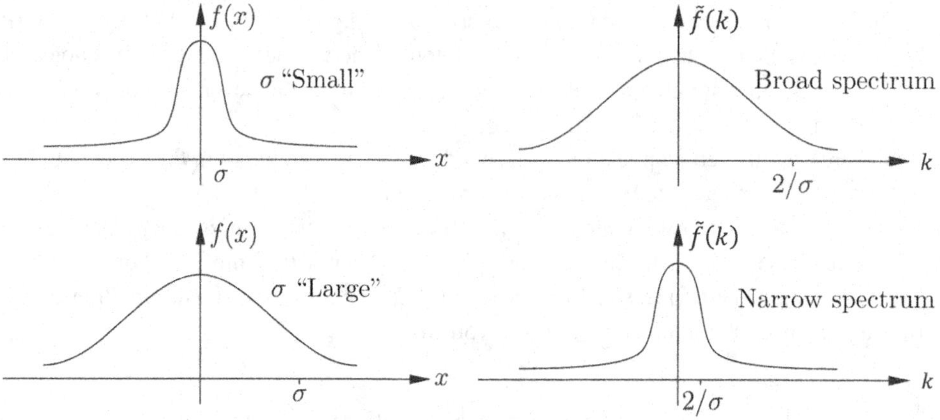

Figure 10.2 The general nature of Fourier transforms is that compact functions in the real domain have a broad spectrum, while broad functions in the real domain have compact support in the spectral domain.

So the spectrum $\tilde{f}(k)$ is also of Gaussian form but with a different dependence on the standard of deviation σ as depicted in Fig. 10.2.

The general rule that we are observing in this example, that is true for all Fourier transforms, is that compact functions in x have broad spectra in k and broad functions in x have compact spectra in k. Note in this example that if we express the "width" of the function in x as $\Delta x = \sigma\sqrt{2}$ and the width of the spectrum in k as $\Delta k = \sqrt{2}/\sigma$, we are observing that $\Delta x \Delta k = 2$, which is sometimes called the "uncertainty relation." This simply means it requires considerable high frequencies (a broad spectrum) to represent a sharply peaked narrow function but requires mainly low frequencies (a narrower spectrum) to represent a spread-out broad function.

Above, we used the definite integral

$$I = \int_{-\infty}^{\infty} e^{-u^2}\, \mathrm{d}u = \sqrt{\pi}.$$

This integral shows up often in applications because Gaussian functions are important in physics (diffusion in particular) and probability. How do we obtain this result? An easy approach is to focus on the square of the integral or

$$\begin{aligned} I^2 &= \left[\int_{-\infty}^{\infty} e^{-x^2}\, \mathrm{d}x\right]\left[\int_{-\infty}^{\infty} e^{-y^2}\, \mathrm{d}y\right] \\ &= \int_{-\infty}^{\infty}\int_{-\infty}^{\infty} \mathrm{d}x\, \mathrm{d}y\, e^{-(x^2+y^2)}. \end{aligned}$$

The integral over the entire x, y plane can be performed in polar coordinates by using $r^2 = x^2 + y^2$ and $\mathrm{d}x\, dy = \mathrm{d}r\, r\, \mathrm{d}\theta$ to obtain

$$I^2 = \int_0^{2\pi} \mathrm{d}\theta \int_0^{\infty} \mathrm{d}r\, r\, e^{-r^2}.$$

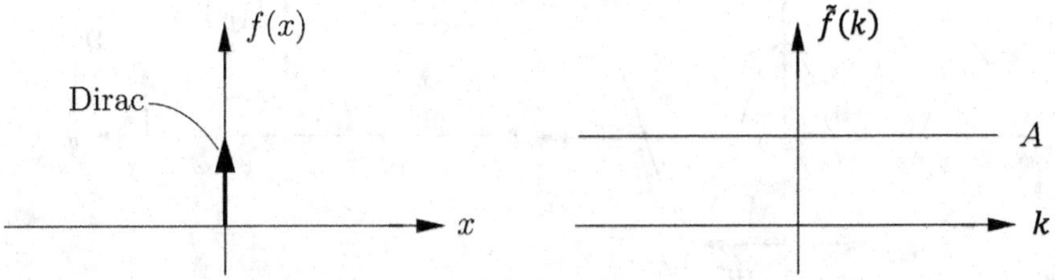

Figure 10.3 The narrowest conceivable function (the Dirac) has the widest conceivable spectrum (a constant).

Making the substitution $u = r^2$ and $\mathrm{d}u = 2r\,\mathrm{d}r$ then gives

$$I^2 = \frac{2\pi}{2}\int_0^\infty \mathrm{d}u\, e^{-u} = \pi\left[-e^{-\infty} + e^0\right] = \pi.$$

We thus have obtained that

$$I = \int_{-\infty}^{\infty} e^{-u^2}\mathrm{d}u = \sqrt{\pi}$$

a result that will be used repeatedly through the remainder of the book.

Returning to the theme of "compact function in the real domain = broad spectrum," let's take that idea to the limit as shown in Fig. 10.3 and let $\sigma \to 0$ in the above Gaussian example so that the Gaussian function becomes a Dirac delta function and the spectrum becomes a constant

$$f(x) = A\,\delta(x) \,\widehat{=}\, \text{Dirac Delta} \quad \text{(the most compact function)} \tag{10.36}$$

$$\tilde{f}(k) = A\int_{-\infty}^{\infty} \delta(x)\, e^{-ikx} = A \quad \text{(the broadest spectrum).} \tag{10.37}$$

The inverse Fourier transform of a constant spectrum A is

$$f(x) = \frac{1}{2\pi}\int_{-\infty}^{\infty} A\, e^{+ikx}\,\mathrm{d}k = A\,\delta(x), \tag{10.38}$$

which is just the completeness relation again. So using the Gaussian function in the limit as $\sigma \to 0$, we have obtained an explicit proof of the completeness relation.

We also have that if

$$f(x) = A \,\widehat{=}\, \text{constant} \quad \text{(the broadest function in } x\text{)} \tag{10.39}$$

$$\tilde{f}(k) = A\int_{-\infty}^{\infty} e^{-ikx}\,\mathrm{d}x = 2\pi\, A\,\delta(k) \quad \text{(the narrowest spectrum in } k\text{).} \tag{10.40}$$

Note that in this last example, we do not satisfy the conditions for existence of the Fourier transform (the function $f(x)$ does not taper off to zero as $|x| \to \infty$) and this is why we end up with a strange spectrum involving a Dirac that diverges (in an integrable way) at $k = 0$.

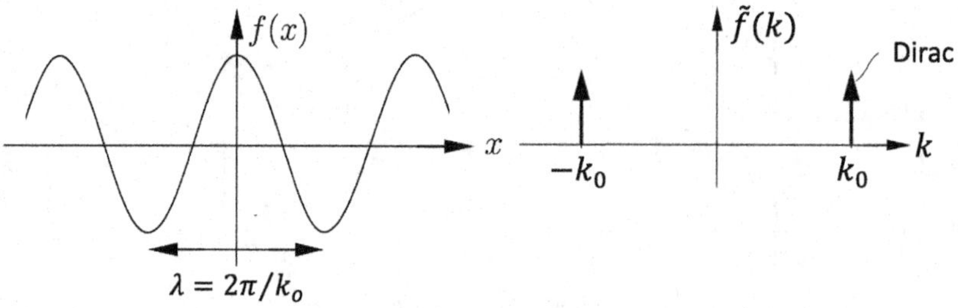

Figure 10.4 The cosine function and its spectrum.

10.2.2 Example 2: The Cosine and Sine Functions

Consider the cosine function $f(x) = A\ \cos k_0 x$. Taking the Fourier transform gives

$$\tilde{f}(k) = A \int_{-\infty}^{\infty} \frac{\left[e^{i\,k_0\,x} + e^{-i\,k_0\,x}\right]}{2}\, e^{-ikx}\,\mathrm{d}x = \pi A \left[\delta(k - k_0) + \delta(k + k_0)\right], \tag{10.41}$$

which is depicted in Fig. 10.4.

Similarly, for the sine function $f(x) = A\ \sin k_0 x$ we have

$$\tilde{f}(k) = A \int_{-\infty}^{\infty} \frac{\left[e^{i\,k_0\,x} - e^{-i\,k_0\,x}\right]}{2i}\, e^{-ikx}\,\mathrm{d}x = i\pi A \left[-\delta(k - k_0) + \delta(k + k_0)\right]. \tag{10.42}$$

We will prove below that, in general, for an even and real function like cosine, the spectrum is purely real and is itself even in k. Similarly, for an odd and real function like sine, the spectrum is purely imaginary and is odd in k. We have shown this explicitly above for the cosine and sine functions.

10.2.3 Example 3: The Boxcar Function

Consider the boxcar function that has the shape given in Fig. 10.5 that has the mathematical description

$$f(x) = \begin{cases} 1 & |x| \le a \\ 0 & |x| > a \end{cases}. \tag{10.43}$$

The Fourier transform of the boxcar is then

$$\tilde{f}(k) = \int_{-a}^{a} e^{-ikx}\,\mathrm{d}x = -\frac{1}{-ik}\left(e^{-ika} - e^{+ika}\right) \tag{10.44}$$

$$= 2a\,\frac{\sin ka}{ka}. \tag{10.45}$$

The function $\sin K/K$ is called the "sinc" function and has the form given in Fig. 10.6. When we work with finite-length data sets that have been recorded, we are implicitly multiplying the actual signal by a boxcar function that corresponds to convolving the Fourier transform of the actual signal with the sinc function. More on this later after we discuss the "convolution theorem" and define the operation of convolution.

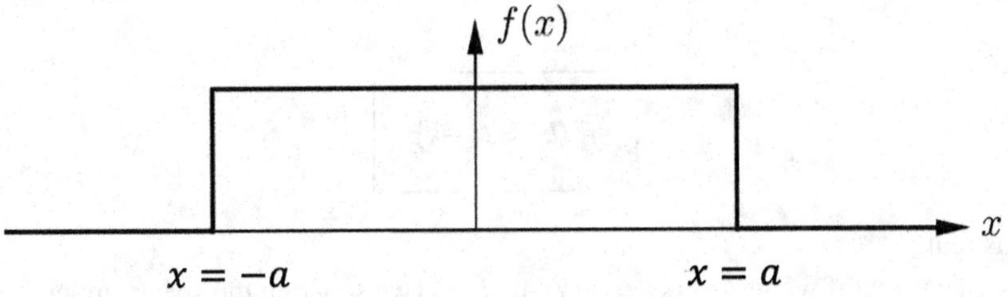

Figure 10.5 The boxcar function.

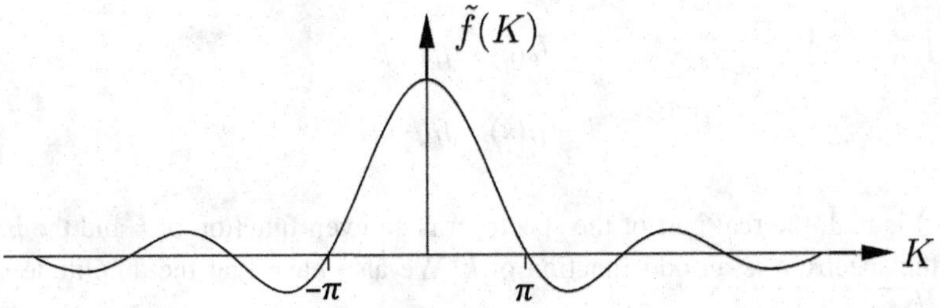

Figure 10.6 The sinc function is the real symmetric spectrum of the real symmetric boxcar function.

10.3 Properties of the Fourier Transform

We have defined the Fourier transform as

$$\tilde{f}(k) = \int_{-\infty}^{\infty} f(x)\, e^{-ikx}\, dx \tag{10.46}$$

and have stated that because $\left|f(x)\, e^{-ikx}\right| \leq |f(x)|$, a condition for the existence of $\tilde{f}(k)$ is that

$$\int_{-\infty}^{\infty} \left|f(x)\right| \text{dx exists.} \tag{10.47}$$

In general, the sufficient condition that $f(x) \to 0$ as $|x| \to \infty$, guarantees the existence of the Fourier transform.

In this section, we consider some general and important properties of the Fourier transform. In what follows, we use * to denote the "complex conjugate" which is the process of replacing each i with $-i$ in a complex expression.

10.3.1 When f(x) Is Real

If the function being transformed $f(x)$ is purely real, then the complex conjugate of the spectrum is

$$\begin{aligned} \tilde{f}^*(k) &= \int_{-\infty}^{\infty} f(x)\, e^{+ikx}\, \text{dx} \\ &= \int_{-\infty}^{\infty} f(x)\, e^{-i(-k)x}\, \text{dx} = \tilde{f}(-k) \end{aligned} \tag{10.48}$$

so that

$$\boxed{\tilde{f}^*(k) = \tilde{f}(-k)} \tag{10.49}$$

if $f(x)$ is real.

This means that if we write $\tilde{f}(k) = \tilde{f}_R(k) + i\tilde{f}_I(k)$ (we separate the spectrum into real and imaginary portions), then

$$\tilde{f}_R(k) = \tilde{f}_R(-k) \tag{10.50}$$

$$-\tilde{f}_I(k) = \tilde{f}_I(-k). \tag{10.51}$$

So if $f(x)$ is real, the real part of the spectrum is an even function of k and the imaginary part of the spectrum is an odd function of k. We also have that the amplitude spectrum $A(k) = \sqrt{\tilde{f}_R(k)^2 + \tilde{f}_I(k)^2}$ is even and the phase spectrum $\theta(k) = \tan^{-1}\left(\tilde{f}_I(k)/\tilde{f}_R(k)\right)$ is odd.

10.3.2 When $f(x)$ Is Real and Even

An even function is of the form $f(x) = f(-x)$. If our function to be transformed is both real and even in x, we have that the complex conjugate of the spectrum is given by

$$\tilde{f}^*(k) = \int_{-\infty}^{\infty} f(x)\, e^{+ikx}\, \mathrm{d}x \tag{10.52}$$

$$= \int_{-\infty}^{\infty} f(x)\, e^{-ik(-x)}\, dx.$$

Let $u = -x$, $f(u) = f(x)$ (even function) and $\mathrm{d}u = -\mathrm{d}x$ so that (note that $\int_a^b f(u)\, \mathrm{d}u = -\int_b^a f(u)\, \mathrm{d}u$)

$$\tilde{f}^*(k) = \int_{-\infty}^{\infty} f(u)\, e^{-iku}\, \mathrm{d}u = \tilde{f}(k), \tag{10.53}$$

which says that the spectrum does not have an imaginary part. Thus we have shown that if $f(x)$ is even and real, then $\tilde{f}(k)$ is purely real. Further, because $f(x)$ is real, we have that the purely real $\tilde{f}(k)$ is also even.

10.3.3 When $f(x)$ Is Real and Odd

If our function to be transformed is both real and odd in x, we have that $\tilde{f}(k)$ is purely imaginary and odd, which is something you can prove in an end-of-chapter exercise.

10.3.4 Shifting Property

Given $\tilde{f}(k) = \text{FT}\{f(x)\}$, what is $\text{FT}\{f(x-a)\}$? We address this by simply using the definition of the Fourier transform to write

$$\text{FT}\{f(x-a)\} = \int_{-\infty}^{\infty} f(x-a)\, e^{-ikx}\, \mathrm{d}x. \tag{10.54}$$

Let $u = x - a$ and $\mathrm{d}u = \mathrm{d}x$ to give

$$\text{FT}\{f(x-a)\} = \int_{-\infty}^{\infty} f(u)\, e^{-iku}\, e^{-ika}\, \mathrm{d}u \tag{10.55}$$

so that

$$\boxed{\text{FT}\{f(x-a)\} = e^{-ika}\tilde{f}(k).} \tag{10.56}$$

This says that if a function is shifted in space by an amount a, a phase shift of ka is introduced into the spectrum. If we write the inverse Fourier transform, this simply means that

$$f(x-a) = \frac{1}{2\pi}\int_{-\infty}^{\infty} \tilde{f}(k)e^{ik(x-a)}\, \mathrm{d}k, \tag{10.57}$$

which again just says that if you introduce a phase shift $-ka$ into the spectrum, you shift the function in space by an amount a.

10.3.5 Stretching or Contracting Property

Consider the function $f(x)$ and a scaled version of it $f_\alpha(x) = \alpha f(\alpha x)$. If $\alpha > 1, f_\alpha(x)$ is contracted in x and enhanced in amplitude relative to $f(x)$, while if $\alpha < 1, f_\alpha(x)$ is stretched in x and diminished in amplitude. The Fourier transform of f_α is then

$$\text{FT}\,\{f_\alpha\} = \tilde{f}_\alpha(k) = \int_{-\infty}^{+\infty} \alpha f(\alpha x)e^{-ikx}\, \mathrm{d}x. \tag{10.58}$$

Substituting $\chi = \alpha x$ then gives

$$\tilde{f}_\alpha(k) = \int_{-\infty}^{+\infty} f(\chi)e^{-ik\chi/\alpha}\, \mathrm{d}\chi = \tilde{f}(k/\alpha). \tag{10.59}$$

Thus, $\tilde{f}_\alpha(k) = \tilde{f}(k/\alpha)$, which says that contracting a function by a factor $\alpha > 1$ in the real domain x has the effect of stretching the spectrum by a factor $1/\alpha$ in the spectral domain k. This is the same message we learned from the example of the Fourier transform of the Gaussian function in Section 10.2.1.

10.3.6 Differentiation Property

We next consider the Fourier transform of the derivative of a function $df(x)/dx$. Calculating the Fourier transform gives

$$\begin{aligned}\mathrm{FT}\left\{\frac{d}{dx}f(x)\right\} &= \int_{-\infty}^{\infty} \frac{df(x)}{dx}\, e^{-ikx}\, dx \\ &= \int_{-\infty}^{\infty} \left\{\frac{d}{dx}\left[f(x)\, e^{-ikx}\right] - f(x)\frac{d}{dx}\, e^{-ikx}\right\} dx \qquad (10.60) \\ &= \cancelto{0}{f(+\infty)}\, e^{-ik\infty} - \cancelto{0}{f(-\infty)}\, e^{+ik\infty} + ik \int_{-\infty}^{\infty} f(x)\, e^{-ikx}\, dx = +ik\tilde{f}(k),\end{aligned}$$

where we used our existence condition that $\lim_{|x|\to\infty} f(x) = 0$. We thus have

$$\mathrm{FT}\left\{\frac{d}{dx}f(x)\right\} = +ik\tilde{f}(k). \qquad (10.61)$$

In words, this says that "differentiation in the space domain = multiplication by $+ik$ in the spectral domain," which is an idea we will use over and over again in solving PDEs. We can also just take the derivative of the inverse Fourier transform to obtain

$$\begin{aligned}\frac{d}{dx}f(x) &= \frac{d}{dx}\left\{\frac{1}{2\pi}\int_{-\infty}^{\infty} \tilde{f}(k)\, e^{+ikx}\, dk\right\} \qquad (10.62) \\ &= \frac{1}{2\pi}\int_{-\infty}^{\infty} \underbrace{ik\tilde{f}(k)}_{\substack{\uparrow \\ \text{spectrum of } df(x)/dx}} \; e^{ikx}\, dk.\end{aligned}$$

Similarly, upon taking any number n of derivatives of the function $f(x)$, we obtain

$$\boxed{\mathrm{FT}\left\{\frac{d^n}{dx^n}f(x)\right\} = (ik)^n\tilde{f}(k).} \qquad (10.63)$$

When you see factors of ik multiplying some spectrum $\tilde{f}(k)$ in the frequency domain, you can perform the inverse transform $\tilde{f}(k) \Rightarrow f(x)$ and then perform one differentiation of the resulting $f(x)$ for each factor of ik that was present in the spectrum. We will use this fact in our applications of the Fourier transform to solving differential equations.

10.3.7 Parseval's Theorem

Parseval's theorem is generally interpreted to mean that the total energy of a function in the space (or time) domain is equal to the total energy in the spectrum. A first version of Parseval's theorem is that any two functions $f(x)$ and $g(x)$ (complex, real, whatever) satisfy

$$\int_{-\infty}^{\infty} f(x)\, g(x)\, dx = \frac{1}{2\pi}\int_{-\infty}^{\infty} \tilde{f}(k)\, \tilde{g}(-k)\, dk. \qquad (10.64)$$

You can prove this statement as an end-of-chapter exercise. Given this, if $f(x)$ is a real function so that $\tilde{f}(-k) = \tilde{f}^*(k)$, we then have a second form of Parseval's theorem:

$$\int_{-\infty}^{\infty} f^2(x)\, \mathrm{d}x = \frac{1}{2\pi} \int_{-\infty}^{\infty} |\tilde{f}(k)|^2\, \mathrm{d}k. \tag{10.65}$$

Note that if we had placed the factor of 2π on the transform pair so that each integral of the pair had a factor of $1/\sqrt{2\pi}$ in front of it, there would be no factor of 2π in these statements of Parseval's theorem.

10.3.8 Convolution Theorem

The idea of the convolution theorem is that "convolution in the real domain = multiplication in the spectral domain." To understand what these words mean, consider two functions $f(x)$ and $g(x)$ along with their Fourier transforms

$$\tilde{g}(k) = \mathrm{FT}\{g(x)\} \tag{10.66}$$

$$\tilde{f}(k) = \mathrm{FT}\{f(x)\}. \tag{10.67}$$

We multiply the two spectra together and then take the inverse Fourier transform of the result to obtain

$$\begin{aligned}
\mathrm{FT}^{-1}\left\{\tilde{g}(k)\tilde{f}(k)\right\} &= \frac{1}{2\pi}\int_{-\infty}^{\infty} \tilde{f}(k)\,\tilde{g}(k)\, e^{+ikx}\, \mathrm{d}k \\
&= \frac{1}{2\pi}\int_{-\infty}^{\infty} \tilde{f}(k)\, e^{ikx}\, \mathrm{d}k \int_{-\infty}^{\infty} g(u)\, e^{-iku}\, \mathrm{d}u \\
&= \frac{1}{2\pi}\int_{-\infty}^{\infty} g(u)\, \mathrm{d}u \int_{-\infty}^{\infty} \tilde{f}(k)\, e^{ik(x-u)}\, \mathrm{d}k.
\end{aligned} \tag{10.68}$$

Identifying the integral over k as the inverse Fourier transform then gives the result

$$\mathrm{FT}^{-1}\left\{\tilde{g}(k)\tilde{f}(k)\right\} = \int_{-\infty}^{\infty} g(u) f(x-u)\, \mathrm{d}u \tag{10.69}$$

$$\hat{=}\; g(x) * f(x) \quad \textit{the convolution of two functions.} \tag{10.70}$$

The integral operation called *convolution* as defined by the right-hand side of Eq. (10.69) plays an important role in filter theory, and we will see in the later section on time-series analysis that the fastest way to filter a function is by using the convolution theorem.

By making the substitution of variables $w = x - u$ so that $\mathrm{d}w = -\mathrm{d}u$, we can prove that convolution is commutative

$$\begin{aligned}
g(x) * f(x) &\hat{=} \int_{-\infty}^{\infty} g(u) f(x-u)\, \mathrm{d}u \\
&= -\int_{\infty}^{-\infty} g(x-w) f(w)\, \mathrm{d}w = \int_{-\infty}^{\infty} f(w) g(x-w)\, \mathrm{d}w \\
&= f(x) * g(x).
\end{aligned} \tag{10.71}$$

So it does not matter which function you think of as the filter and which you think of as the function being filtered. You get the same result either way.

The convolution theorem also works the other way and can be stated "multiplication in the real domain = convolution in the spectral domain (and divided by 2π)." The proof proceeds as above:

$$\mathrm{FT}\{g(x)f(x)\} = \int_{-\infty}^{\infty} f(x)\, g(x)\, e^{-ikx}\, \mathrm{d}x \tag{10.72}$$

$$= \frac{1}{2\pi}\int_{-\infty}^{\infty} f(x)\, e^{-ikx} \int_{-\infty}^{\infty} \tilde{g}(\ell)\, e^{i\ell x}\, \mathrm{d}\ell \tag{10.73}$$

$$= \frac{1}{2\pi}\int_{-\infty}^{\infty} \tilde{g}(\ell)\, \mathrm{d}\ell \int_{-\infty}^{\infty} f(x)\, e^{-i(k-\ell)x}\, \mathrm{d}x \tag{10.74}$$

$$= \frac{1}{2\pi}\int_{-\infty}^{\infty} \tilde{g}(\ell)\tilde{f}(k-\ell)\, \mathrm{d}\ell \tag{10.75}$$

$$= \frac{\tilde{g}(k) * \tilde{f}(k)}{2\pi}. \tag{10.76}$$

So, for example, if we have a very longtime function and want to consider only a truncated version of this function by first multiplying it with a boxcar function of finite length $2a$, if we then consider the Fourier transform of the truncated time function it will be the convolution of the actual spectrum of the long time function with the spectrum of the boxcar function (with the result divided by 2π due to our convention of where we place the 2π in the transform pair). We showed earlier that the spectrum of the boxcar function is the scaled sinc function $2a\sin(ak)/(ak)$. Once we learn about contour integration in Chapter 11, we show in Section 11.1.5 that in the limit as $a \to \infty$, the Dirac delta function can be expressed in terms of the sinc function as $\delta(k) = \lim_{a\to\infty} \sin(ak)/(\pi k)$ so that convolving the true spectrum of a signal with the spectrum of the boxcar function (and divided by 2π) will only provide an accurate description of the true spectrum in the limit as $a \to \infty$ so that the sinc function transitions to a Dirac delta function.

10.3.9 Integral Moments of $f(x)$

Using the Fourier transform $\tilde{f}(k) = \int_{-\infty}^{\infty} \mathrm{d}x\, f(x)\, e^{-ikx}$, it is easy to show that the following integral moments of a function $f(x)$ can be identified in terms of how the spectrum behaves near the origin $k = 0$, that is, perform k differentiation on the Fourier transform, evaluate at $k = 0$, and then read off the following results:

$$\int_{-\infty}^{\infty} f(x)\, \mathrm{d}x = \tilde{f}(k)\Big|_{k=0} \tag{10.77}$$

$$\int_{-\infty}^{\infty} x f(x)\, \mathrm{d}x = i\, \frac{\mathrm{d}\tilde{f}(k)}{\mathrm{d}k}\bigg|_{k=0} \tag{10.78}$$

$$\int_{-\infty}^{\infty} x^n f(x)\, \mathrm{d}x = (i)^n \frac{\mathrm{d}^n \tilde{f}(k)}{\mathrm{d}k^n}\bigg|_{k=0}. \tag{10.79}$$

We will now put these relations to use below in deriving two key results of probability theory.

10.4 Probability Distributions and the Central-Limit Theorem

Let's interpret the independent variable x as a measurable quantity that has random fluctuations that allow it to be distributed between $-\infty$ and $+\infty$. The reason for the fluctuations may be random noise that is recorded while measuring x or it may be random forces that change the inherent value of x such as the location of a colloid particle executing Brownian motion (i.e., the random movement of small colloid particles suspended in water that are buffeted by the random thermal motion of the water molecules). We assume that the fluctuations are characterized by a *probability-distribution function* $f(x)$, or PDF, which means that the probability of finding x in the range $[x, x+\mathrm{d}x]$ is $f(x)\mathrm{d}x$. The PDF $f(x)$ in what follows can be arbitrary but it must result in the distribution of x having both a mean value μ and a variance σ^2 about that mean and be such that the probability of finding x somewhere in the range of $[-\infty, \infty]$ is one. These requirements mean that the PDF $f(x)$ has the following integral moments:

$$1 = \int_{-\infty}^{\infty} f(x)\,\mathrm{d}x = \mathbb{E}[1] \tag{10.80}$$

$$\mu = \int_{-\infty}^{\infty} xf(x)\,\mathrm{d}x = \mathbb{E}[x] \mathrel{\widehat{=}} \text{the } \textit{mean} \text{ of } x \tag{10.81}$$

$$\sigma^2 = \int_{-\infty}^{\infty} (x-\mu)^2 f(x)\,\mathrm{d}x = \mathbb{E}[(x-\mu)^2] \mathrel{\widehat{=}} \text{the } \textit{variance} \text{ of } x$$

$$= \int_{-\infty}^{\infty} (x^2 - 2\mu x + \mu^2) f(x)\,\mathrm{d}x = \int_{-\infty}^{\infty} x^2 f(x)\,\mathrm{d}x - \mu^2 = \mathbb{E}[x^2] - \mu^2, \tag{10.82}$$

where the symbol $\mathbb{E}[\ ]$ is the "expected value" of whatever function of x resides inside the brackets, that is, $\mathbb{E}[g(x)] = \int_{-\infty}^{\infty} g(x)f(x)\,\mathrm{d}x$ for some arbitrary function $g(x)$. Higher integral moments of the PDF give information about the shape of the PDF. For example, $\int_{-\infty}^{\infty}[(x-\mu)/\sigma]^3 f(x)\,\mathrm{d}x$ is called the *skewness* and measures the asymmetry of $f(x)$ about the mean, while $\int_{-\infty}^{\infty}[(x-\mu)/\sigma]^4 f(x)\,\mathrm{d}x$ is called the *kurtosis* and measures the sharpness of the central peak of the PDF about the mean.

Let's next take the Fourier transform of the PDF

$$\tilde{f}(k) = \int_{-\infty}^{\infty} e^{-ikx} f(x)\,\mathrm{d}x = \mathbb{E}\left[e^{-ikx}\right]. \tag{10.83}$$

Combining the results of Section 10.3.9 (integral moments) with the definitions of Eqs (10.80)–(10.82) gives

$$\tilde{f}(0) = 1 \tag{10.84}$$

$$\left.\frac{\mathrm{d}\tilde{f}(k)}{\mathrm{d}k}\right|_{k=0} = -i\mu \tag{10.85}$$

$$\left.\frac{\mathrm{d}^2\tilde{f}(k)}{\mathrm{d}k^2}\right|_{k=0} = -(\sigma^2 + \mu^2). \tag{10.86}$$

Thus, if we represent the Fourier transform of the PDF as a Taylor series, we have

$$\tilde{f}(k) = \tilde{f}(0) + \frac{1}{1!}\left.\frac{\mathrm{d}\tilde{f}(k)}{\mathrm{d}k}\right|_{k=0} k + \frac{1}{2!}\left.\frac{\mathrm{d}^2\tilde{f}(k)}{\mathrm{d}k^2}\right|_{k=0} k^2 + \ldots \tag{10.87}$$

$$= 1 - i\mu k - \frac{(1+\sigma^2/\mu^2)}{2}(\mu k)^2 + O\left[(\mu k)^3\right]. \tag{10.88}$$

So for any PDF, the low-frequency limit of the function's Fourier transform is $1 - i\mu k - (\mu^2 + \sigma^2)k^2/2$, which is entirely independent of the specific functional form of the PDF. It is the higher-frequency part $O[(\mu k)^3]$ that depends on the shape of the PDF being treated.

10.4.1 The Arithmetic Mean and the Normal Distribution

We now put the above to work to derive the *central-limit theorem.* If we define the arithmetic mean of the randomly fluctuating variable x over N measurements to be

$$\overline{x}_a = \frac{1}{N}\sum_{n=1}^{N} x_n \tag{10.89}$$

the central-limit theorem says that as N gets big, the PDF of the arithmetic mean $p(\overline{x}_a)$ is a Gaussian or "normal" distribution regardless of the functional nature of the underlying distribution $f(x_n)$. In the limit as $N \to \infty$, we also expect $\overline{x}_a \to \mu$, which is called the *law of large numbers*.

To prove these assertions, we begin with the definition of both the Fourier transform of the PDF controlling the distribution of $\overline{x}_a$ and the expected value of the product of random variables

$$\tilde{p}(k) = \int_{-\infty}^{\infty} e^{-ik\overline{x}_a} p(\overline{x}_a)\,\mathrm{d}\overline{x}_a = \mathbb{E}\left[e^{-ik\overline{x}_a}\right] = \mathbb{E}\left[e^{-ikx_1/N}e^{-ikx_2/N}\ldots e^{-ikx_N/N}\right] \tag{10.90}$$

$$= \mathbb{E}\left[e^{-ikx_1/N}\right]\mathbb{E}\left[e^{-ikx_2/N}\right]\ldots\mathbb{E}\left[e^{-ikx_N/N}\right] \tag{10.91}$$

$$= \left[\int_{-\infty}^{\infty}\mathrm{d}x_1 f(x_1)e^{-ikx_1/N}\right]\left[\int_{-\infty}^{\infty}\mathrm{d}x_2 f(x_2)e^{-ikx_2/N}\right]\ldots\left[\int_{-\infty}^{\infty}\mathrm{d}x_N f(x_N)e^{-ikx_N/N}\right]. \tag{10.92}$$

If two observables, say y and z, have fluctuations that are uncorrelated to each other so that $p(y, z) = p(y)p(z)$, the expected value of their product is $\mathbb{E}\left[yz\right] = \mathbb{E}\left[y\right]\mathbb{E}\left[z\right]$. Because our observations x_n are varying randomly, they are uncorrelated to each other, which leads from

Eq. (10.90) to Eqs (10.91) and (10.92). Thus using Eq. (10.88) for the Fourier transform of the probability distribution of each observation, we have

$$\tilde{p}(k) = \left[\tilde{f}\left(\frac{k}{N}\right)\right]^N = \left[1 - i\frac{\mu k}{N} - \frac{(1+\sigma^2/\mu^2)}{2}\left(\frac{\mu k}{N}\right)^2 + O\left[\left(\frac{\mu k}{N}\right)^3\right]\right]^N . \tag{10.93}$$

As N gets big, the $O[(\mu k/N)^3]$ terms that depend on the specific shape of the underlying PDF $f(x)$ can be dropped relative to the terms that are kept.

So in this limit of large N, we use the Taylor expansion $(1-y)^N = 1 - Ny/1! + N(N-1)y^2/2! + \ldots$ to write

$$\tilde{p}(k) = \left[1 - i\frac{\mu k}{N} - \frac{(1+\sigma^2/\mu^2)}{2}\left(\frac{\mu k}{N}\right)^2\right]^N \tag{10.94}$$

$$= 1 - N\left[i\frac{\mu k}{N} + \frac{(1+\sigma^2/\mu^2)}{2}\left(\frac{\mu k}{N}\right)^2\right] + \frac{N(N-1)}{2}\left[i\frac{\mu k}{N} + \frac{(1+\sigma^2/\mu^2)}{2}\left(\frac{\mu k}{N}\right)^2\right]^2 + \ldots \tag{10.95}$$

$$= 1 - i\mu k - \frac{(\mu^2+\sigma^2)}{2}\frac{k^2}{N} + \frac{(N^2-N)}{2}\left[-\left(\frac{\mu k}{N}\right)^2 + O\left[\left(\frac{\mu k}{N}\right)^3\right]\right] \tag{10.96}$$

$$= 1 - i\mu k - \left(\frac{\mu^2}{2} + \frac{\sigma^2}{2N}\right)k^2, \tag{10.97}$$

where again terms of $O[(\mu k/N)^3]$ are being dropped relative to terms of $O[(\mu k/N)^2]$ as N gets large. Next an equivalent expansion for the exponential function $e^{-z} = 1 - z/1! + z^2/2! + \ldots$ results in

$$\exp\left\{-N\left[i\frac{\mu k}{N} + \frac{(\sigma/\mu)^2}{2}\left(\frac{\mu k}{N}\right)^2\right]\right\}$$

$$= 1 - N\left[\frac{i\mu k}{N} + \frac{(\sigma/\mu)^2}{2}\left(\frac{\mu k}{N}\right)^2\right] + \frac{N^2}{2}\left[\frac{i\mu k}{N} + \frac{(\sigma/\mu)^2}{2}\left(\frac{\mu k}{N}\right)^2\right]^2 + \ldots \tag{10.98}$$

$$= 1 - i\mu k - \frac{\sigma^2 k^2}{2N} + \frac{N^2}{2}\left[-\left(\frac{\mu k}{N}\right)^2 + O\left[\left(\frac{\mu k}{N}\right)^3\right]\right] \tag{10.99}$$

$$= 1 - i\mu k - \left(\frac{\mu^2}{2} + \frac{\sigma^2}{2N}\right)k^2 \tag{10.100}$$

to the same leading order as N becomes large. Equating Eqs (10.97) and (10.100) then gives

$$\tilde{p}(k) = e^{-i\mu k - \sigma^2 k^2/(2N)} \tag{10.101}$$

as the Fourier transform of the sought after probability distribution for the arithmetic mean of N measurements as N becomes large.

The probability distribution for the arithmetic mean $\overline{x}_a$ is obtained as the inverse Fourier transform of $\tilde{p}(k)$

$$p(\overline{x}_a) = \frac{1}{2\pi}\int_{-\infty}^{\infty} \mathrm{d}k e^{ik\overline{x}_a} e^{-ik\mu} e^{-\sigma^2 k^2/(2N)} \tag{10.102}$$

$$= \frac{1}{2\pi}\int_{-\infty}^{\infty} \mathrm{d}k e^{ik(\overline{x}_a-\mu)} e^{-\sigma^2 k^2/(2N)}. \tag{10.103}$$

We complete the square in the exponent $-\sigma^2k^2/(2N) + ik(\overline{x}_a - \mu) = -[\sigma k/\sqrt{2N} + i(\overline{x}_a - \mu)\sqrt{2N}/(2\sigma)]^2 - (\overline{x}_a - \mu)^2 N/(2\sigma^2)$ and substitute variables $u = \sigma k/\sqrt{2N} + i(\overline{x}_a - \mu)\sqrt{2N}/(2\sigma)$ with $\mathrm{d}k = \mathrm{d}u\sqrt{2N}/\sigma$ to obtain

$$p(\overline{x}_a) = \frac{\sqrt{2N}}{2\pi\sigma} \exp\left[-\frac{1}{2}\left(\frac{\overline{x}_a-\mu}{\sigma/\sqrt{N}}\right)^2\right]\int_{-\infty}^{\infty} e^{-u^2}\,\mathrm{d}u \tag{10.104}$$

or

$$\boxed{p(\overline{x}_a) = \frac{1}{\sigma_a\sqrt{2\pi}} \exp\left[-\frac{1}{2}\left(\frac{\overline{x}_a-\mu}{\sigma_a}\right)^2\right].} \tag{10.105}$$

This is exactly the Gaussian, or normal, distribution with a mean of μ (that of the random variable x) and a standard of deviation

$$\boxed{\sigma_a = \frac{\sigma}{\sqrt{N}}.} \tag{10.106}$$

So the arithmetic mean of a random variable measured N times is distributed as a Gaussian as N gets large, a result known as the *central-limit theorem*. The equally important result here is that the standard of deviation of the arithmetic mean σ_a decreases as $1/\sqrt{N}$ with increasing number N of measurements. So as $N \to \infty$, we have $p(\overline{x}_a) = \delta(\overline{x}_a - \mu)$ or $\overline{x}_a = \mu$, a result known as the *law of large numbers*.

Because so many observables can be thought of as the arithmetic average of a quantity measured repeatedly in the presence of random noise, the normal distribution is widely observed in experimental science.

10.4.2 The Geometric Mean and the Log-Normal Distribution

Another average that has wide application is the geometric mean defined as

$$\overline{x}_g = \prod_{n=1}^{N} x_n^{1/N} = [x_1 x_2 \ldots x_N]^{1/N}. \tag{10.107}$$

For the geometric mean to be real, the x_n must be positive and will typically have some physical units. We often use the geometric mean when the x_n are the positive physical properties of some material having fine-scale heterogeneity. If we first divide both sides of Eq. (10.107) by a positive constant x_o having the physical units of x_n, we can take the logarithm to give

$$\ln\left(\frac{\overline{x}_g}{x_o}\right) = \frac{1}{N}\sum_{n=1}^{N}\ln\left(\frac{x_n}{x_o}\right). \tag{10.108}$$

Because the random variables x_n are distributed over $0 < x_n < \infty$, the $\ln(x_n/x_o)$ are randomly distributed between $-\infty$ and $+\infty$, with a mean (expected) value $\ln(\mu/x_o)$ and a standard-of-deviation $\ln(\sigma/x_o)$, that is, the distribution function $f(\ln(x/x_o))$ must have the same low-frequency spectrum $\tilde{f}(k)$ given by Eq. (10.88) but with $\mu \to \ln(\mu/x_o)$ and $\sigma \to \ln(\sigma/x_o)$.

Given the above, an immediate corollary to the central-limit theorem is that the logarithm of the geometric mean satisfies a normal distribution in the form

$$p_l\left(\ln(\overline{x}_g/x_o)\right) = \frac{1}{\ln(\sigma_g/x_o)\sqrt{2\pi}}\exp\left[-\frac{1}{2}\left(\frac{\ln(\overline{x}_g/x_o) - \ln(\mu/x_o)}{\ln(\sigma_g/x_o)}\right)^2\right], \tag{10.109}$$

where

$$\ln\left(\frac{\sigma_g}{x_o}\right) = \frac{1}{\sqrt{N}}\ln\left(\frac{\sigma}{x_o}\right). \tag{10.110}$$

In terms of the geometric mean $\overline{x}_g$ directly, which is distributed on $[0, \infty]$, and not its logarithm, which is distributed on $[-\infty, \infty]$, we have that $p_l \, \mathrm{d}\ln(\overline{x}_g/x_o) = (p_l/\overline{x}_g)\, \mathrm{d}\overline{x}_g$ so that the desired distribution is $p(\overline{x}_g) = p_l/\overline{x}_g$ or

$$p(\overline{x}_g) = \frac{1}{\overline{x}_g \ln(\sigma_g/x_o)\sqrt{2\pi}}\exp\left[-\frac{1}{2}\left(\frac{\ln(\overline{x}_g/x_o) - \ln(\mu/x_o)}{\ln(\sigma_g/x_o)}\right)^2\right], \tag{10.111}$$

which is called the log-normal distribution.

Taking $x_o = \mu$ is a convenient choice when σ_g is measured as factors of μ. In this case, the log-normal distribution may also be written

$$\boxed{p(\overline{x}_g) = \frac{1}{\overline{x}_g \ln(\sigma_g/\mu)\sqrt{2\pi}}\exp\left[-\frac{1}{2}\left(\frac{\ln(\overline{x}_g/\mu)}{\ln(\sigma_g/\mu)}\right)^2\right],} \tag{10.112}$$

where again $0 < \overline{x}_g < \infty$. For positive material properties with random distributions through space (random heterogeneity), the geometric mean can be a good effective-medium model, that is, the material with random small-scale heterogeneity can be replaced by a homogeneous material having a material property given by the geometric mean of the small-scale heterogeneity. We thus expect that log-normal distributions are well suited for describing the randomness of material properties within a heterogeneous body like the Earth as has been demonstrated experimentally by Masson and Pride (2015) for various types of rocks.

10.4.3 The Harmonic Mean and the Inverse-Normal Distribution

A final average to be considered is the harmonic mean defined as

$$\frac{1}{\overline{x}_h} = \frac{1}{N}\sum_{n=1}^{N}\frac{1}{x_n}. \tag{10.113}$$

We assume the x_n^{-1} are randomly distributed on $[-\infty, +\infty]$, which means that the x_n are distributed between 0_- and $-\infty$ and 0_+ and $+\infty$, and that the distribution of the x_n^{-1} possesses a mean μ^{-1} and standard of deviation σ^{-1}, then again the Fourier transform of the distribution $f(x^{-1})$ will have the low-frequency form of Eq. (10.88) but with $\mu \to \mu^{-1}$ and $\sigma \to \sigma^{-1}$.

An immediate corollary of the central-limit theorem is that the inverse of the harmonic mean satisfies a normal distribution in the form

$$p_h(\overline{x}_h^{-1}) = \frac{\sqrt{N}}{\sigma^{-1}\sqrt{2\pi}}\exp\left[-\frac{N}{2}\left(\frac{\overline{x}_h^{-1} - \mu^{-1}}{\sigma^{-1}}\right)^2\right] \tag{10.114}$$

$$= \frac{\sigma\sqrt{N}}{\sqrt{2\pi}}\exp\left[-\frac{\sigma^2 N}{2}\left(\frac{\overline{x}_h - \mu}{\overline{x}_h\mu}\right)^2\right]. \tag{10.115}$$

Now the expected value of some function of $\overline{x}_h^{-1}$, say $g(\overline{x}_h^{-1})$, is

$$\mathbb{E}[g(\overline{x}_h^{-1})] = \int_{-\infty}^{\infty} \mathrm{d}\overline{x}_h^{-1}\, p_h(\overline{x}_h^{-1}) g(\overline{x}_h^{-1})$$

$$= \int_{0_+}^{\infty} \mathrm{d}\overline{x}_h^{-1}\, p_h(\overline{x}_h^{-1}) g(\overline{x}_h^{-1}) + \int_{-\infty}^{0_-} \mathrm{d}\overline{x}_h^{-1}\, p_h(\overline{x}_h^{-1}) g(\overline{x}_h^{-1}). \tag{10.116}$$

So in terms of the harmonic average itself $\overline{x}_h$ and not its inverse, the expected value of some function $g(\overline{x}_h)$ is

$$\mathbb{E}[g(\overline{x}_h)] = -\int_{\infty}^{0_+} \mathrm{d}\overline{x}_h \frac{p_h(\overline{x}_h^{-1})}{\overline{x}_h^2} g(\overline{x}_h) - \int_{0_-}^{-\infty} \mathrm{d}\overline{x}_h \frac{p_h(\overline{x}_h^{-1})}{\overline{x}_h^2} g(\overline{x}_h) = \int_{-\infty}^{\infty} \mathrm{d}\overline{x}_h \frac{p_h(\overline{x}_h^{-1})}{\overline{x}_h^2} g(\overline{x}_h). \tag{10.117}$$

Thus, the harmonic mean satisfies a probability distribution $p(\overline{x}_h) = \overline{x}_h^{-2} p_h(\overline{x}_h^{-1})$ or

$$\boxed{p(\overline{x}_h) = \frac{\sigma\sqrt{N}}{\overline{x}_h^2\sqrt{2\pi}}\exp\left[-\frac{\sigma^2 N}{2}\left(\frac{\overline{x}_h - \mu}{\overline{x}_h\mu}\right)^2\right],} \tag{10.118}$$

which can be called the inverse-normal distribution.

So in this section we have used the Fourier transform to prove that the arithmetic mean of arbitrarily distributed random variables, that are only required to possess a mean and variance, follows a normal distribution, a result known as the central-limit theorem. Similarly, the geometric mean of such random variables follows a log-normal distribution and the harmonic mean follows an inverse-normal distribution.

10.5 Multidimensional Transform Pairs

We can define multidimensional transform pairs as alluded to earlier. For example, we can write the inverse transform of some function of space as

$$f(\boldsymbol{r}) = f(x, y, z) = \frac{1}{(2\pi)^3} \int_{-\infty}^{+\infty} dk_x\, e^{ik_x x} \int_{-\infty}^{+\infty} dk_y\, e^{ik_y y} \int_{-\infty}^{+\infty} dk_z\, e^{ik_z z} \tilde{f}(k_x, k_y, k_z) \tag{10.119}$$

$$= \frac{1}{(2\pi)^3} \iiint_{-\infty}^{+\infty} dk_x\, dk_y\, dk_z\, e^{i(k_x x + k_y y + k_z z)} \tilde{f}(k_x, k_y, k_z)$$

$$= \frac{1}{(2\pi)^3} \iiint_{-\infty}^{+\infty} d^3\boldsymbol{k}\, e^{i\boldsymbol{k}\cdot\boldsymbol{r}} \tilde{f}(\boldsymbol{k}), \tag{10.120}$$

where to write the last more compact expression we used $\boldsymbol{r} \mathrel{\widehat{=}} x\hat{\boldsymbol{x}} + y\hat{\boldsymbol{y}} + z\hat{\boldsymbol{z}} \mathrel{\widehat{=}} (x, y, z)$, $\boldsymbol{k} \mathrel{\widehat{=}} k_x\hat{\boldsymbol{x}} + k_y\hat{\boldsymbol{y}} + k_z\hat{\boldsymbol{z}} \mathrel{\widehat{=}} (k_x, k_y, k_z)$, $\boldsymbol{k}\cdot\boldsymbol{r} = k_x x + k_y y + k_z z$, and $d^3\boldsymbol{k} \mathrel{\widehat{=}} dk_x\, dk_y\, dk_z$. For the Fourier transform we write

$$\tilde{f}(\boldsymbol{k}) = \tilde{f}(k_x, k_y, k_z) = \int_{-\infty}^{+\infty} dx\, e^{-ik_x x} \int_{-\infty}^{+\infty} dy\, e^{-ik_y y} \int_{-\infty}^{+\infty} dz\, e^{-ik_z z} f(x, y, z) \tag{10.121}$$

$$= \iiint_{-\infty}^{+\infty} dx\, dy\, dz\, e^{-i(k_x x + k_y y + k_z z)} f(x, y, z)$$

$$= \iiint_{-\infty}^{+\infty} d^3\boldsymbol{r}\, e^{-i\boldsymbol{k}\cdot\boldsymbol{r}} f(\boldsymbol{r}), \tag{10.122}$$

where we wrote $d^3\boldsymbol{r} = dx\, dy\, dz$ in the final compact expression. If we include a time transform in addition the three space transforms we have the four-fold transform pairs written compactly as

$$f(\boldsymbol{r}, t) = \frac{1}{(2\pi)^4} \iiiint_{-\infty}^{+\infty} d^3\boldsymbol{k}\, d\omega\, e^{i(\boldsymbol{k}\cdot\boldsymbol{r} - \omega t)} \tilde{\tilde{f}}(\boldsymbol{k}, \omega) \tag{10.123}$$

$$\tilde{\tilde{f}}(\boldsymbol{k}, \omega) = \iiiint_{-\infty}^{+\infty} d^3\boldsymbol{r}\, dt\, e^{-i(\boldsymbol{k}\cdot\boldsymbol{r} - \omega t)} f(\boldsymbol{r}, t). \tag{10.124}$$

But remember that when you see a compact Fourier kernel like $e^{i\boldsymbol{k}\cdot\boldsymbol{r}}$, it just means $e^{i\boldsymbol{k}\cdot\boldsymbol{r}} = e^{i(k_x x + k_y y + k_z z)} = e^{ik_x x}\, e^{ik_y y}\, e^{ik_z z}$ when written out in Cartesian coordinates. We use two tildes on the transformed function, one for the time transform and one for the space transforms.

Consider next the spatial gradient of some function $f(\boldsymbol{r}, t)$ that is expressed as an inverse Fourier transform

$$\nabla f(\boldsymbol{r}, t) = \left(\hat{\boldsymbol{x}}\frac{\partial}{\partial x} + \hat{\boldsymbol{y}}\frac{\partial}{\partial y} + \hat{\boldsymbol{z}}\frac{\partial}{\partial z}\right)\frac{1}{(2\pi)^4}\iiiint_{-\infty}^{+\infty} \mathrm{d}^3\boldsymbol{k}\,\mathrm{d}\omega\, e^{i(k_x x + k_y y + k_z z - \omega t)}\,\tilde{\tilde{f}}(\boldsymbol{k}, \omega) \quad (10.125)$$

$$= \frac{1}{(2\pi)^4}\iiiint_{-\infty}^{+\infty} \mathrm{d}^3\boldsymbol{k}\,\mathrm{d}\omega\, \left(\hat{\boldsymbol{x}}\, ik_x + \hat{\boldsymbol{y}}\, ik_y + \hat{\boldsymbol{z}}\, ik_z\right)\, e^{i(\boldsymbol{k}\cdot\boldsymbol{r} - \omega t)}\,\tilde{\tilde{f}}(\boldsymbol{k}, \omega) \quad (10.126)$$

$$= \frac{1}{(2\pi)^4}\iiiint_{-\infty}^{+\infty} \mathrm{d}^3\boldsymbol{k}\,\mathrm{d}\omega\, e^{i(\boldsymbol{k}\cdot\boldsymbol{r} - \omega t)}\, i\boldsymbol{k}\tilde{\tilde{f}}(\boldsymbol{k}, \omega). \quad (10.127)$$

This demonstrates the important and convenient result that the operator ∇ acting in the space domain simply becomes multiplication by the vector $i\boldsymbol{k}$ in the transform domain. We will use this repeatedly when we apply the Fourier transform to PDEs. For example, if we have the vector differential operation $\nabla \cdot \nabla\nabla\nabla \cdot \boldsymbol{f}(\boldsymbol{r})$ in the space domain and want to know the Fourier transform of this expression, we can replace each ∇ with $i\boldsymbol{k}$ to immediately write the transform as $k^2\boldsymbol{k}\boldsymbol{k} \cdot \tilde{\boldsymbol{f}}(\boldsymbol{k})$ where $k^2 = \boldsymbol{k}\cdot\boldsymbol{k} = k_x^2 + k_y^2 + k_z^2$.

Similarly, for a time derivative we have

$$\frac{\partial}{\partial t} f(\boldsymbol{r}, t) = \frac{1}{(2\pi)^4}\iiiint_{-\infty}^{+\infty} \mathrm{d}^3\boldsymbol{k}\,\mathrm{d}\omega\, e^{i(\boldsymbol{k}\cdot\boldsymbol{r} - \omega t)}\,(-i\omega)\tilde{\tilde{f}}(\boldsymbol{k}, \omega). \quad (10.128)$$

This says that the operator $\partial/\partial t$ acting in the time domain simply becomes multiplication by the scalar $-i\omega$ in the temporal transform domain.

So if we consider the elastodyamic wave equation for elastic waves in an infinite homogeneous solid created by a directed point source at the origin $\boldsymbol{f} = s(t)\delta(\boldsymbol{r})\hat{\boldsymbol{s}}$

$$\rho\frac{\partial^2 \boldsymbol{u}(\boldsymbol{r}, t)}{\partial t^2} = (K + G/3)\,\nabla\nabla\cdot \boldsymbol{u}(\boldsymbol{r}, t) + G\nabla^2\boldsymbol{u}(\boldsymbol{r}, t) + s(t)\delta(\boldsymbol{r})\hat{\boldsymbol{s}}, \quad (10.129)$$

and if we want to know how to write down this PDE in the four-fold transform domain, we use $\nabla \to i\boldsymbol{k}$, $\partial/\partial t \to -i\omega$ and FT $\{\delta(\boldsymbol{r})\} = 1$ to write

$$-\rho\omega^2\tilde{\tilde{\boldsymbol{u}}}(\boldsymbol{k}, \omega) = -(K + G/3)\boldsymbol{k}\boldsymbol{k}\cdot\tilde{\tilde{\boldsymbol{u}}}(\boldsymbol{k}, \omega) - Gk^2\tilde{\tilde{\boldsymbol{u}}}(\boldsymbol{k}, \omega) + \tilde{s}(\omega)\hat{\boldsymbol{s}}, \quad (10.130)$$

which can be rewritten

$$\left[(K + G/3)\boldsymbol{k}\boldsymbol{k} + \left(Gk^2 - \rho\omega^2\right)\boldsymbol{I}\right]\cdot\tilde{\tilde{\boldsymbol{u}}}(\boldsymbol{k}, \omega) = \tilde{s}(\omega)\hat{\boldsymbol{s}}, \quad (10.131)$$

where $\boldsymbol{I}$ is the identity tensor. We show later in Chapter 12 how to solve this algebraic (linear algebra) expression to obtain $\tilde{\tilde{\boldsymbol{u}}}(\boldsymbol{k}, \omega)$ and how to return to the time and space domain through the inverse Fourier transform (this will require contour integration that we have not yet covered). The result is called the elastodynamic Green's function. Solving for the Green's function in the four-fold transform domain is an easy algebraic exercise. The crux of solving PDEs using the Fourier transform is in returning from the transform domain back to the real domain.

In some applications of solving PDEs, we will find it convenient to transform over space, but not over time; in others, we will transform over time, but not over space; and in others, we will transform over both space and time. In all such applications of applying the Fourier transform to PDEs, the domain over which we transform is infinite.

10.6 Diffusion Initial-Value Problems

As our first example of using the Fourier transform in the solution of a boundary-value problem, consider a diffusion initial-value problem for a field $\psi(\boldsymbol{r}, t)$ in a 3D wholespace that satisfies the diffusion equation

$$D\,\nabla^2\,\psi(\boldsymbol{r}, t) - \frac{\partial\psi(\boldsymbol{r}, t)}{\partial t} = 0 \tag{10.132}$$

subject to the BC at infinity that

$$\psi(\boldsymbol{r}, t) \to 0 \text{ as } |\boldsymbol{r}| \to \infty. \tag{10.133}$$

The initial condition at $t = 0$ is that

$$\psi(\boldsymbol{r}, t)|_{t=0} = \psi_0(\boldsymbol{r}). \tag{10.134}$$

To solve this problem, we first Fourier transform over 3D space ($\boldsymbol{r} \to \boldsymbol{k}$ and $\nabla \to i\boldsymbol{k}$) to obtain the diffusion equation in the form

$$D(i\boldsymbol{k}\cdot i\boldsymbol{k})\,\tilde{\psi}(\boldsymbol{k}, t) - \frac{\partial}{\partial t}\,\tilde{\psi}(\boldsymbol{k}, t) = 0 \tag{10.135}$$

with an initial condition given by

$$\tilde{\psi}(\boldsymbol{k}, t)\Big|_{t=0} = \tilde{\psi}_0(\boldsymbol{k}) = \iiint_{-\infty}^{+\infty} \psi_0(\boldsymbol{r})\,e^{-i\boldsymbol{k}\cdot\boldsymbol{r}}\,\mathrm{d}^3\boldsymbol{r}. \tag{10.136}$$

So upon noting that $\boldsymbol{k}\cdot\boldsymbol{k} = k^2 = k_x^2 + k_y^2 + k_z^2$, we write the diffusion equation as

$$-\,D\,k^2\,\tilde{\psi} - \frac{\partial}{\partial t}\tilde{\psi} = 0, \tag{10.137}$$

which can be solved in the time domain to give

$$\boxed{\tilde{\psi}(\boldsymbol{k}, t) = \tilde{\psi}_0(\boldsymbol{k})\,e^{-D\,k^2 t}.} \tag{10.138}$$

Obtaining this solution in the $(\boldsymbol{k}, t)$ domain was straightforward and relatively easy.

The crux of the problem is to return to the space domain using the inverse Fourier transform

$$\psi(\boldsymbol{r},t)=\frac{1}{(2\pi)^3}\iiint_{-\infty}^{+\infty}\mathrm{d}^3\boldsymbol{k}\,e^{i(\boldsymbol{k}\cdot\boldsymbol{r})}\,\tilde{\psi}_0(\boldsymbol{k})\,e^{-Dk^2t}. \tag{10.139}$$

If we write the initial condition in $\boldsymbol{k}$ domain as

$$\tilde{\psi}_0(\boldsymbol{k})=\iiint_{-\infty}^{+\infty}\mathrm{d}^3\boldsymbol{r}'\,e^{-i\boldsymbol{k}\cdot\boldsymbol{r}'}\,\psi_0(\boldsymbol{r}'). \tag{10.140}$$

our solution can be written in the form

$$\psi(\boldsymbol{r},t)=\iiint_{-\infty}^{+\infty}\mathrm{d}^3\boldsymbol{r}'\,\psi_0(\boldsymbol{r}')\underbrace{\iiint_{-\infty}^{+\infty}\frac{\mathrm{d}^3\boldsymbol{k}}{(2\pi)^3}\,e^{i\boldsymbol{k}\cdot(\boldsymbol{r}-\boldsymbol{r}')}\,e^{-k^2Dt}}_{\text{Let's do these integrals first}}. \tag{10.141}$$

The integrals over $\boldsymbol{k}$ separate as

$$\iiint_{-\infty}^{+\infty}\frac{\mathrm{d}k_x\,\mathrm{d}k_y\,\mathrm{d}k_z}{(2\pi)^3}\,e^{-(k_x^2+k_y^2+k_z^2)Dt}\,e^{i[k_x(x-x')+k_y(y-y')+k_z(z-z')]}=$$

$$=\left[\frac{1}{2\pi}\int_{-\infty}^{+\infty}\mathrm{d}k_x\,e^{-k_x^2Dt+ik_x(x-x')}\right]\left[\frac{1}{2\pi}\int_{-\infty}^{+\infty}\mathrm{d}k_y\,e^{-k_y^2Dt+ik_y(y-y')}\right]\left[\frac{1}{2\pi}\int_{-\infty}^{+\infty}\mathrm{d}k_z\,e^{-k_z^2Dt+ik_z(z-z')}\right], \tag{10.142}$$

which is simply the product of three integrals having identical form. Each integral is the inverse Fourier transform of a Gaussian that we performed earlier by completing the square in the exponent:

$$\frac{1}{2\pi}\int_{-\infty}^{+\infty}\mathrm{d}k_x\,e^{-k_x^2Dt+ik_x(x-x')}=\frac{1}{2\pi}\int_{-\infty}^{+\infty}\mathrm{d}k_x\,e^{-\left[\sqrt{Dt}\,k_x-i(x-x')/(2\sqrt{Dt})\right]^2}\,e^{-(x-x')^2/(4Dt)}. \tag{10.143}$$

Let $u=\sqrt{Dt}\,k_x-i(x-x')/(2\sqrt{Dt})$ with $\mathrm{d}u=\sqrt{Dt}\,\mathrm{d}k_x$, so that

$$\frac{1}{2\pi}\int_{-\infty}^{+\infty}\mathrm{d}k_x\,e^{-k_x^2Dt+ik_x(x-x')}=\frac{e^{-(x-x')^2/(4Dt)}}{2\pi\sqrt{Dt}}\underbrace{\left[\int_{-\infty}^{+\infty}\mathrm{d}u\,e^{-u^2}\right]}_{=\sqrt{\pi}}=\frac{e^{-(x-x')^2/(4Dt)}}{2\sqrt{\pi Dt}}. \tag{10.144}$$

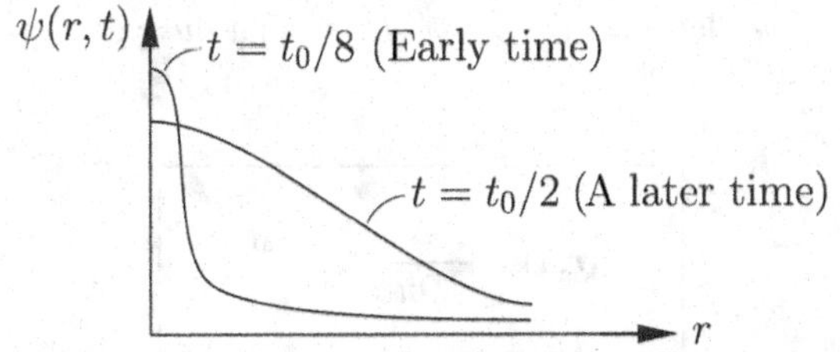

Figure 10.7 Diffusive field evolving through time from an initial field concentrated at the origin $r = 0$.

Thus, because $|\boldsymbol{r} - \boldsymbol{r}'|^2 = (x - x')^2 + (y - y')^2 + (z - z')^2$, we find that for any initial condition $\psi_0(\boldsymbol{r})$ distributed through all of space,

$$\psi(\boldsymbol{r}, t) = \frac{1}{8}\left(\frac{1}{D\pi t}\right)^{3/2} \iiint_{-\infty}^{+\infty} \mathrm{d}^3\boldsymbol{r}' \, \psi_0(\boldsymbol{r}')\, e^{-|\boldsymbol{r}-\boldsymbol{r}'|^2/(4Dt)}. \tag{10.145}$$

So given how a field is distributed in space at $t = 0$, we can calculate how it evolves through time and space for all $t > 0$ by performing this integral. We can use this result in many ways.

10.6.1 A Hot Spot

Let's consider the specific initial condition that the field is concentrated at a point at $t = 0$ so that $\psi_0(\boldsymbol{r}) = A\,\delta(\boldsymbol{r}) = A\,\delta(x)\,\delta(y)\,\delta(z)$. This then gives immediately that

$$\psi(r, t) = \frac{A}{8\,(D\pi\, t)^{3/2}}\, e^{-r^2/(4Dt)}, \tag{10.146}$$

whose nature in time and space is depicted in Fig. 10.7. Mathematically, the distance diffused r_d after a time t_d is controlled by the exponent in Eq. (10.146)

$$r_d^2 \approx 4\,Dt_d \quad \text{or} \quad r_d = \sqrt{4\,Dt_d} \quad \text{or} \quad t_d = \frac{r_d^2}{4D} \tag{10.147}$$

As stated earlier, this simple relation between distance diffused r_d in a time t_d is the most important single thing to remember about diffusion.

10.6.2 A Hot Line

In this case, we will take the initial condition to have the form $\psi_o(\boldsymbol{r}) = A\delta(y)\delta(z)$ to give

$$\psi(\boldsymbol{r}, t) = \frac{A}{8}\left(\frac{1}{D\pi\, t}\right)^{3/2} e^{-r^2/(4Dt)} \int_{-\infty}^{+\infty} \mathrm{d}x'\, e^{-(x-x')^2/(4Dt)}, \tag{10.148}$$

where $r^2 = y^2 + z^2$. If we let $u = (x' - x)/\sqrt{4Dt}$ so that $du = dx'/\sqrt{4Dt}$ we obtain (note $\int_{-\infty}^{\infty} e^{-u^2}\, du = \sqrt{\pi}$)

$$\psi(\mathbf{r}, t) = \frac{A}{4D\pi\, t} e^{-r^2/(4Dt)}. \tag{10.149}$$

Although there is a slightly less rapid fall off of ψ as a function of time compared to an initially hot spot, the relation between r and t for how the diffusive response moves out in cylindrical symmetry from the line is again controlled by the argument in the exponent which is identical to the 3D hot spot.

10.6.3 A Hot Plane

In this case, we will take the initial condition to have the form $\psi_o(\mathbf{r}) = A\delta(z)$ to give

$$\psi(\mathbf{r}, t) = \frac{A}{8}\left(\frac{1}{D\pi\, t}\right)^{3/2} e^{-z^2/(4Dt)} \left[\int_{-\infty}^{+\infty} dx'\, e^{-(x-x')^2/(4\,Dt)}\right]\left[\int_{-\infty}^{+\infty} dy'\, e^{-(y-y')^2/(4\,Dt)}\right]. \tag{10.150}$$

Both integrals in brackets are identical and were performed above so that we obtain

$$\psi(\mathbf{r}, t) = \frac{A}{2\sqrt{D\pi\, t}} e^{-z^2/(4Dt)}. \tag{10.151}$$

This defines how the temperature diffusively moves away from an initially hot plane. It is exactly a Gaussian function in z with a standard of deviation $\sigma = \sqrt{2Dt}$ that grows through time

It is informative to compare the above three boxed expressions for the diffusion away from a point (3D spherical diffusion), a line (2D cylindrical diffusion), and a plane (1D planar diffusion); the space–time relation coming from the exponential function is similar in all three cases but the amplitude variation through time at a given distance from the initial concentration of heat is different.

10.6.4 An Advection–Diffusion Problem

When fluid flow and diffusion are occurring simultaneously there is an additional advection term that must be added to the diffusion equation to create a so-called advection–diffusion or dispersion equation. In Chapter 7, and specifically in Section 7.3.5, we derive the governing equation for the solute concentration $c(z, t)$ down the length z of a cylindrical capillary tube of radius a that is supporting a uniform average flow velocity v_z. The concentration field here $c(z, t)$ is the average concentration across each cross section of the tube and obeys the 1D "Taylor-dispersion" equation

$$\frac{\partial c(z,t)}{\partial t} + v_z \frac{\partial c(z,t)}{\partial z} = D_{\text{eff}} \frac{\partial^2 c(z,t)}{\partial z^2}, \tag{10.152}$$

where the effective solute diffusivity D_{eff} for a given Péclet number $\text{Pe} \,\hat{=}\, av_z/D_c$ takes the form $D_{\text{eff}} = D_c(1 + \text{Pe}^2/48)$ with D_c the diffusivity of the solute when there is no flow in the tube.

The initial condition for the solute concentration can be anything but to create a specific problem to solve we take it to be analogous to the above "hot plane" problem

$$c_0(z) = c(z, t=0) = \Delta z \Delta c\, \delta(z), \tag{10.153}$$

where $\delta(z)$ is again the Dirac delta. So the problem to be solved here involves an infinitely long cylindrical capillary tube that is supporting steady fluid flow of pure solvent. At $t = 0$, there is a one-time injection of solute at the origin $z = 0$ that is applied over a small distance Δz such that the initial solute concentration at $t = 0$ is $c_0(z) = \Delta z \Delta c \delta(z)$.

We solve this problem by taking a Fourier transform over z in both the governing equation (where as usual $\partial/\partial z \to ik$) and the initial condition to give

$$\frac{\partial \tilde{c}(k,t)}{\partial t} = -\left(D_{\text{eff}} k^2 + ikv_z\right) \tilde{c}(k,t) \tag{10.154}$$

subject to $\tilde{c}(k, t=0) = \tilde{c}_0(k) = \Delta z \Delta c$ at $t = 0$. Integrating this equation gives

$$\tilde{c}(k,t) = \tilde{c}_0(k) \exp\left[-\left(D_{\text{eff}} k^2 + ikv_z\right) t\right] \tag{10.155}$$

so that upon applying the inverse Fourier transform, we obtain

$$c(z,t) = \frac{\Delta z \Delta c}{2\pi} \int_{-\infty}^{\infty} \mathrm{d}k\, e^{-D_{\text{eff}} k^2 t} e^{ik(z - v_z t)} \tag{10.156}$$

$$= \boxed{\frac{\Delta z \Delta c}{2\sqrt{\pi D_{\text{eff}} t}} e^{-(z - v_z t)^2/(4 D_{\text{eff}} t)},} \tag{10.157}$$

where the integral over k was performed as above by completing the square in the exponential. So compared to the 1D hot-plane problem solved above when there was no advection, there is a translation through time of the origin $z = v_z t$ and a different diffusivity as derived in Section 7.3.5 on Taylor dispersion given by $D_{\text{eff}} = D_c(1 + \text{Pe}^2/48)$.

10.6.5 An Incorrect Estimate of the Age of the Earth

Lord Kelvin was a great admirer of the work of Fourier and put the Fourier transform to use in the middle of the nineteenth century in an attempt to determine the age of the Earth. Although Kelvin's mathematical treatment of his model is correct, the model does not predict the correct age of the Earth. Let's formulate Kelvin's model and discuss.

Assume that the Earth was initially a hot molten planet of uniform temperature T_0 at the time ($t = 0$) of its birth. Subsequently, the Earth has been gradually freezing from the outside in, losing its heat through simple diffusion across a solid crust that is constantly growing in thickness as the cold temperature at the surface works its way deeper into an otherwise static Earth via diffusion. If through history the surface is maintained at the constant temperature $T = 0$ (if you want this to be a different constant, say T_s, just make the replacement $T \to T + T_s$ in all that follows), the temperature distribution is controlled by the diffusion equation

$$D\,\nabla^2 T(\mathbf{r}, t) - \frac{\partial T(\mathbf{r}, t)}{\partial t} = 0 \tag{10.158}$$

subject to the BC that T (surface) $= 0$ for all t and the initial condition

$$T(\mathbf{r}, t = 0) = T_0. \tag{10.159}$$

Now because the depth that the BC of $T = 0$ has perturbed the initial temperature T_0 throughout the Earth's interior is small compared to the radius of the Earth, we can solve the problem in Cartesian coordinates with z being depth (a plane Earth problem). Neglecting the spherical curvature of the Earth can be shown to not be a significant error.

To allow for the surface boundary condition, Kelvin used the so-called method of images that he had invented earlier in his career to allow for boundary conditions in electrostatic problems. The trick in the present case is to let the initial condition be that

$$T_0(\mathbf{r}) = \begin{cases} +T_0 & \text{for } z > 0 \quad \text{(into the Earth)} \\ -T_0 & \text{for } z < 0 \end{cases} \tag{10.160}$$

throughout all of space. We will see that this maintains the boundary condition $T(z = 0, t) = 0$ for all time.

The solution is then to use what we derived above for diffusion through a whole space

$$T(\mathbf{r}, t) = \iiint_{-\infty}^{+\infty} \mathrm{d}x'\,\mathrm{d}y'\,\mathrm{d}z'\,T_0(z') \frac{e^{-\left[(x-x')^2+(y-y')^2+(z-z')^2\right]/(4Dt)}}{8(D\pi t)^{3/2}}, \tag{10.161}$$

where we are to use the above $T_0(z)$ for $-\infty < z < \infty$ as the initial condition. Integrating over x' and y', consider first the integral

$$I_x = \int_{-\infty}^{+\infty} \mathrm{d}x'\, e^{-(x-x')^2/(4Dt)} \tag{10.162}$$

and let $u = (x - x')/\sqrt{4\,Dt}$ with $\mathrm{d}u = -\mathrm{d}x'/\sqrt{4\,Dt}$, $u(x' = +\infty) = -\infty$, and $u(x' = -\infty) = +\infty$ so that

$$I_x = 2\sqrt{Dt} \int_{-\infty}^{+\infty} e^{-u^2}\,\mathrm{d}u = 2\sqrt{D\pi t}. \tag{10.163}$$

An identical result holds for the integral over y'. We then obtain exactly

$$T(z,t) = \frac{T_0}{2\sqrt{\pi D t}}\left[\int_0^{+\infty} dz'\, e^{-(z-z')^2/(4Dt)} - \int_{-\infty}^{0} dz'\, e^{-(z-z')^2/(4Dt)}\right]$$
$$= \frac{T_0}{2\sqrt{\pi D t}}\int_0^{+\infty} dz'\left[e^{-(z-z')^2/(4Dt)} - e^{-(z+z')^2/(4Dt)}\right], \tag{10.164}$$

which does indeed satisfy the BC that $T = 0$ on $z = 0$. This is Lord Kelvin's solution for how the temperature in the near surface of the Earth has been evolving over Earth's history. If you prefer to work with the actual surface boundary condition that $T = T_s$ at $z = 0$ as measured in say kelvin, then you first solve the above problem to obtain $T(\boldsymbol{r}, t)$ and then add to that result the constant T_s to have the result in kelvin.

Lord Kelvin noted that one thing we can measure on the Earth's surface is the temperature gradient

$$\left.\frac{\partial T}{\partial z}\right|_{z=0}. \tag{10.165}$$

This can be measured in tunnels and mine shafts. Thus, we calculate

$$\frac{\partial T}{\partial z} = \frac{T_0}{2\sqrt{\pi D t}}\int_0^{+\infty} dz'\left[-\frac{2(z-z')}{4Dt}e^{-(z-z')^2/(4Dt)} + \frac{2(z+z')}{4Dt}e^{-(z+z')^2/(4Dt)}\right], \tag{10.166}$$

which then gives at $z = 0$

$$\left.\frac{\partial T}{\partial z}\right|_{z=0} = \frac{T_0}{2\sqrt{\pi D t}}\int_0^{+\infty} dz'\,\frac{z' e^{-z'^2/(4Dt)}}{Dt}. \tag{10.167}$$

Making the substitution $u = z'^2/(4Dt)$ and $du = 2z'\,dz'/4Dt$ yields the tidy result

$$\left.\frac{\partial T}{\partial z}\right|_{z=0} = \frac{T_0}{\sqrt{\pi D t}}. \tag{10.168}$$

By observing $\partial T/\partial z|_{z=0}$ at present, the time t that has resulted in this gradient is, under the assumptions of Kelvin, a good estimate of the age of the Earth:

$$\boxed{t = \frac{1}{\pi D}\left(\frac{T_0}{\partial T/\partial z|_{z=0}}\right)^2 \,\widehat{=}\, \text{age of Earth.}} \tag{10.169}$$

Using the rough estimates $D \approx 10^{-6}$ m^2/s, $T_0 \approx 1{,}000$ °C, and $\partial T/\partial z|_{z=0} = 10^{-2}$ °C/m, one obtains the estimate

$$t \cong \frac{10^{16}}{\pi}\text{s}\left(\frac{1\text{ yr}}{3.15\times 10^7\text{s}}\right) \cong 10^8 \text{ yrs} = 100 \text{ Myr}. \tag{10.170}$$

When Kelvin first made the prediction in the 1860s that Earth could only be about 100 million years old, the geologists of the time struggled to fit all of geological history into that time frame. Today, geologists use radiometric measurements of isotope decay in rocks both from Earth and from meteorites to determine that Earth is about 4,500 Myr old. So what did Lord Kelvin get wrong in the above model?

A few possibilities include: (1) Earth's mantle is convecting beneath a solid crust of roughly constant thickness, which replaces Kelvin's diffusive heat loss at the base of the crust with heat brought from depth by convection, so that if the crustal temperature gradient is decreasing through time, it is doing so much more slowly than in Kelvin's purely diffusive model; (2) as the liquid mantle freezes to solid crust, there is a release of latent heat that partially replaces the heat lost due to diffusive cooling, which keeps the crust from growing as rapidly as in Kelvin's model; or (3) there is radioactivity inside the Earth that is constantly generating heat, which also acts to replace the heat lost due to diffusive cooling. When such mechanisms are quantified, it is the neglect of convection in the mantle that is the primary reason why Kelvin's simple diffusive-cooling model in a passive Earth cannot explain the age of the Earth. As you might imagine, the details are complicated. New oceanic crust is constantly formed at the mid-ocean ridges where hot mantle comes right to the surface. The oceanic crust and uppermost underlying mantle that together form the solid oceanic "lithosphere" thickens over time as it cools diffusively, which is well quantified by Kelvin's model above as the lithosphere moves away from the mid-ocean ridge until it is subducted at a continental margin and remelted. The continental lithosphere does not thicken as rapidly as in the Kelvin model because convection brings heat to the base of a continent that replaces the heat lost there by diffusive cooling. In Earth's overall heat balance, more heat is lost through diffusive cooling than is generated by radioactivity. So although mantle temperatures are going down through time, mantle convection supplies heat to the base of the crust that keeps the crustal temperature gradient from decreasing as rapidly as in Kelvin's passive cooling model.

10.7 Wave Problems

We can also address some types of wave problems using Fourier analysis even prior to learning about contour integration in Chapter 11. We will treat a wave-equation initial-value problem in 3D, a 1D wave problem with a time-dependent boundary condition, and a 1D wave initial-value problem that is most easily treated with Kelvin's "method of images."

10.7.1 A 3D Initial-Value Wave Problem

Let's start by considering an initial-value wave problem in 3D for some wavefield $w(\boldsymbol{r}, t)$, analogous to the diffusion problem treated above. If we take the initial conditions to be $w(\boldsymbol{r}, t=0) = w_0(\boldsymbol{r})$ and, for simplicity, the time derivative of w to be zero at $t=0$, this initial value problem is stated

$$c^2\nabla^2 w(\boldsymbol{r}, t) - \frac{\partial^2 w(\boldsymbol{r}, t)}{\partial t^2} = 0 \tag{10.171}$$

subject to the initial conditions

$$w(\boldsymbol{r}, t=0) = w_0(\boldsymbol{r}) \tag{10.172}$$

$$\left.\frac{\partial w(\boldsymbol{r}, t)}{\partial t}\right|_{t=0} = 0 \tag{10.173}$$

and, for some reasonably "compact-around-the-origin" initial displacement field $w_0(\boldsymbol{r})$, the boundary condition that $w \to 0$ as $|\boldsymbol{r}| \to \infty$.

As with the diffusion initial-value problem, we take three Fourier transforms over the three space dimensions (with $\nabla \to i\boldsymbol{k}$ and $\nabla^2 \to -k^2$) and solve the remaining differential equation in time to obtain the solution

$$\tilde{w}(\boldsymbol{k}, t) = \tilde{w}_0(\boldsymbol{k}) \cos(kct), \tag{10.174}$$

where we choose the time solution $\cos(kct)$ instead of $\sin(kct)$ in order to satisfy the initial condition. Here $\boldsymbol{k} = k_x\hat{\boldsymbol{x}} + k_y\hat{\boldsymbol{y}} + k_z\hat{\boldsymbol{z}}$, $k = \sqrt{k_x^2 + k_y^2 + k_z^2}$ and c is the wave speed. Identical to our handling of the diffusion problem, we then obtain the solution as

$$w(\boldsymbol{r}, t) = \iiint\limits_{-\infty}^{+\infty} \mathrm{d}^3\boldsymbol{r}'\, w_0(\boldsymbol{r}') \iiint\limits_{-\infty}^{+\infty} \frac{\mathrm{d}^3\boldsymbol{k}}{(2\pi)^3}\, e^{i\boldsymbol{k}\cdot(\boldsymbol{r}-\boldsymbol{r}')} \cos\left(\sqrt{k_x^2 + k_y^2 + k_z^2}\, ct\right). \tag{10.175}$$

The challenge now is to perform the three integrals over k_x, k_y, and k_z, which due to the $\sqrt{k_x^2 + k_y^2 + k_z^2}$, do not separate as they did in the diffusive case.

To proceed in this case, we instead express $\boldsymbol{k}$ in spherical-k coordinates (k, θ, ϕ). If we temporarily define $\boldsymbol{r}_o = \boldsymbol{r} - \boldsymbol{r}' = r_o\hat{\boldsymbol{r}}_o$, where $r_o = |\boldsymbol{r} - \boldsymbol{r}'| = |\boldsymbol{r}' - \boldsymbol{r}|$ and measure azimuthal angle θ in k space from the direction $\hat{\boldsymbol{r}}_o$, we have

$$\boldsymbol{k} = k\hat{\boldsymbol{k}}, \tag{10.176}$$

$$\hat{\boldsymbol{k}} \cdot \hat{\boldsymbol{r}}_o = \cos\theta, \tag{10.177}$$

$$\boldsymbol{k} \cdot \boldsymbol{r}_o = kr_o \cos\theta, \tag{10.178}$$

$$\mathrm{d}^3\boldsymbol{k} = k^2\mathrm{d}k\, \sin\theta \mathrm{d}\theta\, \mathrm{d}\phi = -k^2\mathrm{d}k\, \mathrm{d}(\cos\theta)\, \mathrm{d}\phi \tag{10.179}$$

so that the three-fold inverse Fourier transform that gives the spatial dependence is

$$\iiint\limits_{-\infty}^{+\infty} \frac{\mathrm{d}^3\boldsymbol{k}}{(2\pi)^3}\, e^{i\boldsymbol{k}\cdot\boldsymbol{r}_o} \cos(kct) = \frac{1}{(2\pi)^3} \int_0^\infty \mathrm{d}k\, k^2 \cos(kct) \int_{-1}^{1} \mathrm{d}(\cos\theta) e^{ikr_o\cos\theta} \int_0^{2\pi} \mathrm{d}\phi, \tag{10.180}$$

where the integral from $\theta = 0$ to $\theta = \pi$ is written in terms of $\cos\theta$ going from -1 to 1 using the minus sign in the last statement of Eq. (10.179). We then perform the angular integrals in k space to give

$$\iiint\limits_{-\infty}^{+\infty} \frac{\mathrm{d}^3\boldsymbol{k}}{(2\pi)^3}\, e^{i\boldsymbol{k}\cdot\boldsymbol{r}_o} \cos(kct) = \frac{1}{(2\pi)^3} \int_0^\infty \mathrm{d}k\, k^2 \cos(kct) \int_{-1}^{1} \mathrm{d}(\cos\theta) e^{ikr_o\cos\theta} \int_0^{2\pi} \mathrm{d}\phi \tag{10.181}$$

$$= \frac{1}{(2\pi)^2 i r_o} \int_0^\infty \mathrm{d}k\, k \cos(kct) \left(e^{ikr_o} - e^{-ikr_o}\right) \tag{10.182}$$

$$= \frac{2}{(2\pi)^2 r_o} \int_0^\infty \mathrm{d}k\, k \sin(kr_o) \cos(kct) \tag{10.183}$$

$$= \frac{1}{(2\pi)^2 r_o} \int_{-\infty}^\infty \mathrm{d}k\, k \sin(kr_o) \cos(kct), \tag{10.184}$$

where because $k \sin(kr_o)$ is even, the integral can be extended to the range $-\infty$ to ∞ in the last line. Writing out the sines and cosines then gives

$$\iiint_{-\infty}^{\infty} \frac{\mathrm{d}^3\boldsymbol{k}}{(2\pi)^3} e^{i\boldsymbol{k}\cdot\boldsymbol{r}_o} \cos(kct) = \frac{-1}{(2\pi)^2 4 r_o} \int_{-\infty}^\infty \mathrm{d}k\, (ik)$$

$$\times \left(e^{ik(r_o+ct)} - e^{-ik(r_o+ct)} + e^{ik(r_o-ct)} - e^{-ik(r_o-ct)}\right)$$

$$= \frac{-1}{(2\pi)^2 2 r_o} \int_{-\infty}^\infty \mathrm{d}k\, (ik) \left(e^{ik(r_o+ct)} + e^{ik(r_o-ct)}\right) \tag{10.185}$$

$$= \frac{-1}{(2\pi)^2 2 r_o c} \frac{\partial}{\partial t} \int_{-\infty}^\infty \mathrm{d}k \left(e^{ik(r_o+ct)} - e^{ik(r_o-ct)}\right) \tag{10.186}$$

$$= \frac{-1}{4\pi r_o c} \frac{\partial}{\partial t} [\delta(r_o + ct) - \delta(r_o - ct)], \tag{10.187}$$

where in the first line we used $-\int_{-\infty}^{\infty} \mathrm{d}k\,(ik) \exp(-iku) = \int_{-\infty}^{\infty} \mathrm{d}k\,(ik)\exp(+iku)$ by making the substitution $k \to -k$ and in the last line used the completeness relation. Because $r_o = |\boldsymbol{r}' - \boldsymbol{r}|$ and t are both positive, the first Dirac has an argument that is always positive and is thus always zero.

Restoring $r_o = |\boldsymbol{r}' - \boldsymbol{r}|$ we then obtain the solution of this initial-value wave problem as

$$w(\boldsymbol{r}, t) = \frac{1}{4\pi c} \frac{\partial}{\partial t} \iiint_{-\infty}^{+\infty} \mathrm{d}^3\boldsymbol{r}'\, w_0(\boldsymbol{r}') \frac{\delta\left(|\boldsymbol{r}' - \boldsymbol{r}| - ct\right)}{|\boldsymbol{r}' - \boldsymbol{r}|}. \tag{10.188}$$

This corresponds to Dirac-derivative pulses carrying the initial data from points $\boldsymbol{r}'$ where the initial data has support to observation points $\boldsymbol{r}$ and times t. For a $w_0(\boldsymbol{r}')$ having compact support around the origin, the factor of $|\boldsymbol{r}' - \boldsymbol{r}|$ in the denominator causes the wavefield to drop to zero as $|\boldsymbol{r}| \to \infty$. This solution also corresponds to outward propagating waves as $|\boldsymbol{r}| \to \infty$. Note that we could also write the Dirac inside the integral as $\delta\left(|\boldsymbol{r}' - \boldsymbol{r}| - ct\right) = \delta\left[c\left(t - |\boldsymbol{r}' - \boldsymbol{r}|/c\right)\right] = c^{-1}\delta\left(t - |\boldsymbol{r}' - \boldsymbol{r}|/c\right)$.

Mathematicians refer to Eq. (10.188) as *Kirchoff's formula* for the initial-value problem of the scalar wave equation in 3D space. They usually express the result differently however. The Dirac in the integrand requires $|\boldsymbol{r}' - \boldsymbol{r}| = ct$ for the integrand to be nonzero. This means that $w_0(\boldsymbol{r}')$ is being evaluated and used at each point in time on a spherical surface $\partial\Omega_{|\boldsymbol{r}'-\boldsymbol{r}|=ct}$ defined by points $\boldsymbol{r}'$ lying on the spherical surface centered on $\boldsymbol{r}$ as given by the radial distance $|\mathbf{r}' - \mathbf{r}| = ct$ from the center of the sphere. They thus write Kirchoff's formula as

$$w(\mathbf{r}, t) = \frac{1}{4\pi c^2}\frac{\partial}{\partial t}\left(\frac{1}{t}\int_{\partial\Omega_{|\mathbf{r}'-\mathbf{r}|=ct}} w_o(\mathbf{r}')\, dS'\right), \tag{10.189}$$

for what we express as Eq. (10.188).

10.7.2 A Wave Problem with Time-Dependent Boundary Conditions

Another class of wave problems and solution technique is when the force exciting the wave is expressed as a time-dependent boundary condition.

Let's consider the specific 1D boundary-value wave problem in which $u(x, t)$ is say displacement and $\partial u(x, t)/\partial x$ is proportional to stress and the wave problem is

$$c^2\frac{\partial^2 u(x, t)}{\partial x^2} - \frac{\partial^2 u(x, t)}{\partial t^2} = 0 \quad \text{on} \quad 0 \le x \le \infty \tag{10.190}$$

subject to the two boundary conditions that

$$\left.\frac{\partial u(x, t}{\partial x}\right|_{x=0} = s(t) \tag{10.191}$$

$$c\frac{\partial u(x, t)}{\partial x} + \frac{\partial u(x, t)}{\partial t} = 0 \quad \text{as} \quad x \to \infty. \tag{10.192}$$

In the first boundary condition at $x = 0$, the applied-strain (or applied-stress) time function $s(t)$ has either been operating for a long time or turns on at $t = 0$ to excite the wave propagation, with the implicit idea that $u = 0$ and $\partial u/\partial t = 0$ for $t \le 0$. In either case, initial conditions at $t = 0$ are not required for this problem. The second boundary condition says the waves are propagating outward as $x \to \infty$.

To use these boundary conditions, we take a Fourier transform over time with $\partial/\partial t \to -i\omega$, to obtain the spatial problem

$$c^2\frac{\partial^2 \tilde{u}(x, \omega)}{\partial x^2} + \omega^2\tilde{u}(x, \omega) = 0 \quad \text{on} \quad 0 \le x \le \infty, \tag{10.193}$$

$$\left.\frac{\partial \tilde{u}(x, \omega)}{\partial x}\right|_{x=0} = \tilde{s}(\omega) \quad \text{at} \quad x = 0, \tag{10.194}$$

$$c\frac{\partial \tilde{u}(x, \omega)}{\partial x} - i\omega\tilde{u}(x, \omega) = 0 \quad \text{as} \quad x \to \infty. \tag{10.195}$$

The differential equation has solutions $\exp(i\omega x/c)$ and $\exp(-i\omega x/c)$. Either works with the boundary condition at $x = 0$ but we must use the $\exp(i\omega x/c)$ solution to satisfy the outward-radiation condition as $x \to \infty$ so that

$$\tilde{u}(x, \omega) = \frac{c\tilde{s}(\omega)}{i\omega}e^{i\omega x/c}. \tag{10.196}$$

The solution of interest is then given by the inverse Fourier transform

$$u(x, t) = -\frac{c}{2\pi} \int_{-\infty}^{\infty} d\omega \, \frac{\tilde{s}(\omega)}{(-i\omega)} e^{-i\omega(t - x/c)}. \tag{10.197}$$

Using that division by $-i\omega$ in the frequency domain corresponds to time integration in the time domain, we end up with

$$\boxed{u(x, t) = -c \int_{-\infty}^{t} dt' \, s\left(t' - \frac{x}{c}\right).} \tag{10.198}$$

Upon taking a derivative with respect x and noting that $\partial s(t' - x/c)/\partial x = -c^{-1} \partial s(t' - x/c)/\partial t'$, so that at $x = 0$ our solution indeed gives that $\partial u(x, t)/\partial x|_{x=0} = s(t)$ as required.

10.7.3 An Initial-Value Wave Problem on a String

The problem is to find the displacements $u(x, t)$ on a semi-infinite string that satisfy the wave equation

$$c^2 \frac{\partial^2 u(x, t)}{\partial x^2} - \frac{\partial^2 u(x, t)}{\partial t^2} = 0 \quad \text{on} \quad 0 \le x \le \infty \tag{10.199}$$

subject to the boundary condition that the string is clamped at $x = 0$

$$u(0, t) = 0 \quad \text{at} \quad x = 0 \tag{10.200}$$

and subject to the outward-radiation condition as $x \to \infty$

$$c \frac{\partial u(x, t)}{\partial x} + \frac{\partial u(x, t)}{\partial t} = 0 \quad \text{as} \quad x \to \infty. \tag{10.201}$$

The waves are created by the initial condition in which the string is pulled up sharply at the point $x = x_0$

$$u(x, 0) = w(x - x_0) \quad \text{at} \quad t = 0, \tag{10.202}$$

where $w(x)$ is some pulse (or wavelet) function that is peaked at $x = 0$. Possible examples could include an appropriately scaled Gaussian, sinc, or Dirac-delta function. In what follows, however, we will not assume that $w(x)$ is necessarily an even function. The string is pulled up at the point $x = x_0$, where $w(x - x_0)$ is peaked and held like that with $\partial u/\partial t = 0$ until the string is released at $t = 0$. We expect a pair of pulses having the shape of $w(x)$ but with half the amplitude to move away from x_0 for $t > 0$ and for there to be a reflection of one of the two pulses at the clamped end.

One way to solve this problem is to extend the analysis to an infinitely long string that goes out to both $\pm\infty$ and to introduce an additional "image" initial condition $-w(-x - x_0)$, which is peaked at $x = -x_0$ and is a mirror symmetric and negative version of the actual initial condition $w(x - x_0)$ that is peaked at $x = x_0$. This additional image source generates waves such that the clamped boundary condition at $x=0$ is always satisfied. We then only look at the resulting solution for the region $x>0$ that corresponds to our semi-infinite string of interest.

So for an infinite string having an initial condition

$$u(x,0)=w(x-x_0)-w(-x-x_0) \quad \text{at} \quad t=0, \tag{10.203}$$

we take a spatial Fourier transform of both the wave equation and initial conditions to obtain

$$\frac{\partial^2 \tilde{u}(k,t)}{\partial t^2}=-k^2c^2\tilde{u}(k,t), \tag{10.204}$$

$$\tilde{u}(k,0)=\tilde{w}(k)e^{-ikx_0}-\tilde{w}(-k)e^{ikx_0}, \tag{10.205}$$

$$\left.\frac{\partial \tilde{u}(k,t)}{\partial t}\right|_{t=0}=0, \tag{10.206}$$

where $\tilde{w}(k)$ is the Fourier transform of the nonshifted pulse function $w(x)$ and, as you can show in an end-of-chapter exercise, $\tilde{w}(-k)$ is the Fourier transform of $w(-x)$. The solution satisfying both the temporal differential equation and the two initial conditions is

$$\tilde{u}(k,t)=\left[\tilde{w}(k)e^{-ikx_0}-\tilde{w}(-k)e^{ikx_0}\right]\cos(kct) \tag{10.207}$$

$$=\frac{1}{2}\left[\tilde{w}(k)e^{-ikx_0}-\tilde{w}(-k)e^{ikx_0}\right]\left(e^{ikct}+e^{-ikct}\right). \tag{10.208}$$

We now can take the inverse Fourier transform and use the shifting property to obtain

$$u(x,t)=\frac{1}{4\pi}\int_{-\infty}^{\infty}dk\,\left[\tilde{w}(k)\left(e^{ik(x-x_0+ct)}+e^{ik(x-x_0-ct)}\right)-\tilde{w}(-k)\left(e^{ik(x+x_0+ct)}+e^{ik(x+x_0-ct)}\right)\right] \tag{10.209}$$

$$=\frac{1}{2}\left[w(x-x_0+ct)+w(x-x_0-ct)-w(-x-x_0-ct)-w(-x-x_0+ct)\right]. \tag{10.210}$$

This is the solution on the infinite string excited by the two mirror-symmetric initial-condition pulses peaked at $x=\pm x_0$. It corresponds to four pulses propagating at speed c: two pulses propagating to the left $(-x)$ and right $(+x)$ directions from each of the two source positions $x=\pm x_0$. The two pulses traveling to the left satisfy the outward radiation condition as $x\to-\infty$ and the two pulses traveling to the right satisfy the outward radiation condition as $x\to+\infty$. We see that at $x=0$, we always have $u(x,t)=0$ as required.

The solution to the original problem on a semi-infinite string $x\geq 0$ with a clamped boundary condition at $x=0$ and an initial-condition source point located at $x=x_0$ is obtained by agreeing to look at the solution of Eq. (10.210) only for points $x\geq 0$. The pulse $-w(-x-x_0-ct)/2$ is that generated at the image source position and propagating to the left and is thus only in our field of interest $x\geq 0$ if its tail extends into $x\geq 0$ at $t=0$. The pulse $-w(-x-x_0+ct)/2$ is that generated at the image source position and moving to the right and corresponds to the pulse reflected at the clamped boundary $x=0$ in our problem of interest. The pulse $w(x-x_0+ct)/2$ corresponds to the pulse generated at the

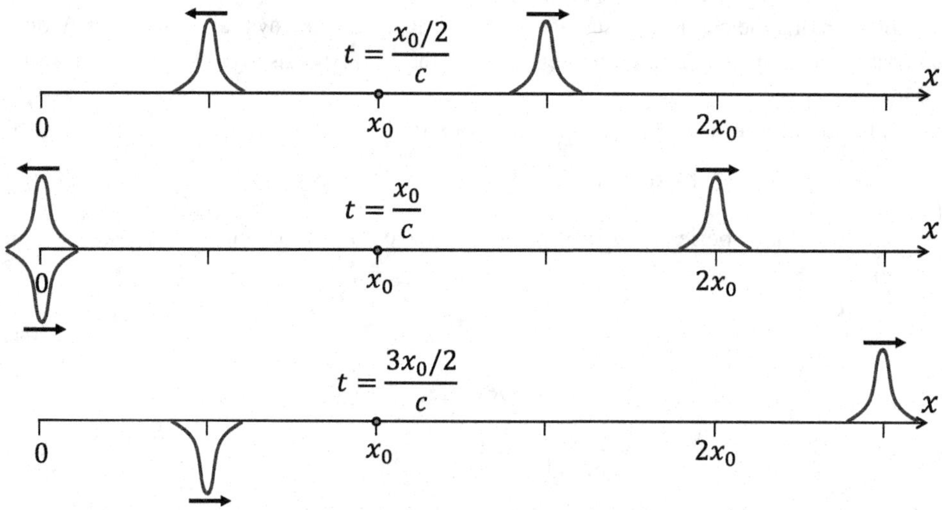

Figure 10.8 Three snapshots of the two pulses moving to the left and right that were generated at the source point $x = x_0$ on the semi-infinite string. Upon reflection at the clamped boundary $x = 0$, the pulse changes sign as it then begins to propagate in the $+x$ direction. All this is captured by including an image source of the opposite sign located at $x = -x_0$ on an infinite string with pulses outward propagating at $\pm\infty$.

actual source position (where the initial-condition is peaked at x_0) and propagating to the left prior to it being reflected at $x = 0$. Last, the pulse generated at the actual source position that is always moving to the right is $w(x - x_0 - ct)/2$. These pulses (or wavelets) and their propagation direction and amplitudes on the semi-infinite string are depicted in Fig. 10.8 at three different points in time.

10.8 Time-Series Analysis

Consider a geophone placed on the Earth's surface and attached via a cable to a recording device called a seismograph. When a source of seismic energy (e.g., an explosion) is excited, we begin to record on the seismograph the vibrations of the Earth that the geophone detects. After some period of time, the recording is stopped and we have a digital record (a time series) of all the vibrations that the geophone sensed during the recording period. Many of these signals are the seismic waves of interest that travelled from the source to the geophone either directly along the Earth's surface (surface waves) or after having refracted and reflected from the layers and structure of the subsurface (head waves and scattered body waves). In addition to recording the vibrations associated with the arrival of all these source-generated seismic waves, the geophone also detects various types of "noise" generated from sources such as car traffic, machine vibrations, or trees blowing in the wind that cause their trunk and root system to generate seismic waves.

The subject of time-series analysis corresponds to the various types of manipulations we may want to perform on the recorded data in order to enhance a desired signal attribute or to decrease the noise. Such manipulation is generally called "filtering," and we will see that Fourier analysis provides an effective way to perform such filtering. Further, the vibrations picked up by the geophone are being recorded digitally with a finite sampling interval. Another aspect of time-series analysis is to understand the effect of such discretization of the signal on the calculated spectrum of the signal.

In what follows, we first give a brief primer on filtering and its connection to the Fourier transform and then discuss discretization effects.

10.8.1 Filtering and Fourier Analysis

You do not need to perform forward- and inverse-Fourier transforms on your data in order to filter it. The filtering may be performed entirely in the time domain as we discuss in the forthcoming subsection. However, it is often more effective to filter in the frequency domain and the connection between time-domain filtering and frequency-domain filtering is provided by the convolution theorem that we derived earlier.

Time-Domain Filtering

To filter a signal means to convolve the signal with a filter, which is another time function that we select so that the output of the convolution operation has some desired attribute. We can state the filtering operation as

$$\underset{\substack{\uparrow \\ \text{Recorded} \\ \text{data}}}{\text{Input}} * \text{filter} = \underset{\substack{\uparrow \\ \text{Filtered} \\ \text{data}}}{\text{output}}$$

where the symbol $*$ denotes the convolution operation. Defining the various time functions as $I(t)$ = input signal to be filtered, $F(t)$ = filter, and $O(t)$ = output or "filtered" signal, the convolution $O(t) = I(t) * F(t)$ is the integral operation defined by

$$O(t) = \int_{-\infty}^{+\infty} I(s)\, F(t-s)\, \mathrm{d}s \mathrel{\hat{=}} I(t) * F(t), \tag{10.211}$$

where s is just the dummy time variable that we integrate over. If you change dummy variables from s to w using $w = t - s$ so that $\mathrm{d}w = -\mathrm{d}s$, you obtain that

$$O(t) = \int_{-\infty}^{+\infty} I(t-w)\, F(w)\, \mathrm{d}w \mathrel{\hat{=}} F(t) * I(t). \tag{10.212}$$

This shows that the convolution operation commutes.

With s the time variable that is being integrated over, the integrand of the convolution integral is the product of the function being filtered $I(s)$ with the filter function after it has been flipped mirror-image-fashion about time zero to generate $F(-s)$ and then slid in the positive s direction by an amount t to create the time-reversed and displaced function $F(t-s)$. So to create the output function $O(t) = I(t) * F(t)$, we slide the time-reversed filter

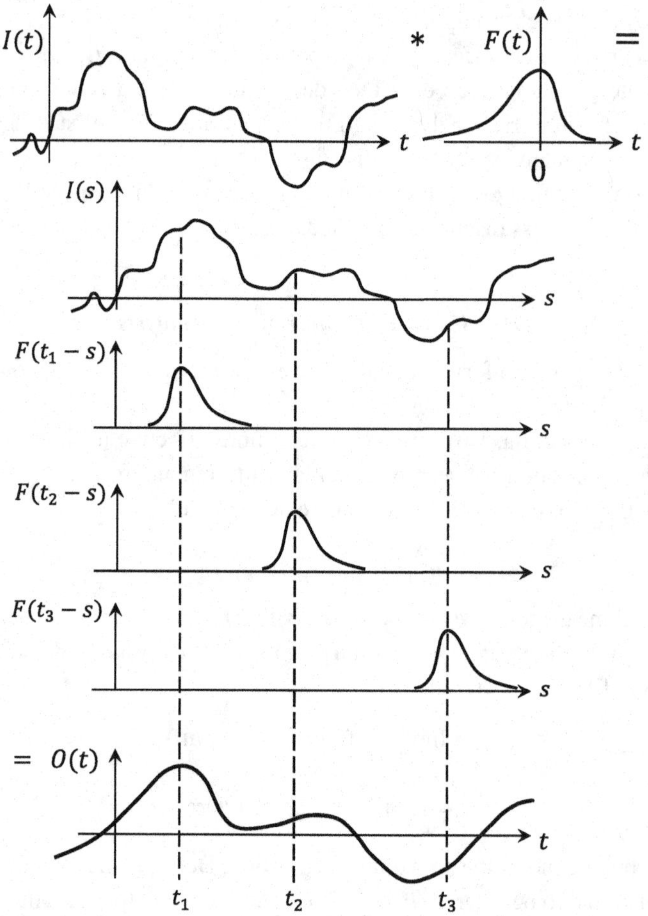

Figure 10.9 The convolution filtering operation: (1) flip your filter in time to create $F(-s)$ and then shift it by an amount t to create the reversed and shifted filter $F(t-s)$, (2) multiply the noisy input signal $I(s)$ with $F(t-s)$, and (3) integrate over s to obtain the output $O(t) = \int_{-\infty}^{\infty} I(s)F(t-s)\,\mathrm{d}s$ at each t. This filter removes the little bumps in $I(t)$ to create the smoothed function $O(t)$ shown at the bottom.

to each new position t, perform the multiplication $I(s)F(t-s)$ and then integrate over s at that particular t to obtain $O(t)$. This operation can be visualized in Fig. 10.9.

Although it does not matter, we generally think of the filter as being the time function that is slid over the input signal even if we understand that convolution commutes so that we can also think of the filter as being the fixed function $F(s)$ with the input signal being flipped about $s = 0$ and slid along with increasing t in the positive s direction as $I(t-s)$. Often, but not always, we think of the filter as being the shorter time function and the input as being the longer time function.

As another way to have a pictorial example of the convolution (or filtering) operation, imagine sending a sharp Dirac-delta time pulse into the Earth as an elastic wave. It bounces

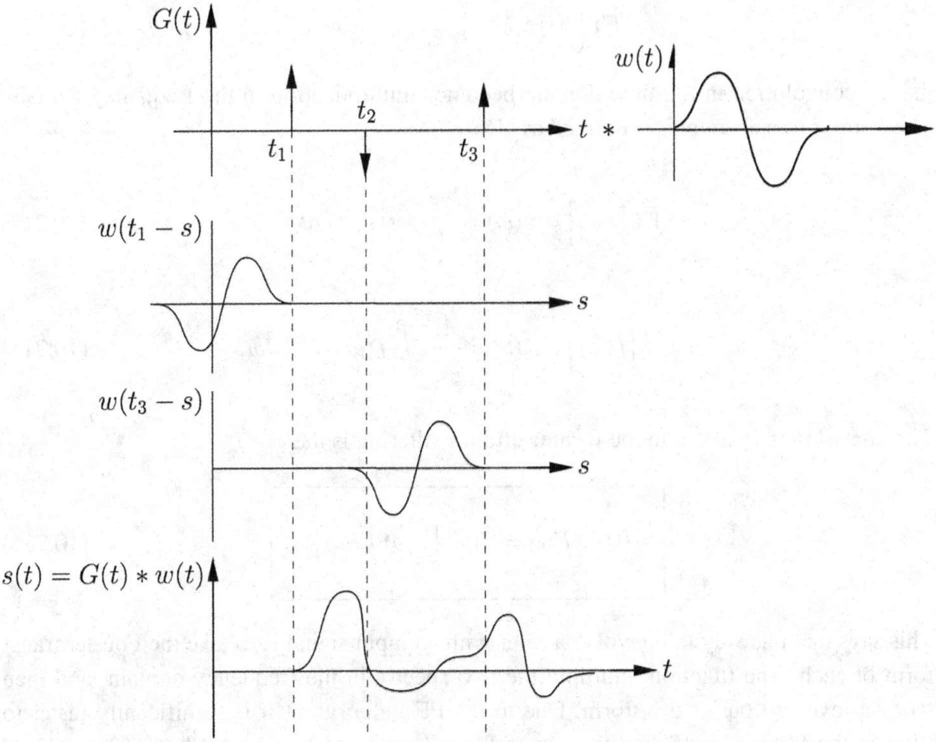

Figure 10.10 Creating a seismic trace by convolving a wavelet $w(t)$ with a Green's function response $G(t)$.

around in the Earth and we record at the surface a series of Dirac pulses (assuming no "attenuation") that we denote as $G(t)$. Such an impulse response is sometimes called the "Green's function" of the Earth (more on Green's functions or "impulse-response functions" in Chapter 12). If we convolve $G(t)$ with a wavelet $w(t)$, we get a model of a seismic trace $s(t)$ that a geophone records, that is, $s(t) = G(t) * w(t)$. Graphically, this is depicted in Fig. 10.10 where we choose to slide the flipped wavelet over the impulse response (though we could have slid the flipped impulse response over the wavelet). Thus the propagation of signals through the Earth's subsurface can itself be understood as a type of filtering operation where an injected wavelet $w(t)$ becomes multiple wavelets, perhaps overlapping and canceling with each other as in the above graphical example, due to the reflection and transmission process at interfaces in the subsurface. The trace recorded at the surface is then modeled as $s(t) = \int_{-\infty}^{\infty} G(t-s)w(s)\,\mathrm{d}s = \int_{-\infty}^{\infty} w(t-s)G(s)\,\mathrm{d}s$.

Convolution Theorem

The fastest way to perform convolution numerically is using the Fourier transform and the *fast-Fourier-transform* (or FFT) numerical algorithm. A nice description of the FFT routine for performing the Fourier transform numerically is given by Press et al. (2007). The basis for filtering using the Fourier transform is the convolution theorem, which we have seen states that

$$\mathrm{FT}\Big\{I(t) * F(t)\Big\} = \tilde{I}(\omega)\, \tilde{F}(\omega), \tag{10.213}$$

that is, convolution in the time domain becomes multiplication in the frequency domain. The Fourier transform pair is defined as always by

$$\mathrm{FT}\Big\{f(t)\Big\} = \tilde{f}(\omega) = \int_{-\infty}^{+\infty} f(t)\, e^{+iwt}\, \mathrm{d}t \tag{10.214}$$

$$\mathrm{FT}^{-1}\Big\{\tilde{f}(\omega)\Big\} = f(t) = \frac{1}{2\pi} \int_{-\infty}^{+\infty} \tilde{f}(\omega)\, e^{-iwt}\, \mathrm{d}\omega. \tag{10.215}$$

The convolution theorem in the form useful for filtering is then

$$\boxed{I(t) * F(t) = \mathrm{FT}^{-1}\Big\{\tilde{I}(\omega)\, \tilde{F}(\omega)\Big\}.} \tag{10.216}$$

This says that one way to convolve a filter with an input signal is to take the Fourier transform of each time function, multiply the two spectra in the frequency domain, and then take an inverse Fourier transform. Due to the FFT algorithm, it is significantly faster to filter in the frequency domain than to perform the convolution integral at each time t of desired output. Note that a convolution integral is not just a single integral. Each time we slide the filter to a new time t, we have to perform another time integral. If the input signal has N time points, each integral at each time position t involves N floating-point operations and so time-domain filtering involves N^2 operations. The FFT algorithm goes as $N \log N$ operations and so filtering by passing through the frequency domain goes as a handful of $N \log N$ operations which is close to N times faster than time-domain convolution. Given that N is typically 1,000 or much more, this is a considerable computational advantage.

In addition to such computational advantages, one goal of filtering, very often, is to eliminate a range of frequencies from a recording $I(t)$. Consider an input signal and its associated amplitude spectrum as depicted in Fig. 10.11. So we might want to eliminate the highest frequencies while keeping the lower frequencies. This can be accomplished by formulating a filter $\tilde{F}(\omega)$ directly in the frequency domain that retains the frequencies over a certain desired band and eliminates the other frequencies as depicted in Fig. 10.12.

The final output is calculated as $O(t) = \mathrm{FT}^{-1}\{\tilde{I}(\omega)\, \tilde{F}(\omega)\} = \mathrm{FT}^{-1}\{\tilde{O}(\omega)\}$ and is depicted in Fig. 10.13. Because we do not have to formulate the filter in the time domain, this operation corresponds to taking a Fourier transform of the input signal, multiplying it with the low-pass filter $\tilde{F}(\omega)$, and then returning to the time domain with an inverse Fourier transform (or inverse FFT). This is the preferred way to eliminate bands of frequencies from an input signal.

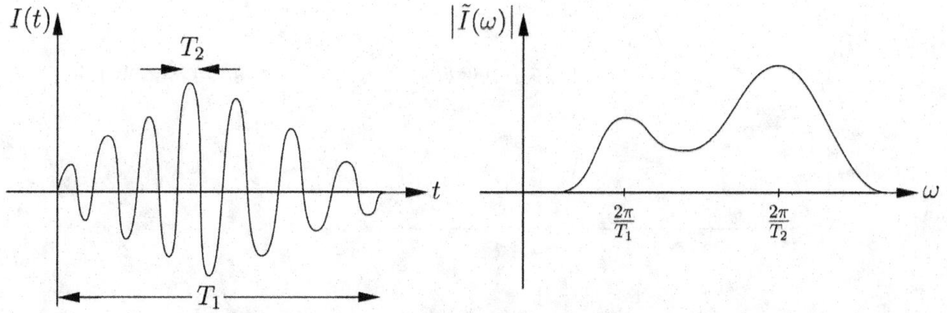

Figure 10.11 An imagined signal and its associated amplitude spectrum.

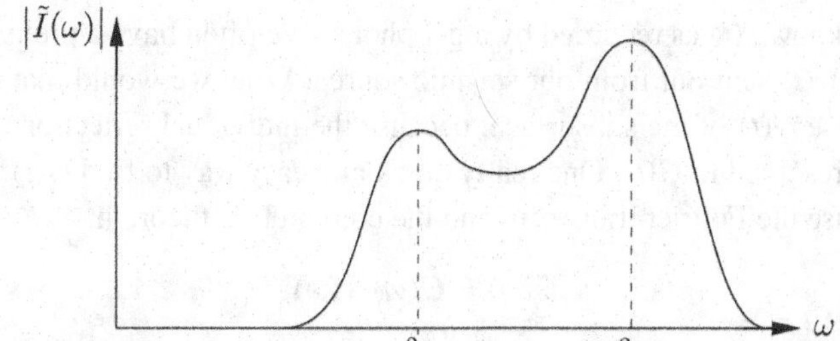

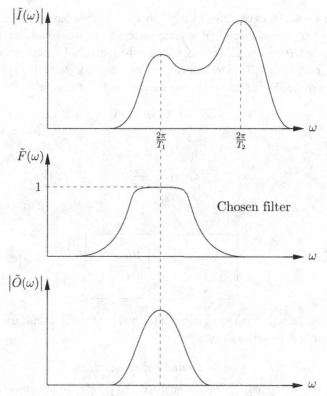

Figure 10.12 A bandpass filter chose to select a certain bandwidth of desired frequencies from the amplitude spectrum and to reject others.

Deconvolution

This is the act of undoing the filtering operation and is easiest when performed in the frequency domain. If we return to the earlier example of a recording of the seismic trace, our signal or seismic recording was modeled as

$$s(t) = G(t) * w(t). \qquad (10.217)$$

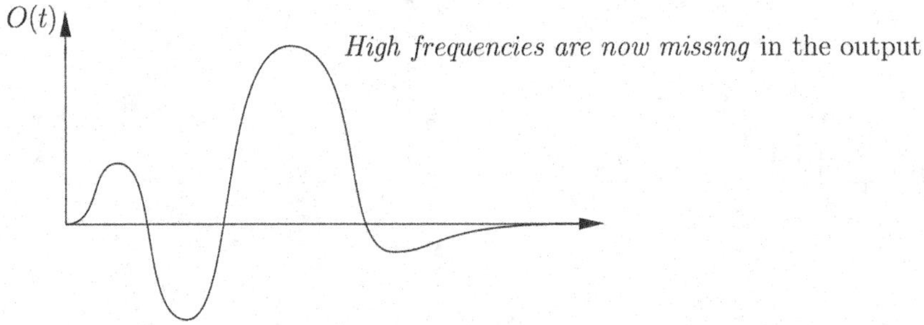

Figure 10.13 The time domain output after using the bandpass filter of Fig. 10.12.

We assume to know $s(t)$ as recorded by a geophone. We often have a pretty good estimate of the wavelet $w(t)$ sent out from our seismic source. What we would like to obtain is the impulse response $G(t)$ of the subsurface, because the individual reflections off the Earth's strata are well resolved in $G(t)$. One really quick and easy way to find $G(t)$ for a given $s(t)$ and $w(t)$ is to use the Fourier transform and the convolution theorem

$$\tilde{s}(\omega) = \tilde{G}(\omega)\,\tilde{w}(\omega) \tag{10.218}$$

so

$$\tilde{G}(\omega) = \frac{\tilde{s}(\omega)}{\tilde{w}(\omega)} \tag{10.219}$$

and

$$\boxed{G(t) = \mathrm{FT}^{-1}\left\{\frac{\tilde{s}(\omega)}{\tilde{w}(\omega)}\right\}.} \tag{10.220}$$

This operation of deconvolving an estimate of the wavelet from a seismic trace is a standard operation in exploration seismology.

Windowing and Fourier Analysis

If we just stop recording a given input signal $I(t)$ at a point in time t_s, this is equivalent to multiplying the actual ongoing signal $I(t)$ by the window function (or "boxcar" function). The window function is

$$w(t) = \begin{cases} 1 & 0 \le t \le t_s \\ 0 & \text{otherwise} \end{cases} \tag{10.221}$$

so the truncated signal $I_s(t)$ is simply

$$\boxed{I_s(t) = w(t)\,I(t).} \tag{10.222}$$

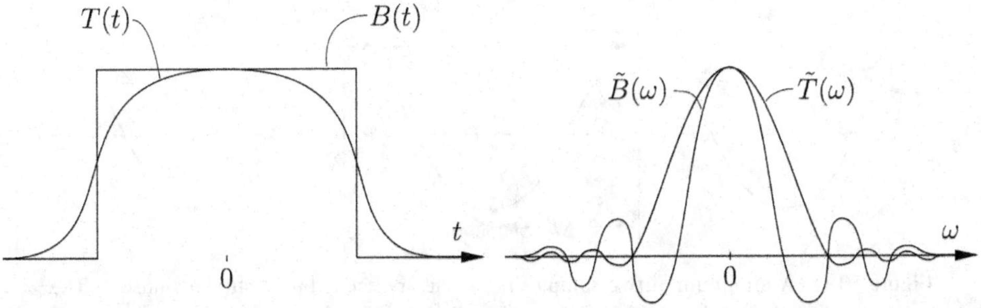

Figure 10.14 The boxcar and tapered window functions in the time domain (left) and frequency domain (right). The goal of a window function is to be as near to a Dirac spike as possible in the frequency domain.

In the frequency domain, we then have

$$\tilde{I}_s(\omega) = \tilde{w}(\omega) * \tilde{I}(\omega) \neq \tilde{I}(\omega) \tag{10.223}$$

because the window function is not a Dirac but is instead

$$\tilde{w}(\omega) = 2\,\frac{\sin\omega\, t_s/2}{\omega}\, e^{iwt_s/2}. \tag{10.224}$$

We want the window function to have a spectrum that is as close as possible to a Dirac function $\delta(\omega)$ so that the truncated signal has a spectrum as close as possible to the actual spectrum of the signal. Usually, a tapered window function causes less total distortion to the spectrum than does the simple boxcar function as is depicted in Fig. 10.14. Spectral sidelobes are reduced using the tapered window $T(t)$, even as the width of the spectral spike in the frequency domain is widened. This is the fundamental tradeoff in choosing window functions. As we sharpen up the central spectral spike in the transform domain, we increase the energy in the sidelobes while if we use tapering to reduce the sidelobes we widen the central spike. Again, the goal is for the transform of a window function to be as close as possible to a perfect Dirac spike in the frequency domain. Typically, a tapered window is preferred for windowing.

10.8.2 Discretization and the FFT

Data recorded over time is sampled at discrete time intervals Δt. In the past, magnetic tape recording allowed for continuous time records but such recordings are no longer used. The highest frequency f_N that can be captured at a discretization interval Δt is that for which we can fit two samples per period $T_N = 1/f_N$. This is depicted in Fig. 10.15. So we need at least two sampling intervals to capture a given frequency $f = 1/T$ and the critical or "maximum" frequency that can be captured at Δt sampling is called the *Nyquist frequency*:

$$\boxed{f_N = \frac{1}{T_N} = \frac{1}{2\,\Delta t}.} \;\widehat{=}\; \text{Nyquist or "highest" frequency in discretized data} \tag{10.225}$$

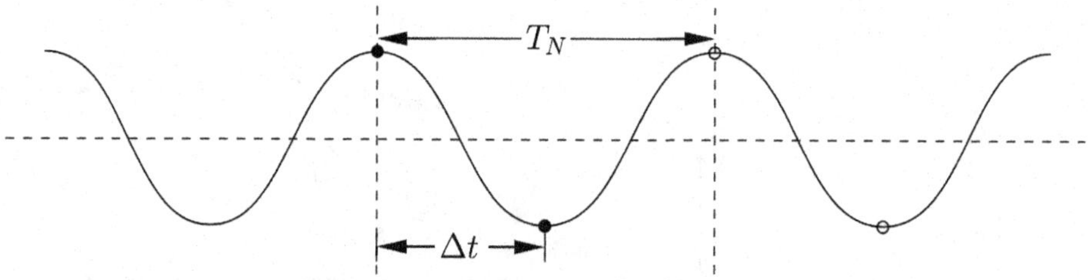

Figure 10.15 A minimum of two samples per cycle are needed to capture a sinusoid. The frequency $f_N = 1/T_N$ where this just occurs is called the Nyquist frequency and is given by $f_N = 1/(2\Delta t)$.

When a time series of length $N\,\Delta t$, where N is the number of samples, is transformed into the Fourier domain, there are N frequency domain samples between $-f_N$ and $+f_N$.

Aliasing

If our signal contains frequencies greater than f_N, then the act of undersampling these signals spuriously folds the energy at $f > f_N$ back into the principal band $-f_N < f < f_N$. This is depicted in Fig. 10.16. The aliased sinusoid does not really exist, but appears to be present, because we have too coarsely sampled the true sinusoid at f. This represents an error in the spectrum of the undersampled signal.

We can see what is happening in the aliased spectrum in Fig. 10.17. All the energy at frequencies greater than the Nyquist frequency is folded back into the principal band $-f_N < f < f_N$ that is the only band present in the discretized signal. All that energy folded back into the principal band (denoted by the gray shading in the above figure) is polluting the actual spectrum of the principal band and is a serious problem when present.

You can remember how energy at frequencies f outside the principal band $(-f_N, f_N)$ folds back into the principal band at the aliased frequency $-f_N < f_a < f_N$ using the pictorial device shown in Fig. 10.18. The only way to eliminate aliasing is to eliminate the signals having $f > f_N$ prior to discretization or to sample in time more finely. Once a signal is

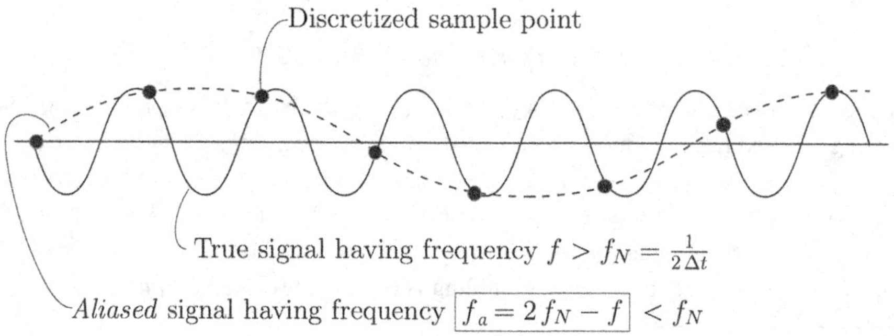

Figure 10.16 A sinusoid of frequency f (solid line) that is under sampled at Δt appears to have a lower *aliased* frequency f_a (dashed line).

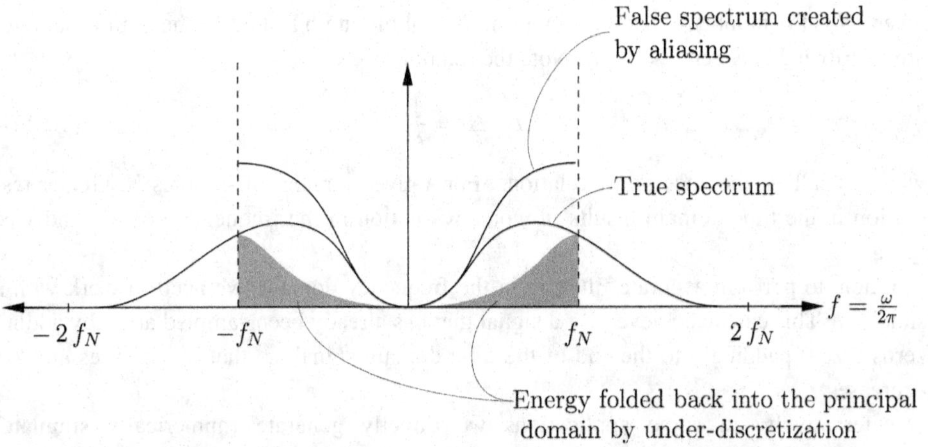

Figure 10.17 If the true spectrum of our time series has energy at frequencies larger than f_N, that energy is folded back into the principal domain as shown. This causes a great distortion of both the spectrum and the time-domain signal when we return to the time domain.

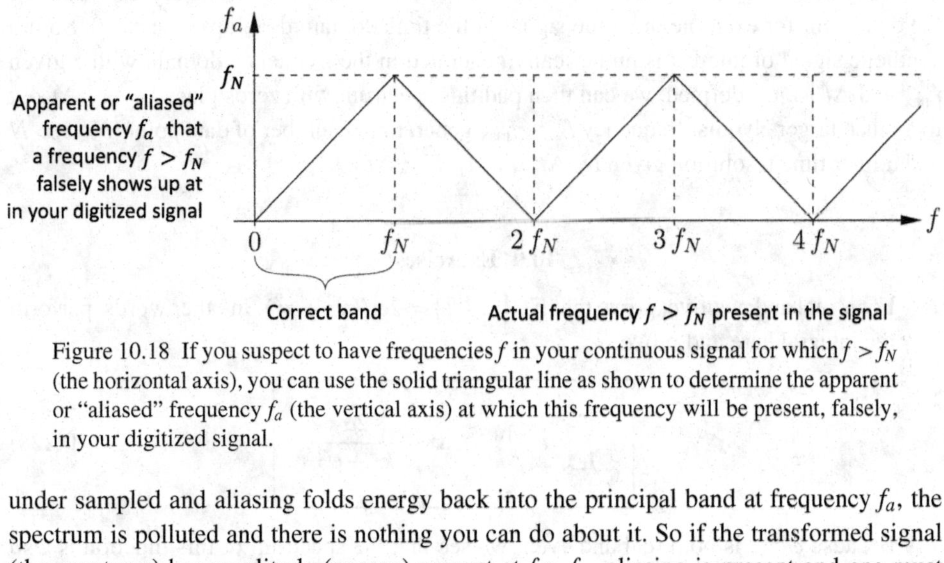

Figure 10.18 If you suspect to have frequencies f in your continuous signal for which $f > f_N$ (the horizontal axis), you can use the solid triangular line as shown to determine the apparent or "aliased" frequency f_a (the vertical axis) at which this frequency will be present, falsely, in your digitized signal.

under sampled and aliasing folds energy back into the principal band at frequency f_a, the spectrum is polluted and there is nothing you can do about it. So if the transformed signal (the spectrum) has amplitude (energy) present at $f > f_N$, aliasing is present and one must rerecord at a finer sampling rate.

Duration of a Sampled Signal

Just as the duration in the frequency domain f_N of a discretized signal is determined by the sampling rate Δt as $f_N = 1/(2\,\Delta t)$ or $\Delta t = 1/(2f_N)$, so does the duration of the time signal t_s determine the frequency sampling interval Δf over the N frequencies distributed between $-f_N < f < +f_N$ as

$$\Delta f \,\hat{=}\, \frac{2f_N}{N} = \frac{1}{N\,\Delta t} = \frac{1}{t_s}, \qquad (10.226)$$

where N is the number of samples in a time signal of length t_s, that is, the N time samples are distributed between $0 < t < t_s$. Note the relation

$$\Delta f \, \Delta t = \frac{1}{N}, \tag{10.227}$$

which is called "the uncertainty relation." For a given number of samples N, greater resolution in the time domain results in worse resolution in the frequency domain and vice versa.

Often, to perform accurate filtering in the frequency domain, we need to work with a small Δf. This can be achieved for a signal that has already been sampled at Δt by adding zeros ("zero padding") to the end of the time-domain signal, so that N increases and Δf decreases.

Alternatively, in some applications we directly generate (numerically simulate) frequency-domain signals using a Δf up to some maximum frequency $f_{\max}$ at which the spectral signal first approaches zero. If we simply identify this $f_{\max}$ as the Nyquist frequency associated with a signal that has $N = 2f_{\max}/\Delta f$ data points, the associated Δt of the inverse transform given by $\Delta t = 1/(N\Delta f)$ may be too coarse of a time resolution to see the details of, for example, how the signal in the time domain develops near $t = 0$. So if a synthetic signal of interest is numerically generated in the frequency domain with a given $f_{\max}$ and Δf as just defined, we can then pad this spectrum with zeros placed every Δf out to a much larger Nyquist frequency $f_N \gg f_{\max}$ generating a number of data points $N_{\text{new}} \gg N$ and a finer time resolution given by $\Delta t_{\text{new}} = 1/(N_{\text{new}} \Delta f)$.

10.9 Exercises

1. For α real and positive, prove that FT $\left\{e^{-\alpha|x|}\right\} = 2\alpha/(\alpha^2 + k^2)$. In other words, perform the integral here and prove

$$\boxed{\int_{-\infty}^{\infty} e^{-\alpha|x|-ikx}\, dx = \frac{2\alpha}{k^2+\alpha^2}.} \tag{10.228}$$

   Because $e^{-\alpha|x|}$ is both real and even, we see that the spectrum of this function is also real and even.

2. Prove that if $f(x)$ is odd in x and real, then the Fourier transform $\tilde{f}(k)$ is pure imaginary and odd in k. HINT: Give the integral representation for the complex conjugate of the spectrum $\tilde{f}^*(k)$ like we did in Section 11.3.1 and then make the substitution $x = -u$ and use the fact that $f(-u) = -f(u)$. Note also that if you can show that for some function $\psi(x) = -\psi(x)$ for all x, then $\psi = 0$.

3a. *Parseval's theorem* (first form): Prove that for two arbitrary functions $f(x)$ and $g(x)$

$$\int_{-\infty}^{\infty} f(x)\, g(x)\, dx = \frac{1}{2\pi} \int_{-\infty}^{\infty} \tilde{f}(k)\tilde{g}(-k)\, dk. \tag{10.229}$$

HINT: Start by writing out explicitly the integral expression for $\tilde{g}(-k)$ (in terms of $g(x)$) on the right-hand side and go from there.

3b. *Parseval's theorem* (second form): Use 3a to prove that if $f(x)$ is real, then

$$\int_{-\infty}^{\infty} f^2(x)\, dx = \frac{1}{2\pi}\int_{-\infty}^{\infty} |\tilde{f}(k)|^2\, dk. \tag{10.230}$$

HINT: Use the result for the symmetry of the Fourier transform of a real function then use 3a.

4. Use Parseval's theorem to show that

$$\boxed{\int_0^{\infty} \frac{du}{(\alpha^2+u^2)(\beta^2+u^2)} = \frac{\pi}{2\alpha\beta(\alpha+\beta)}.} \tag{10.231}$$

HINT: By replacing u with the frequency k and noting that the integrand is even over the domain $-\infty < k < \infty$, use your result from exercise 1 to show this using Parseval's theorem.

5. *Symmetry property of the Fourier transform*: For any function $f(x)$ that has a Fourier transform $\tilde{f}(k)$, show that the Fourier transform of $f(-x)$ is $\tilde{f}(-k)$. HINT: Begin by writing $\tilde{f}(k) = \int_{-\infty}^{\infty} f(x)e^{-ikx}\, dx$ and making the substitution $x \to -x$.

6. If you are given the Fourier transform of some function to be

$$\tilde{f}(k) = ike^{-(bk)^2}, \tag{10.232}$$

where b is some constant, demonstrate that the function $f(x)$ in the real space domain is

$$\boxed{f(x) = \frac{-x}{4\sqrt{\pi}b^3}e^{-[x/(2b)]^2}.} \tag{10.233}$$

HINT: You can and should use the differentiation property that says $d\psi(x)/dx = (2\pi)^{-1}\int_{-\infty}^{\infty} ik\tilde{\psi}(k)\, dk$ where $\tilde{\psi}(k)$ is the Fourier transform of $\psi(x)$, that is, "multiplication by ik in the frequency domain is equivalent to the operation d/dx in the real domain."

Application of the Fourier Transform to Partial-Differential Equations

Although one of our main applications of Fourier analysis in this book is to the solution of PDEs for problems posed in infinite or semi-infinite domains, we cannot fully solve a range of such problems until we learn how to integrate on the complex plane (so-called contour integration) in Chapter 11. So we only treat here a few problems where the inverse Fourier transform can be performed without using contour integration.

7. Consider the equation for a wave $u(x, t)$ on a semi-infinite string $0 \leq x < \infty$

$$\frac{\partial^2 u(x, t)}{\partial x^2} - \frac{1}{c^2}\frac{\partial^2 u(x, t)}{\partial t^2} = 0 \tag{10.234}$$

subject to the Dirichlet condition at $x = 0$ that

$$u(x = 0, t) = f(t), \tag{10.235}$$

where $f(t)$ is some given time function and to the radiation condition for large x that

$$\frac{\partial u(x, t)}{\partial x} + \frac{1}{c}\frac{\partial u(x, t)}{\partial t} = 0 \quad \text{as} \quad x \to \infty. \tag{10.236}$$

We have stated earlier in the class that the solution of this problem is simply $f(t - x/c)$ as was justified by direct substitution into the PDE. Derive this result using the Fourier transform. HINT: Take a Fourier transform over time, solve the resulting Helmholtz differential equation in x to obtain $\tilde{u}(x, \omega)$ that satisfies both boundary conditions (you will need to time transform the radiation condition) and then inverse Fourier transform over ω to obtain the desired result $u(x, t) = f(t - x/c)$. This is the "blind math" approach for figuring out the intuitively obvious solution $f(t - x/c)$.

8. Consider a 1D diffusion initial-value problem given by:

$$\frac{\partial^2 P(x, t)}{\partial x^2} - \frac{1}{D}\frac{\partial P(x, t)}{\partial t} = 0, \tag{10.237}$$

$$P(x, t = 0) = \ell \Delta P\, \delta(x - x_o), \tag{10.238}$$

$$P(x = -\infty, t) = 0 \tag{10.239}$$

$$P(x = \infty, t) = 0, \tag{10.240}$$

where $\ell \Delta P$ is a given constant that characterizes the elevated value of P on the plane at $x = x_o$ and $t = 0$ that then diffuses away for $t > 0$. Starting fresh and not using Eq. (10.151), solve this by first taking a Fourier transform over x to obtain the solution $\tilde{P}(k, t)$ that satisfies the initial condition and then taking the inverse transform to obtain the desired expression for $P(x, t)$. You should find that

$$\boxed{P(x, t) = \frac{\ell \Delta P}{2\sqrt{D\pi\, t}} e^{-x^2/(4Dt)}.} \tag{10.241}$$

9. Let's consider a variation of the previous exercise. We now have a 1D diffusion initial-value problem posed in a half space $0 \leq x \leq \infty$ and given as:

$$\frac{\partial^2 P(x, t)}{\partial x^2} - \frac{1}{D}\frac{\partial P(x, t)}{\partial t} = 0, \tag{10.242}$$

$$P(x, t = 0) = \ell \Delta P\, \delta(x - x_o), \tag{10.243}$$

$$P(x = 0, t) = 0, \tag{10.244}$$

$$P(x = \infty, t) = 0. \tag{10.245}$$

So in this problem, the surface $x = 0$ is maintained at $P = 0$ for all time. This surface $x = 0$ is now a finite distance from the initial plane of elevated P located at $x = x_o$. Use the method of images, analogous to how Lord Kelvin treated the diffusively cooling Earth when the surface is maintained at a constant temperature, to show that

$$\boxed{P(x, t) = \frac{\ell \Delta P}{2\sqrt{D\pi\, t}} \left[e^{-(x-x_o)^2/(4Dt)} - e^{-(x+x_o)^2/(4Dt)} \right] \quad \text{for} \quad 0 \leq x \leq \infty.} \tag{10.246}$$

Show that this solution satisfies the diffusion PDE.

11

Contour-Integration Methods and Applications

Contour-integration methods are approaches for solving certain classes of definite integrals of the general forms

$$\text{(i)} \int_{-\infty}^{+\infty} \frac{P(x)}{Q(x)} f_1(x)\,\mathrm{d}x; \quad \text{(ii)} \int_{-\infty}^{+\infty} f_2(x)\,e^{iax}\,\mathrm{d}x; \quad \text{(iii)} \int_{0}^{\infty} f_3(x)\,\mathrm{d}x; \quad \text{(iv)} \int_{0}^{2\pi} f_4(\theta)\,\mathrm{d}\theta, \tag{11.1}$$

where $P(x)$ and $Q(x)$ are polynomials, $f_1(x)$ is a rather general function, $f_2(x)$ should tend to zero as $|x| \rightarrow 0$ but can be singular at finite values of x, $f_3(x)$, when not even in x, should tend to zero as $x \rightarrow \infty$ and $f_4(\theta) = f_4(\cos\theta, \sin\theta)$ is a function of cosines and sines. In the above integral forms, the integration variable x (or θ) is assumed to be real. The trick of contour-integration methods is to extend analysis onto the complex plane where x is considered a complex variable. The two key theorems used in contour integration are *Cauchy's theorem* and the *residue theorem* that are complemented by a lemma called *Jordan's lemma*.

Perhaps the main motivation for introducing contour integration in this book is to perform the inverse Fourier transform. Specifically, when we want to solve inhomogeneous partial-differential equations (PDEs) that have a source term explicitly present using Fourier transform methods, we will end up with inverse Fourier-transform integrals to perform that are optimally treated using the residue theorem (i.e., contour integration). This topic is treated in Section 11.7 and throughout Chapter 12. But we will also give an example of using the residue theorem and the Fourier transform to derive important relations between the forces and responses of constitutive laws that are called *dispersion* or *Kramers–Krönig* relations that are a manifestation of causality (response cannot occur prior to forcing) when the current response depends on the history of the forcing and not just on the current force as we have assumed up to this point in the book.

In what follows, we want to begin solving definite integrals using contour integration as quickly as possible. As such, we only derive those results that allow us to get going on applications and do not attempt to be exhaustive in presenting all that is known about complex functions on the complex plane.

11.1 Analyticity, Cauchy's Theorem, and Singularities

For our purposes of solving definite integrals using contour integration, the central result that will be needed is called *Cauchy's theorem,* which applies to complex functions that are *analytic* (an equivalent word is *meromorphic*). A complex function is analytic if it satisfies the so-called *Cauchy-Rieman conditions,* which are relations between derivatives of the real and imaginary parts of the function as derived below. Points on the complex plane where a function is not analytic are called *singularities* and singularities play an essential role in contour-integration methods. All of these foundational ideas are introduced in this first section.

11.1.1 Complex Functions on the Complex Plane

Consider a complex function

$$f(z) = f_R(z_R, z_I) + i f_I(z_R, z_I) \tag{11.2}$$

of a complex variable z

$$z = z_R + i z_I. \tag{11.3}$$

All of f_R, f_I, z_R, and z_I are real. We sometimes represent a point z on the complex plane using polar coordinates $z = R\, e^{i\theta} = R\,(\cos\theta + i \sin\theta)$ where $R = \sqrt{zz^*} = \sqrt{z_R^2 + z_I^2}$ is the amplitude of z and $\theta = \tan^{-1}(z_I/z_R)$ is the phase angle (measured in radians). Note that the superscript $*$ denotes the "complex conjugate," which means to replace each i within a complex expression with $-i$ so that, for example, $z^* = z_R - iz_I$. These quantities are depicted in Fig. 11.1 of the complex plane. The vertical axis of the complex plane always represents purely imaginary numbers. Sometimes we label this vertical axis as iz_I which is the formally correct way to do it. But many authors, including me, sometimes just label the vertical axis as z_I with it understood that values along this axis are iz_I. Consider some specific points on the complex plane that have $R = 1$ (are on the unit circle) but have different phase angles θ

$$z = e^{i\pi/4} = \frac{\sqrt{2}}{2}(1 + i) = \sqrt{i} \tag{11.4}$$

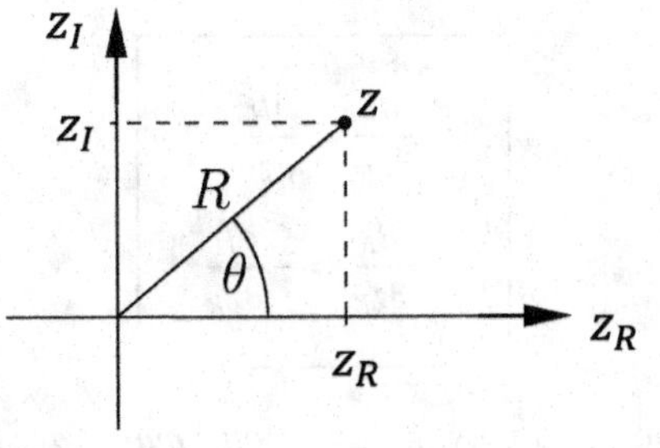

Figure 11.1 A point z in the complex plane. The vertical axis is, implicitly, iz_I.

$$z = e^{i\pi/2} = i \tag{11.5}$$

$$z = e^{i\pi} = -1. \tag{11.6}$$

We will have occasion to use these specific expressions for $\sqrt{i}$, i and -1 in what follows.

11.1.2 Analytic Functions

For a complex function $f(z)$ of a complex variable z to be called "analytic" at a point z, the complex derivative $\mathrm{d}f(z)/\mathrm{d}z$ defined by the limiting process

$$\frac{\mathrm{d}f(z)}{\mathrm{d}z} = \lim_{\Delta\to 0} \frac{f(z+\Delta) - f(z)}{\Delta} \tag{11.7}$$

must be independent of the way that Δ goes to zero on the complex plane, that is, independent of whether the decrease in Δ is along the real z_R axis, the imaginary z_I axis or some other straight or curved path in the neighborhood of z. Such path independence of the complex derivative surrounding some point z, which is the defining requirement for a complex function to be "analytic" at z, is at first glance remarkable and can occur only if there are strict relations between how the real part and imaginary part of the function smoothly vary with z_R and z_I in the neighborhood of z.

So if $\Delta = \Delta_R$ is purely real, we obtain

$$\frac{\partial f(z)}{\partial z_R} = \lim_{\Delta_R\to 0} \frac{f(z+\Delta_R) - f(z)}{\Delta_R} \tag{11.8}$$

while if $\Delta = i\Delta_I$ is purely imaginary we obtain

$$\frac{\partial f(z)}{i\,\partial z_I} = \lim_{\Delta_I\to 0} \frac{f(z+i\Delta_I) - f(z)}{i\Delta_I}. \tag{11.9}$$

By the definition of the complex derivative of an analytic function in the neighborhood of point z, both of these expressions must yield the same limit so upon equating them and inserting $f = f_R + if_I$ we have

$$\frac{\partial f_R}{\partial z_R} + i\frac{\partial f_I}{\partial z_R} = -i\frac{\partial f_R}{\partial z_I} + \frac{\partial f_I}{\partial z_I}, \tag{11.10}$$

which upon equating real and imaginary parts yields the so-called *Cauchy-Rieman* conditions

$$\frac{\partial f_R}{\partial z_R} = \frac{\partial f_I}{\partial z_I} \tag{11.11}$$

$$\frac{\partial f_R}{\partial z_I} = -\frac{\partial f_I}{\partial z_R}. \tag{11.12}$$

So a complex function that is "analytic" at a point must have real and imaginary parts that satisfy the Cauchy-Rieman conditions at that point. Further, the partial derivatives in the Cauchy-Rieman conditions must remain finite at that point. So complex functions that satisfy the Cauchy-Rieman conditions at a point also have real and imaginary parts that are smooth, continuous and well-behaved in the small neighborhood surrounding that point.

To emphasize this last point, note that if we take the z_R derivative of Eq. (11.11) and the z_I derivative of Eq. (11.12) and add the results together, we obtain

$$\frac{\partial^2 f_R}{\partial z_R^2} + \frac{\partial^2 f_R}{\partial z_I^2} = 0. \tag{11.13}$$

Similarly, if we take the z_I derivative of Eq. (11.11) and the z_R derivative of Eq. (11.12) and subtract the results, we obtain

$$\frac{\partial^2 f_I}{\partial z_R^2} + \frac{\partial^2 f_I}{\partial z_I^2} = 0. \tag{11.14}$$

So the real and imaginary parts of a function that is analytic over some domain of the complex plane each satisfy Laplace's equation on this domain which further emphasizes the smooth nature of the real and imaginary parts of an analytic function. If a complex function is analytic at a point in the complex plane, it in fact possesses an infinite number of complex derivatives at that point, none of which blow up. So we can represent a complex function that is analytic at some point as a Taylor series in the neighborhood of that point. These facts will be demonstrated later in Section 11.4.

As some simple examples, consider the smooth function $f(z) = e^z = e^{z_R} e^{iz_I} = e^{z_R}(\cos z_I + i \sin z_I)$ that possesses derivatives of the real and imaginary parts

$$\frac{\partial f_R}{\partial z_R} = e^{z_R} \cos z_I = \frac{\partial f_I}{\partial z_I} \tag{11.15}$$

$$\frac{\partial f_R}{\partial z_I} = -e^{z_R} \sin z_I = -\frac{\partial f_I}{\partial z_R} \tag{11.16}$$

that satisfy the Cauchy-Rieman equations. So $f(z) = e^z$ is an analytic function for all $z < \infty$. Similarly, $f(z) = z^2 = (z_R + iz_i)^2 = z_R^2 - z_I^2 + 2iz_I z_R$ has derivatives of the real and imaginary parts

$$\frac{\partial f_R}{\partial z_R} = 2z_R = \frac{\partial f_I}{\partial z_I} \tag{11.17}$$

$$\frac{\partial f_R}{\partial z_I} = -2z_I = -\frac{\partial f_I}{\partial z_R} \tag{11.18}$$

that satisfy the Cauchy-Rieman equations. So $f(z) = z^2$ is an analytic function at all points of the complex plane. However, note that the purely real function $f(z) = zz^* = z_R^2 + z_I^2$ does not satisfy the Cauchy-Rieman equations and therefore is not analytic (except at the origin $z = 0$).

As another example, consider $f(z) = \ln z = \ln \sqrt{z_R^2 + z_I^2} + i \tan^{-1}(z_I/z_R)$ which has a complex derivative $\mathrm{d}f(z)/\mathrm{d}z = z^{-1}$ that blows up at $z = 0$, which tells us that $\ln z$ is not analytic at the origin of the complex plane. The real and imaginary parts satisfy the Cauchy–Rieman equation

$$\frac{\partial f_R}{\partial z_R} = \frac{z_R}{z_R^2 + z_I^2} = \frac{\partial f_I}{\partial z_I} \tag{11.19}$$

$$\frac{\partial f_R}{\partial z_I} = \frac{z_I}{z_R^2 + z_I^2} = -\frac{\partial f_I}{\partial z_R} \tag{11.20}$$

at all points except at $z = 0$ where the partial derivatives blow up. So we say that $f(z) = \ln z$ is not analytic at the point $z = 0$, that we will later identify as a *branch point*, but is analytic at other points on the complex plane.

11.1.3 Cauchy's Theorem

For a closed contour C on the complex plane that surrounds a domain D as depicted in Fig. 11.2, *Cauchy's theorem* states that

$$\oint_C dz\, f(z) = 0 \quad \text{if } f(z) \text{ is analytic at all points of } D. \tag{11.21}$$

The sense (direction) of a closed contour integral $\oint_C$ is always taken to be counterclockwise as shown in Fig. 11.2. So any function that satisfies Cauchy's theorem for a given closed contour C must satisfy the Cauchy-Rieman conditions at all points of the domain D that resides within C as we now demonstrate.

The proof of Cauchy's theorem employs "Green's theorem" (equivalent to "Stokes theorem") that you proved as an exercise at the end of Chapter 1. Green's theorem states that for any two real differentiable functions $P(x, y)$ and $Q(x, y)$ of two real variables (x, y) defined over a domain D of the $x - y$ plane and bounded by a closed curve C, the following integral relation is true

$$\oint_C (P\, dx + Q\, dy) = \iint_D \left(\frac{\partial Q}{\partial x} - \frac{\partial P}{\partial y} \right) dxdy. \tag{11.22}$$

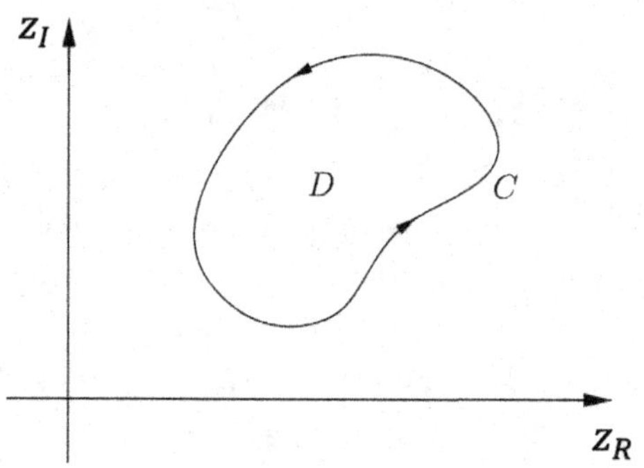

Figure 11.2 A domain D and enclosing contour C in the complex plane.

By separating into real and imaginary parts, we can write the integral of Cauchy's theorem as

$$\oint_C f(z)\,\mathrm{d}z = \oint_C (f_R\,\mathrm{d}z_R - f_I\,\mathrm{d}z_I) + i\oint_C (f_I\,\mathrm{d}z_R + f_R\,\mathrm{d}z_I). \tag{11.23}$$

Applying Green's theorem to the first of these integrals gives

$$\oint_C (f_R\,\mathrm{d}z_R - f_I\,\mathrm{d}z_I) = \iint_D \left(-\frac{\partial f_I}{\partial z_R} - \frac{\partial f_R}{\partial z_I}\right)\mathrm{d}z_R\mathrm{d}z_I = 0 \tag{11.24}$$

which is zero from the second Cauchy-Riemann condition. Applying Green's theorem to the second integral (the imaginary part) gives

$$i\oint_C (f_I\,\mathrm{d}z_R + f_R\,\mathrm{d}z_I) = i\iint_D \left(\frac{\partial f_R}{\partial z_R} - \frac{\partial f_I}{\partial z_I}\right)\mathrm{d}z_R\mathrm{d}z_I = 0 \tag{11.25}$$

which is zero from the first Cauchy-Rieman condition. Thus, given Green's theorem and an application of the Cauchy-Rieman conditions, we obtain Cauchy's theorem $\oint_C f(z)\,\mathrm{d}z = 0$ by combining Eqs (11.23)–(11.25). To summarize, Cauchy's theorem is obeyed by any complex function $f(z)$ having real and imaginary parts that satisfy the Cauchy-Rieman equations and is therefore called "analytic" at all points within and on the closed contour.

We can ask whether $f(z)$ is analytic in some domain by appealing to Cauchy's theorem. Consider $f(z) = \sqrt{z}$. First, if we take the derivative with respect to z in the complex plane, we have $\mathrm{d}f(z)/\mathrm{d}z = 1/(2\sqrt{z})$ which blows up at $z = 0$, so $\sqrt{z}$ is not analytic on domains that include the origin. But we can also see this by calculating the Cauchy integral $\oint_{C_1}\sqrt{z}\,\mathrm{d}z$ for a contour C_1 that is a unit circle encircling the origin on the complex plane. Writing $z = Re^{i\theta}$, $\sqrt{z} = \sqrt{R}e^{i\theta/2}$ and noting that $R = 1$ on the unit circle, we have

$$\oint_{C_1}\sqrt{z}\,\mathrm{d}z = \int_0^{2\pi} e^{i\theta/2}\,\mathrm{d}\theta = \frac{2}{i}\left[e^{i\pi} - e^0\right] = 4i \neq 0. \tag{11.26}$$

So the violation of Cauchy's theorem tells us that $\sqrt{z}$ is not analytic in domains that include the origin.

Using this same argument, we can show, for example, that $f(z) = z^{1.25}$ is not analytic within the unit circle of the complex plane. First note that although the function and it first complex derivative $\mathrm{d}f(z)/\mathrm{d}z = (1.25)z^{0.25}$ are well behaved on the domain within the unit circle, the second complex derivative $\mathrm{d}^2 f(z)/\mathrm{d}z^2 = (1.25)(0.25)/z^{0.75}$ and all higher-order derivatives blow up at $z = 0$. For the Cauchy closed-contour integral on the unit circle C_1, we use $f(z) = f(1, \theta) = e^{i(1.25)\theta}$ and obtain

$$\oint_{C_1} z^{1.25}\,\mathrm{d}z = \int_0^{2\pi} e^{i(1.25)\theta}\,\mathrm{d}\theta = \frac{(e^{i\pi/2} - 1)}{(1.25)i} = \frac{1+i}{1.25} \neq 0. \tag{11.27}$$

Thus, this function and indeed any function of the form z^β with $\beta > 0$ and not an integer does not satisfy Cauchy's theorem and is not analytic within the unit circle. So although

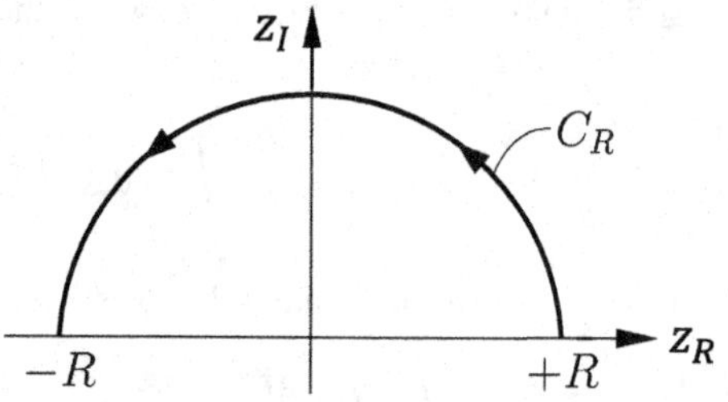

Figure 11.3 The half circle at infinity C_R as $R \to \infty$ that Jordan's lemma applies to when the real parameter a is positive.

$z^{1.25}$ and its first derivative are well behaved within the unit circle, the second and higher order derivatives blow up at the origin and the function is not analytic for any domain containing the origin.

11.1.4 Jordan's Lemma

In order to use Cauchy's theorem to obtain nontrivial results for definite integrals of interest, we will also need a lemma called *Jordan's lemma*, which states that if $g(z)$ is a function that has the behavior $\lim_{|z|\to\infty} g(z) \to 0$ (though it may be nonanalytic at points in the z plane), then

$$I_\infty = \lim_{R\to\infty} \int_{C_R} dz\, g(z)\, e^{iaz} = 0, \tag{11.28}$$

where the contour C_R is the half circle at $R = \infty$ depicted in Fig. 11.3 and where the real parameter a is positive $a > 0$. If $a = 0$, we have the corollary that $I_\infty = 0$ if $\lim_{|z|\to\infty} z\, g(z) \to 0$. This corollary makes good sense because $\int_{C_R} dz = \pi R$ so we need $g(z)$ going to zero faster than $|z|^{-1}$ as $|z| = R \to \infty$ in order for $I_\infty = 0$. When $a > 0$, on the imaginary axis $z = iz_I$ we have $e^{iaz} = e^{-az_I}$, which goes to zero at $z_I = +\infty$ but blows up at $z_I = -\infty$. This is why we take C_R to be the half circle in the upper-half plane when $a > 0$. When $a < 0$, we therefore need to take the half circle C_R to reside in the lower-half z-plane in order for Jordan's lemma to apply.

11.1.5 An Example Definite Integral Performed Using Contour Integration

Given Cauchy's theorem and Jordan's lemma we can already begin evaluating definite integrals that may arise in the course of our analysis of physics problems. As a first integral, let's consider how to obtain

$$I = \int_0^\infty \frac{\sin x}{x}\, dx \tag{11.29}$$

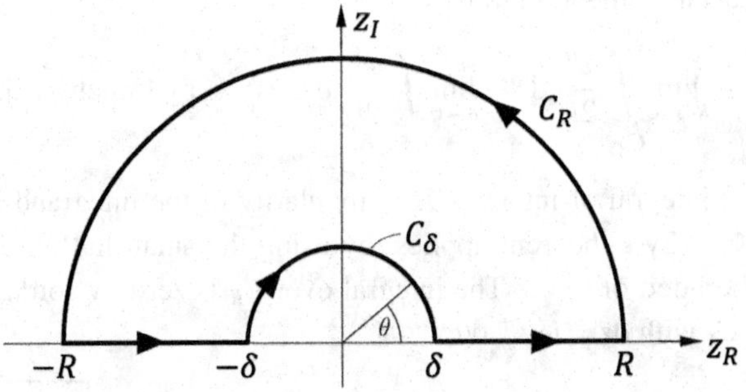

Figure 11.4 The closed contour C used in the analysis of the real integral $\int_0^\infty dx\,(\sin x)/x$.

where x is real. To use Cauchy's theorem and Jordan's lemma to solve this integral, consider the closed contour C in the complex plane depicted in Fig. 11.4. The total closed contour C is being decomposed into the four contributions depicted in the figure

$$\oint_C = \int_\delta^R + \int_{C_R} + \int_{-R}^{-\delta} + \int_{C_\delta} \tag{11.30}$$

in the limits as $R \to \infty$ and $\delta \to 0$. To proceed, we write the sine as a function of the complex variable z where $x = z_R$

$$\sin z = \frac{e^{iz} - e^{-iz}}{2i} \tag{11.31}$$

so that our integral of interest is (let $R \to \infty$)

$$I = \lim_{\delta \to 0} \int_\delta^\infty \frac{e^{iz} - e^{-iz}}{2iz}\, dz \tag{11.32}$$

$$= \lim_{\delta \to 0} \left[\int_\delta^\infty \frac{e^{iz}}{2iz}\, dz + \int_{-\infty}^{-\delta} \frac{e^{iz}}{2iz}\, dz \right] \tag{11.33}$$

$$\widehat{=}\, \mathrm{PV} \int_{-\infty}^{\infty} \frac{e^{iz}}{2iz}\, dz \tag{11.34}$$

where we let $z = -z$ in the second integral of the first line and where the symbol PV means *principal value* and is defined by the two integrals in the second line in the limit as $\delta \to 0$. Even as $\delta \to 0$ and the integrand e^{iz}/z becomes unbounded on either side of $z = 0$, the principal value remains bounded in this example because e^{iz}/z has different signs on either side of $z = 0$ that symmetrically balance each other in their contribution to the integrals as $\delta \to 0$. Further note that the final expression is equivalent to the original integral I as seen by substituting $e^{iz} = \cos z + i \sin z$ and noting that $(\cos z)/z$ is an odd function (and thus integrates to zero over $-\infty \leq z \leq \infty$), while $(\sin z)/z$ is even over $-\infty < z < \infty$. The line integrals involved in these different expressions for I are implicitly along the real z (or x or z_R) axis within the complex z plane.

We now use Cauchy's theorem to write

$$\oint_C \frac{e^{iz}}{2iz}\,\mathrm{d}z = I + \lim_{R\to\infty}\int_{C_R}\frac{e^{iz}}{2iz}\,\mathrm{d}z + \lim_{\delta\to 0}\int_{C_\delta}\frac{e^{iz}}{2iz}\,\mathrm{d}z = 0, \quad \text{by Cauchy's theorem} \tag{11.35}$$

where I is our real integral of interest. The singularity of the integrand at $z=0$ has been avoided, so that Cauchy's theorem applies, by using the small half circle path C_δ in the immediate neighborhood of $z=0$. The integral over C_R is zero by Jordan's lemma. Thus, using $z=\delta e^{i\theta}$ on C_δ with $\mathrm{d}z = i\delta\, e^{i\theta}\,\mathrm{d}\theta$,

$$I = -\lim_{\delta\to 0}\int_{C_\delta}\frac{e^{iz}}{2iz}\,\mathrm{d}z = -\lim_{\delta\to 0}\int_\pi^0 \mathrm{d}\theta\,\frac{i\delta\, e^{i\theta}\, e^{i\,\delta\, e^{i\theta}}}{2i\delta\, e^{i\theta}} = \frac{1}{2}\int_0^\pi \mathrm{d}\theta = \boxed{\frac{\pi}{2}}. \tag{11.36}$$

So by using Cauchy's theorem and Jordan's lemma and some insight on how to define an appropriate closed contour C that avoids the singularity at $z=0$, we have obtained that $I=\int_0^\infty (\sin x/x)\,\mathrm{d}x = \pi/2$ in a relatively painless manner. In this example, although $\sin z/z$ is well behaved at $z=0$ (use l'Hôpital's rule to show that $\sin z/z = 1$ at $z=0$), because we identified

$$I = \int_0^\infty \frac{\sin z}{z}\,\mathrm{d}z = \mathrm{PV}\int_{-\infty}^{+\infty}\frac{e^{iz}}{2iz}\,\mathrm{d}z, \tag{11.37}$$

the second integrand $e^{iz}/(2iz)$ does not behave well at $z=0$ (it blows up there) and this is why we avoided that point when integrating between $-\infty$ and ∞.

Finally, we note that $\sin x/x$ is what we call the "sinc" function. The boxcar function $b(x)$ is defined to be 1 when $-a \le x \le a$ and zero for $|x| > a$. We showed in Chapter 10 that the Fourier transform of the boxcar function is $2a\sin(ak)/(ak)$ where k is the frequency variable. What we have shown above is that $(2\pi)^{-1}\int_{-\infty}^{\infty}\mathrm{d}k\, 2a\sin(ak)/(ak) = 2(2\pi)^{-1}\int_{-\infty}^{\infty}\mathrm{d}u\,(\sin u)/u = 1$. Thus, one definition of the Dirac delta function is $\delta(k) = \lim_{a\to\infty}\sin(ak)/(\pi k)$ as posited in Chapter 1. This is also another separate proof of the completeness relation in the limit as $a\to\infty$ and the boxcar function becomes a constant.

11.1.6 Singularities

To use Cauchy's theorem, the function being integrated around the closed contour C must be analytic at all points within the domain bounded by C. Points where the integrand is not analytic within a domain of the complex plane are called *singularities*. There are two classes of singularities:

(1) *Poles:* the function itself blows up at a pole.

(2) *Branch points:* a function has multiple possible values as you circle around a branch point on the complex z plane.

We briefly describe these two types of singularities using some examples.

Functions Having Poles

Simple examples of functions having poles are :

$$f(z) = \frac{1}{z} \quad \text{which has a "simple" pole at } z = 0, \tag{11.38}$$

$$f(z) = \frac{1}{(z-a)^m} \quad \text{which has a pole of "order m" at } z = a. \tag{11.39}$$

In Section 11.3, we will show how to determine the location of a pole and its order for nonobvious examples such as $f(z) = \tan\left(\pi(z - z_o)^2/b^2\right)$ which blows up whenever $\pi(z - z_0)^2/b^2 = 2\pi n$, where $n = 0, 1, 2, \ldots$ is an integer.

Functions Having Branch Points

Consider the square root function

$$f(z) = \sqrt{z} = f(R, \theta) = \sqrt{R}\, e^{i\theta/2}. \tag{11.40}$$

As we circle around the point $z = 0$, we have $f(R, \theta + 2\pi) = \sqrt{R}\, e^{i\frac{\theta}{2}}\, e^{i\pi} = -\sqrt{R}\, e^{i\theta/2} = -f(R, \theta)$. Thus, instead of $f(R, \theta + 2\pi) = f(R, \theta)$, which would be the case for a single-valued function, we have two possible values for the square root function on the z plane depending on how many times we circulate around the point $z = 0$. We call the point $z = 0$ in this case a *branch point*. Similar reasoning shows that $f(z) = z^\beta$, where β is a positive noninteger exponent is also multivalued. If the function was given as $f(z) = (z - z_0)^\beta$, we would say the branch point is at $z = z_0$.

Another example of a multivalued function is $f(z) = \ln z$. Again, writing $z = Re^{i\theta}$ we have $f(z) = f(R, \theta) = \ln R + i\theta$. So as we circle around the point $z = 0$ in the complex z plane, we have θ increasing by 2π each time we make a full rotation. So, we have for example that $f(R, -\pi) = \ln R - i\pi$ and $f(R, +\pi) = \ln R + i\pi$ so that there is an ambiguity in $\ln z$ of $\pm i2\pi n$ at each point z depending on how many times n we have circulated around the branch point at $z = 0$.

A characteristic of branch points, about which a function is multivalued as we circulate around the branch point, is that the derivatives of multivalued functions blow up at each branch point, at least if we take enough derivatives. So multivalued functions are not analytic in domains that contain their branch points. For example, the function $f(z) = (z - z_0)^{3.2}$ and its first three derivatives do not blow up at $z = z_0$. But its fourth derivative is $\mathrm{d}^4(z - z_0)^{3.2}/\mathrm{d}z^4 = (3.2)(2.2)(1.2)(0.2)/(z - z_0)^{0.8}$, so the fourth and higher-order derivatives of this function all blow up at the branch point $z = z_0$, which means that Cauchy's theorem is not satisfied by this function on domains that include z_0.

We will see in the various integrals being treated in this chapter how to deal with both types of singularities when performing contour integration.

11.2 The Residue Theorem

When an integrand has singularities on the domain bounded by a closed contour, we need to modify Cauchy's theorem. Let's say there is a pole of $f(z)$ at the point $z = z_0$ that lies within our closed contour. We can develop $f(z)$ about the pole $z = z_0$ as a *Laurent series*:

$$f(z) = \sum_{n=-\infty}^{\infty} a_n (z - z_0)^n. \tag{11.41}$$

This can be rewritten into portions for $n \geq 0$ (that we call the analytic part) and for $n < 0$ (that we call the singular part)

$$f(z) = \sum_{n=0}^{\infty} a_n (z - z_0)^n + \sum_{n=1}^{\infty} \frac{a_{-n}}{(z - z_0)^n}. \tag{11.42}$$

If the singularity at $z = z_0$ is a pole of order m, we have $a_{-n} = 0$ for all $n > m$ and can write the above as

$$f(z) = \sum_{n=0}^{\infty} a_n (z - z_0)^n + \sum_{n=1}^{m} \frac{a_{-n}}{(z - z_0)^n} \quad \text{for a pole } z_0 \text{ of order } m \tag{11.43}$$

$$= \frac{a_{-m}}{(z - z_o)^m} + \frac{a_{-(m-1)}}{(z - z_o)^{(m-1)}} + \ldots \frac{a_{-1}}{(z - z_o)} + a_0 + a_1(z - z_o) + a_2(z - z_o)^2 + \ldots. \tag{11.44}$$

The Laurent series is a close relative to the Taylor series. The Taylor series is the power-series development about a point $z = z_0$, where a complex function is analytic and is thus confined to terms in Eq. (11.41) having $n \geq 0$ where the function and its derivatives all exist and never blow up. So in a Taylor series development of a function that is analytic in the neighborhood of z_0, we have $a_n = 0$ for all $n < 0$. The part of the Laurent series for which $n < 0$ is called the singular part of the series because these terms and their derivatives all blow up at $z = z_0$. We will derive the Laurent series in a later section just to be complete and will show how to calculate the a_n for $n < 0$ in terms of $f(z)$ but if you are willing to accept its existence for describing the nonanalytic nature of some function $f(z)$ you can forego that section. We now use the Laurent series to represent a function having one or more poles to derive the residue theorem.

Consider a closed contour C_0 surrounding the pole at z_0 as shown in Fig. 11.5. From Cauchy's theorem, we have that the analytic part ($n \geq 0$) of the Laurent series of Eq. (11.41) satisfies

$$\oint_{C_0} a_n (z - z_0)^n \, dz = 0 \quad \text{when } n \geq 0. \tag{11.45}$$

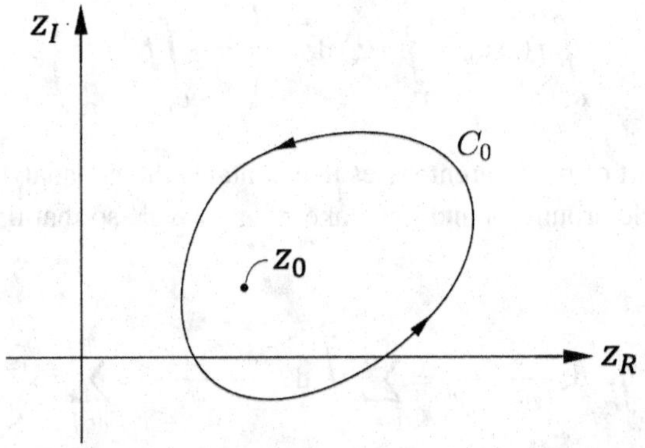

Figure 11.5 A closed contour C_0 surrounding the point z_0 where the integrand has a pole.

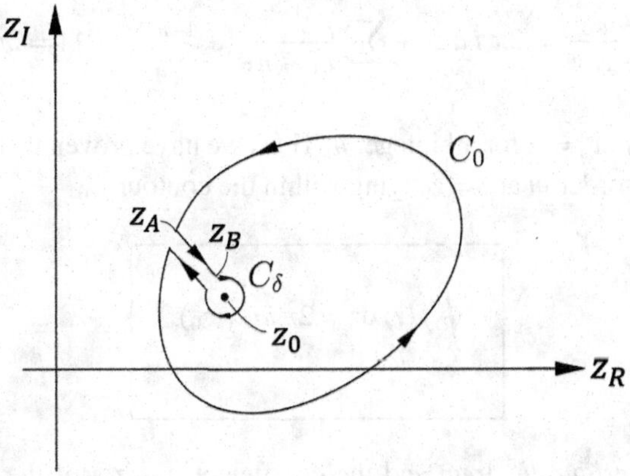

Figure 11.6 A modified contour that excludes the pole at z_0.

To treat the singular part of the Laurent series, we exclude the pole using the contour shown in Fig. 11.6. For this modified contour C_m, we have:

$$\oint_{C_m} f(z)\,dz = 0 \quad \text{(because } z_0 \text{ is not in the interior of } C_m\text{)} \tag{11.46}$$

$$= \int_{z_A}^{z_B} f(z)\,dz + \int_{C_\delta} f(z)\,dz + \int_{z_B}^{z_A} f(z)\,dz + \int_{C_0} f(z)\,dz. \tag{11.47}$$

The two line integrals satisfy $\int_{z_A}^{z_B} f(z)\,dz = -\int_{z_B}^{z_A} f(x)\,dz$ because the line heading from C_0 toward C_δ can be made to lie arbitrarily close to the line heading from C_δ back out to C_0 and in this region the function $f(z)$ is perfectly analytic and single valued. For the same reason, we can assume the integrals over C_0 and C_δ are over closed loops (note that when we use the symbol $\oint_C$ we mean an integral taken in the counter-clockwise direction around the closed loop C), which leads to

$$\oint_{C_0} f(z)\,\mathrm{d}z = \oint_{C_\delta} f(z)\,\mathrm{d}z \quad \left(= -\int_{C_\delta} f(z)\,\mathrm{d}z \right). \tag{11.48}$$

Returning to the part of the Laurent series that contains the nonanalytic singular behavior, we let C_δ be a circle around z_0 and thus take $z - z_0 = \delta\, e^{i\theta}$ so that $\mathrm{d}z = i\delta\, e^{i\theta}\,\mathrm{d}\theta$ along C_δ to give

$$\oint_{C_0} f(z)\,\mathrm{d}z = \sum_{n=1}^{m} \oint_{C_\delta} \mathrm{d}z \frac{a_{-n}}{(z-z_0)^n} = \sum_{n=1}^{m} \int_0^{2\pi} \mathrm{d}\theta\, \frac{i\delta\, e^{i\theta}\, a_{-n}}{\delta^n\, e^{in\theta}} = \sum_{n=1}^{m} i\,\delta^{1-n}\, a_{-n} \int_0^{2\pi} e^{i(1-n)\theta}\,\mathrm{d}\theta. \tag{11.49}$$

Performing the integral for $n = 1$ and $n \geq 2$ separately, we have that

$$\sum_{n=1}^{m} \oint_{C_\delta} \mathrm{d}z \frac{a_{-n}}{(z-z_0)^n} = 2\pi\, i\, a_{-1} + \sum_{n=2}^{m} \frac{\delta^{1-n}}{(1-n)} \left(e^{i(1-n)2\pi} - 1\right) = 2\pi\, i\, a_{-1}, \tag{11.50}$$

that is, $\exp\left[i(1-n)2\pi\right] = 1$ for all integer n. Thus, we have proven that for a function with a single pole of any order m at $z = z_0$ lying within the contour C_0

$$\boxed{\oint_{C_0} f(z)\,\mathrm{d}z = 2\pi i a_{-1}(z_0).} \tag{11.51}$$

This is called the *residue theorem* and the coefficient $a_{-1}(z_0)$ of the Laurent expansion about the point $z = z_0$ is called the *residue*.

If the function $f(z)$ has many poles located at $z = z_s$ with $s = 1, 2, \ldots$ within the closed contour C_0, it is straightforward to deform the contour like we did above so that each pole is surrounded by a small circle defined by $z - z_s = \delta e^{i\theta}$ so that we obtain the generalization to many poles within the contour

$$\boxed{\oint_{C_0} f(z)\,\mathrm{d}z = 2\pi\, i \sum_s a_{-1}(z_s),} \tag{11.52}$$

where the residues $a_{-1}(z_s)$ are determined from the Laurent expansion of $f(z)$ about each pole z_s.

The residues are easy to determine. For a pole of order m located at $z = z_s$, we have the Laurent development

$$f(z) = \frac{a_{-m}}{(z-z_s)^m} + \cdots + \frac{a_{-1}}{z-z_s} + a_0 + a_1(z-z_s) + \cdots \tag{11.53}$$

so to get the residue a_{-1}, we multiply both sides by $(z - z_s)^m$, take $m - 1$ derivatives, and evaluate at $z = z_s$ to obtain

$$a_{-1}(z_s) = \frac{1}{(m-1)!} \lim_{z \to z_s} \frac{d^{m-1}\left[(z - z_s)^m f(z)\right]}{dz^{m-1}}$$
for a pole of order m located at $z = z_s$. (11.54)

For a simple pole of order $m = 1$, this then gives

$$a_{-1}(z_s) = \lim_{z \to z_s} \left[(z - z_s) f(z)\right] \quad \text{for a simple pole of order } m = 1 \text{ at } z = z_s. \tag{11.55}$$

We now go on to consider in greater detail how to determine the location and order of poles and their associated residues.

11.3 The Order of Poles and Calculating Residues

In order to determine the residue using the above Eqs (11.54) or (11.55), we need to know the order of the pole. Often, this is trivial. For example, the function

$$\frac{g(z)}{(z - z_0)^m}, \tag{11.56}$$

where $g(z)$ is a well-behaved analytic function, has a pole at $z = z_0$ of order m. A function like

$$\frac{g(z)}{(z - z_0)^m (z - z_1)^n}, \tag{11.57}$$

where $g(z)$ is again analytic has two poles: one at $z = z_0$ of order m and the other at $z = z_1$ of order n. Each pole in these examples has a residue that is easily determined using Eq. (11.54).

Sometimes, however, it is not obvious by simple inspection to determine the order of the poles. For example, the function

$$\frac{e^{az}}{1 - \cos mz} \tag{11.58}$$

definitely has a pole at $z = 0$ (it blows up there as well as at $mz = 2\pi n$ where $n = 0, 1, 2, \ldots$ is an integer) but what is the order of the pole? You can always determine the order by performing the Laurent expansion around the pole. So in this particular case, we have for our pole at $z = 0$ the Taylor expansions of the denominator and numerator:

$$\cos mz = 1 - \frac{(mz)^2}{2!} + \frac{(mz)^4}{4!} - \cdots \tag{11.59}$$

$$\exp az = 1 + az + \frac{(az)^2}{2!} + \frac{(az)^3}{3!} + \cdots. \tag{11.60}$$

Thus, taking the ratio of these two expansions gives

$$\frac{e^{az}}{1 - \cos mz} = \frac{1 + az + (az)^2/2! + \cdots}{\frac{(mz)^2}{2!}\left[1 - \frac{2!}{4!}(mz)^2 + \frac{2!}{6!}(mz)^4 - \cdots\right]} \tag{11.61}$$

$$= \frac{2}{m^2 z^2}\left(1 + az + \frac{(az)^2}{2!} + \cdots\right)\left(1 + \frac{2!}{4!}(mz)^2 - \cdots\right) \tag{11.62}$$

$$= \frac{2}{m^2}\left[\frac{1}{z^2} + \frac{a}{z} + \left(\frac{a^2}{2!} + \frac{2!}{4!}m^2\right) + O(z)\right], \tag{11.63}$$

where, as discussed in Chapter 1, the "big-O" notation $O(z)$ means terms of order z and higher as $z \to 0$. So this function has a pole at $z = 0$ that is of order 2. Since we have determined the Laurent expansion, we can simply read off the coefficient a_{-1} of the series expanded around the pole at $z_i = 0$ to obtain the residue as

$$\boxed{a_{-1}(z_i = 0) = \frac{2a}{m^2}.} \tag{11.64}$$

Once having determined that the pole is of order 2, we can also try to determine the residue using the residue formula of Eq. (11.54)

$$a_{-1}(z_i = 0) = \lim_{z \to 0} \frac{\mathrm{d}}{\mathrm{d}z}\left[\frac{z^2\, e^{az}}{1 - \cos mz}\right], \tag{11.65}$$

but this is actually more tedious than having developed the Laurent series to obtain the a_{-1} coefficient directly.

Let's consider another function

$$f(z) = \frac{\cos az}{\cos bz} \quad \text{where } a \neq b. \tag{11.66}$$

There is clearly a pole at $z = \pi/(2b)$ (as well as at $z = \pi(2n + 1)/(2b)$ for any integer n). Let's expand $f(z)$ about the pole at $z = \pi/(2b)$. Recall that to expand the analytic function $g(z)$ about $z = z_0$, we have the Taylor series: $g(z) = g(z_0) + g'(z)|_{z=z_0}(z - z_0)/1! + g''(z)|_{z=z_0}(z - z_0)^2/2! + \cdots$. Thus,

$$\begin{aligned} \cos bz &= \cancelto{0}{\cos\frac{\pi}{2}} - b\,\sin\frac{\pi}{2}\left(z - \frac{\pi}{2b}\right) + \frac{b^3}{3!}\,\sin\frac{\pi}{2}\left(z - \frac{\pi}{2b}\right)^3 + \cdots \\ &= -b\left(z - \frac{\pi}{2b}\right)\left[1 - \frac{b^2}{3!}\left(z - \frac{\pi}{2b}\right)^2 + O\left(\left(z - \frac{\pi}{2b}\right)^4\right)\right]. \end{aligned} \tag{11.67}$$

Similarly, we can also expand $\cos az$ around the same pole $z = \pi/(2b)$ as

$$\cos az = \cos\frac{a\pi}{2b} - a\,\sin\frac{a\pi}{2b}\left(z - \frac{\pi}{2b}\right) + \cdots \tag{11.68}$$

and so to leading order in $z - \pi/(2b)$, we have

$$\frac{\cos az}{\cos bz} = -\frac{\cos[a\pi/(2b)]}{b\left[z - \pi/(2b)\right]} + \frac{a}{b}\,\sin\frac{a\pi}{2b} + O\left(z - \frac{\pi}{2b}\right). \tag{11.69}$$

So the pole in this case is order 1 (a simple pole) and the residue is directly read off from the expansion to be

$$\boxed{a_{-1}\left(\frac{\pi}{2b}\right) = -\frac{\cos[a\pi/(2b)]}{b}.} \tag{11.70}$$

In these examples, where it is not obvious by direct inspection how to determine the order of a pole, one has to first expand the function in a Laurent series and then read off the a_{-1} coefficient to obtain the residue.

As a final residue rule that is convenient to know, consider a function $f(z)$ that has a simple pole at $z = z_i$ and is of the form

$$f(z) = \frac{p(z)}{q(z)}, \tag{11.71}$$

where $p(z)$ is analytic near z_i and $q(z) = g(z)(z - z_i)$ where $g(z_i) \neq 0$. An example we will see later is the function $f(z) = \ln z/(z^3 + 1)$ where the zeroes of the polynomial $z_i^3 + 1 = 0$, which are the simple poles of $f(z)$, are located at $z_i^3 = -1 = e^{i(\pi + 2\pi n)}$ so that the three simple poles are at $z_i = e^{i\pi/3}$, $e^{i\pi}$ and $e^{i5\pi/3}$. For this example, you could write $f(z) = \ln z/[(z - z_1)(z - z_2)(z - z_3)]$ and evaluate the residue at say z_1 as $a_{-1}(z_1) = \ln z_1/[(z_1 - z_2)(z_1 - z_3)]$. But there is a better more compact rule for the residues of $f(z) = p(z)/q(z)$ that possess simple poles that comes from expanding both $p(z)$ and $1/q(z)$ as Taylor series and then reading off the a_{-1} coefficient of the subsequent Laurent series

$$f(z) = \frac{p(z)}{q(z)} = \frac{p(z_i) + p'(z_i)(z - z_i) + \ldots}{q(z_i) + q'(z_i)(z - z_i) + q''(z_i)(z - z_i)^2/2 + \ldots}, \tag{11.72}$$

where, for example, $p'(z_i) = \mathrm{d}p(z)/\mathrm{d}z|_{z=z_i}$. Using that $q(z_i) = 0$ we then have the Laurent series

$$f(z) = \frac{p(z_i)}{q'(z_i)(z - z_i)} + \frac{p(z_i)}{q'(z_i)}\left[\frac{p'(z_i)}{p(z_i)} - \frac{q''(z_i)}{2q'(z_i)}\right] + O(z - z_i), \tag{11.73}$$

which gives the residue of the simple pole at $z = z_i$ as

$$\boxed{a_{-1}(z_i) = \frac{p(z_i)}{q'(z_i)}.} \tag{11.74}$$

So for our specific example of $f(z) = \ln z/(z^3 + 1)$ with $p(z) = \ln z$, $q(z) = z^3 + 1$, and $q'(z) = 3z^2$, we find the residues of the three simple poles at $e^{i\pi/3}$, $e^{i\pi}$, and $e^{i5\pi/3}$ as

$$a_{-1}\left(e^{i\pi/3}\right) = \frac{i\pi/3}{3\exp(i2\pi/3)} = -\frac{i\pi}{9}\left(\frac{1}{2} + i\frac{\sqrt{3}}{2}\right), \tag{11.75}$$

$$a_{-1}\left(e^{i\pi}\right) = \frac{i\pi}{3\exp(i2\pi)} = \frac{i\pi}{3}, \tag{11.76}$$

$$a_{-1}\left(e^{i5\pi/3}\right) = \frac{i5\pi/3}{3\exp(i10\pi/3)} = -\frac{i5\pi}{9}\left(\frac{1}{2} - i\frac{\sqrt{3}}{2}\right). \tag{11.77}$$

where we used, for example, that $e^{i\pi/3} = 1/2 + i\sqrt{3}/2$ so that $e^{i2\pi/3} = e^{-i4\pi/3} = -1/2 + i\sqrt{3}/2$. We will use these specific residues in a later integral example.

Although expanding the integrand into a Laurent series and reading off a_{-1} works best for more complicated integrands, our earlier formulas of Eqs (11.54) and (11.55) serve us well when we can directly observe the order of the pole. For example, if we are given the function

$$f(z) = \frac{\sin z}{z - z_0}, \tag{11.78}$$

we can say immediately that it has a pole of order 1 located at $z = z_0$, so that we may calculate

$$a_{-1}(z_0) = \lim_{z\to z_0}(z - z_0)\frac{\sin z}{z - z_0} = \boxed{\sin z_0.} \tag{11.79}$$

As another example, consider

$$f(z) = \frac{\sin z}{z^2 - z_0^2} = \frac{\sin z}{(z - z_0)(z + z_0)}, \tag{11.80}$$

which has two simple poles: a pole of order 1 at $z = z_0$ and another pole of order 1 at $z = -z_0$. So the residue at $z = z_0$, for example, is easily calculated as

$$a_{-1}(z_0) = \lim_{z \to z_0} (z - z_0) \frac{\sin z}{(z - z_0)(z + z_0)} = \boxed{\frac{\sin z_0}{2\, z_0}}. \tag{11.81}$$

But note as well that if we use Eq. (11.74) in this example that involves simple poles, we get the exact same result. As yet another example, consider

$$f(z) = \frac{\sin az}{(z - z_0)^3}, \tag{11.82}$$

which has a third-order pole at $z = z_0$ so that the residue is

$$a_{-1}(z_0) = \frac{1}{2!} \lim_{z \to z_0} \frac{\mathrm{d}^2}{\mathrm{d}z^2} \left[(z - z_0)^3 \frac{\sin az}{(z - z_0)^3} \right] \tag{11.83}$$

$$= \boxed{-\frac{a^2}{2} \sin a\, z_0.} \tag{11.84}$$

Now that we know how to determine the order of a pole and how to calculate residues, we can begin using the residue theorem to perform certain classes of definite integrals.

11.4 The Cauchy Integral Theorem and the Laurent Series

If you are happy with the Laurent series as presented above, you can go straight to Section 11.5 and begin solving definite integrals with the residue theorem. This section simply provides an alternative way to calculate the coefficients a_n of the Laurent series and demonstrates some additional facts about analytic functions.

Again, we use a Laurent series to represent a function on the complex plane as an expansion about the point z_0

$$f(z) = \sum_{n=-\infty}^{\infty} a_n (z - z_0)^n \tag{11.85}$$

when the function $f(z)$ has a singularity at $z = z_0$. The portion of the series having $n < 0$ is representing the singularity at $z = z_0$.

To obtain an alternative way to calculate the coefficients a_n, we begin by using the residue theorem to derive a result called the *Cauchy integral theorem* which says that for an analytic function $f(z)$ on a domain D that is surrounded by the closed contour C, we have

$$f(z) = \frac{1}{2\pi i}\oint_C d\zeta\, \frac{f(\zeta)}{\zeta - z}. \tag{11.86}$$

This result is straightforward to prove: write $2\pi i f(z) = \oint d\zeta f(\zeta)(\zeta - z)^{-1} = 2\pi i a_{-1}(z)$ by the residue theorem, where the residue of the integrand at the simple pole $\zeta = z$ is $a_{-1}(z) = f(z)$, which proves the theorem. If we take a derivative of both sides of this expression with respect to z, we obtain

$$f'(z) = \frac{1}{2\pi i}\oint_C d\zeta\, \frac{f(\zeta)}{(\zeta - z)^2}. \tag{11.87}$$

If we take n derivatives we similarly obtain

$$f^{\{n\}}(z) = \frac{n!}{2\pi i}\oint_C d\zeta\, \frac{f(\zeta)}{(\zeta - z)^{n+1}}. \tag{11.88}$$

This can also be demonstrated by using the residue theorem and using Eq. (11.54) to determine $a_{-1}(z)$ corresponding to the $n+1$ order pole at z.

Before getting to the Laurent series for a function possessing singularities, let's begin with the Taylor series and assume we want to represent a complex function $f(z)$ as an expansion about the point z_0 in the complex plane where this function is analytic (so z_0 is not a pole of $f(z)$). We can obtain this expansion using the above results. We first write the denominator in the Cauchy integral theorem as

$$\frac{1}{\zeta - z} = \frac{1}{\zeta - z_0 - (z - z_0)} \tag{11.89}$$

$$= \frac{1}{\zeta - z_0}\left(1 - \frac{(z - z_0)}{\zeta - z_0}\right)^{-1} \tag{11.90}$$

$$= \frac{1}{\zeta - z_0}\sum_{n=0}^{\infty}\left(\frac{z - z_0}{\zeta - z_0}\right)^n. \tag{11.91}$$

To go from the second line to third line we used the Taylor (or binomial) expansion which is justified because for points ζ on the contour C and for z and z_0 inside the contour, we have $|z - z_0|/|\zeta - z_0| < 1$ which is the condition for convergence of the series. We now insert this expression for $1/(\zeta - z)$ in the Cauchy integral theorem to obtain the series representation of $f(z)$ given by

$$f(z) = \sum_{n=0}^{\infty} a_n (z - z_0)^n, \tag{11.92}$$

where the a_n are read from the Cauchy integral theorem to be

$$a_n = \frac{1}{2\pi i}\oint_C d\zeta\, \frac{f(\zeta)}{(\zeta - z_0)^{n+1}}, \tag{11.93}$$

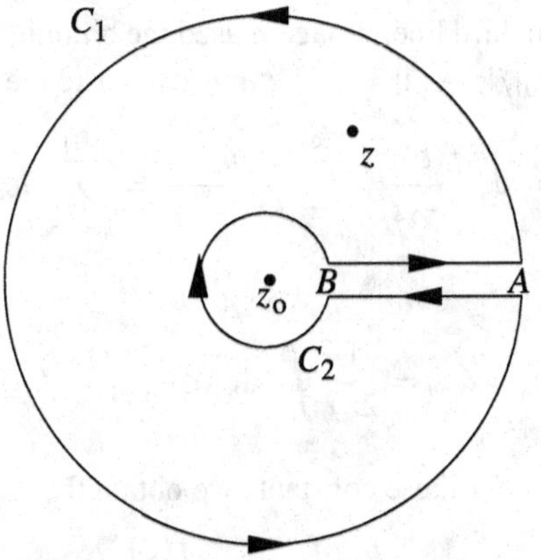

Figure 11.7 A path that excludes the point z_0.

which as was demonstrated above in Eq. (11.88) is also equal to

$$a_n = \frac{1}{n!} f^{\{n\}}(z_0). \tag{11.94}$$

This is just the usual expression for the coefficients a_n of a Taylor series expansion that we have now shown is also applicable to a complex analytic function $f(z)$ on the complex z plane. So in the neighborhood of a point z_0 where a complex function is analytic, the function possesses all of its derivatives (none of them blow up at that point) and we can represent the function as a complex Taylor series in this neighborhood, a result that was stated earlier but has now been demonstrated.

We now allow for the function $f(z)$ to possess singularities at the point z_0. In this case, we consider the contour shown in Fig. 11.7. By excluding the point $z = z_0$ using this contour, our function $f(z)$ is analytic within the annulus between and on the two contours C_1 and C_2 so that we can use the Cauchy integral theorem to write (again, an integral $\oint_C$ is always taken to be in the counter-clockwise direction)

$$f(z) = \frac{1}{2\pi i}\oint_{C_1} \mathrm{d}\zeta \, \frac{f(\zeta)}{\zeta - z} - \frac{1}{2\pi i}\oint_{C_2} \mathrm{d}\zeta \, \frac{f(\zeta)}{\zeta - z}. \tag{11.95}$$

The first integral over C_1 can be used to represent our function which is analytic within the domain bounded by the contour. This will produce a Taylor series just like we did above.

To analyze the second integral that encloses the singularity, we write

$$\frac{1}{\zeta - z} = \frac{1}{\zeta - z_0 - (z - z_0)} \tag{11.96}$$

$$= -\frac{1}{z - z_0}\left(1 - \frac{(\zeta - z_0)}{z - z_0}\right)^{-1} \tag{11.97}$$

$$= -\frac{1}{z - z_0}\sum_{n=0}^{\infty}\left(\frac{\zeta - z_0}{z - z_0}\right)^n. \tag{11.98}$$

In going from the second to third line, we again used the binomial expansion that is justified for ζ on C_2 because $|\zeta - z_0|/|z - z_o| < 1$. We thus can write the second integral over C_2 as

$$-\frac{1}{2\pi i}\oint_{C_2} d\zeta \frac{f(\zeta)}{\zeta - z} = \sum_{n=1}^{\infty} \frac{b_n}{(z-z_0)^n} = \sum_{n=-\infty}^{-1} a_n (z-z_0)^n, \tag{11.99}$$

where

$$b_n = a_{-n} = \frac{1}{2\pi i}\oint_{C_2} d\zeta\, (\zeta - z_0)^{n-1} f(\zeta) \tag{11.100}$$

or if we take $-\infty \leq n \leq -1$ for these constants we obtain the a_n as

$$a_n = \frac{1}{2\pi i}\oint_{C_2} d\zeta \frac{f(\zeta)}{(\zeta - z_0)^{n+1}}, \tag{11.101}$$

which is the same expression as the a_n corresponding to the Taylor series for $0 \leq n \leq +\infty$. If we now add together the two integrals above over C_1 and C_2 and take the limit where C_1 and C_2 approach each other (which we could just call the contour C), we obtain the Laurent series

$$f(z) = \sum_{n=-\infty}^{\infty} a_n (z - z_0)^n, \tag{11.102}$$

where the a_n are given by Eq. (11.101) for the entire range of n from negative to positive and with the contour $C_2 = C$ chosen to be sufficiently small that it surrounds only the one pole z_0. Each coefficient a_n can thus be obtained using the residue theorem for evaluating Eq. (11.101). Note that because the function $f(z)$ is not analytic within the domain, we do not have the Cauchy integral theorem available and we cannot express the a_n in terms of the derivatives of the function evaluated at $z = z_0$ like we could for the Taylor series of an analytic function.

As a somewhat trivial example to see that everything is working as it should, consider expanding the function

$$f(z) = \frac{g(z)}{(z - z_0)^3} \tag{11.103}$$

about the third-order pole z_0 where the function $g(z)$ is perfectly analytic. Before using the Laurent series as just derived, we can obtain this expansion by simply performing a Taylor series of the analytic function $g(z)$ in the numerator and dividing the result by $(z - z_0)^3$ to obtain

$$f(z) = \frac{1}{(z-z_0)^3}\left[g(z_0) + \frac{g'(z_0)}{1!}(z-z_0) + \frac{g''(z_0)}{2!}(z-z_0)^2 + \frac{g'''(z_0)}{3!}(z-z_0)^3 + \dots\right] \tag{11.104}$$

$$= \frac{g(z_0)}{(z-z_0)^3} + \frac{g'(z_0)}{1!(z-z_0)^2} + \frac{g''(z_0)}{2!(z-z_0)} + \frac{g'''(z_0)}{3!} + \dots. \tag{11.105}$$

Let's now verify that we obtain this same result by using the Laurent expansion $f(z) = \sum_{-\infty}^{\infty} a_n(z - z_0)^n$ and calculating the a_n from Eq. (11.101) after inserting $f(z) = g(z)/(z - z_0)^3$ or

$$a_n = \frac{1}{2\pi i} \oint_C \mathrm{d}\zeta \, \frac{g(\zeta)}{(\zeta - z_0)^{n+4}}, \tag{11.106}$$

where the contour C includes the point z_0. For $n = -4$, we then have

$$a_{-4} = \frac{1}{2\pi i} \oint_C \mathrm{d}\zeta \, g(\zeta) = 0 \quad \text{by Cauchy's theorem.} \tag{11.107}$$

We equivalently have that all coefficients a_n for $n < -4$ are also zero. For $n \geq -3$ we then have

$$a_{-3} = \frac{1}{2\pi i} \oint_C \mathrm{d}\zeta \, \frac{g(\zeta)}{\zeta - z_0} = g(z_0) \qquad \text{by the residue theorem} \tag{11.108}$$

$$a_{-2} = \frac{1}{2\pi i} \oint_C \mathrm{d}\zeta \, \frac{g(\zeta)}{(\zeta - z_0)^2} = g'(z_0) \qquad \text{by the residue theorem} \tag{11.109}$$

$$a_{-1} = \frac{1}{2\pi i} \oint_C \mathrm{d}\zeta \, \frac{g(\zeta)}{(\zeta - z_0)^3} = \frac{g''(z_0)}{2!} \qquad \text{by the residue theorem} \tag{11.110}$$

$$a_0 = \frac{1}{2\pi i} \oint_C \mathrm{d}\zeta \, \frac{g(\zeta)}{(\zeta - z_0)^4} = \frac{g'''(z_0)}{3!} \qquad \text{by the residue theorem} \tag{11.111}$$

and so forth. So indeed, the Laurent series gives the proper series with the coefficients determined from Eq. (11.101) with $C_2 = C$ chosen to surround only the one pole z_0. We have further shown by this example that when we calculate the Laurent expansion of a function that has an mth-order pole at z_0 we obtain

$$f(z) = \sum_{n=-m}^{\infty} a_n(z - z_0)^n \quad \text{with} \quad a_n = \frac{1}{2\pi i} \oint_C \mathrm{d}\zeta \, \frac{f(\zeta)}{(\zeta - z_0)^{n+1}}. \tag{11.112}$$

The first term in the series is $a_{-m}/(z - z_0)^m$ which is indeed an mth-order pole.

In practice, the singular part of a function is often explicitly present as the denominator of an expression. As we did in Section 11.3, it is often easier to obtain the series in this case, at least for the negative n portion of the series that contains the singularity, not by using the above integral expression for the a_n that involves using the residue theorem to perform the integral but to expand both the numerator and denominator in separate Taylor series and then place the series in the denominator into the numerator using the binomial expansion (an additional Taylor series). We then calculate the first terms (negative n) through to the residue term $a_{-1}(z)$ which is often all that we are interested in determining for calculating residues and using the residue theorem to solve definite integrals. Now back to solving integrals.

11.5 Evaluating Definite Integrals with the Residue Theorem

We now consider some examples of evaluating definite integrals using the residue theorem. The trick of how to define the contour and how to apply the residue theorem may be different from one integral to the next and it is best to just learn by example.

Consider the definite integral

$$I = \int_0^{+\infty} \frac{\mathrm{d}x}{x^2 + a^2} \quad \text{where } a > 0. \tag{11.113}$$

Use the fact that the integrand is even in x to write this as an integral from $-\infty$ to $+\infty$ and substitute in the complex argument z

$$2I = \int_{-\infty}^{+\infty} \frac{\mathrm{d}z}{z^2 + a^2} = \int_{-\infty}^{+\infty} \frac{\mathrm{d}z}{(z + ia)(z - ia)}. \tag{11.114}$$

There are thus simple poles at $z = ia$ and $z = -ia$. Consider the contour in Fig. 11.8. For this contour, we can write

$$\begin{aligned} \oint_C \frac{\mathrm{d}z}{(z + ia)(z - ia)} &= \int_{C_\infty} \frac{\mathrm{d}z}{z^2 + a^2} + 2I \\ &= 2\pi i\, a_{-1}(ia) \quad \text{(from the residue theorem).} \end{aligned} \tag{11.115}$$

From Jordan's lemma, we have that the integral over C_∞ is zero. The residue is easily calculated from Eq. (11.55) as

$$a_{-1}(ia) = \lim_{z \to ia} \frac{z - ia}{(z - ia)(z + ia)} = \frac{1}{2ia} \tag{11.116}$$

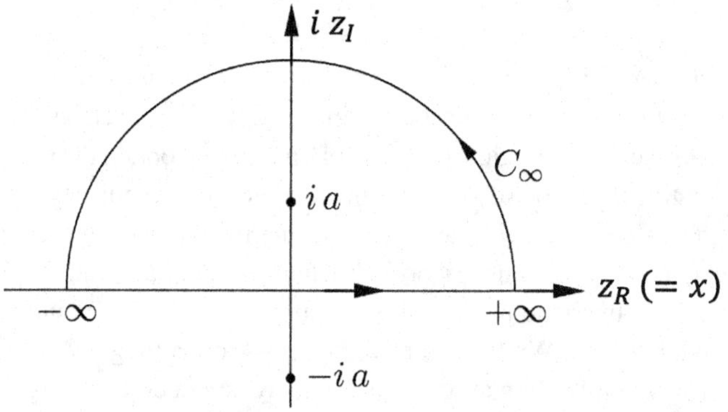

Figure 11.8 The contour used in evaluating $\int_{-\infty}^{\infty} \mathrm{d}x/(x^2 + a^2)$.

so that we obtain the result

$$\boxed{I = \int_0^\infty \frac{dx}{x^2 + a^2} = \frac{\pi}{2a}.} \tag{11.117}$$

Most of the integrals encountered when using the inverse Fourier transform in the solution of a PDE will be integrals from $-\infty$ to $+\infty$ and will employ contours that have a semicircle at infinity to which Jordan's lemma is applicable. However, let's next consider some other examples that do not involve a half circle at infinity.

Consider the integral

$$I = \int_0^{2\pi} \frac{d\theta}{1 - 2p\,\cos\theta + p^2} \quad \text{where } |p| \neq 1. \tag{11.118}$$

Convert this to an integral in the complex plane by setting $z = e^{i\theta}$ (i.e., $|z| = 1$) so that $dz = iz\,d\theta$ and

$$\cos\theta = \frac{1}{2}\left(z + \frac{1}{z}\right) = \frac{z^2+1}{2z}. \tag{11.119}$$

Through this change of variables, the integral can be written directly as

$$I = \oint_{C_1} \frac{dz}{i[z - p(z^2+1) + p^2 z]} \quad \text{where } C_1 = \text{unit circle} \tag{11.120}$$

$$= \oint_{C_1} \frac{dz}{i\,(z-p)(1-pz)} = \oint_{C_1} \frac{i}{p}\frac{dz}{(z-p)(z-1/p)}, \tag{11.121}$$

which thus has simple poles at both $z = p$ and $z = 1/p$. We must make allowances for the two cases where either $p > 1$ or $p < 1$ as indicated in Fig. 11.9. From the residue theorem, we have the two cases:

(i)

$$I = \oint_{C_1} \frac{i\,dz}{p\,(z-p)(z-\frac{1}{p})} = 2\pi i\,a_{-1}\left(\frac{1}{p}\right) = \boxed{\frac{2\pi}{p^2-1} \quad \text{when } p > 1} \tag{11.122}$$

(ii)

$$I = \oint_{C} \frac{i\,dz}{p\,(z-p)(z-\frac{1}{p})} = 2\pi i\,a_{-1}\,(p) = \boxed{\frac{2\pi}{1-p^2} \quad \text{when } p < 1.} \tag{11.123}$$

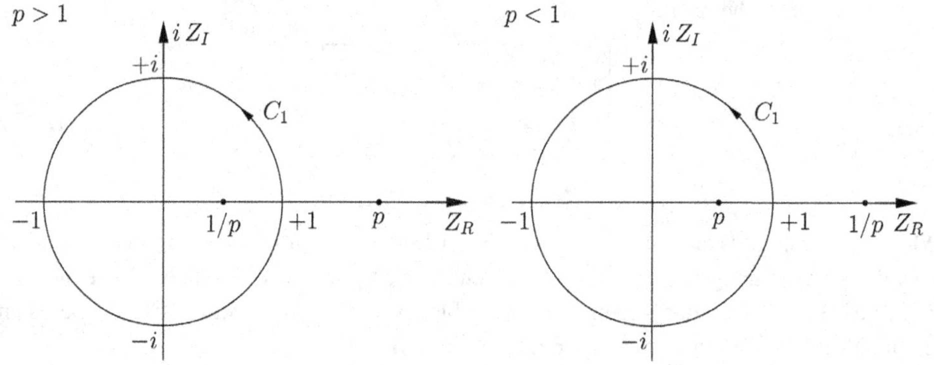

Figure 11.9 The unit-circle contours used in evaluating $\int_0^{2\pi} \mathrm{d}\theta \left[1 - 2p \cos\theta + p^2\right]^{-1}$ for the cases $p > 1$ and $p < 1$.

This style of approach is valid for integrals of the form

$$I = \int_0^{2\pi} \frac{P(\cos\theta, \sin\theta)}{Q(\cos\theta, \sin\theta)} \, \mathrm{d}\theta, \tag{11.124}$$

where P and Q are polynomials in their arguments.

Let's perform another style of integral that does not require (or allow) the use of Jordan's lemma and half circles at infinity. Consider the integral

$$I = \int_{-\infty}^{+\infty} \frac{e^{ax}\, dx}{\cosh \pi x} \quad \text{where } |a| < \pi. \tag{11.125}$$

When we extend analysis to the complex plane, this integrand does not decay to zero along the imaginary axis as $|z_I| \to \infty$. The exponential and hyperbolic cosine are sinusoids along the z_I axis that do not decay with increasing z_I so that we cannot use Jordan's lemma by closing the contour with a half circle at infinity and arguing that the contribution on the half circle is negligible.

To find an appropriate contour, we first note that $\cosh \pi z = 0$ at $z = (2n+1)\, i/2$ for $n = 0, \pm 1, \pm 2, \ldots$ where we used $\cosh iu = \cos u$ for u real. Thus, the integrand has an infinity of poles up and down the z_I axis. Let's consider the rectangular contour as $R \to \infty$ that contains only one of the poles as shown in Fig. 11.10 The residue at $z = i/2$ can be obtained by first expanding $\cosh \pi z$ about $i/2$

$$\cosh \pi z = \cancelto{0}{\cosh \frac{\pi i}{2}} + \pi \sinh \frac{\pi i}{2} \left(z - \frac{i}{2}\right) + \cdots \tag{11.126}$$

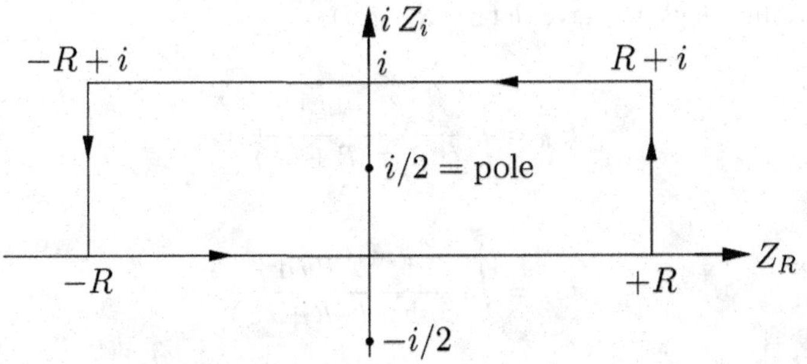

Figure 11.10 A rectangular contour as $R \to \infty$ for the integral $\int_{-\infty}^{+\infty} \mathrm{d}x\, e^{ax}/\cosh(\pi x)$.

and noting that $\sinh(\pi i/2) = i$ and then expanding e^{az} about $i/2$

$$e^{az} = e^{ia/2} + a\,e^{ia/2}\left(z - \frac{i}{2}\right) + \cdots. \tag{11.127}$$

To leading order in $z - i/2$, we then have

$$\frac{e^{az}}{\cosh \pi z} = \frac{e^{ia/2}}{\pi i (z - i/2)} + \cdots, \tag{11.128}$$

where we used that $\sinh \pi i/2 = i\,\sin \pi/2 = i$. We can read off the residue of this first-order pole as

$$a_{-1}\left(\frac{i}{2}\right) = \frac{e^{ia/2}}{\pi i}. \tag{11.129}$$

For the given contour, we have that our integral of interest I is the integral I_R (let $z = z_R$) given by

$$I_R = \int_{-R}^{+R} \frac{e^{az_R}}{\cosh \pi\, z_R}\,\mathrm{d}z_R \quad \text{in the } R \to \infty \text{ limit.} \tag{11.130}$$

The other horizontal integral from $R + i$ to $-R + i$ is the integral I'_R (let $z = z_R + i$) given by

$$I'_R = \int_{R+i}^{-R+i} \frac{e^{a(z_R+i)}}{\cosh \pi\, (z_R + i)}\,\mathrm{d}z_R. \tag{11.131}$$

We have that $\cosh \pi (z_R + i) = \left[e^{\pi z_R}\, e^{i\pi} + e^{-\pi z_R}\, e^{-i\pi}\right]/2 = -\cosh \pi z_R$ so that

$$I'_R = \int_{-R}^{+R} \frac{e^{az_R}\, e^{ia}}{\cosh \pi z_R}\,\mathrm{d}z_R = e^{ia}\, I_R. \tag{11.132}$$

For the two vertical legs, we have (let $z = R + i\, z_I$)

$$J_{+R} = \int_0^1 \frac{e^{a(R+i\,z_I)}\, i\, \mathrm{d}z_I}{\cosh \pi\, (R + i\, z_I)} \tag{11.133}$$

$$J_{-R} = \int_1^0 \frac{e^{a(-R+i\,z_I)}\, i\, \mathrm{d}z_I}{\cosh \pi\, (-R + i\, z_I)}. \tag{11.134}$$

Now, in the limit as $R \to \infty$, we have that $\cosh \pi\, (\pm R + i\, z_I) \to e^{\pi R}/2$ and $e^{a(R+i\, z_I)} \to e^{aR}$. Thus, because $a < \pi$, we have that both integrands $\to 0$ as $R \to \infty$. We can at last bring it all together

$$\oint \frac{e^{az}}{\cosh \pi z} = \frac{2\pi i\, e^{i\,a/2}}{i\pi} \quad \text{(the residue theorem)} \tag{11.135}$$

$$= \lim_{R\to\infty} \left[I_R + I'_R + J_{+R} + J_{-R} \right] \tag{11.136}$$

$$= I \left(1 + e^{ia}\right). \tag{11.137}$$

So, we obtain at last

$$I = \frac{2}{e^{i\,a/2} + e^{-i\,a/2}} = \boxed{\frac{1}{\cos(a/2)}.} \tag{11.138}$$

It would be difficult to perform this integral in any manner other than that outlined here.

Let's next do an integral that involves a branch point

$$I = \int_0^\infty \frac{\sqrt{x}}{x^2 + 1}\, \mathrm{d}x. \tag{11.139}$$

When analysis is extended onto the complex z plane, the $\sqrt{z}$ in the numerator has a branch point at $z = 0$ that we want to exclude. So we consider the closed contour C shown in Fig. 11.11 that excludes the branch point but includes the two simple poles at $z = \pm i$. The residue theorem applied to this contour gives

$$\begin{aligned} \oint_C \frac{\sqrt{z}}{(z+i)(z-i)}\, \mathrm{d}z = \lim_{\delta\to 0} \lim_{R\to\infty} \Bigg[& \int_\delta^R \frac{\sqrt{z}}{(z+i)(z-i)}\, \mathrm{d}z + \int_{C_R} \frac{\sqrt{z}}{(z+i)(z-i)}\, \mathrm{d}z \\ & + \int_R^\delta \frac{\sqrt{z}}{(z+i)(z-i)}\, \mathrm{d}z + \int_{C_\delta} \frac{\sqrt{z}}{(z+i)(z-i)}\, \mathrm{d}z \Bigg] \\ = 2\pi i \left[a_{-1}(+i) + a_{-1}(-i) \right]. \end{aligned} \tag{11.140}$$

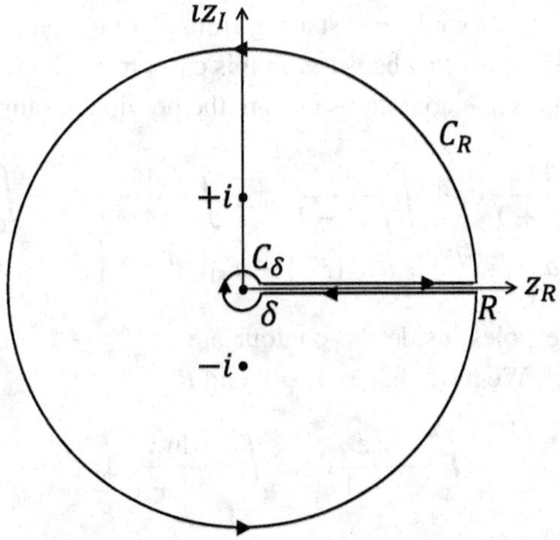

Figure 11.11 Path that excludes the branch point at the origin but includes the simples poles at $z = \pm i$.

As $R \to \infty$, the integral on C_R is zero and similarly as $\delta \to 0$, the integral on C_δ is zero. Along the path from δ to R, we have $\sqrt{z} = \sqrt{z_R e^{i0/2}} = \sqrt{z_R}$, while along the path from R to δ, we have $\sqrt{z} = \sqrt{z_R e^{i2\pi/2}} = -\sqrt{z_R}$. But because $\int_R^\delta = -\int_\delta^R$ we then have

$$2I = 2\pi i \left[a_{-1}(+i) + a_{-1}(-i)\right]. \tag{11.141}$$

The sum of the residues for these two simple poles are

$$a_{-1}(i) + a_{-1}(-i) = \frac{\sqrt{i}}{2i} + \frac{\sqrt{-i}}{(-2i)} = \frac{\sqrt{i}}{2i}(1-i) = \frac{(1+i)(1-i)}{i2\sqrt{2}} = \frac{1}{i\sqrt{2}}. \tag{11.142}$$

We thus obtain

$$\boxed{I = \int_0^\infty \frac{\sqrt{x}}{x^2+1}\,\mathrm{d}x = \frac{\pi}{\sqrt{2}}.} \tag{11.143}$$

As we have seen in each of the above examples, integrals that can be performed using contour integration often have results that are satisfyingly compact.

As a final example that employs a branch point, consider the integral

$$I = \int_0^\infty \frac{\mathrm{d}x}{x^3+1}. \tag{11.144}$$

Because $1/(x^3+1)$ is not even across $x = 0$, we cannot rewrite this as an integral from $-\infty$ to ∞ and close the contour with a half circle at infinity. Instead we employ the trick of inserting a $\ln x$ into the integrand and consider the modified integrand $\ln(x)/(x^3+1)$ on the positive real axis. When analysis of this integrand is extended onto the complex plane, we saw earlier that $\ln z$ has a branch point at $z = 0$ which means that each time n we circulate by an amount 2π around $z=0$ starting from a point x on the real axis with $\theta=0$, we have $\ln\left(xe^{i2\pi n}\right)-\ln(x)=i2\pi n$ because, in this case, $\ln z=\ln(xe^{i\theta})=\ln x+i\theta$.

So if we use the exact same contour as used in the previous example, we have

$$\oint \frac{\ln z}{z^3+1}\,dz=\int_\delta^R \frac{\ln z}{z^3+1}\,dz+\int_{C_R}\frac{\ln z}{z^3+1}\,dz+\int_R^\delta \frac{\ln z}{z^3+1}\,dz+\int_{C_\delta}\frac{\ln z}{z^3+1}\,dz \quad (11.145)$$

$$=2\pi i\left[a_{-1}\left(e^{i\pi/3}\right)+a_{-1}\left(e^{i\pi}\right)+a_{-1}\left(e^{i5\pi/3}\right)\right], \quad (11.146)$$

where the three simple poles inside the contour are at $z_i^3=-1=e^{i\pi(2m-1)}$ or $z_1=e^{i\pi/3}$, $z_2=e^{i\pi}$, and $z_3=e^{i5\pi/3}$. We have that as $\delta\to 0$ and $R\to\infty$

$$\int_\delta^R \frac{\ln z}{z^3+1}\,dz=\int_0^\infty \frac{\ln x}{x^3+1}\,dx \quad (11.147)$$

while in the same limit

$$\int_R^\delta \frac{\ln z}{z^3+1}\,dz=-\int_0^\infty \frac{(\ln x+2\pi i)}{x^3+1}\,dx \quad (11.148)$$

so that upon adding these two integrals together

$$\int_\delta^R \frac{\ln z}{z^3+1}\,dz+\int_R^\delta \frac{\ln z}{z^3+1}\,dz=\int_0^\infty \frac{(\ln x-\ln x-2\pi i)}{x^3+1}\,dx=-2\pi i I, \quad (11.149)$$

where $I=\int_0^\infty dx/(x^3+1)$ is the integral of interest. The integral around the circle at C_R as $R\to\infty$ is zero by Jordan's lemma. The integral over the circle at C_δ as $\delta\to 0$ with $z=\delta e^{i\theta}$ is

$$\lim_{\delta\to 0}\int_{C_\delta}\frac{\ln z}{z^3+1}\,dz=\lim_{\delta\to 0}\int_{2\pi}^0 \frac{i\delta(\ln\delta+i\theta)e^{i\theta}}{1+\delta^3\exp(i3\theta)}\,d\theta=-\lim_{\delta\to 0} i\frac{\ln\delta}{\delta^{-1}}\int_0^{2\pi} d\theta\, e^{i\theta}. \quad (11.150)$$

The prefactor is evaluated by L'Hôpital's rule as

$$\lim_{\delta\to 0}\frac{\ln\delta}{\delta^{-1}}=-\lim_{\delta\to 0}\frac{\delta^{-1}}{\delta^{-2}}\to 0. \quad (11.151)$$

For the sum of the three residues, we have the earlier results given by Eqs (11.75)–(11.77) so that

$$\boxed{I=\int_0^\infty \frac{dx}{x^3+1}=\frac{2\pi}{3\sqrt{3}}.} \quad (11.152)$$

We could continue giving endless examples of solving particular definite integrals using contour-integration techniques but it is best if you, at this point, try solving the integrals given in the exercises at the end of the chapter with some guiding hints on how to either obtain the residues or set up the contours to use for each integral.

11.6 Linear Response, Kramers–Krönig Relations, and Loss Processes

We now put our understanding of Fourier transforms and contour integration to use on physics problems. In this first example, we consider linear constitutive laws in which some force is driving some response. We will demonstrate that if the force is varying in time and if the response depends linearly on past values of the force in addition to the present value, then in the temporal frequency domain, the "modulus" relating force and response will be a complex function of frequency with the imaginary part controlling the energy lost to heat in each sinusoidal time oscillation and with the real part controlling the average energy stored reversibly in each sinusoidal oscillation. In particular, we will show using contour integration that if, for example, we are given the real part of the modulus and its frequency dependence we can then calculate the imaginary part of the modulus and its frequency dependence. These results are very general and apply to all linear constitutive laws when the driving force is varying in time and the current response depends not only on the current force level but the past forcing as well.

11.6.1 Delayed Linear-Response Processes and Kramers–Krönig Relations

Consider some response $r(t)$ generated by some force $f(t)$. We will write the constitutive law between $r(t)$ and $f(t)$ as a relation between scalar quantities, but we can easily generalize to vectorial and tensorial constitutive laws. So in what follows, $r(t)$ might be a component of strain and $f(t)$ a component of stress. Equivalently, $r(t)$ could be a component of dielectric displacement and $f(t)$ a component of the electric field. The particular physics does not matter. However, we do assume that the response is linear in the applied force in writing the general delayed linear-response relation

$$\boxed{r(t) = \int_{-\infty}^{t} \mathrm{d}\tau\, M(t-\tau) f(\tau),} \tag{11.153}$$

where the modulus time function $M(t-\tau)$ (often called the "susceptibility" time function) is telling us how the force at times past is creating a response at the present time. In other words, when a force acts on the material, it sets off a process that does not occur instantaneously. Such delay processes are quantified at the macroscopic level by an $M(t-\tau)$ that is not a perfect Dirac $M_0\delta(t-\tau)$ and result, as we will demonstrate, in energy being irreversibly lost to heat. The chosen limits in the integration is a statement of "causality"; namely, that response at the present time t can only be created by forces at present or past times τ and not by the force at future times.

Let $u = t - \tau$, $\tau = t - u$, and $\mathrm{d}\tau = -\mathrm{d}u$, to obtain

$$r(t) = -\int_{\infty}^{0} \mathrm{d}u\, M(u) f(t-u) = \int_{0}^{\infty} \mathrm{d}u\, M(u) f(t-u) = \int_{-\infty}^{\infty} \mathrm{d}u\, M(u) f(t-u) \tag{11.154}$$

under the condition that $M(u) = 0$ for $u < 0$. The fact that $u > 0$ or, equivalently, that $M(u) = 0$ for $u < 0$, is the causality statement that future force values cannot influence the present response. The present response is thus the convolution of the force history with the modulus time function.

Now, from the convolution theorem, we have that

$$\tilde{r}(\omega) = \tilde{M}(\omega)\tilde{f}(\omega), \tag{11.155}$$

where these expressions are the Fourier transforms of $r(t)$, $M(t)$, and $f(t)$. Combining the inverse Fourier transforms

$$r(t) = \frac{1}{2\pi}\int_{-\infty}^{+\infty} d\omega\, \tilde{r}(\omega)\, e^{-i\omega t} \quad \text{and} \quad f(t-u) = \frac{1}{2\pi}\int_{-\infty}^{+\infty} d\omega \tilde{f}(\omega)\, e^{-i\omega(t-u)} \tag{11.156}$$

with the response created by past forcing

$$r(t) = \int_0^{\infty} du\, M(u) f(t-u) \tag{11.157}$$

allows us to identify

$$\begin{aligned}\int_{-\infty}^{+\infty} d\omega\, \tilde{r}(\omega)\, e^{-i\omega t} &= \int_0^{\infty} du\, M(u) \int_{-\infty}^{+\infty} d\omega \tilde{f}(\omega)\, e^{-i\omega(t-u)} \\ &= \int_{-\infty}^{+\infty} d\omega \left[\int_0^{\infty} du\, M(u) e^{+i\omega u}\right] \tilde{f}(\omega)\, e^{-i\omega t}\end{aligned} \tag{11.158}$$

or

$$\tilde{M}(\omega) = \tilde{M}_R(\omega) + i\tilde{M}_I(\omega) = \int_0^{\infty} du\, M(u)\, e^{i\omega u}. \tag{11.159}$$

Only positive time contributes to the Fourier transform $\tilde{M}(\omega)$, which is, again, the statement of causality. Note that for a process without delay, $M(t) = M_0\,\delta(t)$, where M_0 is a real constant, so that $\tilde{M}(\omega) = M_0$ for all ω. Thus, when $\tilde{M}(\omega)$ is complex, the imaginary part is due to delay processes that allow past force values to influence the present response.

The real functions $\tilde{M}_R(\omega)$ and $\tilde{M}_I(\omega)$ are related to each other, that is, given one of these functions, the other is obtained by performing an integral. To obtain this relation, we begin by integrating

$$\oint_C d\alpha\, \frac{\tilde{M}(\alpha)}{\alpha - \omega} \tag{11.160}$$

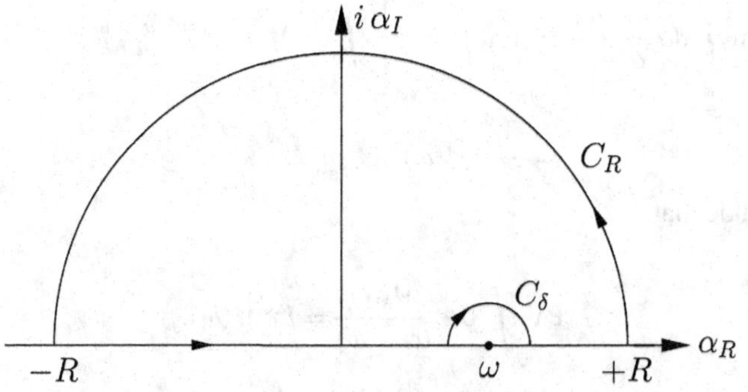

Figure 11.12 Contour used to develop the dispersion or Kramers–Krönig relations.

over the closed contour C shown in Fig. 11.12. Cauchy's theorem gives

$$\oint_C \mathrm{d}\alpha \frac{\tilde{M}(\alpha)}{\alpha-\omega} = 0 = \mathrm{PV}\int_{-\infty}^{+\infty} \mathrm{d}\alpha\, \frac{\tilde{M}(\alpha)}{\alpha-\omega} + \lim_{\delta\to 0}\int_{C_\delta} \mathrm{d}\alpha\, \frac{\tilde{M}(\alpha)}{\alpha-\omega} + \lim_{R\to\infty}\int_{C_R} \mathrm{d}\alpha\, \frac{\tilde{M}(\alpha)}{\alpha-\omega}, \quad (11.161)$$

where PV again means "principal value." It is defined whenever there is a pole in the integrand that lies on the path of integration and is given by

$$\mathrm{PV}\int_{-\infty}^{+\infty} \mathrm{d}\alpha\, \frac{f(\alpha)}{(\alpha-\omega)^m} = \lim_{\delta\to 0}\left\{\int_{-\infty}^{\omega-\delta} \mathrm{d}\alpha\, \frac{f(\alpha)}{(\alpha-\omega)^m} + \int_{\omega+\delta}^{+\infty} \mathrm{d}\alpha\, \frac{f(\alpha)}{(\alpha-\omega)^m}\right\}, \quad (11.162)$$

where $f(\alpha)$ is analytic along α real. In the denominator of the integrals, $(-\delta)^{-m} = -(\delta)^{-m}$ when m is odd, so the PV stays finite as $\delta \to 0$ because the large contributions from either side of $\alpha = \omega$ are of equal amplitude and opposite sign and therefore cancel to produce a finite principal value.

Note that in the Fourier-transform definition of $\tilde{M}(\omega)$, when analysis is extended to the complex frequency plane $\alpha = \alpha_R + i\alpha_I$

$$\lim_{\alpha_I\to+\infty} \tilde{M}(\alpha) = \lim_{\alpha_I\to+\infty}\int_0^{\infty} \mathrm{d}u\, M(u)\, e^{iu\,\alpha_R}\, e^{-u\alpha_I} \to 0, \quad (11.163)$$

that is, the integrand goes to zero as $e^{-u\alpha_I}$ for $u > 0$ in the upper half α plane. So by Jordan's lemma, the integral over C_R in Eq. (11.161) is zero as $R \to \infty$. Again, this result is a consequence of causality, which is the statement that $u > 0$ or "current response can only depend on current or past forcing."

For the small half circle C_δ centered on ω and of radius δ, we take $\alpha - \omega = \delta\, e^{i\theta}$ and $\mathrm{d}\alpha = i\delta\, e^{i\theta}\,\mathrm{d}\theta$ to obtain

$$\lim_{\delta\to 0}\int_{C_\delta} d\alpha\,\frac{\tilde{M}(\alpha)}{\alpha-\omega} = \lim_{\delta\to 0}\int_{\pi}^{0} d\theta\, i\left[\int_0^\infty du\, M(u)\, e^{i\omega u}\, e^{i\delta e^{i\theta}u}\right] \tag{11.164}$$

$$= -i\pi\,\tilde{M}(\omega).$$

We thus conclude that

$$\mathrm{PV}\int_{-\infty}^{+\infty} d\alpha\,\frac{\tilde{M}(\alpha)}{(\alpha-\omega)} = i\pi\,\tilde{M}(\omega). \tag{11.165}$$

Upon putting $\tilde{M} = \tilde{M}_R + i\,\tilde{M}_I$ and taking Re and Im parts of Eq. (11.165), we obtain

$$\tilde{M}_I(\omega) = -\frac{1}{\pi}\,\mathrm{PV}\int_{-\infty}^{+\infty} d\alpha\,\frac{\tilde{M}_R(\alpha)}{\alpha-\omega} \tag{11.166}$$

$$\tilde{M}_R(\omega) = +\frac{1}{\pi}\,\mathrm{PV}\int_{-\infty}^{+\infty} d\alpha\,\frac{\tilde{M}_I(\alpha)}{\alpha-\omega} \tag{11.167}$$

which are called the *Kramers–Krönig* or *dispersion* relations. These say that if we know, for example, the function $\tilde{M}_R(\omega)$ for all frequencies, we can calculate the imaginary part $\tilde{M}_I(\omega)$ by performing the above integral. The mathematical relation between $\tilde{M}_R(\omega)$ and $\tilde{M}_I(\omega)$ given by the Kramers–Krönig relations are also called *Hilbert-transform* pairs. Using these relations in practice can be tricky because they require you know either function $\tilde{M}_R(\omega)$ or $\tilde{M}_I(\omega)$ over the entire band that they have support. If you only have either function over a portion of the band and assume in error that the spectrum is zero outside the band that you have, these relations can give unreliable results.

Note that if $\tilde{M}_R(\omega) = \text{const}$ for all ω, then

$$\mathrm{PV}\int_{-\infty}^{+\infty}\frac{d\alpha}{\alpha-\omega} = \mathrm{PV}\int_{-\infty+\omega}^{+\infty+\omega}\frac{dx}{x} = \mathrm{PV}\int_{-\infty}^{+\infty}\frac{dx}{x} = 0, \tag{11.168}$$

where it was assumed that ω is finite (though possibly quite large). The last integral is zero because x^{-1} is odd. Thus, when $\tilde{M}_R = \text{const}$, we have that $\tilde{M}_I(\omega) = 0$. So we can always add a constant to $\tilde{M}_R(\omega)$ without changing Eq. (11.166).

To see how the Kramers–Krönig relations work in practice, let's consider what is likely the simplest delay model called *Debye relaxation* in which the temporal delay is controlled by an exponential time function

$$M(t) = M_\infty\delta(t) + (M_0 - M_\infty)\omega_r e^{-\omega_r t} S(t), \tag{11.169}$$

where ω_r is called the Debye relaxation frequency and $S(t)$ is the unit step function $S(t) = 0$ for $t < 0$ and $= 1$ for $t > 0$, which ensures that $M(t) = 0$ for $t < 0$ (causality).

The Fourier transform is easily calculated to give the complex modulus associated with Debye relaxation

$$\tilde{M}(\omega) = M_\infty + \frac{M_0 - M_\infty}{1 - i\omega/\omega_r} \tag{11.170}$$

that separates into real and imaginary parts as

$$\tilde{M}(\omega) = \left[M_\infty + \frac{\omega_r^2(M_0 - M_\infty)}{\omega^2 + \omega_r^2}\right] + i\left[\frac{\omega\omega_r(M_0 - M_\infty)}{\omega^2 + \omega_r^2}\right]. \tag{11.171}$$

Let's imagine that by some experimental means (e.g., measuring the wave velocity at different frequencies in a wave application) for a process in which Debye relaxation is operating, we have determined the real part of the modulus to have the frequency dependence

$$\tilde{M}_R(\omega) = M_\infty + \frac{\omega_r^2(M_0 - M_\infty)}{\omega^2 + \omega_r^2}. \tag{11.172}$$

Given this real measured function $\tilde{M}_R(\omega)$, what is the imaginary modulus?

The Kramers–Krönig relation for $\tilde{M}_I(\omega)$ given $\tilde{M}_R(\omega)$ is

$$\tilde{M}_I(\omega) = -\frac{1}{\pi}\,\text{PV}\int_{-\infty}^{+\infty} d\alpha\,\frac{\tilde{M}_R(\alpha)}{\alpha - \omega} = -\frac{1}{\pi}\,\text{PV}\int_{-\infty}^{\infty} d\alpha\,\frac{(M_0 - M_\infty)\omega_r^2}{(\alpha - \omega)(\alpha + i\omega_r)(\alpha - i\omega_r)}. \tag{11.173}$$

This integral is easily performed using contour-integration methods. Consider the contour shown in Fig. 11.13. We exclude the pole on the real α axis at ω using a small half circle C_δ but include the pole at $i\omega_r$ using the half circle at infinity C_∞ that itself makes no contribution by Jordan's lemma. From the residue theorem, we have

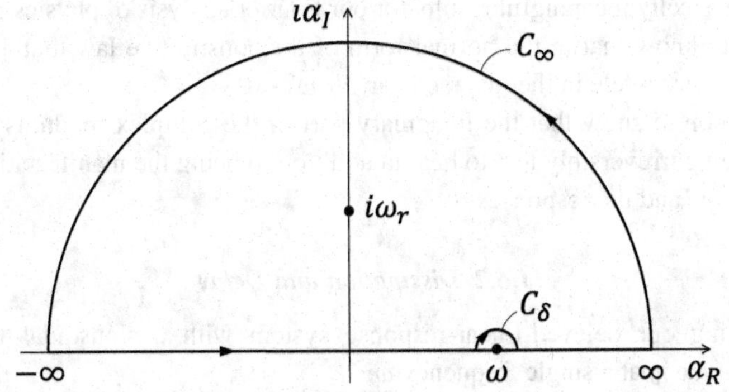

Figure 11.13 The contour used to obtain $\tilde{M}_I(\omega)$ given $\tilde{M}_R(\omega)$ for the Debye-relaxation process using the Kramers–Krönig relation.

$$2\pi i a_{-1}(i\omega_r) = \tilde{M}_I(\omega) + \left[-\frac{1}{\pi} \lim_{\delta \to 0} \int_{C_\delta} \mathrm{d}\alpha \frac{(M_0 - M_\infty)\omega_r^2}{(\alpha - \omega)(\alpha + i\omega_r)(\alpha - i\omega_r)} \right]. \tag{11.174}$$

The residue due to the simple pole of the integrand gives

$$2\pi i a_{-1}(i\omega_r) = \frac{\omega_r(M_0 - M_\infty)}{\omega - i\omega_r} = \frac{\omega_r(M_0 - M_\infty)(\omega + i\omega_r)}{\omega^2 + \omega_r^2}, \tag{11.175}$$

while the integral over the small half circle C_δ with $\alpha - \omega = \delta e^{i\theta}$ and $\mathrm{d}\alpha/(\alpha - \omega) = i\mathrm{d}\theta$ gives

$$\begin{aligned} -\frac{1}{\pi} \lim_{\delta \to 0} \int_{C_\delta} \mathrm{d}\alpha \frac{(M_0 - M_\infty)\omega_r^2}{(\alpha - \omega)(\alpha + i\omega_r)(\alpha - i\omega_r)} &= -\frac{i}{\pi} \frac{(M_0 - M_\infty)\omega_r^2}{(\omega^2 + \omega_r^2)} \int_\pi^0 \mathrm{d}\theta \\ &= \frac{i(M_0 - M_\infty)\omega_r^2}{\omega^2 + \omega_r^2}. \end{aligned} \tag{11.176}$$

We then have that

$$\boxed{\tilde{M}_I(\omega) = \frac{\omega\omega_r(M_0 - M_\infty)}{\omega^2 + \omega_r^2}} \tag{11.177}$$

exactly as required by Eq. (11.171). This exercise, using Debye relaxation as an example, simply demonstrates that the Kramers–Krönig relations work as advertised.

But we also learn something useful from Eq. (11.177). If the real part of the complex modulus at infinite frequency M_∞ is smaller than the real modulus at zero frequency M_0, the imaginary part of the complex modulus is positive for all positive frequencies. While when the real modulus at infinite frequency is larger than the real modulus at zero frequency, the imaginary modulus is negative for all positive frequencies. These two scenarios correspond to the constitutive law written in what we may call "normal" or "positive" form $\tilde{r}(\omega) = \tilde{M}(\omega)\tilde{f}(\omega)$ where $M_\infty < M_0$ and $\tilde{M}_I(\omega) > 0$ for all $\omega > 0$, and in "inverse" or "negative" form $\tilde{f}(\omega) = \tilde{N}(\omega)\tilde{r}(\omega)$ where $\tilde{N}(\omega) = 1/\tilde{M}(\omega)$ with $N_\infty > N_0$ and $\tilde{N}_I(\omega) < 0$ for all $\omega > 0$. We will show in Sections 11.6.2 and 11.6.3, that either form of the constitutive law results in the same physically meaningful results for our Fourier analysis of physics problems, but it is essential to know that in the normal form of the constitutive law that $\tilde{M}_I(\omega) > 0$ for positive frequencies while in the inverse form $\tilde{N}_I(\omega) < 0$.

We now go on to show that the imaginary part of the complex modulus is associated with energy being irreversibly lost to heat in addition to being the manifestation of a delay between the force and the response.

11.6.2 Dissipation and Delay

Consider driving our delayed-linear-response system with a sinusoidal force $f(t)$ of constant amplitude f_0 at a single frequency ω:

$$f(t) = \mathrm{Re}\left\{f_0 e^{-i\omega t}\right\} = \frac{1}{2}\left[f_0 e^{-i\omega t} + f_0^* e^{+i\omega t}\right]. \tag{11.178}$$

The response $r(t)$ in terms of the complex modulus $\tilde{M}(\omega)$ is then:

$$r(t) = \text{Re}\left\{\tilde{M}(\omega) f_0 e^{-i\omega t}\right\} = \frac{1}{2}\left[\tilde{M}(\omega) f_0 e^{-i\omega t} + \tilde{M}^*(\omega) f_0^* e^{+i\omega t}\right]. \tag{11.179}$$

As the force is creating a response, work is performed on the system. Some of this work energy is stored reversibly in the response field and some is lost irreversibly to heat. Our goal here is to see how the real and imaginary parts of the complex modulus $\tilde{M}(\omega)$ are influencing the reversibly stored and irreversibly dissipated energies.

The reversibly stored energy is given by

$$E(t) = \frac{1}{2} r(t) f(t) = \frac{1}{8}\left(\tilde{M} f_0 e^{-i\omega t} + \tilde{M}^* f_0^* e^{+i\omega t}\right)\left(f_0 e^{-i\omega t} + f_0^* e^{+i\omega t}\right) \tag{11.180}$$

$$= \frac{1}{8}\left[\left(\tilde{M} + \tilde{M}^*\right) f_0 f_0^* + \tilde{M} f_0^2 e^{-2i\omega t} + \tilde{M}^* f_0^{*2} e^{2i\omega t}\right]. \tag{11.181}$$

The rate that energy is being reversibly stored is thus given by

$$\frac{\mathrm{d}E(t)}{\mathrm{d}t} = -\frac{i\omega}{4}\left[\tilde{M} f_0^2 e^{-2i\omega t} - \tilde{M}^* f_0^{*2} e^{2i\omega t}\right] \tag{11.182}$$

and is sometimes positive and sometimes negative as we proceed through each forcing cycle. The rate that work is being performed by the force $f(t)$ acting on the system is given by

$$\frac{\mathrm{d}U(t)}{\mathrm{d}t} = \frac{\mathrm{d}r(t)}{\mathrm{d}t} f(t) \tag{11.183}$$

$$= \frac{1}{4}\left(-i\omega \tilde{M} f_0 e^{-i\omega t} + i\omega\, \tilde{M}^* f_0^* e^{+i\omega t}\right)\left(f_0 e^{-i\omega t} + f_0^* e^{i\omega t}\right) \tag{11.184}$$

$$= -\frac{i\omega}{4}\left[\left(\tilde{M} - \tilde{M}^*\right) f_0 f_0^* + \tilde{M} f_0^2 e^{-2i\omega t} - \tilde{M}^* f_0^{*2} e^{2i\omega t}\right]. \tag{11.185}$$

The difference between the work rate and the rate that energy is stored reversibly is the dissipation rate $\mathrm{d}D(t)/\mathrm{d}t$, that is, $\mathrm{d}U(t)/\mathrm{d}t = \mathrm{d}E(t)/\mathrm{d}t + \mathrm{d}D(t)/\mathrm{d}t$. As an aside, this energy relation is simply the first law of thermodynamics in which $\mathrm{d}U/\mathrm{d}t$ is the change in the internal energy of an element that is being acted upon by the force $f(t)$ and $\mathrm{d}D/\mathrm{d}t = T\mathrm{d}S/\mathrm{d}t$ is the change in the heat energy of the element. The energy being stored reversibly $\mathrm{d}E/\mathrm{d}t$ in the element depends on the particular physics represented by $r(t)$ and $f(t)$ in the constitutive law but was identified in the chapter on thermodynamics as either elastic energy or electric- and/or magnetic-polarization energy.

The first law then yields

$$\frac{\mathrm{d}D(t)}{\mathrm{d}t} = \frac{\mathrm{d}U(t)}{\mathrm{d}t} - \frac{\mathrm{d}E(t)}{\mathrm{d}t} = -\frac{i\omega}{4}\left(\tilde{M} - \tilde{M}^*\right) f_0 f_0^* = \frac{\omega}{2} \tilde{M}_I f_0 f_0^*, \tag{11.186}$$

where we used that $\tilde{M} - \tilde{M}^* = 2i\tilde{M}_I$. So energy dissipation is controlled by the imaginary part of the complex modulus. The second law of thermodynamics requires that $\mathrm{d}D/\mathrm{d}t > 0$ so that we always have $\tilde{M}_I(\omega) > 0$ when $\omega > 0$ or $\omega \tilde{M}_I(\omega) > 0$ for all frequencies where

$\tilde{M}_I(\omega) = -\tilde{M}_I(-\omega)$ is an odd function because $M(t)$ is real. The total energy lost to heat in one period $T = 2\pi/\omega$ of time-harmonic forcing is then

$$\Delta D = \int_0^T \mathrm{d}t \frac{\mathrm{d}D(t)}{\mathrm{d}t} = \pi \tilde{M}_I(\omega) f_0 f_0^*. \tag{11.187}$$

So the imaginary part of the complex modulus creates irreversible loss, which is a first key message of this section.

We next use Eq. (11.181) to define the average energy stored in a period $T = 2\pi/\omega$ of the sinusoidal cycling

$$E_{\text{ave}} = \frac{1}{T} \int_0^T \mathrm{d}t\, E(t) = \frac{1}{4} \tilde{M}_R(\omega) f_0 f_o^*, \tag{11.188}$$

where we used that $\tilde{M} + \tilde{M}^* = 2\tilde{M}_R$ and that the integrals involving $e^{\pm 2i\omega t}$ are zero over one period of cycling. The average energy that is stored reversibly during each cycle is thus controlled by $\tilde{M}_R(\omega)$.

If the constitutive law is written in inverse form $\tilde{f}(\omega) = \tilde{N}(\omega)\tilde{r}(\omega)$ we have that

$$\tilde{N}(\omega) \,\hat{=}\, \frac{1}{\tilde{M}_R(\omega) + i\tilde{M}_I(\omega)} = \frac{\tilde{M}_R(\omega) - i\tilde{M}_I(\omega)}{\tilde{M}_R^2(\omega) + \tilde{M}_I^2(\omega)}. \tag{11.189}$$

So the inverse complex modulus has an imaginary part that is negative $\tilde{N}_I(\omega) < 0$ when $\omega > 0$ because we showed above that $\tilde{M}_I(\omega) > 0$ when $\omega > 0$ (second law of thermodynamics).

To identify the $r(t)$ and $f(t)$ in a linear-delay constitutive law for a particular application such as linear electrical polarization or linear elastic response, you look at the form of the change in internal energy in an element of initial volume V_0 that is appropriate to the particular application. So in Chapter 3 we showed that for electrical polarization $V_0^{-1}\mathrm{d}U/\mathrm{d}t = (\mathrm{d}\boldsymbol{D}/\mathrm{d}t) \cdot \boldsymbol{E}$ which says that the dielectric displacement $\boldsymbol{D}$ is the natural "response" variable and the total electric field $\boldsymbol{E}$ is the "force" variable when the constitutive law is written in normal form. The complex form of this constitutive law in an isotropic material is $\tilde{\boldsymbol{D}}(\omega) = \tilde{\varepsilon}(\omega)\tilde{\boldsymbol{E}}(\omega)$ which indeed has a real permittivity at infinite frequency $\varepsilon_\infty = \epsilon_0\kappa_\infty$ that is smaller (in nearly all cases) than the real permittivity at zero frequency $\varepsilon_0 = \epsilon_0\kappa_0$ because there is no time for any polarization to take place at infinite frequency (i.e., the dielectric constant almost always satisfies $\kappa_\infty < \kappa_0$ though certain phases of matter such as an electron plasma can have increasing dielectric constant with increasing frequency). So from Eq. (11.177) of Section 11.6.1, we indeed have that $\tilde{M}_I(\omega) > 0$ for positive frequencies when the electrical-polarization constitutive law is written in normal form, which is also what the second law of thermodynamics showed us above. Similarly, in Chapter 4 we showed that for elastic response $V^{-1}\mathrm{d}U/\mathrm{d}t = (\mathrm{d}\boldsymbol{e}/\mathrm{d}t) : \boldsymbol{\tau}$, which says that the strain tensor is the natural response variable and the stress tensor is the natural force variable when the linearized Hooke's law is written in "normal" form. The complex form of this constitutive law is $\tilde{\boldsymbol{e}}(\omega) = {}_4\tilde{\boldsymbol{S}}(\omega) : \tilde{\boldsymbol{\tau}}(\omega)$ where the elastic compliances ${}_4\tilde{\boldsymbol{S}}(\omega)$ will have real parts at infinite frequency that are smaller than at zero frequency because the delay mechanism in elastic response involves some type of internal viscosity so that the material is stiffer (and therefore less compliant) at high frequencies than at low frequencies. So from Eq. (11.177), we again have that the complex moduli satisfy $\tilde{M}_I(\omega) > 0$ for positive frequencies when Hooke's law is written in "normal" (or compliance) form. These facts are important to remember when using Fourier analysis to address the linear response in physics applications where delay processes are operative.

A convenient measure of loss or dissipation is the real and dimensionless *quality factor* $Q(\omega)$, defined as

$$\frac{1}{Q(\omega)} \,\widehat{=}\, \frac{\text{energy dissipated over one cycle}}{(4\pi)(\text{average energy stored in each cycle})} = \frac{\Delta D}{4\pi\, E_{\text{ave}}} \tag{11.190}$$

or

$$\boxed{\frac{1}{Q(\omega)} = \frac{\tilde{M}_I(\omega)}{\tilde{M}_R(\omega)}.} \tag{11.191}$$

Other definitions of $Q(\omega)$ exist in the literature for quantifying dissipated energy for sinusoidal linear response, but the above is the most useful, convenient, and common.

To show how to measure $Q^{-1}(\omega)$ from the observed $r(t)$ and $f(t)$, let's write $\tilde{M}(\omega)$ in the form

$$\tilde{M}(\omega) = \tilde{M}_R(\omega) + i\,\tilde{M}_I(\omega) \tag{11.192}$$

$$= \sqrt{\tilde{M}_R^2 + \tilde{M}_I^2}\ \exp\left[i\,\tan^{-1}\left(\frac{\tilde{M}_I}{\tilde{M}_R}\right)\right]. \tag{11.193}$$

Now let's return to the time response $r(t)$ generated by a sinusoidal force $f(t)$ at a single frequency ω

$$r(t) = \text{Re}\left\{\tilde{M}(\omega)\, f_0\, e^{-i\omega t}\right\} \tag{11.194}$$

$$= \text{Re}\left\{\sqrt{\tilde{M}_R^2 + \tilde{M}_I^2}\, f_0\ \exp\left[-i\left(\omega t - \tan^{-1}\left(\frac{\tilde{M}_I}{\tilde{M}_R}\right)\right)\right]\right\} \tag{11.195}$$

$$= \text{Re}\left\{A\,\exp\left[-i\omega(t - t_0)\right]\right\}, \tag{11.196}$$

where the last line corresponds to what is observed on, say, an oscilloscope as depicted in Fig. 11.14. In terms of the observed time shift t_0, we have

$$\omega\, t_0 = \tan^{-1}\left(\frac{\tilde{M}_I}{\tilde{M}_R}\right), \tag{11.197}$$

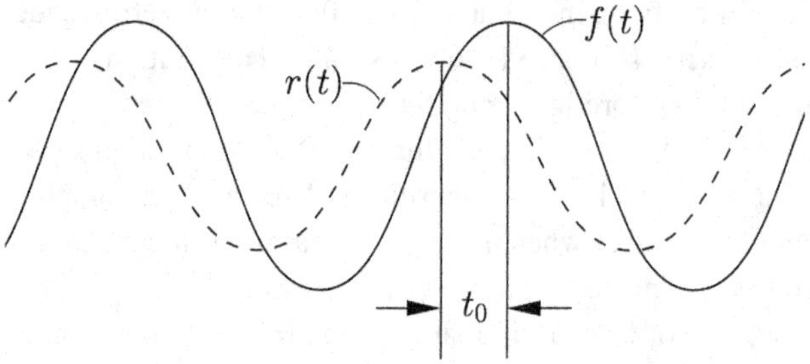

Figure 11.14 A sinusoidal response $r(t)$ delayed relative to the sinusoidal force $f(t)$.

which then gives

$$\boxed{\frac{1}{Q(\omega)} = \tan(\omega\, t_0).} \tag{11.198}$$

The right-hand side of this expression is often called the *loss tangent*. The measured time t_0 that the response lags behind the force is positive due to $\tilde{M}_I/\tilde{M}_R$ being positive at positive frequency ω.

We showed earlier that $\tilde{M}_I$ is responsible for dissipation and the present analysis shows that it also creates delay in the response. An easy way to measure how much loss is taking place in each sinusoidal cycle as quantified by $1/Q(\omega)$ is to record both $r(t)$ and $f(t)$ at a given frequency, observe the time lag t_0 between them and then use Eq. (11.198). Different frequencies will have different time lags which means that $t_0 = t_0(\omega)$ so we have to measure t_0 at each ω over the entire band of interest to obtain the frequency-dependent attenuation function $1/Q(\omega)$.

If the constitutive law is a transport law coming from nonequilibrium thermodynamics, which was not the perspective taken above, when the transport flux is instantaneous in the force driving transport, all the work energy is being lost to heat with no energy being stored reversibly. So when there is a delay between the flux and force in a time-dependent transport law, the delay mechanism is providing a means for work energy to be stored reversibly. If the constitutive law comes from equilibrium thermodynamics, which is the perspective we took in the above, when the response is instantaneous in the force there is no dissipated energy so that if there is a delay between the response and the force it is because there is also an internal-transport mechanism taking place in the delay process that causes energy to be dissipated. In either scenario (equilibrium or nonequilibrium constitutive laws), the modulus relating the response to the force is a complex frequency-dependent function in the frequency domain with the real part and imaginary part of the modulus related to each other as Hilbert-transform pairs.

11.6.3 Dispersive Waves and Group Velocity

Let's now consider a wave example where delayed-linear response is occurring. Hooke's law is the relation between stress and strain. In a simple 1D plane-wave scenario, in which, say, uniaxial stress τ responds instantaneously and linearly to uniaxial strain, we often write Hooke's law in stiffness form as $\tau = C\partial u/\partial x$, where C is the elastic stiffness controlling this 1D stress–strain scenario. Per the discussion of Section 11.6.2, when delay mechanisms are operative, some type of internal viscosity is present such that the stiffness at high frequencies C_∞ is larger than the stiffness at low frequencies C_0, which corresponds to the inverse form of the constitutive law in which the imaginary part of the complex stiffness is negative for positive frequencies. For the complex elastic modulus to have an imaginary part that is positive for positive frequencies, the delayed-linear response is written in the "normal" or "compliance" form of Hooke's law

$$\frac{\partial u(x,t)}{\partial x} = \int_{-\infty}^{t} dt'\, M(t-t')\tau(x,t'). \tag{11.199}$$

Taking an x derivative of both sides and inserting Newton's law of conservation of momentum $\partial\tau(x,t)/\partial x = \rho\partial^2 u(x,t)/\partial t^2$ yields the wave equation in delayed linear-response form

$$\frac{\partial^2 u(x,t)}{\partial x^2} = \int_{-\infty}^{t} dt'\, \rho M(t-t')\frac{\partial^2 u(x,t')}{\partial t'^2}. \tag{11.200}$$

The mass density ρ has no delayed-linear response associated with it. Taking a Fourier transform over time of this delayed-response form of the scalar wave equation and using the convolution theorem gives

$$\frac{\partial^2 \tilde{u}(x,\omega)}{\partial x^2} + \rho\omega^2\tilde{M}(\omega)\tilde{u}(x,\omega) = 0. \tag{11.201}$$

The frequency-dependent elastic compliance $\tilde{M}(\omega)$ is complex (it has an imaginary part) because the real temporal-response function $M(t)$ is not even. The real and imaginary parts of the complex compliance $\tilde{M}$ satisfy the Kramers–Krönig relations, are both smaller at infinite frequency compared to zero frequency, and satisfy the even and odd symmetry properties (due only to the fact that $M(t)$ is real in the time domain) of $\tilde{M}_R(\omega) = \tilde{M}_R(-\omega)$ and $\tilde{M}_I(\omega) = -\tilde{M}_I(-\omega)$. Further, $\omega\tilde{M}_I(\omega) > 0$ for all frequencies per the second law of thermodynamics and the just-quoted odd-symmetry property.

We solve the complex wave equation of Eq. (11.201) for the particular case where a source for the waves is placed at $x=0$, which provides the boundary condition $u(x=0,t) = u_0(t)$ with $u_0(t)$ some given displacement time function. We consider the wave response in the semi-infinite lossy slab (or semi-infinite lossy string) located in $0 \le x \le \infty$ and require that the waves are outgoing as $x \to \infty$.

If we define the complex frequency-dependent "slowness" $\tilde{s}(\omega) = \sqrt{\rho\tilde{M}(\omega)}$ (which has physical units of inverse wave speed), the solution of Eq. (11.201) that satisfies the two boundary conditions is

$$\tilde{u}(x,\omega) = \tilde{u}_0(\omega)e^{i\omega\tilde{s}(\omega)x}. \tag{11.202}$$

Returning to the time domain via the inverse Fourier transform then gives

$$u(x, t) = \frac{1}{2\pi} \int_{-\infty}^{\infty} \tilde{u}_0(\omega) e^{-i[\omega t - \omega \tilde{s}_R(\omega) x]} e^{-\omega \tilde{s}_I(\omega) x}\, d\omega. \tag{11.203}$$

We see that $\omega \tilde{s}_I(\omega)$ controls the exponential loss of amplitude of this 1D plane wave in a material in which loss, or delay, mechanisms are operative. We next show that $\omega \tilde{s}_I(\omega) > 0$ for all frequencies and for the constitutive law written in either normal or inverse form.

The complex slowness $\tilde{s}(\omega)$ has real and imaginary parts given as

$$\begin{aligned} \tilde{s}(\omega) &= \sqrt{\rho\left[\tilde{M}_R(\omega) + i\tilde{M}_I(\omega)\right]} = \tilde{s}_R(\omega) + i\tilde{s}_I(\omega) \\ &= \sqrt{\rho\sqrt{\tilde{M}_R^2(\omega) + \tilde{M}_I^2(\omega)}} \\ &\quad \times \left\{\cos\left[\frac{1}{2}\tan^{-1}\left(\frac{\tilde{M}_I(\omega)}{\tilde{M}_R(\omega)}\right)\right] + i \sin\left[\frac{1}{2}\tan^{-1}\left(\frac{\tilde{M}_I(\omega)}{\tilde{M}_R(\omega)}\right)\right]\right\}. \end{aligned} \tag{11.204}$$

Thus, because $\tan^{-1} u$ is an odd function of u as is $\sin u$, the complex slowness has real and imaginary parts that satisfy the same symmetry properties as do $\tilde{M}_R(\omega)$ and $\tilde{M}_I(\omega)$; namely, $\tilde{s}_R(\omega) = \tilde{s}_R(-\omega)$ and $\tilde{s}_I(\omega) = -\tilde{s}_I(-\omega)$. Further, because the complex compliance has an imaginary part $\tilde{M}_I$ that is positive for positive ω, Eq. (11.204) requires that $\tilde{s}_I(\omega)$ is also positive for positive ω so that $\omega \tilde{s}_I(\omega) > 0$ for all frequencies, a result that is needed for Eq. (11.203) to give meaningful results.

We can express Eq. (11.204) in terms of the real loss parameter $Q^{-1}(\omega) = \tilde{M}_I(\omega)/\tilde{M}_R(\omega)$ that we defined earlier as the ratio of the energy irreversibly lost in each cycle of frequency ω to the energy reversibly stored in each cycle (and divided by 4π). We also define the real *wavenumber* $k(\omega) = \omega \tilde{s}_R(\omega)$ (units of inverse length) and the real *absorption coefficient* $\alpha(\omega) = \omega \tilde{s}_I(\omega)$ (also units of inverse length) from Eq. (11.204) as

$$k(\omega) \widehat{=} \omega \tilde{s}_R(\omega) = \frac{\omega}{c_p(\omega)} = \omega\sqrt{\rho \tilde{M}_R(\omega)}\sqrt{\frac{\sqrt{1 + Q^{-2}(\omega)} + 1}{2}}, \tag{11.205}$$

$$\alpha(\omega) \widehat{=} \omega \tilde{s}_I(\omega) = \frac{\omega}{c_p(\omega)}\sqrt{\frac{\sqrt{1 + Q^{-2}(\omega)} - 1}{\sqrt{1 + Q^{-2}(\omega)} + 1}} > 0. \tag{11.206}$$

The wave's *phase velocity* $c_p(\omega) = \omega/k(\omega) = 1/\tilde{s}_R(\omega)$ is the wave speed at which each frequency ω comprising the wavelet $u_0(t)$ is advancing through space and time. We used standard trigonometric identifies to rewrite the cosine and sine of $(1/2)\tan^{-1} Q^{-1}$ in terms of the square root operation. It is common that the delay mechanism and associated loss process is such that $Q^{-1} \ll 1$ so that a Taylor expansion of Eqs (11.205) and (11.206) yields the useful and widely used approximations

$$k(\omega) = \frac{\omega}{c_p(\omega)} = \omega\sqrt{\rho\tilde{M}_R(\omega)}\left(1 + \frac{Q^{-2}(\omega)}{8} + \dots\right) \approx \omega\sqrt{\rho\tilde{M}_R(\omega)}, \tag{11.207}$$

$$\alpha(\omega) = \frac{\omega Q^{-1}(\omega)}{2c_p(\omega)}\left(1 - \frac{Q^{-2}(\omega)}{4} + \dots\right) \approx \frac{\omega Q^{-1}(\omega)}{2c_p(\omega)}, \tag{11.208}$$

where in this small Q^{-1} limit, the phase velocity becomes simply $c_p(\omega) \approx 1/\sqrt{\rho\tilde{M}_R(\omega)}$. Although the linear relation between the absorption coefficient $\alpha(\omega)$ and $Q^{-1}(\omega)$ as given by Eq. (11.208) is widely used, it is only valid when $Q^{-1} \ll 1$. In terms of the real wave number $k(\omega)$ and real absorption coefficient $\alpha(\omega)$ the wave response of Eq. (11.203) is written

$$u(x, t) = \frac{1}{2\pi}\int_{-\infty}^{\infty} \tilde{u}_0(\omega)e^{-i[\omega t - k(\omega)x]}e^{-\alpha(\omega)x}\, d\omega. \tag{11.209}$$

Wave problems in which each frequency ω comprising a wavelet advances with a different phase velocity $c_p(\omega) = \omega/k(\omega)$ and absorption coefficient $\alpha(\omega)$ are called *dispersive*.

Note as well that if we had worked with the inverse form of Hooke's law, using the inverse modulus (or stiffness) $\tilde{N}(\omega) = 1/\tilde{M}(\omega)$, we would have

$$\begin{aligned}\tilde{s}(\omega) &= \sqrt{\frac{\rho}{\left[\tilde{N}_R(\omega) + i\tilde{N}_I(\omega)\right]}}\\ &= \sqrt{\frac{\rho}{\sqrt{\tilde{N}_R^2(\omega) + \tilde{N}_I^2(\omega)}}}\\ &\quad \times \left\{\cos\left[\frac{1}{2}\tan^{-1}\left(\frac{\tilde{N}_I(\omega)}{\tilde{N}_R(\omega)}\right)\right] - i\sin\left[\frac{1}{2}\tan^{-1}\left(\frac{\tilde{N}_I(\omega)}{\tilde{N}_R(\omega)}\right)\right]\right\}.\end{aligned} \tag{11.210}$$

Because $\tilde{N}_I(\omega) < 0$ for all positive frequencies, we again have that $\tilde{s}_I(\omega) > 0$ for all positive frequencies or $\alpha(\omega) = \omega\tilde{s}_I(\omega) > 0$ for all frequencies. In this case, we define $Q^{-1} = -\tilde{N}_I(\omega)/\tilde{N}_R(\omega)$ and in the small Q^{-1} limit can identify the phase velocity as $c_p(\omega) = \sqrt{\tilde{N}_R(\omega)/\rho}$ and will again have $Q^{-1} = 2c_p(\omega)\alpha(\omega)/\omega$. So Eq. (11.209) gives the same dispersive-wave results whether we work with the constitutive law in normal or inverse form.

We can now consider some special cases of evaluating the integral of Eq. (11.203). If there is no delay mechanism, there is no acoustic attenuation $\tilde{M}_I(\omega)/\tilde{M}_R(\omega) = 0$ and $\tilde{s}_I(\omega) = 0$. The wave slowness is now $\tilde{s}_R(\omega) = 1/c_p$ in which the wave speed c_p is a constant independent of frequency. In this case, the integral of Eq. (11.203) gives

$$u(x, t) = u_0\left(t - \frac{x}{c_p}\right), \tag{11.211}$$

which is the standard solution of the 1D wave equation without a delay and associated loss mechanism that we have seen earlier.

If there is a delay and loss mechanism present and if the source at $x = 0$ (the time-dependent boundary condition) is *monochromatic*, meaning driven at a single circular frequency ω_0, with $u_0(t)=A\cos(\omega_0 t)$ and $\tilde{u}_0(\omega)=A\pi\,[\delta(\omega-\omega_0)+\delta(\omega+\omega_0)]$, then because $c_p(\omega)=c_p(-\omega_0)$ and $\alpha(\omega_0)=\alpha(-\omega_0)$, Eq. (11.203) gives

$$u(x,t)=\frac{A}{2}\left[e^{-i\omega_0[t-x/c_p(\omega_0)]}e^{-\alpha(\omega_0)x}+e^{+i\omega_0[t-x/c_p(\omega_0)]}e^{-\alpha(\omega_0)x}\right] \tag{11.212}$$

or

$$u(x,t)=A\cos\left[\omega_0\left(t-\frac{x}{c_p(\omega_0)}\right)\right]e^{-\alpha(\omega_0)x}. \tag{11.213}$$

At each single frequency ω_0 considered, this corresponds to a sinusoid through all of space with exponential spatial decay as x increases. This monochromatic response cannot really be said to transmit information from the source at $x=0$ at a wave speed $c_p(\omega_0)$ because the source has been operating forever and the attenuated sinusoid is already present through all of space without arrival times.

So let's next consider that the source term $u_0(t)$ corresponds to a finite-duration pulse. This pulse has a Fourier transform consisting of a narrow band of frequencies $\Delta\omega_p$ surrounding a center frequency ω_0 that is called the *carrier* frequency. The associated time function $u_0(t)$ is depicted in Fig. 11.15. Due to the narrow band of frequencies $\Delta\omega_p$ surrounding the carrier frequency, the real dispersive wave number $k(\omega)$ can be expanded about ω_0 to give

$$k(\omega)=k(\omega_0)+k'(\omega_0)(\omega-\omega_0)+\ldots, \tag{11.214}$$

where we are using the notation $k'(\omega_0)=\mathrm{d}k(\omega)/\mathrm{d}\omega|_{\omega=\omega_0}$. Inserting this truncated expansion into Eq. (11.209) gives

$$u(x,t)=\frac{e^{i[k(\omega_0)-\omega_0 k'(\omega_0)]x}}{2\pi}\int_{-\infty}^{\infty}\tilde{u}_0(\omega)e^{-i\omega(t-k'(\omega_0)x)}e^{-\alpha(\omega)x}\,\mathrm{d}\omega. \tag{11.215}$$

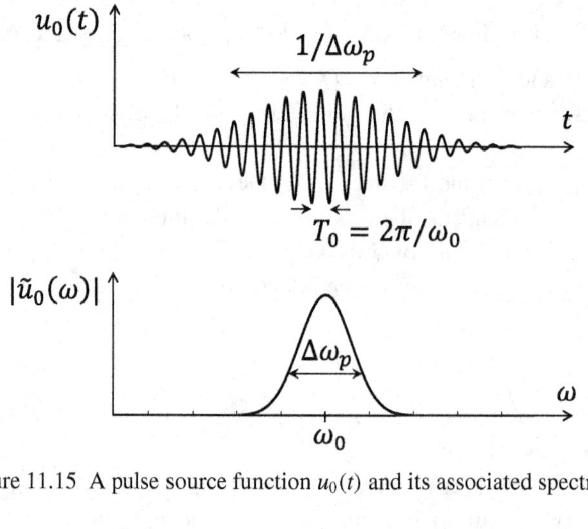

Figure 11.15 A pulse source function $u_0(t)$ and its associated spectrum.

The way in which the pulse is moving through space and time is controlled by the term $\exp\left[-i\omega(t-k'(\omega_0)x)\right]$. This shows that the speed at which the pulse advances, which is called the *group velocity* c_g, is given by

$$c_g = \frac{1}{k'(\omega_0)} = \frac{1}{\tilde{s}_R(\omega_0) + \omega_0 \, d\tilde{s}_R(\omega)/d\omega|_{\omega=\omega_0}}. \tag{11.216}$$

Expressed in terms of the phase velocity $c_p(\omega_0) = 1/\tilde{s}_R(\omega_0)$ this becomes the so-called *Rayleigh dispersion formula*

$$\boxed{c_g = \frac{c_p(\omega_0)}{1 - \omega_0 \left[c_p(\omega_0)\right]^{-1} dc_p(\omega)/d\omega\big|_{\omega=\omega_0}}.} \tag{11.217}$$

For elastic waves, the stiffness of the material is increasing with increasing frequency so that the phase velocity is increasing with increasing frequency and $c_g > c_p$. But depending on the dispersive-wave scenario being considered, the group velocity can either be larger than or less than the phase velocity at the given carrier frequency ω_0. For example, for electromagnetic waves in a plasma characterized by an electrical permittivity $\varepsilon(\omega) = \epsilon_0\left(1 - \omega_p^2/\omega^2\right)$ and for $\omega > \omega_p$ such that the permittivity is getting bigger with increasing frequency, an electromagnetic plasma wave has a phase velocity $c_p(\omega) = 1/\sqrt{\mu_0\varepsilon(\omega)}$ that is decreasing with increasing frequency so that $c_g < c_p$. However, in a more normal polarizing material, the electrical permittivity decreases with increasing frequency (less time for polarization with increasing frequency) and $c_g > c_p$.

At small time t and therefore small distances x propagated, the initial pulse $u_0(t - x/c_g)$ is moving out with the group velocity c_g, which is really the only time where group velocity is clearly defined. As time and the distance propagated increase, the pulse becomes progressively distorted as the higher-frequency components of the pulse move at a different speed than the lower-frequency components and the effects of frequency-dependent attenuation begin to accumulate. At any time or space position, the pulse is therefore distorted by dispersion (often called the "chirped" pulse) but is always given by the integral of Eq. (11.203) for given frequency-dependent $\tilde{s}_R(\omega)$ and $\tilde{s}_I(\omega)$ and given $u_0(t)$.

11.7 Using Fourier Transforms and Contour Integration for Solving PDEs

We now consider another important application of using the Fourier transform and contour integration for obtaining the inverse Fourier transform in the solution of some PDEs. We will solve two 1D wave problems here and a 1D diffusion problem but will continue using the Fourier transform and contour integration in Chapter 12 to obtain 3D Green's functions for a range of physics problems.

11.7.1 Wave Example 1

Consider the 1D wave-equation without delayed and dispersive response defined by

$$\frac{\partial^2 \psi}{\partial x^2} - \frac{1}{c^2}\frac{\partial^2 \psi}{\partial t^2} = N\delta(x)\cos(bt) \quad \text{with outward waves as } |x| \to \infty \tag{11.218}$$

and with the wave speed c (phase velocity) a given constant. This corresponds to a source sitting at the origin $x = 0$ radiating waves in both the $x > 0$ direction and $x < 0$ direction, each of which has the same sinusoidal character. If the given parameter N has units of force (Newtons), ψ has units of energy (Joules) because the Dirac $\delta(x)$ has units of inverse length while the cosine is unitless. Note that the given circular frequency b at which the source is driven has units of radians per second. It is always good to check that the final answer comes out with the proper physical units of the field you are searching for. In this problem, the sinusoidal forcing term $\cos(bt)$ has been operating for a very long time so that initial conditions are not required.

We solve this problem by taking the Fourier transform over both x and t and identifying the transforms:

$$\frac{\partial}{\partial x} \to ik, \quad \frac{\partial}{\partial t} \to -i\omega, \quad \delta(x) \to 1, \quad \text{and} \quad \cos(bt) \to \frac{2\pi}{2}\Big[\delta(\omega - b) + \delta(\omega + b)\Big].$$

The PDE in the transform domain becomes

$$\left(-k^2 + \frac{\omega^2}{c^2}\right)\tilde{\tilde{\psi}}(k, \omega) = N\pi\Big[\delta(\omega - b) + \delta(\omega + b)\Big] \tag{11.219}$$

which has the solution

$$\tilde{\tilde{\psi}}(k, \omega) = -N\pi\frac{[\delta(\omega - b) + \delta(\omega + b)]}{(k^2 - \omega^2/c^2)}. \tag{11.220}$$

Thus, as always, it is straightforward to solve the problem in the transform domain and the main mathematical work is in performing the inverse transforms to obtain $\psi(x, t)$.

Let's do the k integral first (inverse transform to x)

$$\tilde{\psi}(x, \omega) = \frac{1}{2\pi}\int_{-\infty}^{+\infty} \tilde{\tilde{\psi}}(k, \omega)\, e^{+ikx}\, dk = -\frac{N\big[\delta(\omega - b) + \delta(\omega + b)\big]}{2}\int_{-\infty}^{+\infty} \frac{e^{ikx}\, dk}{(k + \omega/c)\,(k - \omega/c)}. \tag{11.221}$$

We have two poles at $k = \omega/c$ and $k = -\omega/c$ that reside directly on the path of integration. The easiest and most physically meaningful way of handling this so that the waves are outgoing as required is to introduce a tiny loss parameter ϵ into the original scalar wave equation that causes the waves to lose amplitude with distance traveled

$$\frac{\partial^2 \psi(x, t)}{\partial x^2} - \frac{1}{c^2}\left(\frac{\partial}{\partial t} + \epsilon\right)^2 \psi(x, t) = \delta(x)\, \cos bt. \tag{11.222}$$

Once we solve the problem with a nonzero positive ϵ (units of inverse time) we then put $\epsilon = 0$ in the final result. To understand that $\epsilon > 0$ introduces attenuation, write out the time operator as

$$\left(\frac{\partial}{\partial t} + \epsilon\right)^2 \psi(x,t) = \frac{\partial^2 \psi(x,t)}{\partial t^2} + 2\epsilon \frac{\partial \psi(x,t)}{\partial t} + \epsilon^2 \psi(x,t) \tag{11.223}$$

and note that for sufficiently small ϵ, the ϵ^2 term is negligible while the $2\epsilon\partial\psi/\partial t$ term corresponds to a small amount of diffusion that always produces a loss in amplitude.

Upon taking the temporal Fourier transform of Eq. (11.222), we have $\partial/\partial t + \epsilon \to -i\omega + \epsilon = -i(\omega + i\epsilon)$. So in Eq. (11.221), we replace ω by $\omega + i\epsilon$ in the denominator of the integrand to obtain

$$\tilde{\psi}(x,\omega) = -\frac{N[\delta(\omega - b) + \delta(\omega + b)]}{2} \int_{-\infty}^{+\infty} \frac{e^{ikx}\, dk}{(k + \omega/c + i\epsilon/c)\,(k - \omega/c - i\epsilon/c)}. \tag{11.224}$$

The contour with the poles displaced by the replacement $\omega/c \to \omega/c + i\epsilon/c$ is depicted in Fig. 11.16. In order to neglect the contribution from the half circle at infinity, we need to close the contour in the upper-half k plane when $x > 0$, and in the lower-half k plane when $x < 0$. We have when $x > 0$

$$\int_{-\infty}^{+\infty} \frac{e^{ikx}\, dk}{(k + \omega/c + i\epsilon/c)\,(k - \omega/c - i\epsilon/c)} = 2\pi i\, a_{-1}\left(\frac{\omega + i\epsilon}{c}\right) \tag{11.225}$$

$$= 2\pi i \frac{\exp(i\omega x/c)\exp(-\epsilon x/c)}{2\,(\omega/c + i\epsilon/c)} \quad \text{when } x > 0, \tag{11.226}$$

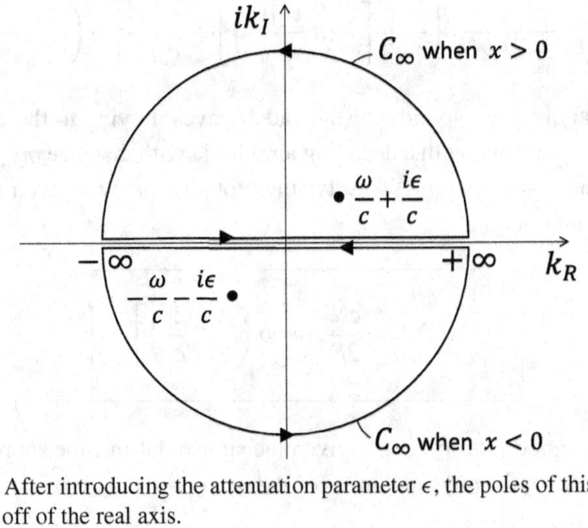

Figure 11.16 After introducing the attenuation parameter ϵ, the poles of this wave problem are displaced off of the real axis.

while when $x < 0$,

$$\int_{-\infty}^{+\infty} \frac{e^{ikx}\,\mathrm{d}k}{(k+\omega/c+i\epsilon/c)\,(k-\omega/c-i\epsilon/c)} = -2\pi i\, a_{-1}\left(-\frac{\omega+i\epsilon}{c}\right) \tag{11.227}$$

$$= 2\pi i\,\frac{\exp(-i\omega x/c)\exp(\epsilon x/c)}{2\,(\omega/c+i\epsilon/c)} \quad \text{when } x<0. \tag{11.228}$$

Note that on the closed counter-clockwise path in the lower-half complex k plane, our integral of interest is in the opposite direction (from $-\infty$ to $+\infty$) compared to the integral on the path (from $+\infty$ to $-\infty$). That is why a negative showed up on the right-hand side of Eq. (11.227). Either of these results (x either positive or negative) can equivalently be expressed $\pi i e^{i\omega|x|/c}\, e^{-\epsilon|x|/c}/(\omega/c+i\epsilon/c)$. As $|x|$ grows larger, the parameter ϵ causes the spatial response to get smaller than it would if $\epsilon=0$. Now that we have returned to the space domain, we could set $\epsilon=0$ but we will keep it around to the very end.

Doing the inverse transform over ω then gives:

$$\psi(x,t) = \frac{Ne^{-\epsilon|x|/c}}{4i}\int_{-\infty}^{+\infty}\mathrm{d}\omega\,\frac{e^{-i\omega(t-|x|/c)}}{(\omega/c+i\epsilon/c)}\Big[\delta(\omega+b)+\delta(\omega-b\Big] \tag{11.229}$$

$$= \frac{N}{4i}\left[\frac{e^{+ib(t-|x|/c)}}{(-b/c+i\epsilon/c)}+\frac{e^{-ib(t-|x|/c)}}{(b/c+i\epsilon/c)}\right]e^{-\epsilon|x|/c}. \tag{11.230}$$

So using $\sin u = (e^{iu}-e^{-iu})/(2i)$ and $\cos u = (e^{iu}+e^{-iu})/2$, we can write the full solution with ϵ as

$$\psi(x,t) = \frac{-N}{2\left(b^2/c^2+\epsilon^2/c^2\right)}\left\{\frac{b}{c}\sin\left[b\left(t-\frac{|x|}{c}\right)\right]+\frac{\epsilon}{c}\cos\left[b\left(t-\frac{|x|}{c}\right)\right]\right\}e^{-\epsilon|x|/c}, \tag{11.231}$$

which is purely real and corresponds to sinusoidal waves moving to the right when $x>0$ and to the left when $x<0$ and with a decaying amplitude with distance propagated as set by the positive attenuation parameter ϵ. To solve the problem that was given at the beginning, we now set $\epsilon=0$ and obtain

$$\boxed{\psi(x,t) = -\frac{cN}{2b}\sin\left[b\left(t-\frac{|x|}{c}\right)\right].} \tag{11.232}$$

This solution makes good intuitive sense given the sinusoidal-in-time source for the waves at the origin $x=0$. Also, the physical units for ψ have come out correctly as energy (force N times distance c/b).

11.7.2 Wave Example 2

Consider the 1D wave problem defined by

$$\frac{\partial^2 \psi}{\partial x^2} - \frac{1}{c^2}\frac{\partial^2 \psi}{\partial t^2} = \delta(x)\,\delta(t) \quad \text{where } \psi = 0 \text{ for } t < 0 \tag{11.233}$$

and where we want outward radiating waves for both $x > 0$ and $x < 0$. In this problem, the physical units of ψ are velocity (length divided by time). Note that the condition that $\psi = 0$ for $t < 0$ is equivalent to the initial condition that ψ and $\partial\psi/\partial t = 0$ at $t = 0^-$ (i.e., just prior to the Dirac $\delta(t)$ going off at $t = 0$).

Solve this problem by transforming over space and time to obtain

$$\left(-k^2 + \frac{\omega^2}{c^2}\right)\tilde{\tilde{\psi}}(k, \omega) = 1, \tag{11.234}$$

which has the solution

$$\tilde{\tilde{\psi}}(k, \omega) = \frac{c^2}{\omega^2 - c^2 k^2}. \tag{11.235}$$

In this example, let's return to the time domain first

$$\tilde{\psi}(k, t) = \frac{c^2}{2\pi}\int_{-\infty}^{+\infty} \frac{e^{-i\omega t}\,\mathrm{d}\omega}{(\omega + ck)(\omega - ck)}. \tag{11.236}$$

We thus have two poles at $\omega = +ck$ and $\omega = -ck$ that lie right on the path of integration as is always the case in pure wave problems. To perform this integral using the residue theorem for $t > 0$, we must close the contour in the lower-half ω plane due to the factor $e^{-i\omega t}$. For $t < 0$, we want this integral to be zero per the problem statement. We therefore do not want any poles in the upper-half plane, so that when we close the contour in the upper-half plane as required when $t < 0$, we get zero. One might think to try the contour given in Fig. 11.17. The problem is that when we close the contour in the upper-half plane

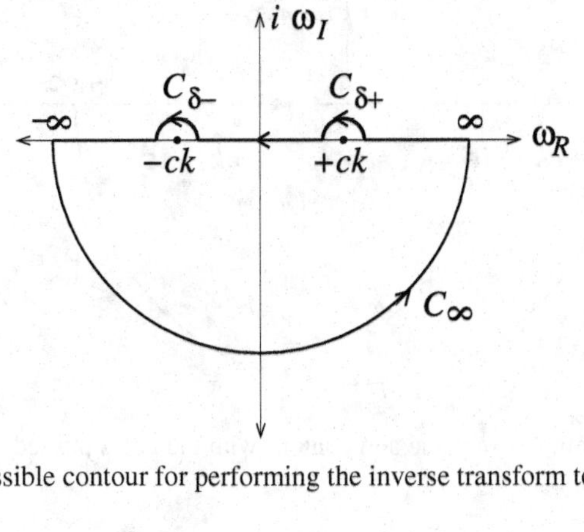

Figure 11.17 Possible contour for performing the inverse transform to the time domain.

for $t < 0$ using the same path along the real axis that puts the poles outside the closed contour, the two small half circles around the poles nonetheless contribute and we obtain nonzero response for $t < 0$.

Again, the way to most easily resolve this problem is to push the poles off the real ω axis by defining the wave equation with a small attenuation parameter ϵ

$$\frac{\partial^2 \psi(x,t)}{\partial x^2} - \frac{1}{c^2}\left(\frac{\partial}{\partial t} + \epsilon\right)^2 \psi(x,t) = \delta(x)\delta(t). \tag{11.237}$$

This corresponds to the replacement $\omega \to \omega + i\epsilon$ in the denominator of the integrand of Eq. (11.236) so that the inverse Fourier transform to the time domain is

$$\tilde{\psi}(k,t) = \lim_{\epsilon\to 0} \frac{c^2}{2\pi} \int_{-\infty}^{+\infty} \frac{e^{-i\omega t}\, d\omega}{(\omega + i\epsilon + ck)(\omega + i\epsilon - ck)}. \tag{11.238}$$

When $t > 0$, the closed contour in the lower-half plane with the displaced poles is depicted in Fig. 11.18. On the half circle at negative infinity, we have $e^{-i\omega t} \to 0$ when $t > 0$ and the integral of Eq. (11.238) is given by the residue theorem. When $t < 0$, a closed contour in the upper-half plane yields $\tilde{\psi}(k,t) = 0$ and we are good there.

The line integral on the real axis given in Fig. 11.18 is from $+\infty$ to $-\infty$ while our inverse transform of Eq. (11.238) is in the opposite direction which accounts for the negative sign in the result

$$\begin{aligned}\tilde{\psi}(k,t) &= \frac{c^2}{2\pi} \int_{-\infty}^{+\infty} \frac{e^{-i\omega t}\, d\omega}{(\omega + i\epsilon + ck)(\omega + i\epsilon - ck)} \\ &= -2\pi i \left[a_{-1}(ck - i\epsilon) + a_{-1}(-ck - i\epsilon)\right]\end{aligned} \tag{11.239}$$

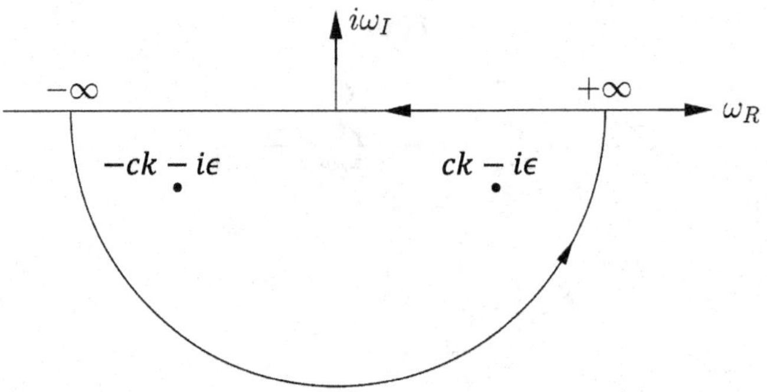

Figure 11.18 The complex ω plane and contour with the poles pushed off the real axis by an amount $-i\epsilon$.

$$= -2\pi i \frac{c^2}{2\pi}\left[\frac{-e^{+ickt}+e^{-ickt}}{2ck}\right]e^{-\epsilon t} \tag{11.240}$$

$$= -\frac{c}{k}\sin(ckt)e^{-\epsilon t}. \tag{11.241}$$

Thus we again see that the effect of the parameter ϵ is to create an attenuation to the wave propagation. We can set $\epsilon = 0$ at this point but will keep it around until the end.

Next, we return to the space domain

$$\psi(x,t) = \left[-\frac{c}{2\pi}\int_{-\infty}^{+\infty} \frac{e^{ikx}\sin ckt}{k}\,dk\right]e^{-\epsilon t}. \tag{11.242}$$

Now, just like $\partial/\partial x \to ik$, we also have $\int_{-\infty}^{x} dx_0 \to (ik)^{-1}$ so that

$$\psi(x,t) = \left[-\frac{c}{2\pi}\int_{-\infty}^{x} dx_0 \int_{-\infty}^{+\infty} i\,e^{ikx_0}\frac{\left(e^{ictk}-e^{-ictk}\right)}{2i}\,dk\right]e^{-\epsilon t} \tag{11.243}$$

$$= \left[-\frac{c}{4\pi}\int_{-\infty}^{x} dx_0 \int_{-\infty}^{+\infty} \left(e^{i(x_0+ct)k}-e^{i(x_0-ct)k}\right)dk\right]e^{-\epsilon t} \tag{11.244}$$

$$= \left[-\frac{c}{4\pi}\int_{-\infty}^{x} dx_0\, 2\pi\left[\delta\left(x_0+ct\right)-\delta\left(x_0-ct\right)\right]\right]e^{-\epsilon t} \tag{11.245}$$

which then results in the solution

$$\psi(x,t) = -\frac{c}{2}\Big[S(x+ct)-S(x-ct)\Big]e^{-\epsilon t}, \tag{11.246}$$

where the unit step function is defined

$$S(u) = \int_{-\infty}^{u}\delta(w)\,dw = \begin{cases} 1 & \text{if } u > 0 \\ 0 & \text{if } u < 0 \end{cases}. \tag{11.247}$$

We can now take $\epsilon = 0$ to obtain the solution to the problem as initially stated. This solution is thus an expanding boxcar function as depicted in Fig. 11.19. Note that $S(u)$ is dimensionless for whatever its argument u, going through the jump from 0 to 1 (unitless) when $u = 0$. With $\epsilon = 0$, we can therefore write this expanding boxcar-function solution as

$$\boxed{\psi(x,t) = -\frac{c}{2}S\left(t-\frac{|x|}{c}\right),} \tag{11.248}$$

which is a more compact statement that is more easily remembered. The physical units of the solution are indeed velocity as required.

This problem defines the Green's function for waves on a string (i.e., in 1D) when the impulse generating the waves occurs at time $t = 0$ (no waves present on the string prior to

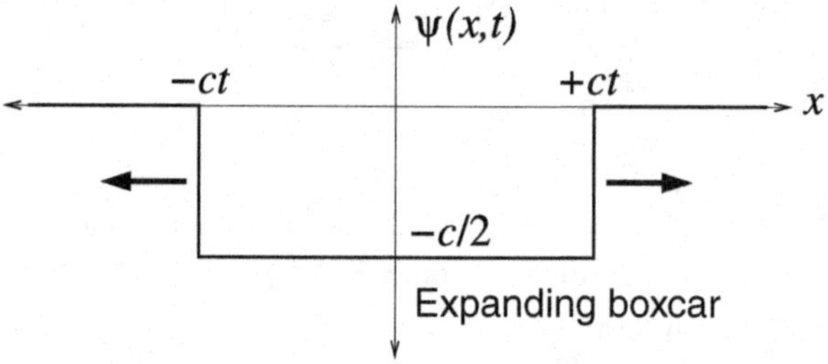

Figure 11.19 The solution of the 1D wave equation for a temporal and spatial impulse (Dirac) applied at $t=0$ and $x=0$; this is the Green's function for the 1D wave equation.

this impulse) and for waves radiating outward from the source located at the origin $x=0$. In Chapter 12, we will obtain the Green's function for the wave equation in 3D.

11.7.3 A Diffusion Example

Let's consider the particular diffusion equation

$$D\frac{\partial^2 P(x,t)}{\partial x^2} - \frac{\partial P(x,t)}{\partial t} = A\delta(x)\cos(bt), \tag{11.249}$$

which has a source term of identical form to the wave problem treated above as Wave Example 1. As in the wave problem, we assume that the sinusoidal injection term $\cos(bt)$ at $x=0$ has been operating for a very long time so initial conditions are not required. The boundary condition is that $P \to 0$ as $|x| \to \infty$.

We solve this problem by taking the Fourier transform over x and t but for this diffusion problem, we will return to the time domain first and then to the space domain second. In performing the inverse transform to the x domain, contour-integration methods are required. The Fourier transform of the governing equation gives

$$(-k^2 D + i\omega)\tilde{\tilde{P}}(k,\omega) = A\pi\,[\delta(\omega+b) + \delta(\omega-b)] \tag{11.250}$$

or

$$\tilde{\tilde{P}}(k,\omega) = \frac{A\pi\,[\delta(\omega+b)+\delta(\omega-b)]}{i\left(\omega + iDk^2\right)}. \tag{11.251}$$

As always, obtaining the solution in the double-transformed domain is almost trivial.

Returning to the time domain using the inverse Fourier transform gives

$$\tilde{P}(k,t) = \frac{1}{2\pi}\int_{-\infty}^{\infty} \frac{A\pi\,[\delta(\omega+b)+\delta(\omega-b)]}{i\left(\omega+iDk^2\right)} e^{-i\omega t}\, d\omega. \tag{11.252}$$

This integral is easy due to the two Diracs with the result

$$\tilde{P}(k,t) = -\frac{A}{2D}\left[\frac{e^{ibt}}{k^2 + ib/D} + \frac{e^{-ibt}}{k^2 - ib/D}\right]. \tag{11.253}$$

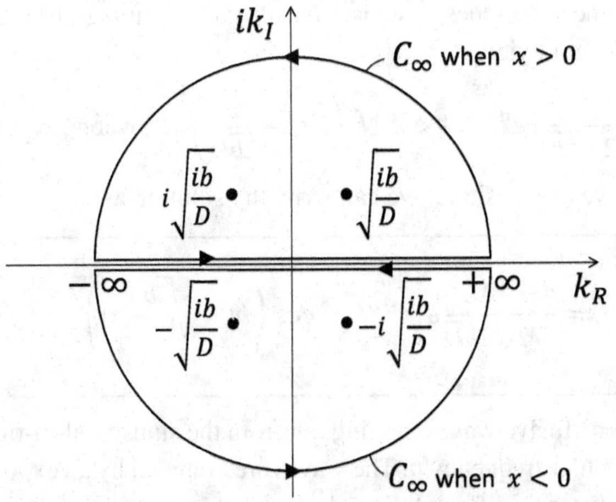

Figure 11.20 The position of the four poles involved in the inverse transform to the space domain of the given diffusion problem involving sinusoidal injection of a diffusing quantity at a temporal frequency b and at spatial position $x = 0$.

We next return to the x domain by performing the inverse Fourier transform over k

$$P(x, t) = -\frac{A}{4\pi D}\int_{-\infty}^{\infty}\left[\frac{e^{ibt}}{(k + i\sqrt{ib/D})(k - i\sqrt{ib/D})} + \frac{e^{-ibt}}{(k + \sqrt{ib/D})(k - \sqrt{ib/D})}\right] e^{ikx}\, dk. \tag{11.254}$$

The position of the four simple (first order) poles is shown in Fig. 11.20. Because of the common factor e^{ikx}, we must close the contour with a half circle C_∞ in the upper-half complex k plane when $x > 0$ and with a half circle in the lower-half complex k plane when $x < 0$.

When $x > 0$, the two integrals contributing to $P(x, t)$ are each solved with the residue theorem giving the result

$$P(x, t) = 2\pi i a_{-1}\left(i\sqrt{\frac{ib}{D}}\right) + 2\pi a_{-1}\left(\sqrt{\frac{ib}{D}}\right), \tag{11.255}$$

where the first residue is due to the simple pole involved in the first integral of Eq. (11.254) and the second residue is due to the simple pole involved in the second integral. Using $\sqrt{i} = (1 + i)/\sqrt{2}$ in the calculation of these residues and doing some straightforward algebra yields the purely real result

$$P(x, t) = -\frac{A}{2\sqrt{2bD}} e^{-xb/\sqrt{2bD}} \cos\left[b\left(t - \frac{x}{\sqrt{2bD}}\right)\right] \quad \text{when} \quad x > 0. \tag{11.256}$$

When $x < 0$, we need to enclose the poles in the lower-half complex k plane so that

$$P(x, t) = -2\pi i a_{-1}\left(-i\sqrt{\frac{ib}{D}}\right) - 2\pi a_{-1}\left(-\sqrt{\frac{ib}{D}}\right). \tag{11.257}$$

Again, calculating these residues associated with the two integrals in Eq. (11.254) and going through the algebra yields

$$P(x,t) = -\frac{A}{2\sqrt{2bD}} e^{xb/\sqrt{2bD}} \cos\left[b\left(t + \frac{x}{\sqrt{2bD}}\right)\right] \quad \text{when} \quad x < 0. \tag{11.258}$$

So for both x negative and positive, we can write the solution as

$$P(x,t) = -\frac{A}{2\sqrt{2bD}} e^{-|x|/\sqrt{2D/b}} \cos\left(bt - \sqrt{\frac{b}{2D}}|x|\right). \tag{11.259}$$

This corresponds to diffusive waves moving out from the sinusoidal-in-time injection point at $x = 0$ driven at circular frequency b. The waves are bounded by an exponentially decreasing envelope with distance diffused. The decay of this envelope is characterized by the skin depth $\delta = \sqrt{2D/b}$ which is common to all time-harmonic diffusion problems. This skin depth associated with the decay of the envelope is proportional to the wavelength as $\lambda = 2\pi\delta = 2\pi\sqrt{2D/b}$, which is another universal characteristic of diffusive processes with time-harmonic sourcing. You can compare this solution to that of Wave Example 1 to see how the time-harmonic response of a diffusion equation is distinct from that of a wave equation when the sourcing terms in each problem have identical form.

11.8 Exercises

1. Using contour-integration methods, show that

$$\int_{-\infty}^{+\infty} \frac{dx}{x^4 + a^4} = \frac{\pi}{\sqrt{2}a^3}.$$

HINTS: You need to find all the roots of the denominator in the integrand (these are the poles of the integrand). You do this by writing $z^4 = -a^4$ and then writing $-1 = e^{i\pi(2n+1)}$ where n is any integer. You find that there are four roots at $ae^{i\pi/4}$, $ae^{i3\pi/4}$, $ae^{i5\pi/4}$, and $ae^{i7\pi/4}$. It's probably a good idea to plot these points on the complex z plane (just use polar coordinates). Note that $z = ae^{i\theta} = a(\cos\theta + i\sin\theta)$.

You could then write the polynomial as

$$z^4 + a^4 = (z - z_1)(z - z_2)(z - z_3)(z - z_4) \tag{11.260}$$

using the above four roots/simple poles. For the poles that lie inside the contour you choose, you could then determine the residues of those poles using your formula for first-order poles. There is a bit of algebra that is not so bad and you are done.

It is a bit faster, however, to find the residues by expanding the integrand as a Laurent series and reading off the a_{-1} coefficient of the series for each simple pole. Earlier in the chapter, we did this for integrands of the form $p(z)/q(z)$ that only have simple poles z_i and showed that the residue for each simple pole is $a_{-1}(z_i) = p(z_i)/q'(z_i)$ where $q'(z_i) = \mathrm{d}q(z)/\mathrm{d}z|_{z=z_i}$. Using this formula for the residues of the simple poles inside the closed contour is the fastest way to perform this integral.

2. Using contour-integration methods, show that

$$\boxed{\int_0^\infty \frac{x^2\,\mathrm{d}x}{x^6+1} = \frac{\pi}{6}.}$$

HINTS: Proceed identically as in the previous problem. Once you find the roots of the denominator (i.e., simple first-order poles of the integrand) by setting $z_i^6 = -1 = e^{i\pi(2n+1)}$ and define an appropriate contour, you find the residues by again using the formula for integrands of the form $p(z)/q(z)$ that have simple poles, that is, $a_{-1}(z_i) = p(z_i)/q'(z_i)$. You then use the residue theorem for an appropriately defined closed contour.

3. Using contour-integration methods, show that

$$\boxed{\int_{-\infty}^\infty \frac{\mathrm{d}x}{(x^2+1)^3} = \frac{3\pi}{8}.}$$

HINTS: This is one of the easier results to obtain. To find the residue associated with the pole inside your contour, you can just use your formula for a_{-1} which will involve taking two derivatives since this involves third-order poles.

Alternatively, you can develop the Laurent series of the integrand by expanding around the poles $z = z_i$ defined from $z_i^2 + 1 = 0$. As practice, let's take this approach. Make the substitution $u = z - z_i$ to obtain

$$\frac{1}{(z^2+1)^3} = \frac{1}{(z_i^2 + 2z_i u + u^2 + 1)^3} \tag{11.261}$$

$$= \frac{1}{(2uz_i)^3[1 + u/(2z_i)]^3} \quad \text{by virtue of } z_i^2 + 1 = 0 \tag{11.262}$$

$$= \frac{1 - 3u/(2z_i) + 6[u/(2z_i)]^2 + \ldots}{8z_i^3 u^3} \tag{11.263}$$

$$= \frac{1}{8z_i^3 u^3} - \frac{3}{16z_i^4 u^2} + \frac{3}{16z_i^5 u} - \ldots \tag{11.264}$$

and then identify the residue as $a_{-1}(z_i) = 3/(16z_i^5)$. But obtaining this expansion is likely a bit more work than just using our residue formula for third-order poles.

4. Using contour-integration methods, show that

$$\int_0^{\infty} \frac{x \sin x \, dx}{x^2+1} = \frac{\pi}{2e}.$$

HINTS: You write $\sin z = (e^{iz} - e^{-iz})/(2i)$ which means this integral will actually involve two integrals. Now, for the integral involving e^{iz}, you will need to close the contour in the upper-half z plane so that this exponential does not blow up at imaginary infinity and for the integral involving e^{-iz} you will need to close it in the lower-half z plane. That is the one thing to be careful about.

5. Using contour-integration methods, show that

$$\int_{-\infty}^{\infty} \frac{\cos kx \, dx}{(x-a)^2+b^2} = \frac{\pi}{b} e^{-kb} \cos ka.$$

HINTS: A good place to start is to make the change of variables to the complex plane using $z = x - a$. Just like in the previous problem, you use the identity $\cos x = (e^{ix} + e^{-ix})/2$ along with $x = z + a$ which again gives two integrals to perform. Again, you will need to be careful to close each contour in the half of the z plane where $e^{\pm iz}$ does not blow up at imaginary infinity.

6. Using contour-integration methods, show that

$$\int_{\infty}^{\infty} \frac{e^{ax} \, dx}{1+e^{bx}} = \frac{\pi}{b \sin(\pi a/b)} \quad \text{where } |a| < |b|.$$

HINTS: In the previous problems, half-circle contours at infinity and Jordan's lemma were invoked. Here, you must use a rectangular contour like used earlier in the chapter. With $z = x + iy$, the four sides of the rectangle are $x = +R$, $x = -R$, $y = 0$, and $y = 2\pi/b$ in the limit $R \to \infty$.

7. An RL electric circuit consists of a resistor having resistance R in series with an inductor (a coil that converts electrical current into a magnetic field and that acquires a voltage drop across it when the current is changing in time) having an inductance L. A sharp voltage impulse is applied at time $t = 0$ and the resulting current in the circuit is controlled by the differential equation

$$L\frac{dI(t)}{dt} + RI(t) = W_0 \delta(t), \tag{11.265}$$

where W_0 has units of volt-seconds and $\delta(t)$ is the Dirac delta function. Using the Fourier transform and residue theorem, show that the current $I(t)$ excited by this impulse voltage is

$$I(t) = \frac{2\pi W_0}{L} e^{-Rt/L}. \tag{11.266}$$

8. For the time function controlling delayed response given by

$$M(t) = M_\infty \delta(t) + (M_0 - M_\infty)\omega_r^2 t e^{-\omega_r t} S(t), \tag{11.267}$$

where $S(t)$ is the unit step function and the relaxation frequency ω_r is a measured constant, the delayed linear-response constitutive law is

$$r(t) = \int_0^\infty M(u) f(t-u)\, du. \tag{11.268}$$

Calculate the complex modulus given by

$$\tilde{M}(\omega) = \int_{-\infty}^{\infty} M(t) e^{i\omega t}\, dt \tag{11.269}$$

and show that it has a real part given by

$$\tilde{M}_R(\omega) = M_\infty + \frac{(M_0 - M_\infty)\omega_r^2(\omega_r^2 - \omega^2)}{(\omega_r^2 + \omega^2)^2}. \tag{11.270}$$

Using this $\tilde{M}_R(\omega)$, calculate the Kramers–Krönig integral of Eq. (11.166) using contour-integration methods (i.e., define an appropriate closed contour and use the residue theorem) and show that this integral gives

$$\tilde{M}_I(\omega) = \frac{2(M_0 - M_\infty)\omega_r^3 \omega}{(\omega_r^2 + \omega^2)^2}. \tag{11.271}$$

Like for all relaxation or time-delay mechanisms, so long as $M_\infty < M_0$ (corresponding to the "normal" form of the constitutive law and not the inverse form), we have $\omega \tilde{M}_I(\omega) > 0$ for all frequencies.

9. Find the solution of the following wave problem:

$$\frac{\partial^2 u}{\partial x^2} - \frac{1}{c^2}\frac{\partial^2 u}{\partial t^2} = A e^{-a|x|}\sin(bt), \tag{11.272}$$

where $|x| = x$ if $x > 0$ and $|x| = -x$ if $x < 0$. The solution is to be valid and nonzero for all of $-\infty < x < \infty$ and $-\infty < t < \infty$ (i.e., the source term has been running for a very long time). The boundary conditions are that the waves are to be propagating outward (away from $x = 0$) as $|x| \to \infty$.

Use the following approach: (1) Take a Fourier transform over both x and t including determining the transforms of the source terms on the right-hand side to obtain the wave equation in the double transform domain (k, ω); (2) return to the space domain first, using the residue theorem and making sure to close the contour properly when $x < 0$ (the Jordan's lemma half circle must be in the lower complex half plane) and

$x > 0$ (the half circle must be in the upper complex half plane) and dealing with the poles on the real k axis as we did in Wave Examples 1 and 2 by introducing the real positive "attenuation" parameter ϵ; and (3) set $\epsilon = 0$ and return to the time domain to give the answer that is valid when $x > 0$ and the answer when $x < 0$. Show that for x either positive or negative:

$$\boxed{u(x,t) = \frac{A}{a^2 + b^2/c^2}\left[e^{-a|x|}\sin(bt) - \frac{ac}{b}\cos\left[b\left(t - \frac{|x|}{c}\right)\right]\right].} \tag{11.273}$$

10. Consider the diffusion equation

$$D\frac{\partial^2 \psi(x,t)}{\partial x^2} - \frac{\partial \psi(x,t)}{\partial t} = Be^{-a|x|}\sin(bt) \tag{11.274}$$

that has the same sourcing term as in the previous wave problem. Again assume that the sinusoidal injection term has been operating for a very long time so initial conditions are not required. The boundary condition is that $\psi \to 0$ as $|x| \to \infty$.

Solve this problem by taking a Fourier transform over x and t, then returning first to the time domain and then to the spatial domain. When returning to the spatial domain, you again must be careful to consider the cases $x > 0$ and $x < 0$ separately (will depend whether the half circle at infinity C_∞ is in the upper-half or lower-half complex k plane). Show that for x either positive or negative:

$$\boxed{\begin{aligned}\psi(x,t) &= \frac{Be^{-a|x|}}{D\left(a^4 + b^2/D^2\right)}\left[a^2\sin(bt) - \frac{b}{D}\cos(bt)\right] \\ &\quad - \frac{aBe^{-|x|\sqrt{b/(2D)}}}{D\sqrt{b/(2D)}\left(a^4 + b^2/D^2\right)}\left[\left(a^2 + \frac{b}{D}\right)\sin\left(bt - \sqrt{\frac{b}{2D}}|x|\right)\right. \\ &\quad \left. - \left(a^2 - \frac{b}{D}\right)\cos\left(bt - \sqrt{\frac{b}{2D}}|x|\right)\right].\end{aligned}}$$

Comment about how the solution of this time-harmonic diffusion problem is distinct from the solution of the previous time-harmonic wave problem.

12
Green's Functions

A *Green's function* is defined as a solution of a partial-differential equation (PDE) that has an inhomogeneous source term in the form of Dirac delta functions in both space and time (i.e., a point source that is also a sharp impulse in time) and that satisfies appropriate homogeneous boundary conditions on the surface surrounding the domain of interest. The Green's function or *impulse-response* function predicts the reaction of a system that is "pinged" at a point. The greatest use of the Green's function comes for linear physics problems where we can obtain the response of a system having an extended source distribution (either in the form of a distributed source term in the PDE itself or in the form of inhomogeneous boundary and/or initial conditions) by convolving the distributed source terms over space and time with the appropriate Green's function. Using the Green's function in this way for problems involving linear physics is called the *Green's function method*. We will begin by showing how the Green's function method works for scalar wave problems. We then perform an equivalent analysis for the scalar diffusion equation. Finally, we go full circle in the book and show how to use and determine Green's tensors for the vectorial response of elastodynamics, viscous flow, and electromagnetics that were the focus of Part I of the book.

12.1 Green's Function and the Scalar Wave Equation

12.1.1 The "Green's Function Method" Applied to the Wave Equation

We begin by using the scalar wave equation to illustrate the ideas surrounding how to use and define a Green's function. We pose a problem within a finite domain Ω having an inhomogeneous source term in the PDE, inhomogeneous boundary conditions on $\partial\Omega$ and inhomogeneous initial conditions at $t=0$. The source term in the PDE and the boundary conditions are possibly time dependent.

Stated mathematically, we want to obtain the wavefield $\psi(\boldsymbol{r}, t)$ that satisfies

$$\nabla^2\psi(\boldsymbol{r}, t) - \frac{1}{c^2}\frac{\partial^2\psi(\boldsymbol{r}, t)}{\partial t^2} = f(\boldsymbol{r}, t), \tag{12.1}$$

where $f(\boldsymbol{r}, t)$ is a given source term throughout Ω and where $\psi(\boldsymbol{r}, t)$ is subject to the boundary condition that either

$$\psi(\boldsymbol{r}, t) = b_0(\boldsymbol{r}, t) \quad \text{for } \boldsymbol{r} \text{ on } \partial\Omega \tag{12.2}$$

or

$$\boldsymbol{n} \cdot \nabla \psi(\boldsymbol{r}, t) = b_1(\boldsymbol{r}, t) \quad \text{for } \boldsymbol{r} \text{ on } \partial\Omega. \tag{12.3}$$

Additionally, to obtain a unique solution, we also have to satisfy the two initial conditions

$$\psi(\boldsymbol{r}, 0) = T_0(\boldsymbol{r}) \tag{12.4}$$

and

$$\left.\frac{\partial \psi(\boldsymbol{r}, t)}{\partial t}\right|_{t=0} = T_1(\boldsymbol{r}). \tag{12.5}$$

Corresponding to this problem we define the so-called Green's function $G(\boldsymbol{r}, t|\boldsymbol{r}', t')$ as obeying the differential equation

$$\nabla^2 G(\boldsymbol{r}, t|\boldsymbol{r}', t') - \frac{1}{c^2}\frac{\partial^2 G(\boldsymbol{r}, t|\boldsymbol{r}', t)}{\partial t^2} = \delta(\boldsymbol{r} - \boldsymbol{r}')\,\delta(t - t') \tag{12.6}$$

subject to the boundary condition that either

$$G(\boldsymbol{r}, t|\boldsymbol{r}', t') = 0 \quad \text{for } \boldsymbol{r} \text{ on } \partial\Omega \tag{12.7}$$

or

$$\boldsymbol{n} \cdot \nabla G(\boldsymbol{r}, t|\boldsymbol{r}', t') = 0 \quad \text{for } \boldsymbol{r} \text{ on } \partial\Omega. \tag{12.8}$$

Whatever BC the field ψ satisfies, the Green's function G satisfies the homogeneous version of the same BC. The Green's function is also subject to the initial condition that

$$G(\boldsymbol{r}, t|\boldsymbol{r}', t') = 0 \quad \text{for } t - t' < 0. \tag{12.9}$$

Physically, $G(\boldsymbol{r}, t\,|\,\boldsymbol{r}', t')$ represents the response (wave response in this case) at an observation point $\boldsymbol{r}, t$ due to a sharp impulse applied at the point $\boldsymbol{r}', t'$.

Before showing how to use G to obtain the solution for ψ, we need to establish the following symmetry property of the Green's function

$$\boxed{G(\boldsymbol{r}, t\,|\,\boldsymbol{r}', t') = G(\boldsymbol{r}', -t'\,|\,\boldsymbol{r}, -t),} \tag{12.10}$$

which says that if you run time backwards and interchange the source and receiver positions, you get the same response. This symmetry condition is sometimes referred to as the *reciprocity* of the Green's function. For the special case of the Green's function in an infinite homogeneous body, we can write the time and space dependencies as $G(|\boldsymbol{r} - \boldsymbol{r}'|, t - t')$, which is seen to satisfy the general reciprocity of Eq. (12.10). But for problems with finite boundaries and either homogeneous or inhomogeneous boundary conditions, we must write the dependencies on the observation and source coordinates in the more general form $G(\boldsymbol{r}, t|\boldsymbol{r}', t')$, which always satisfies the reciprocity of Eq. (12.10).

To establish this symmetry, we first make the substitutions $\boldsymbol{r}' \to \boldsymbol{r}''$, $t \to -t$, and $t' \to -t''$ in Eq. (12.6) to obtain the PDE

$$\nabla^2 G(\boldsymbol{r}, -t \,|\, \boldsymbol{r}'', -t'') - \frac{1}{c^2}\frac{\partial^2 G(\boldsymbol{r}, -t \,|\, \boldsymbol{r}'', -t'')}{\partial t^2} = \delta(\boldsymbol{r} - \boldsymbol{r}'')\,\delta(t - t''), \tag{12.11}$$

where because $\delta(u) = \delta(-u)$ we used that $\delta(t'' - t) = \delta(t - t'')$. If we multiply Eq. (12.6) by $G(\boldsymbol{r}, -t \,|\, \boldsymbol{r}'', -t'')$ and multiply Eq. (12.11) by $G(\boldsymbol{r}, t \,|\, \boldsymbol{r}', t')$, subtract, and integrate overall all time and all of Ω we obtain

$$\begin{aligned} &\int_{-\infty}^{\infty} \mathrm{d}t \int_{\Omega} \mathrm{d}^3\boldsymbol{r} \left\{ G(\boldsymbol{r}, t \,|\, \boldsymbol{r}', t')\nabla^2 G(\boldsymbol{r}, -t \,|\, \boldsymbol{r}'', -t'') - G(\boldsymbol{r}, -t \,|\, \boldsymbol{r}'', -t'')\nabla^2 G(\boldsymbol{r}, t \,|\, \boldsymbol{r}', t') \right. \\ &\quad \left. - \frac{1}{c^2}\frac{\partial}{\partial t}\left[G(\boldsymbol{r}, t \,|\, \boldsymbol{r}', t')\frac{\partial G(\boldsymbol{r}, -t \,|\, \boldsymbol{r}'', -t'')}{\partial t} - G(\boldsymbol{r}, -t \,|\, \boldsymbol{r}'', -t'')\frac{\partial G(\boldsymbol{r}, t \,|\, \boldsymbol{r}', t')}{\partial t} \right] \right\} \\ &= G(\boldsymbol{r}'', t'' \,|\, \boldsymbol{r}', t') - G(\boldsymbol{r}', -t' \,|\, \boldsymbol{r}'', -t''). \end{aligned} \tag{12.12}$$

Next, we use the identity that $\nabla \cdot (G_1 \nabla G_2 - G_2 \nabla G_1) = G_1 \nabla^2 G_2 - G_2 \nabla^2 G_1$ and the divergence theorem to rewrite the above as

$$\begin{aligned} &G(\boldsymbol{r}'', t'' \,|\, \boldsymbol{r}', t') - G(\boldsymbol{r}', -t' \,|\, \boldsymbol{r}'', -t'') \\ &= \int_{-\infty}^{\infty} \mathrm{d}t \int_{\partial\Omega} \mathrm{d}^2\boldsymbol{r}\, \boldsymbol{n} \cdot \left[G(\boldsymbol{r}, t \,|\, \boldsymbol{r}', t')\nabla G(\boldsymbol{r}, -t \,|\, \boldsymbol{r}'', -t'') - G(\boldsymbol{r}, -t \,|\, \boldsymbol{r}'', -t'')\nabla G(\boldsymbol{r}, t \,|\, \boldsymbol{r}', t') \right] \\ &\quad - \frac{1}{c^2}\int_{\Omega} \mathrm{d}^3\boldsymbol{r} \left[G(\boldsymbol{r}, t \,|\, \boldsymbol{r}', t')\frac{\partial G(\boldsymbol{r}, -t \,|\, \boldsymbol{r}'', -t'')}{\partial t} - G(\boldsymbol{r}, -t \,|\, \boldsymbol{r}'', -t'')\frac{\partial G(\boldsymbol{r}, t \,|\, \boldsymbol{r}', t')}{\partial t} \right]_{t=-\infty}^{t=\infty}. \end{aligned} \tag{12.13}$$

The homogeneous BC that G satisfies on $\partial\Omega$ puts the surface integral to zero while the initial condition that $G(\boldsymbol{r}, t = -\infty \,|\, \boldsymbol{r}', t') = 0$ puts the final integral over Ω to zero. Thus, with the right-hand side zero, upon substituting $\boldsymbol{r}'' \to \boldsymbol{r}$ and $t'' \to t$, we obtain the symmetry condition we set out to prove $G(\boldsymbol{r}, t \,|\, \boldsymbol{r}', t') = G(\boldsymbol{r}', -t' \,|\, \boldsymbol{r}, -t)$.

We now use this symmetry property to rewrite the PDE for G. In Eq. (12.6), we make the exchange $\boldsymbol{r} \leftrightarrow \boldsymbol{r}'$ and $t \leftrightarrow -t'$ and use the symmetry property just derived to obtain (note that the derivatives are now acting on the source position coordinates due to this symmetry)

$$\nabla'^2 G(\boldsymbol{r}, t \,|\, \boldsymbol{r}', t') - \frac{1}{c^2}\frac{\partial^2 G(\boldsymbol{r}, t \,|\, \boldsymbol{r}', t')}{\partial t'^2} = \delta(\boldsymbol{r} - \boldsymbol{r}')\delta(t - t'). \tag{12.14}$$

Let's also rewrite the PDE for ψ by making the change of variables $\boldsymbol{r} \to \boldsymbol{r}'$ and $t \to t'$

$$\nabla'^2 \psi(\boldsymbol{r}', t') - \frac{1}{c^2}\frac{\partial^2 \psi(\boldsymbol{r}', t')}{\partial t'^2} = f(\boldsymbol{r}', t'). \tag{12.15}$$

We multiply the equation for ψ by G and the equation for G by ψ and subtract to obtain

$$\nabla' \cdot \left[G\nabla'\psi - \psi\nabla' G \right] - \frac{1}{c^2}\frac{\partial}{\partial t'}\left[G\frac{\partial \psi}{\partial t'} - \psi\frac{\partial G}{\partial t'} \right] = Gf(\boldsymbol{r}', t') - \psi\,\delta(\boldsymbol{r} - \boldsymbol{r}')\,\delta(t - t'). \tag{12.16}$$

Next, we integrate over t' from the initial condition at $t' = 0$ to $t' = \infty$ and integrate over all of Ω and use the divergence theorem to obtain

$$\int_0^\infty \mathrm{d}t' \int_{\partial\Omega} \boldsymbol{n}\cdot\left[G\nabla'\psi - \psi\nabla' G\right]\mathrm{d}^2\boldsymbol{r}' - \frac{1}{c^2}\int_\Omega \mathrm{d}^3\boldsymbol{r}'\left[G\frac{\partial\psi}{\partial t'} - \psi\frac{\partial G}{\partial t'}\right]_{t'=0}^{t'=\infty}$$
$$= \int_0^\infty\int_\Omega G(\boldsymbol{r},t\,|\,\boldsymbol{r}',t')f(\boldsymbol{r}',t')\,d^3\boldsymbol{r}'\,\mathrm{d}t' - \int_0^\infty\int_\Omega \psi(\boldsymbol{r}',t')\,\delta(\boldsymbol{r}'-\boldsymbol{r})\,\delta(t'-t)\,\mathrm{d}^3\boldsymbol{r}'\,\mathrm{d}t'. \tag{12.17}$$

From the fact that $G = 0$ for $t' > t$ (which means that G and its time derivatives are all zero at $t' = \infty$ and that there is no need to integrate beyond $t' = t$), we obtain at last

$$\begin{aligned}\psi(\boldsymbol{r},t) = &\int_0^t\int_\Omega G(\boldsymbol{r},t\,|\,\boldsymbol{r}',t')f(\boldsymbol{r}',t')\,d^3\boldsymbol{r}'\,\mathrm{d}t'\\ &-\int_0^t \mathrm{d}t'\int_{\partial\Omega}\boldsymbol{n}\cdot\Big[G(\boldsymbol{r},t\,|\,\boldsymbol{r}',t')\nabla'\psi(\boldsymbol{r}',t')\\ &-\psi(\boldsymbol{r}',t')\nabla' G(\boldsymbol{r},t\,|\,\boldsymbol{r}',t')\Big]\mathrm{d}^2\boldsymbol{r}'\\ &-\frac{1}{c^2}\int_\Omega \mathrm{d}^3\boldsymbol{r}'\Big[G(\boldsymbol{r},t\,|\,\boldsymbol{r}',t')\big|_{t'=0}\left.\frac{\partial\psi(\boldsymbol{r}',t')}{\partial t'}\right|_{t'=0}\\ &-\psi(\boldsymbol{r}',t')\big|_{t'=0}\left.\frac{\partial G(\boldsymbol{r},t\,|\,\boldsymbol{r}',t')}{\partial t'}\right|_{t'=0}\Big].\end{aligned} \tag{12.18}$$

This expression shows how the Green's function allows us to obtain the solution of the problem for the wavefield ψ. It is sometimes called the *representation theorem* for the wavefield ψ. The various sources for the field ψ are any inhomogeneous (nonzero) contributions to the source term in the PDE, in the BCs, or in the initial conditions. The Green's function produces a wave response at $(\boldsymbol{r}, t)$ from a stimulation at $(\boldsymbol{r}', t')$. Through linear superposition, the total wavefield in Eq. (12.18) is just the sum from all the points of stimulation either inside the domain or on the boundary.

So, for example, if the problem for ψ had an inhomogeneous source term in the PDE but homogeneous boundary and initial conditions, then only the first term (first line) would contribute

$$\psi(\boldsymbol{r},t) = \int_0^t\int_\Omega G(\boldsymbol{r},t\,|\,\boldsymbol{r}',t')f(\boldsymbol{r}',t')\,d^3\boldsymbol{r}'\,\mathrm{d}t'. \tag{12.19}$$

If we had a homogeneous PDE for ψ, an inhomogeneous Dirichlet condition on $\partial\Omega$ given by $\psi(\boldsymbol{r}, t) = b_0(\boldsymbol{r}, t)$ (so that we choose the BC that $G = 0$ on $\partial\Omega$) and a homogeneous initial condition, we then have

$$\psi(\boldsymbol{r}, t) = \int_0^t \mathrm{d}t' \int_{\partial\Omega} b_0(\boldsymbol{r}', t')\, \boldsymbol{n}(\boldsymbol{r}') \cdot \nabla' G(\boldsymbol{r}, t \,|\, \boldsymbol{r}', t')\, \mathrm{d}^2\boldsymbol{r}'. \tag{12.20}$$

If we had a homogeneous PDE for ψ, an inhomogeneous Neumann condition on $\partial\Omega$ given by $\boldsymbol{n} \cdot \nabla\psi(\boldsymbol{r}, t) = b_1(\boldsymbol{r}, t)$ (so that we choose the BC that $\boldsymbol{n} \cdot \nabla G = 0$ on $\partial\Omega$) and a homogeneous initial condition, we then have

$$\psi(\boldsymbol{r}, t) = -\int_0^t \mathrm{d}t' \int_{\partial\Omega} b_1(\boldsymbol{r}', t')\, G(\boldsymbol{r}, t \,|\, \boldsymbol{r}', t')\, \mathrm{d}^2\boldsymbol{r}'. \tag{12.21}$$

Finally, if we had a homogeneous PDE for ψ, a homogeneous Dirichlet or Neumann condition on $\partial\Omega$ for both ψ and G, and an inhomogeneous initial condition given by $\psi(\boldsymbol{r}, t = 0) = T_0(\boldsymbol{r})$ and $\partial\psi/\partial t|_{t=0} = T_1(\boldsymbol{r})$, we obtain

$$\psi(\boldsymbol{r}, t) = -\frac{1}{c^2} \int_\Omega \mathrm{d}^3\boldsymbol{r}' \left[G(\boldsymbol{r}, t \,|\, \boldsymbol{r}', t')\Big|_{t'=0} T_1(\boldsymbol{r}') - \left.\frac{\partial G(\boldsymbol{r}, t \,|\, \boldsymbol{r}', t')}{\partial t'}\right|_{t'=0} T_0(\boldsymbol{r}') \right]. \tag{12.22}$$

We could have any combination of these various ways to generate nonzero wave response for $\psi(\boldsymbol{r}, t)$, that is an inhomogeneous PDE and/or an inhomogeneous BC and/or an inhomogeneous initial condition.

The above defines the *Green's function method* for finding the solution for the wavefield $\psi(\boldsymbol{r}, t)$. The equivalent holds for any linear physics (e.g., diffusion and potential problems).

12.1.2 The Wave Equation Green's Function in a Wholespace

We now obtain the unique, explicit expression for the Green's function of the scalar wave equation as defined above in Eqs (12.6) and (12.9) in an unbounded domain Ω_∞. The wave response is propagating outward from the point source located at $\boldsymbol{r}'$ and falls to zero as $|\boldsymbol{r} - \boldsymbol{r}'| \to \infty$. For convenience in what follows, we can set $\boldsymbol{r}' = 0$ and $t' = 0$, solve for $G(\boldsymbol{r}, t)$ and then simply make the change of variables $\boldsymbol{r} \to \boldsymbol{r} - \boldsymbol{r}'$ and $t \to t - t'$ at the end. This Green's function satisfies the homogeneous boundary conditions that $G(\boldsymbol{r}, t) = 0$ and $\boldsymbol{n} \cdot \nabla G(\boldsymbol{r}, t) = 0$ on $\partial\Omega_\infty$.

To obtain $G(\boldsymbol{r}, t)$, begin by taking a 4D Fourier transform over space and time and noting as usual that

$$G(\boldsymbol{r}, t) \to \tilde{\tilde{G}}(\boldsymbol{k}, \omega), \quad \frac{\partial}{\partial t} \to -i\omega, \quad \nabla \to i\boldsymbol{k}, \quad \delta(t - t') \to 1, \quad \text{and} \quad \delta(\boldsymbol{r} - \boldsymbol{r}') \to 1$$

so that the PDE for G transforms to

$$\left(-k^2 + \frac{\omega^2}{c^2}\right) \tilde{\tilde{G}}(\boldsymbol{k}, \omega) = 1, \tag{12.23}$$

which gives the transformed result

$$\tilde{\tilde{G}}(\boldsymbol{k}, \omega) = \frac{1}{-k^2 + \omega^2/c^2}. \tag{12.24}$$

As always, it is almost trivial to obtain the solution of a PDE in the Fourier domain and the mathematical effort is in performing the inverse transforms in returning to the space and time domain.

We first return to the space domain using the inverse transform over the $\boldsymbol{k}$ coordinates

$$\tilde{G}(\boldsymbol{r},\omega)=\frac{1}{(2\pi)^3}\int \mathrm{d}^3\boldsymbol{k}\, e^{i\boldsymbol{k}\cdot\boldsymbol{r}}\frac{1}{\left(-k^2+\omega^2/c^2\right)}. \tag{12.25}$$

To do the three integrals over all of $\boldsymbol{k}$ space, we work in spherical $\boldsymbol{k}$ coordinates (k,θ,ϕ) with the latitudinal angle θ (where $0\le\theta\le\pi$) is defined to be measured from the direction of $\boldsymbol{r}$:

$$\hat{\boldsymbol{k}}\cdot\hat{\boldsymbol{r}}=\cos\theta, \tag{12.26}$$

$$\mathrm{d}^3\boldsymbol{k}=k^2\,\mathrm{d}k\,\sin\theta\,\mathrm{d}\theta\,\mathrm{d}\phi=-k^2\,\mathrm{d}k\,\mathrm{d}(\cos\theta)\,\mathrm{d}\phi. \tag{12.27}$$

Thus, we obtain three integrals to perform with the integrals over ϕ and $\cos\theta$ (that you can rename $u=\cos\theta$ if you prefer) performed very easily

$$\tilde{G}(\boldsymbol{r},\omega)=\frac{-1}{(2\pi)^3}\int_0^\infty \frac{\mathrm{d}k\,k^2}{\left(k^2-\omega^2/c^2\right)}\int_{-1}^{+1}\mathrm{d}(\cos\theta)\,e^{ikr\cos\theta}\int_0^{2\pi}\mathrm{d}\phi \tag{12.28}$$

$$=\frac{-1}{(2\pi)^2\,ir}\int_0^\infty \frac{\mathrm{d}k\,k}{\left(k^2-\omega^2/c^2\right)}\left(e^{ikr}-e^{-ikr}\right) \tag{12.29}$$

$$=\frac{-1}{(2\pi)^2\,ir}\int_{-\infty}^{+\infty}\frac{\mathrm{d}k\,k\,e^{ikr}}{(k+\omega/c)\,(k-\omega/c)}. \tag{12.30}$$

In the first integral, the integral from $\cos 0=1$ to $\cos\pi=-1$ was rewritten as an integral from -1 to 1 by using the minus sign present in Eq. (12.27). In the second integral involving e^{-ikr} of the second line, we replaced $-k$ with $+k$ as the integration variable which makes that integral go from $-\infty$ to 0 so the two integrals can be combined into a single integral from $-\infty$ to $+\infty$ as displayed in the third line.

As always in pure wave problems, there are poles on the real-k axis of integration and we must specify a way to integrate around them. In order to obtain outward-radiating waves, we again introduce the positive real attenuation parameter ϵ into the defining PDE for the Green's function

$$\nabla^2 G-\frac{1}{c^2}\left(\frac{\partial}{\partial t}+\epsilon\right)^2 G=\delta(\boldsymbol{r})\delta(t), \tag{12.31}$$

to obtain $\tilde{G}(\boldsymbol{r},\omega)$ in the form

$$\tilde{G}(\boldsymbol{r},\omega)=\frac{-1}{(2\pi)^2\,ir}\int_{-\infty}^{+\infty}\frac{\mathrm{d}k\,k\,e^{ikr}}{(k+\omega/c+i\epsilon/c)\,(k-\omega/c-i\epsilon/c)}. \tag{12.32}$$

We can solve this integral and then take $\epsilon=0$ at the end in order to obtain outward-radiating wave response. The closed contour on the complex k plane used in the residue theorem is shown in Fig. 12.1. The residue theorem states that

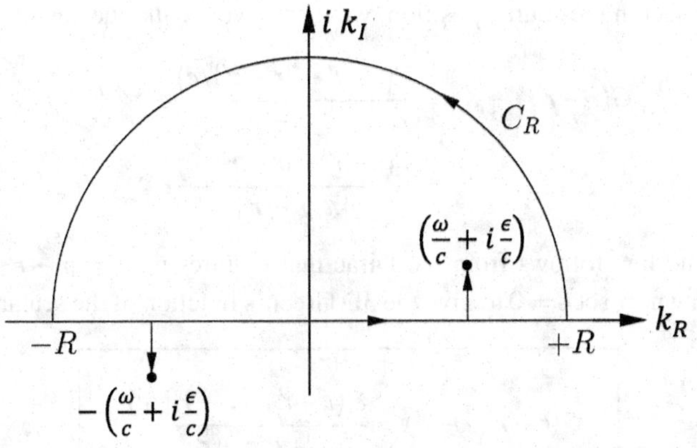

Figure 12.1 The contour with attenuation-displaced poles and as $R \to \infty$ that is used for performing the inverse k transform for the 3D Green's function of the wave equation.

$$\oint \frac{dk\, k\, e^{ikr}}{(k+\omega/c+i\epsilon/c)\,(k-\omega/c-i\epsilon/c)} = 2\pi i\, a_{-1}\,(\omega/c+i\epsilon/c) \tag{12.33}$$

while the closed contour integral is broken into pieces as

$$\oint \frac{dk\, k\, e^{ikr}}{(k+\omega/c+i\epsilon/c)\,(k-\omega/c-i\epsilon/c)} = -(2\pi)^2\, ir\tilde{G}(\mathbf{r},\omega) + \lim_{R\to\infty}\int_{C_R} \frac{dk\, k\, e^{ikr}}{(k+\omega/c+i\epsilon/c)\,(k-\omega/c-i\epsilon/c)}. \tag{12.34}$$

The integral over the half-circle at infinity is zero by Jordan's Lemma. The residue associated with the simple pole at $\omega/c + i\epsilon/c$ is determined as

$$a_{-1}\left(\frac{\omega}{c}+i\frac{\epsilon}{c}\right) = \lim_{k\to\omega/c+i\epsilon/c} \frac{k\, e^{ikr}\left[\cancel{k-(\omega/c+i\epsilon/c)}\right]}{\left[k+\ \omega/c+i\epsilon/c\right]\left[\cancel{k-(\omega/c+i\epsilon/c)}\right]} = \frac{e^{i(\omega/c+i\epsilon/c)r}}{2}, \tag{12.35}$$

which then gives the result

$$\tilde{G}(\mathbf{r},\omega) = \frac{e^{i\omega r/c}\, e^{-\epsilon r/c}}{4\pi r}. \tag{12.36}$$

We can now eliminate the attenuation parameter $\epsilon = 0$ but will keep it around until the very end.

Next, we return to the time domain using the completeness relation $(2\pi)^{-1}\int_{\infty}^{\infty} d\omega e^{-i\omega(t-r/c)} = \delta(t-r/c)$ or

$$G(\mathbf{r},t) = \frac{1}{2\pi}\int_{-\infty}^{+\infty} d\omega\, \frac{e^{-i\omega(t-r/c)}}{4\pi r}\, e^{-\epsilon r/c} = \frac{\delta\,(t-r/c)}{4\pi r}\, e^{-\epsilon r/c}. \tag{12.37}$$

Restoring the offset in the source position, we then have the unique answer that

$$G(\boldsymbol{r}-\boldsymbol{r}', t-t') = \frac{\delta\left(t-t'-|\boldsymbol{r}-\boldsymbol{r}'|/c\right)}{4\pi\,|\boldsymbol{r}-\boldsymbol{r}'|}e^{-\epsilon|\boldsymbol{r}-\boldsymbol{r}'|/c} \tag{12.38}$$

$$= \frac{\delta\left(t-t'-|\boldsymbol{r}-\boldsymbol{r}'|/c\right)}{4\pi\,|\boldsymbol{r}-\boldsymbol{r}'|}e^{-\epsilon(t-t')}, \tag{12.39}$$

where the second line follows from the Dirac that requires $t-t' = |\boldsymbol{r}-\boldsymbol{r}'|/c$ for nonzero response. We can now set $\epsilon = 0$ to give the 3D Green's function of the scalar wave equation

$$\boxed{G(\boldsymbol{r}-\boldsymbol{r}', t-t') = \frac{\delta\left(t-t'-|\boldsymbol{r}-\boldsymbol{r}'|/c\right)}{4\pi\,|\boldsymbol{r}-\boldsymbol{r}'|}.} \tag{12.40}$$

Thus we see that in an infinite homogeneous body, $G(\boldsymbol{r}, t|\boldsymbol{r}', t') = G(\boldsymbol{r}-\boldsymbol{r}', t-t') = G(|\boldsymbol{r}-\boldsymbol{r}'|, t-t')$ as conjectured earlier. An impulsive (Dirac) source of wave energy moves out into space as an impulsive (Dirac) wave with spherical symmetry about the source position $\boldsymbol{r}'$ and a $1/r$ fall off in wave amplitude with distance r from the source position. Wave energy is proportional to G^2 in any wave response, so the wave energy falls off as $1/r^2$ which is a statement of energy conservation, that is, since the area of a spherical surface is proportional to r^2 and the wave energy is proportional to $1/r^2$, the total wave energy passing through a spherical surface at any distance r from the source is always the same. Last, for $t-t' < 0$, the form of our Green's function satisfies $G = 0$ as required.

12.1.3 The Wave Equation Green's Function in 1D Revisited

We showed in Chapter 11 ("Wave Example 2" of Section 11.7.2) that the Green's function for the 1D wave equation

$$\frac{\partial^2 G(x-x', t-t')}{\partial x^2} - \frac{1}{c^2}\frac{\partial^2 G(x-x', t-t')}{\partial t^2} = \delta(x-x')\delta(t-t') \tag{12.41}$$

with $G = 0$ for $t-t' < 0$ and outward radiating waves as $|x-x'| \to \infty$ is given uniquely by

$$G(x-x', t-t') = -\frac{c}{2}S\left(t-t'-\frac{|x-x'|}{c}\right), \tag{12.42}$$

which is an expanding boxcar function with $S(u)$ the step function ($S = 0$ when $u < 0$ and $S = 1$ when $u > 0$). We obtained this result by taking a double Fourier transform over x and t and then returning to the space and time domain using the inverse Fourier transform.

Note that when we introduce the attenuation parameter ϵ in the scalar wave equation

$$\frac{\partial^2 G(x-x', t-t')}{\partial x^2} - \frac{1}{c^2}\left(\frac{\partial}{\partial t}+\epsilon\right)^2 G(x-x', t-t') = \delta(x-x')\delta(t-t'), \tag{12.43}$$

we showed in Chapter 11 that the Green's function is

$$G(x-x', t-t') = -\frac{c}{2} S\left(t - t' - \frac{|x-x'|}{c}\right) e^{-\epsilon(t-t')}. \tag{12.44}$$

As will be seen, it can be convenient to work with this Green's function prior to taking $\epsilon = 0$ in a final result.

In what follows we will use this example to demonstrate some general behavior of the Green's function at the origin. We will also use it to put the Green's function method to use.

Continuity Properties of the Green's Function across the Origin

For simplicity, let's first set $x' = 0$ and $t' = 0$ with no loss in generality (we can just make the substitution $x \to x - x'$ and $t \to t - t'$ later) to obtain

$$\frac{\partial^2 G(x,t)}{\partial x^2} - \frac{1}{c^2}\frac{\partial^2 G(x,t)}{\partial t^2} = \delta(x)\delta(t). \tag{12.45}$$

This equation requires that $G(x,t)$ is continuous across either $x=0$ or $t=0$, that is $G(-\varepsilon_x, t) = G(+\varepsilon_x, t)$ and $G(x, -\varepsilon_t) = G(x, +\varepsilon_t)$ where ε_x is an infinitesimal increment in x and ε_t is an infinitesimal increment in t. If these were not continuous across the origin, then the first derivative at the origin would contain a Dirac and the second derivative would contain terms of either $d\delta(x)/dx$ or $d\delta(t)/dt$ which is inconsistent with the differential equation for G that contains only $\delta(x)$ and $\delta(t)$. For the same reason, it must be that the first derivative contains a jump across the origin. For example, if we take a Fourier transform over x so that $\partial/\partial x \to ik$ we obtain

$$-k^2\tilde{G}(k,t) - \frac{1}{c^2}\frac{\partial^2 \tilde{G}(k,t)}{\partial t^2} = \delta(t). \tag{12.46}$$

Integrating this equation across the origin from $t = -\varepsilon_t$ to $t = +\varepsilon_t$ in the limit as $\varepsilon_t \to 0$ then gives (note that $\tilde{G}(k,t)$ is continuous across the origin as just discussed), we obtain

$$\left.\frac{\partial \tilde{G}(k,t)}{\partial t}\right|_{t=+\varepsilon_t} - \left.\frac{\partial \tilde{G}(k,t)}{\partial t}\right|_{t=-\varepsilon_t} = -c^2. \tag{12.47}$$

Transforming this jump condition back to the space domain then gives

$$\left.\frac{\partial G(x,t)}{\partial t}\right|_{t=+\varepsilon_t} - \left.\frac{\partial G(x,t)}{\partial t}\right|_{t=-\varepsilon_t} = -c^2\delta(x). \tag{12.48}$$

So the Green's function always has a jump in the first derivative at the origin. Similarly, if we had transformed over t so that $\partial/\partial t \to -i\omega$, we obtain the jump condition in x at the origin that

$$\left.\frac{\partial \tilde{G}(x,\omega)}{\partial x}\right|_{x=+\varepsilon_x} - \left.\frac{\partial \tilde{G}(x,\omega)}{\partial x}\right|_{x=-\varepsilon_x} = -1. \tag{12.49}$$

Transforming this jump condition back to the time domain gives

$$\left.\frac{\partial G(x,t)}{\partial x}\right|_{x=+\varepsilon_x} - \left.\frac{\partial G(x,t)}{\partial x}\right|_{x=-\varepsilon_x} = -\delta(t). \tag{12.50}$$

These facts about the behavior of the Green's function at the origin of either space or time (where the Diracs are nonzero) can be used to solve for the Green's function in a slightly different manner.

For example, after taking the Fourier transform over x we can solve Eq. (12.46) above for $t > 0$ as

$$\tilde{G}(k,t) = Ae^{ikct} + Be^{-ikct}. \tag{12.51}$$

If this is to also be the solution for $t \geq 0$, we can use the continuity of $\tilde{G}(k,t)$ across the origin $t=0$ to deduce that the $B=-A$ (because $\tilde{G}(k,-\varepsilon_t)=0$ from the initial condition) and then use the above jump condition in the first derivative to deduce that

$$\lim_{\varepsilon_t\to 0} \left.\frac{\partial \tilde{G}(k,t)}{\partial t}\right|_{x=-\varepsilon_t}^{x=+\varepsilon_t} = -c^2 = 2Aikc. \tag{12.52}$$

We thus have that $A = -c/(2ik)$ which then gives

$$\tilde{G}(k,t) = -\frac{c}{k}\sin(ckt), \tag{12.53}$$

which is exactly the result we obtained in Chapter 11 by performing an inverse transform over ω and leads to Eq. (12.42) when we perform the inverse transform over k.

The point of this is to show a general property of all Green's functions that are solutions of second-order differential equations; namely, that across the origin where the Dirac source terms are nonzero, the Green's function is continuous while the first derivatives of the Green's function (both in space and time) have discontinuities that can be determined by integrating the differential equation across the origin. As we showed here, these facts can be used in the construction of the Green's function. If the differential equation was first order (e.g., the diffusion equation after we have transformed over the space variables to obtain a first order in time-differential equation), the Green's function itself would need to be discontinuous at the origin so that its derivative generates a Dirac. We will use this fact in the upcoming section on diffusion problems.

"Wave Example 1" from Chapter 11 Revisited

In Section 11.7.1 of Chapter 11, we demonstrated that the solution of the PDE

$$\frac{\partial^2\psi(x,t)}{\partial x^2} - \frac{1}{c^2}\frac{\partial^2\psi}{\partial t^2} = \delta(x)\cos bt \tag{12.54}$$

with outward waves as $|x| \to \infty$ was given by

$$\psi(x,t) = -\frac{c}{2b}\sin\left[b\left(t - \frac{|x|}{c}\right)\right]. \tag{12.55}$$

We could also have obtained this result by convolving our 1D wave-equation Green's function $G(x-x', t-t')$ with the source term above to give

$$\psi(x, t) = \int_{-\infty}^{t} \mathrm{d}t' \int_{-\infty}^{\infty} \mathrm{d}x' \, G(x-x', t-t')\delta(x') \cos bt' \tag{12.56}$$

$$= \int_{-\infty}^{t} \mathrm{d}t' \, G(x, t-t') \cos bt'. \tag{12.57}$$

In this problem, the source has been turned on forever and does not satisfy the condition $\psi(x, t) = 0$ for $t < 0$. This is why the time integration is from $t' = -\infty$ to $t' = t$. As always, we have $G(x-x', t-t') = 0$ for $t - t' < 0$ which is why there is no need to integrate beyond $t' = t$. To handle the limit of $t' = -\infty$, it is convenient to include the attenuation factor ϵ in the 1D Green's function given by Eq. (12.44) prior to taking $\epsilon = 0$. As such, we have

$$\psi(x, t) = -\frac{c}{2} \lim_{\epsilon \to 0} \int_{-\infty}^{t} \mathrm{d}t' \, S(t - |x|/c - t') \, e^{-\epsilon(t-t')} \cos bt' \tag{12.58}$$

$$= -\frac{c}{2} \lim_{\epsilon \to 0} e^{-\epsilon t} \int_{-\infty}^{t-|x|/c} \mathrm{d}t' \, e^{\epsilon t'} \cos bt' \tag{12.59}$$

$$= -\frac{c}{2} \lim_{\epsilon \to 0} e^{-\epsilon t} \left[\frac{b}{(b^2+\epsilon^2)} e^{\epsilon t'} \sin bt' + \frac{\epsilon}{(b^2+\epsilon^2)} e^{\epsilon t'} \cos bt' \right]_{t'=-\infty}^{t'=t-|x|/c} \tag{12.60}$$

$$= -\frac{c}{2} \lim_{\epsilon \to 0} \frac{e^{-\epsilon |x|/c}}{(b^2+\epsilon^2)} \left\{ b \sin \left[b \left(t - \frac{|x|}{c} \right) \right] + \epsilon \cos \left[b \left(t - \frac{|x|}{c} \right) \right] \right\} \tag{12.61}$$

$$= -\frac{c}{2b} \sin \left[b \left(t - \frac{|x|}{c} \right) \right], \tag{12.62}$$

which is exactly the result we obtained in Chapter 11 without the use of the Green's function.

12.2 Green's Function and the Scalar Diffusion Equation

We now run through much of the above analysis but as applied to the scalar diffusion equation.

12.2.1 The "Green's Function Method" Applied to the Diffusion Equation

As we did for the scalar wave equation, we now show how the solution of a scalar diffusion problem can be obtained using an appropriately defined Green's function for the diffusion equation.

The diffusion problem we want to solve within some domain Ω can be stated

$$\nabla^2 \psi(\boldsymbol{r}, t) - \frac{1}{D} \frac{\partial \psi(\boldsymbol{r}, t)}{\partial t} = s(\boldsymbol{r}, t) \tag{12.63}$$

with BCs that either

$$\psi(\boldsymbol{r}, t) = b_0(\boldsymbol{r}, t) \quad \text{for } \boldsymbol{r} \text{ on } \partial\Omega \tag{12.64}$$

or

$$\boldsymbol{n} \cdot \nabla\psi(\boldsymbol{r}, t) = b_1(\boldsymbol{r}, t) \quad \text{for } \boldsymbol{r} \text{ on } \partial\Omega. \tag{12.65}$$

Additionally, to obtain a unique solution, we also have to satisfy the initial condition that

$$\psi(\boldsymbol{r}, 0) = T(\boldsymbol{r}). \tag{12.66}$$

Corresponding to this problem we define the Green's function

$$\nabla^2 G(\boldsymbol{r}, t \,|\, \boldsymbol{r}', t') - \frac{1}{D}\frac{\partial\, G(\boldsymbol{r}, t \,|\, \boldsymbol{r}', t')}{\partial t} = \delta(\boldsymbol{r} - \boldsymbol{r}')\, \delta(t - t') \tag{12.67}$$

subject to the boundary condition that either

$$G(\boldsymbol{r}, t) = 0 \quad \text{for } \boldsymbol{r} \text{ on } \partial\Omega \tag{12.68}$$

or

$$\boldsymbol{n} \cdot \nabla G(\boldsymbol{r}, t) = 0 \quad \text{for } \boldsymbol{r} \text{ on } \partial\Omega. \tag{12.69}$$

As with the wave problem, whatever BC the field ψ satisfies, the diffusion Green's function G satisfies the homogeneous version of the same BC. The Green's function is also subject to the initial condition that

$$G(\boldsymbol{r}, t \,|\, \boldsymbol{r}', t') = 0 \quad \text{for } t - t' < 0. \tag{12.70}$$

Physically, $G(\boldsymbol{r}, t \,|\, \boldsymbol{r}', t')$ represents the response (diffusive response in this case) at $\boldsymbol{r}$, t due to an impulse applied at $\boldsymbol{r}'$, t'.

As with the wave problem, to solve for ψ we need to establish again the symmetry property of the Green's function

$$G(\boldsymbol{r}, t \,|\, \boldsymbol{r}', t') = G(\boldsymbol{r}', -t' \,|\, \boldsymbol{r}, -t). \tag{12.71}$$

To obtain this result, we again make the substitutions $\boldsymbol{r}' \to \boldsymbol{r}''$, $t \to -t$, and $t' \to -t''$ in Eq. (12.67) to obtain the PDE

$$\nabla^2 G(\boldsymbol{r}, -t \,|\, \boldsymbol{r}'', -t'') + \frac{1}{D}\frac{\partial\, G(\boldsymbol{r}, -t \,|\, \boldsymbol{r}'', -t'')}{\partial t} = \delta(\boldsymbol{r} - \boldsymbol{r}'')\, \delta(t - t''), \tag{12.72}$$

where we used that $\delta(t'' - t) = \delta(t - t'')$. Note the important change in the sign of the time derivative due to it being only a first-order derivative. If we multiply Eq. (12.72) by $G(\boldsymbol{r}, t \,|\, \boldsymbol{r}', t')$ and multiply Eq. (12.67) by $G(\boldsymbol{r}, -t \,|\, \boldsymbol{r}'', -t'')$, subtract and then integrate overall all time and all of Ω we obtain

$$\int_{-\infty}^{\infty} dt \int_{\Omega} d^3\boldsymbol{r} \left\{ G(\boldsymbol{r}, t \,|\, \boldsymbol{r}', t')\nabla^2 G(\boldsymbol{r}, -t \,|\, \boldsymbol{r}'', -t'') - G(\boldsymbol{r}, -t \,|\, \boldsymbol{r}'', -t'')\nabla^2 G(\boldsymbol{r}, t \,|\, \boldsymbol{r}', t') \right.$$
$$\left. + \frac{1}{D}\frac{\partial}{\partial t}\left[G(\boldsymbol{r}, t \,|\, \boldsymbol{r}', t')G(\boldsymbol{r}, -t \,|\, \boldsymbol{r}'', -t'')\right] \right\}$$
$$= G(\boldsymbol{r}'', t'' \,|\, \boldsymbol{r}', t') - G(\boldsymbol{r}', -t' \,|\, \boldsymbol{r}'', -t''). \tag{12.73}$$

We again use the identity that $\nabla \cdot (G_1 \nabla G_2 - G_2 \nabla G_1) = G_1 \nabla^2 G_2 - G_2 \nabla^2 G_1$ and the divergence theorem to rewrite the above as

$$G(\boldsymbol{r}'', t'' \,|\, \boldsymbol{r}', t') - G(\boldsymbol{r}', -t' \,|\, \boldsymbol{r}'', -t'')$$
$$= \int_{-\infty}^{\infty} dt \int_{\partial\Omega} d^2\boldsymbol{r}\, \boldsymbol{n} \cdot \left[G(\boldsymbol{r}, t \,|\, \boldsymbol{r}', t')\nabla G(\boldsymbol{r}, -t \,|\, \boldsymbol{r}'', -t'') - G(\boldsymbol{r}, -t \,|\, \boldsymbol{r}'', -t'')\nabla G(\boldsymbol{r}, t \,|\, \boldsymbol{r}', t')\right]$$
$$+ \frac{1}{D}\int_{\Omega} d^3\boldsymbol{r}\, G(\boldsymbol{r}, t \,|\, \boldsymbol{r}', t')G(\boldsymbol{r}, -t \,|\, \boldsymbol{r}'', -t'')\Big|_{t=-\infty}^{t=\infty}. \tag{12.74}$$

The homogeneous BC that G satisfies on $\partial\Omega$ puts the surface integral to zero while the initial condition that $G(\boldsymbol{r}, t = -\infty \,|\, \boldsymbol{r}', t') = 0$ puts the final integral over Ω to zero. Thus, with the right-hand side zero, upon substituting $\boldsymbol{r}'' \to \boldsymbol{r}$ and $t'' \to t$, we obtain the symmetry condition we set out to prove: $G(\boldsymbol{r}, t \,|\, \boldsymbol{r}', t') = G(\boldsymbol{r}', -t' \,|\, \boldsymbol{r}, -t)$, which is the same symmetry that the wave-equation Green's function satisfies.

We now use this symmetry property to rewrite the PDE for G. In Eq. (12.67), we make the exchange $\boldsymbol{r} \leftrightarrow \boldsymbol{r}'$ and $t \leftrightarrow -t'$ and use the symmetry property just derived to obtain

$$\nabla'^2 G(\boldsymbol{r}, t \,|\, \boldsymbol{r}', t') + \frac{1}{D}\frac{\partial G(\boldsymbol{r}, t \,|\, \boldsymbol{r}', t')}{\partial t'} = \delta(\boldsymbol{r} - \boldsymbol{r}')\delta(t - t'). \tag{12.75}$$

Note the derivatives here are acting on the position of the source in space and time. Let's also rewrite the PDE for ψ by making the change of variables $\boldsymbol{r} \to \boldsymbol{r}'$ and $t \to t'$

$$\nabla'^2 \psi(\boldsymbol{r}', t') - \frac{1}{D}\frac{\partial \psi(\boldsymbol{r}', t')}{\partial t'} = s(\boldsymbol{r}', t'). \tag{12.76}$$

We multiply the equation for ψ by G and the equation for G by ψ and subtract to obtain

$$\nabla' \cdot \left[G\nabla'\psi - \psi\nabla' G\right] - \frac{1}{D}\frac{\partial}{\partial t'}\left[G(\boldsymbol{r}, t \,|\, \boldsymbol{r}', t')\psi(\boldsymbol{r}', t')\right] = G\, s(\boldsymbol{r}', t') - \psi\, \delta(\boldsymbol{r} - \boldsymbol{r}')\, \delta(t - t'). \tag{12.77}$$

Next, we integrate over t' from the initial condition at $t' = 0$ to $t' = \infty$ and integrate over all of Ω and use the divergence theorem to obtain

$$\int_0^{\infty} dt' \int_{\partial\Omega} \boldsymbol{n} \cdot \left[G\nabla'\psi - \psi\nabla' G\right] d^2\boldsymbol{r}' - \frac{1}{D}\int_{\Omega} d^3\boldsymbol{r}' \left[G(\boldsymbol{r}, t \,|\, \boldsymbol{r}', t')\, \psi(\boldsymbol{r}', t')\right]\Big|_{t'=0}^{t'=\infty}$$
$$= \int_0^{\infty}\int_{\Omega} G(\boldsymbol{r}, t \,|\, \boldsymbol{r}', t')\, s(\boldsymbol{r}', t')\, d^3\boldsymbol{r}'\, dt' - \int_0^{\infty}\int_{\Omega} \psi(\boldsymbol{r}', t')\, \delta(\boldsymbol{r}' - \boldsymbol{r})\, \delta(t' - t)\, d^3\boldsymbol{r}'\, dt'. \tag{12.78}$$

From the fact that $G = 0$ for $t' > t$ (which means that G is zero at $t' = \infty$ and that there is no need to integrate beyond $t' = t$), we obtain at last

$$\boxed{\begin{aligned}\psi(\boldsymbol{r}, t) = &\int_0^t \int_\Omega G(\boldsymbol{r}, t \,|\, \boldsymbol{r}', t')\, s(\boldsymbol{r}', t')\, d^3\boldsymbol{r}'\, dt' \\ &- \int_0^t dt' \int_{\partial\Omega} \boldsymbol{n} \cdot \left[G(\boldsymbol{r}, t \,|\, \boldsymbol{r}', t')\nabla'\psi(\boldsymbol{r}', t') - \psi(\boldsymbol{r}', t')\nabla' G(\boldsymbol{r}, t \,|\, \boldsymbol{r}', t')\right] d^2\boldsymbol{r}' \\ &- \frac{1}{D}\int_\Omega d^3\boldsymbol{r}'\, G(\boldsymbol{r}, t \,|\, \boldsymbol{r}', 0)\, \psi(\boldsymbol{r}', 0).\end{aligned}} \quad (12.79)$$

This representation theorem shows how the Green's function allows us to obtain the solution $\psi(\boldsymbol{r}, t)$ of the diffusion problem. As for the wave equation, the sources for the field ψ are any inhomogeneous (nonzero) contributions to the source term in the PDE, in the BCs, or in the initial condition.

12.2.2 The Diffusion Equation Green's Function in a Wholespace

As throughout this chapter, and with no loss in generality, we can solve for $G(\boldsymbol{r}, t)$ that is the solution of

$$\nabla^2 G(\boldsymbol{r}, t) - \frac{1}{D}\frac{\partial G(\boldsymbol{r}, t)}{\partial t} = \delta(\boldsymbol{r})\delta(t) \quad (12.80)$$

and then substitute $\boldsymbol{r} \to \boldsymbol{r} - \boldsymbol{r}'$ and $t \to t - t'$ at the end to obtain $G(\boldsymbol{r} - \boldsymbol{r}', t - t')$. The Green's function is required to satisfy $G(\boldsymbol{r}, t) = 0$ for all $t < 0$ and will be seen to decay to zero as $|\boldsymbol{r}| \to \infty$.

To obtain the solution, we transform over space but not time to obtain

$$-k^2\tilde{G}(\boldsymbol{k}, t) - \frac{1}{D}\frac{\partial \tilde{G}(\boldsymbol{k}, t)}{\partial t} = \delta(t). \quad (12.81)$$

By integrating this from $t = -\epsilon$ to $t = +\epsilon$ in the limit as $\epsilon \to 0$, we obtain the jump condition at the origin that

$$\tilde{G}(\boldsymbol{k}, +\epsilon) - \tilde{G}(\boldsymbol{k}, -\epsilon) = -D. \quad (12.82)$$

The solution of Eq. (12.81) when $t > 0$ is given by

$$\tilde{G}(\boldsymbol{k}, t) = Ae^{-Dk^2 t}, \quad (12.83)$$

where $k^2 = \boldsymbol{k} \cdot \boldsymbol{k}$ and A is determined from the jump condition at $t = 0$ in order to give the solution that is valid at $t \geq 0$. From the initial condition, we have $\tilde{G}(\boldsymbol{k}, -\epsilon) = 0$ so that the jump condition (as $\epsilon \to 0$) gives that $A = -D$. We thus have found that

$$\tilde{G}(k, t) = -De^{-Dk^2 t} \quad \text{for } t \geq 0 \tag{12.84}$$

while $\tilde{G}(k, t) = 0$ for $t < 0$.

We now perform the triple inverse transform to return to the space domain $\boldsymbol{r} = (x, y, z)$. We have that with $k^2 = k_x^2 + k_y^2 + k_z^2$

$$G(\boldsymbol{r}, t) = -\frac{D}{(2\pi)^3}\left[\int_{-\infty}^{\infty} \mathrm{d}k_x\, e^{ik_x x - Dtk_x^2}\right]\left[\int_{-\infty}^{\infty} \mathrm{d}k_y\, e^{ik_y y - Dtk_y^2}\right]\left[\int_{-\infty}^{\infty} \mathrm{d}k_z\, e^{ik_z z - Dtk_z^2}\right]. \tag{12.85}$$

Each of these integrals is of identical form and is obtained by completing the square in the exponent as was done in the earlier Chapter 10 on Fourier analysis. So, for example, we have

$$\int_{-\infty}^{\infty} \mathrm{d}k_x\, e^{ik_x x - Dtk_x^2} = \frac{e^{-x^2/(4Dt)}}{2\sqrt{\pi Dt}}. \tag{12.86}$$

We then obtain the result (restoring the offset in the source position)

$$\boxed{G(\boldsymbol{r} - \boldsymbol{r}', t - t') = -\frac{D}{8[D\pi(t - t')]^{3/2}} \exp\left(-\frac{|\boldsymbol{r} - \boldsymbol{r}'|^2}{4D(t - t')}\right),} \tag{12.87}$$

where $|\boldsymbol{r} - \boldsymbol{r}'|^2 = (x - x')^2 + (y - y')^2 + (z - z')^2$. This is the 3D Green's function for the diffusion equation when $t - t' \geq 0$. For $t - t' < 0$, we have $G = 0$. The solution does indeed fall off to zero as $|\boldsymbol{r} - \boldsymbol{r}'| \to \infty$ as required in the problem statement.

So if we had a diffusion problem in a wholespace Ω_∞ where the source term in the diffusion equation was zero and there was an initial condition that $\psi(\boldsymbol{r}, 0) = T(\boldsymbol{r})$, we have from Eq. (12.79) that

$$\psi(\boldsymbol{r}, t) = \frac{1}{8(\pi Dt)^{3/2}} \int_{\Omega_\infty} \mathrm{d}^3\boldsymbol{r}'\, T(\boldsymbol{r}') \exp\left(-\frac{|\boldsymbol{r} - \boldsymbol{r}'|^2}{4Dt}\right), \tag{12.88}$$

which is exactly the result we obtained earlier in Chapter 10 on Fourier analysis prior to having defined Green's functions.

12.3 Green's Tensors for Elastodynamic Response

We now use Fourier transforms to obtain explicit Green's tensors for the various types of continuum vector response treated in Part I of the book, namely, elastodynamics, slow viscous flow, and electromagnetics. For each type of vectorial response, we either review or derive how the Green's function method puts the Green's tensors to use in solving boundary-value problems (BVPs). We begin with elastodynamics.

12.3.1 The "Green's Function Method" Applied to Elastodynamics

In Chapter 4, we derived a result that we called the *seismic representation theorem*, which was just the "Green's function method" applied to the special case of linear elastodynamics as we now review.

Let's consider the linear elastodynamic response within a domain Ω that is surrounded by a surface $\partial\Omega$, which corresponds to solutions of

$$\rho \frac{\partial^2 \boldsymbol{u}}{\partial t^2} = \nabla \cdot ({}_4\boldsymbol{C} : \nabla \boldsymbol{u}) + \boldsymbol{F}(\boldsymbol{r}, t) \tag{12.89}$$

for the material displacements $\boldsymbol{u}(\boldsymbol{r}, t)$ throughout the domain Ω. The density $\rho(\boldsymbol{r})$ and elastic stiffnesses ${}_4\boldsymbol{C}(\boldsymbol{r})$ may be heterogeneous throughout Ω but are both independent of time through our assumption of linearity. One possible source for creating displacements is the body force $\boldsymbol{F}(\boldsymbol{r}, t)$ that we assume is zero for $t < 0$ (though this is not essential). Other sources for creating displacements are nonzero boundary values of either

$$\boldsymbol{u}(\boldsymbol{r}, t) = \boldsymbol{b}_u(\boldsymbol{r}, t) \quad \text{or} \quad \boldsymbol{n} \cdot [{}_4\boldsymbol{C} : \nabla \boldsymbol{u}(\boldsymbol{r}, t)] = \boldsymbol{b}_\tau(\boldsymbol{r}, t) \tag{12.90}$$

for points $\boldsymbol{r}$ on $\partial\Omega$, which are also required conditions for the solution to this BVP to be unique. Additional sources for creating displacements are nonzero values of the initial conditions

$$\boldsymbol{u}(\boldsymbol{r}, 0) \quad \text{and} \quad \left.\frac{\partial \boldsymbol{u}(\boldsymbol{r}, t)}{\partial t}\right|_{t=0}, \tag{12.91}$$

which are also required conditions that must be specified throughout Ω to obtain unique solutions. As derived in Chapter 4, the above is a "well-posed" BVP that leads to unique solutions for the displacement field $\boldsymbol{u}(\boldsymbol{r}, t)$ throughout Ω.

Corresponding to the above, we define the *displacement Green's tensor* $\boldsymbol{G}(\boldsymbol{r}, t|\boldsymbol{r}', t')$ as the solution of

$$\rho \frac{\partial^2 \boldsymbol{G}}{\partial t^2} = \nabla \cdot ({}_4\boldsymbol{C} : \nabla \boldsymbol{G}) + \boldsymbol{I}\delta(\boldsymbol{r} - \boldsymbol{r}')\delta(t - t'), \tag{12.92}$$

where $\boldsymbol{I}$ is the second-order identity tensor. This second-order Green's tensor is required to satisfy homogeneous versions of the boundary conditions in Eq. (12.90); namely, either

$$\boldsymbol{G}(\boldsymbol{r}, t|\boldsymbol{r}', t') = 0 \quad \text{or} \quad \boldsymbol{n} \cdot \left[{}_4\boldsymbol{C} : \nabla \boldsymbol{G}(\boldsymbol{r}, t|\boldsymbol{r}', t')\right] = 0 \tag{12.93}$$

at each point $\boldsymbol{r}$ on $\partial\Omega$. The Green's tensor is also required to satisfy the causal initial condition that

$$\boldsymbol{G}(\boldsymbol{r}, t|\boldsymbol{r}', t') = 0 \quad \text{for} \quad t < t'. \tag{12.94}$$

Further, in Chapter 4, we derived the reciprocity condition $\boldsymbol{G}(\boldsymbol{r}', -t'|\boldsymbol{r}, -t) = \boldsymbol{G}^T(\boldsymbol{r}, t|\boldsymbol{r}', t')$.

As shown in Chapter 4, the above suite of equations can be combined in a manner we have been calling the "Green's function method," which results in the following *representation theorem* for elastodynamic response

$$
\begin{aligned}
\boldsymbol{u}(\boldsymbol{r}, t) = & \int_0^t \mathrm{d}t' \int_\Omega \mathrm{d}^3 r' \, \boldsymbol{F}(\boldsymbol{r}', t') \cdot \boldsymbol{G}^T(\boldsymbol{r}, t\,|\boldsymbol{r}', t') \\
& + \int_0^t \mathrm{d}t' \int_{\partial\Omega} \mathrm{d}^2 r' \left[\boldsymbol{n}(\boldsymbol{r}') \cdot {}_4\boldsymbol{C}(\boldsymbol{r}') : \nabla' \boldsymbol{u}(\boldsymbol{r}', t')\right] \cdot \boldsymbol{G}^T(\boldsymbol{r}, t\,|\boldsymbol{r}', t') \\
& - \int_0^t \mathrm{d}t' \int_{\partial\Omega} \mathrm{d}^2 r' \, \boldsymbol{u}(\boldsymbol{r}', t') \cdot \left[\boldsymbol{n}(\boldsymbol{r}') \cdot {}_4\boldsymbol{C}(\boldsymbol{r}') : \nabla' \boldsymbol{G}^T(\boldsymbol{r}, t\,|\boldsymbol{r}', t')\right] \\
& + \int_\Omega \mathrm{d}^3 r' \rho \left. \left(\frac{\partial \boldsymbol{u}(\boldsymbol{r}', t')}{\partial t'} \cdot \boldsymbol{G}^T(\boldsymbol{r}, t|\boldsymbol{r}', t') - \boldsymbol{u}(\boldsymbol{r}', t') \cdot \frac{\boldsymbol{G}^T(\boldsymbol{r}, t|\boldsymbol{r}', t')}{\partial t'} \right) \right|_{t'=0} .
\end{aligned}
\quad (12.95)
$$

Note that at those places $\boldsymbol{r}'$ on the boundary where the tractions $\boldsymbol{n}(\boldsymbol{r}') \cdot {}_4\boldsymbol{C}(\boldsymbol{r}') : \nabla' \boldsymbol{u}(\boldsymbol{r}', t') = \boldsymbol{b}_\tau(\boldsymbol{r}, t)$ are specified through time, the Green's tensor satisfies the homogeneous version of this condition $\boldsymbol{n}(\boldsymbol{r}') \cdot {}_4\boldsymbol{C}(\boldsymbol{r}')\!: \nabla' \boldsymbol{G}^T(\boldsymbol{r}, t\,|\boldsymbol{r}', t') = \boldsymbol{n}(\boldsymbol{r}') \cdot {}_4\boldsymbol{C}(\boldsymbol{r}') : \nabla' \boldsymbol{G}(\boldsymbol{r}', -t'\,|\boldsymbol{r}, t) = 0$. As usual, we integrate from the time $t' = 0$ where the initial conditions are specified to $t' = \infty$ but because the Green's tensor is zero for $t' > t$, we stop the integration at $t' = t$. Note that if the initial conditions are taken to be zero, this means that the boundary values and effective body force generating the displacements are also zero for times less than zero.

So any combination of a nonzero body force, nonzero boundary conditions, or nonzero initial conditions act as source terms for the elastodynamic displacements. If the corresponding Green's tensor is known, then the displacements from these sources are obtained by performing the integrations of the representation theorem, a solution procedure that we call the "Green's function method."

12.3.2 The Green's Tensor of Elastodynamic Response in a Wholespace

To obtain the Green's tensor in a wholespace, consider the elastodynamic wave equation in an isotropic, homogeneous and infinite material

$$\rho \frac{\partial^2 \boldsymbol{u}}{\partial t^2} = \nabla \cdot \boldsymbol{\tau} + \boldsymbol{s}\,\delta(\boldsymbol{r})\,\delta(t) \quad (12.96)$$

$$\boldsymbol{\tau} = K \nabla \cdot \boldsymbol{u} \boldsymbol{I} + G \left[\nabla \boldsymbol{u} + (\nabla \boldsymbol{u})^T - \frac{2}{3} \nabla \cdot \boldsymbol{u} \boldsymbol{I} \right] \quad (12.97)$$

$$\nabla \cdot \boldsymbol{\tau} = \left(K + \frac{G}{3} \right) \nabla \nabla \cdot \boldsymbol{u} + G \nabla^2 \boldsymbol{u}, \quad (12.98)$$

where in the point source, $\boldsymbol{s}$ is a spatially and temporally constant vector that has units of force times time (N s) and denotes the direction of the impulsive force applied at the origin. The boundary condition is that the waves are propagating outward at infinity and the initial condition is that $\boldsymbol{u}=0$ for $t<0$. We can solve this as $\boldsymbol{u}=\boldsymbol{G}\cdot\boldsymbol{s}$ and then insert $\boldsymbol{r}\to\boldsymbol{r}-\boldsymbol{r}'$ and $t\to t-t'$ to obtain the second-order elastodynamic outward-radiating Green's tensor $\boldsymbol{G}(\boldsymbol{r}-\boldsymbol{r}', t-t')$ in a wholespace, which then can be used to obtain the response for any given inhomogeneous source distribution in the wholespace $\boldsymbol{F}(\boldsymbol{r}, t)$ by performing a convolution of the form $\int_0^t \mathrm{d}t' \int_{\Omega_\infty} \mathrm{d}^3r'\, \boldsymbol{G}(\boldsymbol{r}-\boldsymbol{r}', t-t')\cdot\boldsymbol{F}(\boldsymbol{r}', t')$.

We solve for the outward-radiating $\boldsymbol{G}$ by taking the Fourier transform over time $\partial/\partial t \to -i(\omega+i\epsilon)$ where ϵ is the attenuation parameter that we have discussed previously and that results in outward-radiating waves

$$\left(K+\frac{G}{3}\right)\nabla\nabla\cdot\tilde{\boldsymbol{u}}+G\,\nabla^2\tilde{\boldsymbol{u}}+\rho\,(\omega+i\epsilon)^2\tilde{\boldsymbol{u}}=-\boldsymbol{s}\,\delta(\boldsymbol{r}). \tag{12.99}$$

We also take the Fourier transform over space $\nabla\to+i\boldsymbol{k}$ giving

$$\left[\boldsymbol{I}+\frac{K+G/3}{G\,k^2-\rho\,(\omega+i\epsilon)^2}\boldsymbol{k}\boldsymbol{k}\right]\cdot\tilde{\tilde{\boldsymbol{u}}}=\frac{1}{G\,k^2-\rho\,(\omega+i\epsilon)^2}\,\boldsymbol{s}, \tag{12.100}$$

where $\boldsymbol{k}\cdot\boldsymbol{k}=k^2$. To continue, we must be able to multiply through by the inverse of the tensor on the left-hand side.

As discussed in Chapter 1, an algebraic function of a second-order tensor is itself a second-order tensor that is defined in terms of a Taylor expansion. So from the tensorial generalization of the binomial expansion we obtain (note that for scalars the binomial expansion is $(1+\alpha)^{-1}=1-\alpha+\alpha^2-\alpha^3+\ldots$)

$$\begin{aligned}[\boldsymbol{I}+\alpha\,\boldsymbol{x}\boldsymbol{x}]^{-1} &= \boldsymbol{I}-\alpha\,\boldsymbol{x}\boldsymbol{x}+(\alpha\,\boldsymbol{x}\boldsymbol{x})\cdot(\alpha\,\boldsymbol{x}\boldsymbol{x})-(\alpha\,\boldsymbol{x}\boldsymbol{x})\cdot(\alpha\,\boldsymbol{x}\boldsymbol{x})\cdot(\alpha\,\boldsymbol{x}\boldsymbol{x})+\cdots\\ &= \boldsymbol{I}-\alpha\left(1-\alpha x^2+\alpha^2x^4-\cdots\right)\boldsymbol{x}\boldsymbol{x}\\ &= \boldsymbol{I}-\frac{\boldsymbol{x}\boldsymbol{x}}{x^2+1/\alpha}\quad \text{where } x^2=\boldsymbol{x}\cdot\boldsymbol{x}.\end{aligned} \tag{12.101}$$

One may be concerned about the validity of the above expansion when $|\alpha||\boldsymbol{x}||\boldsymbol{x}|>1$. This would be a concern if we left the expansion as a series, but we collapse it and it is easy to confirm by direct multiplication that the product $[\boldsymbol{I}+\alpha\boldsymbol{x}\boldsymbol{x}]^{-1}\cdot[\boldsymbol{I}+\alpha\boldsymbol{x}\boldsymbol{x}]=\boldsymbol{I}$ as required, independent of the amplitude of $|\alpha||\boldsymbol{x}||\boldsymbol{x}|$.

So, using this inverse tensor in the above we obtain

$$\tilde{\tilde{\boldsymbol{u}}}(\boldsymbol{k},\omega)=\frac{1/G}{k^2-[(\omega+i\epsilon)/c_s]^2}\left[\boldsymbol{I}-\left(1-\frac{c_s^2}{c_p^2}\right)\frac{(i\boldsymbol{k})(i\boldsymbol{k})}{k^2-\left[(\omega+i\epsilon)/c_p\right]^2}\right]\cdot\boldsymbol{s}, \tag{12.102}$$

where

$$c_p^2=\frac{K+4G/3}{\rho}\quad\text{and}\quad c_s^2=\frac{G}{\rho} \tag{12.103}$$

are the square of the P-wave and S-wave velocities. This is the solution for the response in the four-fold transform domain.

We then take the inverse Fourier transform in spherical-$\boldsymbol{k}$ coordinates as we did earlier for the scalar wave-equation Green's function, use the residue theorem to perform the radial k integral, let the attenuation parameter $\epsilon \to 0$ and use the fact that $i\boldsymbol{k}$ in the transform domain becomes ∇ in the space domain to obtain

$$\tilde{\boldsymbol{u}}(\boldsymbol{r},\omega) = \frac{1}{G}\left\{\frac{e^{i\omega r/c_s}}{4\pi r}\boldsymbol{I} + \frac{c_s^2}{4\pi\omega^2}\nabla\nabla\left[\frac{e^{i\omega r/c_s}}{r} - \frac{e^{i\omega r/c_p}}{r}\right]\right\}\cdot\boldsymbol{s}. \tag{12.104}$$

Carrying out the spatial derivatives [where $\nabla r = \hat{\boldsymbol{r}} = \boldsymbol{r}/r$ and $\nabla\hat{\boldsymbol{r}} = \left(\boldsymbol{I} - \hat{\boldsymbol{r}}\hat{\boldsymbol{r}}\right)/r$] we obtain

$$\begin{aligned}\tilde{\boldsymbol{u}}(r,\omega) = &\left\{\frac{e^{i\omega r/c_s}}{\rho\, c_s^2\, 4\pi r}\left(\boldsymbol{I} - \hat{\boldsymbol{r}}\hat{\boldsymbol{r}}\right) + \frac{e^{i\omega r/c_p}}{\rho\, c_p^2\, 4\pi r}\hat{\boldsymbol{r}}\hat{\boldsymbol{r}}\right.\\ &\left.+\left[\frac{e^{i\omega r/c_s}}{\rho\, c_s^2\, 4\pi r}\left(\frac{i\, c_s}{\omega r} - \frac{c_s^2}{\omega^2 r^2}\right) - \frac{e^{i\omega r/c_p}}{\rho\, c_p^2\, 4\pi r}\left(\frac{i\, c_p}{\omega r} - \frac{c_p^2}{\omega^2 r^2}\right)\right]\left(\boldsymbol{I} - 3\,\hat{\boldsymbol{r}}\hat{\boldsymbol{r}}\right)\right\}\cdot\boldsymbol{s}.\end{aligned} \tag{12.105}$$

Note that the polarization of $\boldsymbol{I} - \hat{\boldsymbol{r}}\hat{\boldsymbol{r}}$ denotes a "transverse" response (let $\boldsymbol{s} = s_0\,\hat{\boldsymbol{\theta}}$ and $\boldsymbol{s} = s_0\,\hat{\boldsymbol{r}}$ to see this), while $\hat{\boldsymbol{r}}\hat{\boldsymbol{r}}$ denotes a "longitudinal" response (let $\boldsymbol{s} = s_0\,\hat{\boldsymbol{r}}$ to see this). The third type of polarization $\boldsymbol{I} - 3\,\hat{\boldsymbol{r}}\hat{\boldsymbol{r}}$ is a "mixed" polarization associated with the so-called near field that dominates when both

$$r \ll \frac{c_s}{\omega} \quad \text{and} \quad r \ll \frac{c_p}{\omega} \qquad [\text{note that } c/\omega = \lambda/(2\pi) \text{ where } \lambda = \text{wavelength}]. \tag{12.106}$$

So the near-field term is important whenever the distance from the source is less than seismic wavelengths. The response in this near-field case is associated with the topic of *strong-motion seismology*. In the far field where distance from the source is much larger than seismic wavelengths, the near-field term may be neglected and the response corresponds to outward propagating spherical body waves, both compressional and shear.

We finally return to the time domain by noting that

$$\mathrm{FT}^{-1}\left\{e^{i\omega r/c_s}\right\} = \delta\left(t - \frac{r}{c_s}\right) \tag{12.107}$$

$$\mathrm{FT}^{-1}\left\{\frac{c_s}{-i\omega}\,e^{i\omega r/c_s}\right\} = c_s\int_{-\infty}^{t} \mathrm{d}t_0\,\delta\left(t_0 - \frac{r}{c_s}\right) = c_s\,S\left(t - \frac{r}{c_s}\right) \tag{12.108}$$

$$\mathrm{FT}^{-1}\left\{\left(\frac{c_s}{-i\omega}\right)^2 e^{i\omega r/c_s}\right\} = c_s^2\int_{-\infty}^{t} S\left(t_0 - \frac{r}{c_s}\right)\,\mathrm{d}t_0 = c_s^2\left(t - \frac{r}{c_s}\right)\,S\left(t - \frac{r}{c_s}\right), \tag{12.109}$$

where $S(u)$ is again the unit step function that has the behavior that $S(u) = 0$ when $u < 0$ and $S(u) = 1$ (unitless) when $u > 0$. We thus obtain the outward-radiating elastodynamic Green's tensor in the form

$$G(r,t)=\frac{1}{4\pi r}\left[\frac{\delta(t-r/c_s)}{\rho\, c_s^2}(I-\hat{r}\hat{r})+\frac{\delta(t-r/c_p)\hat{r}\hat{r}}{\rho\, c_p^2}\right]$$

far-field radiating response

$$+\frac{t}{4\pi\,\rho\, r^3}\left[S(t-r/c_s)-S(t-r/c_p)\right](I-3\,\hat{r}\hat{r})\,.$$

near-field response (12.110)

Each far-field radiating body wave (either P or S wave) is a sharp Dirac pulse moving in the $+r$ direction as a spherical wave. The near-field response is nonzero in a spherical shell residing between $tc_s < r < tc_p$ that grows in radial size as time increases due to $c_p > c_s$ but whose amplitude drops through time at the radial bounds of the expanding nonzero response shell as t^{-2}.

Last, we can substitute $r=|r-r'|$ and $t=t-t'$ to obtain the tensor $G(r-r',t-t')$ that tells us how an impulse at r', t' generates outward-radiating elastodynamic response at r, t. This Green's tensor satisfies the initial condition that $G=0$ when $t<t'$. Note that $\hat{r}=(r-r')/|r-r'|$ in this case (radial direction away from the point r').

12.3.3 The Elastostatic Green's Tensor in a Wholespace

The elastostatic Green's tensor in a uniform isotropic wholespace is obtained from the solution of

$$\left(K+\frac{G}{3}\right)\nabla\nabla\cdot u+G\nabla^2 u=-s_0\,\delta(r). \tag{12.111}$$

We can take a Fourier transform over space $\nabla\rightarrow ik$ and then use the inverse tensor given by Eq. (12.101) to obtain the transformed solution

$$\tilde{u}(k)=\left[I+\frac{(ik)(ik)}{k^2}\left(\frac{K+G/3}{K+4G/3}\right)\right]\cdot\frac{s_0}{Gk^2}. \tag{12.112}$$

It is then a straightforward exercise to perform the inverse transform in spherical-k coordinates as you can do yourself to obtain Eq. (12.116) as given below.

Alternatively, we can also obtain the desired elastostatic Green's tensor from Eq. (12.105) for $\tilde{u}(r,\omega)$ by carefully taking the $\omega\rightarrow 0$ limit and replacing s with s_0, which now has units of Newtons.

Expanding the exponentials in the near-field terms of Eq. (12.105) as $\omega r/c\rightarrow 0$ we obtain

$$\begin{aligned}\tilde{u}(r,0)=\lim_{\omega\rightarrow 0}\tilde{u}(r,\omega)=\frac{1}{4\pi r}&\left\{\frac{(I-\hat{r}\hat{r})}{G}+\frac{\hat{r}\hat{r}}{K+4G/3}\right.\\ &-\frac{1}{\rho\omega^2 r^2}\left[\left(1+\frac{i\omega r}{c_s}-\frac{\omega^2 r^2}{2!c_s^2}+\ldots\right)\left(1-\frac{i\omega r}{c_s}\right)\right.\\ &\left.\left.-\left(1+\frac{i\omega r}{c_p}-\frac{\omega^2 r^2}{2!c_p^2}+\ldots\right)\left(1-\frac{i\omega r}{c_p}\right)\right](I-3\hat{r}\hat{r})\right\}\cdot s_0\end{aligned} \tag{12.113}$$

$$= \frac{1}{4\pi r}\left[\frac{(I-\hat{r}\hat{r})}{G} + \frac{\hat{r}\hat{r}}{K+4G/3} - \frac{1}{\rho\omega^2 r^2}\left(\frac{\omega^2 r^2}{2c_s^2} - \frac{\omega^2 r^2}{2c_p^2}\right)(I-3\hat{r}\hat{r})\right]\cdot s_0. \tag{12.114}$$

Thus, we have the result

$$\tilde{u}(r,0) = u(r) = \frac{1}{8\pi rG}\left[\left(\frac{K+7G/3}{K+4G/3}\right)I + \left(\frac{K+G/3}{K+4G/3}\right)\hat{r}\hat{r}\right]\cdot s_0, \tag{12.115}$$

which with the replacement $r \to r - r'$ gives the elastostatic Green's tensor in a uniform isotropic wholespace

$$G_0(r-r') = \frac{1}{8\pi|r-r'|G} \times \left[\left(\frac{K+7G/3}{K+4G/3}\right)I + \left(\frac{K+G/3}{K+4G/3}\right)\frac{(r-r')(r-r')}{|r-r'|^2}\right]. \tag{12.116}$$

The exercise performed this way shows that the elastostatic response involves both the radiating and near-field contributions of the elastodynamic response in the $\omega \to 0$ limit and is not made up purely of the near-field contribution alone. For a time-dependent point force $s_0(t)\delta(r)$ applied at the origin, if our distances of observation r are much smaller than all seismic wavelengths generated by this time-varying point force, we obtain the so-called quasi-static response as $u(r,t) = G_0(r)\cdot s_0(t)$ where $G_0(r)$ is given by Eq. (12.116) with $r' = 0$.

12.4 Green's Tensors for Slow Viscous Flow

In Chapter 5, we derived the linear governing equations for slow viscous flow that is called *Stokes flow*. In that chapter, we also derived what flow fields must be specified on the system boundary for the flow field to be unique. However, we did not derive the representation theorem for Stokes flow as we now will do.

12.4.1 The "Green's Function Method" Applied to Stokes Flow

We are interested in the steady-state flow field $v(r)$ within some flow domain Ω that is generated by two possible types of sourcing. The governing equations for such steady, slow, incompressible, viscous flow are the Stokes equations

$$\nabla\cdot\left[\eta\left(\nabla v + (\nabla v)^T\right)\right] - \nabla P + F = 0 \quad \text{and} \quad \nabla\cdot v = 0, \tag{12.117}$$

where the body-force density $F(r)$ acting in Ω is one possible source for the flow field. For generality, we have written the Stokes equation here for the case where η can have spatial variation. We showed in Chapter 5 that unique solutions of this problem are obtained if either

$$v = b_v(r) \quad \text{or} \quad -nP + \eta n\cdot\left[\nabla v + (\nabla v)^T\right] = b_\tau(r) \tag{12.118}$$

are specified at each point $\boldsymbol{r}$ of the system boundary $\partial\Omega$. If either the given flow field $\boldsymbol{b}_v(\boldsymbol{r})$ or given traction vector $\boldsymbol{b}_\tau(\boldsymbol{r})$ on the boundary is nonzero, such a BC provides a second type of sourcing for $\boldsymbol{v}(\boldsymbol{r})$ in Ω.

Corresponding to the above well-posed Stokes-flow problem, we now define the viscous-flow Green's tensor $\boldsymbol{W}(\boldsymbol{r}|\boldsymbol{r}')$ and pressure Green's vector $\boldsymbol{Q}(\boldsymbol{r}|\boldsymbol{r}')$ that are the solutions of

$$\nabla\cdot\left[\eta\left(\nabla\boldsymbol{W})+(\nabla\boldsymbol{W})^{T}_{213}\right)\right]-\nabla\boldsymbol{Q}+\boldsymbol{I}\delta(\boldsymbol{r}-\boldsymbol{r}')=0 \quad \text{and} \quad \nabla\cdot\boldsymbol{W}=0, \tag{12.119}$$

where $\boldsymbol{I}$ is the second-order identity tensor. The boundary conditions are

$$\boldsymbol{W}(\boldsymbol{r}|\boldsymbol{r}')=0 \quad \text{or} \quad -\boldsymbol{n}\boldsymbol{Q}(\boldsymbol{r}|\boldsymbol{r}')+\eta\boldsymbol{n}\cdot\left[\nabla\boldsymbol{W}(\boldsymbol{r}|\boldsymbol{r}')+\left(\nabla\boldsymbol{W}(\boldsymbol{r}|\boldsymbol{r}')\right)^{T}_{213}\right]=0, \tag{12.120}$$

which are the homogeneous versions of the boundary conditions specified in Eq. (12.118).

The above equations are now combined to give the Stokes-flow representation theorem. To do so, we work for convenience in Cartesian coordinates and write Eq. (12.117) for the flow field in component form and make the change of variables $\boldsymbol{r}\to\boldsymbol{r}'$:

$$\frac{\partial}{\partial x_j'}\left[\eta\left(\frac{\partial v_i(\boldsymbol{r}')}{\partial x_j'}+\frac{\partial v_j(\boldsymbol{r}')}{\partial x_i'}\right)-\delta_{ij}P(\boldsymbol{r}')\right]+F_i(\boldsymbol{r}')=0. \tag{12.121}$$

For the corresponding Green's tensor equation written in component form, we also make the exchange of variables $\boldsymbol{r}\leftrightarrow\boldsymbol{r}'$ to obtain

$$\frac{\partial}{\partial x_j'}\left[\eta\left(\frac{\partial W_{ik}(\boldsymbol{r}'|\boldsymbol{r})}{\partial x_j'}+\frac{\partial W_{jk}(\boldsymbol{r}'|\boldsymbol{r})}{\partial x_i'}\right)-\delta_{ij}Q_k(\boldsymbol{r}'|\boldsymbol{r})\right]-\delta_{ik}\delta(\boldsymbol{r}'-\boldsymbol{r})=0. \tag{12.122}$$

Multiply Eq. (12.121) by $W_{ik}(\boldsymbol{r}'|\boldsymbol{r})$ and Eq. (12.122) by $v_i(\boldsymbol{r}')$, subtract, and rearrange to obtain

$$\begin{aligned}\frac{\partial}{\partial x_j'}\Bigg\{&W_{ik}(\boldsymbol{r}'|\boldsymbol{r})\left[\eta\left(\frac{\partial v_i(\boldsymbol{r}')}{\partial x_j'}+\frac{\partial v_j(\boldsymbol{r}')}{\partial x_i'}\right)-\delta_{ij}P(\boldsymbol{r}')\right]\\&-v_i(\boldsymbol{r}')\left[\eta\left(\frac{\partial W_{ik}(\boldsymbol{r}'|\boldsymbol{r})}{\partial x_j'}+\frac{\partial W_{jk}(\boldsymbol{r}'|\boldsymbol{r})}{\partial x_i'}\right)-\delta_{ij}Q_k(\boldsymbol{r}'|\boldsymbol{r})\right]\Bigg\}\\&+F_i(\boldsymbol{r}')W_{ik}(\boldsymbol{r}'|\boldsymbol{r})-v_k(\boldsymbol{r}')\delta(\boldsymbol{r}'-\boldsymbol{r})=0.\end{aligned} \tag{12.123}$$

Integrating $\boldsymbol{r}'$ over Ω, using the fundamental theorem of 3D calculus and returning to the bold-face notation applicable to all coordinate systems yields

$$\boxed{\begin{aligned}\boldsymbol{v}(\boldsymbol{r})=&\int_\Omega d^3r'\,\boldsymbol{F}(\boldsymbol{r}')\cdot\boldsymbol{W}(\boldsymbol{r}'|\boldsymbol{r})\\&+\int_{\partial\Omega}d^2r'\left\{\left[\eta\boldsymbol{n}\cdot\left(\nabla'\boldsymbol{v}(\boldsymbol{r}')+\left(\nabla'\boldsymbol{v}(\boldsymbol{r}')\right)^T\right)-\boldsymbol{n}P(\boldsymbol{r}')\right]\cdot\boldsymbol{W}(\boldsymbol{r}'|\boldsymbol{r})\right.\\&\left.-\boldsymbol{v}(\boldsymbol{r}')\cdot\left[\eta\boldsymbol{n}\cdot\left(\nabla'\boldsymbol{W}(\boldsymbol{r}'|\boldsymbol{r})+\left(\nabla'\boldsymbol{W}(\boldsymbol{r}'|\boldsymbol{r})\right)^{T}_{213}\right)-\boldsymbol{n}\boldsymbol{Q}(\boldsymbol{r}'|\boldsymbol{r})\right]\right\}\end{aligned}} \tag{12.124}$$

as our statement of the *Stokes-flow representation theorem.* So if we have expressions for $\boldsymbol{W}(\boldsymbol{r}|\boldsymbol{r}')$ and $\boldsymbol{Q}(\boldsymbol{r}|\boldsymbol{r}')$ and analytically make the exchange $\boldsymbol{r} \leftrightarrow \boldsymbol{r}'$ in these expressions or if we calculate numerically $\boldsymbol{W}(\boldsymbol{r}'|\boldsymbol{r})$ and $\boldsymbol{Q}(\boldsymbol{r}'|\boldsymbol{r})$ for point sources located at $\boldsymbol{r}$ and observation points located at $\boldsymbol{r}'$, we can determine through integration how given nonzero velocities and/or tractions acting on the system boundary $\partial\Omega$ and/or how a given body-force distribution in Ω creates the flow field $\boldsymbol{v}(\boldsymbol{r})$.

The Stokes-flow Green's tensor possesses the reciprocity that $\boldsymbol{W}(\boldsymbol{r}|\boldsymbol{r}') = \boldsymbol{W}^T(\boldsymbol{r}'|\boldsymbol{r})$. To prove this, we again work in Cartesian coordinates and write the governing equation for the Green's tensor in component form and for two different locations of the point source

$$\frac{\partial}{\partial x_j}\left[\eta\left(\frac{\partial W_{ik}(\boldsymbol{r}|\boldsymbol{r}')}{\partial x_j} + \frac{\partial W_{jk}(\boldsymbol{r}|\boldsymbol{r}')}{\partial x_i}\right) - \delta_{ij}Q_k(\boldsymbol{r}|\boldsymbol{r}')\right] - \delta_{ik}\delta(\boldsymbol{r}-\boldsymbol{r}') = 0 \tag{12.125}$$

and

$$\frac{\partial}{\partial x_j}\left[\eta\left(\frac{\partial W_{il}(\boldsymbol{r}|\boldsymbol{r}'')}{\partial x_j} + \frac{\partial W_{jl}(\boldsymbol{r}|\boldsymbol{r}'')}{\partial x_i}\right) - \delta_{ij}Q_l(\boldsymbol{r}|\boldsymbol{r}'')\right] - \delta_{il}\delta(\boldsymbol{r}-\boldsymbol{r}'') = 0. \tag{12.126}$$

Multiply Eq. (12.125) by $W_{il}(\boldsymbol{r}|\boldsymbol{r}'')$ and Eq. (12.126) by $W_{ik}(\boldsymbol{r}|\boldsymbol{r}')$, subtract, distribute the derivatives, and use the incompressible flow condition to obtain

$$\begin{aligned}
&\frac{\partial}{\partial x_j}\left\{W_{il}(\boldsymbol{r}|\boldsymbol{r}'')\left[\eta\left(\frac{\partial W_{ik}(\boldsymbol{r}|\boldsymbol{r}')}{\partial x_j} + \frac{\partial W_{jk}(\boldsymbol{r}|\boldsymbol{r}')}{\partial x_i}\right) - \delta_{ij}Q_k(\boldsymbol{r}|\boldsymbol{r}')\right]\right.\\
&\qquad\left. - W_{ik}(\boldsymbol{r}|\boldsymbol{r}')\left[\eta\left(\frac{\partial W_{il}(\boldsymbol{r}|\boldsymbol{r}'')}{\partial x_j} + \frac{\partial W_{jl}(\boldsymbol{r}|\boldsymbol{r}'')}{\partial x_i}\right) - \delta_{ij}Q_l(\boldsymbol{r}|\boldsymbol{r}'')\right]\right\}\\
&\qquad - W_{kl}(\boldsymbol{r}|\boldsymbol{r}'')\delta(\boldsymbol{r}-\boldsymbol{r}') + W_{lk}(\boldsymbol{r}|\boldsymbol{r}')\delta(\boldsymbol{r}-\boldsymbol{r}'') = 0.
\end{aligned} \tag{12.127}$$

Next, integrate this equation over all of Ω and use the fundamental theorem of 3D calculus. The integrand of the surface integral is then zero from the boundary conditions of Eq. (12.120), which results in $W_{kl}(\boldsymbol{r}'|\boldsymbol{r}'') = W_{lk}(\boldsymbol{r}''|\boldsymbol{r}')$. Rewriting this result in the bold-face notation applicable to any coordinate system and substituting $\boldsymbol{r}'' \to \boldsymbol{r}$ then gives

$$\boxed{\boldsymbol{W}(\boldsymbol{r}'|\boldsymbol{r}) = \boldsymbol{W}^T(\boldsymbol{r}|\boldsymbol{r}')} \tag{12.128}$$

as the general reciprocity condition describing what happens when the source point and observation point are exchanged.

If viscosity η is uniform in an unbounded domain, we have additional symmetry properties of $\boldsymbol{W}$ and $\boldsymbol{Q}$ that are obtained from the governing equation in this case

$$\eta\nabla^2\boldsymbol{W} - \nabla\boldsymbol{Q} = -\boldsymbol{I}\delta(\boldsymbol{r}-\boldsymbol{r}'). \tag{12.129}$$

Because there are no boundaries or heterogeneity influencing the $\boldsymbol{r}$ and $\boldsymbol{r}'$ dependence in this equation, we have $\boldsymbol{W}(\boldsymbol{r}|\boldsymbol{r}') = \boldsymbol{W}(\boldsymbol{r}-\boldsymbol{r}')$ and $\boldsymbol{Q}(\boldsymbol{r}|\boldsymbol{r}') = \boldsymbol{Q}(\boldsymbol{r}-\boldsymbol{r}')$. Taking the divergence of Eq. (12.129) and using the incompressible flow requirement $\nabla\cdot\boldsymbol{W} = 0$ gives

$$\nabla^2\boldsymbol{Q}(\boldsymbol{r}-\boldsymbol{r}') = \nabla\delta(\boldsymbol{r}-\boldsymbol{r}'). \tag{12.130}$$

Make the substitution $\boldsymbol{r} \leftrightarrow \boldsymbol{r}'$ to give

$$\nabla'^2 \boldsymbol{Q}(\boldsymbol{r}' - \boldsymbol{r}) = \nabla' \delta(\boldsymbol{r} - \boldsymbol{r}'). \tag{12.131}$$

Because $\delta(\boldsymbol{r} - \boldsymbol{r}') = \delta(\boldsymbol{r}' - \boldsymbol{r})$ and $\nabla\delta(\boldsymbol{r} - \boldsymbol{r}') = -\nabla'\delta(\boldsymbol{r} - \boldsymbol{r}')$, we have $\nabla^2 \boldsymbol{Q}(\boldsymbol{r} - \boldsymbol{r}') = -\nabla'^2 \boldsymbol{Q}(\boldsymbol{r}' - \boldsymbol{r})$. Further, $\nabla' \boldsymbol{Q}(\boldsymbol{r} - \boldsymbol{r}') = -\nabla \boldsymbol{Q}(\boldsymbol{r} - \boldsymbol{r}')$ and $\nabla'^2 \boldsymbol{Q}(\boldsymbol{r} - \boldsymbol{r}') = +\nabla^2 \boldsymbol{Q}(\boldsymbol{r} - \boldsymbol{r}')$, so that we obtain the Laplace equation $\nabla'^2 \left[\boldsymbol{Q}(\boldsymbol{r} - \boldsymbol{r}') + \boldsymbol{Q}(\boldsymbol{r}' - \boldsymbol{r})\right] = 0$. Because the fields are zero at infinity in an unbounded medium, there is no nonzero "boundary data" generating nonzero solutions of this Laplace equation so that

$$\boldsymbol{Q}(\boldsymbol{r} - \boldsymbol{r}') = -\boldsymbol{Q}(\boldsymbol{r}' - \boldsymbol{r}). \tag{12.132}$$

Making the same substitution $\boldsymbol{r} \leftrightarrow \boldsymbol{r}'$ in Eq. (12.129), introducing $\boldsymbol{Q}(\boldsymbol{r} - \boldsymbol{r}') = -\boldsymbol{Q}(\boldsymbol{r}' - \boldsymbol{r})$ and going through a similar argument results in $\nabla'^2 \left[\boldsymbol{W}(\boldsymbol{r} - \boldsymbol{r}') - \boldsymbol{W}(\boldsymbol{r}' - \boldsymbol{r})\right] = 0$ so that

$$\boldsymbol{W}(\boldsymbol{r} - \boldsymbol{r}') = \boldsymbol{W}(\boldsymbol{r}' - \boldsymbol{r}). \tag{12.133}$$

From the general reciprocity relation of Eq. (12.128), we then have $\boldsymbol{W}(\boldsymbol{r} - \boldsymbol{r}') = \boldsymbol{W}^T(\boldsymbol{r} - \boldsymbol{r}')$. These various symmetries in an unbounded uniform domain can also be seen by obtaining the explicit functional form for $\boldsymbol{W}(\boldsymbol{r} - \boldsymbol{r}')$ and $\boldsymbol{Q}(\boldsymbol{r} - \boldsymbol{r}')$ as we now will do.

12.4.2 The Green's Tensor of Slow Viscous Flow in a Wholespace

Put a point body force $\boldsymbol{s}\delta(\boldsymbol{r})$ as a source term in the Stokes-flow equations of slow, incompressible, steady flow

$$\eta\nabla^2 \boldsymbol{v}(\boldsymbol{r}) - \nabla P(\boldsymbol{r}) + \boldsymbol{s}\delta(\boldsymbol{r}) = 0, \tag{12.134}$$

$$\nabla \cdot \boldsymbol{v}(\boldsymbol{r}) = 0, \tag{12.135}$$

in a domain of infinite extent so that fields are zero on the domain boundary at infinity. Here, $\boldsymbol{s}$ is a given force vector (units of Newtons). Our goal is to find the solutions $\boldsymbol{v}(\boldsymbol{r}) = \boldsymbol{W}(\boldsymbol{r}) \cdot \boldsymbol{s}$ and $P(\boldsymbol{r}) = \boldsymbol{Q}(\boldsymbol{r}) \cdot \boldsymbol{s}$ of these governing equations. Upon placing the point source at $\boldsymbol{r}'$, we thus obtain our desired Green's tensor $\boldsymbol{W}(\boldsymbol{r}|\boldsymbol{r}') = \boldsymbol{W}(\boldsymbol{r} - \boldsymbol{r}')$ and pressure vector $\boldsymbol{Q}(\boldsymbol{r}|\boldsymbol{r}') = \boldsymbol{Q}(\boldsymbol{r} - \boldsymbol{r}')$. We will use 3D Fourier transformation to obtain these results.

Upon taking the Fourier transform and using the usual facts $\nabla \to i\boldsymbol{k}$, $\nabla^2 \to -k^2$ and $\delta(\boldsymbol{r}) \to 1$, the transformed Stokes-flow equations with a given point force (source) are

$$-\eta k^2 \tilde{\boldsymbol{v}}(\boldsymbol{k}) - i\boldsymbol{k}\tilde{P}(\boldsymbol{k}) + \boldsymbol{s} = 0, \tag{12.136}$$

$$i\boldsymbol{k} \cdot \tilde{\boldsymbol{v}}(\boldsymbol{k}) = 0. \tag{12.137}$$

Dot multiply the transformed Stokes equation with $i\boldsymbol{k}$ and use the transformed incompressibility condition to obtain

$$\tilde{P}(k) = -\frac{i\boldsymbol{k}}{k^2} \cdot \boldsymbol{s}. \tag{12.138}$$

We perform the inverse Fourier transform of this transformed pressure field using spherical-$\boldsymbol{k}$ coordinates for which a volume element of $\boldsymbol{k}$ space is $\mathrm{d}^3\boldsymbol{k} = -k^2 \mathrm{d}k\, \mathrm{d}(\cos\theta)\, \mathrm{d}\phi$

and $e^{i\boldsymbol{k}\cdot\boldsymbol{r}} = e^{ikr\cos\theta}$. Further, multiplication by $i\boldsymbol{k}$ in the frequency domain becomes the gradient operation ∇ in the real-space domain, which gives

$$P(r) = -\nabla\left[\frac{1}{(2\pi)^3}\int_0^\infty dk\frac{k^2}{k^2}\int_{-1}^{1} d(\cos\theta)e^{ikr\cos\theta}\int_0^{2\pi} d\phi\right]\cdot \boldsymbol{s}. \tag{12.139}$$

The term in square brackets is exactly minus the Green's function of the Laplace equation. In the guided Exercise 2 at the end of the chapter, you will prove that the 3D inverse-Fourier transform of k^{-2} is

$$\frac{1}{(2\pi)^3}\int_0^\infty dk\int_{-1}^{1} d(\cos\theta)e^{ikr\cos\theta}\int_0^{2\pi} d\phi = \frac{1}{4\pi r}. \tag{12.140}$$

Thus, we find the real pressure field generated by the point force at the origin to be

$$P(r) = -\nabla\left(\frac{1}{4\pi r}\right)\cdot \boldsymbol{s} = \frac{\hat{\boldsymbol{r}}}{4\pi r^2}\cdot \boldsymbol{s}, \tag{12.141}$$

when there is radial symmetry $\nabla = \hat{\boldsymbol{r}}\partial/\partial r$. If we now place the point source at $\boldsymbol{r}'$ and note that $\hat{\boldsymbol{r}} = \boldsymbol{r}/r$, we obtain the Green's pressure vector in an unbounded, uniform domain

$$\boxed{\boldsymbol{Q}(\boldsymbol{r}-\boldsymbol{r}') = \frac{\boldsymbol{r}-\boldsymbol{r}'}{4\pi|\boldsymbol{r}-\boldsymbol{r}'|^3},} \tag{12.142}$$

where $|\boldsymbol{r}-\boldsymbol{r}'| = \sqrt{(\boldsymbol{r}-\boldsymbol{r}')\cdot(\boldsymbol{r}-\boldsymbol{r}')}$. This result does indeed possess the symmetry $\boldsymbol{Q}(\boldsymbol{r}-\boldsymbol{r}') = -\boldsymbol{Q}(\boldsymbol{r}'-\boldsymbol{r})$ as required.

Using the above expression for $\tilde{P}(k)$ in the transformed Stokes equation gives

$$\tilde{\boldsymbol{v}}(\boldsymbol{k}) = \frac{1}{\eta}\left[\frac{(i\boldsymbol{k})(i\boldsymbol{k})}{k^4} + \frac{\boldsymbol{I}}{k^2}\right]\cdot \boldsymbol{s}. \tag{12.143}$$

Again taking the inverse transform in spherical-$\boldsymbol{k}$ coordinates $d^3\boldsymbol{k} = -k^2 dk\, d(\cos\theta)\, d\phi$ and $e^{i\boldsymbol{k}\cdot\boldsymbol{r}} = e^{ikr\cos\theta}$ as well as $i\boldsymbol{k} \to \nabla$ gives

$$\begin{aligned}\boldsymbol{v}(\boldsymbol{r}) = {} & \frac{1}{\eta}\nabla\nabla\left[\frac{1}{(2\pi)^3}\int_0^\infty dk\frac{k^2}{k^4}\int_{-1}^{1} d(\cos\theta)e^{ikr\cos\theta}\int_0^{2\pi} d\phi\right]\cdot \boldsymbol{s} \\ & + \frac{\boldsymbol{I}}{\eta}\left[\frac{1}{(2\pi)^3}\int_0^\infty dk\frac{k^2}{k^2}\int_{-1}^{1} d(\cos\theta)e^{ikr\cos\theta}\int_0^{2\pi} d\phi\right]\cdot \boldsymbol{s}.\end{aligned} \tag{12.144}$$

As you will prove in the guided Exercise 3 at the end of the chapter, the term in the first square brackets is exactly the Green's function of the biharmonic equation. You will show in Exercise 3 that

$$\frac{1}{(2\pi)^3}\int_0^\infty \frac{dk}{k^2}\int_{-1}^{1} e^{ikr\cos\theta} d(\cos\theta)\int_0^{2\pi} d\phi = -\frac{r}{8\pi}. \tag{12.145}$$

Further, the term in the second square brackets is again minus the Green's function of the Laplace equation. Thus, we have

$$v(r) = \frac{1}{\eta}\nabla\nabla\left[\frac{-r}{8\pi}\right]\cdot s + \frac{I}{\eta}\left[\frac{1}{4\pi r}\right]\cdot s. \tag{12.146}$$

Because $\nabla r = \hat{r} = r/r$ and $\nabla r = I$, we have

$$\nabla\nabla\left(\frac{-r}{8\pi}\right) = -\frac{1}{8\pi}\nabla\left(\frac{r}{r}\right) = -\frac{1}{8\pi}\left(\frac{I}{r} - \frac{rr}{r^2}\right) = -\frac{1}{8\pi}\left(\frac{I}{r} - \frac{\hat{r}\hat{r}}{r}\right), \tag{12.147}$$

so that the Stokes-flow velocity induced by a directed point force at the origin is given by

$$v(r) = \frac{1}{8\pi\eta r}\left(I + \hat{r}\hat{r}\right)\cdot s, \tag{12.148}$$

which is also called the "Stokeslet." Upon moving the source point to r', we then obtain the Stokes-flow Green's tensor in a uniform, unbounded domain as

$$\boxed{W(r - r') = \frac{1}{8\pi\eta|r - r'|}\left(I + \frac{(r - r')(r - r')}{|r - r'|^2}\right),} \tag{12.149}$$

a result that is also called the "Oseen tensor." We do indeed see that this tensor satisfies the required symmetries that $W(r - r') = W(r' - r)$ and $W(r - r') = W^T(r - r')$.

12.5 Green's Tensors for Continuum Electromagnetics

In Chapter 3, we derived the continuum governing equations controlling electromagnetic (EM) response. We also derived the boundary and initial conditions that must be prescribed for some domain so that the EM fields can be determined uniquely. However, we did not derive the representation theorem for EM response as we now will do.

12.5.1 The "Green's Function Method" Applied to Continuum EM

For an isotropic domain Ω having uniform magnetic permeability μ, we are interested here in solutions of

$$\nabla \times E(r,t) = -\mu\frac{\partial H(r,t)}{\partial t}, \tag{12.150}$$

$$\nabla \times H(r,t) = \varepsilon(r)\frac{\partial E(r,t)}{\partial t} + \sigma(r)E(r,t) + J_s(r,t), \tag{12.151}$$

where the electrical conductivity σ and permittivity ε are possibly heterogeneous throughout Ω. As derived in Section 3.11 of Chapter 3, unique solutions require that either

$$n \times E = b_e(r,t) \quad \text{or} \quad n \times \nabla \times E = b_h(r,t) \tag{12.152}$$

are specified at each point of the surface $\partial\Omega$, where b_e or b_h are given vectors that act as source terms when nonzero. The vector field J_s is a given current-density distribution that acts as another source term. We will assume (though this is not essential) that $\boldsymbol{J}_s(\boldsymbol{r}, t) = 0$ for $t < 0$. Unique solutions also require that

$$\boldsymbol{E}(\boldsymbol{r}, 0) \quad \text{and} \quad \left.\frac{\partial \boldsymbol{E}(\boldsymbol{r}, t)}{\partial t}\right|_{t=0} \tag{12.153}$$

are both given throughout Ω at $t = 0$. If these initial conditions are nonzero, they too can act as source terms for the EM response. In the usual way, by taking the curl of Faraday's law and the time-derivative of Ampère's law under the assumption of uniform μ and upon using the identity $\nabla \times \nabla \times \boldsymbol{E} = \nabla\nabla \cdot \boldsymbol{E} - \nabla^2 \boldsymbol{E}$, we obtain the governing equation for the electric field to be

$$\nabla\nabla \cdot \boldsymbol{E} - \nabla^2 \boldsymbol{E} = -\varepsilon\mu \frac{\partial^2 \boldsymbol{E}}{\partial t^2} - \sigma\mu \frac{\partial \boldsymbol{E}}{\partial t} - \mu \frac{\partial \boldsymbol{J}_s}{\partial t}. \tag{12.154}$$

Once this electric field is determined, the magnetic field is obtained by time integrating Faraday's law.

Corresponding to the above well-posed form of an EM BVP, we can define the electric-field Green's tensor from the equation

$$\nabla\nabla \cdot \boldsymbol{G} - \nabla^2 \boldsymbol{G} = -\varepsilon\mu \frac{\partial^2 \boldsymbol{G}}{\partial t^2} - \sigma\mu \frac{\partial \boldsymbol{G}}{\partial t} - \boldsymbol{I}\delta(\boldsymbol{r} - \boldsymbol{r}')\delta(t - t'), \tag{12.155}$$

where $\boldsymbol{I}$ is the second-order identity tensor and the second-order tensor $\boldsymbol{G} = \boldsymbol{G}(\boldsymbol{r}, t|\boldsymbol{r}', t')$ describes how a step current impulse applied at $\boldsymbol{r}'$ and time t' produces an electric field at on observation point $\boldsymbol{r}$ at a later time t. We require the Green's tensor to satisfy the causal initial condition that

$$\boldsymbol{G}(\boldsymbol{r}, t|\boldsymbol{r}', t') = 0 \quad \text{for } t < t'. \tag{12.156}$$

In addition, the Green's tensor is required to satisfy the homogeneous version of whatever boundary conditions are specified in Eq. (12.152), that is,

$$\boldsymbol{n} \times \boldsymbol{G} = 0 \quad \text{or} \quad \boldsymbol{n} \times \nabla \times \boldsymbol{G} = 0 \tag{12.157}$$

for each point on the boundary $\partial\Omega$. Further, this electric-field Green's tensor satisfies the reciprocity relation $\boldsymbol{G}(\boldsymbol{r}', -t'|\boldsymbol{r}, -t) = \boldsymbol{G}^T(\boldsymbol{r}, t|\boldsymbol{r}', t')$, which will be needed in the derivation below of the EM representation theorem.

We begin by deriving this reciprocity relation. To do so, we work for convenience in Cartesian coordinates and write the Green's tensor PDE in component form for the components $G_{im} \,\widehat{=}\, G_{im}(\boldsymbol{r}, t|\boldsymbol{r}', t')$

$$\frac{\partial^2 G_{jm}}{\partial x_i \partial x_j} - \frac{\partial^2 G_{im}}{\partial x_j^2} = -\varepsilon\mu \frac{\partial^2 G_{im}}{\partial t^2} - \sigma\mu \frac{\partial G_{im}}{\partial t} - \delta_{im}\delta(\boldsymbol{r} - \boldsymbol{r}')\delta(t - t'), \tag{12.158}$$

where δ_{im} is the Kronecker delta. We now make the change of variables in this equation $\boldsymbol{r}' \to \boldsymbol{r}''$, $t' \to -t''$, and $t \to -t$ and replace the free index m with n to create a PDE for the components denoted as $G'_{in} \,\widehat{=}\, G_{in}(\boldsymbol{r}, -t|\boldsymbol{r}'', -t'')$

$$\frac{\partial^2 G'_{jn}}{\partial x_i \partial x_j} - \frac{\partial^2 G'_{in}}{\partial x_j^2} = -\varepsilon\mu \frac{\partial^2 G'_{in}}{\partial t^2} + \sigma\mu \frac{\partial G'_{in}}{\partial t} - \delta_{in}\delta(\boldsymbol{r}-\boldsymbol{r}'')\delta(t-t''). \tag{12.159}$$

Note the important sign change in the diffusion term (the term involving σ) due to changing the direction of time. As usual, we multiply Eq. (12.159) by $G_{im} = G_{im}(\boldsymbol{r}, t|\boldsymbol{r}', t')$ and multiply Eq. (12.158) by $G'_{in} = G_{in}(\boldsymbol{r}, -t|\boldsymbol{r}'', -t'')$, distribute the spatial derivatives and subtract to obtain

$$\begin{aligned}\frac{\partial}{\partial x_i}\left(G_{im}\frac{\partial G'_{jn}}{\partial x_j} - G'_{in}\frac{\partial G_{jm}}{\partial x_j}\right) - \frac{\partial}{\partial x_j}\left(G_{im}\frac{\partial G'_{in}}{\partial x_j} - G'_{in}\frac{\partial G_{im}}{\partial x_j}\right)\\ = -\varepsilon\mu\frac{\partial}{\partial t}\left(G_{im}\frac{\partial G'_{in}}{\partial t} - G'_{in}\frac{\partial G_{im}}{\partial t}\right) - \sigma\mu\frac{\partial}{\partial t}\left(G_{im}G'_{in}\right)\\ + G'_{mn}\delta(\boldsymbol{r}-\boldsymbol{r}')\delta(t-t') - G_{nm}\delta(\boldsymbol{r}-\boldsymbol{r}'')\delta(t-t'').\end{aligned} \tag{12.160}$$

Next, we integrate $\boldsymbol{r}$ over Ω and integrate t from $-\infty$ to $+\infty$ and use the fundamental theorem of 3D calculus to obtain

$$\begin{aligned}\int_{\partial\Omega} \mathrm{d}^2\boldsymbol{r}\left[G_{im}\left(n_i\frac{\partial G'_{jn}}{\partial x_j} - n_j\frac{\partial G'_{in}}{\partial x_j}\right) - G'_{in}\left(n_i\frac{\partial G_{jm}}{\partial x_j} - n_j\frac{\partial G_{im}}{\partial x_j}\right)\right]\\ = -\int_{\Omega}\mathrm{d}^3\boldsymbol{r}\left[\varepsilon\mu\left(G_{im}\frac{\partial G'_{in}}{\partial t} - G'_{in}\frac{\partial G_{im}}{\partial t}\right)\Bigg|_{t=-\infty}^{t=\infty} + \sigma\mu\, G_{im}G'_{in}\Big|_{t=-\infty}^{t=\infty}\right]\\ + G_{mn}(\boldsymbol{r}', -t'|\boldsymbol{r}'', -t'') - G_{nm}(\boldsymbol{r}'', t''|\boldsymbol{r}', t').\end{aligned} \tag{12.161}$$

Because $G'_{in}\big|_{t=\infty} = G_{in}(\boldsymbol{r}, t=-\infty|\boldsymbol{r}'', t'') = 0$ and $G_{im}|_{t=-\infty} = G_{im}(\boldsymbol{r}, t=-\infty|\boldsymbol{r}', t') = 0$ from the initial condition on the Green's tensor, the integral over Ω is zero on the right-hand side. If we now return to the bold-face notation representing the tensors in any coordinate system, we have

$$\begin{aligned}\int_{\partial\Omega}\mathrm{d}^2\boldsymbol{r}\,\left\{\boldsymbol{G}^T\cdot\left[\boldsymbol{n}\left(\nabla\cdot\boldsymbol{G}'\right) - \boldsymbol{n}\cdot\nabla\boldsymbol{G}'\right] - (\boldsymbol{G}')^T\cdot\left[\boldsymbol{n}(\nabla\cdot\boldsymbol{G}) - \boldsymbol{n}\cdot\nabla\boldsymbol{G}\right]\right\}\\ = \boldsymbol{G}(\boldsymbol{r}', -t'|\boldsymbol{r}'', -t'') - \boldsymbol{G}^T(\boldsymbol{r}'', t''|\boldsymbol{r}', t').\end{aligned} \tag{12.162}$$

The integrand of the surface integral can be written

$$\begin{aligned}\boldsymbol{G}^T\cdot\left[\boldsymbol{n}(\nabla\cdot\boldsymbol{G}') - \boldsymbol{n}\cdot\nabla\boldsymbol{G}'\right] - (\boldsymbol{G}')^T\cdot\left[\boldsymbol{n}(\nabla\cdot\boldsymbol{G}) - \boldsymbol{n}\cdot\nabla\boldsymbol{G}\right]\\ = \boldsymbol{G}^T\cdot\left(\boldsymbol{n}\times\nabla\times\boldsymbol{G}'\right) - (\boldsymbol{G}')^T\cdot(\boldsymbol{n}\times\nabla\times\boldsymbol{G}),\end{aligned} \tag{12.163}$$

where we used the identity proven in Exercise 8 of Chapter 1 that $\boldsymbol{n}\times\nabla\times\boldsymbol{G} = \boldsymbol{n}(\nabla\cdot\boldsymbol{G}) - \boldsymbol{n}\cdot\nabla\boldsymbol{G}$. Because of the homogeneous boundary conditions of Eq. (12.157) that $\boldsymbol{G}(\boldsymbol{r}, t|\boldsymbol{r}', t')$ and $\boldsymbol{G}' = \boldsymbol{G}(\boldsymbol{r}, -t|\boldsymbol{r}'', -t'')$ both satisfy for points $\boldsymbol{r}$ on $\partial\Omega$, the integrand of the surface integral is zero on $\partial\Omega$. Thus, upon making the substitution $\boldsymbol{r}''\to\boldsymbol{r}$ and $t''\to t$, we arrive at

$$\boxed{\boldsymbol{G}(\boldsymbol{r}', -t'|\boldsymbol{r}, -t) = \boldsymbol{G}^T(\boldsymbol{r}, t|\boldsymbol{r}', t')} \tag{12.164}$$

as the needed statement of EM reciprocity that the electric-field Green's tensor satisfies.

To derive the EM representation theorem for the electric field $\boldsymbol{E}(\boldsymbol{r}, t)$, we again work for convenience in Cartesian coordinates and rewrite Eq. (12.154) in component form after making the substitutions $\boldsymbol{r} \to \boldsymbol{r}'$ and $t \to t'$:

$$\frac{\partial^2 E_j(\boldsymbol{r}', t')}{\partial x_i' \partial x_j'} - \frac{\partial^2 E_i(\boldsymbol{r}', t')}{\partial x_j'^2} = -\varepsilon\mu \frac{\partial^2 E_i(\boldsymbol{r}', t')}{\partial t'^2} - \sigma\mu \frac{\partial E_i(\boldsymbol{r}', t')}{\partial t'} - \mu \frac{\partial J_{si}(\boldsymbol{r}', t')}{\partial t'}. \quad (12.165)$$

In the Green's tensor governing equation of Eq. (12.158), make the substitutions $\boldsymbol{r} \leftrightarrow \boldsymbol{r}'$ and $t \leftrightarrow -t'$ so that G_{im} has the dependence $G_{im}(\boldsymbol{r}', -t'|\boldsymbol{r}, -t)$, which the reciprocity relation then gives as $G_{im}(\boldsymbol{r}', -t'|\boldsymbol{r}, -t) = G_{mi}(\boldsymbol{r}, t|\boldsymbol{r}', t')$. After these substitutions, Eq. (12.158) becomes

$$\frac{\partial^2 G_{mj}(\boldsymbol{r}, t|\boldsymbol{r}', t')}{\partial x_i' \partial x_j'} - \frac{\partial^2 G_{mi}(\boldsymbol{r}, t|\boldsymbol{r}', t')}{\partial x_j'^2} = -\varepsilon\mu \frac{\partial^2 G_{mi}(\boldsymbol{r}, t|\boldsymbol{r}', t')}{\partial t'^2} + \sigma\mu \frac{\partial G_{mi}(\boldsymbol{r}, t|\boldsymbol{r}', t')}{\partial t'}$$
$$- \delta_{im}\delta(\boldsymbol{r}' - \boldsymbol{r})\delta(t' - t). \quad (12.166)$$

We now multiply Eq. (12.165) by $G_{mi}(\boldsymbol{r}, t|\boldsymbol{r}', t')$ and multiply Eq. (12.166) by $E_i(\boldsymbol{r}', t')$, subtract and redistribute derivatives to give

$$\frac{\partial}{\partial x_i'}\left(G_{mi}\frac{\partial E_j}{\partial x_j'} - E_i \frac{\partial G_{mj}}{\partial x_j'}\right) - \frac{\partial}{\partial x_j'}\left(G_{mi}\frac{\partial E_i}{\partial x_j'} - E_i\frac{\partial G_{mi}}{\partial x_j'}\right)$$
$$= -\varepsilon\mu \frac{\partial}{\partial t'}\left(G_{mi}\frac{\partial E_i}{\partial t'} - E_i\frac{\partial G_{mi}}{\partial t'}\right) - \sigma\mu\frac{\partial}{\partial t'}(G_{mi}E_i)$$
$$- \mu G_{mi}\frac{\partial J_{si}}{\partial t'} + E_i\delta_{im}\delta(\boldsymbol{r}' - \boldsymbol{r})\delta(t' - t). \quad (12.167)$$

Next, integrate $\boldsymbol{r}'$ over all of Ω and integrate t' from 0 to ∞. Because $G_{mi}(\boldsymbol{r}, t|\boldsymbol{r}', t') = 0$ for $t' > t$, both $G_{mi}(t' = \infty) = 0$ and $\left(\partial G_{mi}/\partial t'\right)\big|_{t'=\infty} = 0$. For the same reason, it is not necessary to integrate t' beyond t. Using the fundamental theorem of 3D calculus, we then have

$$E_m(\boldsymbol{r}, t) = \mu \int_0^t \mathrm{d}t' \int_\Omega \mathrm{d}^3\boldsymbol{r}' G_{mi}\frac{\partial J_{si}}{\partial t'}$$
$$+ \int_0^t \mathrm{d}t' \int_{\partial\Omega} \mathrm{d}^2\boldsymbol{r}' \left[G_{mi}\left(n_i\frac{\partial E_j}{\partial x_j'} - n_j\frac{\partial E_i}{\partial x_j'}\right) - E_i\left(n_i\frac{\partial G_{mj}}{\partial x_j'} - n_j\frac{\partial G_{mi}}{\partial x_j'}\right)\right]$$
$$- \int_\Omega \mathrm{d}^3\boldsymbol{r}' \left[\varepsilon\mu\left(G_{mi}\frac{\partial E_i}{\partial t'} - E_i\frac{\partial G_{mi}}{\partial t'}\right)\bigg|_{t'=0} + \sigma\mu\,(G_{mi}E_i)|_{t'=0}\right]. \quad (12.168)$$

Finally, we return to the bold-face tensor notation applicable to any coordinate system and use the identity that $\boldsymbol{n} \times \nabla \times \boldsymbol{E} = \boldsymbol{n}\nabla \cdot \boldsymbol{E} - \boldsymbol{n} \cdot \nabla\boldsymbol{E}$ on $\partial\Omega$ to obtain the *electromagnetic representation theorem* in final form:

$$
\begin{aligned}
\boldsymbol{E}(\boldsymbol{r}, t) = \mu \int_0^t \mathrm{d}t' \int_\Omega \mathrm{d}^3r' \frac{\partial \boldsymbol{J}_s(\boldsymbol{r}', t')}{\partial t'} \cdot \boldsymbol{G}(\boldsymbol{r}, t|\boldsymbol{r}', t') \\
+ \int_0^t \mathrm{d}t' \int_{\partial\Omega} \mathrm{d}^2r' \Big[\big(\boldsymbol{n} \times \nabla' \times \boldsymbol{E}(\boldsymbol{r}', t')\big) \cdot \boldsymbol{G}(\boldsymbol{r}, t|\boldsymbol{r}', t') \\
- \boldsymbol{E}(\boldsymbol{r}', t') \cdot \big(\boldsymbol{n} \times \nabla' \times \boldsymbol{G}(\boldsymbol{r}, t|\boldsymbol{r}', t')\big)\Big] \\
- \int_\Omega \mathrm{d}^3r' \Bigg[\varepsilon\mu \left(\frac{\partial \boldsymbol{E}(\boldsymbol{r}', t')}{\partial t'} \cdot \boldsymbol{G}(\boldsymbol{r}, t|\boldsymbol{r}', t') - \boldsymbol{E}(\boldsymbol{r}', t') \cdot \frac{\partial \boldsymbol{G}(\boldsymbol{r}, t|\boldsymbol{r}', t')}{\partial t'} \right)\Bigg|_{t'=0} \\
+ \sigma\mu \boldsymbol{E}(\boldsymbol{r}, 0) \cdot \boldsymbol{G}(\boldsymbol{r}, t|\boldsymbol{r}', 0) \Bigg].
\end{aligned}
\tag{12.169}
$$

This shows that if you have the electric-field Green's tensor $\boldsymbol{G}(\boldsymbol{r}, t|\boldsymbol{r}', t')$ and given nonzero values of a current-density source term in Ampère's law and/or given nonzero boundary values of the electric field on $\partial\Omega$ and/or given nonzero initial conditions for the electric field throughout Ω, the electric field generated by these various types of sources is obtained by performing the integrals of the representation theorem.

12.5.2 The Green's Tensors of Electromagnetic Response in a Wholespace

For a uniform isotropic medium of infinite extent, we seek solutions of the EM governing equations

$$
\nabla \times \boldsymbol{E}(\boldsymbol{r}, t) = -\mu \frac{\partial \boldsymbol{H}(\boldsymbol{r}, t)}{\partial t}, \tag{12.170}
$$

$$
\nabla \times \boldsymbol{H}(\boldsymbol{r}, t) = \varepsilon \frac{\partial \boldsymbol{E}(\boldsymbol{r}, t)}{\partial t} + \sigma \boldsymbol{E}(\boldsymbol{r}, t) + \boldsymbol{s}\, \delta(\boldsymbol{r}) S(t). \tag{12.171}
$$

From the solutions of these linear equations, which are proportional to the constant source vector $\boldsymbol{s}$ that is identified in the next paragraph, we obtain the electric-field Green's tensor from $\boldsymbol{E}(\boldsymbol{r}, t) = \boldsymbol{G}(\boldsymbol{r}, t) \cdot \mu \boldsymbol{s}$ and finish by making the substitution $\boldsymbol{r} \to \boldsymbol{r} - \boldsymbol{r}'$ and $t \to t - t'$.

As we now show, the point source $\boldsymbol{J}_s = \boldsymbol{s}\, \delta(\boldsymbol{r}) S(t)$ in Ampère's law of Eq. (12.171) corresponds to the sudden appearance at $\boldsymbol{r} = 0$ and $t = 0$ of a current dipole, where $S(t)$ is the unit step function. A current dipole corresponds to one electrode embedded in the material injecting a current I and another electrode a distance $\boldsymbol{d}$ from the first extracting a current $-I$. The charge-density distribution of this dipole is $\rho_s(\boldsymbol{r}, t) = q(t)\delta(\boldsymbol{r} - \boldsymbol{d}/2) - q(t)\delta(\boldsymbol{r} + \boldsymbol{d}/2)$, where $q(t)$ is the charge emerging at the injection electrode and $-q(t)$ that disappearing at the extraction electrode. Charge conservation requires that $\nabla \cdot \boldsymbol{J}_s = -\partial \rho_s / \partial t$, which upon inserting $\boldsymbol{J}_s = \boldsymbol{s}\, \delta(\boldsymbol{r})$ for times $t > 0$ yields

$$
\nabla \cdot \boldsymbol{J}_s = \boldsymbol{s} \cdot \nabla \delta(\boldsymbol{r}) = -\frac{\partial}{\partial t} \left\{ q(t) \left[\delta\left(\boldsymbol{r} - \frac{\boldsymbol{d}}{2}\right) - \delta\left(\boldsymbol{r} + \frac{\boldsymbol{d}}{2}\right) \right] \right\} \tag{12.172}
$$

$$= -\frac{\mathrm{d}q(t)}{\mathrm{d}t}\left[\delta(\boldsymbol{r}) - \frac{\boldsymbol{d}}{2}\cdot\nabla\delta(\boldsymbol{r}) + \ldots - \delta(\boldsymbol{r}) - \frac{\boldsymbol{d}}{2}\cdot\nabla\delta(\boldsymbol{r}) - \ldots\right] \tag{12.173}$$

$$= I(t)\,\boldsymbol{d}\cdot\nabla\delta(\boldsymbol{r}), \tag{12.174}$$

where $I(t) = \mathrm{d}q(t)/\mathrm{d}t$ is the injected current and we have expanded the Dirac delta associated with each point charge of the dipole in a Taylor series, assuming that $\boldsymbol{d}$ is small relative to observation distances in order to truncate the expansions. Thus, we have shown that $\boldsymbol{s} = I\boldsymbol{d}$, which we will take to be independent of time for our task at hand of finding the EM Green's tensors for an infinite homogeneous body. The source term $\boldsymbol{J}_s = \boldsymbol{s}\,\delta(\boldsymbol{r})S(t)$ indeed corresponds to turning on at $t = 0$ a current $+I$ at the injection electrode and $-I$ at the extraction electrode that is a small distance $\boldsymbol{d}$ from the injection electrode.

In what follows, we first obtain the solution of these equations valid for any values of ε and σ, which involves a final inverse Fourier transform over temporal frequency ω that cannot be performed analytically in closed form. We then obtain the explicit Green's tensor corresponding to use of the representation theorem when the source has support over only high frequencies and the response is pure unattenuated wave propagation. After that, we obtain the Green's tensor when the response is purely diffusive. You will treat the electrostatic case as a guided end-of-chapter exercise.

By combining Eqs (12.170) and (12.171) in the usual way that eliminates the magnetic field $\boldsymbol{H}$ and using the Chapter 1 vector identity that $\nabla\times\nabla\times\boldsymbol{E} = \nabla\nabla\cdot\boldsymbol{E} - \nabla^2\boldsymbol{E}$, the electric field is controlled by the equation

$$\nabla\nabla\cdot\boldsymbol{E}(\boldsymbol{r},t) - \nabla^2\boldsymbol{E}(\boldsymbol{r},t) = -\varepsilon\mu\frac{\partial^2\boldsymbol{E}(\boldsymbol{r},t)}{\partial t^2} - \sigma\mu\frac{\partial\boldsymbol{E}(\boldsymbol{r},t)}{\partial t} - \delta(\boldsymbol{r})\delta(t)\mu\boldsymbol{s}, \tag{12.175}$$

where we used that $\delta(t) = \mathrm{d}S(t)/\mathrm{d}t$ in the time derivative of the source term. We now take Fourier transforms over space and time and use the usual transforms that $\nabla \to i\boldsymbol{k}$ and $\partial/\partial t \to -i\omega$ to obtain

$$\left[\boldsymbol{I} - \frac{\boldsymbol{k}\boldsymbol{k}}{k^2 - \varepsilon\mu\omega(\omega + i\sigma/\varepsilon)}\right]\cdot\tilde{\tilde{\boldsymbol{E}}}(\boldsymbol{k},\omega) = \frac{-\mu\boldsymbol{s}}{k^2 - \varepsilon\mu\omega(\omega + i\sigma/\varepsilon)}. \tag{12.176}$$

This equation is solved algebraically by multiplying through by the inverse tensor of Eq. (12.101) to obtain

$$\tilde{\tilde{\boldsymbol{E}}}(\boldsymbol{k},\omega) = -\left[\boldsymbol{I} + \frac{(i\boldsymbol{k})(i\boldsymbol{k})}{\varepsilon\mu\omega(\omega + i\sigma/\varepsilon)}\right]\cdot\frac{\mu\boldsymbol{s}}{\left[k + \sqrt{\varepsilon\mu\omega(\omega + i\sigma/\varepsilon)}\right]\left[k - \sqrt{\varepsilon\mu\omega(\omega + i\sigma/\varepsilon)}\right]}. \tag{12.177}$$

So we obtain the transformed electric field relatively easily and now must perform the inverse Fourier transforms.

Doing the inverse transforms over $\boldsymbol{k}$ first we obtain

$$\tilde{\boldsymbol{E}}(\boldsymbol{r},\omega) = -\left[\boldsymbol{I} + \frac{\nabla\nabla f(r,\omega)}{\varepsilon\mu\omega(\omega + i\sigma/\varepsilon)}\right]\cdot\mu\boldsymbol{s} \tag{12.178}$$

where $f(r, \omega)$ is the triple inverse Fourier transform performed in spherical-$\boldsymbol{k}$ coordinates that is given by

$$f(r,\omega) = \frac{1}{(2\pi)^3}\int_0^\infty \frac{\mathrm{d}k\, k^2}{\left[k+\sqrt{\varepsilon\mu\omega(\omega+i\sigma/\varepsilon)}\right]\left[k-\sqrt{\varepsilon\mu\omega(\omega+i\sigma/\varepsilon)}\right]} \times \int_{-1}^{+1} \mathrm{d}(\cos\theta) e^{ikr\cos\theta}\int_0^{2\pi}\mathrm{d}\theta, \tag{12.179}$$

$$= \frac{1}{(2\pi)^2}\int_0^\infty \frac{\mathrm{d}k\, k^2\left(e^{ikr}-e^{-ikr}\right)}{ikr\left[k+\sqrt{\varepsilon\mu\omega(\omega+i\sigma/\varepsilon)}\right]\left[k-\sqrt{\varepsilon\mu\omega(\omega+i\sigma/\varepsilon)}\right]}, \tag{12.180}$$

$$= \frac{-1}{(2\pi)^2 r}\int_{-\infty}^\infty \frac{\mathrm{d}k\, ik\, e^{ikr}}{\left[k+\sqrt{\varepsilon\mu\omega(\omega+i\sigma/\varepsilon)}\right]\left[k-\sqrt{\varepsilon\mu\omega(\omega+i\sigma/\varepsilon)}\right]}, \tag{12.181}$$

$$= \frac{-1}{(2\pi)^2 r}\frac{\partial}{\partial r}\int_{-\infty}^\infty \frac{\mathrm{d}k\, e^{ikr}}{\left[k+\sqrt{\varepsilon\mu\omega(\omega+i\sigma/\varepsilon)}\right]\left[k-\sqrt{\varepsilon\mu\omega(\omega+i\sigma/\varepsilon)}\right]}. \tag{12.182}$$

There are two simple poles on the complex k plane of this integrand: (1) one at $\sqrt{\varepsilon\mu\omega(\omega+i\sigma/\varepsilon)}$ that lies in the first quadrant or upper half of the complex k plane and (2) one at $-\sqrt{\varepsilon\mu\omega(\omega+i\sigma/\varepsilon)}$ that lies in the third quadrant or lower half of the complex k plane. Because we form a closed contour using a half-circle at infinity in the upper half plane (due to the factor of e^{ikr}) and because the half-circle makes no contribution to the contour integral due to Jordan's lemma, the residue theorem then gives that

$$f(r,\omega) = \frac{-1}{(2\pi)^2 r}\frac{\partial}{\partial r}\left[\frac{2\pi i\, e^{ir\sqrt{\varepsilon\mu\omega(\omega+i\sigma/\varepsilon)}}}{2\sqrt{\varepsilon\mu\omega(\omega+i\sigma/\varepsilon)}}\right] = \frac{e^{ir\sqrt{\varepsilon\mu\omega(\omega+i\sigma/\varepsilon)}}}{4\pi r}. \tag{12.183}$$

The two gradient operations of Eq. (12.178) are next performed noting that $\nabla \boldsymbol{r} = \boldsymbol{I}$ and that $\nabla = \hat{\boldsymbol{r}}\partial/\partial r = \boldsymbol{r}r^{-1}\partial/\partial r$ due to the purely radial spatial dependence. Carrying out this exercise gives

$$\tilde{\boldsymbol{E}}(\boldsymbol{r},\omega) = -\frac{e^{ir\omega\sqrt{\varepsilon\mu(1+i\gamma)}}}{4\pi r}\left[\boldsymbol{I}-\hat{\boldsymbol{r}}\hat{\boldsymbol{r}}+\left(\frac{i}{r\omega\sqrt{\varepsilon\mu(1+i\gamma)}}-\frac{1}{r^2\omega^2\varepsilon\mu(1+i\gamma)}\right)(\boldsymbol{I}-3\hat{\boldsymbol{r}}\hat{\boldsymbol{r}})\right]\cdot\mu\boldsymbol{s}, \tag{12.184}$$

where

$$\gamma = \frac{\sigma}{\varepsilon\omega}. \tag{12.185}$$

The first term with the transverse polarization $\boldsymbol{I}-\hat{\boldsymbol{r}}\hat{\boldsymbol{r}}$ corresponds to the far-field EM response and the second term with polarization $\boldsymbol{I}-3\hat{\boldsymbol{r}}\hat{\boldsymbol{r}}$ to the near-field response. Note that as you showed in Exercise 1 of Chapter 3, we can separate the square root in this expression into real and imaginary parts as

$$\sqrt{\varepsilon\mu(1+i\gamma)} = \sqrt{\varepsilon\mu}\sqrt{\frac{1+\sqrt{1+\gamma^2}}{2}} + \frac{i\gamma\sqrt{\varepsilon\mu}}{\sqrt{2(1+\sqrt{1+\gamma^2})}}, \tag{12.186}$$

$$= \frac{1}{c(\omega)} + \frac{i}{\omega\delta_{em}(\omega)}. \tag{12.187}$$

The real wave speed $c(\omega)$ and the real EM skin depth $\delta_{em}(\omega)$ are frequency ω dependent. This decomposition allows us to rewrite $\tilde{\boldsymbol{E}}(\boldsymbol{r}, \omega)$ as

$$\tilde{\boldsymbol{E}}(\boldsymbol{r}, \omega) = \frac{e^{ir\omega/c(\omega)} e^{-r/\delta_{em}(\omega)}}{4\pi r}$$
$$\times \left\{ \boldsymbol{I} - \hat{\boldsymbol{r}}\hat{\boldsymbol{r}} + \left[\frac{ic(\omega)}{r\omega(1+i\gamma/\sqrt{1+\gamma^2})} - \left(\frac{c(\omega)}{r\omega(1+i\gamma/\sqrt{1+\gamma^2})} \right)^2 \right] (\boldsymbol{I} - 3\hat{\boldsymbol{r}}\hat{\boldsymbol{r}}) \right\} \cdot \mu \boldsymbol{s}. \tag{12.188}$$

The real electric field is then obtained by the inverse transform over frequency ω as

$$\boldsymbol{E}(\boldsymbol{r}, t) = -\frac{1}{4\pi r}\frac{1}{2\pi}\int_{-\infty}^{\infty} d\omega\, e^{-i\omega[t-r/c(\omega)]} e^{-r/\delta_{em}(\omega)}$$
$$\times \left[(\boldsymbol{I} - \hat{\boldsymbol{r}}\hat{\boldsymbol{r}}) + \frac{c(\omega)\,(1+\gamma^2)}{(-i\omega)r(1+2\gamma^2)} \right.$$
$$\left. \left(1 - \frac{i\gamma}{\sqrt{1+\gamma^2}} + \frac{c(\omega)\left(1 - i2\gamma\sqrt{1+\gamma^2}\right)}{(-i\omega)r(1+2\gamma^2)} \right) (\boldsymbol{I} - 3\hat{\boldsymbol{r}}\hat{\boldsymbol{r}}) \right] \cdot \mu \boldsymbol{s}. \tag{12.189}$$

Unfortunately, due to the complicated presence of the frequency parameter $\gamma = \sigma/(\varepsilon\omega)$, this final integration over ω cannot be performed in closed form. We first treat the higher-frequency regime where $\gamma \to 0$ and the EM response corresponds to nonattenuated waves. We then treat separately the purely diffusive regime.

The Nonattenuated Electromagnetic-Wave-Propagation Limit

If either the electrical conductivity of the material is exactly zero or if in our application of the Green's tensor to the electrical response through the representation theorem applied to an infinite uniform material, the source term $\boldsymbol{J}_s(\boldsymbol{r}, t)$ has temporal variations with support only over high frequencies where $\gamma = \sigma/(\varepsilon\omega) \to 0$, the appropriate Green's tensor corresponds to taking $\gamma = 0$ in Eq. (12.189) to give

$$\boldsymbol{E}(\boldsymbol{r}, t) = -\frac{1}{4\pi r}\left\{ \frac{1}{2\pi}\int_{-\infty}^{\infty} d\omega\, e^{-i\omega(t-r/c)} (\boldsymbol{I} - \hat{\boldsymbol{r}}\hat{\boldsymbol{r}}) \right.$$
$$\left. + \frac{c}{r}\frac{1}{2\pi}\int_{-\infty}^{\infty} \frac{d\omega}{(-i\omega)} e^{-i\omega(t-r/c)} \left[1 + \frac{c}{(-i\omega)r} \right] (\boldsymbol{I} - 3\hat{\boldsymbol{r}}\hat{\boldsymbol{r}}) \right\} \cdot \mu \boldsymbol{s}. \tag{12.190}$$

The wave speed $c = 1/\sqrt{\varepsilon\mu}$ is a simple constant in this limit. Using the results that

$$\frac{1}{2\pi}\int_{-\infty}^{\infty} d\omega\, e^{-i\omega(t-r/c)} = \delta\left(t - \frac{r}{c}\right), \tag{12.191}$$

$$\frac{1}{2\pi}\int_{-\infty}^{\infty}\frac{d\omega}{(-i\omega)}e^{-i\omega(t-r/c)} = S\left(t-\frac{r}{c}\right), \tag{12.192}$$

$$\frac{1}{2\pi}\int_{-\infty}^{\infty}\frac{d\omega}{(-i\omega)^2}e^{-i\omega(t-r/c)} = \left(t-\frac{r}{c}\right)S\left(t-\frac{r}{c}\right), \tag{12.193}$$

then gives

$$\boldsymbol{E}(\boldsymbol{r},t) = \frac{-1}{4\pi r}\left[\delta\left(t-\frac{r}{c}\right)(\boldsymbol{I}-\hat{\boldsymbol{r}}\hat{\boldsymbol{r}}) + \frac{c^2t}{r^2}S\left(t-\frac{r}{c}\right)(\boldsymbol{I}-3\hat{\boldsymbol{r}}\hat{\boldsymbol{r}})\right]\cdot\mu\boldsymbol{s}. \tag{12.194}$$

As for the elastodynamic response, the polarization tensor $\boldsymbol{I}-\hat{\boldsymbol{r}}\hat{\boldsymbol{r}}$ corresponds to transverse waves propagating in the far field while the polarization tensor $\boldsymbol{I}-3\hat{\boldsymbol{r}}\hat{\boldsymbol{r}}$ corresponds to the near-field response. Upon making the substitutions $t \to t-t'$ and $\boldsymbol{r} \to \boldsymbol{r}-\boldsymbol{r}'$, we then obtain the Green's tensor for use in the representation theorem of a uniform, isotropic wholespace in the wave-propagation regime

$$\boxed{\begin{aligned}\boldsymbol{G}(\boldsymbol{r},t|\boldsymbol{r}',t') = &-\frac{1}{4\pi r}\delta\left(t-t'-\frac{|\boldsymbol{r}-\boldsymbol{r}'|}{c}\right)\left(\boldsymbol{I}-\frac{(\boldsymbol{r}-\boldsymbol{r}')(\boldsymbol{r}-\boldsymbol{r}')}{|\boldsymbol{r}-\boldsymbol{r}'|^2}\right)\\ &-\frac{c^2(t-t')}{4\pi|\boldsymbol{r}-\boldsymbol{r}'|^3}S\left(t-t'-\frac{|\boldsymbol{r}-\boldsymbol{r}'|}{c}\right)\left(\boldsymbol{I}-\frac{3(\boldsymbol{r}-\boldsymbol{r}')(\boldsymbol{r}-\boldsymbol{r}')}{|\boldsymbol{r}-\boldsymbol{r}'|^2}\right).\end{aligned}} \tag{12.195}$$

Due to no boundaries or heterogeneity perturbing the spatial response, this Green's tensor satisfies $\boldsymbol{G}(\boldsymbol{r},t|\boldsymbol{r}',t') = \boldsymbol{G}(\boldsymbol{r}-\boldsymbol{r}',t-t') = \boldsymbol{G}^T(\boldsymbol{r}-\boldsymbol{r}',t-t')$. This Green's tensor has a different sign compared to the elastodynamic Green's tensor due only to the sign we placed on the impulsive point source.

The Electromagnetic-Diffusion Limit

If we want to use the representation theorem to obtain the electric response in the diffusive limit where $\varepsilon|\partial\boldsymbol{E}/\partial t| \ll \sigma|\boldsymbol{E}|$, the Green's tensor to use can be obtained from the impulsive point-source response controlled by

$$\nabla\nabla\cdot\boldsymbol{E}(\boldsymbol{r},t) - \nabla^2\boldsymbol{E}(\boldsymbol{r},t) = -\sigma\mu\frac{\partial\boldsymbol{E}(\boldsymbol{r},t)}{\partial t} - \delta(\boldsymbol{r})\delta(t)\mu\boldsymbol{s}, \tag{12.196}$$

where in the present section we are still focusing on solutions in an infinite wholespace with a uniform conductivity σ. This is a vector diffusion equation with *electromagnetic diffusivity*

$$D = \frac{1}{\sigma\mu}. \tag{12.197}$$

This diffusive electric field is zero for $t < 0$ and approaches zero as $r \to \infty$. The problem is solved by taking a Fourier transform over space but not time, solving the temporal differential equation, and then returning to the space domain using the inverse Fourier transform.

So upon taking a Fourier transform over space $\nabla \to i\boldsymbol{k}$ and dividing by $\sigma\mu = 1/D$, we obtain

$$- D\boldsymbol{kk}\cdot\tilde{\boldsymbol{E}}(\boldsymbol{k},t) + Dk^2\tilde{\boldsymbol{E}}(\boldsymbol{k},t) + \frac{\partial \tilde{\boldsymbol{E}}(\boldsymbol{k},t)}{\partial t} = -\delta(t)D\mu\boldsymbol{s}. \tag{12.198}$$

Multiply through by the integrating factor $\exp(Dk^2t)$ to obtain

$$- D\boldsymbol{kk}\cdot e^{Dk^2t}\tilde{\boldsymbol{E}}(\boldsymbol{k},t) + \frac{\partial}{\partial t}\left[e^{Dk^2t}\tilde{\boldsymbol{E}}(\boldsymbol{k},t)\right] = -\delta(t)De^{Dk^2t}\mu\boldsymbol{s}. \tag{12.199}$$

Now dot multiply by the second-order integrating tensor $\exp(-D\boldsymbol{kk}t)$ to give

$$\frac{\partial}{\partial t}\left[\exp(-D\boldsymbol{kk}t)e^{Dk^2t}\cdot\tilde{\boldsymbol{E}}(\boldsymbol{k},t)\right] = -\delta(t)D\exp(-D\boldsymbol{kk}t)e^{Dk^2t}\cdot\mu\boldsymbol{s}, \tag{12.200}$$

where the second-order tensor $\exp(-D\boldsymbol{kk}t)$ is defined, as usual, through the expansion

$$\exp(-D\boldsymbol{kk}t) = \boldsymbol{I} - \frac{1}{1!}D\boldsymbol{kk}t + \frac{1}{2!}(D\boldsymbol{kk}t)\cdot(D\boldsymbol{kk}t) - \ldots \tag{12.201}$$

$$= \boldsymbol{I} - \frac{\boldsymbol{kk}}{k^2}\left(\frac{Dk^2t}{1!} - \frac{(Dk^2t)^2}{2!} + \ldots\right) \tag{12.202}$$

$$= \boldsymbol{I} + \frac{\boldsymbol{kk}}{k^2}\left(e^{-Dk^2t} - 1\right). \tag{12.203}$$

If we integrate Eq. (12.200) from $-\infty$ to the current time and use that the electric field is zero at $t = -\infty$, we obtain the integrated result

$$\exp(-D\boldsymbol{kk}t)e^{Dk^2t}\cdot\tilde{\boldsymbol{E}}(\boldsymbol{k},t) = -D\mu\boldsymbol{s}. \tag{12.204}$$

Dot multiply both sides by the inverse tensor

$$\exp(D\boldsymbol{kk}t)e^{-Dk^2t} = \left[\boldsymbol{I} + \frac{\boldsymbol{kk}}{k^2}\left(\frac{Dk^2t}{1!} + \frac{(Dk^2t)^2}{2!} + \ldots\right)\right]e^{-Dk^2t} \tag{12.205}$$

$$= \boldsymbol{I}e^{-Dk^2t} + \frac{\boldsymbol{kk}}{k^2}\left(1 - e^{-Dk^2t}\right) \tag{12.206}$$

to obtain

$$\tilde{\boldsymbol{E}}(\boldsymbol{k},t) = -D\left[\boldsymbol{I}e^{-Dk^2t} - \frac{(i\boldsymbol{k})(i\boldsymbol{k})}{k^2}\left(1 - e^{-Dk^2t}\right)\right]\cdot\mu\boldsymbol{s}. \tag{12.207}$$

As an aside, you should verify through explicit multiplication that $\exp(+D\boldsymbol{kk}t)\cdot\exp(-D\boldsymbol{kk}t) = \boldsymbol{I}$.

We now return to the space domain by taking the inverse Fourier transform to give

$$\boldsymbol{E}(\boldsymbol{r},t) = \left[-D\boldsymbol{I}f_1(r,t) + D\nabla\nabla\left(f_2(r) - f_3(r,t)\right)\right]\cdot\mu\boldsymbol{s}. \tag{12.208}$$

The function $f_1(r, t)$ is the 3D inverse Fourier transform of $\exp(-Dk^2t)$ that, for convenience, is performed in spherical-$\boldsymbol{k}$ coordinates with $\mathrm{d}^3\boldsymbol{k} = -k^2\mathrm{d}k\,\mathrm{d}(\cos\theta)\,\mathrm{d}\phi$:

$$f_1(r,t) = \frac{1}{(2\pi)^3}\int_0^\infty \mathrm{d}k\,k^2 e^{-Dk^2t}\int_{-1}^{1}\mathrm{d}(\cos\theta)\,e^{ikr\cos\theta}\int_0^{2\pi}\mathrm{d}\phi \tag{12.209}$$

$$= \frac{1}{(2\pi)^2}\int_{-\infty}^{\infty}\frac{\mathrm{d}k\,k^2 e^{-Dk^2t+ikr}}{ikr} \tag{12.210}$$

$$= -\frac{1}{(2\pi)^2 r}\frac{\partial}{\partial r}\left[\int_{-\infty}^{\infty}\mathrm{d}k\,e^{-Dk^2t+ikr}\right] \tag{12.211}$$

$$= -\frac{1}{(2\pi)^2 r}\sqrt{\frac{\pi}{Dt}}\frac{\partial e^{-r^2/(4Dt)}}{\partial r} \tag{12.212}$$

$$= \frac{e^{-r^2/(4Dt)}}{8(\pi Dt)^{3/2}}. \tag{12.213}$$

Verify that you can go from line to line in this development. You can verify that $-Df_1(r,t)$ is the Green's function of the scalar diffusion equation in a uniform wholespace as given by Eq. (12.87). The function $f_2(r)$ is the 3D inverse Fourier transform of k^{-2} and is thus minus the Green's function of the Laplace equation or

$$f_2(r) = \frac{1}{4\pi r}. \tag{12.214}$$

The function $f_3(r,t)$ is the 3D inverse Fourier transform of e^{-Dk^2t}/k^2 or

$$f_3(r,t) = \frac{1}{(2\pi)^3}\int_0^\infty \mathrm{d}k\,\frac{k^2}{k^2}e^{-Dk^2t}\int_{-1}^{1}\mathrm{d}(\cos\theta)e^{ikr\cos\theta}\int_0^{2\pi}\mathrm{d}\phi \tag{12.215}$$

$$= \frac{1}{(2\pi)^2 r}\int_{-\infty}^{\infty}\mathrm{d}k\,\frac{e^{-Dk^2t+ikr}}{ik}. \tag{12.216}$$

Note that

$$\int_0^r \mathrm{d}r'\int_{-\infty}^{\infty}\mathrm{d}k e^{-Dk^2t+ikr'} = \int_{-\infty}^{\infty}\mathrm{d}k e^{-Dk^2t}\left(\frac{e^{ikr}-1}{ik}\right) = \int_{-\infty}^{\infty}\mathrm{d}k\frac{e^{-Dk^2t+ikr}}{ik} \tag{12.217}$$

where to obtain the last statement, we used that $\int_{-\infty}^{\infty}\mathrm{d}k e^{-Dk^2t}/(ik) = 0$ because the integrand is odd. We thus have

$$f_3(r,t) = \frac{1}{(2\pi)^2 r}\sqrt{\frac{\pi}{Dt}}\int_0^r e^{-r'^2/(4Dt)}\mathrm{d}r' \tag{12.218}$$

$$= \frac{1}{4\pi r}\left[\frac{2}{\sqrt{\pi}}\int_0^{r/\sqrt{4Dt}} e^{-u^2}\mathrm{d}u\right], \tag{12.219}$$

where the function in square brackets is called the error function $\mathrm{erf}(r/\sqrt{4Dt})$. The function needed in our Green's tensor is

$$f_2(r) - f_3(r,t) = \frac{1}{4\pi r}\left[1 - \frac{2}{\sqrt{\pi}}\int_0^{r/\sqrt{4Dt}} e^{-u^2}\,\mathrm{d}u\right], \tag{12.220}$$

where the function in the square brackets is called the complimentary error function $\text{erfc}(r/\sqrt{4Dt}) = 1 - \text{erf}(r/\sqrt{4Dt})$, which is a function that monotonically decreases from 1 to 0 as $r/\sqrt{4Dt}$ increases from 0 to ∞.

As earlier, if $f(r)$ is a purely radial function, we have the identity

$$\nabla\nabla f(r) = \frac{(\boldsymbol{I} - \hat{\boldsymbol{r}}\hat{\boldsymbol{r}})}{r}\frac{\partial f(r)}{\partial r} + \hat{\boldsymbol{r}}\hat{\boldsymbol{r}}\frac{\partial^2 f(r)}{\partial r^2}, \tag{12.221}$$

which when used in Eq. (12.208) yields

$$\boldsymbol{E}(\boldsymbol{r},t) = -\left[\frac{De^{-r^2/(4Dt)}}{8(\pi Dt)^{3/2}}(\boldsymbol{I} - \hat{\boldsymbol{r}}\hat{\boldsymbol{r}}) + \frac{D}{4\pi r^2}\frac{e^{-r^2/(4Dt)}}{\sqrt{\pi Dt}}(\boldsymbol{I} - 2\hat{\boldsymbol{r}}\hat{\boldsymbol{r}}) \right.$$
$$\left. + \frac{D}{4\pi r^3}\left(1 - \frac{2}{\sqrt{\pi}}\int_0^{r/\sqrt{4Dt}} e^{-u^2}\,du\right)(\boldsymbol{I} - 3\hat{\boldsymbol{r}}\hat{\boldsymbol{r}})\right]\cdot \mu\boldsymbol{s}. \tag{12.222}$$

We thus see that for vectorial diffusion, there is no clear separation into far-field and near-field contributions, each with distinct polarization, as there was in the case of waves. Because the diffusion advances through space and time with $r \sim \sqrt{Dt}$, each term here, which has its own distinct polarization, is equally present at each r and t.

Finally, upon making the substitutions $\boldsymbol{r} \to \boldsymbol{r} - \boldsymbol{r}'$ and $t \to t - t'$, we obtain the Green's tensor controlling how the electric field diffuses in an infinite material having uniform $D = 1/(\sigma\mu)$

$$\begin{aligned}\boldsymbol{G}(\boldsymbol{r}-\boldsymbol{r}', t-t') = &-\frac{De^{-|\boldsymbol{r}-\boldsymbol{r}'|^2/[4D(t-t')]}}{8[\pi D(t-t')]^{3/2}}\left(\boldsymbol{I} - \frac{(\boldsymbol{r}-\boldsymbol{r}')(\boldsymbol{r}-\boldsymbol{r}')}{|\boldsymbol{r}-\boldsymbol{r}'|^2}\right)\\ &-\frac{D}{4\pi|\boldsymbol{r}-\boldsymbol{r}'|^2}\frac{e^{-|\boldsymbol{r}-\boldsymbol{r}'|^2/[4D(t-t')]}}{\sqrt{\pi D(t-t')}}\left(\boldsymbol{I} - \frac{2(\boldsymbol{r}-\boldsymbol{r}')(\boldsymbol{r}-\boldsymbol{r}')}{|\boldsymbol{r}-\boldsymbol{r}'|^2}\right)\\ &-\frac{D}{4\pi|\boldsymbol{r}-\boldsymbol{r}'|^3}\left(1 - \frac{2}{\sqrt{\pi}}\int_0^{|\boldsymbol{r}-\boldsymbol{r}'|/\sqrt{4D(t-t')}} e^{-u^2}\,du\right)\\ &\times\left(\boldsymbol{I} - \frac{3(\boldsymbol{r}-\boldsymbol{r}')(\boldsymbol{r}-\boldsymbol{r}')}{|\boldsymbol{r}-\boldsymbol{r}'|^2}\right).\end{aligned} \tag{12.223}$$

If you are able to work through and understand how this result for EM diffusion from a point source is obtained, it means you are comfortable with handling tensors and applying Fourier analysis to physics problems, both of which were (at least minor) pedagogic objectives at the outset of the book.

12.6 Overall Conclusions

The main objective of this book has been for you to see how the rules of macroscopic response emerge from the underlying molecular dynamics. In the theory of continuum physics presented in this book, we do not replace the collection of molecules that make up a body with a "continuous mass" that is infinitely divisible and supporting fields whose quantitative meaning must be guessed at. We instead go to each point in the body and sum over the molecules present to determine their collective average response. The fields and governing equations so obtained are complemented with constitutive laws that derive from thermodynamics. This coherent set of macroscopic rules for the collective, averaged response of the molecules surrounding each point is called "the theory of continuum physics" and was shown to be an exact theory.

I also wanted you to see the mathematical and physical commonalities and distinctions between the various types of possible macroscopic response. The explicit wholespace Green's tensors (i.e., impulsive point-source responses in uniform infinite domains) obtained above for elastodynamics, elastostatics, Stokes flow, EM waves, and EM diffusions provide an excellent opportunity to compare and contrast each type of response. You should now have the mindset that if you have expertise in one field, say elastodynamics (i.e., seismology), you immediately have expertise in another field like electromagnetics or viscous flow. All chapters of macroscopic physics can be written in the same language and follow a consistent plot.

When the physics is linear in the applied forcing, the Green's function method presented in this chapter shows how the response is created by forcing that comes from any of (1) an explicit source term in the PDE, (2) an initial condition, or (3) a boundary condition. The Green's function satisfies the homogeneous version of whatever boundary condition the response field satisfies. For simple boundaries or no boundaries, the needed Green's function can be obtained analytically. But whether the Green's function is obtained numerically or analytically, the response is the integral of the Green's function with the spatial distribution of the three types of forcing in the manner prescribed in this chapter.

12.7 Exercises

1. *The Green's function of the 1D diffusion equation*: Find the 1D diffusion Green's function defined by the PDE

$$D\frac{\partial^2 \psi(x,t)}{\partial x^2} - \frac{\partial \psi(x,t)}{\partial t} = \delta(x)\delta(t) \tag{12.224}$$

such that $\psi(x,t) = 0$ for all $t < 0$ and such that $\psi(x.t) \to 0$ as $|x| \to \infty$. Do so by taking the Fourier transform over x and t ($\partial^2/\partial x^2 \to -k^2$ and $\partial^2/\partial t^2 \to -\omega^2$), returning to the time domain first (there is a single pole in the lower-half ω plane that contributes to inverse transform) and to the space domain second (performing the inverse transform by completing the square in the exponential). The answer is:

$$\boxed{\psi(x,t) = \frac{e^{-x^2/(4Dt)}}{2\sqrt{\pi D t}}.} \tag{12.225}$$

Note that this is exactly a Gaussian function with a standard of deviation σ (width in space) that grows with time as $\sigma = \sqrt{2Dt}$.

2. *The Green's function of the Laplace equation*: Find the 3D Green's function of the Laplace equation in a wholespace

$$\nabla^2 \varphi(\boldsymbol{r}) = \delta(\boldsymbol{r}). \tag{12.226}$$

You will be guided through how to obtain $\varphi(\boldsymbol{r})$ using two different approaches.

First approach: Take Fourier transforms over space and use the standard facts that $\nabla \to i\boldsymbol{k}$, $\nabla^2 \to -k^2$ and $\delta(\boldsymbol{r}) \to 1$ to obtain the Fourier transform as

$$\tilde{\varphi}(k) = -\frac{1}{k^2}. \tag{12.227}$$

As always, it is easy to obtain the solution in the frequency domain and the mathematical work is in returning to the real space domain. Perform the inverse Fourier transform in spherical-$\boldsymbol{k}$ coordinates in which a volume element in $\boldsymbol{k}$ space is $d^3\boldsymbol{k} = -k^2 dk\, d(\cos\theta)\, d\phi$ and $e^{i\boldsymbol{k}\cdot\boldsymbol{r}} = e^{kr\cos\theta}$ to obtain

$$\varphi(r) = -\frac{1}{(2\pi)^3}\int_0^\infty dk \frac{k^2}{k^2}\int_{-1}^{1} d(\cos\theta) e^{ikr\cos\theta}\int_0^{2\pi} d\phi, \tag{12.228}$$

$$= -\frac{1}{(2\pi)^2}\int_0^\infty dk \frac{\left(e^{ikr} - e^{-kr}\right)}{ikr}, \tag{12.229}$$

$$= -\frac{1}{(2\pi)^2}\int_{-\infty}^\infty dk \frac{e^{ikr}}{ikr}. \tag{12.230}$$

You should verify that you can get from the second line here to the third line by making the change of variables $k \to -k$ in the second integral of the second line. Using the same contour as in Fig. 11.4, the integral here is obtained using Cauchy's theorem using the fact that the contribution from the half-circle at infinity is zero by Jordan's lemma

$$\int_{-\infty}^\infty dk \frac{e^{ikr}}{ik} + \int_{C_\delta} dk \frac{e^{ikr}}{ik} = 0, \tag{12.231}$$

where the contour C_δ is as shown in Fig. 11.4 and is a half-circle that excludes the origin in which the radius of the half-circle is $\delta \to 0$. Working in polar coordinates in which $k = \delta e^{i\theta}$ on C_δ, show that

$$\int_{-\infty}^\infty dk \frac{e^{ikr}}{ikr} = -\lim_{\delta\to 0}\int_\pi^0 \frac{\delta e^{i\theta}\, i d\theta\, e^{i\delta \exp(i\theta)}}{i\delta e^{i\theta} r} = \frac{\pi}{r}. \tag{12.232}$$

Thus, you have obtained

$$\boxed{\varphi(r) = -\frac{1}{4\pi r}} \tag{12.233}$$

as the Green's function of the Laplace equation. If the source point had been at $\boldsymbol{r} = \boldsymbol{r}_s$ instead of at the origin of the spherical coordinates, you would make the substitution $r \to |\boldsymbol{r} - \boldsymbol{r}_s|$ in this expression.

Second approach: You can obtain this result using a different approach that will prove useful in the next exercise where you find the Green's function of the biharmonic equation. To do so, begin by rewriting Eq. (12.230) as

$$r\varphi(r) = -\frac{1}{(2\pi)^2}\int_{-\infty}^{\infty} dk \frac{e^{ikr}}{ik} \tag{12.234}$$

and show that the integral here can be rewritten as

$$\int_{-\infty}^{\infty} dk \frac{e^{ikr}}{ik} = \frac{1}{2}\int_{-\infty}^{\infty} dk \left(\frac{e^{ikr} - e^{-ikr}}{ik}\right), \tag{12.235}$$

$$= \frac{1}{2}\int_{-r}^{r} dr_1 \int_{-\infty}^{\infty} dk\, e^{ikr_1}. \tag{12.236}$$

Note that if we had incorrectly performed the integral over r_1 in Eq. (12.236) from 0 to r instead of from $-r$ to r with the multiplying factor of $1/2$ we would obtain

$$\int_0^r dr_1 \int_{-\infty}^{\infty} dk\, e^{ikr_1} = \int_{-\infty}^{\infty} dk \left(\frac{e^{ikr} - 1}{ik}\right) = \int_{-\infty}^{\infty} dk \frac{e^{ikr}}{ik} + i \ln k|_{-\infty}^{\infty}, \tag{12.237}$$

which is not properly defined. Thus, using the completeness relation

$$\frac{1}{2\pi}\int_{-\infty}^{\infty} dk e^{ikr} = \delta(r), \tag{12.238}$$

show that

$$r\varphi(r) = -\frac{1}{2(2\pi)}\int_{-r}^{r} dr_1 \delta(r_1) = -\frac{S(r)}{4\pi}, \tag{12.239}$$

where $S(r)$ is the unit step function. But because, in fact, our r is radial distance with $r \geq 0$, the step function can be set to 1 to give

$$\boxed{\varphi(r) = -\frac{1}{4\pi r}.} \tag{12.240}$$

Although this second approach is a bit less direct in this particular problem, it will be the most direct way to obtain the Green's function of the biharmonic equation as you can now see for yourself.

3. *The Green's function of the biharmonic equation*: Find the 3D Green's function of the biharmonic equation in a wholespace

$$\nabla^4\psi(\boldsymbol{r}) = \delta(\boldsymbol{r}). \tag{12.241}$$

Using the facts that $\nabla \to i\boldsymbol{k}$, $\nabla^4 \to k^4$ and $\delta(\boldsymbol{r}) \to 1$, show that

$$\tilde{\psi}(k) = \frac{1}{k^4}. \tag{12.242}$$

By performing the inverse Fourier transform in spherical-$\boldsymbol{k}$ coordinates with $d^3\boldsymbol{k} = k^2 dk\, d(\cos\theta)\, d\phi$ and $e^{i\boldsymbol{k}\cdot\boldsymbol{r}} = e^{ikr\cos\theta}$, show that

$$\psi(r) = \frac{1}{(2\pi)^3}\int_0^\infty dk\frac{k^2}{k^4}\int_{-1}^{1} d(\cos\theta)e^{ikr\cos\theta}\int_0^{2\pi} d\phi, \tag{12.243}$$

$$= \frac{1}{(2\pi)^2}\int_0^\infty dk\frac{(e^{ikr} - e^{-ikr})}{(ik)(ik)(ik)r} \tag{12.244}$$

or, by changing variables $k \to -k$ in the second integral,

$$r\psi(r) = -\frac{1}{(2\pi)^2}\int_{-\infty}^{\infty} dk\frac{e^{ikr}}{(ik)(ik)(ik)}. \tag{12.245}$$

Next, verify that you can go from line to line in the following development

$$\frac{1}{2}\int_0^r dr_3\int_0^{r_3} dr_2\int_{-r_2}^{r_2} dr_1\int_{-\infty}^{\infty} dk e^{ikr_1}$$

$$= \frac{1}{2}\int_0^r dr_3\int_0^{r_3} dr_2\int_{-\infty}^{\infty} dk\left(\frac{e^{ikr_2} - e^{-ikr_2}}{ik}\right), \tag{12.246}$$

$$= \int_0^r dr_3\int_0^{r_3} dr_2\int_{-\infty}^{\infty} dk\frac{e^{ikr_2}}{ik}, \tag{12.247}$$

$$= \int_0^r dr_3\int_{-\infty}^{\infty} dk\left(\frac{e^{ikr_3} - 1}{(ik)(ik)}\right)$$

$$= \int_0^r dr_3\left[\int_{-\infty}^{\infty} dk\frac{e^{ikr_3}}{(ik)(ik)} - \frac{1}{k}\bigg|_{-\infty}^{\infty}\right], \tag{12.248}$$

$$= \int_0^r dr_3\int_{-\infty}^{\infty} dk\frac{e^{ikr_3}}{(ik)(ik)}, \tag{12.249}$$

$$= \int_{-\infty}^{\infty} dk\left(\frac{e^{ikr} - 1}{(ik)(ik)(ik)}\right) = \int_{-\infty}^{\infty} dk\frac{e^{ikr}}{(ik)(ik)(ik)} - \frac{1}{2ik^2}\bigg|_{-\infty}^{\infty}, \tag{12.250}$$

$$= \int_{-\infty}^{\infty} dk\frac{e^{ikr}}{(ik)(ik)(ik)}. \tag{12.251}$$

So using the completeness relation $\delta(r) = (2\pi)^{-1}\int_{-\infty}^{\infty} dk e^{ikr}$, show that

$$r\psi(r) = -\frac{1}{2(2\pi)}\int_0^r dr_3\int_0^{r_3} dr_2\int_{-r_2}^{r_2} dr_1\delta(r_1), \tag{12.252}$$

$$= -\frac{1}{4\pi}\int_0^r dr_3 r_3 = -\frac{r^2}{8\pi}. \tag{12.253}$$

Thus you have found that

$$\boxed{\psi(r) = -\frac{r}{8\pi}} \tag{12.254}$$

is the Green's function of the biharmonic equation. If the source point had been at $\boldsymbol{r} = \boldsymbol{r}_s$ instead of at the origin of the spherical coordinates, you would make the substitution $r \rightarrow |\boldsymbol{r} - \boldsymbol{r}_s|$ in this expression.

4. *The Green's function method for waves on a string*: Consider the 1D wave equation for displacements $\psi(x, t)$ on a finite-length string $a \leq x \leq b$

$$\frac{\partial^2 \psi(x, t)}{\partial x^2} - \frac{1}{c^2}\frac{\partial^2 \psi(x, t)}{\partial t^2} = f(x, t) \tag{12.255}$$

subject to either of the boundary conditions at $x = b$ that

$$\psi(b, t) = \psi_b(t) \quad \text{or} \quad \left.\frac{\partial \psi(x, t)}{\partial x}\right|_{x=b} = s_b(t) \quad \text{at} \quad x = b \tag{12.256}$$

and to either of the boundary conditions at $x = a$ that

$$\psi(a, t) = \psi_a(t) \quad \text{or} \quad \left.\frac{\partial \psi(x, t)}{\partial x}\right|_{x=a} = s_a(t) \quad \text{at} \quad x = b \tag{12.257}$$

and to both of the initial conditions that

$$\psi(x, 0) = T_0(x) \quad \text{and} \quad \left.\frac{\partial \psi(x, t)}{\partial t}\right|_{t=0} = T_1(x) \quad \text{at} \quad t = 0. \tag{12.258}$$

If either $a \rightarrow -\infty$ or $b \rightarrow \infty$ or both, the wave field is required to satisfy outward radiation conditions as $x \rightarrow \pm\infty$.

The Green's function corresponding to this problem is the solution of

$$\frac{\partial^2 G(x, t|x', t')}{\partial x^2} - \frac{1}{c^2}\frac{\partial^2 G(x, t|x', t')}{\partial t^2} = \delta(x - x')\delta(t - t') \tag{12.259}$$

subject to the homogeneous version of the boundary conditions that $\psi(x, t)$ satisfies at $x = a$ and b and subject to the initial condition that

$$G(x, t|x', t') = 0 \quad \text{for} \quad t - t' < 0. \tag{12.260}$$

(a) Begin by showing that this Green's function for waves on a string satisfies the reciprocity condition that

$$\boxed{G(x, t|x', t') = G(x', -t'|x, -t).} \tag{12.261}$$

(b) Using the above defining equations for $\psi(x,t)$ and $G(x,t|x',t')$, obtain the following representation theorem for waves on a string

$$
\boxed{
\begin{aligned}
\psi(x,t) &= \int_0^t \mathrm{d}t' \int_a^b \mathrm{d}x'\, G(x,t|x',t') f(x',t') \\
&- \int_0^t \mathrm{d}t' \left[G(b,-t'|x,-t) \left. \frac{\partial \psi(x',t')}{\partial x'} \right|_{x'=b} - \psi(b,t') \left. \frac{\partial G(x',-t'|x,-t)}{\partial x'} \right|_{x'=b} \right. \\
&\left. - G(a,-t'|x,-t) \left. \frac{\partial \psi(x',t')}{\partial x'} \right|_{x'=a} + \psi(a,t') \left. \frac{\partial G(x',-t'|x,-t)}{\partial x'} \right|_{x'=a} \right] \\
&- \frac{1}{c^2} \int_a^b \mathrm{d}x' \left[G(x',0|x,-t) \left. \frac{\partial \psi(x',t')}{\partial t'} \right|_{t'=0} \right. \\
&\left. - \psi(x',0) \left. \frac{\partial G(x',-t'|x,-t)}{\partial t'} \right|_{t'=0} \right].
\end{aligned}
}
\tag{12.262}
$$

which shows how the source term in the wave equation, time-dependent boundary conditions or initial conditions can excite waves on a string.

(c) Show that for an infinitely long string, the outward radiation of the Green's function can be expressed

$$
\frac{\partial G(x',-t'|x,-t)}{\partial x'} = +\frac{1}{c} \frac{\partial G(x',-t'|x,-t)}{\partial t'} \quad \text{at } x' = \infty, \tag{12.263}
$$

$$
\frac{\partial G(x',-t'|x,-t)}{\partial x'} = -\frac{1}{c} \frac{\partial G(x',-t'|x,-t)}{\partial t'} \quad \text{at } x' = -\infty, \tag{12.264}
$$

while for the wavefield of interest, the outward radiation conditions are

$$
\frac{\partial \psi(x',t')}{\partial x'} = -\frac{1}{c} \frac{\partial \psi(x',t')}{\partial t'} \quad \text{at } x' = \infty, \tag{12.265}
$$

$$
\frac{\partial \psi(x',t')}{\partial x'} = +\frac{1}{c} \frac{\partial \psi(x',t')}{\partial t'} \quad \text{at } x' = -\infty. \tag{12.266}
$$

Using these statements of outward radiation in an infinitely long string, show that the time integral of the boundary conditions in the above representation theorem becomes

$$
\begin{aligned}
&- \int_0^t \mathrm{d}t' \left[G(\infty,-t'|x,-t) \left. \frac{\partial \psi(x',t')}{\partial x'} \right|_{x'=\infty} - \psi(\infty,t') \left. \frac{\partial G(x',-t'|x,-t)}{\partial x'} \right|_{x'=\infty} \right. \\
&\left. - G(-\infty,-t'|x,-t) \left. \frac{\partial \psi(x',t')}{\partial x'} \right|_{x'=-\infty} + \psi(-\infty,t') \left. \frac{\partial G(x',-t'|x,-t)}{\partial x'} \right|_{x'=-\infty} \right]
\end{aligned}
\tag{12.267}
$$

$$= -\frac{1}{c}\int_0^{\infty} dt' \frac{\partial}{\partial t'}\left[G(\infty, -t'|x, -t)\psi(\infty, t') + G(-\infty, -t'|x, -t)\psi(-\infty, t')\right] \tag{12.268}$$

$$= 0, \tag{12.269}$$

where to get that this boundary-condition contribution is zero, you need to use the initial condition for the Green's function that $G(x', -\infty|x, -t) = 0$ and the condition on the wavefield that $\psi(x', 0) \to 0$ as $x' \to \pm\infty$ if the forcing $f(x', t')$ and initial conditions have finite extent along the infinite string (i.e., do not extend out to $\pm\infty$, which we assume is the case).

Thus, you have shown that the representation theorem for waves on an infinite string is

$$\psi(x, t) = \int_0^t dt' \int_{-\infty}^{\infty} dx'\, G(x, t|x', t')f(x', t') - \frac{1}{c^2}\int_{-\infty}^{\infty} dx' \left[G(x', 0|x, -t)\left.\frac{\partial\psi(x', t')}{\partial t'}\right|_{t'=0} - \psi(x', 0)\left.\frac{\partial G(x', -t'|x, -t)}{\partial t'}\right|_{t'=0}\right], \tag{12.270}$$

which corresponds to waves generated either by the source term f or by nonzero initial conditions on ψ.

(d) Using the Green's function for waves on an infinite string, obtained either by using the Fourier transform or by using earlier results, show that

$$\left.\frac{\partial G(x', -t'|x, -t)}{\partial t'}\right|_{t'=0} = \left.\frac{\partial G(x, t|x', t')}{\partial t'}\right|_{t'=0} = \frac{c^2}{2}\left[\delta(x - x' + ct) + \delta(x - x' - ct)\right], \tag{12.271}$$

which corresponds to a pair of Dirac pulses moving out from x' in the $+x$ and $-x$ directions.

(e) Consider the specific problem where the string is infinitely long with waves outward radiating at infinity and for which $f(x, t) = 0$ and there are the two initial conditions $\psi(x, 0) = T_0(x)$ and $\partial\psi(x, t)/\partial t|_{t=0} = 0$. Show that the solution for the wavefield is obtained from the representation theorem as

$$\psi(x, t) = \frac{1}{c^2}\int_{-\infty}^{\infty} dx' T_0(x') \left.\frac{\partial G(x, t|x', t')}{\partial t'}\right|_{t'=0} \tag{12.272}$$

$$= \frac{1}{2} \int_{-\infty}^{\infty} T_0(x') \left[\delta(x - x' + ct) + \delta(x - x' - ct)\right] \tag{12.273}$$

$$= \boxed{\frac{1}{2} \left[T_o(x + ct) + T_0(x - ct)\right] ,} \tag{12.274}$$

which corresponds to two pulses having the shape of $T_0(x)$ but with half the amplitude (that sum to the full $T_0(x)$ at $t = 0$) that are propagating outward (away from each other) in the $-x$ and $+x$ directions respectively.

References

Atkins, P. W. & Friedman, R. S., 2011. *Molecular Quantum Mechanics*, Oxford University Press, Oxford.

Balluffi, R. W., Allen, S. M., & Carter, W. C., 2005. *Kinetics of Materials*, Wiley-Interscience, Hoboken, NJ.

Batchelor, G. K., 1967. *An Introduction to Fluid Dynamics*, Cambridge University Press, Cambridge.

Bohm, D., 1952. A suggested interpretation of the quantum theory in terms of 'hidden variables.' I, *Physical Review*, **85**, 166–179.

Borg, R. J. & Dienes, G. J., 1988. *An Introduction to Solid State Diffusion*, Academic Press, San Diego.

Breen, S. J., Pride, S. R., Masson, Y., & Manga, M., 2022. Stable drainage in a gravity field, *Advances in Water Resources*, **182**, 104150.

Butkov, E., 1968. *Mathematical Physics*, Addison Wesley, Reading, MA.

Cahn, J. W. & Hilliard, J. E., 1958. Free energy of a nonuniform system. I. Interfacial free energy, *Journal of Chemical Physics*, **28**, 258–267.

Callen, H. B., 1985. *Thermodynamics and an Introduction to Thermostatistics,* 2nd ed., John Wiley & Sons, New York.

Curie, P., 1894. Sur la symétrie dans les phènomènes physiques: symétrie d'un champ électrique et d'un champ magnétique, *Journal de Physique*, **3**, 393–415.

deGroot, S. R., 1951. *Thermodynamics of Irreversible Processes*, North Holland, Amsterdam.

Dussan, E. B. & Davis, S. H., 1974. On the motion of a fluid-fluid interface along a solid surface, *Journal of Fluid Mechanics*, **65**, 71–95.

Feder, J., Flekkoy, E. G., & Hansen, A., 2023. *Physics of Flow in Porous Media*, Cambridge University Press, Cambridge.

Feynman, R. P., 1963. *The Feynman Lectures on Physics*, vol. 1, Addison Wesley, Reading, MA.

Flekkøy, E. G., Pride, S. R., & Toussaint, R., 2017. Onsager symmetry from mesoscopic time-reversibility and the hydrodynamic dispersion tensor for coarse-grained systems, *Physical Review E*, **95**, 022136.

Fourier, J., 1822. *Théorie Analytique de la Chaleur*, Firmin Didot Père & Fils, Paris.

Happel, J. & Brenner, H., 1983. *Low Reynolds Number Hydrodynamics*, Mechanics of Fluids and Transport Processes, Martinus Nijhoff Publishers, The Hague.

Israelachvili, J. N., 2011. *Intermolecular and Surface Forces,* 3rd ed., Academic Press, San Diego.

Jackson, J. D., 1975. *Classical Electrodynamics,* 2nd ed., John Wiley & Sons, New York.

Kong, J. A., 1986. *Electromagnetic Wave Theory*, John Wiley & Sons, New York.

Körner, T. W., 1988. *Fourier Analysis*, Cambridge University Press, Cambridge.

Landau, L. D. & Lifshitz, E. M., 1984. *Electrodynamics of Continuous Media,* 2nd ed., Pergamon, New York.

Landau, L. D. & Lifshitz, E. M., 1986. *Theory of Elasticity,* 3rd ed., Pergamon, New York.

Landau, L. D. & Lifshitz, E. M., 1987. *Fluid Mechanics,* 2nd ed., Pergamon, New York.

Lebedev, L. P., Cloud, M. J., & Eremeyev, V. A., 2010. *Tensor Analysis with Applications in Mechanics*, World Scientific Publishing, Singapore.

Liouville, J. & Sturm, C., 1837. Extrait d'un mémoire sur le développement des fonctions en séries dont les différents termes sont assujettis à satisfaire à une même équation différentielle linéaire, contenant un paramètre variable, *Journal de Mathématiques Pures et Appliquées*, **2**, 220–233.

Lyklema, J., 1991. *Fundamentals of Interface and Colloid Science. Volume 1: Fundamentals*, Academic Press, London.

Masson, Y. J. & Pride, S. R., 2015. Mapping the mechanical properties of rocks using automated microindentation tests, *Journal of Geophysical Research*, **120**, 7138–7155.

Nayfeh, A. H., 1973. *Perturbation Methods*, Wiley-Interscience, Hoboken, NJ.

Nye, J. F., 1957. *Physical Properties of Crystals*, Oxford University Press, Oxford.

Onsager, L., 1931. Reciprocal relations in irreversible processes, *Physical Review*, **37**, 405–426.

Press, W. H., Teukolsky, S. A., Vetterling, W. T., & Flannery, B. P., 2007. *Numerical Recipes: The Art of Scientific Computing*, 3rd ed., Cambridge University Press, Cambridge.

Rowlinson, J. S., 1979. Translation of J. D. van der Waals' "The thermodynamic theory of capillarity under the hypothesis of a continuous variation of density," *Journal of Statistical Physics*, **20**, 197–244.

Shannon, C. E., 1948. A mathematical theory of communication, *The Bell System Technical Journal*, **27**, 379–423.

Stokes, G. G., 1851. On the effect of internal friction of fluids on the motion of pendulums, *Transactions of the Cambridge Philosophical Society*, **9**(8), 8–106.

Stratton, J. A., 1941. *Electromagnetic Theory*, McGraw-Hill, New York.

Sturm, C., 1836. Mémoire sur les équations différentielles linéaires du second ordre, *Journal de Mathématiques Pures et Appliquées*, **1**, 106–186.

Swift, M. R., Orlandini, E., Osborn, W. R., & Yeomans, J. M., 1996. Lattice-Boltzmann simulations of liquid-gas and binary fluid systems, *Physical Review E*, **54**, 5041–5052.

Taylor, G. I., 1953. Dispersion of soluble matter in solvent flowing slowly through a tube, *Proceedings of the Royal Society of London, Series A*, **219**, 186–203.

Van Dyke, M. D., 1964. *Perturbation Methods in Fluid Mechanics*, Academic Press, New York.

Weiner, J. H., 2002. *Statistical Mechanics of Elasticity*, Dover, New York.

Young, T., 1805. An essay on the cohesion of fluids, *Philosophical Transactions of the Royal Society of London*, **95**, 65–87.

Zhang, J. Z. H., 1999. *Theory and Application of Quantum Molecular Dynamics*, World Scientific Publishing, Singapore.

Index

Printed in the United States
by Baker & Taylor Publisher Services